ENCYCLOPEDIA OF STATISTICAL SCIENCES

VOLUME 6

**Multivariate Analysis
to Plackett and Burman Designs**

ENCYCLOPEDIA OF STATISTICAL SCIENCES

VOLUME 6

MULTIVARIATE ANALYSIS
to PLACKETT AND BURMAN DESIGNS

A WILEY-INTERSCIENCE PUBLICATION

John Wiley & Sons

NEW YORK · **CHICHESTER** · **BRISBANE** · **TORONTO** · **SINGAPORE**

Library of Congress Cataloging in Publication Data:
Main entry under title:
Encyclopedia of statistical sciences.

  "A Wiley-Interscience publication."
  Includes bibliographies.
  Contents: v. 1. A to Circular probable
error—v. 3. Faà di Bruno's formula
to Hypothesis testing—[etc.]—
v. 6. Multivariate analysis to Plackett and
Burman designs.
  1. Mathematical statistics—Dictionaries.
2. Statistics—Dictionaries. I. Kotz, Samuel.
II. Johnson, Norman Lloyd. III. Read, Campbell B.
QA276.14.E5   1982   519.5'03'21   81-10353
ISBN 0-471-05553-0 (v. 6)

Printed in the United States of America

10 9 8 7 6 5 4 3 2 1

CONTRIBUTORS

A. Abouchar, *University of Toronto, Toronto, Ontario, Canada.* Paasche–Laspeyres Index Numbers

J. Abrahams, *Carnegie-Mellon University, Pittsburgh, Pennsylvania.* Narrowband Process

A. G. Agresti, *University of Florida, Gainesville, Florida.* Ordinal Data

D. J. Aldous, *University of California, Berkeley, California.* Partial Exchangeability

F. B. Alt, *University of Maryland, College Park, Maryland.* Multivariate Quality Control

P. K. Andersen, *Statistical Research Unit, Copenhagen, Denmark.* Multivariate Cox Regression Model

C. E. Antle, *The Pennsylvania State University, University Park, Pennsylvania.* Pivotal Quantities

B. C. Arnold, *University of California, Riverside, California.* Pareto Distribution

A. A. Assad, *University of Maryland, College Park, Maryland.* Optimization, Statistics in; PERT

K. Atkinson, *University of Iowa, Iowa City, Iowa.* Numerical Analysis

R. A. Bailey, *Rothamsted Experimental Station, Harpenden, Herts, England.* Nesting and Crossing in Design; Partially Balanced Designs

L. J. Bain, *University of Missouri, Rolla, Missouri.* Pivotal Quantities

J. Banks, *Georgia Institute of Technology, Atlanta, Georgia.* Nomograms

G. A. Barnard, *Brightlingsea, Essex, England.* Pivotal Inference

D. E. Barton, *University of London, London, England.* Neyman's and Other Smooth Goodness-of-Fit Tests

D. Bates, *University of Wisconsin, Madison, Wisconsin.* Nonlinear Models; Nonlinear Regression

J. M. Begun, *Glaxo Statistical Services, Research Triangle Park, North Carolina.* Newman–Keuls Procedure

D. A. Berry, *University of Minnesota, Minneapolis, Minnesota.* One- and Two-Armed Bandit Problems

H. W. Block, *University of Pittsburgh, Pittsburgh, Pennsylvania.* Multivariate Exponential Distributions (Excluding Marshall–Olkin Distributions)

L. D. Bodin, *University of Maryland, College Park, Maryland.* Network Analysis

R. Bohrer, *University of Illinois, Urbana, Illinois.* Noncentral Studentized Maximal Distributions

K. O. Bowman, *Union Carbide Corporation, Oak Ridge, Tennessee.* Optimal Sample Size Requirements; Padé and Stieltjes Transformations

E. L. Bradley, *The University of Alabama, Birmingham, Alabama.* Overlapping Coefficient

R. A. Bradley, *University of Georgia, Athens, Georgia.* Paired Comparisons

J. M. Cass, *Imperial Tobacco, Ltd., Bristol, England.* Newton–Spurrell Method

v

C. **Chakrapani,** *Applied Marketing Statistics, Toronto, Canada.* Numeracy

O. B. **Chedzoy,** *University of Bath, Bath, England.* Phi-Coefficient; Phi-Deviation; Phi-Max Coefficient

C. S. **Cheng,** *University of California, Berkeley, California.* Nearly Balanced Designs

J. P. **Cohen,** *University of Kentucky, Lexington, Kentucky.* Normal Extremes

P. J. **Coughlin,** *University of Maryland, College Park, Maryland.* Nash Axioms; Nash Equilibrium; Pareto Optimality

M. **Csörgő,** *Carleton University, Ottawa, Canada.* Multivariate Cramér–Von Mises Statistics

T. **Dalenius,** *Brown University, Providence, Rhode Island.* Optimum Stratification

D. J. **Daley,** *The Australian National University, Canberra, Australia.* Pecking Order Problem

F. N. **David,** *University of California, Berkeley, California.* Pearson, Karl

H. A. **David,** *Iowa State University, Ames, Iowa.* Order Statistics

H. **DeVroom,** *OECD Publications and Information Center, Washington, D.C.* OECD

S. W. **Dharmadhikari,** *Southern Illinois University, Carbondale, Illinois.* Multivariate Unimodality

R. L. **Disney,** *Virginia Polytechnic Institute and State University, Blacksburg, Virginia.* Network of Queues

N. R. **Draper,** *University of Wisconsin, Madison, Wisconsin.* Plackett and Burman Designs

O. J. **Dunn,** *University of California, Los Angeles, California.* One-Way Analysis of Variance

R. L. **Dykstra,** *The University of Iowa, Iowa City, Iowa.* Ordering, Starshaped

A. W. F. **Edwards,** *University of Cambridge, Cambridge, England.* Pascal Triangle

C. **Eisenhart,** *National Bureau of Standards, Washington, D.C.* National Bureau of Standards Journal of Research

R. L. **Eubank,** *Southern Methodist University, Dallas, Texas.* Optimal Spacing Problems

D. H. **Evans,** *Oakland University, Rochester, Michigan.* n-Dimensional Quadrature

W. J. **Ewens,** *University of Pennsylvania, Philadelphia, Pennsylvania.* Mutation Process

K.-T. **Fang,** *Academia Sinica, Beijing, China.* Occupancy Problems

V. **Farewell,** *Fred Hutchinson Cancer Research Center, Seattle, Washington.* Nonparametric Estimation of Standard Errors

S. E. **Fienberg,** *Carnegie-Mellon University, Pittsburgh, Pennsylvania.* Multivariate Directed Graphs

K. R. **Gabriel,** *University of Rochester, Rochester, New York.* Multivariate Graphics

J. **Galambos,** *Temple University, Philadelphia, Pennsylvania.* Multivariate Order Statistics; Multivariate Stable Distributions

P. A. **Games,** *The Pennsylvania State University, University Park, Pennsylvania.* Overall and Woodward Test

J. **Gani,** *University of Kentucky, Lexington, Kentucky.* Newton, Isaac

J. D. **Gibbons,** *The University of Alabama, University, Alabama.* Normal Scores Test; Permutation Tests; Pitman Tests

R. E. **Glaser,** *Lawrence Livermore National Laboratory, Livermore, California.* Multivariate Bartlett Test

B. L. **Golden,** *University of Maryland, College Park, Maryland.* Network Analysis; Optimization, Statistics in

E. D. **Goldfield,** *National Research Council, Washington, D.C.* National Statistics, Committee on

A. S. **Goldman,** *Los Alamos National Laboratory, Los Alamos, New Mexico.* Nuclear Methods Safeguards; Particle Size Statistics

P. J. **Green,** *The Wharton School of The University of Pennsylvania, Philadelphia, Pennsylvania.* Peeling Data

R. F. **Green,** *University of Minnesota, Duluth, Minnesota.* Outlier Prone Distribution

R. C. **Griffiths,** *Monash University, Clayton, Australia.* Orthogonal Expansion

S. S. **Gupta,** *Purdue University, West Lafayette, Indiana.* Optimal Sampling in Selection Problems

K. B. **Haley,** *University of Birmingham, Birmingham, England.* Operational Research Society, Journal of the

P. Hall, *Australian National University, Canberra, Australia.* Near Neighbor Estimator; Optional Sampling, Optional Sampling Theorem

T. J. Harris, *E. I. du Pont de Nemours & Co., Inc., Wilmington, Delaware.* Optimal Stochastic Control

H. Hauptman, *Medical Foundation of Buffalo, Buffalo, New York.* Phase Problems of X-Ray Crystallography, Probabilistic Methods in

D. M. Hawkins, *National Research Institute for Mathematical Sciences, Pretoria, South Africa.* Outliers

T. P. Hettmansperger, *Pennsylvania State University, University Park, Pennsylvania.* Multivariate Location Test

P. S. Horn, *University of Cincinnati, Cincinnati, Ohio.* Pivot and Bipivot-t; Pivot Depth

Y.-S. Hsu, *Georgia State University, Atlanta, Georgia.* Ornstein–Uhlenbeck Process

R. Hultquist, *The Pennsylvania State University, University Park, Pennsylvania.* Partial Regression

W. G. Hunter, *University of Wisconsin, Madison, Wisconsin.* Nonlinear Models

F. K. Hwang, *Bell Laboratories, Murray Hill, New Jersey.* Ordering Procedures

N. T. Jazairi, *York University, Downsview, Ontario, Canada.* Paasche–Laspeyres Index Numbers

D. R. Jensen, *Virginia Polytechnic Institute and State University, Blacksburg, Virginia.* Multivariate Distributions; Multivariate Weibull Distribution

K. Joag-Dev, *University of Illinois, Urbana, Illinois.* Multivariate Unimodality

R. A. Johnson, *University of Wisconsin, Madison, Wisconsin.* Multivariate Analysis (Excluding Bayesian)

K. Kafadar, *Hewlett-Packard, Palo Alto, California.* Notched Box and Whisker Plot; One-Wild Distribution

M. A. Kastenbaum, *The Tobacco Institute, Washington, D.C.* Optimal Sample Size Requirements

R. Kay, *University of Sheffield, Sheffield, England.* Partial Likelihood.

J. P. Keating, *University of Texas, San Antonio, Texas.* Percentiles, Estimation of

N. M. Kiefer, *Cornell University, Ithaca, New York.* Multivariate Probit Model

H. E. Klugh, *Alma College, Alma, Michigan.* Normalized T Scores

S. L. Koffler, *Rutgers University, New Brunswick, New Jersey.* Nonparametric Discrimination

P. R. Krishnaiah, *University of Pittsburgh, Pittsburgh, Pennsylvania.* Multivariate Gamma Distribution; Multivariate Multiple Comparisons

E. E. Leamer, *University of California, Los Angeles, California.* Nonexperimental Inference

E. Lehmann, *University of California, Berkeley, California.* Neyman–Pearson Lemma and Applications

S. Lemeshow, *University of Massachusetts, Amherst, Massachusetts.* Nonresponse (in Sample Surveys)

B. G. Lindsay, *Pennsylvania State University, University Park, Pennsylvania.* Nuisance Parameters

S. M. McKinlay, *American Institutes for Research in the Behavioral Sciences, Cambridge, Massachusetts.* Observational Studies

B. F. J. Manly, *University of Otago, Dunedin, New Zealand.* Multivariate Fitness Functions

A. W. Marshall, *University of British Columbia, Vancouver, Canada.* Multivariate Exponential Distributions

R. J. Martin, *University of Sheffield, Sheffield, England.* Papadakis's Method

P. W. Mielke, Jr., *Colorado State University, Fort Collins, Colorado.* Omega Distribution

B. L. Misra, *J. Nehru Agricultural University, Jabalpur, India.* Partially Balanced Youden Squares

I. W. Molenaar, *Rijksuniversiteit te Groningen, Groningen, The Netherlands.* Normal Approximations to Binomial, Negative Binomial, and Hypergeometric Distributions

C. N. Morris, *University of Texas, Austin, Texas.* Natural Exponential Family

A. C. Mukhopadhyay, *Indian Statistical Institute, Calcutta, India.* Orthogonal Arrays and Their Applications

L. S. Nelson, *Nashua Corporation, Nashua, New Hampshire.* Nelder–Mead Simplex Method

G. Noether, *University of Connecticut, Storrs, Connecticut.* Nonparametric Confidence Intervals; Nonparametric Tolerance Limits

D. Oakes, *University of Rochester, Rochester, New York.* Oakes's Test of Concordance

K. O'Brien, *East Carolina University, Greenville, North Carolina.* Odds Ratio Estimator

H. Oja, *University of Oulu, Oulu, Finland.* Ordering of Distributions, Partial

I. Olkin, *Stanford University, Stanford, California.* Multivariate Exponential Distributions

J. K. Ord, *The Pennsylvania State University, University Park, Pennsylvania.* Nearest Neighbor Method; Pearson System of Distributions; Periodogram Analysis

D. B. Owen, *Southern Methodist University, Dallas, Texas.* Noncentral *t* Distribution; Orthant Probabilities

A. G. Pakes, *University of Western Australia, Nedlands, Western Australia.* Palm Functions

W. J. Park, *Wright State University, Dayton, Ohio.* Ornstein–Uhlenbeck Process

G. P. Patil, *Pennsylvania State University, University Park, Pennsylvania.* Multivariate Logarithmic Series Distributions; Multivariate Power Series Distributions

D. A. Penfield, *Rutgers University, New Brunswick, New Jersey.* Nonparametric Discrimination

R. Picard, *Los Alamos National Laboratories, Los Alamos, New Mexico.* Nuclear Materials Safeguard

K. C. S. Pillai, *Purdue University, West Lafayette, Indiana.* Multivariate Analysis of Variance (MANOVA); Pillai Trace

W. R. Pirie, *Virginia Polytechnic Institute and State University, Blacksburg, Virginia.* Page Test for Ordered Alternatives

R. Popping, *Rijksuniversiteit te Groningen, Groningen, The Netherlands.* Nominal Scale Agreements

S. J. Press, *University of California, Riverside, California.* Multivariate Analysis (Bayesian)

R. E. Quandt, *Princeton University, Princeton, New Jersey.* Peak Test

D. B. Ramey, *Washington, D.C.* Nonparametric Clustering Techniques

A. Raveh, *Hebrew University, Jerusalem, Israel.* Partial Order Scalogram Analysis

C. B. Read, *Southern Methodist University, Dallas, Texas.* Nightingale, Florence; Normal Distribution; Partition of Chi-Square

G. W. Reddien, *Southern Methodist University, Dallas, Texas.* Nonparametric Clustering Techniques

P. Redfern, *Bookhurst Hill, Cranleigh, England.* Office of Population Censuses and Surveys

H. T. Reynolds, *University of Delaware, Newark, Delaware.* Nominal Data

H. Rootzén, *University of Copenhagen, Copenhagen, Denmark.* Persson–Rootzén Estimator

J. R. Rosenblatt, *National Bureau of Standards, Washington, D.C.* National Bureau of Standards, Statistics in

B. Rosner, *Harvard Medical School, Boston, Massachusetts.* Ophthalmology, Statistics in

W. L. Ruble, *Sparks Commodities, Inc., Cranford, New Jersey.* Newton Iteration Extensions

A. Rukhin, *Purdue University, West Lafayette, Indiana.* Multivariate Chernoff Theorem

J. S. Rustagi, *The Ohio State University, Columbus, Ohio.* Optimization in Statistics

P. Schmidt, *Michigan State University, East Lansing, Michigan.* Overidentification

I. Schneider, *Universität München, München, Federal Republic of Germany.* Physics, Statistics in, Early History

H. T. Schreuder, *Rocky Mountain Forest and Range Experiment Station, USDA, Fort Collins, Colorado.* Pascual's Estimator

S. J. Schwager, *Cornell University, Ithaca, New York.* Multivariate Skewness and Kurtosis

N. C. Schwertman, *California State University, Chico, California.* Multivariate Median and Rank Sum Tests

S. L. Sclove, *University of Illinois, Chicago, Illinois.* Pattern Recognition

E. L. Scott, *University of California, Berkeley, California.* Neyman, Jerzy

G. A. F. Seber, *University of Auckland, Auckland, New Zealand.* Orthogonal Decomposition

P. K. Sen, *University of North Carolina, Chapel Hill, North Carolina.* Neyman Structure, Tests With; Optimal $C(\alpha)$ Tests; Permutational Central Limit Theorem

E. Seneta, *University of Sydney, Sydney, Australia.* Pascal, Blaise; Path Analysis

G. Shafer, *University of Kansas, Lawrence, Kansas.* Nonadditive Probabilities

M. Shaked, *The University of Arizona, Tucson, Arizona.* Ordering Distributions by Dispersion; Phase-Type Distributions

J. G. Shanthikumar, *University of California, Berkeley, California.* Phase-Type Distributions

L. R. Shenton, *The University of Georgia, Athens, Georgia.* Padé and Stieltjes Transformations

J. Shipley, *Los Alamos National Laboratories, Los Alamos, New Mexico.* Nuclear Materials Safeguard

D. Siegmund, *Stanford University, Stanford, California.* Optimal Stopping Rules

B. W. Silverman, *University of Bath, Bath, England.* Penalized Maximum Likelihood Estimation

N. J. H. Small, *89 Greyhound Road, London, England.* Multivariate Normality, Testing for

A. F. M. Smith, *The University of Nottingham, Nottingham, England.* Observations, Imaginary

P. J. Smith, *University of Maryland, College Park, Maryland.* Noether and Related Conditions

H. Solomon, *Stanford University, Stanford, California.* Neyman's Test of Uniformity

G. W. Somes, *East Carolina University, Greenville, North Carolina.* Odds Ratio Estimators

M. A. Stephens, *Simon Fraser University, Burnaby, Canada.* Neyman's Test of Uniformity

G. W. Stewart, *University of Maryland, College Park, Maryland.* Orthogonalization of Matrices

S. Stigler, *University of Chicago, Chicago, Illinois.* Newcomb, Simon

H. Strasser, *Universität Bayreuth, Bayreuth, Federal Republic of Germany.* Pitman Estimators

A. H. Stroud, *Texas A & M University, College Station, Texas.* Numerical Integration

T. H. Szatrowski, *Rutgers University, Newark, New Jersey.* Patterned Covariances; Patterned Means

Y. Takane, *McGill University, Montreal, Quebec, Canada.* Nonmetric Data Analysis

G. M. Tallis, *University of Adelaide, Adelaide, South Australia.* Permutation Models

V. K. T. Tang, *Humboldt State University, Arcata, California.* Neyman Allocation

M. Taqqu, *Cornell University, Ithaca, New York.* Orthogonal Processes

W. E. Taylor, *Bell Laboratories, Murray Hill, New Jersey.* Panel Data

M. L. Tiku, *McMaster University, Hamilton, Ontario, Canada.* Noncentral Chi-Square Distribution; Noncentral F Distribution

J. W. Tolle, *University of North Carolina, Chapel Hill, North Carolina.* Nonlinear Programming

R. D. Tortora, *U.S. Department of Agriculture, Washington, D.C.* Nonsampling Errors in Surveys

R. C. Tripathi, *University of Texas, San Antonio, Texas.* Negative Binomial Distributions; Neyman Type A, B, C Distributions; Percentiles, Estimation of

D. H. Wagner, *Daniel H. Wagner Associates, Paoli, Pennsylvania.* Operations Research

D. G. Watts, *Queen's University, Kingston, Ontario, Canada.* Nonlinear Regression

W. A. O'N. Waugh, *University of Toronto, Toronto, Ontario, Canada.* Music, Statistics and Probability in

G. H. Weiss, *National Institutes of Health, Bethesda, Maryland.* Passage Times

L. V. White, *Imperial College of Science and Technology, London, England.* Orthogonal Designs

D. Wichern, *University of Wisconsin, Madison, Wisconsin.* Multivariate Analysis (Excluding Bayesian)

J. C. Wierman, *Johns Hopkins University, Baltimore, Maryland.* Percolation Theory

T. R. Willemain, *Harvard University, Cambridge, Massachusetts.* Nomination Sampling

E. J. Williams, *University of Melbourne, Parkville, Victoria, Australia.* Pitman, E. J. G.

H. O. Wold, *University of Uppsala, Uppsala, Sweden.* Partial Least Squares

D. A. Wolfe, *The Ohio State University, Columbus, Ohio.* Placement Statistics

M. B. Woodroofe, *Rutgers University, New Brunswick, New Jersey.* Nonlinear Renewal Theory

F. T. Wright, *University of Missouri, Rolla, Missouri.* Order Restricted Inferences

Y.-C. Yao, *Colorado State University, Fort Collins, Colorado.* Noise; Nyquist Frequency

P. A. Young, *The Wellcome Research Laboratories, Beckenham, Kent, England.* Pharmaceutical Industry, Statistics in

S. Zacks, *State University of New York, Binghamton, New York.* Pitman Efficiency

M
continued

MULTIVARIATE ANALYSIS

INTRODUCTION

The body of statistical methodology used to analyze simultaneous measurements on many variables is called *multivariate analysis*. Many multivariate methods are based on an underlying probability model known as the multivariate normal (*see* MULTINORMAL DISTRIBUTION). Other methods are ad hoc in nature and are justified by logical or common-sense arguments. Regardless of their origin, multivariate techniques invariably must be implemented on a computer. Recent advances in computer technology have been accompanied by the development of rather sophisticated statistical software* packages, making the implementation step easier.

Multivariate analysis is a "mixed bag." It is difficult to establish a classification scheme for multivariate techniques that is both widely accepted and also indicates the appropriateness of the techniques. One classification distinguishes techniques designed to study interdependent relationships from those designed to study dependent relationships. Another classifies techniques according to the number of populations and the number of sets of variables being studied. This entry is divided into sections according to inferences about treatment means, inferences about covariance structure, and techniques for classification or grouping. This should not, however, be regarded as an attempt to place each method into a slot. Rather, the choice of methods and the types of analyses employed are determined largely by the objectives of the investigation.

The objectives of scientific investigations, for which multivariate methods most naturally lend themselves, include the following:

1. Data reduction or structural simplification.
2. Sorting and grouping.
3. Investigation of the dependence among variables.
4. Prediction.
5. Hypothesis construction and testing.

A HISTORICAL PERSPECTIVE

Many current multivariate statistical procedures were developed during the first half of the twentieth century. A reasonably complete list of the developers would be voluminous. However, a few individuals can be cited as making important initial contributions to the theory and practice of multivariate analysis.

F. Galton* and K. Pearson* did pioneering work in the areas of correlation* and regression analysis. R. A. Fisher's derivation of the exact distribution of the sample correlation coefficient and related quantities provided the impetus for multivariate distribution theory. C. Spearman and K. Pearson were among the first to work in the area of factor analysis*. Significant contributions to multivariate analysis were made during the 1930s by S. S. Wilks* (general procedures for testing certain multivariate hypotheses), H. Hotelling* (Hotelling's T^2*, principal component analysis*, canonical correlation analysis), R. A. Fisher* (discrimination and classification*), and P. C. Mahalanobis* (generalized distance, hypothesis testing). J. Wishart* derived an important joint distribution of sample variances and covariances that bears his name. Later M. Bartlett and G. E. P. Box contributed to the large sample theory associated with certain multivariate test statistics.

Many multivariate methods developed in consort with the development of electronic computers. Specifically, ingenious graphical methods for displaying multivariate data (e.g., Chernoff faces*, Andrews plots*) can only be conveniently implemented on a computer (*see* MULTIVARIATE GRAPHICS). Multidimensional scaling* and many clustering procedures were not feasible before the advent of large fast computers.

Multivariate statistical procedures and their refinements are now available in comprehensive software packages such as BMDP, IMSL, SAS, and SPSS. The computer has made it possible for researchers to apply routinely, for good or ill, multivariate statistical methods.

NOTATION

The description of multivariate data and the computations required for their analysis are greatly facilitated by the use of matrix algebra (*see* LINEAR ALGEBRA, COMPUTATIONAL).

Consequently, the subsequent discussion will rely heavily on the following notation:

$\mathbf{X}$ $p \times 1$ random vector

$\mathbf{x}_j$ $p \times 1$ multivariate observation on $\mathbf{X}$

$$\mathbf{X} = \begin{bmatrix} x_{11} & x_{12} & \cdots & x_{1n} \\ x_{21} & x_{22} & \cdots & x_{2n} \\ \vdots & \vdots & & \vdots \\ x_{p1} & x_{p2} & \cdots & x_{pn} \end{bmatrix}$$

$$= [\mathbf{x}_1, \mathbf{x}_2, \ldots, \mathbf{x}_n],$$

a $p \times n$ matrix. (Each column of $\mathbf{X}$ represents a multivariate observation.)

$$\bar{\mathbf{x}} = [\bar{x}_1, \bar{x}_2, \ldots, \bar{x}_p]'$$

$$= \left\{ \bar{x}_i = \frac{1}{n} \sum_{j=1}^{n} x_{ij};\ i = 1, 2, \ldots, p \right\},$$

a $p \times 1$ vector of sample means.

$$\mathbf{S} = \left\{ s_{ik} = \frac{1}{n-1} \sum_{j=1}^{n} (x_{ij} - \bar{x}_i)(x_{kj} - \bar{x}_k);\ i, k = 1, 2, \ldots, p \right\},$$

a $p \times p$ symmetric matrix of sample variances and covariances.

$$\mathbf{R} = \left\{ r_{ik} = \frac{s_{ik}}{\sqrt{s_{ii}} \sqrt{s_{kk}}} \right\},$$

a $p \times p$ symmetric matrix of sample correlation coefficients.

$$\boldsymbol{\mu} = \{ \mu_i \} = E(\mathbf{X}),$$

a $p \times 1$ vector of population means. [$E(\cdot)$ is the expectation operator.]

$$\boldsymbol{\Sigma} = \{ \sigma_{ij} \} = E(\mathbf{X} - \boldsymbol{\mu})(\mathbf{X} - \boldsymbol{\mu})',$$

a $p \times p$ symmetric matrix of population variances and covariances.

$$\boldsymbol{\rho} = \left\{ \rho_{ij} = \frac{\sigma_{ij}}{\sqrt{\sigma_{ii}} \sqrt{\sigma_{jj}}} \right\},$$

a $p \times p$ symmetric matrix of population correlation coefficients.

$$\mathbf{Z} = \left\{ Z_i = \frac{X_i - \mu_i}{\sqrt{\sigma_{ii}}} \right\},$$

a $p \times 1$ vector of standardized variables.

MULTIVARIATE NORMAL DISTRIBUTION

A generalization of the familiar bell-shaped normal density to several dimensions plays a fundamental role in multivariate analysis (*see* MULTINORMAL DISTRIBUTION). In fact, many multivariate techniques assume that the data were generated from a *multivariate* normal distribution. While real data are never *exactly* multivariate normal, the normal density is often a useful approximation to the "true" population distribution. Thus the normal distribution* serves as a bona fide population model in some instances. Also the sampling distributions of many multivariate statistics are approximately normal, regardless of the form of the parent population, because of a *central limit effect* (*see* MULTIDIMENSIONAL CENTRAL LIMIT THEOREM).

The p-dimensional normal density for the random vector $\mathbf{X} = [X_1, X_2, \ldots, X_p]'$, evaluated at the point $\mathbf{x} = [x_1, x_2, \ldots, x_p]'$, is given by

$$f(\mathbf{x}) = (2\pi)^{-p/2} |\mathbf{\Sigma}|^{-1/2}$$
$$\times \exp\left\{ -\tfrac{1}{2}(\mathbf{x} - \boldsymbol{\mu})'\mathbf{\Sigma}^{-1}(\mathbf{x} - \boldsymbol{\mu}) \right\},$$

where $-\infty < x_i < \infty$, $i = 1, 2, \ldots, p$. Here $\boldsymbol{\mu}$ is the population mean vector, $\mathbf{\Sigma}$ is the population variance-covariance matrix, $|\mathbf{\Sigma}|$ is the determinant of $\mathbf{\Sigma}$, and $\exp(\cdot)$ stands for the exponential function. We denote this p-dimensional normal density by $N_p(\boldsymbol{\mu}, \mathbf{\Sigma})$.

Contours of constant density for the p-dimensional normal distribution are ellipsoids defined by $\mathbf{x}$ such that

$$(\mathbf{x} - \boldsymbol{\mu})'\mathbf{\Sigma}^{-1}(\mathbf{x} - \boldsymbol{\mu}) = c^2.$$

These ellipsoids are centered at $\boldsymbol{\mu}$ and have axes $\pm c\sqrt{\lambda_i}\, \mathbf{e}_i$, where $\mathbf{\Sigma}\mathbf{e}_i = \lambda_i \mathbf{e}_i$, $i = 1, 2, \ldots, p$. That is, $\lambda_i, \mathbf{e}_i$ are the eigenvalue-(normalized) eigenvector pairs associated with $\mathbf{\Sigma}$.

The following are true for a random vector $\mathbf{X}$ having a multivariate normal distribution.

1. Linear combinations of the components of $\mathbf{X}$ are normally distributed.

2. All subsets of the components of $\mathbf{X}$ have a (multivariate) normal distribution.

3. Zero covariance implies that the corresponding components are distributed independently.

4. The conditional distributions of the multivariate components are (multivariate) normal.

(See Rao [16, Sec, 8a] and Anderson [2, Chap. 2] for more discussion of the multivariate normal distribution.) Because these properties make the normal distribution easy to manipulate, it has been overemphasized as a population model.

To some degree, the *quality* of inferences made by some multivariate methods depend on how closely the true parent population resembles the multivariate normal form. It is imperative, then, that procedures exist for detecting cases where the data exhibit moderate to extreme departures from what is expected under multivariate normality.

Sometimes nonnormal data can be made more normal looking by considering *transformations* of the data. Normal theory analyses can then be carried out with the suitably transformed data. (See Gnanadesikan [7, Chap. 5] and Johnson and Wichern [10, Sec. 4.7].) Recent advances in the theory of *discrete* multivariate analysis are contained in Bishop et al. [3]; *see* also MULTIDIMENSIONAL CONTINGENCY TABLES.

SAMPLING DISTRIBUTIONS OF $\overline{\mathbf{X}}$ AND $\mathbf{S}$

The tentative assumption that the columns of the data matrix, treated as random vectors, constitute a random sample from a normal population, with mean $\boldsymbol{\mu}$ and covariance $\mathbf{\Sigma}$, completely determines the sampling distribution of $\overline{\mathbf{X}}$ and $\mathbf{S}$. We now summarize the sampling distribution results.

Let $\mathbf{X}_1, \mathbf{X}_2, \ldots, \mathbf{X}_n$ be a random sample of size n from a p-variate *normal* distribution with mean $\boldsymbol{\mu}$ and covariance matrix $\mathbf{\Sigma}$. Then (see Anderson [1, Secs. 3.3 and 7.2] and Rao [16, Sec. 8b]):

1. $\overline{\mathbf{X}}$ is distributed as $N_p(\boldsymbol{\mu}, (1/n)\mathbf{\Sigma})$.

2. $(n - 1)\mathbf{S}$ is distributed as a Wishart random matrix with $n - 1$ d.f.

3. $\overline{\mathbf{X}}$ and $\mathbf{S}$ are independent.

Because $\mathbf{\Sigma}$ is unknown, the distribution of $\overline{\mathbf{X}}$ cannot be used directly to make inferences about $\boldsymbol{\mu}$. However, $\mathbf{S}$ provides independent information about $\mathbf{\Sigma}$ and the distribution of $\mathbf{S}$ does not depend on $\boldsymbol{\mu}$. This allows one to construct a statistic for making inferences about $\boldsymbol{\mu}$.

MULTIVARIATE METHODS FOR SELECTED PROBLEMS

Inferences About Means

SINGLE POPULATION MEAN VECTOR.   An immediate objective in many multivariate studies is to make statistical inferences about population mean vectors. As an initial example, consider the problem of testing whether a multivariate normal population mean vector has a particular value $\boldsymbol{\mu}_0$.

Let the null and alternative hypotheses be $H_0 : \boldsymbol{\mu} = \boldsymbol{\mu}_0$ and $H_1 : \boldsymbol{\mu} \neq \boldsymbol{\mu}_0$. Once the sample is in hand, a sample mean vector far from $\boldsymbol{\mu}_0$ tends to discredit H_0. A test of H_0 based on the (statistical) distance of $\overline{x}$ from $\boldsymbol{\mu}_0$ (assuming the population covariance matrix $\mathbf{\Sigma}$ is unknown) can be carried out using Hotelling's T^2*,

$$T^2 = n(\overline{x} - \boldsymbol{\mu}_0)'\mathbf{S}^{-1}(\overline{x} - \boldsymbol{\mu}_0).$$

Hotelling's T^2 is a distance measure that takes account of the joint variability of the p measured variables. Under H_0, T^2 is distributed as

$$\left[(n - 1)p/(n - p)\right]F_{p, n - p},$$

where n is the sample size and $F_{p, n - p}$ denotes an F random variable with p and $n - p$ degrees of freedom. Let $F_{p, n - p}(\alpha)$ be the upper 100αth percentage point of this F distribution*. The hypothesis $H_0 : \boldsymbol{\mu} = \boldsymbol{\mu}_0$ is rejected, at the α level of significance if

$$T^2 > \left[(n - 1)p/(n - p)\right]F_{p, n - p}(\alpha).$$

Example 1.   The quality-control department of a microwave oven manufacturer is concerned about the radiation emitted by the ovens. They record radiation measurements with the oven doors opened and closed. Measurements for a random sample of microwave ovens could then be compared with the standards for radiation emission set by the manufacturer.

Other principles of test construction (e.g., likelihood ratio*, union-intersection principle*) also lead to the use of Hotelling's T^2 in this testing situation.

Hotelling's T^2 statistic has numerous other applications. There are multivariate analogs of the paired and two-sample univariate t-statistics. For applications, see Anderson [2, Chap. 5], Johnson and Wichern [10, Chap. 6], Kshirsagar [12, Chap. 5] and Rao [16, Sec. 8d].

SEVERAL POPULATION MEANS, MULTIVARIATE ANALYSIS OF VARIANCE* (MANOVA).   Multivariate analysis of variance (MANOVA) is concerned with inferences about several population means. It is a direct generalization of the analysis of variance* (ANOVA) to the case of more than one response variable. In its simplest form, one-way MANOVA, random samples are collected from each of g populations and arranged as

Population 1	$\mathbf{X}_{11}, \mathbf{X}_{12}, \ldots, \mathbf{X}_{1n_1}$
Population 2	$\mathbf{X}_{21}, \mathbf{X}_{22}, \ldots, \mathbf{X}_{2n_2}$
$\vdots$	$\vdots$
Population g	$\mathbf{X}_{g1}, \mathbf{X}_{g2}, \ldots, \mathbf{X}_{gn_g}$

MANOVA is used first to investigate whether the population mean vectors are the same and, if not which mean components differ significantly. It is assumed that

1. $\mathbf{X}_{l1}, \mathbf{X}_{l2}, \ldots, \mathbf{X}_{ln_l}$ is a random sample of size n_l from a population with mean $\boldsymbol{\mu}_l$, $l = 1, 2, \ldots, g$. The random samples from different populations are independent.

2. All populations have a common covariance matrix $\mathbf{\Sigma}$.

3. Each population is multivariate normal.

Table 1   MANOVA Model for Comparing g Population Mean Vectors

$\mathbf{X}_{lj} = \boldsymbol{\mu} + \boldsymbol{\tau}_l + \mathbf{e}_{lj}$, $\quad j = 1, 2, \ldots, n_l$, and $\quad l = 1, 2, \ldots, g$, where $\mathbf{e}_{lj}$ are independent $N_p(\mathbf{0}, \boldsymbol{\Sigma})$ variables. Here the parameter vector $\boldsymbol{\mu}$ is an overall mean (level) and $\boldsymbol{\tau}_l$ represents the lth treatment effect with, for instance,

$$\sum_{l=1}^{g} n_l \boldsymbol{\tau}_l = \mathbf{0}$$

Condition 3 can be relaxed by appealing to the central limit theorem* when the sample sizes n_l are large. The model states that the mean $\boldsymbol{\mu}_l$ consists of a common part $\boldsymbol{\mu}$ plus an amount $\boldsymbol{\tau}_l$ due to the lth treatment. According to the model, *each component* of the observation vector $\mathbf{X}_{lj}$ satisfies the univariate model. The errors for the components of $\mathbf{X}_{lj}$ are correlated, but the covariance matrix $\boldsymbol{\Sigma}$ is the same for all populations.

A vector of observations may be decomposed as suggested by the model. Thus

$$\mathbf{x}_{lj} = \bar{\mathbf{x}} + (\bar{\mathbf{x}}_l - \bar{\mathbf{x}}) + (\mathbf{x}_{lj} - \bar{\mathbf{x}}_l)$$

$$(\text{observation}) = \begin{bmatrix} \text{overall} \\ \text{sample} \\ \text{mean } \hat{\boldsymbol{\mu}} \end{bmatrix} + \begin{bmatrix} \text{estimated} \\ \text{treatment} \\ \text{effect } \hat{\boldsymbol{\tau}}_l \end{bmatrix}$$

$$+ \begin{pmatrix} \text{residual} \\ \hat{\mathbf{e}}_{lj} \end{pmatrix},$$

which leads to a decomposition of the sum of squares and cross-products matrix

$$\sum_{l=1}^{g} \sum_{j=1}^{n_l} (\mathbf{x}_{lj} - \bar{\mathbf{x}})(\mathbf{x}_{lj} - \bar{\mathbf{x}})'$$

and the MANOVA table (see Table 2). One test of $H_0 : \boldsymbol{\tau}_1 = \boldsymbol{\tau}_2 = \cdots = \boldsymbol{\tau}_g = \mathbf{0}$ involves *generalized variances**. We reject H_0 if the ratio of generalized variances

$$\Lambda = \frac{|\mathbf{W}|}{|\mathbf{B} + \mathbf{W}|}$$

$$= \left| \frac{\sum_{l=1}^{g} \sum_{j=1}^{n_l} (\mathbf{x}_{lj} - \bar{\mathbf{x}}_l)(\mathbf{x}_{lj} - \bar{\mathbf{x}}_l)'}{\sum_{l=1}^{g} \sum_{j=1}^{n_l} (\mathbf{x}_{lj} - \bar{\mathbf{x}})(\mathbf{x}_{lj} - \bar{\mathbf{x}})'} \right|$$

is too small. The quantity $\Lambda = |\mathbf{W}|/|\mathbf{B} + \mathbf{W}|$ is called *Wilks' lambda* after its proposer S. Wilks (*see* LAMBDA CRITERION, WILKS'S).

This test is equivalent to the likelihood ratio test*. However, in MANOVA, there are other reasonable tests besides comparing the determinants of sums-of-squares and

Table 2   MANOVA Table for Comparing Population Mean Vectors

Source of Variation	Matrix of Sum of Square and Cross Products (SSP)	Degrees of Freedom (d.f.)
Treatment	$\mathbf{B} = \sum_{l=1}^{g} n_l (\bar{\mathbf{x}}_l - \bar{\mathbf{x}})(\bar{\mathbf{x}}_l - \bar{\mathbf{x}})'$	$g - 1$
Residual (error)	$\mathbf{W} = \sum_{l=1}^{g} \sum_{j=1}^{n_l} (\mathbf{x}_{lj} - \bar{\mathbf{x}}_l)(\mathbf{x}_{lj} - \bar{\mathbf{x}}_l)'$	$\left(\sum_{l=1}^{g} n_l \right) - g$
Total (corrected for the mean)	$\mathbf{B} + \mathbf{W} = \sum_{l=1}^{g} \sum_{j=1}^{n_l} (\mathbf{x}_{lj} - \bar{\mathbf{x}})(\mathbf{x}_{lj} - \bar{\mathbf{x}})'$	$\left(\sum_{l=1}^{g} n_l \right) - 1$

cross-products matrices; there are tests based on the extreme eigenvalues of $\mathbf{W}^{-1}\mathbf{B}$ and Pillai's trace* criterion.

A comparison of the MANOVA table with the familiar $p = 1$ ANOVA table reveals that they are of the same structure. For the multivariate generalization, squares $(\bar{x}_i - \bar{x})^2$ are replaced by sums-of-squares and cross-products matrices $(\bar{\mathbf{x}}_i - \bar{\mathbf{x}})(\bar{\mathbf{x}}_i - \bar{\mathbf{x}})'$. The same type of replacement holds for any fixed ANOVA, so MANOVA tables can be constructed easily for any of the common designs. (See Anderson [2, Chap. 8] and Johnson and Wichern [10, Chap. 6], e.g.)

SUMMARY REMARKS ON INFERENCE ABOUT MEANS. Multivariate analysis takes into account the joint variation of several responses. One noticeable difference from the univariate situation is that rejection of a null hypothesis $H_0 : \boldsymbol{\mu} = \boldsymbol{\mu}_0$ must be followed by a determination of which component(s) led to the rejection. Technically, it is at least one linear combination $a_1\mu_1 + \cdots + a_p\mu_p = \mathbf{a}'\boldsymbol{\mu}$ that is different from $\mathbf{a}'\boldsymbol{\mu}_0$, but this class typically includes some individual μ_i. As we proceed to several treatments, rejection of the null hypothesis $H_0 : \boldsymbol{\mu}_1 = \boldsymbol{\mu}_2 = \cdots = \boldsymbol{\mu}_g$ must be followed by a comparison of the $\boldsymbol{\mu}_l$, to determine which treatments are different and then to determine which components contribute to the difference.

MULTIVARIATE MULTIPLE REGRESSION

Regression analysis is the statistical methodology for predicting values of one or more *response* (dependent) variables from a collection of *predictor* (independent) variable values. It can also be used for assessing the effects of the predictor variables on the responses. In its simplest form, it applies to the fitting of a straight line to data.

The classical linear regression* model states that the response Y is composed of a mean, which depends in a linear fashion on the predictor variables z_i and random error ϵ which accounts for measurement error* and the effects of other variables not explicitly

considered in the model. The values of the predictor variables recorded from the experiment or set by the investigator are treated as *fixed*. The error (and hence the response) is viewed as a random variable whose behavior is characterized by a set of distributional assumptions.

Specifically, the linear regression model with a single response and n measurements on Y and the associated predictors z_1, $z_2, \ldots, z_r$ can be written in matrix notation as

$$\mathbf{Y} = \mathbf{Z}\boldsymbol{\beta} + \boldsymbol{\epsilon},$$

with $E(\boldsymbol{\epsilon}) = \mathbf{0}$ and $\mathrm{Cov}(\boldsymbol{\epsilon}) = \sigma^2\mathbf{I}$ (*see* GENERAL LINEAR MODEL).

The *least-squares* * estimator of $\boldsymbol{\beta}$ is given by $\hat{\boldsymbol{\beta}} = (\mathbf{Z}'\mathbf{Z})^{-1}\mathbf{Z}'\mathbf{Y}$ and σ^2 is estimated by $(\mathbf{Y} - \mathbf{Z}\hat{\boldsymbol{\beta}})'(\mathbf{Y} - \mathbf{Z}\hat{\boldsymbol{\beta}})/(n - r - 1)$. The literature on multiple linear regression* is vast; see the numerous books on the subject including Draper and Smith (1981) or Seber (1977).

Multivariate multiple regression is the extension of multiple regression to several response variables. Each response is assumed to follow its own regression model but with the same predictors, so that

$$Y_1 = \beta_{01} + \beta_{11}z_1 + \cdots + \beta_{r1}z_r + \epsilon_1$$
$$Y_2 = \beta_{02} + \beta_{12}z_1 + \cdots + \beta_{r2}z_r + \epsilon_2$$
$$\vdots \qquad \vdots$$
$$Y_m = \beta_{0m} + \beta_{1m}z_1 + \cdots + \beta_{rm}z_r + \epsilon_m .$$

The error term $\boldsymbol{\epsilon} = [\epsilon_1, \epsilon_2, \ldots, \epsilon_m]'$ has $E(\boldsymbol{\epsilon}) = \mathbf{0}$ and $\mathrm{Var}(\boldsymbol{\epsilon}) = \boldsymbol{\Sigma}$. Thus the error terms associated with different responses may be correlated.

To establish notation conforming to the classical linear regression model, let $[z_{j0}, z_{j1}, \ldots, z_{jr}]$ denote the values of the predictor variables for the jth trial, $\mathbf{Y}_j = [Y_{j1}, Y_{j2}, \ldots, Y_{jm}]'$ the responses, and $\boldsymbol{\epsilon}_j = [\epsilon_{j1}, \epsilon_{j2}, \ldots, \epsilon_{jm}]'$ the errors. In matrix notation, the design matrix

$$\mathbf{Z}_{(n \times (r+1))} = \begin{bmatrix} z_{10} & z_{11} & \cdots & z_{1r} \\ z_{20} & z_{21} & \cdots & z_{2r} \\ \vdots & \vdots & & \vdots \\ z_{n0} & z_{n1} & \cdots & z_{nr} \end{bmatrix}$$

is the same as that for the single-response regression model. The other matrix quantities have multivariate counterparts. Set

$$
\underset{(n \times m)}{\mathbf{Y}}
$$

$$
= \begin{bmatrix}
Y_{11} & Y_{12} & \cdots & Y_{1m} \\
Y_{21} & Y_{22} & \cdots & Y_{2m} \\
\vdots & \vdots & & \vdots \\
Y_{n1} & Y_{n2} & \cdots & Y_{nm}
\end{bmatrix}
$$

$$
= \begin{bmatrix} \mathbf{Y}_{(1)} & \vdots & \mathbf{Y}_{(2)} & \vdots & \cdots & \vdots & \mathbf{Y}_{(m)} \end{bmatrix},
$$

$$
\underset{((r+1) \times m)}{\boldsymbol{\beta}}
$$

$$
= \begin{bmatrix}
\beta_{01} & \beta_{02} & \cdots & \beta_{0m} \\
\beta_{11} & \beta_{12} & \cdots & \beta_{1m} \\
\vdots & \vdots & & \vdots \\
\beta_{r1} & \beta_{r2} & \cdots & \beta_{rm}
\end{bmatrix}
$$

$$
= \begin{bmatrix} \boldsymbol{\beta}_{(1)} & \vdots & \boldsymbol{\beta}_{(2)} & \vdots & \cdots & \vdots & \boldsymbol{\beta}_{(m)} \end{bmatrix},
$$

and

$$
\underset{(n \times m)}{\boldsymbol{\epsilon}}
$$

$$
= \begin{bmatrix}
\epsilon_{11} & \epsilon_{12} & \cdots & \epsilon_{1m} \\
\epsilon_{21} & \epsilon_{22} & \cdots & \epsilon_{2m} \\
\vdots & \vdots & & \vdots \\
\epsilon_{n1} & \epsilon_{n2} & \cdots & \epsilon_{nm}
\end{bmatrix}
$$

$$
= \begin{bmatrix} \boldsymbol{\epsilon}_{(1)} & \vdots & \boldsymbol{\epsilon}_{(2)} & \vdots & \cdots & \vdots & \boldsymbol{\epsilon}_{(m)} \end{bmatrix}.
$$

Simply stated, the ith response $\mathbf{Y}_{(i)}$ follows the linear regression model

$$
\mathbf{Y}_{(i)} = \mathbf{Z}\boldsymbol{\beta}_{(i)} + \boldsymbol{\epsilon}_{(i)}, \qquad i = 1, 2, \ldots, m,
$$

with $\mathrm{Cov}(\boldsymbol{\epsilon}_{(i)}) = \sigma_{ii}\mathbf{I}$. However, the errors for *different* responses on the *same* trial can be correlated.

Given the outcomes $\mathbf{Y}$ and the values of the predictor variables $\mathbf{Z}$, we determine the least-squares estimates $\hat{\boldsymbol{\beta}}_{(i)}$ exclusively from the observations, $\mathbf{Y}_{(i)}$, on the ith response. Since

$$
\hat{\boldsymbol{\beta}}_{(i)} = (\mathbf{Z}'\mathbf{Z})^{-1}\mathbf{Z}'\mathbf{Y}_{(i)},
$$

$$
\hat{\boldsymbol{\beta}} = \begin{bmatrix} \hat{\boldsymbol{\beta}}_{(1)} & \vdots & \hat{\boldsymbol{\beta}}_{(2)} & \vdots & \cdots & \vdots & \hat{\boldsymbol{\beta}}_{(m)} \end{bmatrix}
$$

$$
= (\mathbf{Z}'\mathbf{Z})^{-1}\mathbf{Z}\mathbf{Y}.
$$

Using the least-squares estimates $\hat{\boldsymbol{\beta}}$, we can form the matrices of

Predicted values $\quad \hat{\mathbf{Y}} = \mathbf{Z}\hat{\boldsymbol{\beta}} = \mathbf{Z}(\mathbf{Z}'\mathbf{Z})^{-1}\mathbf{Z}'\mathbf{Y}$,

Residuals $\quad \hat{\boldsymbol{\epsilon}} = \mathbf{Y} - \hat{\mathbf{Y}} = \mathbf{I} - \mathbf{Z}(\mathbf{Z}'\mathbf{Z})^{-1}\mathbf{Z}'\mathbf{Y}$.

Example 2. Companies considering the purchase of a computer must first assess their future needs in order to determine the proper equipment. Data from several similar company sites can be used to develop a forecast equation of computer-hardware requirements for, say, inventory management. The independent variables might include z_1 = customer orders and z_2 = add-delete items. The multivariate responses might include Y_1 = central processing unit (CPU) time and Y_2 = disc input/output capacity. (See Johnson and Wichern [10, Chap. 7].)

Anderson [2, Sec. 8.7] and Srivastava and Khatri [18, Sec. 6.3] derive test statistics and

Table 3   Multivariate Linear Regression Model

$$
\underset{(n \times m)}{\mathbf{Y}} = \underset{(n \times (r+1))}{\mathbf{Z}} \underset{((r+1) \times m)}{\boldsymbol{\beta}} + \underset{(n \times m)}{\boldsymbol{\epsilon}}
$$

with

$$
E(\boldsymbol{\epsilon}_{(i)}) = \mathbf{0};
$$
$$
\mathrm{Cov}(\boldsymbol{\epsilon}_{(i)}, \boldsymbol{\epsilon}_{(k)}) = \sigma_{ik}\mathbf{I}, \qquad i, k = 1, 2, \ldots, m.
$$

The m observations on the jth trial have covariance matrix $\boldsymbol{\Sigma} = \{\sigma_{ik}\}$, but observations from different trials are uncorrelated. Here $\boldsymbol{\beta}$ and $\boldsymbol{\Sigma}$ are matrices of unknown parameters and the design matrix $\mathbf{Z}$ has jth row $[z_{j0}, \ldots, z_{jr}]$.

discuss the distribution theory for multivariate regression.

By allowing $\mathbf{Z}$ to have less than full rank, all of fixed effects MANOVA can be incorporated into the multivariate multiple regression framework. This unifying concept is also most valuable in connecting ANOVA with the classical multiple linear regression model.

ANALYSIS OF COVARIANCE STRUCTURE

Principal Components

A principal component analysis is concerned with explaining the variance-covariance structure through a few *linear* combinations of the original variables. Its general objectives are (1) data reduction and (2) interpretation.

Although p components are required to reproduce the total system variability, often much of this variability can be accounted for by a small number k of the principal components. If so, there is (almost) as much information in the k components as there is in the original p variables. The k principal components can then replace the initial p variables, and the original data set is reduced to one consisting of n measurements on k principal components (*see* COMPONENT ANALYSIS).

Analyses of principal components are more of a means to an end than an end in themselves because they frequently serve as intermediate steps in much larger investigations. For example, principal components may be inputs to a multiple linear regression*.

Algebraically, principal components are particular linear combinations of the p random variables $X_1, X_2, \ldots, X_p$. Geometrically, these linear combinations represent the selection of a new coordinate system obtained by rotating the original system with $X_1, X_2, \ldots, X_p$ as the coordinate axes. The new axes represent the directions with maximum variability and provide a simpler and more parsimonious description of the covariance structure.

Principal components depend solely on the covariance matrix $\mathbf{\Sigma}$ (or the correlation matrix $\boldsymbol{\rho}$) of $X_1, X_2, \ldots, X_p$. Their development does not require a multivariate normal assumption.

The first principal component is the linear combination with maximum variance. That is, it maximizes $\mathrm{Var}(Y_1) = \ell_1' \mathbf{\Sigma} \ell_1$. It is clear that $\mathrm{Var}(Y_1) = \ell_1' \mathbf{\Sigma} \ell_1$ can be increased by multiplying any ℓ_1 by some constant. To eliminate this indeterminacy, it is convenient to restrict attention to coefficient vectors of unit length. Therefore, we define

first principal component = linear combination $\ell_1' \mathbf{X}$ that maximizes

$$\mathrm{Var}(\ell_1' \mathbf{X}) \quad \text{subject to} \quad \ell_1' \ell_1 = 1.$$

second principal component = linear combination $\ell_2' \mathbf{X}$ that maximizes

$$\mathrm{Var}(\ell_2' \mathbf{X}) \quad \text{subject to} \quad \ell_2' \ell_2 = 1$$

and

$$\mathrm{Cov}(\ell_1' \mathbf{X}, \ell_2' \mathbf{X}) = 0.$$

At the ith step

ith principal component = linear combination $\ell_i' \mathbf{X}$ that maximizes

$$\mathrm{Var}(\ell_i' \mathbf{X}) \quad \text{subject to} \quad \ell_i' \ell_i = 1$$

and

$$\mathrm{Cov}(\ell_i' \mathbf{X}, \ell_k' \mathbf{X}) = 0 \quad \text{for} \quad k < i.$$

Let $\mathbf{\Sigma}$ be the covariance matrix associated with the random vector $\mathbf{X}' = [X_1, X_2, \ldots, X_p]$. Let $\mathbf{\Sigma}$ have the eigenvalue-eigenvector pairs $(\lambda_1, \mathbf{e}_1), (\lambda_2, \mathbf{e}_2), \ldots, (\lambda_p, \mathbf{e}_p)$, where $\lambda_1 \geqslant \lambda_2 \geqslant \cdots \geqslant \lambda_p \geqslant 0$. The *ith principal component* is given by

$$Y_i = \mathbf{e}_i' \mathbf{X} = e_{1i} X_i + e_{2i} X_2 + \cdots + e_{pi} X_p,$$
$$i = 1, 2, \ldots, p.$$

With these choices

$$\mathrm{Var}(Y_i) = \mathbf{e}_i' \mathbf{\Sigma} \mathbf{e}_i = \lambda_i, \quad i = 1, 2, \ldots, p;$$

$$\mathrm{Cov}(Y_i, Y_k) = \mathbf{e}_i' \mathbf{\Sigma} \mathbf{e}_k = 0, \quad i \neq k.$$

Thus the principal components are uncorrelated and have variances equal to the eigenvalues of $\mathbf{\Sigma}$. If some λ_i are equal, the choices of the corresponding coefficient vectors e_i, and hence Y_i are not unique.

How do we summarize the sample variation in n measurements on p variables with a few judiciously chosen linear combinations?

Assume the data $x_1, x_2, \ldots, x_n$ represent n independent drawings from some p-dimensional population with mean vector μ and covariance matrix Σ. These data yield the sample mean vector $\bar{x}$, the sample covariance matrix S, and the sample correlation matrix R. These quantities are substituted for the corresponding population quantities above to get *sample principal components*.

Example 3. The weekly rates of return for five stocks (Allied Chemical, DuPont, Union Carbide, Exxon, and Texaco) listed on the New York Stock Exchange were determined for the period January 1975–December 1976. The weekly rates of return are defined as (current Friday closing price − previous Friday closing price)/(previous Friday closing price) adjusted for stock splits and dividends. The observations in 100 successive weeks appear to be distributed independently, but the rates of return *across* stocks are correlated, since, as one expects, stocks tend to move together in response to general economic conditions.

Let $x_1, x_2, \ldots, x_5$ denote observed weekly rates of return for Allied Chemical, DuPont, Union Carbide, Exxon, and Texaco, respectively. Then

$$\bar{x}' = [0.0054, 0.0048, 0.0057, 0.0063, 0.0037],$$

$$R = \begin{bmatrix} 1.000 & 0.577 & 0.509 & 0.387 & 0.462 \\ 0.577 & 1.000 & 0.599 & 0.389 & 0.322 \\ 0.509 & 0.599 & 1.000 & 0.436 & 0.426 \\ 0.387 & 0.389 & 0.436 & 1.000 & 0.523 \\ 0.462 & 0.322 & 0.426 & 0.523 & 1.000 \end{bmatrix}.$$

We note that R is the covariance matrix of the standardized observations

$$z_1 = \frac{x_1 - \bar{x}_1}{\sqrt{s_{11}}}, \qquad z_2 = \frac{x_2 - \bar{x}_2}{\sqrt{s_{22}}}, \ldots,$$

$$z_5 = \frac{x_5 - \bar{x}_5}{\sqrt{s_{55}}}.$$

The eigenvalues and corresponding normalized eigenvectors of R were determined by a computer and are

$$\hat{\lambda}_1 = 2.857,$$
$$\hat{e}_1' = [0.464, 0.457, 0.470, 0.421, 0.421],$$
$$\hat{\lambda}_2 = 0.809,$$
$$\hat{e}_2' = [0.240, 0.509, 0.260, -0.526, -0.582],$$
$$\hat{\lambda}_3 = 0.540,$$
$$\hat{e}_3' = [-0.612, 0.178, 0.335, 0.541, -0.435],$$
$$\hat{\lambda}_4 = 0.452,$$
$$\hat{e}_4' = [0.387, 0.206, -0.662, 0.472, -0.382],$$
$$\hat{\lambda}_5 = 0.343,$$
$$\hat{e}_5' = [-0.451, 0.676, -0.400, -0.176, 0.385].$$

Using the standardized variables, we obtain the first two sample principal components

$$\hat{y}_1 = \hat{e}_1' z = 0.464 z_1 + 0.457 z_2 + 0.470 z_3 + 0.421 z_4 + 0.421 z_5,$$

$$\hat{y}_2 = \hat{e}_2' z = 0.240 z_1 + 0.509 z_2 + 0.260 z_3 - 0.526 z_4 - 0.582 z_5.$$

These components, which account for

$$\left(\frac{\hat{\lambda}_1 + \hat{\lambda}_2}{p} \right) 100\% = \left(\frac{2.857 + 0.809}{5} \right) 100\%$$
$$= 73\%$$

of the total (standardized) sample variance, have interesting interpretations. The first component is a (roughly) equally weighted sum or "index," of the five stocks. This component might be called a *general stock-market component* or simply a *market component*. (In fact, these five stocks are included in the Dow Jones Industrial Average.)

The second component represents a contrast between the chemical stocks (Allied Chemical, DuPont, and Union Carbide) and the oil stocks (Exxon and Texaco). It might be called an *industry component*. Thus we see that most of the variation in these stock returns is due to market activity and uncorrelated industry activity.

The remaining components are not easy to interpret and, collectively, represent variation that is probably specific to each stock.

In any event, they do not explain much of the total sample variance.

FACTOR ANALYSIS. The essential purpose of factor analysis* is to describe, if possible, the covariance relationships among many variables in terms of a few underlying, but unobservable, random quantities called *factors*. Basically, the factor model is motivated by the following argument. Suppose variables can be grouped by their correlations. That is, all variables within a particular group are highly correlated among themselves, but have relatively small correlations with variables in a different group. It is conceivable that each group of variables represents a single underlying construct or factor that is responsible for the observed correlations. For example, correlations from the group of test scores in classics, French, English, mathematics, and music collected by Spearman suggested an underlying "intelligence" factor. A second group of variables, representing physical-fitness scores, if available might correspond to another factor. It is this type of structure that factor analysis seeks to confirm.

Factor analysis can be considered an extension of principal component analysis. Both can be viewed as attempts to approximate the covariance matrix Σ. The approximation based on the factor analysis model is more elaborate; the primary question is whether the data are consistent with a prescribed structure.

The observable random vector $\mathbf{X}$ with p components has mean μ and covariance matrix Σ. The factor model postulates that $\mathbf{X}$ is linearly dependent on a few unobservable random variables $F_1, F_2, \ldots, F_m$, called *common factors* and p additional sources of variation $\epsilon_1, \epsilon_2, \ldots, \epsilon_p$, called *errors* or, sometimes, *specific factors*. In particular the factor analysis model is

$$X_1 - \mu_1 = l_{11}F_1 + l_{12}F_2 + \cdots + l_{1m}F_m + \epsilon_1$$
$$X_2 - \mu_2 = l_{21}F_1 + l_{22}F_2 + \cdots + l_{2m}F_m + \epsilon_2$$
$$\vdots \qquad\qquad\qquad\qquad \vdots$$
$$X_p - \mu_p = l_{p1}F_1 + l_{p2}F_2 + \cdots + l_{pm}F_m + \epsilon_p$$

or, in matrix notation,

$$\mathop{\mathbf{X} - \mu}_{(p \times 1)} = \mathop{\mathbf{L}}_{(p \times m)} \mathop{\mathbf{F}}_{(m \times 1)} + \mathop{\epsilon}_{(p \times 1)}.$$

The coefficient l_{ij} is the *loading* of the ith variable on the jth factor, so the matrix $\mathbf{L}$ is the *matrix of factor loadings*. Note that the ith specific factor ϵ_i is associated only with the ith response X_i. The p deviations $X_1 - \mu_1, X_2 - \mu_2, \ldots, X_p - \mu_p$ are expressed in terms of $p + m$ variables $F_1, F_2, \ldots, F_m, \epsilon_1, \epsilon_2, \ldots, \epsilon_p$ which are unobservable.

With so many unobservable quantities, a direct verification of the factor model from observations on $X_1, X_2, \ldots, X_p$ is hopeless. However, with some additional assumptions about the random vectors $\mathbf{F}$ and ϵ, the preceding model implies certain covariance relationships that can be checked. It follows immediately from the factor model that

1. $\text{Cov}(\mathbf{X}) = \mathbf{LL'} + \mathbf{\Psi}$
    or
    $$\text{Var}(X_i) = l_{i1}^2 + \cdots + l_{im}^2 + \psi_i$$
    $$\text{Cov}(X_i, X_k) = l_{i1}l_{k1} + \cdots + l_{im}l_{km}$$

Table 4   Orthogonal Factor Model with *m* Common Factors

$$\mathop{\mathbf{X}}_{(p \times 1)} = \mathop{\mu}_{(p \times 1)} + \mathop{\mathbf{L}}_{(p \times m)} \mathop{\mathbf{F}}_{(m \times 1)} + \mathop{\epsilon}_{(p \times 1)}$$

$\mu_i = $ *mean* of variable i
$\epsilon_i = i$th *specific factor*
$F_j = j$th *common factor*
$l_{ij} = $ *loading* of the ith variable on the jth factor

The unobservable random vectors $\mathbf{F}$ and ϵ satisfy:
　　$\mathbf{F}$ and ϵ are independent.
　　$E(\mathbf{F}) = \mathbf{0}, \qquad \text{Cov}(\mathbf{F}) = \mathbf{I}.$
　　$E(\epsilon) = \mathbf{0}, \qquad \text{Cov}(\epsilon) = \mathbf{\Psi}, \qquad$ where $\mathbf{\Psi}$ is a diagonal matrix.

2. $\mathrm{Cov}(\mathbf{X}, \mathbf{F}) = \mathbf{L}$
or
$$\mathrm{Cov}(X_i, F_j) = l_{ij}.$$

That portion of the variance of the ith variable contributed by the m common factors is called the ith *communality*. That portion of $\mathrm{Var}(X_i) = \sigma_{ii}$ due to the specific factor is often called the *uniqueness* or *specific variance*. Denoting the ith communality by h_i^2,

$$\underbrace{\sigma_{ii}}_{\mathrm{Var}(X_i)} = \underbrace{l_{i1}^2 + l_{i2}^2 + \cdots + l_{im}^2}_{\text{communality}} + \underbrace{\psi_i}_{\substack{\text{specific} \\ \text{variance}}}$$

or

$$h_i^2 = l_{il}^2 + l_{i2}^2 + \cdots + l_{im}^2,$$
$$\sigma_{ii} = h_i^2 + \psi_i, \qquad i = 1, 2, \ldots, p.$$

The ith communality is the sum of squares of the loadings of the ith variable on the m common factors.

Given observations $\mathbf{x}_1, \mathbf{x}_2, \ldots, \mathbf{x}_n$ on p generally correlated variables, factor analysis seeks to answer the question, Does the factor model (see Table 4), with a small numer of factors, adequately represent the data? In essence, we tackle this statistical model-building problem by trying to verify covariance relationships 1 and 2.

The sample covariance matrix $\mathbf{S}$ is an estimator of the unknown population covariance matrix $\boldsymbol{\Sigma}$. If the off-diagonal elements of $\mathbf{S}$ are small or those of the sample correlation matrix $\mathbf{R}$ essentially zero, the variables are not related and a factor analysis will not prove useful. In these circumstances, the *specific* factors play the dominant role, whereas the major aim of the factor analysis is to determine a few important *common* factors.

If $\boldsymbol{\Sigma}$ appears to deviate significantly from a diagonal matrix, then a factor model can be entertained and the initial problem is one of estimating the factor loadings l_{ij} and specific variances ψ_i. (Methods of estimation are discussed in Lawley and Maxwell [13, Chaps 4 & 7].)

All factor loadings obtained from the initial loading by an orthogonal transformation have the same ability to reproduce the co-variance (or correlation) matrix. From matrix algebra, we know that an orthogonal transformation corresponds to a rigid rotation (or reflection) of the coordinate axes. For this reason an orthogonal transformation of the factor loadings and the implied orthogonal transformation of the factors is called *factor rotation*.

If $\hat{\mathbf{L}}$ is the $p \times m$ matrix of estimated factor loadings obtained by any method, then

$$\hat{\mathbf{L}}^* = \hat{\mathbf{L}}\mathbf{T}, \qquad \text{where} \quad \mathbf{T}\mathbf{T}' = \mathbf{T}'\mathbf{T} = \mathbf{I}$$

is a $p \times m$ matrix of "rotated" loadings. Moreover, the estimated covariance (or correlation) matrix remains unchanged, since

$$\hat{\mathbf{L}}\hat{\mathbf{L}}' + \hat{\boldsymbol{\Psi}} = \hat{\mathbf{L}}\mathbf{T}\mathbf{T}'\hat{\mathbf{L}} + \hat{\boldsymbol{\Psi}} = \hat{\mathbf{L}}^*\hat{\mathbf{L}}^{*\prime} + \hat{\boldsymbol{\Psi}}.$$

The residual matrix $\mathbf{S} - \hat{\mathbf{L}}\hat{\mathbf{L}}' - \hat{\boldsymbol{\Psi}} = \mathbf{S} - \hat{\mathbf{L}}^*\hat{\mathbf{L}}^{*\prime} - \hat{\boldsymbol{\Psi}}$ also remains unchanged after rotation. Moreover, the specific variances $\hat{\psi}_i$ and hence the communalities $\hat{h}_i^2$ are unaltered. Thus from a mathematical veiwpoint it is immaterial whether $\hat{\mathbf{L}}$ or $\hat{\mathbf{L}}^*$ is obtained.

Since the original loadings may not be readily interpretable, the usual practice is to rotate them until a "simple structure" is achieved. The rationale is very much akin to sharpening the focus of a microscope in order to see the detail more clearly.

Ideally, we should like to see a pattern of loadings such that each variable loads highly on a single factor and has small-to-moderate loadings on the remaining factors. It is not always possible to get this simple structure.

Kaiser [11] has suggested an analytical measure of simple structure known as the *varimax* (or normal varimax) *criterion*. Define $\tilde{l}_{ij}^* = \hat{l}_{ij}^* / \hat{h}_i$ to be the final rotated coefficients scaled by the square root of the communalities. The (normal) varimax* procedure selects the orthogonal transformation T that makes

$$V = \frac{1}{p} \sum_{j=1}^{m} \left[\sum_{i=1}^{p} \tilde{l}_{ij}^{*4} - \left(\sum_{i=1}^{p} \tilde{l}_{ij}^{*2} \right)^2 / p \right]$$

as large as possible.

Effectively maximizing V corresponds to "spreading out" the squares of the loadings on each factor as much as possible. Therefore, we hope to find groups of large and negligible coefficients in any *column* of the rotated loadings matrix **L***.

In factor analysis, interest is usually centered on the parameters in the factor model. However, the estimated values of the common factors, called *factor scores*, may also be required. Often these quantities are used for diagnostic purposes as well as inputs to a subsequent analysis. (See FACTOR ANALYSIS; Johnson and Wichern [10, Sec. 9.5]; Morrison [15, Sec. 9.10]; and Lawley and Maxwell [13, Chap. 9] for further discussion of factor scores.)

Example 4. Beginning with correlations between the scores of the Olympic decathlon events, a factor analysis can be employed to see if the 10 events can be explained in terms of two, three, or four underlying "physical" factors. One study of this kind (see Linden [14]) found that the four factors "explosive arm strength," "explosive leg strength," "running speed," and "running endurance" represented several years of decathlon data quite well.

CANONICAL CORRELATIONS AND VARIABLES. Canonical correlation analysis seeks to identify and quantify the associations between two sets of variables (see CANONICAL ANALYSIS and GENERALIZED CANONICAL CORRELATIONS). H. Hotelling [9], who initially developed the technique, provided the example of relating arithmetic speed and arithmetic power to reading speed and reading power. Other examples include relating governmental policy variables to economic goal variables and relating college "performance" variables with precollege "achievement" variables.

Canonical correlation analysis focuses on the correlation between a linear combination of the variables in one set and a linear combination of the variables in another set. When the association between the two sets is expected to be unidirectional—from one set to the other—we label one set the independent or "predictor" variables and the other set the dependent or "criterion" variables.

The idea of canonical correlation analysis is to first determine the pair of linear combinations having the largest correlation. Next, one determines the pair of predictor set–criterion set linear combinations having the largest correlation among all pairs uncorrelated with the initially selected pair. The process continues by selecting, at each stage, the pair of predictor set–criterion set linear combinations having largest correlation among all pairs that are uncorrelated with the preceding choices. The pairs of linear combinations are the *canonical variables* and their correlations are *canonical correlations*. The following discussion gives the necessary details for obtaining the canonical variables and their correlations. In practice, sample covariance matrices are substituted for the corresponding population quantities yielding sample canonical variables and sample canonical correlations.

Suppose $p \leqslant q$ and let the random vectors

$$\underset{(p \times 1)}{\mathbf{X}_1} \quad \text{and} \quad \underset{(q \times 1)}{\mathbf{X}_2}$$

have

$$\underset{(p \times p)}{\text{Cov}(\mathbf{X}_1) = \Sigma_{11}} , \quad \underset{(q \times q)}{\text{Cov}(\mathbf{X}_2) = \Sigma_{22}}$$

$$\text{and} \quad \underset{(p \times q)}{\text{Cov}(\mathbf{X}_1 \mathbf{X}_2) = \Sigma_{12}} .$$

For coefficient vectors

$$\underset{(p \times 1)}{\mathbf{a}} \quad \text{and} \quad \underset{(q \times 1)}{\mathbf{b}} ,$$

form the linear combinations $U = \mathbf{a}'\mathbf{X}_1$ and $V = \mathbf{b}'\mathbf{X}_2$. Then

$$\underset{\mathbf{a},\mathbf{b}}{\text{Max}}\, \text{Corr}(U, V) = \rho_1^* ,$$

attained by the linear combination (first canonical variate pair)

$$U_1 = \mathbf{a}_1'\mathbf{X}_1 \quad \text{and} \quad V_1 = \mathbf{b}_1'\mathbf{X}_2$$

The kth pair of canonical variates, $k = 2, 3, \ldots, p$,

$$U_k = \mathbf{a}_k'\mathbf{X}_1 \quad \text{and} \quad V_k = \mathbf{b}_k'\mathbf{X}_2$$

maximize

$$\text{Corr}(U_k, V_k) = \rho_k^*$$

among those linear combinations uncorrelated with the preceding $1, 2, \ldots, k-1$ canonical variables.

The canonical variates have the properties:

$$\mathrm{Var}(U_k) = \mathrm{Var}(V_k) = 1,$$

$$\mathrm{Cov}(U_k, U_l) = \mathrm{Corr}(U_k, U_l) = 0, \qquad k \neq l,$$

$$\mathrm{Cov}(V_k, V_l) = \mathrm{Corr}(V_k, V_l) = 0, \qquad k \neq l,$$

$$\mathrm{Cov}(U_k, V_l) = \mathrm{Corr}(U_k, V_l) = 0, \qquad k \neq l,$$

for $k, l = 1, 2, \ldots, p$.

In general canonical variables are artificial; they have no physical meaning. If the original variables $\mathbf{X}_1$ and $\mathbf{X}_2$ are used, the canonical coefficients $\mathbf{a}$ and $\mathbf{b}$ have units proportional to those of the $\mathbf{X}_1$ and $\mathbf{X}_2$ sets. If the original variables are standardized to have zero means and unit variances, the canonical coefficients have no units of measurement, and they must be interpreted in terms of the standardized variables.

Methods for determining the coefficient vectors $\mathbf{a}$ and $\mathbf{b}$ and examples of canonical analyses are given in CANONICAL ANALYSIS.

Classification and Grouping Techniques

DISCRIMINANT ANALYSIS* AND CLASSIFICATION*. Discriminant analysis and classification are multivariate techniques concerned with *separating* distinct sets of objects (or observations) and with *allocating* new objects (observations) to previously defined groups. Discriminant analysis is rather exploratory in nature. As a separating procedure, it is often employed on a one-time basis in order to investigate observed differences when causal relationships are not well understood. Classification procedures are less exploratory in the sense that they lead to well-defined rules, which can be used for assigning new objects. Classification ordinarily requires more problem structure than discrimination.

Thus, the immediate goals of discrimination and classification, respectively, are

1. To describe either graphically (in three or fewer dimensions) or algebraically, the differential features of objects (observations) from several known collections (populations). We try to find "discriminants" whose numerical values are such that the collections are separated as much as possible.

2. To sort objects (observations) into two or more labeled classes. The emphasis is on deriving a rule that can be used to assign a *new* object to the labeled classes optimally.

To fix ideas, we will list situations where one may be interested in (1) separating two classes of objects, or (2) assigning a new object to one of the two classes (or both). It is convenient to label the classes π_1 and π_2. The objects are ordinarily separated or classified on the basis of measurements on, for instance, p associated random variables $\mathbf{X}' = [X_1, X_2, \ldots, X_p]$. The observed values of $\mathbf{X}$ differ to some extent from one class to the other. We can think of the totality of values from the first class as being the population of $\mathbf{x}$ values for π_1 and those from the second class as the population of $\mathbf{x}$ values for π_2. These two populations can then be described by probability density functions $f_1(\mathbf{x})$ and $f_2(\mathbf{x})$, and, consequently, we can talk of assigning observations to populations or objects to classes interchangeably.

Populations π_1 and π_2	Measured Variables $\mathbf{X}$
Solvent and distressed property-liability insurance companies.	Total assets, cost of stocks and bonds, market value of stocks and bonds, loss expenses, surplus, amount of premiums written.
Federalist papers written by James Madison and those written by Alexander Hamilton.	Frequencies of different words and length of sentences.
Purchasers of a new product and laggards (those "slow" to purchase).	Education, income, family size, amount of previous brand switching.

You may wonder at this point how it is we *know* some observations belong to a particular population but we are unsure about others. (This, of course, is what makes classification a problem!) There are several conditions that can give rise to this apparent anomaly.

Incomplete knowledge of future performance.

"Perfect" information requires destroying object.

Unavailable or expensive information.

It should be clear that classification rules cannot usually provide an error-free method of assignment. This is because there may not be a clear distinction between the measured characteristics of the populations; that is, the groups may overlap. It is then possible, for example, to incorrectly classify a π_2 object as belonging to π_1 or a π_1 object as belonging to π_2.

For a discussion of discriminant analysis and subsequent classification procedures, *see* DISCRIMINANT ANALYSIS. (See also Anderson [2, Chap 6] and Johnson and Wichern [10, Chap. 10].) Since the literature on this subject is large, we simply display below the "best" allocation rule for two multivariate normal populations with a common covariance matrix Σ. In practice, sample quantities replace the corresponding population quantities.

Let μ_1 and μ_2 be the two population mean vectors, $c(1|2)$ the cost of incorrectly assigning a population 2 observation to population 1, $c(2|1)$ the cost of incorrectly assigning a population 1 observation to population 2, p_1 the "prior" probability of population 1 and p_2 the "prior" probability of population 2, then we

Allocate $\mathbf{x}$ to π_1 if

$$(\mu_1 - \mu_2)'\Sigma^{-1}\mathbf{x} - \tfrac{1}{2}(\mu_1 - \mu_2)'\Sigma^{-1}(\mu_1 + \mu_2)$$

$$\geqslant \ln\left[\frac{c(1|2)}{c(2|1)}\left(\frac{p_2}{p_1}\right)\right].$$

Allocate $\mathbf{x}$ to π_2 otherwise.

The first term, $y = (\mu_1 - \mu_2)'\Sigma^{-1}\mathbf{x}$, above is *Fisher's linear discriminant function* (Fisher actually developed the sample version $\hat{y} = (\bar{\mathbf{x}}_1 - \bar{\mathbf{x}}_2)'\mathbf{S}_{\text{pooled}}^{-1}\mathbf{x}$, where $\mathbf{S}_{\text{pooled}}$ is the pooled sample covariance matrix.) Assuming a common population covariance matrix, it is the linear function $\ell'\mathbf{x}$ with $\ell \propto (\mu_1 - \mu_2)'\Sigma^{-1}$ that maximizes the separation between the two populations as measured by

$$\frac{(\mu_{1y} - \mu_{2y})^2}{\sigma_y^2} = \frac{(\ell'\mu_1 - \ell'\mu_2)^2}{\ell'\Sigma\ell}.$$

One important way of judging the performance of any classification procedure is to calculate its "error rates," or misclassification probabilities. When the forms of the parent populations are known completely, misclassification probabilities can be calculated with relative ease. Because parent populations are rarely known, one must concentrate on the error rates associated with the sample classification function.

Finally, it should be intuitive that good classification (low error rates) will depend on the separation of the populations. The farther apart the groups, the more likely it is that a *useful* classification rule can be developed.

Fisher also proposed a several-population extension of his discriminant method. The motivation behind Fisher discriminant analysis is the need to obtain a reasonable representation of the populations that involves only a *few* linear combinations of the observations, such as $\ell_1'\mathbf{x}$, $\ell_2'\mathbf{x}$, and $\ell_3'\mathbf{x}$. His approach has several advantages when one is interested in *separating* several populations for visual inspection or graphical descriptive purposes. It allows for:

1. Convenient representations of the g populations that reduce the dimension from a very large number of characteristics to a relatively few linear combinations. Of course, some information—needed for optimal classification—may be lost unless the population means lie complelely in the lower-dimensional space selected.

2. Plotting of the means of the first two or

three linear combinations (discriminants). This helps display the relationships and possible groupings of the populations.

3. Scatterplots of the sample values of the first two discriminants, which can indicate outliers* or other abnormalities in the data. (See Gnanadesikan [7, Chap. 2] and Johnson and Wichern [10, Sec. 10.8] for examples of low-dimensional representations.)

CLUSTERING AND GRAPHICAL PROCEDURES. Rudimentary, exploratory procedures are often quite helpful in understanding the complex nature of multivariate relationships. Searching the data for a structure of "natural" groupings is an important exploratory technique. Groupings can provide an informal means for assessing dimensionality, identifying outliers, and suggesting interesting hypotheses concerning relationships.

Grouping, or clustering, is distinct from the classification methods discussed earlier. Classification pertains to a *known* number of groups, and the operational objective is to assign new observations to one of these groups. Cluster analysis is a more primitive technique in that no assumptions are made concerning the number of groups or the group structure. Grouping is done on the basis of similarities of distances (dissimilarities). The inputs required are similarity measures or data from which similarities can be computed (*see* MEASURES OF SIMILARITY, DISSIMILARITY AND DISTANCE). Good references on clustering include Everitt [5, 6], Anderberg [1], Gnanadesikan [7, Sec. 4.3], and Hartigan [8].

Even without the precise notion of a natural grouping, we are often able to cluster objects in two- or three-dimensional scatter plots by eye. To take advantage of the mind's ability to group similar objects, several graphical procedures have been developed recently for depicting high-dimensional observations in two dimensions. Stars*, Andrews plots*, and Chernoff faces* seem to be useful graphical techniques for representing multivariate data.

References

[1] Anderberg, M. R. (1973). *Cluster Analysis for Applications*. Academic Press, New York.

[2] Anderson, T. W. (1958). *An Introduction to Multivariate Statistical Analysis*. Wiley, New York.

[3] Bishop, Y. M. M., Fienberg, S. E., and Holland, P. W. (1975). *Discrete Multivariate Analysis: Theory and Practice*. MIT Press, Cambridge, MA.

[4] Draper, N. R. and Smith, H. (1981). *Applied Regression Analysis*, 2nd ed. Wiley, New York.

[5] Everitt, B. (1974). *Cluster Analysis*. Heinemann, London.

[6] Everitt, B. (1978). *Graphical Techniques for Multivariate Data*. North-Holland, New York.

[7] Gnanadesikan, R. (1977). *Methods for Statistical Data Analysis of Multivariate Observations*. Wiley, New York.

[8] Hartigan, J. A. (1975). *Clustering Algorithms*. Wiley, New York.

[9] Hotelling, H. (1936). *Biometrika*, **28**, 321–377.

[10] Johnson, R. A. and Wichern, D. W. (1982). *Applied Multivariate Statistical Analysis*. Prentice-Hall, Englewood Cliffs, NJ.

[11] Kaiser, H. F. (1958). *Psychometrika*, **23**, 187–200.

[12] Kshirsagar, A. M. (1972). *Multivariate Analysis*. Marcel Dekker, New York.

[13] Lawley, D. N. and Maxwell, A. E. (1971). *Factor Analysis As a Statistical Method*, 2nd ed. American Elsevier, New York.

[14] Linden, M. (1977). *Res. Quart.*, **48**, 562–568.

[15] Morrison, D. F. (1976). *Multivariate Statistical Methods*, 2nd ed. McGraw-Hill, New York.

[16] Rao, C. R. (1973). *Linear Statistical Inference and Its Applications*, 2nd ed. Wiley, New York.

[17] Seber, G. A. F. (1977). *Linear Regression Analysis*. Wiley, New York.

[18] Srivastava, M. S. and Khatri, C. G. (1979). *An Introduction to Multivariate Statistics*. North-Holland, New York.

Bibliography

Eaton, M. L. (1983). *Multivariate Statistics*. Wiley, New York.

Gnanadesikan, R. and Kettenring, J. R. (1984). In *Statistics: An Appraisal*, H. A. David and H. T. David, eds. Iowa State University Press, Ames, IA.

(CANONICAL ANALYSIS
CLASSIFICATION
COMPONENT ANALYSIS
DISCRIMINANT ANALYSIS
FACTOR ANALYSIS
GENERALIZED CANONICAL
CORRELATIONS

RICHARD A. JOHNSON
DEAN WICHERN

MULTIVARIATE ANALYSIS (BAYESIAN)

Multivariate analysis is that branch of statistics devoted to the study of random variables that are not necessarily independent. Where inference is concerned several (generally correlated) measurements are made on every observed subject. Bayesian multivariate analysis involves the study of such random variables that arise in connection with the use of Bayes' theorem* (see the following subsection). In this brief article, we discuss only the basic concepts of the subject, including the likelihood principle*, multivariate prior and posterior distributions, and the use of computer programs to implement multivariate Bayesian procedures.

BAYES' THEOREM, POSTERIOR DISTRIBUTIONS, AND INFERENCE

Multivariate Bayesian inference is based on Bayes' theorem for correlated random variables. The theorem asserts that the joint density of several correlated, jointly continuous, but unobservable random variables, given observations on one or more observ-

able random variables, is proportional to the product of the likelihood function for the observable random variables and the density function for the unobservable random variables. (If the unobservable random variables are jointly discrete, we use the joint probability mass function instead of the joint density in Bayes' theorem.)

Symbolically, let Θ denote a collection (vector) of k unobservable random variables, $\mathbf{X}$ a collection (vector) of p observable random variables and $f(\cdot)$, $g(\cdot)$, and $h(\cdot)$, densities (probability mass functions) of their arguments. Bayes' theorem asserts that

$$h(\theta \mid \mathbf{x}) = \frac{1}{k} f(\mathbf{x} \mid \theta) \cdot g(\theta),$$

where $\theta, \mathbf{x}$ denote fixed values of $\Theta, \mathbf{X}$, respectively, and k denotes a constant (depending on $\mathbf{x}$, but not on θ), which is given by

$$k = \int f(\mathbf{x} \mid \theta) g(\theta) \, d\theta.$$

The integration is taken over all possible values in k-dimensional space, and the notation $f(\mathbf{x} \mid \theta)$ should be understood to mean the density of the conditional distribution of $\mathbf{X}$ given $\Theta = \theta$. $f(\mathbf{x} \mid \theta)$ is the *likelihood function*; $g(\theta)$ is the *prior density* of θ since it is the density of θ prior to having observed $\mathbf{X}$ (it is a density if the variables in the θ array are continuous, and it is a probability mass function if they are discrete); $h(\theta \mid \mathbf{x})$ is the *posterior density* (probability mass function) of θ since it is the distribution of θ "subsequent" to having observed $\mathbf{X}$.

Bayesian inference* in multivariate distributions is based on the posterior distribution* of the unobservable random variables, say Θ, given the observable data (the unobservable random variable may be a vector or a matrix). A measure of location of the posterior distribution, such as the mean, median, or mode*, generally is used as a Bayesian estimator of Θ. For example, if there tends to be an underlying "quadratic loss" penalty function in an estimation problem, the mean of the posterior distribution is optimal as an estimator since it minimizes the expected loss (penalty).

To obtain the marginal posterior density of Θ given the data, it is often necessary to integrate the joint posterior density over spaces of other unobservable random variables that are jointly distributed with Θ. For example, if the sampling distribution of X given (Θ, Σ) is $N(\Theta, \Sigma)$, the marginal posterior density of Θ is obtainable by integrating the joint posterior density of (Θ, Σ) over all Σ that are positive definite. Bayesian confidence intervals or regions (called *credibility intervals* or *regions*, respectively) are obtainable for any preassigned level of confidence directly from the cumulative distribution function of the posterior distribution. Predictions about a data vector not yet observed are carried out by averaging the likelihood for the future observation vector over the posterior distribution. Hypotheses may be tested by comparing the posterior probabilities of all competing hypotheses and selecting the hypothesis with the largest posterior probability. These notions are identical with those in univariate Bayesian analysis.

Likelihood Principle

The likelihood function is not uniquely defined. It may be taken to be any constant multiple of the ordinary sampling, or frequency, function (probability mass function) of the joint distribution of all of the observable random variables given the unobservable ones. The *likelihood principle** asserts that all relevant information about Θ obtainable from the observable data is found in the likelihood function.

Prior Distributions

None of the variables in a collection of unobservables, Θ, is ever known. The function $g(\boldsymbol{\theta})$ is used to denote the degrees of belief* about $\boldsymbol{\theta}$ held by the analyst (*see* PRIOR DISTRIBUTIONS.)

For example, suppose Θ is bivariate ($k = 2$), so that there are two unobservable, one-dimensional random variables Θ_1 and Θ_2. Suppose further (for simplicity) that Θ_1 and Θ_2 are discrete random variables, and

Table 1

θ_2 \ θ_1	0	1
0	0.2	0.1
1	0.3	0.4

let $g(\theta_1, \theta_2)$ denote the joint probability mass function for $\Theta = (\Theta_1, \Theta_2)$. Suppose Θ_1 and Θ_2 can each assume only two values, 0 and 1, and the analyst believes the probable values to be given by those in Table 1. Thus, for example, the analyst believes that the chances that Θ_1 and Θ_2 are both 1 is 0.4, i.e.,

$$P\{\Theta_1 = 1, \Theta_2 = 1\} = g(1, 1) = 0.4.$$

Multivariate prior distributions are sometimes difficult to generate due to the complexities of thinking in many dimensions simultaneously. While there are already satisfactory methods for generating one-dimensional marginal prior distributions, the distribution of a person's joint beliefs about several random variables simultaneously is more difficult to develop. The higher the dimension of the problem, the more this difficulty is exacerbated. One type of solution proposed is to use regression on observables to develop the multidimensional prior distribution (see Kadane et al. [6]). Another proposal has been to use homogeneous, informed groups of experts (see Press [8]). The process of developing a prior distribution to express the beliefs of the analyst about the likely values of a collection of unobservables is called *subjective probability assessment*. The ability of people to assess correlation coefficients was studied by Gokhale and Press [3].

Vague Priors

In some situations the analyst does not feel at all knowledgeable about the likely values of unknown, unobservable variables. In such cases he will probably resort to use of a "vague" (sometimes called "diffuse") prior distribution. Let Θ denote a collection of k continuous variables each defined on

$(-\infty, +\infty)$; $g(\boldsymbol{\theta})$ is a vague prior density if the variables are mutually independent and if the probability mass of each variable is diffused evenly over all possible values. We write the (improper) prior density for $\boldsymbol{\Theta}$ as

$$g(\boldsymbol{\Theta}) \propto \text{constant},$$

where $\propto$ denotes proportionality.

If an unobservable variable were strictly positive, such as a variance σ^2, we could adopt a vague prior for σ^2 by considering $(\log \sigma^2)$ as a new variable [defined on $(-\infty, \infty)$] and taking a vague prior on the variable $(\log \sigma^2)$, as above. Thus

$$g(\log \sigma^2) \propto \text{constant}.$$

But by a change of variable this implies an (improper) prior for σ^2, that is,

$$g^*(\sigma^2) \propto \sigma^{-2}.$$

The notion of "positive," one-dimensional random variables, extends, in a multivariate context, to "positive definite," when we consider an array (a matrix) of variables. Thus if $\boldsymbol{\Sigma}$ denotes a k-dimensional square and symmetric matrix of variances and covariances, and $\boldsymbol{\Sigma}$ is a positive definite matrix, a vague prior on $\boldsymbol{\Sigma}$ is given by

$$g(\boldsymbol{\Sigma}) \propto |\boldsymbol{\Sigma}|^{-(k+1)/2},$$

where $|\boldsymbol{\Sigma}|$ denotes the determinant of the matrix $\boldsymbol{\Sigma}$. (For an elaboration of these priors, see Press, [8, Sections 3.6 and 3.8].) For invariance arguments relating to these priors see Hartigan [4], Jeffreys [5], and Villegas [11]. The formulation presented here was first given by Geisser and Cornfield [2].

For discussions of controversial issues relating to multivariate vague priors, see Stein [10] and Dawid et al. [1].

Natural Conjugate Priors

It is sometimes convenient for an analyst to describe his prior information about some unobservable, say $\boldsymbol{\Theta}$, by adapting the prior information to some preassigned family of distributions. The family most often used is called the *natural conjugate family* of prior distributions (the term and concept is due to

Raiffa and Schlaifer [9]. It is obtained by interchanging the roles of the observable and unobservable random variables in the likelihood functions and "enriching" the parameters. (*See* CONJUGATE FAMILIES OF DISTRIBUTIONS.)

For example, if $\mathscr{L}(\mathbf{X} \mid \boldsymbol{\theta}) = N(\boldsymbol{\theta}, \mathbf{I}_k)$, where $\mathbf{I}_k$ denotes the k-dimensional identity matrix, and $N(\boldsymbol{\theta}, \mathbf{I}_k)$ denotes the normal distribution with mean vector $\boldsymbol{\theta}$, and covariance matrix $\mathbf{I}_k$, $\mathscr{L}(\boldsymbol{\theta}) = N(\boldsymbol{\phi}, \mathbf{A})$ is a natural conjugate prior distribution for $\boldsymbol{\theta}$. $(\boldsymbol{\phi}, \mathbf{A})$, the parameters that index the natural conjugate prior, are called *hyperparameters*. This result is obtained by writing out the density of $(\mathbf{X} \mid \boldsymbol{\theta})$ and noting that if the same density is viewed as a density of $(\boldsymbol{\Theta} \mid \mathbf{x})$, the density is still that of a normal distribution. So we adopt a normal distribution as a prior for $\boldsymbol{\Theta}$. We then "enrich" the parameters by adopting completely general parameters for this prior, namely, $\boldsymbol{\phi}$ and $\mathbf{A}$ (and this way, the hyperparameters do not depend upon the sample data). Next, we use our prior beliefs about $\boldsymbol{\Theta}$ to assess the hyperparameters $(\boldsymbol{\phi}, \mathbf{A})$.

Exchangeability

A multivariate distribution that does not depend on the order in which the random variables appear is sometimes referred to as exchangeable. The corresponding populations are also said to be exchangeable; (*see* EXCHANGEABILITY). Suppose, for example, that $(\Theta_1, \ldots, \Theta_k, \ldots)$ are one-dimensional random variables any k of which follow the joint distribution $N(a\mathbf{e}, \mathbf{H})$, where $\mathbf{e}$ denotes a k-dimensional vector of ones, a denotes any scalar, and $\mathbf{H}$ denotes a covariance matrix with equal diagonal elements, and equal off-diagonal elements. If the Θ_i's are permuted, the joint distribution does not change, so it is called exchangeable. The original concept was applied to Bernoulli sequences of trials (infinite sequences) and has now been extended.

In some situations in Bayesian multivariate analysis it is useful to adopt an exchangeable prior distribution to express ignorance. For instance, suppose we have observations from three multivariate normal populations with equal covariance matrices, and we wish to carry out Bayesian inference on the mean vectors to compare the closeness of the three populations (multivariate analysis of variance). In many situations like this it would not be unreasonable to take the prior distributions for each of the mean vectors to be the same; i.e., to assume, a priori, that the populations are exchangeable (in the absence of any information to the contrary). Thus, if Θ, Φ, η denote the mean vectors for the three normal populations, we could adopt the joint prior distribution for their mean vectors,

$$f(\Theta, \Phi, \eta) = g(\Theta)\, g(\Phi)\, g(\eta),$$

where the distribution of Θ (or Φ, or η) is $N(\mu, \Sigma)$, and the hyperparameters (μ, Σ) must be assessed.

Computer-Assisted Statistical Inference

It is often the case in Bayesian multivariate analysis that posterior distributions are sufficiently complicated that numerical procedures and computers are required to effect posterior inferences. Fortunately, computer programs have already been written for many of the known multivariate Bayesian inference procedures (see Press [7] for a recent compilation and accompanying descriptions).

References

[1] Dawid, A. P., Stone, M. and J. V. Zidek (1973). *J. R. Statist. Soc. Ser. B*, **35**, 189–233.

[2] Geisser, S. and Cornfield, J. (1963). *J. R. Statist. Soc. Ser. B*, **25**, 368–376.

[3] Gokhale, D. V. and Press, S. James (1982). *J. R. Statist. Soc. Ser. A*, **145**, 237–249.

[4] Hartigan, J. (1964). *Ann. Math. Statist.* **35**, 836–845.

[5] Jeffreys, H. (1961 and 1966). *Theory of Probability*, 3rd ed. Clarendon, Oxford.

[6] Kadane, J. B., Dickey, J. M., Winkler, R. L., Smith, W. S., and Peters, S. C. (1980). *J. Amer. Statist. Ass.*, **75**, 845–854.

[7] Press, S. J. (1980). In *Bayesian Analysis in Econometrics and Statistics*, ed. by A. Zellner, North-Holland, New York, Chap. 27.

[8] ———. (1982). *Applied Multivariate Analysis: Using Bayesian and Frequentist Methods of Inference*, 2nd ed. (revised) Krieger, Melbourne, FL.

[9] Raiffa, H. and Schlaifer, R. (1961). *Applied Statistical Decision Theory*. Harvard University Press, Boston.

[10] Stein, C. (1956). *Proc. 3rd Berkeley Symp. Math. Statist. Prob.*, **1**. Berkeley: University of California Press, pp. 197–206.

[11] Villegas, C (1969). *Ann. Math. Statist.*, **40**, 1098–1099.

Bibliography

Box, George E. P. and George C. Taio (1973). *Bayesian Inference in Statistical Analysis*. Addison-Wesley, Reading, MA.

de Finetti, B. (1937). *Ann. Inst. Henri Poincaré*, **7**, 1–68 (reprinted in English translation (1964). In *Studies in Subjective Probability*, H. E. Kyburg, Jr. and H. E. Smokler, eds. Wiley, New York.

———. (1974) *Theory of Probability*, Vols. 1 and 2. Wiley, New York (especially Section 11.4).

Geisser, Seymour (1965). *Ann. Math. Statist.* **36**, 150–159.

Leamer, E. E. (1978). *Specification Searches*. Wiley, New York.

Lindley, D. V. (1965). *Introduction to Probability and Statistics*, Vols. 1 and 2, Cambridge University Press.

———. (1972). *Bayesian Statistics: A Review*, SIAM, Philadelphia.

——— and Novick, M. R. (1972). *Ann. Statist.*, **9**, 45–58.

Press, S. J. (1983). *Technol. Forecasting Soc. Change*, **28**, 247–259.

———. (1984). In *Bayesian Statistics 2*, J. M. Bernardo, M. H. de Groot, D. V. Lindley, and A. F. M. Smith, eds. North-Holland, Amsterdam.

Villegas, C. (1977). *J. Amer. Statist. Ass.*, **72**, 453–458.

———. (1977). *J. Amer. Statist. Ass.*, **72**, 651–654.

———. (1981). *Ann. Statist.*, **9**, 768–776.

Zellner, A. (1971). *An Introduction to Bayesian Inference in Econometrics*, Wiley, New York.

(BAYESIAN INFERENCE
BELIEF FUNCTIONS
CONJUGATE FAMILIES OF
  DISTRIBUTIONS
DEGREES OF BELIEF
EXCHANGEABILITY
LIKELIHOOD PRINCIPLE
MULTIVARIATE ANALYSIS
POSTERIOR DISTRIBUTIONS

PRIOR DISTRIBUTIONS
SUBJECTIVE PROBABILITIES)

S. JAMES PRESS

MULTIVARIATE ANALYSIS OF VARIANCE (MANOVA)

Consider the test of the following hypothesis: $H_0 : \mu_1 = \cdots = \mu_l$ vs. H_1: not all μ_h's in H_0 equal, in l p-variate normal populations, $N_p(\mu_h, \Sigma)$, $h = 1, \ldots, l, \Sigma$ symmetric positive definite (pd) and unknown. In order to test this hypothesis, one uses *multivariate analysis of variance*, which generalizes the analysis of variance* for the test of equality of means of l univariate normal populations having a common unknown variance. The l populations are assumed to be normal in this test. If this assumption cannot be met, nonparametric methods may be sought (see Sen [20]) for the test of H_0. However, proper choices of test statistics in the normal MANOVA approach that are robust against non-normality (see below, as well as PILLAI'S TRACE) would justify the use of the MANOVA model even when the populations are not strictly normal.

MULTIVARIATE GENERAL LINEAR HYPOTHESIS (MGLH)

The hypothesis H_0 is a special case of the multivariate general linear hypothesis (*see also* GENERAL LINEAR MODEL) which, therefore, is considered first and is described here in the light of the union-intersection* approach of S. N. Roy leading to his largest root test statistic (discussed later; *see also* ROY'S CHARACTERISTIC ROOT STATISTIC) [Roy (1957), Morrison (1976)].

In order to introduce the MGLH, it is natural to start with the univariate model

$$\mathbf{x}_{N \times 1} = \mathbf{A}_{N \times m}\boldsymbol{\xi}_{m \times 1} + \boldsymbol{\epsilon}_{N \times 1},$$

where $\mathbf{x}$ is a vector of uncorrelated random (response) variables, $\mathbf{A}$ is a given (design)

matrix of rank $r \leqslant m \leqslant N$, $\boldsymbol{\xi}$, a vector of unknown parameters, and $\boldsymbol{\epsilon}$, a vector of random (error) variables with $E(\boldsymbol{\epsilon}) = \mathbf{0}$ and $E(\boldsymbol{\epsilon}\boldsymbol{\epsilon}') = \sigma^2\mathbf{I}$. Also, let a linear compound $\mathbf{c}'\boldsymbol{\xi}$, for given vector $\mathbf{c}_{m \times 1}$, be considered estimable if an estimate $\mathbf{b}'\mathbf{x}$ exists such that $E\mathbf{b}'\mathbf{x} = \mathbf{c}'\boldsymbol{\xi}$. Then $\mathbf{c}'\boldsymbol{\xi}$ in the univariate model is estimable only if

$$\mathbf{c}_2' = \mathbf{c}_1'(\mathbf{A}_1'\mathbf{A}_1)^{-1}\mathbf{A}_1'\mathbf{A}_2', \quad \mathbf{c}' = \left(\underset{r}{\mathbf{c}_1'}, \underset{m-r}{\mathbf{c}_2'}\right),$$

$$\mathbf{A} = \left(\underset{r}{\mathbf{A}_1}, \underset{m-r}{\mathbf{A}_2}\right),$$

where $\mathbf{A}_1$ is taken as a basis. Now if $\boldsymbol{\epsilon} \sim N_N(\mathbf{0}, \sigma^2\mathbf{I})$, the F-statistic for testing the hypothesis: $\mathbf{C}_{d \times m}\boldsymbol{\xi}_{m \times 1} = \mathbf{0}$ vs. $\mathbf{C}\boldsymbol{\xi} \neq \mathbf{0}$ where $\mathbf{C}$ is a given matrix of rank $t \leqslant \min(r, d)$ is given by

$$F_{t, N-r} = \frac{(N-r)\mathbf{x}'\mathbf{T}_1\mathbf{x}}{t\mathbf{x}'\mathbf{T}_2\mathbf{x}},$$

$$\mathbf{C} = \underset{d-t}{\overset{t}{\left(\begin{matrix} \mathbf{C}_{11} & \mathbf{C}_{12} \\ \mathbf{C}_{21} & \mathbf{C}_{22} \end{matrix}\right)}}, \underset{r \quad m-r}{}$$

$$\mathbf{T}_1 = \mathbf{A}_1(\mathbf{A}_1'\mathbf{A}_1)^{-1}\mathbf{C}_{11}'\big(\mathbf{C}_{11}(\mathbf{A}_1'\mathbf{A}_1)^{-1}\mathbf{C}_{11}'\big)^{-1}$$

$$\times \mathbf{C}_{11}(\mathbf{A}_1'\mathbf{A}_1)^{-1}\mathbf{A}_1',$$

$$\mathbf{T}_2 = \mathbf{I} - \mathbf{A}_1(\mathbf{A}_1'\mathbf{A}_1)^{-1}\mathbf{A}_1',$$

and in view of the estimability conditions, $\mathbf{C}_{12} = \mathbf{C}_{11}(\mathbf{A}_1'\mathbf{A}_1)^{-1}\mathbf{A}_1'\mathbf{A}_2$. For a test of level α, reject $\mathbf{C}\boldsymbol{\xi} = \mathbf{0}$ if $F_{t, N-r} > F_{t, N-r, 1-\alpha}$; accept otherwise.

The extension to p-variate response variables could be achieved as follows. Let $\mathbf{X}_{p \times N}$ be N p-variate column vectors and consider the model

$$\mathbf{X}_{N \times p}' = \mathbf{A}_{N \times m}\boldsymbol{\xi}_{m \times p} + \boldsymbol{\epsilon}_{N \times p}',$$

where the columns of $\boldsymbol{\epsilon}$ are mutually independently distributed, $\boldsymbol{\epsilon}_{k \ p \times 1} \sim N_p(\mathbf{0}, \Sigma)$, $k = 1, \ldots, N$, where Σ is symmetric pd, $\mathbf{A}$ is a given design matrix, and $\boldsymbol{\xi}$ is a matrix of unknown parameters. In view of these assumptions, the columns of $\mathbf{X}$ are mutually independently distributed, $\mathbf{x}_{k \ p \times 1} \sim N_p(E\mathbf{X}_k, \Sigma)$. Now, the MGLH is given by

$$\mathbf{C}_{d \times m}\boldsymbol{\xi}_{m \times p}\mathbf{M}_{p \times u} = \mathbf{0}_{d \times u}$$

with the alternative $\mathbf{C\xi M} \neq \mathbf{0}$, where the rank of $\mathbf{C}$ as before is t and $u \leqslant p$. In order to test this hypothesis, consider the union-intersection* approach of Roy (1957). (*See also* HOTELLING'S T^2 and HOTELLING'S TRACE.) The hypothesis $\mathbf{C\xi M} = \mathbf{0}$ is true if and only if $\mathbf{C\xi Ma} = \mathbf{0}$ for all non-null $\mathbf{a}_{u \times 1}$. If $\mathbf{a}$ is fixed, the latter hypothesis could be tested using the F-statistic

$$F_{t, N-r}(\mathbf{a}) = \frac{(N-r)\mathbf{a}'\mathbf{M}'\mathbf{X}\mathbf{T}_1\mathbf{X}'\mathbf{Ma}}{t\mathbf{a}'\mathbf{M}'\mathbf{X}\mathbf{T}_2\mathbf{X}'\mathbf{Ma}},$$

replacing $\mathbf{x}$ by $\mathbf{X}'\mathbf{Ma}$ in the F-statistic obtained earlier. For a level α' test, reject the hypothesis if $F_{t, N-r}(\mathbf{a}) > F_{t, N-r, 1-\alpha'}$; otherwise accept. According to the union-intersection principle, hypothesis $\mathbf{C\xi M} = \mathbf{0}$ is rejected at some level α ($> \alpha'$) over

$$\bigcup_{\mathbf{a}} F_{t, N-r}(\mathbf{a}) > F_{t, N-r, 1-\alpha'},$$

for all non-null $\mathbf{a}$.

This leads to the test: Reject the hypothesis $\mathbf{C\xi M} = \mathbf{0}$ if $f_s > f_{s, 1-\alpha}$; accept otherwise, where $0 < f_1 < \cdots < f_s < \infty$ are almost everywhere (ae) positive characteristic roots of $|\mathbf{S}^* - f\mathbf{S}| = 0$ and where

$$\mathbf{S}^* = \mathbf{M}'\mathbf{X}\mathbf{T}_1\mathbf{X}'\mathbf{M}, \qquad \mathbf{S} = \mathbf{M}'\mathbf{X}\mathbf{T}_2\mathbf{X}'\mathbf{M},$$

$$s = \min(u, t).$$

SPECIAL CASES

One-way Fixed-Effects Model

Let

$$N = N_1 + \cdots + N_l, \qquad m = l + 1,$$

$$\mathbf{A} = N(\mathbf{A}_1, \mathbf{A}_2),$$

where

$$\mathbf{A}_{1\, N \times l} = \mathrm{diag}(\mathbf{e}_{N_1}, \mathbf{e}_{N_2}, \ldots, \mathbf{e}_{N_l}),$$

$$\mathbf{A}_2 = \mathbf{e}_N,$$

$$\mathbf{\xi}_{(l+1) \times p} = (\mathbf{\tau}_1, \ldots, \mathbf{\tau}_l, \mathbf{\mu})',$$

where $\mathbf{e}'_{j\, 1 \times j} = (1, \ldots, 1)$ and $\mathbf{\tau}'_{h\, 1 \times p} = (\tau_{1h}, \ldots, \tau_{ph})$, $h = 1, \ldots, l$ and $\mathbf{\mu}'_{1 \times p} = (\mu_1, \ldots, \mu_p)$. Then the one-way fixed effects

model is given by

$$\mathbf{X}'_{N \times p} = \mathbf{A\xi} + \mathbf{\epsilon}'_{N \times p},$$

$$\mathbf{X}_{p \times N} = (\mathbf{X}_1, \ldots, \mathbf{X}_l),$$

$$\mathbf{X}_{h\, p \times N_h} = \begin{bmatrix} x_{11}^{(h)} & \cdots & x_{1 N_h}^{(h)} \\ \cdots\cdots\cdots\cdots\cdots \\ x_{p1}^{(h)} & \cdots & x_{p N_h}^{(h)} \end{bmatrix},$$

($h = 1, \ldots, l$), the hth treatment sample of size N_h distributed independently of other treatment samples; $\mathbf{\epsilon}'$ is defined as earlier with its rows following the sample order. Now, the hypothesis: $\mathbf{\tau}_1 = \cdots = \mathbf{\tau}_l$ can be rewritten as

$$\mathbf{C\xi} = \mathbf{0},$$

$$\mathbf{C}_{(l-1) \times (l+1)} = (\mathbf{C}_1, \mathbf{C}_2),$$

$$\mathbf{C}_{1\,(l-1) \times l} = \begin{bmatrix} 1 & 0 & \cdots & 0 & -1 \\ 0 & 1 & \cdots & 0 & -1 \\ \cdot & \cdot & \cdots & \cdot & \cdot \\ \cdot & \cdot & \cdots & \cdot & \cdot \\ 0 & 0 & \cdots & 1 & -1 \end{bmatrix};$$

$\mathbf{C}_{2\,(l-1) \times 1}$ is a null column vector. (Note that $\mathbf{C}_1$ stands for $\mathbf{C}_{11}$ and $\mathbf{C}_2$ for $\mathbf{C}_{12}$ in the absence of $\mathbf{C}_{21}$ and $\mathbf{C}_{22}$.) Here $\mathbf{M} = \mathbf{I}(p)$ and the alternative hypothesis is $\mathbf{C\xi} \neq \mathbf{0}$. Under these assumptions $\mathbf{S}^*$ is the between-SP (sum of products) matrix given by $\mathbf{S}^*_{p \times p} = (s^*_{ij})$, where

$$s^*_{ij} = \sum_{h=1}^{l} N_h \left(\bar{x}_i^{(h)} - \bar{x}_i \right)\left(\bar{x}_j^{(h)} - \bar{x}_j \right),$$

where $\bar{x}_i^{(h)}$ is the mean of the ith response variable in the hth treatment sample and $\bar{x}_i$ the mean of the ith response variable from all N observations. Further, $\mathbf{S}$ is the within-SP matrix given by $\mathbf{S}_{p \times p} = (s_{ij})$, where

$$s_{ij} = \sum_{h=1}^{l} \sum_{k=1}^{N_h} \left(x_{ik}^{(h)} - \bar{x}_i^{(h)} \right)\left(x_{jk}^{(h)} - \bar{x}_j^{(h)} \right).$$

Here $\mathbf{S}^*$ has $l - 1$ degrees of freedom and $\mathbf{S}$ has $N - l$.

It may be appropriate to give an analysis of variance table for the one-way fixed-effect model as in Table 1. (*See* ONE-WAY CLASSIFICATION; FIXED-, RANDOM- AND MIXED-EFFECTS MODELS.)

For $H_0 : \mathbf{C\xi M} = \mathbf{0}$ vs. $H_1 : \mathbf{C\xi M} \neq \mathbf{0}, \mathbf{S}^*$ in the preceding table will be replaced by

Table 1   MANOVA

Source	df	SS/SP Matrix
Hypothesis	$l - 1$	$\mathbf{S^*} = \mathbf{X A_1 (A_1' A_1)}^{-1} \mathbf{C_1' (C_1 (A_1' A_1)}^{-1} \mathbf{C_1')}^{-1} \mathbf{C_1 (A_1' A_1)}^{-1} \mathbf{A_1' X'}$
Error	$N - l$	$\mathbf{S} = \mathbf{X (I - A_1 (A_1' A_1)}^{-1} \mathbf{A_1') X'}$
Total	$N - 1$	$\mathbf{S^* + S} = \mathbf{S_0}$ (see discussion)

$\mathbf{M' S^* M}$ and $\mathbf{S}$ by $\mathbf{M' SM}$. Note that $\mathbf{S^*}$ is the between-SP matrix and $\mathbf{S}$ the within-SP matrix defined earlier. In the special case of $H_0 : \boldsymbol{\xi M} = \mathbf{0}$ vs. $H_1 : \boldsymbol{\xi M} \neq \mathbf{0}$, $\mathbf{S^* + S} = \mathbf{XX'}$.

For the various standard test statistics useful for the test of the one-way fixed-effects model hypothesis, see the next section.

Multivariate Multisample Problem

Consider the model

$$\mathbf{X'}_{N \times p} = \mathbf{A}_{1\ N \times l} \boldsymbol{\mu'}_{l \times p} + \boldsymbol{\epsilon'}_{N \times p},$$

$$\boldsymbol{\mu}_{p \times l} = (\boldsymbol{\mu}_1, \ldots, \boldsymbol{\mu}_l), \qquad \boldsymbol{\mu}_h = \boldsymbol{\tau}_h + \boldsymbol{\mu};$$

$\mathbf{X}_{h\ p \times N_h}$ is a random sample from $N_p(\boldsymbol{\mu}_h, \boldsymbol{\Sigma})$, $h = 1, \ldots, l$, independent of other samples, and $\mathbf{A}_1$ and $\boldsymbol{\epsilon'}$ are defined as before (see GENERAL LINEAR MODEL). Now in the test H_0 vs. H_1 discussed in the Introduction, H_0 can be rewritten as $\mathbf{C}_{1\ (l-1) \times l} \boldsymbol{\mu'}_{l \times p} = \mathbf{0}$ and the between-SP matrix $\mathbf{S^*}$ and within-SP matrix $\mathbf{S}$ that arise in this test are the same as those in the One-way Fixed-Effects section. The largest root test of level α rejects H_0 if $f_s > f_{s,1-\alpha}$ where $s = \min(l - 1, p)$.

The Likelihood Ratio Principle

The likelihood ratio* for testing H_0 vs. H_1 is given by

$$\lambda = (|\mathbf{S}|/|\mathbf{S}_0|)^{N/2},$$

where $\mathbf{S}$ is the within-SP matrix and $\mathbf{S}_0$ is the SP matrix of the N observations given by $\mathbf{S}_{0\ p \times p} = (s_{0ij})$, where

$$s_{0ij} = \sum_{h=1}^{l} \sum_{k=1}^{N_h} \left(x_{ik}^{(h)} - \bar{x}_i \right) \left(x_{jk}^{(h)} - \bar{x}_j \right).$$

$\mathbf{S}_0 = \mathbf{S^* + S}$, which involves the analysis of sums of squares and sums of products, called multivariate analysis of variance

(MANOVA). Now

$$\lambda^{N/2} = \frac{|\mathbf{S}|}{|\mathbf{S^* + S}|} = \prod_{i=1}^{s} (1 + f_i)^{-1} = W^{(s)}$$

(say) where $s = \min(l - 1, p)$. The statistic $W^{(s)}$ proposed by S. S. Wilks (1932) is known as Wilks' criterion or Wilks' Λ for the test of H_0 vs. H_1. H_0 is rejected at level α if $W^{(s)} < W_\alpha^{(s)}$. (See also LAMBDA CRITERION, WILKS'S.)

Another statistic for test of H_{0_s} vs. H_1 is Hotelling's trace*,

$$U^{(s)} = \operatorname{tr} \mathbf{S^* S}^{-1} = \sum_{i=1}^{s} f_i,$$

whose union-intersection character also has been demonstrated. A fourth statistic that has several optimal properties including robustness* against non-normality and heteroscedasticity* is Pillai's trace*,

$$V^{(s)} = \operatorname{tr} \mathbf{S^* (S^* + S)}^{-1}$$

$$= \sum_{i=1}^{s} \frac{f_i}{1 + f_i} = \sum_{i=1}^{s} b_i, \qquad b_i = \frac{f_i}{1 + f_i}.$$

Note that $0 < b_1 < \cdots < b_s \leqslant 1$. For both $U^{(s)}$ and $V^{(s)}$, rejection regions for level α tests are taken greater than the respective $100(1 - \alpha)$ percentiles.

Application

To illustrate the test procedure for MANOVA two numerical examples are given.

Example 1. A study was made [Ventura (1957), Pillai (1960)] for a MANOVA test of hypothesis with four variables based on measurements of (a) height (inches), (b) weight (pounds), (c) chest (inches), and (d) waist (inches) of male reserve officers in civilian status of the armed forces of the Philippines, hailing from six different regions of the Philippine Islands but all within the age interval 29–31. The sample contained 25 officers from each region (the assumption of equality of covariance matrices was found to be justified in view of earlier tests).

Here $H_0: \mu_1 = \cdots = \mu_6$ vs. H_1: not all μ_h's equal in $N_4(\mu_h, \Sigma)$, $h = 1, \ldots, 6$. The S^* and S matrices are discussed in HOTELLING'S TRACE. The degrees of freedom for S^*, denoted by $v_1 = l - 1 = 5$ and that for S, $v_2 = N - l = 144$, and the two arguments m and n needed to read many of the tables of percentiles, are given by $m = \frac{1}{2}(|l - 1 - p| - 1) = 0$ and $n = \frac{1}{2}(N - l - p - 1) = 69.5$. $s = \min(l - 1, p) = 4$. The characteristic roots of S^*S^{-1} are as follows: $f_1 = 0.0068$, $f_2 = 0.0240$, $f_3 = 0.0438$, and $f_4 = 0.1207$.

LARGEST ROOT TEST. $b_4 = f_4/(1 + f_4) = 0.1077 < b_{4,0.95}$. Do not reject H_0. [See Pillai (1960, 1964, 1965, 1967, 1970), Pillai and Bantegui (1959), Foster and Rees (1957), Foster (1957, 1958), Morrison (1976), Pearson and Hartley, Vol. 2 (1972), Krishnaiah (1980 and ref. 8).]

LIKELIHOOD RATIO TEST. $W^{(4)} = 0.8292 > W_{0.05}^{(4)}$ and hence do not reject H_0. [See Schatzoff (1966), Pillai and Gupta (1969), Pearson and Hartley, Vol. 2 (1972), and Lee (1972).]

HOTELLING'S TRACE TEST. $U^{(4)} = \text{tr}\, S^*S^{-1} = 0.1953 < U_{0.95}^{(4)}$ as shown in HOTELLING'S TRACE. Do not reject H_0. [See Davis (1970 and ref. 2) and Pillai (1960).]

PILLAI'S TRACE TEST. $V^{(4)} = \text{tr}\, S^*(S^* + S)^{-1} = 0.1799 < V_{0.95}^{(4)}$. Hence do not reject H_0. [See Pillai (1960), Mijares (1964), Timm (1975), and Krishnaiah (1980 and ref. 8).]

Thus all the four tests agree in the conclusion not to reject H_0 based on $\alpha = 0.05$.

Example 2.   Rao (1952, p. 263) has considered a test of equality of mean vectors of 6 three-variate populations using measurements of head length, height, and weight of 140 schoolboys of almost the same age belonging to six different schools in an Indian city. The 3×3 matrices, S^* and S have been computed [see Rao (1972) and Pillai and Samson (1959)]. Consider the test of $H_0: \mu_1 = \cdots = \mu_6$ vs. H_1: not all μ_h's equal with $s = 3$, $v_1 = 5$, $v_2 = 134$, $m = 0.5$, and $n = 65$.

LIKELIHOOD RATIO TEST.   Rao (1952) has shown that $W^{(3)} = 0.8239 < W_{0.05}^{(3)}$ and $> W_{0.01}^{(3)}$, showing significance at level $\alpha = 0.05$ but not at $\alpha = 0.01$.

LARGEST ROOT TEST.   Foster (1957) has computed the largest root statistic in this problem and $b_3 = 0.10055$, which he has shown to be significant only at the 15% level.

HOTELLING'S TRACE.   Pillai and Samson (1959) have obtained $U^{(3)} = 0.2016 > U_{0.95}^{(3)}$ but $< U_{0.99}^{(3)}$ and have shown that the results in this case agree with those of the likelihood ratio test, unlike those of the largest root.

PILLAI'S TRACE.   Now $V^{(3)} = 0.1863 > V_{0.95}^{(3)}$ but $< V_{0.99}^{(3)}$, which again agrees with the findings for $U^{(3)}$ and $W^{(3)}$ but not those of the largest root.

The Examples 1 and 2 bring to the fore the need for using all the four tests in drawing inferences from sample data. While in the first example, all four tests agreed in their findings not to reject H_0 at the 5% level, in the second one, all the overall tests (i.e., $U^{(3)}$, $V^{(3)}$, and $W^{(3)}$) concluded rejection of H_0 at the 5% but not at the 1% level, unlike the largest root, which was significant only at the 15% level. This behavior of the largest root test is explained partially by the following power* and robustness* considerations.

POWER AND ROBUSTNESS COMPARISONS

The optimum properties of the four tests defined above for MANOVA have been discussed elsewhere (see HOTELLING'S T^2; MAHALANOBIS D^2; HOTELLING'S TRACE; PILLAI'S TRACE; LAMBDA CRITERION, etc.). Here a comparison will be made of the powers of the four tests as well as their robustness aspects.

Power Comparisons

Let $\omega_1, \ldots, \omega_p$, be the characteristic roots of $\Omega = \Sigma^* \Sigma^{-1}$, where

$$\Sigma^*_{p \times p} = (\sigma^*_{ij}),$$

where

$$\sigma^*_{ij} = \sum_{h=1}^{l} N_h (\mu_{ih} - \bar{\mu}_i)(\mu_{jh} - \bar{\mu}_j),$$

$$\bar{\mu}_i = \sum_{h=1}^{l} \frac{N_h \mu_{ih}}{N}.$$

Since the joint density of the sample characteristic roots, $f_1, \ldots, f_p$, $(b_1, \ldots, b_p)$, $p \leqslant v_1, v_2$, involves as parameters only the population characteristic roots, $\omega_1, \ldots, \omega_p$, power studies of the four tests have been carried out with respect to each population root. The joint density of $f_1, \ldots, f_s$, $(b_1, \ldots, b_s)$ for $v_1 < p \leqslant v_2$ can be obtained from the following density for $p \leqslant v_1, v_2$ by making the following changes:

$$(p, v_1, v_2) \to (v_1, p, v_2 - p + v_1):$$

$$f(f_1, \ldots, f_p)$$
$$= C(p, v_1, v_2) e^{-\operatorname{tr} \Omega/2} |\mathbf{F}|^{(v_1 - p - 1)/2}$$
$$\times |\mathbf{I} + \mathbf{F}|^{-(v_1 + v_2)/2} \prod_{i>j} (f_i - f_j)$$
$$\times {}_1F_1\left(\tfrac{1}{2}(v_1 + v_2); \tfrac{1}{2}v_1;\right.$$
$$\left.\tfrac{1}{2}\Omega, \mathbf{F}(\mathbf{I} + \mathbf{F})^{-1}\right),$$
$$0 < f_1 < \cdots < f_p < \infty,$$

where $\mathbf{F} = \mathbf{S}_2^{-1/2} \mathbf{S}_1 \mathbf{S}_2^{-1/2}$, $C(p, v_1, v_2)$ is a constant and ${}_1F_1(a; b; \mathbf{T}, \mathbf{S})$ is the hypergeometric function of two matrix arguments defined by Constantine (1963), who also de-

rived the preceding distribution. See also James (1964), Pillai [15], Muirhead [13]. All the expressions in the density given above can be expressed fully in terms of the characteristic roots of $\mathbf{F}$ and Ω. For obtaining the joint density of $b_1, \ldots, b_p$, one need only transform $\mathbf{F} = (\mathbf{I} - \mathbf{B})^{-1} - \mathbf{I}$ or $\mathbf{B} = \mathbf{I} - (\mathbf{I} + \mathbf{F})^{-1}$ in the above density. The result is

$$g(b_1, \ldots, b_p)$$
$$= C(p, v_1, v_2) e^{-\operatorname{tr} \Omega/2} |\mathbf{B}|^{(v_1 - p - 1)/2}$$
$$\times |\mathbf{I} - \mathbf{B}|^{(v_2 - p - 1)/2} \prod_{i>j} (b_i - b_j)$$
$$\times {}_1F_1\left(\tfrac{1}{2}(v_1 + v_2); \tfrac{1}{2}v_1; \tfrac{1}{2}\Omega, \mathbf{B}\right),$$
$$0 < b_1 < \ldots < b_p < 1.$$

Pillai and Jayachandran (1967, 1968) obtained the distributions of $U^{(2)}, V^{(2)}, W^{(2)}$ and the largest root b_2, using these joint densities of the characteristic roots. They also carried out power studies of these four tests and made power comparisons based on tabulations of their exact powers for selected values of the parameter vector (ω_1, ω_2). Their findings are as follows:

1. For small deviations from the hypothesis, the order of the power is given by $V^{(2)} > W^{(2)} > U^{(2)}$.

2. For large deviations from the hypothesis; when the values of the roots are far apart, the order of the power is $U^{(2)} > W^{(2)} > V^{(2)}$; when the roots are close, $V^{(2)} > W^{(2)} > U^{(2)}$.

3. When there is only one nonzero deviation and large $b_2 > U^{(2)} > W^{(2)} > V^{(2)}$. Otherwise the power of b_2 is far inferior to those of the other three. For example, for $m = 0$, $n = 30$, $\omega_1 = 4$, $\omega_2 = 4$, the power of b_2 is below that of $V^{(2)}$ ($= 0.505$) by 0.082, that of $W^{(2)}$ by 0.075 and of $U^{(2)}$ by 0.072. This is a typical illustration, and in this manner the largest root test stands aloof from the other three (see ROY'S CHARACTERISTIC ROOT STATISTIC).

4. For $\omega_1 + \omega_2 =$ constant, powers of $V^{(2)}$ and $W^{(2)}$ increase as the roots tend to

be equal, while the powers of $U^{(2)}$ and b_2 decrease.

Schatzoff (1966) carried out a Monte Carlo study for comparison of powers and also for larger number of roots; his findings are similar to 1–4. Fujikoshi (1970) computed some approximate powers for $U^{(3)}$, $V^{(3)}$, and $W^{(3)}$; Lee (1971) obtained some approximate powers for $p = 3$ and 4.

The noncentral distributions in the general case are available for $W^{(p)}$ and the largest root but only partial results have been obtained for $U^{(p)}$ and $V^{(p)}$ (*see* HO-TELLING'S TRACE *and* PILLAI'S TRACE). Pillai et al. (1969) obtained the exact noncentral distribution of $W^{(p)}$ using the inverse Mellin transform in terms of Meijer's G-functions (*see* INTEGRAL TRANSFORMS). Pillai and Nagarsenker (1972) derived the distributions of a statistic $\prod_{i=1}^{p} b_i^a (1 - b_i)^b$, of which the Pillai–Al-Ani–Jouris results are special cases. Pillai and Sudjana (1975) extended these results starting from Pillai's (1975) distribution of the characteristic roots of $\mathbf{S}_1 \mathbf{S}_2^{-1}$ under violations. As for the largest root, Hayakawa (1967) and independently Khatri and Pillai (1968) obtained the distribution of b_p in a beta function series with coefficients involving zonal polynomials*. Further, Pillai and Sugiyama (1969) obtained the density of b_p in power-series form simpler than that derived before. For approximate and asymptotic distributions of all four test statistics see Pillai (1976 and ref. 15).

Robustness Comparisons

Robustness* aspects here are of two types: against non-normality and against heteroscedasticity* in the sense of violation of the assumption of equality of covariance matrices. Mardia [11] has shown, based on certain permutation distributions*, that the $V^{(s)}$-test is robust against moderate non-normality. Olson [14] has made a Monte Carlo study concerning robustness of six MANOVA tests, including the four discussed earlier. For general protection against departures from normality and from homo-geneity of covariance matrices in the fixed-effects model, he has recommended the $V^{(s)}$-test as the most robust of the MANOVA tests with adequate power against a variety of alternatives. His specific findings are that (a) the largest root test, which produces excessive rejections of H_0 under both kurtosis and heteroscedasticity, may be dropped from consideration, (b) for protection against kurtosis the $V^{(s)}$-test is generally better than others in terms of type I error rates but $U^{(s)}$ and $W^{(s)}$ are in some cases more powerful, and (c) for protection against heteroscedasticity, $U^{(s)}$ and $W^{(s)}$ should be avoided since their behavior is similar to that of the largest root in this case, but the $V^{(s)}$-test is robust against this type of violation although its type I error rate is somewhat high.

Pillai and Sudjana (1975) have carried out an exact robustness study in the two-roots case based on Pillai's distribution of the characteristic roots of $\mathbf{S}_1 \mathbf{S}_2^{-1}$ under violations [Pillai (1975)]. Based on the numerical values of the ratio $e = (p_1 - p_0)/(p_0 - \alpha)$, where p_1 = power under violations, p_0 = power without violation, and $\alpha = 0.05$, the $V^{(2)}$-test has been observed to be most robust among all the four tests against heteroscedasticity. In fact, the order of robustness was observed to be $V^{(2)} > W^{(2)} > U^{(2)} > b_2$. Davis (1980b, [3], and ref. 4) has studied the effects of non-normality on $W^{(s)}$ and the largest root test in multivariate Edgeworth populations, expanding the distributions to terms of the first order. These first-order approximations under non-normality are shown to involve Mardia's measures of multivariate skewness and kurtosis [10] together with a supplementary skewness measure. Both for $W^{(s)}$ and b_s increasing kurtosis lowers type I error while increasing skewness raises it. The effect of skewness is considerably more serious for lower v_2, more so as l increases, but kurtosis becomes significant for large v_2. For lower v_2, the skewness and kurtosis terms have larger magnitude for $W^{(s)}$ than for b_s for small departures from normality. However, as v_2 increases, the b_s-test becomes more sensitive to non-normality. (For more details on robustness, see HOTELLING'S TRACE; PILLAI'S TRACE *and* LAMBDA CRITERION; *see also* MARDIA'S TEST FOR MULTIVARIATE NORMALITY *and* MULTIVARIATE NORMALITY, TESTING FOR.)

TEST FOR DIMENSIONALITY

Consider l p-variate normal populations $N_p(\boldsymbol{\mu}_h, \boldsymbol{\Sigma})$, $h = 1, \ldots, l$, where $\boldsymbol{\Sigma}$ is pd. Given independent random samples of sizes $N_1, \ldots, N_l$, from the respective populations, one may wish to test the hypothesis: $\boldsymbol{\mu}_h$ lie in an r-dimensional hyperplane, $r \leqslant s = \min(l - 1, p)$, $h = 1, \ldots, l$, vs. the alternative: $\boldsymbol{\mu}_h$'s are unrestricted. If $\boldsymbol{\Sigma}$ is known, $-2\log\lambda = g_1 + \cdots + g_{p-r}$, where $0 < g_1 < \cdots < g_p < \infty$, are the characteristic roots of $\mathbf{S}^*\boldsymbol{\Sigma}^{-1}$ and λ is the likelihood ratio criterion for the test of the null hypothesis with $\boldsymbol{\Sigma}$ known. For large values of $N_1, \ldots, N_l$, $-2\log\lambda = g_1 + \cdots + g_{p-r}$ is distributed as χ_v^2, i.e., chi-square where the degrees of freedom $v = (p - r)(l - r - 1)$. If $\boldsymbol{\Sigma}$ is unknown, one may estimate $\boldsymbol{\Sigma}$ by $\mathbf{S}/v_2$; then, asymptotically for large v_2, $v_2(f_1 + \cdots + f_{p-r})$ is distributed as χ_v^2, where $0 < f_1 < \cdots < f_p < \infty$ are the characteristic roots of $\mathbf{S}^*\mathbf{S}^{-1}$. Bartlett (1947) suggested

$$-\left[N - 1 - \tfrac{1}{2}(p + l) \right] \sum_{i=1}^{p-r} \log(1 - b_i) \sim \chi_v^2$$

as an improved approximation. The rejection region is taken greater than $\chi_{v,1-\alpha}^2$. The test of dimensionality may now be performed sequentially for $r = 0, 1, \ldots, s$. [For more details, *see* Rao (1973), Mardia et al. (ref. 12), Kshirsagar (1972).]

CANONICAL VARIATES AND DISCRIMINANT FUNCTIONS*

From the preceding test for dimensionality or otherwise, consider that the dimension of the plane spanned by the true group means is r. If $\boldsymbol{\Sigma}$ is unknown, let $\mathbf{l}_i$ be the characteristic vector of $\mathbf{S}^{-1}\mathbf{S}^*$ corresponding to f_i

normalized by $\mathbf{l}_i'[\mathbf{S}/(N - l)]\mathbf{l}_i = 1$, $i = p - r + 1, \ldots p$. These $\mathbf{l}_i$'s can be used to estimate the plane of the true group means. Consider the r-dimensional canonical coordinates $(\mathbf{l}_{p-r+1}'\mathbf{x}, \ldots, \mathbf{l}_p'\mathbf{x})$, the projection of a point $\mathbf{x}$ onto the estimated plane. The canonical means of the l groups, $\mathbf{m}_h = (\mathbf{l}_{p-r+1}'\bar{\mathbf{x}}^{(h)}, \ldots, \mathbf{l}_p'\bar{\mathbf{x}}^{(h)})'$, $h = 1, \ldots, l$, represent the projection of the group means onto this plane which could be used to study the group differences. Let $y_i = \mathbf{l}_i'\mathbf{x}$. $\mathbf{l}_i$ is the canonical vector for the ith canonical variable y_i and the canonical variables are *optimal discriminant functions* in view of the fact that for the l-samples data matrix the ith canonical variable is that linear function that maximizes the between-group variance relative to the within-subject to the constraint that it is uncorrelated with canonical variables numbered $i + 1, \ldots, p$. In view of this, for any value $r \leqslant s$, the y_i's are linear functions that separate the l sample means as much as possible.

For $r = 1$ or 2, a graph of the canonical means can give some idea of the strength of separation between groups. In order to discuss the accuracy of each of the canonical means a rough $100(1 - \alpha)\%$ confidence region for the hth true canonical mean $\boldsymbol{\mu}_h^* = (\mathbf{l}_{p-r+1}'\boldsymbol{\mu}_h, \ldots, \mathbf{l}_p'\boldsymbol{\mu}_h)'$ is given by the disc of radius $n_h^{-1/2}\sqrt{\chi_{r,1-\alpha}^2}$ about the sample canonical mean $\mathbf{m}_h$.

Unlike the principal components (*see* COMPONENT ANALYSIS), canonical coordinates are invariant under changes of scale of the original variables in view of the fact that $\mathbf{S}/(N - l)$, an estimate of $\boldsymbol{\Sigma}$, is taken into account in the development here. [See ref. 12 for details and examples, and Kshirsagar (1972) for more aspects of multigroup discrimination.]

MULTIVARIATE REGRESSION ANALYSIS

Let $\mathbf{X}$ be a $p \times N$ matrix of N independent normally distributed column vectors with covariance matrix $\boldsymbol{\Sigma}$, which is pd and $E\mathbf{X} = \boldsymbol{\beta}_{0\,p \times N} + \boldsymbol{\beta}_{p \times q}\mathbf{U}_{q \times N}$, where $\boldsymbol{\beta}_0$ has identical columns of unknown parameters, $\boldsymbol{\beta}$ is a matrix of unknown parameters, $\mathbf{U}$ is a matrix of rank q of given observations of concomitant variables with each row sum assumed to be zero without any loss of generality. Alternatively, one can write $\mathbf{X}' = \boldsymbol{\beta}_0' + \mathbf{U}'\boldsymbol{\beta}' + \boldsymbol{\epsilon}'$, the p-variable regression model, where the columns of $\boldsymbol{\epsilon}_{p \times N}$ are independently distributed $N_p(\mathbf{0}, \boldsymbol{\Sigma})$. The minimum variance unbiased estimates of the regression parameters are given by $(\bar{x}_1, \ldots, \bar{x}_p)'$ for the columns of $\boldsymbol{\beta}_0$ and $\mathbf{B}' = (\mathbf{U}\mathbf{U}')^{-1}\mathbf{U}\mathbf{X}'$ for parameters of $\boldsymbol{\beta}$. For testing the hypothesis $\boldsymbol{\beta} = \mathbf{0}$ vs. $\boldsymbol{\beta} \neq \mathbf{0}$,

$$\mathbf{S}^* = \mathbf{X}\mathbf{U}'(\mathbf{U}\mathbf{U}')^{-1}\mathbf{U}\mathbf{X}',$$

$$\mathbf{S} = \mathbf{X}\left(\mathbf{I} - \frac{\mathbf{e}_N \mathbf{e}_N'}{N} - \mathbf{U}'(\mathbf{U}\mathbf{U}')^{-1}\mathbf{U}\right)\mathbf{X}'.$$

$$s = \min(q, p),$$

$$m = \tfrac{1}{2}(|q - p| - 1), \quad n = \tfrac{1}{2}(N - q - p - 2).$$

Here $\mathbf{A} = (\mathbf{e}_N, \mathbf{U}')$ and $\mathbf{C} = (\mathbf{0}, \mathbf{I}(q))$. The test could be carried out using the four statistics (discussed earlier) based on the characteristic roots of $\mathbf{S}^*\mathbf{S}^{-1}$. [For details, see Roy (1957) and Morrison (1976).]

MULTIVARIATE ANALYSIS OF COVARIANCE

The linear model in this case can be written

$$\mathbf{X}_{N \times p}' = \mathbf{A}_{N \times m}\boldsymbol{\xi}_{m \times p} + \mathbf{U}_{N \times q}'\boldsymbol{\beta}_{q \times p}' + \boldsymbol{\epsilon}_{N \times p}'$$

$$= (\mathbf{A}\mathbf{U}')\begin{pmatrix} \boldsymbol{\xi} \\ \boldsymbol{\beta}' \end{pmatrix} + \boldsymbol{\epsilon}',$$

where $\mathbf{A}$ is a design matrix of rank r, $\boldsymbol{\xi}$ and $\boldsymbol{\epsilon}'$ are as defined earlier for MGLH, $\mathbf{U}'$ is a matrix of concomitant variable observations, and $\boldsymbol{\beta}'$ a matrix of regression parameters. $\mathbf{A}_{1 \, N \times r}$ is taken as a basis in $\mathbf{A}$. For testing $\mathbf{C}_1\boldsymbol{\xi} = \mathbf{0}$ vs. $\mathbf{C}_1\boldsymbol{\xi} \neq \mathbf{0}$, where $\mathbf{C}_1$ is a $t \times m$ matrix of rank t,

$$\mathbf{S}^* = \mathbf{S}_{xx}^* - \mathbf{S}_{0,xu}\mathbf{S}_{0,uu}^{-1}\mathbf{S}_{0,xu}' + \mathbf{S}_{xu}\mathbf{S}_{uu}^{-1}\mathbf{S}_{xu}',$$

$$\mathbf{S} = \mathbf{S}_{xx} - \mathbf{S}_{xu}\mathbf{S}_{uu}^{-1}\mathbf{S}_{xu}',$$

for

$$\mathbf{S}_{xu}^* = \mathbf{X}\mathbf{T}_1\mathbf{U}', \qquad \mathbf{S}_{xu} = \mathbf{X}\mathbf{T}_2\mathbf{U}',$$

$$\mathbf{S}_{0,xu} = \mathbf{S}_{xu}^* + \mathbf{S}_{xu},$$

where $\mathbf{T}_1$ and $\mathbf{T}_2$ are defined as in the $F_{t,N-r}$ in MGLH, noting the appropriate definitions for $\mathbf{A}_1$ here, and $\mathbf{C}_{11}$ being replaced by $\mathbf{C}_1$. Under the hypothesis, $s = \min(t, p)$, $m = \tfrac{1}{2}(|t - p| - 1)$ and $n = \tfrac{1}{2}(N - r - q - p - 1)$, and the four tests described could be used for testing the hypothesis using the characteristic roots of $\mathbf{S}^*\mathbf{S}^{-1}$. Further, if the hypothesis is $\boldsymbol{\beta} = \mathbf{0}$ vs. $\boldsymbol{\beta} \neq \mathbf{0}$, the appropriate $\mathbf{S}^* = \mathbf{S}_{xu}\mathbf{S}_{uu}^{-1}\mathbf{S}_{xu}'$ with $\mathbf{S}$ and n as before, but $s = \min(q, p)$ and $m = \tfrac{1}{2}(|q - p| - 1)$. [See Morrison (1976) and Timm (1975) for examples.]

MULTIPLE COMPARISONS IN MGLH

In the MGLH, $100(1 - \alpha)\%$ simultaneous confidence bounds on all functions $\mathbf{b}'\mathbf{C}\boldsymbol{\xi}\mathbf{M}\mathbf{a} = \mathbf{b}'(\mathbf{C}_{11}\mathbf{C}_{12})\boldsymbol{\xi}\mathbf{M}\mathbf{A}$ have been obtained in the form [see Bose and Roy (1953) and Roy (1957)] $h \pm \sqrt{d}$, where

$$h = \mathbf{b}'\mathbf{C}_{11}(\mathbf{A}_1'\mathbf{A}_1)^{-1}\mathbf{A}_1'\mathbf{X}'\mathbf{M}\mathbf{a},$$

$$d = f_{s,1-\alpha}\mathbf{a}'\mathbf{M}'\mathbf{X}\mathbf{T}_2\mathbf{X}'\mathbf{M}\mathbf{a}$$

for all non-null $\mathbf{a}$ and $\mathbf{b}$ subject to $\mathbf{b}'\mathbf{C}_{11}(\mathbf{A}_1' \mathbf{A}_1)^{-1}\mathbf{C}_{11}'\mathbf{b} = 1$. [See Roy (1957) and Morrison (1976) for special cases.]

GROWTH CURVE ANALYSIS

Potthoff and Roy [16] introduced the growth curve* model, which may be written as $\mathbf{X}_{p \times N} = \mathbf{B}\boldsymbol{\xi}\mathbf{A} + \boldsymbol{\epsilon}$ [see refs. 7 and 21 and Morrison (1976)], where $\mathbf{B}_{p \times q}$ is a known non-random matrix of full rank $q \leqslant p$; $\boldsymbol{\xi}_{q \times m}$, a matrix of unknown parameters; $\mathbf{A}_{m \times N}$, a design matrix of rank $m < N$ (if of rank $< m$, a basis $\mathbf{A}_1$ could be chosen and similarly for $\mathbf{B}$, or use of some generalized inverse could be made (see Srivastava and Khatri [21]); $\boldsymbol{\epsilon}_{p \times N}$, random error matrix, the columns being independently distributed $N_p(\mathbf{0}, \boldsymbol{\Sigma})$, where $\boldsymbol{\Sigma}$ is pd. For example, if x_{ijt} denotes a growth measurement of the ith individual in the jth group at time t, then

$$x_{ijt} = \beta_{0j} + \beta_{1j}t + \cdots + \beta_{q-1,j}t^{q-1} + \epsilon_{ijt},$$

$$i = 1, \ldots, N_j, \quad j = 1, \ldots l, \quad t = 1, \ldots, T.$$

(Note that here $N_1 + \cdots + N_l = N$, $T = p$, $l = m$.) Potthoff and Roy [16] gave an analysis of the model involving an arbitrary matrix. Rao [18] and Khatri [7] independently (see also Grizzle and Allen [6]) evolved a conditional model approach for estimation and tests that was not affected by the arbitrary matrix. Khatri [7] obtained the maximum likelihood estimate of ξ in the form

$$\hat{\xi} = (\mathbf{B'E^{-1}B})^{-1}\mathbf{B'E^{-1}XA'(AA')^{-1}},$$

$$\mathbf{E} = \mathbf{X}(\mathbf{I} - \mathbf{A'(AA')^{-1}A})\mathbf{X'},$$

which could be considered proportional to an estimate of Σ. For testing the hypothesis $\mathbf{H}\xi\mathbf{D} = \mathbf{0}$ vs. $\mathbf{H}\xi\mathbf{D} \neq \mathbf{0}$, where $\mathbf{H}_{c \times q}$ has rank $c \leqslant q$ and $\mathbf{D}_{m \times v}$ has rank v,

$$\mathbf{S^*} = \mathbf{H}\hat{\xi}\mathbf{D}(\mathbf{D'TD})^{-1}\mathbf{D'}\hat{\xi}'\mathbf{H'},$$

$$\mathbf{S} = \mathbf{H}(\mathbf{B'E^{-1}B})^{-1}\mathbf{H'},$$

$$\mathbf{T} = (\mathbf{AA'})^{-1} + (\mathbf{AA'})^{-1}\mathbf{AX'E^{-1}XA'(AA')^{-1}}$$
$$- \hat{\xi}'(\mathbf{B'E^{-1}B})\hat{\xi}.$$

Hence tests could be carried out using the four preceding statistics based on the characteristic roots of $\mathbf{S^*S^{-1}}$. Note that $s = \min(c, v)$, $m = \frac{1}{2}(|c - v| - 1)$ and $n = \frac{1}{2}(N - m - p + q - c - 1)$. [For further details and examples, see refs. 21 and 22 and Morrison (1976); for a Bayesian approach, see ref. 5].

Other Topics

For further applications of the MGLH (e.g., the two-way layout and others) see Morrison (1976), Timm (1975), Mardia et al. [12], Press [17], Roy et al. [19], Arnold [1], and references therein. Also see Krishnaiah [8] for a variety of papers on MANOVA.

Further Reading

For references not listed here, see Pillai [15], which is also annotated in HOTELLING'S TRACE.

See the new book (published after this article was prepared) by M. E. Eaton, *Multivariate Statistics* (Wiley, New York, 1983)

for a vector space approach to MANOVA and for invariance considerations to describe the structure of the linear models for inference for means and to suggest testing procedures. Maximum likelihood estimators are considered and likelihood ratio tests are obtained for most of the examples discussed. The problems treated include: (a) the MGLH, the characteristic roots, f_i's (b_i's) as the maximal invariant for the test of the linear hypothesis under appropriate group transformation, and the derivation of the likelihood ratio statistic (Wilks' Λ); (b) the MANOVA problem with block diagonal covariance structure in which the likelihood ratio statistic is a product of individual likelihood ratios that arise in view of the fact that the original linear model decomposes into independent component MANOVA models; (c) an intraclass (compound symmetric) covariance structure reducing the problem of test for means to two linear models individually under group transformation; (d) an example involving cyclic covariances; and (e) complex covariance structures.

References

[1] Arnold, S. F., (1981). *The Theory of Linear Models in Multivariate Analysis*. Wiley, New York.

[2] Davis, A. W. (1980). *Commun. Statist. Simul. Comp.*, **B9**, 321–336.

[3] Davis, A. W. (1982). *Biometrika*, **63**, 661–670.

[4] Davis, A. W. (1982). *J. Amer. Statist. Ass.*, **77**, 896–900.

[5] Geisser, S. (1980). In [8], pp. 89–115.

[6] Grizzle, J. E. and Allen, D. M. (1969). *Biometrika*, **25**, 357–381.

[7] Khatri, C. G., (1966). *Ann. Inst. Statist. Math.*, **18**, 75–86.

[8] Krishnaiah, P. R. (1980). *Handbook of Statistics*, Vol. 1: *Analysis of Variance*, P. R. Krishnaiah, ed. North-Holland, New York.

[9] Krishnaiah, P. R. In [8], pp. 745–971.

[10] Mardia, K. V. (1970). *Biometrika*, **57**, 519–530.

[11] Mardia, K. V. (1971). *Biometrika*, **58**, 105–127.

[12] Mardia, K. V., Kent, J. T., and Bibby, J. M. (1979). *Multivariate Analysis*. Academic Press, New York.

[13] Muirhead, R. J. (1982). *Aspects of Multivariate Statistical Theory*. Wiley, New York.

[14] Olson, C. L., (1974). *J. Amer. Statist. Ass.*, **69**, 894–908.

[15] Pillai, K. C. S. (1977). *Canad. J. Statist.*, **5**, 1–62.

[16] Potthoff, R. F. and Roy, S. N. (1964). *Biometrika*, **51**, 313–326.

[17] Press, S. J. (1972). *Applied Multivariate Analysis.* Holt, Rinehart, and Winston, New York.

[18] Rao, C. R. (1965). *Biometrika*, **52**, 447–458.

[19] Roy, S. N., Gnanadesikan, R., and Srivastava, J. N. (1971). *Analysis and Design of Certain Quantitative Multiresponse Experiments.* Pergamon, New York.

[20] Sen, P. K. (1980). In [8], pp. 673–702.

[21] Srivastava, M. S. and Khatri, C. G. (1979). *An Introduction to Multivariate Statistics.* North-Holland, New York.

[22] Timm, N. H. (1980). In [8], pp. 41–87.

(ANALYSIS OF COVARIANCE
ANALYSIS OF VARIANCE
GENERAL LINEAR MODEL
GROWTH CURVES
HOTELLING TRACE
LAMBDA CRITERION, WILKS'S
MULTINORMAL DISTRIBUTION
MULTIVARIATE ANALYSIS
PILLAI'S TRACE
ROY'S CHARACTERISTIC ROOT
 STATISTIC
UNION-INTERSECTION PRINCIPLE)

K. C. S. Pillai

MULTIVARIATE BARTLETT TEST

The multivariate Bartlett test [5] is a test of homogeneity* of covariance matrices. At issue is the question of whether $k \geqslant 2$ populations of ($p \geqslant 2$)-dimensional multivariate normal random variables have the same covariance matrix. The mean vectors are assumed unknown and not necessarily equal. The univariate Bartlett test* [4] may be generalized in this context to a multivariate version involving determinants of sample covariance matrices. A principal application is in multivariate analysis of variance* (MANOVA) situations, where covariance matrix homogeneity is commonly an underlying assumption. The multivariate Bartlett test may be used to corroborate this assumption. In addition, the multivariate Bartlett test statistic is used as a factor in an omnibus test of homogeneity of several multivariate normal distributions (see Anderson [2, Sec. 10.3], Giri [8, Sec. 8.5.1], or Kendall [9, Chap. 9]).

METHODOLOGY

Consider k p-dimensional multivariate normally distributed populations with unknown mean vectors $\boldsymbol{\mu}_j$ and unknown covariance matrices $\boldsymbol{\Sigma}_j$, $j = 1, \ldots, k$. Independent random samples are taken, one from each population. Let n_j denote the sample size for the jth population and $\{\mathbf{X}_{ji}\}$, $i = 1, \ldots, n_j$, the sample. Introduce the unbiased sample covariance matrices,

$$\mathbf{S}_j = \nu_j^{-1} \sum_{i=1}^{n_j} \left(\mathbf{X}_{ji} - \overline{\mathbf{X}}_{j\cdot}\right)\left(\mathbf{X}_{ji} - \overline{\mathbf{X}}_{j\cdot}\right)',$$

$$\nu_j = n_j - 1$$

and

$$\overline{\mathbf{X}}_{j\cdot} = n_j^{-1} \sum_{i=1}^{n_j} \mathbf{X}_{ji}.$$

The objective is to test $H_0: \boldsymbol{\Sigma}_1 = \cdots = \boldsymbol{\Sigma}_k$ against $H_1: \boldsymbol{\Sigma}_t \neq \boldsymbol{\Sigma}_u$ for some $t \neq u$. The likelihood ratio* procedure obtained by Wilks [12] has critical region $0 < L_1 < A_1$, where $L_1 = \prod_{j=1}^k |\hat{\boldsymbol{\Sigma}}_j|^{n_j/2} / |\hat{\boldsymbol{\Sigma}}|^{n/2}$ is the test statistic, $\hat{\boldsymbol{\Sigma}}_j = (\nu_j/n_j)\mathbf{S}_j$ is the maximum likelihood* estimator of $\boldsymbol{\Sigma}_j$ under $H_0 \cup H_1$, $\hat{\boldsymbol{\Sigma}} = \sum_{j=1}^k (n_j/n)\hat{\boldsymbol{\Sigma}}_j$, with $n = \sum_{j=1}^k n_j$, is the maximum likelihood estimator of the common covariance matrix under H_0, and A_1 is the critical value determined by $P_{H_0}\{0 < L_1 < A_1\} = \alpha$, α being the size of the test. Based on the feeling that populations having relatively small sample sizes are weighted too heavily in L_1, the Bartlett approach makes corrective adjustments that result in the modified test statistic,

$$L = \prod_{j=1}^k \frac{|\mathbf{S}_j|^{\nu_j/2}}{|\mathbf{S}|^{\nu/2}},$$

where $\mathbf{S} = \sum_{j=1}^k (\nu_j/\nu)\mathbf{S}_j$, with $\nu = \sum_{j=1}^k \nu_j$.

The size α multivariate Bartlett critical region is $0 < L < A$, where A is determined by $P_{H_0}\{0 < L < A\} = \alpha$.

Exact computation of the critical value A requires a tractable expression for the exact null distribution of L. Unfortunately, such expressions exist only for the case $k = 2$ (see Anderson [2, Sec. 10.6] for a representation in terms of beta integrals for $p = 2$ and Khatri and Srivastava [10] for a representation in terms of the H-function* for general p). For $k > 2$ populations, asymptotic expressions that provide approximate critical values for large sample sizes are available. Such approximations involve expressions of the null distribution of $M = -2\ln L$ (Bartlett's M) in terms of chi-square (χ^2) factors. Starting from the moments of L (see Anderson [2, Sec. 10.4]) and using an expansion due to Barnes [3] for the logarithm of the gamma function, Box [6] derived series expressions for the null CDF of ρM having prescribed orders of accuracy. The coefficient ρ is a constant chosen to simplify computation and/or improve accuracy.

The size α Bartlett test criterion, put in terms of $U = \rho M$, is to reject H_0 if U exceeds $u_{1-\alpha}$, the $(1 - \alpha)$-quantile of the null distribution of U. The Box approximation of the null CDF of U with error of order $\nu^{-(m+1)}$ is

$$P_{H_0}\{U \leqslant u\} = \sum_{r=0}^{m} Q_r(u) + O(\nu^{-(m+1)}),$$

$$Q_r(u) = \sum_{j=0}^{r} t_j^{(r)}\mathrm{Pr}\{\chi_{f+2j}^2 \leqslant u\},$$

$$\text{where} \quad f = \tfrac{1}{2}(k-1)p(p+1),$$

and the coefficients $t_j^{(r)}$ come from the polynomials, $T_r(x) = \sum_{j=0}^{r} t_j^{(r)}x^j$, defined as follows: Let

$$\omega_r = \frac{(-1)^{r+1}}{r(r+1)}\left(\frac{2}{\rho}\right)^r$$

$$\times \left\{ \sum_{j=1}^{k} \frac{1}{\nu_j^r} \sum_{i=1}^{p} B_{r+1}\left(\tfrac{1}{2}(1-\rho)\nu_j + \tfrac{1}{2}(1-i)\right) \right.$$

$$\left. - \frac{1}{\nu^r} \sum_{i=1}^{p} B_{r+1}\left(\tfrac{1}{2}(1-\rho)\nu + \tfrac{1}{2}(1-i)\right) \right\},$$

where $B_{r+1}(\cdot)$ denotes the Bernoulli polynomial* of degree $r + 1$. $B_1(h) = h - \tfrac{1}{2}$, $B_2(h) = h^2 - h + \tfrac{1}{6}$, etc. (See Abramowitz and Stegun [1].) Then

$$T_0(x) \equiv 1, \qquad T_1(x) = \omega_1(x - 1),$$

$$T_2(x) = \omega_1^2\left(\tfrac{1}{2}x^2 - x + \tfrac{1}{2}\right) + \omega_2(x^2 - 1),$$

$$T_3(x) = \omega_1^3\left(\tfrac{1}{6}x^3 - \tfrac{1}{2}x^2 + \tfrac{1}{2}x - \tfrac{1}{6}\right)$$

$$+ \omega_1\omega_2(x^3 - x^2 - x + 1)$$

$$+ \omega_3(x^3 - 1),$$

$$T_4(x) = \omega_1^4\left(\tfrac{1}{24}x^4 - \tfrac{1}{6}x^3 + \tfrac{1}{4}x^2 - \tfrac{1}{6}x + \tfrac{1}{24}\right)$$

$$+ \omega_1^2\omega_2\left(\tfrac{1}{2}x^4 - \tfrac{1}{2}\right)$$

$$+ \omega_1\omega_3(x^4 - x^3 - x + 1)$$

$$+ \omega_2^2\left(\tfrac{1}{2}x^4 - x^2 + \tfrac{1}{2}\right) + \omega_4(x^4 - 1),$$

and so on. (In general, $T_r(x)$ is a degree r polynomial of order ν^{-r} defined to be the sum of all terms in

$$\left\{ \sum_{j=1}^{m} \left(\sum_{i=0}^{\infty} (\omega_j x^j)^i / i! \right) \right\} \left\{ \sum_{j=1}^{m} \sum_{i=0}^{\infty} (-\omega_j)^i / i! \right\}$$

whose involvement with the ω's is a factor of the form $\omega_1^{d_1} \ldots \omega_m^{d_m}$, where $\sum_{j=1}^{m} jd_j = r$.) The order $\nu^{-(m+1)}$ Box approximation of the critical value $u_{1-\alpha}$ is then the value $u = u_{1-\alpha}^{(m)}$ satisfying $\sum_{r=0}^{m} Q_r(u) = 1 - \alpha$. Equivalently, the Box criterion rejects H_0 if $\sum_{r=0}^{m} Q_r(u_0) > 1 - \alpha$, where u_0 denotes the observed value of $U = \rho M$.

Computational complications are reduced by setting

$$\rho = \rho_0$$

$$= 1 - \frac{2p^2 + 3p - 1}{6(p+1)(k-1)}\left(\sum_{j=1}^{k} \nu_j^{-1} - \nu^{-1} \right).$$

Then $\omega_1 = 0$, $Q_1(u) \equiv 0$, and subsequent $Q_r(u)$ terms are greatly simplified. In fact, for this value of ρ, $U = \rho_0 M$ is distributed, to order ν^{-2}, as χ_f^2 under H_0, giving the simplest Box approximation, $u_{1-\alpha}^{(1)} = \chi_{f;1-\alpha}^2$. (In contrast, use of $\rho \neq \rho_0$, e.g., $\rho = 1$, gives $U = \rho M$ as χ_f^2 to order only ν^{-1}.) Similarly, the order ν^{-3} approximation based on $\rho = \rho_0$ is $u = u^{(2)}_{1-\alpha}$, where

$$1 - \alpha = \Pr\{\chi^2_f \leqslant u\}$$
$$+ \omega_2\Big[\Pr\{\chi^2_{f+4} \leqslant u\} - \Pr\{\chi^2_f \leqslant u\}\Big],$$

and

$$\omega_2 = \frac{p(p+1)}{48\rho_0^2}$$
$$\times\left[(p-1)(p+2)\left(\sum_{j=1}^{k}\nu_j^{-2} - \nu^{-2}\right)\right.$$
$$\left. - 6(k-1)(1-\rho_0)^2\right].$$

An alternative order ν^{-3} approximation offered by Box [6] and based on the F distribution* is considered superior for smaller samples ($\nu/k < 20$). Define

$$\tau = \frac{(p-1)(p+2)}{6(k-1)}\left(\sum_{j=1}^{k}\nu_j^{-2} - \nu^{-2}\right),$$
$$g = (f+2)\Big/\left[\tau - (1-\rho_0)^2\right],$$
$$\gamma = (\rho_0 - f/g)/f.$$

Then the the null distribution of $V = \gamma M$ is $F_{f,g}$ to order ν^{-3}, with f and g degrees of freedom, respectively, and the corresponding test criterion rejects H_0 if $V > F_{f,g;1-\alpha}$.

In applications, the selection of an appropriate order of accuracy will depend on the data. Enough terms in the series approximation need to be taken to establish clearly, on which side of $1 - \alpha$, $\Sigma Q_r(u_0)$ falls. Examples are provided by Box [6], Anderson [2, Sec. 10.5], and Kendall [9, Chap. 9].

PROPERTIES

Properties of the power function of the multivariate Bartlett test are as yet rather undeveloped. It is known (see Sugiura and Nagao [11]) that the test is unbiased if $k = 2$ (and for unequal sample sizes, the likelihood ratio test is not). Moreover, the power function is calculable for the case $k = 2$ from a form of

the non-null distribution of L derived by Khatri and Srivastava [10]. For arbitrary k, under orthogonality conditions placed on the covariance matrices, certain admissibility* and unbiasedness* results are available (see Giri [8, Sec. 8.5]). The severe sensitivity to the assumption of normality, well documented in the univariate case, is apparently present also in the multivariate context (see Box [7]). Thus use of the procedure is discouraged in settings where approximate multinormality is not justified.

References

[1] Abramowitz, M., and Stegun, I. A., eds. (1970). *Handbook of Mathematical Functions with Formulas, Graphs, and Mathematical Tables*. National Bureau of Standards, Washington, DC.

[2] Anderson, T. W. (1958). *An Introduction to Multivariate Statistical Analysis*. Wiley, New York. (Excellent exposition of Box's approximation.)

[3] Barnes, E. W. (1899). *Mess. Math.*, **29**, 64–128.

[4] Bartlett, M. S. (1937). *Proc. R. Soc. Lond. Ser. A*, **160**, 268–282.

[5] Bartlett, M. S. (1938). *Proc. Camb. Philos. Soc.*, **34**, 33–40.

[6] Box, G. E. P. (1949). *Biometrika*, **36**, 317–346.

[7] Box, G. E. P. (1953). *Biometrika*, **40**, 318–335.

[8] Giri, N. C. (1977). *Multivariate Statistical Inference*. Academic Press, New York. (Detailed description of Bartlett and related tests with many references.)

[9] Kendall, M. (1980). *Multivariate Analysis*, 2nd ed. Macmillan, New York. (Excellent collection of examples.)

[10] Khatri, C. G. and Srivastava, M. S. (1971). *Sankhyā*, **33**, 201–206.

[11] Sugiura, N., and Nagao, H. (1968). *Ann. Math. Statist.*, **39**, 1689–1692.

[12] Wilks, S. S. (1932). *Biometrika*, **24**, 471–494.

(ANALYSIS OF VARIANCE
BARTLETT'S TEST OF HOMOGENEITY
    OF VARIANCES
HOMOGENEITY AND TESTS OF
    HOMOGENEITY
LIKELIHOOD RATIO TESTS
MULTIVARIATE ANALYSIS OF
    VARIANCE (MANOVA))

R. E. GLASER

MULTIVARIATE CHERNOFF THEOREM

Let $X_1, X_2, \ldots$ be a sequence of independent identically distributed random vectors taking values in a space V. When V is a Euclidean space, then according to a version of the multivariate Chernoff theorem, for any open convex set U

$$s(U) = \lim_{n \to \infty} \left[n^{-1} \log \Pr\{ \overline{X}_n \in U \} \right]$$

$$= \sup\{ \rho(u) : u \in U \},$$

where $\overline{X}_n = (X_1 + \cdots + X_n)/n$ and the so-called Chernoff function $\rho(u)$ is defined by the formula

$$\rho(u) = \inf_t \left[-t'u + \log \phi(t) \right],$$

$$\phi(t) = E \exp(t'X_1).$$

In the case when the expected value $\mu = E(X_1)$ exists and does not belong to U this theorem shows that the probabilities of large deviations* $\Pr\{ \overline{X}_n \in U \}$ tend to zero exponentially fast and gives the exact rate of this convergence. The Chernoff theorem* implies, for instance, that

$$\rho(u) = \lim_{\epsilon \to 0} s(U_\epsilon),$$

where U_ϵ is the sphere of radius ϵ with the center at u.

The multivariate Chernoff theorem extends this result to a broad class of (locally convex) topological vector spaces V and some open sets U. In this situation, in the definition of Chernoff's function t is an element of a dual vector space (i.e., t is a continuous linear functional, and $t'X_1$ denotes its value on X_1.)

For a convex open set U, the multivariate Chernoff theorem has the following interpretation: there exists a supporting hyperplane H to the set U such that $s(U) = s(H)$.

Chernoff's theorem is intimately related to convex analysis. Indeed the Chernoff function is essentially the convex conjugate (Fenchel transform) of the logarithm of the moment generating function* $\phi(t)$, which is convex.

An equivalent formulation of the multivariate Chernoff theorem arises in the case when $\overline{X}_n$ is replaced by P_n, where P_n is the empirical measure corresponding to a random sample $Y_1, \ldots, Y_n$. If X_i denotes the distribution function degenerate at Y_i, then $X_1, X_2, \ldots$ is a sequence of independent and identically distributed random elements taking values in the vector space V of functions of bounded variation on the real line, and $\overline{X}_n = P_n$. Thus one comes to a formulation of the so-called *Sanov's problem* concerning the limiting behavior of the probability that the empirical distribution function belongs to a given set U of distribution functions. Similar results can be proved for the probabilities that a continuous piecewise linear function $S_n(t)$ with nodes at the points $(k/n, k\overline{X}_k/n)$, $k = 1, \ldots, n$, belongs to a set of continuous functions.

The Chernoff theorem plays a significant role in mathematical statistics, where it is used for the asymptotical study of tests and estimators. In this application, the sum of the X's typically corresponds to the log-likelihood ratio $\sum_j \log[f(Y_j, \eta)/f(Y_j, \theta)]$ for two parametric values η and θ. Chernoff's theorem implies

$$\lim_{n \to \infty} \left[n^{-1} \log \Pr_\theta \left\{ \prod_1^n f(Y_j, \eta) \right. \right.$$

$$\left. \left. > \prod_1^n f(Y_j, \theta) \right\} \right]$$

$$= \inf_{t > 0} \left(\log E_\theta [f(Y_1, \eta)/f(Y_1, \theta)]' \right).$$

If $K(Q, P) = E^Q \log dQ/dP$ is the Kullback–Liebler information number for probability measures Q and P, then under mild regularity assumptions

$$\inf_{t > 0} \left[\log \{ E_\theta [f(Y_1, \eta)/f(Y_1, \theta)]^t \} \right]$$

$$= - \inf_Q \left[\{ K(Q, F_\theta), \right.$$

$$E^Q \log[f(Y, \eta)/f(Y, \theta)] \geqslant 0 \} \right]$$

$$\sim - (\eta - \theta) I(\theta)(\eta - \theta)' \quad \text{as} \quad \eta \to \theta,$$

where $I(\theta)$ is the Fisher information* matrix for the family $\{ F_\theta \}$.

Thus Chernoff's theorem is related to two important quantities of information theory*. Analogous results obtained with the help of the multivariate Chernoff theorem are used in the study of adaptive methods* in statistical problems with finite decision spaces and a nuisance parameter that takes a number of values equal to the dimension of V. Another application of this theorem in statistics is to determine the asymptotic Bahadur efficiency* of various statistics that are functions of the empirical distribution P_n.

The multivariate Chernoff theorem is also a very useful tool in many other applications, e.g., in statistical mechanics*, where it allows us to evaluate the entropy* $s(U)$ for some sets U and in statistical communication theory, where it is needed to obtain upper bounds for error probabilities of some codes.

Bibliography

Bahadur, R. R., and Zabell, S. L. (1979). *Ann. Prob.*, **7**, 587–621. (A fundamental paper contains a proof of Chernoff's theorem for topological vector spaces, conditions for the existence of $s(U)$, and different formulas for this function.)

Borovkov, A. A., and Mogulski, A. A. (1978, 1980). *Siberian Math. J.*, **19**, 679–683; **21**, 653–663. (A slight generalization of the main theorem; extension to the case of the continuous curve S_n.)

Groeneboom, P., Oosterhoff, J., and Ruymgaart, F. H. (1979). *Ann. Prob.*, **7**, 553–586. (The most general results on Sanov's problem. Chernoff's theorem for random vectors obtained as a corollary.)

Jelinek, F. (1968). *Probabilistic Information Theory, Discrete and Memoryless Models*, McGraw-Hill, New York (Chap. 5, in particular, exercise 5.12, p. 125). (Application of Chernoff's theorem to the error bounds for source block codes.)

Lanford, O. E., (1971). In Statistical Mechanics and Mathematical Problems, *Lecture Notes in Physics*, **20**, 1–113. (The finite-dimensional Chernoff theorem as related to statistical mechanics, in particular to the notion of entropy.)

Rukhin, A. L. (1982). *Ann. Statist.*, **10**, 1148–1162. (The multivariate Chernoff theorem is used to establish the existence of adaptive procedures in multiple decision problems.)

Sievers, G. L. (1975). *Ann. Statist.*, **3**, 897–905. (Contains expressions for $s(U)$ as the supremum of $s(B)$ for "rectangular" sets B and as the limit of a sequence of density functions.)

Steinebach, J. (1978). *Ann. Prob.*, **6**, 751–759. (Sufficient conditions for the existence of $s(U)$.)

(ADAPTIVE METHODS
CHERNOFF THEOREM
EDF STATISTICS
INFORMATION THEORY AND CODING
  THEORY
LARGE DEVIATIONS AND APPLICATIONS
LARGE SAMPLE THEORY
STATISTICAL MECHANICS)

ANDREW L. RUKHIN

MULTIVARIATE COX REGRESSION MODEL

Cox's regression model* for survival* data [5] specifies the hazard function $\lambda(t)$ of the lifetime T of an individual with covariates $\mathbf{z} = (z_1, \ldots, z_p)$ to have the form

$$\lambda(t; \mathbf{z}) = \lambda_0(t)e^{\beta'\mathbf{z}}, \qquad (1)$$

where β is a p-vector of unknown regression parameters and $\lambda_0(t)$ an unknown and unspecified hazard function for individuals with $\mathbf{z} = \mathbf{0}$. This model is widely used in, e.g., medical contexts when studying the effect of concomitant variables* on survival.

MULTIVARIATE COUNTING PROCESSES

Often a similar model is needed when studying the occurrence of a recurrent* phenomenon rather than an ultimate event, such as death. Examples are admissions to a hospital or the event of getting unemployed. Such models can be studied within the framework of multivariate counting processes. A counting process $\mathbf{N}(t)$ is a collection of, say, n univariate counting processes

$$\mathbf{N}(t) = (N_1(t), \ldots, N_n(t)), \qquad t \geq 0,$$

where N_i can be thought of as counting observed events for individual i, $i = 1, \ldots, n$. To each component N_i a random intensity process λ_i corresponds [6, 7], so that the processes M_i defined by

$$M_i(t) = N_i(t) - \int_0^t \lambda_i(u)\,du,$$

$$t \geqslant 0, \quad i = 1, \ldots, n,$$

are orthogonal local square integrable martingales*. One possible statistical model for a multivariate counting process is the multiplicative intensity model of Aalen [1], which specifies that the intensity process must have the form

$$\lambda_i(t) = \alpha_i(t)Y_i(t),$$

where $Y_i(t)$ is an observable stochastic process and $\alpha_i(t)$ an unknown function. In the multivariate Cox regression model [2] $\alpha_i(t)$ is further specified as

$$\alpha_i(t) = \lambda_0(t)e^{\boldsymbol{\beta}'\mathbf{Z}_i(t)}, \qquad (2)$$

where $\boldsymbol{\beta}$ and $\lambda_0(t)$ are as in (1) and $\mathbf{Z}_i(t)$ is a p-vector of stochastic processes* observable of individual i, $i = 1, \ldots, n$. Most often $Y_i(t)$ will be the indicator of individual i being "at risk" at t for experiencing an event of the type under study. Thus the Cox regression model for survival data arises as a special case of the multivariate Cox regression model, where each of the components N_i counts at most once and where $Y_i(t)$ is the indicator of individual i being alive and uncensored at t.

In an example where the processes N_i count admissions to psychiatric hospitals for women giving birth and women having induced abortion [3], $\mathbf{Z}_i(t)$ contained both time-independent demographic characteristics of woman i and information on prior admissions to psychiatric hospitals. In a model describing labor market dynamics [4], $\mathbf{Z}_i(t)$ could contain information on the length of employment and income during employment periods.

ESTIMATION

In the nonparametric model, where the "underlying hazard function" $\lambda_0(t)$ exists and is unspecified, maximum likelihood estimation* cannot be performed [9], but Johansen [9] demonstrated that in an extended model where the absolutely continuous measure

$$\Lambda_0(t) = \int_0^t \lambda_0(u)\,du$$

is replaced by an arbitrary measure $\Lambda(t)$, maximum likelihood estimation amounts to estimating $\boldsymbol{\beta}$ by the value $\hat{\boldsymbol{\beta}}$ that maximizes the generalized Cox partial likelihood* function

$$L(\boldsymbol{\beta}) = \prod_{t \geqslant 0} \prod_{i=1}^n \left(e^{\boldsymbol{\beta}'\mathbf{Z}_i(t)} S_n(t, \boldsymbol{\beta}) \right)^{dN_i(t)}, \quad (3)$$

where $dN_i(t) = N_i(t) - N_i(t-)$ and $S_n(t, \boldsymbol{\beta}) = (\sum_{j=1}^n Y_j(t)e^{\boldsymbol{\beta}'\mathbf{Z}_j(t)})^{-1}$ and to estimating $\Lambda(t)$ by

$$\hat{\Lambda}(t) = \int_0^t S_n(u, \hat{\boldsymbol{\beta}})\,d\overline{N}(u), \qquad (4)$$

where $\overline{N} = N_1 + \cdots + N_n$. (For a different derivation of (3), see [12].) Jacobsen [8] studied another extension of the absolutely continuous model and derived slightly different, computationally more difficult but asymptotically equivalent estimators of $\boldsymbol{\beta}$ and $\Lambda(t)$.

PROPERTIES OF ESTIMATORS

The asymptotic properties of $\hat{\boldsymbol{\beta}}$ and $\hat{\Lambda}(t)$ when $n \to \infty$ were derived by Andersen and Gill [2] using martingale* results, thus extending the results of Tsiatis [13] valid for the Cox regression model for survival data with time-independent covariates. The key step in proving a central limit theorem* for $\hat{\boldsymbol{\beta}}$ is to note that the score statistic*

$$U(\boldsymbol{\beta}) = \frac{\partial}{\partial \boldsymbol{\beta}} \log L(\boldsymbol{\beta})$$

considered as a stochastic process in t is a local square integrable martingale when the true value $\boldsymbol{\beta} = \boldsymbol{\beta}_0$ is inserted. Thus conditions can be found using martingale central limit theorems [11] ensuring that $\hat{\boldsymbol{\beta}}$ is consistent* and that the distribution of $n^{1/2}(\hat{\boldsymbol{\beta}} - \boldsymbol{\beta}_0)$ converges to a p-variate normal distribution with mean $\mathbf{0}$ and a covariance matrix $\boldsymbol{\Sigma}$, where $\boldsymbol{\Sigma}^{-1}$ can be estimated consistently by

$$-\frac{1}{n}\frac{\partial^2}{\partial \boldsymbol{\beta}^2} \log L(\boldsymbol{\beta}),$$

where $\beta = \hat{\beta}$ is inserted. Furthermore, it can be proved that the process $n^{1/2}(\hat{\Lambda} - \Lambda_0)$ converges weakly on any compact interval $[0, \tau]$ to a mean zero Gaussian process* whose variance function can be estimated consistently by

$$\int_0^t S_n(u, \hat{\beta}) \, d\overline{N}(u) + \mathbf{H}(\hat{\beta}, t)' \hat{\Sigma} \mathbf{H}(\hat{\beta}, t),$$

$$\mathbf{H}(\beta, t) = -\int_0^t \sum_{i=1}^n Y_i(u) \mathbf{Z}_i(u) e^{\beta' \mathbf{Z}_i(u)}$$

$$\times S_n((u, \beta))^2 \, d\overline{N}(u).$$

Here the key step is to note that the process

$$\int_0^t S_n(u, \beta_0) \, d\overline{N}(u) - \Lambda_0(t)$$

is a local square integrable martingale. From the estimate $\hat{\Lambda}(t)$ of the integrated underlying hazard, an estimate of $\lambda_0(t)$ can be obtained using kernel function smoothing [10]; thus $\lambda_0(t)$ can be estimated by

$$\hat{\lambda}_0(t) = \frac{1}{b} \int_0^t K\left(\frac{t - u}{b}\right) d\hat{\Lambda}(u),$$

where the kernel function K is nonnegative with support on $[-1, 1]$ and integral 1 and the window b is a positive parameter.

HYPOTHESIS TESTING AND MODEL CHECKING

The preceding results indicate that when analyzing a multivariate Cox regression model, the usual large-sample tests for maximum likelihood estimates such as the Wald test or the likelihood ratio test* can be applied to $\hat{\beta}$.

Also graphical checks of the assumption of the covariates having a multiplicative effect on the intensity can be performed using the techniques known from the analysis of Cox's regression model for survival data.

References

[1] Aalen, O. O. (1978). *Ann. Statist.*, **6**, 701–726. (The fundamental paper on nonparametric inference in counting processes.)

[2] Andersen, P. K., and Gill, R. D. (1982). *Ann. Statist.*, **10**, 1100–1120. (Presents the multivariate Cox regression model and proofs for asymptotic properties of estimators.)

[3] Andersen, P. K., and Rasmussen, N. K. (1982). *Research Report 82/6.* Statistical Research Unit, Copenhagen. (Presents an analysis of a set of data using the multivariate Cox regression model.)

[4] Andersen, P. K. (1985). In *Longitudinal Analysis of Labor Market Data*, J. J. Heckman and B. Singer, eds. Cambridge University Press, New York. (Discusses statistical models using multivariate counting processes for labor market dynamics.)

[5] Cox, D. R. (1972). *J. R. Statist. Soc. B*, **34**, 187–220. (The fundamental paper where Cox's regression model for survival data is presented.)

[6] Dolivo, F. G. (1974). *Counting Processes and Integrated Conditional Rates: A Martingale Approach with Application to Detection Theory.* Ph.D. thesis, University of Michigan, Ann Arbor, MI.

[7] Gill, R. D. (1980). In *Mathematical Centre Tracts*, Vol. 124: *Censoring and Stochastic Integrals.* Mathematisch Centrum, Amsterdam. (Gives a rigorous treatment of the application of counting processes to survival data.)

[8] Jacobsen, M. (1982). In *Springer Lecture Notes in Statistics*, Vol. 12: *Statistical Analysis of Counting Processes.* Springer Verlag, New York. (A self-contained textbook particularly suitable for graduate courses.)

[9] Johansen, S. (1983). *Int. Statist. Rev.*, **51**, 258–262. (Discusses MLE in the multivariate Cox regression model.)

[10] Ramlau-Hansen, H. (1983). *Ann. Statist.*, **11**, 453–466.

[11] Rebolledo, R. (1980). *Wahrscheinlichkeitsth.*, **51**, 269–286. (Presents CLT for local martingales.)

[12] Self, S. G. and Prentice, R. L. (1982). *Ann. Statist.*, **10**, 1121–1124.

[13] Tsiatis, A. A. (1981). *Ann. Statist.*, **9**, 93–108. (Proves large-sample results in Cox's regression model for survival data.)

(COX'S REGRESSION MODEL LIMIT THEOREM, CENTRAL MARTINGALES SURVIVAL ANALYSIS)

PER KRAGH ANDERSEN

MULTIVARIATE CRAMÉR–VON MISES STATISTICS

Let $Y_1, \ldots, Y_n$ be independent random variables (rvs) uniformly distributed over the d-dimensional unit cube I^d ($d \geq 1$), and let

$E_n(y)$ be the empirical distribution function of these d-dimensional rvs, i.e., for $y = (y_1, \ldots, y_d) \in I^d$, $E_n(y)$ is the proportion of $Y_j = (Y_{j1}, \ldots, Y_{jd})$, $j = 1, \ldots, n$, whose components are less than or equal to the corresponding components of y, conveniently written as

$$E_n(y) = E_n(y_1, \ldots, y_d)$$
$$= n^{-1} \sum_{j=1}^{n} \prod_{i=1}^{d} I_{[0, y_i]}(Y_{ji}), \quad (1)$$

where, for any subset B of the line,

$$I_B(u) = \begin{cases} 1 & \text{if } u \in B, \\ 0 & \text{if } u \notin B. \end{cases} \quad (2)$$

The corresponding *uniform empirical process* α_n is defined by

$$\alpha_n(y) = n^{1/2}\{E_n(y) - \lambda(y)\},$$
$$y \in I^d, \quad d \geqslant 1, \quad (3)$$

where

$$\lambda(y) = \prod_{i=1}^{d} y_i.$$

In the context of continuous distribution functions F on d-dimensional Euclidean space $\mathbb{R}^d$ ($d \geqslant 1$), this process occurs in the following way. Let $\mathcal{F}$ be the class of continuous distribution functions on $\mathbb{R}^d$, and let $\mathcal{F}_0$ the subclass consisting of every member of $\mathcal{F}$ that is a product of its associated one-dimensional marginal distribution functions. Let $X_1, \ldots, X_n$ be independent random d-vectors with a common distribution function $F \in \mathcal{F}$, and let $F_n(x)$ be the empirical distribution function of these d-dimensional rvs, i.e., for $x = (x_1, \ldots, x_d) \in \mathbb{R}^d$ (cf. (1))

$$F_n(x) = F_n(x_1, \ldots, x_d)$$
$$= n^{-1} \sum_{j=1}^{n} \prod_{i=1}^{d} I_{(-\infty, x_i]}(X_{ji}). \quad (4)$$

Consider now the *empirical process*

$$\beta_n(x) = n^{1/2}\{F_n(x) - F(x)\},$$
$$x \in \mathbb{R}^d, \quad d \geqslant 1. \quad (5)$$

Let $y_i = F_{(i)}(x_i)$ $(i = 1, \ldots, d)$ be the ith marginal distribution of $F \in \mathcal{F}$ and let $F_{(i)}^{-1}(y_i) = \inf\{x_i \in \mathbb{R}^1 : F_{(i)}(x_i) \geqslant y_i\}$ be its inverse. Define the mapping $L^{-1} : I^d \to \mathbb{R}^d$ by

$$L^{-1}(y_1, \ldots, y_d)$$
$$= \left(F_{(1)}^{-1}(y_1), \ldots, F_{(d)}^{-1}(y_d)\right),$$
$$y = (y_1, \ldots, y_d) \in I^d \quad (d \geqslant 1). \quad (6)$$

Then (cf. (3) and (5)), whenever $F \in \mathcal{F}_0$,

$$\alpha_n(y) = \beta_n\left(L^{-1}(y)\right),$$
$$y = (y_1, \ldots, y_d) \in I^d \quad (d \geqslant 1), \quad (7)$$

i.e., if $F \in \mathcal{F}_0$, then the empirical process β_n is distribution-free (does not depend on the distribution function F). In statistical terminology, we say that when we are testing the independence null hypothesis

$$H_0 : F \in \mathcal{F}_0$$

against the alternative

$$H_1 : F \in \mathcal{F} - \mathcal{F}_0 \quad (d \geqslant 2), \quad (8)$$

then the null distribution of $\beta_n(L^{-1}(y))$ is that of $\alpha_n(y)$, i.e., the same for all $F \in \mathcal{F}_0$ and for $d = 1$ with F simply continuous. Otherwise (i.e., if H_1 obtains), the empirical process β_n is a function of F and so will be also its distribution.

When testing the null hypothesis H_0 of (8), one of the frequently used statistics is the Cramér–von Mises statistic* $\omega_{n,d}^2$, defined by

$$\omega_{n,d}^2 = \int_{\mathbb{R}^d} \beta_n^2(x) \prod_{i=1}^{d} dF_{(i)}(x_i)$$
$$= \int_{I^d} \alpha_n^2(y) \prod_{i=1}^{d} dy_i$$
$$= n^{-1} \sum_{k=1, j=1}^{n} \left\{ \prod_{i=1}^{d} \left(1 - (y_{ki} \vee y_{ji})\right) \right.$$
$$- \prod_{i=1}^{d} \tfrac{1}{2}(1 - y_{ki}^2) - \prod_{i=1}^{d} \tfrac{1}{2}(1 - y_{ji}^2)$$
$$\left. + 3^{-d} \right\}, \quad (9)$$

$d \geqslant 1$, where $(y_{j1}, \ldots, y_{jd})_{j=1}^{n}$ with $y_{ji} = F_{(i)}(X_{ji})$, $(i = 1, \ldots, d)$, are the observed values of the random sample $X_j = (X_{j1}, \ldots, X_{jd})$, $j = 1, 2, \ldots, n$. One rejects H_0 of (8) if, for a given random sample $X_1, \ldots, X_n$ on F, the computed value of $\omega_{n,d}^2$ is too large for a given level of significance (fixed-size type I error). Naturally, in order to be able to compute the value of $\omega_{n,d}^2$ for a sample, H_0 of (8), i.e., the marginals of F, will have to be completely specified (simple statistical hypothesis). Although it is true that the distribution of $\omega_{n,d}^2$ will not depend on the specific form of these marginals [cf. (7)], the problem of finding and tabulating this distribution is not an easy task.

Let $V_{n,d}(x)$ be the distribution function of the rv $\omega_{n,d}^2$, i.e.,

$$V_{n,d}(x) = \mathbb{P}\{\omega_{n,d}^2 \leqslant x\}, \qquad 0 < x < \infty.$$

$$(10)$$

Csörgő and Stachó (1979) gave a recursion formula for the exact distribution function $V_{n,1}$ of the rv $\omega_{n,1}^2$. The latter in principle is applicable to tabulating $V_{n,1}$ exactly for any given n. Naturally, much work has already been done to compile tables for $V_{n,1}$. A survey and comparison of these can be found in Knott [14], whose results prove to be the most accurate so far. All these results and tables are based on some kind of an approximation of $V_{n,1}$. As to higher dimensions, $d \geqslant 2$, no analytic results appear to be known about the exact distribution function $V_{n,d}$. Hence asymptotic results for the latter are especially important to have around.

Let $\{B(y); y \in I^d\}$ be a Brownian bridge, i.e., a separable Gaussian process* with $EB(y) = 0$ and $EB(x)B(y) = \prod_{i=1}^d (x_i \wedge y_i) - (\prod_{i=1}^d x_i)(\prod_{i=1}^d y_i)$. From invariance principle considerations (cf., e.g., (1.11) and (1.12) in Cotterill and Csörgő [4])

$$\lim_{n \to \infty} V_{n,d}(x) = \mathbb{P}\{\omega_d^2 \leqslant x\} = V_d(x),$$

$$0 < x < \infty, \quad d \geqslant 1, \quad (11)$$

where $\omega_d^2 = \int_{I^d} B^2(y)\,dy$ with $dy = \prod_{i=1}^d dy_i$ from now on.

For the sake of describing the speed of convergence of the distribution functions

$\{V_{n,d}\}_{n=1}^\infty$ to the distribution function V_d of ω_d^2 [cf. (11)], we define

$$\Delta_{n,d} = \sup_{0 < x < \infty} |V_{n,d}(x) - V_d(x)|. \quad (12)$$

Csörgő [6] showed that

$$\Delta_{n,1} = O(n^{-1/2}\log n)$$

and, on the basis of his complete asymptotic expansion for the Laplace transform of the rv $\omega_{n,1}^2$ [cf. (9)], he conjectured that $\Delta_{n,1}$ is of order $1/n$. Indeed, the latter turned out to be correct (cf. Corollary 1: $\Delta_{n,1} = O(n^{-1})$, in Cotterill and Csörgő [4]) and can be deduced from the ground breaking work of Götze [12]. The latter work, when combined with Dugue [10] and Bhattacharya and Ghosh [1], implies an asymptotic expansion of arbitrary order for the distribution function $V_{n,d}$ of (10) and also that $\Delta_{n,d} = O(n^{-1})$, $d \geqslant 1$ (cf. Section 2 and Corollary 3 in Cotterill and Csörgő [3]).

An extensive tabulation of the distribution function V_1 [cf. (11)] can be found in the monograph of Martynov [17], where the theory and applications of a wide range of univariate Cramér–von Mises-type statistics are also surveyed.

There appear to be no tables available for the distribution function $V_{n,d}$ ($d \geqslant 2$) [cf. (10)]. Hence, and in the light of the just quoted result $\Delta_{n,d} = O(n^{-1})$ ($d \geqslant 1$), tables for the distribution function V_d ($d \geqslant 2$) of (11) are of special interest. Durbin [11] tabulated V_d for $d = 2$, and Krivyakova et al. [16] for $d = 3$. Using the characteristic function* of V_d (cf. Dugué [10] and Durbin [11]), Cotterill and M. Csörgő obtained a recursive equation for the cumulants* of the rv ω_d^2 and, using the first six of these cumulants in the Cornish–Fisher* asymptotic expansion, tabulated its critical values for $d = 2, 3, \ldots, 50$ at various levels of significance. These critical values are within 3% of Durbin's values for $d = 2$ and those of Krivyakova et al. for $d = 3$. We note also that errors in the said tables for higher dimensions should be further reduced, because the cumulants of ω_d^2 are $O(e^{-d})$ (cf. Corollary 7 and Remark 3.2 in Cotterill and

Csörgő [4]). As far as we know, there exist no other tables for $d \geq 4$.

As mentioned already, for the sake of computing the value of $\omega_{n,d}^2$ for a sample, H_0 of (8) will have to be completely specified. Hoeffding [13] and Blum et al. [2] suggested an alternative route to testing H_0 of (8). Their approach does not require the specification of the marginals of F under H_0

Let F_{ni} be the marginal empirical distribution function of the ith component of x_j $(j = 1, \ldots, n)$, i.e.,

$$F_{ni}(x_i) = n^{-1} \sum_{j=1}^{n} I_{(-\infty, x_i]}(X_{ji}),$$

$$i = 1, \ldots, d, \quad (13)$$

and define

$$T_n(x) = T_n(x_1, \ldots, x_d)$$

$$= n^{1/2} \left\{ F_n(x) - \prod_{i=1}^{d} F_{ni}(x_i) \right\},$$

$$d \geq 2, \quad (14)$$

where F_n is as in (4). In terms of the mapping L^{-1} of (6), we define t_n, the uniform version of T_n, by

$$t_n(y) : T_n(L^{-1}(y))$$

$$= n^{1/2} \left\{ F_n(F_{(1)}^{-1}(y_1), \ldots, F_{(d)}^{-1}(y_d)) \right.$$

$$\left. - \prod_{i=1}^{d} F_{ni}(F_{(i)}^{-1}(y_i)) \right\}$$

$$= n^{1/2} \left\{ E_n(y) - \prod_{i=1}^{d} E_{ni}(y_i) \right\}, \quad (15)$$

where $E_{ni}(y_i)$ $(i = 1, \ldots, d)$ is the ith uniform empirical distribution function of the ith component of $L(X_j) : (F_{(1)}(X_{j1}), \ldots, F_{(d)}(X_{jd}))$ $(j = 1, \ldots, n)$, and L is the inverse of L^{-1} of (6).

Consequently, $H_0 : F \in \mathscr{F}_0$ of (8) is equivalent to $H_0 : F(L^{-1}(y)) = \prod_{i=1}^{d} y_i = \lambda(y)$, i.e., given H_0, T_n is distribution-free. Hence, in order to study the distribution of T_n under H_0, we may take F to be the uniform distribution on I^d $(d \geq 2)$ and then study the distribution of t_n instead. Blum et al. [2] proposed the following Cramér–von

Mises-type test statistic for H_0:

$$C_{n,d} = \int_{\mathbb{R}^d} T_n^2(x) \prod_{i=1}^{d} dF_{(i)}(x_i) = \int_{I^d} t_n^2(y) \, dy,$$

$$d \geq 2. \quad (16)$$

One rejects H_0 of (8) if for a given random sample $X_1, \ldots, X_n$ on F the computed value of $C_{n,d}$ is too large for a given level of significance.

Let

$$\{ T(y); y \in I^d \, (d \geq 2) \}$$

$$= \left\{ B(y) - \sum_{i=1}^{d} B(\underline{1}, y_i, \underline{1}) \prod_{j \neq i} y_i \right. ;$$

$$\left. y \in I^d \, (d \geq 2) \right\}, \quad (17)$$

where $\{ B(y); y \in I^d \, (d \geq 2) \}$ is a Brownian bridge over the d-dimensional unit interval I^d. Let $\Gamma_{n,d}$ be the distribution of the rv $C_{n,d}$, and let Γ_d be that of the rv $C_d = \int_{I^d} T^2(y) \, dy$. From invariance principle considerations (cf., e.g., (1.21) in Cotterill and Csörgő [4])

$$\lim_{n \to \infty} \Gamma_{n,d}(x) = \Gamma_d(x),$$

$$0 < x < \infty, \quad d \geq 2. \quad (18)$$

Nothing seems to be known about the exact distribution function $\Gamma_{n,d}$ of the rv $C_{n,d}$. As to the speed of convergence in (18), we have (cf. Theorem 1 in Cotterill and Csörgő [3])

$$\nabla_{n,d} : \sup_{0 < x < \infty} |\Gamma_{n,d}(x) - \Gamma_d(x)|$$

$$= \begin{cases} O\left(n^{-(d+1)^{-1}/2}(\log n)^{3/2}\right) & \text{if } d \geq 3, \\ O\left(n^{-1/2} \log^2 n\right) & \text{if } d = 2. \end{cases}$$

$$(19)$$

As far as we know, the rates of (19) for $\nabla_{n,d}$ are the only ones available so far.

As to tables for the distribution function Γ_d, for $d = 2$, Blum et al. [2] obtained the characteristic function* of the distribution function Γ_d of the rv c_d and tabulated its distribution via numerical inversion. The statistic $C_{n,d}$ of (16) itself cannot be computed unless $F \in \mathscr{F}_0$ of H_0 is completely specified. Hoeffding [13] and Blum et al. [2] suggested, for the critical region for H_0 of (8), large values of

$$\hat{C}_{n,d} = \int_{\mathbb{R}^d} T_n^2(x) \, dF_n(x), \qquad d \geqslant 2, \quad (20)$$

or

$$\tilde{C}_{n,d} = \int_{\mathbb{R}^d} T_n^2(x) \prod_{i=1}^{d} dF_{ni}(x_i), \qquad d \geqslant 2.$$

$$(21)$$

These two statistics are equivalent to $C_{n,d}$ in that both converge in distribution to the rv C_d. This was already noted by Blum et al. [2]; for a detailed proof of this statement see Section 4 in Cotterill and Csörgő [3]. Recently DeWet [9] studied a version of (21) in the case of $d = 2$ with some nonnegative weight functions multiplying the integrand T_n^2 of $\tilde{C}_{n,d}$. Koziol and Nemec [15] studied $\hat{C}_{n,d}$ of (20) and its power properties in testing for independence with bivariate normal random observations.

Concerning tables for the distribution function Γ_d for $d \geqslant 2$, Cotterill and Csörgő [3, Section 4] find an expression for the characteristic function of the rv C_d, $d \geqslant 2$, utilizing the representation of the stochastic process $T(y)$ of (17) in terms of Brownian bridges. This enables them to find the first five cumulants of C_d; using these in the Cornish–Fisher* asymptotic expansion, they tabulated its critical values for $d = 2$, ..., 20 at the "usual" levels of significance. These tables and details of the calculations for approximate critical values of the rv C_d for all $d \geqslant 2$ are given in Sections 5 and 6 of Cotterill and Csörgő [3]. Compared with the figures of Blum et al. [2] for $d = 2$, the Cornish–Fisher approximation seems to work quite well. For $d > 2$, we do not know of any other tables for the rv C_d.

Another approach to this problem was suggested by Deheuvels [8], who showed that the Gaussian process $T(y)$ of (17) which approximates $t_n(y)$ of (15) (cf. Theorem 1 in Csörgő [5]) can be decomposed into $2^d - d - 1$ independent Gaussian processes whose covariance functions are of the same structure for all $d \geqslant 2$ as that of $T(y)$ for $d = 2$. If tables for the Cramér–von Mises functionals of these $2^d - d - 1$ inde-

pendent rvs were available, then one could test asymptotically independently whether there are dependence relationships within each subset of the coordinates of $X \in \mathbb{R}^d$, $d \geqslant 2$.

References

[1] Bhattacharya, R. N. and Ghosh, J. K. (1978). *Ann. Statist.*, **6**, 434–451.

[2] Blum, J. R., Kiefer, J., and Rosenblatt, M. (1961). *Ann. Math. Statist.*, **32**, 485–498.

[3] Cotterill, D. S., and Csörgő, M. (1980). "On the Limiting Distribution of and Critical Values for the Hoeffding–Blum–Kiefer–Rosenblatt Independence Criterion." *Carleton Mathematics Lecture Note No. 24*, Carleton University, Ottawa.

[4] Cotterill, D. S., and Csörgő, M. (1982). *Ann. Statist.*, **10**, 233–244.

[5] Csörgő, M., (1979). *J. Multivariate Anal.*, **9**, 84–100.

[6] Csörgő, S., (1976). *Acta Sci. Math. (Szeged)*, **38**, 45–67.

[7] Csörgő, S. and Stachó, L. (1979). *Coll. Math. J. Bolyai* **21**; *Analytic Function Methods in Probability Theory*, B. Gyires, ed. North-Holland, New York, pp. 53–65.

[8] Deheuvels, P. (1981). *J. Multivariate Anal.*, **11**, 102–113.

[9] DeWet, T. (1980). *J. Multivariate Anal.*, **10**, 38–50.

[10] Dugué, D. (1969). *Multivariate Analysis*, Vol. 2, P. R. Krishnaiah, ed. Academic Press, New York pp. 289–301.

[11] Durbin, J. (1970). *Nonparametric Techniques in Statistical Inference*, M. L. Puri, ed. Cambridge University Press, pp. 435–451.

[12] Götze, F. (1979). *Z. Wahrscheinlichkeitsth. Verw. Geb.*, **50**, 333–355.

[13] Hoeffding, W., (1948), *Ann. Math. Statist.*, **19**, 546–557.

[14] Knott, M., (1974), *J. R. Statist. Soc. Ser. B*, **36**, 430–438.

[15] Koziol, J. A. and Nemec, A. F. (1979). *Can. J. Statist.*, **7**, 43–52.

[16] Krivyakova, E. N., Martynow, G. V., and Tyurin, Yu. N. (1977). *Theory Probab. Appl.*, **22**, 406–410.

[17] Martynov, G. V. (1978). *Omega-Square Criteria* (in Russian). Nauka, Moscow.

(CRAMÉR–VON MISES STATISTIC PROCESSES, EMPIRICAL)

MIKLÓS CSÖRGŐ

MULTIVARIATE DIRECTED GRAPHS IN STATISTICS

UNIVARIATE DIRECTED GRAPHS

A directed graph consists of a set of *g nodes* and a set of *directed arcs* connecting pairs of nodes. Such graphs are natural mathematical representations of biological and social networks. They are also used in various other applications, such as statistical geography and transportation networks, and in the study of disease contagion (acquaintance networks). In a social network the nodes of a graph may represent individuals, groups, or even organizations, and the arcs correspond to relationships or choices broadly interpreted to represent any type of binary relationship. It is customary (e.g. see Harary et al. [12]) to use an incidence matrix representation of directed graphs. Thus, corresponding to each graph is an *adjacency matrix* $\mathbf{x} = (x_{ij})$, such that

$$x_{ij} = \begin{cases} 1 & \text{if } i \text{ chooses } j \\ 0 & \text{otherwise,} \end{cases} \quad (1)$$

where $x_{ii} = 0$.

Holland and Leinhardt [13, 14] and Frank [8, 9] summarize the historical development of *random graphs**, for which the observed adjacency matrix is treated as the realization of a matrix random variable $\mathbf{X}$ that has a probability distribution on the set of all directed graphs with g nodes. Typically, the observed features of an empirically constructed directed graph are compared with the distribution of features that is generated by some random graph. This basic idea can be traced back in the social science literature to Moreno [17].

One of the more interesting developments in the modeling of directed graphs is due to Holland and Leinhardt [1981], who begin by assuming independence of relationships amongst *pairs* of nodes or *dyads*. Their basic model can be represented in the form

$$\log \Pr\left[(1 - X_{ij})(1 - X_{ji}) = 1\right] = \lambda_{ij},$$

$$\log \Pr\left[X_{ij}(1 - X_{ji}) = 1\right] = \lambda_{ij} + a_i + \beta_j + \theta,$$
$$(2)$$

$$\log \Pr\left[(1 - X_{ij})X_{ji} = 1\right] = \lambda_{ij} + a_j + \beta_i + \theta,$$

$$\log \Pr\left[X_{ij}X_{ji} = 1\right]$$
$$= \lambda_{ij} + a_i + a_j + \beta_i + \beta_j + 2\theta + \rho.$$

The parameter λ_{ij} is required for normalization purposes (each dyad must be in one of the four possible states), $\{a_i\}$ and $\{\beta_j\}$ are effects that measure the productivity and attractiveness of the nodes, θ is a choice parameter, and ρ is a measure of "reciprocity." Note that model (2) is log-linear in structure. Holland and Leinhardt present iterative methods for maximum likelihood estimation* of the parameters in this model (see the discussion of the estimation of parameters in loglinear models for categorical data in CONTINGENCY TABLES), and Fienberg and Wasserman [6] provide an alternative approach based on a simple transformation of the data and the use of the method of *iterative proportional fitting**. They also suggest several generalizations of the Holland–Leinhardt model where, for example, the parameter ρ in expression (2) is replaced by

$$\rho_{ij} = \rho + \rho_i + \rho_j, \quad (3)$$

where $\sum \rho_i = 0$, and demonstrate how the parameters of this model can also be estimated by iterative proportional fitting.

Two outstanding theoretical statistical problems in connection with the Holland and Leinhardt univariate model and its generalizations are (a) the lack of an appropriate asymptotic framework for inference (see the discussions in Fienberg and Wasserman [7], and Haberman [11], which is needed to carry out goodness-of-fit* tests, and (b) the need for alternative models that allow for dyadic dependence and include the Holland–Leinhardt model as a special case.

MULTIVARIATE DIRECTED GRAPHS

A *multivariate directed graph* is a collection of univariate directed graphs with the same g nodes. (The term *multigraph* is also in widespread use.) If there are R such univariate graphs, then we represent the multivariate graph by the collection of adjacency matrices for the R univariate graphs, $\{\mathbf{x}_1,$

$x_2, \ldots, x_R$}. We may think of the R graphs as representing either R different types of relationships among the g nodes or the same relationship at R different points in time. In either case, we wish to think of an observed multivariate graph as a realization of a random multivariate graph $X = \{X_1, X_2, \ldots, X_R\}$.

In the univariate situation, we saw that each dyad had four possible realizations:

$(1, 1)$:　　arc in both directions,

$(1, 0)$ or $(0, 1)$:　　arc in one direction,

$(0, 0)$:　　no arc.

Now each dyad has 2^{2R} possible realizations.

Fienberg et al. [5] have proposed a class of log-linear models for random multivariate directed graphs that generalize some aspects of the Holland–Leinhardt model to the multivariate case. By sacrificing the node-level parameters $\{a_i\}$ and $\{\beta_j\}$ associated with each univariate graph, these models incorporate not only *reciprocity* effects for dyadic patterns of the form

$$X \underset{\text{Relation } r}{\longleftrightarrow} Y,$$

but also *exchange* effects for patterns of the form

$$X \quad \overset{\text{Relation } r_1}{\underset{\text{Relation } r_2}{\overset{\longrightarrow}{\longleftarrow}}} \quad Y,$$

and *multiplex* choice effects for patterns of the form

$$X \quad \overset{\text{Relation } r_1}{\underset{\text{Relation } r_2}{\overset{\longrightarrow}{\longrightarrow}}} \quad Y,$$

as well as multivariate generalizations of these effects.

Although there are 2^{2R} possible dyadic realizations, we only get to observe

$$2^R + 2^R(2^R - 1)/2$$

states when the nodes lose their individual identities. We can still summarize the data in the adjacency matrices of the multivariate graph by counting every dyad twice, once from the perspective of each node. We end up with a 2^{2R} table of counts, with the 2^R cells corresponding to reciprocal arcs on each relation (both present or both absent) containing double the actual number of dyads, and each of the other $2^{R-1}(2^R - 1)$ patterns yielding two symmetrically placed duplicate counts in the table.

Fienberg et al. [5] show how fitting a simple affine translation of a loglinear model for the $2^R + 2^{R-1}(2^R - 1)$ counts corresponds to fitting standard loglinear models to the 2^{2R} table of duplicated and doubled counts.

In an attempt at combining the univariate approach, involving individual-level parameters, with the multivariate approach involving multirelational parameters, Fienberg et al. [5a] develop an even more elaborate loglinear model approach to the analysis of multivariate directed graphs.

TWO EXAMPLES

Holland and Leinhardt [15] illustrate their univariate model on data collected by Sampson [15], who spent a year observing monks in an American monastery. Sampson measured both negative and positive relationships on four dimensions at five different points in time. The same 18 monks were interviewed at three of these time points. Thus the data can be represented in the form of an $R = 4 \times 2 \times 3 = 24$-variate directed graph involving 18 nodes. Holland and Leinhardt analyze only a single relationship from this data set. Fienberg et al. [5a] analyze simultaneously the data on all 8 relations at one of the 3 intervals.

Galaskiewicz and Marsden [10] and Fienberg et al. [5] describe data from a study of the formal organizations in a small midwest U.S. community of 32,000 persons referred to by the pseudonym Towertown. They focus their analyses on a subset of 73 organizations and their links on three relations: (1) information, (2) money, and (3) support. Thus the original data take the form of three 73×73 adjacency matrices, but the analyses focus on a summary of these in the form of a 2^6 table of counts of pairs of organizations. The full adjacency matrices are available in Fienberg and Galaskiewicz [4]. The most substantial estimated effects in the loglinear models fitted by Fienberg et al. [5] are associated with choices (θ's), reciprocity (ρ's), and a multiplex-reciprocity effect associated with the dyadic pattern:

$$\begin{array}{c} \text{Information} \\ \longleftrightarrow \\ X \qquad\qquad Y \\ \longleftrightarrow \\ \text{Support} \end{array}$$

SOME RELATED STATISTICAL APPROACHES

In a pair of related papers, White et al. [19] and Boorman and White [2] proposed a method, labeled *block modeling*, for the analysis of data in the form of multivariate directed graphs. A *block model* for a network consists of a partition of the nodes into blocks of structurally equivalent nodes (i.e., ones that relate in the same way to all other nodes in the network) and corresponds to a deterministic rather than a stochastic model. Unfortunately, few directed graphs yield exactly to such block models, and substantive social science theory does not always suggest appropriate partitions. Thus White et al. [19] suggested the use of an approach similar to the statistical approach to search for an "acceptable" block model of a particular form, and they demonstrate their approach on Sampson's monastery data [15].

Breiger et al. [3] describe a more general search procedure for a block model structure, based on *hierarchical clustering* * methods and apply their method to a study of directorship interlocks in American industry. These methods are closely related to other exploratory statistical procedures for row-column permutations of a matrix, such as nonmetric *multidimensional scaling* * (see Arabie et al. [1]).

The major drawbacks of block-model methods include their inexplicit use of formal parametric models, the use of arbitrary criterion functions for the choice of partitions, and the inability to distinguish actual structure from chance variation. Their major advantage is that they provide an explicit model for the pattern of responses that many sociometricians find very useful for thinking about sociological theory (see Light and Mullins [16]).

References

[1] Arabie, P., Boorman, S. A., and Levitt, P. R. (1978). *J. Math. Psychol.*, **17**, 21–63.

[2] Boorman, S. A. and White, H. C. (1976). *Amer. J. Sociol.*, **81**, 1384–1446.

[3] Breiger, R. L., Boorman, S. A., and Arabie, P. (1975). *J. Math. Psychol.*, **12**, 328–383.

[4] Fienberg, S. E. and Galaskiewicz, J. (1984). In *Data*, D. Andrews and A. Herzberg, eds. Springer-Verlag, New York, in press.

[5] Fienberg, S. E., Meyer, M. M., and Wasserman, S. S. (1981). In *Interpreting Multivariate Data*, V. Barnett, ed. Wiley, New York, pp. 289–306.

[5a] Fienberg, S. E., Meyer, M. M., and Wasserman, S. S. (1985). *J. Amer. Statist. Ass.*, **80**, in press.

[6] Fienberg, S. E. and Wasserman, S. S. (1981). *Sociol. Methodol.*, **1981**, 156–192.

[7] Fienberg, S. E. and Wasserman, S. S. (1981). *J. Amer. Statist. Ass.*, **76**, 54–57.

[8] Frank, O. (1971). *Statistical Inference in Graphs*. Swedish Research Institute of National Defense, Stockholm.

[9] Frank, O. (1981). *Sociol. Methodol.*, **1981**, 110–155.

[10] Galaskiewicz, J. and Marsden, P. V. (1978). *Soc. Sci. Res.*, **7**, 89–107.

[11] Haberman, S. J. (1981). *J. Amer. Statist. Ass.*, **76**, 60–62.

[12] Harary, F., Norman, R. Z., and Cartwright, D. (1965). *Structural Models: An Introduction to the Theory of Directed Graphs*. Wiley, New York.

[13] Holland, P. W. and Leinhardt, S. (1976). *Sociol. Methodol.*, **1976**, 1–45.

[14] Holland, P. W. and Leinhardt, S. (1979). In *Perspectives on Social Network Research*, P. W. Holland and S. Leinhardt, eds., pp. 63–83.

[15] Holland, P. W. and Leinhardt, S. (1981). *J. Amer. Statist. Ass.*, **76**, 33–50.

[16] Light, J. M. and Mullins, N. C. (1979). In *Perspectives on Social Network Research*, P. W. Holland and S. Leinhardt, eds., pp. 85–118.

[17] Moreno, J. L. (1934). *Who Shall Survive?* Nervous and Mental Disease Publishing Co., Washington, DC.

[18] Sampson, S. F. (1969). *Crisis in a Cloister*. Ph.D. thesis, Dept. of Sociology, Cornell University, Ithaca, NY.

[19] White, H. C., Boorman, S. A., and Breiger, R. L. (1976). *Amer. J. Sociol.*, **81**, 730–780.

Acknowledgment

The preparation of this entry was supported in part by the Office of Naval Research Contract N00014-80-C-0637 to Carnegie–Mellon University.

(CONTINGENCY TABLES
HIERARCHICAL CLUSTERING
ITERATIVE PROPORTIONAL FITTING
MULTIDIMENSIONAL SCALING
RANDOM GRAPHS)

STEPHEN E. FIENBERG

MULTIVARIATE DISTRIBUTIONS

BASIC CONCEPTS

Origins and Uses

Multivariate distributions are defined on finite-dimensional spaces. They serve to model dependent outcomes of random experiments. They derive from other distributions through transformations, projections, convolutions, extreme values, mixing, compounding, truncating, and censoring. From them derive the marginal distributions of various statistics. They characterize random processes through finite-dimensional projections. On occasion, they support probabilistic proofs for mathematical theorems. In short, these distributions arise throughout statistics and applied probability and are essential to understanding those fields.

Origins of this topic are found in the work of Adrian [1], Laplace* [28], Plana [34], Gauss* [10], and Bravais [2] on bivariate normal distributions*. Early developments on the use of multivariate distributions in empirical research include those of Galton* [9], Pearson* [33], Spearman [36], and Student (*see* GOSSET, WILLIAM SEALY) [38].

In practice, the choice of model is often critical. Some problems have a clear mandate; for others, the choice must be guided by experience, conjecture, and empirical validation. To aid in this choice, numerous discrete and continuous multivariate distributions are known. The fitting of frequency curves*, as used in the univariate case, does not extend readily to higher dimensions (*see* FREQUENCY SURFACES, SYSTEMS OF). A contemporary approach to modeling, which stipulates classes of distributions rather than a particular one, is basic in nonparametrics, in studying stochastic dependence*, in modeling systems' reliability* through type of failure rate, and in validating many statistical procedures for use with other than standard distributions. This point of view is developed in the Continuous Distributions section of this article.

The objectives here are to set forth basic concepts, to survey the principal multivariate distributions of continuous and discrete types, and to give some insight regarding their use. Entries on many of these topics are provided elsewhere in this encyclopedia as noted.

The Basic Tools

Let $\mathscr{X}$ be a set and $(\Omega, \mathscr{B}, P)$ a probability space with Ω an event set, $\mathscr{B}$ a field of subsets of Ω, and P a probability measure*. An $\mathscr{X}$-valued random element is a measurable mapping $X(\omega)$ from Ω to $\mathscr{X}$ which, when $\mathscr{X}$ is finite-dimensional such as the Euclidean space $\mathscr{R}^n$, is multivariate. The cumulative distribution function (CDF) of $\mathbf{X}(\omega) = [X_1(\omega), \ldots, X_n(\omega)]$ on $\mathscr{R}^n$ is

$$F(x_1, \ldots, x_n)$$

$$= P\big(\omega : X_1(\omega) \leqslant x_1, \ldots, X_n(\omega) \leqslant x_n\big)$$

$$(1)$$

with values in the unit interval. Corresponding to each CDF is a probability measure*

P_X and, conversely, giving the model $(\mathscr{R}^n, \mathscr{B}_n, P_X)$.

The study of multivariate distributions draws on the calculus of $\mathscr{R}^n$, on integral transforms* of Fourier, Laplace, and Mellin, including characteristic functions* on $\mathscr{R}^n$, and on imbedding other finite-dimensional spaces in $\mathscr{R}^n$. Distributions often can be derived through change of variables and integral transforms, and their properties through inverse images. Generating functions* for joint moments*, cumulants*, factorial moments*, and probabilities are used. Projection methods apply, as the distribution of $\mathbf{X}(\omega)$ on $\mathscr{R}^n$ is characterized by the one-dimensional distributions of every linear transformation of it.

Some multivariate distribution functions admit simple expressions. Others require series expansions subject to problems with convergence and stability. Limit theorems often suggest simple approximations. *See* APPROXIMATIONS TO DISTRIBUTIONS, ASYMPTOTIC EXPANSIONS OF DISTRIBUTIONS, *and* CORNISH–FISHER AND EDGEWORTH EXPANSIONS*.

Types of Distributions

Discrete distributions arise with counting data such as numbers of adult and larval insects. Continuous distributions typically associate with measurements. A formal statement follows.

Suppose $\mathscr{X}$ is finite-dimensional. Each probability measure* P on $(\mathscr{X}, \mathscr{B}, \cdot)$ is decomposable as a mixture

$$P = \alpha_1 P_1 + \alpha_2 P_2 + \alpha_3 P_3;$$
$$\alpha_i \geqslant 0, \quad \alpha_1 + \alpha_2 + \alpha_3 = 1, \quad (2)$$

such that P_1 assigns positive probability to the mass points of P, P_2 has absolute continuity* with respect to Lebesgue (i.e., volume) measure on $(\mathscr{X}, \mathscr{B}, \cdot)$, and P_3 is purely singular on a set in $\mathscr{X}$ having Lebesgue measure zero, often a linear subspace of $\mathscr{X}$.

Corresponding to P_1, P_2, and P_3 on $(\mathscr{R}^n, \mathscr{B}_n, P)$ are CDFs F_1, F_2, and F_3, respectively, as in (1). $F_2(x_1, \ldots, x_n)$ has a corresponding probability density function

(PDF) given by

$$f_2(x_1, \ldots, x_n)$$
$$= \frac{\partial^n}{\partial x_1 \ldots \partial x_n} F_2(x_1, \ldots, x_n) \quad (3)$$

for almost all $\{x_1, \ldots, x_n\}$, and $F_1(x_1, \ldots, x_n)$ has a corresponding probability mass function (PMF)

$$p(x_1, \ldots, x_n) = P(X_1 = x_1, \ldots, X_n = x_n),$$
$$(4)$$

giving the jumps of $F_1(x_1, \ldots, x_n)$ at its mass points.

The study of multivariate distributions is concerned mainly with functions of the continuous [equation (3)] and discrete [equation (4)] types; the principal distributions of these types are surveyed under the Continuous Distributions and Discrete Distributions sections. In practice, P_3 typically is a degenerate distribution, often concentrated on a subspace of $\mathscr{R}^n$ and absolutely continuous there. These pure types may be combined by mixture as in (2).

To fix notation, $\mathscr{R}^n_+$, $\mathscr{F}_{n \times m}$, $\mathscr{S}_m$, and $\mathscr{S}^+_m$ are the positive orthant of $\mathscr{R}^n$, the $(n \times m)$ real matrices, the real symmetric $(m \times m)$ matrices, and the positive definite varieties, respectively. The array $\mathbf{a} \in \mathscr{R}^n$ is a column vector, $\mathbf{a}' = [a_1, \ldots, a_n]$ its transpose, and $\mathbf{I}_n$ the identity of order n. The transpose, inverse, trace, and determinant of $\mathbf{A} \in \mathscr{F}_{n \times n}$ are $\mathbf{A}'$, $\mathbf{A}^{-1}$, $\mathrm{tr}\,\mathbf{A}$, and $|\mathbf{A}|$, respectively. If $\mathbf{y} \in \mathscr{R}^n$ is random, $E(\mathbf{y}) \in \mathscr{R}^n$ and $V(\mathbf{y}) \in \mathscr{S}^+_n$ denote the vector of expected values and the dispersion (or covariance) matrix of $\mathbf{y}$ when defined. If $\mathbf{Y} = [\mathbf{y}_1, \ldots, \mathbf{y}_n]' \in \mathscr{F}_{n \times m}$, conventions for the corresponding arrays are $E(\mathbf{Y}) = [E(Y_{ij})] \in \mathscr{F}_{n \times m}$ and $V(\mathbf{Y}) = V(\mathbf{y}) \in \mathscr{S}^+_{nm}$ such that $\mathbf{y} = [\mathbf{y}'_1, \ldots, \mathbf{y}'_n]' \in \mathscr{R}^{nm}$. The law of distribution of the random element $X(\omega)$ is denoted by $\mathscr{L}(X)$.

CONTINUOUS DISTRIBUTIONS

Some distributions arise from multidimensional central limit theorems*, many serve as models for random experiments, and others are of interest primarily as derived distributions. Prominent among limit distributions and those from which others derive is the multinormal distribution*. Some distributions derived from it are known to be the same for all parent distributions in a class containing symmetric multivariate stable distributions*. These facts bear on robustness* and the validity of normal-theory procedures for use with nonnormal data, including multivariate analysis*, multivariate analysis of variance*, and the multivariate Bartlett test*. Symmetric distributions are surveyed next, invariance properties of distributions derived from them are noted, and, subsequently, derived distributions having these properties are identified. The principal reference for the Continuous Distributions is Johnson and Kotz [25].

Symmetric Distributions

Often multivariate data are scattered more heavily away from the center of location than are multinormal data. Symmetric distributions that are either more or less scattered than multinormal are described here, where symmetry is *invariance under a group of transformations*. For many purposes, these supplant multinormal distributions as models for the outcomes of random experiments.

DISTRIBUTIONS ON $\mathscr{R}^n$. Let $S_n(\boldsymbol{\theta}, \boldsymbol{\Sigma})$ be the class of *ellipsoidal distributions* on $\mathscr{R}^n$ having the typical PDF

$$f(\mathbf{y}) = |\boldsymbol{\Sigma}|^{-1/2} \psi\big((\mathbf{y} - \boldsymbol{\theta})' \boldsymbol{\Sigma}^{-1}(\mathbf{y} - \boldsymbol{\theta})\big) \quad (5)$$

with $\psi(\cdot)$ a suitable function on $[0, \infty)$. $S_n(\mathbf{0}, \mathbf{I}_n)$ contains *isotropic distributions** for which $\mathbf{z}$ and $\mathbf{Qz}$ have the same distribution for every real orthogonal matrix $\mathbf{Q}(n \times n)$. Examples of ellipsoidal distributions are given in Table 1, where Student's t, Cauchy, and stable laws with $\alpha < 2$ have heavier tails than multinormal distributions. It is known that $E(\mathbf{y}) = \boldsymbol{\theta}$ and $V(\mathbf{y}) = \gamma \boldsymbol{\Sigma}$ with $\gamma > 0$ when these moments exist; further properties are given in review articles by Devlin et al. [6] and Chmielewski [3].

DISTRIBUTIONS ON $\mathscr{F}_{n \times m}$. Let $S_{n,m}(\boldsymbol{\Theta}, \boldsymbol{\Gamma} \times \boldsymbol{\Sigma})$, with $\boldsymbol{\Gamma} \times \boldsymbol{\Sigma} = [\gamma_{ij}\boldsymbol{\Sigma}]$, be the class of distributions on $\mathscr{F}_{n \times m}$ having the typical PDF

$$f(\mathbf{Y}) = |\boldsymbol{\Gamma}|^{-m/2} |\boldsymbol{\Sigma}|^{-n/2}$$
$$\times \psi\big(\text{tr}(\mathbf{Y} - \boldsymbol{\Theta})' \boldsymbol{\Gamma}^{-1}(\mathbf{Y} - \boldsymbol{\Theta}) \boldsymbol{\Sigma}^{-1}\big)$$

$$(6)$$

with $\psi(\cdot)$ a function on $[0, \infty)$. This class contains symmetric stable distributions* including the matrix normal distribution (*see* MATRIX-VALUED DISTRIBUTIONS) with $E(\mathbf{Y}) = \boldsymbol{\Theta}$ and $V(\mathbf{Y}) = \boldsymbol{\Gamma} \times \boldsymbol{\Sigma}$, as well as matrix versions of other examples from Table 1. Independence of the rows of $\mathbf{Y} = [\mathbf{y}_1, \ldots, \mathbf{y}_n]'$ and multinormality are linked: If $\mathscr{L}(\mathbf{Y}) \in S_{n,m}(\boldsymbol{\Theta}, \mathbf{I}_n \times \boldsymbol{\Sigma})$, then $\{\mathbf{y}_1, \ldots, \mathbf{y}_n\}$ are mutually independent if and only if $\mathbf{Y}$ is matrix normal on $\mathscr{F}_{n \times m}$.

Table 1   Examples of Spherical Distributions on $\mathscr{R}^n$ Having Probability Density Functions $f(\mathbf{x})$ or Characteristic Functions $\xi(\mathbf{t})$

Type	Description
Multinormal	$f(\mathbf{x}) = c_1 \exp(-\mathbf{x}'\mathbf{x}/2)$
Logistic	$f(\mathbf{x}) = c_2 \exp(-\mathbf{x}'\mathbf{x})/[1 + \exp(-\mathbf{x}'\mathbf{x})]^2$
Pearson type II	$f(\mathbf{x}) = c_3(1 - \mathbf{x}'\mathbf{x})^{\gamma-1}, \gamma > 1$
Pearson type VII	$f(\mathbf{x}) = c_4(1 + \mathbf{x}'\mathbf{x})^{-\gamma}, \gamma > n/2$
Student's t	$f(\mathbf{x}) = c_5(1 + \mathbf{x}'\mathbf{x})^{-(\nu+n)/2}, \nu$ a positive integer
Cauchy	$f(\mathbf{x}) = c_6(1 + \mathbf{x}'\mathbf{x})^{-(n+1)/2}$
Scale mixtures	$f(\mathbf{x}) = c_7 \int_0^\infty t^{-n/2} \exp(-\mathbf{x}'\mathbf{x}/2t) \, dG(t), G(t)$ a CDF
Stable laws	$\xi(\mathbf{t}) = \exp[\gamma(\mathbf{t}'\mathbf{t})^{\alpha/2}], 0 < \alpha \leqslant 2$ the index

More generally, $L_{n,m}(\Theta, \Gamma, \Sigma)$ is the class of matrix distributions having the typical PDF

$$f(\mathbf{Y}) = |\Gamma|^{-m/2}|\Sigma|^{-n/2}\phi(\mathbf{D}'\Gamma^{-1}\mathbf{D})$$

$$\text{with} \quad \mathbf{D} = (\mathbf{Y} - \Theta)\Sigma^{-1/2} \quad (7)$$

with $\phi(\cdot)$ a function on $\mathscr{S}_m^+$ and $\Sigma^{1/2}$ a factor of Σ. A subclass of these is $S_{n,m}(\Theta, \Gamma \times \Sigma)$. Distributions in $L_{n,m}(\Theta, \mathbf{I}_n, \Sigma)$ have the property that $\mathbf{Y}$ and $\mathbf{QY}$ have the same distribution for every real orthogonal matrix $\mathbf{Q}(n \times n)$ (cf. Dempster [5] and Dawid [4]). For a treatment of the class $L_{n,m}(\Theta, \Gamma, \Sigma)$, see ref. 22.

INVARIANCE PROPERTIES. Basic distributions derived from these classes are invariant. Let $\mathscr{M}$ be a subspace of $\mathscr{R}^n$ or $\mathscr{F}_{n \times m}$ as appropriate; let T be a mapping to a finite-dimensional space $\mathscr{T}$, and consider parametric families generated by $S_n(\theta, \Sigma)$, $S_{n,m}(\Theta, \Gamma \times \Sigma)$ and $L_{n,m}(\Theta, \Gamma, \Sigma)$ as their parameters are varied. The following summary is from ref. 22, where proofs are provided. These facts apply to the possible invariance of derived distributions not considered here.

Property 1. If $T(c(\mathbf{y} + \mathbf{m})) = T(\mathbf{y})$ for each $c > 0$ and $\mathbf{m} \in \mathscr{M} \subset \mathscr{R}^n$, then the distribution of $T(\mathbf{y})$ is invariant for all distributions $\mathscr{L}(\mathbf{y})$ in the class $\{ S_n(\theta, \Sigma); \ \theta \in \mathscr{M}, \ \Sigma \in \mathscr{S}_n^+ \}$ of families on $\mathscr{R}^n$.

Property 2. If $T(c(\mathbf{Y} + \mathbf{M})) = T(\mathbf{Y})$ for each $c > 0$ and $\mathbf{M} \in \mathscr{M} \subset \mathscr{F}_{n \times m}$, then $\mathscr{L}(T(\mathbf{Y}))$ is invariant for all distributions $\mathscr{L}(\mathbf{Y})$ in the class $\{ S_{n,m}(\Theta, \Gamma \times \Sigma); \ \Theta \in \mathscr{M}, \ \Gamma \times \Sigma \in \mathscr{S}_{nm}^+ \}$ of families on $\mathscr{F}_{n \times m}$.

Property 3. If $T((\mathbf{Y} + \mathbf{M})\mathbf{B}) = T(\mathbf{Y})$ for each $\mathbf{M} \in \mathscr{M} \subset \mathscr{F}_{n \times m}$ and each nonsingular $\mathbf{B}(m \times m)$, then $\mathscr{L}(T(\mathbf{Y}))$ is invariant for all distributions $\mathscr{L}(\mathbf{Y})$ in $\{ L_{n,m}(\Theta, \Gamma, \Sigma); \ \Theta \in \mathscr{M}, \ \Gamma \in \mathscr{S}_n^+, \ \Sigma \in \mathscr{S}_m^+ \}$.

The principal multivariate continuous distributions are surveyed next by name, although terminology is not yet standard. The multinormal members of $S_n(\theta, \Sigma)$ and $S_{n,m}(\Theta, \Gamma \times \Sigma)$ are denoted by $N_n(\theta, \Sigma)$ and $N_{n,m}(\Theta, \Gamma \times \Sigma)$, respectively. $\chi^2(\nu, \lambda)$ denotes the noncentral chi-square distribution* having ν degrees of freedom and noncentrality parameter λ, and the central case is abbreviated to $\chi^2(\nu)$.

Gamma Distributions

Matrix and vector generalizations of the gamma* and chi-square* distributions are considered.

MATRIX DISTRIBUTIONS. Suppose $\Sigma(m \times m)$ is positive definite, $\mathbf{W}$ is random with values in $\mathscr{S}_m^+$, and $K(\cdot)$ is a constant. The PDF with $\lambda > 0$, $\mathbf{W} \in \mathscr{S}_m^+$ given by

$$f(\mathbf{W}) = K(\lambda, \Sigma)|\mathbf{W}|^{\lambda - 1}\exp(-\operatorname{tr}\mathbf{W}\Sigma^{-1}),$$

$$(8)$$

$f(\mathbf{W}) = 0$ otherwise, is that of a *matric gamma distribution* [29, pp. 40 ff.]. If $\mathbf{W} = \mathbf{Y}'\mathbf{Y}$ with $\mathscr{L}(\mathbf{Y}) \in L_{n,m}(\mathbf{0}, \mathbf{I}_n, \Sigma)$ as in (7) and $n \geqslant m$, then the PDF of $\mathbf{W}$ is

$$f(\mathbf{W}) = K(n, m, \Sigma)|\mathbf{W}|^{(n-m-1)/2}$$

$$\times \phi(\Sigma^{-1/2}\mathbf{W}\Sigma^{-1/2}) \quad (9)$$

for $\mathbf{W} \in \mathscr{S}_m^+$, $f(\mathbf{W}) = 0$ otherwise, a result of Hsu [15]. Also, $\mathscr{L}(\mathbf{Y}) = N_{n,m}(\mathbf{M}, \mathbf{I}_n \times \Sigma)$ with $n \geqslant m$ and if $\mathbf{W} = \mathbf{Y}'\mathbf{Y}$, then $\mathbf{W}$ has a *noncentral Wishart distribution*, denoted by $W_m(n, \Sigma, \Lambda)$, with noncentrality $\Lambda = \mathbf{M}'\mathbf{M}$. The central version is $W_m(n, \Sigma)$; its PDF is a special case of (8) and (9); the noncentral PDF has a series expansion in special polynomials [25, pp. 170 ff.].

Wishart matrices arise in multinormal sampling, e.g., as the sample dispersion matrix, and otherwise in multivariate distribution theory. Parallel remarks apply to (9) and the class $L_{n,m}(\mathbf{M}, \mathbf{I}_n, \Sigma)$. The noncentral Wishart distribution, intractable numerically, has approximations based on the following. As $n \to \infty$, the limit distribution is multinormal for standardized central and noncentral matrices, and for fixed n it is asymptotically multinormal as the noncentrality parameters grow in a specified manner [19].

DISTRIBUTIONS ON $\mathscr{R}_+^m$. The diagonal elements of $\mathbf{W} = [W_{ij}]$ arise in the analysis of variance*, time-series* analyses, multiple comparisons*, the analysis of multidimensional contingency tables*, extensions of Friedman's chi-square test*, and elsewhere in statistical methodology. There is a multivariate gamma distribution* for case (8), a multivariate chi-square distribution when $\mathbf{W}$ is Wishart, and a multivariate exponential distribution* in the central case with $n = 2$. The joint distribution of $\{W_{11}^{1/2}, W_{22}^{1/2}, \ldots, W_{mm}^{1/2}\}$, a multivariate Rayleigh distribution, arises in the detection of signals from noise [30]. More general Rayleigh distributions* are known [17] as are more general multivariate chi-square distributions with differing marginal degrees of freedom [18].

Densities of these distributions are intractable, apart from special cases. However, as $n \to \infty$, the standardized chi-square and Rayleigh distributions in the limit are multinormal for both central and noncentral cases, and for fixed n, the limits again are multinormal as the noncentrality parameters grow [16]. Another approximation is based on normalizing transformations [20].

Student Distributions

Vector and matrix versions of Student's statistic are considered. Central versions of their distributions are invariant given the symmetry of the parent distribution.

DISTRIBUTIONS ON $\mathscr{R}^m$. There are two basic types. Suppose $[X_1, \ldots, X_m]$ is multinormal with means $[\mu_1, \ldots, \mu_m]$, unit variances, and correlation matrix $\mathbf{R}(m \times m)$. A type I distribution (see MULTIVARIATE t DISTRIBUTION) is that of $\{t_j = X_j / S, \ j = 1, \ldots, m\}$ such that the distribution of νS^2 is $\chi^2(\nu)$ independently of $[X_1, \ldots, X_m]$. A type II distribution is that of $\{t_j = X_j / S_{jj}^{1/2}, \ j = 1, \ldots, m\}$ such that $\nu[S_{jj}]$ is Wishart, that is, $W_m(\nu, \mathbf{R})$, independently of $[X_1, \ldots, X_m]$. Both types are central if $\mu_1 = \cdots = \mu_m = 0$ and are noncentral otherwise. These arise in multiple comparisons* procedures, in the construction of rectangular confidence sets for means, in the Bayesian analysis of multinormal data (see MULTIVARIATE ANALYSIS

(BAYESIAN)), and in various multistage procedures. See Johnson and Kotz [25, Chap. 37].

More generally, if $\mathscr{L}(X_1, \ldots, X_m, Z_1, \ldots, Z_\nu)$ is in the class $S_n(\boldsymbol{\theta}, \boldsymbol{\Gamma})$ with $\boldsymbol{\theta}' = [\mu_1, \ldots, \mu_m, 0, \ldots, 0]$ and $\boldsymbol{\Gamma} = \text{diag}(\mathbf{R}, \mathbf{I}_\nu)$, a block-diagonal matrix, then with $\nu S^2 = Z_1^2 + \cdots + Z_\nu^2$, the central distribution of $\{t_j = X_j / S; \ j = 1, \ldots, m\}$ is type I multivariate t for all distributions in $S_n(\boldsymbol{\theta}, \boldsymbol{\Gamma})$ having the required structure. This follows from Property 1, so that normal-theory multiple comparisons using $\{t_1, \ldots, t_m\}$ are exact in linear models having spherical errors [21].

Similarly, if $\mathscr{L}(\mathbf{Y}) \in S_{n,m}(\boldsymbol{\Theta}, \mathbf{I}_n \times \boldsymbol{\Sigma})$ with parameters $\boldsymbol{\Theta} = [\boldsymbol{\theta}, \ldots, \boldsymbol{\theta}]'$, $\boldsymbol{\theta} \in \mathscr{R}^m$, if $X_j = n^{1/2} \overline{Y}_j$ with $\overline{Y}_j = (Y_{1j} + \cdots + Y_{nj})/n$, $j = 1, \ldots, m$, and if $\mathbf{S}$ is the sample dispersion matrix, then Property 2 asserts that the central distribution of $\{t_j = X_j / S_{jj}^{1/2}; \ j = 1, \ldots, m\}$ is type II multivariate t for every $\mathscr{L}(\mathbf{Y})$ in $S_{n,m}(\mathbf{0}, \mathbf{I}_n \times \boldsymbol{\Sigma})$. Noncentral distributions generally depend on the particular distribution in $S_{n,m}(\boldsymbol{\Theta}, \mathbf{I}_n \times \boldsymbol{\Sigma})$.

MATRIC t DISTRIBUTIONS*. Let $\mathbf{Y}$ and $\mathbf{W}$ be independent with $\mathscr{L}(\mathbf{Y}) = N_{k,m}(\mathbf{0}, \mathbf{I}_k \times \boldsymbol{\Sigma})$ and $\mathscr{L}(\mathbf{W}) = W_m(\nu, \boldsymbol{\Sigma})$ such that $\nu \geqslant m$, and let $\mathbf{T} = \mathbf{Y}\mathbf{W}^{-1/2}$ using any factorization $\mathbf{U}'\mathbf{U}$ of $\mathbf{W}$ with $\mathbf{W}^{1/2} = \mathbf{U}$. Then $\mathbf{T}$ has a matric t distribution; for origins, uses, properties, extensions, and references see the appropriate entry.

Alternatively, consider $\mathbf{X} = [\mathbf{Y}', \mathbf{Z}']'$ with distribution in $S_{n,m}(\mathbf{0}, \mathbf{I}_n \times \boldsymbol{\Sigma})$ such that $n = k + \nu$ and $\nu \geqslant m$, and again let $\mathbf{T} = \mathbf{Y}\mathbf{W}^{-1/2}$ with $\mathbf{W} = \mathbf{Z}'\mathbf{Z}$. These variables arise from distributions in $S_{n,m}(\mathbf{0}, \mathbf{I}_n \times \boldsymbol{\Sigma})$ in the same manner as for the multinormal case. From Property 2, $\mathbf{T}$ has a matric t distribution* for every distribution $\mathscr{L}(\mathbf{Y})$ in $S_{n,m}(\mathbf{0}, \mathbf{I}_n \times \boldsymbol{\Sigma})$. This invariance property of $\mathscr{L}(\mathbf{T})$ transfers directly to the scaled distribution $\mathscr{L}(\mathbf{ATB})$ considered by Dickey [7] with $\mathbf{A}$ and $\mathbf{B}$ nonsingular.

Beta and F Distributions

If X and Y are independent gamma* variates having common scale, then $U = X/(X + Y)$ has a beta distribution* and

$V = X / Y$ has an inverted beta distribution*, with the Snedecor–Fisher F distribution* as a special case. This section treats vector and matrix versions of these.

DIRICHLET DISTRIBUTIONS*. If $\{Z_0, Z_1, \ldots, Z_k\}$ are independent gamma variates having common scale and shape parameters $\{\alpha_0, \alpha_1, \ldots, \alpha_k\}$ and if $T = Z_0 + Z_1 + \cdots + Z_k$, then the joint distribution of $\{U_j = Z_j / T; \ j = 1, \ldots, k\}$ is the *k-dimensional Dirichlet distribution* $\mathscr{D}(\alpha_0, \alpha_1, \ldots, \alpha_k)$. An important case is that $\{\alpha_j = \nu_j / 2; \ j = 0, 1, \ldots, k\}$ with $\{\nu_0, \nu_1, \ldots, \nu_k\}$ as positive integers and $\{Z_0, Z_1, \ldots, Z_k\}$ as independent chi-square* variates. However, in this case neither independence nor chi-square distributions are required. For if $\mathbf{y} = [\mathbf{y}_0', \mathbf{y}_1', \ldots, \mathbf{y}_k']' \in \mathscr{R}^n$ with $\{\mathbf{y}_j \in \mathscr{R}^{\nu_j}; \ j = 0, 1, \ldots, k\}$ and $n = \nu_0 + \nu_1 + \cdots + \nu_k$ such that $\mathscr{L}(\mathbf{y}) \in S_n(\mathbf{0}, \mathbf{I}_n)$, then Property **1** assures that $\{U_j = \mathbf{y}_j' \mathbf{y}_j / T; \ j = 1, \ldots, k\}$, with $T = \mathbf{y}_0' \mathbf{y}_0 + \mathbf{y}_1' \mathbf{y}_1 + \cdots + \mathbf{y}_k' \mathbf{y}_k$, has the distribution $\mathscr{D}(\nu_0 / 2, \nu_1 / 2, \ldots, \nu_k / 2)$.

A *matric Dirichlet distribution* is known [31] for which $\{\mathbf{S}_0, \mathbf{S}_1, \ldots, \mathbf{S}_k\}$ are independent Wishart matrices with $\{\mathscr{L}(\mathbf{S}_j) = W_m(\nu_j, \mathbf{\Sigma}), \nu_j \geqslant m; j = 0, 1, \ldots, k\}$. If

$$\mathbf{W}_j = \left(\sum_{j=0}^{k} \mathbf{S}_j \right)^{-1/2} \mathbf{S}_j \left(\sum_{j=0}^{k} \mathbf{S}_j \right)^{-1/2},$$

$$j = 1, \ldots, k, \quad (10)$$

then for any choice of square root their joint P.D.F. is

$$f(\mathbf{W}_1, \ldots, \mathbf{W}_k) = K \left(\prod_{j=1}^{k} |\mathbf{W}_j|^{(\nu_j - m - 1)/2} \right)$$

$$\times \left| \mathbf{I}_m - \sum_{j=1}^{k} \mathbf{W}_j \right|^{(\nu_0 - m - 1)/2}$$

$$(11)$$

for $\mathbf{W}_j$ and $(\mathbf{I}_m - \sum_{j=1}^{k} \mathbf{W}_j)$ positive definite; $f(\mathbf{W}_1, \ldots, \mathbf{W}_k) = 0$, otherwise (see ref. 25, p. 234). As before, neither independence nor Wishart distributions are required. For if $\mathbf{Y} = [\mathbf{Y}_0', \mathbf{Y}_1', \ldots, \mathbf{Y}_k']' \in \mathscr{F}_{n \times m}$ with $n = \nu_0 + \nu_1 + \cdots + \nu_k$, such that $\nu_j \geqslant m$ and $\mathscr{L}(\mathbf{Y}) \in S_{n,m}(\mathbf{0}, \mathbf{I}_n \times \mathbf{\Sigma})$, then Property

2 assures that the joint P.D.F. of $\{\mathbf{W}_1, \ldots, \mathbf{W}_k\}$, with $\{\mathbf{S}_j = \mathbf{Y}_j' \mathbf{Y}_j; \ j = 0, 1, \ldots, k\}$, is identical to (11) for every distribution $\mathscr{L}(\mathbf{Y})$ in $S_{n,m}(\mathbf{0}, \mathbf{I}_n \times \mathbf{\Sigma})$.

Connections among these distributions follow. When $m = 1$, equation (11) is Dirichlet. The ratios of quadratic forms*

$$U_j(\mathbf{a}) = \mathbf{a}' \mathbf{S}_j \mathbf{a} / \mathbf{a}' (\mathbf{S}_0 + \mathbf{S}_1 + \cdots + \mathbf{S}_k) \mathbf{a},$$

$$j = 1, \ldots, k \quad (12)$$

for fixed $\mathbf{a} \in \mathscr{R}^m$ and the ratios of traces,

$$U_j = \operatorname{tr} \mathbf{S}_j / \operatorname{tr}(\mathbf{S}_0 + \mathbf{S}_1 + \cdots + \mathbf{S}_k),$$

$$j = 1, \ldots, k, \quad (13)$$

are both Dirichlet. The special case of (11) with $k = 1$, sometimes called a type I multivariate beta distribution (*see* MATRIX-VARIATE BETA DISTRIBUTION).

INVERTED DIRICHLET AND F DISTRIBUTIONS. The inverted Dirichlet distribution* is that of $\{V_j = Z_j / Z_0; \ j = 1, \ldots, k\}$ when $\{Z_0, Z_1, \ldots, Z_k\}$ are independent gamma variates having common scale and shape parameters $\{\alpha_0, \alpha_1, \ldots, \alpha_k\}$ (see ref. 25, p. 238). The scaled variates $\{V_j^* = \nu_0 Z_j / \nu_j Z_0; \ j = 1, \ldots, k\}$ have a *multivariate F distribution* when $\{\alpha_j = \nu_j / 2; \ j = 0, 1, \ldots, k\}$ with $\{\nu_0, \nu_1, \ldots, \nu_k\}$ as positive integers. This arises in the analysis of variance for ratios of independent mean squares to a common denominator [8]. As before, neither independence nor multinormality are required; take $\{V_j^* = \nu_0 \mathbf{y}_j' \mathbf{y}_j / \nu_j \mathbf{y}_0' \mathbf{y}_0; \ j = 1, \ldots, k\}$ with $\mathscr{L}(\mathbf{y}) \in S_n(\mathbf{0}, \mathbf{I}_n)$ as stipulated for Dirichlet distributions.

An inverted matric Dirichlet distribution is known [31] with $\{\mathbf{S}_0, \mathbf{S}_1, \ldots, \mathbf{S}_k\}$ as before and $\{\mathbf{V}_j = \mathbf{S}_0^{-1/2} \mathbf{S}_j \mathbf{S}_0^{-1/2}; j = 1, \ldots, k\}$ using the symmetric root of $\mathbf{S}_0$. The PDF $f(\mathbf{V}_1, \ldots, \mathbf{V}_k)$ is given in the entry MATRIX-VALUED DISTRIBUTIONS allowing $\mathbf{S}_0$ to be noncentral. The special case with $k = 1$ is sometimes called a *type II multivariate beta distribution*. Neither the independence nor the Wishart distribution is required in the central case; take $\{\mathbf{S}_j = \mathbf{Y}_j' \mathbf{Y}_j; \ j = 0, 1, \ldots, k\}$ as for matric Dirichlet distributions with $\mathbf{Y} = [\mathbf{Y}_0', \mathbf{Y}_1', \ldots, \mathbf{Y}_k']'$, and conclude that $f(\mathbf{V}_1, \ldots, \mathbf{V}_k)$ is invariant for every $\mathscr{L}(\mathbf{Y})$ in $S_{n,m}(\mathbf{0}, \mathbf{I}_n \times \boldsymbol{\Sigma})$.

Some connections among distributions follow. When $m = 1$, $f(\mathbf{V}_1, \ldots, \mathbf{V}_k)$ is the P.D.F. of the inverted Dirichlet distribution. The collections of ratios $\{ V_j(\mathbf{a}) = \mathbf{a}'\mathbf{S}_j\mathbf{a}/\mathbf{a}'\mathbf{S}_0\mathbf{a}; \ j = 1, \ldots, k \}$, for fixed $\mathbf{a} \in \mathscr{R}^m$, and $\{ V_j = \operatorname{tr}\mathbf{S}_j/\operatorname{tr}\mathbf{S}_0; \ j = 1, \ldots, k \}$, both have inverted Dirichlet distributions.

Other distributions of these types are known. Multivariate F distributions having correlated numerators have been found as ratios of multivariate chi-square variates to a common denominator (see ref. 25, p. 240 ff.).

DISTRIBUTIONS OF LATENT ROOTS*. Many problems entail latent roots of random matrices, particularly in multivariate analysis* and in studies of energy levels of physical systems (*see also* MULTIVARIATE ANALYSIS OF VARIANCE). If $\mathbf{S}_0$ and $\mathbf{S}_1$ are independent with $W_m(\nu_0, \boldsymbol{\Sigma})$ and $W_m(\nu_1, \boldsymbol{\Sigma}, \boldsymbol{\Lambda})$ distributions, then central $(\boldsymbol{\Lambda} = \mathbf{0})$ and noncentral joint distributions of the roots of

$$|\mathbf{S}_1 - \ell\, \mathbf{S}_0| = 0 \qquad (14)$$

are known [25, pp. 181–188]: these are the latent roots of $\mathbf{W}_1$ at (10) when $k = 1$. In the central case, an invariance property holds. If $\mathbf{Y} = [\mathbf{Y}_0', \mathbf{Y}_1']$ with $n = \nu_0 + \nu_1$, $\mathbf{S}_0 = \mathbf{Y}_0'\mathbf{Y}_0$ and $\mathbf{S}_1 = \mathbf{Y}_1'\mathbf{Y}_1$, then, by Property 3, the latent root distribution is the same for all $\mathscr{L}(\mathbf{Y})$ in $L_{n,m}(\mathbf{0}, \mathbf{I}_n, \boldsymbol{\Sigma})$. For a fuller discussion and references, *see* LATENT ROOT DISTRIBUTIONS.

The roots of (14) with $\mathbf{S}_0$ replaced by $\boldsymbol{\Sigma}$ arise in tests for hypotheses about dispersion parameters. Sometimes the ratios of roots are required [25, p. 205], in which case there is an invariance property for central distributions. For if $\mathbf{S} = \mathbf{Y}'\mathbf{Y}$, then the joint distributions of ratios of various roots of the equation $|\mathbf{S} - \ell\, \boldsymbol{\Sigma}| = 0$ are invariant for all $\mathscr{L}(\mathbf{Y}) \in S_{\nu,m}(\mathbf{0}, \mathbf{I}_\nu \times \boldsymbol{\Sigma})$ by Property 2.

Other Distributions

Numerous other continuous multivariate distributions are known. Multivariate versions of *Burr distributions** arise through gamma mixtures of independent Weibull distributions* [25, pp. 288–291]. *Multivariate exponential distributions** of various types are treated in this encyclopedia, as are *multivariate stable distributions** and other distributions mentioned earlier. *Multivariate extreme-value distributions* are treated in ref. 25 (pp. 249–260) with emphasis on the bivariate case. The *Beta–Stacy distribution* [25, pp. 273–284] yields a *multivariate Weibull distribution** as a special case. *Multivariate Pareto distributions* [25, pp. 285–288] have their origins in econometrics*. The *multivariate logistic distribution* [25, pp. 291–294] is used to model binary data in the analysis of quantal responses. Kibble [26] used properties of characteristic functions* to obtain a bivariate distribution having normal* and gamma* marginals.

DISCRETE DISTRIBUTIONS

Many discrete distributions* have multivariate extensions. These serve as building blocks for other distributions through *compounding*, in which distributions are assigned to some or all parameters of a family. Here the principal distributions are surveyed and connections among them noted. Generic names are used for distributions including "negative" and "inverse" types. The principal references are Chapter 11 of Johnson and Kotz [24] and selections from Patil and Joshi [32]. Additional references are cited here, including the inequalities of Jogdeo and Patil [23] for a number of discrete multivariate distributions.

Binomial Distributions

The number of successes in n independent Bernoulli trials, each having the probability π of success, has the binomial distribution* $B(n, \pi)$. The number of trials to k successes has a negative binomial distribution*. Some extensions follow.

MULTIVARIATE BINOMIAL DISTRIBUTIONS. The outcome of a random experiment is classified as having or not having each of s attributes $\{A_1, \ldots, A_s\}$. If $\{X_1, \ldots, X_s\}$ are the numbers having these attributes in n independent trials, then theirs is an s-dimensional binomial distribution with parameters

$$\pi_i = P(A_i), \qquad i = 1, \ldots, s$$
$$\pi_{ij} = P(A_iA_j), \quad i \neq j, i, j = 1, \ldots, s$$
$$\vdots \qquad\qquad (15)$$
$$\pi_{12\ldots s} = P(A_1A_2 \ldots A_s).$$

The marginal distribution of X_i is $B(n, \pi_i)$, all having the same index n for $i = 1, \ldots, s$. Bivariate distributions having different indices are treated in refs. 11 and 13.

For sequences of identical experiments, the limiting standardized distribution is multinormal* as $n \to \infty$. For nonidentical sequences such that $\pi_i \to 0$ as $n \to \infty$, $i = 1, \ldots, s$, the limit is a multivariate Poisson distribution under conditions given later. For further developments, see ref. 32, p. 81.

MULTIVARIATE PASCAL DISTRIBUTIONS. Independent trials of the preceding type are continued until exactly k trials exhibit none of the s attributes. The joint distribution of the numbers $\{Y_1, \ldots, Y_s\}$ of occurrences of $\{A_1, \ldots, A_s\}$ during these trials is an s-dimensional Pascal distribution [32, p. 83].

MULTIVARIATE NEGATIVE BINOMIAL DISTRIBUTIONS. The result of using a gamma* variate with parameters (α, k) to scale the parameters Λ of an s-variate Poisson distribution and mixing is an s-variate negative binomial distribution [32, p. 83], its marginals negative binomial. It reduces to the multivariate Pascal distribution when k is an integer and to the negative multinomial distribution on mixing multiple Poisson distributions. *See* NEGATIVE BINOMIAL DISTRIBUTIONS.

Multinomial Distributions

Let $\{A_0, A_1, \ldots, A_s\}$ be exclusive and exhaustive outcomes having probabilities $\{\pi_0, \pi_1, \ldots, \pi_s\}$ with $0 < \pi_i < 1$ and $\pi_0 +$ $\pi_1 + \cdots + \pi_s = 1$. The numbers $\{X_1, \ldots, X_s\}$ of occurrences of $\{A_1, \ldots, A_s\}$ in n independent trials has the multinomial distribution* with parameters $(n, \pi_1, \ldots, \pi_s)$.

NEGATIVE MULTINOMIAL DISTRIBUTIONS. If independent trials are repeated until A_0 occurs exactly k times, the numbers of occurrences of $\{A_1, \ldots, A_s\}$ during these trials have a negative multinomial distribution with parameters $(k, \pi_1, \ldots, \pi_s)$. This distribution arises through mixtures: first as a gamma* mixture of multiple Poisson distributions as noted, second as a negative binomial mixture on n of multinomials (Property 9 of MULTINOMIAL DISTRIBUTIONS). As $k \to \infty$ and $\pi_i \to 0$ such that $k\pi_i \to \lambda_i$, $0 < \lambda_i < \infty$, $i = 1, \ldots, s$, the negative multinomial distribution with parameters $(k, \pi_1, \ldots, \pi_s)$ converges to the multiple Poisson distribution with parameters $(\lambda_1, \ldots, \lambda_s)$. Further properties are given in refs. 24 (p. 292) and 32 (p. 70).

MULTIVARIATE MULTINOMIAL DISTRIBUTIONS. These are the joint distributions of marginal sums in multidimensional contingency tables*. Classify an outcome according to each of k criteria having the exclusive and exhaustive classes $\{A_{i0}, A_{i1}, \ldots, A_{is_i}\}$ for $i = 1, \ldots, k$. If in n independent trials $\{X_{i1}, \ldots, X_{is_i}; i = 1, \ldots, k\}$ are the numbers occurring in $\{A_{i1}, A_{i2}, \ldots, A_{is_i}, i = 1, \ldots, k\}$, then their joint distribution is called a *multivariate* (also *multivector*) *multinomial distribution*, including the k-variate binomial distribution when $s_1 = s_2 = \cdots = s_k = 1$. Further developments are given in refs. 24 (p. 312) and 32 (p. 86).

MULTIVARIATE NEGATIVE MULTINOMIAL DISTRIBUTIONS. Continue independent trials of the preceding type until exactly t trials are classified in all of $\{A_{10}, A_{20}, \ldots, A_{k0}\}$. The numbers occurring in $\{A_{i1}, \ldots, A_{is_i}, i = 1, \ldots, k\}$ during these trials have a *multivariate negative multinomial distribution*, reducing to the negative multinomial distribution when $k = 1$ and to the multivariate Pascal distribution when $s_1 = s_2 = \cdots = s_k$

= 1. For further discussion see ref. 24 (p. 314).

Hypergeometric Distributions*

A collection of N items consists of $s + 1$ types: N_0 of type A_0; N_1 of type $A_1, \ldots$; N_s of type A_s, with $N = N_0 + N_1 + \cdots + N_s$. Random samples are taken from this collection.

MULTIVARIATE HYPERGEOMETRIC DISTRIBUTIONS. In a random sample of n items drawn without replacement, the joint distribution of the numbers of items of types $\{A_1, \ldots, A_s\}$ is an s-dimensional hypergeometric distribution (see HYPERGEOMETRIC DISTRIBUTIONS) with parameters $(n, N, N_1, \ldots, N_s)$. With replacement, their distribution is multinomial with parameters $(n, N_1/N, \ldots, N_s/N)$. As $N \to \infty$ and $N_i \to \infty$ such that $N_i/N \to \pi_i$ with $0 < \pi_i < 1$ and $\pi_1 + \cdots + \pi_s < 1$, the hypergeometric converges to the multinomial distribution with parameters $(n, \pi_1, \ldots, \pi_s)$. If instead $N \to \infty$, $N_i \to \infty$ and $n \to \infty$ such that $N_i/N \to 0$ and $nN_i/N \to \lambda_i$ with $0 < \lambda_i < \infty$, $i = 1, \ldots, s$, then the limit distribution is multiple Poisson with parameters $(\lambda_1, \ldots, \lambda_s)$. For further properties, see refs. 24 (p. 200) and 32 (p. 76); for extensions, see GENERALIZED HYPERGEOMETRIC DISTRIBUTIONS.

MULTIVARIATE INVERSE HYPERGEOMETRIC DISTRIBUTIONS. If successive items are drawn without replacement until exactly k items of type A_0 are drawn, then the numbers of types $\{A_1, \ldots, A_s\}$ thus drawn have an s-variate inverse hypergeometric distribution with parameters $(k, N, N_1, \ldots, N_s)$. As $N \to \infty$, $N_i \to \infty$ such that $N_i/N \to \pi_i$ with $0 < \pi_i < 1$ and $\pi_1 + \cdots + \pi_s < 1$, this distribution converges to the s-variate negative multinomial distribution with parameters $(k, \pi_1, \ldots, \pi_s)$. If instead $N \to \infty$, $N_i \to \infty$, and $k \to \infty$ such that $N_i/N \to 0$ and $kN_i/N \to \lambda_i$ with $0 < \lambda_i < \infty$, $i = 1, \ldots, s$, then the multivariate inverse hypergeometric converges to the multiple Poisson distribution with parameters $(\lambda_1, \ldots, \lambda_s)$. See ref. 32 (p. 76).

MULTIVARIATE NEGATIVE HYPERGEOMETRIC DISTRIBUTIONS. Sampling proceeds in two stages. In the first stage, m items are drawn without replacement, giving $\{x_1, \ldots, x_s\}$ items of types $\{A_1, \ldots, A_s\}$. Without replacing the first sample, n additional items are drawn without replacement at the second stage, giving $\{Y_1, \ldots, Y_s\}$ items of types $\{A_1, \ldots, A_s\}$. The conditional distribution of $\{Y_1, \ldots, Y_s\}$, given that $\{X_1 = x_1, \ldots, X_x = x_s\}$, is a multivariate negative hypergeometric distribution. It arises on compounding the multinomial distribution, with parameters $(n, \pi_1, \ldots, \pi_s)$, by assigning to $(\pi_1, \ldots, \pi_s)$ the s-dimensional Dirichlet* distribution and then mixing. Under alternative conditions, this distribution converges either to the multinomial distribution or to the product of negative binomial distributions. See ref. 32 (p. 77) for further details.

Poisson Distributions

MULTIPLE POISSON DISTRIBUTIONS. If $\{X_1, \ldots, X_s\}$ are independent Poisson* random variables with parameters $\{\lambda_1, \ldots, \lambda_s\}$, their joint distribution is a *multiple Poisson distribution* with parameters $(\lambda_1, \ldots, \lambda_s)$.

MULTIVARIATE POISSON DISTRIBUTIONS. Let $\{X_1, \ldots, X_s\}$ have the multivariate binomial distribution with the parameters as in equation (15), and suppose that $n \to \infty$, $\pi_i \to 0$, $i = 1, \ldots, s$, such that

$$n\left\{ \pi_i - \sum_j \pi_{ij} + \sum_{j<k} \pi_{ijk} - \cdots \right.$$
$$\left. + (-1)^{s-1}\pi_{12\ldots s} \right\} \to \lambda_i,$$

$$n\left\{ \pi_{ij} - \sum_k \pi_{ijk} + \sum_{k<l} \pi_{ijkl} - \cdots \right.$$
$$\left. + (-1)^{s-2}\pi_{12\ldots s} \right\} \to \lambda_{ij},$$

$$\vdots$$

$$n\pi_{12\ldots s} \to \lambda_{12\ldots s}. \qquad (16)$$

Then the limiting distribution of $\{X_1,$

$\ldots, X_s\}$ is a *multivariate Poisson distribution* with parameters given by (16). This distribution also can be derived as the joint distribution of various partial sums of 2^{s-1} independent Poisson random variables with parameters as appropriate. See Johnson and Kotz [24, p. 297] and Patil and Joshi [32, p. 82] for references and further details.

Multivariate Series Distributions

Further classes of discrete multivariate distributions are identified by types of their PMFs.

MULTIVARIATE LOGARITHMIC SERIES DISTRIBUTIONS. These distributions arise through truncation and limits. If $[X_1, \ldots, X_s]$ has the s-variate negative multinomial distribution with parameters $(k, \pi_1, \ldots, \pi_s)$, then the conditional distribution of $[X_1, \ldots, X_s]$, given that $[X_1, \ldots, X_s] \neq [0, \ldots, 0]$, converges to the s-*variate logarithmic series distribution* with parameters $(\theta_1, \ldots, \theta_s)$ as $k \to 0$, where $\theta_i = 1 - \pi_i$, $i = 1, \ldots, s$. See ref. 32 (p. 71) for details. A modified multivariate logarithmic series distribution (see Property 8 of MULTINOMIAL DISTRIBUTIONS) arises as a mixture, on n, of the multinomial distribution with parameters $(n, \pi_1, \ldots, \pi_s)$, where the mixing distribution is a logarithmic series distribution* (see ref. 32, p. 73).

MULTIVARIATE POWER SERIES DISTRIBUTIONS. A class of distributions with parameters $(\theta_1, \ldots, \theta_s) \in \Theta$, derived from convergent power series, has PMFs of the form

$$p(x_1, \ldots, x_s)$$
$$= \frac{a(x_1, \ldots, x_s)\theta_1^{x_1} \cdots \theta_s^{x_s}}{f(\theta_1, \ldots, \theta_s)}, \quad (17)$$

for $\{x_i = 0, 1, 2, \ldots; \ i = 1, \ldots, s\}$, $p(x_1, \ldots, x_s) = 0$, otherwise. Such distributions, called *multivariate power series distributions*, have been studied as a class. This class contains the s-variate multinomial distribution with parameters $(n, \pi_1, \ldots, \pi_s)$, the s-variate logarithmic series distribution with parameters $(\theta_1, \ldots, \theta_s)$; the s-variate

negative multinomial distribution with parameters $(k, \pi_1, \ldots, \pi_s)$; and others. See Patil and Joshi [32, p. 74] and MULTIVARIATE POWER SERIES DISTRIBUTIONS for further properties.

A nonexhaustive sampling of other discrete multivariate distributions is given here. In many applications, compound distributions arise from random environments, where the mixing distribution describes the variation of parameters over the possible environments.

BIVARIATE BOREL–TANNER DISTRIBUTIONS. The Borel–Tanner distribution* is the distribution of the number of customers served before a queue vanishes for the first time. If service in a single-server queue begins with r customers of type I and s of type II with different arrival rates and service needs for each type, then the joint distribution of the numbers served is the *bivariate Borel–Tanner distribution* studied by Shenton and Consul [35].

COMPOUND MULTIVARIATE DISTRIBUTIONS. Numerous discrete bivariate and multivariate distributions have been found as compound distributions. Many of these are well motivated by the structure of the problem at hand. Some examples are listed in Table 2, together with references and brief comments.

References

An asterisk denotes a reference of historical interest.

[1] *Adrian, R. (1808). *The Analyst or Mathematical Museum*, Vol. 1.

[2] *Bravais, A. (1846). *Mém. Inst. Fr.*, **9**, 255–332.

[3] Chmielewski, M. A. (1981). *Int. Statist. Rev.*, **49**, 67–74. (Excellent survey article on ellipsoidal distributions.)

[4] Dawid, A. P. (1977). *J. R. Statist. Soc. B*, **39**, 254–261. (Technical source paper on the structure of distributions.)

[5] Dempster, A. P. (1969). *Elements of Continuous Multivariate Analysis*. Addison-Wesley, London. (General reference featuring a geometric approach.)

Table 2 Some Discrete Multivariate Compound Distributions

Basic Distribution	Mixing Parameters	Mixing Distribution	References	Comments
Bivariate binomial $(n, \pi_{01}, \pi_{10}, \pi_{11})$	n	Poisson	14	Gives bivariate Poisson distribution.
Multinomial $(n, \pi_1, \ldots, \pi_s)$	$(\pi_1, \ldots, \pi_s)$	Dirichlet	24 (p. 309)	Gives s-variate negative hypergeometric distribution.
Multinomial $(n, \pi_1, \ldots, \pi_s)$	n	Logarithmic series	32 (p. 69)	Gives s-variate modified logarithmic series distribution.
Multinomial $(n, \pi_1, \ldots, \pi_s)$	n	Negative binomial	32 (p. 68)	Gives s-variate negative multinomial distribution.
Multinomial $(n, \pi_1, \ldots, \pi_s)$	n	Poisson	32 (p. 69)	Gives multiple Poisson distribution.
Multiple Poisson $(u\lambda_1, \ldots, u\lambda_s)$	u	Gamma	32 (p. 70)	Gives s-variate negative multinomial distribution.
Multiple Poisson $(\lambda_1, \ldots, \lambda_s)$	$(\lambda_1, \ldots, \lambda_s)$	Multinormal	37	Gives s-variate Poisson-normal distribution.
Multiple Poisson $(\lambda, \ldots, \lambda)$ $\lambda = \alpha + (\beta - \alpha)u$	u	Rectangular on (0, 1)	32 (p. 80)	Gives s-variate Poisson-rectangular distribution.
Multivariate Poisson $(u\lambda_i, u\lambda_{ij}, \ldots, u\lambda_{12\ldots s})$	u	Gamma	12 (p. 82)	Gives s-variate negative binomial distribution.
Negative multinomial $(k, \pi_1, \ldots, \pi_s)$	$(\pi_1, \ldots, \pi_s)$	Dirichlet	24 (p. 311) 32 (p. 80)	Gives s-variate negative multinomial-Dirichlet distribution.
Convolution of multinomials $(\gamma_1, \ldots, \gamma_{2^k}, \theta_1, \ldots, \theta_s)$	$(\gamma_1, \ldots, \gamma_{2^k})$	Multivariate hypergeometric	27	Gives the distribution of numbers judged defective of k types in lot inspection.

[6] Devlin, S. J., Gnanadesikan, R., and Kettenring, J. R. (1976). In *Essays in Probability and Statistics*, S. Ikeda et al., eds. Shinko Tsusho, Tokyo. (Excellent survey article on ellipsoidal distributions.)

[7] Dickey, J. M. (1967). *Ann. Math. Statist.*, **38**, 511–518. (Source paper on matric *t* distributions and their applications.)

[8] Finney, D. J. (1941). *Ann. Eugen.*, **11**, 136–140. (Source paper on dependent *F* ratios in the analysis of variance.)

[9] *Galton, F. (1889). *Natural Inheritance*. MacMillan, London.

[10] *Gauss, K. F. (1823). *Theoria Combinationis Observationum Erroribus Minimis Obnoxiae*. Muster-Schmidt, Göttingen.

[11] Hamdan, M. A. (1972). *Int. Statist. Rev.*, **40**, 277–280. (Source paper on bivariate binomial distributions.)

[12] Hamdan, M. A. and Al-Bayyati, H. A. (1971). *J. Amer. Statist. Ass.*, **66**, 390–393. (Source paper on a compound bivariate Poisson distribution.)

[13] Hamdan, M. A. and Jensen, D. R. (1976). *Aust. J. Statist.*, **18**, 163–169. (Source paper on bivariate binomial distributions.)

[14] Hamdan, M. A. and Tsokos, C. P. (1971). *Int. Statist. Rev.*, **39**, 60–63. (Source paper on bivariate compound Poisson distributions.)

[15] Hsu, P. L. (1940). *Proc. Edinburgh Math. Soc. Ser. 2*, **6**, 185–189. (Source paper on generalizations of Wishart's distribution.)

[16] Jensen, D. R. (1969). *SIAM J. Appl. Math.*, **17**, 802–814. (Source paper on limits of certain noncentral distributions.)

[17] Jensen, D. R. (1970). *Sankhyā A*, **32**, 193–208. (Source paper on generalizations of Rayleigh distributions.)

[18] Jensen, D. R. (1970). *Ann. Math. Statist.*, **41**, 133–145. (Source paper on multivariate chi-square and *F* distributions.)

[19] Jensen, D. R. (1972). *Aust. J. Statist.*, **14**, 10–16. (Source paper on limits of noncentral Wishart distributions.)

[20] Jensen, D. R. (1976). *J. Statist. Comp. Simul.*, **4**, 259–268. (Source paper on normalizing bivariate transformations.)

[21] Jensen, D. R. (1979). *Biometrika*, **66**, 611–617. (Source paper on linear models under symmetric errors.)

[22] Jensen, D. R. and Good, I. J. (1981). *J. R. Statist. Soc. B*, **43**, 327–332. (Source paper on invariance of derived distributions under symmetry.)

[23] Jogdeo, K. and Patil, G. P. (1975). *Indian J. Statist.*, **37**, 158–164. (Source paper on probability inequalities for discrete multivariate distributions.)

[24] Johnson, N. L. and Kotz, S. (1969). *Distributions in Statistics: Discrete Distributions*. Wiley, New York. (An excellent primary source with extensive bibliography.)

[25] Johnson, N. L. and Kotz, S. (1972). *Distributions in Statistics: Continuous Multivariate Distributions*. Wiley, New York. (An excellent primary source with extensive bibliography.)

[26] Kibble, W. F. (1941). *Sankhyā*, **5**, 137–150. (Source paper on expansions of bivariate distributions.)

[27] Kotz, S. and Johnson, N. L. (1983). *Commun. Statist. Theor. Meth.*, **12**, 2809–2821.

[28] *Laplace, P. S. (1811). *Mémoires de l'Institut Impérial de France, Année 1810*, 279–347.

[29] Lukacs, E. and Laha, R. G. (1964). *Applications of Characteristic Functions*. Hafner, New York. (Excellent reference with emphasis on multivariate distributions.)

[30] Miller, K. S. (1975). *Multivariate Distributions*. Krieger, Huntington, NY. (An excellent reference with emphasis on problems in engineering and communications theory.)

[31] Olkin, I. and Rubin, H. (1964). *Ann. Math. Statist.*, **35**, 261–269. (Source paper on matric Dirichlet, beta, inverted beta, and related distributions.)

[32] Patil, G. P. and Joshi, S. W. (1968). *A Dictionary and Bibliography of Discrete Distributions*. Hafner, New York. (An excellent primary source with extensive bibliography.)

[33] *Pearson, K. (1905). *Drapers Co. Res. Mem. Biom. Ser.*, **2**, 1–54.

[34] *Plana, G. A. A. (1813). *Mém. Acad. Impériale de Turin, 1811–1812*, **20**, 355–498.

[35] Shenton, L. R. and Consul, P. C. (1973). *Sankhyā A*, **35**, 229–236. (Source paper on bivariate Lagrange and Borel–Tanner distributions and their applications.)

[36] *Spearman, C. (1904). *Amer. J. Psychol.*, **15**, 72–101.

[37] Steyn, H. S. (1976). *J. Amer. Statist. Ass.*, **71**, 233–236. (Source paper on multivariate Poisson-normal distributions.)

[38] *Student (1908). *Biometrika*, **6**, 1–25.

D. R. JENSEN

MULTIVARIATE EXPONENTIAL DISTRIBUTION

A distribution is said to be *multivariate exponential* if its univariate marginals are exponential. A similar property defines other multivariate distributions* discussed here. Many distributions exist with univariate exponential marginals; we shall discuss some of the more important.

A similar situation holds for most other multivariate distributions where the marginals are all members of a well-known univariate family of distributions. The exception is the multivariate normal distribution (*see* MULTINORMAL DISTRIBUTION). Even though there are many multivariate distributions with univariate normal marginals, we generally speak of *the* multivariate normal distribution. Since this distribution has so many desirable properties, it is distinguished from all others having univariate normal marginals.

Some of the earliest examples of multivariate exponential distributions occurred implicitly as special cases of multivariate chi-square and gamma distributions. These can be obtained by choosing parameters such that the univariate marginals reduce to exponentials. One such bivariate example is given by Johnson and Kotz [21, p. 260]. Others can be similarly obtained from the other multivariate gamma distributions in Johnson and Kotz [21, Chap. 40]. (*See also* Chmelynski [9].) In this entry, multivariate expo-

nential distributions that arise in their own right will be the main focus. One of the most cited, Marshall and Olkin [26], will only be briefly treated, since MULTIVARIATE EXPONENTIAL DISTRIBUTION, MARSHALL–OLKIN is devoted to it.

We first discuss bivariate exponential and related bivariate distributions. Following this, the general multivariate case will be considered. Besides the basic reference work of Johnson and Kotz [21], other recent references dealing with multivariate exponential distributions are Basu and Block [3], Arnold [2], Block et al. [7] (whose results are summarized in Block [5]) and Block and Savits [8].

BIVARIATE EXPONENTIALS

One of the first to explicitly discuss bivariate exponential distributions was Gumbel [18]. The joint cumulative distribution functions of two of these are given by

$$F(x, y) = 1 - e^{-x} - e^{-y} + e^{-x-y-\delta xy},$$

$$x \geqslant 0, \quad y \geqslant 0, \quad 0 \leqslant \delta \leqslant 1;$$

$$(1)$$

$$F(x, y) = (1 - e^{-x})(1 - e^{-y})(1 + \alpha e^{-x-y}),$$

$$x \geqslant 0, \quad y \geqslant 0, \quad -1 \leqslant \alpha \leqslant 1.$$

The latter was a special case of a family considered by Morgenstern (*see* FARLIE–GUMBEL–MORGENSTERN DISTRIBUTION). See Johnson and Kotz [21, pp. 23 and 263] for details and references. A third bivariate exponential was considered briefly also.

A distribution that did not have exponential marginals but was one of the first multivariate distributions to be based on a model involving the exponential distribution was that of Freund [13]. This model involves a two-component system where the failure of one component affects the lifetime of the other component. Initially the unaffected lifetimes are exponential; the affected lifetimes are also exponential. (See Block and Savits [8] for an interpretation.) The joint distribution has joint survival distribution function $\Pr\{X_1 > x_1, X_2 > x_2\}$ given by

$$S_{x_1, x_2}$$

$$= \begin{cases} \gamma_2^{-1}\left\{\alpha_1 e^{-\gamma_2 x_1 - \alpha_2' x_2} \right. \\ \qquad \left. + (\alpha_2 - \alpha_2')e^{-(\alpha_1 + \alpha_2)x_2}\right\} \\ \qquad\qquad\qquad\qquad x_1 < x_2, \\ \gamma_1^{-1}\left\{\alpha_2 e^{-\alpha_1' x_1 - \gamma_1 x_2} \right. \\ \qquad \left. + (\alpha_1 - \alpha_1')e^{-(\alpha_1 + \alpha_2)x_1}\right\} \\ \qquad\qquad\qquad\qquad x_1 > x_2, \end{cases}$$

where $\gamma_j = \alpha_1 + \alpha_2 - \alpha_j'$ $(j = 1, 2)$ (2)

for $\alpha_1, \alpha_1', \alpha_2, \alpha_2' > 0$. Properties of this distribution are given in Johnson and Kotz [21]. Since a marginal distribution is not exponential this is called a bivariate exponential extension rather than a bivariate exponential distribution.

An important bivariate exponential distribution was derived by Marshall and Olkin [26]. It has joint survival distribution function

$$S(x_1 x_2)$$

$$= \exp(-\lambda_1 x_1 - \lambda_2 x_2 - \lambda_{12}\max(x_1, x_2))$$
$$\text{for} \quad x_1 > 0, \quad x_2 > 0$$

for $\lambda_1, \lambda_2, \lambda_{12} > 0$. This distribution is derivable from: (a) a fatal shock model; (b) a nonfatal shock model; and (c) a loss-of-memory model. For details on these models and for properties of this distribution, *see* MULTIVARIATE EXPONENTIAL DISTRIBUTION, MARSHALL–OLKIN.

Block and Basu [6] have studied a distribution closely related to the Marshall and Olkin and Freund distributions. It can be obtained by considering a special case of the model (2). The choices

$$\alpha_1 = \lambda_1 + \lambda_{12}\lambda_1(\lambda_1 + \lambda_2)^{-1},$$
$$\alpha_1' = \lambda_1 + \lambda_{12},$$
$$\alpha_2 = \lambda_2 + \lambda_{12}\lambda_2(\lambda_1 + \lambda_2)^{-1}$$
$$\alpha_2' = \lambda_2 + \lambda_{12}$$

in (2) yield the survival function of this distribution, which is also derivable from a loss-of-memory model similar to that of Marshall and Olkin. Furthermore, it is the absolutely continuous part of the Marshall

and Olkin distribution. See Block and Basu [6] for details. The lifetime of the two organ systems of Gross et al. [16] and also of the two organ subsystems of Gross [15] are special cases of the maximum lifetime of the two-component system that has the distribution under discussion. Estimation and testing have been done by Mehrotra and Michalek [27] and by Gross and Lam [17]. In the latter paper, an application of this distribution to bivariate relief-times for patients receiving different treatments is considered.

Proschan and Sullo [31] suggest a model that contains both the Marshall and Olkin and the Freund distributions as special cases. Friday and Patil [14] have pursued the idea of a distribution containing these two distributions still further. They have developed a similar but more general distribution than that of Proschan and Sullo that is derivable from a threshold model, a gestation model, and a warmup model. See ref. 14 for details or Block and Savits [8] for a summary.

The distribution of Downton [10], given in (3), is a special case of a classical bivariate gamma distribution due to Wicksell [33] and to Kibble [22] (see Krishnaiah and Rao [24] for a discussion and references.) Downton [10] developed a model that gave rise to this bivariate exponential distribution and proposed its use in the setting of reliability theory*.

An interpretation by Arnold [2] leads to the following distribution. In a two-component system each component is subjected to nonfatal shocks that occur according to two independent Poisson processes*. If each component fails after a random number of shocks, where these random numbers follow a certain correlated geometric distribution, the joint density for the times to failure of the two components is

$$f(y_1, y_2) = \frac{\mu_1 \mu_2}{1 - \rho} \exp\left(-\frac{\mu_1 y_1 + \mu_2 y_2}{1 - \rho}\right)$$
$$\times I_0\left\{\frac{2\sqrt{\rho\mu_1\mu_2 y_1 y_2}}{1 - \rho}\right\}$$
$$\text{for} \quad y_1 > 0, \quad y_2 > 0 \quad (3)$$

where I_0 is the modified Bessel function* of the first kind of order 0. This is the bivariate exponential distribution of Downton [10]. For a detailed discussion of this interpretation, see Block and Savits [8].

As mentioned initially the preceding distribution (derived by Wicksell [33] and Kibble [22]) can be obtained as follows as the special case of a particular bivariate gamma distribution. Let $(X_1, Y_1), (X_2, Y_2), \dots, (X_n, Y_n)$ have independent identically distributed (iid) bivariate standard normal distributions with correlation ω. Then

$$\left(\sum_{i=1}^{n} X_i^2, \sum_{i=1}^{n} Y_i^2 \right)$$

has a correlated bivariate gamma (chi-square) distribution. The characteristic function and the joint density are given by (10.1) and (10.2), respectively, of Chapter 40 of ref. 21 with $\alpha = n$. For the case $n = 2$, the joint density is of the form of equation (3).

The bivariate exponential distribution of Hawkes [19] is obtained from the same model as that of Downton. The difference is in the choice of a more general bivariate geometric distribution for the random number of shocks to failure for the two components. Hawkes [19] derives a particular geometric distribution. The same distribution was derived independently by Arnold [1] and by Esary and Marshall [11], whose derivation utilizes a discrete nonfatal shock model. See Block and Savits [8] or Hawkes [19] for details.

Paulson [28] derives a bivariate exponential distribution through a characteristic function equation which generalizes a one-dimensional characteristic function equation that arises from a compartment model* (see Paulson and Uppuluri [29]). A generalization of the compartment model also leads to the bivariate equation. It can be shown that the distribution arising from this equation is exactly the Hawkes distribution. We refer to this as the Hawkes–Paulson distribution; its form is given in Paulson [28] and is fairly involved. For properties, see Paulson [28] and Hawkes [19].

Classes of bivariate exponential distributions have been developed by Arnold [1]

using a generalized bivariate geometric distribution. This is a reparametrized version of the distribution used by Esary and Marshall [11] and by Hawkes [19]. Let (N_1, N_2) have this distribution. Then Arnold's bivariate classes $\epsilon_n^{(2)}$ consist of the random variables

$$(Y_1, Y_2) = \left(\sum_{i=1}^{N_1} X_{i1}, \sum_{i=1}^{N_2} X_{i2} \right),$$

where (X_{i1}, X_{i2}) for $i = 1, 2, \dots$ are bivariate iid random vectors, where X_{i1} and X_{i2} are not in general independent, with distributions in $\epsilon_{n-1}^{(2)}$ for $n > 1$, and where $\epsilon_0^{(2)}$ consists of (X, X) where X is exponential. It follows that $\epsilon_1^{(2)}$ contains the independent exponentials and the Marshall and Olkin distribution. Furthermore, $\epsilon_2^{(2)}$ contains the Downton and the Hawkes–Paulson distributions. See Arnold [1] or Block and Savits [8] for details.

The Arnold classes of distributions have been described in Block et al. [7] using the characteristic function equation approach of Paulson and Uppuluri [29] and Paulson [27]. In ref. 7, the characteristic function equation approach has been used to derive properties of the distributions in this class including descriptions of the standard distributions in the class, the infinite divisibility* of the distributions, moment properties, and asymptotic properties. These results are summarized, without proof, in Block [4], in which it is also shown how the distributions in the class lead to multivariate shock models of the type studied in the univariate case by Esary et al. [12]. See also [6a].

MULTIVARIATE EXPONENTIAL AND RELATED DISTRIBUTIONS

Most of the bivariate models in the preceding section have multivariate $(n > 3)$ analogs. In general, the ideas are similar to the bivariate case, but the notational complexity is greatly increased.

The Freund distribution has been generalized to the multivariate case by Weinman [31], but only for identically distributed marginals (see Johnson and Kotz [21] for details). Block [5] considered a generalization of the Freund distribution for the case when the marginals need not be distributed identically and also of the Block and Basu [6] and the Proschan and Sullo [31] models.

Generalizations of the Downton [10], Hawkes [19], and Paulson [27] distributions exist implicitly within the framework of the general multivariate gamma distribution* of Krishnamoorthy and Parthasarathy [25] (see also Krishnaiah and Rao [24] and Krishnaiah [23]) and also within the framework of the Arnold classes. A specific parametric form has been given in Hsu et al. [20]. See also Chmelynski [9], who discusses many multivariate gamma distributions.

References

[1]   Arnold, B. C. (1975). *Sankhyā A*, **37**, 164–173.

[2]   Arnold, B. C. (1975). *J. Appl. Prob.*, **12**, 142–147. (A broad class of multivariate exponential distributions is discussed.)

[3]   Basu, A. P. and Block, H. (1975). In *Statistical Distributions in Scientific Work*, Vol. 3, G. P. Patil, S. Kotz, and J. K. Ord, eds. D. Reidel, Dordrecht, The Netherlands. (A basic reference work on multivariate exponential distributions and their characterizations.)

[4]   Block, H. W. (1975). In *Reliability and Fault Tree Analysis*, R. E. Barlow, J. B. Fussell, and N. D. Singpurwalla, eds. SIAM, Philadelphia. (A multivariate generalization of the distributions in refs. 6 and 13.)

[5]   Block, H. W. (1977). In *Theory and Applications of Reliability*, Vol. I, C. P. Tsokos and I. Shimi, eds. Academic Press, New York. (A summary of the results in ref. 7.)

[6]   Block, H. W. and Basu, A. P. (1974). *J. Amer. Statist. Ass.*, **69**, 1031–1037. (A bivariate distribution related to those in refs. 13 and 28.)

[6a] Block, H. W. and Paulson, A. S. (1984). *Sankhyā A*, **46**, 102–109. (Proofs of results stated in [4].)

[7]   Block, H. W., Paulson, A. S., and Kohberger, R. C. (1976). "Some Bivariate Exponential Distributions: Syntheses and Properties." Unpublished report. (A characteristic function approach to describing the distributions in ref. 2.)

[8]   Block, H. W. and Savits, T. H. (1981). In *Statistical Distributions in Scientific Work*, Vol. 5, C. Taillie, G. P. Patil, and B. A. Baldessari, eds., D. Reidel, Dordrecht, The Netherlands. (A recent survey of parametric and nonparametric distributions useful in reliability theory.)

[9]   Chmelynski, H. (1982). "A New Multivariate Error Structure for Multiple Regression Based on a Bayesian Analysis of the Gamma Process." Ph.D.

thesis, Carnegie–Mellon University. (Contains an up-to-date treatment of multivariate gamma distributions.)

[10] Downton, F. (1970). *J. R. Statist. Soc. B*, **32**, 408–417. (A fundamental bivariate exponential distribution.)

[11] Esary, J. D. and Marshall, A. W. (1973). "Multivariate Geometric Distributions Generated by a Cumulative Damage Process." *Naval Postgraduate School Rept.* NP55EY73041A.

[12] Esary, J. D., Marshall, A. W., and Proschan, F. (1973). *Ann. Prob.*, **1**, 627–649.

[13] Freund, J. (1961). *J. Amer. Statist. Ass.*, **56**, 971–977. (A fundamental bivariate exponential extension.)

[14] Friday, D. S. and Patil, G. P. (1977). In *Theory and Applications of Reliability*, Vol. I., C. P. Tsokos and I. Shimi, eds. Academic Press, New York. (A general multivariate exponential extension.)

[15] Gross, A. J. (1973). *IEEE Trans. Rel.*, **R-22**, 24–27.

[16] Gross, A. J., Clark, V. A. and Liu, V. (1971). *Biometrics*, **27**, 369–377.

[17] Gross, A. J. and Lam, C. F. (1981). *Biometrics*, **37**, 505–512. (An application of the distribution in ref. 6.)

[18] Gumbel, E. J. (1960). *J. Amer. Statist. Ass.*, **55**, 698–707. (Several early bivariate exponential distributions.)

[19] Hawkes, A. G. (1972). *J. R. Statist. Soc. B*, **34**, 129–131. (A fundamental bivariate exponential distribution.)

[20] Hsu, C. L., Shaw, L., and Tyan, S. G. (1977). "Reliability Applications of Multivariate Exponential Distributions." *Polytechnic Institute of New York Report POLY-EE-77-036.*

[21] Johnson, N. L. and Kotz, S. (1972). *Distributions in Statistics: Continuous Multivariate Distributions.* Wiley, New York. (A fundamental reference on multivariate distributions.)

[22] Kibble, W. F. (1941). *Sankhyā*, **5**, 137–150 (One of the first bivariate gammas.)

[23] Krishnaiah, P. R. (1977). In *Theory and Applications of Reliability*, Vol. I, C. P. Tsokos and I. Shimi, eds. Academic Press, New York. (A review of various multivariate gamma distributions.)

[24] Krishnaiah, P. R. and Rao, M. M. (1961). *Amer. Math. Monthly*, **68**, 342–346. (Properties of a particular multivariate gamma distribution.)

[25] Krishnamoorthy, A. S. and Parthasarathy, M. (1951). *Ann. Math. Statist.*, **22**, 549–557. (A multivariate gamma distribution.)

[26] Marshall, A. W. and Olkin, I. (1967). *J. Amer. Statist. Ass.*, **62**, 30–44. (A fundamental multivariate exponential distribution.)

[27] Mehrotra, K. G. and Michalek, J. E. (1976). "Estimation of Parameters and Tests of Indepen-

dence in a Continuous Bivariate Exponential Distribution. Unpublished manuscript. (Estimation for the distribution in ref. 6.)

[28] Paulson, A. S. (1973). *Sankhyā A*, **35**, 69–78. (A fundamental bivariate exponential distribution.)

[29] Paulson, A. S. and Uppuluri, V. R. R. (1972). *Math. Biosci.* **13**, 325–333. (A compartment model for the distributions in refs. 28 and 30.)

[30] Paulson, A. S. and Uppuluri, V. R. R. (1972). *Sankhyā A*, **34**, 88–91. (A bivariate geometric distribution.)

[31] Proschan, F. and Sullo, P. (1974). In *Reliability and Biometry*, F. Proschan and R. J. Serfling, eds. SIAM, Philadelphia. (A general multivariate exponential extension.)

[32] Weinman, D. G. (1966). "A Multivariate Extension of the Exponential Distribution." Ph.D. thesis, Arizona State University, Tempe, AZ. (A particular multivariate generalization of the distribution in ref. 13.)

[33] Wicksell, S. D. (1933). *Biometrika*, **25**, 121–136. (One of the first bivariate gammas.)

Acknowledgment

The work of Henry W. Block has been supported by ONR Contract N00014-76-C-0839. Reproduction in whole or in part is permitted for any purpose of the U.S. government.

(EXPONENTIAL DISTRIBUTION
MULTIVARIATE EXPONENTIAL
DISTRIBUTIONS, MARSHALL–OLKIN
MULTIVARIATE GAMMA DISTRIBUTION)

H. W. BLOCK

MULTIVARIATE EXPONENTIAL DISTRIBUTIONS, MARSHALL–OLKIN

The bivariate exponential distribution (BVE) introduced by Marshall and Olkin [18] is given by

$$\bar{F}(x, y)$$
$$\equiv P[X > x, Y > y]$$
$$= \exp\{-\lambda_1 x - \lambda_2 y - \lambda_{12}\max(x, y)\},$$
$$x, y \geqslant 0, \quad (1)$$

where $\lambda_1, \lambda_2, \lambda_{12}$ are nonnegative parameters such that $\lambda_1 + \lambda_{12} > 0$, $\lambda_2 + \lambda_{12} > 0$. This distribution has the following origins.

A "Fatal Shock" Model

Independent Poisson processes* $Z_1(t; \lambda_1)$, $Z_2(t; \lambda_2)$, and $Z_{12}(t; \lambda_{12})$ govern the occurrence of fatal shocks, respectively, to component 1, to component 2, and to components 1 and 2 simultaneously. If X and Y denote the respective life lengths of components 1 and 2, their joint distribution is given by (1).

Minima

If

$$X = \min(U, W), \qquad Y = \min(V, W), \quad (2)$$

where U, V, and W are independently and exponentially distributed with respective parameters λ_1, λ_2, and λ_{12}, then the joint distribution of X and Y is given by (1). This fact is essentially the same as that given in the "Fatal Shock" Model section.

A "Nonfatal Shock" Model

Events in the Poisson process $Z(t; \theta)$ cause failure to the ith component (but not the other) with probability p_i, $i = 1, 2$, and they cause failure to both components with probability p_{12}, where $1 - p_1 - p_2 - p_{12} \geqslant 0$. If $\lambda_i = p_i\theta$, $i = 1, 2$, and $\lambda_{12} = p_{12}\theta$, then the respective times to failure X and Y of components 1 and 2 have a joint distribution given by (1).

A Random Sums Model

Let $X = U_1 + \cdots + U_N$, $Y = U_1 + \cdots + U_M$, where $\{U_i\}$ are independently and exponentially distributed with parameter θ and where (N, M) have the bivariate geometric distribution described by Hawkes [15]:

$$P[N > n, M > m]$$
$$= \begin{cases} p_{00}^n(p_{01} + p_{11})^{m-n}, & \text{if } n \leqslant m, \\ p_{00}^m(p_{10} + p_{11})^{n-m}, & \text{if } n \geqslant m. \end{cases}$$

If $\{U_i\}$ and (N, M) are independent, $\lambda_1 = (p_{10} + p_{11})\theta$, $\lambda_2 = (p_{01} + p_{11})\theta$, and $\lambda_{12} = p_{11}\theta$, then X and Y have a joint distribution given by (1). This fact, the same as that given in the "Nonfatal Shock" Model section, was noticed by Esary and Marshall [7] and by Arnold [2]. The modification $X = U_1 + \cdots + U_N$ and $Y = V_1 + \cdots + V_M$, where $\{U_i\}$ and $\{V_i\}$ are independently and exponentially distributed, leads to the bivariate exponential distribution of Hawkes [15], which generalizes that of Downton [6].

Lack of Memory Property

The univariate exponential distribution* is characterized by the functional equation

$$\overline{F}(s + t) = \overline{F}(s)\overline{F}(t),$$

$$s, t \geqslant 0, \quad \text{where } \overline{F}(x) = P[X > x]. \quad (3)$$

The same equation in vectors s, t leads to the case of independence. The less stringent functional equation

$$\overline{F}(s_1 + \delta, s_2 + \delta) = \overline{F}(s_1, s_2)\overline{F}(\delta, \delta),$$

$$s_1, s_2, \delta \geqslant 0 \quad (4)$$

has many solutions, but the only solutions with exponential marginals are given by (1). Other solutions of (4) include the bivariate distribution of Friday and Patil [10], and Freund [9], and the distribution F_a defined by (6) and studied by Block and Basu [5]. The functional equation (4) has also been studied by Block [4], and Fermann [8]. The class of all solutions of (4) is characterized by Ghurye and Marshall [13].

PROPERTIES

From (1) or the form in the "Fatal Shock" Model section, it is clear that $\lambda_{12} > 0$ implies $P[X = Y] > 0$, so that the distribution (1) is not absolutely continuous. The singular and absolutely continuous parts F_s and F_a are given by

$$\overline{F}_s(x, y) = \exp\{-\lambda \max(x, y)\}, \quad (5)$$

$$\overline{F}_a(x, y) = [\lambda/(\lambda_1 + \lambda_2)]\overline{F}(x, y)$$
$$- [\lambda_{12}/(\lambda_1 + \lambda_2)]\overline{F}_s(x, y), \quad (6)$$

where $\lambda = \lambda_1 + \lambda_2 + \lambda_{12}$.

The moment-generating function* ψ is given by

$$\psi(s, t)$$
$$= \frac{(\lambda_1 + \lambda_{12})(\lambda_2 + \lambda_{12})(\lambda + s + t) + st\lambda_{12}}{(\lambda_1 + \lambda_{12} + s)(\lambda_2 + \lambda_{12} + t)(\lambda + s + t)}. \quad (7)$$

It follows that $\text{Cov}(X, Y) = \lambda_{12}/[\lambda(\lambda_1 + \lambda_{12})(\lambda_2 + \lambda_{12})]$, $\text{Corr}(X, Y) = \lambda_{12}/\lambda \geqslant 0$. The nonnegativity of the correlation follows from the fact that (2) implies that X and Y are associated.

STATISTICS

In a sample $(X_1, Y_1), \ldots, (X_n, Y_n)$ from the bivariate exponential distribution (1), let $N_{10}, N_{01}, N_{11} = n - N_{10} - N_{01}$ be, respectively, the number of $X_i > Y_i$, of $X_i < Y_i$, and of $X_i = Y_i$. A sufficient statistic is $[N_{10}, N_{01}, \Sigma X_i, \Sigma Y_i, \Sigma \min(X_i, Y_i)]$.

To estimate the parameters of the distribution, an iterative method for solving the nonlinear maximum likelihood* equations is given by Proschan and Sullo, who also show that the MLEs are consistent. Method of moments* estimators are obtained by Bemis et al. [3], and an asymptotically more efficient intuitive estimator is proposed and studied by Proschan and Sullo [23].

An estimator of the correlation ρ is proposed by Bemis et al. [3], who also propose a test of the hypothesis $\rho = 0$ (equivalently, $P[X = Y] = 0$, or $\lambda_{12} = 0$, or X and Y are independent). They show that for $\lambda_1 = \lambda_2$, the test is uniformly most powerful against the alternative $\rho > 0$.

If only $T_i = \min(X_i, Y_i)$, $i = 1, \ldots, n$ and N_{10}, N_{01} are observed, then MLEs of the parameters are obtained explicitly by George [12], who also studies some tests of hypotheses for this type of data.

APPLICATIONS

Applications of the BVE appear most notably in the literature on nuclear reactor safety (*see* NUCLEAR MATERIALS SAFEGUARDS), competing risks*, and reliability*. These applications are tied together by the possibility of common causes of failure, an intrinsic feature of the BVE. For references on competing risks and on life lengths, see Gail [11], Prentice et al. [22], Tolly et al. [25], Langberg et al. [17]. For references in the context of nuclear risk, see Vesely [26] and Hagan [14], and in the context of reliability, see Sarkar [24] and Apostolakis [1].

THE MULTIVARIATE CASE

Let $\{U_J, J \in \mathscr{J}\}$ be a set of independent exponentially distributed random variables indexed by the set $\mathscr{J}$ of all nonempty subsets of $\{1, \ldots, n\}$ and let U_J have parameter $\lambda_J \geqslant 0$, $J \in \mathscr{J}$. Assume that $\sum_{i \in J} \lambda_J > 0$ and as an extension of (2), let $X_i = \min_{i \in J} U_J$, $i = 1, \ldots, K$. Then $(X_1, \ldots, X_k)$ has the multivariate exponential distribution given by

$$\overline{F}(x_1, \ldots, x_k)$$
$$\equiv P[X_1 > x_1, \ldots, X_k > x_k]$$
$$= \exp\left\{-\sum_{J \in \mathscr{J}} \lambda_J \left(\max_{i \in J} x_i\right)\right\}, \quad (8)$$

$x_1, \ldots, x_k \geqslant 0$. The $(k-1)$-dimensional marginal distributions of (8) have the same structure, and the two-dimensional marginal distributions are BVE of the farm (1). Moreover, the functional equation

$$\overline{F}(s_1 + \delta, \ldots, s_k + \delta)$$
$$= \overline{F}(s_1, \ldots, s_k)\overline{F}(\delta, \ldots, \delta) \quad (9)$$

is satisfied, and the only distributions with exponential marginal distributions that satisfy (9) have the form (8).

GENERALIZATIONS

Random variables with the BVE distributions (1) can be regarded as joint waiting times in a bivariate Poisson process. A more general definition of "waiting times" leads to the more general BVE distribution of Marshall and Olkin [19].

The multivariate exponential distributions given by (8) are examples of distributions with exponential minima, i.e., $\min_{i \in I} X_i$ has an exponential distribution for each nonempty $I \subset \{1, \ldots, k\}$. Pickands [21] has shown that distributions with exponential minima have the form

$$\log \overline{F}(x_1, \ldots, x_k) = \int_S \left(\max_{1 \leqslant i \leqslant k} q_i x_i\right) \mu(q),$$
$$(10)$$

where μ is a finite measure on the unit simplex $S = \{q : q_i \geqslant 0, \ i = 1, \ldots, k, \ \Sigma q_i = 1\}$. If $(X_1, \ldots, X_n)$ has a distribution given by (10), then there exist random variables $Y_1, \ldots, Y_k$ with a distribution of the form (8) such that $\min_{i \in I} X_i$ has the same distribution as $\min_{i \in i} Y_i$ for all nonempty $I \in \{1, \ldots, K\}$. Thus, for some purposes, notably in reliability theory, one can assume (8) whenever (10) holds (Esary and Marshall, [7] and Langberg et al. [16].

Distributions of the form (8) arise as boundary cases in certain nonparametric classes of life distributions important in reliability theory (see e.g., Marshall and Shaked [20]).

References

[1] Apostolakis, G. E. (1976). *Nucl. Eng. Des.*, **36**, 123–133.

[2] Arnold, B. C. (1975). *J. Appl. Prob.*, **12**, 142–147.

[3] Bemis, B. M., Bain, L. J., and Higgins, J. J. (1972). *J. Amer. Statist. Ass.*, **67**, 927–929.

[4] Block, H. W. (1977). *Ann. Statist.*, **5**, 803–812.

[5] Block, H. W. and Basu, A. L. (1974). *J. Amer. Statist. Ass.*, **69**, 1031–1037.

[6] Downton, F. (1970). *J. R. Statist. Soc. B*, **32**, 403–417.

[7] Esary, J. D. and Marshall, A. W. (1974). *Ann. Statist.*, **2**, 84–98.

[8] Fermann, P. (1981). (Russian). *Vestn. Moskov. Univ. Ser. I Mat. Meh.*, 44–47.

[9] Freund, J. E. (1961). *J. Amer. Statist. Ass.*, **56**, 971–977.

[10] Friday, D. S. and Patil, G. P. (1977). In *The Theory and Applications of Reliability*, Vol. I, C. P. Tsokos and I. N. Shimi, eds. Academic Press, New York, pp. 527–549.

[11] Gail, M. (1975). *Biometrics*, **31**, 209–222.

[12] George, L. L. (1977). *IEEE Trans. Rel.*, **R-26**, 270–272.

[13] Ghurye, S. G. and Marshall, A. W. (1982). "Shock Processes with Aftereffects and Multivariate Lack of Memory." *Tech. Rep.*, Department of Statistics, Stanford University, Stanford, CA.

[14] Hagen, E. W. (1980). *Ann. Nucl. Energy*, **7**, 509–517.

[15] Hawkes, A. G. (1972). *J. R. Statist. Soc. B*, **34**, 129–131.

[16] Langberg, N., Proschan, F., and Quinzi, A. J. (1978). *Ann. Prob.*, **6**, 174–181.

[17] Langberg, N., Proschan, F., and Quinzi, A. J. (1981). *Ann. Statist.*, **9**, 157–167.

[18] Marshall, A. W. and Olkin, I. (1967). *J. Amer. Statist. Ass.*, **62**, 30–44.

[19] Marshall, A. W. and Olkin, I. (1967). *J. Appl. Prob.*, **4**, 291–302.

[20] Marshall, A. W. and Shaked, M. (1982). "Multivariate New Better Than Used Distributions." *Technical Report*, Department of Statistics, Stanford University, Stanford, CA.

[21] Pickands, J. (1982). "Multivariate Negative Exponential and Extreme Value Distributions." (Unpublished manuscript.)

[22] Prentice, R. L., Kalbfleisch, J. D., Peterson, A. V., Jr., Flournoy, N., Farewell, V. T., and Breslow, N. E. (1978). *Biometrics*, **34**, 541–554.

[23] Proschan, F. and Sullo, P. (1976). *J. Amer. Statist. Ass.*, **71**, 465–472.

[24] Sarkar, T. K. (1971). *Technometrics*, **13**, 535–546.

[25] Tolley, H. D., Manton, K. G., and Poss, S. S. (1978). *Biometrics*, **34**, 581–591.

[26] Vesely, W. E. (1977). In *Nuclear Systems Reliability Engineering and Risk Assessment*, J. B. Fussell and G. R. Burdick eds. SIAM, Philadelphia, pp. 314–341.

(EXPONENTIAL DISTRIBUTION
MULTIVARIATE EXPONENTIAL
  DISTRIBUTION)

A. W. Marshall
I. Olkin

MULTIVARIATE FITNESS FUNCTIONS

Suppose that the individuals in a large population are subjected to some form of selection related to certain variables $X_1, X_2, \ldots, X_p$, such that for every individual in the population at time zero with the particular values $x_1, x_2, \ldots, x_p$ for the X's, there are $w_t(x_1, x_2, \ldots, x_p)$ individuals in the population at some later time t. Then w_t is a *multivariate fitness function*, and it can be estimated on the basis of changes in the distribution of the X's between time zero and time t. More generally, if a series of s samples is taken from a population at times $t_1, t_2, \ldots, t_s$, then a fitness function can be estimated for the time period covered by the samples on the basis of the changes in the distribution of the X's from sample to sample.

The biological concept of the *survival of the fittest* goes back to Darwin, although in practice it is difficult to define fitness exactly (Cook [2]). The idea of relating fitness to measurable characters of individuals is also quite old. For example, Weldon [12] investigated the question of how the survival of snails is related to shell dimensions.

The general use of fitness functions in biology was considered at length by O'Donald [8–10]. He discussed the case of one variable and one time period, where the fitness of an individual is determined by one of the fitness functions

$$w(x) = 1 - \alpha - K(\theta - x)^2 \qquad (1)$$

or

$$w(x) = (1 - \alpha)\exp\left\{ -K(\theta - x)^2 \right\}. \qquad (2)$$

In both these functions, K is assumed to be a positive constant and α to be less than one. Hence an individual with $X = \theta$ has the maximum fitness of $1 - \alpha$.

The main advantage of the quadratic fitness function (1) is that it is easy to fit to data using the method of moments*. However, it has the considerable disadvantage of giving negative fitness values for individuals with extreme values of X. The function (2) is called the nor-optimal fitness function, following Cavalli-Sforza and Bodmer [1]. This is easy to fit to data if X has a normal distribution before and after selection.

There have been numerous applications of single variate fitness functions to biological data involving one sample before selection and one sample after selection. These include selection related to tooth widths of rats

(O'Donald [8]), bristle numbers of *Drosophila* (O'Donald [9]), birth weight for human babies (Karn and Penrose [5], Cavalli-Sforza and Bodmer [1]), and shell size for snails (Cook and O'Donald [3]).

Manly [6] generalized the estimation of a univariate fitness function to the case where there are more than two samples with a nor-optimal fitness function. He also considered selection on gamma* and beta* distributions.

In a later paper Manly [7] proposed two methods for estimating fitness functions in the completely general situation where there are p variables $X_1, X_2, \ldots, X_p$ and s samples taken at times $t_1, t_2, \ldots, t_s$. For the first of these two methods, it is necessary to assume that the probability density function for the X's before selection (at time zero) is multivariate normal with mean vector μ_0 and covariance matrix V_0. If the fitness function takes the form

$$w_t(x) = \exp\{(L'x + x'Mx)t\}, \qquad (3)$$

where L is a vector of constants and M is a $p \times p$ symmetric matrix of constants, then at a later time t the distribution is still multivariate normal, with mean vector

$$\mu_t = V_t(Lt + V_0^{-1}\mu_0) \qquad (4)$$

and covariance matrix

$$V_t = \{V_0^{-1} - 2Mt\}.^{-1} \qquad (5)$$

Equations (4) and (5) form the basis of a regression method for estimating the vector L and the matrix M. They were given in a slightly different form by Felsenstein [4]. They were first derived by Pearson [11] in an important paper that has been neglected until recently.

For his second method of estimating a multivariate fitness function, Manly [7] proposes setting up a likelihood function where the probability associated with a sampled individual is the probability of that individual appearing in the sample that it was in, given that it was in one of the samples taken. Any sensible form of fitness function can be assumed, and maximum likelihood estimates* of the parameters of this function

can then be determined by maximizing the likelihood function numerically. Manly illustrates his two methods of estimation using data on Egyptian skulls in samples covering the time period from 4000 B.C. to A.D. 150.

References

[1] Cavalli-Sforza, L. L. and Bodmer, W. F. (1972). *The Genetics of Human Populations*, W. H. Freeman, San Francisco.

[2] Cook, L. M. (1971). *Coefficients of Natural Selection*. Hutchinson University Library, London.

[3] Cook, L. M. and O'Donald, P. (1971). In *Ecological Genetics and Evolution*, R. Creed, ed. Blackwell, Oxford, England.

[4] Felsenstein, J. (1977). *Proceedings of the International Conference on Quantitative Genetics*, E. Pollak, O. Kempthorne, and T. B. Bailey, eds. Iowa State University, Ames, IA.

[5] Karn, M. N. and Penrose, L. S. (1951). *Ann. Eugen.*, **16**, 145–164.

[6] Manly, B. F. J. (1977). *Biom. J.*, **19**, 391–401.

[7] Manly, B. F. J. (1981). *Biom. J.*, **23**, 267–281.

[8] O'Donald, P. (1968). *Nature*, **220**, 197–198.

[9] O'Donald, P. (1970). *Theor. Popul. Biol.*, **1**, 219–232.

[10] O'Donald, P. (1971). *Heredity*, **27**, 137–153.

[11] Pearson, K. (1903). *Phil. Trans. Roy. Soc. Lond.*, **A200**, 1–66.

[12] Weldon, W. F. R. (1901). *Biometrika*, **1**, 109–124.

(ANTHROPOLOGY, STATISTICS IN
ECOLOGICAL STATISTICS
GENETICS, STATISTICS IN
HUMAN GENETICS, STATISTICS IN
MATHEMATICAL THEORY OF
  POPULATION)

B. F. J. MANLY

MULTIVARIATE GAMMA DISTRIBUTIONS

Multivariate gamma distributions play an important role in various areas such as hydrology*, meteorology*, pattern recognition*, point processes*, reliabillity*, simultaneous test procedures (*see* MULTIPLE COMPARISONS), etc. For example, in the area of reliability, it is realistic to assume that the joint distribution of failure times of components or the intervals between the shocks on a component are jointly distributed as a multivariate gamma distribution. For various applications in reliability, see Krishnaiah [10, 13]. An application in hydrology was discussed in Prekopa and Szantai [19]. Applications in simultaneous test procedures are discussed in Krishnaiah [12].

Here we give an overview of some developments on multivariate gamma distributions. Given that the marginal distributions are gamma, one can construct multivariate gamma distributions in various ways; we will emphasize reviews on the multivariate gamma distributions considered by Krishnamoorthy and Parthasarathy [15] and by Krishnaiah and Rao [8].

A MULTIVARIATE GAMMA DISTRIBUTION

Let $\mathbf{x}'_j = (x_{1j}, \ldots, x_{pj})$, $(j = 1, 2, \ldots, n)$, be distributed independently and identically as a multivariate normal with mean vector $\boldsymbol{\mu}'= (\mu_1, \ldots, \mu_p)$ and covariance matrix $\boldsymbol{\Sigma}= (\sigma_{ij})$. Also, let $z_i = \frac{1}{2}\sum_{j=1}^n x_{ij}^2$ for $i = 1, 2, \ldots, p$. Then the joint distribution of $(z_1, \ldots, z_p)$ is a *central* or *noncentral multivariate gamma distribution* with $n/2$ as shape parameter and with $\boldsymbol{\Sigma}$ as the covariance matrix of the accompanying multivariate normal. The joint distribution of $(2z_1, \ldots, 2z_p)$ is a *multivariate chi-square distribution* with n degrees of freedom, whereas the joint distribution of $(2z_1)^{1/2}, \ldots, (2z_p)^{1/2}$ is a *multivariate chi distribution* with n degrees of freedom, also known as the generalized Rayleigh distribution in the literature. The multivariate gamma distribution with shape parameter one is a *multivariate exponential distribution*. The distributions just defined are central or noncentral according to whether $\boldsymbol{\mu} = \mathbf{0}$ or $\boldsymbol{\mu} \neq \mathbf{0}$. Unless stated otherwise, we consider central distributions only in the sequel.

Bose [1] derived an expression for the bivariate chi distribution and Krishnaiah et al. [7] studied its properties. The probability density function (P.D.F.) of the bivariate chi-square distribution with n degrees of freedom and with $\boldsymbol{\Sigma} = (\rho_{ij})$, $\rho_{12} = \rho_{21} = \rho$, $\rho_{ii} = 1$ as the correlation matrix of the accompanying bivariate normal, is known to be

$$f(y_1, y_2)$$
$$= (1 - \rho^2)^{n/2} \cdot \sum_{i=0}^{\infty} \left[\frac{(\frac{1}{2}n)^{[i]}}{i!} \right.$$
$$\left. \cdot e^{2i} \prod_{j=1}^2 \frac{y_j^{n/2+i-1} \exp\{-\frac{1}{2}y_j/(1-\rho^2)\}}{2^{n/2+i}\Gamma(\frac{1}{2}n+i)(1-\rho^2)^{n/2+i}} \right]$$

where $a^{[b]} = a(a+1) \ldots (a+b-1)$. This expression can be obtained from Bose's expression by simple transformation. Moran [17] discussed tests of the hypothesis that $\rho = 0$. Kibble [6] gave the following alternative expression for the bivariate gamma distribution with $\alpha = n/2$ as shape parameter:

$$f(y_1, y_2)$$
$$= g(y_1; \alpha) g(y_2; \alpha)$$
$$\times \left[1 + \sum_{j=1}^{\infty} \frac{\rho^{2j} L_j(y_1; \alpha) L_j(y_2; \alpha)}{j! \, \alpha^{[j]}} \right],$$
$$(1)$$

where

$$L_r(x; \alpha) = \frac{1}{g(x; \alpha)} \frac{d^r}{dx^r} \left[(-x)^r g(x; \alpha) \right],$$
$$g(x; \alpha) = \frac{\exp(-x) x^{\alpha-1}}{\Gamma(\alpha)}.$$

Since the preceding bivariate gamma distribution is infinitely divisible (*see* INFINITE DIVISIBILITY), the right side of (1) is a PDF for any real $\alpha > 0$. Krishnamoorthy and Parthasarathy [15] expressed the multivariate gamma distribution as an infinite series involving products of Laguerre polynomials (*see* LAGUERRE SERIES). The characteristic function* of the multivariate gamma distribution is known to be

$$\phi(t_1, \ldots, t_p) = |\mathbf{I}_p - i\mathbf{T}\boldsymbol{\Sigma}|^{-n/2},$$

where $\mathbf{T} = \text{diag}(t_1, \ldots, t_p)$. Here we note that the multivariate chi-square distribution is the joint density of the diagonal elements of the Wishart* matrix. Moran and Vere-Jones [18] showed that the multivariate gamma distribution is infinitely divisible when Σ is of the form $\sigma^2(\rho_{ij})$, $\rho_{ii} = 1$ and $\rho_{ij} = \rho$ ($i \neq j$). When $p = 3$ and $\rho_{ij} = \rho^{|i-j|}$, they showed that the preceding distribution is infinitely divisible. Griffiths [4] established necessary and sufficient conditions for the infinite divisibility of the trivariate gamma distribution. Exact percentage points of the bivariate chi-square distribution and approximate percentage points of the multivariate chi-square distribution are available (see Krishnaiah [11]).

A generalization of the multivariate gamma distribution is the joint distribution of the correlated quadratic forms considered by Khatri et al. [5]. Krishnaiah and Waikar [9] derived the distribution of the linear combination of correlated quadratic forms.

We will now discuss some alternative bivariate gamma distributions. Cheriyan [2] and David and Fix [3] considered the joint distribution of u and v, where $u = x + y$ and $v = x + z$; here x, y, and z are distributed independently as gamma variables.

Next, let y_1 and y_2 be nonnegative random variables. Then the joint density of y_1 and y_2 can be expressed as

$$f(y_1, y_2)$$
$$= g(y_1; \alpha) g(y_2; \alpha)$$
$$\times \left\{ 1 + \sum_{k=1}^{\infty} c_k L_k(y_1; \alpha) L_k(y_2; \alpha) \right\},$$
$$(2)$$

where $\{c_k\}$ is some sequence of nonnegative numbers such that $\sum_{k=1}^{\infty} c_k^2 < \infty$. Sarmanov [21] showed that the necessary and sufficient condition for the expansion (2) to be valid is that $\{c_k\}$ form a moment sequence of some distribution concentrated in $[0, 1)$. Kibble's expression for the bivariate gamma distribution is a special case of the preceding bivariate gamma distribution. For a discussion of alternative multivariate gamma distributions, see Krishnaiah [13].

MULTIVARIATE GAMMA-WEIBULL AND GAMMA-NORMAL DISTRIBUTIONS

In this section, we discuss the multivariate gamma-Weibull and gamma-normal distributions considered by Krishnaiah [13].

Let $(X_{1j}, \ldots, X_{pj}, y_{1j}, \ldots, y_{qj})$, ($j = 1, 2, \ldots, n$), be distributed independently and identically as multivariate normal with mean vector $\mathbf{0}$ and covariance matrix Σ. Then the joint distribution of $(z_1, \ldots, z_p, \bar{y}_1, \ldots, \bar{y}_q)$ is a *multivariate gamma-normal distribution*, where

$$z_i = \frac{1}{2} \sum_{j=1}^{n} x_{ij}^2, \qquad n\bar{y}_t = \sum_{j=1}^{n} y_{tj}.$$

The joint distribution of $(z_1, \ldots, z_p, v_1, \ldots, v_q)$ is a *multivariate gamma–Weibull distribution*, where

$$v_t = \left(\frac{1}{2} \sum_{j=1}^{2} y_{tj}^2 \right)^{1/\beta}.$$

When $\beta = 1$, this distribution is a *multivariate gamma-exponential distribution*. The joint characteristic function of $(z_1, \ldots, z_p, v_1^{\beta}, \ldots, v_q^{\beta})$ is given by

$$\phi(t_1, \ldots, t_{p+q}) = D_1 D_2,$$
$$\mathbf{T} = \operatorname{diag}(t_1, \ldots, t_{p+q}),$$
$$\mathbf{T}_{11} = \operatorname{diag}(t_1, \ldots, t_p)$$
$$D_1 = |\mathbf{I} - i\mathbf{T}\Sigma|^{-1},$$
$$D_2 = |\mathbf{I} - i\mathbf{T}_{11}\Sigma_{11}|^{-(n-2)/2}$$
$$\Sigma = \begin{pmatrix} \Sigma_{11} & \Sigma_{12} \\ \Sigma_{21} & \Sigma_{22} \end{pmatrix}.$$

The characteristic function of the multivariate gamma-normal distribution is given by

$$\phi(t_1, \ldots, t_{p+q})$$
$$= |\mathbf{I} - 2i\Sigma\mathbf{T}|^{-n/2}$$
$$\times \exp\left\{ (n/2)\mathbf{t}'(\Sigma^{-1} - 2i\mathbf{T})^{-1}\mathbf{t} \right\},$$
$$\mathbf{T} = \operatorname{diag}(t_1/2, \ldots, t_p/2, 0, \ldots, 0),$$
$$\mathbf{t}' = (0, \ldots, 0, t_{p+1}/n, \ldots, t_{p+q}/n).$$

Acknowledgment

This work was sponsored by the Air Force Office of Scientific Research under Contract F49620-82-K-0001.

References

[1]   Bose, S. (1935). *Sankhyā*, **2**, 65.

[2]   Cheriyan, K. C. (1941). *J. Indian Math. Soc.*, **5**, 133.

[3]   David, F. N. and Fix, E. (1961). *Proc. of the 4th Berkeley Symp.*, **1**, 177.

[4]   Griffiths, R. C. (1970). *Sankhyā A*, **32**, 393.

[5]   Khatri, C. G., Krishnaiah, P. R., and Sen, P. K. (1977). *J. Statist. Plann. Inf.*, **1**, 299.

[6]   Kibble, W. F. (1941). *Sankhyā*, **5**, 137.

[7]   Krishnaiah, P. R., Hagis, P., Jr., and Steinberg, L. (1963). *SIAM Rev.*, **5**, 140.

[8]   Krishnaiah, P. R. and Rao, M. M. (1961). *Amer. Math. Monthly*, **68**, 342.

[9]   Krishnaiah, P. R. and Waikar, V. B. (1973). *Commun. Statist.*, **1**, 371.

[10] Krishnaiah, P. R. (1977). *Proceedings of the Conference on the Theory and Applications of Reliability with Bayesian and Nonparametric Methods*, Vol. 1, C. P. Tsokos and I. N. Shimi, eds. Academic Press, New York, p. 475.

[11] Krishnaiah, P. R. (1980). *Handbook of Statistics*, Vol. 1. North-Holland, Amsterdam, p. 745. (Computations of some multivariate distributions.)

[12] Krishnaiah, P. R. ed. (1979). *Developments in Statistics*, Vol. 1. p. 157. (Some developments on simultaneous test procedures.)

[13] Krishnaiah, P. R. (1983). "Multivariate Gamma Distributions and Their Applications in Reliability." *Tech. Rep. No. 83-09*, Center for Multivariate Analysis, University of Pittsburgh, PA.

[14] Krishnaiah, P. R. and Sarkar, S. (1983). *Multivariate Analysis VI*, P. R. Krishnaiah, ed. North-Holland, Amsterdam. (Nonparametric estimation of density using orthogonal polynomials and its applications in pattern recognition and reliability.) (To appear)

[15] Krishnamoorthy, A. S. and Parthasarathy, M. (1951). *Ann. Math. Statist.*, **22**, 549; erratum **31**, 229.

[16] Linhart, H. (1970). *S. Afr. Statist. J.*, **4**, 1.

[17] Moran, P. A. P. (1967). *Biometrika*, **54**, 385.

[18] Moran, P. A. P. and Vere-Jones, D. (1969). *Sankhyā A*, **31**, 191.

[19] Prekopa, A. and Szantai, T. (1978). *Water Resour. Res.*, **14**, 19.

[20] Ramabhadran, V. K. (1951). *Sankhyā*, **11**, 45.

[21] Sarmanov, I. O. (1968). *Sov. Math. Dokl.*, **9**, 547.

(CHI-SQUARE DISTRIBUTION
GAMMA DISTRIBUTION
LAGUERRE SERIES
MULTINORMAL DISTRIBUTION
MULTIVARIATE DISTRIBUTIONS
MULTIVARIATE EXPONENTIAL
    DISTRIBUTION
MULTIVARIATE WEIBULL
    DISTRIBUTION)

P. R. KRISHNAIAH

MULTIVARIATE GRAPHICS

Multivariate graphics display data of several variables observed on a batch of units. The display is to show viewers aspects of the data that perusal of the numbers would not reveal —or that would require more time and effort to discover. Some viewers need guidance with very simple aspects of the data and may benefit from expository displays, such as the several single univariate histograms* or the profiles of means and standard deviations. This article is concerned with the needs of viewers who require tools to explore multivariate data in more detail, to seek transformations, to locate special effects and/or outliers*, to identify patterns and clusters, to diagnose models, and generally to hunt for novel and unexpected phenomena. Such graphical tools are not needed as substitutes for mathematical methods of multivariate analysis when the latter are appropriate: their real role is to explore the unkown and reveal the unexpected, such as special interactions of treatments, new forms of models, outliers, and repeated irregularities that suggest a change of model. (*See also* GRAPHICAL REPRESENTATION OF DATA.)

The flexibility of the display-viewer interplay is what makes graphics important— especially in the interactive mode—not the ability to mimic graphically what can be done mathematically. For example, linear discrimination can be carried out analytically in an optimal way, but graphics may be helpful in checking the underlying assumptions and diagnosing nonlinearity when it exists. (For summary discussions of graphics of multivariate data see refs. 9, 14, 24, 31,

34, and 43 and especially Chambers et al. [7, Chap. 5].)

The choice of centering, double centering, scaling and/or reexpression is obviously important, but it is preliminary to analysis and display and will not be discussed here (*see* EXPLORATORY DATA ANALYSIS).

Many of the displays used in multivariate analysis are not in themselves *multidimensional*, since they only show single dimensions of the multivariate data. Thus the separate histograms for each of the variates are *multiple univariate displays* rather than multivariate ones, since they do not show any *joint* variation. Scatter plots, on the other hand, are by their nature multivariate. Even a plot of multivariate statistics such as that of the successive eigenvalues of a multivariate analysis of variance is essentially unidimensional. Thus Gnanadesikan [24] plots expectations of eigenvalues against observed eigenvalues from a multivariate analysis: this may indicate the number of dimensions whose variability is essentially "noise," and which may therefore be neglected in interpreting the data. That has been explored in greater detail by Wachter [45].

There is evidently much room for such *univariate* and *unidimensional displays* in exploring multivariate data. However, these displays do not differ in their graphical features from univariate displays; therefore, they will not be dealt with here. The only univariate technique mentioned in this article will be Q-Q and P-P plotting. In multivariate analysis Q-Q (quantile vs. quantile) plots permit checking normality of marginal distributions [7, Chap. 6]. They can also be used to plot summary measures [4] as, for example, when multivariate distances of units from the batch centroid are gamma-plotted to check the multivariate normality of the batch's distribution [47]. Such plots may also indicate the existence of nonlinear regularities in the units' scatter. Q-Q plotting can further be used to compare the quantiles of two samples, and this idea may be extended to the multivariate case by defining a ranking along a *minimal spanning tree* [16]. P-P plotting of the two samples' empirical cdf's against each other is more easily extended to multivariate data [23], though its usefulness may be more limited.

CLASSES OF MULTIVARIATE DISPLAYS

Some classification of multivariate and multidimensional graphics will be helpful for the following discussion. Thus graphics will be classified as being either *static* or *dynamic*, as using Euclidean, pictorial, or mixed *mappings* of the data, and as having a certain *mode* of choice of displays.

Static displays may be hand-drawn or computer-plotted, whereas dynamic displays appear on computer-driven CRT screens. Either of these can be used in different *modes*, depending on when and how the choice of displays is made, i.e., whether the display is prechosen by the viewer, driven by the data, or interactive between the data and the viewer. (*see also* GRAPHICAL REPRESENTATION, COMPUTER-AIDED).

When an analyst decides to present the results of a multivariate analysis by, say, all the bivariate scatters, he/she operates in the *prechosen* mode. When the computer is instructed to compute the plane of best fit and then plot the point scatter in that plane, the display is said to be in the *data-driven* mode because that particular plane is chosen to accord with the data's own features. Finally, the mode of display is said to be *interactive* when the investigator sits at the terminal, inspects a view of the data, and then asks for more detail or for a complementary aspect by, say, zooming in on the display or rotating it about an axis.

This article concentrates on multidimensional displays of *p*-variate data ($p \geqslant 2$) on a batch of *n* units or on several batches. Any such display requires a *mapping* of the *data* into the display. The most common mapping is Euclidean: each unit is represented by a point or other marker whose coordinates correspond to the values the variable assumes for the different units; each variable is mapped into a point whose coordinates correspond to the *n* units' observations on that variable. These two alternative Euclidean geometries are well known in the forms of, respectively, *scatter plots* and *variable configurations* (the latter are familiar from geometric illustrations in texts on the general linear hypothesis). A third geometric display merges the two in a *biplot** that carries markers for both the units and the variables.

Alternatively, mappings may be *pictorial*. Thus each unit and/or variable may be represented by a picture whose separate features correspond to the characteristics of the unit and/or variable. The simplest form of such a display is that in which every unit is represented by a profile showing the value of each of the variables recorded on the unit. More sophisticated pictorial mappings use stars [5], metroglyphs [1], directed rays [35], function plots [2], shadings [3], trees and castles [31], faces [8], etc. Some of these will be described later. (See also Chambers et al. [7, Chap. 5].)

Mixed mappings use two of the variables to locate the units' markers in a plane and then display the remaining $p - 2$ variables pictorially. Thus, in a demographic example, each cohort might be located by its birth date (abscissa) and total fertility (ordinate), and a profile at that location might then show the $p - 2$ age-specific fertility rates. A special case of these mixed mappings is *cartographic* representation, in which the picture for each unit is placed on a map according to the longitude and latitude of its geographic location and the remaining $p - 2$ variables determine the features of the picture.

USE OF COMPUTERS

The role of the computer is obviously central to modern-day graphical displays, but its function depends very much on the mode of representation. Any amount of multivariate analysis* depends on the use of a computer, and all but the simplest prechosen static displays need computer plots and dynamic graphics on a CRT. A computer is needed for the data-driven mode—the choice of the

best-fitting plane, for example, cannot really be obtained otherwise. Plotting hardware is crucial for almost all multivariate graphics, be they pictorial displays or scatters of multivariate observations. Interactive facilities are crucial only in the interactive mode, in which observation of one view on the CRT leads the viewer to select and implement the next one. (See Newman and Sproull [33].)

No attempt is made here to describe the state of the art in this field of computing, since the technology is developing so fast that anything said about available hardware and software is likely to be outdated by the time it sees print. (See, however, refs. 12 and 15.) Buyers of equipment for multivariate graphics would do well to anticipate the actual uses to which they want to put it and be guided by their needs rather than by the more spectacular features of some new device. Thus, for example, if graphics must be distributed widely or published, a CRT display may not be helpful and color may result in exorbitant costs of reproduction. The user might be better served by straightforward printouts and simple pen plotters.

EUCLIDEAN MAPPINGS OF UNIT SCATTERS

Displays of Euclidean mappings of multivariate data will be discussed first in terms of the representation of units with respect to variables' axes, i.e., the multivariate scatter. The discussion extends, *mutatis mutandis*, to the variables' configuration and to the biplot.

One of the purposes of such displays is to view the distribution of the units with respect to the variables. If the density contours are reasonably close to elliptic, a multivariate normal distribution may describe the data well; if a high density of points in one corner is accompanied by rapidly falling-off densities in several directions, then a highly skewed distribution is diagnosed and a transformation may be indicated; if a few points lie well away from the main swarm, their identification may lead either to the correction of faulty data or to the finding of un-usual events. (Calling them "outliers" is convenient, but studying them carefully may be more productive of fresh insights. We are more likely to obtain new ideas from the unexpected than from that which follows the established pattern.)

Another use of graphical displays of scatters is to see if a single batch's units have a reasonably homogeneous distribution or whether they cluster into identifiable subsets or special patterns. Alternatively, if the units are known to come in several batches, one would want to see if the units' scatter is differentiated by these batches or whether the batches' scatters overlap.

A third important use of displays is to diagnose models that may fit the data. Obvious examples are data lying on or near hyperplanes or other low-dimensional manifolds*. A view of the units' scatter may reveal such patterns clearly and suggest a mathematical formulation to model the phenomenon studied. The importance of this use of graphics cannot be overemphasized, especially since the eye does not seem to have a mathematical counterpart in the ability to grasp unexpected patterns. And modeling is often primary to the formulation of underlying laws—after all, Kepler's orbital ellipses did precede Newton's laws of gravity.

How then can a p-dimensional scatter be viewed? If $p = 2$, the answer is obvious—a scatter plot is readily inspected (though the eye may be misled by the scales or forms of display [10]). If $p = 3$, it is more difficult to obtain a good view without the expense and delay involved in constructing a solid model. Visual simulation of three-dimensional views can be attempted in a number of ways: stereograms and analglyphs (using two differently polarized projections or two colors) can be highly effective, but they work mainly in the prechosen mode since they need advance preparation and use of projection equipment unless a stereo viewer is attached to the CRT displaying the pair of stereo views. Less satisfactory simulation of

stereographics can be obtained from perspective views—in which the size of markers and/or their labels indicate the third dimension. These are relatively cheap and can be readily produced in print and shown on an on-line screen. Dynamic rotation of the display that is sufficiently fast to simulate continuous movement is highly effective in producing the semblance of three-dimensional perception [14]. However, such displays are still too expensive to be used routinely for inspection of multivariate data [16]. Expressing the third dimension by a succession of colors and hues is probably the easiest way to simulate depth on modern equipment, but its effectiveness in simulating three-dimensional views is not clear [41]. Other techniques, such as light intensity or shading, are readily available on some computer systems and may turn out to be useful. The effectiveness, and cost, of these various simulations of three-dimensional views depends strongly on the system used, and since this type of equipment is undergoing rapid development at this time, it is difficult to predict what will turn out to be usable and cost-effective.

In many cases one will attempt to elucidate trivariate scatters by viewing them in several planar projections. These are easy to plot and print or to view on a screen, whether in prechosen or data-driven mode or interactively. The most obvious prechosen display is the *draftsman's view*, i.e., the three orthogonal projections onto the planes whose axes represent pairs of the original variables, corresponding of three sides of a cube. Figure 1 shows the draftsman's view of three of the variables of Anderson's well-known iris data—the three species being indicated by different markers. (This display, as well as Figs. 2–4, are reproduced courtesy of Paul A. Tukey.) More prechosen views could be obtained by projecting orthogonally onto six sides of a regular dodecahedron—see Fig. 2—or onto ten sides of a regular icosahedron. Other planes of view might be selected interactively by rotating the scatter in the pursuit of interesting as-

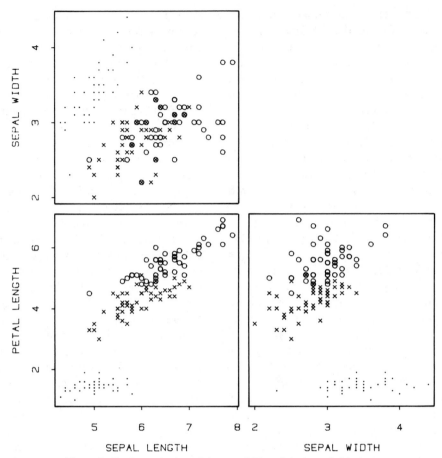

Figure 1 Draftsman's view of three variables of Anderson's iris data.

ANDERSON IRIS DATA

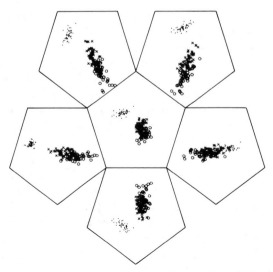

Figure 2 Dodecahedral views of three variables of the Anderson iris data.

pects. A number of computer systems allow the viewer to do this on the screen (Tsiano et al. [40]).

Alternatively, one might go to the data-driven mode and let the computer pursue the particular projection that shows the scatter in some optimal manner. (This would formalize some of what the viewer presumably would be attempting to do in the interactive mode.) An obvious choice of plane is the one that maximizes the squared fit (i.e., minimizes the squared residuals). This choice leads to the first two principal axes and corresponds to Pearson's [36] formulation of principal components (*see* COMPONENT ANALYSIS). It fits closer to the scatter, in the least-squares sense, than any other plane, but this does not ensure that it will always yield the most revealing view. For example, an outlier might dominate one of the first two dimensions and result in poor fit for many other units. A *resistant* choice of principal axes might do better in that it would fit more closely to the majority of units while ignoring outliers. (The resulting view, on the other hand, would not be help-

ANDERSON IRIS DATA

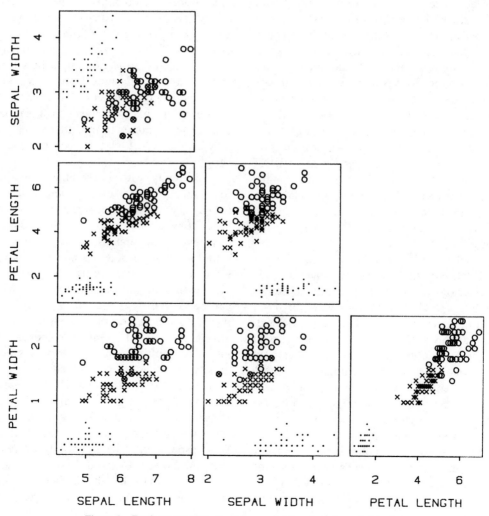

Figure 3   Draftsman's view of all four variables of Anderson's iris data.

ful in identifying the outlier, though the computations used in fitting should flag it— Gabriel and Odoroff [22].) Yet another criterion for fitting a plane would be the majorization of some measure of clottedness, which is " . . . the degree to which data points are concentrated locally . . . while at the same time expanded globally. Experience suggests that such projections tend to be unusually interesting to researchers" (Friedman and Tukey [18]). Such criteria have been discussed and implemented under the general name of *projection pursuit*.

All these displays of three-variable data view *projections* of the scatter onto planes, i.e., show *marginal bivariate scatters*. An alternative method is to show *partial* or *conditional* bivariate scatters for given intervals of the third variable. One can think of this as slicing the tridimensional space through the third variable's axis in order to show partial planar views of the first and second variables' scatter. That is unlikely to give a feeling of depth, but might reveal interesting aspects of the data.

A similar idea is that of masking*. Here the display in the x_1, x_2 plane may be restricted to points within certain intervals of another variable, x_3, the remaining points being masked. This is obviously related to conditional scatter plots, since the masking of x_3 determines the conditioning of that variable.

Display of *four-dimensional* data needs a little more ingenuity. The three planes of the draftsman's view for $p = 3$ can readily be extended to the draftsman's view for $p = 4$ by arranging the required six bivariate scatters suitably [44]. Figure 3 shows this view of the four variables of the iris data. Again, in the data-driven mode, one might project the four-dimensional scatter onto planes going through the principal axes (see Fig. 5 below) or onto planes chosen by some other projection pursuit criterion.

When it comes to conditional, or partial, bivariate scatters, one needs scatters for x_1, x_2, say, separately for given intervals of x_3, x_4 (or masked for both of them). An inge-

nious method of showing both marginal and conditional scatter is Tukey and Tukey's [43, 44] *casement displays*. (See also ref. 7.) Here the x_3, x_4 plane is cellulated (i.e., sliced on both x_3 and x_4), and the partial x_1, x_2 scatter is displayed within each cell. An illustration is given in Fig. 4 for Anderson's iris data.

Marginal and conditional displays may both be of interest; a reasonable strategy is first to inspect the marginal draftsman's views (transforming if necessary) and then select which two variables shall determine the cellulation into multiple windows and which two variables define the scatters within these windows. (That, of course, introduces an element of interactive mode.)

A useful supplement to scatter plots and other Euclidean mappings is the selection of *clusters*, *outliers*, or *patterns* noted on one plot and displaying them *distinctively* on another plot. Thus, for example, an outlier in the x_1, x_2 plane projection might be selected to reappear as a flashing point in the x_3, x_4 plane projection, etc., or a cluster in the former plane may reappear as a specially colored set of points in the latter plane. This may help to reveal relations between pairs of variables.

A difficulty with interpreting the above four-variable scatter displays and indeed all displays in which $p > 3$, is that most viewers have considerable difficulty in visualizing anything in four dimensions, let alone in five or more dimensions. Graphical display of scatters of necessity consists of projections onto planes or three-spaces, each of which can be viewed by the methods discussed for $p = 3$. As p increases, the complexity of visualizing the general picture becomes greater, and the multiplicity of views augments greatly; thus, for $p = 10$, there are 45 bivariate, 120 trivariate, and 210 four-variate scatters in the original variables. The prechosen mode can thus be effective only if the investigator has some a priori clue as to what aspects may be interesting or revealing; otherwise, it would flood the viewer with more than he/she can absorb. The interactive mode may be helpful here, letting

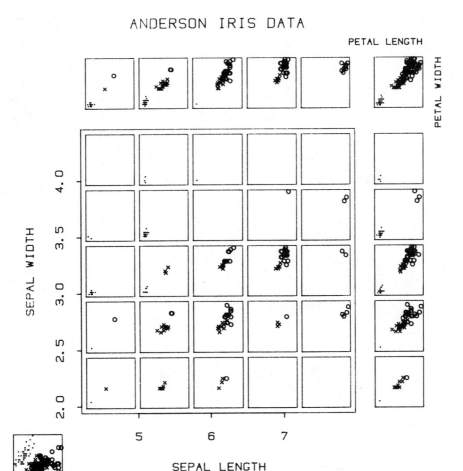

Figure 4 Casement display of Anderson's iris data.

the viewer progress from one view to another, being guided by each view in the selection of the next. Programs such as PRIM-9 and its descendants allow such interactive inspection by permitting real-time rotation of any three variables, exchange of any displayed variable for another, zooming, etc. [12, 15, 42]. However, the practical usefulness of this elaborate technology is limited by the viewers' difficulties with visualizing anything in hyperspace. The viewers' geometric experience is in a three-dimensional world, and this limits their ability to grasp higher-dimensional data, however clearly or cleverly displayed. But perhaps with experience we will get better at it!

The answer to viewers' difficulties with

hyperspace may be to try and reduce as much as possible of the data to three dimensions or planes. The role of *lower-dimensional approximation* seems to be crucial here, and the success of graphical displays of data may depend on how well the data's features can be recaptured in one of several planes or three-spaces. Clearly, this makes the data-driven mode central to the exploration of higher-dimensional scatters. The data need to be projected onto two- or three-dimensional displays chosen by least-squares* or resistant principal component methods or according to some criterion of interest such as the clottedness index in projection pursuit. Methods of two- and three-dimensional display can then be used to

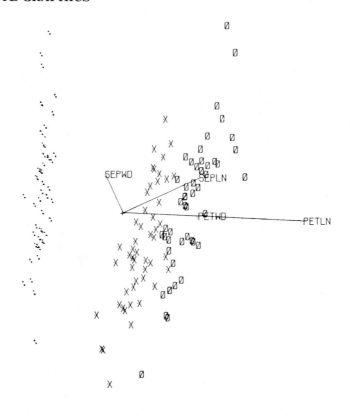

Figure 5 Biplot of Anderson's iris data (·, *setosa*; ×, *versicolor*; 0, *virginica*).

explore these projections. It is, of course, important that the perpendiculars to these projections, i.e., the residuals*, not be ignored either, but be subjected to further multivariate graphical display and analysis or at least to the recording of salient features such as outliers. Figure 5 is a two-dimensional biplot of the iris data along its principal axes: variables are displayed as lines, units as markers indicating the species. Figure 6 shows a three-dimensional biplot of the same data after rotation to a direction highlighting species differentiation. In the latter biplot, the unit markers for each species are replaced by a concentration ellipse.

OTHER EUCLIDEAN MAPPINGS

The foregoing detailed discussion considered the multivariate scatter of units with respect

to axes along which the variables were measured. The *dual* of this display, the *configuration of variables*, is geometrically analogous in that variables are displayed with respect to axes for the units of observation. Inspection of the displayed configuration of markers for the variables is likely to be more concerned with correlations and sheaves (clusters) of highly correlated variables. The methods of display, however, are analogous; again, high-dimensional configurations are likely to need projection onto lower-dimensional subspaces, and methods such as principal components may serve that purpose.

A third type of display is that of the biplot* [11, 19, 21], in which *biplot* markers for units and for variables appear jointly. Inspection of biplots allows simultaneous examination of units' scatter and variables'

CØLS   XL=-2.15E+02   XU= 2.15E+02   YL=-2.15E+02   YU= 2.15E+02   ZL=-2.15E+02   ZU= 2.15E+02

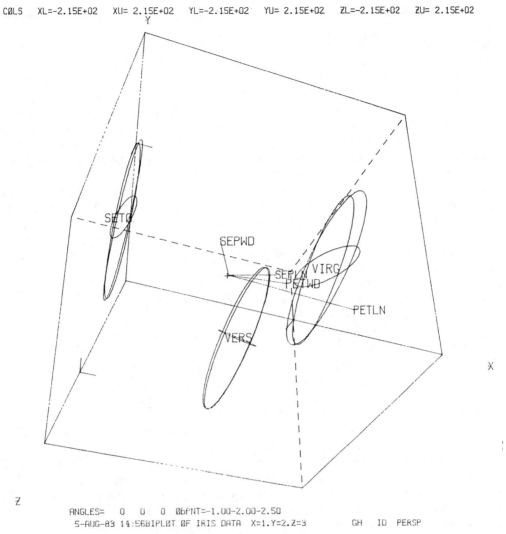

ANGLES=   0   0   0 ØbPNT=-1.00-2.00-2.50
5-AUG-83 14:56BIPLØT ØF IRIS DATA X=1.Y=2.Z=3   GH   ID PERSP

Figure 6   Rotated three-dimensional biplot of Anderson's iris data.

configuration as well as of the relation between the two—see Figs. 5 and 6. It is therefore likely to be more revealing than inspection of either units' scatters alone or the variables' configuration alone. In particular, it is especially useful for diagnosing models to fit the data (Bradu and Gabriel, [6]).

Biplots are constructed by superimposition of the principal axes for the units' scatter upon those for the variables' configuration. The data values are then represented by inner products of unit marker vectors and variable marker vectors—no reference to axes is necessary. The exact mapping of the data is higher-dimensional, but the actual displays show orthogonal projections onto two or three axes, usually those of the first several principal components. The techniques for displaying these biplots, interactively and otherwise, are analogous to those discussed earlier. (A special display package, BGRAPH, facilitates this [40]; Figs. 5 and 6 were produced by it.)

A closely related technique, *correspondence analysis** [5, 27], also superimposes the scatter and configuration, but it does not permit as simple a mapping of the data as the biplot does.

EUCLIDEAN MAPPING OF SIMILARITIES

The above techniques map the observations directly into a display. Another class of techniques maps the units (or variables) into a Euclidean space in terms of interunit similarities, e.g., some monotonically decreasing function of Mahalanobis distances (or correlations between variables). Such methods go under the general name of *multidimensional scaling** and result in a display of units by means of markers whose distances vary inversely with the units' similarities (or variables by means of markers whose proximities vary with the variables' correlations); thus close markers represent similar units (or highly correlated variables). A scaling technique that is closely related to principal component analysis is *principal coordinate analysis* [26]. In practice these methods, as well as the biplot and correspondence analysis, often produce quite similar graphical representations of interunit similarities. (An analogous statement holds for the variables' configuration.)

In general, multidimensional scaling techniques [28, 32, 37, 38, 39] are *nonmetric* and do not use orthogonal projections. Instead they fit displayed distances to dissimilarities largely by attempting to preserve their rank ordering. This makes them more flexible than Euclidean techniques and allows them to fit nonlinear patterns better. As regards actual display, reduction of dimensionality, etc., these techniques are very similar to the Euclidean ones already discussed [20].

DENDROGRAMS AND CLUSTERING

A different type of display of multivariate data uses a tree structure and is useful for clustering. Thus a *dendrogram** is a tree that branches out from one trunk to several large *branches*, and these further branch out to smaller and smaller branches until they culminate in leaves representing the units (or variables). The tree is constructed in such a manner that if any branch—of whatever size —were lopped off, the units (variables) attached to it and its subbranches would be more similar to each other than to other units. Descriptions and algorithms for constructing such dendrograms, and for using them in *clustering*, can be found in many texts [29].

At times, a *minimal spanning tree* is constructed to minimize the total of the statistical distances (i.e., dissimilarities) along its branches that yet allows joining all units (or variables). This may be of interest in itself, but it can also supplement other Euclidean displays in showing whether their planar views reproduce the higher-dimensional structure of the data as given by this tree [13].

PICTORIAL MAPPINGS

A brief description of pictorial mapping is given next. (The reader is directed to the References for technical detail.) The principle is straightforward: some picture, or *glyph*, is chosen to represent each unit, and p features of each unit's glyph are drawn so as to represent the unit's observations of the p variables. Thus, with $p = 5$, unit i might be pictured by a polygonal profile line joining five points whose heights correspond to the five variables' observations on unit i. Similarly, star-shaped profiles would draw these observations as the lengths of five rays of a star. A variety of other pictorial mappings of this kind have been proposed from time to time, the most notable of which are *Andrews function plots** [2] and *Chernoff faces** [8]. Andrews represented the ith unit's observations by a harmonic function of angle t, so that each unit is plotted as a sinusoidal curve, the height of which at different t's expresses the relative contribution of the different variables. Chernoff, on the other hand, represented each unit by a humanoid face, the individual features of which are associated with the variables. For example, one variable might determine breadth of face, a second size of eyes; a third, slant of eyes; a fourth, length of nose; a fifth, curvature of the mouth, etc. The actual computations of Andrews' function plots and Chernoff's faces can be done by a computer that also plots the pictures. Hence the methods are feasible so long as there are not too many units to be conveniently plotted and inspected.

Examples of the use of mapping into faces include data on air quality in which observations at each hour of the day and night were displayed as a face in which, for example, ozone concentration was shown by the mouth's curvature [43]. Changing trends in air quality during the day were easily discernible from this array of faces. This may be a highly effective means of conveying multivariate information to a relatively unsophisticated audience, especially if the features are cued to the content, as, for example, a downward curving mouth to poor air quality. It has also been shown to be quite effective in allowing unsophisticated viewers to cluster multivariate data into fairly homogeneous subbatches and to discern turning points in sequences of observations [8]. But that does not mean that these methods are more effective than available mathematical

algorithms that can trace trends and cluster units from large batches at high speed. Such elementary applications are therefore unlikely to be of much use to the serious data analyst, nor is it likely that pictorial mappings will help the viewer to grasp linear correlations and perhaps more subtle relationships that are not readily apparent from scatter plots and other Euclidean mappings. (But will faces overcome the viewer's difficulty with visualizing hyperspace?)

The usefulness of such pictorial mappings depends on the ability of the viewer to discern and rapidly absorb the multivariate data displayed. Most of these methods are likely to be quite adequate for classifying two very different subgroups of units by clustering together similar profiles, or similar function plots, or similar faces. Various studies have shown this to be possible on relatively small batches, though it depends somewhat on the particular assignment of features to variables. (See, however, Kleiner and Hartigan's [31] suggested use of clustering to define an objective assignment; see also ref. 30.) Among the different pictures,

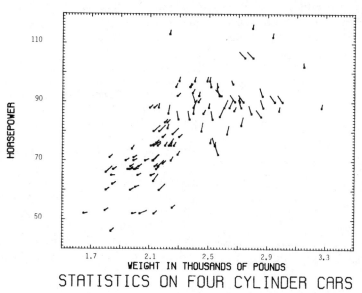

Figure 7 Enhanced four-dimensional scatter plot of data on four-cylinder cars. (Courtesy of W. L. Nicholson.)

the most effective ones are presumably the ones that appeal to human perception most readily, such as faces whose recognition we learn from an early age and simple rays whose direction and length map the third and fourth dimension [35]. An illustration is shown in Fig. 7.

At the other extreme, sinusoidal functions are probably too sophisticated mathematically—and too cluttered in most displays—to be interpreted easily by most viewers.

The real value of the pictorial representations is likely to be in *mixed mappings*, where some of the variables are used to locate the pictures and the others are represented by features of the pictures [46]. Thus a map of the United States with 50 faces displaying a set of economic variables for each state might be helpful in showing the relation of economic phenomena to each other and to their geographical location. Again, a bivariate scatter plot on the first two principal axes, say, might be usefully supplemented by plotting each unit as a glyph showing its *p*-variate values. Such a plot might reveal linear and other relationships between the various variables. If there are too many points in the scatter, the bivariate plane might be cellulated and a glyph drawn in each cell to represent the averages of the various variables for the units in that cell. Clearly, such mixed mappings need to be experimented with; they may prove an important tool in multivariate data analysis.

References

[1] Anderson, E. (1960). *Technometrics*, **2**, 387–391.

[2] Andrews, D. F. (1972). *Biometrics*, **28**, 125–136.

[3] Bachi, R. (1968). *Graphical Rational Patterns: A New Approach to Graphical Representation of Statistics*. Israel Universities Press, Jerusalem.

[4] Barnett, V. (1976). *J. R. Statist. Soc., A*, **139**, 318–355.

[5] Benzecri, J.-P. (1973). *L'Analyse des Données* (vol. 2: *L'Analyse des Correspondances*.) Dunod, Paris.

[6] Bradu, D. and Gabriel, K. R. (1978). *Technometrics*, **20**, 47–68.

[7] Chambers, J. M., Cleveland, W. S., Kleiner, B., and Tukey, P. A. (1983). *Graphical Methods for Data Analysis*. Wadsworth, Belmont, CA; Duxbury Press, Boston.

[8] Chernoff, H. (1973). *J. Amer. Statist. Ass.* **68**, 361–368.

[9] Chernoff, H. (1978). In *Graphical Representation of Multivariate Data*, P. C. C. Wang, ed. Academic Press, New York, pp. 1–11.

[10] Cleveland, W. S., Diaconis, P., and McGill, R. (1982). *Science*, **216**, 1138–1141.

[11] Cox, C. and Gabriel, K. R. (1982). In *Modern Data Analysis*, R. L. Launer and A. F. Siegel, eds. Academic Press, New York, pp. 45–82.

[12] Donoho, D. L., Huber, P. J., Ramos, E., and Thoma, H. M. (1982). *Proceedings of the Third Annual Conference of the National Computer Graphics Association*, Vol. 1, pp. 393–400.

[13] Everitt, B. S. (1978). *Graphical Techniques for Multivariate Data*. Heineman, London.

[14] Fienberg, S. E. (1979). *Amer. Statist.*, **33**, 165–178.

[15] Friedman, H. P., Farell, E. S., Goldwyn, R. M., Miller, M., and Siegel, J. M. (1972). *Proceedings of the Sixth Interface Symposium on Computer Science and Statistics*. University of California, Berkeley, pp. 56–59.

[16] Friedman, J. H. and Rafsky, L. C. (1981). *J. Amer. Statist. Ass.*, **76**, 277–287, with discussion on pp. 287–295.

[17] Friedman, J. H. and Stuetzle, W. (1982). In *Modern Data Analysis*, R. L. Launer and A. F. Siegel, eds. Academic Press. New York, pp. 123–148.

[18] Friedman, J. H. and Tukey, J. W. (1974). *IEEE Trans. Comp.* **C-23**, 881–890.

[19] Gabriel, K. R. (1971). *Biometrika*, **58**, 453–467.

[20] Gabriel, K. R. (1978). In *Theory Construction and Data Analysis in the Social Sciences*, S. Shye, ed. Jossey-Bass, San Francisco, pp. 350–370.

[21] Gabriel, K. R. (1981). In *Interpreting Multivariate Data*, Vic Barnett, ed. Wiley, Chichester, England, 167–173.

[22] Gabriel, K. R. and Odoroff, C. L. (1983). "Resistant Lower Rank Approximation of Matrices." *Statistics Technical Report 83/02*, University of Rochester, NY.

[23] Gentleman, J. F. (1981). *J. Amer. Statist. Ass.* **76**, 289–290.

[24] Gnanadesikan, R. (1973). *Bull. Intern. Statist. Inst.*, **45**(4), 195–206.

[25] Gnanadesikan, R. (1977). *Methods for Statistical Data Analysis of Multivariate Observations*. Wiley, New York.

[26] Gower, J. C. (1966). *Psychometrika*, **53**, 325–338.

[27] Greenacre, M. J. (1981). In *Interpreting Multivariate Data*, Vic Barnett, ed. Wiley, Chichester, UK, pp. 119–146.

[28] Guttman, L. and Lingoes, J. C. (1967). *Multivariate Behav. Res.*, **2**, 485–505.

[29] Hartigan, J. A. (1975). *Clustering Algorithms*. Wiley, New York.

[30] Jacob, R. J. K. (1981). *J. Amer. Statist. Ass.* **76**, 270–272.

[31] Kleiner, B. and Hartigan, J. A. (1981). *J. Amer. Statist. Ass.*, **76**, 260–269.

[32] Kruskal, J. B. (1964). *Psychometrika*, **27**, 1–27.

[33] Newman, W. M. and Sproull, R. F. (1979). *Principles of Interactive Computer Graphics*. McGraw-Hill, New York.

[34] Newton, C. M. (1978). In *Graphical Representation of Multivariate Data*, P. C. C. Wang, ed. Academic Press, New York, pp. 59–92.

[35] Nicholson, W. L. and Littlefield, R. J. (1983). *Computer Science and Statistics Proceedings of the 14th Symposium on the Interface*. Springer-Verlag, New York.

[36] Pearson, K. (1901). *Philos. Mag. 6th Ser.*, **2**, 559–572.

[37] Shepard, R. N. (1962). *Psychometrika*, **27**, 125–139 and 219–246.

[38] Shepard, R. N. and Carroll, J. D. (1966). In *Multivariate Analysis*, P. R. Krishnaiah, ed. Academic Press, New York, pp. 561–592.

[39] Shepard, R. N., Romney, A. K. and Nerlove, S. B., eds. (1973). *Multidimensional Scaling*. Seminar Press, New York.

[40] Tsianco, M. C., Odoroff, C. L., Plumb, S., and Gabriel, K. R. (1981). "BGRAPH—A Program for Biplot Multivariate Graphics. User's Guide." *Tec. Rep. 81/20*, Dep. of Statistics, University of Rochester, NY.

[41] Trumbo, B. E. (1981). *Amer. Statist.*, **35**, 220–226.

[42] Tukey, J. W., Friedman, J. F., and Fisherkeller, M. A. (1976). *Proceedings of the 4th International Congress for Stereology*, Gaithersburg, Maryland, cf. p. 15.

[43] Tukey, P. A. and Tukey, J. W. (1981). In *Interpreting Multivariate Data*, Vic Barnett, ed. Wiley, Chichester, England, Chaps. 10–12.

[44] Tukey, J. W. and Tukey, P. A. (1983). *Computer Science and Statistics: Proceedings of the Fourteenth Symposium on the Interface*, K. W. Heiner and J. W. Wilkinson, eds. Springer-Verlag, New York, pp. 60–66.

[45] Wachter, K. W. (1975). "User's Guide and Tables to Probability Plotting for Large-Scale Principal Component Analysis." Report W-75-1, Harvard University Research, Cambridge, MA.

[46] Wainer, H. (1981). *J. Amer. Statist. Ass.*, **76**, 272–275.

[47] Wilk, M. B., Gnanadesikan, R., and Huyett, M. J. (1962). *Technometrics*, **4**, 1–20.

(ANDREWS FUNCTION PLOTS
BIPLOT
CHERNOFF FACES
DENDROGRAMS
GEOGRAPHY, STATISTICS IN
GRAPHICAL REPRESENTATION, COMPUTER-AIDED
GRAPHICAL REPRESENTATION OF DATA
MULTIDIMENSIONAL SCALING
STATISTICAL SOFTWARE)

K. Ruben Gabriel

MULTIVARIATE KURTOSIS   *See* MULTI-VARIATE SKEWNESS AND KURTOSIS

MULTIVARIATE LOCATION TESTS

ONE-SAMPLE TESTS

Suppose we are presented with a set of p measurements on each of n subjects. Let $\mathbf{X}_i$, $i = 1, \ldots, n$ denote a column vector of p measurements on the ith subject. The data can be written as a $p \times n$ array:

$$\begin{matrix} X_{11} & \cdots & X_{1n} \\ \vdots & & \vdots \\ X_{p1} & \cdots & X_{pn} \end{matrix}$$

where the ith column represents the p measurements on the ith subject and the jth row is a set of n independent measurements on the jth measurement variable.

If we further suppose that each column is an independent observation from a multivariate normal population with mean vector μ and covariance matrix Σ, then we can ask for a test of the hypothesis $H_0 : \mu = \mu_0$ vs. $H_A : \mu \neq \mu_0$, where μ_0 is a specified vector.

Under the normality assumption, Hotelling's T^{2*} statistic is the basis for the test. Let

$$T^2 = (\overline{\mathbf{X}} - \mu_0)' \left[n^{-1} \mathbf{S} \right]^{-1} (\overline{\mathbf{X}} - \mu_0),$$

where $\overline{\mathbf{X}}$ is the $p \times 1$ vector of sample means, with ith component $\overline{X}_i = n^{-1} \sum_{t=1}^{n} X_{it}$, and $\mathbf{S}$ is the $p \times p$ sample covariance matrix, with (i, j)th component

$$(n-1)^{-1} \sum_{t=1}^{n} (X_{it} - \overline{X}_i)(X_{jt} - \overline{X}_j).$$

The test rejects $H_0 : \mu = \mu_0$ at significance

level α if

$$T^2 > [(n-1)p/(n-p)]F(\alpha, p, n-p),$$

where $F(\alpha, p, n-p)$ is the upper α critical point from an F-distribution* with p and $n - p$ degrees of freedom. The form of T^2 generalizes, in the natural way, the square of the one-sample univariate t-statistic in which $\mathbf{S}$ represents the unbiased estimator of the population variance.

The normality assumption may be too strong. If we suppose that a multivariate density satisfies $f(x_1 - \theta_1, \ldots, x_p - \theta_p) = f(-x_1 + \theta_1, \ldots, -x_p + \theta_p)$, said to be *diagonally symmetric* about $\mathbf{\theta}$, then the multivariate rank test can be constructed. The statistic is a generalization to the multivariate setting of the univariate one-sample Wilcoxon signed rank* statistic. Let

$$W_i = \sum_{j=1}^{n} R(|X_{ij} - \theta_i|)\text{sgn}(X_{ij} - \theta_i),$$

$i = 1, \ldots, p$, denote the Wilcoxon signed rank statistic computed on the n observations in the ith component and $R(|X_{ij} - \theta_i|)$ the rank of $|X_{ij} - \theta_i|$ among $|X_{i1} - \theta_i|$, $\ldots, |X_{in} - \theta_i|$. Let $\mathbf{W}$ be the $p \times 1$ vector with W_i as the ith component. Following the pattern for the construction of Hotelling's statistic, we have $W^2 = \mathbf{W}'\mathbf{V}^{-1}\mathbf{W}$, where $n^{-3}\mathbf{V}$ is a consistent estimate of the asymptotic covariance matrix of $n^{-3/2}\mathbf{W}$. Let $\mathbf{V} = ((v_{ij})); i, j = 1, \ldots, p$, then

$$v_{ii} = n(n+1)(2n+1)/6,$$

$$v_{ij} = \sum_{t=1}^{n} R(|X_{it} - \theta_i|)R(|X_{jt} - \theta_j|)$$
$$\times \text{sgn}(X_{it} - \theta_i)\text{sgn}(X_{jt} - \theta_j).$$

The limiting distribution of $n^{-3/2}\mathbf{W}$ is multivariate normal with $\mathbf{0}$ mean vector, and the limiting distribution of W^2 is chi-square* with p degrees of freedom. To test H_0: $\mathbf{\theta} = \mathbf{\theta}_0$ vs. H_A: $\mathbf{\theta} \neq \mathbf{\theta}_0$ at approximate significance level α, we reject H_0 if $W^2 > \chi^2(\alpha, p)$, the upper chi-square critical value. Note that W^2 is computed by inserting the known, hypothesized values $\theta_{10}, \ldots, \theta_{p0}$ into the

ranking formulas for W_i, $i = 1, \ldots, p$ and the covariance formula v_{ij}, $i, j = 1, \ldots, p$.

Finally, it should be noted that under the null hypothesis, W^2 is not distribution-free in finite samples, in contrast to the univariate Wilcoxon signed rank test. Conditional on the absolute values, the 2^n sign configurations are equally likely, and the test based on W^2 is conditionally distribution-free. In practice, the conditional randomization test is not easy to compute for moderate sample sizes, so the (asymptotically distribution-free) chi-square approximation is used.

If the diagonal symmetry assumption is not appropriate, then the multivariate extension of the simple univariate sign test* is available. In this case, no shape assumption is imposed on $f(x_1 - \theta_1, \ldots, x_p - \theta_p)$. The parameter $\mathbf{\theta}$ will be taken as the vector of population medians. Let $\mathbf{S}$ be the $p \times 1$ vector of sign statistics computed for each component. Hence the ith component of $\mathbf{S}$ is $S_i = \sum_{t=1}^{n}\text{sgn}(X_{it} - \theta_i)$. Similar to T^2 and W^2, define $S^2 = \mathbf{S}'\mathbf{V}^{-1}\mathbf{S}$, where $n^{-1}\mathbf{V}$ is a consistent estimate of the asymptotic covariance matrix of $n^{-1/2}\mathbf{S}$. Let $\mathbf{V} = ((v_{ij}))$, $i, j = 1, \ldots, p$; then

$$v_{ii} = n,$$

$$v_{ij} = \sum_{t=1}^{n} \text{sgn}(X_{it} - \theta_i)\text{sgn}(X_{jt} - \theta_j).$$

Again, to test H_0: $\mathbf{\theta} = \mathbf{\theta}_0$ vs. H_A: $\mathbf{\theta} \neq \mathbf{\theta}_0$ at approximate level α, we reject H_0 if $S^2 > \chi^2(\alpha, p)$, the upper chi-square critical point.

When $p = 2$, the bivariate case, S^2 has a particularly simple form:

$$S^2 = \frac{(C_1 - C_2)^2}{C_1 + C_2} + \frac{(D_1 - D_2)^2}{D_1 + D_2},$$

$$C_1 = \#(X_{1i} \leqslant \theta_{01}, X_{2i} \leqslant \theta_{02}),$$
$$C_2 = \#(X_{1i} > \theta_{01}, X_{2i} > \theta_{02}),$$
$$D_1 = \#(X_{1i} \leqslant \theta_{01}, X_{2i} > \theta_{02}),$$
$$D_2 = \#(X_{1i} > \theta_{01}, X_{2i} \leqslant \theta_{02}),$$
$$i = 1, \ldots, n; \quad \mathbf{\theta}' = (\theta_{01}, \theta_{02}).$$

The test based on S^2 is conditionally distribution-free, but not unconditionally distribution-free, in contrast to the univariate sign test. The test is asymptotically distribution-free, and, in practice, the chi-square distribution is used to approximate the critical values of the test.

For additional discussion of the multivariate rank and sign tests see Maritz [4, Chap. 7] and Puri and Sen [5, Chap. 4]. The former reference deals extensively with applied issues while the latter reference contains the theoretical development. Bickel [2, p. 172] concludes that W^2 and S^2 are superior to T^2 when there are gross errors present in the data. However, they should be used with caution in situations where considerable degeneracy is present.

Example. We now illustrate the calculations on a small data set. We will consider a portion of the data in Johnson and Wichern [3, Table 5.1, p. 182]. We have the first 10 measurements on sodium content (X_1) and potassium content (X_2) of the perspiration of healthy female subjects. We will test the hypotheses $H_0 : \mu' = (50, 10)$ vs. $H_A : \mu' \neq (50, 10)$. The data is shown in Table 1.

From the data, we find $\bar{X} = (46.53, 9.94)$ and

$$S^{-1} = \begin{pmatrix} 0.007 & 0.009 \\ 0.009 & 0.235 \end{pmatrix}.$$

Hence $T^2 = 0.89$ and will fail to reject H_0: $\mu' = (50, 10)$ at any reasonable level. The Wilcoxon statistic W^2 is easily computed from the table. By summing the fourth and seventh columns we find $\mathbf{W}' = (-9, -5)$

and again using these columns

$$\mathbf{V}^{-1} = \begin{pmatrix} 0.0026 & -0.0002 \\ -0.0002 & 0.0026 \end{pmatrix}.$$

Then $W^2 = \mathbf{W}'\mathbf{V}^{-1}\mathbf{W} = 0.26$ and also fails to reject $H_0 : \mu' = (50, 10)$ at any reasonable level. Finally, we can illustrate the multivariate sign test using the simple formula for S^2 when $p = 2$. A quick count from the table yields $C_1 = 4$, $C_2 = 2$, $D_1 = D_2 = 2$; hence $S^2 = \frac{2}{3}$ and will also fail to reject the null hypothesis.

TWO-SAMPLE TESTS

Tests in the two-sample multivariate location problem can be developed along the same lines. We will briefly outline the three cases: Hotelling's test, a multivariate Mann–Whitney–Wilcoxon test, * and a multivariate Mood test. These statistics can be used to test hypotheses on the difference in location vectors of two populations. The location vector may be the vector of means or medians, depending on the model.

If we have two samples, sizes n_1 and n_2, from two p-variate normal populations, then to test $H_0 : \mu_1 = \mu_2$ vs. $H_A : \mu_1 \neq \mu_2$, we would use the two-sample version of Hotelling's T^2 statistic. The test statistic is written as

$$T^2 = \left[\bar{X}_1 - \bar{X}_2 \right]' \left[\left(\frac{1}{n_1} + \frac{1}{n_2} \right) S_p \right]^{-1} \times \left[\bar{X}_1 - \bar{X}_2 \right],$$

Table 1   Perspiration Data

Subject	X_1	$X_1 - 50$	$R(\lvert X_1 - 50 \rvert) \times$ $\text{sgn}(X_1 - 50)$	X_2	$X_2 - 10$	$R(\lvert X_2 - 10 \rvert) \times$ $\text{sgn}(X_2 - 10)$
1	48.5	-1.5	1	9.3	-0.7	-2
2	65.1	15.1	8	8.0	-2	-6
3	47.2	-2.8	-3	10.9	0.9	3
4	53.2	3.2	4	12.2	2.2	8
5	55.5	5.5	6	9.7	-0.3	-1
6	36.1	-13.9	-7	7.9	-2.1	-7
7	24.8	-25.2	-10	14.0	4	10
8	33.1	-16.9	-9	7.6	-2.4	-9
9	47.4	-2.6	-2	8.5	-1.5	-5
10	54.1	4.1	5	11.3	1.3	4

where $\overline{\mathbf{X}}_1, \overline{\mathbf{X}}_2$ are the vectors of sample means and

$$\mathbf{S}_p = \left[(n_1 - 1)/(n_1 + n_2 - 2) \right] \mathbf{S}_1$$
$$+ \left[(n_2 - 1)/(n_1 + n_2 - 2) \right] \mathbf{S}_2,$$

with $\mathbf{S}_1, \mathbf{S}_2$ the sample covariance matrices. The test rejects $H_0: \boldsymbol{\mu}_1 = \boldsymbol{\mu}_2$ at significance level α if

$$T^2 > \left[(n_1 + n_2 - 2)p/(n_1 + n_2 - p - 1) \right]$$
$$\times F(\alpha, p, n_1 + n_2 - p - 1),$$

where $F(\alpha, p, n_1 + n_2 - p - 1)$ is the upper $100\alpha\%$ point of an F distribution with p and $n_1 + n_2 - p - 1$ degrees of freedom.

If we do not wish to make the normality assumption, then the multivariate version of the Mann–Whitney–Wilcoxon* rank sum test is available. We will assume that the two sample population distributions differ at most in their location vectors. It is not necessary to assume symmetry in the two-sample problem. Consider the ith component with $N = n_1 + n_2$ and let $W_i = \sum_{t=1}^{n_1} R_{it} - n_1(N + 1)/2$, the centered sum of ranks of the observations in the ith component of the first sample when ranked together with the observations in the ith component of the second sample. Let $\mathbf{W}$ be the $p \times 1$ vector whose ith component is W_i.

To test $H_0: \boldsymbol{\theta}_1 = \boldsymbol{\theta}_2$ vs. $H_A: \boldsymbol{\theta}_1 \neq \boldsymbol{\theta}_2$, we will use $W^2 = \mathbf{W}'\mathbf{V}^{-1}\mathbf{W}$, where $N^{-3}\mathbf{V}$ is a consistent estimate of the asymptotic covariance matrix of $N^{-3/2}\mathbf{W}$. Let $\mathbf{V} = ((v_{ij}))$, $i, j = 1, \ldots, p$; then

$$v_{ii} = n_1 n_2 (N + 1)/12,$$

$$v_{ij} = \left[n_1 n_2 / \{ (N - 1)N \} \right]$$
$$\times \left[\sum_{t=1}^{N} R_{it} R_{jt} - \tfrac{1}{4} N(N + 1)^2 \right]$$

The limiting distribution of W^2 is chi-square with p degrees of freedom, so the test rejects $H_0: \boldsymbol{\theta}_1 = \boldsymbol{\theta}_2$ at approximate significance level α if $W^2 > \chi^2(\alpha, p)$.

Mood's median test, which generalizes the one-sample sign test (*see* BROWN–MOOD ME-

DIAN TEST) can also be extended to the multivariate two-sample problem. Consider the ith component and let M_i be the number of observations in the first sample that are less than the median of the combined observations in the two samples minus $n_1/2$. Then M_i is the centered Mood's statistic. We will suppose that N is even. Then let $\mathbf{M}$ be the $p \times 1$ vector of Mood's statistics. The test statistic is $M^2 = \mathbf{M}'\mathbf{V}^{-1}\mathbf{M}$, where $N^{-1}\mathbf{V}$ is a consistent estimate of the asymptotic covariance matrix of $N^{-1/2}\mathbf{M}$. Let $\mathbf{V} = ((v_{ij}))$, $i, j = 1, \ldots, p$, then

$$v_{ii} = n_1 n_2 / \left[4(N - 1) \right],$$

$$v_{ij} = \left[n_1 n_2 / (N - 1) \right] \left[(N_{ij}/N) - \tfrac{1}{4} \right],$$

where N_{ij} is the number of pairs in the combined data such that both observations are less than their respective combined sample medians.

The limiting distribution of M^2 is chi-square with p degrees of freedom. Hence, to test $H_0: \boldsymbol{\theta}_1 = \boldsymbol{\theta}_2$ vs. $H_A: \boldsymbol{\theta}_1 \neq \boldsymbol{\theta}_2$, reject H_0 at approximate significance level α if $M^2 > \chi^2(\alpha, p)$.

Neither W^2 nor M^2 are distribution-free for finite samples. However, they are conditionally and asymptotically distribution-free. For further discussion, see Maritz [4, Chap. 7] for applied issues and Puri and Sen [5, Chap. 5] for theoretical issues.

For further discussion of Hotelling's one- and two-sample tests, see Johnson and Wichern [3] for applied issues and Arnold [1] for theoretical issues.

References

[1] Arnold, S. F. (1981). *The Theory of Linear Models and Multivariate Analysis*. Wiley, New York.

[2] Bickel, P. J. (1965). *Ann. Math. Statist.*, **36**, 160–173. (Discusses the asymptotic distribution theory and efficiency of the rank tests.)

[3] Johnson, R. A. and Wichern, D. W. (1982). *Applied Multivariate Statistical Analysis*. Prentice-Hall, Englewood Cliffs, NJ.

[4] Maritz, J. S. (1981). *Distribution-Free Statistical Methods*. Chapman and Hall, New York.

[5] Puri, M. L. and Sen, P. K. (1971). *Nonparametric Methods in Multivariate Analysis*. Wiley, New York.

(BROWN–MOOD MEDIAN TEST
DISTRIBUTION-FREE METHODS
MANN–WHITNEY–WILCOXON TEST
HOTELLING'S T^2
MULTIVARIATE ANALYSIS
WILCOXON SIGNED-RANK STATISTIC)

T. P. HETTMANSPERGER

MULTIVARIATE LOGARITHMIC SERIES DISTRIBUTION

INTRODUCTION

A growing interest has been witnessed in recent years in multivariate discrete probability models. See, for example, Patil et al. [8]. The multivariate logarithmic series distribution (LSD) is a multivariate analog of the univariate LSD. It was introduced by Khatri [4] as an illustrative example of the multivariate power series distribution*. In the context of a problem in population and community ecology, Patil and Bildikar [7] found the multivariate LSD fruitful and studied some of its structural and inferential properties. They applied the multivariate LSD model to a data set in human ecology presented by Clark et al. [2], whereas Taillie et al. [12] have discussed the model for bivariate species frequency data in aquatic ecology. Chatfield et al. [1] touch upon its potential in the field of stationary purchasing behavior, whereas Kemp [3], Phillippou and Roussas [11], and Wani [13] provide some properties and procedures.

DEFINITIONS AND PROPERTIES

Definition 1. A random vector rv $\mathbf{x} = (x_1, x_2, \ldots, x_s)$ is said to have the s-variate LSD with the parameter vector (pv) $\boldsymbol{\theta} = (\theta_1, \theta_2, \ldots, \theta_s)$ if its probability function (pf) is given by

$$f(\mathbf{x}, \boldsymbol{\theta}) = \frac{\Gamma(x_1 + x_2 + \cdots + x_s)}{x_1! \, x_2! \cdots x_s!}$$
$$\cdot \frac{\theta_1^{x_1} \theta_2^{x_2} \cdots \theta_s^{x_s}}{-\log(1 - \theta_1 - \theta_2 \cdots - \theta_s)},$$

(1)

where for $i = 1, 2, \ldots, s$, $x_i = 0, 1, 2, \ldots, \infty$, and $0 < \theta_i < 1$, such that $x_1 + x_2 + \cdots + x_s > 0$ and $\theta_1 + \theta_2 + \cdots + \theta_s < 1$.

Analogous to the univariate case, the s-variate LSD is available as the limit of the origin-truncated s-variate NBD (negative multinomial) when the parameter $k \to 0$. (See Patil [5], Taillie et al. [12], Patil et al. [8], and MULTINOMIAL DISTRIBUTIONS.)

Clearly, a multivariate LSD is a multivariate power series distribution,* with series function $f(\boldsymbol{\theta}) = -\log(1 - \theta_1 - \theta_2 - \cdots - \theta_s)$, and it inherits the series function's properties. (*See* MULTIVARIATE POWER SERIES DISTRIBUTIONS.)

Definition 2. A rv $\mathbf{x}$ is said to have the s-variate modified LSD (MLSD) with the pv $(\delta; \boldsymbol{\theta})$ if its pf is given by

$$f(\mathbf{x}; \delta, \boldsymbol{\theta}) = \begin{cases} \delta, & \mathbf{x} = \mathbf{0} \\ (1 - \delta) f(\mathbf{x}; \boldsymbol{\theta}), & \mathbf{x} \neq \mathbf{0}, \end{cases}$$

(2)

where $0 \leqslant \delta < 1$ and $f(\mathbf{x}, \boldsymbol{\theta})$ is defined by (1). (*See also* MODIFIED POWER SERIES DISTRIBUTION.)

Analogous to the univariate case, the multivariate MLSD arises as a compound multinomial distribution when the multinomial parameter n follows the LSD.

Property 1. The multivariate LSD possesses a rather interesting modal property. The number of modes of an s-variate LSD with pv $\boldsymbol{\theta}$ is equal to the number of maximal components of the parameter vector. Further, a mode occurs at the ith s-dimensional standard basis vector if θ_i is the maximal component of the pv $\boldsymbol{\theta}$.

Property 2. The components of the mean vector $\boldsymbol{\mu}$ of the multivariate LSD (1) are given by

$$\mu_i = \theta_i/(\gamma L), \qquad i = 1, 2, \ldots, s, \quad (3)$$

where $\gamma = 1 - \theta_1 - \theta_2 \ldots - \theta_s$ and $L = -\log \gamma$.

Further, the covariances are given by

$$\sigma_{ij} = \mu_i \big[(\theta_j/\gamma) - \delta_{ij}(\mu_i - 1) - (1 - \delta_{ij})\mu_j \big], \quad (4)$$

where δ_{ij} is the Kronecker delta.

Property 3. The crude factorial moments of order $\mathbf{r} = (r_1, r_2, \ldots, r_s)$ defined by $m_{(\mathbf{r})} = E[x_1^{(r_1)} x_2^{(r_2)} \ldots x_s^{(r_s)}]$, where $z^{(k)} = z(z - 1) \ldots (z - k + 1)$ is the descending factorial of order k, are given by

$$m_{(\mathbf{r})} = \frac{1}{L} \left(\frac{\theta_1}{\gamma}\right)^{r_1} \left(\frac{\theta_2}{\gamma}\right)^{r_2} \cdots \left(\frac{\theta_s}{\gamma}\right)^{r_s}. \quad (5)$$

Property 4. The distribution of the k-dimensional vector $\mathbf{s} = (s_1, s_2, \ldots, s_k)$ of the disjoint partial sums defined by $s_i = \sum x_j$, $r_{i-1} < j \leqslant r_i$, with $1 \leqslant r_1 < r_2 \cdots < r_k = s$, is the k-variate LSD with pv $\boldsymbol{\lambda} = (\lambda_1, \lambda_2, \ldots, \lambda_k)$, where $\lambda_i = \sum \theta_j$, $r_{i-1} < j \leqslant r_i$.

Further, in the conditional distribution of $\mathbf{x}$ given $\mathbf{s}$, the k conditional random vectors $\mathbf{y}_i = \{x_j\}$, $r_{i-1} < j \leqslant r_j$, $i = 1, 2, \ldots, k$, are statistically independent of each other, and the conditional distribution of $\mathbf{y}_i$ given s_i is the singular multinomial distribution with pv $(s_i; \{p_j\})$, $y_{i-1} < j \leqslant r_i$, where $p_j = \theta_j/\lambda_i$.

Property 5. The conditional distribution of $(x_1, x_2, \ldots, x_r)$ given $(x_{r+1}, x_{r+2}, \ldots, x_s)$ depends only on the sum $x_{r+1} + x_{r+2} + \cdots + x_s$, and not the individual components. Further, if $x_{r+1} + x_{r+2} + \cdots + x_s = 0$, the conditional distribution is the r-variate LSD with pv $(\theta_1, \theta_2, \ldots, \theta_r)$, whereas if $x_{r+1} + x_{r+2} + \cdots + x_s = k > 0$, it is the r-variate negative multinomial with pv $(k; \theta_1, \theta_2, \ldots, \theta_r)$.

In view of this property, we can study without loss of generality the regression of x_1 on $x_{r+1} + x_{r+2} + \cdots + x_s$ in order to study the regression of x_1 on $x_{r+1}, x_{r+2}, \ldots, x_s$. Thus, in order to predict the value of x_1, no weighing of the individual components of the given conditioning vector $(x_{r+1}, x_{r+2}, \ldots, x_s)$ is necessary.

Property 6. The multiple correlation coefficient* $\rho_{1.23 \ldots s}$ of x_1 on $x_2, x_3, \ldots, x_s$ is equal to the ordinary correlation coefficient of x_1 and $x_2 + x_3 + \cdots + x_s$ and is given by

$$\rho_{1.23 \ldots s}^2$$

$$= \frac{\theta_1(\theta - \theta_1)(L - 1)^2}{\{\theta_1(L-1) + \gamma L\}\{(\theta - \theta_1)(L-1) + \gamma L\}}, \quad (6)$$

where $\theta = \theta_1 + \theta_2 + \cdots + \theta_s$ and γ and L are defined in (3).

ESTIMATION RESULTS

Suppose that we have a random sample of size n consisting of x_{ij}; $i = 1, 2, \ldots, s$; $j = 1, 2, \ldots, n$, drawn from the s-variate LSD defined by (1).

Maximum Likelihood Estimation*

The equations for the maximum likelihood estimates $\hat{\theta}_i$ are given by

$$\hat{\theta}_i = (\bar{x}_i/\bar{x})\hat{\varphi}, \qquad i = 1, 2, \ldots, s, \quad (7)$$

where $\bar{x}_i = (1/n)\sum_j x_{ij}$ is the sample mean of the ith component x_i and $\bar{x} = \bar{x}_1 + \bar{x}_2 + \cdots + \bar{x}_s$ with $\hat{\varphi} = \hat{\theta}_1 + \hat{\theta}_2 + \cdots + \hat{\theta}_s$. The value of $\hat{\varphi}$ is obtained from the equation

$$\bar{x} = \hat{\varphi} / \{(1 - \hat{\varphi})[-\log(1 - \hat{\varphi})]\} \quad (8)$$

which is nothing but the likelihood equation for the univariate LSD, the solution of which has been extensively tabulated, e.g., in Patil and Wani [10].

The information matrix is given by $\mathbf{I} = (I_{ij})$, where $I_{ij} = n\sigma_{ij}/(\theta_i\theta_j)$ for σ_{ij} given by (4). The asymptotic variance-covariance matrix of the maximum likelihood estimators $\hat{\theta}_1, \hat{\theta}_2, \ldots, \hat{\theta}_s$ is then given by $\mathbf{V} = \mathbf{I}^{-1}$.

Minimum Variance Unbiased Estimation

With symbols carrying their usual meaning, the MVU estimate of θ_i is given by

$$\tilde{\theta}_i = (\bar{x}_i / \bar{x})\tilde{\varphi}. \tag{9}$$

The value of $\tilde{\varphi}$ is obtained from the tabulation of the MVU estimation of the parameter φ of the univariate LSD given in Patil and Bildikar [6].

It is interesting that both the maximum likelihood estimates and the best unbiased estimates of $\theta_1, \theta_2, \ldots, \theta_s$ are proportional to the corresponding sample means $\bar{x}_1$, $\bar{x}_2, \ldots, \bar{x}_s$.

Concluding Remarks

For explicit statements of chance mechanisms for multivariate LSDs analogous to those for univariate LSDs, see Kemp [3] and LOGARITHMIC SERIES DISTRIBUTION.

References

[1] Chatfield, C., Ehrenberg, A. S., and Goodhardt, G. J. (1966). *J. R. Statist. Soc. A.*, **129**, 317–367.

[2] Clark, P. J., Eckstrom, P. T., and Linden, L. C. (1964). *Ecology*, **45**, 367–372.

[3] Kemp, A. W. (1981). In *Statistical Distributions in Scientific Work*, Vol. 5, C. Taillie, G. P. Patil, and B. Baldessari, eds. D. Reidel, Dordrecht and Boston, pp. 57–73.

[4] Khatri, C. G. (1959). *Biometrika*, **46**, 486–490.

[5] Patil, G. P. (1968). *Sankhyā B*, **30**, 355–366.

[6] Patil, G. P. and Bildikar, S. (1966). *Sankhyā A*, **28**, 239–250.

[7] Patil, G. P. and Bildikar, S. (1967). *J. Amer. Statist. Ass.*, **62**, 655–674.

[8] Patil, G. P., Boswell, M. T., Joshi, S. W., and Ratnaparkhi, M. V. (1984). *A Modern Dictionary and Classified Bibliography of Statistical Distributions, Vol. 3: Discrete Models*. International Cooperative Publishing House, Fairland, MD.

[9] Patil, G. P. and Joshi, S. W. (1968). *A Dictionary and Bibliography of Discrete Distributions*. Oliver and Boyd, Edinburgh, Scotland.

[10] Patil, G. P. and Wani, J. K. (1965). In *Classical and Contagious Discrete Distributions*, G. P. Patil, ed. Statistical Publishing Society, Calcutta and Pergamon Press, New York, pp. 398–409. Also in *Sankhyā A*, **27**, 281–291.

[11] Philippou, A. N. and Roussas, G. G. (1974). *Commun. Statist.*, **3**, 469–472.

[12] Taillie, C., Ord, J. K., Mosimann, J. E., and Patil, G. P. (1979). In *Statistical Distributions in Ecological Work*, J. K. Ord, G. P. Patil, and C. Taillie, eds. International Cooperative Publishing House, Fairland, MD, pp. 157–191.

[13] Wani, J. K. (1970). *Skand. Aktuarietidskr.* 1–5.

(LOGARITHMIC SERIES DISTRIBUTION
MODIFIED POWER SERIES
  DISTRIBUTIONS
MULTIVARIATE POWER SERIES
  DISTRIBUTIONS
POWER SERIES DISTRIBUTIONS)

G. P. PATIL

MULTIVARIATE MEDIAN AND RANK SUM TESTS

Nonparametric methods have especially broad applications in the analysis of data since they are not bound by restrictions on the population distribution. Nonparametric methods are robust, and hence are appropriate for analyzing data sets from populations with a common but general continuous distribution function. A continuous distribution function is required theoretically to ensure that there are no ties.

Consider the problem of testing the equality of c ($c \geqslant 2$) continuous distribution functions, $F_1, F_2, \ldots, F_c$, that have the same general form but that may have different parameters. Two common univariate nonparametric procedures for testing the equality of the distribution functions are the median test (*see* BROWN–MOOD MEDIAN TEST) and the Kruskal–Wallis test,* which uses the sum of the ranks. When there are only two populations to compare ($c = 2$) the Kruskal–Wallis test is equivalent to the Mann–Whitney–Wilcoxon test.* Puri and Sen [2], adapting both the median and the rank sum test for multivariate data analysis, proposed the *multivariate multisample median test* (MMMT) and the *multivariate multisample rank sum test* (MMRST). The

MMRST is equivalent to the Kruskal–Wallis test when there is only one response variable ($p = 1$).

The statistic for testing the equality of the continuous distribution functions using either the MMMT or the MMRST is the L_N statistic, which is a weighted sum of quadratic forms. Specifically, the L_N statistic is

$$L_N = \sum_{i=1}^{c} n_i (\mathbf{T}_i - \bar{\mathbf{T}}.)' \mathbf{V}^{-1} (\mathbf{T}_i - \bar{\mathbf{T}}.),$$

where $\bar{\mathbf{T}}. = \sum_{i=1}^{c} n_i \mathbf{T}_i / (\sum_{i=1}^{c} n_i)$.

To calculate the L_N for the MMRST test, initially for each of the p multivariate responses, the combined data from all the samples is ordered from smallest to largest with rank 1 assigned to the smallest, rank 2 to the next smallest, etc. These ranks then replace the original data. Theoretically, since the distribution functions are continuous, the probability of a tie is zero. In practice, ties do occur occasionally, and then the average of their rankings is used (see the Example). In computing L_N, $\mathbf{T}_i$ is the p-vector of average ranks for each of the p multivariate responses in the ith sample, c is the number of populations sampled, n_i is the number of data vectors from the ith population, $\mathbf{V}$ is the dispersion matrix of the rank vector and $\bar{\mathbf{T}}.$ is the vector of the average ranks for the combined data from all samples. The (k, l) element of $\mathbf{V}$ is given by

$$\mathbf{V}(k, l) = \frac{1}{N} \sum_{i=1}^{c} \sum_{j=1}^{n_i} \left(r_{ij}^{(k)} - \bar{r}_{..}^{(k)} \right)\left(r_{ij}^{(l)} - \bar{r}_{..}^{(l)} \right),$$

where $r_{ij}^{(\alpha)}$ is the rank of the jth experimental unit in the ith sample, αth multivariate response, $\bar{r}_{..}^{(\alpha)}$ is the average rank of the αth multivariate response for all samples combined, and N is the sum of the n_i's.

To calculate the L_N statistic for the MMMT test, initially, for each of the p multivariate responses, the medians of the combined data from all the samples are determined. The original data is replaced by $E_{ij}^{(\alpha)}$ which is either 0 or 1: one if the observed value is less than or equal the median for the particular multivariate response; zero, otherwise. Then in computing L_N, $\mathbf{T}_i$ is

the p-vector of proportions of observations less than or equal to the median of the combined samples for each of the p multivariate responses from the ith sample, $\mathbf{V}$ is the dispersion matrix of the proportions, c is the number of populations sampled, n_i is the number of data vectors from the ith population and $\bar{\mathbf{T}}.$ is the vector of proportions of observations less than or equal to the median of the combined samples. The (k, l) element of $\mathbf{V}$ is given by

$$\mathbf{V}(k, l)$$
$$= \frac{1}{N} \sum_{i=1}^{c} \sum_{j=1}^{n_i} \left(E_{ij}^{(k)} - \bar{E}_{..}^{(k)} \right)\left(E_{ij}^{(l)} - \bar{E}_{..}^{(l)} \right),$$

where $E_{ij}^{(\alpha)}$ has been previously defined. $\bar{E}_{..}^{(\alpha)}$ is the proportion of the combined data from all samples that are less than or equal to the median for the αth multivariate response, and N is the sum of the n_i's.

When the null hypothesis that the distribution functions $F_1, F_2, \ldots, F_c$ are identical is true, one would expect for the MMRST that the average of the ranks from the various samples would be approximately equal, and similarly for the MMMT, one would expect that the proportions less than or equal to the median from the various samples would be approximately equal.

Substantial deviation from equality is reflected in $(\mathbf{T}_i - \bar{\mathbf{T}}.)$. The weighting factor n_i gives increased importance to those deviations based on large samples. Hence large values of $|\mathbf{T}_i - \bar{\mathbf{T}}.|$ lead to large values of L_N and rejection of the hypothesis of identical distribution functions.

When there are only two populations to compare ($c = 2$), the L_N statistic becomes

$$L_N = \frac{n_1(n_1 + n_2)}{n_2} (\mathbf{T}_1 - \bar{\mathbf{T}}.)' \mathbf{V}^{-1} (\mathbf{T}_1 - \bar{\mathbf{T}}.)$$

The L_N statistic is asymptotically distributed as chi-square with $p(c - 1)$ degrees of freedom. Currently no tables are available for L_N for small samples. In a Monte Carlo* study, Schwertman [3] investigated the use of the asymptotic chi-square critical values for small samples of size 5 and 10 from each

Table 1

Dog	Data			Indicator Data (MMMT)			Rank Transformations (MMRST)		
	1	7	13	1	7	13	1	7	13
1	4.0	3.6	3.1	0	1	1	9.5	5	2
2	4.2	3.9	4.9	0	0	0	13.5	12.5	11
3	3.7	4.8	4.2	1	0	0	7	11	8.5
4	3.4	3.1	3.3	1	1	1	3.5	2.5	4
5	3.0	3.0	3.1	1	1	1	1	1	2
6	3.8	3.9	3.4	0	1	1	8	6	5.5
7	4.2	4.1	4.0	0	0	1	13.5	8	7
8	4.1	4.3	4.2	0	0	0	11.5	10	8.5
9	3.1	3.1	3.1	1	1	1	2	2.5	2
10	3.5	5.4	5.6	1	0	0	5.5	14	13
11	3.4	3.3	3.4	1	1	1	3.5	4	5.5
12	4.0	4.9	5.8	0	0	0	9.5	12.5	14
13	4.1	4.0	4.7	0	1	0	11.5	7	10
14	3.5	4.2	5.0	1	0	0	5.5	9	12

of two populations. A variety of dispersion structures using normal and nonnormal variables were simulated, and while the procedure was less than ideal in some cases, it was generally conservative with respect to significance level for data with four or more multivariate responses.

A program to compute the L_N statistic for either the MMMT or the MMRST is available (see Schwertman [4]). As a final note, the maximum value of the L_N statistic for the MMMT is Np and for the MMRST it is

$$ncp\left(1 - \frac{n^2 - 1}{n^2 c^2 - 1}\right),$$

when the sample size n for each population is the same.

Example. To demonstrate the computation of the L_N statistic for both the MMMT and MMRST, consider a portion of the Grizzle–Allen [1] data on the coronary sinus potassium level on dogs observed at several time periods. Only a portion of the data is used and only at time 1, 7, and 13 minutes to reduce the computation. The data, the corre-

sponding indicators ($E_{ik}^{(\alpha)}$'s) and rank transformations are given in Table 1.

The first three dogs were from population 1, the next five from population 2, the next two from population 3, and the last four from population 4, i.e., $n_1 = 3$, $n_2 = 5$, $n_3 = 2$, and $n_4 = 4$.

For the MMRST:

$\mathbf{T}_1 = (10, 9.5, 7.1667)$, $\mathbf{T}_2 = (7.5, 5.5, 5.4)$,

$\mathbf{T}_3 = (3.75, 8.25, 7.5)$, $\mathbf{T}_4 = (7.5, 8.125, 10.375)$,

$\bar{\mathbf{T}}. = (7.5, 7.5, 7.5)$;

$$\mathbf{V} = \frac{1}{14}\begin{bmatrix} 225 & 131.25 & 100.5 \\ & 226.5 & 201.5 \\ \text{symmetric} & & 224.5 \end{bmatrix},$$

$$\mathbf{V}^{-1} = 14\begin{bmatrix} 0.0069882, & -0.0062843, & 0.0025124 \\ & 0.0275602, & -0.0219234 \\ \text{symmetric} & & 0.0230071 \end{bmatrix},$$

$$L_N = 3(1.66) + 5(0.386) + 2(2.09) + 4(1.71)$$
$$= 17.93.$$

If the asymptotic chi-square distribution of the L_N statistic is used to determine the critical value, the degrees of freedom are $p(c - 1) = 3(4 - 1) = 9$, and the critical value at $\alpha = 0.05$ is 16.919. Therefore, the L_N computed is significant at the 0.05 level.

For the MMMT:

$$T_1 = (\tfrac{1}{3}, \tfrac{1}{3}, \tfrac{1}{3}), \qquad T_2 = (\tfrac{2}{5}, \tfrac{3}{5}, \tfrac{4}{5}),$$

$$T_3 = (1, \tfrac{1}{2}, \tfrac{1}{2}), \qquad T_4 = (\tfrac{1}{2}, \tfrac{1}{2}, \tfrac{1}{4}),$$

$$\bar{T}. = (\tfrac{1}{2}, \tfrac{1}{2}, \tfrac{1}{2})$$

$$V = \frac{1}{14} \begin{bmatrix} \frac{14}{4} & \frac{1}{2} & \frac{1}{2} \\ & \frac{14}{4} & \frac{10}{4} \\ \text{symmetric} & & \frac{14}{4} \end{bmatrix},$$

$$V^{-1} = 14 \begin{bmatrix} 0.2926829, & -0.0243902, & -0.0243902 \\ & 0.5853659, & -0.4146341 \\ & & 0.5853659 \end{bmatrix},$$

$$L_N = 3(0.2087) + 5(0.5395) + 2(1.0244)$$
$$+ 4(0.5122)$$
$$= 7.43.$$

If the asymptotic chi-square distribution of the L_N statistic is used to determine the critical value, the degrees of freedom are $p(c - 1) = 3(4 - 1) = 9$, and the critical value at $\alpha = 0.05$ is 16.919. Therefore, the L_N computed is not significant at the 0.05 level.

The different values for the L_N statistic for the MMMT and MMRST test may be due to the MMRST retaining more information from the original data.

References

[1] Grizzle, J. E. and Allen, D. M. (1969). *Biometrics*, **25**, 359.

[2] Puri, M. L. and Sen P. S. (1971). *Nonparametric Methods in Multivariate Analysis*, Wiley, New York. (Quite difficult to read.)

[3] Schwertman, N. C. (1982). *Commun. Statist. B*, **11**, 667–676.

[4] Schwertman, N. C. (1982). *J. R. Statist. Soc. C*, **31**, 80–85.

(BROWN–MOOD MEDIAN TEST
MANN–WHITNEY–WILCOXON TEST
KRUSKAL–WALLIS TEST
RANK TESTS)

NEIL C. SCHWERTMAN

MULTIVARIATE MULTIPLE COMPARISONS

Multivariate multiple comparison tests are useful in finding out which populations dif-

fer from each other when there are significant differences between them. Some data analysts advocate the use of ANOVA* multiple comparison* tests on each variable separately. But separate tests do not take into account the correlations between the variables, and so some information is not utilized. Sometimes, the experimenter may be interested in making inferences on each variable separately. But, even then, it would be of interest sometimes to eliminate the effect of other variables in order to make multiple comparisons of means of various groups on any given variable. Also, in a number of situations, it is of interest to test for the equality of the means on various variables simultaneously in order to draw inferences on the populations. By using multivariate analysis of variance* (MANOVA) multiple comparison procedures, we can specify the overall type I error, which we cannot do if we use ANOVA multiple comparison procedures. Another advantage of the MANOVA procedures is that we can draw conclusions as to whether certain linear combinations of the means of various variables for different populations are equal, whereas separate ANOVA tests on each variable are not helpful for these purposes.

Here we review some procedures for multivariate normal populations with special emphasis on multiple comparisons of mean vectors. Reviews of the literature on some aspects of parametric multiple comparison procedures have been given in Krishnaiah [8] and Krishnaiah et al. [12]. For a review of the literature on nonparametric multiple comparison procedures see Sen [15].

FIT FOR MULTIPLE COMPARISONS OF MEANS

In any practical situation, the experimenter is interested in testing the hypotheses on only a finite number of contrasts on means of normal populations. Scheffé's simultaneous comparison procedure* is equivalent to testing an infinite number of contrasts simultaneously and the lengths of confidence intervals associated with this procedure are quite large. So Krishnaiah proposed procedures for simultaneously testing the hypotheses on a finite number of linear combinations of means. These are known as the finite intersection tests (FIT), were proposed in an unpublished report in 1960, and later published in Krishnaiah [5, 6]. A brief description of them in the univariate case follows.

Consider k univariate normal populations with means $\mu_1, \ldots, \mu_k$ and a common variance σ^2. Also, let $H_i : \lambda_i = 0$ and $A_i : \lambda_i \neq 0$, $(i = 1, 2, \ldots, q)$, where $\lambda_i = c_{i1}\mu_1 + \cdots + c_{ik}\mu_k$ and the c_{ij}'s are some known constants. In addition, let $\bar{x}_i$ and n_i, respectively, denote the mean and size of the sample from the ith population; s^2 is the usual unbiased estimate of σ^2, $F_i = \hat{\lambda}_i^2/(s^2 d_i)$, $d_i = \sum_{j=1}^{n_i} c_{ij}^2/n_i$. If we are testing the hypotheses $H_1, \ldots, H_q$ simultaneously against $A_1, \ldots, A_q$, we accept or reject H_i according as

$$F_i \lessgtr F_\alpha,$$

where

$$P[\, F_i \leq F_\alpha ; i = 1, 2, \ldots, q \,|\, H\,] = 1 - \alpha,$$

$$(1)$$

and $H = \bigcap_{i=1}^q H_i$. The simultaneous confidence intervals associated with the FIT are given by

$$\hat{\lambda}_i - \sqrt{F_\alpha d_i s^2} \leq \lambda_i \leq \hat{\lambda}_i + \sqrt{F_\alpha d_i s^2} . \quad (2)$$

The joint distribution of $F_1, \ldots, F_q$ is the multivariate F distribution (in the sense of Krishnaiah [6, 7]) with $(1, \nu)$ degrees of freedom and with a certain correlation matrix of the "accompanying" multivariate normal, where $\nu = n - k$ and n is the total sample size. Values of F_α for given values of α were given for some cases by Krishnaiah and Armitage [10, 11] and reproduced in Krishnaiah et al. [12]. For situations where exact percentage points are not available, various bounds available in the literature can be used to approximate F_α. Empirical investigations on the sharpness of some of these bounds were made in Krishnaiah and Armitage [10, 11] and Cox et al. [3]. The

simultaneous confidence intervals on contrasts associated with the FIT yield lengths shorter than those yielded by Scheffé's method. As the number of hypotheses tested becomes smaller, the lengths of the confidence intervals associated with the FIT become shorter.

When the sample sizes are equal, Krishnaiah's FIT is equivalent to Tukey's test for pairwise comparisons of means. But Krishnaiah's FIT can be applied even when the sample sizes are not equal. In this case, confidence intervals are given by

$$(\bar{x}_i - \bar{x}_j) - \sqrt{F_\alpha \left(\frac{1}{n_i} + \frac{1}{n_j} \right) s^2}$$

$$\leq \mu_i - \mu_j$$

$$\leq (\bar{x}_i - \bar{x}_j) + \sqrt{F_\alpha \left(\frac{1}{n_i} + \frac{1}{n_j} \right) s^2} . \quad (3)$$

In (3) F_α can be approximated by taking advantage of one of the known inequalities for the probability integral of the multivariate F distribution, with $(1, \nu)$ degrees of freedom. In particular, if we make use of the bound by Khatri [4] and Sidak [16], we can approximate $\sqrt{F_\alpha}$ with upper $100\alpha\%$ point of the distribution of the studentized maximum modulus.* Percentage points of the square of this distribution were given in Armitage and Krishnaiah [1] and later reproduced in Krishnaiah [9]. The computer program of Armitage and Krishnaiah can be used to extend the preceding percentage points. Thus Krishnaiah's FIT for pairwise comparisons of means for unequal sample sizes can be implemented by replacing F_α with approximate values. Cox et al. [2] developed a comprehensive computer program for implementation of Krishnaiah's FIT for simultaneous testing of hypotheses on linear combinations of means. Besides pairwise comparisons of means, the problems of testing the following hypotheses can be treated as special cases of the FIT: (a) testing $\mu_i - \mu_k = 0$ simultaneously for $i = 1, 2, \ldots, (k - 1)$, and (b) testing $\mu_i - \mu_j = 0$ $(i \neq j)$ and $\mu_i = 0$ $(i = 1, 2, \ldots, k)$. For some alternative procedures for multiple comparisons of means of normal populations, *see* MULTIPLE RANGE AND ASSOCIATED TEST PROCEDURES; k-RATIO t-TESTS, t-INTERVALS AND POINT ESTIMATES, etc., and MULTIPLE COMPARISONS. A discussion of the multiple comparison procedures for variances is given in Krishnaiah [8].

FIT FOR REGRESSION COEFFICIENTS

Consider the classical univariate regression model (*see* MULTIPLE LINEAR REGRESSION)

$$y_j = \beta_0 + \beta_1 x_{1j} + \cdots + \beta_k x_{kj} + \epsilon_j, \quad (4)$$

where the ϵ_j's $(j = 1, 2, \ldots, n)$ are distributed independently and normally with mean 0 and variance σ^2. Let $H_i : \lambda_i = 0$ and $A_i : \lambda_i \neq 0$, where $\lambda_i = \mathbf{c}_i' \boldsymbol{\beta}$, $\mathbf{c}_i' = (c_{i0}, \ldots, c_{ik})$, and $\boldsymbol{\beta}' = (\beta_0, \ldots, \beta_k)$; here $\mathbf{c}_i$ $(i = 1, 2, \ldots, q)$ are known. The least-squares estimate of $\boldsymbol{\beta}$ is given by

$$\hat{\boldsymbol{\beta}} = (\mathbf{X}'\mathbf{X})^{-1}\mathbf{X}'\mathbf{y},$$

$$\mathbf{X}' = (x_{ij}), \quad \mathbf{y}' = (y_1, \ldots, y_n).$$

In addition, let

$$F_i = t_i^2, \quad t_i = \hat{\lambda}_i / \left\{ \mathbf{c}_i'(\mathbf{X}'\mathbf{X})^{-1}\mathbf{c}_i s^2 \right\}^{1/2},$$

$$\hat{\lambda}_i = \mathbf{c}_i'\hat{\boldsymbol{\beta}},$$

$$(n - k - 1)s^2 = \mathbf{y}'\left[\mathbf{I} - \mathbf{X}(\mathbf{X}'\mathbf{X})^{-1}\mathbf{X}'\right]\mathbf{y}.$$

If we use Krishnaiah's finite intersection test, we accept or reject H_i according as

$$F_i \lessgtr F_{\alpha 1},$$

where

$$P\left[F_i \leqslant F_{\alpha 1}; i = 1, 2, \ldots, q \mid \bigcap_{i=1}^{q} H_i \right]$$

$$= 1 - \alpha. \quad (5)$$

When $\bigcap_{i=1}^{q} H_i$ is true, the joint distribution of $F_1, \ldots, F_q$ is the multivariate F distribution (in the sense of Krishnaiah [5, 6] with $(1, n - k - 1)$ degrees of freedom. As before, we can replace F_α with approximate values. The simultaneous confidence intervals asso-

ciated with this test are

$$\hat{\lambda}_i - \sqrt{F_{\alpha 1} \mathbf{c}_i'(\mathbf{X}'\mathbf{X})^{-1}\mathbf{c}_i s^2}$$

$$\leqslant \lambda_i \leqslant \hat{\lambda}_i + \sqrt{F_{\alpha 1} \mathbf{c}_i'(\mathbf{X}'\mathbf{X})^{-1}\mathbf{c}_i s^2}. \quad (6)$$

This procedure was proposed in Krishnaiah [6]. The lengths of the confidence intervals on λ_i associated with Krishnaiah's FIT are shorter than the lengths of the corresponding confidence intervals associated with the well-known overall F test. In (6), if we put $\mathbf{c}_1' = (1, x_1, x_2, \ldots, x_k)$, then we get the confidence interval on $\beta_0 + \beta_1 x_1 + \cdots + \beta_k x_k$. For a brief review of the literature on simultaneous tests on linear combinations of the regression coefficients, see Krishnaiah [8].

Next, consider the model

$$y_j = \beta_0 + \beta_1 x_{1j}^* + \cdots + \beta_k x_{kj}^* + \epsilon_j, \quad (7)$$

where $x_{ij}^* = x_{ij} - \bar{x}_i$ and ϵ_j's are as defined in model (4). Then the procedure for testing the hypotheses $H_1, \ldots, H_q$ simultaneously and the associated confidence intervals are the same as before when x_i is replaced with x_i^*. The simultaneous confidence intervals for λ_i $(i = 1, 2, \ldots, q)$ in this case are given by

$$\hat{\lambda}_i - \sqrt{F_{\alpha_1} \mathbf{c}_i'(\mathbf{X}^{*\prime}\mathbf{X}^*)^{-1}\mathbf{c}_i s^2}$$

$$\leqslant \lambda_i \leqslant \hat{\lambda}_i + \sqrt{F_{\alpha_1} \mathbf{c}_i'(\mathbf{X}^{*\prime}\mathbf{X}^*)^{-1}\mathbf{c}_i s^2}, \quad (8)$$

$$\mathbf{X}^{*\prime} = (x_{ij}^*),$$

$$(n - k - 1)s^2 = \mathbf{y}'\left[\mathbf{I} - \mathbf{X}^*(\mathbf{X}^{*\prime}\mathbf{X}^*)^{-1}\mathbf{X}^{*\prime}\right]\mathbf{y},$$

$$F_i = \frac{(\mathbf{c}_i'\hat{\boldsymbol{\beta}})^2}{\mathbf{c}_i'(\mathbf{X}^{*\prime}\mathbf{X}^*)^{-1}\mathbf{c}_i s^2},$$

$$\hat{\boldsymbol{\beta}} = (\mathbf{X}^{*\prime}\mathbf{X}^*)^{-1}\mathbf{X}^{*\prime}\mathbf{y},$$

$$P\left[F_i \leqslant F_{\alpha_1}; i = 1, 2, \ldots, q \mid \bigcap_{i=1}^{q} H_i \right]$$

$$= 1 - \alpha.$$

We will now discuss some procedures for multiple comparisons of mean vectors of multivariate normal populations.

ROY'S LARGEST ROOT TEST

For $i = 1, 2, \ldots, k$, let $\mathbf{x}_{i1}, \ldots, \mathbf{x}_{in_i}$ be a sample from a p-variate normal population with mean vector $\boldsymbol{\mu}_i$ and covariance matrix $\boldsymbol{\Sigma}$. Also, let $H_i : \boldsymbol{\lambda}_i = \mathbf{0}$, $H : \boldsymbol{\mu}_1 = \cdots = \boldsymbol{\mu}_k$ and $A_i : \boldsymbol{\lambda}_i \neq \mathbf{0}$ $(i = 1, 2, \ldots, q)$ where $\boldsymbol{\lambda}_i = c_{i1}\boldsymbol{\mu}_1 + \cdots + c_{ik}\boldsymbol{\mu}_k$ and the c_{ij}'s are known constants subject to the restrictions $c_{i1} + \cdots + c_{ik} = 0$. We first discuss Roy's largest root test for testing $H_1, \ldots, H_q$ simultaneously (see ROY'S CHARACTERISTIC ROOT STATISTIC).

Let $\mathbf{S}_H = (s_{1ij})$ and $\mathbf{S} = (s_{2ij})$ where

$$\mathbf{S}_H = \sum_{i=1}^{k} n_i (\bar{\mathbf{x}}_{i\cdot} - \mathbf{x}_{\cdot\cdot})(\bar{\mathbf{x}}_{i\cdot} - \bar{\mathbf{x}}_{\cdot\cdot})'$$

$$\mathbf{S} = \sum\sum_{ij} (\mathbf{x}_{ij} - \bar{\mathbf{x}}_{i\cdot})(\mathbf{x}_{ij} - \bar{\mathbf{x}}_{i\cdot})',$$

$$n_i \bar{\mathbf{x}}_{i\cdot} = \sum_{j=1}^{n_i} \mathbf{x}_{ij}, \qquad n\bar{\mathbf{x}}_{\cdot\cdot} = \sum\sum_{ij} \mathbf{x}_{ij},$$

and $n = n_1 + \cdots + n_k$. According to the largest root test, we accept or reject H according to whether

$$c_L(\mathbf{S}_H \mathbf{S}^{-1}) \lessgtr c_\alpha, \tag{9}$$

where

$$P\left[\frac{(n-k)}{(k-1)} c_L(\mathbf{S}_H \mathbf{S}^{-1}) \leqslant c_\alpha \mid H \right] = 1 - \alpha, \tag{10}$$

and $c_L(A)$ denotes the largest eigenvalue of A. Here $\mathbf{S}_H$ is distributed independent of $\mathbf{S}$ as central (noncentral) Wishart distribution* with $(k-1)$ degrees of freedom when H is true (not true), whereas $\mathbf{S}$ is distributed as central Wishart with $(n-k)$ degrees of freedom and $E(\mathbf{S}) = (n-k)\boldsymbol{\Sigma}$ when H is true. Tables for c_α are given in Krishnaiah [9] for some values of the parameters. If the total hypothesis H is rejected, we accept or reject H_i according to whether

$$T_i^2 \lessgtr (k-1)c_\alpha, \tag{11}$$

where

$$T_i^2 = (n-k)\hat{\boldsymbol{\lambda}}_i' \mathbf{S}^{-1} \hat{\boldsymbol{\lambda}}_i \Big/ \left\{ \sum_{j=1}^{k} n_j^{-1} c_{ij}^2 \right\}.$$

Roy's largest root can be interpreted as follows. Let

$$H = \bigcap_{\mathbf{a} \neq \mathbf{0}} \bigcap_{\mathbf{b} \neq \mathbf{0}} H_{\mathbf{a},\mathbf{b}} \tag{12}$$

where $H_{\mathbf{a},\mathbf{b}} : \mathbf{a}'\boldsymbol{\eta}\mathbf{b} = 0$, $\mathbf{b}'\mathbf{1} = 0$, $\boldsymbol{\eta} = (\boldsymbol{\mu}_1, \ldots, \boldsymbol{\mu}_k)$ and $\mathbf{1}$ denotes a $k \times 1$ vector whose elements are all equal to unity. For illustration, let $\mathbf{a}' = (1, -1, 0, \ldots, 0)$, $\mathbf{b}' = (0, 1, 0, -1, 0, \ldots, 0)$. Then $H_{\mathbf{a},\mathbf{b}}$ indicates the hypothesis that the difference between the means of the first two variables is the same for the second and fourth populations. Now let

$$F_{\mathbf{a},\mathbf{b}} = \frac{\{\mathbf{a}'\hat{\boldsymbol{\zeta}}\mathbf{b}\}^2 (n-k)}{\mathbf{a}'\mathbf{S}\mathbf{a}d}, \tag{13}$$

$$\boldsymbol{\zeta} = (\bar{\mathbf{x}}_1, \ldots, \bar{\mathbf{x}}_k.),$$

$$d = \frac{b_1}{n_1} + \cdots + \frac{b_k}{n_k}.$$

While the classical ANOVA F test is equivalent to testing an infinite number of contrasts on means simultaneously, Roy's largest root is equivalent to testing $\mathbf{a}'\boldsymbol{\eta}\mathbf{b} = 0$ simultaneously for all $\mathbf{a} \neq \mathbf{0}$, $\mathbf{b} \neq \mathbf{0}$ where $\mathbf{b}$ is subject to the restriction $\mathbf{b}'\mathbf{1} = 0$. When $p = 1$, Roy's largest root test and the ANOVA F test are equivalent. The $100(1 - \alpha)\%$ simultaneous confidence intervals associated with the largest root test are

$$\mathbf{a}'\boldsymbol{\zeta}\mathbf{b} - \sqrt{c_\alpha^* \mathbf{a}'\mathbf{S}\mathbf{a}} \leqslant \mathbf{a}'\boldsymbol{\zeta}\mathbf{b} \leqslant \mathbf{a}'\hat{\boldsymbol{\zeta}}\mathbf{b} + \sqrt{c_\alpha^* \mathbf{a}'\mathbf{S}\mathbf{a}} \tag{14}$$

for all $\mathbf{a} \neq \mathbf{0}$ and $\mathbf{b} \neq \mathbf{0}$, where the $\mathbf{b}$'s are subject to the restriction $\mathbf{b}'\mathbf{b} = 1$ and $c_\alpha^* = d(k-1)c_\alpha/(n-k)$.

$T_{\max}^2$ TEST

The hypothesis H_i can be decomposed as $H_i = \bigcap_{\mathbf{a} \neq \mathbf{0}} H_{i\mathbf{a}}$ where

$$H_{i\mathbf{a}} : c_{i1}(\mathbf{a}'\boldsymbol{\mu}_1) + \cdots + c_{ik}(\mathbf{a}'\boldsymbol{\mu}_k) = 0.$$

When $H_{i\mathbf{a}}$ $(\mathbf{a} \neq \mathbf{0}, i = 1, \ldots, q)$ are tested simultaneously for all $\mathbf{a} \neq \mathbf{0}$ and i, we accept or reject $H_{i\mathbf{a}}$ for given i and $\mathbf{a}$ according to whether

$$F_{i\mathbf{a}} \lessgtr T_\alpha^2,$$

where

$$P\left[F_{ia} \leqslant T_{\alpha}^2 \text{ for all } i \text{ and } \mathbf{a} \neq \mathbf{0} \mid H \right]$$
$$= 1 - \alpha, \tag{15}$$

$$F_{ia} = \frac{(a'\hat{\boldsymbol{\lambda}}_i)^2 (n - k)}{\mathbf{a}' S \mathbf{a} d_i}, \tag{16}$$

$$\hat{\boldsymbol{\lambda}}_i = c_{i1}\bar{\mathbf{x}}_1. + \cdots + c_{ik}\bar{\mathbf{x}}_k.$$

But (16) is equivalent to

$$P\left[T_i^2 \leqslant T_{\alpha}^2 ; i = 1, \ldots, q \mid H \right]$$
$$= 1 - \alpha, \tag{17}$$

where T_i is as defined in (11). The hypothesis H_i, for any given i, is accepted or rejected according to whether $T_i^2 \lessgtr T_{\alpha}^2$ (see HOTELLING'S T^2). The total hypothesis H is accepted if and only if $H_1, \ldots, H_q$ are accepted (see UNION-INTERSECTION PRINCIPLE). The $100(1 - \alpha)\%$ simultaneous confidence intervals associated with the $T_{\max}^2$ test are

$$\mathbf{a}'\hat{\boldsymbol{\lambda}}_i - \left\{ ((d_i/(n - k)))T_{\alpha}^2 \mathbf{a}' S \mathbf{a} \right\}^{1/2}$$
$$\leqslant \mathbf{a}'\boldsymbol{\lambda}_i$$
$$\leqslant \mathbf{a}'\hat{\boldsymbol{\lambda}}_i + \left\{ ((d_i/(n - k)))T_{\alpha}^2 \mathbf{a}' S \mathbf{a} \right\}^{1/2} \tag{18}$$

for all $\mathbf{a} \neq \mathbf{0}$. The $T_{\max}^2$ test is equivalent to testing H_{ia} simultaneously for $i = 1, 2, \ldots, q$ and for all nonnull vectors $\mathbf{a}$. The lengths of the confidence intervals (8) on $\mathbf{a}'\boldsymbol{\lambda}_i$ associated with the $T_{\max}^2$ test are shorter than the lengths of the corresponding confidence intervals associated with the largest root test. Approximate percentage points associated with the $T_{\max}^2$ test were given in Siotani [17, 18] for some special cases and are produced in Krishnaiah [9]. Roy and Bose [14] formulated the $T_{\max}^2$ test for pairwise comparisons of mean vectors. For other details of the $T_{\max}^2$ test, see Krishnaiah [8a].

FIT FOR MULTIPLE COMPARISONS OF MEAN VECTORS

In the FIT, we test $H_{\mathbf{a}_j, \mathbf{b}_i}$ simultaneously (or in a certain sequential manner) for $j = 1, 2, \ldots, p$ and $i = 1, 2, \ldots, q$, where the $\mathbf{b}_i$'s

are subject to the restrictions that $\mathbf{b}_i' \mathbf{1} = 0$. Here

$$\mathbf{a}_1' = (1, 0, \ldots, 0), \quad \mathbf{a}_2' = (-\boldsymbol{\beta}_1', 1, 0, \ldots, 0),$$
$$\mathbf{a}_3' = (-\boldsymbol{\beta}_2', 1, 0, \ldots, 0), \quad \mathbf{a}_p' = (-\boldsymbol{\beta}_{p-1}', 1),$$

$$\boldsymbol{\beta}_i = \boldsymbol{\Sigma}_i^{-1} \begin{bmatrix} \sigma_{1,i+1} \\ \vdots \\ \sigma_{i,i+1} \end{bmatrix}, \tag{19}$$

and $\boldsymbol{\Sigma}_i$ is the top $i \times i$ left-hand corner of $\boldsymbol{\Sigma}$.

The MANOVA model discussed earlier can be written as

$$E(\mathbf{X}) = \mathbf{A}\boldsymbol{\theta}, \tag{20}$$

$$\mathbf{X} = [\mathbf{x}_1, \ldots, \mathbf{x}_p], \quad \boldsymbol{\theta}' = [\boldsymbol{\mu}_1, \ldots, \boldsymbol{\mu}_k],$$
$$\mathbf{x}_t' = (x_{11t}, \ldots, x_{1n_1t}, \ldots, x_{k1t}, \ldots, x_{kn_kt}),$$

where x_{ijt} denotes the observation of the jth individual in the ith group and for the tth variable. Also

$$\mathbf{A}_{n \times k} = \begin{bmatrix} \mathbf{e}_1 & \mathbf{0} & \mathbf{0} & \ldots & \mathbf{0} \\ \mathbf{0} & \mathbf{e}_2 & \mathbf{0} & \ldots & \mathbf{0} \\ \cdot & \cdot & \cdot & \ldots & \cdot \\ \cdot & \cdot & \cdot & \ldots & \cdot \\ \cdot & \cdot & \cdot & \ldots & \cdot \\ \mathbf{0} & \mathbf{0} & \cdot & \ldots & \mathbf{e}_k \end{bmatrix},$$

where $\mathbf{l}_i'$ is the $1 \times n_i$ vector with all its elements equal to unity. In addition, let $\mathbf{X}_j = [\mathbf{x}_1, \ldots, \mathbf{x}_j]$ and $\boldsymbol{\eta}_{j+1} = \boldsymbol{\theta}_{j+1} - \boldsymbol{\phi}_j \boldsymbol{\beta}_j$ where $\boldsymbol{\phi}_j = [\boldsymbol{\theta}_1, \ldots, \boldsymbol{\theta}_j]$ and $\boldsymbol{\theta}_j$ is the jth column of $\boldsymbol{\theta}$. The conditional distribution of x_{j+1} given $\mathbf{X}_j$ is multivariate normal with covariance matrix $\sigma_{j+1}^2 \mathbf{I}$ and mean vector

$$E_c(\mathbf{x}_{j+1}) = \mathbf{A}\boldsymbol{\eta}_{j+1} + \mathbf{X}_j \boldsymbol{\beta}_j \tag{21}$$

for $j = 1, 2, \ldots, p - 1$. Also, let

$$E(\mathbf{x}_1) = \mathbf{A}\boldsymbol{\eta}_1 + \mathbf{X}_1 \boldsymbol{\beta}_1, \tag{22}$$

$$F_{gj} = (\mathbf{c}_g' \hat{\boldsymbol{\eta}}_j)^2 (n - k - j + 1) / s_j^2 d_{gj}, \tag{23}$$

where $\boldsymbol{\eta}_j' = (\hat{\eta}_{1j}, \ldots, \hat{\eta}_{kj})$ is the least-squares estimate of $\boldsymbol{\eta}_j'$ under model (22)–(23), s_j^2 is the error sum of squares under the model, and $\mathrm{var}(\hat{c}_g \hat{\eta}_j) = d_{gj} \sigma_j^2$.

The hypothesis $H_g : c_{g1}\boldsymbol{\mu}_1 + \cdots + c_{gk}\boldsymbol{\mu}_k = 0$ can be decomposed as $H_g = \bigcap_{j=1}^p H_{gj}$, where $H_{gj} : c_{g1}\eta_{1j} + \cdots + c_{gk}\eta_{kj} = 0$. Motivated by this, Krishnaiah [6, 7] proposed the following procedures. The hypothesis H_g

$(g = 1, 2, \ldots, q)$ is accepted if and only if

$$F_{gj} \leqslant F_{j\alpha} \quad \text{for} \quad j = 1, 2, \ldots, p$$

and rejected otherwise, where

$$P\big[\, F_{gj} \leqslant F_{j\alpha} \,;\, g = 1, \ldots, q,$$

$$j = 1, 2, \ldots, p \,|\, H \,\big]$$

$$= \prod_{j=1}^{p} P\big[\, F_{gj} \leqslant F_{j\alpha} \,;\, g = 1, 2, \ldots, q \,|\, H \,\big]$$

$$= 1 - \alpha. \tag{24}$$

The simultaneous confidence intervals associated with this procedure are

$$\mathbf{c}'_g \hat{\boldsymbol{\eta}}_j - \left\{ F_{j\alpha} s_j^2 d_{gj} / (n - k - j + 1) \right\}^{1/2}$$

$$\leqslant \mathbf{c}'_g \boldsymbol{\eta}_j$$

$$\leqslant \mathbf{c}'_g \hat{\boldsymbol{\eta}}_j + \left\{ F_{j\alpha} s_j^2 d_{gj} / (n - k - j + 1) \right\}^{1/2}. \tag{25}$$

A computer program has been written by Cox et al. [2] for implementation of Krishnaiah's finite intersection tests.

In the step-down procedure proposed by Roy [13], we test the hypotheses H_j ($j = 1, 2, \ldots, p$) simultaneously, using the classical F tests* under the model (21)–(22). Inference on subhypotheses H_{ij} can be made by examining the simultaneous confidence intervals on $\mathbf{c}'_i \boldsymbol{\eta}_j$. Krishnaiah's finite intersection tests yield shorter confidence intervals on $\mathbf{c}'_i \boldsymbol{\eta}_j$ than the step-down procedure of Roy.

Example. The Air Force Flight Dynamics Laboratory conducted a simulated experiment to study pilot performance when three different densities of threat symbols (for airplanes, surface-to-air missiles, and antiaircraft artillery) are displayed on the CRT of the multifunction keyboard in an aircraft cockpit. The experiment was conducted with densities 10, 20, and 30 of these symbols. Data was collected on 18 pilots on the keyboard operation time to complete their assigned tasks. We will illustrate the usefulness of some special cases of Krishnaiah's tests for multiple comparisons of means using this data.

We expect some correlations between the scores of any given pilot when different densities of symbols are used. For our example, we assume that the correlations are zero; to illustrate the tests for unequal sample sizes, we omitted some data. We refer to the keyboard operation times associated with symbol densities 10, 20, and 30 as belonging to populations 1, 2, and 3, respectively.

Let x_{ij} denote the score on the jth pilot ($i = 1, 2, 3; j = 1, 2, \ldots, n_i$) at the ith symbol density. For $i = 1, 2, 3$, we assume that $x_{i1}, \ldots, x_{in_i}$ is a sample from a normal population with mean μ_i and variance σ^2. Also let

$$n_i \bar{x}_{i\cdot} = \sum_{j=1}^{n_i} x_{ij}, \qquad s^2 = \sum\sum_{ij} \frac{\left(x_{ij} - \bar{x}_{i\cdot} \right)^2}{n - k},$$

$$n = \sum_{i=1}^{k} n_i.$$

Here $k = 3$, $n_1 = 10$, $n_2 = 12$, $n_3 = 14$, $\bar{x}_{1\cdot} = 6.0176$, $\bar{x}_{2\cdot} = 8.1041$, $\bar{x}_{3\cdot} = 9.2459$, and $s^2 = 4.4813$. We first discuss certain approximations used in the application of FIT.

Let $F_1, \ldots, F_q$ and H_i be as defined in Fit for Regression Coefficients section. Let c_1, c_2, c_3, and c_4 be defined as follows when $\bigcap_{i=1}^{q} H_i$ is true:

$$P\big[\, F_i \leqslant F_\alpha \,;\, i = 1, 2, \ldots, q \,\big] = 1 - \alpha, \tag{26}$$

$$1 - \sum_{i=1}^{q} P\big[\, F_i \geqslant c_1 \,\big] = 1 - \alpha, \tag{27}$$

$$1 - \sum_{i=1}^{q} P\big[\, F_i \geqslant c_2 \,\big]$$

$$+ \sum_{i<j} P\big[\, F_i \geqslant c_2 \,;\, F_j \geqslant c_2 \,\big]$$

$$= 1 - \alpha, \tag{28}$$

$$\prod_{i=1}^{q} P\big[\, F_1 \leqslant c_3 \,\big] = 1 - \alpha \tag{29}$$

$$P\big[\, \left(y_i^2 / s^2 \right) \leqslant c_4 \,;\, i = 1, 2, \ldots, q \,\big]$$

$$= 1 - \alpha, \tag{30}$$

where $y_1, \ldots, y_q$ are distributed independently as normal with zero mean and variance σ^2, and s^2 is distributed independent of the y_i's. The constants c_1, c_3, and c_4 are upper bounds on the exact critical value F_α whereas c_2 is a lower bound on F_α; c_3 is a sharper upper bound than c_1. Let

$$F_{ij} = \frac{(\bar{x}_{i\cdot} - \bar{x}_{j\cdot})^2}{s^2 \left(\frac{1}{n_i} + \frac{1}{n_j} \right)} . \qquad (31)$$

We will first discuss the problem of testing H_{12}, H_{13}, and H_{23}, respectively, against A_{12}, A_{13}, and A_{23}. The values of the test statistics are given by $F_{12} = 5.299$, $F_{13} = 13.567$, and $F_{23} = 1.880$. Various approximations to the critical vaulue F_α in this case are given by $c_1 = 6.362$, $c_2 = 6.017$, $c_3 = 6.326$, and $c_4 = 6.299$. Since F_{12} and F_{23} are both less than the preceding approximations to F_α, we accept H_{12} and H_{23}. But we reject H_{13} since F_{13} is greater than all of the c_i's.

Consider the problem of testing the hypotheses H_{12} and H_{23} simultaneously using the FIT. In this case, $F_{12} = 5.299$, $F_{23} = 1.880$, $c_1 = 5.515$, $c_2 = 5.342$, $c_3 = 5.488$, and $c_4 = 5.474$. Since F_{12} and F_{23} are less than any of the preceding approximations, we accept H_{12} and H_{23}.

In testing H_{13} and H_{23} simultaneously, $F_{13} = 13.567$, $F_{23} = 1.880$, $c_1 = 5.515$, $c_2 = 5.372$, $c_3 = 5.488$, and $c_4 = 5.474$. Since F_{13} is greater than the c_i's, we reject H_{13}. But we accept H_{23} since F_{23} is less than any c_i.

We can similarly test the hypotheses H_{12}, H_{23}, H_{13}, H_1^*, H_2^*, and H_3^* simultaneously. H_1^*, H_2^*, and H_3^* can be tested simultaneously by using the test statistics F_1^*, F_2^*, and F_3^*, respectively, where

$$F_i^* = \frac{\bar{x}_{i\cdot}^2}{s^2/n_i} .$$

For the data on pilot performance described earlier, we have $F_1^* = 80.806$, $F_2^* = 175.866$, $F_3^* = 267.072$, $c_1 = 6.362$, $c_2 = 6.298$, $c_3 = 6.326$, and $c_4 = 6.299$. Since F_1^*, F_2^*, and F_3^* are all greater than the critical values, we reject H_1^*, H_2^*, and H_3^*.

In situations where the F statistics have values between the lower and upper bounds of the critical values, c_2 or $(c_2 + c_4)/2$ is recommended as the critical value unless the

error degrees of freedom are large (say, greater than 40), in which case one may use c_3 as the critical value.

A computer program is available for implementation of Krishnaiah's finite intersection tests for multiple comparisons of means. It computes the values of F_i and c_i and gives confidence intervals on various linear combinations of means. The program implements the FIT for multiple comparisons of mean vectors also. As the number of hypotheses tested increases, the lengths of the confidence intervals increase.

The work of Krishnaiah was sponsored by the Air Force Office of Scientific Research under Contract F49620-79-0161. Reproduction in whole or in part is permitted for any purpose of the United States Government.

References

[1] Armitage, J. V. and Krishnaiah, P. R. (1964). "Tables for the Studentized Largest Chi-square Distribution and Their Applications." *Tech. Rep. No. ARL 64-188*, Wright-Patterson Air Force Base, Ohio.

[2] Cox, C. M., Fang, C., and Boudreau, R. (1981). "Computer Program for Krishnaiah's Finite Intersection Tests for Multiple Comparisons of Mean Vectors." *CPSS Tech. Rep. No. 81-2*, Institute for Statistics and Applications, University of Pittsburgh, Pittsburgh, PA.

[3] Cox, C. M., Krishnaiah, P. R., Lee, T. C., Reising, J., and Schuurmann, F. J. (1980). *Multivariate Analysis* Vol. V, P. R. Krishnaiah, ed. Academic Press, New York, pp. 435–4761. (A study on finite intersection tests of multiple comparisons of means.)

[4] Khatri, C. G. (1967). *Ann. Math. Statist.*, **38**, 1853–1867. (On certain inequalities for normal distributions and their applications to simultaneous confidence bounds.)

[5] Krishnaiah, P. R. (1963). "Simultaneous Tests and the Efficiency of Generalized Incomplete Block Designs." *Tech. Rep. No. ARL 63-174*, Wright-Patterson Air Force Base, Ohio.

[6] Krishnaiah, P. R. (1965). *Ann. Inst. Statist. Math*, **17**, 35–53. (On the simultaneous ANOVA and MANOVA tests.)

[7] Krishnaiah, P. R. (1965). *Sankhyā A*, **27**, 31–36. (Multiple comparisons in multiresponse experiments.)

[8] Krishnaiah, P. R., ed. (1969). *Developments in Statistics*, Vol. 2. Academic Press, New York, pp.

157–201. (Some developments on simultaneous test procedures.)

[8a] Krishnaiah, P. R. (1969). In *Multivariate Analysis*, P. R. Krishnaiah, Ed. Vol. 2. Academic Press, New York, pp. 121–143. (Simultaneous test procedures under general MANOVA models.)

[9] Krishnaiah, P. R., ed. (1980). *Handbook of Statistics*, Vol. 1. North-Holland, Amsterdam, pp. 745–971. (Computations of some multivariate distributions.)

[10] Krishnaiah, P. R. and Armitage, J. V. (1965). "Probability Integrals of the Multivariate *F* Distribution, with Tables and Applications." *Tech. Rep. No. ARL* 65-236, Wright-Patterson Air Force Base, Ohio.

[11] Krishnaiah, P. R. and Armitage, J. V. (1970). *Essays in Probability and Statistics*, R. C. Bose, ed. University of North Carolina Press, Chapel Hill, pp. 439–468. (On a multivariate *F* distribution.)

[12] Krishnaiah, P. R., Mudholkar, G. S., and Subbaiah, P. (1980). *Handbook of Statistics*, Vol. 1, P. R. Krishnaiah, ed. North-Holland, Amsterdam, pp. 631–671. (Simultaneous test procedures for mean vectors and covariance matrices.)

[13] Roy, J. (1958). *Ann. Math. Statist.*, **29**, 1177–1187. (Step-down procedure in multivariate analysis.)

[14] Roy, S. N. and Bose, R. C. (1953). *Ann. Math. Statist.*, **24**, 513–536. (Simultaneous confidence intervals estimation.)

[15] Sen, P. K. (1980). *Handbook of Statistics*, Vol. 1, P. R. Krishnaiah, ed. North-Holland, Amsterdam, 673–702. (Nonparametric simultaneous inference for some MANOVA models.)

[16] Sidak, Z. (1967). *J. Amer. Statist. Ass.*, **62**, 626–633. (Rectangular confidence regions for the means of multivariate normal distributions.)

[17] Siotani, M. (1959). *Ann. Inst. Statist. Math.*, **10**, 183–203. (The extreme value of the generalized distances of the individual points in the multivariate normal sample.)

[18] Siotani, M. (1960). *Ann. Inst. Statist. Math.*, **11**, 167–182. (Notes on the multivariate confidence bounds.)

P. R. KRISHNAIAH
J. M. REISING

MULTIVARIATE NORMAL DISTRIBUTION *See* MULTINORMAL DISTRIBUTION

MULTIVARIATE NORMALITY, TESTING FOR

INTRODUCTION

Although methods of multivariate analysis, since their inception at the end of the nineteenth century, have for the most part been based on the assumption that the data have a multivariate normal distribution, the first published practical quantitative test of multivariate normality was that proposed in 1970 by Mardia [8]. (*See* MARDIA'S TEST OF MULTINORMALITY.) Relative to the univariate case, there are few alternative distributions that have been considered, and virtually all of these have marginal distributions belonging to a common class. Other than when one of these alternatives is clearly appropriate, it is likely that multivariate data (possibly after an obvious transformation of one or more of the variates) will be regarded as being approximately multivariate normal, and some standard analysis made.

The importance of any particular kind of departure from multivariate normality will depend on what the analysis is. Many methods of multivariate analysis depend on covariance matrices, and, consequently, any appreciable nonlinearity of dependence is likely to be serious. The reason for this is that in such a case the covariance of two variates is a poor indication of their association; nonnormality of the marginal distributions, as such, does not have this consequence. There are clearly many possibilities for departure from multivariate normality, implying the need for a variety of techniques for its detection, as omnibus tests will have low power; there is no single best method, and choice should be guided by what departure might be expected a priori or would have the most serious consequences.

It is valuable if, when a test reveals departure from multivariate normality, it indicates a normalizing transformation or suggests alternative methods of analysis. There is a choice between tests which are invariant under arbitrary nonsingular linear transformations of the variates and those which are not. If such transformations of the coordinate system of the observations are of no interest, then tests in the former class are of limited value, since any indication of nonnormality is likely to be difficult to interpret.

Before analysis, the data should be plotted, whether using simple scatter diagrams of pairs of variates or a more sophisticated technique (*see* MULTIVARIATE GRAPHICS). This may give indications of nonnormality or reveal outliers* that, if included in the analysis, might give rise to spurious indications of departure from normality or perhaps conceal real departure from normality. The need for robust estimation* of parameters of the distribution may be suggested.

When a test is carried out on the full set of variates, serious nonnormality confined to a subset of the variates may not be apparent owing to the "diluting" effect of the other variates. To detect nonnormality in such a case requires that subsets of the variates be subject to the test. The subsets may be selected by prior considerations or, perhaps, if computationally feasible, all subsets may be considered; in either case the true significance level of the test will be difficult to estimate.

The requirement of computational feasibility may, indeed, prevent the inclusion of all variates in a test, owing to their number, so that the testing of subsets of the variates is forced upon one. In such a case, the subsets may be selected from prior knowledge of the variates, or, if possible, so that there is little dependence between subsets.

Provided that disjoint subsets are selected, the subtests are approximately independent, and there is no difficulty in assessing the significance level of the overall test. The following tests are arranged in three groups:

1. Tests of marginal normality.
2. Univariate tests of joint normality or of normality in a particular direction.
3. Multivariate techniques for testing joint normality.

NOTATION

Let $x_1, x_2, \ldots, x_n$ be n observations of a random vector $\mathbf{X}$ with p components $X_1, X_2, \ldots, X_p$. Let $\bar{x}$ and $\mathbf{S}$ be the mean and covariance matrix of the sample, with $\boldsymbol{\mu}$ and $\boldsymbol{\Sigma}$ the corresponding population parameters. The null hypothesis is that $\mathbf{X}$ is multivariate normal.

The squared radius or *Mahalanobis distance* of $\mathbf{x}_i$ from $\bar{x}$ is defined as $r_i^2 = (\mathbf{x}_i - \bar{\mathbf{x}})' \mathbf{S}^{-1} (\mathbf{x}_i - \bar{\mathbf{x}})$, and the *Mahalanobis angle* between the vectors $\mathbf{x}_i - \bar{\mathbf{x}}$ and $\mathbf{x}_j - \bar{\mathbf{x}}$ is $r_{ij} = (\mathbf{x}_i - \bar{\mathbf{x}})' \mathbf{S}^{-1} (\mathbf{x}_j - \bar{\mathbf{x}})$. The *scaled residuals* are $\mathbf{y}_i = \mathbf{S}^{-1/2} (\mathbf{x}_i - \bar{\mathbf{x}})$.

EVALUATING MARGINAL NORMALITY

Although it should be remembered that marginal normality does not imply joint normality, the converse is true, and many types of nonnormality can most simply be detected from their marginal effects.

The simplest technique in this class is to test the univariate normality of the marginal distributions and to evaluate the overall significance level according to the usual rules for simultaneous testing (*see* MULTIPLE COMPARISONS). The univariate tests are described elsewhere (*see* DEPARTURE FROM NORMALITY, TESTS FOR and TRANSFORMATION TESTS), and attention here is focused on techniques to combine test statistics for the marginal distributions to produce a single test statistic.

Let $\mathbf{v}_1$ and $\mathbf{v}_2$ be the $p \times 1$ vectors of the marginal coefficients of skewness and kurtosis, respectively. Small [11] shows how Johnson's S_U transformation (*see* JOHNSON'S SYSTEM OF DISTRIBUTIONS) can be applied individually to the components of these vectors to yield vectors $\mathbf{w}_1$ and $\mathbf{w}_2$, the components of which have approximately standard normal distributions. The covariance matrices of $\mathbf{w}_1$ and $\mathbf{w}_2$, say $\mathbf{U}_1$ and $\mathbf{U}_2$, therefore have the elements of their main diagonals all equal to unity; it can be shown that the off-diagonal elements are asymptotically ρ_{ij}^3 and ρ_{ij}^4, respectively, where ρ_{ij} is corr(X_i, X_j), estimated in practice by the sample correlations. The test statistics $Q_1 = \mathbf{w}_1' \mathbf{U}_1^{-1} \mathbf{w}_1$ and $Q_2 = \mathbf{w}_2' \mathbf{U}_2^{-1} \mathbf{w}_2$ are nearly independent, and each has a null distribution approximately χ_p^2 (chi-square with p degrees of freedom). In a similar manner, Royston [9] combines marginal values of the Shapiro–Wilk W statistic*.

These combined tests avoid the difficulty of attempting to evaluate the true significance level in simultaneous testing of marginal statistics and might detect smaller deviations from normality that affect two or more marginal distributions in a similar manner; against this, they run the risk of obscuring nonnormality in a single marginal distribution, so marginal test statistics should always be examined.

TESTS BASED ON UNIDIMENSIONAL VIEWS

A simple but useful test of multivariate normality is obtained by plotting the ordered squared radii r_i^2 against the expected order statistics of their null distribution, which Gnanadesikan and Kettenring [6] show to be proportional to that of a beta* variable of the first kind. They show, too, that this distribution may be approximated by a χ_2^2 distribution for the case $p = 2$ and $n \geqslant 25$. Small [10] observes that approximation by a χ_p^2 distribution for $p > 2$ is inadequate and describes how to estimate the expected order statistics of the beta distribution more directly. Since the null distribution of the r_i^2 is

known, a quantitative test can be made by converting to normal scores* and applying a test of univariate normality; however, such a test will be equally influenced by values of r_i^2 for points near to and distant from the mean, whereas, usually, the latter are the main interest. (An algorithm for computing the incomplete beta integral is given by Cran et al. [5].)

Another basis for tests is the characterizaton of the multivariate normal distribution by the univariate normality of all linear combinations of the variates (*see* MULTINORMAL DISTRIBUTION). Malkovich and Afifi [7] apply this fact to derive three tests based on finding the linear combination giving the largest values of skewness or kurtosis and the smallest value of the Shapiro–Wilk W statistic, respectively. Clearly, for all but small values of p, it will require considerable computational effort to discover the extremal linear combination. It would be computationally simpler to apply any desired test of univariate normality to the projection of the data in a chosen direction. Andrews et al. [1], in the context of transformation tests, suggest that one select a suitable direction by considering the weighted sum of the scaled residuals $\mathbf{y}_i$ defined by

$$\mathbf{d}_\alpha = \sum_{i=1}^n w_i \mathbf{y}_i \bigg/ \left\| \sum_{i=1}^n w_i \mathbf{y}_i \right\|,$$

where $w_i = \|\mathbf{y}_i\|^\alpha$, $\|\mathbf{z}\|$ denotes the length of $\mathbf{z}$, and α is a constant to be chosen. For $\alpha > 0$, $\mathbf{d}_\alpha$ will reflect the direction of any clustering remote from the mean, whereas for $\alpha < 0$, it will point to clustering close to the mean. For $\alpha = -1$, $\mathbf{d}_\alpha$ depends only on the orientations of the $\mathbf{y}_i$. In the space of observations, the direction corresponding to $\mathbf{d}_\alpha$ is $\mathbf{d}_\alpha^* = \mathbf{S}^{1/2} \mathbf{d}_\alpha$. Note that when the direction is chosen in a manner dependent on the data, the formal significance levels of tests are not applicable.

The final test in this section takes a two-dimensional view of the data, but its test statistic is the maximum of a function of a single linear combination of the variates. Cox and Small [4] suggest that two linear combinations of the variates be found such that one of these has maximum curvature when regressed on the other. Let the linear combinations V and W be defined by

$$V = \mathbf{a}'\mathbf{x}, \qquad W = \mathbf{b}'\mathbf{x},$$
$$\text{where} \quad \mathbf{a}'\Sigma\mathbf{a} = \mathbf{b}'\Sigma\mathbf{b} = 1,$$

so that V and W have zero mean and unit variance. Define

$$\eta^2(\mathbf{a}, \mathbf{b}) = \gamma^2 \Big/ \Big[E(W^4) - 1 - \big\{ E(W^3) \big\}^2 \Big],$$

where γ is the least-squares regression coefficient of V on W^2 adjusting for linear regression on W. This expression is the sum of squares accounted for by the quadratic term. For fixed $\mathbf{b}$ this expression can be maximized analytically with respect to $\mathbf{a}$, to give $\eta^2(\mathbf{b})$. The maximum of the sample value of this must be found numerically, giving $\hat{\eta}^2_{\max}$. Simulation shows that under the null hypothesis, for $n \geqslant 50$. $p \leqslant 6$, $\log \hat{\eta}^2_{\max}$ is normally distributed with mean $\log(5p^2/8n)$ and standard deviation $\log(0.53 + 3.87/p)$.

EVALUATING JOINT NORMALITY

A graphical approach is possible by converting the scaled residuals $\mathbf{y}_i$ to polar coordinates with squared radii, giving p coordinates $r_i^2 = \mathbf{y}_i'\mathbf{y}_i$ plus $(p - 1)$ independent angles. The plotting of the r_i^2 was considered in the preceding section. One of the angles will be uniformly distributed on $[0, 2\pi)$, so it is readily plotted. For $p > 2$, the remaining angles have distributions with densities proportional to $\sin^{j-1}\theta$ $(0 \leqslant \theta \leqslant \pi, j = 2, \ldots, p - 1)$.

Mardia [8] proposed the following statistics as multivariate measures of skewness and kurtosis, respectively:

$$b_{1,p} = \frac{1}{n^2} \sum_{i=1}^{n} \sum_{j=1}^{n} r_{ij}^3 \quad \text{and} \quad b_{2,p} = \frac{1}{n} \sum_{i=1}^{n} r_i^4.$$

Asymptotically, $nb_{1,p}/6$ is distributed as χ^2 with $p(p+1)(p+2)/6$ degrees of freedom, and $b_{2,p}$ is asymptotically normal with mean $p(p+2)$ and variance $8p(p+2)/n$. (For details, *see* MARDIA'S TEST OF MULTINORMALITY.)

The extension to multivariate distributions by Andrews et al. [1] of the method for transformation to univariate normality of Box and Cox [3] can be applied to give a likelihood ratio test of multivariate normality. The transformations considered are of the type:

$$x_{ij}^{(\lambda_j)} = \begin{cases} (x_{ij}^{\lambda_j} - 1)/\lambda_j, & \lambda_j \neq 0, \\ \log x_{ij}, & \lambda_j = 0, \end{cases}$$

where x_{ij} is the jth component of $\mathbf{x}_i$. The null hypothesis is then that $\boldsymbol{\lambda} = (\lambda_1\lambda_2, \ldots, \lambda_p) = \mathbf{1}$. The log-likelihood function for $\boldsymbol{\lambda}$ is

$$L^*(\boldsymbol{\lambda}) = -\frac{n}{2}\log|\hat{\boldsymbol{\Sigma}}|$$
$$+ \sum_{j=1}^{p} \left\{ (\lambda_j - 1) \sum_{i=1}^{n} \log x_{ij} \right\},$$

where $\hat{\boldsymbol{\Sigma}}$ is the maximum likelihood estimate of $\boldsymbol{\Sigma}$. Let this be maximized by $\boldsymbol{\lambda} = \hat{\boldsymbol{\lambda}}$. Then the null hypothesis is tested by considering $2\{L^*(\hat{\boldsymbol{\lambda}}) - L^*(\mathbf{1})\}$, which is asymptotically distributed as χ_p^2. One limitation of a transformation test that considers only a restricted set of transformations is that there is no certainty that any member of the set will significantly improve normality, so that failure to reject the null hypothesis does not guarantee normality of the data; on the other hand, should nonnormality be detected, the method provides a normalizing transformation.

Bera and John [2] have derived a test for multivariate normality by considering alternatives with a multivariate Pearson distribution*, for which the PDF, $f(\mathbf{x})$, satisfies

$$\frac{\partial \log f}{\partial x_i} = \frac{\alpha_{i00} + \sum_{r=1}^{p}\alpha_{ir0}x_r}{\begin{array}{c}\beta_{i00} + \sum_{r=1}^{p}\beta_{ir0}x_r \\ + \sum_{r=1}^{p}\sum_{s=1}^{p}\beta_{irs}x_r x_s\end{array}},$$

$$i = 1, 2, \ldots, p,$$

where the α's and β's are parameters and $\beta_{irs} = \beta_{isr}, i, r, s = 1, 2, \ldots, p$. The multivariate normal distribution corresponds to the case where all β's except β_{i00}, $i = 1, 2, \ldots, p$, are zero. By transforming to a normalized variate and making maximum likelihood estimates of the parameters that vanish if the parent distribution is multivariate normal, Bera and John [2] obtain test statistics

$$T_i = \sum_{t=1}^{n} \frac{y_{ti}^3}{n}, \qquad T_{ii} = \sum_{t=1}^{n} \frac{y_{ti}}{n},$$

$$T_{ij} = \sum_{t=1}^{n} \frac{y_{ti}^2 y_{tj}^2}{n},$$

$i, j = 1, 2, \ldots, p$, $i \neq j$, where y_{ti} is the ith component of the scaled residual $\mathbf{y}_t$. Under the null hypothesis, the T's are asymptotically uncorrelated and normally distributed with

$$E(T_i) = 0, \qquad V(T_i) = 6/n,$$

$$E(T_{ii}) = 3, \qquad V(T_{ii}) = 24/n,$$

$$E(T_{ij}) = 1, \qquad V(T_{ij}) = 4/n, \qquad i \neq j.$$

The T_i's may be tested simultaneously using

$$C_1 = n \sum_{i=1}^{p} \frac{T_i^2}{6},$$

which is asymptotically χ_p^2. If this is not significant, then

$$C_2 = n \left\{ \frac{1}{24} \sum_{i=1}^{p} (T_{ii} - 3)^2 \right.$$

$$\left. + \frac{1}{4} \sum_{i=2}^{p} \sum_{j=1}^{i-1} (T_{ij} - 1)^2 \right\}$$

may be tested as approximately $\chi^2_{p(p+1)/2}$, the overall significance level following from the usual rules for simultaneous testing. Alternatively,

$$C_3 = n \left\{ \frac{1}{6} \sum_{i=1}^{p} T_i^2 + \frac{1}{24} \sum_{i=1}^{p} (T_{ii} - 3)^2 \right\}$$

or

$$C_4 = C_1 + C_2$$

may be tested as χ_p^2 or $\chi^2_{p(p+3)/2}$, respectively. All these combined tests are conservative, and Bera and John give a table of the ratios of empirical to asymptotic mean values of C_1, C_2, C_3, and C_4.

A coordinate-dependent method for testing multivariate normality by investigating

nonlinearity of regression has been developed by Cox and Small [4]. Starting with the case $p = 2$, they define $Q_{2,1}$ as the standard Student t statistic for the significance of the regression coefficient of X_2 on X_1^2, when X_2 is regressed on X_1 and X_1^2. (In special circumstances, nonlinear functions other than X_1^2, for example $1/X_1$, could be used.) $Q_{1,2}$ is defined similarly. Under the null hypothesis, the joint distribution of $(Q_{2,1}, Q_{1,2})$ is asymptotically bivariate normal with zero mean, unit variance, and correlation $\rho(2 - 3\rho^2)$ where $\rho = \text{corr}(X_1, X_2)$. Then, if required, the composite test statistic $\max(|Q_{2,1}|, |Q_{1,2}|)$ can be tested from tables of the bivariate normal distribution*, or the quadratic form

$$[Q_{2,1} \; Q_{1,2}] \begin{bmatrix} 1 & r(2-3r^2) \\ r(2-3r^2) & 1 \end{bmatrix}^{-1} \begin{bmatrix} Q_{2,1} \\ Q_{1,2} \end{bmatrix},$$

where r is the sample estimate of ρ and is asymptotically χ_2^2. A variety of extensions to $p > 2$ is suggested. In particular, $Q_{r,t}^{(p)}$ is defined as the Student t statistic for the quadratic contribution when regressing X_r on all other X_s and on X_t^2. The Q's ($Q_{r,s}^{(p)}$, $r, s = 1, 2, \ldots, p$, $r \neq s$) may be ordered and plotted against expected normal order statistics, provided that the sample size is such that the Student t-distribution is effectively normal. For a more quantitative approach, it is natural to consider the Q's as a square array and to examine row and column sums $Q_{r,.}^{(p)}$, $Q_{.,t}^{(p)}$ or sums of squares

$$S_{r,.}^{(p)} = \sum_{t \neq r} \left\{ Q_{r,t}^{(p)} \right\}^2, \quad S_{.,t}^{(p)} = \sum_{r \neq t} \left\{ Q_{r,t}^{(p)} \right\}^2.$$

The last two statistics have, approximately, means $(p - 1)$ and variances

$$2(p - 1)\{1 + 2(p - 2)/n\}.$$

References

[1] Andrews, D. F., Gnanadesikan, R., and Warner, J. L. (1971). *Biometrics*, **27**, 825–840. (Gives transformations to normality corresponding to each of the classes of tests described.)

[2] Bera, A. and John, S. (1983). *Commun. Statist. Theor. Meth.*, **12**, 103–117.

[3] Box, G. E. P. and Cox, D. R. (1964). *J. R. Statist. Soc. B*, **26**, 211–252.

[4] Cox, D. R. and Small, N. J. H. (1978). *Biometrika*, **65**, 263–272.

[5] Cran, G. W., Martin, K. J., and Thomas, G. E. (1977). *Appl. Statist.*, **26**, 111–114.

[6] Gnanadesikan, R., and Kettenring, J. R. (1972). *Biometrics*, **28**, 81–124.

[7] Malkovich, J. F. and Afifi, A. A. (1973). *J. Amer. Statist. Ass.*, **68**, 176–179.

[8] Mardia, K. V. (1970). *Biometrika*, **57**, 519–530.

[9] Royston, J. P. (1983). *Appl. Statist.*, **32**, 121–133.

[10] Small, N. J. H. (1978). *Biometrika*, **65**, 657–658.

[11] Small, N. J. H. (1980). *Appl. Statist.*, **29**, 85–87.

Bibliography

More detailed reviews of tests of multivariate normality, together with extensive bibliographies, are to be found in:

Gnanadesikan, R. (1977). *Methods for Statistical Data Analysis of Multivariate Observations*. Wiley, New York, pp. 161–195.

Mardia, K. V. (1980). In *Handbook of Statistics*, Vol. 1, P. R. Krishnaiah, ed. North-Holland, Amsterdam, pp. 310–320.

(DEPARTURES FROM NORMALITY,
   TESTS FOR
MARDIA'S TEST OF MULTINORMALITY
MULTINORMAL DISTRIBUTION
MULTIVARIATE ANALYSIS
ROBUST ESTIMATION)

N. J. H. Small

MULTIVARIATE NORMAL-WISHART DISTRIBUTION

Let $X_1, \ldots, X_n$ be a random sample from a k-variate multinormal distribution* with unknown mean vector M and unknown covariance matrix R. It is also assumed that the conditional prior distribution of M given $R = r$ is a multivariate normal distribution with mean μ and covariance matrix νr where $\nu > 0$, while the marginal distribution of R is a Wishart distribution* with α degrees of freedom and covariance matrix Σ. (It is assumed that $\alpha > k - 1$.) The posterior joint

distribution of M and R given $X_i = x_i$ ($i = 1, \ldots, n$) has then a density of the form

$$f(m, r \mid x_i, i = 1, \ldots, n)$$
$$\propto \left\{ |r|^{1/2} \exp\left[-\frac{\nu + n}{2} (m - \mu^*)' \right.\right.$$
$$\left.\left. \times r(m - \mu^*) \right] \right\}$$
$$\cdot \left\{ |r|^{(\alpha + n - k - 1)/2} \exp\left[-\tfrac{1}{2} \operatorname{tr}(\Sigma^* r) \right] \right\}$$

where

$$\mu^* = \frac{\nu\mu + n\bar{x}}{\nu + n},$$

$$\Sigma^* = \Sigma + \sum_{i=1}^{n} (x_i - \bar{x})(x_i - \bar{x})'$$
$$+ \frac{\nu n}{\nu + n}(\mu - \bar{x})(\mu - \bar{x})'.$$

This is a multivariate normal-Wishart distribution. For more details see, e.g., De-Groot [1].

Reference

[1] DeGroot, M. H. (1970). *Optimal Statistical Decisions*. McGraw-Hill, New York.

(BAYESIAN INFERENCE
MULTINORMAL DISTRIBUTION
WISHART DISTRIBUTION)

MULTIVARIATE ORDER STATISTICS

INTRODUCTION

There is no obvious extension of the concept of order statistics* of univariate random variables* to the multivariate case. In fact, one cannot hope for a meaningful ordering of multivariate data without some kind of transformation to the univariate case nor can the large variety of applications of order statistics of univariate data in statistical inference* be imitated in a multivariate setting. There is, however, one aspect, i.e., model building, in which the significant role of the theory of order statistics directly extends from the univariate case to the multivariate one.

If one is considering building a dam at each of two locations on the same river, then *flood* or *drought* levels (highest or lowest water levels) at each of these two locations must be predicted. If the water level on this river on the ith day is X_i at location C_1 and Y_i at C_2, then from the observations (X_i, Y_i), one is led to two types of "extremal vectors": one is the joint flood level $(\max X_i, \max Y_i)$, and the other type is $(\max X_i, Y_{[n]})$, where $Y_{[n]}$ is the Y-measurement when the X values have achieved their maximum. Additional examples when "order statistics" of multivariate data play an important role can be taken from reliability theory* (e.g., for a k-out-of-n system, if X_i is the random life length of the ith component and Y_i is another of its random characteristics, then $(X_{n-k+1:n}, Y_{n:n})$ is the random life length of the system, combined with the highest Y value (here and in what follows, $U_{1:n} \leqslant U_{2:n} \leqslant \cdots \leqslant U_{n:n}$ denote the order statistics of the (univariate) random variables $U_1, U_2, \ldots, U_n$); from sociology (if a random characteristic for the father is X and for his son is Y, then, in a population of size n, $(X_{k:n}, Y_{[k:n]})$ represents the measurement for the son of that father who is kth in the population, can be used to predict $Y_{[k:n]}$ on the base of $X_{k:n}$) and from other areas. Evidently, these bivariate examples immediately extend to arbitrary dimensions.

Several other attempts to introduce ordering of multivariate data have been made. Most of these are associated with a single statistical method and thus do not have general appeal. A good analysis can be found in the survey of Barnett [1].

The discussion in the rest of this article will be limited to the two types of ordering introduced in our examples. For these, the following terms will be used. If $\mathbf{X} = (X^{(1)}, X^{(2)}, \ldots, X^{(k)})$ and if $\mathbf{X}_1, \mathbf{X}_2, \ldots, \mathbf{X}_n$ are independent copies of $\mathbf{X}$, then the vectors $(X_{t_1:n}^{(1)}, X_{t_2:n}^{(2)}, \ldots, X_{t_k:n}^{(k)})$ will be called *multivariate order statistics*. When only one component, the first one say, is ordered, then in the vector $(X_{[t:n]}^{(1)}, X_{[t:n]}^{(2)}, \ldots, X_{[t:n]}^{(k)})$,

where the subscript $[t:n] = j$ whenever $X_{t:n}^{(1)} = X_j^{(1)}$, the components $X_{[t:n]}^{(s)}$, $2 \leqslant s \leqslant k$, will be called *concomitants of $X_{t:n}^{(1)}$*, or, as a general reference, *concomitants of order statistics*.

DISTRIBUTION THEORY OF MULTIVARIATE EXTREMES

Let $\mathbf{X} = (X^{(1)}, X^{(2)}, \ldots, X^{(k)})$ and let $\mathbf{X}_1, \mathbf{X}_2, \ldots, \mathbf{X}_n$ be independent copies of $\mathbf{X}$. We call the vectors $\mathbf{Z}_n = (X_{n:n}^{(1)}, X_{n:n}^{(2)}, \ldots, X_{n:n}^{(k)})$ and $\mathbf{W}_n = (X_{1:n}^{(1)}, X_{1:n}^{(2)}, \ldots, X_{1:n}^{(k)})$ multivariate extremes. Just as in the univariate case, $\mathbf{Z}_n$ and $\mathbf{W}_n$ can be transformed into each other by taking $-\mathbf{X}$ for $\mathbf{X}$ in the definitions. Therefore, it suffices to discuss the distribution theory for $\mathbf{Z}_n$ only.

Let $F(x_1, x_2, \ldots, x_k)$ be the cumulative distribution function of $\mathbf{X}$. Then, by the assumptions on the $\mathbf{X}_j$, the cumulative distribution function of $\mathbf{Z}_n$ is given by

$$H_n(x_1, x_2, \ldots, x_k)$$
$$= \Pr\left(\bigcap_{j=1}^{k} \left\{ X_{n:n}^{(j)} \leqslant x_j \right\} \right)$$
$$= F^n(x_1, x_2, \ldots, x_k).$$

Now, if F is known, so in principle, is H_n (its computation may still be difficult if it is so for F). However, if F can only be estimated, then we may get no information on H_n through the preceding formula because a high power is very sensitive to even slight errors in the base (compare $0.992^{400} = 0.040$ and $0.999^{400} = 0.670$). In such cases, an asymptotic formula for $F^n(x_1, x_2, \ldots, x_k)$, where each x_j also depends on n, may give the appropriate value for $H_n(x_1, x_2, \ldots, x_k)$.

The marginal distributions $H_j(x_j) = \Pr(X_{n:n} \leqslant x_j^{(j)})$ follow the univariate theory of *extremes*. Hence when x_j is of the form $x_j = a_{j,n} + b_{j,n} z_j$, the existence of a limiting distribution for $H_j(x_j)$ as well as the choice of $a_{j,n}$ and $b_{j,n} > 0$ can be determined by the rules of the univariate theory. (See Galambos [13, Chap. 2] and EXTREME-VALUE DISTRIBUTIONS.)

In order to express the dependence of a multivariate distribution $F(x_1, x_2, \ldots, x_k)$ on its marginals $F_j(x_j)$, it is convenient to introduce the concept of a dependence function: $D(u_1, u_2, \ldots, u_k)$ is said to be a *dependence function* of the distribution function $F(x_1, x_2, \ldots, x_k)$ if it is a continuous distribution function whose marginals are uniform on the interval $(0, 1)$ such that, for each point $(x_1, x_2, \ldots, x_k)$ at which the marginals $F_j(x_j)$, $1 \leqslant j \leqslant k$, are continuous,

$$F(x_1, x_2, \ldots, x_k)$$
$$= D(F_1(x_1), F_2(x_2), \ldots, F_k(x_k)).$$

The fact that such a dependence function exists is not evident, but as a tool it had been applied widely in the literature. The first rigorous proof of its existence was given by Deheuvels [10] (*see also* DEPENDENCE, CONCEPTS OF). The importance of dependence functions in the theory of multivariate extremes is expressed by the following basic theorems.

Theorem 1. With the previous notations and assumptions, there are constants $a_{j,n}$ and $b_{j,n} > 0$, $1 \leqslant j \leqslant n$, such that $H_n(a_{1,n} + b_{1,n} z_1, a_{2,n} + b_{2,n} z_2, \ldots, a_{k,n} + b_{k,n} z_k)$ converges weakly to a distribution function $H(z_1, z_2, \ldots, z_k)$, whose univariate marginals are nondegenerate if and only if each marginal of H_n converges to a nondegenerate univariate extreme-value distribution* (for the maxima), and if $D_F^n(u_1^{1/n}, u_2^{1/n}, \ldots, u_k^{1/n})$ converges to a dependence function, where D_F is the dependence function of the population distribution F.

Theorem 2. A distribution function $H(z_1, z_2, \ldots, z_k)$ is a limiting distribution function in Theorem 1 if and only if each of its univariate marginals is a univariate *extreme value distribution* for the maxima and if its dependence function D_H satisfies

$$D_H^t(u_1^{1/t}, u_2^{1/t}, \ldots, u_k^{1/t})$$
$$= D_H(u_1, u_2, \ldots, u_k),$$

where $t > 1$ is an arbitrary integer.

Theorem 3. Let $H(z_1, z_2, \ldots, z_k)$ be a limiting distribution function in Theorem 1. Assume that each of its univariate marginals is of the form $H_j(z_j) = \exp(-\exp(-z_j - A_j))$, where $A_j > 0$ is a constant. Then with some finite measure $U(p_1, p_2, \ldots, p_k)$ on the k-dimensional unit simplex $\mathbf{S}$, H has the representation

$$H(z_1, z_2, \ldots, z_k)$$
$$= \exp\left\{ - \int_{\mathbf{S}} \max_{1 \leqslant t \leqslant k} (p_t e^{-z_t}) \, dU(p_1, \ldots, p_k) \right\}.$$

Although Theorems 2 and 3 are equivalent, the form of Theorem 2 is more convenient when one wants to decide whether a distribution function H is an *asymptotic distribution* of normalized maxima, while Theorem 3 provides a method of construction of multivariate extreme-value distributions.

Several inequalities involving the limiting distribution H and its marginals are known. One is the important inequality

$$H(z_1, z_2, \ldots, z_k)$$
$$\geqslant H_1(z_1) H_2(z_2) \ldots H_k(z_k),$$

from which it follows that the correlation* between any two components is positive. Another set of inequalities implies that the complete independence of the components of a vector whose distribution is H is equivalent to pairwise independence, which in turn is equivalent to zero correlation. This remarkable similarity with the multivariate normal distributions should lead to a successful theory of regressions when the underlying distribution is an H-function*.

There is a detailed analysis of the literature prior to 1978 in Galambos [13, pp. 271–273]. The theory presented in this section started with independent and extensive studies for the bivariate case by Geffroy [15], Tiago de Oliveira [30], and Sibuya [28]. A very simple representation was established by Tiago de Oliveira [31] for $H(z_1, z_2)$ when its density exists. Extensions of the bivariate case to arbitrary dimension, leading to Theorems 1–3, were studied in Galambos [12], Pickands [24], de Haan and Resnick [9], Galambos [13, Chap. 5], and Deheuvels [10]. Theorem 3 is due to Pickands [24].

A population distribution F is said to be in the *domain of attraction* of a limiting distribution H if Theorem 1 is satisfied by them. The first criteria for a bivariate F to belong to the domain of attraction of a given H were given by Nair [23]; these criteria, with some restrictions on H, were extended to higher dimension in Galambos [14]. The most general result in this direction is given in Marshall and Olkin [21]. (See also Galambos [13, p. 263]).

DISTRIBUTION THEORY OF MULTIVARIATE ORDER STATISTICS

By multivariate extensions of the classical Bonferroni identities and inequalities*, one can calculate or approximate the cumulative distribution function of the multivariate order statistic $(X_{t_1:n}^{(1)} X_{t_2:n}^{(2)}, \ldots, X_{t_k:n}^{(k)})$. The details of such an approach are given in Galambos [12], where asymptotic results are also established for bounded $s_j = n - t_j$, $1 \leqslant j \leqslant k$. In this case, the same normalizing constants can be used as for maxima, and the dependence functions obtained for maxima can be utilized for constructing the dependence functions of these limits. The conditions for asymptotic independence of the components is the same as for maxima; this was proved independently by Mikhailov [22] and Galambos [12], who extended the results of Mardia [20] and Srivastava [29] for the bivariate case.

STATISTICAL INFERENCE

The paper by Posner et al. [26] demonstrates well the effectiveness of the theory of bivariate extremes. Assume that coded messages are transmitted through two random frequencies $X^{(1)}$ and $X^{(2)}$. If $X^{(1)}$ is above a given level x, then a wrong code is recorded, while a part of the message is erased if $X^{(2)}$ exceeds another level y. If transmissions at different times are independent, then the probability of receiving a message in n transmissions free of error is $H_n(x, y) = \Pr(X_{n:n}^{(1)} \leqslant x, X_{n:n}^{(2)} \leqslant y)$. The authors utilize the asymptotic theory of maxima (Theorems 1–3), they choose a one-parameter family of de-

pendence functions from the set of differentiable models (Tiago de Oliveira [31]), and develop and compute estimators of the parameters (those of the marginals and the one for the dependence function) from actual data. Earlier, Gumbel and Mustafi [17] used a similar, but somewhat less efficient method in another context.

Tiago de Oliveira, in a number of papers, developed estimators*, proposed test statistics* for parameters, and analyzed the regression* of one component on another for some special bivariate extremal models. Some of these are summarized in Taigo de Oliveira [32], and further references are given therein. Pickands [25] recommends a nonparametric estimator of the dependence function of H of Theorem 3.

CONCOMITANTS OF ORDER STATISTICS

We use the definition and notation of the last paragraph of the Introduction. The concept of *concomitants* of order statistics was first used by David [5] and, independently, under the name of *induced order statistics*, by Bhattacharya [3]. In these papers as well as in David and Galambos [7], Sen [27], Yang [33], and Galambos [13, p. 269], different aspects of the asymptotic theory of concomitants of order statistics are developed. Distribution theory for finite sample sizes is discussed in David et al. [8] and Yang [33]. Interesting applications of concomitants of order statistics, either as an underlying model or as a tool in statistical inference, can be found in Barnett et al. [2], David et al. [8], Yang [33], Galambos [13, p. 268], Gomes [16], and Kaminsky [19]. In David [6, Chaps. 5 and 9] a systematic development of this theory can be found. A more recent comprehensive work on concomitants of order statistics is that of Bhattacharya [4].

Further Reading

All aspects of the theory introduced here are comparatively new, hence there is not a large selection of books or expositions for consultation. For the theory of multivariate order statistics, see Galambos [13, Chap. 5]. For applications and for statistical methods, see the papers mentioned in the section on Statistical Inference as well as Johnson and Kotz [18, Chap. 41]. David [6, Chaps. 5 and 9] and Bhattacharya [4] can be consulted for the theory of concomitants. Point processes in which the dependence structure of the exponential interarrivals is that of multivariate extremes are studied by Deheuvels [11].

References

[1] Barnett, V. (1976). *J. R. Statist. Soc. A*, **139**, 318–354.

[2] Barnett, V., Green, P. J., and Robinson, A. (1976). *Biometrika*, **63**, 323–328.

[3] Bhattacharya, P. K. (1974). *Ann. Statist.*, **2**, 1034–1039.

[4] Bhattacharya, P. K. (1984). In *Handbook of Statistics*, Vol. 4, P. R. Krishnaiah and P. K. Sen, eds. North-Holland, Amsterdam.

[5] David, H. A. (1973). *Bull. Int. Statist. Inst.*, **45**(1), 295–300.

[6] David, H. A. (1981). *Order Statistics*. 2nd ed. Wiley, New York.

[7] David, H. A. and Galambos, J. (1974). *J. Appl. Prob.*, **11**, 762–770.

[8] David, H. A., O'Connell, M. J., and Yang, S. S. (1977). *Ann. Statist.*, **5**, 216–223.

[9] de Haan, L. and Resnick, S. I. (1977). *Zeit. Wahrscheinlichkeitsth. verw. Geb.* **40**, 317–337.

[10] Deheuvels, P. (1978). *Publ. Inst. Statist. Univ. Paris*, **23**, 1–36.

[11] Deheuvels, P. (1982). "Point Processes and Multivariate Extreme Values." *Tech. Rep.*, University of Paris.

[12] Galambos, J. (1975). *J. Amer. Statist. Ass.*, **70**, 674–680.

[13] Galambos, J. (1978). *The Asymptotic Theory of Extreme Order Statistics*. Wiley, New York.

[14] Galambos, J. (1981). "Multivariate Extreme Value Distributions." *Tech. Rep.*, Temple University, Philadelphia, PA.

[15] Geffroy, J. (1958/59). *Publ. Inst. Statist. Univ. Paris*, 7–8, 37–185.

[16] Gomes, M. I. (1981). In *Statistical Distributions in Scientific Work*, Vol. 6, C. Taillie et al., eds., D. Reidel, Dordrecht, pp. 389–410.

[17] Gumbel, E. J. and Mustafi, C. K. (1967). *J. Amer. Statist. Ass.*, **62**, 569–588.

[18] Johnson, N. L. and Kotz, S. (1972). *Continuous Multivariate Distributions*. Wiley, New York.

[19] Kaminsky, A. S. (1981). *43rd session ISI*, Contrib. Papers Vol., pp. 161–164.

[20] Mardia, K. V. (1964). *Calcutta Statist. Ass. Bull.*, **13**, 172–178.

[21] Marshall, A. W. and Olkin, I. (1983). *Ann. Prob.*, **13**, 168–177.

[22] Mikhailov, V. G. (1974). *Teor. veroyatn. ee primen.*, **19**, 817–821.

[23] Nair, K. A. (1976). *Commun. Statist.*, **5**, 575–581.

[24] Pickands, J., III (1977). "Multivariate Extreme Value Distributions." *Tech. Rep.*, University of Pennsylvania, Philadelphia, PA.

[25] Pickands, J., III (1981). *Bull. Int. Statist. Inst.*, **49**(2), 859–878.

[26] Posner, E. C., Rodenich, E. R., Ashlock, J. C., and Lurie, S. (1969). *J. Amer. Statist. Ass.*, **64**, 1403–1415.

[27] Sen, P. K. (1976). *Ann. Prob.*, **4**, 474–479.

[28] Sibuya, M. (1960). *Ann. Inst. Statist. Math.*, **11**, 195–210.

[29] Srivastava, O. P. (1967). *Sankhyā A*, **29**, 175–182.

[30] Tiago de Oliveira, J. (1958). *Rev. Fac. Cienc. Univ. Lisboa*, **A7**, 215–227.

[31] Tiago de Oliveira, J. (1962/63). *Estud. Mat. Estat. Econ.*, **7**, 165–195.

[32] Tiago de Oliveira, J. (1980). In *Proceedings of the Fifth International Symposium on Multivariate Analysis*, P. R. Krishnaiah, ed. North-Holland, Amsterdam.

[33] Yang, S. S. (1977). *Ann. Statist.*, **5**, 996–1002.

(DEPENDENCE, CONCEPTS OF
EXTREME-VALUE DISTRIBUTIONS
LARGE-SAMPLE THEORY
LIMIT THEOREMS
ORDER STATISTICS)

JANOS GALAMBOS

MULTIVARIATE PHASE-TYPE DISTRIBUTION   *See* PHASE-TYPE DISTRIBUTIONS

MULTIVARIATE POWER SERIES DISTRIBUTIONS

The family of multivariate power series distributions is a discrete version of the multivariate linear exponential family*. It provides a perceptive formulation of several multivariate classical discrete distributions.

Variants of multivariate power series distributions have been introduced and studied by Khatri [10] and by Patil in a series of papers [14–16]. For some details and references, see also Johnson and Kotz [6], Ord [12], Patil and Joshi [18], and Patil et al. [17].

DEFINITIONS AND PROPERTIES

Let T be a subset of the set I of nonnegative integers. Define $f(\theta) = \sum a(x)\theta^x$ where the summation extends over T and $a(x) > 0$, $\theta \geqslant \Theta$, the parameter space, such that $f(\theta)$ is finite and differentiable. One has $\Theta = \{\theta : 0 \leqslant \theta < \rho\}$, where ρ is the radius of convergence of the power series of $f(\theta)$. Then a random variable X with probability function (pf)

$$\Pr\{X = x\} = p(x) = p(x; \theta)$$
$$= a(x)\theta^x / f(\theta),$$
$$x \in T \quad (1)$$

is said to have the (generalized) power series distribution* (PSD) with *range* T and the *series function* $f(\theta)$. θ is the *series parameter*, whereas $a(x)$ is the *coefficient function* (Patil [13]).

For the multivariate analog, let T be a subset of the s-fold Cartesian product of the set I of nonnegative integers. Define $f(\theta) = \sum a(\mathbf{x})\theta_1^{x_1}\theta_2^{x_2} \ldots \theta_s^{x_s}$, where the summation extends for $\mathbf{x} = (x_1, x_2, \ldots, x_s)$ over T and $a(\mathbf{x}) > 0$, with $\boldsymbol{\theta} \in \Omega$, the parameter space, itself a subset of the first orthant of an s-dimensional space, such that $f(\boldsymbol{\theta})$ is finite and differentiable. Then an s-dimensional random variable $\mathbf{X}$ with probability function (pf)

$$p(\mathbf{x}) = p(\mathbf{x}; \boldsymbol{\theta}) = a(\mathbf{x})\theta_1^{x_1}\theta_2^{x_2} \ldots \theta_s^{x_s} / f(\boldsymbol{\theta}),$$
$$\mathbf{x} \in T \quad (2)$$

is said to have the multivariate (generalized) power series distribution (MPSD) with *range* T and *series function* (sf) $f(\boldsymbol{\theta})$. $\boldsymbol{\theta}$ is the *series parameter*, which takes values in the region of convergence of the power series of $f(\boldsymbol{\theta})$ in powers of θ_i's. The coefficient $a(\mathbf{x})$ is the *coefficient function*.

Property 1. A conditional distribution of $\mathbf{X}$ over a subset of its range T is a *truncated*

version of the distribution of $\mathbf{x}$. Then a truncated MPSD is an MPSD in its own right, and the general properties that hold for an MPSD hold automatically for the truncated MPSD.

Property 2. The r-dimensional random variable $(x_{i_1}, x_{i_2}, \ldots, x_{i_r})$, $1 \leqslant r \leqslant s$ follows an MPSD with sf $f(\boldsymbol{\theta})$ as expanded in powers of $\theta_{i_1}, \theta_{i_2}, \ldots \theta_{i_r}$, other θ's being treated as constants.

Property 3. The means, variances, and covariances are given by

$$E[X_i] = \mu_i = \theta_i \frac{\partial}{\partial \theta_i} \log f(\boldsymbol{\theta}),$$
$$i = 1, 2, \ldots, s,$$

$$V(X_i) = \sigma_{ii} = \theta_i \frac{\partial \mu_i}{\partial \theta_i}, \qquad i = 1, 2, \ldots, s,$$

$$\mathrm{cov}(X_i, X_j) = \sigma_{ij} = \theta_j \frac{\partial \mu_i}{\partial \theta_j},$$
$$i, j = 1, 2, \ldots, s.$$

Further, the MPSD is uniquely determined by its vector of means and the variance-covariance matrix given as functions of the parameters $\theta_1, \theta_2, \ldots, \theta_s$.

Property 4. Let $\mathbf{r} = (r_1, r_2, \ldots, r_s)$ and let $\mathbf{e}_i$ be the ith standard basis vector having unit component in the ith place with zero components elsewhere. The recurrence relations for different kinds of moments here are as follows:

For crude moments,

$$m_{(\mathbf{r})} : m_{(\mathbf{r}+\mathbf{e}_i)} = \theta_i \frac{\partial m_{(\mathbf{r})}}{\partial \theta_i} + \mu_i m_{(\mathbf{r})}.$$

For factorial moments*,

$$m_{[\mathbf{r}]} : m_{[\mathbf{r}+\mathbf{e}_i]} = \theta_i \frac{\partial m_{[\mathbf{r}]}}{\partial \theta_i} + (\mu_i - r_i)m_{[\mathbf{r}]}.$$

For central moments,

$$\mu_{(\mathbf{r})} : \mu_{(\mathbf{r}+\mathbf{e}_i)} = \theta_i \frac{\partial \mu_{(\mathbf{r})}}{\partial \theta_i} + \sum_{j=1}^{ks} r_j \sigma_{ij} \mu_{(\mathbf{r}-\mathbf{e}_i)}.$$

Property 5. The recurrence relation that the cumulants satisfy is

$$k_{(\mathbf{r}+\mathbf{e}_i)} = \theta_i \frac{\partial k_{(r)}}{\partial \theta_i} .$$

This recurrence relation characterizes an MPSD among multivariate discrete distributions. (See Patil [14].)

Property 6. The multinomial*, singular multinomial, multiple Poisson, negative multinomial, and the multivariate logarithmic series distributions* are special cases of MPSDs as follows:

$$f(\boldsymbol{\theta}) = (1 + \theta_1 + \theta_2 + \cdots + \theta_s)^n$$

n a positive integer, for multinomial;

$$f(\boldsymbol{\theta}) = (\theta_1 + \theta_2 + \cdots + \theta_s)^n$$

n a positive integer, for singular multinomial;

$$f(\boldsymbol{\theta}) = \exp(\theta_1 + \theta_2 + \cdots + \theta_s)$$

for multiple Poisson;

$$f(\boldsymbol{\theta}) = (1 - \theta_1 - \theta_2 - \cdots - \theta_s)^k, \quad k > 0$$

for negative multinomial;

$$f(\boldsymbol{\theta}) = -\log(1 - \theta_1 - \theta_2 - \cdots - \theta_s)$$

for multivariate log series.

ESTIMATION

Result 1. To estimate the parameters θ_1, $\theta_2, \ldots, \theta_s$ by the method of maximum likelihood* on the basis of a random sample of size n with x_{ij}, $i = 1, 2, \ldots s$; $j = 1, 2, \ldots, n$, drawn from the MPSD given by (1), the likelihood equations are

$$\bar{x}_i = \mu_i(\theta_1, \theta_2, \ldots, \theta_s),$$

where $\bar{x}_i = (1/n)\sum_{j=1}^{n} x_{ij}$ is the sample mean of the ith component. The likelihood equations are also the equations provided by the method of moments*. The information matrix (*see* FISHER INFORMATION) is given by $\mathbf{I} = (I_{ij})$, where

$$I_{ij} = n\sigma_{ij}/\theta_i\theta_j .$$

Result 2. Analogous to the univariate power series distributions, results for minimum variance unbiased estimation* are available for an MPSD. Here again the MVU estimability of $\boldsymbol{\theta}$ depends only on the structure of the range T and has nothing to do with the specific form of an MPSD as determined by its coefficient function $a(\mathbf{x})$. For details, see Patil [14].

RADIAL MPSDs AND THEIR MIXTURES

While studying problems of light and severe accidents, Bates and Neyman [2] introduced an interesting multivariate discrete probability model by adopting an ingenious fundamental hypothesis in the formulation of a theory of accident proneness. Motivated by their approach, Patil [15] introduced radial MPSDs and their mixtures. Some of the highlights are as follows.

Definitions

The s-dimensional space Ω is a *radial parameter space* relative to $\mathbf{a} = (a_1, a_2, \ldots, a_s)$ if its parameter vectors $\boldsymbol{\theta}$ are nonnegative multiples of the fixed vector $\mathbf{a}$, i.e., $\tilde{\boldsymbol{\theta}} = \lambda\mathbf{a}$, where $\lambda \geqslant 0$ and $\boldsymbol{\theta} \in \Omega$.

The parameter space Ω is a *unit parameter space* if the vector of units $(1, 1, \ldots, 1)$ belongs to Ω.

An MPSD $(\lambda\mathbf{a})$ is defined to be an MPSD having a radial parameter space relative to $\mathbf{a}$.

The distribution of $\mathbf{x}$ is a λ-*mixture* of an MPSD $(\lambda\mathbf{a})$ if the conditional distribution of $\mathbf{X}$ given λ is the MPSD $(\lambda\mathbf{a})$ and if λ is a random variable.

The distribution function, say $F(\lambda)$, of λ is the *mixing distribution*, whereas the MPSD $(\lambda\mathbf{a})$ is the *mixed distribution*. Clearly, the pf of the λ-mixture is available as

$$p(\mathbf{x}) = \int_0^{\infty} p(\mathbf{x}; \lambda\mathbf{a}) \, dF(\lambda)$$

$$= a(\mathbf{x})a_1^{x_1}a_2^{x_2} \ldots a_s^{x_s} \int_0^{\infty} \frac{\lambda^z \, dF(\lambda)}{f(\lambda\mathbf{a})},$$

where $z = x_1 + x_2 + \cdots + x_s$ and $p(\mathbf{x}, \boldsymbol{\theta})$ is defined by (1).

Properties

1. If the distribution of $\mathbf{x}$ is a λ-mixture of MPSD $(\lambda\mathbf{a})$, then the conditional distribution of $\mathbf{x}$ given the disjoint partial sums $\mathbf{z} = (z_1, z_2, \ldots, z_k)$ defined by

$$z_i = x_{r_{i-1}+1} + x_{r_{i-2}+2} + \cdots + x_{r_i}$$

with $\quad 1 = r_0 \leqslant r_1 < r_2 < \cdots < r_k = s$

is only the truncation of the mixed MPSD $(\lambda\mathbf{a})$. Further, it is not independent of the mixing distribution $F(\lambda)$ of λ. For example, if the distribution of $\mathbf{x}$ is a λ-mixture of multiple Poisson $(\lambda\mathbf{a})$, then the conditional distribution of $(x_1, x_2, \ldots, x_{s-1})$, given $x_1 + x_2 + \cdots + x_s = z$, is multinomial $(z; p_1, p_2, \ldots, p_{s-1})$ where $p_i = a_i / \sum a_i$ for $i = 1, 2, \ldots, s - 1$, which does not depend on z. Curiously, the converse is also true.

2. For every λ-mixture of radial MPSD $(\lambda\mathbf{a})$ with $T = I_s = I \times I \times \cdots \times I$ having unit parameter space and independent components, the conditional distribution of x_1, $x_2, \ldots, x_r$ given $x_{r+1}, x_{r+2}, \ldots, x_s$ depends only on the sum $x_{r+1} + x_{r+2} + \cdots + x_s$, and not on the individual values.

3. The only λ-mixture of multiple Poisson $(\lambda\mathbf{a})$ which admits positive linear regression of x_1 on $y_1 = x_{r+1} + x_{r+2} + \cdots + x_s$ for some $r > 1$ is the negative multinomial distribution (*see* MULTINOMIAL DISTRIBUTIONS).

4. A λ-mixture of radial MPSD $(\lambda\mathbf{a})$ with range $T_n = \{\mathbf{x} : x_1 + x_2 + \cdots + x_s = z, z = 0, 1, 2, \ldots, n\}$ is identifiable if and only if $n = \infty$ (*see* IDENTIFIABILITY).

For more results, see Patil [15] and Ghosh and Sinha [5].

MULTIVARIATE FACTORIAL SERIES DISTRIBUTIONS

While studying problems of finite population sampling*, capture-recapture* sampling, sampling from frames with duplicates, and sampling from symmetric trials, Berg [3] introduced an interesting family of multivariate factorial series distributions (FSD)

which is structurally somewhat analogous to the family of multivariate power series distributions (*see* FACTORIAL SERIES DISTRIBUTIONS). The similarity becomes evident when the pf of the multivariate FSD is written as

$$p(\mathbf{x}) = p(\mathbf{x}, \mathbf{N})$$
$$= a(\mathbf{x}) N_1^{(x_1)} N_2^{(x_2)} \cdots N_s^{(x_s)} / f(\mathbf{N}),$$
$$\mathbf{x} \in T;$$

$$f(\mathbf{N}) = \sum_{\mathbf{x}} a(\mathbf{x}) N_1^{(x_1)} N_2^{(x_2)} \cdots N_s^{(x_s)},$$

is the Newton expansion of $f(\mathbf{N})$ in factorial powers of $N_1, N_2, \ldots, N_s$. The coefficient $a(\mathbf{x})$ can be obtained in terms of mixed differences of order $\mathbf{x}$ of $f(\mathbf{N})$ evaluated at the origin. In the case of PSDs, the mixed derivatives of the sf at the origin define the coefficient function of the PSD in a similar way. Berg has derived several interesting structural properties providing means, variances, covariances, binomial moments, likelihood estimates, and minimum variance unbiased estimates, etc. He has also demonstrated their similarity to those of the multivariate power series distributions.

Concluding Remarks

See the entry on univariate power series distributions* written by this author for specific results and related details; see also SUM-SYMMETRIC POWER SERIES DISTRIBUTIONS, which is an elegant subclass of MPSDs.

References

[1] Barndorff-Nielsen, O. (1977). *Information and Exponential Families*. Wiley, New York, pp. 205–208.

[2] Bates, G. E. and Neyman, J. (1952). *Univ. Calif. Publ. Statist.*, **1**, 215–254.

[3] Berg, S. (1977). *Scand. J. Statist.*, **4**, 25–30.

[4] Bildikar, S. and Patil, G. P. (1968). *Ann. Math. Statist.*, **39**, 1316–1326.

[5] Ghosh, J. K., Sinha, B. and Sinha, B. K. (1977). *J. Multivariate Anal.*, **7**, 397–408.

[6] Johnson, N. L. and Kotz, S. (1969). *Discrete Distributions*. Wiley, New York.

[7] Joshi, S. W. and Patil, G. P. (1970). In *Random Counts in Scientific Work*, Vol. 2, G. P. Patil, ed.

Pennsylvania State University Press, University Park, pp. 189–204.

[8] Joshi, S. W. and Patil, G. P. (1971). *Sankhyā A*, **33**, 175–184.

[9] Joshi, S. W. and Patil, G. P. (1972). *Sankhyā A*, **34**, 377–386. Also in *Theory Prob. Appl. Moscow*, **19**, (1974).

[10] Khatri, C. G. (1959). *Biometrika*, **46**, 486–490.

[11] Neyman, J. (1965). In *Classical and Contagious Discrete Distributions*. G. P. Patil, ed. Statistical Publishing Society, Calcutta and Pergamon, New York.

[12] Ord, J. K. (1972). *Families of Frequency Distributions*. Hafner, New York.

[13] Patil, G. P. (1959). "Contributions to Estimation in a Class of Discrete Distributions." Ph.D. thesis, University of Michigan, Ann Arbor.

[14] Patil, G. P. (1965). In *Classical and Contagious Discrete Distributions*, G. P. Patil, ed. Statistical Publishing Society, Calcutta and Pergamon, New York, pp. 183–194. Also in *Sankhyā A*, **28**, 225–237.

[15] Patil, G. P. (1965). *Sankhyā A*, **27**, 259–270.

[16] Patil, G. P. (1968). *Sankhyā B*, **30**, 355–366.

[17] Patil, G. P., Boswell, M. T., Joshi, S. W., and Ratnaparkhi, M. V. (1984). In *A Modern Dictionary and Classified Bibliography of Statistical Distributions*, Vol. 3: *Discrete Models*. International Cooperative Publishing House, Fairland, MD.

[18] Patil, G. P. and Joshi, S. W. (1968). *Bibliography and Dictionary of Discrete Distributions*. Oliver and Boyd, Edinburgh, Scotland.

[19] Patil, G. P. and Joshi, S. W. (1970). *Ann. Math. Statist.*, **41**, 567–575.

(FACTORIAL SERIES DISTRIBUTIONS
MODIFIED POWER SERIES
  DISTRIBUTIONS
MULTINOMIAL DISTRIBUTIONS
MULTIVARIATE DISTRIBUTIONS
MULTIVARIATE LOGARITHMIC SERIES
  DISTRIBUTIONS
POISSON DISTRIBUTIONS
POWER SERIES DISTRIBUTIONS)

G. P. PATIL

MULTIVARIATE PROBIT

Multivariate probit is a method for analyzing several quantal variables. For example, probabilities of positive responses to the survey questions, Do you own a house? and Do you have children? might be modeled within the multivariate probit framework as functions of prices, income, and other variables. In an experiment, a sample of organisms might be given stimuli at various levels and several quantal responses (e.g., corresponding to effects on different physiological systems) could be noted. Sequences of quantal responses over time provide another source of application of the multivariate probit specification.

Let D_{it} be equal to one if the tth quantal response* for the ith unit of observation is positive; equal to zero if the response is negative. A probit model specifies

$$\Pr(D_{it} = d_{it})$$
$$= \left\{ \Phi(\mathbf{x}'_{it}\boldsymbol{\beta}_t) \right\}^{d_{it}} \left[1 - \Phi(\mathbf{x}'_{it}\boldsymbol{\beta}_t) \right]^{1 - d_{it}}$$

with $\Phi(\cdot)$ the standard normal* cdf, $\mathbf{x}_{it}$ a $k \times 1$ vector of covariates and $\boldsymbol{\beta}_t$ a $k \times 1$ vector of parameters. The multivariate probit specification for the $T \times 1$ vector $\mathbf{d}_i = (d_{i1}, d_{i2}, \ldots, d_{iT})'$ is

$$P_i = \Pr(D_{i1} = d_{i1}, \ldots, D_{iT} = d_{iT})$$
$$= \int \ldots \int_A f(u_1, u_2, \ldots, u_T : \mathbf{R}) \, du_1 \ldots du_T$$

with $f(\cdot; \mathbf{R})$ the T-variate normal density with means zero, unit variances, and correlation matrix $\mathbf{R}$. The tth integral is from $-\infty$ to $\mathbf{x}'_{it}\boldsymbol{\beta}_t$ if $d_{it} = 1$ (i.e., if the tth quantal response is positive) and from $\mathbf{x}'_{it}\boldsymbol{\beta}_t$ to ∞ if $d_{it} = 0$. This defines the set A.

In this specification, marginal probabilities do not depend on unknown elements of R but conditional probabilities and probabilities of combinations of events do. Consequently, estimates of R will be important for many prediction applications.

ESTIMATION

The likelihood function under random sampling is given by $L = \prod_i P_i$. Maximization of L with respect to the parameters $\beta_1, \ldots, \beta_t$ and the unknown elements of $\mathbf{R}$ is complicated due to the presence of the multinormal integral. Series approximations (*see* NUMERICAL ANALYSIS) or numerical integration* techniques can be used when T is not too large. In some applications it is appropriate to give **R** a special structure, e.g., the one-factor structure in which **R** is the sum of a diagonal and a rank-one matrix. In that case, the problem can be simplified substantially, and efficient computational algorithms are available.

In general, the $\boldsymbol{\beta}_t$ can be estimated by univariate probit. These estimates will not be fully efficient but they are inexpensive to obtain. Constraints (e.g., $\boldsymbol{\beta}_t = \boldsymbol{\beta}$ for all t) can be imposed by taking weighted averages of the univariate probit estimates. All unknown elements of **R** can be obtained by considering pairs of responses involving only bivariate normal integrals. In this way all parameters of the model can be estimated consistently using only univariate and bivariate integration.

A Bayesian* analysis of the probit model is given by Zellner and Rossi [6].

MULTINOMIAL PROBIT: QUALITATIVE CHOICE

Specializing the notion of a general quantal response to that of an explicit choice leads to a model with more structure. Consider the problem of a decision maker choosing among T alternatives on the basis of the utility u_t, $t = 1, \ldots, T$, associated with each. If the utilities are random, then the probability that alternative t is selected is $\Pr(u_t - u_{t'} > 0, \text{ all } t' \neq t)$. If the utilities are jointly normally distributed the model is termed *multinomial probit** (*see* MULTINOMIAL LOGIT, MULTINOMIAL PROBIT). This formulation is a special case of the general multivariate probit. Define $D_{tt'} = 1$ if $u_t - u_{t'} > 0$, $D_{tt'} = 0$ if $u_t - u_{t'} \leqslant 0$ for $t < t'$, $t' = 2, \ldots, T$. For normally distributed utilities, the random variables $D_{tt'}$ could be modeled by a $T(T-1)/2$-variate probit. However, an internal consistency is desirable, and an observed choice is not informative on all the utility comparisons. Each observation, in fact, provides information on a $(T-1)$-variate margin. For example, the proba-

bility that item 1 is chosen is $\Pr(D_{1t'} = 1, t' = 2, \ldots, T)$.

A convenient specification for the utility of option t to individual i is $u_{it} = \mathbf{x}'_{it}\boldsymbol{\beta}_i + \epsilon_{it}$ with $\boldsymbol{\beta}_i$ normally distributed (across individuals, not choices) with mean $\bar{\boldsymbol{\beta}}$, variance $\boldsymbol{\Sigma}_\beta$, and $\boldsymbol{\epsilon}_i = (\epsilon_{i1}, \ldots, \epsilon_{iT})'$ normally distributed, independently of $\boldsymbol{\beta}_i$, with mean zero and variance $\boldsymbol{\Sigma}$. Here $\mathbf{x}_{it}$ is a $k \times 1$ vector of attributes of the tth potential choice (e.g., mileage when cars are being considered for purchase).

With this specification, the probability $\Pr_i(t)$ that individual i makes choice t is a complicated integral of the multinormal distribution. Under random sampling, the likelihood function is given by $L = \prod_i \Pr_i(t_i)$ with $\Pr_i(t_i)$ evaluated at the actual choice made by individual i. The parameters to be estimated are the unknown elements of $\bar{\boldsymbol{\beta}}$, $\boldsymbol{\Sigma}_\beta$, and $\boldsymbol{\Sigma}$ (a normalization is necessary). Hausman and Wise [3] apply this model to a problem with $T = 3$ using series approximations to evaluate the multinormal probabilities. A simple approximation to the distribution of the maximum of a set of normally distributed variates (the Clark approximation) is considered by Manski [4], whose results indicate that the multinomial probit specification can be computationally and economically feasible in applied work.

References

[1] Amemiya, T. (1981). *J. Econ. Lit.*, **19**, 1483–1536.

[2] Ashford, J. R. and Sowden, R. R. (1970). *Biometrics*, **26**, 535–546.

[3] Hausman, J. A. and Wise, D. (1978). *Econometrica*, **46**, 403–436.

[4] Manski, C. and McFadden, D., eds. (1981). *Structural Analysis of Discrete Data*. MIT Press, Cambridge, MA. (A very useful collection of papers. Especially relevant are Manski and McFadden and Cosslett on estimation theory, McFadden on probabilistic choice theory, and Lerman and Manski on estimation practice.)

[5] McFadden, D. (1976). *Ann. Econ. Soc. Meas.*, **4**, 363–390.

[6] Zellner, A. and Rossi, P. E. Proceedings of the 1982 Business and Economic Statistics Section of the American Statistical Association (to be published).

Bibliography

Kolakowski, D. and Bock, R. D. (1981). *Biometrics*, **37**, 541–551. (This presents a probit analysis of several quantal responses under the control of a single continuous variable.)

(BIOASSAY, STATISTICAL METHODS IN MULTINOMIAL LOGIT; MULTINOMIAL PROBIT
NORMAL EQUIVALENT DEVIATE
QUANTAL RESPONSE ANALYSIS)

NICHOLAS M. KIEFER

MULTIVARIATE QUALITY CONTROL

The general multivariate statistical quality control problem considers a repetitive process where each item is characterized by p quality characteristics, $X_1, X_2, \ldots, X_p$. Because of the chance causes inherent in the process, these quality characteristics are random variables. Because of the interdependency between the characteristics, the random variables are correlated. The problem thus requires a multivariate approach. The underlying probability distribution of the p quality characteristics is assumed to be multivariate normal with mean vector μ and covariance matrix Σ.

As a simple example, consider a lumber manufacturing plant where the variables of interest for a particular grade of lumber may be stiffness (X_1) and bending strength (X_2), both measured in pounds/(inches)^2. Because these characteristics are correlated, it is inappropriate to use *individual* Shewhart charts to detect changes in each process average from prespecified values.

This entry is closely related to several others. Many concepts discussed therein are utilized below with appropriate modifications (*see* QUALITY CONTROL, STATISTICAL; CONTROL CHARTS).

The multivariate approach to quality control was first widely publicized in 1947 and 1951 by Hotelling [18, 19] in the testing of bombsights. In his procedure, two sights were randomly selected from each lot of 20 sights. Each sight was then tested by dropping four bombs on each of two flights. The range error (measured in the flight direction of the airplane) and the deflection error (measured perpendicular to the direction of flight), which are correlated, were used to monitor the quality of each bomb dropped at target. Hotelling introduced the T^2-control chart as a technique for monitoring the overall quality of a flight, sight, or lot by summing over the appropriate number of bombs involved. The field remained relatively dormant until the late 1950s with the increasing availability of computers.

In a set of related papers, Jackson [20, 21] and Jackson and Morris [26] used an elliptical control region and extended Hotelling's procedure for use with principal components (*see* COMPONENT ANALYSIS) in monitoring a photographic process. By using a suitable base period during which no apparent trouble had occurred, the multivariate data were analyzed for lack of control. Ghare and Torgersen [14], Montgomery and Wadsworth [33], Alt [1, 4], and Alt et al. [8] examined the simultaneous control of several related variables when the data is in the form of rational subgroups. Montgomery and Klatt [31, 32] employed Hotelling's T^{2*} control charts in determining the optimal sample size, interval between samples and in obtaining the control chart constant. Some of the conclusions and results presented in the papers by Jackson [20, 21] and those discussed in Ghare and Torgersen [14], Montgomery and Wadsworth [33], and Montgomery and Klatt [31, 32] require certain alterations. Specific details are in Alt [2, 4]. Alt and Deutsch [5] determined the appropriate sample size and control chart constant by extending the univariate scheme of Page [34] to multivariate data. Alt et al. [6] developed control charts for when there is correlation across the data vectors as well as within each vector. In the univariate case, the process dispersion is monitored by sigma charts or range charts. Alt [1] and Alt et al.

[7] develop and present the multivariate counterparts. Few textbooks discuss the use of T^2-control charts; a notable exception is the textbook by Johnson and Leone [27].

Multivariate sampling plans*, as distinguished from multivariate control charts, were treated by Jackson and Bradley [22–25], who employed multivariate sequential procedures for monitoring the means of ballistic missile data. Shakun [36] also discusses the multivariate acceptance sampling* problem by considering an ellipsoidal specification region when the variables of interest have a multivariate normal distribution with known covariance matrix. Chapman et al. [12] have developed a model to determine the expected total cost of quality control per inspection lot for multivariate acceptance sampling. Patel [35] has developed quality control methods for when the observation vectors emanate from a multivariate binomial or multivariate Poisson population. When the quality of a product is measured by standards on several variables, Berger [11] proposes a method of combining the acceptance sampling procedures for the individual parameters. Additional aspects of the literature will be considered throughout this paper.

There are two distinct phases of control chart practice. Phase I consists of using the charts for (1) retrospectively testing whether the process was in control when the first subgroups were being drawn and (2) testing whether the process remains in control when future subgroups are drawn. These are two separate and distinct stages of analysis. Phase II consists of using the control chart to detect any departure of the underlying process from standard values (μ_0, Σ_0). Phase II charts will be considered first since this facilitates the development of Phase I charts.

CONTROL CHARTS FOR THE MEAN (PHASE II)

When there is only one quality characteristic X, where $X \sim N(\mu_0, \sigma_0^2)$, a Shewhart control chart for the mean has the following control limits:

$$\left.\begin{array}{c} \text{UCL} = \mu_0 + z_{\alpha/2}(\sigma_0/\sqrt{n}), \\ \text{CL} = \mu_0, \\ \text{LCL} = \mu_0 - z_{\alpha/2}(\sigma_0/\sqrt{n}), \end{array}\right\} \quad (1)$$

where $z_{\alpha/2}$ is the corresponding z-percentile. When $\alpha = 0.0027$, $z_{0.00135} = 3.0$, and the upper and lower control limits are written as $\mu_0 \pm A\sigma_0$. Tables of A are given in Duncan [13] for $n = 2, \ldots, 25$.

For successive random samples of size n, this control chart can be viewed as repeated tests of significance of the form $H_0: \mu = \mu_0$ vs. $H_1: \mu \neq \mu_0$. The regions above the UCL and below the LCL thus correspond to the likelihood ratio test* rejection region. This hypothesis testing* viewpoint lays the necessary groundwork for extending the univariate to the multivariate case.

Assume that the p-quality characteristics are jointly distributed as a p-variate normal and that a random sample of size n is available from the process. The likelihood ratio test (see ref. 10) of $H_0: \mu = \mu_0$ vs. $H_1: \mu \neq \mu_0$ specifies that the null hypothesis be rejected if

$$n(\bar{x} - \mu_0)'\Sigma_0^{-1}(\bar{x} - \mu_0) > \chi_{p,\alpha}^2,$$

where $\bar{x}$ denotes the $(p \times 1)$ vector of sample means and $\chi_{p,\alpha}^2$* is the corresponding χ^2-percentile. The control chart is formed by letting

$$\left.\begin{array}{c} \text{UCL} = \chi_{p,\alpha}^2, \\ \text{LCL} = 0, \end{array}\right\} \quad (2)$$

and plotting the values of $n(\bar{x} - \mu_0)'\Sigma_0^{-1}(\bar{x} - \mu_0)$. If this statistic plots above the upper control limit, the process mean is deemed out of control and assignable causes of variation are sought. The control chart is essentially a visual representation of the likelihood ratio test for successive samples. Hereafter, this will be referred to as a χ^2-chart. When there are two quality characteristics, an elliptical control region, centered at μ_0, can be used in place of the χ^2-chart. While the χ^2-chart provides a comprehensive summary over time, the elliptical control region is useful in indicating which characteristic led to an out-of-control condition.

Because of the equivalence between this control chart and a significance test, there is an operating characteristic curve or, equivalently, a power curve associated with this chart. The power* is given by

$$P\left(\chi_{p,\lambda}^{\prime 2} > \chi_{p,\alpha}^2\right),$$

where $\chi_{p,\lambda}^{\prime 2}$ denotes the noncentral chi-square* random variable with p degrees of freedom and noncentrality parameter $\lambda = n(\mu - \mu_0)'\Sigma_0^{-1}(\mu - \mu_0)$. However, this is the power for a single sample only, i.e., the one currently being analyzed. Alt et al. [9] have investigated the power of the test on the mean vector when there are two quality characteristics and one of the process standard deviations, denoted by σ_1, can be ad-

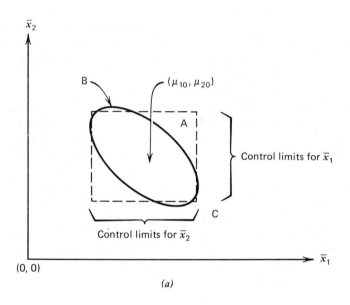

(a)

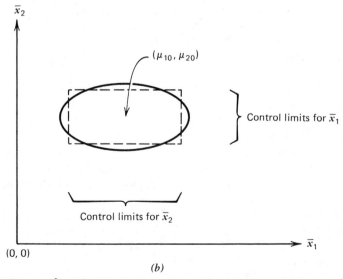

(b)

Figure 1   The elliptical χ^2 control region and the Bonferroni adjusted rectangular control region in the presence of (*a*) high correlation between X_1 and X_2; (*b*) no correlation between X_1 and X_2.

justed. When the correlation between the two characteristics is positive, the power is not a monotonically decreasing function of σ_1 as it is in the univariate case.

When the control chart indicates that the process is out of control, Bonferroni's inequality* can be used to investigate which of the p quality characteristics are responsible for this (*see* BONFERRONI INEQUALITIES AND INTERVALS). To accomplish this, let A_i denote the event that the sample mean for the ith quality characteristic, $i = 1, 2, \ldots, p$, plots within the control limits specified in (1) with $z_{\alpha/2}$ replaced by $z_{\alpha/2p}$. Then Bonferroni's inequality states that $P(A_1 \cap A_2 \cap \cdots \cap A_p) \geqslant 1 - \alpha$. Thus p individual charts would be constructed, each with type I error equal to α/p. However, this does not imply that p univariate charts should be used in place of the χ^2-chart. The danger of this approach is illustrated in Fig. 1 when there are two quality characteristics. Both sample means may plot outside the elliptical

control region. However, if the individual charts are used, it is possible to conclude that both process means are at their nominal values (Region A), one process mean is out of control while the other is not (Region B), and the process is in control (Region C). The size of these regions and their corresponding errors depend on the degree of correlation, or lack thereof, between the quality characteristics (see Fig. 1). Furthermore, when the process mean (μ) equals the nominal value (μ_0), the probability that the vector of sample mean plots within the elliptical control region is *exactly* equal to $1 - \alpha$, whereas this probability is *at least* $1 - \alpha$ when the rectangular control region is used.

Table 1 presents simulated data for a lumber manufacturing plant, as reported in Alt et al. [7]. The data represent measurements on stiffness (X_1) and bending strength (X_2) in units of pounds/(inches)^2 for a particular grade of lumber. The standard values, either derived from a large amount of past data or selected by management to attain certain objectives, are: $\mu_{01} = 265$, $\mu_{02} = 470$, $\sigma_{01} = 10$, $\sigma_{02} = 11$, and $\rho_0 = 0.6$. It follows that

$$\mu_0 = \begin{bmatrix} 265 \\ 470 \end{bmatrix}, \qquad \Sigma_0 = \begin{bmatrix} 100 & 66 \\ 66 & 121 \end{bmatrix}.$$

To simulate an out-of-control process, μ_{01} was increased by $\sigma_{01}/2$ and μ_{02} was decreased by $\sigma_{02}/2$ for samples 16–20. The sample size is 10. The value of $n(\bar{x} - \mu_0)' \Sigma_0^{-1}(\bar{x} - \mu_0)$ is denoted by χ_0^2. If type I error is set equal to 0.0054, then UCL $= \chi_{2,0.0054}^2 = 10.44$. In the corresponding control chart (Fig. 2), it is seen that sample numbers 16, 18, and 20 plot out of control.

With $\alpha = 0.0027$ for the individual control charts (à la Bonferroni), the limits are

$$\text{UCL}_1 = 274.49, \qquad \text{UCL}_2 = 480.44$$
$$\text{CL}_1 = 265.00, \qquad \text{CL}_2 = 470.00$$
$$\text{LCL}_1 = 255.51, \qquad \text{LCL}_2 = 459.56$$

With respect to stiffness, only for sample number 20 does $\bar{x}_1$ plot out of control; with respect to breaking strength, no sample means plot out of control. Thus, for this example, the individual control charts performed poorly in detecting shifts in the pro-

Table 1 Statistics for Control Charts for the Mean (Phase II)

Sample Number	$\bar{x}_1$	$\bar{x}_2$	χ_0^2
1	263.52	469.05	0.22
2	265.62	470.37	0.03
3	265.87	474.68	2.27
4	261.72	470.66	2.13
5	263.51	466.90	0.80
6	261.42	469.76	1.86
7	266.14	470.73	0.13
8	268.85	469.10	3.02
9	258.11	459.74	8.97
10	261.52	465.17	2.05
11	262.28	466.25	1.23
12	267.86	474.19	1.50
13	263.32	473.24	2.72
14	263.08	474.02	3.98
15	262.19	474.76	6.44
16	269.99	465.22	10.91[a]
17	268.21	468.46	2.77
18	272.85	467.62	13.53[a]
19	269.93	466.16	8.92
20	278.79	474.16	22.17[a]

UCL $= \chi_{2,0.0054}^2 = 10.44$

[a]Samples 16, 18, and 20 plot out of control.

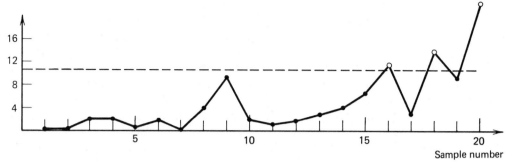

Figure 2 A χ^2-control chart (Phase II).

cess means. This illustrates the danger in relying on such individual charts to detect shifts in the process means.

In the development of the χ^2-control chart, it was assumed that the observation vectors constitute a random sample. Alt et al. [6] have developed a control chart for when the vectors of observations are correlated with each other. Although the UCL is still $\chi^2_{p,\alpha}$, the test statistic differs from that previously discussed.

When there is one quality characteristic, Page [34] determined the control chart constant and the sample size needed to minimize the average run length* of an out-of-control process for a large fixed value of the average run length of an in-control process. Alt and Deutsch [5] extended Page's scheme to the multivariate case and determined that the sample size needed to detect shifts in the process means does not always decrease as the magnitude of the shifts increase. For a relatively large positive correlation, a larger sample size is needed to detect large positive shifts than to detect small positive shifts.

Although Hotelling [18] discussed the use of a χ^2-control chart, he did not actually implement it, since Σ_0 was not available.

CONTROL CHARTS FOR PROCESS DISPERSION (PHASE II)

In the first section, it was assumed that the process dispersion remained constant at Σ_0. This assumption must be validated in practice; methods for investigating it are presented in this section. Several different con-

trol charts for process dispersion will be presented since different statistics can be used to describe variability. To lay the groundwork for the development of control charts for multivariate data, first consider the case of one quality characteristic.

Let S^2 denote the (unbiased) sample variance for a random sample of size n from a process. When the process variance is σ_0^2, then $(n-1)S^2/\sigma_0^2 \sim \chi^2_{n-1}$, and an S^2-chart is obtained by pivoting on this expression. The control limits are presented as Case 1a in Table 2. For successive random samples of size n, this control chart can also be viewed as repeated tests of significance of the form $H_0: \sigma^2 = \sigma_0^2$ vs. $H_1: \sigma^2 \neq \sigma_0^2$. A control chart for S is obtained by taking the square root of these limits (Table 2, Case 1b). Both charts are presented in Guttman and Wilks [16].

The next two process dispersion charts do not utilize the complete distributional properties of the sample statistic but only the first two moments. It can be shown that $E(S) = \sigma_0 c_2'$ and $\text{Var}(S) = \sigma_0^2(1 - c_2'^2)$, where tables of c_2' for $n = 2, \ldots, 25$ are given in Johnson and Leone [27]. Since most of the probability distribution of S is contained in the interval $E(S) \pm 3\sqrt{V(S)}$, it seems reasonable that a control chart for S would have control limits corresponding to this interval (Table 2, Case 2). The lower control limit is replaced by 0 for $n < 6$. Since S is not normally distributed, these control limits cannot be thought of as probability limits [13]. The same rationale applies in the development of what is tradi-

Table 2 Univariate Dispersion Control Charts (Phase II)

	Statistic	Lower Control Limit (LCL)	CL	Upper Control Limit (UCL)
1a.	$S^2 = (n-1)^{-1}\sum_{i=1}^n (X_i - \bar{X})^2$	LCL $= \sigma_0^2 \chi_{n-1,1-(\alpha/2)}^2/(n-1)$	—	UCL $= \sigma_0^2 \chi_{n-1,\alpha/2}^2/(n-1)$
1b.	$S = \sqrt{S^2}$	LCL $= \sigma_0\sqrt{\chi_{n-1,1-(\alpha/2)}^2/(n-1)}$	—	UCL $= \sigma_0\sqrt{\chi_{n-1,\alpha/2}^2/(n-1)}$
2.	$S = \sqrt{S^2}$	LCL $= \max\{0, \sigma_0 c_2' - 3(1-c_2'^2)^{1/2}\}$	CL $= \sigma_0 c_2'$	UCL $= \sigma_0[c_2' + 3(1-c_2'^2)^{1/2}]$
3.	$V = \sqrt{n^{-1}\sum_{i=1}^n (X_i - \bar{X})^2}$	LCL $= \max\{0, \sigma_0[c_2 - 3n^{-1/2}(n-1-nc_2^2)^{1/2}]\}$ $= \max\{0, \sigma_0 B_1\}$	CL $= \sigma_0 c_2$	UCL $= \sigma_0[c_2 + 3n^{-1/2}(n-1-nc_2^2)^{1/2}]$ $= \sigma_0 B_2$

Table 3 Multivariate Dispersion Control Charts (Phase II)

	Statistic	Lower Control Limit (LCL)	Center Line	Upper Control Limit (UCL)
1.	W Equation (3)	—	—	UCL $= \chi_{p(p+1)/2,\alpha}^2$ (Approximate)
2.	$|S|$	LCL $= |\Sigma_0|(\chi_{2n-4,1-(\alpha/2)}^2)^2/[4(n-1)^2]$	—	UCL $= |\Sigma_0|(\chi_{2n-4,\alpha/2}^2)^2/[4(n-1)^2]$
3.	$|\Sigma|$	LCL $= |\Sigma_0|(b_1 - 3b_2^{1/2})$	CL $= |\Sigma_0|b_1$	UCL $= |\Sigma_0|(b_1 + 3b_2^{1/2})$

tionally referred to as the sigma chart, except that V is used in place of S, where $V = \{\sum_{i=1}^{n}(X_i - \bar{X})^2/n\}^{1/2}$ (Table 2, Case 3). Tables of c_2 and the B_1 and B_2 factors are available in Duncan [13].

Although the range chart and the standardized range chart are used widely to monitor univariate process dispersion, they will not be discussed here since the multivariate analog is intractable (*see* CONTROL CHARTS).

For multivariate data, the first chart to be considered is the analog of the S^2-chart (Table 2, Case 1a), which is equivalent to repeated tests of significance of the form H_0: $\sigma^2 = \sigma_0^2$ vs. $H_1 : \sigma^2 \neq \sigma_0^2$. Here $H_0 : \Sigma = \Sigma_0$ vs. $H_1 : \Sigma \neq \Sigma_0$. Using the asymptotic likelihood ratio test result [10], one would compute the following statistic for each random sample:

$$W = -pn + pn \ln n$$
$$- n \ln(|A|/|\Sigma_0|) + \text{tr}(\Sigma_0^{-1}A), \quad (3)$$

where A denotes the sum of squares and cross-products matrix and tr is the trace operator. Note that $A = (n-1)S$, where S is the $(p \times p)$ sample variance-covariance matrix. If the value of this test statistic plots above the UCL $= \chi^2_{p(p+1)/2, \alpha}$, the process is deemed to be out of control. Refer to Table 3, Case 1. Korin [29] found that the asymptotic chi-square approximation, slightly modified, is quite good even for moderate n. He proposes an F approximation that appears to be better.

A widely used measure of multivariate dispersion is the sample generalized variance*, denoted by $|S|$, where S is the $(p \times p)$ sample variance-covariance matrix. The sample generalized variance is the basis for the other multivariate dispersion charts to be considered. However, Johnson and Wichern [28] present three sample covariance matrices for bivariate data that all have the same generalized variance and yet have distinctly different correlations, $r = 0.8, 0.0,$ and -0.8. "Consequently, it is often desirable to provide more than the single number $|S|$ as a summary of S." Thus a control chart for $|S|$

should be used in conjunction with the univariate dispersion charts.

One approach in developing a $|S|$-control chart is to utilize its distributional properties. Suppose there are two quality characteristics. Since $2(n-1)|S|^{1/2}/|\Sigma_0|^{1/2}$ is distributed as χ^2_{2n-4}, it follows that

$$P\big[\,|\Sigma_0|^{1/2}\chi^2_{2n-4,1-(\alpha/2)}/(2(n-1))$$
$$\leqslant |S|^{1/2} \leqslant |\Sigma_0|^{1/2}\chi^2_{2n-4,\alpha/2}/(2(n-1))\big]$$
$$= 1 - \alpha. \qquad (4)$$

This yields the control chart limits in Table 3, Case 2. When there are more than two quality characteristics, one may employ Anderson's asymptotic normal approximation [10] or the approximation suggested by Gnanadesikan and Gupta [15].

Another $|S|$-control chart can be constructed using only the first two moments of $|S|$ and the property that most of the probability distribution of $|S|$ is contained in the interval $E(|S|) \pm 3\sqrt{V(|S|)}$, where $E(|S|) = b_1|\Sigma_0|$ and $V(|S|) = |\Sigma_0|^2 b_2$ with

$$\left.\begin{array}{l} b_1 = (n-1)^{-p} \prod_{i=1}^{p} (n-i), \\[2mm] b_2 = (n-1)^{-2p} \prod_{i=1}^{p} (n-i) \\[2mm] \qquad \times \left[\prod_{j=1}^{p} (n-j+2) - \prod_{j=1}^{p} (n-j) \right]. \end{array}\right\}$$
$$(5)$$

However, since S is positive definite with probability one, $|S| > 0$ and it is not meaningful to have a lower control limit that is negative. If this should occur, set LCL = 0 for Table 3, Case 3.

To illustrate the use of these process dispersion control charts, consider the hypothetical lumber manufacturing plant data of Alt et al. [7] referred to earlier, where the variables of interest are stiffness and bending strength. To simulate an out-of-control process, both process standard deviations were increased by 40% for sample numbers 16–20. The sample statistics measuring dis-

Table 4   Statistics for Univariate and Multivariate Dispersion Charts (Phase II)

Sample Number	s_1	s_2	s_{12}	$\lvert\mathbf{S}\rvert$	W
1	8.29	12.38	50.42	7982.25	1.53
2	6.59	10.10	47.86	2141.95	5.58
3	14.80	14.58	169.37	17867.19	3.27
4	12.21	7.74	81.75	2242.26	9.87
5	8.89	11.01	60.31	5946.96	0.70
6	10.05	10.49	52.38	8376.14	0.29
7	12.07	6.96	62.13	3196.27	7.54
8	8.41	7.77	39.75	2682.44	3.56
9	8.17	6.12	25.08	1873.17	6.20
10	8.07	7.96	48.93	1733.95	6.09
11	11.56	13.20	98.99	13489.77	0.42
12	10.98	10.75	75.77	8190.77	0.31
13	7.67	13.31	84.98	3192.69	6.79
14	6.12	7.33	28.33	1205.71	7.86
15	11.79	9.70	54.64	10092.65	1.55
16	13.79	8.37	82.31	6541.44	6.03
17	15.57	16.79	203.23	27007.14	5.28
18	12.87	9.54	25.82	14405.95	5.80
19	12.59	19.58	156.37	36250.03	9.50
20	14.94	16.76	106.54	51314.10	10.87

persion are presented in Table 4. The numerical values of the univariate control limits described in Table 2 are presented in Table 5. Inspection of the entries in Table 5a reveals that the increased variability was detected only for the second characteristic and only for sample 19. The data is also examined using the multivariate charts described in Table 3. The statistics for these charts are presented in Table 4, and a summary of the results is in Table 5. For the percentage point charts, type I error was set equal to 0.0054. Although the control chart based on the likelihood ratio test (Table 3, Case 1) failed to detect the increased variability, this was not so for the other two charts.

Additional properties of these multivariate dispersion charts is presented in Alt [1].

CONTROL CHARTS FOR THE MEAN (PHASE I)

Recall that in the *first stage* of Phase I, the control chart is used for retrospectively testing whether the process was in control when the data for the first m rational subgroups was collected. Since the standard values in equation (1) are not known, the ad hoc and customary procedure is to use unbiased estimates in their place (*see* CONTROL CHARTS, Table 1). Since there are several sample measures of process variability, one has the option of which control chart to implement.

When there is only one quality characteristic ($p = 1$) and the number of rational subgroups is small, Hillier [17] and Yang and Hillier [37] developed control charts based on small sample probability limits. This was extended to the case of multiple quality characteristics by Alt et al. [8], and it is this multivariate situation that will be presented.

Let $\mathbf{X}_i$ denote the ($p \times n$) data matrix for subgroup i, $i = 1, \ldots, m$. For each data matrix, the ($p \times 1$) vector of sample means ($\bar{\mathbf{x}}_i$) and the ($p \times p$) matrix of sample variances–covariances ($\mathbf{S}_i$) can be calculated. Unbiased estimates of $\boldsymbol{\mu}$ (the process mean) and $\boldsymbol{\Sigma}$ (the process variance-covariance matrix) are

Table 5 Control Chart Limits for Process Dispersion (Phase II)

Statistic	Lower Control Limit	Samples Out	Centerline	Upper Control Limit	Samples Out
(a)	Univariate Dispersion Control Limits (Phase II)				
1a. S_2	$LCL_1 = 13.79$	None	—	$UCL_1 = 301.03$	None
	$LCL_2 = 16.69$	None	—	$UCL_2 = 364.25$	19
1b. S	$LCL_1 = 3.71$	None	—	$UCL_1 = 17.35$	None
	$LCL_2 = 4.09$	None	—	$UCL_2 = 19.09$	19
2. S	$LCL_1 = 2.81$	None	$CL_1 = 9.73$	$UCL_1 = 16.65$	None
	$LCL_2 = 3.09$	None	$CL_2 = 10.70$	$UCL_2 = 18.32$	19
3. V	$LCL_1 = 2.62$	None	$CL_1 = 9.23$	$UCL_1 = 15.84$	None
	$LCL_2 = 2.88$	None	$CL_2 = 10.15$	$UCL_2 = 17.42$	19
(b)	Control Limits for Multivariate Dispersion Charts (Phase II)				
1. W	—	—	—	$UCL = 12.67$	None
2. $|S|$	$LCL = 512.87$	None	—	$UCL = 31,348.95$	19, 20
3. $|S|$	$LCL = 0.00$	None	6884.42	$UCL = 21,885.92$	17, 19, 20

given by $\bar{\mathbf{x}} = (1/m)\sum_{i=1}^{m}\bar{\mathbf{x}}_i$ and $\bar{\mathbf{S}} = (1/m)$ $\sum_{i=1}^{m}\mathbf{S}_i$. Alt et al. [8] show that $(\bar{\mathbf{X}}_i - \bar{\bar{\mathbf{X}}})'$ $\bar{\mathbf{S}}^{-1}(\bar{\mathbf{X}}_i - \bar{\bar{\mathbf{X}}})$ is distributed as $c(m, n, p)$ $\times F_{p,mn-m-p+1}$, where

$$c(m, n, p) = \frac{p(m-1)(n-1)}{n(mn-m-p+1)}. \quad (6)$$

Thus the stability of the process during Stage 1 is tested by plotting the value of $(\bar{\mathbf{x}}_i - \bar{\bar{\mathbf{x}}})'\bar{\mathbf{S}}^{-1}(\bar{\mathbf{x}}_i - \bar{\bar{\mathbf{x}}})$, for $i = 1, \ldots, m$, on a control chart with

$$\text{UCL} = c(m, n, p)F_{p,mn-m-p+1,\alpha} \quad \text{and}$$

$$\text{LCL} = 0.$$

$F_{v_1,v_2,\alpha}$ denotes the F-percentile* with v_1 and v_2 degrees of freedom. If one or more of the initial m subgroups is out of control, the Stage I control limits are recalculated on the basis of the remaining subgroups. This process is repeated as often as necessary until the test statistics for the remaining subgroups fall within the control limits. When this occurs, a second-stage control chart is introduced for determining whether the process remains in control when *future* subgroups are analyzed.

Let $\bar{\mathbf{X}}_f$ denote the $(p \times 1)$ vector of sample means for a future subgroup. Then we have that $(\bar{\mathbf{X}}_f - \bar{\bar{\mathbf{X}}})'\bar{\mathbf{S}}^{-1}(\bar{\mathbf{X}}_f - \bar{\bar{\mathbf{X}}})$ is distributed as $c^*(m, n, p)F_{p,mn-m-p+1}$, where

$$c^*(m, n, p) = \frac{p(m+1)(n-1)}{n(mn-m-p+1)}. \quad (7)$$

Consequently, the control of the process mean during Stage 2 is tested by plotting the values of $(\bar{\mathbf{x}}_f - \bar{\bar{\mathbf{x}}})'\bar{\mathbf{S}}^{-1}(\bar{\mathbf{x}}_f - \bar{\bar{\mathbf{x}}})$ for future subgroups on a control chart with UCL = $c^*(m, n, p)F_{p,mn-m-p+1,\alpha}$ and LCL = 0. A suggested policy is to update the Stage 2 control limits and, of course, $\bar{\mathbf{x}}$ and $\bar{\mathbf{S}}$ fairly often in the beginning, with less updating after the process has stabilized.

The use of these Stage 1 and Stage 2 procedures is illustrated via the example presented in Alt et al. [7]. Refer to Table 6. This data could represent measurements on opacity and gloss for a hypothetical plastic film extrusion process. For Stage I, it was decided to let $m = 5$ and $\alpha = 0.001$. Since sam-

ples of size 10 were used, $c(m, n, p) = 0.1636$ and UCL = 1.3268. The entries in Table 6 reveal that the statistic for subgroup 4 is above the UCL. An assignable cause was found, subgroup 4 was eliminated, and the Stage 1 control limits were revised. The test statistics for the remaining four initial subgroups plot in control. In the analysis of future subgroups (Stage 2), α was set equal to 0.005. Since $m = 4$, $n = 10$, and $p = 2$, UCL = 1.5906. It is now apparent that the test statistic for the third future subgroup ($f = 3$, $i = 8$) plots out of control. Again, an assignable cause was sought. After 10 test statistics plotted in control, the Stage 2 control limits, together with $\bar{\mathbf{x}}$ and $\bar{\mathbf{S}}$, were updated.

The Stage 1 control charts for the mean presented in this section, and variations of them, are sometimes referred to as (Hotelling's) T^2-charts (*see* HOTELLING'S T^2). Other uses of the T^2 statistic in a control chart setting are presented in Alt [4].

CONTROL CHARTS FOR PROCESS DISPERSION (PHASE I)

First consider the case of one quality characteristic. Suppose data is available from m rational subgroups of n observations each. To determine if process variability was the same across the m subgroups, a control chart can be developed by using the reasonable procedure of replacing the standard value σ_0 in Phase II control charts by an unbiased estimate. Consider the sigma chart (Table 2, Case 3). Define $\bar{V} = (1/m)\sum_{i=1}^{m}V_i$, where V_i is defined in Table 2. Since $\bar{V}/c_2$ is an unbiased estimator of σ_0, we replace σ_0 by $\bar{V}/c_2$. Thus LCL = $\max\{0, \bar{V}B_3\}$ and UCL = $\bar{V}B_4$. Tables of the B_3 and B_4 factors are in Duncan [13].

The multivariate Phase II dispersion charts were presented in Table 3. Since the standard value $|\Sigma_0|$ is now unknown, an unbiased estimate is used in its place. Define $|\mathbf{S}^*| = (1/m)\sum_{i=1}^{m}|\mathbf{S}_i|$, where $\mathbf{S}_i$ is the sample variance-covariance matrix for subgroup i, $i = 1, \ldots, m$. Since $E(|\mathbf{S}_i|) = b_1|\Sigma_0|$, an

Table 6   An Example of Control Charts for the Mean (Phase I)

(a)   Stage 1

i	$\bar{x}_i$	S_i		$\bar{\bar{x}}$ and $\bar{S}$ based on subgroups 1, 2, 3, 4, and 5		
1	3.028	0.009	0.011	$\bar{\bar{x}}$	$\bar{S}$	
	7.978	0.011	0.034			
2	3.025	0.012	0.011	3.036	0.009	0.007
	8.089	0.011	0.017	7.989	0.007	0.017
3	3.031	0.013	0.008			
	8.022	0.008	0.011	$\bar{\bar{x}}$ and $\bar{S}$ based only on subgroups 1, 2, 3, and 5.		
4	3.072	0.003	−0.0003	$\bar{\bar{x}}$	$\bar{S}$	
	7.796	−0.0003	0.007			
5	3.025	0.009	0.007	3.027	0.011	0.009
	8.059	0.007	0.015	8.037	0.009	0.019

Subgroup Number i	Value of Test Statistic $(\bar{x}_i - \bar{\bar{x}})'\bar{S}^{-1}(\bar{x}_i - \bar{\bar{x}})$	Stage 1 UCL	Revised Value of Test Statistic	Revised Stage 1 UCL
1	0.009	1.3268	0.315	0.9546
2	1.102	1.3268	0.258	0.9546
3	0.130	1.3268	0.031	0.9546
4	4.617*	1.3268	Subgroup 4 is not used.	
5	0.586	1.3268	0.051	0.9546

(b)   Stage 2

f	i	$\bar{x}_f$	S_f		$\bar{\bar{x}}$ and $\bar{S}$ based on subgroups 1, 2, 3, 5, 6, 7, 9, 10, 11, and 12		
1	6	3.019	0.014	0.011	$\bar{\bar{x}}$	$\bar{S}$	
		8.094	0.011	0.015			
2	7	3.003	0.017	0.016	3.016	0.011	0.009
		7.977	0.016	0.021	8.022	0.009	0.018
3	8	2.887	0.015	0.018			
		8.085	0.018	0.035			
4	9	3.033	0.005	0.003			
		7.998	0.003	0.009			
5	10	3.070	0.009	0.007			
		8.052	0.007	0.016			
6	11	2.964	0.009	0.010			
		8.001	0.010	0.026			
7	12	2.967	0.013	0.009			
		7.948	0.009	0.012			

Subgroup Number i	Value of Test Statistic $(\bar{x}_f - \bar{\bar{x}})'\bar{S}^{-1}(\bar{x}_f - \bar{\bar{x}})$	Stage 2 UCL
6	0.370	1.5906
7	0.190	1.5906
8	4.350*	1.5906
9	0.174	1.5906
10	0.212	1.5906
11	0.402	1.5906
12	0.460	1.5906

unbiased estimate of $|\Sigma_0|$ is $|S^*| \div b_1$, where b_1 is defined in equation (5). It seems reasonable to replace $|\Sigma_0|$ in Table 3 by $(|S^*| \div b_1)$. This substitution yields the corresponding Phase I multivariate dispersion control charts.

RESEARCH DIRECTIONS

Research in multivariate quality control can extend in many directions. For example, the cumulative sum control chart* (CSCC) is being used widely and appears to have certain advantages over the Shewhart chart, especially in the detection of a small shift. Although the properties of the CSCC have been thoroughly investigated for one quality characteristic, their development and properties for multivariate data is one important goal.

In the multivariate case, separate control charts were used to monitor the process mean and dispersion. This parallels the treatment for one quality characteristic. The development of one control chart for the *simultaneous* monitoring of both location and dispersion is needed. Determination of the economic factors (control chart constant, sampling interval, and sample size) for this chart is yet another direction for research.

References

[1] Alt, F. B. (1973). "Aspects of Multivariate Control Charts." M.S. thesis, Georgia Institute of Technology, Atlanta.

[2] Alt, F. B. (1976). *Manag. Sci.*, **22**, 1167–1168.

[3] Alt, F. B. (1977). "Economic Design of Control Charts for Correlated, Multivariate Observations." Ph.D. dissertation, Georgia Institute of Technology, Atlanta, GA.

[4] Alt, F. B. (1982). *Ann. Tech. Conf. Trans. ASQC*, 886–893.

[5] Alt, F. B. and Deutsch, S. J. (1978). *Proc. Seventh Ann. Meeting, Northeast Regional Conf., Amer. Inst. Decision Sci.*, 109–112.

[6] Alt, F. B., Deutsch, S. J., and Walker, J. W. (1977). *Ann. Tech. Conf. Trans., ASQC*, 360–369.

[7] Alt, F. B., Goode, J. J., Montgomery, D. C., and Wadsworth, H. M. (1973). "Variables Control Charts for Multivariable Data." *Proceedings of the 133rd Annual Meeting of the American Statistical Association.*

[8] Alt, F. B., Goode, J. J., and Wadsworth, H. M. (1976). *Ann. Tech. Conf. Trans., ASQC*, 170–176.

[9] Alt, F. B., Walker, J. W., and Goode, J. J. (1980). *Ann. Tech. Conf. Trans., ASQC*, 754–759.

[10] Anderson, T. W. (1958). *An Introduction to Multivariate Statistical Analysis*. Wiley, New York.

[11] Berger, R. L. (1982). *Technometrics*, **24**, 295–300.

[12] Chapman, S. C., Schmidt, J. W., and Bennett, G. K. (1978). *Naval Res. Logist. Quart.*, **25**, 633–651.

[13] Duncan, A. J. (1974). *Quality Control and Industrial Statistics*, 4th ed. Richard D. Irwin, Homewood, IL.

[14] Ghare, P. M. and Torgersen, P. E. (1968). *J. Ind. Eng.*, **19**, 269–272.

[15] Gnanadesikan, M. and Gupta, S. S. (1970). *Technometrics*, **12**, 103–117.

[16] Guttman, I. and Wilks, S. S. (1965). *Introductory Engineering Statistics*, Wiley, New York.

[17] Hillier, F. S. (1969). *J. Qual. Tech.*, **1**, 17–26.

[18] Hotelling, H. (1947). In *Techniques of Statistical Analysis*, C. Eisenhart, H. Hastay, and W. A. Wallis, eds. McGraw-Hill, New York, pp. 111–184.

[19] Hotelling, H. (1951). *Proceedings of the Second Berkeley Symposium on Mathematical Statistics and Probability*, University of California Press, Berkeley, CA pp. 23–41.

[20] Jackson, J. E. (1956). *Ind. Qual. Control*, **12**(7) 2–6.

[21] Jackson, J. E. (1959). *Technometrics*, **1**, 359–377.

[22] Jackson, J. E. and Bradley, R. A. (1959). "The Development of Statistical Methods for Experimental Design in Quality Control and Surveillance Testing." *Tech. Rep. No. 10*, Virginia Polytechnic Institute, Blacksburg, VA.

[23] Jackson, J. E. and Bradley, R. A. (1961). *Technometrics*, **3**, 519–534.

[24] Jackson, J. E. and Bradley, R. A. (1961). *Ann. Math. Statist.*, **32**, 1063–1077.

[25] Jackson, J. E. and Bradley, R. A. (1966). In *Multivariate Analysis*, Vol. I, P. R. Krishnaiah, ed. Academic Press, New York, pp. 507–518.

[26] Jackson, J. E. and Morris, R. H. (1957). *J. Amer. Statist. Ass.*, **52**, 186–199.

[27] Johnson, N. L. and Leone, F. C. (1977). *Statistics and Experimental Design in Engineering and the Physical Sciences*, Vol. 1, 2nd ed., Wiley, New York.

[28] Johnson, R. A. and Wichern, D. W. (1982). *Applied Multivariate Statistical Analysis*. Prentice-Hall, Englewood Cliffs, NJ.

[29] Korin, B. P. (1968). *Biometrika*, **55**, 171–178.

[30] Miller, R. G. (1966). *Simultaneous Statistical Inference*. McGraw-Hill, New York.

[31] Montgomery, D. C. and Klatt, P. J. (1972). *Manag. Sci.*, **19**, 76–89.

[32] Montgomery, D. C. and Klatt, P. J. (1972). *AIIE Trans.*, **4**, 103–110.

[33] Montgomery, D. C. and Wadsworth, H. M. (1972). *Ann. Tech. Conf. Trans.*, ASQC, 427–435.

[34] Page, E. S. (1954). *J. R. Statist. Soc. B*, **16**, 131–135.

[35] Patel, H. I. (1973). *Technometrics*, **15**, 103–112.

[36] Shakun, M. F. (1965). *J. Amer. Statist. Ass.*, **60**, 905–913.

[37] Yang, C.-H. and Hillier, F. S. (1970). *J. Qual. Tech.*, **2**, 9–16.

(ACCEPTANCE SAMPLING
CONTROL CHARTS
QUALITY CONTROL, STATISTICAL
SAMPLING PLANS)

FRANK B. ALT

MULTIVARIATE RATIO ESTIMATORS

See RATIO ESTIMATORS

MULTIVARIATE SKEWNESS AND KURTOSIS

Multivariate skewness and kurtosis are generalizations of (univariate) skewness* and kurtosis*, the standardized third and fourth moments, to multivariate distributions and samples. Other measures of univariate asymmetry, such as Pearson's (mean − mode)$/\sigma$, are also referred to as skewness (*see* MEAN, MEDIAN, MODE AND SKEWNESS), and can be generalized similarly.

Let F denote an arbitrary p-dimensional distribution, μ its $p \times 1$ mean vector, and Σ its $p \times p$ covariance matrix. Let $X_1, \ldots, X_n$ denote a set of $p \times 1$ observations whose sample mean vector and covariance matrix are

$$\overline{X} = \frac{1}{n} \sum_{i=1}^{n} X_i,$$

$$S = \frac{1}{n} \sum_{i=1}^{n} (X_i - \overline{X})(X_i - \overline{X})'.$$

In many of the following computations, n must exceed p. All expectations $E(\cdot)$ appearing are assumed to exist.

Mardia [5, 6] defined the *multivariate skewness* $\beta_{1,p}$ and *kurtosis* $\beta_{2,p}$ of distribution F as

$$\beta_{1,p} = E\left\{ \left[(X - \mu)' \Sigma^{-1}(Y - \mu) \right]^3 \right\},$$

$$\beta_{2,p} = E\left\{ \left[(X - \mu)' \Sigma^{-1}(X - \mu) \right]^2 \right\}$$

where X and Y are independent $p \times 1$ random vectors with this distribution. He defined the *multivariate sample skewness* $b_{1,p}$ and *kurtosis* $b_{2,p}$ of the set of observations $X_1, \ldots, X_n$ as

$$b_{1,p} = \frac{1}{n^2} \sum_{i,j=1}^{n} \left[(X_i - \overline{X})' S^{-1}(X_j - \overline{X}) \right]^3,$$

$$b_{2,p} = \frac{1}{n} \sum_{i=1}^{n} \left[(X_i - \overline{X})' S^{-1}(X_i - \overline{X}) \right]^2.$$

For any nonsingular $p \times p$ matrix A and any $p \times 1$ vector D, $b_{1,p}$, and $b_{2,p}$ are invariant* under the affine transformation $AX + D$ of the sample; $\beta_{1,p}$ and $\beta_{2,p}$ are also invariant under this transformation. When the dimension p is 1, $\beta_{1,p}$ and $b_{1,p}$ reduce to the squares of the usual univariate population skewness $\sqrt{\beta_1}$ and sample skewness $\sqrt{b_1}$. In addition, $\beta_{2,p}$ and $b_{2,p}$ reduce to the usual univariate population kurtosis β_2 and sample kurtosis b_2. When $p = 2$ and $\Sigma = I$,

$$\beta_{1,p} = \mu_{30}^2 + \mu_{03}^2 + 3\mu_{12}^2 + 3\mu_{21}^2,$$

$$\beta_{2,p} = \mu_{04} + \mu_{40} + 2\mu_{22}.$$

Similar but more complex expressions are available for $p > 2$, $\Sigma \neq I$ (see ref. 5).

The skewness of any distribution symmetric about its mean is $\beta_{1,p} = 0$. Thus the p-dimensional multivariate normal $N(\mu, \Sigma)$ distribution has skewness $\beta_{1,p} = 0$. For a random sample from this distribution, the statistic $nb_{1,p}/6$ has an asymptotic chi-square distribution* with $p(p + 1)(p + 2)/6$ degrees of freedom. An improved version of this result follows from the mean of $b_{1,p}$ under normal sampling [6].

The p-dimensional multivariate normal $N(\mu, \Sigma)$ distribution has kurtosis $\beta_{2,p} = p(p + 2)$. For a random sample from this distribution, the statistic

$$\frac{b_{2,p} - p(p+2)(n-1)(n+1)^{-1}}{\left\{8p(p+2)n^{-1}\right\}^{1/2}}$$

is asymptotically normal $(0, 1)$. (*See* ASYMP-TOTIC NORMALITY.) An improved version of this result follows from the mean and variance of $b_{2,p}$ under normal sampling [6].

Mardia [5–7] advocated using his multivariate sample skewness and multivariate sample kurtosis to test for normality. A test based on skewness is given by rejecting the hypothesis of multivariate normality if $b_{1,p}$ is very large. A test based on kurtosis may be performed by rejecting the hypothesis of multivariate normality if $b_{2,p}$ is either very large or very small. To perform these tests, tables of critical points of the distributions of $b_{1,p}$ and $b_{2,p}$ under normal sampling are necessary for small to moderately large values of n. Tables for $p = 2$ and selected values of n from 10 to 5000, produced by Monte Carlo simulations* and smoothing, appear in Mardia [6], where recommendations for the case of $p > 2$ are also found. For extremely large n, critical points of $b_{1,p}$ and $b_{2,p}$ can be approximated from their asymptotic behavior. Tests of normality that use both $b_{1,p}$ and $b_{2,p}$ have also been suggested. (*See* MARDIA'S TEST OF MULTINORMALITY.) Schwager and Margolin [10] showed that rejecting the null hypothesis whenever $b_{2,p}$ is sufficiently large gives the locally best* invariant test of H_0: The data are a multivariate normal random sample vs. H_1: There are some outliers resulting from mean slippage*.

The magnitude of the multivariate skewness $\beta_{1,p}$ and the extent to which the multivariate kurtosis $\beta_{2,p}$ differs from $p(p+2)$ are measures of the nonnormality of a distribution. These can be useful in robustness* studies. Nonnormality reflected by $\beta_{1,p}$ affects the size of Hotelling's T^2* test, while nonnormality reflected by $\beta_{2,p}$ does not appear to have much impact on the size of this test. In contrast, nonnormality reflected by $\beta_{2,p}$ affects the size of the normal theory likelihood ratio test* for equal covariance

matrices in several populations, which does not seem to be influenced much by nonnormality reflected in $\beta_{1,p}$ [5–7].

Algorithms for computing $b_{1,p}$ and $b_{2,p}$ were given by Mardia and Zemroch [8]. Gnanadesikan [1, Chap. 5] used $b_{1,p}$ and $b_{2,p}$ in analyzing several data sets.

Malkovich and Afifi [4] introduced different definitions of multivariate skewness and kurtosis, based on Roy's union-intersection principle*. If X has distribution F, then for any nonzero $p \times 1$ vector C, the scalar variable $C'X$ has squared skewness

$$\beta_1(C) = \left\{ E\left[(C'X - C'\mu)^3 \right] \right\}^2 / (C'\Sigma C)^3.$$

The multivariate skewness of the distribution of X is defined as

$$\beta_1^M = \max_C \beta_1(C),$$

the largest squared skewness produced by any projection of the p-dimensional distribution onto a line. Similarly, the scalar variable $C'X$ has kurtosis

$$\beta_2(C) = E\left[(C'X - C'\mu)^4 \right] / (C'\Sigma C)^2.$$

The multivariate kurtosis of the distribution of X is defined as

$$\beta_2^M = \max_C |\beta_2(C) - 3|;$$

this is the greatest deviation from 3, the kurtosis of the univariate normal distribution, produced by any projection of the p-dimensional distribution onto a line. The multivariate normal $N(\mu, \Sigma)$ distribution has $\beta_1^M = 0$ and $\beta_2^M = 0$, since every scalar variable $C'X$ is univariate normal, so $\beta_1(C) = 0$ and $\beta_2(C) = 3$ for every C.

For a sample $X_1, \ldots, X_n$ and any nonzero $p \times 1$ vector C, the square of the sample skewness of the scalars $C'X_1, \ldots, C'X_n$ is

$$b_1(C) = \frac{n\left[\sum_{i=1}^n (C'X_i - C'\overline{X})^3 \right]^2}{\left[\sum_{i=1}^n (C'X_i - C'\overline{X})^2 \right]^3}.$$

The multivariate sample skewness of $X_1, \ldots, X_n$ is [4]

$$b_1^M = \max_C b_1(C).$$

Similarly, the sample kurtosis of $C'X_1, \ldots,$

$\mathbf{C}'\mathbf{X}_n$ is

$$b_2(\mathbf{C}) = \frac{n\sum_{i=1}^{n}\left(\mathbf{C}'\mathbf{X}_i - \mathbf{C}'\overline{\mathbf{X}}\right)^4}{\left[\sum_{i=1}^{n}\left(\mathbf{C}'\mathbf{X}_i - \mathbf{C}'\overline{\mathbf{X}}\right)^2\right]^2}.$$

A union-intersection test* of multivariate normality based on kurtosis is given by rejecting the hypothesis of normal random sampling whenever $b_2(\mathbf{C})$ is far from 3 for any $\mathbf{C}$. The multivariate sample kurtosis of $\mathbf{X}_1, \ldots, \mathbf{X}_n$ is [4]

$$b_2^M = \max_{\mathbf{C}}|b_2(\mathbf{C}) - K|,$$

where the constant K is chosen to equalize, under the hypothesis of multivariate normality, the probabilities of rejecting this hypothesis because $\min_{\mathbf{C}}b_2(\mathbf{C})$ is very small and because $\max_{\mathbf{C}}b_2(\mathbf{C})$ is very large. As n increases, $K \to 3$.

For any p, b_1^M and b_2^M are invariant under nonsingular affine transformations $\mathbf{AX} + \mathbf{D}$, as are β_1^M and β_2^M. When the dimension p is 1, β_1^M and b_1^M reduce to the squares β_1 and b_1 of the usual population and sample skewness. Also, $\beta_2^M = |\beta_2 - 3|$ and $b_2^M = |b_2 - K|$, where β_2 and b_2 are the usual population and sample kurtosis.

Malkovich and Afifi proposed using their multivariate sample skewness and kurtosis to test for multivariate normality. A union-intersection test based on skewness is given by rejecting the null hypothesis of normal random sampling whenever b_1^M is very large. The analogous test based on kurtosis was discussed in defining b_2^M. The maximization and evaluation of K required to calculate b_1^M and b_2^M involve computations whose difficulty increases with p.

Isogai [2] extended Pearson's measure of univariate skewness, $(\text{mean} - \text{mode})/\sigma$, to multivariate distributions and samples. He defined a measure τ_p of the multivariate skewness of the distribution F,

$$\tau_p = (\mu - \theta)'\omega^{-1}(\Sigma)(\mu - \theta),$$

where θ is the mode of F and $\omega(\Sigma)$ is a specified $p \times p$ function of Σ, possibly equal to Σ itself. The multivariate sample skewness of $\mathbf{X}_1, \ldots, \mathbf{X}_n$ is

$$t_p = (\overline{\mathbf{X}} - \hat{\theta})'\omega^{-1}(\mathbf{S})(\overline{\mathbf{X}} - \hat{\theta}),$$

where the sample mode $\hat{\theta}$ is obtained by density estimation* with an appropriate kernel function. When $\mathbf{X}_1, \ldots, \mathbf{X}_n$ are a random sample from a multivariate normal distribution, t_p is asymptotically distributed as a linear combination of p independent χ_1^2 variables. Isogai suggested using t_p to test for multivariate normality.

Neither t_p nor τ_p is invariant under nonsingular affine transformations. If the dimension p is 1 and ω is the identity function, then τ_p reduces to the square of Pearson's measure of population skewness, and t_p to its sample analog. These are not directly related to β_1 and b_1 in general; however, a relationship among τ_p ($p = 1$), β_1, and β_2 holds for any distribution in the Pearson system* [3, pp. 85 and 149].

Oja [9] defined multivariate skewness and kurtosis by considering the volume of the simplex in p-dimensional space determined by $p + 1$ points $\mathbf{x}_1, \ldots, \mathbf{x}_{p+1}$. Let $\mathbf{x}_i = (x_{i1}, x_{i2}, \ldots, x_{ip})'$ for $i = 1, \ldots, p + 1$. The volume of this simplex is

$$\Delta(\mathbf{x}_1, \ldots, \mathbf{x}_{p+1})$$

$$= \frac{1}{p!}\left|\det\begin{bmatrix} 1 & 1 & \cdots & 1 \\ x_{11} & x_{21} & \cdots & x_{p+1,1} \\ \vdots & \vdots & & \vdots \\ x_{1p} & x_{2p} & \cdots & x_{p+1,p} \end{bmatrix}\right|.$$

A $p \times 1$ measure of location for the distribution F is given by μ_α ($0 < \alpha < \infty$) satisfying

$$E\left\{\Delta(\mathbf{X}_1, \ldots, \mathbf{X}_p, \mu_\alpha)^\alpha\right\}$$
$$= \min_\lambda E\left\{\Delta(\mathbf{X}_1, \ldots, \mathbf{X}_p, \lambda)^\alpha\right\},$$

where $\mathbf{X}_1, \ldots, \mathbf{X}_p$ are independent $p \times 1$ random vectors with this distribution. The mean μ of F equals μ_2, and μ_1 can be used to define the generalized median of F. Oja defined the distribution's multivariate skewness as

$$\eta_p = \frac{E\left\{\Delta(\mathbf{X}_1, \ldots, \mathbf{X}_{p-1}, \mu_1, \mu_2)\right\}}{\left(E\left\{\Delta(\mathbf{X}_1, \ldots, \mathbf{X}_p, \mu_2)^2\right\}\right)^{1/2}}$$

and its multivariate kurtosis as

$$\beta_{2,p}^* = \frac{E\left\{\Delta(\mathbf{X}_1, \ldots, \mathbf{X}_p, \mu_2)^4\right\}}{\left(E\left\{\Delta(\mathbf{X}_1, \ldots, \mathbf{X}_p, \mu_2)^2\right\}\right)^2},$$

where $\mathbf{X}_1, \ldots, \mathbf{X}_p$ are independent vectors with this distribution.

For a sample $\mathbf{X}_1, \ldots, \mathbf{X}_n$, any $\boldsymbol{\mu}_\alpha$ $(0 < \alpha < \infty)$ may be estimated by solving the equation

$$\sum \left[\Delta\left(\mathbf{X}_{i_1}, \ldots, \mathbf{X}_{i_p}, \hat{\boldsymbol{\mu}}_\alpha\right)\right]^\alpha$$
$$= \min_\lambda \sum \left[\Delta\left(\mathbf{X}_{i_1}, \ldots, \mathbf{X}_{i_p}, \lambda\right)\right]^\alpha,$$

where summation is over $1 \leqslant i_1 < i_2 < \cdots < i_p \leqslant n$. This gives the sample mean $\overline{\mathbf{X}}$ as $\hat{\boldsymbol{\mu}}_2$. The sample median $\hat{\boldsymbol{\mu}}_1$ may be a single point, but may be a convex set from which the median can be chosen. Oja defined the multivariate sample skewness of $\mathbf{X}_1, \ldots, \mathbf{X}_n$ as

$$h_p = \left(\begin{array}{c} n \\ p-1 \end{array}\right)^{-1}$$
$$\times \sum \frac{\Delta\left(\mathbf{X}_{i_1}, \ldots, \mathbf{X}_{i_{p-1}}, \hat{\boldsymbol{\mu}}_1, \overline{\mathbf{X}}\right)}{v_2^{1/2}},$$

and the multivariate sample kurtosis as

$$b_{2,p}^* = v_4 / v_2^2,$$

where

$$v_j = \left(\begin{array}{c} n \\ p \end{array}\right)^{-1} \sum \left[\Delta\left(\mathbf{X}_{i_1}, \ldots, \mathbf{X}_{i_p}, \overline{\mathbf{X}}\right)\right]^j,$$

with summation over $1 \leqslant i_1 < i_2 < \cdots < i_{p-1} \leqslant n$ in the numerator of h_p and over $1 \leqslant i_1 < \cdots < i_p \leqslant n$ for each v_j.

For any dimension p, h_p and $b_{2,p}^*$ are invariant under nonsingular affine transformations, as are η_p and $\beta_{2,p}^*$. When p is 1, $\beta_{2,p}^* = \beta_2$ and $b_{2,p}^* = b_2$; in addition, η_p reduces to $|\mu - \mu_1|/\sigma$, where μ, μ_1, and σ are the mean, median, and standard deviation of the distribution F, and h_p reduces to the sample analog of this quantity. These are not directly related to β_1 and b_1.

References

[1] Gnanadesikan, R. (1977). *Methods for Statistical Data Analysis of Multivariate Observations*. Wiley, New York.

[2] Isogai, T. (1982). *Ann. Inst. Statist. Math.*, **34**, A, 531–541.

[3] Kendall, M. G. and Stuart, A. (1969). *The Advanced Theory of Statistics*, 3rd ed., Vol. 1. Hafner, New York.

[4] Malkovich, J. F. and Afifi, A. A. (1973). *J. Amer. Statist. Ass.*, **68**, 176–179.

[5] Mardia, K. V. (1970). *Biometrika*, **57**, 519–530.

[6] Mardia, K. V. (1974). *Sankhyā B*, **36**, 115–128.

[7] Mardia, K. V. (1975). *Appl. Statist.*, **24**, 163–171.

[8] Mardia, K. V. and Zemroch, P. J. (1975). *Appl. Statist.*, **24**, 262–265.

[9] Oja, H. (1983). *Statist. Prob. Lett.*, **1**, 327–332.

[10] Schwager, S. J. and Margolin, B. H. (1982). *Ann. Statist.*, **10**, 943–954.

(DEPARTURES FROM NORMALITY,
 TESTS FOR
KURTOSIS
MARDIA'S TEST OF MULTINORMALITY
MULTIVARIATE NORMALITY,
 TESTING FOR SKEWNESS)

STEVEN J. SCHWAGER

MULTIVARIATE STABLE DISTRIBUTIONS

The term stable distribution is usually associated with limiting distributions when one operation such as summation or taking extremes, and others, is applied to a sequence of random variables, including vector variables. The present article is limited to the component by component summation of random vectors. For multivariate extremal stable distributions, *see* MULTIVARIATE ORDER STATISTICS.

Let $\mathbf{X}_j = (X_j^{(1)}, X_j^{(2)}, \ldots, X_j^{(d)})$, $1 \leqslant j \leqslant n$, be independent and identically distributed d-dimensional random vectors. Put $\mathbf{S}_n = (S_n^{(1)}, S_n^{(2)}, \ldots, S_n^{(d)})$, where $S_n^{(t)} = X_1^{(t)} + X_2^{(t)} + \cdots + X_n^{(t)}$, $1 \leqslant t \leqslant d$.

Definition 1. Assume that there are constant vectors $\mathbf{a}_n = (a_n^{(1)}, a_n^{(2)}, \ldots, a_n^{(d)})$ and $\mathbf{b}_n = (b_n^{(1)}, b_n^{(2)}, \ldots, b_n^{(d)})$ with $b_n^{(t)} > 0$, $1 \leqslant t \leqslant d$, such that the limit $F(x_1, x_2, \ldots, x_d)$ of the distribution of $(\mathbf{S}_n - \mathbf{a}_n)/\mathbf{b}_n$, where arithmetic operations are component by component, exists at each continuity point of F, and F is not concentrated at a single point. Then we call F a *d-dimensional stable distribution*.

By standard techniques of the theory of (weak) convergence* of distributions, it easily follows that Definition 1 is equivalent to the following property:

Definition 2. The common distribution function $F(x_1, x_2, \ldots, x_d)$ of the vectors $\mathbf{X}_j$, $1 \leqslant j \leqslant n$, is stable if it is not concentrated at a single point and if, for every $n \geqslant 1$, there are constant vectors $\mathbf{a}_n$ and $\mathbf{b}_n$ with $b_n^{(t)} > 0$, $1 \leqslant t \leqslant d$, such that the distribution function of $(\mathbf{S}_n - \mathbf{a}_n)/\mathbf{b}_n$ is F itself.

Definition 1 implies that every univariate marginal of a stable distribution is (univariate) stable (if not degenerated at a single point) (*see* STABLE DISTRIBUTIONS). Another consequence of Definition 1 is that the family of d-dimensional stable distributions includes the d-dimensional multinormal distributions*. That is, when, in Definition 1, the variance V_t of each component $X_1^{(t)}$, $1 \leqslant t \leqslant d$, is finite, the well-known multivariate extension of the classical central limit theorem* yields that the asymptotic distribution of $(\mathbf{S}_n - \mathbf{a}_n)/\mathbf{b}_n$ is normal, where one can choose $a_n^{(t)} = nE(X_1^{(t)})$ and $b_n^{(t)} = (nV_t)^{1/2}$. This simple form of asymptotic normality* of sums, in turn, has a significant implication for stable distributions: a multivariate stable distribution is either normal, and thus all moments* of each univariate marginal are finite, or such that at least one univariate marginal does not have a finite second moment*. This is why the very appealing properties of the normal distribution do not extend to the larger family of stable distributions. (The present article does not deal with properties special to the *normal distribution**, even though the latter belongs to the class of stable distributions.)

There is a large number of applied fields in which multivariate stable distributions are used as either the exact or approximate underlying distribution. In fact, when one faces a component by component sum of vectors with a large number of terms, then, in view of Definition 1, an approximation to its distribution is either normal, or if no normal distribution would fit the data, then it is necessarily another member of the family of stable distributions. For an example of concluding from Definition 2 that the exact underlying distribution is stable, let us look at the following problem of portfolio* management.

Example. Let X and Y be the random prices of the stocks of companies C_1 and C_2, respectively, on a stock exchange. If the price of company C_1 is X_0 at a given time, and if the changes of X over successive periods of time are $\Delta X_1, \Delta X_2, \ldots$, then, after n time units, $X = X_0 + \Delta X_1 + \cdots + \Delta X_n$. Assuming that $\Delta X_1, \Delta X_2, \ldots, \Delta X_n$ are independent and identically distributed, the distribution of $X - X_0$ is the n-fold convolution* of the distribution F of the ΔX_j, but the distribution of $X - X_0$ should be similar to F, since both $X - X_0$ and ΔX_j represent random fluctuation of the same price (over different time units). Hence, by arguing similarly with the vector (X, Y), we get from Definition 2 that the distribution $F(x_1, x_2)$ of (X, Y) is stable. Since empirical studies show that, for several companies, the marginals of $F(x_1, x_2)$ are skewed, and thus not normal, here we face the practical appearance of nonnormal stable distributions. The development of this idea in several papers, particularly in Mandelbrot [11] and Fama [3], increased interest in stable distributions; a good summary of related results can be found in the book by Press [14] (see particularly Chaps. 6 and 12). The mathematical foundations for the multivariate case were laid down by Lévy [10] and Rvaceva [16]. Gleser [4] studied conditions under which the limit in Definition 1 remains stable if the sample size n is a random variable.

DISTRIBUTION THEORY

There are two major difficulties in applying multivariate nonnormal stable distributions. One mentioned earlier, is that second moments are infinite. The other difficulty is that there is hardly any multivariate stable distribution (or density) which is known in closed form. Their characteristic functions*, however, are known in explicit form, from which, through inversion formulas, values of distributions can be computed as well as approximations to densities can be deduced. Let us state the following form of stable characteristic functions. Let the distribution function of the vector $\mathbf{X} = (X^{(1)}, X^{(2)}, \ldots, X^{(d)})$ be stable. Then the characteristic function $(\mathbf{t} = (t_1, t_2, \ldots, t_d))$

$$\phi(\mathbf{t}) = E\{\exp[i(\mathbf{t}, \mathbf{X})]\}$$

has the form

$$\phi(\mathbf{t}) = \exp\{\psi(\mathbf{t})\}$$

with

$$\psi(\mathbf{t}) = i(\mathbf{t}, \mathbf{b}) - c\|\mathbf{t}\|^\alpha \int_{C_d} g_\alpha(\mathbf{t}; \mathbf{z}) \, dG(\mathbf{z}),$$

where $\mathbf{z} = (z_1, z_2, \ldots, z_d)$

where $\mathbf{b} = (b_1, b_2, \ldots, b_d)$ is an arbitrary vector, $c > 0$, $0 < \alpha \leqslant 2$, C_d is the unit d-dimensional cube, G is a measure on C_d, and g_α is an explicitly known function. Here $(\mathbf{x}, \mathbf{y}) = x_1 y_1 + x_2 y_2 + \cdots + x_d y_d$ is the so-called dot product of vectors, and $\|\mathbf{t}\| = (t_1^2 + t_2^2 + \cdots + t_d^2)^{1/2}$; $\mathbf{b}$ is a *location* parameter, c a *scale* parameter, and α the *index* (sometimes the *characteristic exponent*) of the distribution. The value $\alpha = 2$ corresponds to the normal case.

The preceding representation of the characteristic function is essentially due to Lévy [10]. Press [14, Chap. 6 Section 5] transforms the preceding form into one that more closely resembles the familiar form of a normal characteristic function; the quoted form, on the other hand, is more familiar in the light of univariate stable characteristic functions. A more systematic study of the convergence in Definition 1 is given in Rvaceva [16], who characterized all those population distributions (so-called domain of attraction*), for which the limiting distribution of $(\mathbf{S}_n - \mathbf{a}_n)/\mathbf{b}_n$ is a given stable distribution F. In this regard, see also de Haan and Resnick [2].

Kalinauskaite [8] and Press [14] observe that, since $|\phi(\mathbf{t})|$ is integrable, all multivariate stable distributions have a density that is continuously differentiable of all orders. Kalinauskaite ([8] and [9]) gave infinite se-

ries expansions of some multivariate stable densities and observed that a density of Cauchy type has a closed form. Press [14, Chap. 6] also produced a class of densities of Cauchy type as a special case of stable densities. As a matter of fact, he showed that the density

$$f(\mathbf{x}) = K\big[1 + (\mathbf{x} - \mathbf{a})'\Sigma^{-1}(\mathbf{x} - \mathbf{a})\big]^{-(d+1)/2}$$

is stable for every vector $\mathbf{a}$, where K is a suitable constant and Σ is a positive definite matrix. Press calls this density *multivariate Cauchy of order one*. Higher-order Cauchy distributions are obtained through convolutions of Cauchy distributions of order one with differing location vectors $\mathbf{a}$ and scale matrices Σ. Besides these Cauchy densities and the multinormal densities, no multivariate stable density is known in closed form.

We return to discussion of the literature. Paulauskas [13] studied conditions under which the representation for characteristic functions quoted earlier takes special forms. Interestingly, some questions posed by Paulauskas were solved by Holmes et al. [6] by turning to a more abstract class of distributions, termed *operator stable distributions* (the concept is very similar to Definition 1: simply write $(\mathbf{S}_n - \mathbf{a}_n)/\mathbf{b}_n$ as $\mathbf{B}_n\mathbf{S}_n - \mathbf{A}_n$, let the terms X_j of $\mathbf{S}_n$ belong to an abstract vector space, and let $\mathbf{B}_n$ be a linear operator). While the theory of operator stable distributions is fast developing, no statistical application has been found yet (see Brockett [1], Holmes et al. [6], and their references).

One class of multivariate stable distributions can be generated as follows. Let Y_1, $Y_2, \ldots, Y_k$ be independent and identically distributed random variables with (univariate) stable distribution*. Put $X^{(t)} = \sum_{j=1}^k c_{jt} Y_j$, $1 \leqslant t \leqslant d$, where c_{jt} are real numbers. A direct substitution into the quoted form of stable characteristic functions (or an appeal to Definition 1) yields that the distribution of $(X^{(1)}, X^{(2)}, \ldots, X^{(d)})$ is stable. Since so-called moving averages* are linear functions like the preceding ones, this model is widely used in connection with random price levels; it is also fundamental in developing an integral representation of stable random variables in Schilder [17].

STATISTICAL INFERENCE

Three major categories of inference have been treated in the literature: parameter estimation*, goodness-of-fit* tests, and linear regression*. All require special methods due to the lack of closed forms of distribution and, mainly in the case of regression, the lack of finiteness of second moments. These difficulties have not adequately been overcome by the presently available results.

The location parameter $(b_1, b_2, \ldots, b_d)$ (see the preceding section) can be estimated through known univariate methods for the marginals. Such methods are developed by Press [15]; his own methods and others are well treated in [14, Chap. 12]. A critical view of estimating location through the means* is presented in Utts and Hettmansperger [19], and they propose an alternate estimator in terms of rank statistics which is shown to be superior to means, trimmed means, and other robust estimators* in the case of so-called heavy-tailed* marginals (which applies to stable distributions). When estimating the other parameters, and in goodness-of-fit* tests, the multivariate character of the data is not exploited adequately; several methods are simply limited to the marginals only. Exceptions are nonparametric tests and other generally applicable methods such as those suggested by Tanaka [18]. For methods that are particular to stable distributions, see Press [14 Chap. 12].

Turning to linear regression, consider the model in the last paragraph of the preceding section. That is, let $Y_1, Y_2, \ldots, Y_k$ be independent and identically distributed univariate stable random variables. With constants a_j and c_j, define

$$X_1 = a_1 Y_1 + a_2 Y_2 + \cdots + a_k Y_k,$$
$$X_2 = c_1 Y_1 + c_2 Y_2 + \cdots + c_k Y_k.$$

Then (X_1, X_2) is bivariate stable. Problems like estimating u in $E(X_1 \mid X_2) = uX_2$ have received considerable attention in the literature. "Usual" methods of linear regression do not work, because, when the variables are not normal, second moments are infinite. For a theory of linear regression with infinite variances, see Mandelbrot [12], Granger and Orr [5], and Kanter and Steiger [7].

References

[1] Brockett, P. L. (1977). *Ann. Prob.*, **5**, 1012–1017. (A very well-written theoretical rather than statistical paper on infinitely divisible measures on Hilbert space. When the space is limited to *d*-dimensional Euclidean space and infinite divisibility to stability, the author reobtains and extends several results on multivariate stable distributions. See also the references of the paper.)

[2] de Haan, L. and Resnick, S. (1979). *Stoch. Proc. Appl.*, **8**, 349–355. (A newer study of so-called domains of attraction of multivariate stable distributions.)

[3] Fama, E. F. (1965). *J. Bus.*, **38**, 34–105. (A basic paper on analyzing stock prices in the light of stable distributions.)

[4] Gleser, L. J. (1969). *Ann. Math. Statist.*, **40**, 935–941. (Studies conditions under which the limiting distribution of normalized sums of random vectors remains stable when the sample size is a random variable.)

[5] Granger, C. and Orr, D. (1972). *J. Amer. Statist. Ass.*, **67**, 275–285. (Deals with linear regression when the variances are infinite.)

[6] Holmes, J. P., Hudson, W. N., and Mason, J. D. (1982). *Ann. Prob.*, **10**, 602–612. (The structure of operator stable laws is studied. See also the references of the paper for earlier results on this subject matter.)

[7] Kanter, M. and Steiger, W. L. (1974). *Adv. Appl. Prob.*, **6**, 768–783. (An estimator is established for the regression coefficient in a linear model when second moments are infinite.)

[8] Kalinauskaite, N. (1970). *Litov. Mat. Sb.*, **10**, 491–495.

[9] Kalinauskaite, N. (1970). *Litov. Mat. Sb.*, **10**, 727–732. (This, together with the work in ref. 8 is a two-part study of infinite series expansions of multidimensional stable densities.)

[10] Lévy, P. (1937). *Théorie de l'Addition des Variables Aléatoires*. Gauthier-Villars, Paris. (The foundations of the theory of stable distributions, including the multivariate case, are laid down in this monograph. In particular, an integral representation of stable characteristic functions is given.)

[11] Mandelbrot, B. (1963). *J. Bus.*, **36**, 394–419. (This is one of the earliest and most significant papers devoted to the justification of utilizing stable distributions as population distributions. This, and

subsequent papers by this author, are responsible for the active research in connection with stable distributions.)

[12] Mandelbrot, B. (1972). *Ann. Econ. Soc. Measurement*, **1**, 259–290. (This paper can be viewed as the foundation of regression theory when the variables have infinite second moments.)

[13] Paulauskas, V. J. (1976). *J. Multivariate Anal.*, **6**, 356–368. (Analyzes the structure of some multivariate stable distributions.)

[14] Press, S. J. (1972). *Applied Multivariate Analysis*. Holt, Reinhart and Winston, New York. (Its second edition, under the same title, was published by Krieger in 1982. Chapters 6 and 12 summarize the work of the author and others on multivariate stable distributions and their use in the portfolio problem. The section on estimation of parameters is quite extensive and thorough. The lists of references of both editions are very useful.)

[15] Press, S. J. (1972). *J. Amer. Statist. Ass.*, **67**, 842–846. (Deals with the problem of estimation of the parameters of stable distributions.)

[16] Rvaceva, E. L. (1954). *Uch. Zap. Lvov. Gos. In.-ta im. I. Franko Ser. Mech.-Mat.*, **29**, 5–44. Translated into English in *Select. Translat. Math. Statist. Prob.*, **2** (1962), 183–207. (This is the most extensive study of multivariate stable distributions. In particular, the solutions to the problem of domains of attraction are quite general and thorough.)

[17] Schilder, M. (1970). *Ann. Math. Statist.*, **41**, 412–421. (In establishing an integral representation of stable random variables, a special model of multivariate stable distributions is utilized.)

[18] Tanaka, M. (1970). *Ann. Math. Statist.*, **41**, 1999–2020. (Challenges the assumption of asymptotic normality in stochastic models and develops general methods of constructing confidence intervals for multivariate population distributions as well as goodness-of-fit tests on the basis of empirical distribution functions.)

[19] Utts, J. M. and Hettmansperger, T. P. (1980). *J. Amer. Statist. Ass.*, **75**, 939–946. (An estimator of location is proposed in terms of rank statistics, which is shown to be superior to means and trimmed means in the case of heavy-tailed marginals, which includes the stable distributions.)

Bibliography

DeSilva, B. M. (1978). *J. Multivariate Anal.*, **8**, 335–345.

Hendricks, W. J. (1973). *Ann. Prob.*, **1**, 849–853.

Kuelbs, J. (1973). *Zeit. Wahrscheinlichkeitsth. verw. Geb.*, **26**, 259–271.

Miller, G. (1978). *J. Multivariate Anal.*, **8**, 346–360.

Zolotarev, V. M. (1981). In *Contributions to Probability and Statistics*, J. Gani et al., eds. Academic Press, New York, pp. 283–305.

(INFINITE DIVISIBILITY
MULTIVARIATE ORDER STATISTICS
STABLE DISTRIBUTIONS)

JANOS GALAMBOS

MULTIVARIATE STUDENT DISTRIBUTION   *See* MULTIVARIATE *t*-DISTRIBUTION

MULTIVARIATE *t*-DISTRIBUTION

There are several forms of multivariate *t*-distribution* (see Johnson and Kotz [2] for more details). We describe the most common one. A random vector $\mathbf{X} = (X_1, \ldots, X_m)$ has an *m-dimensional t-distribution* (also called *m-dimensional Student distribution*) with ν degrees of freedom, mean vector $\boldsymbol{\mu}$, and *precision matrix* $\mathbf{T}$ if

$$f(\mathbf{x}) = \frac{\Gamma\left(\frac{m+\nu}{2}\right)\sqrt{|\mathbf{T}|}}{\Gamma(\nu/2)\sqrt{2\pi m}}$$
$$\times \left[1 + \frac{1}{\nu}(\mathbf{x} - \boldsymbol{\mu})'\mathbf{T}(\mathbf{x} - \boldsymbol{\mu})\right]^{(-m+\nu)/2},$$

where $\mathbf{T}$ is a positive definite matrix, and $\mathbf{x}'$ is the transpose of vector $\mathbf{x}$. The moments of first and second order of this distribution are

$$\mathbf{E}(\mathbf{X}) = \boldsymbol{\mu}; \qquad \text{Cov}(\mathbf{X}) = \frac{n}{n-2}\mathbf{T}^{-1},$$
$$n > 2.$$

If $\mathbf{Z}$ is an *m*-dimensional multinormal variable with zero mean vector and a nonsingular variance-covariance matrix* $\boldsymbol{\Sigma} = \mathbf{T}^{-1}$, Y is a χ^2 (chi-squared*) random variable with n degrees of freedom, then $\mathbf{X} = (X_1, \ldots, X_m)'$, where $X_i = (\sqrt{n}\, Z_i/\sqrt{Y}) + \mu_i$ $(i = 1, \ldots, m)$ has an *m*-dimensional *t*-distribution with n degrees of freedom, mean vector $\boldsymbol{\mu} = (\mu_1, \ldots, \mu_m)$, and precision matrix $\mathbf{T}$.

If $\mathbf{X}$ is an *m*-dimensional *t* random variable with n degrees of freedom, mean vector $\boldsymbol{\mu}$, and precision matrix $\mathbf{T}$, then the random variable

$$U = (\mathbf{X} - \boldsymbol{\mu})'\mathbf{T}(\mathbf{X} - \boldsymbol{\mu})/m$$

possesses an *F*-distribution* with *m* and *n* degrees of freedom. For an account of related distributional problems, see Nagarsenkar [3]. Chen [1] gives some recent tables of percentage points.

References

[1] Chen, H. J. (1979). *Biom. J.*, **21**, 347–360.

[2] Johnson, N. L. and Kotz, S. (1972). *Distributions in Statistics: Continuous Multivariate Distributions.* Wiley, New York.

[3] Nagarsenkar, B. N. (1975). *Metron*, **33**, 66–74.

(MATRIC *t*-DISTRIBUTION
t-DISTRIBUTION)

MULTIVARIATE UNIMODALITY

Unimodality of distributions is important because it indicates which values of a variable are typical and which ones are significantly different from the norm. For distributions on the line, there is basically just one natural definition of unimodality. In terms of the density *g*, a distribution is called unimodal about a mode *M* if $g(x)$ is nonincreasing as *x* goes away from *M*. A unimodal distribution may have several modes. To make the class of distributions smooth, one also includes limits of distributions of the type mentioned. It was shown by Khintchine [9] that every unimodal distribution on *R* is a limit of mixtures of uniform distributions.

DEFINITIONS AND BASIC RELATIONS

Let *S* be a subset of R^n. Then *S* is said to be *star-shaped* about a point **a** if, for every point **x** in *S*, the line segment [**a**, **x**] is completely contained in *S*. Further, *S* is said to be *centrally symmetric* if $\mathbf{x} \in S \Rightarrow (-\mathbf{x}) \in S$.

Let *P* be the probability distribution of a random vector $\mathbf{X} = (X_1, X_2, \ldots, X_n)$. Assume for simplicity that *P* has a density *f*. One can define the unimodality of *P* (about the origin) in several different ways by requiring that:

1. *f* is maximum at the origin and the restriction of *f* to every line is univariate unimodal, or, equivalently, for every $c > 0$, the set $\{\mathbf{x} : f(\mathbf{x}) \geqslant c\}$ is convex.

2. *f* is maximum at the origin and the restriction of *f* to every line passing through the origin is univariate unimodal, or, equivalently, for every $c > 0$, the set $\{\mathbf{x} : f(\mathbf{x}) \geqslant c\}$ is star-shaped about the origin.

3. The distribution of every linear function $\Sigma a_i X_i$ is univariate unimodal about zero.

In the univariate case, these definitions are equivalent. In higher dimensions $1 \Rightarrow 2$, but there are no other implication relationships among the definitions. Definitions 1, 2, and 3 identify, respectively, *convex unimodal*, *star unimodal*, and *linear unimodal* distributions. Definition 1 was given by Anderson [1], 2 is based on the work of Olshen and Savage [13], and 3 was proposed by Ghosh [8].

Again let *P* and **X** be as earlier. The distribution *P* is called *centrally symmetric* if **X** and $(-\mathbf{X})$ have the same distribution, or, equivalently, if $P(S) = P(-S)$ for all (Borel) subsets *S* or R^n. For centrally symmetric distributions, Definitions 1–3 can be modified in an obvious way and two additional definitions can be based on a paper by Sherman [16]:

4. Call a distribution *central convex unimodal* if it is a limit of mixtures (belongs to the closed convex hull) of the set of all uniform distributions on centrally symmetric convex bodies.

5. Call a distribution *monotone unimodal* if $P(C + k\mathbf{y})$ is univariate unimodal in *k* for every centrally symmetric convex set *C* and every nonzero **y**.

For a unimodal distribution on the line, the probability carried by an interval of fixed length is also a unimodal function of the position of the interval. Further, when this probability becomes a maximum, the interval contains a mode. A multivariate version of this property appears in Definition 5 and forms the basic motivation for Definitions 1 and 4. The multivariate normal distribution with mean vector 0 clearly satisfies definitions 1–4. In a pioneering paper, Anderson [1] proved the following result.

Anderson's Theorem. *Every centrally symmetric convex unimodal distribution is monotone unimodal.*

This theorem was generalized by Sherman [16], who showed that every central convex unimodal distribution is monotone unimodal. A detailed analysis of other implication relationships and preservation of unimodality properties under convolution and marginalization is given by Dharmadhikari and Jogdeo [7].

A notion of ordering closely related to the concept of unimodality is that of peakedness. On the line, a distribution P_1 is said to be *more peaked* (about 0) than a distribution P_2 if P_1 assigns more mass than P_2 to any interval symmetric about 0. With this definition, $N(0, \sigma_1^2)$ is more peaked than $N(0, \sigma_2^2)$ whenever $\sigma_1^2 < \sigma_2^2$. Generalizing this to R^n, Anderson [1] defined P_1 to be *more peaked* than P_2 (about **0**) if $P_1(C) \geqslant P_2(C)$ for every centrally symmetric convex set C. He proved that $N(\mathbf{0}, \boldsymbol{\Sigma}_1)$ is more peaked than $N(\mathbf{0}, \boldsymbol{\Sigma}_2)$ if $\boldsymbol{\Sigma}_2 - \boldsymbol{\Sigma}_1$ is positive definite. Sherman's generalization [16] of an earlier result of Birnbaum [3] implies that for central convex unimodal distributions, ordering by peakedness is preserved under convolutions. That is, if P_1, P_2, Q_1, and Q_2 are central convex unimodal distributions and P_i is more peaked than Q_i, $i = 1, 2$, then $P_1^* P_2$ is more peaked than $Q_1^* Q_2$.

APPLICATIONS

An important application of the concept of unimodality is in proving unbiasedness* of certain tests and the monotonicity of their power functions. In the simplest case, if a random variable X has density $f(x - \theta)$

where f is symmetric and unimodal about 0, then a natural test of $H : \theta = 0$ against $K : \theta \neq 0$ has acceptance region $(-c, c)$. The power function of this test is monotone in $|\theta|$, and, consequently, the test is unbiased. Anderson's theorem enables us to prove similar results in multivariate situations. For example, suppose we have a random sample from $N(\boldsymbol{\mu}, \boldsymbol{\Sigma})$, and the problem is to test $H : \boldsymbol{\mu} = \mathbf{0}$ against $K : \boldsymbol{\mu} \neq \mathbf{0}$. Many tests have been developed for this problem. Since the multivariate normal distribution is centrally symmetric and convex unimodal, Anderson's theorem can be used to prove the monotonicity of power functions of some of these tests whenever their acceptance regions are suitably convex. Following this method, Das Gupta et al. [6] have proved the monotonicity of the power functions of Roy's maximum root test, the Hotelling–Lawley trace* test, the likelihood ratio test, and Pillai's trace* test for the preceding problem. Similar monotonicity results for tests of independence and tests for the homogeneity of covariance matrices have been given, respectively, by Anderson and Das Gupta [2] and by Perlman [14].

Unimodality of distributions has also been used to derive certain probability inequalities. For example, let $(X_1, \ldots, X_n)$ be $N(\mathbf{0}, \boldsymbol{\Sigma})$. Šidák [17] proved that

$$P[\, |X_i| \leqslant c_i, i = 1, \ldots, n\,]$$
$$\geqslant \prod_{i=1}^{n} P[\, |X_i| \leqslant c_i\,],$$

for every $(c_1, \ldots, c_n)$. This result has been used to obtain conservative confidence sets for the mean vector of a normal distribution.

RELATED NOTIONS AND RESULTS

Applications of unimodality result from the important role played by the notions of convexity and central symmetry. Other notions of convexity and symmetry can be used to obtain variants of the concept of unimodality. For example, in the bivariate case, a density $f(x, y)$ is called *Schur concave* if $f(x, y) = f(y, x)$ and $f(c + x, c - x)$ is unimodal in x for every c (see MAJORIZATION). This definition has an n-variate version which requires that, wherever any $(n - 2)$ variables are fixed, the resulting bivariate function should satisfy the conditions stated earlier. A recent book by Marshall and Olkin [11] studies the concept of Schur convexity and its applications. A more general definition of symmetry has been used by Mudholkar [12] to prove an Anderson-type theorem.

A density f on R^n is *log concave* if $\log f$ is concave. Log concave densities are convex unimodal. Moreover, convolutions of log concave densities are again log concave [15]. Convex unimodality and log concavity are special cases of the concept of *s-unimodality*, which requires that

$$f(\Sigma \phi_i x_i) \geqslant \left\{ \Sigma \phi_i \left[f(x_i) \right]^s \right\}^{1/s},$$

where $\phi_i \geqslant 0$ and $\Sigma \phi_i = 1$. Results on s-unimodal densities are given by Borell [4] and Das Gupta [5].

In an important paper, Olshen and Savage [13] define an index of unimodality for distributions on vector spaces, thereby giving a hierarchy of classes of unimodal laws. A density f on R^n is called α-*unimodal* if $t^{n-\alpha} f(t\mathbf{x})$ is nonincreasing in $t > 0$ for every nonzero $\mathbf{x}$. Star unimodality as defined earlier corresponds to the case $\alpha = n$. Olshen and Savage have proved a theorem similar to Anderson's for α-unimodal laws.

Finally, the uniqueness of the maximum likelihood estimate is often established through the unimodality of the likelihood function. For example, for compound multinomial distributions, such a result has been given by Levin and Reeds [10].

References

[1] Anderson, T. W. (1955). *Proc. Amer. Math. Soc.*, **6**, 170–176.

[2] Anderson, T. W. and Das Gupta, S. (1964). *Ann. Math. Statist.*, **35**, 206–208.

[3] Birnbaum, Z. W. (1948). *Ann. Math. Statist.*, **19**, 76–81.

[4] Borell, C. (1975). *Period. Math. Hung.*, **6**, 111–136.

[5] Das Gupta, S. (1977). *Sankhyā B*, **38**, 301–314.

[6] Das Gupta, S., Anderson, T. W., and Mudholkar, G. S. (1964). *Ann. Math. Statist.*, **35**, 200–205.

[7] Dharmadhikari, S. W. and Jogdeo, K. (1976). *Ann. Statist.*, **4**, 607–613.

[8] Ghosh, P. (1974). *Commun. Statist.*, **3**, 567–580.

[9] Khintchine, A. Y. (1938). *Izv. Nauchno-Issled. Inst. Mat. Mech. Tomsk. Gosuniv.*, **2**, 1–7.

[10] Levin, B. and Reeds, J. (1977). *Ann. Statist.*, **5**, 79–87.

[11] Marshall, A. W. and Olkin, I. (1979). *Inequalities: The Theory of Majorization with Applications to Combinatorics, Probability, Statistics and Matrix Theory*. Academic Press, New York.

[12] Mudholkar, G. S. (1966). *Proc. Amer. Math. Soc.*, **17**, 1327–1333.

[13] Olshen, R. A. and Savage, L. J. (1970). *J. Appl. Prob.*, **7**, 21–34.

[14] Perlman, M. D. (1980). *Ann. Statist.*, **8**, 247–263.

[15] Prékopa, A. (1973). *Acta Sci. Math. Szeged.*, **34**, 335–343.

[16] Sherman, S. (1955). *Ann. Math. Statist.*, **26**, 763–766.

[17] Šidák, Z. (1967). *J. Amer. Statist. Ass.*, **62**, 626–633.

(MAJORIZATION
UNBIASEDNESS
UNIMODALITY)

S. W. DHARMADHIKARI
KUMAR JOAG-DEV

MULTIVARIATE WEIBULL DISTRIBUTIONS

Multivariate Weibull distributions are distributions of possibly dependent random variables whose marginals are Weibull distributions*. Applications are found in such diverse fields as engineering* and biology, often in connection with the life lengths of dependent components in systems reliability*.

Many multivariate distributions* have Weibull marginals. Five classes are noted here, including the class $C1$ of distributions of independent Weibull variates. A characteristic property of the one-dimensional Weibull family is its closure under the operation of taking minima. That is, the minimum

$X_{(m)} = \min\{X_1, \ldots, X_n\}$ of independent, identically distributed variates has a Weibull distribution if and only if X_i does (see Property c under WEIBULL DISTRIBUTION). This yields the class $C2$ of joint distributions of minima of various overlapping subsets of a collection of independent Weibull variates. Because $W = (bX^{1/c} + a)$ has a Weibull distribution when the distribution of X is standard exponential (see Property b in WEIBULL DISTRIBUTION), this yields the class $C5$ of distributions generated from $\{X_1^{1/c_1}, \ldots, X_k^{1/c_k}\}$ when $\{X_1, \ldots, X_k\}$ have a multivariate exponential distribution* of any type (cf. Johnson and Kotz [5, p. 269]). See also multivariate gamma distributions* and appropriate sections under multivariate distributions*.

Other joint distributions having Weibull properties are the class $C3$ of multivariate distributions having Weibull minima after arbitrary scaling (i.e., $\min\{a_1X_1, \ldots, a_kX_k\}$ is Weibull for arbitrary positive weights $\{a_1, \ldots, a_k\}$), and the class $C4$ having Weibull minima with unit scaling for all nonempty subsets of $\{X_1, \ldots, X_k\}$. Lee [8] studied these classes of distributions, proposed the classification scheme adopted here, and showed that the inclusions $C1 \subset C2 \subset C3 \subset C4 \subset C5$ are strict.

Bivariate and multivariate Weibull distributions of various types are studied in refs. 1–3 and 5–11. Recent comprehensive treatments are found in [2] and [8]. Related work in [12] deals with properties of joint distributions of minima over sets of random size. Connections are known between particular multivariate Weibull distributions and the multivariate extreme-value distributions* of Gumbel [4] (see also Johnson and Kotz [5, p. 249 ff.]).

References

[1] Arnold, B. C. (1967). *J. Amer. Statist. Ass.*, **62**, 1460–1461.

[2] Block, H. W. and Savits, T. H. (1980). *Ann. Prob.*, **8**, 793–801.

[3] David, H. A. (1974). In *Reliability and Biometry*, F. Proschan and R. J. Serfling, eds. SIAM, Philadelphia, PA, pp. 275–290.

[4] Gumbel, E. J. (1958). *Statistics of Extremes*, 2nd ed. Columbia University Press, New York.

[5] Johnson, N. L. and Kotz, S. (1972). *Distributions in Statistics: Continuous Multivariate Distributions*. Wiley, New York.

[6] Johnson, N. L. and Kotz, S. (1975). *J. Multivariate Anal.*, **5**, 53–66.

[7] Krishnaiah, P. R. (1977). In *The Theory and Applications of Reliability*, Vol. 1, C. P. Tsokos and I. N. Shimi, eds. Academic, New York, pp. 475–494.

[8] Lee, L. (1979). *J. Multivariate Anal.*, **9**, 267–277.

[9] Lee, L. and Thompson, W. A. (1974). In *Reliability and Biometry*, F. Proschan and R. J. Serfling, eds. SIAM, Philadelphia, PA, pp. 291–302.

[10] Marshall, A. W. and Olkin, I. (1967). *J. Amer. Statist. Ass.*, **62**, 30–44.

[11] Moeschberger, M. L. (1974). *Technometrics*, **16**, 39–47.

[12] Shaked, M. (1974). In *The Theory and Applications of Reliability*, Vol. 1, C. P. Tsokos and I. N. Shimi, eds. Academic, New York, pp. 227–242.

(EXPONENTIAL DISTRIBUTIONS
WEIBULL DISTRIBUTIONS)

D. R. JENSEN

MULTIWAY CONTINGENCY TABLES
See MULTIDIMENSIONAL CONTINGENCY TABLES

MURTHY ESTIMATOR

This is a product estimator used in sample survey* methods. Suppose that there are N units in the population, that the character under study Y takes the value y_i for the ith unit, and that an auxiliary character X takes the value x_i for the ith unit in the population ($i = 1, 2, \ldots, N$), where X and Y are negatively correlated. It is assumed that the population mean $\bar{X} = (\sum_{i=1}^{N} x_i)/N$ of X is known in advance, and the objective is to estimate the population mean $\bar{Y} = (\sum_{i=1}^{N} y_i)/N$ of Y.

A simple random sample is made without replacement from the population; the measurements on x and y are obtained for each of the selected sample units. The *Murthy estimator* of $\overline{Y}$ is then given [1] by

$$\hat{y}_p = \bar{y}\bar{x}/\overline{X}, \qquad (1)$$

where $\bar{y}$ and $\bar{x}$ are the sample mean values for Y and X, respectively, in the sample.

Then $\hat{y}_p$ is biased, but is more efficient than the ratio estimator* $\bar{y}\overline{X}/\bar{x}$ or the unbiased estimator $\bar{y}$, provided that the correlation coefficient $\rho(x, y)$ between X and Y lies between certain limits, both negative.

A similar estimator was discussed by Srivastava [4].

For even sample sizes, Shukla [3] proposed a modification of Murthy's estimator, splitting the sample into two subsamples of equal size, and using Quenouille's method of bias reduction [2] (*see* JACKKNIFE METHODS) to construct the estimator

$$\hat{y} = w\hat{y}_p^{(1)} + w\hat{y}_p^{(2)} + (1 - 2w)\hat{y}_p^{(3)},$$

where $\hat{y}_p^{(1)}$, $\hat{y}_p^{(2)}$, and $\hat{y}_p^{(3)}$ are the Murthy estimators based on the two subsamples and the entire sample, respectively, and w is an appropriate weight. Shukla proposed several choices of w which yield (almost) unbiased estimators of $\overline{Y}$ to the first degree of approximation.

References

[1] Murthy, M. N. (1964). *Sankhyā A*, **26**, 69–74.

[2] Quenouille, M. H. (1956). *Biometrika*, **43**, 353–360.

[3] Shukla, N. D. (1976). *Metrika*, **23**, 127–133.

[4] Srivastava, S. K. (1966). *J. Indian Statist. Ass.*, **4**, 29–37.

(RATIO ESTIMATORS
SURVEY SAMPLING)

MUSIC, PROBABILITY, AND STATISTICS

In the main stream of Western classical music, almost every performance is based on a score. There may be variations of interpretation, but basically the performance is fixed by what the composer has written. In other types of music, such as jazz, improvisation plays a greater role, and performances are less predictable.

Some composers have, however, deliberately set out to write works that will be unpredictable in performance. This has occurred particularly among those described as avant-garde or experimental. They have used randomization*, and the terms *aleatoric*, *chance*, and *stochastic* appear in the literature of musicology. Two journals in which theoretical articles on these developments will be found are *Die Reihe* (now discontinued) and *Perspectives of New Music*.

There are much earlier examples of randomization in the Western tradition: one example, often attributed to Mozart, consists of a set of musical phrases in the style of a minuet, together with instructions for determining the order in which they are to be played by rolling dice. [3, p. 100] It is doubtful that Mozart was the composer, but the composition of such musical games of chance was a popular pastime in his day.

Composers in the period since 1950 have also introduced unpredictability into pieces of music by dividing them into subunits that are selected and ordered randomly at each performance. In his *Elytres*, the American composer Lukas Foss has written a score for an ensemble divided into four "forces" (solo flute, violins, pitchless percussion, and pitched percussion). The total score consists of 12 phrases in a fixed order. With the score, Foss provides a diagram that is, in effect, a 4×12 incidence matrix* (actually a 4×12 grid in which some squares are black and the rest white; see Fig. 1).

In preparing to play, two assignments are made. A "force" is assigned to each row. One of the phrases is assigned to the first column, its successor to the second, and so on cyclically. The composer directs that these assignments should be made arbitrarily by the conductor and that "to repeat the identical version is to violate the intent of these compositions." Since the row assignment can be made in 24 ways, and the column assignment in 12, there will be 288

PERFORMANCE SCHEME
Aufführungsplan

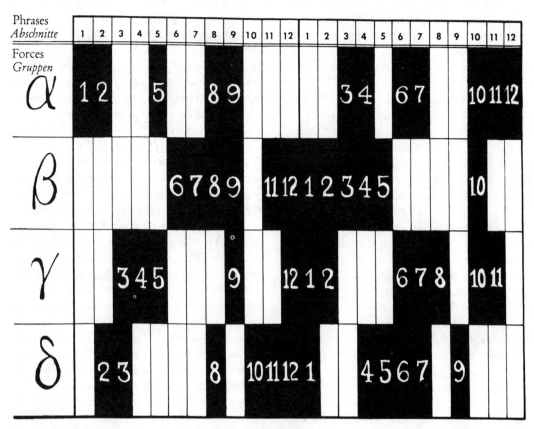

Figure 1 Diagram at the beginning of Lukas Foss's *Elytres* score. (Copyright © 1965 by Carl Fischer, Inc. New York. Reprinted by permission.)

possible variations among which the arbitrary assignments will select one randomly, though not, apparently, in an unbiased way.

A performance consists of playing the 12 phrases in cyclical order, starting with the phrase assigned to column 1, but when the *j*th phrase is being played, the "force" assigned row *i* only plays if entry (i, j) in the diagram is a black square, otherwise they remain silent. The 12 phrases are then played a second time, with each "force" playing those phrases in which it previously remained silent and vice versa. The incidence matrix has been structured to render

the fluctuations among the "forces" playing complex yet ultimately balanced.

Lukas Foss has written another work (*Fragments of Archilochos*) involving selection and ordering of much greater complexity, and the same method has been adopted by several other composers, notably Xenakis, Boulez, Stockhausen, and John Cage [3].

Turning to other ways of randomizing, we note particularly the experiments of Xenakis, who has described his own compositional procedures in a book [7]. Xenakis' approach to his work is extremely cerebral, and the book deals with the application of a wide range of mathematical ideas to composition of which those of probability are only part. One of his methods is to make a musical performance follow a game with chance elements in it: for example, his *Duel* and *Stratégie* involve guessing games between the conductors of two competing orchestral groups, for which the game matrices are given, and discussed at length, in Xenakis [7]. Xanakis also introduces probability in an essentially different way, in other works, as part of the process of composition. The result in this case is a fixed score, not an unpredictable performance, but the composer uses procedures based on various stochastic processes* to determine what notes are to be played by what instruments, the lengths and durations of glissandi, and numerous other components of the whole composition. A variety of stochastic processes are employed, such as the Poisson process* and various Markov chains (*see* MARKOV PROCESSES); the interested reader is referred to Xenakis [7] for a discussion of the methods and their results that is full and sometimes distinctly recondite.

Another process intended to generate unpredictability is the introduction of a large number of simultaneous processes, so that the interest of the piece derives from the complexity resulting from their superposition. Two composers who have employed this technique in a number of works are Ligeti and John Cage. The transition from separate, possibly quite simply structured time series* to "chaos" as increasing numbers of series are superimposed has its counterpart in Dobrushin's theorem on the limiting Poissonian nature of the superposition of time series (see Dobrushin [1] and Stone [6]). At a performance at the Institut de Recherche et Coordination Acoustique/Musique in Paris in 1979, Cage employed 64 simultaneously running tape tracks. Cage, who mixes speculative philosophy with musical theory in his public statements, shows great interest in the nature of probability itself and is influenced in his composition by two aspects of probability that are somewhat re-

moved from the frequentist one: (1) the occurrence of an event "at hazard" as a result of simultaneous events in two series, themselves possibly deterministic but apparently far too widely separated for any interconnection to be traced; (2) the treatment of events as if they were random when they derive from processes that could possibly be considered deterministic, but are far too complex for the details to be worked out.

The generation of music by random processes has been investigated by Voss and Clarke of IBM from a technical point of view. Their work is well summarized by Martin Gardner in a *Scientific American* article [2] together with interesting connections with the concept of fractal curves due to Mandelbrot (*see* FRACTALS), and further references will be found there. They generated "music" in the form of random sequences of notes, one type of sequence being a discretized version of white noise and another of Brownian motion*. When played, "white music" proved too disjointed to be accepted as music by the ordinary listener, while "brown music" sounded quite different, being strongly correlated, but also failed to appeal. Noting that the spectral density of Brownian noise is $1/f^2$, while that of white noise* is $1/f^0$, they generated "tunes" based on noise ("flicker noise") with spectral density $1/f$ and discovered that listeners were much more able to accept them as music. While there is clearly a great gap between this process and the craft of the composer, some insight into the mixture of order and surprise that the listener appreciates in real music may be gained from it.

Use of information theory* to characterize the music of various styles (e.g., cowboy songs and American children's songs) and various composers (e.g., Schubert, Mendelssohn, and Schumann) are described by Yaglom and Yaglom [8].

Information theory is also applied to music, among other arts, by Moles [5], who is concerned with its relation to aesthetic quality. Combining the information-theoretic characterizations with simpler statistical measures such as the frequency of occurrence of particular notes and two-note combinations, attempts have been made to generate synthetic music that would sound similar to the music characterized, and this has influenced computer-generated music.

References

[1] Dobrushin, R. L. (1956). *Ukr. Math. Z.*, **8**, 127–134 (in Russian).

[2] Gardner, M. (1978). *Sci. Amer.*, **238**(4), 16–33.

[3] Kostelanetz, R. ed. (1970). *John Cage*. Praeger, New York.

[4] Mandelbrot, B. (1977). *Fractals: Form, Chance and Dimension*. W. H. Freeman, San Francisco.

[5] Moles, A. A. (1972). *Theorie de l'Information et Perception Esthétique*. Denoël/Gonthier, Paris.

[6] Stone, C. (1968). *Ann. Math. Statist.*, **39**, 1391–1401.

[7] Xenakis, I. (1971). *Formalized Music: Thought and Mathematics in Composition*. Indiana University Press, Bloomington, IN.

[8] Yaglom, A. M., and Yaglom, I. M. (1960). Translated (with revisions by the authors) as *Wahrscheinlichkeit und Information*, Deutscher Verlag der Wissenschaften, Berlin, 1967.

(RANDOMIZATION
STOCHASTIC PROCESSES
WHITE NOISE)

W. A. O'N. WAUGH

MUTATION PROCESSES

A *mutation* is a heritable change in the genetic material. The consequences of mutation have been studied in great detail from the mathematical point of view in the subject of population genetics, and the theory of mutation processes has found its greatest application in genetic and evolutionary areas. Nevertheless, mutation processes can be described in abstract terms and this allows an application of the theory beyond genetics* and biological evolution.

The essential elements in the structure of a mutation process are a population of individuals (in genetics, genes), each individual being of one or other of a set of types (in genetics, alleles), a well-defined model describing the formation of one generation of individuals from the parental generation, and a mutation structure describing the probability that a mutant offspring of a parent of given type should be of any other type. The population is normally assumed to be of large and fixed size and the number of possible types is in some models a fixed finite number and in other models, infinite. In biological evolutionary theory, one also allows the possibility of selection, that is, of differential reproduction rates of different types, but here we do not consider this generalization (*see* GENETICS, STATISTICS IN). Attention is paid to time-dependent and also to stationary properties of the process, and also to properties of samples of individuals taken from the population, in particular at stationarity (*see* STATIONARY PROCESSES).

Properties of mutation processes may be studied either retrospectively, by considering properties of the ancestor sequence of any sample of individuals in the current population (see, in particular, Kingman [5]) or prospectively, by considering lines of descent from any such sample (see, in particular, Griffiths [1, 2]). The two approaches can be unified largely through the concept of *time reversibility* (see Tavaré [7] for a review of these and associated matters).

When there exists a finite number m of types, with symmetric mutation structure, the stationary distribution of the frequencies $x_1, \ldots, x_m$ of the m types is (in large populations) in the Dirichlet* form

$$f(x_1, \ldots, x_m) = \text{const} \prod_i x_i^{\theta/(m-1)-1},$$

$$\Sigma x_i = 1, \qquad x_i \geq 0.$$

Here $\theta = cNu$, c being a constant depending on the model assumed for the formation of each new generation (often $c = 1, 2$, or 4); N is the population size, and u the mutation rate for each individual. There exists no nontrivial limit for the distribution of each frequency as $m \to \infty$. Nevertheless, a limiting concept does exist if we focus on the order statistics* $x_{(1)}, x_{(2)}, \ldots$; for any fixed j, there exists a nondegenerate limiting distribution for the first j order statistics*. This is the marginal distribution of the first j components of the so-called Poisson–Dirichlet distribution with parameter θ introduced by Kingman [4]. From this distribution, one may find properties of the *infinite-type process* and thus the m-type and the infinite-type models may be related through standard convergence arguments as $m \to \infty$.

It is also possible, and often simpler, to proceed directly to the infinite-type process. Here all mutants are regarded as being of an entirely novel type and the concept of stationarity refers to patterns of type frequencies rather than the frequency of any specific type. Stationarity properties may be found by a retrospective analysis using [5] the concept of the N-*coalescent*. This is a Markov chain of equivalence relations in which, for any $i, j = 1, \ldots, N$ $(i \neq j)$, we have $i \sim j$ at step s of the chain if individuals i and j at time 0 have a common ancestor at time $-s$. A backward Kolmogorov argument, watching the equivalence classes formed by a sample of n individuals during each ancestor generation until a common ancestor for all n individuals is reached, shows that the probability that in the sample there exist k different types $(k = 1, 2, \ldots, n)$, in such a way that β_i types are represented by exactly i individuals $(\Sigma \beta_i = k, \Sigma i\beta_i = n)$, is

$$n! \, \theta^k / \left\{ \theta^{[n]} \prod (\beta_i! \, i^{\beta_i}) \right\}$$

where $\theta^{[n]} = \theta(\theta + 1) \ldots (\theta + n - 1)$.

$$(1)$$

This is the Ewens–Karlin–McGregor–Kingman–Watterson sampling formula, which has many applications in population genetics theory. Note that, given a sample of n individuals yielding an observed value for the vector $(\beta_1, \beta_2, \ldots, \beta_n)$, the statistic k is sufficient for the parameter θ, and that the conditional distribution of $(\beta_1, \beta_2, \ldots, \beta_n)$,

given k, is of the form

$$\text{const} \times \left\{ \prod \beta_i! \, i^{\beta_i} \right\}^{-1}.$$

From this it follows that, although the mutation process is symmetric and the selection of individuals to be parents is random, the most likely observed configurations of $(\beta_1, \ldots, \beta_n)$, given k, are those where one type predominates, together with a small number of types at low frequency. This unexpected conclusion may be explained by considering the times at which the various types in the sample first arose in the population.

The retrospective analysis also yields results on the "ages" of types represented in a sample. For example, if from the sample of n individuals, we take a subsample of m, the probability that the oldest type present in the sample is represented in the subsample is $m(n + \theta)/\{n(m + \theta)\}$. The probability distribution of the frequency of the oldest type in the sample is [3]

$$\Pr(j \text{ individuals of oldest type})$$
$$= \theta \binom{n - 1}{j - 1} \bigg/ \left\{ n \binom{n + \theta - 1}{j} \right\},$$
$$j = 1, 2, \ldots, n.$$

The probability that there exist m types in the population older than the oldest type in the sample is [6]

$$\Pr(m \text{ older types}) = \frac{n}{n + \theta} \left(\frac{\theta}{n + \theta} \right)^m,$$
$$m = 0, 1, 2, \ldots .$$

A large variety of similar results are described in the references.

The prospective, as opposed to the retrospective, properties of the infinite-type process have been found by Griffiths [1, 2]. These include time-dependent analogs of the sampling formula (1) as well as properties of samples taken from the population t generations apart. The latter include the distribution of the number and frequencies of types common to both samples. An important adaptation of these results concerns properties of samples taken from two different populations that split from a common stock $\frac{1}{2}t$

generations in the past. A sufficient statistic* for t is the set of type frequencies in common in the two populations. By time reversibility arguments, the properties of the age of the oldest type are identical to corresponding properties of the time that the current types present survive. Given j types in the population at any time, the distribution of the time until the first type is lost may be found. This time has exponential distribution* with mean $2[j(j + \theta - 1)]^{-1}$.

A final question in the infinite-type model is to find the probability that the most frequent type in the population is also the oldest. This is neatly answered by a time reversal, since the probability in question is identical to the probability that, of the current types, the most frequent will survive the longest and is thus the mean frequency of the most frequent type. This may be found from the Poisson–Dirichlet distribution of $x_{(1)}$. Details are given by Watterson and Guess [8], who also provide further similar examples.

References

[1] Griffiths, R. C. (1979). *Adv. Appl. Prob.*, **11**, 310–325.

[2] Griffiths, R. C. (1980). *Theor. Pop. Biol.*, **17**, 37–50.

[3] Kelly, F. (1977). *J. Appl. Prob.*, **13**, 127–131.

[4] Kingman, J. F. C. (1975). *J. R. Statist. Soc. B*, **37**, 1–22.

[5] Kingman, J. F. C. (1982). *J. Appl. Prob.*, **19A**, 27–43.

[6] Saunders, I., Tavaré, S., and Watterson, G. A. (1984). *Adv. Appl. Prob.*, **16**, 471–491.

[7] Tavaré, S. (1984). *Theoret. Pop. Biol.*, **26**, 119–164.

[8] Watterson, G. A. and Guess, H. A. (1977). *Theor. Pop. Biol.*, **11**, 141–160.

(GENETICS, STATISTICS IN HUMAN GENETICS, STATISTICS IN STATIONARY PROCESSES)

W. J. Ewens

MUTUALLY EXCLUSIVE EVENTS

Events that cannot occur simultaneously are *mutually exclusive*. For such events $E_1, \ldots, E_k$, the probability of the joint event $\bigcap_{i=1}^{k} E_i$ is zero. The converse is not necessarily true, for two reasons.

1. We can have $\Pr[\bigcap_{i=1}^{k} E_k] = 0$ $(k > 2)$, although for some (or all) pairs of events E_i, E_j, $\Pr[E_i \cap E_j] > 0$. As a simple example, if a card is drawn from a standard pack of 52 playing cards, we can define the three events:

    E_1 : card is red; E_2 : card is a 2;

    E_3 : card is not a red 2.

    The three events are not mutually exclusive, although

    $$\Pr[E_1 \cap E_2 \cap E_3] = 0.$$

2. Either or both of events E_1, E_2 may have zero probability. Then we have $\Pr[E_1 \cap E_2] = 0$, although $E_1 \cap E_2$ may not be mutually exclusive. As an example, suppose X_1 and X_2 are independent continuous random variables and define E_1, E_2 as the events $X_1 = 0$, $X_2 = 0$, respectively.

(ADDITION THEOREM)

N

NAIVE ESTIMATOR *See* DENSITY ESTIMATION

NARROWBAND PROCESS

A narrowband process is a random process whose power spectral density function differs from zero only in some narrow frequency band centered around a given frequency that is large compared with the width of the power spectral density. These processes arise in modeling noise* in amplitude- and frequency-modulation communications systems. The fundamental work on narrowband processes is due to Rice [3]. Discussions of Rice's work appear in most probability textbooks directed toward a communications engineering audience as well as in most statistical communications theory* textbooks; for example, see Papoulis [2] or Thomas [4]. Cramer and Leadbetter's [1] discussion presupposes no communications background; see *envelope* in their index.

Let us concentrate on the case of narrowband Gaussian processes, since it is in the context of Gaussian processes* that the narrowband concept generally arises. Let $\{X(t), -\infty < t < \infty\}$ be a stationary* zero-mean Gaussian process with autocorrelation function $R(\tau) \overset{\Delta}{=} EX(t)X(t + \tau)$ and suppose that the Fourier transform of $R(\tau)$ (*see* INTEGRAL TRANSFORMS), the power spectral density function $S(\omega)$, exists. If $S(\omega)$ is nonzero only for $|\omega| \in [\omega_0, \omega_1]$, where for $\Delta\omega \overset{\Delta}{=} \omega_1 - \omega_0$ and some $\omega_0 \leqslant \omega_c \leqslant \omega_1$, we have that $\Delta\omega \ll \omega_c$, we say that $X(t)$ is a narrowband Gaussian process. In order to visualize the typical sample path behavior of a narrowband Gaussian process, it is helpful to consider the random sinusoid process given by

$$Y(t) = A \cos(\omega_0 t + \theta)$$
$$= Y_c \cos(\omega_0 t) - Y_s \sin(\omega_0 t),$$

where A and θ are independent random variables, A has a Rayleigh distribution*, and θ is uniform on $[0, 2\pi)$, and where $Y_c = A \cos\theta$ and $Y_s = A \sin\theta$ are independent, identically distributed Gaussian random variables. Since $Y(t)$ has an autocorrelation function proportional to $\cos(\omega_0\tau)$, the power spectral density of $Y(t)$ would consist of Dirac delta functions* located at $\pm\omega_0$. The same paths of $Y(t)$ are sinusoids of random amplitude and phase and of frequency ω_0.

In the narrowband case, the representations of the process $X(t)$ in terms of related processes $A(t), \theta(t)$, $X_c(t) = A(t)\cos(\theta(t))$, and $X_s(t) = A(t)\sin(\theta(t))$ are particularly illuminating. Although in general a random process $X(t)$ can be represented by

$$X(t) = A(t)\cos(\omega_c t + \theta(t))$$

$$= X_c(t)\cos(\omega_c t) - X_s(t)\sin(\omega_c t),$$

this representation has special significance in terms of sample path behavior in the narrowband case. For narrowband processes, typical sample paths look like sinusoids of slowly varying amplitude or envelope described by the random process $A(t)$, and slowly varying phase described by the random process $\theta(t)$, about the nominal frequency ω_c. "Slow" is with respect to the rapidly varying $\cos \omega_c t$.

For fixed t, $A(t)$ and $\theta(t)$ are independent random variables, Rayleigh and uniform respectively, although these are not independent random processes. Likewise, $X_c(t)$ and $X_s(t)$ are marginally independent Gaussian random variables, although they are only approximately independent random processes. Additional results concerning these processes can be found in the references.

References

[1] Cramer, H. and Leadbetter, M. R. (1967). *Stationary and Related Stochastic Processes*. Wiley, New York.

[2] Papoulis, A. (1984). *Probability, Random Variables, and Stochastic Processes*, 2nd ed. McGraw-Hill, New York.

[3] Rice, S. O. (1944). *Bell System Tech. J.*, **23**, 282–332; (1945). *ibid*, **24**, 46–156. [Reprinted in Wax, N., ed. (1954). *Selected Papers on Noise and Stochastic Processes*. Dover, New York.]

[4] Thomas, J. B. (1969). *An Introduction to Statistical Communication Theory*. Wiley, New York.

(COMMUNICATION THEORY,
  STATISTICAL
GAUSSIAN PROCESSES
NOISE)

JULIA ABRAHAMS

NASH AXIOMS

Bargaining is a widespread phenomenon. It occurs between individuals (e.g., in haggling over a price or over the details of an exchange of goods and services), between organizations in the same country (e.g., a union and a company that its members work for), and between nations (e.g., in negotiations over trade restrictions). In all of these situations, bargaining is undertaken because of the fact that there is an opportunity for a binding agreement to be reached that would benefit both parties. In most of these situations, however, there is also some difference of opinion as to what the final agreement should be.

One way in which two bargainers (or an arbitrator acting on their behalf) can narrow down the set of possible final agreements is by limiting the characteristics that the gains to the two bargainers can have. A list of axioms that can be used for this purpose has been provided by Nash [3]. In his view, they are ones that bargainers should be willing to accept. The most basic bargaining situations that Nash's axioms can be used for are the ones that fit the following description.

There are two bargainers, $i = 1, 2$. There is a set A of possible final agreements. It is also possible for them to not reach an agreement. This possibility will be denoted by a_*. The set of possible states of the world that can occur subsequent to the bargainers having an opportunity to reach an agreement will be denoted by $Z = \{z^1, \ldots, z^m\}$ (where m is finite). The set of all possible probability distributions on Z will be denoted by

$$p = \left\{ (p^1, \ldots, p^m) \in R^m : p^k \geq 0, \right.$$

$$\left. \forall k = 1, \ldots, m \text{ and } \sum_{k=1}^m p^k = 1 \right\}.$$

For each $a \in A$, there exists a corresponding probability distribution $\rho(a) = (\rho^1(a), \ldots, \rho^m(a)) \in P$ that summarizes the likelihoods of the elements in Z if they both agree to a. The probability distribution on Z

in the absence of an agreement will be denoted by $\rho(a_*)$. For each probability distribution $p \in P - \rho(a_*)$, there is a binding agreement $a \in A$ such that $\rho(a) = p$ (i.e., the function $\rho : A \cup \{a_*\} \to P$ is onto).

Each bargainer i has *complete preferences* on the set $A \cup \{a_*\}$. That is, for each pair $a', a'' \in A \cup \{a_*\}$, either a' is at least as good as a'' for i or a'' is at least as good as a' for i (or both). In addition, for each i, there is also a scalar-valued function $u_i(z)$ on Z such that his preferences on $A \cup \{a_*\}$ can be *represented* by the function

$$Eu_i(a) = \sum_{k=1}^{m} u_i(z^k)\rho^k(a),$$

i.e., by the function that assigns the expected value of the random variable $u_i(z^k | \rho(a))$ to a. That is, for each pair $a', a'' \in A \cup \{a_*\}$, we have $Eu_i(a') \geqslant Eu_i(a'')$ if and only if a' is at least as good as a'' for i. Specific axioms on preferences that lead to the existence of such functions have been provided by von Neumann and Morgenstern [7] and others. (For a more detailed discussion of the assumptions that have been made on preferences here, *see* UTILITY THEORY.) Finally, there is at least one $a \in A$ such that $Eu_1(a) > Eu_1(a_*)$ and $Eu_2(a) > Eu_2(a_*)$.

The set of all bargaining situations that satisfy the above description will be denoted by B.

For any given $b \in B$, a particular pair of utility values $(s_1, s_2) \in R^2$ is *feasible* if and only if there exists an $a \in A \cup \{a_*\}$ such that $Eu_1(a) = s_1$ and $Eu_2(a) = s_2$. The set of all feasible pairs of utility values for the two bargainers (in the given bargaining situation) will be denoted by S. From the assumptions already stated, it follows that S is closed, convex, and bounded. The particular pair of utility values that will occur if there is no agreement, $(Eu_1(a_*), Eu_2(a_*)) \in S$, is called the *status quo*. It will be denoted, more simply, as (s_1^*, s_2^*) when this is convenient.

Any final agreement a that is reached will have a corresponding pair of gains for the two bargainers: $g_1 = Eu_1(a) - s_1^*$ and g_2

$= Eu_2(a) - s_2^*$. As a consequence, the basic problem for the two bargainers can be viewed as the selection of a feasible pair of utility gains—or, equivalently, (since s_1^* and s_2^* are known) as the selection of a feasible pair of utilities $(\bar{s}_1, \bar{s}_2) = (s_1^* + g_1, s_2^* + g_2)$ for the two bargainers. Indeed, once such a (unique and mutually agreeable) pair of utility levels has been selected, the bargainers will no longer have any difference of opinion about whether a particular agreement is acceptable. Rather, an agreement will be acceptable to both of them if and only if it leads to the selected pair of utilities. Thus, once S has been narrowed down to a unique acceptable pair (s_1, s_2), any $a \in A$ with $(Eu_1(a), Eu_2(a)) = (\bar{s}_1, \bar{s}_2)$ can be selected as the final agreement. Because of this, the pair $(\bar{s}_1, \bar{s}_2)$ is called the *bargaining solution* (or *solution*) for the given bargaining situation.

The Nash axioms place specific restrictions on the utility pairs that can be chosen when solutions for all of the possible bargaining situations in B are to be selected at one time. More specifically, they restrict any function $\psi : B \to R^2$ that assigns a feasible bargaining solution $(\bar{s}_1, \bar{s}_2) = \psi(b)$ to each $b \in B$. The first axiom applies to each bargaining situation. For a given $b \in B$, $(s_1, s_2) \in S$ *Pareto dominates* $(t_1, t_2) \in S$ if and only if $s_1 > t_1$ and $s_2 > t_2$ (*see* PARETO OPTIMALITY).

A1 For each $b \in B$, there is *no* $(s_1, s_2) \in S$ that Pareto dominates $(\bar{s}_1, \bar{s}_2) = \psi(b)$.

A bargaining situation is *symmetric* if and only if (a) the set of feasible pairs of utility for the two bargainers is symmetric around the line $s_1 = s_2$ (i.e., $(s, t) \in S$ implies $(t, s) \in S$) and (b) the utilities of the two bargainers are equal at the status quo (i.e., $s_1^* = s_2^*$). The second axiom applies to all bargaining situations that are symmetric:

A2 If $b \in B$ is symmetric, then $\psi(b) = (\bar{s}_1, \bar{s}_2)$ has $\bar{s}_1 = \bar{s}_2$.

The remaining 2 axioms apply to certain specific pairs of bargaining situations.

Two bargaining situations $b', b'' \in B$ *differ from each other only by affine transformations of the bargainers utility functions* if and only if (a) $A' = A''$, (b) $a'_* = a''_*$, (c) $Z' = Z''$, (d) $\rho' = \rho''$, and (e) there exist scalars α and γ and positive scalars β and δ such that

$$u'_1(z_k) = \alpha + \beta u''_1(z_k),$$
$$u'_2(z_k) = \gamma + \delta u''_2(z_k),$$

at each $z_k \in Z' = Z''$. The bargaining solutions $\psi(b') = (\bar{s}'_1, \bar{s}'_2)$ and $\psi(b'') = (\bar{s}''_1, \bar{s}''_2)$ *differ by the same transformations* if and only if $\bar{s}'_1 = \alpha + \beta \bar{s}''_1$ and $\bar{s}'_2 = \gamma + \delta \bar{s}''_2$. If two bargaining situations $b', b'' \in B$ differ from each other only by affine transformations of the bargainers' utility functions, then (using the preceding scalars)

$$Eu'_1(a) = \alpha + \beta Eu''_1(a),$$
$$Eu'_2(a) = \gamma + \delta Eu''_2(a),$$

at each $a \in A' \cup \{a'_*\} = A'' \cup \{a''_*\}$. Therefore the bargainers' preferences on $A' \cup \{a'_*\}$ in b' are exactly the same as the bargainers' preferences on $A'' \cup \{a''_*\}$ in b''. Similarly, if the bargaining solutions $\psi(b')$ and $\psi(b'')$ differ by the same transformations, then the set of "acceptable" agreements in b' will be exactly the same as the set of "acceptable" agreements in b'' (i.e., $\{a \in A : Eu'_1(a) = \bar{s}'_1$ and $Eu'_2(a) = \bar{s}'_2\} = \{a \in A : Eu''_1(a) = \bar{s}''_1$ and $Eu''_2(a) = \bar{s}''_2\}$).

A3 If $b', b'' \in B$ differ from each other only by affine transformations of the bargainers' utility functions, then $\psi(b')$ and $\psi(b'')$ differ by the same transformations.

One bargaining situation $b' \in B$ is a *contraction* of a second bargaining situation $b'' \in B$ if and only if (a) the two bargaining situations have the same status quo [i.e., $(Eu'_1(a'_*), Eu'_2(a'_*)) = (Eu''_1(a''_*), Eu''_2(a''_*))$] and (b) the set of feasible pairs of utility values in b' is a subset of the set of feasible pairs of utility values in b' (i.e., $S' \subseteq S''$).

A4 If $b' \in B$ is a contraction of $b'' \in B$ and $\psi(b'')$ is feasible in b', then $\psi(b') = \psi(b'')$.

From Nash [3], we have:

Theorem. ψ satisfies **A1–A4** if and only if, for each $b \in B$, $\psi(b)$ is the unique pair that maximizes the function $N(s_1, s_2) = (s_1 - s_1^*)(s_2 - s_2^*)$ over the set $H = \{(s_1, s_2) \in S : s_1 \geq s_1^*$ and $s_2 \geq s_2^*\}$.

The particular pair $(\bar{s}_1, \bar{s}_2)$ that maximizes $N(s_1, s_2)$ over the set H for a particular bargaining situation is, accordingly, called the *Nash bargaining solution* for that situation. For each $b \in B$, the fact that S is compact and convex implies that the Nash bargaining solution exists and is unique. More general versions of the preceding theorem and further discussion of the axioms can be found in Nash [3, 4] and in the recent texts of Friedman [1], Harsanyi [2], Owen [5], and Roth [6].

References

[1] Friedman, J. (1977), *Oligopoly and the Theory of Games*. North-Holland, Amsterdam.

[2] Harsanyi, J. (1977). *Rational Behavior and Bargaining Equilibrium in Games and Social Situations*. Cambridge University Press, Cambridge.

[3] Nash, J. (1950). *Econometrica*, **18**, 155–162.

[4] Nash, J. (1953). *Econometrica*, **21**, 128–140.

[5] Owen, G. (1982). *Game Theory*, 2nd ed. Academic Press, New York.

[6] Roth, A. (1979). *Axiomatic Models of Bargaining*. Springer-Verlag, Berlin.

[7] von Neumann, J. and O. Morgenstern (1944), *Theory of Games and Economic Behavior*, 2nd ed. Princeton University Press, Princeton, NJ.

(DECISION THEORY
GAME THEORY
OPTIMIZATION
PARETO OPTIMALITY
UTILITY THEORY)

PETER J. COUGHLIN

NASH EQUILIBRIUM

A group of individuals can easily find themselves in a situation where (a) each individual in the group has to make a decision, (b) for each individual, the relevant consequences depend not only on his decision, but also on the decisions of the other individuals, (c) each individual will act in his own self-interest, and (d) the individuals cannot or will not reach any binding agreements with each other. Any such situation is a *noncooperative game* (*see* GAME THEORY). The absence of binding agreements, in particular, may arise because of the presence of strictly opposing interests, because of the absence of communication or for some other reason (such as an antitrust law).

Many noncooperative games can be modeled by using a *normal form.* Let $N = \{1, \ldots, n\}$ be an index set for the n individuals (or players) in a game. Then a normal form is (1) a set of possible *decisions* (or *strategies*), X_i, for each $i \in N$ and (2) a *payoff function,* $\pi_i : X_N \rightarrow R^1$ for each $i \in N$ —where $X_N = X_1 \times X_2 \times \cdots \times X_n$ is the set of all possible n-tuples of strategies for the players. (In what follows, X_N will be referred to more succinctly as the set of possible *outcomes*). For a specific normal form to be an appropriate model for a particular game, it must be true that, for each $i \in N$, (I) X_i contains all of the decisions available to i and nothing else, (II) i has complete *preferences* on the set of possible outcomes—that is, for each pair $\mathbf{x}, \mathbf{y} \in X_N$, either $\mathbf{x}$ is at least as good as $\mathbf{y}$ for i *or* $\mathbf{y}$ is at least as good as $\mathbf{x}$ for i, and (III) for each pair $\mathbf{x}, \mathbf{y} \in X_N$, we have $\pi_i(\mathbf{x}) \geqslant \pi_i(\mathbf{y})$ if and only if $\mathbf{x}$ is at least as good as $\mathbf{y}$ for i—that is, $\pi_i(\mathbf{x})$ is a *utility function* that summarizes i's preferences on X_N (*see* UTILITY THEORY).

Since in a noncooperative game, there is no possibility of the individuals in the game making binding agreements, analyses of specific noncooperative games are usually concerned with the question: Are there decisions that the individuals can make such that once each person has learned what everyone else has decided, nobody will want to change his mind? If the answer to the question is no, then we cannot identify any particular decisions as the ones that can be expected—since no matter what ones may be selected initially, there will always be at least one individual who will "upset" the outcome (i.e., by changing his decision on finding out what everyone else has decided). On the other hand, if the answer to the question is yes, then whenever such decisions occur they will tend to persist (or "stick around"). At the same time, all other decisions by the individuals will tend to disappear (or "be replaced"). Because of this, researchers have (quite naturally) investigated the question: Which noncooperative games are ones in which such "equilibrium" decisions exist?

In any such game, before the existence of equilibrium decisions can be established, an exact definition of an equilibrium must be specified. Any such definition, in turn, involves a specification of how each individual assesses whether or not he should change an initial decision (after learning everyone else's decision). One such specification is: Each player first makes the assumption that the other players will not change their decisions (from the initial decisions), even if he changes his decision, then the player asks himself, "Given this assumption, is the outcome that occurs with my initial decision at least as good (for me) as the outcome that would result from any other decision that I could make?" If the answer is yes, then i doesn't change his decision. If the answer is no, then i changes his initial decision to one which results in an outcome that is better for him. Thus an individual will leave an initial decision x_i at the outcome $x \in X_N$ unchanged if and only if x_i is a "best reply" to the initial decisions that have been made by the others. That is, (letting $(y_i, \hat{x}_i)$ be the vector that is obtained by replacing x_i by y_i in $\mathbf{x}$) if and only if $\pi_i(\mathbf{x}) = \max_{y_i \in X_i} \pi_i(y_i, \hat{x}_i)$. In any such game, any outcome where no individual can gain by *unilaterally* changing his strategy (i.e., by altering it while all of the other players' strategies are held fixed) will have the following important property:

No player in the game will cause the outcome to change and hence no change will occur. Since these outcomes have this property, each such outcome is called an *equilibrium point*, or, more specifically (in honor of Nash [5, 6]) a *Nash equilibrium*.

Definition. An outcome $\mathbf{x} = (x_1, \ldots, x_n)$ $\in X_N$, is a *Nash equilibrium* if and only if for each $i \in N$ we have

$$\pi_i(\mathbf{x}) \geqslant \pi_i(\mathbf{y})$$

for each $\mathbf{y} \in X_N$ such that $y_i \in X_i$ and $y_j = x_j$, $\forall j \neq i$. (That is, if and only if $\mathbf{x}$ is such that, for each $i \in N$, $\pi_i(\mathbf{x}) = \max_{y_i \in X_i} \pi_i(y_i, \hat{x}_i)$.)

Note that, while the preceding rationale for considering Nash equilibria is based on certain assumptions that involve each player carrying out a specific assessment of whether he should change his decision, the definition itself applies to any normal form—even if these assumptions are not explicitly satisfied in the game being modeled.

Some Examples

The following examples illustrate some important facts about Nash equilibria. Consider an election in which we have two candidates (indexed by $i = 1$ and $i = 2$) and one issue (whether or not a certain tax should be increased). Suppose that each candidate must decide either to come out in favor of increasing the tax (f) or to come out against increasing the tax (a). That is, each candidates strategy set is $X_i = \{f, a\}$. Suppose, further, that each candidate knows: if one of them chooses a and the other one chooses f then the former will win; if they both choose the same strategy, then they will tie.

Example 1. Consider the case in which each candidate cares only about whether he wins, ties, or loses *and* prefers winning to tying to losing. Then (letting the rows in the following matrices correspond to strategies for candidate 1 and letting the columns correspond to strategies for candidate 2), the

specification of the normal form for this game can be completed by using the payoff functions

$$\begin{array}{c} & \begin{array}{cc} f & a \end{array} \\ \begin{array}{c} f \\ a \end{array} & \left[\begin{array}{cc} 0 & -1 \\ +1 & 0 \end{array} \right] \end{array} = w_1(x_1, x_2),$$

$$\begin{array}{c} & \begin{array}{cc} f & a \end{array} \\ \begin{array}{c} f \\ a \end{array} & \left[\begin{array}{cc} 0 & +1 \\ -1 & 0 \end{array} \right] \end{array} = w_2(x_1, x_2).$$

The unique Nash equilibrium in this game is (a, a). The payoff functions in this example are *zero-sum* (i.e., $w_1(x_1, x_2) + w_2(x_1, x_2) = 0 \ \forall(x_1, x_2) \in X_1 \times X_2$) and (a, a) is a *saddle point* for the game (i.e., $\pi_1(x_1, a) \leqslant \pi_1(a, a) \leqslant \pi_1(a, x_2)$, $\forall x_1 \in X_1$, $\forall x_2 \in X_2$) (see GAME THEORY). This illustrates the fact that the definition of a Nash equilibrium includes the definition of a saddle point (von Neumann and Morgenstern [7]) as a special case—specifically, in the case of a two-person zero-sum game.

Example 2. Consider the case in which each candidate has the following preferences: Winning outright is best; If one can't win, the next best thing would be a tie at (f, f)—since f is personally preferred; if one isn't going to win outright (i.e., will either tie or lose) *and* knows that a is going to be the policy proposal of the person who is elected, then one would rather choose f and go down to defeat than choose a and only tie with the other candidate. That is,

$$\begin{array}{c} & \begin{array}{cc} f & a \end{array} \\ \begin{array}{c} f \\ a \end{array} & \left[\begin{array}{cc} 3 & 2 \\ 4 & 1 \end{array} \right] \end{array} = u_1(x_1, x_2),$$

$$\begin{array}{c} & \begin{array}{cc} f & a \end{array} \\ \begin{array}{c} f \\ a \end{array} & \left[\begin{array}{cc} 3 & 4 \\ 2 & 1 \end{array} \right] \end{array} = u_2(x_1, x_2).$$

Note that this game cannot be appropriately modeled by any normal form that has payoff functions that are zero-sum (since $u_1(f, f) > u_1(a, a)$ and $u_2(f, f) > u_2(a, a)$). There are two Nash equilibria in this game: (f, a) and (a, f). This illustrates the fact that a Nash equilibrium need not be unique. It also illustrates the fact that (unlike with saddle pairs in two-person zero-sum games), the payoffs at two different equilibria need not be the same and, as a consequence, different players may prefer different Nash equilibria [here 1 prefers (a, f) while 2 prefers (f, a)]. For each candidate, the strategy f is a *maximin* strategy [i.e., $\min_{x_2} \pi_1(f, x_2) = \max_{x_1} \min_{x_2} \pi_1(x_1, x_2)$ and $\min_{x_1} \pi_2(x_1, f) = \max_{x_2} \min_{x_1} \pi_2(x_1, x_2)$]. Furthermore, for each candidate, f is the only maximin strategy. Therefore, this example also illustrates that (unlike in a saddle point in a two-person, zero-sum game) the strategies used in a Nash equilibrium need not be maximin strategies and, in addition, a game can have an outcome that is a Nash equilibrium and also be such that none of the outcomes where all of the players use maximin strategies are Nash equilibria.

Example 3. Consider the case in which candidate 1 has the same preferences as in Example 2, but candidate 2 has the following preferences: f, being the policy proposed by the person elected, is most important; if f isn't going to be the policy proposed by the person who is elected, then candidate 2 prefers winning to tying to losing. That is,

$$
\begin{array}{c}
\quad\ f \quad a \\
\begin{array}{c} f \\ a \end{array}
\begin{bmatrix} 3 & 2 \\ 4 & 1 \end{bmatrix} = z_1(x_1, x_2),
\end{array}
$$

$$
\begin{array}{c}
\quad\ f \quad a \\
\begin{array}{c} f \\ a \end{array}
\begin{bmatrix} 4 & 3 \\ 1 & 2 \end{bmatrix} = z_2(x_1, x_2).
\end{array}
$$

There is no Nash equilibrium in this game. This illustrates the fact that (as with saddle pairs in two-person zero-sum games) not every nonzero-sum game has a Nash equilibrium.

Nash's Theorem

Nash [5, 6] developed an important theorem that identifies a large class of noncooperative games in normal form in which Nash equilibria exist. These games can be described as follows:

There are n individuals (n is finite), indexed by the elements in the set $N = $ $\{1, \ldots, n\}$. Each $i \in N$ has a finite set Z_i of possible final choices. The elements in this set will be denoted by $z_i^1, \ldots, z_i^{m(i)}$, where $m(i)$ is the number of elements in Z_i. Each i can select either a specific $z_i^k \in Z_i$ or a nondegenerate probability distribution on Z_i that will be used to determine his or her final choice. That is, the set of possible strategies for i is the set of possible probability distributions over Z_i—i.e.,

$$
P_i = \left\{ \left(p_i^1, \ldots, p_i^{m(i)} \right) \in R^{m(i)} : p_i^k \geq 0, \right.
$$

$$
\left. \forall k = 1, \ldots, m(i), \text{ and } \sum_{k=1}^{m(i)} p_i^k = 1 \right\},
$$

where p_i^k is the probability that z_i^k is i's final choice. (It should be noted that the decision to select a certain final choice z_i^h for sure is the strategy given by the degenerate probability distribution $p_i \in P_i$ with $p_i^h = 1$ and $p_i^k = 0$, $\forall k \neq h$). Since the individuals are playing noncooperatively, the strategies selected by different players are always independent probability distributions.

It is the case that each individual in the game, $i \in N$, has *complete preferences* over the set $P_N = P_1 \times \cdots \times P_n$. That is, for each pair $\mathbf{p}', \mathbf{p}'' \in P_N$, either $\mathbf{p}'$ is at least as good as $\mathbf{p}''$ for i or $\mathbf{p}''$ is at least as good as $\mathbf{p}'$ for i (or both). In addition, for each individual i there is a scalar-valued function $u_i(z)$ on $Z_N = Z_1 \times \cdots \times Z_n$ such that i's preferences on P_N can be *represented* by the function

$$
Eu_i(\mathbf{p})
$$

$$
= \sum_{k(1)=1}^{m(1)} \cdots \sum_{k(n)=1}^{m(n)} u_i\left(z_1^{k(1)}, \ldots, z_n^{k(n)} \right)
$$

$$
\times p_1^{k(1)} \cdots p_n^{k(n)},
$$

that is, by the function that assigns the expected value of the random variable $u_i(\mathbf{z} \mid \mathbf{p})$ to $\mathbf{p} = (p_1, \ldots, p_n) \in P_N$. Thus, for each pair $\mathbf{p}', \mathbf{p}'' \in P_N$, $Eu_i(\mathbf{p}') \geq Eu_i(\mathbf{p}'')$ if and only if $\mathbf{p}'$ is at least as good as $\mathbf{p}''$ for i. Specific axioms that imply the existence of such functions have been provided by von Neumann and Morgenstern [7] and others. (*See* UTILITY THEORY for a more detailed discussion of the assumptions made about individuals' preferences here.)

Nash's theorem [5, 6] is:

Theorem 1. The noncooperative game $(P_i, Eu_i; i \in N)$ has at least one Nash equilibrium.

While Nash originally developed his theorem for the games described earlier, the logic of his proof (see Nash [6] or Vorob'ev [10]) is not dependent on this interpretation. Rather, the logic is based on the fact that these games have three particular characteristics. First, for each $i \in N$, the strategy set is the unit simplex in a Euclidean space. That is, for each $i \in N$, there exists a positive integer $m(i)$ such that i's strategy set is

$$X_i = \left\{ \left(x_i^1, \ldots, x_i^{m(i)} \right) \in R^{m(i)} : x_i^k \geqslant 0, \right.$$

$$\left. \forall k = 1, \ldots, m(i), \sum_{k=1}^{m(i)} x_i^k = 1 \right\}.$$

Second, each individual's payoff function is a continuous function of $\mathbf{x} \in X_N = X_1 \times \cdots \times X_n$. Third, each individual's payoff function is linear in his strategies. That is, for each $i \in N$,

$$\pi_i(\lambda v_i + (1 - \lambda)w_i, \hat{x}_i)$$
$$= \lambda \pi_i(v_i, \hat{x}_i) + (1 - \lambda)\pi_i(w_i, \hat{x}_i),$$
$$\forall x \in X_N, \quad \forall v_i, w_i \in X_i, \quad \forall \lambda \in [0, 1].$$

As a consequence, his proof also implies that any noncoopertive game that has those three characteristics also has at least one Nash equilibrium.

Subsequent generalizations of Nash's theorem have identified many more noncooperative games in normal form that have Nash equilibria. One that is particularly useful in applications has also been dubbed "Nash's theorem" (in Aubin [1] and Moulin [4]). The generalization is specifically for games in which each player's payoff function is concave in his strategies (i.e., for each $i \in N$,

$$\pi_i(\lambda v_i + (1 - \lambda)w_i, x_i)$$
$$\geqslant \lambda \pi(v_i, x_i) + (1 - \lambda)\pi_i(w_i, x_i),$$
$$\forall \mathbf{x} \in X_N, \quad v_i, w_i \in X_i, \quad \forall \lambda \in [0, 1]).$$

(Where, in particular, $\geqslant$ has replaced $=$ in the equation given above.) It is:

Theorem 2. Suppose $G = (X_i, \pi_i, i \in N)$ is a noncooperative game in which, for each $i \in N$, X_i is a compact, convex subset of a topological space; π_i is a continuous function of $\mathbf{x} \in X_N$; and π_i is concave in x_i. Then G has at least one Nash equilibrium.

For a proof of this theorem and further generalizations, see Aubin [1], Friedman [2], or Moulin [4]. For additional information about Nash equilibria, see also Luce and Raiffa [3], Owen [8], Shubik [9], and/or Vorob'ev [10].

References

[1] Aubin, J. (1979). *Mathematical Methods of Game and Economic Theory*. North-Holland, Amsterdam.

[2] Friedman, J. (1977). *Oligopoly and the Theory of Games*. North-Holland, Amsterdam.

[3] Luce, R. D. and Raiffa, H. (1957). *Games and Decisions*. Wiley, New York.

[4] Moulin, H. (1982). *Game Theory for the Social Sciences*. New York University Press, New York.

[5] Nash, J. (1950). *Proc. Natl. Acad. Sci. (U.S.)*, **36**, 48–49.

[6] Nash, J. (1951). *Ann. Math.*, **54**, 286–295.

[7] von Neumann, J. and O. Morgenstern (1947). *Theory of Games and Economic Behavior*, 2nd ed. Princeton University Press, Princeton, NJ.

[8] Owen, G. (1982). *Game Theory*, 2nd ed. Academic Press, New York.

[9] Shubik, M. (1982), *Game Theory in the Social Sciences*. MIT Press, Cambridge, MA.

[10] Vorob'ev, N. (1977). *Game Theory: Lectures for Economists and System Scientists*, translated by S. Kotz. Springer-Verlag, New York.

(DECISION THEORY
GAME THEORY
OPTIMIZATION
UTILITY THEORY)

PETER J. COUGHLIN

NAT

When the base of logarithms used in the definition of the entropy* function is chosen to be *e*, the unit for numerical values of entropy is *nat*.

(ENTROPY
INFORMATION THEORY)

NATIONAL BUREAU OF STANDARDS

The National Bureau of Standards (NBS) is the oldest national laboratory in the United States, established in 1901 as the central reference laboratory for measurements in the physical sciences and engineering. (*See NATIONAL BUREAU OF STANDARDS, JOURNAL OF RESEARCH OF THE* for additional history.) Attached to the Department of Commerce since 1903, the NBS also provides scientific and technological advice to other agencies of the federal government. The Center for Applied Mathematics is one of eleven technical centers, grouped into two major units, the National Measurement Laboratory and the National Engineering Laboratory.

The NBS program in statistics began shortly after World War II in recognition of the value of statistical methodology in "the achievement of objectivity in experimentation, in testing, and in the settting of more exacting standards for drawing scientific conclusions" (A. V. Astin, cited in ref. 3).

The primary focus of statistical work at the National Bureau of Standards is the Statistical Engineering Division in the NBS Center for Applied Mathematics. "Statistical engineering is the name given to that phase of scientific research in which statisticians advise, guide, and assist other scientists in the conduct of experiments and tests," wrote Churchill Eisenhart, describing the role of the group he founded in 1946 and headed until 1963.

The Statistical Engineering Division:

Provides consulting services in application of mathematical statistics to physical science experiments and engineering tests, and

Collaborates in the development and implementation of statistical quality control* procedures for measurement operations and of statistical sampling procedures for monitoring and field inspection activities based on physical measurements and test methods.

This work draws on and contributes to the development of a wide variety of statistical methods, especially experiment design and analysis, reliability*, statistical computing, and statistical quality control.

The Statistical Engineering Division provides supporting services at the two NBS laboratory sites in Gaithersburg, Maryland, and Boulder, Colorado, with a professional staff of approximately 15 people. Other NBS units have employed individual statisticians from time to time, notably John Mandel, statistical consultant in the National Measurement Laboratory. Eisenhart's successors as Chief of Statistical Engineering are J. M. Cameron (1963–69), J. R. Rosenblatt (1969–78), and H. H. Ku (1978–). Headquarters of the Division are in Maryland. The mailing address:

A337 Administration Building
National Bureau of Standards
Gaithersburg, MD 20899.

MEASUREMENT PROCESSES

A method of laboratory measurement can be thought of as a production process in which the output is a series of observed numerical values of some physical quantity. This idea applies with special force to the operation of calibration* services for physical standards and instruments by which NBS provides the central basis for the national measurement system.

It is an NBS goal to establish and maintain a state of statistical control in the Bureau's measurements and in the procedures for transferring the NBS reference values to those measurement processes whose results are reported relative to designated national or international standards (e.g., length, mass, frequency, and derived physical quantities).

Eisenhart [5] gives a definitive treatment of the interpretation of modern probability and statistics as a rigorous foundation for the theory of errors (*see* ERROR ANALYSIS *and* LAWS OF ERROR) and its application to the operational definition and practical evaluation of the precision and accuracy of measurement processes (*see* MEASUREMENT ERRORS). The concept of statistical control (in the simplest case, a sequence of measurements is represented by a sequence of independent identically distributed random variables) permits the application of the Strong Law of Large Numbers* to define the *limiting mean* of a stable measurement process. When applied to the Bureau's work in the national measurement system, operational meaning is given to statements of uncertainty of measurements relative to reference standards maintained by NBS.

On the basis of measurement process concepts, J. M. Cameron collaborated with P. E. Pontius (head of the mass and volume laboratory) in a broadened approach to measurement services, supplementing the traditional NBS calibration services. The NBS began in 1966 to offer "measurement assurance program" (MAP) services, cooperating with customers to place primary emphasis on measurement processes and their adequacy for their intended purpose rather than on the properties of standards and instruments.

The measurement assurance concept entails use of designs and control procedures by both NBS and the participating laboratories to accomplish the transfer from NBS to calibration customers of the capability to characterize the uncertainty of measurement results relative to national standards.

TEST METHOD EVALUATION

A substantial part of the statistical work at the NBS contributes to the characterization of the uncertainties of results obtained from physical, chemical, and engineering tests.

The NBS issues standard reference materials (SRM) for use in calibration of testing instruments or analytical procedures.

Development and certification of an SRM usually involves comparative experiments with several different analytical techniques and evaluation of the homogeneity of the material from which specimens are made. An SRM is issued with a certificate that states the properties of interest and describes the uncertainty of the reported values, usually giving all-inclusive "uncertainty limits." These are based, as appropriate, on the standard errors of the NBS determinations, upper bounds for the effects of systematic errors (*see* ERROR ANALYSIS), and statistical tolerance limits (*see* TOLERANCE REGIONS, STATISTICAL) for the distribution of the certified properties in the inventory of samples [2].

The NBS also participates in the development and standardization of test methods by collaborating in the work of committees that develop voluntary standards under the auspices of the American National Standards Institue (ANSI), the American Society for Testing and Materials (ASTM), and many other similar technical organizations and their international counterparts. (Unlike other nations, the United States relies on private organizations for engineering standards. Federal and local government agencies generally issue standards only when required for regulatory use.) NBS statisticians collaborate with NBS members of voluntary standards committees in the planning and analysis of interlaboratory experiments that are required for the validation of test method standards and, in particular, serve on committees such as the ASTM Committee E-11 on Statistical Methods, which develops generic statistical standards.

W. J. Youden and John Mandel introduced innovations in experiment design and statistical methodology for interlaboratory experiments ([12]; see Mandel [8] for bibliography). Recent methodological developments for test method studies include robust estimation* methods and techniques for fitting and using calibration curves.

EVALUATED SCIENTIFIC DATA

The NBS administers the National Standard Reference Data System, which publishes definitive compilations of physical and chemical data. For this program and also as a result of other research into measurement methods, the NBS produces data ranging from fundamental physical constants through atomic and molecular data, from thermochemical and thermophysical properties of matter to empirical properties of materials. The experimental work that produces carefully characterized data calls on a wide range of statistical models, experiment designs, and estimation techniques. See, e.g., Youden [13].

NBS scientists are served both by statistical consulting* and by the provision of statistical computing tools. NBS statisticians pioneered in the development of computer program packages for statistical and graphical analysis of small- to moderate-sized sets of data (*see* STATISTICAL SOFTWARE). Hilsenrath [6] included a comprehensive collection of these statistical procedures in OMNITAB, a precursor of MINITAB*. Recent developments focus on interactive nonlinear modeling and graphics (*see* GRAPHICAL REPRESENTATION, COMPUTER-AIDED).

Special tasks are assigned to the NBS in its role as a source of technical services to the federal government. For example, see Rosenblatt and Filliben [10].

Literature

In addition to contributing to the general statistics literature, NBS statisticians have published in the *Journal of Research of the NBS*, (*see* NATIONAL BUREAU OF STANDARDS, JOURNAL OF RESEARCH OF THE) and in other NBS publications series (handbooks, monographs, and technical notes). The NBS applied mathematics series includes mathematical tables of interest to statisticians. Selected titles from these series are included in the list of references [1, 2, 4, 6, 7, 9, 11].

References

[1] Abramowitz, M. and Stegun, I. A., eds. (1964). *Handbook of Mathematical Functions*, NBS Appl. Math. Series **55**.

[2] Cali, J. P. and Ku, H. H., et al. (1975). *The Role of Standard Reference Materials in Measurement Systems*, NBS Monograph **148**.

[3] Cameron, J. M. (1969). *Technometrics*, **11**, 247–254.

[4] Clatworthy, W. H. (1973). *Tables of Two-Associate-Class Partially Balanced Designs*, NBS Appl. Math. Series **63**.

[5] Eisenhart, C. (1963). *J. Res. Nat. Bur. Std.*, **67C**, 161–187.

[6] Hilsenrath, J., et al. (1966). *OMNITAB*, NBS Handbook **101**.

[7] Ku, H. H. ed. (1969). *Precision Measurement and Calibration: Statistical Concepts and Procedures*, NBS Special Publication **300**, Vol. 1.

[8] Mandel, J. (1976). *J. Qual. Tech.*, **8**, 86–97.

[9] Natrella, M. G. (1963). *Experimental Statistics*, NBS Handbook **101**.

[10] Rosenblatt, J. R. and Filliben, J. J. (1971). *Science*, **171**, 306–308.

[11] Statistical Engineering Laboratory (1957). *Fractional Factorial Designs for Factors at Two Levels*, NBS Appl. Math. Series **48**.

[12] Youden, W. J. (1967). *Statistical Techniques for Collaborative Tests*, Association of Official Analytical Chemists, Washington, DC.

[13] Youden, W. J. (1972). *Technometrics*, **14**, 1–11.

References [1], [2], [4], [6], [7], [9], and [11] are all published by the U.S. Government Printing Office, Washington, D.C.

(CALIBRATION
ERROR ANALYSIS
MEASUREMENT ERRORS
NATIONAL BUREAU OF STANDARDS,
JOURNAL OF RESEARCH OF THE)

JOAN R. ROSENBLATT

NATIONAL BUREAU OF STANDARDS, JOURNAL OF RESEARCH OF THE

The *Journal of Research of the National Bureau of Standards* is the primary scientific journal of the U.S. National Bureau of Standards* (NBS). It is published bimonthly for the Bureau by the U.S. Government Printing Office (GPO), Washington DC 20234. The number of annual subscriptions fluctuates around 1300. In addition, the Journal is sent by the GPO to 735 U.S. government depository libraries in the United States and by the NBS to an official mailing list of 581 of which over two-thirds are foreign professional and technical organizations. The Contents page of each issue is reproduced in the corresponding issues of *Current Contents— Physical & Chemical Sciences* and *Current Contents—Engineering Technology and Applied Sciences*. The Journal's papers on statistical theory and methods, and applications thereof are indexed in *Current Index to Statistics**.

HISTORY

The U.S. National Bureau of Standards was established by Act of Congress of March 3, 1901 (Public Law 177—56th Congress) to provide custody, maintenance, and development of the national standards of measurement, means and methods of making measurements consistent with those standards, solution of problems arising in connection with standards, and determinations of physical constants and properties of materials of great importance to science and industry when values of sufficient accuracy are not obtainable elsewhere. Subsequent legislation expanded the functions of the Bureau gradually to include development of methods of testing materials, mechanisms and structures, programs of research in engineering, mathematics, the physical sciences, radio propagation, etc.

The Bureau's research papers in science and technology were published initially in two series, scientific papers (1904–1928) and technologic papers (1910–1928). These were combined into a single series of research papers in July 1928 and issued monthy under the title *Bureau of Standards Journal of Research*. In 1934, the name of the journal was changed to *Journal of Research of the National Bureau of Standards*, beginning with volume 13, July–December 1934.

With volume 63, July–December 1959, the journal was reorganized into four separately published sections: (A) *Physics and Chemistry* (bimonthly); (B) *Mathematics and Mathematical Physics* (quarterly); (C) *Engineering and Instrumentation* (quarterly); (D) *Radio Propagation* (bimonthly). The caption of section D was changed to *Radio Science*, and it became a monthly publication with the January 1964 issue (volume 68, no. 1). Its publication was assumed by the new Environmental Science Services Administration of the Department of Commerce following transfer of the Bureau's Central Radio Propagation Laboratory to that agency in January 1966. Section C expired with the July–December 1970 issue (volume 74C, nos. 3 and 4). Sections A and B were combined in a single bimonthly publication beginning with the July–August 1977 issue, designated volume 82, no. 1, of the *Journal of Research of the National Bureau of Standards*. It is restricted to refereed papers from within the Bureau, plus an occasional invited paper from the outside that is very timely and makes a unique contribution to a current research program of the Bureau.

The *Journal of Research* has served from its beginning as a medium of publication for complete accounts of Bureau research and development work, both theoretical and experimental, in physics, chemistry, and engineering, and the results of test and instrumentation activities. There were no papers on mathematics per se and none on or incorporating statistical theory and methodology —other than routine applications of the method of least squares*, calculation of measures of imprecision (probable errors* or standard deviations), tabular and graphical presentation of measurement results—until after the establishment of the Applied Mathematics Division (predecessor to the present Center for Applied Mathematics, 1978–) on July 1, 1947.

Table 1 Selected Statistical Papers 1947–82

Year	Volume(s)	Author(s) and Title of Paper
1947	39	J. H. Curtiss. "Acceptance Sampling by Variables, with Special Reference to the Case in Which Quality Is Measured by Average or Dispersion."
1949	43	U. Chand. "Formulas for the Percentage Points of the Distribution of the Arithmetic Mean in Random Samples from Certain Symmetrical Universes."
1951	47	U. Chand. "On the Derivation and Accuracy of Certain Formulas for Sample Sizes and Operating Characteristics of Nonsequential Sampling Procedures."
1952	48	R. P. Peterson. "Uniformly Best Constant Risk and Minimax Point Estimates."
1952	48	J. Lieblein. "Properties of Certain Statistics Involving the Closest Pair in a Sample of Three Observations."
1953	50	K. L. Chung. "Contributions to the Theory of Markov Chains."
1954	53	W. J. Youden and W. S. Connor. "New Experimental Designs for Paired Observations."
1954	53	M. Zelen. "Bounds on a Distribution Function that Are Functions of Moments to Order Four."
1955	54	W. H. Clatworthy. "Partially Balanced Incomplete Block Designs with Two Associate Classes and Two Treatments per Block."
1957	59	J. Lieblein and H. E. Salzer. "Table of the First Moment of Ranked Extremes."
1958	60	F. Scheid. "Radial Distribution of the Center of Gravity of Random Points on a Unit Circle."
1960–63	64B–67B	L. S. Deming. "Selected Bibliography of Statistical Literature, 1930 to 1960."
1961	65B	F. A. Haight. "Index to the Distributions of Mathematical Statistics."
1961	65B	I. R. Savage. "Probability Inequalities of the Tchebycheff Type."
1962	66B	G. H. Weiss. "Reliability of a System in Which Spare Parts Deteriorate in Storage."
1962	66D	M. M. Siddiqui. "Some Problems Connected with Rayleigh Distributions."
1963	67B	G. H. Weiss. "An Analysis of Pedestrian Queueing."
1964	68B	H. Ruben. "An Asymptotic Expansion for the Multivariate Normal Distribution and Mills' Ratio."

1964	68B	T. A. Willke. "General Application of Youden's Rank Sum Test for Outliers and Tables of One-sided Percentage Points."
1964	68B	L. Weiss. "On the Asymptotic Joint Normality of Quantiles from a Multivariate Distribution."
1964	68B	J. Van Dyke. "Fitting $y = \beta x$ when the Variance Depends on x."
1964	68B	A. J. Goldman and M. Zelen. "Weak Generalized Inverses and Minimum Variance Linear Unbiased Estimation."
1965	69B	R. C. Bose and J. M. Cameron. "The Bridge Tournament Problem and Calibration Designs for Comparing Pairs of Objects."
1966	70B	F. J. Anscombe and B. A. Barron. "Treatment of Outliers in Samples of Size Three."
1967	71B	R. C. Bose and J. M. Cameron. "Calibration Designs Based on Solutions to the Tournament Problems."
1968	72B	D. Hogben. "The Distribution of a Sample Correlation Coefficient with One Variable Fixed."
1968	72B	H. H. Ku and S. Kullback. "Interaction in Multidimensional Contingency Tables: An Information Theoretic Approach."
1969	73B	J. Mandel. "The Partitioning of Interaction in Analysis of Variance."
1969	73C	B. L. Joiner. "Student-t Deviate Corresponding to a Given Normal Deviate."
1970	74B	J. Mandel. "The Distribution of Eigenvalues of Covariance Matrices of Residuals in Analysis of Variance."
1972	76B	P. V. Tryon. "Covariances of Two Sample Rank Sum Statistics."
1972	76B	J. A. Lechner. "Efficient Techniques for Unbiasing a Bernoulli Generator."
1972	76B	J. J. Filliben and J. E. McKinney. "Confidence Limits for the Abscissa of Intersection of Two Linear Regressions."
1973	77B	D. R. Holt and E. L. Crow. "Tables and Graphs of the Stable Probability Density Functions."
1980	85	C. H. Spiegelman and W. J. Studden. "Design Aspects of Scheffé's Calibration Theory Using Linear Splines."
1982	87	C. H. Spiegelman. "A Note on the Behavior of Least-Squares Regression Estimates When Both Variables Are Subject to Error."
1982	87	M. C. Croarkin and G. Yang. "Acceptance Probabilities for Sampling Procedure Based on the Mean and an Order Statistics."

PROBABILITY AND STATISTICAL THEORY AND METHODS

About 90 of the papers in the journal from 1947–82 contain material on probability theory, statistical theory and methodology, or applications thereof of potential interest and value to statisticians and teachers of statistics. Those published in the Journal's Section B were written in most cases for mathematician or statistician readers and their titles tend to indicate adequately their substance. A few in the pre-1959 unpartitioned journal and several in the reconsolidated journal (1977–) have these characteristics too. Table 1 gives the author(s), titles, year, and volume of publication for a selection of such papers.

In contrast, essentially all of the papers addressed to readers in a field of application have titles that do not reveal their statistical methodology content of potential interest to statistician readers. For example, in "A New Method of Radioactive Standard Calibration" by H. H. Seliger (in volume 48), a Latin square* experiment design involving three mutually orthogonal Latin squares was employed at the suggestion of W. J. Youden and J. M. Cameron to reduce the influence of intermittent extraneous disturbances on Geiger-counter measurements of four activity sources and provide three independent estimates of the experimental error variance for comparison with the theoretical value for purely Poisson variation. Using the appropriate modification of the χ^2 dispersion index for Poisson distributed data, 333 sets of such error variance estimates were found to be individually and collectively in good agreement with the variances expected for Poisson-distributed disintegrations, showing that the Latin-square arrangement had been completely successful in eliminating extraneous non-Poisson variation in the measurements and that the counting time could be cut in half yet yield measurements of the activities of the source of greater precision than by the previous method.

The balanced incomplete block designs* advocated by Youden and Connor in vol-

ume 53 were given a trial in volume 54 by B. L. Page of the NBS Length Section. He evaluated the corrections to the nominal lengths of 10 meter-bar line standards of length using 15, 25, 30, and all 45 of the differences observed directly in the 1953 series of intercomparisons and concluded that in comparison of basic standards, direct observation of all differences between bars may be justified but in calibration of secondary standards or others submitted for calibration, a lesser number of direct comparisons should be sufficient. In a subsequent comparison of the dimensional stabilities of 2 standard and 13 other stainless-steel decimeter bars (reported in volume 58), by intercomparing the bars in three groups of six bars, and one group of four, he reduced the number of directly observed differences from 105 to 29.

Some more general partially balanced* incomplete block designs were tabulated explicitly and used to intercompare gamma-ray point sources measured at different positions on the wheel of an automatic calibrator, in a paper by S. B. Garfinkel, W. B. Mann, and W. J. Youden, in volume 70C (1966).

In the developmental work on gage blocks of superior stability reported by M. R. Meyerson et al. in volume 64C (1960), the scheduling of the successive measurements was such as to nullify completely the effects of a linear trend with time, and to nullify approximately the effects of a moderately nonlinear component (i.e., a *trend-elimination design*).

In "Statistical Investigation of the Fatigue Life of Deep-Groove Ball Bearings," by J. Lieblein and M. Zelen, in volume 57 (1956), linear functions of order statistics* were developed that provide minimum variance unbiased estimators* of the *rating life* L_{10} (exceeded with probability 0.90) and *median life* L_{50} in millions of revolutions (or hours) when the fatigue life data are modeled by a Weibull distribution*.

"Variability of Spectral Tristimulus Values" by I. Nimeroff, J. R. Rosenblatt, and M. C. Danemiller, in volume 65A (1961), provided an exposition of the application of multivariate analysis* to the interpretation of spectral tristimulus values, x, y, z, obtained in a color-mixture experiment.

The concept of measurement as a production process was developed in "Realistic Evaluation of the Precision and Accuracy of Instrument Calibration Systems," by C. Eisenhart, in volume 67C (1963), with probabilistic and statistical aspects of the quality control of measurement processes spelled out in detail. This paper provided the groundwork for the development of the Bureau's measurement assurance programs (MAPs). Reference [1] provides a complete exposition of one of the more recent MAPs.

"Determinations Based on Duplicate Readings" by J. A. Speckman, in volume 68B (1964), provided a statistical evaluation of an ASTM procedure that prescribed taking successive circumferential measurements of certain types of cylindrical tanks with a graduated tape until two identical readings were obtained.

NBS research on radiowave propagation fostered a variety of publications on "Rayleigh Distributions" in Section D. One by M. M. Siddiqui is listed in Table 1.

Since the NBS has had an active Operations Research group for over two decades, many mathematical papers in or related to this field appeared in Section B of the journal. Two by G. H. Weiss are included in Table 1.

The Applied Mathematics Division was active from its beginning in developing and exploring solutions of systems of linear equations and methods of matrix inversion. Papers on these topics have appeared in the journal. Some should be useful to statisticians, e.g., "Solving Equations Exactly," by Morris Newman, in volume 71B (1967) and his "How To Determine the Accuracy of the Output of a Matrix Inversion Program" in volume 78B (1974).

Reference

[1] Croarkin, M. C., Beers, J., and Tucker, C. (1979). "Measurement Assurance for Gage Blocks," *Natl.*

Bur. Stand. (U.S.) Mono. **163**. U.S. GPO, Washington, DC. (Introduction by J. M. Cameron.)

(NATIONAL BUREAU OF STANDARDS)

CHURCHILL EISENHART

NATIONAL STATISTICS, COMMITTEE ON

The Committee on National Statistics (U.S.) was established by the National Research Council of the National Academy of Sciences (U.S.) in January 1972, after the creation of such a body was recommended by the President's Commission on Federal Statistics. The Commission's report (1971) stated that:

> . . . a need exists for continuous review of federal statistical activities, on a selective basis, by a group of broadly representative professionals without direct relationships with the federal government.

> Such a body could monitor the implemention of Commission recommendations and, even more important, conduct special studies on statistical questions it deemed important because their favorable resolution would contribute to the continuing effectiveness of the federal system. The body would need to have the independence that is a prerequisite to completely objective review. With independence, with a policy of publishing its findings, and with the leadership and resources required for the active, continuing review we have in mind, it is our opinion that the purely advisory services of this group would make a critically important contribution to the continued success and vitality of the federal statistical system. . . . The quasi-governmental National Academy of Sciences–National Research Council provides an appropriate status and sponsorship for the committee.

Created by a congressional charter signed by President Lincoln in 1863, the National Academy of Sciences is a private, honorary society of scientists and engineers dedicated to the furtherance of science and its use for the general welfare. It responds to requests by departments of the U.S. government to examine and report on subjects of science and technology.

Within the National Research Council, which is the operating agency of the National Academy of Sciences, the Committee and its staff, initially located in the Assembly of Mathematical and Physical Sciences, were transferred in 1977 to the Assembly of Behavioral and Social Sciences. Although most of its major studies are in the latter fields, the Committee's interests cover all fields in which there are statistical issues important to the public—important in the sense that public decisions or understanding may be affected by the need for relevant and accurate information or by the need for good statistical methodology. Thus the Committee is concerned not only with statistical activities of government, but also with the application of statistics elsewhere in public affairs, in science, and in private decision making.

The first chairman of the Committee was William H. Kruskal, who had been a member of the president's commission. He was succeeded as chairman by Conrad Taeuber in 1978 and Stephen E. Fienberg in 1981. Members are appointed by the NAS, usually for three-year terms. The staff of the Committee was initially headed by executive director Margaret E. Martin, who was succeeded in 1978 by Edwin D. Goldfield. Miron L. Straf serves as research director.

Funding for the Committee was originally provided by the Russell Sage Foundation. Currently, core funding is mainly provided by annual contributions from a consortium of federal agencies with some additional foundation support. Major studies are funded separately, usually by a federal agency or group of agencies. For each of these, the Committee establishes a panel of experts and a project staff. The panel membership usually includes a Committee member.

Ongoing panel studies are on the subjects of immigration statistics, statistical assessments as evidence in the courts, natural gas statistics, cognitive aspects of survey methodology, and statistical methodology for the U.S. decennial census* of population and housing. Recently completed panel studies with reports in preparation for publication are on the subjects of survey-based measures of subjective phenomena (e.g., attitudes and opinions) and the treatment of incomplete data in surveys. A number of other major panel studies are in the early stages of development. Some studies are conducted by the Committee itself with staff assistance, without a separate panel. These currently include a study of the problems of sharing research data, a review of developments in federal statistics* in the decade following the report of the president's commission, industry coding, and statistical uses of administrative records. The Committee has also, on request of federal agencies and commissions, reviewed draft reports on statistical matters.

The following major panel reports have been published and are available from the National Academy Press, National Academy of Sciences, 2101 Constitution Avenue, NW, Washington, DC 20418:

Estimates of Increases in Skin Cancer Due to Increases in Ultraviolet Radiation Caused by Reducing Stratospheric Ozone. Appendix C of Environmental Impact of Stratospheric Flight (1975)

Surveying Crime (1977)

Environmental Monitoring (1977)

Counting the People in 1980: An Appraisal of Census Plans (1978)

Privacy and Confidentiality as Factors in Survey Response (1979)

Measurement and Interpretation of Productivity (1979)

Estimating Population and Income of Small Areas (1980)

Rural America in Passage: Statistics for Policy (1981)

Surveying Subjective Phenomena: Summary Report (1981)

Single copies of the following reports are currently available free of charge from the Committee on National Statistics, National

Research Council, 2101 Constitution Avenue, NW, Washington, D.C. 20418:

Setting Statistical Priorities (panel report, 1976)

Planning and Coordination of the Federal Statistics System (staff paper, 1977)

Subnational Statistics and Federal-State Cooperative Systems (staff paper, 1977)

Statistical Data Requirements in Legislation (staff paper, 1977)

Report on the Conference on Immigration Statistics (1980)

Report on the Conference on Indicators of Equity in Education (1981)

A Review of the Statistical Program of the Bureau of Mines (1982)

Family Assistance and Poverty: An Assessment of Statistical Needs (1983)

Bibliography

Assessment of Federal Support for the Committee on National Statistics. (1981). *Statistical Reporter No. 81-11*, pp. 437–451. (Includes descriptions of Committee studies.)

Commission on Federal Statistics. (1971). *Federal Statistics: Report of the President's Commission.* U.S. GPO, Washington, DC, pp. 175–176.

Kruskal, W. (1973). *Science*, **180**, 1256–1258.

Martin, M. E. (1974). *Amer. Statist.*, **28** (3), 104–107.

National Academy of Sciences. (1979). National Statistics. In *The National Research Council in 1979: Current Issues and Studies*. Washington, DC, pp. 56–63.

(FEDERAL STATISTICS)

EDWIN D. GOLDFIELD

NATURAL CONJUGATE PRIORS *See* CONJUGATE FAMILIES OF DISTRIBUTIONS

NATURAL EXPONENTIAL FAMILIES

Univariate one-parameter natural exponential families (NEF) of distributions are the subclass of all exponential families* (EF) of distributions, or of the Koopman–Darmois family of distributions if the random variable X has density function $f(x,\theta) \equiv \exp(x\theta - \psi(\theta))$ and probabilities

$$P_\theta(X \in A) = \int_A \exp(x\theta - \psi(\theta))\, dH(x),$$

$$\tag{1}$$

H being an increasing function, or distribution function, not depending on θ. The set Θ of all θ, the *natural parameters*, for which (1) is finite is an interval assumed to have positive length. In (1), X is the *natural observation*. The terms *natural parameters* and *natural observation* are referred to as the "canonical parameter" and the "canonical statistic" in the entry EXPONENTIAL FAMILIES. Other univariate exponential families, not NEFs, are given by one-to-one transformations $Y = t(X)$ of X in (1).

Examples of one-parameter NEFs are the normal*, Poisson*, gamma* (scale parameter), binomial*, and negative binomial* distributions. The lognormal* and beta* distributions are examples of EFs that are not NEFs. In the statistical literature, authors may mean either NEF or EF when they discuss "exponential familes." The modifier *natural* has only recently been introduced to distinguish these cases.

NEFs are special EFs because convolutions* $\sum X_i$ of independent identically distributed (iid) members of a NEF also belong to a NEF, and the derivatives of $\psi(\theta)$ yield the cumulants of X, but this is untrue of other EFs.

The natural observation in (1) has mean $EX \equiv \mu = \psi'(\theta)$ and $\mathrm{Var}(X) = \psi''(\theta) = V(\mu)$, the *variance function* (VF), expressing the variance as a function of the mean. One can write $X \sim \mathrm{NEF}(\mu, V(\mu))$ unambiguously because the variance function characterizes the NEF. $\psi(\theta)$ is called the *cumulant function* because the rth cumulant* C_r of X is $\psi^{(r)}(\theta)$, the rth derivative with respect to θ. Because $d\mu/d\theta = V(\mu)$, we have $C_{r+1}(\mu) = V(\mu)\, dC_r(\mu)/d\mu$ with $C_2(\mu) = V(\mu)$.

Exactly six distinct classes of NEF distributions have quadratic variance functions (QVF)

$$V(\mu) = v_0 + v_1 + v_2\mu^2. \tag{2}$$

These include the five most important NEFs:

(a) The normal, $N(\mu, \sigma^2)$ with $V(\mu) = \sigma^2$ (constant VF).

(b) The Poisson, Poiss(μ) with $V(\mu) = \mu$ (linear VF).

(c) The gamma, Gam(r, λ), $\mu = r\lambda$, $V(\mu) = \mu^2/r$.

(d) The binomial, Bin(n, p), $\mu = np$, $V(\mu) = -\mu^2/n + \mu$.

(e) The negative binomial, NB(n, p), $\mu = np/(1-p)$, $V(\mu) = \mu^2/n + \mu$.

The exponential, chi-square, Rayleigh, Bernoulli, and geometric distributions are special cases of these. There also is the NEF generated by the generalized hyperbolic secant distribution, NEF–GHS, having $V(\mu) = \mu^2/n + n$, $\mu > 0$.

Linear transformations $Y = \Sigma(X_i - b)/c$ preserve both the NEF and the QVF properties if $X_1, \ldots, X_n$ are iid as NEF–QVF. Then Y has mean $\mu^* = n(\mu - b)/c$ and VF $V^*(\mu^*)$ with coefficients

$$v_0^* = nV(b)/c^2, \qquad v_1^* = V'(b)/c,$$
$$v_2^*/n = v_2/n. \tag{3}$$

The convolution parameter n actually can be any positive number if the NEV–QVF is infinitely divisible (*see* INFINITE DIVISIBILITY), which only excludes the binomial distribution.

One can prove many facts for these six families together by using both the NEF and the QVF properties. For example, in NEF–QVF distributions, even-numbered cumulants C_{2m} are polynomials in $V(\mu)$ of degree m with a similar formula for odd cumulants. The orthogonal polynomials are

$$P_m(X, \mu)$$
$$= V^m(\mu)\{d^m f(x, \theta)/d\theta^m\}/f(x, \theta) \tag{4}$$

with $f(x, \theta) = \exp(x\theta - \psi(\theta))$, and satisfy $P_0 = 1$, $P_1 = x - \mu$, and for $m \geqslant 1$:

$$P_{m+1} = (x - \mu - mV'(\mu))P_m$$
$$- m\{1 + (m-1)v_2\}V(\mu)P_{m-1}. \tag{5}$$

These are known individually as the Hermite

(normal distribution) (*see* CHEBYSHEV–HERMITE POLYNOMIALS), Poisson–Charlier (Poisson distribution), generalized Laguerre (gamma) (*see* LAGUERRE SERIES), Krawtchouk* (binomial), Meixner* (negative binomial), and Pollaczek (NEF–GHS) polynomials. These and other properties appear in Morris [1].

Other results appearing in Morris [2] for NEF–QVF families build on previous NEF–QVF results, particularly the orthogonal polynomials. For example, an analytic function $g(\mu)$ has a uniformly minimum variance unbiased estimate* if and only if

$$\sum_1^\infty \{g^i(\mu)\}^2 V^i(\mu)/(i!\,b_i) \tag{6}$$

is convergent, $b_i \equiv \prod_1^i\{1 + (i-1)v_2\}$, and then the estimate is

$$\hat{g}(X) = \sum_0^\infty g^i(\mu_0)P_i(X, \mu_0)/(i!\,b_i) \tag{7}$$

for some μ_0, P_i being the ith orthogonal polynomial (4). Then (6) is Var($\hat{g}(X)$), and the partial sums of (6) are Bhattacharyya's lower bounds for the variance of an unbiased estimator (*see* BHATTACHARYYA BOUNDS).

If X_1, X_2 are independent NEF distributions with the same θ, the conditional distributions of X_1 given $Y = X_1 + X_2$ have quadratic variance in Y, Var($X_1 \mid Y$) is quadratic in Y, if and only if the NEF has QVF. The NEF-QVF conditional distributions include the normal, binomial, beta, hypergeometric, and geometric, which are not all NEV-QVF distributions. NEF conjugate prior distributions (*see* CONJUGATE FAMILIES OF DISTRIBUTIONS) on μ are in the Pearson system* of distributions if the NEF has QVF. These prior distributions include the normal, gamma, reciprocal gamma, beta, F- and t-distributions. Other well-known distributions are marginal distributions for X if X given μ has a NEF-QVF distribution and μ has a conjugate prior distribution. General formulas for moments of conditional distributions, conjugate prior distributions, posterior distributions, and marginal distributions are available in terms of the variance function for NEV-QVF distributions [2].

References

[1] Morris, C. (1982). *Ann. Statist.*, **10**, 65–80.

[2] Morris, C. (1982). *Natural Exponential Families with Quadratic Variance Functions: Statistical Theory*. Dept. of Mathematics, Institute of Statistics, University of Texas, Austin, TX.

(EXPONENTIAL FAMILIES
NORMAL DISTRIBUTION
POISSON DISTRIBUTIONS
GAMMA DISTRIBUTIONS
BINOMIAL DISTRIBUTION
NEGATIVE BINOMIAL DISTRIBUTION
KOOPMAN–DARMOIS–PITMAN
   FAMILIES)

CARL N. MORRIS

NATURALISTIC SAMPLING

This is a method of sampling that does not prespecify any frequencies except the overall total. It is also called *cross-sectional* or *multinomial* sampling. It is applicable in statistical methodology of rates* and proportions and contingency tables*. See, e.g., Fleiss [1] for more detail.

Reference

[1] Fleiss, J. (1973). *Statistical Methods for Rates and Proportions*, Wiley, New York.

(CONTINGENCY TABLES)

NATURAL PARAMETER *See* CONJUGATE FAMILES OF DISTRIBUTIONS

NAVAL RESEARCH LOGISTICS QUARTERLY

Until 1982, this journal was published by the Office of Naval Research of the United States Department of the Navy. Volume 1 appeared in 1954, the journal continuing

with one volume per year. Its purpose, as stated in the first issue, remained unchanged over the years:

> The *Naval Research Logistics Quarterly* is devoted to the dissemination of scientific information in logistics and will publish research and expository papers, including those in certain areas of mathematics, statistics and economics, relevant to the overall effort to improve the efficiency and effectiveness of logistics operations.

The managing editor for volume 1, and later for volume 9, was Jack Laderman. The managing editors of volumes 2–8 were M. E. Rose, M. I. Rosenberg, H. P. Jones, and H. D. Moore. In 1963, Seymour M. Selig became managing editor and remained in that position until 1982, with the assistance of an editorial board of four and several associate editors.

With volume 30 in 1983 came a change in structure. The position of managing editor was replaced by that of an editor-in-chief, and John Wiley & Sons became the publishers of the journal. There is an editorial advisory board of three persons. Manuscripts should be submitted to Dr. Herbert S. Solomon, editor-in-chief, *Naval Research Logistics Quarterly*, Statistics Department, Sequoia Hall, Stanford University, Stanford, CA 94305.

The journal was established by the Office of Naval Research because an outlet was needed for articles dealing with a scientific and theoretical approach to logistics. For further discussion, *see* MILITARY STATISTICS.

n-DIMENSIONAL QUADRATURE

n-dimensional quadrature is concerned with the numerical approximation of integrals in two or more variables by the use of weighted sums of evaluations of the integrand at selected points. It is not a simple extension of one-dimensional quadrature (*see* NUMERICAL INTEGRATION); the diversity of possible regions of integration and of the singularities possible for *n*-dimensional functions are daunting. General methods have been developed only for regions of integrations that are quite simple; even the case where the region is bounded by general hyperplanes is too difficult. This article will be restricted to the determination of the expected value of a function of several independent random variables and all functions will be assumed to be well-behaved. Thus the desired approximate relationship is:

$$\int \int \cdots \int_S g(x_1, x_2, \ldots, x_n) \prod_{i=1}^{n} f_{X_i}(x_i) \, dx_i$$

$$\approx \sum_{j=1}^{N} w_j g(P_j), \qquad (1)$$

where f_{X_i} is the probability density function for X_i, $i = 1, 2, \ldots, n$; $S = S_1 \times S_2 \times \cdots \times S_n$ is the Cartesian product of the sample spaces S_i for the random variables; $P_j = (x_1^{(j)}, x_2^{(j)}, \ldots, x_n^{(j)})$ is a point, preferably in S, and w_j, the weight associated with P_j, is preferably a positive number, both the P_j and w_j, $j = 1, 2, \ldots, N$, depending only on the f_{X_i}; and $g(P_j)$ is the evaluation of g, an arbitrary function, at P_j. Note that under the restrictions the integral is *n* iterated integrals, each over an S_i, and we prefer that for each P_j, $f_{X_i}(x_i^{(j)}) > 0$ for all i, j. This is not always true, and it is not clear what should be done when it is not. The reason it is desirable that $w_j > 0$ for all j is for the sake of stability. To see this, set $g = 1$, so that $\sum_j w_j = 1$. Thus, if the magnitudes of the w_j can be large, precision is lost.

Problems involving the propagation of error in science and engineering commonly lead to *n*-dimensional integrals such as those in (1) being evaluated numerically. Suppose there is a system with response $y = h(x_1, x_2, \ldots, x_n)$, which is known, where the x_i are system parameters and/or component values. Further suppose that x_i are subject to error: let them be random variables, independent with known densities $f_{X_i}(x_i)$. Then the system response is a random variable Y, and typically the question of interest is its distribution. Quadrature enters when this problem is intractable. We can characterize

the distribution of Y by its moments, expressing them as *n*-dimensional integrals and use (1) to write

$$\mu_k' = E(Y^k)$$

$$= \int \int \cdots \int [h(\mathbf{x})]^k \Pi_i f_{X_i}(x_i) \, dx_i$$

$$\approx \sum_j w_j [h(P_j)]^k, \qquad (2)$$

where $\mathbf{x} = (x_1, x_2, \ldots, x_n)$ and k ranges through the orders of the lower moments desired. It is helpful in practice to redefine the integrand so that

$$g(\mathbf{x}) = [h(\mathbf{x}) - h(\boldsymbol{\mu})]^k, \qquad \mu_i = E(X_i),$$

where $\boldsymbol{\mu} = (\mu_1, \mu_2, \ldots, \mu_n)$, in order to reduce round-off error; the redefined moments are, of course, now around $h(\mu_1, \mu_2, \ldots, \mu_n)$. The generalization to the vector case $y_j = h_j(x_1, x_2, \ldots, x_n), j = 1, 2, \ldots, m$, is straightforward; one simply redefines g in (1) to generate the desired moments.

GENERAL RESULTS

Currently available results by numerical analysts are found in Davis and Rabinowitz [1], Stroud [5], and Haber [4]. There are no general results akin to the ones obtainable in one-dimensional quadrature based on orthogonal polynomials. There is a general theorem that states that if g in (1) is a sum of monomials $x_1^\alpha x_2^\beta \ldots x_n^\gamma$, for $\alpha, \beta, \ldots, \gamma$ nonnegative integers, there exist points $\mathbf{P}_j$ in the region of integration and corresponding weights w_j that are positive so that (1) is exact for monomials of degree less than or equal to $d = \alpha + \beta + \cdots + \gamma$ and N is minimal for d. No general way of generating them efficiently, however, is now known. Some optimum rules have been found, but generally they have no intuitive probabilistic interpretation and are for low-degree monomials.

PRODUCT RULES

In spite of the aforementioned problems with the rules, there are two general classes of rules which, although they do not have all the features desired, are nevertheless useful for applications in probability and statistics. One class is the set of product rules, useful for small *n*, the other is the set of rules for symmetric regions, useful for moderate *n*; while Monte Carlo* is always an alternative method for any size *n*; for large *n* it is the only applicable technique.

Let us first consider product rules, specializing in those based on one-dimensional Gauss quadrature rules since these are the most broadly useful. The basic idea is to treat (1) as an iterated integral and use one-dimensional quadrature at each iteration. Thus, suppose that for f_{X_i} the corresponding k-point Gauss rule weights and abscissas are $w_{i1}, w_{i2}, \ldots, w_{iK}$ and $x_{i1}, x_{i2}, \ldots, x_{iK}$, respectively. Then (1) becomes

$$\int \int \cdots \int g(x_1, x_2, \ldots, x_n) \Pi_i f_{X_i}(x_i)\, dx_i$$

$$= \sum_{a=1}^{K} \sum_{b=1}^{K} \cdots \sum_{c=1}^{K} w_{1a} w_{2b} \cdots w_{nc}$$

$$\times g(x_{1a}, x_{2b}, \ldots, x_{nc}). \tag{3}$$

This rule is exact for *g* consisting of sums of monomials of degree $d \leqslant 2K - 1$, a result that is the same as for the one-dimensional Gauss rule and derives directly from it. The problem in *n* dimensions is that the number of evaluations of *g* is K^n, which grows rapidly with *n*; this explains why numerical analysts suggest using this rule for *n* no greater than 5 or 6. Note that the X_i need not be identically distributed and, if desired, the rule for each X_i can be a K_i-point rule.

There is a special case common in statistical applications for which the limitation on *n* can be relaxed, that is, when the moments of a statistic *Y* are approximated. In this case $Y = h(X_1, X_2, \ldots, X_n)$ is a symmetric function and only one evaluation is needed for all points P_j, which are permutations of one another. It can be shown that the total number of evaluations required is $\binom{n+K-1}{K-1}$, which is significantly smaller than K^n, so that calculations for sample sizes of order 10 are reasonable. Indeed, if there is additional symmetry, the number may yet be smaller.

SYMMETRIC RULES

The second general quadrature rule relies on the symmetry argument that, since *g* in (1) is an arbitrary function, there is no reason for treating any of its components differently from the others. To write the quadrature formula in a perspicuous format let $E(X_i) = 0$ and $\mathrm{var}(X_i) = \sigma_i^2$, by transforming if necessary; further, for shorthand, let only the nonzero components of the function *h* be displayed explicitly. Thus set

$$h_0 \equiv h(0, 0, \ldots, 0)$$

for all components at their means,

$$h(x_i) \equiv h(0, \ldots, 0, x_i, 0, \ldots, 0)$$

for only component *i* off-mean,

$$h(x_i, x_j) \equiv h(0, \ldots, 0, x_i, 0, \ldots, 0,$$
$$x_j, 0, \ldots, 0)$$

for only components *i* and *j* off-mean.

Now postulate the quadrature formula

$$E\left[(h - h_0)^k\right]$$

$$= \int \int \cdots \int [h(\mathbf{x}) - h_0]^k \Pi_i f_{X_i}(x_i)\, dx_i$$

$$\approx Q\left[(h - h_0)^k\right]$$

$$= \sum_k \left\{ H_k^+ \left[h(a_k^+ \sigma_k) - h_0\right]^k \right.$$

$$\left. + H_k^- \left[h(a_k^- \sigma_k) - h_0\right]^k \right\}$$

$$+ \sum_{k<j} \left\{ P_{kj}^{++} \left[h(b_k^+ \sigma_k, b_j^+ \sigma_j) - h_0\right]^k \right.$$

$$+ P_{kj}^{+-} \left[h(b_k^+ \sigma_k, b_j^- \sigma_j) - h_0\right]^k$$

$$+ P_{kj}^{-+} \left[h(b_k^- \sigma_k, b_j^+ \sigma_j) - h_0\right]^k$$

$$\left. + P_{kj}^{--} \left[h(b_k^- \sigma_k, b_j^- \sigma_j) - h_0\right]^k \right\},$$

$$\tag{4}$$

where $H_k^{\pm}, P_{kj}^{\pm\pm}, a_k^{\pm}, b_k^{\pm}$ are constants to be determined (see Evans, ref. 2, 1972). They are determined by successively setting *h* to all monomials of degree *d* for $d = 1, 2, \ldots$ up to the highest degree for which equality in (4) is possible. It turns out that the lowest-degree monomials for which equality is impossible for general X_i is the fifth degree, and therefore in (4)

$$E\left[(h - h_0)^k\right] = Q\left[(h - h_0)^k\right] + O(\sigma^5),$$

i.e., the error is $O(\sigma^5)$, where σ is representative of the σ_i. In the special case in which all X_i are symmetric, the error is $O(\sigma^6)$. Further analysis of (4) shows that $\mu'_k = O(\sigma^k)$ so that as k is increased, the approximation loses accuracy; this general property is exhibited by all the quadrature rules considered although the exact behavior varies. The total number of evaluations of h required in (4) is $N = 2n^2 + 1$. A Fortran computer program is available [3] to do the calculation shown; it will also handle the vector case, $y_i = h_i(x_1, x_2, \ldots, x_n)$, $i = 1, \ldots, m$. A description of one feature of the program may help in comprehending the benefit of n-dimensional quadrature methods: To use the program, the user must supply a computer subroutine called RESP; it specializes the computer program to the particular response $y = h(x_1, x_2, \ldots, x_n)$ of interest. The main program provides RESP with a $P_j = (x_1^{(j)}, x_2^{(j)}, \ldots, x_n^{(j)})$ and the subroutine must return the corresponding value for y. How the subroutine does this is of no consequence; the functional relationship can be explicit or implicit, or the calculation can require using an algorithm, or indeed an experiment can be performed to determine y, etc. The only constraints are that the effort involved in obtaining the $2n^2 + 1$ evaluations by RESP be practicable and that the error be acceptable.

References

[1] Davis, P. J. and Rabinowitz, P. (1975). *Methods of Numerical Integration*. Academic Press, New York. (An excellent general reference.)

[2] Evans, D. H. (1967, 1971, 1972). *Technometrics*, **9**, 441–456; **13**, 315–324; **14**, 22–25. (A series on the application of n-dimensional numerical integrations to statistical tolerancing. The last contains the general results cited here; the second discusses the error.)

[3] Evans, D. H. and Falkenburg, D. R. (1976). *J. Quality Tech.*, **8**, 108–114. (Contains listing of computer programs and examples of usage.)

[4] Haber, S. (1970). *SIAM Rev.* **12**, 481–526.

[5] Stroud, A. H. (1971). *Approximate Calculation of Multiple Integrals*. Prentice-Hall, Englewood Cliffs, NJ. (Has extensive list of known n-dimensional rules.)

(NUMERICAL ANALYSIS
NUMERICAL INTEGRATION)

D. H. EVANS

NEAREST-NEIGHBOR METHODS

The notion that "near things are more related than distant things" is an attractive one and a variety of analyses have been developed that give substance to the idea. First of all, we must distinguish between point processes*, where the location of the objects is of primary interest and other processes that are conditioned upon the set of observed locations. In this context, a *location* may be a position in one-, two-, or three-dimensional Euclidean space or in any general space where measurements can be recorded.

POINT PROCESSES

Much of the work on nearest neighbors has been carried out for processes in the plane, and our discussion follows this path. However, the extension to higher-dimensional spaces usually follows without difficulty.

In modeling a point process, it seems plausible to begin with the assumptions that: (1) objects locate independently in the plane and (2) the probability that an object is located in any small area (δA) is proportional to δA.

These two assumptions suffice to establish that, when objects occur with *intensity* λ, the number of objects located in any area of size A follows the Poisson distribution* with mean λA. Indeed, the scheme is known as a Poisson (point) process*. This is an oft-rediscovered result that Holgate [8] traced back to 1890. A natural attraction of the Poisson model is that the probability density function describing the incidence of objects in any *given* area is uniform. The relevance of this to nearest neighbors is that the square of the distance X_1^2, say, from a randomly selected point in the plane (or a randomly selected object) to the nearest object follows an exponential distribution* with mean $1/(\pi\lambda)$. X_1 is known as the (first) nearest-neighbor distance. Likewise, X_k^2, the square of the kth nearest-neighbor distance, follows a gamma distribution* with index k and mean $k/(\pi\lambda)$. These results give rise to a variety of tests of *randomness** (i.e., of the Poisson process hypothesis) based on nearest-neighbor distances (see Cormack [4]).

Various attempts have been made to formulate estimators for the intensity λ based on nearest-neighbor distances. Such estimators depend critically on the Poisson assumption (Persson [11]) although estimators that are much less sensitive to departures from randomness have been constructed in recent years (Diggle [5]). Estimators based on areal sampling remain unbiased for any spatial pattern, but may be more time-consuming to perform in practice. Sampling designs for the study of nearest-neighbor distances are discussed by Diggle and Matern [6].

The nearest-neighbor distributions for more complex point processes have been discussed by Warren [13] and Bartlett [1]. The use of nearest-neighbor methods in modeling spatial patterns is fully described by Ripley [12]. Models for spatial processes* conditioned on the set of observed locations typically rely on the dependence of neighboring locations (see Bartlett [1] and Cliff and Ord [3]).

Models that describe the development of a point process in both space and time (such as an epidemic) may use nearest-neighbor contact distributions to describe spread of a disease or other phenomenon (see Mollison [9]).

OTHER NEAREST-NEIGHBOR METHODS

In agricultural experiments, the different blocks in a design may be physically contiguous. Papadakis [10] first suggested the use of neighboring plot values as covariates to adjust for variation in fertility and other location-dependent effects. Bartlett [2] revisited this problem and shows that the benefits may be substantial when the number of treatments is large. Freeman [7] gives some two-dimensional Latin square* designs that are balanced for nearest neighbors.

Nearest-neighbor techniques for density estimation* and their application to classification* and discrimination* are described in NEAR-NEIGHBOR ESTIMATION.

References

[1] Bartlett, M. S. (1975). *The Statistical Analysis of Spatial Pattern*. Chapman and Hall, London; Halsted, New York. (A concise description of stochastic models for spatial processes including nearest-neighbor systems and distributions for nearest neighbors.)

[2] Bartlett, M. S. (1978). *J. R. Statist. Soc. B*, **40**, 147–158. (Examines the use of neighboring plots as "covariates" in the analysis of designed experiments.)

[3] Cliff, A. D. and Ord, J. K. (1981). *Spatial Processes: Models and Applications*. Pion, London; Methuen, New York. (Provides a detailed account of tests and estimation methods for spatial dependence including nearest-neighbor analysis. Also includes a full bibliography.)

[4] Cormack, R. M. (1979). In *Spatial and Temporal Analysis in Ecology*, R. M. Cormack and J. K. Ord, eds. International Co-operative Publishing House, Fairland, MD, pp. 151–211. (Describes different aspects of competition among nearest neighbors in space and related statistical methods. Extensive bibliography.)

[5] Diggle, P. (1979). In *Spatial and Temporal Analysis in Ecology*, R. M. Cormack and J. K. Ord, eds. International Co-operative Publishing House, Fairland, MD. (A review of statistical methods for the analysis of spatial point patterns, including a variety of applications. Extensive bibliography.)

[6] Diggle, P. J. and Matern, B. (1980). *Scand. J. Statist.*, **7**, 80–84. (Discusses sampling designs for the study of point-to-object distance distributions.)

[7] Freeman, G. H. (1979). *J. R. Statist. Soc.*, **B41**, 88–95. (Develops Latin-square designs balanced for nearest neighbors.)

[8] Holgate, P. (1972). In *Stochastic Point Processes*, P. A. W. Lewis, ed. Wiley, New York, pp. 122–135. (A review of spatial Poisson point processes and tests of "randomness." Extensive bibliography.)

[9] Mollison, D. (1977). *J. R. Statist. Soc.*, **B39**, 283–306. (Considers nearest-neighbor and related models for ecological and epidemic spread.)

[10] Papadakis, J. S. (1937). *Bull. Inst. Amelior. Plant. Salonique*, **23**. (The original source of nearest-neighbor methods for the analysis of designed experiments; see ref. 2.)

[11] Persson, O. (1971). In *Statistical Ecology*, Vol. 2, G. P. Patil, E. C. Pielou, and W. E. Waters, eds. Pennsylvania State University Press, University Park, PA, pp. 175–190. (Examines the robustness of distance-based estimates of the intensity of a point process.)

[12] Ripley, B. D. (1981). *Spatial Statistics*. Wiley, New York. (Provides a review of stochastic models and statistical analysis for spatial data. In particular, covers the use of nearest-neighbor methods for mapped data. Contains a comprehensive bibliography.)

[13] Warren, W. G. (1971). In *Statistical Ecology*, Vol. 2, G. P. Patil, E. C. Pielou, and W. E. Waters, eds. Pennsylvania State University Press, University Park, PA, pp. 87–116. (Gives the density functions for nearest neighbor distributions for several non-Poissonian processes.)

(NEAR-NEIGHBOR ESTIMATION PROCESSES, POINT
POISSON PROCESSES
SPATIAL PROCESSES)

J. K. ORD

NEARLY BALANCED DESIGNS

In designs for a comparative experiment, *balance* usually refers to an equal and impartial treatment, e.g., equal replications, equal numbers of meetings, etc., of the objects (varieties of grains, different fertilizers, or drugs, etc.) under investigation. The highly symmetric structures of balanced designs lead to very simple analyses and quite often are also responsible for their high efficiencies (or optimalities) (*see* BALANCE IN EXPERIMENTAL DESIGN and GENERAL BALANCE). The stringent conditions of symmetry, however, make balanced designs rather sparse; often they do not exist. Designs that are "nearly balanced" are then useful and recommended.

Literally, nearly balanced designs are those that are close to being balanced. One could make this vague concept precise by introducing a measure of *imbalance*. Balanced designs are those with zero imbalance; nearly balanced designs then are the designs with minimum imbalance when balance is not attainable. We shall present a general theory of the quantification of balance together with an illustration on block designs. The application of course is not limited to this particular setting.

Consider a situation where there are available certain resources to run an experiment with N observations. Suppose there are n unknown parameters $\theta_1, \theta_2, \ldots, \theta_n$ in the system and from each design arise N observations $y_1, y_2, \ldots, y_N$ such that

$$\mathbf{y} = \mathbf{X}\boldsymbol{\theta} + \boldsymbol{\epsilon}, \qquad (1)$$

where $\mathbf{y} = (y_1, \ldots, y_N)'$, $\boldsymbol{\theta} = (\theta_1, \ldots, \theta_n)'$, $\mathbf{X}$ is the design matrix*, and $\boldsymbol{\epsilon}$ is a random vector such that $E(\boldsymbol{\epsilon}) = \mathbf{0}$ and $\mathrm{Cov}(\boldsymbol{\epsilon}) = \sigma^2 \mathbf{I}_N$ (*see* GENERAL LINEAR MODEL). Suppose $\boldsymbol{\theta} = (\boldsymbol{\theta}_1', \boldsymbol{\theta}_2')'$ and $\mathbf{X} = (\mathbf{X}_1, \mathbf{X}_2)$, where $\boldsymbol{\theta}_1$ is $p \times 1$ and $\mathbf{X}_1$ is $N \times p$, and one is only interested in $\boldsymbol{\theta}_1$; then the least-squares* estimate of $\boldsymbol{\theta}_1$ can be obtained by solving the reduced normal equations

$$\mathbf{X}_1'\left(\mathbf{I} - \mathbf{X}_2(\mathbf{X}_2'\mathbf{X}_2)^{-}\mathbf{X}_2'\right)\mathbf{X}_1\hat{\boldsymbol{\theta}}_1$$

$$= \mathbf{X}_1'\left(\mathbf{I} - \mathbf{X}_2(\mathbf{X}_2'\mathbf{X}_2)^{-}\mathbf{X}_2'\right)\mathbf{y}, \qquad (2)$$

where $'$ and $^{-}$ represent transpose and generalized inverse*, respectively.

The symmetric and nonnegative definite matrix $\mathbf{X}_1'(\mathbf{I} - \mathbf{X}_2(\mathbf{X}_2'\mathbf{X}_2)^{-}\mathbf{X}_2')\mathbf{X}_1$ will be called the *information matrix* (for $\boldsymbol{\theta}_1$). We shall denote the information matrix associated with design d by $\mathbf{C}_d$. It is well known that a linear function $\mathbf{a}'\boldsymbol{\theta}_1$ is estimable if and only if $\mathbf{a}$ belongs to $\mathscr{C}(\mathbf{C}_d)$, the space generated by the column vectors of $\mathbf{C}_d$ (*see* ESTIMABILITY).

In some settings, $\mathbf{C}_d$ is singular for all d, i.e., not all the parameters in $\boldsymbol{\theta}_1$ are estimable. Suppose $\mathscr{C}(\mathbf{C}_d) \subset \mathscr{L}$ for all d, where $\mathscr{L}$ is a q-dimensional space with $q \leqslant p$, and we are interested in all the linear functions $\mathbf{a}'\boldsymbol{\theta}_1$ with $\mathbf{a} \in \mathscr{L}$. Then each $\mathbf{C}_d$ has at most q nonzero eigenvalues, and we will only consider designs with $\mathscr{C}(\mathbf{C}_d) = \mathscr{L}$. Equation (2) can be solved by calculating a generalized inverse of $\mathbf{C}_d$. The computation requires minimum effort if C_d has q nonzero eigenvalues that are equal. Such a design will be called a *balanced design*.

For example, suppose v varieties are to be compared by the use of an incomplete block design* with b blocks of size k. Under the usual additive homoscedastic uncorrelated model, for any design d, the information matrix for the variety effects is

$$C_d = \text{diag}(r_{d1}, \ldots, r_{dv}) - k^{-1}N_d N_d', \quad (3)$$

where r_{di} is the number of units assigned to variety i and $N_d = (n_{dij})_{v \times b}$ is the variety-block incidence matrix*, i.e., n_{dij} is the number of times variety i appears in block j.

In the present setting, C_d always has zero row sums, since only the variety contrasts are estimable. If we are interested in estimating all the variety contrasts, then $q = v - 1$ and a balanced design is one such that C_d has constant diagonal elements and constant off-diagonal elements. This amounts to saying that all the pairwise comparisons between the varieties are estimated with the same variance. We can also characterize this in terms of combinatorial symmetry of the design. One important balanced design is the well-known balanced incomplete block design (BIBD) (*see* BLOCKS, BALANCED INCOMPLETE). Recall that a BIBD is a design such that:

(a) each variety appears in each block at most once,

(b) all the varieties appear in the same number of blocks,

(c) any two varieties appear together in the same number of blocks.

Certainly, BIBDs are not the only balanced designs. It is conditions (b) and (c) that make a BIBD balanced. Condition (a) distinguishes a BIBD from other balanced designs. In the literature, a design satisfying (a) is called a *binary design*. The following theorem by Kiefer [8] clearly reveals the roles played by these three conditions.

Theorem. In the general setting (1), let $\mathscr{D}$ be the collection of all the competing designs. For each $d \in \mathscr{D}$, let $\mu_{d1} \geqslant \mu_{d2} \geqslant \cdots$

$\geqslant \mu_{dq} \geqslant \mu_{d,q+1} = \mu_{d,q+2} = \cdots = \mu_{dp} = 0$ be the eigenvalues of C_d. If there exists a balanced design $d^* \in \mathscr{D}$ such that d^* maximizes $\text{tr}\,C_d$ over $\mathscr{D}$, then d^* minimizes $\Phi_f(C_d) = f(\mu_{d1}, \ldots, \mu_{dq})$ over $\mathscr{D}$ for any permutation-invariant real-valued function f that is convex and nonincreasing in each component.

For the sake of simplicity, the theorem is not stated in its most general form. Let $a_1'\theta_1$, $a_2'\theta_1, \ldots, a_q'\theta_1$ be q mutually orthogonal linear functions of θ_1 such that $a_i \in \mathscr{L}$ and $a_i'a_i = 1$ for all i. Then the choice $f(\mu_1, \ldots, \mu_q) = \sum_{i=1}^{q} \mu_i^{-1}$, $-\sum_{i=1}^{q} \log \mu_i$, or μ_q^{-1} leads to the so called *A-*, *D-*, or *E-* criterion which minimizes $\text{tr}\,V_d$, $\det V_d$, or the maximum eigenvalue of V_d, respectively, where V_d is the covariance matrix of the least-squares estimates of $a_1'\theta_1, \ldots$ and $a_q'\theta_1$. Thus a balanced design has strong optimality properties if it also maximizes $\text{tr}\,C_d$. In the setting of block designs, it can be shown that $\text{tr}\,C_d$ is maximized by binary designs; a BIBD is optimal because it is balanced *and* binary.

When an optimum balanced design does not exist, one can try to use a design closest to it in an appropriate sense. Shah [10] proposed the use of "Euclidean distance." For a balanced design with $\text{tr}\,C_d = A$, all the eigenvalues $\mu_{d1}, \mu_{d2}, \ldots$ and μ_{dq} are equal to A/q. This suggests that one can use the Euclidean distance between $(\mu_{d1}, \mu_{d2}, \ldots, \mu_{dq})$ and $(A/q, A/q, \ldots, A/q)$, i.e.,

$$IB(d) \equiv \left[\sum_{i=1}^{q} \mu_{di}^2 - \left(\sum_{i=1}^{q} \mu_{di} \right)^2 / q \right]^{1/2},$$

as a measure of imbalance. In view of Kiefer's theorem, it seems sensible to maximize $\text{tr}\,C_d$ first and then choose a design that minimizes $IB(d)$ among those maximizing $\text{tr}\,C_d$, which is the same as minimizing $\text{tr}\,C_d^2$ among the designs that maximize $\text{tr}\,C_d$. This is exactly the (M, S)-criterion introduced by Eccleston and Hedayat [5]. Indeed, for a continuous function f, one would expect the value $\Phi_f(C_d)$ of an (M, S)-optimal design to be very close to the ideal optimum. We shall consider (M, S)-*optimal design* and *nearly balanced design* synonymous.

In the block design setting, as discussed earlier, *balance* can be characterized in terms of combinatorial symmetry of the design. It turns out that nearly balanced designs can also be characterized combinatorially. Since $\text{tr}\,\mathbf{C}_d$ is maximized by binary designs, let us consider the minimization of $IB(d)$ (or $\text{tr}\,\mathbf{C}_d^2$) among the binary designs. From (3)

$$\text{tr}\,\mathbf{C}_d^2 = k^{-2}\left\{(k-1)^2\sum_{i=1}^{v}r_{di}^2 + \sum\sum_{i\neq j}\lambda_{dij}^2\right\},$$

where λ_{dij} is the number of blocks in which varieties i and j appear together. Since $\sum_{i=1}^{v}r_{di} = bk$ and $\sum\sum_{i\neq j}\lambda_{dij} = bk(k-1)$ are constants, $\text{tr}\,\mathbf{C}_d^2$ is minimized by choosing the r_{di}'s as well as the λ_{dij}'s as equal as possible.

One important kind of nearly balanced design is the class of regular graph designs* introduced by John and Mitchell [7]. These are binary designs with all the r_{di}'s equal and $|\lambda_{dij} - \lambda_{di'j'}| \leqslant 1$ for all $i \neq j$ and $i' \neq j'$, i.e., there are only two possible values of λ_{dij}, that is, λ or $\lambda + 1$. If we consider the v varieties as the vertices of a graph in which there is a line between vertices i and j if and only if $\lambda_{dij} = \lambda$, then the resulting graph is a regular graph, i.e., all the vertices are adjacent to the same number of other vertices.

The (M, S)-criterion is a very crude criterion that does not produce a unique design. Indeed, there are many regular graphs with the same number of vertices and lines. Further comparison among regular graph designs* requires the examination of the structure of their corresponding graphs. For example, some optimality properties had been proved for group-divisible designs* with $\lambda_2 = \lambda_1 + 1$ (see Takeuchi [11], Conniffe and Stone [4], and Cheng [1]). These are regular graph designs whose corresponding graphs are disjoint unions of complete graphs. A BIBD, in fact, can be viewed as a regular graph design corresponding to a complete graph. Recent researches by the author and G. M. Constantine considered other graphs yielding optimum designs that include some triangular and L_2-type partially balanced* incomplete block designs with $\lambda_2 = \lambda_1 + 1$. But in any case, as long as the number of

blocks is not too small, one would expect little difference among the regular graph designs; any of them is highly efficient if not optimal. The results of Cheng [1] can be used to study the efficiencies of regular graph designs; see discussions in Cheng [2]. Using the relation to graphs with the help of a computer, John and Mitchell [7] compiled a table of A-, D-, and E-optimal regular graph designs for parameter values in practical range. This is a useful source of efficient incomplete block designs.

When bk, the total number of experimental units, is not a multiple of v, a binary design satisfying the following conditions is (M, S)-optimal:

(a) $|r_{di} - r_{di'}| \leqslant 1$ for all $i \neq i'$.
(b)   For any fixed i, $|\lambda_{dij} - \lambda_{dij'}| \leqslant 1$ for all $j, j' \neq i$.

This clearly is a generalization of regular graph designs. Such designs are related to graphs with, at most, two different degrees, just as regular graph designs are related to regular graphs. For details, see Cheng and Wu [3], who also discussed the existence, construction, and efficiencies of these designs. Among other results, Jacroux and Seely [6] showed the (M, S)-optimality of the preceding designs.

Although our discussion of near balance was focused on block designs, this simple and useful idea also appears in other types of problems. For example, a simple and efficient algorithm was developed by Wu [12] to achieve nearly balanced assignment of treatments to experimental units with categorical covariate information. He also defined a criterion to measure the distance between an unbalanced assignment and the ideally balanced assignment.

We close with a remark about partially balanced incomplete block (PBIB) designs. Our definition of nearly balanced designs was based on a consideration of statistical efficiency. The analysis of an arbitrary nearly balanced design requires much more effort than that of a BIBD. Of course this may not be a serious problem in the era of

h-speed computing. Partially balanced in-
nplete block designs do not necessarily
·e the λ_{dij}'s as equal as possible, but their
pler structures (especially those with two
ociate classes) leads to easier analysis.
der a PBIB design with two associate
sses, the pairwise comparisons between
varieties are estimated with two different
iances. In a broader sense (in terms of
icture), this can also be considered a kind
near balance. Thus for practical purposes,
IB designs with two associate classes and
= $\lambda_1 \pm 1$ are particularly recommended.
ey are nearly balanced in terms of both
iciency and structure. A different but reed notion of nearly balanced designs may
found in Nigam [9], who discussed a
thod to construct binary designs with un-
al block sizes under which the pairwise
iety comparisons are also estimated with
) different (but very close) variances.

ferences

Cheng, C. S. (1978). *Ann. Statist.*, **6**, 1239–1261.
(The optimality of group-divisible designs with
two groups and $\lambda_2 = \lambda_1 + 1$ with respect to a large
class of criteria is established.)

Cheng, C. S. (1978). *Commun. Statist.*, **A7**, 1327–
1338. [Discusses (M, S)-optimality and computes
the efficiencies of regular graph design.]

Cheng, C. S. and Wu, C. F. (1981). *Biometrika*,
68, 493–500. (Discusses nearly balanced incomplete block designs with unequal replications.)

Conniffe, D. and Stone, J. (1975). *Biometrika*, **62**,
685–686. (Shows the A-optimality of a group-
divisible design with two groups and $\lambda_2 = \lambda_1 + 1$.)

Eccleston, J. A. and Hedayat, A. (1974). *Ann.
Statist.*, **2**, 1238–1255. [Defines the concept of
(M, S)-optimality.]

Jacroux, M. and Seely, J. (1980). *J. Statist. Plan.
Inf.*, **4**, 3–11. [Discusses (M, S)-optimal designs
with unequal replications.]

John, J. A. and Mitchell, T. J. (1977). *J. R.
Statist. Soc. Ser. B*, **39**, 39–43. (Introduces regular
graph designs.)

Kiefer, J. (1975). In *A Survey of Statistical Design
and Linear Models*, J. N. Srivastava, ed. North-
Holland, Amsterdam, pp. 333–353. (This is an
important paper on the optimality of balanced
designs.)

Nigam, A. K. (1976). *Sankhyā B*, **38**, 195–198.

[10] Shah, K. R. (1960). *Ann. Math. Statist.*, **31**, 791–
794. (Introduces the criterion of minimizing
$\operatorname{tr} \mathbf{C}_d^2$.)

[11] Takeuchi, K. (1961). *Rep. Statist. Appl. Res. Union
of Japan Sci. Eng.*, **8**, 140–145. (Shows the E-
optimality of a group-divisible design with λ_2
= $\lambda_1 + 1$.)

[12] Wu, C. F. (1981). *Technometrics*, **23**, 37–44. (Discusses nearly balanced assignment of treatments
to experimental units.)

(BALANCE IN EXPERIMENTAL DESIGN
BLOCKS, BALANCED INCOMPLETE
ESTIMABILITY
GENERAL BALANCE
GROUP-DIVISIBLE DESIGNS
IMBALANCE FUNCTIONS
PARTIALLY BALANCED DESIGNS
REGULAR GRAPH DESIGNS)

CHING-SHUI CHENG

NEARLY BALANCED INCOMPLETE BLOCK DESIGNS *See* REGULAR GRAPH DESIGNS

NEAR-NEIGHBOR ESTIMATION

Near-neighbor methods, introduced by Fix
and Hodges [5], comprise a nonparametric
tool for use in decision theory* or density
estimation* and may be employed with either continuous or discrete data. One of
their main applications is to problems of
classification* or discrimination (*see* DIS-
CRIMINANT ANALYSIS). In their simplest form
they rely on information contained in a single observation near a certain point (the
so-called nearest neighbor or kth nearest-
neighbor rules). More generally, they may
combine information from several neighbors.

Perhaps the simplest near-neighbor procedure is the *nearest neighbor decision rule* or
NN rule. Suppose we have m populations
$\Pi_1, \ldots, \Pi_m$ and a training sample X_1,
$\ldots, X_n$ of correctly classified observations
from these populations. A new observation
X is classified as coming from population Π_i
if the value of X_j which minimizes $|X - X_j|$

comes from Π_i. Cover and Hart [3] showed that in an asymptotic sense, this simple procedure has less than twice the probability of error of the Bayes classification rule in the case of a simple zero-one loss function. Thus at least half the classification information in a large sample is contained in the nearest neighbor. The nearest-neighbor rule has received considerable attention in the context of pattern recognition* and information theory*; see Cover [3], Fritz [6], Fukunaga and Hostetler [7], and the authors cited therein.

A more sophisticated approach to a nonparametric classification is to construct a likelihood estimate for each population and classify new observations on the basis of relative likelihood. This idea motivated Loftsgaarden and Quesenberry [10] to suggest the following nearest-neighbor estimator of a continuous, multivariate density function. Given a d-dimensional random sample $X_1, \ldots, X_n$, and a point x in d-dimensional space, let r denote the distance from x to the sample value kth nearest to x. The volume of the d-dimensional ball centered on x of radius r is given by $v = 2r^d \pi^{d/2} / \{d\Gamma(d/2)\}$, and the quantity $(k-1)/n$ estimates the amount of probability inside this ball. Hence $(k-1)/(nv)$ estimates the value of the underlying probability density at the point x. This estimator is weakly consistent if k and n/k diverge to infinity with n. Devroye and Wagner [4] established strong uniform consistency under the additional restriction that $k/\log n \to \infty$.

Loftsgaarden and Quesenberry's estimator may be regarded as a nonparametric density estimator of the kernel type in which the kernel is the density of a uniform distribution* and the window size or smoothing parameter is taken equal to a nearest-neighbor distance. It has been studied in this context by several authors; references may be found in Mack and Rosenblatt [11]. The value of k that minimizes the mean square error* is asymptotically equivalent to a multiple of $n^{4/(d+4)}$, and the minimum mean square error is asymptotically equivalent to a multiple of $n^{-4/(d+4)}$.

The nearest-neighbor density estimator is self-adjusting to some extent, since the window size is calculated by a data-based method. This property is part of its appeal. However, the self-adjusting property sometimes works to the detriment of the estimator, particularly out in the tails. Mack and Rosenblatt [11] have shown that if the value of the density at x is small, then the bias of the estimator can be excessively large, rendering the nearest-neighbor estimator uncompetitive with the kernel estimator for such values of x.

A problem related to density estimation is that of estimating the Bayes risk (*see* BAYESIAN INFERENCE) associated with a pattern recognition* or classification problem. In the case where the loss function takes only the values zero and one, an upper bound to the Bayes risk may be estimated very easily by counting the errors committed by a nearest-neighbor rule in classifying a test set of data; see Cover [2, Section III.2]. Fukunaga and Hostetler [7] and the authors cited therein have developed more sophisticated near-neighbor methods for estimating Bayes risk.

Near-neighbor methods have important applications in nonparametric regression (*see* REGRESSION, DISTRIBUTION-FREE METHODS IN). A large class of estimators of $E[Y | X = x]$ based on a random sample $(X_1, Y_1), \ldots, (X_n, Y_n)$, may be expressed in the form

$$\hat{E}[Y | X = x] = \sum_{i=1}^{n} W_{ni}(x) Y_i,$$

where the weights $W_{ni}(x)$ depend on $X_1, \ldots, X_n$. Often the W_{ni}'s are chosen as simple functions of nearest-neighbor distances within the X sample. Regression estimators of this type are examined by Stone [12] and in the discussion and references of this article.

We turn now to the case of discrete data, for which the simplest density estimator is the cell proportion estimator. This estimator can become unworkable in the context of classification if the new observation takes a value not previously observed in the training sample. The difficulty can be alleviated by using an estimator based on weighted near neighbors. We shall describe this procedure by considering the case of data on the d-dimensional binary space $B = \{0, 1\}^d$.

The space B admits a natural metric $|\cdot|$, defined by

$$\left| (x_1, \ldots, x_d) - (y_1, \ldots, y_d) \right|$$
$$= \sum_1^d |x_i - y_i|.$$

Two vectors $\mathbf{x}$ and $\mathbf{y}$ in B are said to be distance j apart if $|\mathbf{x} - \mathbf{y}| = j$, where $0 \leqslant j \leqslant d$. If $X_1, \ldots, X_n$ is a random sample from a distribution with density p on B, an estimate of the probability of being distance j from the point x in B is given by

$$\hat{p}_j(\mathbf{x}) = n^{-1}(\text{no. of } X_i\text{'s with } |X_i - x| = j).$$

The usual cell proportion estimator is equal to $\hat{p}_0(\mathbf{x})$. A near-neighbor estimator takes the form $\hat{p}(\mathbf{x}) = \sum_j w_j \hat{p}_j(\mathbf{x})$, for weights w_j depending on n. In the special case where $w_0 = 1$ and $w_j = 0$ for $j \geqslant 1$, $\hat{p}(\mathbf{x})$ is equivalent to the cell proportion estimator. Indeed, the condition for consistency is that the weights converge to those of the cell proportion estimator: $w_j \to \delta_{0j}$ (the Kronecker delta) as $n \to \infty$. Note that $\sum_x \hat{p}(x) = 1$ if the weights w_j are constrained by the relation $\sum_j w_j \binom{d}{j} = 1$. Various versions of this estimator have been suggested by Hall [8] and Hills [9] and the authors cited therein. Examples are given in these references.

The notion of *near-neighbor estimator* has found application in several other areas of statistics, although in contexts disjoint from those already considered. In particular, if there is a correlation between the yields of adjacent plots in a field experiment, the precision of estimates of treatment effects can be improved considerably by using the neighboring residuals as concomitant variables. This problem and others are discussed by Bartlett [1] and the authors cited therein.

References

[1] Bartlett, M. S. (1978). *J. R. Statist. Soc. B*, **40**, 147–174. (Discusses near-neighbor methods in field experiments.)

[2] Cover, T. M. (1969). In *Methodologies of Pattern Recognition*, S. Watanabe, ed., Academic Press, New York, pp. 111–132. (Discusses nearest-neighbor methods in pattern recognition.)

[3] Cover, T. M. and Hart, P. E. (1967). *IEEE Trans. Inf. Theory*, **IT-13**, 21–27.

[4] Devroye, L. P. and Wagner, T. J. (1977). *Ann. Statist.*, **5**, 536–540; (1978) addendum ibid. **6**, 935.

[5] Fix, E. and Hodges, J. L. (1951). "USAF School of Aviation Medicine," Randolph Field, Texas, *Project 21-49-004, Rep. 4, Contract AF41(128)-31*. (Earliest description of near-neighbor methods.)

[6] Fritz, J. (1975). *IEEE Trans. Inf. Theory*, **IT-21**, 552–557.

[7] Fukunaga, K. and Hostetler, L. D. (1975). *IEEE Trans. Inf. Theory*, **IT-21**, 285–293.

[8] Hall, P. (1981). *Biometrika*, **68**, 572–575. (Discusses near-neighbor methods for discrete data.)

[9] Hills, M. (1967). *Appl. Statist.*, **16**, 237–250. (Discrete data case.)

[10] Loftsgaarden, D. O. and Quesenberry, C. P. (1965). *Ann. Math. Statist.*, **36**, 1049–1051. (Introduces nearest-neighbor density estimator.)

[11] Mack, Y. P. and Rosenblatt, M. (1979). *J. Multivariate Anal.*, **9**, 1–15.

[12] Stone, C. J. (1977). *Ann. Statist.*, **5**, 595–645. (With discussion. Describes near-neighbor methods in nonparametric regression.)

(CLASSIFICATION
DENSITY ESTIMATION
DISCRIMINANT ANALYSIS
INFORMATION THEORY AND CODING THEORY
NEAREST-NEIGHBOR METHODS
PATTERN RECOGNITION
REGRESSION, DISTRIBUTION-FREE METHODS IN)

PETER HALL

NEGATIVE BINOMIAL DISTRIBUTION

The negative binomial distribution (NBD) has been used in many disciplines involving count data, such as accident statistics [3, 24], biological sciences [10, 47], ecology [66], epidemiology of noncommunicable events [21], market research [15], medical research [30, 69], and psychology [65]. Bartko [4] has given a summary of the applications and properties of NBD. Kemp [44] has presented an excellent historic review of how NBD and some other discrete distributions arise in accident proneness.

A formulation of the NBD as the distribution of the number of tosses of a coin necessary to achieve a fixed number of heads was published by Montmort* in 1714, although its special forms had been discussed earlier by Pascal* [54] in 1679. On analyzing the effects of various departures from the conditions that lead to the Poisson distribution* for the occurrence of individuals in divisions of space or time, Student [67] concluded that if different divisions have different chances of containing individuals, the NBD provides a better fit than does the Poisson. He also concluded that "if the presence of one individual in a division increases the chance of other individuals falling in that division, a negative binomial will fit best. . . . " This effect, which arises whenever each favorable event increases or decreases the chance of future favorable events, is called *true contagion*. Eggenberger and Pólya's derivation of the NBD from Pólya's urn scheme (*see* URN MODEL DISTRIBUTIONS) renders it an example of true contagion. Greenwood and Yule [24] arrived at the NBD by way of *apparent contagion*, which is the result of heterogeneity arising from the distributions of parameters involved in a population. For example, the NBD arises when the number of accidents sustained by individuals in a time interval is Poisson distributed with parameter λ and λ is assumed to be gamma distributed [3]. *See also* CONTAGIOUS DISTRIBUTIONS.

The NBD is defined in terms of the series expansion $(Q - P)^{-k}$, $k > 0$, $P > 0$, $Q = 1 + P$. Its probability generating function* (pgf) and the probability function (pf) are

$$G(z) = (Q - Pz)^{-k}, \qquad (1)$$

$$P_x = \binom{k + x - 1}{k - 1}\left(\frac{P}{Q}\right)^x\left(1 - \frac{P}{Q}\right)^k,$$

$$x = 0, 1, 2, \ldots, \qquad (2)$$

respectively. The mean and the variance are given by $\mu = kP$ and $\sigma^2 = kP(1 + P)$ with

$\sigma^2 > \mu$. Various authors have used differe forms for the pf of NBD; for a compari of related characteristics, see Shenton Meyers [64].

CHANCE MECHANISMS GENERATING THE NBD [12]

a. **As a Waiting Time* Distribution.**   I sequence of Bernoulli trials with pro bility p of success, let Y be the num of failures before the first success anc the number of failures before the success. Then Y has a geometric di bution* with pgf $(Q - Pz)^{-1}$, wh $Q = p^{-1}$ and $P = (1 - p)p^{-1}$. Since the sum of k independent geometric dom variables (rvs), it has a NBD w pgf (1).

b. **As a Poisson Sum of Logarithmic Seri rv's.**   Let $Y = X_1 + X_2 + \cdots +$ where X's are independent identic distributed (iid) logarithmic rv's with $\ln(1 - \theta z)/\ln(1 - \theta)$, $\theta > 0$. Let N b Poisson rv with parameter λ, indep dent of the X's. Then Y has a NBD w the pgf (1), where $k = -\lambda/\ln(1 - P = \theta/(1 - \theta)$ (see Quenouille [5 Thus the NBD arises as a generali Poisson distribution denoted

Poisson $\vee$ logarithmic

and offers an appropriate physi model for random distribution colonies [30]. Douglas [18] calls t form of NBD *Poisson-stopped logar mic*.

c. **As a Poisson Mixture with Gamma M ing Distribution.**   Let X, given Λ, hav Poisson distribution with mean Λ a let Λ have a gamma distribution*. Th unconditionally, X has a NBD. This proach to NBD has been used to mo accident proneness [3] and purchases nondurable consumer goods [15].

d. **As a Limit of Pólya's Distributi** From an urn containing N balls which a fraction p are white and $1 -$ black, a random sample of size n

taken. After each draw the ball drawn is replaced along with $c = \beta N$ balls of the same color. Let X be the number of white balls in the sample. Then

$$P(X = x)$$

$$= \binom{n}{x}\left(\frac{p}{\beta}\right)^{[x]}\left(\frac{q}{\beta}\right)^{[n-x]} \bigg/ \left(\frac{1}{\beta}\right)^{[n]},$$

$$x = 0, 1, \ldots, n, \quad (3)$$

where $a^{[b]} = a(a+1)\ldots(a+b-1)$. If we let $n \to \infty$, $p \to 0$, $\beta \to 0$, with $np = \lambda$ and $n\delta = \eta$ fixed, then (3) approaches the pf of the NBD with $k = \lambda/\eta$, $P = \eta$, and $Q = 1 + \eta$.

e NBD can also arise

As a Limit of a Binomial Mixture with the Beta Distribution*.

As a Pólya Process.

From a Population Growth with Immigration*.

As the Equilibrium Case of a Markov Chain.

Based on Randomly Distributed Parents and Normally Distributed Progeny, and

From a Queueing* Process.

e zero truncated NBD can arise

As a Group Size Distribution and

As a Zero Truncated Poisson Mixture.

r a detailed discussion, see Boswell and til [12].

**ME GENERALIZATIONS;
LATIONSHIP TO OTHER FAMILIES**

e NBD belongs to many classes of distritions. It is a member of the class satisfy-, the difference equation [4]

$$P_{x+1}/P_x = (\alpha + \beta x)/(x + 1),$$

$$\alpha > 0, \quad 0 < \beta < 1, \quad x = 0, 1, \ldots . \quad (4)$$

rland and Tripathi [32] extended the dif-

ference equation (4) to

$$P_{x+1}/P_x = (\alpha + \beta x)/(x + \lambda),$$

$$\lambda > 0, \quad \beta < 1, \quad x = 0, 1, 2, \ldots \quad (5)$$

The class defined by (5) contains that defined by (4) for $\lambda = 1$. For $0 < \beta < 1$, (5) gives a three-parameter extension of the NBD (see also Tripathi and Gurland [70]). The systems defined by (4) and (5) are contained in a wide class considered by Kemp [43] (see KEMP FAMILIES OF DISTRIBUTIONS).

Gupta [26] defined the modified power series distributions* (MPSD) as

$$P_x = a(x)\{g(\theta)\}^x/f(\theta), \quad (6)$$

where $f(\theta) = \Sigma a(x)\{g(\theta)\}^x$. The NBD belongs to this class for $f(\theta) = (1 - \theta)^{-k}$ and $g(\theta) = \theta, \theta > 0$. It also belongs to the power series distributions* (PSD) of Noak [50] and the generalized power series distributions (GPSD) of Patil [56]. The PSD and the GPSD themselves belong to the MPSD class when $g(\theta)$ is a one-one function of θ.

Jain and Consul [37] defined a class of discrete distributions by utilizing Lagrange's formula

$$\phi(z) = \phi(0) + \sum_{j=1}^{\infty} \left\{ \frac{1}{j!} \frac{d^{j-1}}{dz^{j-1}} (f(z))^j \phi'(z) \right\}\bigg|_{z=0}$$

$$\times \left\{ \frac{z}{f(z)} \right\}^j, \quad (7)$$

where $f(z)$ and $\phi(z)$ are both pgf's. A generalized NBD with pf

$$P_x = \frac{k\Gamma(k + \beta x)}{x!\,\Gamma(k + \beta x - x + 1)}$$

$$\times \theta^x(1 - \theta)^{k+\beta x-x},$$

$$k > 0, \quad x = 0, 1, \ldots \quad (8)$$

is obtained by a proper choice of $\phi(z)$ and $f(z)$ in (7). The NBD is a special case of (8) for $\beta = 1$ (see LAGRANGE AND RELATED PROBABILITY DISTRIBUTIONS). Bhalerao and Gurland [7] developed a class of distributions called Poisson $\vee$ POLPAB with the pgf $\exp\{\lambda(g(z) - 1)\}$, where $g(z) = \{1 - \beta(z - 1)/(1 - \beta)\}^{-\alpha/\beta}$ is the pgf of the Katz system of distribution* given by (4). The NBD is a limiting case of the Poisson $\vee$ POLPAB for $\alpha \to 0$, $\lambda \to \infty$ with $\alpha\lambda/\beta = k$, $0 < \beta < 1$.

MODES, MOMENTS, AND COMPUTATIONAL FORMULAS

Since for the NBD,

$$P_{x+1}/P_x = (k+x)P/\{(x+1)Q\},$$

$P_{x+1} \lessgtr P_x$ according as $x \lessgtr kP - Q$. If $m_1 = kP - Q$ is an integer, the NBD is bimodal with modes at m_1 and $m_1 + 1$. If m_1 is not an integer, the NBD is unimodal with the mode at $[m_1 + 1]$. If $kP < Q$, the mode is at zero [38, 39].

The descending jth factorial moment* of the NBD is

$$\mu_{(j)} = (k+j-1)^{(j)}P^j, \qquad j = 1, 2, \ldots,$$

$$a^{(j)} = a(a-1) \ldots (a-j+1).$$

The recurrence relation for the cumulants is $\kappa_{r+1} = PQ(\partial\kappa_r/\partial Q)$, $r \geqslant 1$, from which the cumulants can be derived. Gupta and Singh [29] derived the expressions for the moments and the factorial moments of the generalized NBD as

$$\mu_j' = \sum_{y=0}^{\infty} \sum_{i=0}^{\infty} \frac{k\Gamma(k+\beta(y+i))}{\Gamma(k+\beta(y+i)-(y+i)+1)}$$

$$\times \frac{\left\{\theta(1-\theta)^{\beta-1}\right\}^{y+i}}{i!(1-\theta)^{-k}} s(j, y), \qquad (9)$$

$$\mu_{(j)} = \sum_{i=0}^{\infty} \frac{k\Gamma(k+\beta(j+i))}{\Gamma(k+\beta(j+1)-(j+i)+1)}$$

$$\times \frac{\left\{\theta(1-\theta)^{\beta-1}\right\}^{j+i}}{i!(1-\theta)^{-k}}, \qquad (10)$$

where the $s(j, y)$'s are the Stirling numbers* of the second kind. When $\beta = 1$, (9) and (10) given the moments and the factorial moments of the NBD. These can also be obtained from the corresponding recurrence relations given by Gupta [26] for the MPSD, with the proper choice of $g(\theta)$.

Computing Formulae and Approximations

Let

$$F(r, k, P)$$

$$= \sum_{x=0}^{r} \binom{k+x-1}{k-1} \left(\frac{Q}{P}\right)^x \left(\frac{1-Q}{P}\right)^k$$

denote the cumulative distribution function of the NBD. Williamson and Bretherton [73] have tabulated values of P_x for many combinations of (k, P), from which $F(r, k, P)$ can be computed. When (k, P) are beyond their table, one needs alternatives. When k is an integer, Patil [55] suggests the formula $F(r, k, P) = 1 - B(k-1, Q^{-1}, r+k)$ where

$$B(c, Q^{-1}, n)$$

$$= \sum_{x=0}^{c} \binom{n}{x}(Q^{-1})^x(1-Q^{-1})^{n-x}.$$

When k is not an integer, $F(r, k, P) = I_{(1/Q)}(k, r+1)$, the incomplete beta function. Thus one can compute $F(r, k, P)$ from tables of binomial probabilities if k is an integer or from the tables of the incomplete beta function otherwise.

Bartko [5] proposed and compared five approximations to the NBD. Two of his most useful are:

A corrected (Gram–Charlier) Poisson approximation.

The Camp–Paulson approximation (see Johnson and Kotz [39]).

The Camp–Paulson approximation is remarkably good, but somewhat complicated. Guenther [25] proposed an approximation based on the incomplete gamma function given by

$$F(r, k, P)$$

$$\approx \left(\int_0^{r_0} t^{N-1}\exp(-\tfrac{1}{2}t)\,dt\right)\Big/\left(2^N\Gamma(N)\right),$$

where $N = kP/Q$ and $r_0 = (2r+1)Q^{-1}$. The extensive tables of the incomplete gamma function can be used for this computation (for references on incomplete gamma function tables, see Guenther [25]).

STATISTICAL INFERENCE

Estimation

For estimation, we consider two cases:

Case 1. k known and P unknown. Roy and Mitra [60] gave the uniform minimum variance unbiased estimator* (UMVUE) of P as $\hat{P} = \hat{\theta}/(1 - \hat{\theta})$ where $\hat{\theta} = T/(nk + T - 1)$ and T is the sample sum based on a sample of size n. The UMVUE and the maximum likelihood estimator* (MLE) of P for the NBD and the generalized NBD can be obtained from the work of Gupta [27, 28] for the mpsd. Maynard and Chow [46] gave an approximate Pitman-type "close" estimator of P as $\hat{p} = k/(k + X + 1)$ for small values of n and P where $p = 1 - P/Q$ (*see* CLOSE-NESS OF ESTIMATORS). Scheaffer [62] proposed some methods for constructing confidence intervals for $p = 1 - P/Q$. Based on an empirical study, he recommends the method that utilizes Anscombe's variance stabilizing transformation (*see* EQUALIZA-TION OF VARIANCES). For sequential estimation* of the mean of the NBD when k is known, see Binns [9] and Gerrard and Cook [22].

Case 2. k and P both unknown. Some of the well-known methods of estimation in this case are:

METHOD 1: METHOD OF MOMENTS*. This gives

$$\hat{k} = \bar{x}^2/(s^2 - \bar{x}), \qquad \hat{P} = s^2/\bar{x} - 1,$$

where $\bar{x}$ is the sample mean and s^2 the sample variance. When $s^2 < \bar{x}$, both $\hat{k}$ and $\hat{P}$ turn out to be negative; in such a case, the NBD should be regarded as inappropriate for the data set.

METHOD 2: FIRST MOMENT AND FREQUENCY OF ZEROS. Equating the population mean and the zero frequency to the sample mean and the sample zero frequency gives

$$\hat{k}\hat{P} = \bar{x}, \qquad (1 + \hat{P})^{-\hat{k}} = f_0,$$

where f_0 is the sample zero relative frequency. These give

$$\hat{P}/\ln(1 + \hat{P}) = \bar{x}/(-\ln f_0),$$

which can be solved iteratively for $\hat{P}$. If $\bar{x} > -\ln(f_0)$, a unique solution always exists.

METHOD 3: MAXIMUM LIKELIHOOD. The maximum likelihood estimators (MLE) sat-isfy the equations

$$\hat{k}\hat{P} = \bar{x}, \quad \ln(1 + \hat{P}) = \sum_{j=1}^{\infty} (\hat{k} + j - 1)^{-1} F_j,$$

where F_j is the proportion of X's in the sample that are greater than or equal to j. On utilizing the second of these equations an estimate of k is obtained through iteration, starting with the moment estimate of k as the initial value. If $s^2 > \bar{x}$, there must be at least one solution $k > 0$. If $s^2 \leq \bar{x}$, the NBD may not be appropriate.

METHOD 4: THE DIGAMMA FUNCTION ESTIMA-TOR. For large values of k and kP, Ans-combe [1, 2] suggested an iterative method of estimating k based on the transformation $y = 2\sinh^{-1}[(x + \frac{3}{8})/(k - \frac{3}{4})]^{1/2}$. The vari-ance of y is approximately $\psi'(k)$, where ψ and ψ' are digamma* and trigamma func-tions. For α large, a good approximation for $\psi'(\alpha)$ is $\psi'(\alpha) \approx (\alpha - \frac{1}{2})^{-1}$. Starting with an initial estimator k_0 of k, the method involves computing the y's from the x's and then computing s_y^2. Successive estimates of k are obtained from $s_y^2 = \psi'(\hat{k})$ until desired accu-racy is achieved.

METHOD 5: GENERALIZED MINIMUM CHI-SQUARE* (GMCS) ESTIMATORS. Methods 2–4 each involve an iterative process to obtain the estimates. The GMCS method yields highly efficient estimators obtainable by solving linear equations [31, 34]. Let $\boldsymbol{\eta}' = (\eta_1, \eta_2, \ldots, \eta_s)$ be functions of the mo-ments and/or frequencies such that $\boldsymbol{\eta} = \mathbf{w}\boldsymbol{\theta}$, where $\mathbf{w}$ is an $s \times r$ matrix of known con-stants and $\boldsymbol{\theta}$ is an $r \times 1$ parametric vector. Let $\mathbf{h}$ be a sample counterpart of $\boldsymbol{\eta}$ and $\hat{\boldsymbol{\Sigma}}_h$ a consistent estimate of the asymptotic covari-ance matrix $\boldsymbol{\Sigma}_h$ of $\mathbf{h}$. A GMCS estimator of $\boldsymbol{\theta}$ is obtained by minimizing the quadratic form

$$Q = (\mathbf{h} - \mathbf{w}\boldsymbol{\theta})'\hat{\boldsymbol{\Sigma}}_h^{-1}(\mathbf{h} - \mathbf{w}\boldsymbol{\theta})$$

with respect to $\boldsymbol{\theta}$ and is given by $\hat{\boldsymbol{\theta}} = (\mathbf{w}'\hat{\boldsymbol{\Sigma}}_h^{-1}\mathbf{w})^{-1}(\mathbf{w}'\hat{\boldsymbol{\Sigma}}_h^{-1}\mathbf{h})$.

For recommendations regarding some of these estimators with respect to efficiency, bias, and other considerations see Anscombe [2] and Shenton and Meyers [64]. Pieters et al. [58] made small-sample comparisons by simulation of Methods 2–4. The authors recommend that for small samples one estimates P and k by Method 1. If $\hat{P} < \hat{k}$, the process is terminated, otherwise ML estimates should be obtained by utilizing the moment estimates as initial values.

In an efficiency comparison of several GMCS estimators, Katti and Gurland [41] concluded that the estimators based on the first two factorial cumulants and logarithm of zero frequency were highly efficient. However, these estimators were obtainable only by solving nonlinear equations. Gurland [31] and Hinz and Gurland [34] compared the efficiency of GMCS estimators relative to that of the MLE and concluded that the estimators based on factorial cumulants and a certain function of zero frequency were highly efficient.

Anscombe [2] discussed the estimation of assumed common k from several negative binomial populations (see Johnson and Kotz [39]). A detailed discussion along with a good bibliography appears in Bliss and Owen [11].

Test of Hypotheses

Hinz and Gurland [36] utilized the statistic $\hat{Q} = (\mathbf{h} - \mathbf{w}\hat{\boldsymbol{\theta}})'\boldsymbol{\Sigma}_h^{-1}(\mathbf{h} - \mathbf{w}\hat{\boldsymbol{\theta}})$, the minimum value of Q, for testing the fit of the NBD and other contagious distributions*. The asymptotic null distribution of $\hat{Q}$ is chi-square with $s - r$ degrees of freedom. They developed methods for testing linear hypotheses regarding the means of several NBD and other contagious distributions (see Hinz and Gurland [35] and Tripathi and Gurland [71]). These procedures do not need any transformation of the data to achieve constant variance and normality. However, $\hat{\boldsymbol{\Sigma}}_h$ in $\hat{Q}$ is obtained by replacing the population moments in $\boldsymbol{\Sigma}_h$ by the corresponding sample moments. Since higher sample moments are subject to large sampling fluctuations, Bhalerao et al. [8] used a statistic similar to $\hat{Q}$ in which $\hat{\boldsymbol{\Sigma}}_h$ is obtained by replacing the parameters involved with their consistent estimators. This yields tests with high power.

Chi [16] gave a locally most power[ful] similar test for testing homogeneity of [sev]eral negative binomial populations.

Graphical Methods for Model Selection

Various methods have been used to iden[tify] an appropriate model from among the [po]tential models for the data at hand. S[ome] methods utilize ratios of factorial cumula[nts] [34], ratios of factorial moments [53], pr[ob]ability-ratio cumulants [31], and a func[tion] of ratios $(x + 1)P_{x+1}/P_x$ of successive pr[ob]abilities [51, 72]. Grimm [23] suggest[s a] method that uses the graph of empir[ical] sum-percent curve plotted on Poisson pr[oba]bility paper. These methods suggest w[hen] the NBD or some other distribution may [be] appropriate.

TRUNCATED (DECAPITATED) NEGATIV[E] BINOMIAL DISTRIBUTION

In many applications of group size distri[bu]tions [33] such as the number of anim[als] born in a litter, the number of cars invol[ved] in an accident, the number of passenger[s in] a vehicle, an NBD with the zero class tr[un]cated is appropriate. The pf of the ze[ro] truncated NBD is

$$P_x = \left(1 - Q^{-k}\right)^{-1}\binom{k + x - 1}{k - 1}$$
$$\times \left(\frac{P}{Q}\right)^x\left(1 - \frac{P}{Q}\right)^k, \quad x = 1, 2,, \ldots$$

Its moments are $\left(1 - Q^{-k}\right)^{-1}$ times the [mo]ments of the full NBD. Thus

$$\mu = kP\left(1 - Q^{-k}\right)^{-1},$$
$$\sigma^2 = kPQ\left(1 - Q^{-k}\right)^{-1}$$
$$\times \left[1 - kPQ^{-1}\left\{\left(1 - Q^{-k}\right)^{-1} - 1\right\}\right]$$

Estimators based on equating the sam[ple] mean and the sample variance to the co[rre]sponding population mean and the varia[nce] do not have simple explicit solutions. San[d]ford [61] suggested a trial-error method ing these equations. David and Johnson used the first three sample moments, ch gave an explicit solution, but these mates were very inefficient. Brass [13] osed estimates based on the frequency nes in the sample, the sample mean, and sample variance. Brass concluded that estimates are more efficient than those of pford for $k \leqslant 5$ and not much less efficit when $k > 5$. In a modification of the , Brass suggested replacing $k\hat{P}(1 - k)^{-1}\hat{Q}^{-(\hat{k}+1)}$ in the equation for $\hat{Q}$ by he proportion of ones in the sample. This s to a simpler equation without any subtial loss in efficiency. Pichon et al. [57] osed a method that uses only the first ple moment and the first sample frency. Schenzle [63] examined the efficy of the estimates of Sampford [61], ss [13], and of Pichon et al. [57] for small es of k. He observed that for small values of k, the estimates of Pichon et al. e slightly higher efficiency than the other estimates. However, in this case all the e estimates have low efficiency; hence enzle recommends the ML estimates. For IVU estimates of the truncated NBD, see oullos and Chamberlides [14].

ARIATE AND MULTIVARIATE ;ATIVE BINOMIAL DISTRIBUTIONS

multivariate NBD, sometimes also ed the negative multinomial distribution MULTINOMIAL DISTRIBUTIONS for details), been used to model the joint distribution he number of accidents suffered by an vidual in k separate periods [6]. For e applications, see Neyman [49].

he bivariate NBD has been used to del accidents in two separate time periods Arbous and Kerrich [3], Edwards and land [19], and Fitzpatrick [20]. The biate model as set forth by Arbous and rich is derived by assuming independent sson distributions for the number of accits in the two intervals with the parame-$\delta_1\lambda$ and $\delta_2\lambda$, respectively; λ is assumed ave a gamma distribution. The pgf of the

resulting distribution is

$$g_1(z_1, z_2) = (A + Bz_1 + B_2z_2)^{-k},$$

$$A > 0, \quad B_1, B_2 > 0, \quad k > 0.$$

Edwards and Gurland [19] extended this model by taking the joint distribution conditional on λ as a correlated bivariate Poisson distribution. Then, assuming λ to have a gamma distribution, the pgf of the extended proneness model is

$$g_2(z_1, z_2)$$

$$= (A' + B_1'z_1 + B_2'z_2 + B_{12}'z_1z_2)^{-k},$$

$$B_1', B_2', B_{12}' < 0, \quad A' > 0, \quad k > 0.$$

If $B_{12} = 0$, g_2 reduces to g_1.

Arbous and Kerrich fitted the distribution represented by g_1 utilizing the method of moments. Edwards and Gurland also fitted their extended model utilizing the method of moments. For this model, Subrahmaniam and Subrahmaniam [68] compared the efficiency of the method of moments and of the method based on zero-zero cell frequency relative to the MLE. They recommend the ML estimates as the other two methods yield estimates with low efficiency.

Conclusions

Martin and Katti [45] fitted the NBD and some other widely used distributions to 35 sets of data published and analyzed by many authors. It turns out that the NBD and the Neyman type-A distribution* have wide applicability. One difficulty with the NBD is that it can be arrived at in different ways such as by *true contagion* and by *apparent contagion*, etc. If the NBD is found to be empirically appropriate for a data set the experimenter has to decide which interpretation is more appropriate. For this purpose, the experimenter has to have a deeper understanding of the mechanism that generates the data so that an appropriate interpretation may be adopted. Some graphical techniques may be helpful in the preliminary selection of an appropriate model.

References

[1] Anscombe, F. J. (1948). *Biometrika*, **35**, 246–254.

[2] Anscombe, F. J. (1950). *Biometrika*, **37**, 358–382.

[3] Arbous, A. G. and Kerrich, J. E. (1951). *Biometrics*, **7**, 340–429.

[4] Bartko, J. J. (1961). *Va. J. Sci.*, **12**, 18–37.

[5] Bartko, J. J. (1966). *Technometrics*, **8**, 345–350.

[6] Bates, G. E. and Neyman, J. (1952). *Univ. Calif. Publ. Statist.*, **1**, 215–276.

[7] Bhalerao, N. R. and Gurland, J. (1977). *Tech. Report 399*, University of Wisconsin, Madison, WI.

[8] Bhalerao, N. R., Gurland, J., and Tripathi, R. C. (1980). *J. Amer. Statist. Ass.*, **75**, 934–938.

[9] Binns, M. R. (1975). *Biometrika*, **62**, 433–440.

[10] Bliss, G. I. and Fisher, R. A. (1953). *Biometrics*, **9**, 176–200.

[11] Bliss, G. T. and Owen, A. R. C. (1958). *Biometrika*, **45**, 37–58.

[12] Boswell, M. T. and Patil, G. P. (1970). *Random Counts in Scientific Work*, Vol. 1, G. P. Patil, ed. Pennsylvania State University Press, University Park, PA, pp. 3–21. (An excellent discussion of the chance mechanisms giving rise to the negative binomial.)

[13] Brass, W. (1958). *Biometrika*, **45**, 59–68.

[14] Cacoullos, T. and Chamberlides, C. A. (1975). *Ann. Inst. Statist. Math.*, **27**, 235–244.

[15] Chatfield, C. (1970). *Random Counts in Scientific Work*, Vol. 3, G. P. Patil, ed. Pennsylvania State University Press, University Park, PA, pp. 163–181.

[16] Chi, P. Y. (1980). *Biometrika*, **67**, 252–254.

[17] David, F. N. and Johnson, N. L. (1952). *Biometrics*, **8**, 275–285.

[18] Douglas, J. B. (1980). *Analysis with Standard Contagious Distributions*. International Co-operative Publishing House, Burtonsville, MD. (An excellent book on properties and applications of contagious distributions. It also has an extensive bibliography.)

[19] Edwards, C. B. and Gurland, J. (1961). *J. Amer. Statist. Ass.*, **56**, 503–517.

[20] Fitzpatrick, R. (1958). *Biometrics*, **14**, 50–66.

[21] Froggatt, P. (1970). *Random Counts in Scientific Work*, Vol. 2, G. P. Patil, ed. Pennsylvania State University Press, University Park, PA, pp. 15–40.

[22] Gerrard, D. J. and Cook, R. D. (1972). *Biometrics*, **28**, 971–980.

[23] Grimm, H. (1970). *Random Counts in Scientific Work*, Vol. 1, G. P. Patil, ed. Pennsylvania State University Press, University Park, PA, pp. 193–206.

[24] Greenwood, M. and Yule, G. U. (1920). *J. R. Statist. Soc.*, **83**, 255–279.

[25] Guenther, W. C. (1972). *Technometrics*, **14**, 385–389.

[26] Gupta, R. C. (1974). *Sankhyā B*, **36**, 288–296.

[27] Gupta, R. C. (1975). *Commun. Statist.*, **A4**, 689–697.

[28] Gupta, R. C. (1977). *Commun. Statist.*, **A6**, 977–991.

[29] Gupta, P. L. and Singh, J. (1981). *Statistical Distributions in Scientific Work*, Vol. 4, C. Taillie, G. P. Patil, and B. Baldessari, eds. D. Reidel, Dordrecht and Boston, pp. 189–195.

[30] Gurland, J. (1957). *Amer. J. Public Health*, **49**, 1388–1399. (An excellent historical review, including interpretation and applications of the negative binomial and other contagious distributions.)

[31] Gurland, J. (1965). *Classical and Contagious Discrete Distributions*, G. P. Patil, ed. Statistical Publishing Society, Calcutta, pp. 141–158.

[32] Gurland, J. and Tripathi, R. C. (1975). *Statistical Distributions in Scientific Work*, Vol. 1, G. P. Patil, S. Kotz, and J. K. Ord, eds. D. Reidel, Dordrecht and Boston, pp. 59–82.

[33] Haight, F. A. (1970). *Random Counts in Scientific Work*, Vol. 3, G. P. Patil, ed. Pennsylvania State University Press, University Park, PA, pp. 95–105.

[34] Hinz, P. N. and Gurland, J. (1967). *Biometrika*, **54**, 555–566.

[35] Hinz, P. N. and Gurland, J. (1968). *Biometrika*, **55**, 315–322.

[36] Hinz, P. N. and Gurland, J. (1970). *J. Amer. Statist. Ass.*, **65**, 887–903.

[37] Jain, G. C. and Consul, P. C. (1971). *SIAM J. Appl. Math.*, **21**, 501–513.

[38] Janardan, K. G. and Patil, G. P. (1970). *Random Counts in Scientific Work*, Vol. 1, G. P. Patil, ed. Pennsylvania State University Press, University Park, PA, pp. 57–75.

[39] Johnson, N. L. and Kotz, S. (1969). *Distributions in Statistics: Discrete Distributions*. Wiley, New York. (This book has a detailed review of the literature on discrete distributions up to 1968. It has an excellent bibliography at the end of each chapter.)

[40] Johnson, N. L. and Kotz, S. (1980). *Int. Statist. Rev.*, **50**, 70–101. (This article has a brief discussion of the developments in discrete distributions during 1969–1980. It has an extensive bibliography on discrete distributions.)

[41] Katti, S. K. and Gurland, J. (1962). *Biometrika*, **49**, 215–226.

[42] Katz, L. (1965). *Classical and Contagious Discrete Distributions*, G. P. Patil, ed. Statistical Publishing Society, Calcutta, pp. 175–182.

[43] Kemp, A. W. (1968). *Sankhyā A*, **30**, 401–410.

[44] Kemp, C. D. (1970). *Random Counts in Scientific Work*, Vol. 2, G. P. Patil, ed. Pennsylvania State University Press, University Park, PA, pp. 41–65. (This article gives an excellent historical review of accident proneness and discrete distributions. It also has an excellent bibliography on the subject.)

[45] Martin, D. C. and Katti, S. K. (1965). *Biometrics*, **21**, 34–48. (An extensive comparison of fits on a collection of 35 data sets by some of the most common contagious distributions.)

[46] Maynard, J. M. and Chow, B. (1972). *Technometrics*, **14**, 77–88.

[47] McGuire, J. V., Brindley, T. A., and Bancroft, T. A. (1957). *Biometrics*, **13**, 65–78.

[48] Montmort, P. R. (1714). "Essai d'analyse sur les jeux de hasards." Paris.

[49] Neyman, J. (1965). *Classical and Contagious Discrete Distributions*, G. P. Patil, ed. Statistical Publishing Society, Calcutta, pp. 4–14.

[50] Noak, A. (1950). *Ann. Math. Statist.*, **21**, 127–132.

[51] Ord, J. K. (1967). *J. R. Statist. Soc. A.*, **130**, 232–238.

[52] Ord, J. K. (1972). *Families of Frequency Distributions*. Hafner, New York.

[53] Ottestad, P. (1939). *Skand. Actu.*, **22**, 22–31.

[54] Pascal, B. (1679). *Varia Opera Mathematica*. D. Pettri de Fermat, Tolossae.

[55] Patil, G. P. (1960). *Technometrics*, **2**, 501–505.

[56] Patil, G. P. (1962). *Ann. Inst. Statist. Math.*, **14**, 179–182.

[57] Pichon, G., Merlin, M., Fagneaux, G., Riviere, F., and Laigret, J. (1976). *Tech. Rep.*, Institut de Recherches Medicale "Louis Malarde," Papeete, Tahiti.

[58] Pieters, E. P., Gates, C. E., Matis, J. H., and Sterling, W. L. (1977). *Biometrics*, **33**, 718–723.

[59] Quenouille, M. H. (1949). *Biometrics*, **5**, 718–723.

[60] Roy, J. and Mitra, S. K. (1957). *Sankhyā A*, **18**, 371–378.

[61] Sampford, M. R. (1955). *Biometrika*, **42**, 58–69.

[62] Scheaffer, R. L. (1976). *Commun. Statist.*, **A5**, 149–158.

[63] Schenzle, D. (1979). *Biometrics*, **35**, 637–640.

[64] Shenton, L. R. and Meyers, R. (1965). *Classical and Contagious Discrete Distributions*, G. P. Patil, ed. Statistical Publishing Society, Calcutta, pp. 241–262. (An excellent comparison of various estimators with respect to bias and efficiency. Also includes a comparison of different forms of NBD.)

[65] Sichel, H. S. (1951). *Psychometrika*, **16**, 107–127.

[66] Skellam, J. G. (1952). *Biometrika*, **39**, 346–382.

[67] Student (1919). *Biometrika*, **12**, 211–215.

[68] Subrahmaniam, K. and Subrahmaniam, K. (1973). *J. R. Statist. Soc. B*, **35**, 131–146.

[69] Talwarkar, S. (1975). *Statistical Distributions in Scientific Work*, Vol. 2, G. P. Patil, S. Kotz, and J. K. Ord, eds. D. Reidel, Dordrecht and Boston, pp. 263–274.

[70] Tripathi, R. C. and Gurland, J. (1977). *J. R. Statist. Soc. B*, **39**, 349–356.

[71] Tripathi, R. C. and Gurland, J. (1978). *Bull. Greek Math. Soc.*, **19**, 217–239.

[72] Tripathi, R. C. and Gurland, J. (1979). *Commun. Statist.*, **A8**, 855–869.

[73] Williamson, E. and Bretherton, M. H. (1963). *Tables of the Negative Binomial Probability Distribution*. Wiley, New York.

(CONTAGIOUS DISTRIBUTIONS)
KEMP FAMILIES OF DISTRIBUTIONS
LAGRANGE AND RELATED
    PROBABILITY DISTRIBUTIONS
MODIFIED POWER SERIES
    DISTRIBUTION
MULTINOMIAL DISTRIBUTIONS
NEYMAN'S TYPE A, B, AND C
    DISTRIBUTIONS
POWER SERIES DISTRIBUTIONS)

RAM C. TRIPATHI

NEGATIVE EXPONENTIAL DISTRIBUTION *See* EXPONENTIAL DISTRIBUTION

NEGATIVE HYPERGEOMETRIC DISTRIBUTION *See* HYPERGEOMETRIC DISTRIBUTIONS; GENERALIZED HYPERGEOMETRIC DISTRIBUTIONS

NEGATIVE MOMENTS

These are simply moments of negative (usually integral) order. The rth negative moment of X is

$$\mu'_{-r}(X) = E[X^{-r}].$$

It is, of course, also the regular rth moment

of X^{-1}:

$$\mu_r'(X^{-1}) = E\left[\left(X^{-1}\right)^r\right]$$
$$= E\left[X^{-r}\right] = \mu_{-r}'(X).$$

(MOMENTS)

NEGATIVE MULTINOMIAL DISTRIBUTION *See* MULTINOMIAL DISTRIBUTIONS

NEIGHBOR DESIGNS

These were introduced by Rees [4], who applied them to problems in serology.

These designs are described as an arrangement of v symbols in b circles (blocks or plates) such that

1. Every circle has k symbols not necessarily all distinct,
2. Each symbol appears r times in the design, not necessarily on r different circles, and
3. Every symbol is a neighbor of every other symbol precisely λ times.

Lawless [3] studied the relationship between balanced incomplete block designs* (BIB) and neighbor designs and devised a necessary condition for a BIB to be a neighbor design. Hwang [2] constructed classes of neighbor designs with $\lambda = 1$ (by repeating these designs t times, neighbor designs with $\lambda = t$ arise). Das and Laha [1] generalized neighbor designs by stipulating that the frequency of occurrence (λ) of every symbol as a neighbor of every other may be one or more than one. They also developed constructions of complete block (i.e., $k = v$) neighbor designs as well as incomplete block (i.e., $k < v$) neighbor designs for even values of v. (The case of neighbor designs for odd values of v is covered in Rees [4].)

References

[1] Das, A. D. and Laha, G. M. (1976). *Bull. Calcutta Statist. Ass.*, **25**, 151–163.

[2] Hwang, F. K. (1973). *Ann. Statist.*, **1**, 786–79?

[3] Lawless, J. F. (1971). *Ann. Math. Statist.*, **42**, 1 1441.

[4] Rees, D. H. (1967). *Biometrics*, **23**, 779–791.

(BLOCKS, BALANCED INCOMPLETE)

NEI'S DIVERSITY MEASURES

Nei (1978) proposed three measures of di sity (with simple genetic interpretatic called minimum, standard, and maxim genetic distances, respectively. These m sures were analyzed and extended by ? [2], who converted them to strict distar functions (satisfying the triangular ineq ity).

References

[1] Nei, M. (1978). *Japan J. Hum. Genet.*, **23**, 341

[2] Rao, C. R. (1982). *Theor. Popul. Biol.*, **21**, 24–

(DISTANCE FUNCTIONS
DIVERSITY INDICES
GENETICS, STATISTICS IN
MEASURES OF SIMILARITY,
 DISSIMILARITY, AND DISTANCE)

NELDER–MEAD SIMPLEX METHOD

Many computer algorithms have been vised to search for the minimum of a fu tion (referred to as the objective function several variables. See, for example, var: chapters in Fletcher [9]. See also *The C puter Journal*, where many articles on subject have appeared, and the entry MA? MATICAL PROGRAMMING. Optimization o function is a more general term and refer either minimization or maximization. should be noted that maximization probl can be converted to minimization probl by changing the sign of the objective fu tion.

There are two broad classes of algorith (1) those that use information about slope or shape of the response of the ob function and (2) those whose rules of ~~~ement are independent of the shape of ~~~response surface. The latter are referred ~~~s direct methods. The Nelder–Mead ~~~) simplex method [4] belongs to the ~~~nd class.

~~~he simplex method was originated by ~~~dley et al. [10] who suggested the use of ~~~rimental designs in the shape of geomet-~~~implexes for the optimization of physical ~~~ems and, in particular, for automating ~~~technique of evolutionary operation*.

~~~purposes of numerical minimization, ~~~ler and Mead extended the method by ~~~nging for the simplex to adapt its form ~~~tinually to the response surface, "elon-~~~g down long inclined planes, changing ~~~ction on encountering a valley at an ~~~e, and contracting in the neighborhood ~~~ minimum" [4]. With these modifica-~~~s, the technique becomes an effective ~~~, in particular, robust procedure for ~~~tion minimization as reported in detail ~~~Nelder and Mead [4], Nelson [5], and ~~~on and Nelson [7]. The main ideas can ~~~be appreciated by considering the mini-~~~ation of a function of two variables, as ~~~trated in Fig. 1. The simplex to be used ~~~have $k + 1$ vertices, where k is the num-~~~of variables present in the objective ~~~tion. For two variables, the simplex will ~~~triangle.

As with all such procedures, choices of starting values and initial step sizes must be made. Subsequent step sizes are chosen by the algorithm. In Fig. 1, let point A represent the starting coordinates and points B and C represent the results of taking an initial step in the X_1 and X_2 directions, respectively. Let $f(A)$ be the value of the objective function at the point A.

If $f(B)$ is greater than both $f(A)$ and $f(C)$, then point A is replaced by its *reflection* E on the line BF that passes through D, the centroid of A and C. If $f(E)$ is not less than $f(A)$ and $f(C)$, then *contraction* occurs to G or H (depending on whether $f(B)$ or $f(E)$ is lower), respectively. If $f(G)$ is greater than $f(B)$ or $f(H)$ is greater than $f(E)$, then *shrinkage* occurs, giving a new simplex $AA'D$ or $CC'D$ depending on whether $f(A)$ or $f(C)$ is the lower, respectively. If $f(E)$ is less than both $f(A)$ and $f(C)$, then *extension* to point F occurs, and either E or F is chosen to replace A, depending on whether $f(E)$ or $f(F)$ is the lower, respectively. The process continues with the newly formed simplex.

Reflection (DE/BD), contraction ($DH/BD = DG/BD$) and extension (DF/BD) coefficients are usually taken as $\alpha = 1$, $\beta = 1/2$, and $\gamma = 2$, respectively. Shrinkage of the simplex involves a present vertex and the centroids of the remaining sides. Rather than attempting to move as directly as possible toward the minimum, the procedure involves moving away from the point having the largest value of the objective function in a direction toward a lower objective function value. Lack of dependence on shape information slows the process, but contributes to its robustness*.

Other than the coefficients mentioned in the preceding paragraph, the only freedom of choice in the simplex method is the selection of the initial simplex. This involves selection of the starting values and step sizes for each of the k independent variables. Good starting values, i.e., ones close to the final values, are always desirable and sometimes critical. The simplex method is no exception here. While it is usually stated that

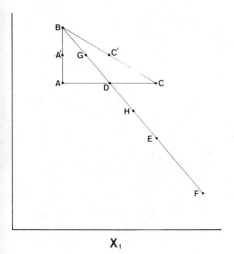

Figure 1   Operation of the simplex method.

the simplex method is limited to problems with no more than six to eight independent variables, it has been used successfully with as many as ten variables when good starting variables were available.

The initial step sizes, i.e., the size of the initial simplex, frequently have only a minor effect on the speed of convergence. However, quite large step sizes can occasionally make up for poor starting values. Also it has sometimes been found useful after a minimum has been reached to restart the procedure at that point but with step sizes significantly larger (possibly a hundredfold) than initially used. In this way, one may move from a local minimum to a global minimum.

A particular advantage of the N-M simplex method is that it is just as easily applied to complex objective functions as to simple ones. For example, in fitting an empirical function to a set of table values, the appropriate objective function is the maximum absolute deviation. This analytically difficult function presents no procedural problems for the N-M simplex method. Response surfaces representing objective functions containing absolute values frequently have been found to have numerous local minima. This difficulty can be circumvented by first using the least-squares* criterion to provide good starting values for the absolute criterion.

Consideration of stopping criteria is the same for the N-M simplex method as for any optimization technique. It seems reasonable to stop iterating when the computer cannot detect any differences either in successive values of the objective function or in changes in any of the parameter values. Consideration of constraints is also the same as for other optimization procedures.

Three methods of handling constraints have been used. The first involves the penalty approach in which the value of the objective function is replaced with a very large number (say, 10^{38}) whenever a constraining boundary is passed. This will keep the search in the required region. The second method transforms the bounded space into an unbounded one so that the constraining value is moved to infinity. A logarithmic transformation for use with both singly and doubly bounded spaces is given by Atwood and Foster [1]. Both of these methods suffer from the fact that an exact result cannot be obtained in the search if the best value of any variable lies on a boundary. If it is known that the solution lies in the plane of the constraint, then a third method is appropriate. This is recommended by Spendley [9] and involves setting equal to the boundary value any parameter that violates the constraint with continuous monitoring to detect when all simplex points arrive at the constraint. The least favorable point is then eliminated, and the search continues in one less dimension.

In statistical problems it may be of interest to obtain the variances and covariances of the final estimates. It is shown by Spendley [9] and Spendley et al. [10] that, for least-squares fitting, the function values at the $k + 1$ points of a k-dimensional simplex will define the required quadratic surface. It is, of course, necessary to choose this simplex, which will enclose the minimum value of the objective function, in such a way that it is neither so small that rounding errors are significant nor so large that the quadratic approximation is poor.

Summary

The N-M simplex method is a simple and robust procedure for seeking the minimum of a function of up to, say, 10 variables. It has been used to solve problems involving (1) nonlinear least squares, (2) curve fitting* with various minimization criteria, (3) estimation using minimum chi-square*, (4) direct maximization of likelihood functions, (5) solution of nonlinear simultaneous equations, and so on. Examples of the first three are given by Nelson [5] and of the last two, by Olsson and Nelson [7]. It has also been used to find optimum values for the parameters of partial sequential binomial tests as reported e.g., by Billard [2] and to find the minimum operating cost for a production scheduling problem with 10 variables as reported by Fan et al. [3].

The method is based neither on gradients (first-order derivatives) nor on quadratic forms (second-order derivatives). No assumptions are made about the response surface other than that it is continuous and has a unique ("global") minimum. The fact that no account is kept of past positions and that most of the steps in the algorithm involve additions, subtractions, or comparisons make it attractive for implementation on microcomputers. FORTRAN programs are given by Olsson [6] and O'Neill [8].

References

[1] Atwood, G. R. and Foster, W. W. (1973). *Ind. Eng. Chem. Process Des. Dev.*, **12**, 485–486.

[2] Billard, L. (1977). *J. Amer. Statist. Ass.*, **72**, 197–201.

[3] Fan, L. T., Hwang, C. L., and Tillman, F. A. (1969). *AIIE Trans.*, **1**, 267–273.

[4] Nelder, J. A. and Mead, R. (1965). *Comput. J.*, **7**, 308–313.

[5] Nelson, L. S. (1973). *Twenty-seventh Annual Technical Conference Transactions*. American Society for Quality Control, pp. 107–117.

[6] Olsson, D. M. (1974). *J. Quality Tech.*, **6**, 53–57.

[7] Olsson, D. M. and Nelson, L. S. (1975). *Technometrics*, **17**, 45–51.

[8] O'Neill, R. (1971). *Appl. Statist.*, **20**, 338–345.

[9] Spendley, W. (1969). In *Optimization*, R. Fletcher, ed. Academic Press, London, pp. 259–270.

[10] Spendley, W., Hext, G. R., and Himsworth, F. R. (1962). *Technometrics*, **4**, 441–461.

(MATHEMATICAL PROGRAMMING
NUMERICAL ANALYSIS
SADDLE-POINT APPROXIMATIONS)

LLOYD S. NELSON

NELSON–AALEN ESTIMATOR   *See*
SURVIVAL ANALYSIS

NESTED ROW-COLUMN DESIGNS

Nested row-column designs make up a class of designs that are arrangements of treatments in several sets of rows and columns such that within each set, row vs. column classification is orthogonal. These designs were introduced by Singh and Dey [2], where the formal definition is presented, and studied by Agrawal and Prasad [1]. These designs are a generalization of lattice square designs.

References

[1] Agrawal, H. L. and Prasad, S. (1982), *Calcutta Statist. Ass. Bull.*, **31**, 131–136.

[2] Singh, M. and Dey, A. (1979). *Biometrika*, **66**, 321–326.

(LATTICE DESIGNS
ROW AND COLUMN DESIGNS)

NESTING AND CROSSING IN DESIGN

NESTED FACTORS

In many experiments, the plots, or experimental units, cannot be assumed to be similar. If possible, the experimenter should partition the set of plots into blocks (or recognize an existing partition) in such a way that the plots within each block may reasonably be assumed to be more alike than plots in different blocks (*see* CONFOUNDING; BLOCKS, BALANCED INCOMPLETE; BLOCKS, RANDOMIZED COMPLETE and DESIGN OF EXPERIMENTS). The plots may then be symbolically listed as u_{ij} for suitable values of i and j; here u_{ij} denotes the jth plot in the ith block. Thus plots with the same value of i are in the same block, but plots with the same value of j are in no special relationship (unless they also have the same value of i, when they are the same plot). The plots in any block may be renumbered without affecting the structure; the subscript j need not even restart at 1 within each block nor does every block need to have the same number of plots. Table 1 shows three different ways of labeling the plots in the *same* structure: in two blocks, each consisting of three plots.

Table 1

(a)	Block number (i)	1	1	1	2	2	2
	Plot number (j)	1	2	3	1	2	3
(b)	Block number (i)	1	1	1	2	2	2
	Plot number (j)	1	2	3	1	3	2
(c)	Block number (i)	1	1	1	2	2	2
	Plot number (j)	1	2	3	4	5	6

The plots are classified into blocks by a *block factor*, whose *levels* are the values of i. They are also classified, in a trivial way, simply into plots. This is a hierarchical classification, because each plot is contained in exactly one block. It is convenient to introduce a *plot factor*, whose levels are the values of j. We say that blocks *nest* plots or that plots *are nested in* blocks to indicate that plots are a division *within* blocks. Equivalently, a block can be distinguished just by the level of the block factor, while the levels of both factors may be needed to identify a plot.

Contrast this structure with that of a drug trial in which twelve patients receive drugs for six months, each patient being given a new drug each month. Here the experimental units are the patient-months, each of which may be specified by the levels of a *patient factor* and a *month factor*. However, neither patients nor months are nested in the other. If u_{ij} denotes the ith patient during the jth month, then the equality of *either* subscript of different units has some physical meaning. The two factors are said to be *crossed* (see Block Structures section).

Although *nesting* is a useful way of describing the blocks-split-into-plots structure, the concept is even more useful when there are three or more factors. A common example with three factors is the split-plot* structure with blocks. Here each block is divided into whole plots (as already shown), and each whole plot is further divided into subplots. Thus blocks nest whole plots, blocks nest subplots, and whole plots nest subplots. In principle, any number of factors may be nested in a chain in this way. Some authors refer to *levels* of nesting. In the split example, blocks are the first level of nes whole plots the second, and so on. (This of the word *level*, implying an order, is ferent from the use in *level of a factor*.)

BLOCK STRUCTURES

The general term for factors that classify plots before application of treatment *block factors*. (This is perhaps unfortu because it is common for the grouping one classification to be called blocks an one of the factors is called *the* block fac A *block structure* on a set of plots is def by one or more block factors and the tions between them. The *nesting* relatic as described earlier. The other common tion is *crossing*.

Unfortunately, at least four different nitions of crossing occur in the litera For definiteness, suppose that two block tors classify the plots into rows and colu According to the different definitions, and columns are *crossed* if:

(a) Neither nests the other.

(b) Each row meets each column, tha the intersection of each row and column contains at least one plot.

(c) The row-column intersections all tain the same number (necessarily zero) of plots (e.g., the rows and umns of a Latin square*).

(d) Condition (c) holds whenever ne restrictions make it possible (e.g rows and columns are both neste blocks, then a row from one block not meet a column from ano block).

Condition (**c**) is stronger than condition which is stronger than condition (**a**); cc tion (**d**) is a sensible modification of cc tion (**c**).

Block structures in which each block tor has equal-sized blocks and each pa block factors is related by nesting or crossing in sense (**d**) are called *simple or* *onal blocks structures*. These were defined by Nelder [11]; several examples are described informally by Bailey [2]. Some common designs whose underlying block structure is simple orthogonal are: completely randomized designs, randomized block designs, Latin squares, split-plot designs (with or without blocks), crisscross designs (*see* STRIP PLOTS), and lattice squares*.

NESTED TREATMENT FACTORS

In a factorial experiment* the treatment factors may also be nested. For example, in an experiment to compare different ways of restricting the energy intake of broilers, there were ten feeding treatments, as follows:

1. Control (i.e., usual diet).
2. Restricting the quantity of feed by three different amounts during the brooding phase.
3. Restricting the quantity of the diet but making compensating increases in the protein content in three different ways during the finishing phase.
4. Changing from the usual diet to a low energy–high protein diet at three different times.

It is sensible to regard the ten treatments as being classified by the four different *methods* of restriction **1–4**. There is thus a four-level method factor, and methods nest treatments. Note that methods **2–4** each have three treatments nested within them, while method **1** has only one. This does not matter, but it makes analysis a little harder.

DESIGNS FOR NESTED STRUCTURES

If a factorial experiment is conducted on a nested block structure, it is quite common to apply levels of each treatment factor to whole levels of one of the block factors. For example, in the simplest classical split-plot designs there are two treatment factors: levels of the first are applied to whole plots,

and levels of the second to subplots. When the numbers of levels of the various factors are suitable, levels of both treatment factors can be applied to subplots in such a way that their interaction*—or some component of it—is confounded with whole plots (*see* CONFOUNDING). Both these types of factorial design for nested structures may be easily constructed using Patterson's *design key*, introduced by Patterson [12] and described more simply by Patterson and Bailey [13].

If the treatments have no factorial structure, one may require that the design have certain desirable properties with respect to each of the nested block factors separately. For example, with three levels of nesting, blocks nesting subblocks nesting plots, it may be possible to allocate treatments in such a way that treatments and blocks form a balanced incomplete block design (BIBD) if subblocks are ignored, and treatments and subblocks form a BIBD if blocks are ignored. Preece [14] termed such designs *nested BIBDs* and gave many examples. Table 2 shows a nested BIBD for seven treatments in seven blocks of two subblocks of three plots. (*See also* BLOCKS, BALANCED INCOMPLETE.)

Kleczkowsi [6] used an interesting design for an experiment with four levels of nesting. The experimental material consisted of the half-leaves of each of the two primary leaves of 28 bean plants. The treatments were eight dilutions of an inoculum of tobacco necrosis virus. The plants were grouped into blocks of two so that each treatment could occur once in each block. Thus blocks nested plants, plants nested leaves, and leaves nested half-leaves. The treatments were allocated to the 112 half-leaves in such a way

Table 2

1 2 4	2 3 5	3 4 6	4 5 0	5 6 1	6 0 2	0 1 3
3 5 6	4 6 0	5 0 1	6 1 2	0 2 3	1 3 4	2 4 5

that: treatments and blocks formed a complete block design; treatments and plants formed a BIBD; and treatments and leaves formed a BIBD.

With three levels of nesting, *nested partially balanced incomplete blocks designs* are similarly defined: If either blocks or subblocks are ignored, the design should be partially balanced with respect to the same association scheme (*see* PARTIALLY BALANCED DESIGNS for a definition of association scheme). Robinson [15] and Homel and Robinson [4, 5] give examples and constructions.

There are similar definitions of balanced and partially balanced designs when the block structure is given by three block factors, called blocks, rows, and columns, with blocks nesting both rows and columns. See Preece [14], Singh and Dey [16], Street [17], and Agrawal and Prasad [1] for some examples and constructions.

ANOTHER MEANING OF NESTING IN DESIGN

Federer [3] has used the term *nested design* in a quite different sense. The sense used elsewhere in this article is that of finer and coarser *partitions*, so that if blocks nest subblocks then each block is a disjoint union of subblocks. Federer's use of the term means that the subblocks do not cover the whole experiment; each block contains *one* subblock, and the remaining plots of that block are not contained in any subblock. The purpose of this partial classification is that two analyses may be made of the results of the experiment, one using all the results and one restricted to those plots that lie in subblocks. If the conclusions from the two analyses do not differ very much, one may perhaps infer that similar future experiments can reasonably be conducted on the smaller number of plots. A *nested design* is an allocation of treatments to plots in such a way that treatments and blocks (ignoring subblocks) form a BIBD, and treatments and subblocks (i.e., ignoring all plots that are not in subblocks)

form a BIBD. Longyear [9, 10] gives constructions and examples.

SURVEYS

The concept of nesting also arises in sampling*. The structure of blocks nesting plots corresponds to cluster sampling*, more levels of nesting to multistage sampling (*see* STRATIFIED MULTISTAGE SAMPLING).

MULTIDIMENSIONAL SCALING

Further different vocabulary is used to describe partitions. Sometimes data values are not comparable across all plots. For example, if the data are several people's rankings of several items, then it is probably not sensible to compare data from two different people directly, and different explanatory models may need to be fitted for each person. In this case, the data would be partitioned into subsets corresponding to people. In the computer package KYST-2 [8], this is called *splitting by people*. Equivalent terms used in multidimensional scaling* are *people-conditioned*, *local order scaling*, and *within people* [7, 18].

Further Reading

The bibliography for DESIGN OF EXPERIMENTS lists textbooks on that subject that deal with nesting. In none of the books is it easy to read the material on nesting out of context.

References

[1] Agrawal, H. L. and Prasad, J. (1982). *Biometrika*, **69**, 481–483.

[2] Bailey, R. A. (1982). In *Applications of Combinatorics*, R. J. Wilson, ed. Shiva, Nantwich, England, pp. 1–18.

[3] Federer, W. T. (1972). In *Statistical Papers in Honor of George W. Snedecor*, T. A. Bancroft, ed., Iowa State University Press, pp. 91–114.

[4] Homel, R. J. and Robinson, J. (1972). *Proc. First Austral. Conf. Combinatorial Math.*, J. S. Wallis

and W. D. Wallis, eds. TUNRA, Newcastle, N.S.W., Australia, pp. 203–206.

[5] Homel, R. J. and Robinson, J. (1975). *Sankhyā B*, **37**, 201–210.

[6] Kleczkowski, A. (1950). *J. Gen. Microbiol.*, **4**, 53–69.

[7] Kruskal, J. B. and Wish, M. (1977). In *Sage University Series on Quantitative Applications in the Social Sciences*, 07-011, Sage Publications, Beverley Hills, CA.

[8] Kruskal, J. B., Young, F. W., and Seery, J. B. (1978). *How to Use KYST-2, a Very Flexible Program to Do Multidimensional Scaling and Unfolding*. Bell Laboratories, Murray Hill, NJ.

[9] Longyear, J. Q. (1981). *J. Statist. Plan. Inf.*, **5**, 181–187.

[10] Longyear, J. Q. (1984). *J. Statist. Plan. Inf.*, **10**, 227–239.

[11] Nelder, J. A. (1965). *Proc. R. Soc. Lond. A*, **283**, 147–178.

[12] Patterson, H. D. (1965). *J. Agric. Sci.*, **65**, 171–182.

[13] Patterson, H. D. and Bailey, R. A. (1978). *Appl. Statist.*, **27**, 335–343.

[14] Preece, D. A. (1967). *Biometrika*, **54**, 479–486.

[15] Robinson, J. (1970). *Biometrika*, **57**, 347–350.

[16] Singh, M. and Dey, A. (1979). *Biometrika*, **66**, 321–326.

[17] Street, D. J. (1981). In *Combinatorial Mathematics VIII*, K. L. McAvaney, ed., *Lect. Notes Math. No. 884*. Springer-Verlag, Berlin, pp. 304–313.

[18] Young, F. W. and Hamer, R. M., *Multidimensional Scaling: Theory and Application* (in preparation).

(BLOCKS, BALANCED INCOMPLETE
BLOCKS, RANDOMIZED COMPLETE
CLASSIFICATION
CLUSTER SAMPLING
CONFOUNDING
DESIGN OF EXPERIMENTS
FACTORIAL EXPERIMENTS
GENERAL BALANCE
HIERARCHICAL CLASSIFICATION
INTERACTION
LATIN SQUARES
MULTIDIMENSIONAL SCALING
ONE-WAY CLASSIFICATION
PARTIALLY BALANCED DESIGNS
SAMPLING
SPLIT PLOTS
STRIP PLOTS)

R. A. BAILEY

NETWORK ANALYSIS

INTRODUCTION

Network analysis is a field of applied mathematics rich in diverse applications. Examples include the transportation of people and goods, the design of computer and pipeline systems, the analysis of financial cash flow management alternatives, the assignment of people to jobs, the routing and scheduling of vehicles and crews, project and production planning, and the grouping of ordered data.

Network problems are important and of interest for the following reasons:

1. Networks provide accurate representations of many real-world problems.

2. Since network models are more visually informative and intuitively appealing than other models, there is a greater likelihood that the models will be implemented and used.

3. Many network problems are also linear programs. Special-purpose network codes are capable of solving certain types of these problems on the order of 100 times as fast as general-purpose linear programming* codes.

4. Network analysis is linked with numerous branches of pure and applied mathematics. Its study has led to significant theoretical developments that have in turn proved useful in the development of these branches of mathematics. This trend is expected to continue.

In a certain sense, the field of network analysis dates back to Leonhard Euler's 1736 paper concerning the so-called Konigsberg bridges problem (*see* GRAPH THEORY). Network models have been *extensively* studied since the advent of the electronic computer during World War II. Some earlier and important excursions into network analysis were made by Leonid Kantorovich and Tjalling Koopmans (corecipients of the 1975 Nobel Prize in Economics). In the last 20 years, however, an enormous amount of research attention has been focused on this field.

Since World War II, an active interface between network analysis and the field of statistics has emerged. Topics from this interface include network reliability, cluster analysis, probabilistic and statistical analysis of network algorithms, PERT*, and, more generally, stochastic network optimization.

Our intent in this article is to introduce appropriate terminology, to describe some important and well-known network analysis problems, and to illustrate the exciting and fertile interface between network analysis and statistics.

Basic Definitions

To begin, we must define what we mean by a network and by network analysis. A *network* $[N, A, W]$ consists of a set N of *nodes* (or *points*), a set A of *arcs* (or *lines*) each of which joins two nodes, and a set W of *weights* (or *costs*) each of which is associated with an arc. An example of a network is provided in Fig. 1. Note that arcs may be directed or undirected. Since undirected arcs such as the one joining b and c in Fig. 1 may be traversed in either direction, we record two instances of the arc, (b, c) and (c, b), to take this into account. This notation, although not standard, is used here to simplify the exposition. The arc weights are prominently displayed in Fig. 1 [e.g., $w(a, b) = 4$].

A *path* joining nodes i and j in a network is a nonempty sequence of arcs of the form $(i, i_1), (i_1, i_2), \ldots, (i_k, j)$. For simplicity, we use the abbreviated notation $i - i_1 - i_2 - \cdots - i_k - j$ to denote a path. A *cycle* is a path joining node i to itself. The *weight of a path p* is given by

$$W(p) \equiv \sum_{(i,j) \in p} w(i, j).$$

A minimization (maximization) problem in network analysis asks for a configuration over $[N, A, W]$ (e.g., a path from node s to node t) such that the total weight associated with the configuration is minimal (maximal). Since there are a finite number of configurations in any finite network problem we discuss, total enumeration is always a candidate solution procedure. However, as we do wish to solve problems within a reasonable amount of computer time, more effective procedures need to be explored. This point serves as motivation for the design and analysis of efficient network algorithms, a major direction in past and current network research.

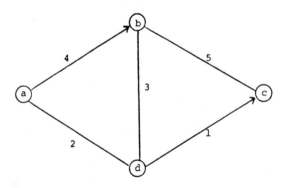

N = {a,b,c,d}

A = {(a,b),(d,c),(b,c),(c,b),(a,d),(d,a),(b,d),(d,b)}

W = {4,1,5,5,2,2,3,3}

Figure 1 An example of network.

Some Well-Studied Network Problems

In this section, we identify a small number of extremely important and widely applicable network analysis problems in order to indicate their diversity in type and difficulty.

Four network analysis problems are sketched briefly. Algorithms for solving them are not presented here, but the reference list points to state-of-the-art solution procedures for each specific network problem. We do, however, make an effort to indicate the computational complexity of each problem by stating worst-case running time as a function of $n = |N|$, where $|N|$ is the cardinality of N or, equivalently, the number of nodes in the network.

An algorithm is said to *run in polynomial time* if there exists an upper bound on the number of operations that is a polynomial in n. An algorithm runs *in exponential time* if the upper bound is $2^{P(n)}$, where $P(n)$ is a polynomial. Polynomial-time algorithms are "good" algorithms, and their order is determined by the highest power of n in the polynomial expression. Clearly, an algorithm of order n^2 is preferable to one of order n^4, and exponential-time algorithms are to be avoided whenever possible.

SHORTEST PATH PROBLEM. For a given network $[N, A, W]$ with arc costs $W = [w(i, j)]$, find the shortest path from a specific origin s to a specific destination t. In Fig. 2, the weights are placed near the corresponding arcs, and the shortest path is clearly indicated. If all the arc weights are nonnegative, an algorithm of order n^2 will solve the prob-

lem. If some arc weights are negative, but there are no cycles with a total weight that is negative, then an algorithm of order n^3 is available. When negative cycles exist, there is no shortest path from s to t.

MINIMAL SPANNING TREE PROBLEM. Given an undirected network $[N, A, W]$, determine a set of arcs $A' \subset A$ connecting all nodes in N for which the sum of the arc weights

$$\sum_{\substack{(i,j) \in A' \\ i < j}} w(i, j)$$

is minimal. The arcs chosen will always form a tree (that is, a network of n nodes and no cycles). In Fig. 3, a minimal spanning tree is displayed. There is a straightforward algorithm (closely related to a shortest path algorithm) of order n^2 for solving this problem.

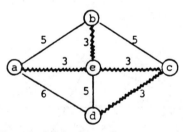

Figure 3   Minimal spanning tree example.

MAXIMUM FLOW PROBLEM. Given a directed network $[N, A, W]$, where $w(i, j)$ is the nonnegative capacity of arc (i, j), find the maximal amount of flow that can be sent from node s to node t. The capacity can be thought of as the maximum amount of some commodity that can "flow" through the arc per unit time in a steady-state situation. For example, let the commodity be fuel oil, let the arcs represent pipelines, and let the arc capacity be a function of pipeline diameter. The idea is illustrated in Fig. 4. There is a recent algorithm for this problem that requires on the order of n^3 computations in the worst case.

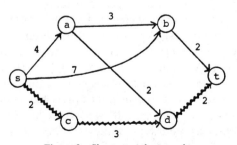

Figure 2   Shortest path example.

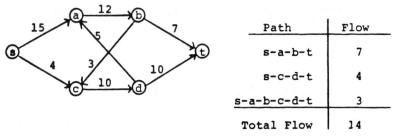

Path	Flow
s-a-b-t	7
s-c-d-t	4
s-a-b-c-d-t	3
Total Flow	**14**

Figure 4   Maximum flow example.

TRAVELING-SALESMAN PROBLEM*. Given an undirected network $[N, A, W]$ where finite arc weights or, in this setting, lengths are defined between every pair of nodes, find the cycle passing through each node $i \in N$ exactly once that minimizes the total length traveled (see Fig. 5 for amplification). The only known algorithms for obtaining an optimal solution to the traveling-salesman problem are exponential-time algorithms. Perhaps the outstanding question in computer science today is whether a polynomially bounded exact algorithm exists for this notorious problem.

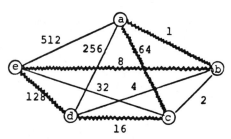

Figure 5   Traveling salesman example.

Interface with Statistics

In this section we illustrate several aspects of the interface between network analysis and statistics. The examples have been chosen for their simplicity but, nonetheless, they should convince the reader of valuable connections between these two important fields.

THE GROUPING OF ORDERED DATA. A problem of interest to statisticians and social scientists alike is how to form a small number of categories from a set of ordered data

that originally consists of many categories. The objective is to lose as little information as possible. Suppose there are n categories in the original data set. Let

$$w(i, j + 1)$$

$$= \begin{cases} \text{the loss in information if categories} \\ i, i+1, \ldots, j \text{ for } i < j \text{ are} \\ \text{combined, and} \\ \infty \qquad \text{if} \quad i \geqslant j. \end{cases}$$

For example, $w(1, n + 1)$ is the information loss when all original categories are combined. Note that on a network with node set $\{1, 2, \ldots, n + 1\}$ and known $(n + 1) \times (n + 1)$ cost matrix $\mathbf{W} = [w(i, j)]$, a path from node 1 to node $n + 1$, which contains at most k arcs, defines a clustering of the original data set into no more than k categories. If the information loss is additive, then the optimal stratification of the original data set into at most k categories is given by the shortest path from node 1 to node $n + 1$ in k or fewer arcs. This statistical problem is a variant of the shortest path problem in the preceding subsection.

HIERARCHICAL SINGLE-LINKAGE CLUSTER ANALYSIS. Hierarchical clustering methods have received a great deal of attention in the statistical literature (*see* HIERARCHICAL CLUSTER ANALYSIS). As an illustration, suppose n data points are to be clustered into k partitions. A similarity matrix $\mathbf{S} = [s(i, j)]$ is given that provides a measure of likeness between every pair of data points based on Euclidean distance. If we view the data points as nodes in a network and define $w(i, j) \equiv -s(i, j)$ for all i and j, then a minimal spanning tree algorithm can be uti-

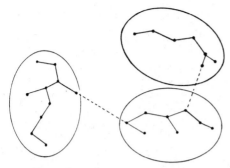

Figure 6   A cluster analysis example.

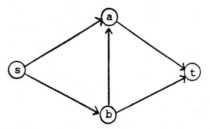

Figure 7   A network reliability example.

lized to perform the clustering. The arcs in the tree will tend to have large similarity or, equivalently, small dissimilarity values. The $k - 1$ arcs with smallest similarity values can then be removed to yield k partitions of the n data points. An example in two-dimensional space with three partitions is exhibited in Fig. 6.

NETWORK RELIABILITY.   An important problem in computer-communication networks is to determine the *reliability* of a network. Suppose that every arc has a probability p of failure and that all the failures are independent events. The probability that nodes s and t are disconnected is given by

$$\Pr(s, t) = \sum_{k=1}^{|A|} A_k p^k (1 - p)^{|A|-k},$$

where A is the set of arcs in the network and A_k is the number of subsets of k arcs in A that disconnect s from t. For the network in Fig. 7, the reader can verify that $A_2 = 3$, $A_3 = 8$, $A_4 = 5$, and $A_5 = 1$. For extremely reliable networks, where p is very small, $\Pr(s, t)$ can be approximated by

$$A_l p^l (1 - p)^{|A|-l},$$

where l is the smallest integer for which $A_l > 0$. Instead of having to enumerate subsets of arcs, which for large networks can become quite burdensome, we can employ a well-known and efficient algorithm for solving the maximum flow problem in order to determine l. Conveniently, it turns out that the dual of the linear programming formulation of the maximum flow problem

can be interpreted as the problem of determining l. This result is known as the *max-flow min-cut theorem*. If we let $p = 0.02$ in Fig. 7, then $\Pr(s, t) \approx 0.0012$ and the approximation yields 0.00113. As the arc reliability decreases (p increases) the approximation begins to lose its validity and the problem of evaluating network reliability becomes much more difficult.

ANALYSIS OF HEURISTICS.   As mentioned earlier, there is strong evidence to suggest that the traveling salesman problem and other combinatorial optimization problems cannot be solved optimally with an algorithm that is guaranteed to run in polynomial time. With this in mind, fast heuristic algorithms (algorithms that tend to produce near-optimal solutions) with polynomial time bounds become a necessity. Several types of analysis have been used to evaluate the accuracy of heuristic solutions. These include *empirical analysis*, *worst case analysis*, *probabilistic analysis*, and *statistical analysis*. The last two categories are of primary interest to us here.

In performing probabilistic analysis on heuristics, we consider algorithms that are guaranteed to provide optimal or near-optimal solutions on almost all problem instances. This requires that a probability distribution over the set of problem instances of each size (e.g., number of nodes) be specified. A result of the following form is then derived:

$$\Pr\left\{ \frac{\text{heuristic solution}}{\text{optimal solution}} \leqslant 1 + \epsilon \right\} \geqslant 1 - \delta,$$

where $\epsilon > 0$ and $\delta \to 0$ as $n \to \infty$. Several successful derivations of this type have been carried out on the traveling-salesman problem in the preceding subsection. For the Euclidean traveling-salesman problem, a problem instance of size n is specified by drawing n points independently from a uniform distribution over the unit square.

The basis for the statistical analysis approach is in the following observation. For a problem such as the traveling-salesman problem, if we can devise a systematic procedure for generating independent heuristic solutions, we should be able to apply techniques from statistical inference and extreme-value theory in order to estimate the value of the optimal solution. Researchers have been successful in computing accurate point and interval estimates for a variety of problems including the aforementioned traveling-salesman problem. In most cases, the confidence intervals are quite narrow and thus informative. Theoretical research in this direction is ongoing.

Concluding Remarks

There are, of course, many other and more profound connections between network analysis and statistics. One important problem that we have not discussed is that of estimating project completion time in a PERT network. The reference list refers the interested reader to sources of information on this and related subject areas. The list is intended to supplement and complement the presentation of topics in this article.

General Bibliography

Bradley, G. (1975). *AIIE Trans.*, **7**, 222–234.

Busacker, R. and Saaty, T. (1965). *Finite Graphs and Networks*. McGraw-Hill, New York.

Christofides, N. (1975). *Graph Theory: An Algorithmic Approach*. Academic Press, New York.

Golden, B. and Magnanti, T. (1977). *Networks*, **7**, 149–183.

Jensen, P. and Barnes, J. (1980). *Network Flow Programming*. Wiley, New York.

Mandl, C. (1979). *Applied Network Optimization*. Academic Press, New York.

Minieka, E. (1978). *Optimization Algorithms for Networks and Graphs*. Marcel Dekker, New York.

Historical Bibliography

Biggs, N., Lloyd, E. and Wilson, R. (1976). *Graph Theory: 1736–1936*. Clarendon Press, Oxford, England.

Ford, L. and Fulkerson, D. R., (1962). *Flows in Networks*. Princeton University Press, Princeton, NJ.

Kantorovich, L. (1960). *Manag. Sci.*, **6**, 366–422.

Koopmans, T. C. (1947). In *Proceedings of the International Statistical Conference*, Washington, DC.

Advanced Texts

Even, S. (1979). *Graph Algorithms*. Computer Science Press.

Frank, H. and Frisch, I. (1971). *Communication, Transmission, and Transportation Networks*. Addison-Wesley, Reading, MA.

Kennington, J. and Helgason, R. (1980). *Algorithms for Network Programming*. Wiley, New York.

Lawler, E. (1976). *Combinatorial Optimization: Networks and Matroids*. Holt, Rinehart and Winston, New York.

Further Reading

Bellmore, M. and Nemhauser, G. (1968). *Oper. Res.*, **16**, 538–558.

Bodin, L. (1972). *Networks*, **2**, 307–310.

Davies, D. and Barber, D. (1973). *Communication Networks for Computers*. Wiley, New York.

Dreyfus, S. (1969). *Oper. Res.*, **17**, 395–412.

Evans, J. (1976). *Networks*, **6**, 161–183.

Frank, H. (1969). *Oper. Res.*, **17**, 583–599.

Garfinkel, R. and Gilbert, K. (1978). *J. ACM*, **25**, 435–448.

Golden, B. and Alt, F. (1979). *Naval Res. Logist. Quart.*, **26**, 69–77.

Golden, B., Bodin, L., Doyle, T., and Stewart, W. (1980). *Oper. Res.*, **28**, 694–711.

Gower, J. and Ross, G. (1969). *Appl. Statist.*, **18**, 54–64.

Hartley, H. O. and Wortham, A. (1966). *Manag. Sci.*, **12**, B469–B481.

Hubert, L. (1974). *Psychometrika*, **39**, 283–309.

Kao, E. (1978). *Oper. Res.*, **26**, 1033–1045.

Karp, R. (1977). *Math. Oper. Res.*, **2**, 209–224.

Lenstra, J. and Rinnooy Kan, A. (1975). *Oper. Res. Quart.*, **26**, 717–733.

Mirchandani, P. (1976). *Comput. Oper. Res.*, **3**, 347–355.

Robillard, P. and Trahan, M. (1977). *Oper. Res.*, **25**, 15–29.

Shogan, A. (1976). *Oper. Res.*, **24**, 1027–1044.

Sigal, C., Pritsker, A., and Solberg, J. (1980). *Oper. Res.*, **28**, 1122–1129.

Thomas, R. (1976). *Networks*, **6**, 287–305.

Van Slyke, R. (1963). *Oper. Res.*, **11**, 839–860.

Zahn, C. (1971). *IEEE Trans. Computers*, **C-20**, 68–86.

(FLOWGRAPH ANALYSIS
GRAPH THEORY
HIERARCHICAL CLUSTER ANALYSIS
PERT
TRAVELING SALESMAN PROBLEM)

BRUCE L. GOLDEN
LAWRENCE D. BODIN

NETWORKS OF QUEUES

The theory of queueing* and a review of multiserver queues* have been presented elsewhere in this encyclopedia. Here we extend these concepts to the area of *queueing networks*.

A queueing network is a finite collection of J systems of servers each with its service time process, queue capacity, and queue discipline, which we will call a *node*. Nodes are connected by paths over which customers travel. The collection of nodes and paths is called a *queueing network*.

There may be arrival processes to several (or all) of the nodes. These are often assumed to be Poisson processes* (but see Kelly [28]), independent of each other and of the queueing network (but see Jackson [25]).

Each arrival proceeds through the network according to some *routing scheme*. The usual assumption is that customers departing from one node choose the next node to visit as a multinomial*, $(J + 1)$-dimensional random variable where the next "node" might be a dummy node Δ, called "the outside." Under this assumption, the routing process is a Markov chain (*see* MARKOV PROCESSES) whose states are the nodes of the network with Δ added. Δ is usually taken to be an absorbing state and all other states are transient. Melamed [41] provides some exceptions to these assumptions. Kelly [28] provides for a fixed routing scheme and identifies types of customers with the routes

taken. Jansen and König [26] provides for yet more general routing.

Research has been concerned primarily with the vector-valued queue length process (the vector of the queue lengths at each node). Few results are available for the sojourn time process (time to traverse the network). Reference 28 is basic to the study of the queue length processes. Sojourn times are reviewed by Disney [15] and Melamed [42].

SOME BACKGROUND

A need for a theory of queueing networks probably arose early in the design of telephone systems. Syski [52] attributes to R. I. Wilkerson the remark: "The problem of interconnections was born with the completion of the third commercial telephone instrument. . . . " To connect, directly, each telephone instrument to every other instrument in a system of, say 10,000, users there would have to be on the order of 10^7 connections made to each instrument, a number that is physically and economically infeasible. Such an infeasibility raises the need to design networks to provide a switching system that will handle many, but not necessarily all, subscribers at once. The inability to handle all possible calls that could arise then brings with it the problems of congestion, waiting, queueing, lost calls, and the others that have formed the area of queueing theory and queueing network theory.

But it was recognized early in fields distinct from telephone systems that related problems of congestion and queueing arise also. Early work in production systems design recognized the need for inventories (*see* INVENTORY THEORY), work-in-process storage, raw material storage, and the like at various stages in the production network. See Muth and Yeralan [44] and Solberg [51], for recent work in this area. Road and highway design has recognized the problems of congestion and queueing in networks. See Haberman [22], for example. Sea traffic [34] and air traffic [36] encountered the same class of networking problems. One can argue, not unconvincingly, that any traffic system is inherently a queueing network with many of the same problems as those first recognized formally in telephone networking.

Research in queueing network theory received additional impetus in the early 1960s when computer scientists considered the interconnections of individual computers into a computer network. Once again the problems of congestion and queueing appeared (see Kleinrock [32]). Today, with a need to transmit both voice and data information across telephone lines, microwave systems, and satellite systems, a need for knowledge of how congestion and queueing occur, how they can be mitigated, how systems can be designed economically so as to avoid the more dire consequences of congestion (system "crashes") has pushed queueing network theory into a prominent position in applied probability*. Reference 16 gives a state-of-the-art look at many of the topics now being considered in this field and gives an idea of the problems and approaches being taken to solve them.

SOME RESULTS

Early papers concerning queueing problems in telephone networks tended to be concerned with specific networks and specific problems therein. Syski [52] provides an introduction to the history of the development of queueing theory and queueing network theory in telephony. Thus, while one can find a considerable amount of research on what now could be called queueing networks in the telephone literature of the 1930s, it is usually argued that queueing network theory had its genesis in the paper by J. R. Jackson called "Networks of Waiting Lines" [24] and the followup paper by the same author [25].

In Jackson [24], there is one arrival process to the network. It is a Poisson process (λ). Each node consists of a single server (or multiple servers, although we only discuss

the single server case); queue capacities are unlimited; and the queue discipline is unspecified. We will assume a first come–first served discipline. The routing scheme is a Markov chain, with Δ the only absorbing state. The set of nodes forms a single transient class. Every arrival to the network eventually leaves.

Then $\mathbf{N}(t) = (N_1(t), N_2(t), \ldots, N_J(t))$ is a vector-valued Markov process*. $N_j(t)$ is the number of customers at node j at time t.

Under these conditions, one finds that

$$\lim_{t \to \infty} \Pr\left[\mathbf{N}(t) = \mathbf{k}\right] = B_J \rho_1^{k_1} \rho_2^{k_2} \ldots \rho_J^{k_J},$$

a *product form solution*. The quantities ρ_j are constants depending on the arrival rate and service rate at node j. The constant $B_J > 0$ is a normalizing constant that, when it exists, ensures that the limiting values are probabilities. In the specific example given, when $B_J > 0$, $B_J = (1 - \rho_1)(1 - \rho_2) \ldots (1 - \rho_J)$, implying that in the limit the queue lengths at the J nodes are independent. Furthermore, this result seems to say that these queue length processes are each generated by $M/M/1$ queues (*see* QUEUEING THEORY). However, while the first implication is correct, the second is not. This striking anomaly has been discussed in detail by many authors (e.g., see Kelly [28]).

In the preceding model, one has customers entering the network from "outside" and leaving the network for the "outside." These conditions define an *open* queueing network. An important question arises in open networks as the existence of $B_J > 0$. Thus any study of open queueing networks must establish conditions on the parameters of the network that ensure this existence. In *closed* queueing networks such a constant always exists, for in such a network one assumes that there is no "outside" but merely a network in which a finite number N of customers flows endlessly. Of course, the independence of the J queue length processes is no longer valid in these cases. The constant term does not factor correctly since $k_1 + k_2 + \cdots + k_J = N$. Closed networks have been used to study equipment repair systems

[21] and fleet operations [34] as well as several computer programs [16]. They are discussed in more detail by Gordon and Newell [19]. A summary of results is given by Koenigsberg [35].

These basic models have been extended to include: mixed networks (e.g., see Basket et al. [2]: A network is mixed if it closed for some customers but not closed for others); more than one type of customer (see, e.g., Kelly [28]); fixed routing schemes depending on customer types (e.g., see Kelly [28]); more general arrival processes and more general service processes (see, e.g., Jansen and König [26] and Kelly [28]). In fact, Kelly [28] starts with a Markov process $X(t)$, whose states need not be scalar-valued, but from which one can ultimately obtain queue length distributions. If $X(t)$ is such a Markov process with the added properties: (a) for fixed t_0, the arrival process after t_0 is independent of $X(t_0)$; (b) for fixed t_0, the departure process before t_0 is independent of $X(t_0)$, then $X(t)$ is called a *quasi-reversible* Markov process. If each node of a queueing network behaves in isolation as a quasi-reversible Markov process, then the network is a quasi-reversible network. These networks exhibit many of the same properties as the network in Jackson [24]. In particular, a similar product form of solution is obtained. The result extends significantly in that several queue disciplines (e.g., processor sharing) may produce the requisite quasi-reversibility, even though service times are not exponentially distributed. See refs. 17, 28, and 43, for example.

These results can be extended yet further to give conditions under which the queue length process distribution (for the stationary case) depends on the service time distribution only through the mean service time. See Syski [52] for the early history of this result. This property, called *insensitivity*, is quite useful, for with it, one need never explore the queue length process for systems whose service times are more general than exponentially distributed. The equilibrium queue length distribution depends only on expected service times and not on distribu-tional assumptions of these service times. In such networks, one need only consider cases in which $\mathbf{N}(t)$ is a Markov process on a countable state space.

For those networks lacking the insensitivity property, $\mathbf{N}(t)$ may not be a Markov process (but see, e.g., Disney [13]). In such cases one must work with non-Markov processes or one must augment the process $\mathbf{N}(t)$ with such random variables as the expended service time, for every nonexponential service time server in the network (e.g., the method of inclusion of supplemental variables must be applied to every nonexponential server). In either case, the state space of the process may be quite large, perhaps uncountable. The analysis of such systems is difficult. See refs. 1, 6, 16, 26, and 28, for example.

MORE GENERAL NETWORKS

There are queueing networks that have neither the product form for their stationary queue length probabilities nor the quasi-reversibility property nor the insensitivity property. These networks are under active investigation, but have yet to be formalized generally; the simplest example is the *overflow network*, first studied by Palm. See Khinchin [29] and Syski [52] for further discussion of this particular network.

In Palm's overflow system with two servers $\mathbf{N}(t) = (N_1(t), N_2(t))$, the vector of queue lengths at server 1 and 2, respectively, is a finite, irreducible Markov process. It has four states. Its stationary distribution does not have a product form. The network is not quasi-reversible, and the insensitivity property does not hold. Networks lacking these properties are important and deserve considerable attention.

SOJOURN TIMES IN QUEUEING NETWORKS

Sojourn time is one of the two major topics in queueing theory (the queue length process is the other). Yet there is little known about this process in queueing networks [15]. It is known (see Lemoine [38] and Melamed [42] for a more extended discussion) that for some networks (e.g., an $M/M/1$ queue in series with a queue having exponentially distributed service times, infinite queue capacities and first come–first served queueing discipline), the sequence of sojourn times for each fixed customer at the nodes it visits is a sequence of independent random variables.

However, if there are multiple paths joining two nodes of the network, then customers flowing along one path may be *overtaken* by customers flowing on others. Effectively what occurs in networks with overtaking is that a given customer who leaves a given node and follows one of the multiple paths to a later node finds, upon arrival at the later node, customers who were behind it at the former node, but who took one of the alternate routes. Sojourn times depend in part on the number of customers a given customer encounters upon arrival to a node. Thus the given customer may have to wait at the subsequent queue a longer time (due to customers who overtook it on the alternate route). The number of customers who could get to the subsequent node ahead of the given customer depends, in part, on the length of time the given customer spent at the former node. In this way, the sojourn time at the former node and that at the subsequent node, for a given customer, will be dependent random variables.

This overtaking phenomenon occurs only in networks. Little is currently known about these dependencies other than the fact that they exist. See, e.g., refs. 5, 48, and 54. Correlations in these sojourn time processes for a given customer seem to be small [30], but the joint distribution of one customer's sojourn times at the queues it visits is difficult to obtain even in special cases [31].

In some cases (the overflow queue described earlier, for example) the sojourn time of a given customer is quite easy to obtain, but the sequence of sojourn times of the sequence of customers is difficult [15]. This entire area is in need of considerably more attention.

CUSTOMER FLOW IN NETWORKS

It is apparent that queueing phenomena in networks come about because of the interaction of arrival processes to nodes, service processes, and network configurations. In most studies the service processes and network configurations are considered as given, as is the arrival process to the network. However, the flow of customers inside the network determines the queueing properties at the nodes. These flows may be significant transformations of the arrival process to the networks due to their interactions with the service time processes and routing scheme of the network. To properly understand queueing network phenomena, one must understand how arrival processes to the network are transformed by service time processes and routing schemes, how the several customer flow processes are related (e.g., cross-correlation), internally related (e.g., autocorrelation), and what effects these have on queue lengths, sojourn times, and other measures of effectiveness.

The study of internal flows requires the study of marked point processes where the mark space may not be countable. In this study one can identify three operations on random point processes* (customer flows) that the point process literature call thinning, stretching, and superposing. See, e.g., Çinlar [8] and Daley and Vere-Jones [12].

One can consider the arrival process to the network as a random point process or, more generally, as a random marked point process. Inside the network, rules (called routing schemes earlier) select points from this process to form arrival processes to each of the J nodes. There are a large number of useful rules [14]; they are called *decomposition rules* in queueing network theory and the corresponding operation on a point process is called *thinning*. Customers enter a queueing system, are served, then they depart. In this way, one can view the departure process as a stretched version of the arrival process. The *stretching* process is the sequence of sojourn times of the customers. See Daley [11] for a survey of many results on queueing departure processes. Also in queueing networks, several streams of customers may be recomposed (merged) to form a single stream. In the most studied case [8], one assumes that independent point or marked point processes are superposed. In most queueing networks, the operation of thinning creates dependencies between flow processes inside the network. See Chandramohan [7] for some results concerning the cross-correlations of customer flows in queueing networks.

OTHER APPROACHES AND TOPICS

Queueing network theory is still in a state of development. Consequently, one finds a diversity of approaches, unsolved problems, areas under active investigation, incomplete results, and gaps of knowledge yet to be filled. Some of these area are described briefly below.

Simulation

For the practitioner faced with a queueing network problem, the most commonly occurring approach is to use a digital computer simulation (Monte-Carlo*). There now exist several computer packages that can be used as building blocks so as to prepare programs to simulate quite complex networks.

Statistical analysis of data output by a computer simulation* is a rapidly developing field that holds promise of making computer simulation a more useful tool for the person confronted with an ongoing problem. See Kobayashi [33], for example, for an indication of current interest in simulations and their statistical analysis. An indication of the level of current interest in the simulation of queueing networks and the statistical analysis of such problems is given in ref. 4, where

one finds over a dozen $1\frac{1}{2}$ hour sessions devoted to this topic.

Approximations

At least three areas of interest can be collected under this title. These include diffusion approximations, numerical approximations, and stochastic process approximations.

DIFFUSION APPROXIMATIONS.   The basic idea here is quite old and has been used to derive equations such as the Fokker–Planck equations* of diffusion from random walks*. See, e.g., Kac [27]. In queueing network theory, one replaces the differential-difference equations (the Kolmogorov differential equations for the Markov queue length process) with a partial differential equation. The resulting partial differential equation is a diffusion equation in J dimensions (*see also* DIFFUSION PROCESSES).

As usual, boundary conditions can be troublesome. To avoid some of these difficulties it is often assumed that the network is in *heavy traffic*. That is, flow rates are such that queues in the network rarely become empty. Basic work here can be found in Gaver [18]. Reiman [46] gives a discussion of the approach, its uses in studying the sojourn time problem in queueing networks, and a bibliography for further reading.

NUMERICAL APPROXIMATIONS.   For large queueing networks where $N(t)$ is a Markov processes and especially for closed networks, there are major problems involved in computing. In closed networks, one must find the normalizing constant B_J, which depends on all of the parameters of the network. But in large queueing network models, with a few exceptions, computational problems abound. Considerable research has gone into this class of problems. References 37, 40, 47, 49, and 50 propose computational algorithms on various aspects of the problem. Each of these papers has a bibliography.

In some networks in which $\mathbf{N}(t)$ is a Markov process where the equilibrium distribution of the queue length process may not have a product form of solution (e.g., the overflow problem), one finds that by a judicious ordering of the states, the generator of the queue length process can be partitioned into a useful block form (e.g., see Disney [13]). In fact, by generating interarrival time or service time distributions as first passage times in a subsidiary Markov process, one can generalize the class of networks with Poisson arrivals and exponential servers to those whose corresponding processes are now called *phase-type processes*. The corresponding queue length process generator can then be partitioned into a block form, and this form can be exploited to obtain some rather powerful computational methods. The basic idea here has been developed and extended. See Neuts [45].

STOCHASTIC APPROXIMATIONS. As has been noted, flows of customers inside of queueing networks, in part, determine queueing properties of customers at each of the nodes. For networks in which $N(t)$ is a Markov process but which do not have a product form for the equilibrium queue length distribution and for the study of a particular node even in some of the networks having such a product form (e.g., for the study of the marginal distribution of the waiting times at a node), the analysis of these customer flows is of importance. For networks in which $N(t)$ is not a Markov process and for other non-Markov processes (e.g., sojourn times), then analysis of customer flow processes may be among the few approaches available. However, as noted, these flows are seldom of simple form. The question then arises: Can one approximate these more general processes with simpler stochastic processes (e.g., renewal* processes)? More cogently, if one does approximate these more general processes with simpler processes, what are the consequences to the queueing properties?

One of the earliest uses of approximating one random process with another in a queueing network, is the *equivalent random*

method (e.g., see Cooper [10]). Here, one replaces the stream of overflowing traffic with a process whose resource utilization in the overflow system has the same first two moments. Holtzman [23] discusses the effects of this approximation on various aspects of the queueing behavior of the network. More recently, Whitt [56, 57] studies the problems of approximating a point process with a renewal process. This area of investigation promises to deliver us from the stringent requirements of networks with product forms of solutions.

Other Models

Most of the work in queueing network theory has evolved through a theory of Markov processes for the study of queue length processes and through the study of point processes for the study of customer flow processes. More recently research is appearing using martingale* methods to study customer flows in networks (see Walrand and Varaiya [55]) and for the study of queueing properties themselves (see Brémaud [3]). Where these new approaches will lead is still an open question. Results so far appear to be promising for the study of customer flow processes by providing an elegance and a formal unity for many existing results.

A promising new approach to queueing networks in which $N(t)$ is a Markov process, including the possibility of estimating transient behavior using operator-theoretic methods and semigroup properties implied by the Chapman–Kolmogorov equations*, can be found in Massey [39].

Literature

For the reader seriously interested in queueing network theory, some of its applications, and some of its problems, Disney [16] is essential. For a study of its application to the design of a large complex system, Kleinrock [32], especially Volume 2, Chapters 5 and 6, is quite worthwhile. For the reader interested in the genesis of queueing network theory in telephony, Syski [52] is important, especially Chapters 7–10. The two Jackson papers [24, 25] are the basis for a considerable amount of research work currently under way. Kelly [28] provides a unification and extension to the papers of Jackson and may be the best beginning place for the reader to catch up as quickly as possible with the state of the art in the study of queue length processes, product forms, insensitivity, and related topics. Its bibliography allows one to go deeper. Franken et al. [17] provide a deep background for understanding the mathematical underpinnings of these networks. This reference does suppose a rather advanced knowledge about point process theory, however. Several textbooks (including Kelly [28]) cover various aspects of queueing networks and some of their problems (e.g., refs. 10, 20, 32, and 33). Cohen [9] is a source for single server queueing results.

Flow process results in networks have not been pulled together in any one place. Disney [14] is a first attempt, now rather out of date. Çinlar [8] surveys results of superposing point processes; Daley [11] surveys results of departure processes. So far, there appears to be no survey of decomposition results.

Martingale methods are discussed in detail by Brémaud [3], but other topics (e.g., statistical analysis) are still under active investigation and have not been summarized.

For the reader trying to stay up (or catch up) with the field, there is no one best place to look for results. Results appear in most applied probability journals* of the world as well as in most computer science journals and various proceedings of conferences, technical reports, and the like. Papers on various applications also are spread over most major engineering journals as well as journals of the sciences.

References

[1] Barbour, A. D. and Schassberger, R. (1981). *Adv. Appl. Prob.*, **13**, 720–736.

[2] Baskett, F., Chandy, M., Muntz, R., and Palacios, J. (1975). *J. Ass. Computer Mach.*, **22**, 248–260.

[3] Brémaud, P. (1980). *Point Processes and Queues: Martingale Dynamics*. Springer-Verlag, Berlin.

[4] *Bulletin of the Joint National Meeting of TIMS/ORSA*, April 25–27, 1983.

[5] Burke, P. J. (1969). *Operat. Res.*, **17**, 754–755.

[6] Burman, D. Y. (1981). *Adv. Appl. Prob.*, **13**, 846–859.

[7] Chandramohan, J., Foley, R. D., and Disney, R. L. (1985). *J. Appl. Prob.* (to appear).

[8] Çinlar, E. (1972). In *Stochastic Point Processes: Statistical Analysis, Theory and Applications*, P. A. W. Lewis, ed. Wiley, New York, pp. 549–606.

[9] Cohen, J. W. (1969). *The Single Server Queue*. North-Holland, Amsterdam.

[10] Cooper, R. B. (1981). *Introduction to Queueing Theory*, 2nd ed. North-Holland, New York.

[11] Daley, D. J. (1976). *Adv. Appl. Prob.*, **8**, 395–415.

[12] Daley, D. J. and Vere-Jones, D. (1972). In *Stochastic Point Processes: Statistical Analysis, Theory and Application*. P. A. W. Lewis, ed. Wiley, New York, pp. 299–383.

[13] Disney, R. L. (1972). *Manag. Sci.*, **19**, 254–265.

[14] Disney, R. L. (1975). *Trans. AIIE*, **7**, 268–288.

[15] Disney, R. L. (1982). *Proc. 1982 IEEE Int. Large-Scale Sys. Symp.*, 104–106.

[16] Disney, R. L. and Ott, T. (eds.) (1982). *Applied Probability—Computer Science: The Interface*. Birkhauser, Boston.

[17] Franken, P., König, D., Arndt, U., and Schmidt, V. (1981). *Queues and Point Processes*. Akademie-Verlag, Berlin.

[18] Gaver, D. (1968). *J. Appl. Prob.*, **5**, 607–623.

[19] Gordon, W. J. and Newell, G. F. (1967). *Operat. Res.*, **15**, 266–278.

[20] Gross, D. and Harris, C. M. (1974). *Fundamentals of Queueing Theory*. Wiley, New York.

[21] Gross, D. and Ince, J. F. (1978). *Trans. AIIE*, **10**, 307–314.

[22] Haberman, R. (1977). *Mathematical Models, Mechanical Vibrations, Population Dynamics and Traffic Flow*. Prentice-Hall, Englewood Cliffs, NJ, pp. 259–394.

[23] Holtzman, J. M. (1973). *Bell Syst. Tech. J.*, **52**, 1673–1679.

[24] Jackson, J. R. (1957). *Operat. Res.*, **5**, 518–521.

[25] Jackson, J. R. (1963). *Manag. Sci.*, **10**, 131–142.

[26] Jansen, U. and König, D. (1980). *Elektron. Informationsverarb. Kybernetik.*, **16**, 385–397.

[27] Kac, M. (1954). In *Selected Papers on Noise and Stochastic Processes*, N. Wax, ed. Dover, New York, pp. 295–337.

[28] Kelly, F. P. (1979). *Reversibility and Stochastic Networks.* Wiley, New York.

[29] Khinchin, A. Y. (1960). *Mathematical Methods in the Theory of Queueing.* Griffin, London.

[30] Kiessler, P. K. (1980). "A Simulation Analysis of Sojourn Times in a Jackson Queueing Network," *Tech. Rep. No. VTR 8016,* Dept. of Industrial Engineering and Operations Research, Virginia Polytechnic Institute and State University, Blacksburg, VA.

[31] Kiessler, P. K. and Disney, R. L. (1982). "The Sojourn Time in a Three Node Acyclic, Jackson Queueing Network." *Tech. Rep. No. VTR 8203,* Department of Industrial Engineering and Operations Research, Virginia Polytechnic Institute and State University, Blacksburg, VA.

[32] Kleinrock, L. (1975/1976). *Queueing Systems,* Vol. 1—theory and Vol. 2—computer applications. Wiley, New York.

[33] Kobayaski, H. (1978). *Modeling and Analysis: An Introduction to System Performance Evaluation Methodology.* Addison-Wesley, Reading, MA.

[34] Koenigsberg, E. and Lam, R. C. (1976). *Operat. Res.,* **24**, 516–529.

[35] Koenigsberg, E. (1982). *J. Opns. Res. Soc.,* **33**, 605–619.

[36] Koopman, B. D. (1972). *Operat. Res.,* **20**, 1089–1114.

[37] Lavenberg, S. S. (1982). In *Applied Probability—Computer Science: The Interface.* R. L. Disney and T. Ott, eds. Birkhauser, Boston, pp. I-219–252.

[38] Lemoine, A. (1970). *Manag. Sci.,* **25**, 1034–1035.

[39] Massey, W. A. (1984). *Adv. Appl. Prob.,* **16**, 176–201.

[40] McKenna, J. and Mitra, D. (1981). In *Proceedings of the Eighth International Symposium on Computer Performance, Modeling, Measurement, and Evaluation.*

[41] Melamed, B. (1979). *Adv. Appl. Prob.,* **11**, 422–438.

[42] Melamed, B. (1982). *Math. Operat. Res.,* **7**, 223–244.

[43] Muntz, R. R. (1972). "Poisson Departure Processes and Queueing Networks," *IBM Res. Rep. RC4145,* T. J. Watson Research Center, Yorktown Heights, NY.

[44] Muth, E. and Yeralan, S. (1981). In *Proc. of the 20th IEEE Conference on Decision and Control,* pp. 643–648.

[45] Neuts, M. F. (1981). *Matrix-Geometric Solutions in Stochastic Models: An Algorithmic Approach.* Johns Hopkins University Press, Baltimore, MD.

[46] Rieman, M. (1982). In *Applied Probability—Computer Science: The Interface.* R. L. Disney and T. Ott, eds. Birkhauser, Boston, pp. II-409–421.

[47] Reiser, M. (1982). In *Applied Probability—Computer Science: The Interface.* R. L. Disney and T. Ott, eds. Birkahuser, Boston, pp. I-253–274.

[48] Simon, B. and Foley, R. D. (1979). *Manag. Sci.,* **25**, 1027–1034.

[49] Sauer, C. (1982). In *Applied Probability—Computer Science: The Interface.* R. L. Disney and T. Ott, eds. Birkhauser, Boston, pp. I-211–218.

[50] Schwetman, H. (1982). In *Applied Probability—Computer Science: The Interface.* R. L. Disney and T. Ott, eds. Birkhauser, Boston, pp. I-135–155.

[51] Solberg, J. J. (1981). *Trans. AIIE,* **13**, 116–122.

[52] Syski, R. (1960). *Introduction to Congestion Theory in Telephone Systems.* Oliver and Boyd, Edinburgh.

[53] Takács, L. (1962). *Introduction to the Theory of Queues.* Oxford University Press, New York.

[54] Walrand, J. and Varaiya, P. (1980). *Adv. Appl. Prob.,* **12**, 1000–1078.

[55] Walrand, J. and Varaiya, P. (1981). *Math. Operat. Res.,* **6**, 387–404.

[56] Whitt, W. (1981). *Manag. Sci.,* **27**, 619–636.

[57] Whitt, W. (1982). *Operat. Res.,* **30**, 125–146.

(MULTISERVER QUEUES
QUEUEING THEORY
TRAFFIC FLOW PROBLEMS)

RALPH L. DISNEY

NEW BETTER (WORSE) THAN USED

See HAZARD RATE AND OTHER CLASSIFICATIONS OF DISTRIBUTIONS

NEWCOMB, SIMON

Born: March 12, 1835, in Wallace, Nova Scotia, Canada.

Died: July 11, 1909, in Washington, D.C.

Contributed to: robust estimation, outlier theory, statistics in astronomy.

Simon Newcomb was the dominant figure in nineteenth-century American astronomy, an intelligent and prolific economist, and a creative and influential statistician. He was born March 12, 1835, in Wallace, Nova Scotia, son of a schoolteacher and a church organist. At the age of 16, he was apprenticed to a county doctor, but his plans for a career in medicine slowly dissolved as he learned that the doctor was a quack. After two years in what was effectively an indentured servitude, Newcomb made a dramatic escape, and until the age of 21, he supported himself as an itinerant teacher [8].

Newcomb's ancestors had come from New England, and his travels took him back there. In January 1857, he obtained employment at the Nautical Almanac office in Cambridge, Massachusetts, and simultaneously began studies at Harvard. He received a bachelor's degree a year later (his first mathematical paper was published in April 1858). In 1861, Newcomb was appointed professor of mathematics at the Naval Observatory in Washington, a post he held until his retirement in 1906, at which time he was given the rank of rear admiral. He died in Washington on July 11, 1909.

Simon Newcomb's scientific career was marked by a far-ranging intellectual curiosity and prodigious energy. His bibliography [1] lists 541 titles, including 318 on astronomy, 35 on mathematics, 42 on economics, and 146 on a diverse miscellany (including political issues of the day, the metric system, and skeptical comments on psychic research and rainmaking). Yet even those large numbers understate his output, as, for example, his 71 articles in the 1901 *Universal Cyclopedia* are counted as just a single item. For twenty years he directed the preparation of the *Nautical Almanac*, and in 1878 he cofounded (and for many years helped edit) the *American Journal of Mathematics*.

Newcomb was an able mathematician and mathematical astronomer (though perhaps not the equal of his colleague G. W. Hill), but his major work was in the organization and analysis of masses of observational data. Today some of Newcomb's determinations of the fundamental constants of astronomy are still accepted as the international standard. In the course of that work he was naturally led to the statistical methods of his day, and he brought to those methods the same probing intelligence that characterized all his scientific work.

Newcomb's best known contribution to statistics was in what we would now call robust estimation*. It was apparent to Newcomb from his analyses of observations that real data tended to be more disperse than can be represented by the normal distribution. He objected to the common practice of rejecting as outliers* those observations with too large residuals, since that practice rendered the resulting estimate a discontinuous function of the observations: as a measurement crossed the cutoff value, the estimate jumps, as from $\frac{1}{3}(a + b + c)$ to $\frac{1}{2}(a + b)$. In 1886 Newcomb published a paper that presented an alternative way of dealing with this problem. He would model the data as a sample from a mixture of normal distributions with different standard deviations and then take the posterior mean as his estimate with respect to a uniform prior. The investigation was framed in terms not much different from modern decision theory*, though he wrote of *evil* instead of *loss*. As mathematical statistics, it was an elegant piece of work, but it was too computationally cumbersome to be practical at the time. It required a trial-and-error type of iteration in the specification of the mixture. Outside of the worked examples Newcomb presented, it may have never been used, although the paper did attract the attention of European statisticians such as F. Y. Edgeworth.

Newcomb's other work in statistics is mostly buried in larger astronomical papers, but some of his separate works are worth note. He seems to have been the first to put forth the logarithmic distribution as the distribution of leading digits of haphazardly encountered data. The distribution is today sometimes referred to as Benford's law (*see* FIRST DIGIT PROBLEM, Newcomb [6], and Raimi [10]). And in an early series of "Notes on the Theory of Probabilities" [4], he gave a clear statement of the idea of sufficiency* in a particular instance, the "serial number problem" of estimating the number of tickets in a bag [13]. He also suggested the application of the Poisson distribution* to data for the first time; in the problem of testing whether or not the stars are randomly distributed in the sky [5]. Newcomb's personal papers have been deposited at the Library of Congress, and they contain portions of unfinished books on least squares*, demography*, and probability.

References

[1]   Archibald, R. C. (1924). *Mem. Nat. Acad. Sci.*, **17**, 19–69. (A complete bibliography of Newcomb's work.)

[2]   Campbell, W. W. (1924). *Mem. Nat. Acad. Sci.*, **17**, 1–18. (Biographical sketch.)

[3]   Marsden, B. G. (1981). *Dictionary of Scientific Biography*, Vol. 10. Scribner's, New York, pp. 33–36. (A recent biographical sketch emphasizing Newcomb's work in astronomy.)

[4]   Newcomb, S. (1859–61). *Mathematical Monthly*, **1–3**. (Appeared in several parts. Reprinted in Stigler [14].)

[5]   Newcomb, S. (1860). *Proc. Amer. Acad. Arts Sci.*, **4**, 433–440. (A slightly different version of the portion of this concerning the Poisson distribution appeared in Newcomb [4].)

[6]   Newcomb, S. (1881). *Amer. J. Math.*, **4**, 39–40. (Reprinted in Stigler [14].)

[7]   Newcomb, S. (1886). *Amer. J. Math.*, **8**, 343–366. (Reprinted in Stigler [14].)

[8]   Newcomb, S. (1903). *Reminiscences of an Astronomer.* Harper, London and New York. (Reprinted in its entirety in Stigler [14].)

[9]   Norberg, Arthur L. (1978). *Isis*, **69**, 209–225. (Discusses Newcomb's astronomical career up to 1870.)

[10] Raimi, R. (1976). *Amer. Math. Monthly*, **83**, 521–538. (A review article on the leading digit distribution.)

[11] Rubin, E. (1967). *Amer. Statist.*, October, 45–48. (Discusses Newcomb's work on the sex ratio at birth.)

[12] Stigler, S. M. (1973). *J. Amer. Statist. Ass.*, **68**, 872–879. (Discusses Newcomb's work on robust estimation. Reprinted in Stigler [14].)

[13] Stigler, S. M. (1977). *Ann. Statist.*, **6**, 239–265. (Quotes Newcomb on sufficiency and discusses his place in early American work. Reprinted in Stigler [14].)

[14] Stigler, S. M., ed. (1980). *American Contributions to Mathematical Statistics in the Nineteenth Cen-*

tury, two volumes. Arno Press, New York. (Includes photographic reprints of several of Newcomb's works in statistics as well as the whole of his autobiography.)

(STATISTICS IN ASTRONOMY)

STEPHEN M. STIGLER

NEWMAN–KEULS PROCEDURE

The Newman–Keuls multiple comparisons procedure is a stepwise technique for simultaneously testing the equality of the means of k different normal distributions. Specifically, consider an analysis of variance situation with k independent sample means $\bar{Y}_i$, $i = 1, \ldots, k$, each based on n independent observations from a normal population $N(\mu_i, \sigma^2)$. The problem is to decide which of the k population means differ from each other. To handle this testing problem, Newman [8], and later Keuls [6], adapted a method originally due to Student [9].

To apply the stepwise Newman–Keuls procedure, arrange the sample means in descending order; let s^2 denote the pooled estimate of σ^2 with v degrees of freedom (d.f.); let $q_{k,v,1-\alpha}$ denote the $(1 - \alpha)100$th percentile of the distribution of the Studentized range of k means with v d.f.; and set $W_p = q_{p,v,1-\alpha}\sqrt{s^2/n}$, for $p = 2, \ldots, k$. The Newman–Keuls procedure begins with the range of all k sample means and proceeds stepwise to the differences of adjacent sample means. At each step, if the range of the p adjacent means is smaller than W_p ($p = 2, \ldots, k$), that sample range is called nonsignificant, the corresponding p population means are declared equal, and there is no further testing among these means. Otherwise, the test proceeds to the next step. Testing stops when all remaining ranges are declared nonsignificant. For a numerical example of this simple computational method, see the second section of MULTIPLE RANGE AND ASSOCIATED TEST PROCEDURES or Chapter 2 of Miller [7]. For tables of the Studentized range distribution, see Harter [4].

When the sample sizes from each population are unequal, the Newman–Keuls procedure may be implemented using a sum of squares statistic instead of a range statistic. Suppose the sample mean $\overline{Y}_i$ ($i = 1, \ldots, k$) is based on n_i independent observations. Simply replace the range of p ($p = 2, \ldots, k$) adjacent sample means by the sum of squares of those sample means, that is,

$$\sum_{i \in P} n_i \overline{Y}_i^2 - \left(\left(\sum_{i \in P} n_i \overline{Y}_i \right)^2 \Big/ \sum_{i \in P} n_i \right),$$

where P denotes the set of indices of the p adjacent sample means; then redefine W_p by

$$W_p = (p - 1)F_{p-1, \nu, 1-\alpha}s^2,$$

where $F_{k, \nu, 1-\alpha}$ denotes the $(1 - \alpha)100$th percentile of the F distribution* with k and ν d.f.

To examine the type I error rates (ERs) of the Newman–Keuls procedure, fix the value of α and suppose that the number of populations k is larger than three. The simulation results of Boardman and Moffitt [2] indicate that in terms of the comparison, ER may be appreciably smaller than α. In terms of the experiment, ER is controlled at level α whenever all the population means are equal [5]. However, if the population means cluster in equal pairs and the pairs are very different from each other, then the experimental ER exceeds α. (See Einot and Gabriel [3] and Hartley [5].)

To bound the experimental ER by α, Peritz (in an unpublished paper) suggested modifying the Newman–Keuls procedure by closing (under intersection) the set of hypotheses to be tested. (See Einot and Gabriel [3, p. 578].) This *closure procedure* is a mixture of the Newman–Keuls and Ryan procedures that: gives fewer rejections than the Newman–Keuls procedure, controls the experimentwise ER at level α, and gives more rejections and hence more power than Ryan's procedure [3]. Although Einot and Gabriel [3] argued analytically that this closure procedure is more powerful than either Ryan's, Duncan's, or Tukey's procedure, their Monte Carlo study indicated that these power differences are small, and hence, they

did not recommend use of the closure procedure since it "involves impractically complicated procedures" [3, p. 582].

More recently, Begun and Gabriel [1] gave a practical stepwise algorithm for performing the closure procedure. To start, perform both the Newman–Keuls and the Ryan stepwise procedures. Accept all sets of means that are accepted by Newman–Keuls and hence Ryan, reject all sets that are rejected by Ryan and hence Newman–Keuls, and label as *contentious* the remaining sets that will be rejected by Newman–Keuls and accepted by Ryan. Next, proceed stepwise from the largest to the smallest of the contentious sets. Accept a contentious set if either: it is contained in another accepted contentious set or using Ryan's significance levels, it is nonsignificant and some set in its complement is also nonsignificant. Otherwise, reject the contentious set.

References

[1] Begun, J. M. and Gabriel, K. R. (1981). *J. Amer. Statist. Ass.*, **76**, 241–245.

[2] Boardman, T. J. and Moffitt, D. R. (1971). *Biometrics*, **27**, 738–744.

[3] Einot, I. and Gabriel, K. R. (1975). *J. Amer. Statist. Ass.*, **70**, 574–583.

[4] Harter, H. L. (1960). *Ann. Math. Statist.*, **31**, 1122–1147.

[5] Hartley, H. O. (1955). *Commun. Pure Appl. Math.*, **8**, 47–72.

[6] Keuls, M. (1952). *Euphytica*, **1**, 112–122.

[7] Miller, R. G., Jr. (1981). *Simultaneous Statistical Inference*, 2nd ed. Springer, New York.

[8] Newman, D. (1939). *Biometrika*, **31**, 20–30.

[9] Student (1927). *Biometrika*, **19**, 151–152.

(DUNCAN'S MULTIPLE RANGE TEST
MULTIPLE COMPARISONS
MULTIPLE RANGE AND ASSOCIATED
 TEST PROCEDURES
RYAN'S MULTIPLE COMPARISONS
 PROCEDURE
SCHEFFÉ'S SIMULTANEOUS
 COMPARISONS PROCEDURE
TUKEY'S SIMULTANEOUS TEST
 PROCEDURE)

JANET M. BEGUN

NEWTON, ISAAC

Born: December 25, 1642, in Woolsthorpe, Lincolnshire, England.

Died: March 20, 1727, in London, England.

Contributed to: algebra, astronomy, infinitesimal calculus, numerical methods, mathematical and experimental physics.

INTRODUCTION

To probabilists and statisticians, Isaac Newton is known as an outstanding mathematician, the discoverer, together with G. W. Leibniz (1646–1716), of the infinitesimal calculus and the originator of the law of universal gravitation. While he does not appear to have taken an active part in the development of probability and statistics in the late seventeenth and early eighteenth centuries, it is clear that he was familiar with the probability calculus of the times and had encountered the problem of the variability of sample means.

A brief biography may prove useful (see e.g., Youschkevitch [13]). Newton was born at Woolsthorpe in Lincolnshire in 1642; after attending school at nearby Grantham, he went up to Trinity College, Cambridge, in 1661, was granted his BA in 1665 and his MA in 1668. In 1669, at the age of 26, he was appointed Lucasian professor, succeeding Isaac Barrow, the first incumbent of this chair at the University of Cambridge. In 1672, he became a Fellow of the Royal Society of London. His lectures, deposited at the University Library, contained new work on optics that appeared in his *Opticks* (1704), on arithmetic and algebra, and on elements of the infinitesimal calculus, later published as Book I of the *Principia* in 1687. In this, his major work, he laid out some basic mathematical principles and rules for limits in Book I, considered the laws of motion of bodies in resisting media in Book

II, and lastly in Book III outlined the laws of celestial mechanics and of universal gravitation. In 1696, he was appointed Warden of the Mint, and moved to London; he became Master of the Mint in 1699 and was knighted by Queen Anne in 1705. He was elected president of the Royal Society in 1703 and is reputed to have ruled it with an iron hand until his death in 1727. During his London period, despite his responsibilities at the Mint and the Royal Society, Newton maintained his scientific interests, published his *Arithmetica Universalis* in 1707, a second edition of the *Principia* in 1713, and an enlarged version of the *Opticks* in 1717.

Newton had worked on the binomial theorem in 1655 (see Whiteside [12]), and would have understood the uses of the binomial distribution*. Such results were well known among mathematicians in Europe at the time, following the publication of Huygens' [7] work in 1657, and the results of James Bernoulli (1654–1705) which appeared in print posthumously [1] in 1713 (*see* HUYGENS, CHRISTIAAN and BERNOULLIS, THE). Although Newton made no original contributions to the theory of probability, a series of letters (see Turnbull [11]) exchanged with Samuel Pepys (1633–1703) in 1693 attest to his familiarity with contemporary probability calculations. Accounts of this correspondence, summarized in the following section, may be found in David [4, 5], Schell [9], Chaundy and Bullard [2], and Gani [6].

After his appointment to the Mint in 1696, Newton must have become familiar with the Trial of the Pyx, a sampling inspection scheme for coinage based on the aggregate weighing of a large number of coins selected at random (see Craig [3]). The concept of the Trial is similar to that of the modern sampling test procedure for means. Stigler [10] has presented some evidence that Newton, through his experience at the Mint and his studies of chronology, may well have had an understanding of the decrease in variability of means as the number of measurements averaged is increased. This is also briefly outlined in the section on statistics.

PROBABILITY: NEWTON'S SOLUTION OF A DICING PROBLEM

On November 22, 1693, Samuel Pepys addressed a letter to Isaac Newton at Cambridge, introducing its bearer Mr. John Smith, the Writing Master of Christ's Hospital School, as one who desired Newton's opinion on a question of dicing. The enquiry may well have resulted from the interest shown in lotteries at that time. Pepys formulated the dicing problem as follows:

The Question.

A–has 6 dice in a Box, wth wch he is to fling a 6.

B–has in another Box 12 Dice, wth wch he is to fling 2 Sixes.

C–has in another Box 18 Dice, wth wch he is to fling 3 Sixes.

Q. whether *B* & *C* have not as easy a Taske as *A*, at even luck?

Newton in his reply to Pepys of November 26, 1693 wrote that the problem was "ill-stated," and took

... the Question to be the same as if it had been put thus upon single throws.

What is ye expectation or hope of *A* to throw every time one six at least wth six dyes?

What is ye expectation or hope of *B* to throw every time two sixes at least wth 12 dyes?

What is ye expectation or hope of *C* to throw every time three sixes or more than three wth 18 dyes?

He then stated "it appears by an easy computation that the expectation of *A* is greater that that of *B* or *C*," without giving any details. After further correspondence, Newton gave Pepys the details of his calculations on December 16, 1693. These were based on the following simple binomial re-

sults:

Pr{1 or more sixes in 1 throw of 6 dice}

$$= 1 - \left(\frac{5}{6}\right)^6 = 1 - a.$$

Pr{2 or more sixes in 1 throw of 12 dice}

$$= 1 - 12\left(\frac{5}{6}\right)^{11}\left(\frac{1}{6}\right) - \left(\frac{5}{6}\right)^{12}$$

$$= 1 - b.$$

Pr{3 or more sixes in 1 throw of 18 dice}

$$= 1 - \frac{18 \cdot 17}{1 \cdot 2}\left(\frac{5}{6}\right)^{16}\left(\frac{1}{6}\right)^2$$

$$- 18\left(\frac{5}{6}\right)^{17}\left(\frac{1}{6}\right) - \left(\frac{5}{6}\right)^{18}$$

$$= 1 - c,$$

where

$$a = \left(\frac{5}{6}\right)^6, \qquad b = \left(\frac{5}{6}\right)^{12}\left(1 + \frac{12}{5}\right),$$

$$c = \left(\frac{5}{6}\right)^{18}\left(1 + \frac{18}{5} + \frac{18 \cdot 17}{2 \cdot 5^2}\right).$$

The values of $1 - a$ and $1 - b$ were found to be

$$1 - a = \frac{31,031}{46,656}, \quad 1 - b = \frac{1,346,704,211}{2,176,782,336},$$

but Newton did not give figures for the 18-dice case. His method of calculation would, however, have led to

$$1 - c = \frac{60,666,401,980,916}{101,559,956,668,416}.$$

In effect, *A* would have the most favorable throw, as $1 - a > 1 - b > 1 - c$.

STATISTICS: THE TRIAL OF THE PYX AND NEWTON'S CHRONOLOGY

Newton's position at the Mint clearly involved familiarity with the Trial of the Pyx. This sampling inspection scheme (*see* SAMPLING PLANS), which had been in operation since the thirteenth century, consisted of taking one gold coin out of roughly every 15 pounds of gold minted or one silver coin out of every 60 pounds of silver (one day's production) at random, over a period of time, and placing them in a box called the Pyx (after the Greek πυξισ for box). At irregular intervals of between one and several years, a Trial of the Pyx would be declared with an adjudicating jury selected from among established goldsmiths. At the Trial, the Pyx would be opened and its contents counted, weighed, and assayed in bulk to ensure that the gold and silver coins were within the allowed tolerances.

The aggregated weight of the sample of n coins was expected not to exceed n times the required tolerance for any single coin. This procedure was equivalent to carrying out a rudimentary two-sided test, where the tolerances were in fact set so that only about 5% of a representative collection of coins would fail to satisfy them. Newton underwent one such Trial in 1710, when he successfully survived the charge that his gold coinage was below standard.

Statistical theory indicates that $\sqrt{n}$ times the tolerance for a single coin would have been a more appropriate measure of tolerance for the aggregate. Newton may possibly have had some understanding of this point. He is known to have emphasized a reduction in the variability of individual coins from the Mint, but the Trial of the Pyx must have raised in his mind the question of the variability of sample means.

Some circumstantial evidence for this is contained in Newton's last work "The Chronology of Ancient Kingdoms Amended," published posthumously in London in 1728. In this, Newton estimated the mean length of a king's reign "at about eighteen or twenty years a-piece." Stigler [10] points out that he repeated this phrase three times without ever quoting nineteen as the mean length; this mean was in fact 19.10 years, with a standard deviation of 1.01. Newton's interval of 18–20 corresponds to a band of one standard deviation about the mean or roughly a 65% confidence interval*.

While we cannot be certain that Newton had in fact pondered the problem of significance tests, he was implicitly providing some form of interval estimate for the length of a king's reign. On this premise, Stigler argues

that Newton "had at least an approximate intuitive understanding of the manner in which the variability of means decreased as the number of measurements averaged increased."

CONCLUDING REMARKS

The correspondence with Pepys, outlined earlier, provides convincing evidence that Newton was conversant with the calculus of probabilities of his time. In this, he was not alone; both Schell [9] and Chaundy and Bullard [2] refer to Pepys' simultaneous request to George Tollet, who obtained the same results as Newton.

The problem itself is of some intrinsic interest; it has been generalized by Chaundy and Bullard [2] in 1960 to take account of an s-faced die, $s \geqslant 2$. The authors study the asymptotic behavior of the probability

$$f(sn, n) = \Pr\{n \text{ or more of a selected face}$$

$$\text{in 1 throw of } sn \text{ dice}\}$$

$$= \sum_{j=n}^{sn} \binom{sn}{j}\left(\frac{1}{s}\right)^j\left(1 - \frac{1}{s}\right)^{sn-j}.$$

Gani [6] has also recently considered de Méré's problem in this more general context. Here, one is concerned with the different question of the number n of repeated throws of $r = 1, 2, \ldots$ six-sided dice, which is necessary to achieve a successful throw of r sixes. In particular, de Méré was interested in the number n of throws required for

$$\Pr\{2 \text{ sixes in } n \text{ throws of 2 dice}\} > \tfrac{1}{2};$$

this Pascal* showed to be $n = 25$.

The evidence for Newton's understanding of the statistical principles involved in the variability of sample means is less secure, but his familiarity with the Trial of the Pyx and his treatment of the mean length of a king's reign indicate that he must at least have considered the problem.

It is interesting to speculate whether in other circumstances, Newton the mathematician might have become more active as a probabilist. Perhaps the simple answer is that in a choice between investigating a "System of the World" and problems which in his time were often related to gambling*, Newton's fundamental seriousness would almost inevitably have caused him to select the first.

No account of Newton's contributions is complete without a mention of two approximation methods in mathematics that bear his name and that are also used in a statistical context. These are the Newton–Raphson method* for approximating the roots of $f(x) = 0$, and the Gauss–Newton method for the replacement of a nonlinear function $g(x_1, \ldots, x_k)$ by its linear approximation

$$g(a_1, \ldots, a_k) + \sum_{i=1}^{k} (x_i - a_i) \frac{\partial g(a_1, \ldots, a_k)}{\partial a_i}$$

(see NEWTON ITERATION EXTENSIONS); for details see Ortega and Rheinboldt [8].

References

[1]   Bernoulli, James (1713). *Ars Conjectandi*.

[2]   Chaundy, T. W. and Bullard, J. E. (1960). *Math. Gaz.*, **44**, 253–260.

[3]   Craig, J. (1953). *The Mint*. Cambridge University Press, Cambridge.

[4]   David, F. N. (1957). *Ann. Sci.*, **13**, 137–147.

[5]   David, F. N. (1962). *Games, Gods and Gambling*. Griffin, London.

[6]   Gani, J. (1982). *Math. Sci.*, **7**, 61–66.

[7]   Huygens, C. (1657). *De Ratiociniis in Ludo Aleae*.

[8]   Ortega, J. M. and Rheinboldt, W. C. (1970). *Iterative Solution of Nonlinear Equations in Several Variables*. Academic Press, New York.

[9]   Schell, E. D. (1960). *Amer. Statist.*, **14**(4), 27–30.

[10] Stigler, S. M. (1977). *J. Amer. Statist. Ass.*, **72**, 493–500.

[11] Turnbull, H. W., ed. (1961). *The Correspondence of Isaac Newton 1668–1694*, Vol. III. Cambridge University Press, London.

[12] Whiteside, D. T. (1961). *Math. Gaz.*, **45**, 175–180.

[13] Youschkevitch, A. P. (1974). Isaac Newton. *Dictionary of Scientific Biography*, Vol. X. Scribner's, New York.

(BERNOULLIS, THE
DE MOIVRE, ABRAHAM
HUYGENS, CHRISTIAAN
PASCAL, BLAISE)

J. GANI

NEWTON ITERATION EXTENSIONS

Newton iteration is a powerful method for estimating a set of parameters that maximize a function when the parameters are related to the function nonlinearly. Two general applications of Newton iteration are nonlinear regression (least squares), often referred to as Gauss–Newton iteration, and maximum likelihood estimation*. Since the method is very general, it is usable in many other estimation procedures as well. References containing numerical examples are noted in the bibliography.

Newton iteration is very powerful for many problems and is the most powerful of the gradient procedures, given certain assumptions. (See Crockett and Chernoff [1] and Greenstadt [4].) For other problems, the method has not converged to a maximum or minimum. In addition to describing Newton iteration, this article gives a procedure which can be used to detect troublesome problems and automatically switch to a Newton extension that performs well for a larger class of problems.

NEWTON ITERATION

To apply Newton iteration, the first and second partial derivatives (or at least an approximation to them) are calculated at each iteration. More specifically, for a function $f(\mathbf{X}:\mathbf{b})$ to be maximized, the m-dimensional vector $\mathbf{b}^*$, which maximizes $f(\mathbf{X}:\mathbf{b})$, is calculated by going through a series of iterations with calculated values of $\mathbf{b}$ (i.e., $\mathbf{b}_{(1)}, \mathbf{b}_{(2)}, \ldots$) until $\mathbf{b}^*$ is found. To minimize a function, merely maximize the negative of that function.

If a local maximum separate from the global maximum exists in a region near any of the $\mathbf{b}_{(i)}$, then convergence is likely to be to the local maximum. Saddle points may be readily handled by the extensions to Newton iteration given further on. Since the focus here is on selecting the $\mathbf{b}^*$ that maximizes $f(\mathbf{X}:\mathbf{b})$, in what follows we will simplify notation by dropping the specific recognition of the matrix of variables, if any, in the function to be maximized and write the function as $f(\mathbf{b})$.

Newton iteration is a gradient method of maximization; that is, at each iteration, the next point, $\mathbf{b}_{(i+1)}$, is chosen in the direction of the steepest ascent from the present point, $\mathbf{b}_{(i)}$. The particular concept of distance used in determining steepest ascent is the Newton metric—the $m \times m$ matrix of second partial derivatives of $f(\mathbf{b})$.

The formulas for Newton iteration may be derived by writing out the first three terms of a Taylor expansion about an m-dimensional point $\mathbf{b}_{(i)}$, taking the first partial derivative of $f(\mathbf{b})$, setting the first partial derivative to 0, and solving for $\mathbf{b}$. (See Crockett and Chernoff [1].) The following is obtained:

$$\mathbf{b}_{(i+1)} = \mathbf{b}_{(i)} + \mathcal{L}_{(i)}^{-1} \mathcal{l}_{(i)},$$

where $\mathcal{l}_{(i)}$ is the m-dimensional vector of first partial derivatives of $f(\mathbf{b})$ evaluated at $\mathbf{b}_{(i)}$ [i.e., with the m values of $\mathbf{b}_{(i)}$ substituted into the formula for the first partial derivative of $f(\mathbf{b})$] and $-\mathcal{L}_{(i)}$ is the $m \times m$ matrix of second partial derivatives of $f(\mathbf{b})$ evaluated at $\mathbf{b}_{(i)}$.

If the first three terms of the Taylor expansion were sufficiently close to $f(\mathbf{b})$ and if $\mathcal{L}_{(i)}$ were positive definite (i.e., $-\mathcal{L}_{(i)}$ were negative definite) then $\mathbf{b}^*$ would be the maximum of $f(\mathbf{b})$ and this article would be almost complete, saving time for everyone. Since the first three terms do not sufficiently represent $f(\mathbf{b})$, the $\mathbf{b}^*$ which maximizes $f(\mathbf{b})$ must be computed by a series of iterations.

The preceding formula suggests $\mathbf{d}_{(i)} = \mathcal{L}_{(i)}^{-1} \mathcal{l}_{(i)}$ as the direction to take at each iteration.

In Newton iteration the length of movement in direction $\mathbf{d}_{(i)}$ is usually generalized so that instead of a step size of one, a step size of $h_{(i)}$ (a scalar) is used, with $h_{(i)}$ varying with each iteration. Thus $\mathbf{b}_{(i+1)}$ with $f(\mathbf{b}_{(i+1)}) > f(\mathbf{b}_{(i)})$ is calculated by the formula

$$\mathbf{b}_{(i+1)} = h_{(i)} \mathbf{d}_{(i)}.$$

$\mathbf{b}_{(i+2)}$ is calculated as $\mathbf{b}_{(i+1)} + h_{(i+1)} \mathbf{d}_{(i+1)}$, etc., until after a series of iterations, $\mathbf{b}^*$, the $\mathbf{b}_{(i)}$ which maximizes $f(\mathbf{b})$, is reached. An algorithm for computing the step size for each iteration is given further on.

Although any positive definite matrix $\mathcal{M}_{(i)}$ could be substituted for $\mathcal{L}_{(i)}$ and $f(\mathbf{b})$ will increase provided $\mathcal{L}_{(i)}$ is positive definite and $h_{(i)}$ is chosen sufficiently small, that $\mathcal{L}_{(i)}$ is a much more efficient metric than most is shown in Greenstadt [4], Crockett and Chernoff [1], and Fiacco and McCormick [3].

GREENSTADT EXTENSION

Although $\mathcal{L}_{(i)}$ may not be positive definite, $\mathcal{L}_{(i)}$ is symmetric, hence (provided $\mathcal{L}_{(i)}$ is nonsingular) it may be forced to be positive definite by deriving eigenvalues* and their corresponding eigenvectors* of $\mathcal{L}_{(i)}$, setting all negative eigenvalues positive, and forming $\mathcal{L}_{(i)}$ to use as a metric [4]. In particular, let

$$\mathcal{L}_{(i)}^* = \mathbf{E}\mathbf{G}^*\mathbf{E}'$$

be the positive definite matrix formed from $\mathcal{L}_{(i)}$ with $\mathbf{G}$ being an $m \times m$ diagonal matrix with the eigenvalues of $\mathcal{L}$ on the diagonal, $\mathbf{G}^*$ being the corresponding matrix with the absolute values of the eigenvalues on the diagonal, and $\mathbf{E}$ being the $m \times m$ matrix with the corresponding eigenvectors as columns of $\mathbf{E}$. $'$ denotes transpose. Since $\mathbf{E}^{-1} = \mathbf{E}'$, $(\mathcal{L}_{(i)}^*)^{-1}$ may be formed directly as $\mathbf{E}(\mathbf{G}^*)^{-1}\mathbf{E}'$. Further, since $\mathbf{G}^*$ is diagonal, the k, lth element of $(\mathcal{L}_{(i)})^{-1}$ can be formed more directly as

$$\sum_{j=1}^{m} \frac{1}{g_j^*} e_{kj} e_{lj},$$

where g_j^* is the absolute value of the jth eigenvalue of $\mathcal{L}_{(i)}$ and e_{kj} is the k, jth element of $\mathbf{E}$.

EIGENVALUE APPROXIMATELY ZERO

Whereas negative eigenvalues are encountered frequently, approximately zero eigenvalues are rarely encountered except for data problems such as those that follow. This section is included for completeness.

If one or more eigenvalues are approximately zero, then the $\mathscr{L}$ matrix is nearly singular and neither $\mathscr{L}^{-1}$ nor $(\mathscr{L}^*)^{-1}$ can be formed. Usually this will occur when the variables in a statistical problem are not linearly independent (remember $f(\mathbf{X}:\mathbf{b})$ was shortened to $f(\mathbf{b})$ for simplicity, since the matrix of variables $\mathbf{X}$ is fixed for a given problem). As an example, such singularity will occur in nonlinear regression* when one variable is a linear combination of others, often due to multicollinearity* among variables or to an insufficient number of observations. In this case, since there is no unique maximum, one or more parameters must be eliminated and the problem recalculated.

On the other hand, a problem could be encountered in which one eigenvalue is approximately zero in a given region, but all are positive at the maximum. In this case, one can set the approximately zero eigenvalue to one and use the Greenstadt extension given earlier in the hope that the difficult region will be moved through in a helpful direction. Then when the eigenvalue becomes not approximately zero, again, hopefully, the Greenstadt extension will move to the maximum without continually returning to the offending region.

INCREASING COMPUTATIONAL EFFICIENCY

When all eigenvalues are positive (i.e., $\mathscr{L}_{(i)}$ is positive definite), considerably fewer computer operations are required if $\mathbf{d}_{(i)} = \mathscr{L}_{(i)}^{-1}\ell_{(i)}$ is directly formed by Gaussian (Gauss–Jordan*) elimination rather than by getting eigenvalues and eigenvectors of $\mathscr{L}_{(i)}$ to form $\mathscr{L}_{(i)}^{-1}\ell_{(i)}$. Luckily, Gaussian elimination gives a means of determining whether any eigenvalue is negative or approximately zero. If during Gaussian elimination, all diagonal elements (pivots) are greater than zero, the resulting $\mathbf{d}_{(i)}$ can be used directly. If, on the other hand, a nonpositive diagonal

element is encountered, then an immediate switch should be made to the Greenstadt extension for the iteration.

MATTHEWS AND DAVIES EXTENSION

Matthews and Davies [6] suggest that if nonpositive diagonal elements of $\mathscr{L}$ are encountered during Gaussian elimination, they be set positive if not approximately zero or set to one if approximately zero, and the Gaussian elimination continued. This is equivalent to using a slightly different positive definite metric than the $\mathscr{L}^*$ metric previously discussed. It seems desirable to use all positive diagonal elements as pivots first, then the negative diagonal elements, and finally the approximately zero diagonal elements.

Computations using the Matthews and Davies extension require substantially less computer time per iteration, less computer space, and (usually) less programming to develop the initial computer routine. Of course, if for particular problems many more iterations are required, or worse, if convergence to a maximum is not obtained, the Greenstadt extension may be superior. The author expects that in general the greater the number of parameters being simultaneously estimated and the more complex the function, the more likely the Greenstadt extension will perform well relative to the Matthews and Davies extension. For a given type of problem, the Matthews and Davies extension could, of course, be tried and if performance is not satisfactory, a switch could be made to the Greenstadt extension.

INITIAL STARTING ESTIMATES

The m initial starting estimates to use for $\mathbf{b}_{(1)}$ are arbitrary. For simple, well-conditioned problems with no local maxima separate from the global maximum, the particular starting estimates will make little difference. For large problems with multiple local maxima, selection of good estimates for $\mathbf{b}_{(1)}$ may make a difference in the answer obtained as well as in the speed of convergence. The values of the parameters that the user most nearly expects in the final result are usually the best starting estimates. The values may come from previous studies or from examination of the function and data.

STEP SIZE

So far only the direction to move from $\mathbf{b}_{(i)}$ to form $\mathbf{b}_{(i+1)}$ has been covered. How far to go in that direction can also be important. Assume that $\mathbf{b}$ consists of only two parameters b_1 and b_2 and consider the situation given in Fig. 1.

If direction $\mathbf{d}_{(1)}$ is taken with step size $h_{(1)}$ to move to $\mathbf{b}_{(2)}$ the first iteration, direction $\mathbf{d}_{(2)}$ is taken with step size $h_{(2)}$ to move to $\mathbf{b}_{(3)}$ the second iteration, etc., due to each step size being too large, many iterations are required to move up the long narrow ridge. This is in spite of an "optimal" local direction for each iteration. Similarly, a series of too small steps may cause little movement up the ridge. Contrast this with Fig. 2, where direction is chosen in the same manner as Fig. 1 but the step size is varied to land at

the top of the ridge on each iteration. As a result, optimal local direction leads up the ridge after the first step. The payoff for a more optimal step size becomes even more important as the number of parameters being estimated increases.

Following is an algorithm that has performed well in determining step size for an iteration:

For each iteration an initial step size of h is tried, i.e., $f(\mathbf{b}_{(i)} + h\mathbf{d}_{(i)})$ is calculated. (The choice of h is somewhat arbitary, since it will be adjusted to a more optimal value by this algorithm. The step size of one implied by the Taylor expansion usually performs well.) Then

1. If $f(\mathbf{b}_{(i)} + h\mathbf{d}_{(i)}) \geqslant f(\mathbf{b}_{(i)})$, $f(\mathbf{b}_{(i)} + 2h\mathbf{d}_{(i)})$ is calculated. The step size is doubled, redoubled, etc. until at the jth trial, for some j,

$$f(\mathbf{b}_{(i)} + 2^j\mathbf{d}_{(i)}) < f(\mathbf{b}_{(i)} + 2^{j-1}h\mathbf{d}_{(i)}).$$

At that time a quadratic approximation (described below) is used to calculate a step size h_2. If $f(\mathbf{b}_{(i)} + h_2\mathbf{d}_{(i)}) > f(\mathbf{b}_{(i)} + 2^{j-1}h\mathbf{d}_{(i)})$, h_2 is used as the step size for the iteration. Otherwise $2^{j-1}h$ is used.

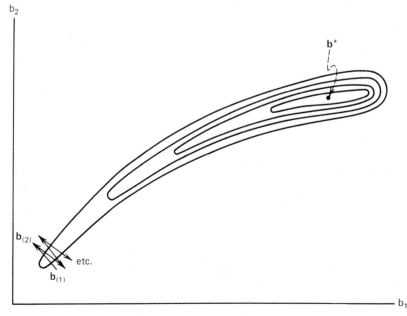

Figure 1 Each step size is too large.

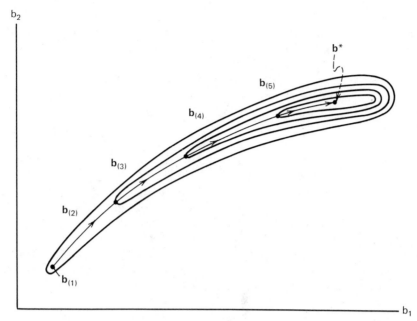

Figure 2 Step size is varied.

2. If $f(\mathbf{b}_{(i)} + h\mathbf{d}) < f(\mathbf{b}_{(i)})$, a trial step half as large is tried. This halving process is continued until $f(\mathbf{b}_{(i)} + 2^{-j}n\mathbf{d}_{(i)}) < f(\mathbf{b}_{(i)})$, at which time a quadratic approximation is applied to calculate a step size h_2. If $f(\mathbf{b}_{(i)} + h_2\mathbf{d}_{(i)}) > f(\mathbf{b}_{(i)} + 2^{-j}h\mathbf{d}_{(i)})$, then h_2 is used as the step size for the iteration. Otherwise $2^{-j}h$ is used.

QUADRATIC APPROXIMATION

Given three n-dimensional points $\mathbf{a}$, $\mathbf{a}^*$ and $\mathbf{a}^{**}$ with $\mathbf{a}^* = \mathbf{a} + k\mathbf{d}$, $\mathbf{a}^{**} = \mathbf{a} + 2k\mathbf{d}$, $f(\mathbf{a}^*) > f(\mathbf{a})$ and $f(\mathbf{a}^*) > f(\mathbf{a}^{**})$, then the maximum of a quadratic function through the points $(\mathbf{a}, f(\mathbf{a}))$, $(\mathbf{a}^*, f(\mathbf{a}^*))$, and $(\mathbf{a}^{**}, f(\mathbf{a}^{**}))$ is given by $\mathbf{a} + k_2\mathbf{d}$, where

$$k_2 = k\left[1 + \frac{f(\mathbf{a}) - f(\mathbf{a}^{**})}{2\{f(\mathbf{a}^{**}) + f(\mathbf{a}) - 2f(\mathbf{a}^*)\}}\right]$$

The quadratic approximation may or may not calculate a more optimal step size for a particular iteration depending on how closely a quadratic function approximates $f(\mathbf{b})$ for the particular point and direction. But computation is fast and leads to an improvement for many iterations.

CONVERGENCE CRITERION

Since $\ell = 0$ and $-\mathscr{L}$ is negative definite at a maximum, iteration continues until all elements of $\ell_{(i)}$ are approximately zero and all diagonal elements of $\mathscr{L}_{(i)}$ are greater than zero during Gaussian elimination.

IMPROVING ACCURACY

Accuracy will be considerably enhanced if the following is done during computation:

1. The moment matrix of variables is normalized so that each diagonal element is one at the start of computation. (This may be done by dividing each row and column of the matrix by the square root of the diagonal element of the row or column.) Normalizing the variables also normalizes the parameters being estimated, making them independent of other potential variable scalings. After convergence to the maximum, the estimated parameters and their related statistics must, of course, be denormalized.

2. During Gaussian elimination or inversion, the largest diagonal element is selected at each step for the pivot rather than the next one in order. Most Gaussian elimination and inversion routines now automatically do this.

3. If available, use double precision rather than single precision throughout computation.

References

[1] Crockett, J. B. and Chernoff, H. (1955). *Pac. J. Math.*, **5**, 33–50. (The basic reference on gradient methods of maximization.)

[2] Eisenpress, H. and Greenstadt, J. (1966). *Econometrica*, **34**, 851–861. (Shows detailed application of Newton iteration in full-information maximum likelihood estimation. No numeric examples other than the final result for an example.)

[3] Fiacco, A. V. and McCormick, G. P. (1968). *Nonlinear Programming: Sequential Unconstrained Minimization Techniques.* Wiley, New York, 209 pp. (Contains examples of application of the Newton metric and additional Newton iteration extensions.)

[4] Greenstadt, J. (1967). *Math. Comp.*, **21**, 360–367. (Contains in-depth persuasive arguments for use of the Newton metric in iteration, especially with the Greenstadt extension.)

[5] Hartley, H. O. (1961). *Technometrics*, **3**, 269–280. (Contains a detailed, easily followed numeric application of Newton iteration to nonlinear regression.)

[6] Matthews, A. and Davies, D. (1971). *Computer J.*, **14**, 293–294. (Has two simple numeric examples of results of Newton iteration as extended by Greenstadt and themselves.)

(GAUSS–JORDAN ELIMINATION
GAUSS–SEIDEL ITERATION
MATHEMATICAL PROGRAMMING
NEWTON–RAPHSON METHOD
NONLINEAR PROGRAMMING
NONLINEAR REGRESSION)

WILLIAM L. RUBLE

NEWTON–KANTOROVICH THEOREM

See NEWTON–RAPHSON METHODS

NEWTON–RAPHSON METHODS

The *Newton–Raphson method*, or simply *Newton's method*, is a basic numerical algorithm for finding approximations to the solutions of nonlinear equations. It can be applied to n simultaneous equations in n unknowns, and it can also be applied to nonlinear operator equations in Banach spaces. In the latter setting, the method is sometimes referred to as *quasi-linearization* [3].

ONE EQUATION, ONE UNKNOWN

In the case of one equation in one unknown, i.e., $f(x) = 0$, the algorithm takes the form

$$x_{i+1} = x_i - f(x_i)/f'(x_i), \qquad (1)$$

where $f'(x) = df/dx$. The algorithm needs both an initial guess x_0 and also a stopping criterion such as requiring a small residual, e.g., $|f(x_i)| \leqslant$ tolerance. If $f(x^*) = 0$, $f'(x^*) \neq 0$, and f has two continuous derivatives, then $x_i \to x^*$ as $i \to \infty$ provided x_0 is chosen sufficiently close to the solution x^*. This last condition is a disadvantage of Newton's method. However, convergence under the preceding assumptions will be quadratic, i.e.,

$$\lim_{i \to \infty} \frac{x_{i+1} - x^*}{(x_i - x^*)^2} = \frac{f''(x^*)}{2f'(x^*)} .$$

This rate means that the number of correct decimal digits is approximately doubled at each iteration as the error goes to zero. This speed is very important if f is expensive to evaluate at each step.

It is possible to avoid the cost of evaluating f' by leaving $f'(x_i)$ fixed at $f'(x_0)$ or by replacing $f'(x_i)$ with $(f(x_i) - f(x_{i-1}))/(x_i - x_{i-1})$. These two modifications are called the *chord* and *secant methods*, repectively. There are many other possibilities [5]. Unfortunately, the chord method converges only linearly, but the secant method error satisfies

$$(x_{i+1} - x^*)/(x_{i+1} - x_i^*)^p \to \text{const},$$

where $p = (\sqrt{5} + 1)/2$. This rate is generally satisfactory in practice, and since the ability to avoid computation of f' is useful, the secant method is often preferred to Newton's method. The secant method is used in the standard library routine ZEROIN [2]. The problem of determining a satisfactory initial guess for the secant method in

ZEROIN is handled by using bisection search. It should also be noted that the Newton and secant methods can be used in the complex plane without any change in the theorems.

Newton's method for $f(x) = 0$ is derived by simply repeatedly linearizing f at the current iterate, i.e., replacing $f(x) = 0$ by $f(x_i) + f'(x_i)(x - x_i) = 0$. Geometrically, this amounts to finding the root of the tangent line to $y = f(x)$ at $(x_i, f(x_i))$. The idea of linearization can be used to extend Newton's method to systems of equations and also to operator equations.

THE SOLUTION OF SIMULTANEOUS EQUATIONS

Given a system of n simultaneous nonlinear algebraic or transcendental equations in n unknowns written as

$$f_1(\mathbf{x}) = 0,$$
$$f_2(\mathbf{x}) = 0, \qquad (2)$$
$$f_n(\mathbf{x}) = 0,$$

where $\mathbf{x}$ is in R^n, one defines Newton's method as follows. First let $\mathbf{F} = (f_1, \ldots, f_n)^T$ so that (2) may be written $\mathbf{F}(\mathbf{x}) = \mathbf{0}$. Then the linearization of $\mathbf{F}$ at $\mathbf{x}_i$ has the form $\mathbf{F}(\mathbf{x}_i) + \mathbf{J}(\mathbf{x}_i)(\mathbf{x} - \mathbf{x}_i)$, where $\mathbf{J}$ is the Jacobian* matrix $(D_j f_s)$ for $\mathbf{F}$. Here D_j denotes the derivative operator with respect to the jth variable. Newton's method for (2) then becomes

$$\mathbf{x}_{i+1} = \mathbf{x}_i - \mathbf{J}(\mathbf{x}_i)^{-1} \mathbf{F}(\mathbf{x}_i). \qquad (3)$$

Although the notation in (3) is commonly used, an implementation of (3) should focus on solving the system $\mathbf{J}(\mathbf{x}_i)(\mathbf{x}_{i+1} - \mathbf{x}_i) = -\mathbf{F}(\mathbf{x}_i)$ and not on explicitly inverting $\mathbf{J}(\mathbf{x}_i)$. The basic theorem for the convergence of Newton's method in the one-variable case extends to the system in (2), and so the method is theoretically attractive. However, it can be difficult to use in practice because of the need to evaluate the n^2 partial derivatives in $\mathbf{J}$ at each step. There are many modifications to the method that avoid the explicit computation of $\mathbf{J}$.

A simple and effective approach is to replace the partial derivatives by divided difference approximations. Denoting $\mathbf{x} = (\xi_1, \ldots, \xi_n)$, one could replace $\partial f_i / \partial \xi_j$ by

$$\{ f_i(\mathbf{x} + \epsilon \mathbf{e}j) - f_i(\mathbf{x}) \} / \epsilon,$$

where $\mathbf{e}_j$ is the jth coordinate vector. This discretized Newton's method will have, in general, only a linear rate of convergence, and in order to achieve the usual fast convergence of Newton's method, it is necessary to replace ϵ by a sequence of values converging to zero as the number of iterations increases.

In a sense, many fast methods for solving (2) are variants of Newton's method. Any sequence $\{\mathbf{x}_i\}$ is said to *converge superlinearly* to $\mathbf{x}^*$ if $\lim_{i \to \infty}[\|\mathbf{x}_{i+1} - \mathbf{x}^*\| / \|\mathbf{x}_i - \mathbf{x}^*\|] = 0$. Many methods for (2) take the form

$$\mathbf{x}_{i+1} = \mathbf{x}_i + \mathbf{p}_i,$$
$$\mathbf{B}_i \mathbf{p}_i = -\mathbf{F}(\mathbf{x}_i), \qquad (4)$$

and the matrix $\mathbf{B}_i$ is an approximate Jacobian matrix for $\mathbf{F}$. If these iterates $\{\mathbf{x}_i\}$ converge linearly, then convergence will be superlinear if and only if the sequence $\{\mathbf{B}_i\}$ satisfies [1]

$$\lim_{i \to \infty} \left[\frac{\|(\mathbf{B}_i - \mathbf{J}(\mathbf{x}^*))(\mathbf{x}_{i+1} - \mathbf{x}_i)\|}{\|\mathbf{x}_{i+1} - \mathbf{x}_i\|} \right] = 0. \qquad (5)$$

It is not necessary for the approximating Jacobian to converge to the true Jacobian to achieve superlinear convergence. However, from (5) it follows that superlinear convergence must occur in the correction directions. This property (5) can be verified under suitable conditions for basic *quasi-Newton methods* such as the well-known Davidson–Fletcher–Powell [5] method. Quasi-Newton methods have the form (4) and satisfy the *quasi-Newton condition* $\mathbf{B}_{i+1}(\mathbf{x}_{i+1} - \mathbf{x}_i) = \mathbf{F}(\mathbf{x}_{i+1}) - \mathbf{F}(\mathbf{x}_i)$.

If $\mathbf{F}$ maps a normed linear space X into a normed linear space Y, then Newton's method can be extended to the equation $\mathbf{F}(\mathbf{x}) = \mathbf{0}$ by using the idea of linearization. Notationally the algorithm remains the same as (3), but the Jacobian is replaced by the Fréchet derivative of F (*see* STATISTICAL FUNCTIONALS). Applications in the setting of differential equations may be found in Kalaba [3].

It is possible to give error estimates for Newton's method that also establish that solutions exist. This is the *Newton–Kantorovich theorem*, which we state in the finite dimensional setting.

Newton–Kantorovich Theorem. Assume that $\mathbf{F}: R^n \to R^n$ is differentiable on a convex set D and that

$$|\mathbf{J}(\mathbf{x}) - \mathbf{J}(\mathbf{y})| \leqslant \gamma|\mathbf{x} - \mathbf{y}|$$

for all $\mathbf{x}, \mathbf{y}$ in D where the absolute value denotes any norm on R^n and also the induced matrix norm over R^n. Suppose there is an $\mathbf{x}_0$ in D such that

$$|\mathbf{J}(\mathbf{x}_0)^{-1}| \leqslant \beta, \qquad |\mathbf{J}(\mathbf{x}_0)^{-1}\mathbf{F}(\mathbf{x}_0)| = \eta,$$

and

$$\alpha \equiv \beta\gamma\eta < \tfrac{1}{2}.$$

Assume that $s \equiv \{\mathbf{x}: |\mathbf{x} - \mathbf{x}_0| \leqslant t^*\}$ is in D, where $t^* = (\beta\gamma)^{-1}(1 - (1 - 2\alpha)^{1/2})$. Then the Newton iterates $\mathbf{x}_{i+1} = \mathbf{x}_i - \mathbf{J}(\mathbf{x}_i)^{-1}\mathbf{F}(\mathbf{x}_i)$, $i = 0, 1, \ldots$, are well-defined and converge to a solution $\mathbf{x}^*$ of $\mathbf{F}(\mathbf{x}) = \mathbf{0}$ in S. Moreover, one has the estimate

$$\|\mathbf{x}^* - \mathbf{x}_i\| \leqslant (2\alpha)^{2^i}/(\beta\gamma 2^i).$$

A proof of this theorem may be found in [5]. Thus, not only do the Newton iterates converge under the conditions of the previous theorem, but also a solution exists. However, it is difficult to use the theorem in practice to establish existence, and it is used more often to give a posteriori error estimates. This follows using the estimate $\|\mathbf{x}_1 - \mathbf{x}^*\| \leqslant 2\beta\gamma\|\mathbf{x}_1 - \mathbf{x}_0\|^2$, which is a consequence of the theorem. For an example of the use of the Newton–Kantorovich theorem to give existence, see Moser [4].

References

[1] Dennis, J. E., Jr. and Moré, J. J. (1974). *Math. Comp.*, **28**, 549–560.

[2] Forsythe, G. E., Malcolm, M. A., and Molen, C. B. (1977), *Computer Methods for Mathematical Computations*. Prentice-Hall, Englewood Cliffs, NJ.

[3] Kalaba, R. (1963). In *Nonlinear Differential Equations and Nonlinear Mechanics*. Academic Press, New York, pp. 135–146.

[4] Moser, J. (1966). *Ann. Sc. Norm. Supl. Pisa*, **20**, 265–315, 499–535.

[5] Ortega, J. M. and Rheinboldt, W. C. (1970). *Iterative Solution of Nonlinear Equations in Several Variables*. Academic Press, New York. (This book includes some historical background.)

(GAUSS–SEIDEL ITERATION
MATHEMATICAL PROGRAMMING
NOMOGRAMS
NONLINEAR PROGRAMMING)

G. W. REDDIEN

NEWTON'S INTERPOLATION FORMULAE

Suppose that a function $f(\cdot)$ has been tabulated at equidistant points $x_i = x_0 + ih$; $i = 0, 1, 2, \ldots$, and it is desired to interpolate $f(x)$ by a polynomial $p_n(x)$ of degree n. Denote by x_0, the tabulated entry nearest to x, and let

$$u = (x - x_0)/h, \qquad y_i = f(x_i).$$

Then Newton's formula interpolates (i.e., approximates) $f(x)$ by

$$p_n(x) = y_0 + u\Delta y_0 + \frac{u(u-1)}{2!}\Delta^2 y_0 + \cdots$$
$$+ \frac{u(u-1)\cdots(u-n+1)}{n!}\Delta^n y_0,$$

$$(1)$$

where

$$\Delta y_0 = y_1 - y_0,$$
$$\Delta^2 y_0 = \Delta y_1 - \Delta y_0,$$
$$\vdots \qquad \vdots$$
$$\Delta^r y_0 = \Delta^{r-1} y_1 - \Delta^{r-1} y_0,$$

the first-, second-, and rth-order forward differences*, respectively. For an illustrative example, see Chakravarti et al. [1, Section 2.2]. Note that only tabulated entries at and on one side of x_0 are used. For interpolation formulas based on equidistant points above and below x_0, *see* BESSEL'S INTERPOLATION

FORMULA and STIRLING'S INTERPOLATION FORMULA; *see also* EVERETT'S CENTRAL DIFFERENCE INTERPOLATION FORMULA.

If $f(\cdot)$ is tabulated at points $x_0, x_1, \ldots, x_n$ that are not equidistant, the same approach as in (1) can be used, where now the finite differences* $\Delta, \Delta^2, \ldots$ are replaced by divided differences. These, together with Newton's divided difference formula for interpolating $f(x)$, are fully discussed in the entry INTERPOLATION and in Section 2.2 of ref. 1.

Reference

[1] Chakravarti, I. M., Laha, R. G., and Roy, J. (1967). *Handbook of Methods of Applied Statistics*, Vol. I. Wiley, New York.

NEWTON–SPURRELL METHOD

THE METHOD OF ELEMENTS FOR SELECTING THE BEST SUBSET OF VARIABLES IN MULTIPLE REGRESSION

In the late 1960s, R. G. Newton and D. J. Spurrell were studying industrial processes by means of multiple regression (*see* MULTIPLE LINEAR REGRESSION). Together they developed a technique for partitioning the regression sum of squares (RSS) into "elements [1, 2]." Subsets of these could be added together to give the RSS for any group of variables drawn from the original set.

This alone considerably reduced the computation necessary to formulate all possible regressions for a given number of variables. However, greater understanding of the way in which elements should be interpreted coupled with practical experience in application of the technique led to formulation of a set of guidelines that could be followed to rapidly obtain the "best" regression. For a given number of variables, the best subset can be defined as the one that *maximizes the regression sum of squares* but element analysis goes beyond this and helps in obtaining a compromise between the number of variables included and the magnitude of the RSS while ensuring that variables selected be as effective as possible for both control and predictive purposes. (*See also* COMMONALITY ANALYSIS.)

MATHEMATICAL BACKGROUND

The traditional model for multiple linear regression is

$$y_i = b_1 x_{i1} + b_2 x_{i2} + \cdots + b_p x_{ip} + e_i,$$

where the p parameters b_j are to be estimated from n sets of observations, y_i and x_{ij}, on the dependent and regressor variables, respectively. In particular we are interested in finding the best subset of variables for any given set of data.

If variable x_i is left out of the regression, the RSS will decrease by b_i^2 / c_{ii}, where c_{ii} is the ith diagonal element of the covariance matrix. This quantity is defined as the *primary element* associated with variable x_i. Note that the sum of all p primary elements does *not* add up to the RSS when all variables are present.

After one variable has been dropped, new primary elements for the $p - 1$ remaining variables can be calculated. These will, in general, be different from their primary elements in the original variable regression, and the change in primary element is called a *secondary element*. Unlike primary elements, which are sums of squares and hence always positive, secondary elements may be positive or negative, showing that the removal of one variable may enhance or diminish the apparent importance of another variable in a regression.

A knowledge of secondary elements at any stage in a regression therefore allows the analyst to look ahead and see the effects of removing next any pair of variables. As more variables are removed, elements of higher orders can be obtained, but experience has shown that knowledge of primary and secondary elements is usually sufficient to quickly find an optimum regression. It can also reveal useful information about the variables under consideration.

INTERPRETATION OF ELEMENTS

With the development of element analysis came the recognition that variables can be of two different kinds: *operationally effective* and *operationally ineffective*, although the boundary between them is not always sharply defined. Operationally effective (OE) variables have a direct effect on the variable of interest, while operationally ineffective (OI) variables are linked through associations with other variables. It is desirable to include only OE variables in a regression equation, because their relationships with the dependent variable are likely to be well determined and stable. Other OI variables may appear to be very good predictors, but there is always a danger in their use that measured relationships between these and the dependent variable are subject to change as conditions present when data were collected change.

The ideal OE variable has a large primary element and relatively small secondary elements. It will always be significant, whatever groups of variables are selected.

Where two variables are linked by a large *positive* secondary element, only one should be retained in the regression. The choice has to be made on scientific rather than statistical grounds, and the variable that is believed to have the most direct influence on the dependent variable should be chosen.

The typical OI variable has large secondary elements relative to its primary, and although it is not significant in the regression containing all variables, it may be highly significant when some of the variables to which it is linked are not included. Often, such a variable will be highly correlated with the dependent variable and will appear to be a *best predictor*. It should not, however, be included in the final regression equation.

Two variables can also be linked by a large *negative* second–order element; removal of one variable will then greatly reduce the significance of the other, and both variables need to be retained, as they form an operationally effective pair. Sometimes such variables are related in practice by a functional relationship. Alternatively it can be indicative of nonlinearity.

Summary of Rules for Determining the Best Subset

Given a set of variables on which multiple regression is to be performed, start with a regression containing all variables. At each stage it is only necessary to compute primary elements for each variable and secondary elements for each pair of variables. This can easily be programmed for a computer, when elements can be printed at each stage of regression and conveniently expressed as a percentage of the total sum of squares. The following rules may then be applied:

1. Remove the variable with the smallest primary element, provided that it is not significant according to the usual t-test.

2. If at any stage a variable is removed that is coupled to another variable (or variables) by a large secondary element, this should be noted, and the effect of removing the other variables in turn at this stage investigated later. This may lead to alternative equations that require assessment on technological grounds to ensure selection of the most OE variables.

3. Continue to apply rules **1** and **2**, removing only one variable at a time, until only variables with significant t values remain. Note the presence of large negative secondary elements linking any variables. These should be investigated further.

Several good examples of how such investigations have led in practice to better understanding of real-life situations are given in the references.

References

[1] Newton, R. G. and Spurrell, D. J. (1967). *Appl. Statist.*, **2**, 51–64.

[2] Newton, R. G. and Spurrell, D. J. (1967). *Appl. Statist.*, **2**, 165–172.

(BACKWARD ELIMINATION SELECTION
  PROCEDURE
COMMONALITY ANALYSIS
ELIMINATION OF VARIABLES
MULTIPLE LINEAR REGRESSION
STEPWISE REGRESSION)

JOHN M. CASS

NEYMAN, JERZY

Born: April 16, 1894, in Bendery, Russia.

Died: August 5, 1981, in Berkeley, California.

Contributed to: mathematical statistics, probability theory, testing hypotheses, confidence intervals, generalized chi-square, stochastic models, statistics in substantive fields.

Jerzy Neyman was one of the great founders of modern statistics. He made fundamental contributions to theoretical statistics and also to the innovative yet precise use of statistics in a wide spectrum of substantive fields ranging from agriculture and astronomy through biology and social insurance to weather modification*. He tackled new areas with great interest and enthusiasm, especially when the problem was of societal importance, because he wanted to "find out" and because he knew that real problems are the source of interesting mathematical-statistical questions. It was not just finding the answer that attracted him, his attention centered on "how to find what we need to know" and even on how to pose the question of what we want to study.

Jerzy Neyman was born of Polish parents in Bendery, Monrovia, then in Russia and now part of the Soviet Union. Both his grandfathers were landowning gentry (as reflected in his full surname, Spława-Neyman) who participated in the Polish uprising of 1863 and thereby had their lands confiscated and their families exiled to Siberia and to Tashkent. But Jerzy Neyman's father Czeslaw was only five years old at this time, and was allowed to stay in the Ukraine on condition that he was not to live near Warsaw. He became a lawyer and was a judge when Jerzy was born. However, Czeslaw had a greater interest in archaeology, and Jerzy remembered going with his father on the digs in the Crimea.

Jerzy Neyman's early education was provided by governesses, alternately French and German, who contributed to his proficiency in many languages. He attended school in Simferopol and then in Kharkov, where his family moved after the death of his father when Jerzy Neyman was twelve. In 1912, he entered the University of Kharkov to study physics and mathematics. One of his lecturers was the Russian probabilist S. N. Bernstein who called his attention to the ideas in Karl Pearson's *Grammar of Science* [28], which Neyman described as influencing his development. Nevertheless, his major interest was in new research in measure theory* of Lebesgue, and this was the area of his early papers. From 1917 to 1921, Neyman was studying graduate mathematics, tutoring, and teaching. Life was very difficult in Kharkov in these years of war and revolution. When reestablished Poland and Russia started fighting over their boundary, Neyman found himself apprehended as an enemy alien, together with the rector of the University of Kharkov, who also happened to be a Pole. After some weeks in jail, they were released because their teaching was needed in the university. In 1921, in an exchange of prisoners of war, Neyman went to Poland for the first time, at the age of 27. Raised and educated in "eastern Poland," Neyman was always fond of being Polish and appreciative of his heritage, although sometimes critical of his country's governance.

In Warsaw, Neyman visited the Polish mathematician W. Sierpinski who was interested in his research and helped him get some of his papers published (those where Neyman had not been anticipated by others while he was isolated in Russia). A position was tentatively promised Neyman in Warsaw for fall, so now the problem was to earn his living during the summer. With the help of Sierpinski, Neyman obtained a position as a statistician at the National Institute of Agriculture in Bydgoszcz. One of his duties was making meteorological observations, but his main duty was to assist in agricultural trials. Neyman applied himself, obtaining funds for a month of study in Berlin since these problems were unknown in Poland, and started to publish. His 1923 doctoral thesis from the University of Warsaw was on probabilistic problems in agricultural experimentation. Beginning in 1923, Neyman lectured at the Central College of Agriculture in Warsaw and at the Universities of Warsaw and Cracow, commuting by train each week.

In 1924 he obtained a post doctoral fellowship to study under Karl Pearson* at University College, London. As Neyman described the situation, the mathematicians in Poland wanted to know whether the statistics he was writing made any sense, that is, whether Karl Pearson would publish it in *Biometrika**. Actually, some of Neyman's statistical papers had been translated into English and were known abroad. During this early London period, Neyman had contacts with Pearson and his son Egon and several other statisticians including R. A. Fisher* and W. S. Gosset*, who turned out to be the first statistician Neyman met at University College. Attired in formal morning coat, Neyman had gone at noon to call on Professor Pearson at his office—the correct dress and the correct hour for a Polish gentleman—but Karl Pearson was out to lunch. Gosset rescued Neyman and invited him to lunch. Thus their friendship started,

before Neyman even knew Gosset's identity —"Student." In retrospect, these early contacts do not seem to have been deep. Neyman's papers, accepted by Karl Pearson, were similar to what he had been doing in Poland. His interest in set theory remained. With the collapse of the Polish currency, Neyman's government fellowship was essentially worthless. He obtained a Rockefeller fellowship and spent the next year in Paris hearing lectures from Lebesgue and Hadamard and meeting Borel.

From 1928 to 1934, Neyman was busy in Poland. His activities were broad, not only agricultural experimentation but also problems in the chemical industry, problems in social health insurance and other socioeconomic questions, and especially statistical research in the behavior of bacteria and viruses leading to his appointment as head of the Statistical Laboratory of the Nencki Institute of Experimental Biology. He was also working in mathematical statistics with his students and with Egon Pearson*. His collaboration with Pearson started in 1925 and at first was carried out largely by correspondence except for a few brief meetings in France. By 1934 Karl Pearson had retired and his department was divided between his son and Fisher. Egon invited Neyman to return to University College, first as senior lecturer and then as reader (associate professor); now cooperation in person was possible and Neyman had a secure position that allowed him time to develop his own research.

The decade 1928–1938 was a fruitful period for Neyman. In their fundamental 1928 paper, Neyman and Pearson [17] put the theory of testing hypotheses* on a firm basis, supplying the logical foundation and mathematical rigor that were missing in the early methodology. An important component was the formulation of statistical hypotheses by the construction of careful stochastic models of the phenomena under question. Whether a proposed statistical test is optimum or even whether it is valid for testing a hypothetical model must be judged in comparison with the other possible models. This idea was resisted by some of the English authorities, especially by R. A. Fisher. The 1928 paper puts forward the fundamental idea and studies the performance of several often used tests and of likelihood ratio tests*. In two papers in 1933, Neyman and Pearson [18, 19] and then, in 1934, Neyman [3] took up the problem of finding tests that are optimal in the sense that they maximize the probability of rejecting a false hypothesis (maximize the power*) while controlling below a preassigned level the probability of rejecting a correct hypothesis. The optimizing is not always easy; in some simple yet very useful cases there is an optimal solution, but in other cases, further restrictions are needed in order to provide a solution. For example, we may also ask that the test be unbiased, that is, not reject a false hypothesis less often than it rejects the true hypothesis. Several papers on these subjects appeared in the *Statistical Research Memoirs* set up by Neyman and Pearson.

The impact of the Neyman–Pearson* ideas spread throughout statistics and into all fields of applications. The concept of level of significance adjusted to the importance of avoiding making an error made sense. Making sure that the experiment was large enough, that enough observations would be collected, to have a reasonable chance of noticing that the hypotheses under study were false when a specified alternative hypothesis was true instead also made sense. Soon the techniques of experimental design included the consideration of the power of the proposed design. Practically, this determination controls the size (and thus the cost) of the experiment or data collection*. Theoretically, many interesting studies evolved, first, as efforts to construct powerful tests and, also, as efforts to extend the concepts. As an example, the theory of statistical decision functions due to A. Wald* [30] has the Neyman–Pearson theory as its basis. In addition to being widely used, the Neyman–Pearson theory was also widely taught, appearing in almost all textbooks. It soon was called the *classical* theory with reference to

Neyman or Pearson omitted. Neyman noticed this rather curious phenomenon occurring so soon within his own lifetime, but he did not complain.

Perhaps this quick adoption was the result of a deeply felt need for a logical basis and straightforward procedure for reaching decisions in the face of uncertainty. However, as noted already, some English statisticians attacked the Neyman–Pearson concepts. They tend to take as a case for ridicule an example that is not a Neyman–Pearson test procedure. One would then presume that any difficulties with this procedure are not reflections on the Neyman–Pearson theory, but this is not always the chain of reasoning of the critics (cf. Hacking [2]). Neyman realized that further extensions are needed to accommodate complex problems, and he continued working on such problems throughout his long life.

In 1934 Neyman created the theory of survey sampling*, thereby initiating the subject of optimum design. Here was the theoretical basis for using probability sampling for cluster samples with a method for estimating the accompanying variances for clustered samples and a statement of the advantages of prior stratification*. As has often been pointed out, Neyman's theoretical work was both based on and aimed toward empirical research. His theoretical results on sampling human populations were obtained for use in a sampling survey of Polish labor conducted by the Institute of Social Problems in Warsaw. The methodology was originally published (1933) in Polish. His fundamental paper is in English, appearing in 1934 in the *Journal of the Royal Statistical Society* [3]. In 1937, Neyman gave a series of invited lectures on sampling for the U.S. Department of Agriculture. The revised and enlarged second edition of these *Lectures and Conferences on Mathematical Statistics and Probability Theory* [7] provided an interesting account of many of Neyman's ideas.

In 1937 Neyman published a third fundamental paper, initiating the theory of estimation* by confidence sets [5]. Estimation by intervals that had end points determined by the observations and had size dependent on the probability that the interval will cover the true point had long been a problem in statistics. Using the relations with the theory of testing hypotheses, Neyman adapted the results from testing theory to the problem of finding a confidence interval* such that he could guarantee that the probability of covering the true value of the parameter to be estimated was at least equal to a preassigned value called the confidence coefficient. Further, the probability of covering a false value, given the first property for the true value, was minimized. Thus the confidence interval was shortest conditionally in a probability sense. Later statisticians constructed confidence intervals that are shortest conditionally in a geometric sense. The concepts were quickly extended to confidence sets of several dimensions.

At first, Neyman thought that confidence sets would coincide with the fiducial intervals of Fisher if these intervals were also required to be shortest in the same sense. In most examples, the two kinds of intervals do coincide, and one can understand this if care is taken to distinguish between random variables and particular values that it may take on and also to distinguish between the random variable and the true unknown parameter to which the random variable is related. Once the observations are made, the interval is no longer random, and there is no longer any probability involved. Either the observed interval covers the true value or it does not. If a series of confidence intervals are constructed by the Neyman method, for the same or different problems and by the same or different observers, the proportion of confidence intervals that cover correctly will tend to the confidence coefficient. There are cases where the confidence interval and the fiducial interval produce different results; the Behrens–Fisher problem* is an example [9].

The uses of confidence intervals soon appeared in many textbooks and works on statistical methodology. Again there usually would be no reference to Neyman as originator. The logic and the rigor behind confidence intervals was appealing as was the easy relation between confidence interval estimation and testing hypotheses. In his lectures, Neyman used to say that this relation "brought order out of chaos."

The characteristic pattern of Neyman's research is evident in all three of these fundamental research efforts to take a rather vague statistical question and make of it a precisely stated mathematical problem for which one can search for an optimal solution. His papers established a new climate in statistical research and new directions in statistical methodology. Neyman, his students, and others vastly extended the new theories. Neyman's ideas quickly went into many fields. As David Kendall reports, "We have all learned to speak statistics with a Polish accent."

Neyman made a six-week tour of the United States in 1937 to lecture at the Department of Agriculture in Washington, and at several universities. He was offered a position as professor at one of these universities and also at the University of California at Berkeley, where professors in several departments had joined in an effort to have a strong statistics unit. Many persons have asked Jerzy Neyman why he moved from a secure position at University College, then the statistical center of the universe, to the faraway University of California, where he had never been and where there was no statistics unit. Two points always appeared in his response: one point is the fact that Neyman was being asked to build a statistical unit at Berkeley where there was none already. The other point was that he realized that World War II was coming fast and he feared what Hitler might do to Poland and to the rest of Europe. He visualized that Poland would be destroyed again and that he would be interned in Britain as an enemy alien. Neyman had survived a war and revolution with vivid difficulties; he wanted to move himself and his family as far away as possible. There was no suggestion of pressure from English statisticians, in particular, from R. A. Fisher. It seems unjust to both individuals to suggest pressure. In any case, it is not in Neyman's character to yield to pressure. Neyman was aware of Fisher's attacks but generally did not respond [12]: "It would not be appropriate." They met and talked at international meetings and at Cambridge where Neyman spent a sabbatical year in 1957–58. In 1961, there was a response [12], however, at the suggestion of Japanese statisticians after a strong series of attacks; it was entitled "Silver Jubilee of my dispute with Fisher."

Jerzy Neyman arrived in Berkeley in August 1938 to spend nearly half of his life at the University of California. When he reached the age of 67, he continued as Director of the Statistical Laboratory and as Professor of Statistics, being recalled each year to active duty. He was going forward all the time. He liked Berkeley and worked with great enthusiasm to build the best department of statistics and to vitalize the university toward stronger yet more flexible programs.

The Statistical Laboratory was founded during his first year at Berkeley, which provided some autonomy, but Neyman felt that a completely independent Department of Statistics was important (cf. Reid [29]). Reaching out to the other departments and to other centers, Neyman established yearly summer sessions and then regular (every five years) Berkeley Symposia on Mathematical Statistics and Probability with participants from all over the world, the leaders in each new direction. Much efforts was involved and many persons had to be convinced, especially in the early years, but Neyman persisted, with the result that Berkeley soon became an exciting place to be and the Berkeley Statistics Department came into existence in 1955, blossoming out as the best. Neyman struggled to obtain more university positions for statistics and then to bring the best possible faculty and students and to support them academically and financially. The University of California has a large measure of faculty governance, but this is very time-consuming, especially during times of stress such as the Year of the Oath (1950, when the Regents imposed a special loyalty oath that interfered with academic freedom) and the student movements in the sixties when more academic governance was demanded by students. Neyman was at the forefront of protecting the rights of the protestors.

With his arrival in Berkeley, Neyman turned more toward the use of statistics in large problems. During World War II, research on directed topics, typically multiple bomb aiming, absorbed all of his time outside of a heavy teaching schedule. As the war was ending, the first Berkeley symposium took place. Neyman presented an important paper [11] which was written before 1945, using a class of procedures that he called best asymptotically normal (BAN) estimates, for estimation and also for testing. Neyman showed that by minimizing various appropriate expressions, one can obtain classes of estimates that are asymptotically equivalent, all producing limiting normal distributions that have minimum variance. The wide choice in the expressions to be minimized, ranging through likelihood ratio* and minimum chi-square*, each with possible modifications, accompanied with a wide choice of estimators for the unknown parameters, provided flexibility that allowed simpler expressions and much easier computation.

Important and useful concepts also appeared in this seminal paper, including what is now called restricted chi-square tests. In order to show that the various test procedures are asymptotically optimal and also asymptotically equivalent, Neyman considered the large-sample performance of these tests by considering "nearby alternatives" that approach the hypothesis at the rate $1/n^{1/2}$. This method is now widely used in asymptotic studies; it was introduced by Neyman in 1937 in an unusual paper on the "smooth" test* for goodness of fit* [6]. The theory of BAN estimation is now widely used in a manner similar to the use of least squares*, as its asymptotic equivalent.

Neyman intensified his interest in how

one studies stochastic problems. His method consisted in constructing a stochastic model consistent with the knowledge available and checking the consequences of the model against possible observations. Usually, certain aspects were of particular interest; he would determine the kinds of observations that would lead to sensitive responses and try to arrange for a cooperative study whenever new observations were needed. The range of topics was broad indeed. He extended [22] his clustering models beyond what are now called Neyman type A*, type B, and so forth, which originally described how the larvae crawl out from where the eggs are clustered, to other entities in other fields, e.g., physical particles and clusters of galaxies [8]. These clustering models were widely adopted in many areas of biology, ecology, and physics by researchers trying to get a better understanding of the processes involved rather than applying interpolation formulas. Neyman, mostly with Scott and with Shane, published a long series of articles on the clustering of galaxies in space as derived from the apparent clustering on the photographic plate [21]. Several cosmological questions are of continuing interest, such as the evolution of galactic types from elliptical to spiral or conversely.

Neyman studied many other stochastic processes, including the mechanism and detection of accident proneness, with Bates, where more realistic and more general models were set up and tested, thus deriving new information about the process [1]. The same vein underlies all of these studies of stochastic processes: catching schools of fish [11], spread of epidemics [24], carcinogenesis in a long series of papers [13, 25] including recent studies where the mechanism inducing the cancer involves high-energy radiation [20]. In his studies of the relation between pollutants and health, the direction had to be more diverse as it also had to be in a long series of studies on the effectiveness of weather modification on which he was working at the time of his death and which he started at the request of the state of California, where lack of rainfall is often a serious difficulty [14, 27].

Jerzy Neyman saw that the models he was using were too complex for the application of his optimum tests developed in the thirties. He turned to asymptotic methods [10], such as his BAN estimates, and developed tests that are locally optimal in the presence of nuisance parameters. These are called C-alpha tests*, in honor of H. Cramér and because they are asymptotically similar, even though no similar test exists. These optimal tests are not difficult to construct [30] from the logarithmic derivatives of the densities; they solve a large class of problems that cannot be solved by straight substitution of observed values for unknown parameters (*see* OPTIMAL $C(\alpha)$-TESTS).

There were other situations where the conflicting hypotheses are not identifiable or the probability of making a correct decision was very tiny. These difficulties arise not only in widely used normal theory [4], but also in applied problems such as competing risks*, when one wants to disentangle which of the possible causes of death (or which combination) actually is the cause, or competition between species, where one wants to predict which of two competing species will be the survivor when both will survive if kept separate [16]. What additional information must be supplied to allow the study of models of relapse and recovery, for example, to be complete?

Jerzy Neyman wanted to ensure that science was not obscured by political expedience or by commercialism. He turned to the scientific societies to which he belonged, of which there were many, for help in enforcing a strict code of ethics. Several organizations have taken action; others are moving. The problems are not easy, but Neyman had the courage to speak out for honesty [15, 23].

Neyman was a superb teacher using his version of the Socratic method. It was not easy to be sent to the chalkboard but the training was invaluable. Also, he was an inspiring research leader who shared his ideas and granted full credit to his students and young colleagues, always in a strict alphabetical order! He was always a winner in number and quality of doctoral students. He maintained a steady interest in all of their activities and liked to follow their progress through the years. He dedicated the 1952 edition of *Lectures and Conferences* to the memory of his Polish students and early co-workers lost in World War II, listing for each the grisly cause of death—nearly all of his Polish students were killed. But he kept his faith in the new students who came from all over the world, helping them in every possible way: "I like young people!"

Neyman strove to strengthen statistics and to build statistical groups. One reason was to make it easier to publish good papers without prejudiced editing, a difficulty that he and other members of his generation had faced. He worked to strengthen the Institute of Mathematical Statistics*, organizing many sessions and serving on the council and as president. He felt that the International Statistical Institute* should be widened to open the program and the elected membership to academic statisticians and to young researchers. He found a way to do this by establishing, with the help of many colleagues, a new section of the ISI, the Bernoulli Society*, whose members could attend ISI sessions. The Society had freedom to organize part of the ISI program. Never mind that in 1958 Neyman agreed to accept the name International Association for Statistics in the Physical Sciences. The name was not a restriction on its activities, and it was very active not only at ISI sessions but also in organizing satellite symposia on a wide range of topics. By 1975 the name Bernoulli Society became official, with a strengthening of its world-wide role. Neyman became an honorary president of the ISI.

Jerzy Neyman was interested in having stronger science and more influence of scientists in general decision making. He spoke out strongly against inequities and worked hard year after year to establish fair treatment, especially for black people. Perhaps his scientific hero was Copernicus. He gave a special talk at the 450th anniversary of Copernicus while he was still in Bydgoszcz. At Copernicus' 500th anniversary in Berkeley, Neyman was even more active. In addition to talks and articles, he edited an un-

usual book, *The Heritage of Copernicus: Theories "Pleasing to the Mind"*, in which scientists from many fields described neo-Copernican revolutions in thought. Neyman received a shower of Copernican medals and plaques.

Jerzy Neyman's achievements received wide recognition. In addition to being elected to the International Statistical Institute and the International Astronomical Union, he was elected to the U.S. National Academy of Sciences, and made a foreign member of the Swedish and Polish Academies of Science and of the Royal Society. He received many medals, including the Guy Medal in gold from the Royal Statistical Society* (London) and, in 1968, the U.S. National Medal of Science. He received honorary doctorates from the University of Chicago, the University of California, Berkeley, the University of Stockholm, the University of Warsaw, and the Indian Statistical Institute*.

Neyman was greatly esteemed, and he was greatly loved. He gave of his own affection, his warmth, and his talents in such a way that they became a part of science to be held and handed on.

References

[1] Bates, G. E. and Neyman, J. (1952). *Univ. Calif. Publ. Statist.*, **1**, 215–254; 255–276. (Theory of accident proneness; true or false contagion.)

[2] Hacking, I. (1965). *Logic of Statistical Inference*. Cambridge University Press, Cambridge, England.

[3] Neyman, J. (1934). *J. R. Statist. Soc.*, **97**, 558–625. Also in *A Selection of Early Statistical Papers of J. Neyman*. (1967). University of California Press, Berkeley, No. 10. [Spanish version appeared in *Estadistica* (1959), **17**, 587–651.] (Fundamental paper on sampling, optimal design, confidence intervals.)

[4] Neyman, J. (1935). *J. R. Statist. Soc. Suppl.*, **2**, 235–242. Also in *A Selection of Early Statistical Papers of J. Neyman*. (1967). University of California Press, Berkeley, No. 15. (Difficulties in interpretation of complex experiments.)

[5] Neyman, J. (1937). *Philos. Trans. R. Soc. Lond. A*, **236**, 333–380. Also in *A Selection of Early Statistical Papers by J. Neyman*. (1967). University of California Press, Berkeley, No. 20. (Fundamental

paper on theory of estimation by confidence sets. See also ref. 7.)

[6] Neyman, J. (1937). *Skand. Aktuarietidskr.*, **20**, 149–199. Also in *A Selection of Early Statistical Papers of J. Neyman.* (1967). University of California Press, Berkeley, No. 21. (Unusual paper on "smooth" test for goodness of fit.)

[7] Neyman, J. (1938, 1952). *Lectures and Conferences on Mathematical Statistics and Probability*, Graduate School. U.S. Dept. of Agriculture, Washington, DC. (Spanish version published by Inter-American Statistical Institute (1967).) (The revised and enlarged second edition provides an interesting account of many of Neyman's ideas.)

[8] Neyman, J. (1939). *Ann. Math. Statist.*, **10**, 35–57. (First model of "contagious" distributions, including Neyman type A clustering.) Also in *A Selection of Early Statistical Papers of J. Neyman.* (1967). University of California Press, Berkeley, No. 25.

[9] Neyman, J. (1941). *Biometrika*, **32**, 128–150. Also in *A Selection of Early Statistical Papers of J. Neyman.* (1967). University of California Press, Berkeley, No. 26. (Investigation of relation between confidence intervals and Fisher's fiducial theory.)

[10] Neyman, J. (1949). *Proc. Berkeley Symp. Math. Statist. Prob.*, (of 1945), 239–273. Also in *A Selection of Early Statistical Papers of J. Neyman.* (1967). University of California Press, Berkeley, No. 28. (Seminal paper on restricted chi-square tests, BAN estimation, and asymptotically optimal and asymptotically equivalent procedures.)

[11] Neyman, J. (1949). *Univ. Calif. Publ. Statist.*, **1**, 21–36. (Catching schools of fish, a study of the decrease in sardine catches.)

[12] Neyman, J. (1961). *J. Operat. Res. Soc. Jpn.*, **3**, 145–154. (Neyman conducts himself "not inappropriately" in controversy with R. A. Fisher.)

[13] Neyman, J. (1961). *Bull. Int. Inst. Statist.*, **38**, 123–135. (Modeling for a better understanding of carcinogenesis. Summary paper in a series on carcinogenesis.)

[14] Neyman, J. (1977). *Proc. Natl. Acad. Sci. (U.S.)*, **74**, 4714–4721. (Invited review paper on a statistician's view of weather modification technology.)

[15] Neyman, J. (1980). *Statistical Analysis of Weather Modification Experiments*, E. Wegman and D. DePriest, eds. Marcel Dekker, New York, pp. 131–137. (Comments on scientific honesty in certain experiments and operations.)

[16] Neyman, J., Park, T. and Scott, E. L. (1956). *Proc. 3rd Berkeley Symp. Math. Statist. Prob.*, **4**, 41–79. (Struggle for existence: Tribolium model.)

[17] Neyman, J. and Pearson, E. S. (1928). *Biometrika*, **20-A**, 175–240 and 263–294. Also in *Joint Statistical Papers of J. Neyman and E. S. Pearson.* (1967). University of California Press, Berkeley, Nos. 1

and 2. (Fundamental paper on testing hypotheses, in two parts.)

[18] Neyman, J. and Pearson, E. S. (1933). *Philos. Trans. R. Soc. Lond. A*, **231**, 289–337. Also in *Joint Statistical Papers of J. Neyman and E. S. Pearson.* (1967). University of California Press, Berkeley, No. 6.

[19] Neyman, J. and Pearson, E. S. (1933). *Proc. Camb. Philos. Soc.*, **29**, 492–510. Also in *Joint Statistical Papers of J. Neyman and E. S. Pearson.* (1967). University of California Press, Berkeley, No. 7.

[20] Neyman, J. and Puri, P. S. (1982). *Proc. R. Soc. Lond. B*, **213**, 139–160. (Models of carcinogenesis for different types of radiation.)

[21] Neyman, J. and Scott, E. L. (1952). *Astrophys. J.*, **116**, 144–163. (Theory of spatial distribution of galaxies; first paper in a long series.)

[22] Neyman, J. and Scott, E. L. (1957). *Proc. Cold Spring Harbor Symp. Quant. Biol.*, **22**, 109–120. (Summary paper on populations as conglomerations of clusters. See also the paper (1959) in *Science*, **130**, 303–308.)

[23] Neyman, J. and Scott, E. L. (1960). *Ann. Math. Statist.*, **31**, 643–655. (Correction of bias introduced by transformation of variables.)

[24] Neyman, J. and Scott, E. L. (1964). *Stochastic Models in Medicine and Biology*, J. Gurland, ed., University of Wisconsin Press, Madison, pp. 45–83. (Stochastic models of epidemics.)

[25] Neyman, J. and Scott, E. L. (1967). *Proc. 5th Berkeley Symp. Math. Statist. Prob.*, **4**, 745–776. (Construction and test of two-stage model of carcinogenesis; summary.)

[26] Neyman, J. and Scott, E. L. (1967). *Bull. Int. Inst. Statist.*, **41**, 477–496. (Use of C(alpha) optimal tests of composite hypotheses. Summary paper with examples.)

[27] Neyman, J. and Scott, E. L. (1967). *Proc. 5th Berkeley Symp. Math. Statist. Prob.*, **5**, 327–350. (Timely summary paper on statistical analysis of weather modification experiments; one of 48 published papers and many reports in this field.)

[28] Pearson, K. (1892, 1937). *The Grammar of Science*, 3rd ed., revised enlarged. E. P. Dutton, New York. (A paperback edition was published in 1957 by Meridian.)

[29] Reid, C. (1982). *Neyman—From Life*. Springer-Verlag, New York. (A sensitive and knowledgeable biography, beautifully rendered.)

[30] Wald, A. (1950). *Statistical Decision Functions*. Wiley, New York. (Fundamental book on decision functions.)

Bibliography

Kendall, D. G., Bartlett, M. S., and Page, T. L. (1982). *Biogr. Mem. Fellows R. Soc. Lond.*, **28**, 378–412. (This interesting and extensive biography is in three parts and contains as complete bibliography through 1982.)

Klonecki, W., and Urbanik, K., (1982). *Prob. Math. Statist.*, Polish Acad. Sci., **2**, I–III. (Neyman viewed by his countrymen.)

LeCam, L. and Lehmann, E. L. (1974). *Ann. Statist.*, **2**, vii–xiii. (Review of Neyman's scientific work, on the occasion of his eightieth birthday.)

(ASYMPTOTIC NORMALITY
CONFIDENCE INTERVALS
  AND REGIONS
ESTIMATION, POINT
HYPOTHESIS TESTING
LARGE-SAMPLE THEORY
MILITARY STATISTICS
NEYMAN ALLOCATION
NEYMAN–PEARSON LEMMA
NEYMAN'S AND OTHER SMOOTH
  GOODNESS-OF-FIT TESTS
NEYMAN'S TEST FOR UNIFORMITY
NEYMAN STRUCTURE, TESTS WITH
OPTIMAL $C(\alpha)$ TESTS
PEARSON, EGON SHARPE
PEARSON, KARL
STATISTICS IN ASTRONOMY
UNBIASEDNESS
WEATHER MODIFICATION,
  STATISTICS IN)

ELIZABETH L. SCOTT

NEYMAN ACCURACY *See* ACCURACY, NEYMAN

NEYMAN ALLOCATION

Consider a finite population Π of M elements. The population Π is divided into k strata:

$$\Pi_1, \Pi_2, \ldots, \Pi_k,$$

with M_i elements in the ith stratum, $i = 1, 2, \ldots, k$, so that

$$\sum_{i=1}^{k} M_i = M.$$

Suppose we are interested in estimating the mean value $\bar{x}_M$ of a certain characteristic X of the population Π by using a stratified random sample of size m in such a way that

m_i elements will be selected from the ith stratum, $i = 1, 2, \ldots, k$, and

$$\sum_{i=1}^{k} m_i = m.$$

An unbiased estimator $\bar{x}_{st}$ of the mean value $\bar{x}_M$ is given by

$$\bar{x}_{st} = \sum_{i=1}^{k} W_i \bar{x}_{m_i},$$

where $W_i = M_i / M$ is the stratum weight and $\bar{x}_{m_i}$ is the mean value of the random sample of size m_i of the ith stratum. Since the efficiency of the estimator $\bar{x}_{st}$ depends on the sample sizes m_i, the problem, then, is how to distribute the total fixed sample size m among strata so as to minimize the variance of the estimator $\bar{x}_{st}$. This variance can be expressed as

$$\text{var}(\bar{x}_{st}) = \sum_{i=1}^{k} \frac{W_i^2 (M_i - m_i) S_i^2}{M_i m_i}.$$

where S_i^2 is the mean square of the M_i elements in the ith stratum. [If σ_i^2 is the variance of the ith stratum then $S_i^2 = M_i \sigma_i^2 / (M_i - 1)$.] Simple algebra shows that the variance can also be written in the form

$$\text{var}(\bar{x}_{st}) = A + B - C,$$

where

$$A = \left(\frac{M}{m} - 1 \right) \sum_{i=1}^{k} M_i S_i^2,$$

$$B = \sum_{i=1}^{k} m_i \left(\frac{M_i S_i}{m_i} - \sum_{i=1}^{k} \frac{M_i S_i}{m} \right)^2,$$

$$C = \frac{M}{m} \sum_{i=1}^{k} M_i \left(S_i - \sum_{i=1}^{k} \frac{M_i S_i}{M} \right)^2,$$

and A, B, and C, are nonnegative. If we select the m_i's so that they are proportional to M_i, then $B = C$ and $\text{var}(\bar{x}_{st}) = A$. This allocation is known as *proportional allocation**. If, however, m_i's are proportional to $M_i S_i$, then $B = 0$ and $\text{var}(\bar{x}_{st}) = A - C$, which is the optimum value of $\text{var}(\bar{x}_{st})$. This allocation was first introduced by Neyman [2]; hence it is named the Neyman allocation in his honor.

An alternative approach to this problem (see Cochran [1]) is to allocate the stratified sample of m elements among the k strata so that either var($\bar{x}_{st}$) is minimized for a specified cost or the cost is minimized for a fixed var($\bar{x}_{st}$). Suppose the cost function is linear, that is,

$$\text{cost} = c_0 + \sum_{i=1}^{k} c_i m_i,$$

where c_0 is the fixed cost and c_i is the within-stratum sampling cost per unit in the ith stratum. The solution is that the m_i's are proportional to $M_i S_i / \sqrt{c_i}$. When the per unit cost is the same throughout all k strata, i.e., $c_i = c$ for $i = 1, 2, \ldots k$, the allocation reduces to Neyman allocation, which therefore is an optimum allocation if the cost per unit is uniform.

It was discovered later than this result had been derived earlier by Tschuprow [3] (*see* CHUPROV, ALEXANDER).

References

[1]   Cochran, W. G. (1977). *Sampling Techniques*, 3rd ed. Wiley, New York.

[2]   Neyman, J. (1934). *J. R. Statist. Soc.*, **97**, 558–606.

[3]   Tschuprow, A. A. (1923). *Metron*, **2**, 461–493 and 646–683.

(FINITE POPULATIONS, SAMPLING FROM
SAMPLING PLANS
STRATIFIED SAMPLING)

VICTOR K. T. TANG

NEYMAN–PEARSON LEMMA, THE

STATEMENT OF THE LEMMA

This lemma, which is due to Neyman and Pearson [13], is the central tool of the theory of hypothesis testing* and plays a crucial role in much of mathematical statistics. It is a mathematical result that in the simplest case is a solution of the following problem.

Let $O_1, \ldots, O_N$ be N objects, each of which has a certain worth, for example its weight, its gold content, or its resale value. Let $V(O_i)$ denote the worth of the ith object and $P(O_i)$ its price. The problem is to select a number of these objects in such a way that their total worth is as large as can be obtained for the price paid for them.

To solve this problem, order the objects according to their worth per dollar, $r(O_i) = V(O_i)/P(O_i)$, and make the selection according to the value of this ratio; take those with highest $r(O_i)$ first, then those with second highest, and so on. If a certain number of objects is selected according to this rule, say those O_i for which i lies in some set I_0, and if their total price is $\Sigma_{i \in I_0} P(O_i) = \alpha$, then, the lemma states, no other set of objects whose total price is $\leqslant \alpha$ can have a larger worth, i.e., I_0 is the set I that maximizes $\Sigma_{i \in I} V(O_i)$ subject to $\Sigma_{i \in I} P(O_i) \leqslant \alpha$. This result is independent of how many objects are selected, that is, of the value of α.

In statistical applications, the objects are the sets of values taken on by random variables $X = (X_1, \ldots, X_n)$, which may be represented as points in an n-dimensional space (the sample space*). The worth V and price P associated with each point x is the value assigned to x by two possible probability distributions P_0 and P_1 of X. It is desired to select a set S_0 of sample points x in such a way that, if $P_0(S_0) = \Sigma_{x \in S_0} P_0(x) = \alpha$, then for any set S satisfying $P(S) = \Sigma_{x \in S} P_0(x) \leqslant \alpha$, one has $P_1(S) \leqslant P_1(S_0)$. The solution is to order the points x according to the value of the *likelihood ratio* $r(x) = P_1(x)/P_0(x)$ and to include in S_0 those points for which $r(x)$ is sufficiently large, say $r(x) > C$. The lemma states that the set $S_0 = \{x : r(x) > C\}$ is a solution of the stated problem, and that this is true for every value of C or of

$$\alpha = P_0[r(X) > C]. \qquad (1)$$

If X is not discrete, but has probability density either p_0 or p_1, the result in (1) remains valid with $P_1(x)/P_0(x)$ replaced by $r(x) = p_1(x)/p_0(x)$. Throughout this article, $p_i(x)$ will be used to denote either discrete probability distributions so that $p_i(x) = P_i(X = x)$ or probability densities (or quite generally, densities with respect to a common measure μ).

Example 1. Let $X_1, \ldots, X_n$ be independently distributed according to the normal distribution $N(\xi_0, 1)$ or $N(\xi_1, 1)$, $\xi_0 < \xi_1$. Then

$$r(x) = p_1(x)/p_0(x)$$

$$= e^{-\sum(x_i - \xi_1)^2/2 + \sum(x_i - \xi_0)^2/2}$$

$$= e^{(\xi_1 - \xi_0)\sum x_i - (n/2)(\xi_1^2 - \xi_0^2)}.$$

For fixed $\xi_0 < \xi_1$, this is a strictly increasing function of $\sum x_i$ and the sample points are therefore ordered according to the value of $\sum x_i$. The larger this sum, the more useful the point is for the purpose at hand.

Example 2. Let X be the number of successes in n binomial trials with success probability either p_0 or p_1 ($p_0 < p_1$). Then the points x are ordered according to

$$r(x) = \binom{n}{x}p_1^x q_1^{n-x} \bigg/ \left[\binom{n}{x}p_0^x q_0^{n-x}\right]$$

$$= \left(\frac{p_1 q_0}{p_0 q_1}\right)^x \left(\frac{q_1}{q_0}\right)^n,$$

or equivalently according to the value of x.

RANDOMIZED SELECTION RULES

The problem stated at the beginning of the first section often arises in the slightly different form that the objects are to be selected subject to a given limited budget, say under the restriction

$$\sum_{i \in I} P(O_i) \leq \alpha. \qquad (2)$$

Subject to this restriction, one wishes to maximize $\sum_{i \in I} V(O_i)$. If there exists a value C for which the price of the set

$$I_0 = \{i : r(O_i) > C\}$$

is exactly α, it follows from the lemma as stated in the first Section that I_0 is a solution of the selection problem subject to (2). If such a value of C does not exist, the prob-

lem has no simple solution. Algorithms for solving it are provided by the theory of linear programming*.

To see why the solution cannot be simple in the present formulation and how the essence of the lemma can be preserved by broadening the set of possible solutions, suppose that a number of objects have been selected and that the budget is nearly but not quite exhausted. Specifically, suppose that the next object that would be selected according to the ordering of the lemma, say O_k, would exceed the budget (and therefore is not eligible), but that there is an object further down in the ordering, say O_1, which would not. To maximize the total value subject to (2) one would then have to violate the selection rule postulated by the lemma. (It is, incidently, not necessarily true that selecting O_1 would solve the problem. An entirely different set of objects may provide a better choice.)

This difficulty can be avoided if it is possible to select part of (or a share in) an object. One could then select that portion of O_k that would expend the budget to the last penny. The lemma (in this extended form) asserts that the proposed selection rule is a solution to the limited budget problem.

In the statistical setting of the first Section when the objects are the possible sample points x, partial selection of a point is possible through *randomization**. Each point x can be selected or not according to stated probabilities $\phi(x)$ and $1 - \phi(x)$. (For example, if a point x is to be selected with probability $\frac{1}{3}$, roll a fair die and select x if and only if the die shows 1 or 2 points.) If x is definitely to be selected, put $\phi(x) = 1$, and put $\phi(x) = 0$ if x is definitely not to be selected. In general, a *selection function* is any function ϕ defined over the sample space and satisfying $0 \leq \phi(x) \leq 1$. The earlier formulation of a nonrandomized selection rule is the special case in which ϕ takes on only the values 1 and 0. The set of points x for which $\phi(x) = 1$ is then the set S of the first section.

The probability that the selection rule will select a random point X, which in the nonrandomized case was given by $P_i(X \in S)$ when the distribution of X is P_i, now becomes the expectation $E_i\phi(X)$, and the problem becomes that of determining the selection function ϕ which, subject to $E_0\phi(X) \leqslant \alpha$, maximizes $E_1\phi(X)$. The lemma in its final form states that ϕ is a solution of this problem if it satisfies

$$\phi(x) = \begin{cases} 1 & \text{when} \quad r(x) > C \\ 0 & \text{when} \quad r(x) < C, \end{cases} \quad (3)$$

where the *critical value* C is determined so that

$$E_0\phi(X) = \alpha. \quad (4)$$

The lemma further asserts that ϕ and C satisfying these conditions always exist. Note, however, that the value of ϕ on the boundary $r(x) = C$ is not specified, except possibly indirectly by (4).

In statistical practice, randomization is usually not acceptable (see, e.g., Kempthorne and Doerfler [10]), but neither is the breaking of the r order so as to obtain the best nonrandomized rule. Instead, the most common practice is to choose a slightly different α for which the best rule is nonrandomized.

THE ROLE OF THE LEMMA IN HYPOTHESIS TESTING

Testing a Simple Hypothesis Against a Simple Alternative

Consider the problem of testing a simple hypothesis $H : p = p_0$ against the simple alternative $p = p_1$. A nonrandomized test is defined by a set S in the sample space* such that H is rejected when $X \in S$. The probability $P_0(S) = P_0(X \in S)$ is the probability of rejecting H when it is true; this is not to exceed a preassigned *significance level* α, so that S is to satisfy

$$P_0(S) \leqslant \alpha. \quad (5)$$

Subject to this condition, one wishes to maximize the power of the test, i.e., the probability of rejection under the alternative that is

given by $P_1(S)$. By the version of the lemma in the Randomized Selection Rules section, the solution of this problem, the *most powerful* level α test, is given by $S = \{x : r(x) > C\}$, provided that there exists a C that satisfies (1). In the contrary case, a randomized rejection rule ϕ is required that rejects H with probability $\phi(x)$ when x is observed and that is given by (3) and (4).

Uniformly Most Powerful Tests

Consider a one-parameter family of probability distributions $P_\theta(X = x) = P_\theta(x)$ or of densities $p_\theta(x)$, and the problem of testing a simple hypothesis $H : \theta = \theta_0$ against the one-sided alternatives $\theta > \theta_0$. For testing H against a simple alternative $\theta_1 > \theta_0$ at level α, the solution is given by the preceding subsection and will typically depend on the alternative θ_1 chosen. There are, however, important cases in which the same test maximizes the power simultaneously for all alternatives $\theta_1 > \theta_0$ and is therefore uniformly most powerful (UMP). From the preceding sections it is seen that this will occur when each pair $\theta_0 < \theta_1$ induces the same ordering of the sample points. This possibility is illustrated by Example 1 where for each pair $\xi_0 < \xi_1$ the points are ordered according to the value of $\sum x_i$, and by Example 2 where for each $p_0 < p_1$ the points are ordered according to the value of x. When a family $\{p_\theta\}$ has the property that the same ordering is induced by all pairs (θ_0, θ_1) with $\theta_0 < \theta_1$, it is said to have *monotone likelihood ratio**. An important class possessing this property is the class of one-parameter exponential families*.

For a family with monotone likelihood ratio, a UMP test of $H : \theta = \theta_0$ against $\theta > \theta_0$ exists at all significance levels α and for each θ_0. This test turns out to be UMP also for testing the more realistic hypothesis $H' : \theta \leqslant \theta_0$ against the alternatives $\theta > \theta_0$.

UMP Unbiased Tests

Consider the problem of testing $H : \theta = \theta_0$ against $\theta > \theta_0$ in a multiparameter exponential family $\{p_\theta, \vartheta\}$ depending not only on the parameter θ being tested but also on nuisance parameters* $\vartheta = (\vartheta_1, \ldots, \vartheta_s)$. A UMP test typically does not exist in this case, and one may then wish to impose the condition of unbiasedness* that the probability of rejection is $\geqslant \alpha$ for all alternatives, i.e., for all (θ, ϑ) with $\theta > \theta_0$. If the density of the exponential family is

$$P_{\theta, \vartheta}(x) = C(\theta, \vartheta)e^{\theta U(x) + \Sigma \vartheta_i T_i(x)}h(x),$$

unbiasedness implies that

$$P_{\theta_0}\left[\text{reject } H \mid T(x) = t\right] = \alpha \qquad \text{for all } t, \tag{6}$$

(where $T(x) = (T_1(x), \ldots, T_s(x))$ and $t = (t_1, \ldots, t_s)$), and it can be shown that a test is most powerful against an alternative (θ_1, ϑ) subject to (6) if it maximizes

$$P_{\theta_1}\left[\text{reject } H \mid T(x) = t\right]. \tag{7}$$

The problem is thus reduced to the application of the lemma to the conditional distribution given t, which depends only on θ, not on ϑ. The test obtained by this application of the lemma turns out to be UMP unbiased.

Example 3. Let Y, Z be independent Poisson variables with expectation λ and μ, and consider the problem of testing $H : \mu = \lambda$ against $\mu > \lambda$. The joint distribution of Y, Z is a two-parameter exponential family with $\theta = \log(\mu/\lambda)$, $\vartheta = \log \lambda$, $U = Z$, $T = Y + Z$. The conditional distribution of U given $T = t$ is the binomial distribution $b(p, t)$ with $p = \mu/(\lambda + \mu)$, and the most powerful conditional test is given by Example 1.

UMP Invariant Tests

When a problem remains invariant under certain transformations of the observations, the principle of invariance permits restriction to a so-called maximal invariant statistic (*see* INVARIANCE CONCEPTS IN STATISTICS). If this has monotone likelihood ratio, a UMP invariant test will exist, as is illustrated in Example 5 of the article HYPOTHESIS TESTING. (For a detailed discussion of UMP un-

biased and UMP invariant tests, see Lehmann [11].)

A GENERALIZATION OF THE LEMMA

The lemma, as stated in the Randomized Selection Rules section, determines the function ϕ $(0 \leqslant \phi(x) \leqslant 1)$ which, subject to

$$\int \phi f \, d\mu \leqslant \alpha \tag{8}$$

or

$$\int \phi f \, d\mu = \alpha \tag{9}$$

maximizes

$$\int \phi g \, d\mu, \tag{10}$$

where f and g are two given probability densities. The lemma remains true if the two probability densities are replaced by arbitrary integrable functions f and g and has an important extension to the case in which side condition (8) or (9) is replaced by a finite set of such side conditions

$$\int \phi f_i \, du \leqslant c_i, \qquad i = 1, \ldots, m \tag{11}$$

or

$$\int \phi f_i \, d\mu = c_i, \qquad i = 1, \ldots, m. \tag{12}$$

A sufficient condition for ϕ to maximize (10) subject to (12) is that it satisfies (12) and

$$\phi(x) = \begin{cases} 1 & \text{if } g(x) > \sum k_i f_i(x) \\ 0 & \text{if } g(x) < \sum k_i f_i(x) \end{cases} \tag{13}$$

for some constants $k_1, \ldots, k_m$. In this result, due to Neyman and Pearson [14], the k's can be viewed as undetermined multipliers whose values are determined by the side conditions (12). Necessity of the form (13) was investigated by Dantzig and Wald [2]. If (12) is replaced by (11), the structure (13) continues to be a sufficient condition for ϕ to maximize (10), provided the k's are nonnegative and (12) holds for all values of i for which $k_i > 0$. (For proofs, see, e.g., Lehmann [11, pp. 87 and 114].)

As a simple application, consider the problem of determining the locally most powerful* unbiased level α test of $H : \theta = \theta_0$, i.e., of maximizing

$$\frac{d^2}{d\theta^2} E_\theta \phi(X) \qquad (14)$$

subject to

$$E_{\theta_0} \phi(X) = \alpha, \qquad \frac{d}{d\theta} E_\theta \phi(X) \Big|_{\theta = \theta_0} = 0. \qquad (15)$$

Under suitable regularity conditions (Neyman and Pearson [14]), the generalized lemma then shows the solution to be given by the test ϕ that rejects H when

$$\frac{d^2}{d\theta^2} p_\theta(x) \geq k_1 \frac{d}{d\theta} p_\theta(x) \Big|_{\theta = \theta_0} + k_2 p_{\theta_0}(x),$$

where $p_\theta(x)$ denotes the probability density of X and where k_1 and k_2 are determined by (15).

TESTS OF COMPOSITE HYPOTHESES AND MINIMAX TESTS

Consider the problem of testing a composite hypothesis $H : \{ p_\theta, \theta \in \omega \}$ against a simple alternative $K : g$. Then a level α test must satisfy

$$E_\theta \phi(X) \leq \alpha \qquad \text{for all} \quad \theta \in \omega. \quad (16)$$

Subject to (16), one wishes to maximize $E_g \phi(X)$.

For the case that ω is finite, this problem is solved by the extended lemma of the preceding section. In the general case (16), one may try replacing the composite hypothesis H by a simple one of the form

$$H' : f_\Lambda(x) = \int_\omega p_\theta(x) \, d\Lambda(\theta)$$

and testing f_Λ against g at level α. If the resulting test given by (3) and (4) satisfies (16), it maximizes $E_g \phi(X)$ subject to (8). The principal difficulty in applying this result is the determination of Λ, which plays the role of the undetermined multipliers in the finite case. In searching for Λ, it is helpful to realize that Λ is *least favorable* in the sense that if β_Λ is the power of the most powerful level α test for testing f_Λ against g, then

$\beta_\Lambda \leq \beta_{\Lambda'}$ for all Λ'. (For illustrations of this result, see Lehmann [11].)

The preceding generalization of the lemma can be extended to the problem of testing a composite hypothesis $H : \{ p_\theta, \theta \in \omega \}$ against a composite alternative $K : \{ p_\theta, \theta \in \omega' \}$, where subject to (16) it is desired to maximize the minimum power over K. A solution to this *maximin problem* (see MINIMAX TESTS) can be obtained by replacing H and K by

$$f(x) = \int_\omega p_\theta(x) \, d\Lambda(\theta) \quad \text{and}$$

$$g(x) = \int_{\omega'} p_\theta(x) \, d\Lambda'(\theta)$$

and testing f against g at level α. If the resulting test satisfies (16) and

$$\inf_{\omega'} E_\theta \phi(X) = E_g \phi(X), \qquad (17)$$

it is a solution of the original problem, and the pair (Λ, Λ') is least favorable.

APPROXIMATE AND SEQUENTIAL VERSIONS OF THE LEMMA

When testing a simple hypothesis p_0 against a simpler alternative p_1 it may happen that p_0 and p_1 are known only approximately. For example, under standard conditions, the data are distributed according to p_0 or p_1, but occasionally something goes wrong in the experiment or data collection. One may then instead wish to test $\mathscr{P}_0$ against $\mathscr{P}_1$, where $\mathscr{P}_i$ is the family of *contaminated distributions*

$$\mathscr{P}_i = \{ Q : Q = (1 - \epsilon_i) P_i + \epsilon_i H_i \} \quad (18)$$

with ϵ_0, ϵ_1 given and the H_i arbitrary unknown distributions. The test that maximizes the minimum power against $\mathscr{P}_1$ can be obtained by the method of the last section.

It was shown by Huber [6] (see also Huber and Strassen [7]) that for sufficiently small ϵ_i the least favorable pair of distributions Λ_0, Λ_1 assigns probability 1 to distributions $Q_0 \in \mathscr{P}_0$, $Q_1 \in \mathscr{P}_1$ and that the maximin test rejects when

$$r(x) = q_1(x) / q_0(x) > C,$$

where

$$r(x) = \begin{cases} ka & \text{when} & p_1(x)/p_0(x) \leqslant a \\ kp_1(x)/p_0(x) & \text{when} & a < p_1(x)/p_0(x) < b \\ kb & \text{when} & p_1(x)/p_0(x) \geqslant b. \end{cases}$$

with $k = (1 - \epsilon_1)/(1 - \epsilon_0)$. The maximin test thus replaces the original probability ratio test of P_0 against P_1 with a censored version.

A quite different generalization of the lemma is to the problem of testing p_0 against p_1 when the sample size N is determined sequentially (see SEQUENTIAL ANALYSIS). This makes it possible to stop early when the observations indicate a clear preference for p_0 or p_1 and to take many observations when they don't, thereby requiring fewer observations on the average. Since by the lemma of the first section, the preference of the observations for p_1 over p_0 after n observations is indicated by $p_{1n}/p_{0n} = p_1(x_1) \ldots p_1(x_n)/p_0(x_1) \ldots p_0(x_n)$, this suggests to take observations as long as $A < p_{1n}/p_{0n} < B$ and to accept or reject H if at the first violation $p_{1n}/p_{0n} \leqslant A$ or $\geqslant B$, respectively.

This *sequential probability ratio test** is due to Wald [16]. If A and B are such that $P_i(\text{rejection}) = \alpha_i$, it was shown by Wald and Wolfowitz [18] that among all tests with the two error probabilities $\leqslant \alpha_i$, the SPRT minimizes the expected sample size both under P_0 and under P_1. (See also Wijsman [19].)

Extensions of this result to stochastic processes* that are observed continuously are given by Dvoretzky et al. [13].

RELATION TO DECISION THEORY

As was pointed out by Wald [17, p. 127], the lemma given by (2) and (3) can be viewed as a complete-class theorem (see ADMISSIBILITY). In particular, it is the simplest special case of the fact that under mild restrictions on the loss function, Bayes solutions constitute a complete class of decision procedures when the parameter space is finite (see DECISION THEORY). There are many generaliza-

tions of the lemma to the more general class of decision problems with a finite number, say s, of actions. A randomized decision procedure for such a problem is a vector-valued function $\phi(x) = (\phi_1(x), \ldots, \phi_s(x))$, $0 \leqslant \phi_i(x) \leqslant 1$, $\sum \phi_i(x) = 1$, where $\phi_i(x)$ denotes the probability with which action i is taken when the observation is x.

The following are some typical examples from the long list of such results.

A form of the lemma generalizing the complete class aspect is given for families with monotone likelihood ratio by Karlin and Rubin [8].

The Bayes solutions for maximizing the probability of a correct decision when deciding among a finite number of distributions is obtained by Hoel and Peterson [5].

Bayes solutions and complete-class theorems for slippage problems (see MEAN SLIPPAGE TESTS) are given by Karlin and Truax [9].

A lemma for determining a minimax solution for certain selection problems is provided by Lehmann [12] and a corresponding lemma for multiple comparison* problems by Spjøtvoll [15]. A lemma for a different type of selection problem is due to Birnbaum and Chapman [1].

References

[1] Birnbaum, Z. W. and Chapman, D. G. (1950). *Ann. Math. Statist.*, **21**, 443–447.

[2] Dantzig, G. B. and Wald, A. (1951). *Ann. Math. Statist.*, **22**, 87–93.

[3] Dvoretzky, A., Kiefer, J., and Wolfowitz, J. (1953). *Ann. Math. Statist.*, **24**, 254–264.

[4] Hall, I. J. and Kudo, A. (1968). *Ann. Math. Statist.*, **39**, 1693–1699.

[5] Hoel, P. G. and Peterson, R. P. (1949). *Ann. Math. Statist.*, **20**, 433–438.

[6] Huber, P. J. (1965). *Ann. Math. Statist.*, **36**, 1753–1758.

[7] Huber, P. J. and Strassen, V. (1973). *Ann. Statist.*, **1**, 251–263.

[8] Karlin, S. and Rubin, H. (1956). *Ann. Math. Statist.*, **27**, 272–300.

[9] Karlin, S. and Truax, D. (1960). *Ann. Math. Statist.*, **31**, 296–324.

[10] Kempthorne, O. and Doerfler, T. E. (1969). *Biometrika*, **56**, 231–248.

[11] Lehmann, E. L. (1959). *Testing Statistical Hypotheses*. Wiley, New York.

[12] Lehmann, E. L. (1961). *Ann. Math. Statist.*, **32**, 990–1012.

[13] Neyman, J. and Pearson, E. S. (1933). *Philos. Trans. R. Soc. Lond. A*, **231**, 289–337.

[14] Neyman, J. and Pearson, E. S. (1936). *Statist. Res. Mem.*, **1**, 1–37.

[15] Spjøtvoll, E. (1972). *Ann. Math. Statist.*, **43**, 398–411.

[16] Wald, A. (1947). *Sequential Analysis*. Wiley, New York.

[17] Wald, A. (1950). *Statistical Decision Functions*. Wiley, New York.

[18] Wald, A. and Wolfowitz, J. (1948). *Ann. Math. Statist.*, **19**, 326–339.

[19] Wijsman, R. A. (1963). *Ann. Math. Statist.*, **34**, 1541–1548.

(ADMISSIBILITY
BAYESIAN INFERENCE
COMPOSITE HYPOTHESIS
DECISION THEORY
EXPONENTIAL FAMILIES
HYPOTHESIS TESTING
INVARIANCE CONCEPTS
  IN STATISTICS
LEAST FAVORABLE DISTRIBUTIONS
LIKELIHOOD PRINCIPLE
LIKELIHOOD RATIO TESTS
LINEAR PROGRAMMING
MEAN SLIPPAGE TESTS
MINIMAX TESTS
MONOTONE LIKELIHOOD RATIO
MULTIPLE COMPARISONS
RANDOMIZATION
ROBUSTNESS OF TESTS
SELECTION PROCEDURES
SEQUENTIAL ANALYSIS
SEQUENTIAL PROBABILITY RATIO TEST
SIMPLE HYPOTHESIS
UNBIASEDNESS)

E. L. Lehmann

NEYMAN–PEARSON THEORY   *See* HYPOTHESIS TESTING

NEYMAN – PEARSON – WALD APPROACH   *See* ADMISSIBILITY

NEYMAN'S AND OTHER SMOOTH GOODNESS-OF-FIT TESTS

Smooth tests of fit were proposed by Neyman* [7] to remedy the perceived weakness of the Pearson chi-squared test in an important class of cases where there may be a serious loss of power because the Pearson statistic X^2 does not make use of the essential ordering of groups (*see* CHI-SQUARE TESTS).

The class of cases concerned is that where a sample of independent random variables $Y_1, \ldots, Y_n$ is drawn from a population and the question at issue is whether the population cumulative distribution function (cdf) of Y has some particular form $F_0(y)$ (the null hypothesis H_0), or whether it is some other form. The cdf $F_0(y)$ may depend on a number q of parameters; when these are unknown H_0 is composite and, in order to calculate probabilities, it is necessary to replace them by estimators (preferably efficient).

To calculate X^2, it is further necessary to group the data. The range of y is partitioned into a finite number k of intervals $\{(\eta_{j-1}, \eta_j), j = 1, \ldots, k\}$ which are indexed in ascending order: (i.e., $-\infty = \eta_0 < \eta_1 < \cdots < \eta_k = +\infty$), the $\{\eta_j\}$ having been chosen in advance of examining the data. The process is called *grouping* since observations lying in the same interval are subsequently grouped together and counted. Denoting the number lying in the jth interval by N_j and its expected value by nf_j, then

$$\mathscr{E} N_j = nf_j = n\{F_0(\eta_j) - F_0(\eta_{j-1})\},$$

when H_0 is true and

$$X^2 = \sum_{j=1}^{k} \frac{(N_j - nf_j)^2}{nf_j}.$$

When parameters have been estimated, the $\{f_j\}$ are estimated.

When H_0 is false and $\mathscr{E}(N_j) = nf_j(1 + d_j)$, then $\lambda^2\sqrt{n} = n\sum_{j=1}^{k} f_j d_j^2$ measures the excess of $\mathscr{E} X^2$ above its value under H_0. When H_0

is simple, the asymptotic power function of the Pearson test is monotone increasing with λ^2 (assumed constant as n increases). More generally, λ^2 is the predominant measure of detectable departure from H_0 and, relative to the chosen grouping, H_0 is expressed by: $d_j = 0$ for all j. Although the standardized deviations $\{d_j\sqrt{f_j}\}$ are indexed in ascending order of y, any rearrangement of their values gives the same value of λ^2. Thus λ^2 does not reflect the pattern of deviations. When goodness of fit* is being assessed, the alternative cdf's commonly contemplated are systematic departures from $F_0(y)$, for example, a shift, a change of dispersion, skewness*, or kurtosis*. These are called *smooth alternatives*. The $\{d_j\}$ correspondingly form a sequence of values of a smooth function of j; for example, they may be values of a monotonically increasing function when the cdf is shifted to the right. The "omnibus" nature of the Pearson chi-squared test does not allow it to make the necessary discrimination, whereas goodness-of-fit* tests designed to detect smooth departures from H_0 may be expected to be much more powerful than X^2 yet they retain the generality of reference required of tests of fit.

Neyman designed his smooth tests for continuous variables and simple H_0. In this case, Y may be replaced by its probability integral transform, $U = F_0(Y)$, where U is uniformly distributed over $(0, 1)$ when H_0 is true. He argued that a low-order polynomial in u would be adequate to describe the logarithm of the probability density function (pdf) of smooth departures from uniformity. Explicitly he chose a parametric family of pdf's of form $\propto \exp[\sum_{r=1}^k \Theta_r \pi_r(u)]$, where $\{\Theta_r\}$ are the parameters and $\{\pi_r(u)\}$ are the normalized Legendre polynomials, which are orthogonal to integration over $(0, 1)$. Writing

$$U_j = F_0(Y_j), \qquad v_r = \sum_{j=1}^n \frac{\pi_r(U_j)}{\sqrt{n}}$$

the $\{v_r\}$ are standardized mean values that have, under H_0, zero means and covariances with unit variances. Neyman's smooth test statistic $\psi_K^2 = \sum_{r=1}^K v_r^2$ has a pdf closely approximated by χ^2 with K d.f. under H_0, when n is of only moderate size. Typically, K would be set at 3 or 4, and the test would have substantially greater power than X^2 for smooth alternatives.

Using David and Johnson's [5] extension of the probability integral transformation*, Barton [2] generalized Neyman's test to cover grouped and discrete variables by defining discrete analogs of the Legendre polynomials with corresponding standardized means $\{V_r\}$. The test function $\Psi_K^2 = \sum_{r=1}^K V_r^2$ has the property that $\sum_{r=1}^K V_r^2 = X^2$, so that it may be regarded both as a grouped form of ψ_K^2 and a "partition of chi squared." Both ψ_K^2 and Ψ_K^2 have an asymptotic χ^2 distribution with K degrees of freedom (d.f.) under H_0, when H_0 is simple. Unfortunately, testing for goodness of fit is commonly required when H_0 is composite. Under these circumstances X^2 simply "loses q degrees of freedom" when the q parameters are efficiently estimated. With both ψ_K^2 and Ψ_K^2, there is an incomplete loss of degrees of freedom. Fuller details are given in Barton [3] but, briefly, the asymptotic distribution of either statistic (under H_0 and with efficient estimation) is as the sum of a χ^2 variable with $K - q$ degrees of freedom and a small independently distributed increment. This increment is a weighted sum of q independent χ^2 variables each of 1 d.f. The weights are small in general, and for moderate values of $K - q$ may be taken as zero approximately. The same incomplete loss of degrees of freedom occurs with X^2 when there is hyperefficient estimation, as there often is, owing to estimation being carried out on the ungrouped data (see Chernoff and Lehmann [4] and Watson [8]). The power of both ψ_K^2 and Ψ_K^2 against the Neyman system of smooth alternatives is essentially unaffected by fitting the q parameters of a composite H_0 relative to the degrees of freedom of the appropriate χ^2 distribution.

Neyman put forward ψ_K^2 as a test statistic to remedy a deficiency of the Pearson chi squared in certain circumstances, and he did so in the light of a family of smooth distributions that modeled those circumstances. His test is at the same time a nonparametric test of goodness of fit based on the probability integral transform, which does not suffer from another feature for which the Pearson chi-squared test of fit has been criticized, i.e., the arbitrariness inherent in the grouping of continuous variables. These properties it shares with the Kolmogorov–Smirnov* and Cramér–von Mises* tests of goodness of fit (see Kendall and Stuart [6] for a comparative discussion). Those tests were proposed a few years earlier than Neyman's and without explicit reference to any alternative hypotheses. It would seem that both compare best with X^2 when the cdf of Y shows a jagged local deviation from $F_0(y)$, so they do not rate as smooth tests. By contrast, such tests as that based on the skewness coefficient (which provides perhaps the best test of normality) are undoubtedly smooth, but only test for the fit of one particular cdf and so do not have the generality to be termed tests of goodness of fit.

References

[1] Barton, D. E. (1953). *Skand. Aktuarietidskr.*, **36**, 24–63.

[2] Barton, D. E. (1956). *Skand. Aktuarietidskr.*, **39**, 1–17.

[3] Barton, D. E. (1957). *Skand. Aktuarietidskr.*, **40**, 216–245.

[4] Chernoff, H. and Lehmann, E. L. (1954). *Ann. Math. Statist.*, **25**, 579–586.

[5] David, F. N. and Johnson, N. L. (1950). *Biometrika*, **37**, 43–49.

[6] Kendall, M. G. and Stuart, A. (1973). *Advanced Theory of Statistics*, 4th ed., Vol. 2, Hafner, New York, Chap. 30.

[7] Neyman, J. (1937). *Skand. Aktuarietidskr.*, **20**, 150–199.

[8] Watson, G. S. (1958). *J. R. Statist. Soc. B*, **20**, 44–72.

(CHI-SQUARE TESTS
CRAMÉR–VON MISES STATISTIC
GOODNESS OF FIT
KOLMOGOROV–SMIRNOV-TYPE
  TESTS OF FIT

NEYMAN'S TEST FOR UNIFORMITY
UNIFORMITY, TESTS FOR)

D. E. BARTON

NEYMAN'S TEST FOR UNIFORMITY

In 1937, Jerzy Neyman* introduced an original method for testing goodness-of-fit* based on the theory of testing hypotheses then recently developed by Neyman and E. S. Pearson*. In one of his last articles, Neyman [10] gives an interesting account of his motivations in introducing the test; among other factors, he was concerned, as have been so many authors, at the inadequacy of the Pearson chi-square test* to take account of the sign of the difference between observed and expected values in the cells. Neyman decided to exploit the probability integral transformation*; if random variable X has continuous density $f(x)$, then random variable Z defined by $Z = \int_{-\infty}^{X} f(x)\,dx$ has the uniform distribution between 0 and 1, written $U(0, 1)$. In a goodness-of-fit test, let H_0 be the null hypothesis that $f(x)$ is the tested distribution (fully specified) for X, for which a random sample $X_1, X_2, \ldots, X_n$ is given; then values Z_i obtained by the preceding transformation, with X_i replacing X, will be a random sample which should be uniform on H_0. Neyman suggested that one should find a test statistic sensitive to smooth departures of the Z distribution from uniformity; such smooth departures he defined by means of Legendre polynomials. Specifically, Neyman proposed that the alternative density can be expressed as

$$f(z) = c \exp\left\{ 1 + \sum_{j=1}^{k} \theta_j l_j(z) \right\},$$

$$0 < z < 1, \quad k = 1, 2, \ldots, \quad (1)$$

where $l_1(z)$, $l_2(z)$, . . . are Legendre polynomials, $\theta_1, \theta_2, \ldots, \theta_k$ are parameters, and c, a function of $\theta_1, \theta_2, \ldots, \theta_k$, is a normalizing constant. When $\theta_j = 0$, for all $j \geq 1$, $f(z)$ is the uniform density $f(z) = 1$. The Legendre polynomials are orthogonal on the interval $(0, 1)$, and, by varying k, $f(z)$ may be made to approximate any given alternative. As the θ_j increase, the density $f(z)$ varies smoothly from the uniform distribution; thus the test for uniformity for Z can be put in the form of a test on the parameter values, i.e., a test of

$$H_0 : \sum_{j=1}^{k} \theta_j^2 = 0.$$

By likelihood ratio* methods, Neyman found an appropriate statistic for testing H_0. For given k, the test statistic is N_k, calculated as follows:
Compute

$$V_j = \frac{1}{\sqrt{n}} \sum_{i=1}^{n} l_j(Z_i), \qquad j = 1, \ldots, k;$$

then

$$N_k = \sum_{j=1}^{k} V_j^2.$$

In these calculations, $l_j(z)$ is best expressed in terms of $y = z - 0.5$. For the first four polynomials,

$$l_1(z) = 2\sqrt{3}\, y; \qquad l_2(z) = \sqrt{5}\, (6y^2 - 0.5);$$

$$l_3(z) = \sqrt{7}\, (20y^3 - 3y);$$

$$l_4(z) = 3(70y^4 - 15y^2 + 0.375).$$

In general, H_0 will be rejected for large values of N_k. Note that N_1 is equivalent to $\overline{X}$, the mean of the X_i. In fact $N_1^2 = V_1^2$ and $V_1 = (12n)^{1/2}(\overline{x} - 0.5)$. Then let t_α be the upper tail percentage point for N_1 at significance level α, and let $Z_{\alpha U}, Z_{\alpha L}$ be the upper and lower tail percentage points at level α for $\overline{X}$; we have $Z_{\alpha U} = 1 - Z_{\alpha L}$, and $t_{2\alpha} = 12n(Z_{\alpha U} - 0.5)^2 = 12n(0.5 - Z_{\alpha L})^2$. Thus significance points for N_1 can be found from significance points for $\overline{X}$; a table of such points is available, for example, in Stephens [14, Table 1]. Further, v_2 derives from $\sum(X_i - 0.5)^2/n = S^2$, a form of sample variance, and so N_2 is a combination of both $\overline{X}$ and S^2.

Neyman showed that, as $n \to \infty$, the V_j are independent, and V_j is normally distributed with mean $\theta_j \sqrt{n}$ and variance 1. Thus the asymptotic null distribution of N_k is χ_k^2, and for the alternative family (1) the asymptotic distribution is noncentral χ_k^2, with parameter $\lambda = n\sum_{j=1}^{k}\theta_j^2$. The tests based on N_k are consistent and unbiased.

Barton [1] considered a slightly different class of alternatives for Z given by

$$f(z) = \sum_{j=0}^{k} \theta_j l_j(z), \quad 0 \leqslant X \leqslant 1, k = 0, 1, \ldots$$

with θ_0 equal to 1 (see NEYMAN'S AND OTHER SMOOTH GOODNESS-OF-FIT TESTS). A restriction must now be placed on the θ_j to ensure that the density is always positive. The same statistics N_k may again be used to test for uniformity against this alternative. Although it is the alternative distribution that is smooth, the name *smooth test* has become attached to the test based on N_k. This hypallage is now fixed in the literature of this subject.

The terms V_j mentioned earlier can be regarded as *components* of N_k; Neyman's statistic appears to have been one of the first to be presented as a sum of components, although in recent years EDF statistics*, for example, have been examined in this way. An important question in making a test based on N_k is to decide the *order* k of N_k, that is, how many components to include. Too few components will not be useful against a wide family of alternatives; on the other hand, the inclusion of too many components can weaken the overall power of N_k against many alternatives. The question was considered early by David [4], who felt that N_2 was all that was needed in most applications. The fact that N_2 uses both sample mean and sample variance makes it plausible that it will detect many types of nonuniformity, and these simple statistics also have a natural appeal.

Since Neyman's early work, many tests for uniformity have been developed, and the statistic N_k has been somewhat overlooked.

Table 1 Upper Tail Percentage Points for N_2: Significance Level α

n \ α	0.1	0.05	0.025	0.01
4	4.116	5.566	7.287	9.643
6	4.316	5.618	7.148	9.384
8	4.421	5.683	7.110	9.276
10	4.476	5.775	7.167	9.265
12	4.486	5.822	7.198	9.170
16	4.527	5.908	7.319	9.233
20	4.542	5.925	7.332	9.234
25	4.554	5.937	7.341	9.230
30	4.562	5.947	7.348	9.230
40	4.573	5.958	7.357	9.230
50	4.579	5.964	7.360	9.223
100	4.592	5.979	7.370	9.220
∞	4.605	5.991	7.378	9.210

In that era before computers, these statistics also required much computation, as was pointed out by David [4]. However, recent studies on tests for uniformity, for example, those by Locke and Spurrier [6] and Miller and Quesenberry [7], together with others by the present authors (*see* UNIFORMITY, TESTS FOR) support the view that N_2 is a good overall statistic, although Miller and Quesenberry advocate N_4 for some alternatives. Monte Carlo points for N_2 have been given by Miller and Quesenberry [7] and by Solomon and Stephens [13]; Solomon and Stephens also obtain points by fitting Pearson curves to the distributions for larger sample sizes n, using the moments first given by David [4]. An abridged set of percentage points for N_2 appears in Table 1. Miller and Quesenberry [7] also give Monte Carlo points for N_1, N_3, and N_4. For $k \leqslant 4$, the null distributions approach the asymptotic χ_k^2 quite rapidly, and the asymptotic percentage points can be used, for, say, $n \geqslant 10$ with little error in significance levels. Typically this will be the situation in practice.

Example. A useful example of a test for uniformity comes from a test of H_0: a sequence of events is occurring randomly in time. It is well known that the intervals between such events will be exponentially distributed. Suppose the events are observed

at times $t_1, t_2, \ldots, t_n$, (giving intervals t_1, $t_2 - t_1$, $t_3 - t_2$, etc). This sequence may be divided by t_n to give values $U_i = t_i/t_n$, and, on H_0, the $n - 1$ values U_i, $i = 1, 2, \ldots$, $n - 1$ will be distributed as the order statistics of a uniform sample of size $n - 1$ drawn from $U(0, 1)$. Suppose 12 intervals are (in hours) 16, 0.5, 21, 13, 35, 19, 35, 10, 17, 56, 35, and 20. The corresponding 11 values u_i (observe that $u_{12} \equiv 1$) are then 0.058, 0.059, 0.135, 0.182, 0.308, 0.377, 0.503, 0.539, 0.600, 0.802, and 0.928. The first four components V_j are then -1.055, 0.182, 0.053, and 0.016, giving $N_1 = 1.113$, $N_2 = 1.146$, $N_3 = 1.149$, and $N_4 = 1.149$. The fairly large negative first component suggests a low mean (the mean of the U_i is 0.408) with the interpretation that the intervals are becoming longer with time. However, none of the N_j is near significance at the 10% level, so there is not strong evidence to reject H_0. The steadiness in the values of N_2, N_3, and N_4 shows how the addition of further components can weaken the power of a test statistic, since these must be compared (approximately) to critical values of χ_2^2, χ_3^2, and χ_4^2, respectively.

TESTS FOR DISCRETE OR GROUPED DATA AND FOR COMPOSITE HYPOTHESES

Neyman's test can be adapted for testing for a fully specified discrete distribution. An adaptation was first suggested by Scott [11], in connection with grouped data* arising from an interesting problem in astrophysics. Scott's method applies to groupings or discrete data with equiprobable classes, and it can be used with up to four components; Barton [2, 3] later gave a more general procedure that can be used with classes of differing probabilities. However, these procedures have not come into general use.

Another important problem is to adapt the Neyman test to the case where unknown parameters in the tested distribution must first be estimated from the data set itself. Again, Barton [3] was an early worker; his procedure takes account of the behavior of the Z_i when estimates of parameters are used in $f(x)$, a question considered by David and Johnson [5]. The resulting test statistic is a linear combination of weighted χ^2 variables, the distribution of which is difficult to find. Since then, analytic results have been given, and good methods of approximation have been found (see Solomon and Stephens [12] for demonstrations and further references). Neyman [9] also considered the test for composite hypotheses. More recently, Thomas and Pierce [15] devised a procedure closely related to Barton's. They first express the density of Z in the form (1), but with powers z^j replacing the polynomials $l_j(z)$; this permits easier calculations later, and the final test statistic of Thomas and Pierce, for the null hypotheses $\{H_0$: the distribution tested is of the correct form except for unknown parameters$\}$, has an asymptotic χ^2 distribution. Specific adaptations of the test criterion are given for testing for the normal, exponential, or Weibull distributions, and a numerical example is shown of the normal test. Again, the authors find the second-order test, the modified form of N_2, to be effective.

References

[1] Barton, D. E. (1953). *Skand. Aktuarietidskr.*, **36**, 24–63.

[2] Barton, D. E. (1955). *Skand. Aktuarietidskr.*, **39**, 1–17.

[3] Barton, D. E. (1956). *Skand. Aktuarietidskr.*, **39**, 216–245.

[4] David, F. N. (1939). *Biometrika*, **31**, 191–199.

[5] David, F. N. and Johnson, N. L. (1948). *Biometrika*, **35**, 182–190.

[6] Locke, C. and Spurrier, J. D. (1978). *Commun. Statist. Theor. Meth.*, **A7**, 241–258.

[7] Miller, R. L., Jr. and Quesenberry, C. P. (1979). *Commun. Statist. B*, **8**, 271–290.

[8] Neyman, J. (1937). *Skand. Aktuarietidskr.*, **20**, 149–199.

[9] Neyman, J. (1959). *Probability and Statistics: The Harald Cramér Volume*. Wiley, New York, pp. 213–234.

[10] Neyman, J. (1980). *Asymptotic Theory of Statistical Tests and Estimation*, I. M. Chakravarti, ed. Academic Press, New York.

[11] Scott, E. L. (1949). *Astrophys. J.*, **109**, 194–207.

[12] Solomon, H. and Stephens, M. A. (1977). *J. Amer. Statist. Ass.*, **72**, 881–885.

[13] Solomon, H. and Stephens, M. A. (1983). *Commun. Statist. B*, **12**, 127–134.

[14] Stephens, M. A. (1966). *Biometrika*, **53**, 235–239.

[15] Thomas, D. R. and Pierce, D. A. (1979). *J. Amer. Statist. Ass.*, **74**, 441–445.

(CHI-SQUARE TESTS
GOODNESS OF FIT
NEYMAN'S AND OTHER SMOOTH
    GOODNESS-OF-FIT TESTS
UNIFORMITY, TESTS FOR)

H. SOLOMON
M. A. STEPHENS

NEYMAN STRUCTURE, TESTS WITH

A *statistical hypothesis* (H) is termed *simple* or *composite* according as it specifies completely or not the probability law (P_θ^x) of X, (a set of) random variables under consideration. A *test* for a *null hypothesis* (H_0) consists in a decomposition of the *sample space* (S) into an *acceptance region* (A) and, its complement, a *critical (rejection) region* ($W = S \backslash A$), such that H_0 is accepted or rejected according as X belongs to A or W (*see* HYPOTHESIS TESTING). Though, ideally, one would like to choose W in such a way that both the probability of *type I error* (i.e., $P\{X \in W \mid H_0$ is true$\}$) and of *type II error* (i.e., $P\{X \in A \mid H_0$ is not true$\}$) are minimized, in reality, this may not be possible, and, in accordance with the classical *Neyman–Pearson theory*, one attempts to maximize the *power* of the test, i.e.,

$$P\{X \in W \mid H_0 \text{ is not true}\}$$
$$(= 1 - P\{\text{type II error}\}) \quad (1)$$

subject to

$$P\{X \in W \mid H_0 \text{ is true}\} = \alpha \quad (0 < \alpha < 1),$$
$$(2)$$

where α (preassigned) is the *level of significance* (or *size*) of the test. When the null and alternative hypotheses are both simple, the *Neyman–Pearson fundamental lemma** provides the desired solution to (1) and (2).

Various modifications and extensions are known to have been worked out to accommodate the case of composite alternatives against a simple H_0. Most of this work is contained in the set of papers in *Statistical Research Memoirs**, Volumes 1 and 2. The situation is somewhat different when H_0 is a composite hypothesis, so that P_θ^x is not completely specified under H_0. It may be convenient to describe the situation in terms of a family P $(= \{ P_\theta^x : \theta \in \Omega \})$ of probability measures of F, where Ω is the *parameter space* and F is a *countably additive family* of sets in S. The null hypothesis H_0 may be stated as $H_0: P_\theta^x \in P_\omega = \{ P_\theta^x : \theta \in \omega \}$, for some subset ω of Ω. Naturally, one would like to have, in place of (2),

$$P_\theta^x(W) = P\{ X \in W \mid P_\theta^x \} = \alpha$$

$$\text{for every } \theta \in \omega. \quad (3)$$

Then, W is said to be a *similar region* of size α ($0 < \alpha < 1$) for the family P_ω, if it satisfies (3) [and hence (2)]. The condition of similarity in (3) is a minimal requirement for the administration of a test of a composite hypothesis in the Neyman–Pearson setup.

Early researchers (viz., Fisher [4], Bartlett [1], Neyman and Pearson [11]) were naturally tempted to uncover the basic role of *sufficient statistics* (*see* SUFFICIENCY) in the construction of similar regions having some desirable or optimal properties. Let T $(= t(x))$, not necessarily real-valued, be a sufficient statistic for the family P_ω. Then T is a measurable transformation from (S, F) into a measurable space (τ, F'); we denote the probability law for T by P_θ^t, $\theta \in \omega$; $P_\omega^t = \{ P_\theta^t : \theta \in \omega \}$. In the preceding setup, we assume the existence of such a T, which need not be a *minimal sufficient statistic*. If a minimal sufficient statistic exists for P_ω, we may be tempted to use that. Neyman [9] noted that if T be sufficient for P_ω and if W has the property that

$$P_\theta^x\{ W \mid T = t \} = \alpha \quad \text{a.e. } P_\omega^t, \quad (4)$$

then W is a similar region of size α for P_ω.

We shall say that the set W in F has the *Neyman structure* with respect to the sufficient statistic T if it satisfies (4), Note that

$$P_\theta^x(W) = \int P_\theta^x(W \mid T = t) \, dP_\theta^t, \qquad \theta \in \omega,$$

$$(5)$$

so that (4) ensures (3) [and hence (2)]. A test with a similar region W satisfying (4) is termed a *test with Neyman structure*.

Significant later contributions to this theory are due to Lehmann and Stein [5], Lehmann and Scheffé [6, 7], Watson [15], among others. These workers were able to identify (and characterize) extended domains of (parametric as well as nonparametric) hypothesis testing problems where tests with Neyman structure exist. The concepts of *completeness** and *bounded completeness* play a vital role in this context. For convenience, we introduce a *critical function* $\phi(x)$ which is a (F-) measurable function of x for which $0 \leqslant \phi(x) \leqslant 1$. Note that by letting $\phi(x) = 1$ ($x \in W$), we are able to write (2) as $E\{ \phi(X) \mid H_0 \} = \alpha$, (3) as $E_\theta^x \phi(X) = \alpha$, for all $\theta \in \omega$, and (4) as $E_\theta^x \{ \phi(X) \mid T = t \} = \alpha$ (a.e. P_ω^t). Since ϕ is bounded, so is $E_\theta^x \{ \phi(X) \mid T = t \} - \alpha$ ($= g(T)$, say), which does not depend on θ due to the sufficiency of T. Thus (3), (4) and the bounded completeness of T lead us to the following result (Lehmann and Scheffé [6]):

If T is a sufficient statistic for P_ω, a necessary and sufficient condition for all similar critical functions for P_ω to have Neyman structure with respect to T is that $P_\omega^t = \{ P_\theta^t ; \theta \in \omega \}$ is boundedly complete.

The preceding characterization in terms of bounded completeness of P_ω^t, instead of minimal sufficient statistics or completeness alone, is a great step: Although the completeness of a sufficient statistic implies its minimality, a minimal sufficient statistic is not necessarily a complete one. Nevertheless, if there exist both a boundedly complete sufficient statistic T and a minimal sufficient statistic U, then T and U are equivalent (see Lehmann and Scheffé [6, 7]). Thus, for the construction of tests with Neyman structure, one needs to construct the minimal sufficient statistics (T) and verify the "bounded completeness" criterion for P_ω^t. This is indeed possible for a variety of parametric as well as nonparametric testing problems.

To make this point clear, we consider the following data transformation technique. Basu [2] has characterized a class of statistics that is independent of the minimal sufficient statistic, whenever the latter is properly defined (*see* BASU THEOREMS). Thus, under $H_0(P_\omega)$, whenever the minimal sufficient statistic T exists and P_ω^t is boundedly complete, it is possible to induce a transformation

$$X \rightarrow (T, Z), \qquad (6)$$

where Z is independent of T (a.e. P_ω^t) and is termed a *noise*. It is always possible, in the preceding setup, to locate a maximal statistic independent of T, which is termed a *maximal noise*, so that without any loss of generality Z in (6) is taken as a maximal noise. In the context of *invariant tests*, the concept of maximal noise coincides with that of *maximal invariants* (*see* INVARIANCE CONCEPTS IN STATISTICS). Note that the conditional distribution of Z, given $T = t$, does not depend on t, and, further, T being a sufficient statistic, this conditional distribution is free from P_θ^x (a.e. P_ω). Thus, if we consider a critical function ϕ depending on X through Z only (i.e., $\phi(X) \equiv \phi(Z)$), then, the conditional distribution of $\phi(Z)$, given $T = t$, is independent of P_θ^t (a.e. P_ω^t), so that a critical region W, or a critical function $\phi(X)$, depending on X through Z alone, can always be selected such that (4) holds. Consequently, in the presence of boundedly complete sufficient statistics, tests based on the maximal noise can be characterized as having Neyman structure.

As illustrations, we consider the following:

Example 1. Let $X = (X_1, \ldots, X_n)$ with the X_i independent, having a common normal distribution with mean μ and variance σ^2, and suppose that $H_0: \sigma = \sigma_0$ and $H_1: \sigma = \sigma_1 > \sigma_0$, with μ as a nuisance parameter*. Under H_0, $\theta = (\mu, \sigma) \in \omega = \{(\mu, \sigma_0); -\infty < \mu < \infty\}$, and a minimal sufficient statistic is

$(\overline{X}_n, \sigma_0)$, where $\overline{X}_n = n^{-1}\sum_{i=1}^{n} X_i$. Since the testing problem is sought to remain invariant under translation, the maximal invariant in this case is $S_n^2 = \sum_{i=1}^{n}(X_i - \overline{X}_n)^2$, where under H_0, S_n^2/σ_0^2 has the chi-square distribution* with $n - 1$ degrees of freedom, independently of $\overline{X}_n$. Thus the test based on S_n^2/σ_0^2 has Neyman structure. A similar picture holds for $H_0: \mu = 0$ vs. $H_1: \mu > 0$, with σ as a nuisance parameter, where the test based on the Student t-statistic $\sqrt{n(n-1)}\,(\overline{X}_n/S_n)$ also has Neyman structure.

Example 2. $X = (X_1, \ldots, X_n)$ where the X_i are independent random variables with unknown distributions F_i, $1 \leqslant i \leqslant n$; $H_0: F_1 = \cdots = F_n = F$ (unknown), against alternatives that the F_i are not all the same. The two or several sample location/scale problem as well as the simple regression problem relate to this model with more structured alternative hypotheses. Assume that F is absolutely continuous, and let $T = \{X_{n:1} < \cdots < X_{n:n}\}$ be the vector of order statistics of X and $Z = (R_1, \ldots, R_n)$ be the vector of ranks of the X_i among themselves (i.e., $X_i = X_{n:R_i}, 1 \leqslant i \leqslant n$). Then, under H_0, T is a minimal sufficient statistic for F and T and Z are independent, with Z having a discrete uniform distribution over the set of $n!$ permutations of $(1, \ldots, n)$. Hence the (rank) tests based on the maximal noise Z have Neyman structure.

Example 3. $X = (X_1, \ldots, X_n)$ where the X_i are independent and have a common distribution F; $H_0: F$ is symmetric about 0. Let T be the vector of the order statistics for the $|X_i|$ and let $Z = (\text{sgn } X_1, \ldots, \text{sgn } X_n, R_1^+, \ldots, R_n^+)$, where R_i^+ is the rank of $|X_i|$ among $|X_1|, \ldots, |X_n|$, $i = 1, \ldots, n$. For absolutely continuous F, T is a minimal sufficient statistic and Z is a maximal noise, so that the (signed rank) tests based on Z have Neyman structure.

In the same manner, the permutation tests for Examples 2 and 3 also have Neyman structure. Similar results hold for the bivariate independence problem (cf. Puri and Sen [13]). For various (parametric as well as nonparametric) tests related to some special families of stochastic processes, Bell [3] has provided a nice account of tests with Neyman structure. Additional references are also cited there.

Within the class of tests having Neyman structure, an optimal (or desirable) test may be located in other ways. For example, for one-sided alternatives, a best (most powerful) test may be obtained within this class. For some multisided alternatives, maximin power tests (*see* MINIMAX TESTS) may be obtained within this class, and so on. For the exponential family* of densities, this can always be done and the best test having Neyman structure remains asymptotically optimal within the entire class of tests with asymptotic size α [see Michel [8]].

References

[1] Bartlett, M. S. (1937). *Pr. R. Soc. Lond. A*, **160**, 268–282.

[2] Basu, D. (1955). *Sankhyā*, **15**, 377–380.

[3] Bell, C. B. (1975). In *Statistical Inference and Related Topics*, Vol. 2, M. L. Puri, ed. Academic Press, New York, pp. 275–290.

[4] Fisher, R. A. (1934). *Statistical Methods for Research Workers*. Oliver and Boyd, Edinburgh.

[5] Lehmann, E. L. and Stein, C. (1949). *Ann. Math. Statist.*, **20**, 28–45.

[6] Lehmann, E. L. and Scheffé, H. (1950). *Sankhyā*, **10**, 305–340.

[7] Lehmann, E. L. and Scheffé, H. (1955). *Sankhyā*, **15**, 219–236.

[8] Michel, R. (1979). *Ann. Statist.*, **7**, 1256–1263.

[9] Neyman, J. (1937). *Philos. Trans. R. Soc. Lond. A*, **236**, 333–380.

[10] Neyman, J. and Pearson, E. S. (1933). *Philos. Trans. R. Soc. Lond. A*, **231**, 289–337.

[11] Neyman, J. and Pearson, E. S. (1936). *Statist. Res. Mem.*, **1**, 1–37.

[12] Neyman, J. and Pearson, E. S. (1938). *Statist. Res. Mem.*, **2**, 25–57.

[13] Puri, M. L. and Sen, P. K. (1971). *Nonparametric Methods in Multivariate Analysis*. Wiley, New York.

[14] Scheffé, H. (1943). *Ann. Math. Statist.*, **14**, 227–233.

[15] Watson, G. S. (1957). *J. R. Statist. Soc. B*, **19**, 262–267.

(BASU THEOREMS
COMPLETENESS
HYPOTHESIS TESTING
INVARIANCE CONCEPTS IN STATISTICS
NEYMAN–PEARSON LEMMA
STATISTICAL RESEARCH MEMOIRS
SUFFICIENCY)

P. K. SEN

NEYMAN'S TYPE A, B, AND C DISTRIBUTIONS

The Poisson distribution*, which describes homogeneous and mutually independent events, is often inadequate for describing heterogeneous and "spotty" counts encountered in the studies of bacteria, larvae, and plants. Under certain assumptions regarding the movements of larvae hatched from the egg masses in a field, Neyman [29] developed three distributions that are appropriate for describing such data. He called them *contagious* because they are suitable for modeling populations in which "the presence of one larva within an experimental plot increases the chance of there being some more larvae." These distributions are now well known as the Neyman Type A (NTA), the Neyman Type B (NTB), and the Neyman Type C (NTC); *see also* CONTAGIOUS DISTRIBUTIONS. The NTA has been the most used distribution. It has been used with great success in various disciplines such as bacteriology [29], ecology (*see* ECOLOGICAL STATISTICS) [1, 11, 30, 34], entomology [3, 4, 28], busy period of queues (*see* QUEUEING THEORY) [22], and clustering of retail food stores [31].

In a general treatment of contagious distributions, Feller [12] derived Neyman's distributions as compound Poisson distributions. This interpretation makes them suitable for modeling heterogeneous populations and renders them examples of "apparent contagion." Neyman's distributions also arise as generalized Poisson if the number of larvae observed at any plot are assumed to be hatched from Poisson distributed egg masses found in the neighboring plots, and the number of larvae hatched from each egg mass have some other discrete distribution. For an excellent discussion of these distributions, see Douglas [10]. For developments between 1969–80 and for an excellent source of references, see Johnson and Kotz [24].

DERIVATION OF NTA, NTB, AND NTC DISTRIBUTIONS

Consider an experimental field F and an experimental plot A of F at which the larvae are to be counted. The area of A is assumed to be small relative to that of F. Let there be $M = m$ egg masses in F, a proportion π, $0 < \pi < 1$, of which is represented at A. Let $N = n$ be the total number of larvae hatched at a typical egg mass with distribution function (df) $F_N(n)$. Let the number of larvae found at the plot A out of the $N = n$ larvae hatched at a typical egg mass have a binomial distribution with probability generating function* (pgf) $(q + pz)^n$, $q = 1 - p$, $0 < p < 1$, where $P = p$ is the probability for each larva to be at the plot A with the df $F_P(p)$. Then the pgf of the number of larvae observed at A from a randomly selected egg mass is

$$h(z) = \int_0^1 \int_0^\infty (q + pz)^n \, dF_N(n) \, dF_P(p).$$

$$(1)$$

Hence the pgf of the number of larvae from a randomly selected egg mass at a randomly selected plot is $1 - \pi + \pi h(z)$, and from m randomly selected egg masses the pgf is $g_m(z) = (1 - \pi + \pi h(z))^m$. On taking the limit of $g_m(z)$ as $\pi \to 0$ and $m \to \infty$ such that $\lambda = m\pi$, $g_m(z)$ approaches $g(z) = \exp\{\lambda(h(z) - 1)\}$, the pgf of a generalized Poisson distribution.

On taking $F_P(p)$ to be degenerate at p_0 and $dF_N(n) = e^{-\theta}\theta^n/n!$, $\theta > 0$, $n = 0, 1,$

$\ldots$, $g(z)$ reduces to the pgf of a two-parameter NTA distribution given by

$$G_A(z) = exp\{\lambda(e^{\phi(z-1)} - 1)\}, \quad (2)$$

where $\phi = \theta p_0$.

The NTB, NTC, and some other variants of NTA can easily be derived from the following alternative formulation given by Feller [12]. Assume that all the egg masses are represented at the plot A (i.e., $\pi = 1$). Let $F_M(m)$ be the df of the egg masses in the field F. Then the pgf of the number of larvae found at the plot A is

$$G(z) = \int_0^\infty \{h(z)\}^m \, dF_M(m). \quad (3)$$

If M has a Poisson distribution with mean λ, then $G(z)$ becomes the pgf of a generalized Poisson distribution. Some variants of the NTA and the NTB, NTC distributions can now be derived from (3). These include:

NTA with Two Parameters

Take

$$dF_M(m) = e^{-\lambda}\lambda^m/m!, \quad m = 0, 1, \ldots,$$
$$dF_P(p_0) = 1,$$
$$dF_N(n) = e^{-\theta}\theta^n/n!, \quad n = 0, 1, \ldots.$$

Then, from (3), we obtain $G_A(z)$ given by (2), the pgf of NTA with two parameters.

NTA with Three Parameters

Take everything as above, except

$$dF_P(p_1) = dF_P(p_2) = \tfrac{1}{2},$$

then (3) reduces to

$$G_A^*(z) = [\lambda\{\exp(\phi_1(z-1))$$
$$+ \exp(\phi_2(z-1))\}/2 - \lambda] \quad (4)$$

with $\phi_i = \theta p_i$, $i = 1, 2$, which is the pgf of a three-parameter NTA distribution. A $(k + 1)$-parameter extension of NTA can be obtained by taking

$$dF_P(p_i) = w_i, \quad i = 1, 2, \ldots, k,$$

with $\sum w_i = 1$ (see Douglas [10]).

For an application of the NTA distribution with three parameters, see Douglas [10], where insects of two kinds lay eggs in masses, each according to a Poisson distribution with mean $\lambda/2$ and the number of eggs per mass for the insect of type i follows a Poisson distribution with mean ϕ_i, $i = 1, 2$.

Neyman Type B Distribution

On assuming a uniform distribution for P with $dF_P(p) = 1$, $0 < p < 1$, and keeping everything else as before, we get the pgf of the NTB distribution:

$$G_B(z) = \exp\left[\frac{\lambda\{\exp(\phi(z-1)) - 1\}}{\phi(z-1) - 1}\right]$$

(5)

with $\phi = \theta$.

Neyman Type C Distribution

If we take $dF_P(p) = 2(1 - p)$, $0 < p < 1$, and letting everything else remain as before, we obtain the pgf of the NTC distribution as

$$G_C(z) = \exp\left[\frac{\lambda\{2\exp(\phi(z-1)) - \phi(z-1) - 1\}}{\{\phi(z-1)\}^2 - 1}\right],$$

(6)

where $\phi = \theta$.

MOMENTS, MODES, RECURRENCE RELATIONS, AND APPROXIMATIONS

NTA Distribution with Two Parameters

MOMENTS. The rth factorial cumulant of NTA is $\kappa_{(r)} = \lambda\phi^r$, $r = 1, 2, \ldots$, from which the higher moments of the distribution can be derived. It can be seen that $\mu = \lambda\phi$ and $\sigma^2 = \lambda\phi(1 + \phi)$. Shenton [32] has given a recurrence relation for the cumulants as

$$\kappa_{r+1} = \phi\left(\kappa_r + \frac{\partial\kappa_r}{\partial\lambda}\right)$$

from which the cumulants and hence the moments of the distribution can be derived.

Johnson and Kotz [23] have tabulated

some values of the ratio $(\beta_2 - 3)/\beta_1$, which depends only on ϕ, for integral values of ϕ between 1 and 20, where $\beta_1 = \mu_3^2/\mu_2^3$ and $\beta_2 = \mu_4/\mu_2^2$. The range of this ratio is very narrow; hence the field of applicability of this distribution is restricted. The NTB, NTC, and some of their generalizations extend the range of this ratio and hence the field of applicability of this family of distributions.

MODES. Neyman [29] pointed out that the observed frequency distributions of biological phenomena were multimodal and that the NTA distribution is capable of being multimodal. Barton [2] studied the modality of the NTA distribution in detail. He noted that it is possible for this distribution to have three or more modes (including the one at zero). The modal values of the random variable occur approximately at integral multiples of ϕ. Barton [2] and Shenton and Bowman [33] give diagrams of the parameter space showing the boundaries of the multimodal regions (see Douglas [10]).

RECURRENCE RELATIONS AND APPROXIMATIONS. For the NTA distribution with the pgf (2), the probability function (pf) is

$$P_x = e^{-\lambda}\sum_{j=1}^{\infty}(\lambda^j/j!)e^{-j\phi}(j\phi)^x/x!$$

$$x = 1, 2, \ldots \quad (7)$$

and

$$P_0 = \exp\{\lambda(e^{-\phi} - 1)\}.$$

Beall [3] derived recurrence relations for P_x in terms of the first $x - 1$ probabilities as

$$P_x = (\lambda\phi/x)\exp(-\lambda)\sum_{j=0}^{x-1}(\lambda^j/j!)P_{x-j-1}.$$

(8)

Douglas [8] derived an expression for P_x in terms of the moments about the origin μ_x' of a Poisson distribution with mean $\lambda e^{-\phi}$, i.e.,

$$P_x = (\phi^x/x!)\mu_x'P_0.$$

(9)

To facilitate the computations further, he derived the expression $P_{x+1} = \phi P_x p_x/(x+1)$ from (9), where $p_x = \mu_{x+1}'/\mu_x'$, and tabulated the values of p_x for different values of $\lambda e^{-\phi}$. However, these recurrence relations are subject to round-off errors* because they either depend on all the probabilities of lower order or on some tables that may not be detailed enough for most applications. We now consider some approximations that may simplify the calculations.

Approximations and Tables

Martin and Katti [26] suggest some approximations for the NTA distribution when the parameters take extreme values:

1. For large λ: When λ is large and ϕ is not too small, the approximate distribution of $Y = (X - \lambda\phi)/\sqrt{\lambda\phi(1 + \phi)}$ is standard normal, where $E(X) = \lambda\phi$ and $\mathrm{var}(X) = \lambda\phi(1 + \phi)$.
2. For small λ: If λ is small, then the NTA is approximately distributed as "Poisson with zeros." Thus $P_0 \approx (1 - \lambda) + \lambda e^{-\phi}$ and $P_x = \lambda\phi^x e^{-\phi}/x!$, $x = 1, 2, \ldots$.
3. For small ϕ: If ϕ is small, then the NTA is approximately distributed as a Poisson rv with mean $\lambda\phi$.

They have described regions in the parameter space where these approximations are good; goodness is measured by the Euclidean distance $\xi^2 = \sum_x(P_x - P_x^*)^2$ or the chi-square distance $\xi^2 = \sum_x(P_x - P_x^*)^2/P_x$ between the NTA and the approximating probabilities P_x and P_x^*, respectively.

Douglas [9] gave the following approximation on applying the steepest descent formula to the pgf (2):

$$P_x \approx \frac{e^{-\lambda}\phi^x \exp(x/g(x))}{\sqrt{2\pi}\,[\,g(x)\,]^x[\,x(1 + g(x))\,]^{1/2}},$$

where $g(x)\exp(g(x)) = x(\lambda e^{-\phi})^{-1}$. For other approximations, see Bowman and Shenton [7].

Grimm [14] has tabulated values of P_x for different values of $\lambda\phi$ and ϕ up to five decimal places. For the recurrence relations and moments of the NTA distribution with three parameters, see Neyman [29].

NTB Distribution

For the NTB distribution with the pgf (5), the mean and the variance are given by $\mu = \lambda\phi/2$ and $\sigma^2 = \lambda\phi(1 + 2\phi/3)/2$, respectively. The recurrence relation for probabilities is [3, 29]

$$P_{x+1} = \frac{\lambda}{(x + 1)\phi} \sum_{j=0}^{x} (j + 1)$$

$$\times \left(1 - e^{-\phi} \sum_{i=0}^{j+1} \frac{\phi^i}{i!} \right) P_{x-j}. \quad (10)$$

Beall [3] used this relation for fitting NTB to some data.

NTC Distribution

For the NTC distribution with the pgf (7), the mean and the variance are given by $\mu = \lambda\phi/3$ and $\sigma^2 = \lambda\phi(1 + \phi/2)/3$, respectively. The recurrence relation for the probabilities is [3, 29]

$$P_{x+1} = \frac{2\lambda e^{-\phi}}{(x + 1)\phi^2} \sum_{j=0}^{x} (j + 1)$$

$$\times \left\{ \phi\left(e^{\phi} - \sum_{i=0}^{j} \frac{\phi^i}{i!} \right) \right.$$

$$\left. - (j + 2)\left(e^{\phi} - \sum_{i=0}^{j+1} \frac{\phi^i}{i!} \right) \right\} P_{x-j}.$$

$$(11)$$

STATISTICAL INFERENCE FOR THE NTA DISTRIBUTION

Estimation

The following methods are available for estimating the parameters λ and ϕ of the NTA distribution. $\hat{\lambda}$ and $\hat{\phi}$ will denote the estimators of λ and ϕ, and it will be clear from the context which method they are derived from.

Method 1. Method of Moments*. We equate the first two population moments to their sample counterparts. This gives $\hat{\lambda} = \bar{x}/\hat{\phi}$ and $\hat{\phi} = (s^2 - \bar{x})/\bar{x}$, where $\bar{x}$ is the sample mean and s^2 the sample variance. If $s^2 < \bar{x}$, the NTA should be regarded as inappropriate for the data set.

Method 2. First Moment and Zero Frequency.

For this method, we equate the population mean and the population zero frequency (P_0) to their sample counterparts. This gives

$$\hat{\lambda} = \bar{x}/\hat{\phi} \quad \text{and}$$

$$\hat{\phi}\{1 - \exp(-\hat{\phi})\}^{-1} = \bar{x}/(-\ln f_0).$$

Method 3. Mean and Ratio of Frequencies of Zeroes and Ones.

This method involves equating the population mean and the ratio of first two population frequencies to their corresponding sample counterparts. This gives $\hat{\lambda} = \bar{x}/\hat{\phi}$ and $\hat{\phi} = \ln(\bar{x}f_0/f_1)$.

Method 4. Method of Maximum Likelihood*.

The maximum likelihood estimators (MLEs) of λ and ϕ are obtained by solving the equations

$$\hat{\lambda}\hat{\phi} = \bar{x}, \qquad \sum_{i=1}^{n} \frac{(x_i + 1)P_{x_i+1}(\hat{\lambda}, \hat{\phi})}{P_{x_i}(\hat{\lambda}, \hat{\phi})} = n\bar{x}$$

for $\hat{\lambda}$ and $\hat{\phi}$. In the preceding equation $P_{x_i}(\hat{\lambda}, \hat{\phi})$ denotes that each P_x involved is a function of $\hat{\lambda}$ and $\hat{\phi}$. Shenton [32] solved these equations iteratively using the Newton–Raphson method*. Some tables appearing in Douglas [8] for p_x, which is related to equation (9), make the calculations easy. Also, a two-dimensional iterative Newton–Raphson method has been developed by Douglas [10].

Method 5. Generalized Minimum Chi-square* Estimator.

For a description of this method, see Method 5 of NEGATIVE BINOMIAL DISTRIBUTION. The generalized minimum chi-square (GMCS) estimators are obtainable by solving linear equations as opposed to the estimators given by methods 2–4, which are obtainable by solving nonlinear equations. These estimators for the NTA

were developed by Gurland [17] and Hinz and Gurland [18].

Several authors have analyzed estimator performance. Shenton [32] tabulated the efficiency of the method of moments estimators. Bowman and Shenton [7] and Katti and Gurland [25] give contours of efficiency of the moment estimators with respect to the MLE. They also give contours of the efficiency of methods 2 and 3 with respect to the moment estimators. Hinz and Gurland [18] conclude that a GMCS estimator based on the first few moments and a function of the zero frequency has very high efficiency relative to the MLE. For a detailed discussion, see also Douglas [10] and Johnson and Kotz [23].

Douglas has developed estimators of a common λ or ϕ utilizing methods 1, 4, and 5 based on samples from k NTA populations.

Test of Hypothesis

As discussed in the section on test of hypotheses of NEGATIVE BINOMIAL DISTRIBUTION, Hinz and Gurland [20] have used the statistic $\hat{Q}$ for testing goodness of fit* of the NTA and other contagious distributions. The modification of $\hat{Q}$ by Bhalerao et al. [6] (discussed in the aforementioned article) yields a test with high power.

Hinz and Gurland [19] proposed tests of linear hypotheses regarding k NTA and other contagious distributions. The procedure is based on the generalized minimum chi-square method and does not need any transformation of the data.

Graphical Methods for Model Selection

Gurland [17] developed a graphical method based on the ratios of probability-ratio cumulants for selecting an appropriate model out of NTA, negative binomial, and other contagious distributions. Hinz and Gurland [18] developed a similar procedure for these distributions based on ratios of factorial cumulants. Grimm [15] suggested a method that uses graphs of empirical sum-percent curve plotted on Poisson probability paper. These methods help in selecting an appropriate model for a given data set.

GENERALIZATIONS OF NEYMAN'S DISTRIBUTIONS

Beall and Rescia's Generalization [4]

Beall and Rescia noted that for some data Neyman's distributions NTA, NTB, and NTC provided progressively better fits, but not better enough. By extending the pgf's (4)–(6) of Neyman's distributions, they presented a generalized family with the pgf (BR for Beall and Rescia)

$$G_{BR}(z)$$

$$= \exp\left[\lambda\left\{\Gamma(\beta+1)\sum_{i=0}^{\infty}\frac{\phi^i(z-1)^i}{\Gamma(\beta+1+i)} - 1\right\}\right],$$

$$(12)$$

where $\beta > 0$, $\lambda, \phi > 0$. This pgf can also be derived from (3) by taking $dF_P(p) = \beta(1-p)^{\beta-1}$, $0 < p < 1$. For $\beta = 0, 1$, and 2, $G_{BR}(z)$ reduces to $G_A(z)$, $G_B(z)$, and $G_C(z)$, respectively. It provides higher members of Neyman's family for $\beta > 2$. For $0 < \beta < 1$, $G_{BR}(z)$ includes those members of Neyman's class that lie between the NTA and the NTB, and for $1 < \beta < 2$, it includes those that lie between the NTB and the NTC.

For the BR family,

$$\mu = \lambda\phi/(\beta+1),$$

$$\sigma^2 = \lambda\phi(1+\beta)^{-1}(1+2\phi(2+\beta)^{-1}).$$

In order to fit this family to data, Beall and Rescia suggest first fixing β and then estimating λ and ϕ by the method of moments to give

$$\hat{\phi} = (\beta+2)(s^2 - \bar{x})/(2\bar{x}),$$

$$\hat{\lambda} = (\beta+1)\bar{x}/s^2.$$

To select an appropriate value of β, the distribution is fitted for different values of β. The value of the χ^2 statistic is calculated for each fit. The value of β for which χ^2 is a minimum is regarded as appropriate. For recurrence relations, see Beall and Rescia [4], Johnson and Kotz [23], and also the following recurrence relations for Gurland's distributions, which reduce to BR relations when $\alpha = 1$.

Gurland's Generalization [16]

Gurland presented an extension of Neyman's distributions from (3) by taking

$$dF_P(p) = (1/B(\alpha, \beta))p^{\alpha-1}(1-p)^{\beta-1},$$

$$0 < p < 1,$$

the beta distribution*, where $B(\alpha, \beta)$ is the beta function. This gives the pgf (subscript G for Gurland) of Gurland's family as

$$G_G(z) = \exp\{\lambda_1 F_1(\alpha, \alpha+\beta, \phi(z-1)) - \lambda\},$$

$$(13)$$

where $_1F_1(\alpha; \alpha+\beta; \phi(z-1))$ is the confluent hypergeometric function* defined as

$$_1F_1(u; v; w) = 1 + \frac{u}{v}w + \frac{u(u+1)}{v(v+1)}\frac{w^2}{2!}$$

$$+ \cdots.$$

Its mean and variance are

$$\mu = \lambda\phi\alpha(\alpha+\beta)^{-1},$$

$$\sigma^2 = \lambda\phi\alpha(\alpha+\beta)^{-1}$$

$$\times\left[1 + \phi(\alpha+1)(\alpha+\beta+1)^{-1}\right],$$

respectively. The recurrence relations for the probabilities are

$$P_{x+1} = \lambda(x+1)^{-1}\sum_{j=0}^{x}F_jP_{x-j}, \quad (14)$$

where

$$F_j = \phi^{j+1}(j!)^{-1}(\alpha)_{j+1}\{(\alpha+\beta)_{j+1}\}^{-1}$$

$$\times {_1F_1}(\alpha+j+1; \alpha+\beta+j+1; -\phi)$$

$$(15)$$

with $(\alpha)_j = \alpha(\alpha+1)\ldots(\alpha+j-1)$. The F_j's can also be computed from the relations

$$F_j = (\phi+\alpha+\beta+j-1)(j)^{-1}F_{j-1}$$

$$- \phi(\alpha+j-1)\{j(j-1)\}^{-1}F_{j-2}.$$

$$(16)$$

For fixed α and β, λ and ϕ are estimated by the method of moments. For estimating all four parameters, Gurland suggests equating the first two population moments and the first two probabilities to the corresponding sample quantities. Relations (14)–(16) can also be used for BR distributions with $\alpha = 1$ and hence for the NTB and the NTC distributions as well.

Relationship to Other Distributions

The NTA distribution arises as a limiting case of many distributions. It is a limiting case of the Poisson-binomial distribution with the pgf $\exp\{\lambda(q + pz)^n - \lambda\}$, $\lambda > 0$, $0 < p < 1$, $q + p = 1$, as $n \to \infty$ and $p \to 0$ such that $np = \phi$. It also arises as a limiting case of the Poisson–Pascal distribution with pgf $\exp\{\lambda(Q - Pz)^{-k} - \lambda\}$, $k > 0$, $P > 0$, $Q = 1 + P$, as $k \to \infty$, $P \to 0$ such that $Pk = \phi$. The NTA distribution also arises as a limiting case of the Poisson V POLPAB family of Bhalerao and Gurland [5] with pgf

$$\exp\left[\lambda\{1 - \beta(z - 1)/(1 - \beta)\}^{-\alpha/\beta} - \lambda\right],$$
$$\beta < 1, \quad \alpha > 0 \quad \text{as} \quad \beta \to 0.$$

Gurland's distributions with pgf (13) approach the NTA as $\alpha \to \infty$ under various conditions; see Gurland [16]. This family also approaches the generalized Polya–Aeppli distribution [23, Chap. 9] when $\beta \to \infty$ and α is fixed or when α and the first two moments are fixed.

BIVARIATE NTA DISTRIBUTIONS

Holgate [21] considered three types of bivariate NTA distributions.

Type I

In (3), if we replace $h(z)$ with the pgf
$$h(z_1, z_2) = \exp\{\phi_1(z_1 - 1) + \phi_2(z_2 - 1) + \xi(z_1 z_2 - 1)\} \tag{17}$$
of a bivariate Poisson distribution, denoted

by Poisson (ϕ_1, ϕ_2, ξ), and take $dF_M(m) = e^{-\lambda}\lambda^m/m!$, we get the pgf of the bivariate NTA distribution of type I, given by

$$G_I(z_1, z_2)$$
$$= \exp\left[\lambda(\exp\{\phi_1(z_1 - 1) + \phi_2(z_2 - 1) + \xi(z_1 z_2 - 1)\} - 1)\right]. \tag{18}$$

Each marginal distribution corresponding to (18) is univariate NTA.

Type II

In (3), replace $h(z)$ with the pgf
$$h(z_1, z_2) = \exp\{\phi_1(z_1 - 1) + \phi_2(z_2 - 1)\}$$
of a bivariate Poisson distribution and replace $dF_M(m)$ by the joint pf of a Poisson $(\lambda_1, \lambda_2, \lambda)$. The resulting pgf of bivariate NTA of type II is

$$G_{II}(z_1, z_2)$$
$$= \exp\left[\lambda_1\exp\{\phi_1(z_1 - 1)\}\right.$$
$$+ \lambda_2\exp\{\phi_2(z_2 - 1)\}$$
$$+ \lambda\exp\{\phi_1(z_1 - 1) + \phi_2(z_2 - 1)\}$$
$$\left. - \lambda_1 - \lambda_2 - \lambda\right]. \tag{19}$$

The marginal distributions corresponding to (19) are univariate NTA with parameters $\lambda + \lambda_j$ and ϕ_j, $j = 1, 2$.

Type III

Let U, V_1, and V_2 have univariate NTA distributions with parameters (λ, ϕ), (λ_1, ϕ), and (λ_2, ϕ), respectively. Let $X_1 = U + V_1$ and $X_2 = U + V_2$. Then (X_1, X_2) has a bivariate NTA distribution of type III with pgf

$$G_{III}(z_1, z_2) = \exp\left[\lambda_1\exp\{\phi(z_1 - 1)\}\right.$$
$$+ \lambda_2\exp\{\phi(z_2 - 1)\}$$
$$+ \lambda\exp\{\phi(z_1 z_2 - 1)\}$$
$$\left. - \lambda_1 - \lambda_2 - \lambda\right]. \tag{20}$$

For recurrence relations of probabilities and other characteristics, see Gillings [13] (see also Holgate [21] and Johnson and Kotz

[23]). Holgate fitted these distributions to a set of botanical data by the method of moments. Gillings [13] developed maximum likelihood and minimum chi-square estimates for the type I distribution.

CONCLUSIONS

The NTA is the most used Neyman's distribution among all those discussed here. For this distribution, MLE, are also developed. However, other members of this family and its extensions have not been widely used because they are tedious to handle. For them, the estimators used are based on other simpler methods.

The negative binomial distribution* and the NTA have often been considered competitors. After fitting some important contagious distributions to 35 sets of data, Martin and Katti [27] concluded that both the NTA and the negative binomial models have wide applicability. Douglas [10] compared these two distributions with respect to the first four factorial moments and the expansions of their zero probability, assuming that they both have the same first two moments. Both the distributions had exactly the same expressions for the expansion of P_0 up to the fifth-degree term. The expressions for the first four factorial moments were also very close. This should explain why they are close competitors. If both these distributions give good and comparable fits to a data set, the experimenter should choose that model which provides a better understanding of the mechanism generating the data.

References

[1] Archibald, E. E. A. (1948). *Ann. Bot.*, **47**, 221–235.

[2] Barton, D. E. (1957). *Trab. Estadíst.*, **8**, 13–32. (An excellent study regarding modes of Neyman Type A distribution.)

[3] Beall, G. (1940). *Ecology*, **21**, 460–474. (Probably one of the earliest articles dealing with the applications of Neyman's distributions, especially Type B and Type C.)

[4] Beall, G. and Rescia, R. (1953). *Biometrics*, **9**, 354–386. (This article gives an extension of Neyman's family of distributions and presents insights on the practical aspects of these distributions as applied to biological data. Intermediate level.)

[5] Bhalerao, N. R. and Gurland, J. (1977). *Tech. Rep. No. 399*, University of Wisconsin, Madison. (A family of generalized Poisson distributions that contains Neyman Type A as a limiting case.)

[6] Bhalerao, N. R., Gurland, J., and Tripathi, R. C. (1980). *J. Amer. Statist. Ass.*, **75**, 934–938.

[7] Bowman, K. O. and Shenton, L. R. (1967). *Rep. No. ORNL-4102*, Oak Ridge National Laboratory, Oak Ridge, TN.

[8] Douglas, J. B. (1955). *Biometrics*, **11**, 149–173. (Deals with a simplified method of fitting Neyman Type A by maximum likelihood using a table developed in the article.)

[9] Douglas, J. B. (1965). *Classical and Contagious Discrete Distributions*, G. P. Patil, ed. Statistical Publishing Society, Calcutta, pp. 291–302.

[10] Douglas, J. B. (1980). *Analysis with Standard Contagious Distributions*. International Cooperative Publishing House, Fairland, MD. (An excellent source of detailed material on Neyman's distributions in particular and contagious distributions in general. Very rich bibliography. Intermediate level.)

[11] Evans, D. A. (1953). *Biometrika*, **40**, 186–211.

[12] Feller, W. (1943). *Ann. Math. Statist.*, **14**, 389–400. (Presents a simplified way to obtain Neyman's distributions in particular and contagious distributions in general; worthwhile reading. Intermediate level.)

[13] Gillings, D. B. (1974). *Biometrics*, **30**, 619–628.

[14] Grimm, H. (1964). *Biom. Zeit.*, **6**, 10–23.

[15] Grimm, H. (1970). *Random Counts in Scientific Work*, Vol. 1, G. P. Patil, ed. Pennsylvania State University Press, University Park, PA, pp. 193–206.

[16] Gurland, J. (1958). *Biometrics*, **14**, 229–249. (Presents an extension of Neyman's, and Beall and Rescia's family of distributions. Gives good foundation for generating many contagious distributions. Intermediate level.)

[17] Gurland, J. (1965). *Classical and Contagious Discrete Distributions*, G. P. Patil, ed. Statistical Publishing Society, Calcutta, pp. 141–158.

[18] Hinz, P. N. and Gurland, J. (1967). *Biometrika*, **54**, 555–566.

[19] Hinz, P. N. and Gurland, J. (1968). *Biometrika*, **55**, 315–322. (Methods for analyzing data that do not need transformation from various contagious populations are developed.)

[20] Hinz, P. N. and Gurland, J. (1970). *J. Amer. Statist. Ass.*, **65**, 887–903.

[21] Holgate, P. (1966). *Biometrika*, **53**, 241–244. (Bivariate versions of the Type A distribution are presented and fitted by the method of moments).

[22] Jain, G. C. (1975). *Commun. Statist.*, **A4**, 1065–1071.

[23] Johnson, N. L. and Kotz, S. (1969). *Distributions in Statistics: Discrete Distributions*. Wiley, New York. (An excellent source of information on discrete distributions in general and Neyman's distributions in particular. A rich bibliography at the end of each chapter. Intermediate level.)

[24] Johnson, N. L. and Kotz, S. (1980). *Int. Statist. Rev.*, **50**, 71–101. (A brief discussion of developments in discrete distributions during 1969–1980. An enormous source of references on discrete distributions during this period.)

[25] Katti, S. K. and Gurland, J. (1962). *Biometrika*, **49**, 215–226.

[26] Martin, D. C. and Katti, S. K. (1962). *Biometrics*, **18**, 354–364.

[27] Martin, D. C. and Katti, S. K. (1965). *Biometrics*, **21**, 34–48. (An excellent comparison of important contagious distributions by fitting them to a large collection of data sets.)

[28] McGuire, J. V., Brindley, T. A., and Bancroft, T. A. (1957). *Biometrics*, **13**, 65–78.

[29] Neyman, J. (1939). *Ann. Math. Statist.*, **10**, 35–57. (Original derivation of Neyman's Type A, B, and C distributions based on mathematical modeling of biological phenomena of the insect population; recommended reading. Intermediate level.)

[30] Pielou, E. C. (1957). *J. Ecol.*, **45**, 31–47.

[31] Rogers, A. (1969). *Environ. Plan.*, **1**, 47–80 and 155–171.

[32] Shenton, L. R. (1949). *Biometrika*, **36**, 450–454. (Development of maximum likelihood for the Type A distribution.)

[33] Shenton, L. R. and Bowman, K. O. (1967). *Technometrics*, **9**, 587–598.

[34] Skellam, J. G. (1948). *J. R. Statist. Soc. B*, **10**, 257–261.

(CONTAGIOUS DISTRIBUTIONS
KEMP FAMILIES OF DISTRIBUTIONS
LAGRANGE AND RELATED
    PROBABILITY DISTRIBUTIONS
MODIFIED POWER SERIES
    DISTRIBUTIONS
NEGATIVE BINOMIAL DISTRIBUTION
POWER SERIES DISTRIBUTIONS)

RAM C. TRIPATHI

NEYMAN–WALD ASSESSMENT *See*
CONDITIONAL INFERENCE

NIGHTINGALE, FLORENCE

Born: May 12, 1820, in Florence, Italy.

Died: August 13, 1910, in London, England.

Contributed to: social reform, nursing, demography, epidemiological studies, vital statistics.

Florence Nightingale is mainly known for improving the squalid hospital conditions at Scutari during the Crimean War and for her subsequent campaigns to reform the health and living conditions of the British army, the sanitary conditions and administration of hospitals, and the nursing profession. Energy, enthusiasm, and a crusading spirit apparently drove her to physical breakdown; for the last 20 years of her working life, she directed a flow of letters and reports from a couch in the Burlington Hotel and (later) from her house in London, whither those in authority came to consult her.

To a small circle the Lady of the Lamp is known as the Passionate Statistician [1], a name linking the eloquence Florence Nightingale mustered with the massive amounts of data she compiled to convince prime ministers, viceroys, secretaries, undersecretaries, and parliamentary commissions of the truths of her cause. Thus she wrote [4, p. 249]:

> We hear with horror of the loss of 400 men on board the Birkenhead by carelessness at sea; but what should we feel, if we were told that 1,100 men are annually doomed to death in our Army at home by causes which might be prevented?

A timely combination of factors makes her place in the development of applied statistics noteworthy. The first is her flair for collecting, arranging, and presenting facts and figures, developed along with her sense of vocation against immense family opposition. Cecil Woodham-Smith writes of the early 1840s [8, Chap. 4]:

> Notebook after notebook was filled with a mass of facts, compared, indexed and tabu-

lated . . . . In the cold dark mornings she laid the foundation of the vast and detailed knowledge of sanitary conditions which was to make her the first expert in Europe. Then the breakfast bell rang, and she came down to be the Daughter of the Home.

She became aware that mortality statistics should be age-specific and that crude death rates can be misleading ([5, p. 55]; *see also* VITAL STATISTICS):

In comparing the deaths of one hospital with those of another, any statistics are justly considered absolutely valueless which do not give the ages, the sexes and the diseases of all the cases. . . . There can be no comparison between old men with dropsies and young women with consumptions.

Thus in 1859 she articulated the need to allow for confounding effects in observational studies*. Later she was to be influenced by Adolphe Quetelet* and his approach to quantifying social behavior (see Diamond and Stone [2]).

The second factor is the contemporary state of statistics. In 1850 no scientific system of tabulating or reporting mortality or morbidity* statistics could be said to exist. When Miss Nightingale arrived at Scutari in 1854, no proper records were being kept, but she introduced there a uniform system of recording mortality rates without which she would have been unable to plead her case successfully. Her campaigns for reform included proposals to standardize and improve the collection and recording of health statistics. In this she shares credit with her adviser William Farr (1807–1883), who brought a scientific approach to vital statistics* in the United Kingdom and, as registrar-general, put it on a sound footing.

She had influential friends like Sidney Herbert, the Secretary of State at War, but the opposition to reform was entrenched in the army bureaucracy and in political high places, and it was determined to defeat her. Her single most effective weapon, the presentation of sound statistical data, is the third and most important factor to be considered. No major national cause had ever been championed primarily with such a weapon. She showed, for example [3], that "those who fell before Sebastopol by disease were above seven times the number who fell by the enemy." The opposition lost because her statistics were unanswerable (see Tables 1 and 2) and their publication led to an outcry.

The monumental privately printed work [4] that comprised her main reform effort contains an enormous number of statistical tabulations and some elaborate pie charts*. It demonstrated the appalling conditions in Scutari and at the front in the Crimea and the gross neglect of the health of ordinary soldiers living in barracks in peacetime.

Table 1   Florence Nightingale's Statistics of Mortality at Different Periods during the Crimean War[a]

Month	Year	Deaths per 1000 (Living) per Annum
January	1855	$1173\frac{1}{2}$
	1856	$21\frac{1}{2}$
May	1855	203
	1856	8
January–May	1855	628
	1856	$11\frac{1}{2}$
Crimea, May 1856		8
Line at home		18.7
Guard at home		20.4

[a] Ref. 3, p. 295.

Table 2   Florence Nightingale's Relative Mortality Statistics of the Army at Home and of the English Male Population at Corresponding Ages[a]

Ages		Deaths Annually to 1000 Living
20–25	Englishmen	8.4
	English soldiers	17.0
25–30	Englishmen	9.2
	English soldiers	18.3
30–35	Englishmen	10.2
	English soldiers	18.4
35–40	Englishmen	11.6
	English soldiers	19.3

[a] Ref. 3, p. 253.

There are suggestions of attempts by army administrators such as her chief antagonist, Dr. John Hall, Chief of Medical Staff in the Crimea, to cover up the extent of the tragedy with misleading statistics (see ref. 8, Chap. 10). Miss Nightingale explained how some of these were devised ([4, Section X, p. XXV]; *see also* LIFE TABLES):

In constructing a Table of Mortality we take 100 men, eight die the first year, there are left 92—2 die the second year, there are left 90. The usual method of stating this mortality would be to *take the hundred over again* and strike the difference, thus—

$$100 + 100 = 200 \qquad 8 + 2 = 10 \qquad 2)10(5$$

Therefore, it is a mortality of 5 per cent., per annum. Now, this is manifestly wrong, and gives the Secretary of State no idea of his *accumulated* loss.

If this method is applied to the data in Table 1 [4, p. 295], the extent of the horror in the Crimea becomes glossed over. The rates show numbers of deaths per 1000 living rather than per thousand out of the pool of manpower in the forces. In modern terms, the mortality for January 1855 at the height of the catastrophe can be measured roughly as 1173.5/(1000 + 1173.5) or 54 percent. Table 1 also indicates the dramatic reduction in death rates within 18 months.

The last three rows of Table 1 suggest that soldiers in the Crimea in May 1856 had a better chance of survival than those in the insanitary barracks in England. Table 2 [4, p. 253] provides the crux of the case for army reform at home. Miss Nightingale was ahead of her time as an epidemiologist, comparing cases (soldiers in barracks) with civilian controls. In one study [8, Chap. 13] she compared these in the same boroughs, thus controlling for any "district" effect. The confounding effect in the comparisons in Table 2 worked to reinforce her conclusions [4, p. 253]:

The Army are picked lives and the inferior lives are thus thrown back. . . .

The general population includes, besides those thus rejected . . . vagrants, paupers, intemperate persons, the dregs of the race, over whose habits we have little or no control. The food, clothing, lodging, employment, and all that concerns the sanitary state of the soldier, are absolutely under our control and may be regulated to the smallest particular.

Yet with all this, the mortality of the Army, from which the injured lives are *subtracted*, is double that of the whole population, to which the injured lives are added.

Of this high mortality, she wrote [4, p. 251]: "The cause is made sufficiently plain by looking at their Barrack accommodation and their mode of life."

In this age, when the collection, editing, and presentation of data are commonplace in government affairs, it is not easy to appreciate the overwhelming difficulties involved in the statistical aspect of Miss Nightingale's enterprise. Nor should we overlook the prejudices against women in Victorian England. She never appeared in a public forum; her case was always presented for her or submitted in writing. At the *International Statistical Congress* in 1860, for example, a letter from her was read and adopted; it called upon governments to publish more extensive and frequent statistical abstracts [3]. A recent study of her life and work [7] presents a negative picture of her character, but it is too early to assess its implications on her reputation as a statistician.

Florence Nightingale was instrumental in the founding of a statistical department in the army. She was ahead of her time in advocating that censuses* should include data on the sick and infirm and on the housing of the population [3] and in advocating the founding of a professorship in applied statistics at a British university. For her correspondence with Francis Galton* on the latter, see Pearson [6, pp. 414–424].

In 1858 she was elected a Fellow of the Royal Statistical Society*, and in 1874 an honorary member of the American Statistical Association*.

References

[1] Cook, E. (1913). *The Life of Florence Nightingale*, 2 vols. MacMillan, London.

[2] Diamond, M. and Stone, M. (1981). *J. R. Statist. Soc. A*, **144**, 66–79; 176–214; 332–351.

[3] Kopf, E. W. (1916). *J. Amer. Statist. Ass.*, **15**, 388–404. (An account of Miss Nightingale's achievements as a statistician, reprinted in *Studies in the History of Statistics and Probability*, Vol. 2, M. G. Kendall and R. L. Plackett, eds.)

[4] Nightingale, F. (1858). *Notes on Matters Affecting the Health, Efficiency, and Hospital Administration of the British Army. Founded Chiefly on the Experience of the Late War*. Harrison, London. (Printed privately.)

[5] Nightingale, F. (1859). *Notes on Nursing: What It Is and What It Is Not*. Harrison, London.

[6] Pearson, K. (1924). *The Life, Letters and Labours of Francis Galton*, Vol. II. Cambridge University Press, Cambridge, England.

[7] Smith, F. B. (1982). *Florence Nightingale: Reputation and Power*. Croom Helm, London (also St. Martin's Press, New York).

[8] Woodham-Smith, C. (1951). *Florence Nightingale*. McGraw-Hill, New York. (An eminently readable and enthralling biography that barely touches on the statistical aspects of her life.)

Bibliography

Cohen, I. B. (1984). *Scientific Amer.*, **250** (3), 128–137. (A fascinating and informative article on Florence Nightingale as a pioneer in the uses of social statistics. Some of the pie charts, bar charts, and tables from ref. 4 and from the Royal Commission report on sanitary conditions in the army are reproduced.)

Eyler, J. M. (1979). *Victorian Social Medicine: The Ideas and Methods of William Farr*. The Johns Hopkins University Press, Baltimore, MD.

(DEMOGRAPHY
EPIDEMIOLOGICAL STATISTICS
LIFE TABLES
VITAL STATISTICS)

CAMPBELL B. READ

NIKODYM DERIVATIVE *See* RADON–NIKODYM DERIVATIVE

NOETHER AND RELATED CONDITIONS

The asymptotic normality* of statistics based on ranks* is an important topic in nonparametric inference (*see* DISTRIBUTION-FREE METHODS). In particular, simple linear rank statistics* of the form

$$S_N = \sum_{i=1}^{N} c_{Ni} a_N(R_i)$$

are the basis of locally most powerful* rank tests [5]. Here $R_1, \ldots, R_N$ are the ranks of the observations $X_1, \ldots, X_N$. That is, $R_i = r$ if X_i is the rth smallest of the X's. The vector $\mathbf{c}_N = (c_{N1}, \ldots, c_{NN})$ is the *vector of regression constants* and $\mathbf{a}_N = (a_N(1), \ldots, a_N(N))$ is the *vector of scores*.

LIMITING DISTRIBUTION OF S_N UNDER RANDOMNESS

Under the randomness* hypothesis, $X_1, \ldots, X_N$ are independent and identically distributed with unknown continuous CDFs. Hence $R_1, \ldots, R_N$ form a random permutation of $1, \ldots, N$. Noether's condition [8] is the weakest of several conditions imposed on $\mathbf{a}_N$ and/or $\mathbf{c}_N$ to guarantee that $\{S_N - E[S_N]\}/\{\operatorname{var}(S_N)\}^{1/2}$ has a limiting standard normal distribution as $N \to \infty$. The condition can be stated as follows:

$$\lim_{N \to \infty} \left[\frac{\max_{1 \le i \le N}(c_{Ni} - \bar{c}_{N\cdot})}{\sum_{i=1}^{N}(c_{Ni} - \bar{c}_{N\cdot})^2} \right] = 0,$$

$$\text{where} \quad \bar{c}_{N\cdot} = N^{-1}\sum_{i=1}^{N} c_{Ni}. \quad (1)$$

As already stated (1) is imposed on the $\mathbf{c}_N$ sequence, but it and any of the other conditions defined here may be imposed on the $\mathbf{a}_N$ sequence in the obvious way.

If $\mathbf{a}_N$ and $\mathbf{c}_N$ both satisfy (1), then $\{S_N - E[S_N]\}/\{\operatorname{var}(S_N)\}^{1/2}$ has a limiting standard normal distribution if and only if [2] for any $\epsilon > 0$,

$$\lim_{N \to \infty} \left\{ N^{-1} \sum_{|\delta_{Nij}| > \epsilon} \delta_{Nij}^2 \right\} = 0, \quad (2)$$

where

$$\delta_{Nij} = \frac{(c_{Ni} - \bar{c}_{N\cdot})(a_N(j) - \bar{a}_N(\cdot))}{N^{1/2}\sigma(\mathbf{a}_N)\sigma(\mathbf{c}_N)}$$

$$(1 \le i, j \le N),$$

$$\sigma^2(\mathbf{a}_N) = N^{-1} \sum_{i=1}^{N} (a_N(i) - \bar{a}_N(\cdot))^2$$

and

$$\sigma^2(\mathbf{c}_N) = N^{-1} \sum_{i=1}^{N} (c_N(i) - \bar{c}_{N\cdot})^2.$$

TWO-SAMPLE PROBLEM

By choosing $c_{Ni} = 1$, $i = 1, \ldots, m$ and $c_{Ni} = 0$ otherwise, S_N is a simple linear rank statistic for the two-sample problem, where $X_1, \ldots, X_m$ and $X_{m+1}, \ldots, X_N$ are samples from two CDFs whose equality is to be tested. In this case, if $\mathbf{c}_N$ satisfies (1), then $\min\{m, N - m\} \to \infty$. The familiar Wilcoxon statistic (see MANN–WHITNEY–WILCOXON TEST) is based on scores $a_N(r) = r$. Other common choices for scores are normal

scores* $(a_N(r) = E[Z_{r:N}]$, where the Z's are independent standard normal variables) or Savage scores $(a_N(r) = \sum_{j=N-r+1}^{N} 1/j)$. All of these satisfy (1) (Hájek and Šidák [5]).

NONPARAMETRIC REGRESSION AND CORRELATION

In the model

$$X_i = \alpha + \beta c_{Ni} + E_i, \quad i = 1, \ldots, N,$$

α and β unknown, tests based on S_N are locally most powerful rank tests for the hypothesis $\beta = 0$. This optimality property also applied to the two-sample case by taking $c_{Ni} = 1$ or 0 as above. For the special choice $c_{Ni} = i$, $a_N(r) = r$, S_N is a linear function of the Spearman rank correlation* statistic. Statistics like S_N also arise in tests of independence of two variables X and Y, in which $\mathbf{c}_N$ is a vector of scores corresponding to the ranks of the Y sample.

HISTORY AND RELATED CONDITIONS

The limiting behavior of S_N was first considered by Wald and Wolfowitz [10], who proved asymptotic normality assuming both $\mathbf{a}_N$ and $\mathbf{c}_N$ satisfy

$$\frac{N^{-1}\sum_{i=1}^{N}(c_{Ni} - \bar{c}_{N\cdot})^r}{\left[N^{-1}\sum_{i=1}^{N}(c_{Ni} - \bar{c}_{N\cdot})^2\right]^{r/2}} = O(1)$$

$$\text{as} \quad N \to \infty, \quad r = 3, 4, \ldots . \quad (3)$$

Noether [8] showed that the asymptotic normality still holds if $\mathbf{c}_N$ satisfies (3) and the scores $\mathbf{a}_N$ satisfy only the weaker condition (1). Noether actually formulated his condition as

$$\lim_{N \to \infty} \frac{\sum_{i=1}^{N}(a_N(i) - \bar{a}_N(\cdot))^r}{\left[N\sigma^2(\mathbf{a}_N)\right]^{r/2}} = 0,$$

$$r = 3, 4, \ldots . \quad (1')$$

Hoeffding [6] proved the equivalence of (1) and (1') and generalized Noether's results by showing the asymptotic normality of S_N

holds if

$$\frac{\left[\sum_{i=1}^{N}(c_{Ni} - \bar{c}_{N.})^{r}\right]\left[\sum_{i=1}^{N}(a_{N}(i) - \bar{a}_{N}(\cdot))^{r}\right]}{N^{(1/2)r+1}\sigma(\mathbf{a}_{N})\sigma(\mathbf{c}_{N})}$$

$$\to 0 \quad \text{as } N \to \infty, \text{ for } r = 3, 4, \dots .$$

(4)

All of the preceding results are based on the method of moments* and can be extended to permutation tests (*see* RANDOMIZATION TESTS) in which the $a_{N}(i)$ are realizations of independent random variables [6, 8, 10].

The relationships among the various conditions were established by Hájek [2], who also introduced the generalized Noether condition

$$\lim_{N \to \infty} (K_{N}/N) = 0,$$

(5)

$$\lim_{N \to \infty}\left[\frac{\max\sum_{\alpha=1}^{K_{n}}(a_{N}(i_{\alpha}) - \bar{a}_{N}(\cdot))^{2}}{\sigma^{2}(\mathbf{a}_{N})}\right] = 0,$$

where the maximum is taken over all subsets of $\{1, \dots, N\}$ such that $1 \leqslant i_{1} < \cdots < i_{K_{n}} \leqslant N$. Observe that (2) and (4) deal with the joint behavior of the score vector $\mathbf{a}_{N}$ and the regression constant vector $\mathbf{c}_{N}$, while (1), (3) and (5) apply only to a single sequence of vectors. If $\mathbf{c}_{N}$ satisifies (1) and $\mathbf{a}_{N}$ satisfies (5) simultaneously, this joint condition will be denoted (1)(5). The following implications hold [2]:

$$(3) \Rightarrow (5) \Rightarrow (1)$$
$$(1)(5)$$
$$\nearrow \qquad \searrow$$
$$(3)(3) \Rightarrow (1)(3) \qquad (2) \Rightarrow (1)(1)$$
$$\searrow \qquad \nearrow$$
$$(4)$$

MATHEMATICAL NOTES
AND EXTENSIONS

Hájek's necessary and sufficient conditions for asymptotic normality are based on the method of projection (see Hájek [2] and Hájek and Šidák [5]): S_{N} is approximated by

$$T_{N} = \sum_{i=1}^{N}(c_{Ni} - \bar{c}_{N.})Y_{i},$$

where Y_{i} is a function of X_{i}. The variable T_{N} is asymptotically normal if its summands are

uniformly asymptotically negligible and if the Lindeberg condition* is satisfied. The asymptotic negligibility condition on the terms in T_{N} hold because $\mathbf{c}_{N}$ satisfies (1), and the Lindeberg condition on T_{N} is equivalent to (2). Noether's condition (1), applied to both $\mathbf{a}_{N}$ and $\mathbf{c}_{N}$, is necessary and very nearly sufficient for asymptotic normality. However, counterexamples can be constructed to show that the preceding implications are not reversible [5, pp. 195–197].

The joint asymptotic normality of several simple linear rank statistics is easily derived using the Cramér–Wold device [2]. This result is important in the nonparametric treatment of the multisample problem.

The theory can also be extended to the wider class of linear rank statistics

$$L_{N} = \sum_{i=1}^{N} a_{n}(i, R_{i}),$$

where $a_{N}(i, j)$ is an $N \times N$ array [2, 6, 7]. Asymptotic normality of L_{N} can be shown under conditions analogous to (4) and (2). However, necessary and sufficient conditions have not yet been proven for the asymptotic normality of L_{N}.

In the one-sample symmetry problem, signed rank score statistics of the form

$$S_{N}^{*} = \sum_{i=1}^{N} \operatorname{sgn}(X_{i})a_{N}(R_{i}^{*})$$

are used to test whether the distribution of X is continuous and symmetric about 0. Here $\operatorname{sgn}(x) = 1, 0, -1$ if $x > 0$, $x = 0$ or $x < 0$, respectively, and R_{i}^{*} is the rank of $|X_{i}|$. In particular, if $a_{N}(r) = r$, one obtains the Wilcoxon signed rank statistic*. Under the symmetry hypothesis, $\operatorname{sgn} X = \pm 1$ with probabilities $\frac{1}{2}$ each and the ranks and signs are independent. Therefore $S_{N}^{*}/(\operatorname{var}(X))^{1/2}$ is asymptotically standard normal if and only if the scores satisfy $\sum a_{N}^{2}(r)/\max a_{N}^{2}(r) \to \infty$ [5, p. 153].

ASYMPTOTIC NORMALITY UNDER
ALTERNATIVES TO RANDOMNESS

The asymptotic normality of S_{N} when $X_{1}, \dots, X_{N}$ are independent with possibly different continuous CDFs is much more difficult to derive, and stringent conditions on the score vector $\mathbf{a}_N$ are needed. However, (1) or the somewhat stronger boundedness condition [1, 4]

$$\frac{N \max_{1 \leqslant i \leqslant N}\left(c_{Ni} - \bar{c}_{N\cdot}\right)^2}{\sum_{i=1}^{N}\left(c_{Ni} - \bar{c}_{N\cdot}\right)^2}$$

$$= O(1) \quad \text{as} \quad N \to \infty \qquad (6)$$

are strong enough conditions on $\mathbf{c}_N$ to obtain the desired results. (6) implies (3).

The scores $a_N(i)$ usually satisfy $a_N(i) \cong \phi(i/[N+1])$, where the score function ϕ is square integrable and satisfies various regularity conditions. The conditions on ϕ and the accuracy of the approximate score formula may be weakened if one imposes additional restrictions on the F_i. This line of research was initiated by Hájek [1, 4, 5] and relies heavily on the projection technique, whereby S_N is approximated by $\sum l_i(U_i)$, where $U_1, \ldots, U_N$ are independent uniform variables and $\text{var}[l_i(U_i)] < \infty$.

References

[1] Dupač, V. and Hájek, J. (1969). *Ann. Math. Statist.*, **40**, 1992–2017. (Derives asymptotic normality when the score function is discontinuous.)

[2] Hájek, J. (1961). *Ann. Math. Statist.*, **32**, 506–523. (Finds necessary and sufficient conditions for asymptotic normality under the randomness hypothesis, relates various conditions and introduces the projection technique.)

[3] Hájek, J. (1962). *Ann. Math. Statist.*, **33**, 1124–1147. (Proves asymptotic normality under "local" alternatives.)

[4] Hájek, J. (1968). *Ann. Math. Statist.*, **39**, 325–346. (Proves asymptotic normality for differentiable score functions under general alternatives.)

[5] Hájek, J. and Šidák, Z. (1967). *Theory of Rank Tests*. Academic Press, New York. (An advanced survey of ranking methods stressing the asymptotic theory.)

[6] Hoeffding, W. (1951). *Ann. Math Statist.*, **22**, 558–566.

[7] Motoo, M. (1957). *Ann. Inst. Statist. Math.*, **8**, 145–154.

[8] Noether, G. E. (1949). *Ann. Math. Statist.*, **20**, 455–458. (First formulation of the Noether condition.)

[9] Serfling, R. J. (1980). *Approximation Theorems of Mathematical Statistics*, Wiley, New York. (General survey of large-sample theory with extensive references. Surveys various approaches to limiting distributions of linear rank statistics with extensions to rates of convergence, laws of large numbers, etc.)

[10] Wald, A. and Wolfowitz, J. (1944). *Ann. Math. Statist.*, **15**, 358–372.

(ASYMPTOTIC NORMALITY
DISTRIBUTION-FREE METHODS
LINDEBERG CONDITION
LINEAR RANK TESTS
LOCALLY OPTIMUM STATISTICAL TESTS
RANKING PROCEDURES
RANK ORDER STATISTICS
RANK TESTS)

PAUL J. SMITH

NOISE

The term *noise* was first used in communication engineering as a result of the undesired acoustic effects accompanying spontaneous electric fluctuations in receivers. (*See* COMMUNICATION THEORY, STATISTICAL). Ever since the advent of electric communications, engineers have searched to reduce the electric noise in communications systems. Nowadays the idea of noise has been used widely in many other fields where no acoustic effect is involved. Roughly speaking, noise means something that interferes with the desired signal. There are a variety of sources; for example, the fundamental source of electric noise is the quantized electric charge. Its graininess causes the voltage fluctuations in electrical circuits. Another common example is the computer-generated noise due to round-off error*. In many statistical methods such as regression analysis and categorical data* analysis, the noise is the randomly fluctuating part (e.g., due to sampling), and the signal is the unknown deterministic parameter. To understand the noise structure and to extract the signal from the noisy environment are the most important objectives in many studies.

Usually noise is treated as a stochastic process* of irregular fluctuations. In applications, the noise process is often assumed to be stationary and ergodic. That is, its statistical properties can be characterized completely by just one sample over a long period. Many tools used in time-series* analysis [4, Part II] are very useful in analyzing noise. For example, the fast Fourier transform (FFT) [10, Chap. 6] is a powerful computational tool in estimating noise power spectra. The analysis of autocorrelation and partial autocorrelation functions* is useful in fitting autoregressive–moving average (ARMA) models* to the noise process.

GAUSSIAN NOISE

In many cases, noise is best described as a Gaussian process* (e.g., thermal noise and shot noise in electronic systems). That is, the joint distribution of noise random variables at any set of time points is multivariate normal. In other words, the noise process can be characterized completely by its autocorrelation function or, equivalently, by its spectral distribution if it is stationary. The assumed normality can be justified by the central limit theorem* if the noise is composed of many small independent (or weakly dependent) random effects.

The main advantage of using the Gaussian assumption is that the best linear estimator of the signal is optimal under the criterion of mean squared error*. (Of course, the signal needs to be Gaussian, too.) That is to say, there is no need to consider nonlinear theory in signal estimation in Gaussian systems. In signal detection*, the likelihood ratio statistic is an optimal test statistic, and many results have been established on absolute continuity* and the Radon–Nikodym derivative* between the two measures induced by pure noise and signal plus noise, when both measures are Gaussian [7, Chap. 3; 13].

For example, Feldman and Hájek derived the dichotomy theorem that two Gaussian measures are either singular or equivalent.

Rice [11, 12] studied the behavior of sample paths of stationary Gaussian noise with zero mean in continuous time. Suppose that $X(t)$ is stationary Gaussian with $EX(t) = 0$ and $E[X(t)X(t + s)] = R(s)$. The expected number of zeros per second in sample paths is

$$2\left[\int_0^\infty f^2 w(f)\,df \Big/ \int_0^\infty w(f)\,df\right]^{1/2},$$

$$w(f) = 4\int_0^\infty R(t)\cos 2\pi ft\,dt,$$

where $w(f)$ is the power spectral density and f is frequency in cycles per second. However, not much is known about the distribution of the distance between two successive zeros. The expected number of local maxima per second in a sample path is

$$\left[\int_0^\infty f^4 w(f)\,df \Big/ \int_0^\infty f^2 w(f)\,df\right]^{1/2}.$$

When the noise $X(t)$ is narrowband (i.e., the spectrum $w(f)$ vanishes except in a small region), the envelope of $X(t)$ has the Rayleigh distribution*, i.e., that of the square root of a random variable with the chi-squared distribution of degree 2 (see also NARROWBAND PROCESS).

Even though optimal statistics can be obtained under the Gaussian condition, robust statistics* are desired so that decisions based on statistics are less sensitive to the Gaussian assumption. Various kinds of non-Gaussian noise take place in different situations. Some are generated from Gaussian noise through nonlinear devices. For example, the output voltage of noise has the Rayleigh distribution when narrowband Gaussian thermal noise is applied to an envelope detector [3]. Another example is quantization noise, which has the uniform distribution* from $-d/2$ to $d/2$, where d is a unit quanitization step. Quantization noise occurs when analog signals are converted to digital form. One more example is impulse noise, which is a generalized random process composed of short bursts occurring at random time points with random amplitudes. In the analysis of EEG wave recordings of brain activity, the EEG recordings are a Gaussian signal process of brain activity plus a Poisson process* of impulse noise due to muscle contractions by nerve impulses [8].

WHITE NOISE

White noise is a stationary stochastic process with constant spectral density (*see* STATIONARY PROCESSES). The term *white* is borrowed from optics, where *white light* has been used to signify uniform energy distribution among the colors. (Actually the analogy is not correct since in optics the uniform energy distribution of white light is based on wavelength (the reciprocal of frequency) rather than frequency [3, p. 14].) Discrete-time white noise is simply an uncorrelated wide-sense stationary time series with zero mean. ARMA processes are derived by discrete-time white noise. Continuous-time white noise $X(t)$ satisfies formally $EX(t) = 0$ and $E[X(t)X(s)] = \sigma^2 \delta(t - s)$, where $\delta(\cdot)$ is the Dirac delta function*. In the following, we only consider continuous-time white noise.

Since an entirely flat spectral density distribution implies infinite power, white noise does not exist in practice. Nevertheless, the use of a white noise model is justified in many aspects. Many real data sets, such as aircraft flight test data, radar return data, and passive sonar detection data, involve an additive random noise that has large bandwidth compared to that of the signal. In some other problems, noise may be best described as a linear transformation of white noise. More importantly, it is often much easier analytically and computationally to deal with white noise (e.g., the Kalman filter*). A crucial problem of robustness* arises: Can the results based on a white noise model effectively approximate those based on a real case where the noise involved has large but finite bandwidth? When only linear operations on the data are considered, the answer is yes, i.e., the results based on a white noise model are consistent with the asymptotic case where the noise

bandwidth tends to infinity in any way desired. However, serious difficulties arise in interpretation when nonlinear operations on the data have to be considered [2]. The problem of robustness has to be further investigated in the nonlinear case.

Let us take a close look at the following nonlinear filtering problem. Let $Z(t)$ and $X(t)$ ($t \geqslant 0$) be the mutually independent signal and white noise, respectively. Let $Y(t) = Z(t) + X(t)$ be the observation process. The classic filtering problem is to calculate the conditional expectation (or distribution) of $Z(t)$ given $\{Y(s): 0 \leqslant s \leqslant t\}$. Traditionally, white noise $X(t)$ is treated as the formal derivative of the Wiener process* $W(t) = \int_0^t X(s)\, ds$. The Wiener process is well defined in terms of the Wiener measure on $C[0, T]$ (the set of all continuous real-valued functions on $[0, T]$), where T is the time span. Now, the filtering problem can be solved using the Itô integral (*see* STOCHASTIC INTEGRALS), which allows nonlinear operations. The conditional expectation of $Z(t)$ given $\{Y(s): 0 \leqslant s \leqslant t\}$ is uniquely determined up to a null subset of sample paths. Unfortunately, the Wiener process sample paths are of bounded variation with probability zero and all the physical sample paths are of bounded variation. Thus nonlinear filtering theory cannot be applied in practice unless we are able to choose a particular version of the conditional expectation that is defined everywhere (not just almost everywhere) and is continuous in the sample path with respect to the supremum norm on $C[0, T]$. This is another robustness problem that has been taken up by several authors [5, 6, 9].

Balakrishnan [2] took a different functional approach that avoids this problem. He developed nonlinear white noise theory in which white noise is defined on $L_2[0, T]$ (the set of all square integrable real-valued functions on $[0, T]$). Specifically, let C be the class of cylinder sets in $L_2[0, T]$ with Borel bases in finite dimensional subspaces. A weak distribution is a finitely additive probability measure on C that is countably additive on any class of cylinder sets with bases in the same finite dimensional subspace. Now, white noise is defined to be the triple $(L_2[0, T], C, \mu_G)$, where μ_G is the weak distribution (Gaussian measure) defined by the characteristic function*

$$\int_{L_2[0,T]} \exp\left[i \int_0^T h(t) x(t) dt \right] d\mu_G(x)$$

$$= \exp\left[-\frac{1}{2} \int_0^T h^2(t) dt \right]$$

for all $h \in L_2[0, T]$. Based on this construction of white noise, some results on likelihood ratio, innovation process, and conditional density are derived [1, Chap. 2; 6].

SIGNAL PROCESSING IN THE PRESENCE OF NOISE

The effect of noise on information transmission is the main subject of statistical communication theory. Information theory*, signal detection, and signal estimation (filtering and smoothing) are three important topics. *See* COMMUNICATION THEORY, STATISTICAL for a detailed exposition. Spline approximation (*see* SPLINE FUNCTIONS) is another subject related to signal estimation in the presence of noise. It is a method of recovering a smooth (signal) function $f(t)$ $(0 \leqslant t \leqslant T)$ when only discrete, noisy measurements $y_i = f(t_i) + \epsilon(t_i)$ $(i = 1, 2, \ldots, n)$ are available. What is known about $f(t)$ is that it is smooth, e.g., $f''(t) \in L_2[0, T]$. The spline estimator of f is the spline function that minimizes

$$n^{-1} \sum_{i=1}^n \left(f(t_i) - y_i \right)^2 + \lambda \int_0^T \left(f''(s) \right)^2 ds,$$

where the smoothing parameter λ is determined by the generalized cross-validation method [14]. This estimator may be regarded as a low-pass smoother in a general sense. The estimator has the remarkable property that the higher-order derivatives of it are good estimators of those of f.

References

[1] Balakrishnan, A. V. (1976). *Applied Functional Analysis*. Springer-Verlag, New York.

[2] Balakrishnan, A. V. (1980). *Multivariate Analysis*, Vol. 5. P. R. Krishnaiah, ed., North-Holland, Amsterdam, pp. 97–109.

[3] Bennett, W. R. (1960). *Electrical Noise*. McGraw-Hill, New York.

[4] Box, G. E. P. and Jenkins, G. M. (1976). *Time Series Analysis: Forecasting and Control*. Holden-Day, San Francisco.

[5] Clark, J. M. C. (1978). In *NATO Advanced Study Institute Series: Communication Systems and Random Process Theory*, J. K. Skwirzynski, ed., Sijthoff and Noordhoff, Alphen aan den Rijn, The Netherlands, pp. 721–734.

[6] Davis, M. H. A. (1982). *Theory Prob. Appl.*, **27**, 167–175.

[7] Ibragimov, I. A. and Rozanov, Y. A. (1978). *Gaussian Random Processes*. Springer-Verlag, New York.

[8] Johnson, T. L., Feldman, R. G., and Sax, D. S. (1973). *Amer. J. EEG Tech.*, **13**, 13–35.

[9] Kushner, H. J. (1979). *Stochastics*, **3**, 75–83.

[10] Oppenheim, A. V. and Schafer, R. W. (1975). *Digital Signal Processing*. Prentice-Hall, Englewood Cliffs, NJ.

[11] Rice, S. O. (1944). *Bell Syst. Tech. J.*, **23**, 283–332.

[12] Rice, S. O. (1945). *Bell Syst. Tech. J.*, **24**, 46–156.

[13] Shepp, L. A. (1966). *Ann. Math. Statist.*, **37**, 321–354.

[14] Wahba, G. (1977). *SIAM J. Numer. Anal.*, **14**, 651–667.

(AUTOREGRESSIVE–MOVING AVERAGE (ARMA) MODELS
COMMUNICATION THEORY, STATISTICAL
GAUSSIAN PROCESSES
KALMAN FILTERING
SIGNAL DETECTION
SPLINE FUNCTIONS
STATIONARY PROCESSES
TIME SERIES
WIENER PROCESS)

YI-CHING YAO

NOMINAL DATA

If measurement is the assignment of numbers to objects according to a rule, then nominal scales represent the simplest form of measurement, for the scale values of these variables consist only of category labels. Thus, a nominal variable is perhaps best viewed as a form of classification rather than a measurement scale [14].

Ideally, the categories are constructed and labeled so that all objects in the universe are assigned to one and only one class. If the categories are mutually exclusive and exhaustive, the members of any one of them will be homogeneous with respect to the trait being measured or classified. Achieving such a set of equivalence classes is not an easy matter, as taxonomists have always realized. Indeed, the practical success of nominal data analysis resides in the labels; the operational denotations must be sufficiently precise to ensure that the objects are correctly assigned, i.e., that the "true" or underlying attribute is represented in such a way that accurate and meaningful inferences about the subjects can be made.

For these reasons and because there is no mathematical order among the categories, nominal variables, sometimes called *attribute data*, have always been considered the weakest form of measurement. Furthermore, until the last three decades there has been relatively little statistical theory and few methods for analyzing them. In the past, scientists turned to this type of data only reluctantly because it was felt that theory construction could not progress very far without better measurement.

In some instances, of course, no particular problems arise since there is a natural isomorphism between a set of observations and a group of equivalence classes as in the classification of people by sex. In other instances, the classification is more tentative and problematic. As examples, responses on public opinion surveys are labeled agree and disagree or occupations are designated blue collar and white collar. However convenient these names are for pigeonholing respondents, they may conceal as much information as they reveal. Most citizens do not simply agree or disagree with matters as complicated as public policy, but have varying degrees of feelings about them. These nuances may be important for drawing conclusions about their behavior. Similarly, the labels blue and white collar leave one wondering where to put farmers, doctors, housewives, and students.

Still, nominal data cannot be avoided, particularly in the behavioral sciences where there are few common units of measurement such as liters or grams, where concepts are often unavoidably ambiguous, and where attitudes and behavior are not easily quantified. Quite naturally then, a huge body of literature on the construction of nominal scales has emerged. Concomitantly, there has also been a recent growth of techniques for analyzing these data. According to many statisticians, there are today ways to estimate and test complex systems of variables and hypotheses that seem to rival even the most advanced multivariate techniques.

UNIVARIATE AND BIVARIATE PROCEDURES

The usual strategy for dealing with a single nominal variable is to propose an underlying model that might account for the observed distribution. Consider, for example, a polytomous response having I levels or categories. The goal might be to account for the observed distribution by testing various hypotheses about the probabilities P_i; P_i is the probability that a randomly drawn object with be in the ith class. One might be interested in the model of equiprobability or of symmetry (e.g., $P_1 = P_I$, $P_2 = P_{I-1}, \ldots$).

These hypotheses are frequently cast as log-linear models that express the logarithms of expected cell frequencies as linear functions of various parameters (*see* CONTINGENCY TABLES). This approach has become popular because the models have meaningful substantive interpretations, because a wide variety of hypotheses can be encompassed within a single analytic framework, and because algorithms for estimating expected values (usually based on maximum likelihood* methods) are widely available. Haberman [11] describes many of the possibilities.

Although numerous other procedures exist (see Fleiss [5]), many of them are less commonly used than before or are being performed within a larger context such as log-linear models. Of much greater interest now are cross-classifications of nominal data.

Most scientists have as a minimal objective the explanation of one variable by reference to another. It is probably more interesting, for instance, to explain categories of mental illness in terms of social class than to analyze separately the distributions of these variables. Consequently, statisticians and practitioners alike have toiled for years developing ways to measure the associations among nominal variables. Their efforts have led in three distinct but related directions.

Test of Statistical Independence

One is the familiar *test of statistical independence* between two cross-classified variables. This method, which is standard fare in elementary statistics textbooks, has been greatly expanded in the past decade to cover a variety of meaningful problems. Goodman [7], for example, extends the usual definition of independence ($P_{ij} = A_i B_j$, where P_{ij} is the population proportion of cases in the ijth cell of an $I \times J$ table and A_i and B_j are positive constants) to the notion of *quasi-independence** in which the independence condition holds only for a subset of cells in the table. Goodman uses quasi-independence to answer a variety of questions. Consider a typical case: A social mobility table shows the relationship between, say, parents' and children's occupations. One can test the hypothesis of quasi-independence in the off-diagonal cells in order to measure status inheritance and disin-

heritance. Other methods of extending and refining the usual chi-square test* of independence are described by Cochran [4], Castellan [3], and Iverson [12], among others. These techniques are useful for partitioning the overall chi-square into additive components, each of which tests a specific hypothesis pertaining to a subtable of the larger array; *see* PARTITION OF CHI-SQUARE.

Measurement of the Degree of Association

A second activity is the *measurement of the degree of association* between two nominal variables. A measure of association* is a numerical index that describes the strength and nature of the relationship. Many measures have clearly defined upper and lower bounds, making them easier to interpret. They equal zero, for example, if the variables are statistically independent (the converse may not hold) and 1.0 if there is "perfect" association. The meaning of "perfect" depends on one's standard or definition [15]. Presumably if there is a one-to-one correspondence between sets of categories, or if the variables have different numbers of categories as nearly one to one as possible, the relationship is perfect and the index equals 1.0 (*see* ASSOCIATION, MEASURES OF and NOMINAL SCALE AGREEMENT).

Many measures are constructed so that values lying between the two bounds have an operational or at least intuitive meaning. Goodman and Kruskal [9] introduced a set of measures that rests on a "proportional-reduction-in-error" [PRE] logic. One imagines guessing to which categories of a nominal variable a sample of cases belongs. The guesses are made in two ways: first, without knowledge of how the objects are classified on any other variable, and second, with information about an additional variable taken into account. Presumably the second method leads to fewer prediction errors than the first, especially if the classes of the two variables are strongly related, and one can thus calculate the estimated proportional reduction in error. This family of indices have become common in applied statistics.

Variables in Regressionlike Equations

As commonplace as they are, however, single measures of association are rapidly giving ground to a third, more general method that expresses cell frequencies or functions of cell frequencies (e.g., odds ratios) as *dependent variables* in *regressionlike equations*. Instead of a single index, the investigator estimates an equation that characterizes the relationship by means of sets of parameters. This procedure, it is claimed, permits a more detailed specification of interactions. Another advantage is that this type of formulation is a special case of multivariate models.

MULTIVARIATE PROCEDURES

Scientists and statisticians normally want to analyze more than two-variable relationships. Two variables viewed in isolation usually reveal little about the world from which they are drawn. Stating that social class and mental health are related probably raises as many questions as it answers. Hence, as in any scientific undertaking, the analysis of nominal data is most rewarding when many variables are considered simultaneously.

The form of multivariate analysis* obviously depends on the nature of the data and the hypothesized model. If the predictors are quantitative whereas the dependent variable is nominal, probit or logit* models may be used [13]. Alternatively, the goal might be to discriminate between two or more groups on the basis of a collection of criterion variables as when one wants to compare, say, business people with artists in terms of scores on a personality test. On the other hand, if nominal scales are used to predict a numerical dependent variable, the choice is usually some form of dummy variable regression. Yet perhaps the most interesting situations arise when all of the variables are nominal.

There has been a veritable explosion in methodologies for the multivariate analysis of multiway tables of nominal data (*see* MULTIDIMENSIONAL CONTINGENCY TABLES). Most fall under the headings of: log-linear models or weighted least-squares analysis [10]. Although the proponents of each quarrel among themselves over various points, the two procedures overlap and frequently lead to the same substantive conclusions.

Table 1, taken from Forthofer and Lehnen [6] illustrates the kinds of data and problems to which these techniques are addressed. The data pertain to two factors, cigarette smoking and level of exposure to lead, and a single response variable, pulmonary function test (PFT). Presented in this

Table 1   Relationship between Pulmonary Function Test (PFT) Results, Smoking Habits, and Lead Level in Ambient Air

		PFT		Logit
Lead Level	Smoking	Normal	Not Normal	(Normal)
High lead	Never	33	3	2.40
	Former	12	2	1.79
	Light	27	2	2.35
	Heavy	16	3	1.67
Low lead	Never	160	4	3.69
	Former	49	6	2.10
	Light	75	6	2.53
	Heavy	84	3	3.33

Source.   Reference 6.

fashion, the figures raise a number of general questions:

1. What is the (conditional) probability of a normal PFT?
2. Is there an association between the level of lead at an individual's residence and the probability of a normal PFT?
3. Does the association still hold when smoking habits are taken into account?
4. What is the nature of the joint effect of smoking and exposure to lead on the PFT results? (For example, is there a multiplicative effect?)
5. If interaction* exists, in which subpopulations of smokers is the relationship strongest? In which is it the weakest?

The list can be extended by examining different functions of cell probabilities or proportions or by considering various types of models. One might consider, for example, linear, logarithmic, exponential, or more complex transformations of the cell entries. The range of models might include those involving interaction parameters, trend terms, or the a priori deletion of certain cells.

In the common form of *log-linear analysis* the logarithms of expected values are expressed as functions of various main and interaction effects, parameters akin to the terms of regular analysis of variance.* Statements about different parameters are sometimes equivalent to hypotheses about equiprobability, independence, and conditional independence; many of these models can be partitioned into submodels.

General log-linear models do not distinguish between dependent and independent variables. The objective instead is to search for interrelationships among the entire set of variables. It is possible, however, to create models in which one or more variables are designated as dependent or response and to explain them as a function of a set of independent or *factor variables*. They have been used to estimate so-called path or causal and latent structure* diagrams, which has put the treatment of nominal data on roughly the same footing as the analysis of recursive and nonrecursive equations [8]. (*See also* CONTINGENCY TABLES for a more complete discussion of log-linear models.)

Using functions of cell probabilities or proportions as the dependent variable, the second approach, *weighted least squares*, employs design matrices to code the categories of the variables. If **F** is a vector of probabilities (or a function of them), **B** is a vector of unknown coefficients, and **X** is a design matrix*, a model **F** = **BX** is estimated with weighted least squares.

Forthofer and Lehnen [6], for example, develop and test a linear model for **F** consisting of logits:

$$\frac{P_{ij1}}{1 - P_{ij1}} = \frac{P_{ij1}}{P_{ij2}} .$$

Table 2   Comparison of Weighted Least-Squares and Maximum Likelihood Estimates for Data in Table 1

Parameter	WLS Estimates	ML Estimates
Constant	2.42	2.46
Lead level		
High	− 0.44	− 0.42
Low	0.44	0.42
Smoking		
Never	0.66	0.64
Former	− 0.62	− 0.63
Light	− 0.16	− 0.15
Heavy	0.12	0.14

Source.   Reference 6.

Using a straightforward design matrix, they estimate a model consisting of a constant term and the main effects of lead level and smoking. The results are summarized in Table 2, which also compares the weighted least-squares estimates with those obtained from maximum-likelihood estimation*. The results suggest that in this type of model both procedures lead to the same conclusion, but generally speaking weighted least-squares deals with a broader class of problems [10]. By no means the only options (the subject has been barely scratched), these two strategies are among the most commonly employed today by applied statisticians. (For a more complete discussion, see Bishop et al. [1].)

THE FUTURE AND PROBLEMS OF NOMINAL DATA ANALYSIS

Nominal data analysis has progressed enormously since the days when manifold contingency tables were scanned visually for interactions and estimation and testing were carried out mainly with percents and chi-square tests*. The newer methods have apparently thrown open the doors to social scientists and others who wanted to achieve the same rigor and precision as their colleagues in the natural sciences but whose data were nonquantitative. These investigators can now deal with multivariate data, test complex hypotheses, estimate simultaneous confidence intervals, express relationships in terms of parameters, etc.

Yet serious problems remain, not simply of refining specific statistics, of discovering sampling distributions, of simplifying calculations, of proving theorems, or of finding more efficient estimating algorithms. Many of *these* hurdles should soon be overcome.

The major difficulties lie in the nature of the data. Apart from circumstances when the trait in question unambiguously specifies the category labels, nominal scales raise a host of questions about measurement—issues that are all too frequently glossed over by the glamour of techniques like weighted least-squares or maximum likelihood estimation (*see* MEASUREMENT STRUCTURES AND STATISTICS). For if a variable misclassifies objects, any statistical procedure, however advanced or elegant, surely will produce nonsense.

Unfortunately the possibilities for error are endless. Mistakes can arise from the use of too few categories or the wrong choice of cutpoints. Take something as prosaic as measuring preferences. Statisticians and social scientists alike sometimes classify opinions as pro or con and then hurriedly put the scores in a complex equation, hoping, no doubt, to gain a better understanding of attitudes. They do not consider seriously the meaningfulness of the categorization process. But it is likely that some people will feel strongly about the matter while others will care not at all. The individuals in the "pro" group probably attach varying meanings to both the issue and questions about it, have different levels of information, and even vary among themselves as much as they differ from the "against" people. All of these nuances tend to get lost when one falls under the sway of a sophisticated, complex multivariate procedure advertised for nominal data.

It is disconcerting but true that substantive conclusions are affected by factors as mundane as the number of categories, the definition of class boundaries, or the collapsing or combining of adjacent groups into one; conclusions can change along with changes in the categorization of the variables.

Such an eventuality might result from poorly defined operational indicators or from inattention in designing the research. More ominously, it could arise from gaps in the theoretical understanding of the underlying constructs, a gap that prevents their being adequately measured. Statistics, after all, has to walk hand in hand with substantive knowledge; it can never replace it. If variables are inadequately conceptualized and measured, it will be difficult to make sense of them no matter what methods are used. Thus the advent of modern multivariate nominal data analysis may, as Blalock [2] suggests, inadvertently handicap as well as help scientific advancement if it leads researchers to remain content with crudely measured data. In this sense, one might view the tools described as temporary expedients, useful only until better measurement arrives, rather than panaceas for our lack of knowledge.

References

[1]   Bishop, Y. M. M., Fienberg, S. E., and Holland, P. W. (1975). *Discrete Multivariate Analysis: Theory and Practice*. MIT Press, Cambridge, MA.

[2]   Blalock, H. M. (1979). *Social Statistics*, rev. 2nd ed. McGraw-Hill, New York.

[3]   Castellan, J. N. (1965). *Psychol. Bull.*, **64**, 330–338.

[4]   Cochran, W. G. (1954). *Biometrics*, **10**, 417–451.

[5]   Fleiss, J. L. (1973). *Statistical Methods for Rates and Proportions*. Wiley, New York.

[6]   Forthofer, R. N. and Lehnen, R. G. (1981). *Public Program Analysis: A New Categorical Data Approach*. Wadsworth, Belmont, CA.

[7]   Goodman, L. A. (1968). *J. Amer. Statist. Ass.*, **63**, 1091–1131.

[8]   Goodman, L. A. (1979). *Analyzing Qualitative/Categorical Data*. Abt Associates, Cambridge, MA.

[9]   Goodman, L. A. and Kruskal, W. (1979). *Measures of Association for Cross-Classifications*. Springer-Verlag, New York.

[10] Grizzle, J. E., Starmer, F., and Koch, G. C. (1969). *Biometrics*, **25**, 489–504.

[11] Haberman, S. J. (1978). *The Analysis of Qualitative Data*. Academic Press, New York.

[12] Iverson, G. E. (1979). *Sociol. Meth. Res.*, **8**, 143–157.

[13] Nerlove, M. and Press, J. (1973). "Univariate and Multivariate Log-linear and Logistic Models." *Tech. Rep. No. R-1306-EDA/NIH*, Rand Corporation, Santa Monica, CA.

[14] Stevens, S. S. (1946). *Science*, **103**, 46–52.

[15] Weisberg, H. (1974). *Amer. Polit. Sci. Rev.*, **68**, 1638–1655.

(ASSOCIATION, MEASURES OF
CATEGORICAL DATA
CONTINGENCY TABLES
DEPENDENCE, TESTS FOR
KAPPA COEFFICIENT
MEASUREMENT STRUCTURES AND
     STATISTICS
MULTIDIMENSIONAL CONTINGENCY
     TABLES
NOMINAL SCALE AGREEMENT
QUASI-INDEPENDENCE
WEIGHTED LEAST SQUARES)

H. T. Reynolds

NOMINAL SCALE AGREEMENT

Nominal scale agreement concerns the determination of the reliability (i.e., the consistency with which a measure assesses a given trait) between two or more classifications of the same data at a nominal level of measurement. In many applications the classifications are made by judges, and then the term *interjudge reliability* is often used, but it is not confined to this situation. In can be used where agreement between two characteristics or subjects is investigated [19] or to select the best judges (or coders) from a pool of judges on the basis of optimal agreement [25]. Moreover, it can be used for the assessment of agreement between the classifications according to different scale criteria or to help in making research design decisions [14]. There is also criticism on the use of agreement indices [21]. With minor exceptions, the nominal scale should consist of a finite number of mutually exclusive and exhaustive categories. To assess the degree of reli-ability, an agreement index is needed with certain desirable properties. Nominal scale agreement does not directly assess the extent to which a measure actually covers the trait for which it was intended (validity). It is confined to the replicability of the nominal measurement. It is important for fundamental research and for applications to publish such ratings of the quality of the classification process and to use them for improving this process.

AGREEMENT VERSUS ASSOCIATION

Agreement is often considered a special kind of association*. There are differences between agreement and association, however. In general, with regard to agreement, most important is the similarity of the content of behavior (in a broad sense) between judges with the goal of determining the degree of identity of this behavior. The behavior of one judge does not have to be predicted from that of another. In the case of association, the strength of the linear relationship between the variables is investigated. Here the goal is to predict the values of one variable from those of the other.

REQUIREMENTS FOR MEASURES OF AGREEMENT

Requirements of agreement indices for nominal data are (partly taken from Galtung [11]): The index should have a fixed maximum value (usually 1) for all cases of complete agreement, and a fixed value (usually 0) given an independence model (this does not exclude occasional negative values). The index should be independent of the number of observations and of the number of categories; it should not be influenced by permutations of the categories. Each judge agrees perfectly with itself, and perfect agreement is a transitive relation. The index should be symmetric (unless one of the judges is a standard). The sampling distribution of the index should be known, and the index should be robust. (*See also* MEASURES OF AGREEMENT.)

HISTORY

Against the background of the difference between association and agreement, it is to be understood that at first the percentage of cases in which both judges agree in classifying observations was used as an agreement index. The disadvantage of this index, however, is that in cases with few categories the probability of equal assignments is greater than in cases with many categories, especially when one takes into account the agreement that might be expected by chance.

Therefore, Bennett et al. [1] proposed a correction for the number of categories. They assume that all categories have equal probability to be used. An index in which this often unrealistic assumption is removed was proposed by Scott [24]. He assumes that the distribution of the observations over the categories is known and that each judge will assign the same number of observations to each category as any other judge. This reasoning was criticized by Cohen [5]. In his view the number of assignments per category will be different in most cases and will be equal merely by coincidence. Disagreement arises also because it is the tendency of the judges to distribute the observations differently over the categories. Cohen proposed an index, kappa, in which this is also taken into account (*see* KAPPA COEFFICIENT). The index is defined as the proportion of agreement after chance agreement is removed from consideration. The coefficient satisfies the requirements mentioned before. Currently it is used most, and it is extended in several directions. A rule of thumb for interpreting kappa was given by Landis and Koch [18]. The sampling characteristics of kappa have been derived by Fleiss et al. [10].

The coefficients mentioned involve assumptions about the way in which chance factors may operate. Indices in which chance is assumed to operate in a purely random way have been proposed by Holley and Guilford [12] and by Maxwell [20] for dichotomous judgements. Their indices differ in the extension to more than two categories.

Also a number of indices have been derived from similarity measures (*see* MEASURES OF SIMILARITY, DISSIMILARITY, AND DISTANCE). The most important one is defined as the ratio between twice the number of observations on which both judges agree and the total sum of assignments by both judges. Extensions are dealt with by Rogot and Goldberg [23]. It was shown by Fleiss [8], however, that most of these indices are special cases of kappa. New types of indices are based on quasi-independence* models [2] and on latent structure* models [4].

Krippendorff [15] proposed an index to be used in content analysis, where coding according to a multilevel decision process is applied.

EXTENSIONS TO VARIOUS EMPIRICAL SITUATIONS

The most common application of agreement indices is in situations where observations are classified twice, i.e., by two judges. However, in empirical research one will meet other situations as well.

First, there can be more than two judges. When it is known which observations are classified by which judges, the investigator may choose pairwise or simultaneous agreement. In a case where it is not known which observations are classified by which judges, the agreement statistic must be modified (see Conger [6]).

Depending on the goal of the investigation, it may be desirable to consider a specific judge as a standard and further, to incorporate the seriousness of various misclassifications. One also might be interested in the amount of agreement within a specific category, if necessary, conditionalized on one of the judges or between some categories. Finally, it may be that observations are classified in more than one category or that there are missing values. For nearly all these situations, extensions of the kappa statistic are available [3, 9, 13, 16, 17, 25]; *see also* HIERARCHICAL KAPPA STATISTICS.

FREE CATEGORIES

In certain research situations a nominal scale or category system is not yet defined when coders start assigning the data. The coders then have two tasks; to stipulate categories and to classify answers to them. This is often the case when content analysis or (participant) observation is used. The scales that will be developed by the coders, often on the nominal level, may then differ in the numbers of categories, and the categories

can refer to a different content. Formal agreement indices can then be based on registration for each pair of observations whether coders classify them in the same category or not (verbal labels of categories are possibly unreliable and will be disregarded). A chance-corrected agreement index $D2$ for this situation was proposed by Popping [22]. If $f(i, j)$ is the number of observations assigned by one rater to category i and by the other rater to category j, and the marginal totals are respectively $f(i, \cdot)$ and $f(\cdot, j)$, and the total number of observations is N, then Popping [22] proposes

$$D2 = (\text{dot} - \text{dexp})/(\text{dmax} - \text{dexp}),$$

where

$$\text{dot} = \sum_i \sum_j f(i, j) * \{f(i, j) - 1\},$$

$$\text{dmax} = \max\left[\sum_i f(i, \cdot) * \{f(i, \cdot) - 1\}, \right.$$

$$\left. \times \sum_j f(\cdot, j) * \{f(\cdot, j) - 1\}\right],$$

$$\text{dexp} = \sum_i \sum_j \text{entier}\{h(i, j)\}$$

$$* \left[2 * h(i, j) - \text{entier}\{h(i, j)\} - 1\right],$$

where $h(i, j) = f(i, \cdot) * f(\cdot, j)/N$. Because $h(i, j)$ is not an integer, an interpolation is used. The index satisfies the requirements for an agreement index given earlier.

References

[1] Bennett, E. M., Alpert, R., and Goldstein, A. C. (1954). *Publ. Opin. Quart.*, **18**, 303–308.

[2] Bergan, J. R. (1980). *J. Educ. Meas.*, **5**, 363–376.

[3] Bishop, Y. M. M., Fienberg, S. E., and Holland, P. W. (1975). *Discrete Multivariate Analysis*. MIT Press, Cambridge, MA.

[4] Clogg, C. C. (1979). *Soc. Sci. Res.* **8**, 287–301.

[5] Cohen, J. (1960). *Educ. Psychol. Meas.*, **20**, 37–46.

[6] Conger, A. J. (1980). *Psychol. Bull.*, **88**, 322–328.

[7] Fleiss, J. L. (1971). *Psychol. Bull.*, **76**, 378–382.

[8] Fleiss, J. L. (1975). *Biometrics*, **31**, 651–659.

[9] Fleiss, J. L. (1981). *Statistical Methods for Rates and Proportions*, 2nd ed. Wiley, New York. (Contains overview of agreement indices and discusses extensions of kappa.)

[10] Fleiss, J. L., Cohen, J., and Everitt, B. S. (1969). *Psychol. Bull.*, **72**, 323–327.

[11] Galtung, J. (1967). *Theory and Methods of Social Research*. Allen and Unwin, London.

[12] Holley, J. W. and Guilford, J. P. (1964). *Educ. Psychol. Meas.*, **24**, 749–753.

[13] Hubert, L. J. (1977). *Psychol. Bull.*, **84**, 289–297.

[14] Kraemer, H. C. (1979). *Psychometrika*, **44**, 467–472.

[15] Krippendorff, K. (1971). *Behav. Sci.*, **16**, 228–235.

[16] Landis, J. R. and Koch, G. G. (1975). *Statist. Neerlandica*, **29**, 101–123. (Overview article, also discusses indices for higher levels of measurement.)

[17] Landis, J. R. and Koch, G. G. (1975). *Statist. Neerlandica*, **29**, 151–161. (Overview article, discusses extensions of kappa.)

[18] Landis, J. R. and Koch, G. G. (1977). *Biometrics*, **33**, 159–174.

[19] Light, R. J. (1971). *Psychol. Bull.*, **76**, 365–377.

[20] Maxwell, A. E. (1977). *Brit. J. Psychiatry*, **130**, 79–83.

[21] Mellenbergh, G. J. and Van der Linden, W. J. (1979). *Appl. Psychol. Meas.*, **3**, 257–273.

[22] Popping, R. (1983). *Quality and Quantity*, **17**, 1–18.

[23] Rogot, E. and Goldberg, I. D. (1966). *J. Chronic Disorders*, **19**, 991–1006.

[24] Scott, W. A. (1955). *Publ. Opin. Quart.*, **19**, 321–325.

[25] Schouten, H. J. A. (1982). *Statist. Neerlandica*, **36**, 45–61. (Discusses several extensions of kappa.)

(ASSOCIATION, MEASURES OF
HIERARCHICAL KAPPA STATISTICS
KAPPA COEFFICIENT
MEASURES OF AGREEMENT
MEASURES OF SIMILARITY,
   DISSIMILARITY, AND DISTANCE
NOMINAL DATA)

ROEL POPPING

NOMINATION SAMPLING

Nomination sampling is an alternative to simple random sampling when obtaining data to estimate a population fractile, such as a median. Estimation from the conventional simple random sample is a one-step process in which the sample values are used directly to estimate the population fractile

(*see* QUANTILE ESTIMATION). In contrast, estimation from a nomination sample is a two-step process. The first step is identifying the *nominees*, i.e., the largest (or smallest) values in each of several simple random samples from the population. The second step estimates the population fractile from the values of the nominees. In many cases, the selection of nominees is performed by a censoring process, as when several runners compete in foot races but only winners' times are recorded. The method derives its name from a scheme to enlist the cooperation of many human service providers in a survey of client costs; the providers have the opportunity to participate actively in the survey by nominating certain of their clients for detailed study. Such participation may decrease survey nonresponse*.

The method can be understood formally as follows. Consider a simple random sample of S values of a random variable X in a population. Let X_i be the ith order statistic of the sample $X_1 \leqslant X_2 \leqslant X_3 \leqslant \cdots \leqslant X_S$. The value of X_i estimates the $i/(S + 1)$ fractile of the population distribution. For instance, in a simple random sample of seven values, the fourth order statistic estimates the population median.

Now consider S simple random samples each of size N. Let X_{ij} be the ith order statistic in the jth sample ($1 \leqslant i \leqslant N$, $1 \leqslant j \leqslant S$). The case with the largest value X_{Nj} is the nominee from the jth sample. Let U_K be the Kth order statistic among the nominees. The value of U_K estimates the $K/(S + 1)$ fractile of the distribution of nominees. Since the nominees are the largest values in samples of N cases, the $K/(S + 1)$ fractile among the nominees corresponds to the $[K/(S + 1)]^{1/n}$ fractile in the population. (If the nominees had the smallest values in their sample, they would estimate the $1 - [K/(S + 1)]^{1/n}$ fractile.) If two values of K can be found to bracket the desired population fractile, an estimate can be made by linear interpolation.

For example, suppose we know four simple random samples of size two and wish to estimate the population median. Table 1 il-

Table 1   The Use of Nomination Sampling to Estimate a Population Median[a]

Sample	Value of Case 1	Value of Case 2	Value of Nominee	Rank among Nominees	Estimated Fractile of Distribution of Nominees	Estimated Fractile of Distribution of Population
1	2.56	0.74	2.56	4	$\frac{4}{5} = 0.80$	$\sqrt{0.80} = 0.89$
2	1.25	1.45	1.45	2	$\frac{2}{5} = 0.40$	$\sqrt{0.40} = 0.63$
3	1.41	0.03	1.41	1	$\frac{1}{5} = 0.20$	$\sqrt{0.20} = 0.45$
4	1.23	1.81	1.81	3	$\frac{3}{5} = 0.60$	$\sqrt{0.60} = 0.77$

[a]The population median is estimated to lie between 1.41 and 1.45. Linear interpolation provides an estimate of 1.42.

lustrates the data and calculations. In each sample, the larger of the cases is the nominee. Among the nominees, since $S = 4$ and $N = 2$, the kth order statistic corresponds to the $(k/5)^{1/2}$ fractile of the population distribution. Linear interpolation produces an estimate of 1.42 for the population median. Note that the median value of the nominees does not provide a good estimate of the population median since the nomination process biases their values upward.

Bibliography

Gumbel, E. (1960). *Statistics of Extremes.* Columbia University Press, New York.

Mosteller, F. and Rourke, R. (1973). *Sturdy Statistics: Nonparametric and Order Statistics.* Addison-Wesley, Reading, MA.

Willemain, T. (1980). *J. Amer. Statist. Ass.*, **75**, 908–911.

(QUANTILE ESTIMATION
RANKED SET SAMPLING)

THOMAS R. WILLEMAIN

NOMOGRAMS

A nomogram or alignment chart consists of three scales not necessarily rectilinear. When a straight line, or isopleth, is drawn across the diagram, it intersects the three numbers, which are related, according to some prescribed equation. The task of the nomographer is to construct the scales and locate their positions. (The words *nomogram* and *nomograph* are virtually interchangeable. We will use the term *nomogram* when talking about the chart and *nomographer* when talking about the developer of the chart.)

The nomogram is attributed to Professor Maurice d'Ocagne whose *Traité de Nomographie* published in Paris in 1889 [3] is still considered a landmark work. The nomogram's early development was largely European in nature, chiefly attributed to French and German mathematicians. The first comprehensive work in English seems to be that originally authored by the Britishers Allcock and Jones in 1932. Later joined by Michel, their book, *The Nomogram: The Theory and Practical Construction of Computation Charts* [2], has been through numerous editions.

The nomogram was very useful in its early days as a device to reduce computational effort. Prior to the hand-held calculator's appearance in the early 1970s, logarithms, square roots, and trigonometric functions were obtained from tables or from a slide rule. A complex mathematical equation could be resolved with a slide rule, but operator error was a potential hazard. The nomogram alleviated the error due to computational mistakes and provided a continuously variable output by rotating the isopleth around a fixed point. The nomogram still provides the latter advantage, the display of numerous policy options, by simply moving the straightedge. The one disadvantage of a nomogram is that accuracy is limited by the fineness of the grid. However, in most practical situations, exact numerical precision is not required.

Technology has not rendered the nomogram useless to applied statisticians. Examination of recent copies of the *Journal of Quality Technology** will indicate the continuing interest and increasingly complex problems for which nomograms are being developed.

We will show two methods for constructing the very simplest nomogram, present some nomogram forms, and give a few examples of statistical applications.

Before proceeding, a sort of disclaimer is needed. The number of nomograms developed for statistical applications is vast. Any attempt to list a substantial portion of them would result in complaints that dozens had been omitted. Suffice it to say that if you think a particular computation looks amenable to a nomogram, you are probably right, and there's probably a nomogram that has been constructed for the purpose. If you cannot locate the nomogram, pick up Levens' easy-to-read text [6] and involve yourself in this very interesting applications area. Texts by Adams [1], Epstein [4] and Fasal [5] can also be consulted for a different emphasis and approach.

CONSTRUCTION

Two construction methods will be described briefly. These are the geometric and determinant methods. We will solve the same problem using each method. The problem is one of addition given by

$$u + v = w.$$

This problem follows the form

$$f_1(u) + f_2(v) = f_3(w).$$

Let $2 \leqslant u \leqslant 8$ and $4 \leqslant v \leqslant 16$ and, further, let the length of the scales be 12 cm.

Geometric Method

The following rules are used:

1. Place the parallel scales u and v a convenient distance apart.

2. Graduate the scales in accordance with their scale equations

$$x_u = m_u f_1(u) \quad \text{and} \quad x_v = m_v f_2(v).$$

3. Locate the scale for w so that its distance from the u scale is to its distance from the v scale as m_u is to m_v.

4. Graduate the w scale from its scale equation

$$x_w = \frac{m_u m_v}{m_u + m_v} \cdot f_3(w).$$

5. If it is desirable for the scales to begin at u_1, v_1, and w_1, satisfying $f_1(u_1) + f_2(v_1) = f_3(w_1)$, the scale equations will be

$$x_u = m_u \big[f_1(u_2) - f_1(u_1) \big],$$
$$x_v = m_v \big[f_2(v_2) - f_2(v_1) \big],$$
$$x_w = m_w \big[f_3(w_2) - f_3(w_1) \big],$$

where u_2, v_2, and w_2 are the upper limits of the range of each scale.

For the problem at hand, the scales u and v are located conveniently in Fig. 1. Then proceeding, we obtain x_u, x_v, and x_w from rule 5 as

$$m_u = \frac{12}{8-2} = 2 \text{ cm}, \qquad x_u = 2(u-2),$$

$$m_v = \frac{12}{16-4} = 1 \text{ cm}, \qquad x_v = 1(v-4),$$

$$m_w = \frac{(2)(1)}{2+1} = \tfrac{2}{3} \text{ cm}, \qquad x_w = \tfrac{2}{3}(w-6).$$

The value of w_1 is obtained from the equation $w_1 = u_1 + v_1 = 2 + 4 = 6$.

Note that the scale for u begins at 2 (or $u - 2 = 0$) in Fig. 1 and continues with graduations of 2 cm to a height of 12 cm; the v and w scales behave similarly. The w scale is located $\tfrac{2}{3}$ units from the u scale, and $\tfrac{1}{3}$ unit from the v scale according to the $2:1$ ratio m_u / m_v given by rule 3.

Determinant Method

The following rules are used:

1. Place the parallel scales u and v a convenient distance apart.

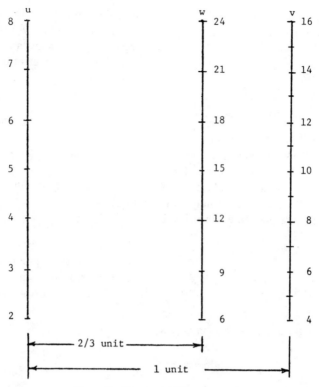

Figure 1   Nomogram for $u + v = w$.

2. State the equations for the nomogram as $f(u, v, w) = 0$.

3. Using elementary row transformations only, develop the "design determinant."

4. If it is desirable for the scales to begin at u_1, v_1, and w_1, satisfying $f_1(u_1) + f_2(v_1) = f_3(w_1)$, modify the design determinant accordingly.

Assume rule **1** has been applied; next, apply rule **2**. Letting $u = x$ and $y = v$, the equations to be satisfied are

$$x - u = 0,$$
$$y - v = 0,$$
$$x + y - 2 = 0.$$

But u and v have to be scaled in the nomogram, so

$$y - m_u u = 0,$$
$$y - m_v v = 0,$$

and $u + v - w = 0$ becomes

$$\frac{x}{m_u} + \frac{y}{m_v} - w = 0.$$

In matrix notation,

$$\begin{bmatrix} 1 & 0 & -m_u u \\ 0 & 1 & -m_v v \\ \dfrac{1}{m_u} & \dfrac{1}{m_v} & -w \end{bmatrix} \begin{bmatrix} x \\ y \\ 1 \end{bmatrix} = \begin{bmatrix} 0 \\ 0 \\ 0 \end{bmatrix}.$$

Now, the determinant of the first array must equal zero to satisfy $f(u, v, w) = 0$ or

$$\begin{vmatrix} 1 & 0 & -m_u u \\ 0 & 1 & -m_v v \\ \dfrac{1}{m_u} & \dfrac{1}{m_v} & -w \end{vmatrix} = 0.$$

Through elementary row transformations only, the design determinant of rule **3** can be developed as

$$\begin{vmatrix} 0 & m_u u & 1 \\ 1 & m_v v & 1 \\ \dfrac{m_u}{m_u + m_v} & \dfrac{m_u m_v}{m_u + m_v} & 1 \end{vmatrix} = 0.$$

As previously, $m_u = 2$ cm and $m_v = 1$ cm.

Further, $u_1 = 2$, $v_1 = 4$, and $w_1 = 6$, so the design determinant is modified as mentioned in rule **4** to yield

$$\begin{vmatrix} 0 & 2(u-2) & 1 \\ 1 & 1(v-4) & 1 \\ \frac{2}{3} & \frac{2}{3}(w-6) & 1 \end{vmatrix} = 0.$$

The third row, first column element, $\frac{2}{3}$, represents the unit distance from u to w. The reader can observe the exact correspondence of the elements in the second column with the result obtained using the geometric approach. The results are the same as in Fig. 1.

Type Forms

1. $f_1(u) + f_2(v) = f_3(w)$.

This type form was seen earlier in the construction of nomograms using the geometric and determinant methods. The type form is greatly extended by using the concept of logarithms. For instance, the sample variance and its logarithm are given, respectively, by

$$S^2 = \sum_{i=1}^{n} \frac{(x_i - \bar{x})^2}{(n-1)},$$

$$2 \log S = \log \left\{ \sum_{i=1}^{n} (x_i - \bar{x})^2 \right\} - \log(n-1),$$

which last corresponds to the preceding type form.

2. $f_1(u) = f_2(v) \cdot f_3(w)$.

Type form **2** is known as the *Z chart*. In standard linear regression* analysis the standard error of the estimate σ_{est} in predicting y scores from x scores is given by

$$\sigma_{est} = \sigma_y \sqrt{1 - r_{xy}^2},$$

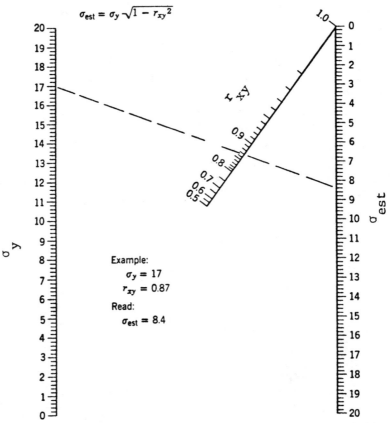

$$\sigma_{est} = \sigma_y \sqrt{1 - r_{xy}^2}$$

Example:
$\sigma_y = 17$
$r_{xy} = 0.87$
Read:
$\sigma_{est} = 8.4$

Figure 2   Standard error of estimate in predicting y scores from x scores. Reproduced with permission from Levens [6].

where r_{xy} is the correlation of x and y scores and σ_y is the standard deviation of y scores. Figure 2 is a Z-chart constructed according to the procedure for this type form. The circular nomogram may be designed for equation type form **2**; it is quite advantageous in dealing with equations whose functions are trigonometric.

$$\textbf{3.} \quad f_1(u) + f_2(v) = f_2(v)/f_3(w).$$

Type form **3** results in a Z-chart very similar to that of the Type form **2**.

$$\textbf{4.} \quad \frac{1}{f_1(u)} + \frac{1}{f_2(v)} = \frac{1}{f_3(w)}.$$

Type form **4** has greater application in optics (the lens formula) and electricity (com-

bination of resistances) than in applied statistics.

$$\textbf{5.} \quad f_1(u) + f_2(v) + f_3(w) + \cdots = f_4(q).$$

Form **5** has application when there are many variables. Its use is facilitated by one or more index lines. A nomogram for the correlation coefficient r follows this form where

$$r = \frac{S_{xy}}{\left[S_{xx} S_{yy} \right]^{1/2}} \; .$$

Here $S_{xx} = \sum_{i=1}^{n}(x_i - \bar{x})^2$, $S_{yy} = \sum_{i=1}^{n}(y_i - \bar{y})^2$, and S_{xy}, the corrected sum of cross-products x_i and y_i, is

$$S_{xy} = \sum_{i=1}^{n} y_i(x_i - \bar{x}).$$

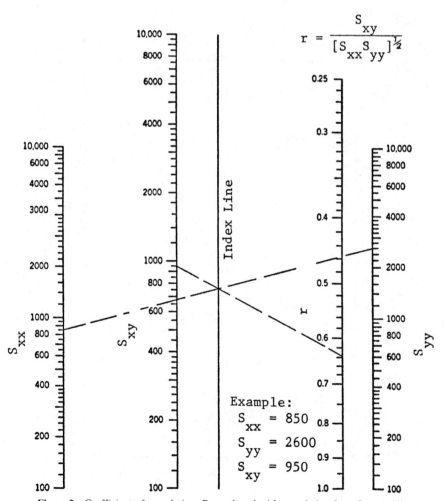

Figure 3 Coefficient of correlation. Reproduced with permission from Levens [6].

The resulting nomogram is shown in Fig. 3. For $S_{xx} = 850$ and $S_{yy} = 2600$, an isopleth crosses the index line as shown. Then another isopleth connecting $S_{xy} = 950$ and the previous mark on the isopleth yields $r = 0.64$.

6. $$\frac{f_1(u)}{f_2(v)} = \frac{f_3(w)}{f_4(q)}.$$

By taking the logarithm of type form **6**, we obtain

$$\log f_1(u) - \log f_2(v) = \log f_3(w) - \log f_4(q),$$

which could be solved using type form **5**. But, in many cases the *proportional chart*, as the resulting nomogram is called, can be

constructed with linear (rather than logarithmic) scales to provide greater accuracy.

Predicted y scores from a regression line can be obtained from

$$y = \left[\frac{S_{xy}}{S_{xx}} \right] x,$$

which, with a slight manipulation, becomes type form (6). Figure 4 is an example of a nomogram for predicting y scores. If $S_{xy} = 100,000$, $S_{xx} = 150,000$, and $x = 50$, the resulting $y = 33.3$.

7. $f_1(u) + f_2(v) = f_3(w)/f_4(q).$

This equation can be solved by a combination of type form **1**,

$$f_1(u) + f_2(v) = K,$$

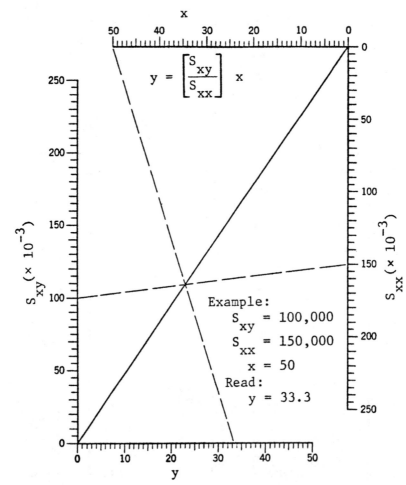

Figure 4 Regression line for predicting scores. Reproduced with permission from Levens [6].

and a Z-chart of type form **2**,

$$K = f_3(w)/f_4(q).$$

However, there are procedures for completion of the nomogram as a unit.

Several other type forms are described in the references. Also described are *net charts*, which make it possible for a nomogram to be constructed such that the solution for a four-variable problem can be solved with one isopleth. No general rules can be stated for constructing nomograms containing a large (more than four) number of variables. However, the procedure is the same in every case. First, resolve the original equation into a number of component equations. Dummy variables that become index scales are created. Usually, there are several ways that a dummy variable can be formed, so the procedure ultimately depends on the nomographer and the nomogram desired.

References

[1] Adams, D. P. (1964). *Nomography: Theory and Application*. Archon, Hamden, CT.

[2] Allcock, H. J., Jones, J. R., and Michel, J. G. L. (1962). *The Nomogram: The Theory and Practical Construction of Computation Charts*, 5th ed. Pitman, New York.

[3] d'Ocagne, M. (1889). *Traité de Nomographie*. Gauthier-Villars, Paris. (A second edition was published in 1921.)

[4] Epstein, L. I. (1958). *Nomography*. Interscience, New York.

[5] Fasal, J. H. (1968). *Nomography*. Ungar, New York.

[6] Levens, A. S. (1959). *Nomography*, 2nd ed. Wiley, New York.

(GAUSS–SEIDEL ITERATION
NEWTON–RAPHSON METHODS
NONLINEAR PROGRAMMING)

JERRY BANKS

NONADDITIVE PROBABILITY

The axiom of additivity is fundamental to the theory of objective probability (*see* AXI-OMS OF PROBABILITY). If $\Pr[E]$ denotes the objective probability with which an event E happens, and if A and B are mutually exclusive events, then

$$\Pr[A \cup B] = \Pr[A] + \Pr[B]. \quad (1)$$

Should we also adopt this axiom in a theory of subjective probability? Bayesians have traditionally contended we should. But in recent decades a number of scholars have developed theories in which subjective probabilities are not required to be additive.

One source of inspiration for the study of nonadditive probability is the betting interpretation of subjective probability. Another is the desire for more flexible modes of probability judgment. After studying the restrictions on nonadditive probabilities suggested by the betting interpretation, we will examine two theories of probability judgment that respect these restrictions: the theory of lower probabilities and the theory of belief functions*.

BETTING INTERPRETATIONS

What does it mean for a person to bet on an event E at odds $p:(1-p)$? It means that both the bettor and an opponent pay money into a bank, the amounts being in the ratio $p:(1-p)$; if E happens the bettor receives the total, whereas if the complementary event $\bar{E}$ happens, the opponent receives the total. If the bettor pays \$p$, then the opponent pays \$(1-p)$, and the total is \$1. So betting \$p$ on E at odds $p:(1-p)$ amounts to paying \$p$ for a contract that returns \$1 if E happens and nothing otherwise.

Let us assume that the happening of E or $\bar{E}$ and the subsequent settling up take place immediately after the bet is made. This means a person will be willing to bet on the sure event at odds $1:0$; to make such a bet is merely to pay money and have it immediately returned. Let us also assume that a person's willingness to bet is independent of the size of the stakes. Let Ω denote the set of possibilities we are considering, so that an event E can be identified with a subset of Ω.

The Basic Betting Interpretation

One way to interpret subjective probabilities as betting rates is to say that having a subjective probability $\Pr[E]$ equal to p means being willing to bet on E at odds $p : (1 - p)$ but at no greater odds—being willing, that is to say, to pay up to $\$p$, but no more, for a contract that returns $\$1$ if E happens. Let us call this the *basic betting interpretation*. What rules does it suggest for the set function $\Pr[\cdot]$?

It suggests first of all that $\Pr[\varnothing] = 0$ and $\Pr[\Omega] = 1$. A person who does not want to give money away will pay nothing for a contract that is sure to return nothing and will pay $\$1$ and no more for a contract that immediately returns $\$1$.

It also suggests that if $A \cap B = \varnothing$, then $\Pr[A \cup B] \geqslant \Pr[A] + \Pr[B]$. Indeed, when a person with subjective probabilities $\Pr[A] = p$ and $\Pr[B] = q$ has paid $\$p$ for a contract that returns $\$1$ if A happens and $\$q$ for a contract that returns $\$1$ if B happens, he has altogether paid $\$(p + q)$ for a contract that pays $\$1$ if $A \cup B$ happens and has thus revealed that his subjective probability for $A \cup B$ is at least $p + q$.

By a slightly more subtle argument, if $A \cup B = \Omega$, then $\Pr[A \cap B] \geqslant \Pr[A] + \Pr[B] - 1$. Indeed, when a person with subjective probabilities $\Pr[A] = p$ and $\Pr[B] = q$ has bet $\$p$ on A and $\$q$ on B, he has altogether paid $\$(p + q)$ for a contract that returns $\$2$ if $A \cap B$ happens and $\$1$ otherwise. Since $\$1$ of the payment is certain to be immediately returned (we may assume that $p + q \geqslant 1$, for otherwise our inequality follows from the fact that $\Pr[A \cap B] \geqslant 0$), we might just as well think of it as not having been paid at all. So we have a payment of $\$(p + q - 1)$ for a contract that returns $\$1$ if $A \cap B$ happens and nothing otherwise, which reveals that the person's subjective probability for $A \cap B$ is at least $p + q - 1$.

The most general conclusion that can be obtained from arguments like these is the following: If by compounding the bets authorized by a person's subjective probabili-

ties $\Pr[A_1], \ldots, \Pr[A_n]$, we find that he is willing to pay $\$r$ for a contract that returns no more than $\$(1 + a)$ if E happens and no more than $\$a$ otherwise, where $a \geqslant 0$, then the person's subjective probability $\Pr[E]$ is at least $r - a$. This can be said more mathematically using indicator functions.

$$\left. \begin{array}{l} \text{If } I_E + aI_\Omega \geqslant \sum a_i I_{A_i}, \\[4pt] \text{where } E, A_1, \ldots, A_n \text{ are subsets} \\[4pt] \text{of } \Omega \text{ and } a, a_1, \ldots, a_n \\[4pt] \text{are nonnegative real numbers,} \\[4pt] \text{then } \Pr[E] \geqslant \sum a_i \Pr[A_i] - a. \end{array} \right\}$$

$$(2)$$

Here is a remarkable fact, independently discovered by P. M. Williams and P. J. Huber: a set function $\Pr[\cdot]$ defined for all subsets of a finite set Ω satisfies (2) and has values $\Pr[\varnothing] = 0$ and $\Pr[\Omega] = 1$ if and only if there exists a nonempty set $\mathcal{M}$ of additive probability measures on Ω such that

$$\Pr[E] = \inf\{\mu[E] \mid \mu \in \mathcal{M}\} \qquad (3)$$

for every subset E of Ω.

It is customary to call a set function of the form (3) a *lower probability function* and to denote it by $P_*[\cdot]$ rather than by $\Pr[\cdot]$. An additive probability measure qualifies, of course, as a lower probability function; this is the case where the set $\mathcal{M}$ has only a single element. But if $\mathcal{M}$ contains more than one element, then $P_*[\cdot]$ will be nonadditive; there will be subsets A and B of Ω such that

$$P_*(A) + P_*(B) < P_*(A \cup B).$$

In particular, there will be at least one subset E of Ω such that $P_*[E] + P_*[\overline{E}] < 1$.

Notice that when $P_*[E] + P_*[\overline{E}] < 1$, there is an interval of values of p, $P_*[E] < p < 1 - P_*[\overline{E}]$, such that the person is unwilling to take either side of a bet of E at odds $p : (1 - p)$. This unwillingness to bet can also be described by saying that the person is willing neither to buy for $\$p$ nor to sell for $\$p$ a contract that pays $\$1$ if E happens. He is unwilling to pay more than $\$P_*[E]$ for the contract, but also unwilling to sell it for less than the larger amount $\$(1 - P_*[E])$.

The Bayesian Betting Interpretation

The Bayesian approach to subjective probability is based on the assumption that a person's preferences form a complete ordering. In terms of betting, this means that if a person is given a choice between $p and a contract that pays $1 if E happens, then he should either prefer the contract (be willing to pay $p for it), prefer the $p (be willing to sell the contract for $p), or be indifferent between the two (be willing to buy or to sell the contract at the price $p). Thus there can be no interval of prices at which he is willing neither to buy nor to sell; his subjective probabilities must satisfy $\Pr[E] = 1 - \Pr[\bar{E}]$ or $\Pr[E] + \Pr[\bar{E}] = 1$. As it turns out, they must also satisfy the more general rule of additivity (1) (*see* AXIOMS OF PROBABILITY).

Though many students of subjective probability have found the Bayesian ideal of completely ordered preferences attractive, the incomplete ordering associated with nonadditive probabilities may sometimes better reflect our limited ability to assign definite monetary values to contracts involving uncertainty*. But where might such nonadditive probabilities come from? How, indeed, are we to make any numerical probability judgments, either additive or nonadditive?

THEORIES OF PROBABILITY JUDGMENT

In order to make probability judgments, we need a standard of comparison. Even when we make merely qualitative probability judgments—practically certain, very probable, fairly probable, etc.—we need more than these mere words; we need a scale of canonical examples, a scale of examples where it is agreed which words are appropriate. Such a scale is necessary to give definite meaning to the words—i.e., to calibrate our use of them.

To make numerical probability judgments, we need a scale of canonical examples where numerical probabilities are agreed on. Given such a scale, we can make probability judgments by comparing our ac-

tual evidence in a particular problem to the scale and choosing from the scale the canonical example that best matches the strength of that evidence.

Different scales of canonical examples produce, of course, different theories of probability judgment.

Lower Probabilities

One way of understanding the Bayesian theory is to say that it uses a scale of canonical examples where the truth is determined according to known chances. For when a Bayesian adopts an additive probability measure $\Pr[\cdot]$, he is saying that his evidence is comparable in strength and significance to knowledge that the truth has been determined by chance*, with the chances governing this determination being given by $\Pr[\cdot]$. An interesting generalization of the Bayesian theory arises when we enlarge this scale of canonical examples to include examples where the truth is determined according to only partially known chances.

We may call this more general theory the *theory of lower probabilities*. Its scale of canonical examples consists of an example for every nonempty set $\mathcal{M}$ of additive probability measures on our set of possibilities Ω. In the example for a given $\mathcal{M}$, it is supposed that we know that the truth has been determined by chance and that the chances are given by one (we do not know which one) of the measures in $\mathcal{M}$. It seems natural in such a case to adopt

$$P_*(E) = \inf\{\,\mu[\,E\,]\,|\,\mu \in \mathcal{M}\,\} \qquad (4)$$

as our subjective probability that the truth is in a given subset E of Ω. Notice that $P_*[\cdot]$ is a lower probability function and hence is compatible with the basic betting interpretation.

In practice the theory of lower probabilities uses a device first proposed by I. J. Good: we make qualitative probability judgments and interpret these judgments as conditions for membership in $\mathcal{M}$. We may, for example, judge that A is more probable than B on our evidence, interpret this by saying that our evidence is like knowing the chance of A is greater than the chance of B, and so require that all the measures in $\mathscr{M}$ give greater probability to A than to B. A qualitative judgment that A would be more probable than B if we knew C for certain may similarly lead to the requirement that all the measures in $\mathscr{M}$ give a greater conditional probability, given C, to A than to B. We may make many qualitative judgments, and $\mathscr{M}$ will consist of all the additive probability measures on Ω that satisfy the resulting conditions.

Since a wide variety of qualitative judgments can be translated into conditions on an additive probability measure, lower probability analyses can take many different forms. But in some cases a lower probability analysis will follow closely the pattern of a Bayesian analysis, and in such a case it will have the character of a check on the sensitivity or robustness of the Bayesian analysis. In the case of an analysis by Bayes' theorem*, for example, the lower probability analysis might consist merely in substituting a range of values for each prior probability and each likelihood, resulting in similar ranges of the posterior probabilities.

Belief Functions

The theory of belief functions* uses canonical examples where chance, instead of determining the truth, determines auxiliary events that affect the meaning or reliability of a message that constitutes our evidence.

Consider, for simplicity, the case where our set of possibilities Ω consists of only two elements: $\Omega = \{\omega_1, \omega_2\}$. We call a set function Bel[$\cdot$] defined for subsets of Ω a *belief function* if Bel[$\emptyset$] = 0, Bel[Ω] = 1, Bel[$\{\omega_1\}$] = p, and Bel[$\{\omega_2\}$] = q, where p and q satisfy $p, q \geqslant 0$ and $p + q \leqslant 1$. We can develop canonical examples for these belief functions by imagining a witness who falls into different modes of behavior randomly and with known chances. Suppose there is a known chance p that he will fall into a mode where he testifies honestly and reliably, a known chance q that he will fall into a mode where

he testifies mendaciously, and a known chance $1 - p - q$ that he will fall into a mode where we do not know how to make any prediction about his behavior. Given testimony from the witness, we might say that there is a chance p that the testimony means what it says, a chance q that it means the opposite, and chance $1 - p - q$ that it means nothing. If the testimony is that ω_1 is true, then this suggests nonadditive probabilities Bel[$\{\omega_1\}$] = p and Bel[$\{\omega_2\}$] = q.

More generally, consider a message of uncertain meaning that bears on a finite set of possibilities Ω that may have more than two elements. For each subset A of Ω, we suppose that there is a chance $m[A]$ that the message's exact true meaning is A, so that for each subset E,

$$\text{Bel}[E] = \sum \{m[A] \mid A \subseteq E\} \quad (5)$$

is the total chance that the message's true meaning implies E. We call any function Bel[$\cdot$] of the form (5) a *belief function*. (We assume that $0 \leqslant m[A] \leqslant 1$, $m[\emptyset] = 0$, and $\sum \{m[A] \mid A \subseteq \Omega\} = 1$.)

Belief functions qualify mathematically as lower probability functions and hence are compatible with the basic betting interpretation. Notice also that if $m[A] = 0$ whenever A contains more than one element, then the belief function Bel[$\cdot$] will be an additive probability function. Thus the theory of belief functions, like the theory of lower probabilities, is a generalization of Bayesian theory.

Central to the theory of belief functions is Dempster's rule for combining belief functions based on independent sources of evidence so as to obtain a belief function based on the total evidence. This rule is obtained by thinking of the chances that affect the meaning or reliability of the messages provided by the different sources of evidence as independent. Consider, for example, two independent witnesses whose reliabilities are assessed as (p_1, q_1) and (p_2, q_2), respectively. If the chances affecting their testimonies are independent, then there is a chance $p_1 + p_2 - p_1 p_2$ that at least one will testify honestly and reliably, a chance $q_1 + q_2 - q_1 q_2$

that at least one will testify mendaciously, and a chance $p_1q_2 + p_2q_1$ that one will testify honestly and reliably while the other testifies mendaciously. Suppose both witnesses testify in favor of ω_1. Their agreement tells us that the event that one testified honestly and reliably and the other mendaciously did not happen, and we must condition on this fact. Thus we obtain conditional chances $c[p_1 + p_2 - p_1p_2 - (p_1q_2 + p_2q_1)]$ for the event that at least one of the testimonies is honest and reliable and $c[q_1 + q_2 - q_1q_2 - (p_1q_2 + p_2q_1)]$ for the event that at least one is mendacious, where $c^{-1} = [1 - (p_1q_2 + p_2q_1)]$. Our belief function based on the total evidence will take these conditional chances as our subjective probabilities for ω_1 and ω_2, respectively. This rule was first given by J. H. Lambert* (1728–1777), though special cases had been given earlier by James Bernoulli* (1654–1705). It was generalized to the case of arbitrary belief functions by A. P. Dempster in 1967.

An important special case of Dempster's rule occurs when one belief function is combined with a second belief function that has $m[A] = 1$ for some particular subset A of Ω. Since the equation $m[A] = 1$ indicates a message that means A with certainty, such combination can be thought of as "conditioning" on A. This approach to conditioning differs from the natural approach for the theory of lower probabilities, which is to condition each of the additive probability measures μ in (4). This difference leads to different approaches to statistical inference.

Literature

The basic betting interpretation for subjective probabilities was first developed by C. A. B. Smith in his "Consistency in Statistical Inference and Decision," (*J. R. Statist. Soc. B*, 1961, Vol. 23, pp. 1–25). For a very elegant exposition, see Peter M. Williams, "Indeterminate Probabilities," in *Formal Methods in the Methodology of Empirical Sciences*, M. Przelecki, K. Szaniawski, and R. Wojciki, eds. (Ossolineum and D. Reidel, Dordrecht, The Netherlands, 1976, pp. 229–

246). See also Peter J. Huber, *Robust Statistics* (Wiley, New York, 1981, Chap. 10), and Robin Giles, "A Logic for Subjective Belief," in *Foundations of Probability Theory, Statistical Inference, and Statistical Theories of Science*, Vol. I, W. L. Harper and C. A. Hooker, eds. (D. Reidel, Dordrecht, The Netherlands, 1976, pp. 41–72).

Twentieth-century work on lower probabilities seems to have begun with Bernard O. Koopman, "The Bases of Probability" (*Bull. Amer. Math. Soc.*, Vol. 46, pp. 763–774, 1940). I. J. Good's most important contributions are in "The Measure of a Nonmeasurable Set," in *Logic, Methodology and Philosophy of Science*, Ernest Nagel, Patrick Suppes, and Alfred Tarski, eds. (Stanford University Press, Stanford, CA, 1962, pp. 319–329). For the role of lower probabilities in checking the robustness of Bayesian analyses, see Huber's book and "Bayesian Inference Using Intervals of Measures," by Lorraine DeRobertis and J. A. Hartigan (*Ann. Statist.*, 1981, Vol. 9, pp. 235–244). For more on the idea of canonical examples, see Glenn Shafer, "Constructive Probability" (*Synthese*, 1981, Vol. 48, pp. 1–60). For related ideas and further references, see Rudolf Beran's "Upper and Lower Risks and Minimax Procedures," in *Proceedings of the Sixth Berkeley Symposium on Mathematical Statistics and Probability*, Vol. I, Lucien M. LeCam, Jerzy Neyman, and Elizabeth Scott, eds. (University of California Press, 1972, pp. 1–16), Patrick Suppes, "The Measurement of Belief" (*J. R. Statist. Soc. B*, 1974, Vol. 36, pp. 160–175), Henry E. Kyburg, Jr., *The Logical Foundations of Statistical Inference* (D. Reidel, Dordrecht, The Netherlands, 1974), Isaac Levi, *The Enterprise of Knowledge* (MIT Press, Cambridge, MA, 1980), and F. J. Giron and S. Rios, "Quasi-Bayesian Behaviour: A More Realistic Approach to Decision Making?," in *Bayesian Statistics*, J. M. Bernardo, M. H. DeGroot, D. V. Lindley, and A. F. M. Smith, eds. (University Press, Valencia, Spain, 1980, pp. 17–37).

The theory of belief functions is developed in Glenn Shafer, *A Mathematical The-*

ory of Evidence (Princeton, University Press, Princeton, NJ, 1976). A summary of A. P. Dempster's earlier formulation is given in his "A Generalization of Bayesian Inference" (*J. R. Statist. Soc. B*, 1968, Vol. 30, pp. 205–247). For references to authors who have raised objections to Dempster's rule, see Shafer's "Constructive Probability," cited earlier. For further historical references, see Shafer's "Non-Additive Probabilities in the Work of Bernoulli and Lambert" (*Arch. History Exact Sci.*, 1978, Vol. 19, pp. 309–370).

Approaches to nonadditive probability not discussed in this article include the theory of "possibility distributions" developed by Lotfi Zadeh [see his "Fuzzy Sets as a Basis for a Theory of Possibility" (*Fuzzy Sets Syst.*, 1978, Vol. 1, pp. 3–28) and FUZZY SET THEORY] and the "Baconian" theory developed by I. J. Cohen in *The Probable and the Provable* (Oxford University Press, London, 1977).

(AXIOMS OF PROBABILITY
BELIEF FUNCTIONS
CHANCE
DEGREES OF BELIEF
FOUNDATIONS OF PROBABILITY
GAMBLING, STATISTICS IN
LAMBERT, JOHANN HEINRICH
UNCERTAINTY
WEIGHT OF EVIDENCE)

GLENN SHAFER

NONCENTRAL BETA DISTRIBUTION

In this article, χ_ν^2 denotes the chi-square distribution* with ν degrees of freedom.

If X_1, X_2 are independent random variables distributed as $\chi_{\nu_1}^2, \chi_{\nu_2}^2$ respectively, then $X_1/(X_1 + X_2)$ has a (central) beta distribution* with parameters $\frac{1}{2}\nu_1, \frac{1}{2}\nu_2$. If X_1 has a noncentral $\chi_{\nu_1}^2$ distribution, with noncentrality parameter λ_1, then the distribution of $X_1/(X_1 + X_2)$ is *noncentral beta* with param-

eters $\frac{1}{2}\nu_1, \frac{1}{2}\nu_2$ and (noncentrality) λ_1. In this case (see NONCENTRAL CHI-SQUARE DISTRIBUTION), X_1 is distributed as a mixture of $\chi_{\nu_1 + 2j}^2$ distributions $(j = 0, 1, \dots)$ with Poisson weights $e^{-\lambda_1/2}(\frac{1}{2}\lambda_1)^j/j!$. Hence the noncentral beta distribution is a mixture of (central) beta distributions with parameters $(\frac{1}{2}\nu_1 + j, \frac{1}{2}\nu_2)$ and the same weights.

If X_2, also, has a noncentral $\chi_{\nu_2}^2$ distribution with noncentrality parameter λ_2, we have a *doubly noncentral beta distribution* with parameters $\frac{1}{2}\nu_1 + j, \frac{1}{2}\nu_2$ and (noncentralities) λ_1, λ_2. This is a doubly infinite mixture of (central) beta distributions with parameters $(\frac{1}{2}\nu_1 + j_1, \frac{1}{2}\nu_2 + j_2)$ and product Poisson weights

$$e^{-(\lambda_1 + \lambda_2)/2}(\tfrac{1}{2}\lambda_1)^{j_1}(\tfrac{1}{2}\lambda_2)^{j_2}/(j_1! \, j_2!).$$

See Hodges [1] and Seber [2] for additional details.

References

[1] Hodges, J. L., Jr. (1955). *Ann. Math. Statist.*, **26**, 648–653.

[2] Seber, G. A. F. (1963). *Biometrika*, **50**, 542–544.

(NONCENTRAL CHI-SQUARE
   DISTRIBUTION
NONCENTRAL *F*-DISTRIBUTION)

NONCENTRAL CHI-SQUARE DISTRIBUTION

Noncentral Chi-square Variate

If $Z_1, Z_2, \dots, Z_\nu$ are ν independently normally distributed random variables, each having zero mean and unit standard deviation, and if $a_1, a_2, \dots, a_\nu$ are ν constants with

$$\lambda = \sum_{i=1}^{\nu} a_i^2, \tag{1}$$

then

$$\chi'^2 = \sum_{i=1}^{\nu} (Z_i + a_i)^2, \tag{2}$$

is a *noncentral chi-square variate*. The probability density function of χ'^2 is the *noncentral chi-square distribution* having ν degrees of freedom and noncentrality parameter λ. The probability density function of χ'^2 has been obtained in a number of different ways. Kerridge's [10] probabilistic method (perhaps the easiest) is in the form

$$p(\chi'^2) = e^{-\lambda/2} \sum_{r=0}^{\infty} \frac{(\tfrac{1}{2}\lambda)^r e^{-\chi'^2/2}(\chi'^2)^{\nu/2+r-1}}{2^{\nu/2+r}r!\,\Gamma(\tfrac{1}{2}\nu+r)}$$

$$0 < \chi'^2 < \infty. \quad (3)$$

An alternative expression with some computational advantages was obtained by Tiku [17] in terms of Laguerre polynomials (*see* LAGUERRE SERIES):

$$p(\chi'^2) = \left\{ \sum_{r=0}^{\infty} \frac{\Gamma(m)}{\Gamma(m+r)} \left(-\tfrac{1}{2}\lambda\right)^r L_r^{(m)}(X) \right\}$$

$$\times p_m(X),$$

$$X = \tfrac{1}{2}\chi'^2, \quad m = \tfrac{1}{2}\nu, \quad (4)$$

where $p_m(X) = \{1/\Gamma(m)\}e^{-X}X^{m-1}$, $0 < X < \infty$, is the gamma distribution and

$$L_r^{(m)}(X) = \frac{1}{r!} \sum_{j=0}^{r} (-1)^j \binom{r}{j} \frac{\Gamma(m+r)}{\Gamma(m+j)} X^j,$$

$$r = 0, 1, 2, \ldots, \quad (5)$$

are the Laguerre polynomials [15] associated with the weight function $p_m(X)$.

Moment-Generating Function

The moment-generating function* of χ'^2 can be easily obtained from (3) or (4) and is given by [12, p. 206; 17, p. 416]

$$M_\theta(\chi'^2) = (1 - 2\theta)^{-\nu/2}\exp\left\{ \frac{\lambda\theta}{1 - 2\theta} \right\}. \quad (6)$$

The mean and the second, third, and fourth central moments of χ'^2 are given by

$$\mu_1' = \nu + \lambda,$$
$$\mu_2 = 2(\nu + 2\lambda),$$
$$\mu_3 = 8(\nu + 3\lambda), \quad (7)$$
$$\mu_4 = 48(\nu + 4\lambda) + 4(\nu + 2\lambda)^2.$$

The jth cumulant of χ'^2 is given by

$$\kappa_j(\chi'^2) = 2^{j-1}(\nu + j\lambda)(j - 1)! \quad (8)$$

Reproductivity

The noncentral chi-square distribution is reproductive under convolution*, that is, if $\chi_i'^2$, $i = 1, 2, \ldots, k$, are k independent noncentral chi-square variates having degrees of freedom ν_i and noncentrality parameters λ_i, respectively, then

$$\chi'^2 = \sum_{i=1}^{k} \chi_i'^2 \quad (9)$$

has a noncentral chi-square distribution having $\nu = \sum_{i=1}^{k}\nu_i$ degrees of freedom and noncentrality parameter $\lambda = \sum_{i=1}^{k}\lambda_i$; this follows from (6). Note that the degrees of freedom and the noncentrality parameters are additive under convolution.

Decomposition

The necessary and sufficient condition for the k noncentral chi-square variates on the right-hand side of (9) to be distributed independently is that $\nu = \sum_{i=1}^{k}\nu_i$ and $\lambda = \sum_{i=1}^{k}\lambda_i$ (see, e.g., [16]). Note that $E(\chi_i'^2) = \nu_i + \lambda_i$, $i = 1, 2, \ldots, k$.

Probability Integral

The probability integral $P(\chi_0'^2) = P(\chi'^2 \geqslant \chi_0'^2)$ is given by

$$P(\chi_0'^2) = e^{-\lambda/2} \sum_{r=0}^{\infty} \frac{(\tfrac{1}{2}\lambda)^r}{r!} Q(\chi_0'^2 \mid \nu + 2r);$$

$$(10)$$

$$Q(\chi_0'^2 \mid \nu + 2r)$$

$$= \frac{1}{2^{\nu/2+r}\Gamma(\tfrac{1}{2}\nu + r)} \int_{\chi_0'^2}^{\infty} e^{-u}u^{\nu/2+r-1}\,du$$

is Hartley and Pearson's [3] central-χ^2 probability integral. The computation of (10) is somewhat involved. However, an alternative expression for the probability integral of χ'^2

is given by [17, p. 417]:

$$P(\chi_0'^2) = P_m(X_0) + \sum_{r=1}^{\infty} P_{(r)}(X_0), \quad (11)$$

$$P_m(X_0) = Q(\chi_0^2 \mid \nu); \quad \nu = 2m, \chi_0^2 = 2X_0.$$

Expression (11) is easy to compute, since

$$P_{(0)}(X_0) = 0,$$

$$P_{(1)}(X_0) = \left(\tfrac{1}{2}\lambda\right)\left\{ e^{-X_0}[X_0^m/\Gamma(m+1)]\right\}, \quad (12)$$

$$P_{(r)}(X_0)$$

$$= -\frac{\tfrac{1}{2}\lambda(m+2r-3-X_0)P_{(r-1)}(X_0)}{r(m+r-1)}$$

$$+ \frac{\left(\tfrac{1}{2}\lambda\right)^2(r-2)P_{(r-2)}(X_0)}{r(r-1)(m+r-1)} \quad r = 2, 3, \ldots$$

and the series (11) converges to the true values rapidly, especially for small values of λ. The first term $P_m(X_0)$ is the type I error α of the chi-square test* and is prespecified in most applications, usually as $\alpha = 0.01$ or 0.05.

Numerous tables of the values of the power $1 - \beta = P(\chi_0'^2)$ of the chi-square test are available; the most extensive is due to Haynam et al. [4], who also provide tables of the values of the noncentrality parameter λ for numerous values of ν, α and $1 - \beta$.

A number of approximations to the noncentral chi-square distribution are available (see Johnson and Kotz [6, pp. 139–143]). The more prominent ones perhaps are the following.

Cube-Root Normal Approximation

From a cube-root transformation of a central chi-square variate, Abdel-Aty [1] obtained the normal approximation

$$P(\chi'^2 \geqslant \chi_0'^2) \simeq \frac{1}{\sqrt{2\pi}} \int_{z_0}^{\infty} e^{-z^2/2}dz, \quad (13)$$

where

$$z_0 = \frac{\left(\dfrac{\chi_0'^2}{\nu + \lambda}\right)^{1/3} + \dfrac{2(\nu + 2\lambda)}{9(\nu + \lambda)^2} - 1}{\sqrt{\left\{\dfrac{2(\nu + 2\lambda)}{9(\nu + \lambda)^2}\right\}}}. \quad (14)$$

Sankaran [14] has given a number of approximations similar to (13) and the one based on $\{\chi'^2/(\nu + \lambda)\}^h$ is particularly accurate, although not as accurate as Pearson's [13] three-moment approximation.

Two-Moment Approximation

By equating the first two moments on both sides of $\chi^2 = \chi'^2/\rho$ (χ^2 being a central chi-square variate having a degree of freedom), Patnaik [12] obtained the approximation

$$P(\chi'^2 \geqslant \chi_0'^2) \simeq \frac{1}{\Gamma(f)} \int_{X_0}^{\infty} e^{-x}x^{f-1}dx,$$

$$f = \tfrac{1}{2}a, \quad X_0 = \tfrac{1}{2}\chi'^2/\rho, \quad (15)$$

$$a = (\nu + \lambda)^2/(\nu + 2\lambda),$$

$$\rho = (\nu + 2\lambda)/(\nu + \lambda).$$

Three-Moment Approximation

By equating the first three moments on both sides of $\chi^2 = (\chi'^2 + c)/h$ (χ^2 being a central chi-square variate having b degrees of freedom), Pearson [13] obtained the approximation

$$P(\chi'^2 \geqslant \chi_0'^2) \simeq \frac{1}{\Gamma(g)} \int_{y_0}^{\infty} e^{-y}y^{g-1}dy,$$

$$g = \tfrac{1}{2}b, \quad y_0 = \tfrac{1}{2}(\chi_0'^2 + c)/h, \quad (16)$$

$$b = (\nu + 2\lambda)^3/(\nu + 3\lambda)^2, \quad c = \lambda^2/(\nu + 3\lambda),$$

$$h = (\nu + 3\lambda)/(\nu + 2\lambda).$$

Expressions (14), (15), or (16) can also be used to obtain approximations to the percentage points of χ'^2.

Comparisons between the preceding approximations and others have been carried out (see, e.g. Johnson [5], Pearson [13], Johnson and Kotz [6, p. 142]); the three-moment approximation turns out to be the most accurate. Approximations more accurate than (16) can, of course, be obtained by fitting Pearson curves (see PEARSON'S DISTRIBUTIONS) to the first four moments of χ'^2; the details and tables of percentage points are given by Johnson et al. [8].

APPLICATIONS

The noncentral chi-square distribution has many applications, in analysis of variance*, in testing hypotheses, etc. We quote only three examples here; see also Boyle [2].

Example 1. In the one-way classification* for analysis of variance, the mathematical model is

$$x_{ij} = \mu + b_i + e_{ij},$$

$$j = 1, 2, \ldots, n, \quad i = 1, 2, \ldots, k, \quad (17)$$

$$\sum_{i=1}^{k} b_i = 0.$$

e_{ij} are assumed to be independently and identically distributed normal $N(0, 1)$. The total sum of squares has the decomposition

$$\sum_{i=1}^{k} \sum_{j=1}^{n} (x_{ij} - \bar{x}..)^2 = n \sum_{i=1}^{k} (\bar{x}_{i.} - \bar{x}..)^2$$

$$+ \sum_{i=1}^{k} \sum_{j=1}^{n} (x_{ij} - \bar{x}_{i.})^2,$$

$$(18)$$

$$\left(\bar{x}_{i.} = \sum_{j=1}^{n} x_{ij}/n, \bar{x}.. = \sum_{i=1}^{k} \sum_{j=1}^{n} x_{ij}/nk \right).$$

These sums of squares are all essentially of the same form as (2), the last one having its noncentrality parameter zero (a central chi-square variate). Their expected values are $(nk - 1) + n\sum_i (b_i/\sigma)^2$, $(k - 1) + n\sum_i (b_i/\sigma)^2$, and $k(n - 1)$, respectively. Since the degrees of freedom and the noncentrality parameters in the last two expected values add up to the corresponding values in the first one, the two components on the right-hand side of (18) are independently distributed.

Example 2. In the multinomial distribution* with probability mass

$$\frac{n!}{n_1! \, n_2! \ldots n_k!} \pi_1^{n_1} \pi_2^{n_2} \ldots \pi_k^{n_k},$$

$$(19)$$

$$\sum_{i=1}^{k} \pi_i = 1, \qquad \sum_{i=1}^{k} n_i = n,$$

suppose that one wants to test the null hypothesis $H_0 : \pi_i = \pi_{i0}$, $i = 1, 2, \ldots, k$; $\sum_{i=1}^{k} \pi_{i0} = 1$. The test usually employed is the chi-square test* based on the statistic (Kendall and Stuart [9, p. 97])

$$\chi^2 = \sum_{i=1}^{k} \frac{(n_i - n\pi_{i0})^2}{n\pi_{i0}}. \qquad (20)$$

The null distribution of χ^2 is approximately central chi-square having degrees of freedom $v = k - 1$. Large values of χ^2 lead to the rejection of H_0. The nonnull (H_0 not true) distribution of χ^2 is noncentral chi-square having degrees of freedom $v = k - 1$ and noncentrality parameter

$$\lambda = n \sum_{i=1}^{k} \frac{(\pi_i - \pi_{i0})^2}{\pi_{i0}}. \qquad (21)$$

The power of the preceding χ^2-test is, therefore, given by the noncentral chi-square distribution.

Example 3. Let $N_p(\mu, \Sigma)$ denote a p-variate multinormal distribution* with mean vector μ and variance–covariance matrix Σ. Given a random sample X_i, $i = 1, 2, \ldots, n$, of size n from this distribution, one wants to test the null hypothesis $H_0 : \mu = 0$. The test usually employed is based on the well-known Hotelling T^{2*} statistic

$$T^2 = n\bar{X}' \mathscr{S}^{-1}\bar{X}, \qquad (22)$$

where $\bar{X}'$ is the sample mean vector and $\mathscr{S}$ is the sample variance–covariance matrix. Large values of T^2 lead to the rejection of H_0. The asymptotic (n tends to infinity) distribution of T^2 is noncentral chi-square with degrees of freedom p and noncentrality parameter $n\delta^2$; $\delta^2 = \mu'\Sigma^{-1}\mu$.

There are situations when one is interested in working out sample estimates of the noncentrality parameter λ. Several such estimates are given and discussed by Neff and Strawderman [11].

The noncentral chi-square distribution generalizes to bivariate and multivariate noncentral chi-square distributions; see Johnson and Kotz [7] for example.

References

[1] Abdel-Aty, S. H. (1954). *Biometrika*, **41**, 538–40. (Gives various Cornish–Fisher-type approximations).

[2] Boyle, P. (1978). *Scand. Actuarial J.*, 108–11. (Establishes a connection between the Poisson exponential model and the noncentral chi-square.)

[3] Hartley, H. O. and Pearson, E. S. (1950). *Biometrika*, **37**, 313–25. (Gives an extensive tabulation of the probability integral of central chi-square.)

[4] Haynam, G. E., Govindarajulu, Z., and Leone, F. C. (1973). *Selected Tables in Mathematical Statistics*, Vol. I. American Mathematical Society, Providence, RI, pp. 2–78. (Gives an extensive tabulation of the noncentral chi-square probability integral.)

[5] Johnson, N. L. (1959). *Biometrika*, **46**, 352–363. (Establishes a connection between the noncentral chi-square and the Poisson and compares various approximations.)

[6] Johnson, N. L. and Kotz, S. (1970). *Continuous Univariate Distributions*, Vol. 2. Wiley, New York. (An exceptionally fine reference source.)

[7] Johnson, N. L. and Kotz, S. (1972). *Continuous Multivariate Distributions*. Wiley, New York. (Another excellent reference source.)

[8] Johnson, N. L., Nixon, E., Amos, D. E., and Pearson, E. S. (1963). *Biometrika*, **50**, 459–98. (Gives an extensive tabulation of the percentage points of Pearson family of distributions.)

[9] Kendall, M. G. and Stuart, A. (1973). *The Advanced Theory of Statistics*, Vol. 2. Charles Griffin, London. (An excellent treatise on statistical inference.)

[10] Kerridge, D. (1965). *Aust. J. Statist.*, **7**, 37–39. (Gives a very interesting probabilistic derivation.)

[11] Neff, N. and Strawderman, E. (1976). *Commun. Statist. A*, **5**, 65–76. (Evaluates the mean square errors of various estimators of the noncentrality parameter.)

[12] Patnaik, P. B. (1949). *Biometrika*, **36**, 202–232. (Points out some interesting geometrical features.)

[13] Pearson, E. S. (1959). *Biometrika*, **46**, 364. (Studies the accuracy of the three-moment chi-square approximation.)

[14] Sankaran, M. (1963). *Biometrika*, **50**, 199–204. (Gives various power-transformations for normalizing the noncentral chi-square.)

[15] Szegő, G. (1959). *Orthogonal Polynomials*. American Mathematical Society Colloquium Publications, New York. (An excellent treatise on orthogonal polynomials.)

[16] Tan, W. Y. (1977). *Canad. J. Statist.*, **5**, 241–250. (Contains some interesting characterizations.)

[17] Tiku, M. L. (1965). *Biometrika*, **52**, 415–427. (Uses Laguerre polynomials to represent the noncentral chi-square distribution.)

(CHI-SQUARE DISTRIBUTION
CHI-SQUARE TESTS
LAGUERRE SERIES)

M. Tiku

NONCENTRAL *F*-DISTRIBUTION

Noncentral *F* Ratio

If $Z_1, Z_2, \ldots, Z_{v_1 + v_2}$ are $v_1 + v_2$ independently normally distributed random variables, each having zero mean and unit standard deviation and if $a_1, a_2, \ldots, a_{v_1}$ are v_1 constants with

$$\lambda = \sum_{i=1}^{v_1} a_i^2,$$

then the probability density function of

$$\chi_1'^2 = \sum_{i=1}^{v_1} (Z_i + a_i)^2 \qquad (1)$$

is the *noncentral chi-square* distribution* having v_1 degrees of freedom and noncentrality parameter λ. The noncentral *F* ratio is defined, for $0 < F' < \infty$, as

$$F' = \frac{v_2 \sum_{i=1}^{v_1}(Z_i + a_i)^2}{v_1 \sum_{i=v_1+1}^{v_1+v_2} Z_i^2} = \frac{\chi_1'^2 / v_1}{\chi_2^2 / v_2}. \qquad (2)$$

$(v_1/v_2)F'$ is the ratio of a noncentral chi-square variate to an independent chi-square variate. The probability density function $p(F')$ of F' is the *noncentral F-distribution* with degrees of freedom v_1 and v_2 and noncentrality parameter λ and is given by [2]

$$p(F') = e^{-\lambda/2} \sum_{r=0}^{\infty} \frac{\dfrac{\left(\frac{1}{2}\lambda\right)^r}{r!} \left(\dfrac{v_1}{v_2}\right)^{v_1/2 + r}}{B\left(\frac{1}{2}v_1 + r, \frac{1}{2}v_2\right)}$$

$$\times \frac{(F')^{v_1/2 + r - 1}}{\left(1 + \dfrac{v_1}{v_2}F'\right)^{(v_1+v_2)/2 + r}}. $$

$$(3)$$

An alternative expression with some computational advantages was derived by Tiku [21] from a Laguerre series* expansion of the noncentral chi-square distribution:

$$p(F')$$

$$= \sum_{r=0}^{\infty} \frac{(-\frac{1}{2}\lambda)^r}{r!} \left\{ \sum_{j=0}^{r} (-1)^j \binom{r}{j} p(F'; \tfrac{1}{2}v_1 + j, \tfrac{1}{2}v_2) \right\}, \tag{4}$$

$$p(F'; \tfrac{1}{2}v_1 + j, \tfrac{1}{2}v_2)$$

$$= \frac{\left(\frac{v_1}{v_2}\right)^{v_1/2+j}}{B(\tfrac{1}{2}v_1 + j, \tfrac{1}{2}v_2)} \frac{(F')^{v_1/2+j-1}}{\left(1 + \frac{v_1}{v_2} F'\right)^{(v_1+v_2)/2+j}} .$$

Moments of *F'*

The first four moments of F' are given explicitly by Pearson and Tiku [14, pp. 177–178]; if $l = \lambda/v_1$, the mean and variance are

$$\mu_1' = \{v_2/(v_2 - 2)\}(1 + l),$$

$$\mu_2 = \frac{2v_2^2(v_1 + v_2 - 2)}{v_1(v_2 - 2)^2(v_2 - 4)}$$

$$\cdot \left(1 + 2l + \frac{v_1 l^2}{v_1 + v_2 - 2}\right).$$

Probability Integral

The probability integral $P(F_0') = P(F' \geqslant F_0') = \int_{F_0'}^{\infty} p(F') \, dF'$ is given by [20]

$$P(F_0') = e^{-\lambda/2} \sum_{r=0}^{\infty} \frac{(\frac{1}{2}\lambda)^r}{r!} I_{u_0}(\tfrac{1}{2}v_2, \tfrac{1}{2}v_1 + r),$$

$$u_0 = 1 \Big/ \left\{ 1 + \frac{v_1}{v_2} F_0' \right\}; \tag{5}$$

$$I_u(a, b) = \int_0^u x^{a-1}(1 - x)^{b-1} \, dx / B(a, b)$$

is Karl Pearson's [12] incomplete beta-function*. The computation of (5) is rather involved; see Tang [20] and Patnaik [11]. However, an alternative expression for $P(F_0')$ is [21, p. 421]

$$P(F_0') = I_{u_0}(\tfrac{1}{2}v_2, \tfrac{1}{2}v_1) + \sum_{r=1}^{\infty} P_r(u_0); \tag{6}$$

$$P_r(u_0) = \frac{1}{r!} I(\tfrac{1}{2}\lambda)^{r-1} \Delta^{r-1}(T_1),$$

$$r = 1, 2, 3, \ldots; \tag{7}$$

$$I = (\lambda/v_1)\left\{ u_0^{v_2/2}(1 - u_0)^{v_1/2} / B(\tfrac{1}{2}v_2, \tfrac{1}{2}v_1) \right\}$$

is a multiple of the beta ordinate; and

$$T_1 = 1, \qquad T_2 = T_1 \frac{v_1 + v_2}{v_1 + 2}(1 - u_0),$$

$$T_3 = T_2 \frac{v_1 + v_2 + 2}{v_1 + 4}(1 - u_0), \ldots; \tag{8}$$

Δ is the usual difference operator (*see* FINITE DIFFERENCES, CALCULUS OF), i.e., $\Delta(T_i) = T_{i+1} - T_i$. The computation of (6) is straightforward; one has simply to prepare a difference table with $T_1, T_2, T_3, \ldots$ as entries, pick out the leading differences, multiply them by a multiple of the constant factor I, and add them to $I_{u_0}(\tfrac{1}{2}v_2, \tfrac{1}{2}v_1)$. This process of computation can be carried out very easily on a computer and the series (6) converges to the true values rapidly, especially for small values of λ. In the context of analysis of variance*, $I_{u_0}(\tfrac{1}{2}v_2, \tfrac{1}{2}v_1)$ is the significance level (type I error) α of the F-test* and is prespecified, usually $\alpha = I_{u_0}(\tfrac{1}{2}v_2, \tfrac{1}{2}v_1) = 0.05$ or 0.01.

For $k = \tfrac{1}{2}v_2$, an integer (i.e., v_2 an even integer) $P(F_0')$ can be expressed in a closed form [17; 21, p. 426]:

$$P(F_0')$$

$$= 1 - e^{-\lambda u_0/2}(1 - u_0)^m \sum_{j=0}^{k-1} u_0^j L_j^{(m)}(-w_0),$$

$$m = \tfrac{1}{2}v_1, \tag{9}$$

in terms of Laguerre polynomials: $w_0 = \tfrac{1}{2}\lambda(1 - u_0)$. Since $L_0^{(m)}(x) = 1$, $L_1^{(m)}(x) = m - x$, and [16, p. 29]

$$jL_j^{(m)}(x) = (m + 2j - 2 - x)L_{j-1}^{(m)}(x)$$

$$- (m + j - 2)L_{j-2}^{(m)}(x), \tag{10}$$

the expression (9) is easy to compute.

Amos [1] gives a different set of equations for $P(F_0')$, but these seem to be somewhat involved computationally.

From (5), (6), or (9), several authors have tabulated the values of $1 - \beta = P(F_0')$; in the context of analysis of variance, $1 - \beta$ is the power (β is the type II error) of the *F*-test. Fox [3], Pearson and Hartley [13], and Odeh and Fox [10] give charts of the power $1 - \beta$ of the *F*-test for numerous values of α and a wide range of values of the degrees of freedom v_1 and v_2 and the noncentrality parameter λ. For a description of these tables and charts, see Johnson and Kotz [7, pp. 193–194], or Odeh and Fox [10, p. 7]

Several approximations that provide reasonably accurate values for $P(F_0')$ and are not difficult to compute have been developed. Some of the prominent ones are the following:

Normal Approximation

From a cube-root transformation of a chi-square variate, Laubscher [8] and Severo and Zelen [18] obtained the normal approximation

$$P(F_0') \simeq \frac{1}{\sqrt{2\pi}} \int_{z_0}^{\infty} \exp\left\{-\tfrac{1}{2}z^2\right\} dz, \quad (11)$$

where

$$z_0 = \frac{\left(\dfrac{v_1 F_0'}{v_1 + \lambda}\right)^{1/3}\left(1 - \dfrac{2}{9v_2}\right) - \left\{1 - \dfrac{2(v_1 + 2\lambda)}{9(v_1 + \lambda)^2}\right\}}{\left\{\dfrac{2(v_1 + 2\lambda)}{9(v_1 + \lambda)^2} + \dfrac{2}{9v_2}\left(\dfrac{v_1 F_0'}{v_1 + \lambda}\right)^{2/3}\right\}^{1/2}}$$

Two-Moment Approximation

By equating the first two moments on both sides of $F = F'/\rho$ (F being a central F variate having degrees of freedom a and v_2), Patnaik [11] obtained the approximation

$$P(F_0') \simeq I_{x_0}\left(\tfrac{1}{2}v_2, \tfrac{1}{2}a\right), \quad (12)$$

where $x_0 = 1/\{1 + (a/v_2)(F_0'/\rho)\}$, $\rho = (v_1 + \lambda)/v_1$, and $a = (v_1 + \lambda)^2/(v_1 + 2\lambda)$.

Three-Moment Approximation

By equating the first three moments on both sides of $F = (F' + c)/h$ (F being a central F variate having degrees of freedom b and v_2), Tiku [21] obtained the approximation

$$P(F_0') \simeq I_{y_0}\left(\tfrac{1}{2}v_2, \tfrac{1}{2}b\right), \quad (13)$$

where $y_0 = 1/\{1 + (b/v_2)(F_0' + c)/h\}$. The values of the constants b, h, and c are given by

$$b = \frac{1}{2}(v_2 - 2)\left\{\sqrt{\left(\frac{E}{E - 4}\right) - 1}\right\},$$

$$h = \frac{b}{v_1}\frac{1}{(2b + b_2 - 2)}\frac{H}{K},$$

$$c = \frac{v_2}{(v_2 - 2)}\left\{h - \frac{v_1 + \lambda}{v_1}\right\};$$

$$H = 2(v_1 + \lambda)^3 + 3(v_1 + \lambda)(v_1 + 2\lambda)(v_2 - 2)$$
$$+ (v_1 + 3\lambda)(v_2 - 2)^2,$$

$$K = (v_1 + \lambda)^2 + (v_2 - 2)(v_1 + 2\lambda),$$

$$E = H^2/K_3.$$

Tiku [22] compared various approximations and concluded that the three-moment approximation (13) is the most accurate. However, Tiku and Yip [25] have given a four-moment approximation similar to (13) and have shown that although considerably more difficult to compute than (13), it is more accurate. For example, they have given the following approximate values of $\beta = 1 - P(F_0')$, $\alpha = 0.01$, $v_1 = 3$, and $\phi = \sqrt{\{\lambda/(v_1 + 1)\}} = 1.0$:

v_2	True	Four-Moment	Three-Moment
12	0.9093	0.9094	0.9096
20	0.8868	0.8870	0.8875
30	0.8725	0.8728	0.8735
60	0.8558	0.8562	0.8571

Mudholkar and Chaubey [9] give Edgeworth-type approximations for $P(F_0')$.

The noncentral *F*-distribution has many applications; see Odeh and Fox [10], Genizi and Soller [4], and Ito [6]. We provide only two examples here.

Example 1. In a certain experiment $k(= 6)$ treatments are being tested and the assumed model is $x_{ij} = \mu + t_i + e_{ij}$ $(i = 1, 2, \ldots, k; j = 1, 2, \ldots, n)$; the treatment effects t_i are constants and $\sum_i t_i = 0$. The random errors e_{ij} are assumed to be independent and normal $N(0, \sigma^2)$. To test the null hypothesis $H_0 : t_i = 0$ for all $i = 1, 2, \ldots, k$, the test usually employed is based on the statistic

$$F = \mathscr{S}^2_{\text{treat}} / \mathscr{S}^2_{\text{error}} ; \tag{14}$$

$$\mathscr{S}^2_{\text{treat}} = n\sum_i (\bar{x}_{i.} - \bar{x}_{..})^2 / (k - 1),$$

$$\mathscr{S}^2_{\text{error}} = \sum_i \sum_j (x_{ij} - \bar{x}_{i.})^2 / \{k(n - 1)\}.$$

Large values of F lead to the rejection of H_0. The null distribution of F is central-F with degrees of freedom $\nu_1 = k - 1$ $(k = 6)$ and $\nu_2 = k(n - 1)$. The nonnull (H_0 not true) distribution of F is noncentral F with degrees of freedom $\nu_1 = 5$, $\nu_2 = 6(n - 1)$, and noncentrality parameter $\lambda = n\sum_i (t_i / \sigma)^2$; the power of the test is $1 - \beta = P(F'_0)$. Suppose that the type I error $\alpha = 0.01$ and one wants to determine n such that the power $1 - \beta$ is at least 0.975 when $\sum_i (t_i / \sigma)^2 = 4$, that is, when the mean of the squares of the effects of the treatments is two-thirds of the variance of the random errors. Here $\phi = \sqrt{\{\lambda / (\nu_1 + 1)\}} = \sqrt{(2n/3)}$. From Tiku's [23, p. 530] tables, one obtains the following values of $1 - \beta$ by two-way linear interpolation ($\alpha = 0.01$):

n	ν_2	ϕ	$1 - \beta$
8	42	2.309	0.9546
9	48	2.449	0.9824

We want $1 - \beta \geqslant 0.975$; by linear interpolation

$$n = 8 + (0.0204 / 0.0278)(9 - 8) = 8.734$$

for the power $1 - \beta$ to be equal to 0.975. Therefore, n has to be greater than or equal to 9; see also Guenther [5] who uses approximations (12) and (13) to determine the sample size n and arrives at the same value.

In Example 1, we assumed that the treatment effects t_i are constants. However, there are situations when t_i are random variables. Tan and Wong [19] have studied the distribution of the corresponding F ratio.

Example 2. Let $\mathbf{X}_i$, $i = 1, 2, \ldots, n$, be a random sample from a p-variate multinormal distribution* $N_p(\boldsymbol{\mu}, \boldsymbol{\Sigma})$; $\boldsymbol{\mu}$ and $\boldsymbol{\Sigma}$ being the mean vector and the variance–covariance matrix, respectively. To test the null hypothesis $\boldsymbol{\mu} = \mathbf{0}$, the test usually employed is based on the Hotelling T^{2*} statistic

$$T^2 = n\bar{\mathbf{X}}' \mathscr{S}^{-1} \bar{\mathbf{X}} ; \tag{15}$$

$\bar{\mathbf{X}}'$ is the sample mean vector and $\mathscr{S}$ is the sample variance–covariance matrix. Large values of T^2 lead to the rejection of the null hypothesis. The nonnull ($\boldsymbol{\mu} \neq \mathbf{0}$) distribution of $\{(n - p)/[p(n - p + 1)]\} T^2$ is noncentral F with degrees of freedom $\nu_1 = p$ and $\nu_2 = n - p$ and noncentrality parameter $n\delta^2$; $\delta^2 = \boldsymbol{\mu}' \boldsymbol{\Sigma}^{-1} \boldsymbol{\mu}$. The power of this test is, therefore, given by the probability integral of the noncentral F-distribution.

The noncentral F generalizes into *doubly noncentral F*, and the latter is defined as

$$F'' = \left(\frac{\chi_1'^2}{\nu_1} \right) \Big/ \left(\frac{\chi_2'^2}{\nu_2} \right) ; \tag{16}$$

$\chi_1'^2$ is a noncentral chi-square variate with degrees of freedom ν_1 and noncentrality parameter λ_1, and $\chi_2'^2$ is an independent noncentral chi-square variate with degrees of freedom ν_2 and noncentrality parameter λ_2. The distribution of F'' is the *doubly noncentral F-distribution* and is discussed in detail by Price [15] and Tiku [24]; it has numerous applications in the context of analysis of variance, information theory*, etc. [24]. Tables of the probability integral of doubly noncentral F are given by Tiku [24].

References

[1] Amos, D. E. (1976). *Commun. Statist. A*, **5**, 261–281. (Gives recurrence relations in terms of a special doubly noncentral *t*-distribution.)

[2] Fisher, R. A. (1928). *Proc. R. Soc. Lond. A*, **121**, 654–673. (Offers the derivation of the distribution for the first time.)

[3] Fox, M. (1956). *Ann. Math. Statist.*, **27**, 484–497. (Provides charts of the power of *F*-test.)

[4] Genizi, A. and Soller, M. (1979). *J. Statist. Plan. Infer.*, **3**, 127–134. (Gives interesting applications of a mixture of two noncentral *F*-distributions.)

[5] Guenther, W. C. (1979). *Amer. Statist.*, **33**, 209–210. (Illustrates the usefulness of the Patnaik and Tiku approximations.)

[6] Itô, P. K. (1980). *Handbook of Statistics*, Vol. 1, P. R. Krishnaiah, ed., North-Holland, Amsterdam. (An excellent review paper on ANOVA and MANOVA procedures, which often give rise to the noncentral *F*.)

[7] Johnson, N. L. and Kotz, S. (1970). *Continuous Univariate Distributions*, Vol. 2. Houghton Mifflin, Boston. (An exceptionally fine reference source.)

[8] Laubscher, N. F. (1960). *Ann. Math. Statist.*, **31**, 1105–1112. (Discusses several normalizing transformations.)

[9] Mudholkar, G. S. and Chaubey, Y. P. (1976). *Technometrics*, **18**, 351–358. (Develops an Edgeworth series expansion based on a cube-root normalizing transformation.)

[10] Odeh, R. E. and Fox, M. (1975). *Sample Size Choice*. Marcel Dekker, New York. (Contains charts and numerous illuminating examples.)

[11] Patnaik, P. B. (1949). *Biometrika*, **36**, 202–232. (Derives the first four moments about the origin and investigates the accuracy of a two-moment approximation.)

[12] Pearson, Karl. (1934). *Tables of the Incomplete Beta Function*. Cambridge University Press, Cambridge, England. (Has an extensive tabulation of the incomplete beta.)

[13] Pearson, E. S. and Hartley, H. O. (1972). *Biometrika Tables for Statisticians*, Vol. II. Cambridge University Press, Cambridge, England. (A collection of very useful tables.)

[14] Pearson, E. S. and Tiku, M. L. (1970). *Biometrika*, **57**, 175–179. (Studies the relationship between central and noncentral *F*-distributions.)

[15] Price, R. (1964). *Biometrika*, **51**, 107–122. (Studies the doubly noncentral *F*-distribution.)

[16] Sansone, G. (1959). *Orthogonal Functions*. Wiley, Interscience, New York. (An excellent treatise on orthogonal polynomials.)

[17] Seber, G. A. (1963). *Biometrika*, **50**, 542–544. (Studies the noncentral beta distribution and gives some interesting formulas.)

[18] Severo, N. C. and Zelen, M. (1960). *Biometrika*, **47**, 411–416. (Derives a useful normal approximation and investigates its accuracy.)

[19] Tan, W. Y. and Wong, S. P. (1980). *J. Amer. Statist. Ass.*, **75**, 655–662. (Studies the distribution under the random-effects model.)

[20] Tang, P. C. (1938). *Statist. Res. Mem.*, **2**, 126–149. (Studies and tabulates the probability integral.)

[21] Tiku, M. L. (1965). *Biometrika*, **52**, 415–427. (Develops series expansions, based on Laguerre polynomials.)

[22] Tiku, M. L. (1966). *Biometrika*, **53**, 606–610. (Investigates the accuracy of various approximations.)

[23] Tiku, M. L. (1967). *J. Amer. Statist. Ass.*, **62**, 525–539 (errata, **63**, 1551; additions, **67**, 709–710). (Extensive tables.)

[24] Tiku, M. L. (1974). *Selected Tables in Mathematical Statistics*, Vol. 2. pp. 139–176. American Mathematical Society, Providence, RI. (Formulas, applications, and tables of the probability integral of doubly noncentral *F*.)

[25] Tiku, M. L. and Yip, Y. N. (1978). *Aust. J. Statist.*, **20**, 257–261. (Contains remarkably accurate four-moment approximation.)

(ANALYSIS OF VARIANCE
F-DISTRIBUTION
F-TESTS
NONCENTRAL CHI-SQUARE
    DISTRIBUTION
VARIANCE COMPONENTS)

M. Tiku

NONCENTRAL MATRIX *F*-DISTRIBU TION   *See* MATRIX-VALUED DISTRIBUTIONS

NONCENTRAL STUDENTIZED MAXIMAL DISTRIBUTIONS

NONCENTRAL STUDENTIZED MAXIMAL DISTRIBUTIONS (STUDENTIZED MAXIMUM AND STUDENTIZED MAXIMUM MODULUS)

Let X_i be independent normal random variables that are independent of S and have mean vector μ and common variance σ^2, where $\nu S^2/\sigma^2$ has the chi-square distribution* with ν degrees of freedom. Define $M = \max(X_i)/S$ and $M^{\ddagger} = \max|X_i|/S$; $i = 1, \ldots, k$.

The noncentral maximum distribution with parameters k, μ and ν is the distribution of M, while the noncentral Studentized maximum modulus distribution is the distribution of $M^{\ddagger}$. Special cases are:

(a) The Studentized maximum and Studentized maximum modulus distributions, being the central cases wherein $\mu = 0$.

(b) The maximum and maximum modulus distributions, being the known variance cases wherein S is replaced by the known value for σ.

The Studentized maximal distributions arise naturally in problems of multiple comparisons*. Whenever k parameters β_j have independent, normally distributed estimators $\hat{\beta}_j$ with standard deviations $d_j\sigma$ for known values d_j, then simultaneous two-sided confidence intervals with exact confidence coefficient $1 - \alpha$ for all the β_j are given by

$$\hat{\beta}_j - M^{\ddagger}_{\alpha;k,\nu}d_j S \leqslant \beta_j \leqslant \hat{\beta}_j + M^{\ddagger}_{\alpha;k,\nu}d_j S. \quad (1)$$

Here $M^{\ddagger}_{\alpha;k,\nu}$ is the upper α point of the (central) Studentized maximum modulus distribution with parameters k and ν, where S is an independent estimator for σ, as in the definition.

Usually a size α test of the hypothesis that $\beta_j = \beta_j^*$ simultaneously for all $j = 1(1)k$ means that the hypothesis must be rejected if the hypothesized parameter value falls outside the set (1). The power of this test is $P(M^{\ddagger} > M^{\ddagger}_{\alpha;k,\nu})$, where $M^{\ddagger}$ has the noncentral Studentized maximum modulus distribution with parameters k, ν, and μ, with $\mu_j = \beta_j/(d_j\sigma)$, and where these are the true parameter values.

Cases abound in the theory and practice of the general linear model* under usual normality assumptions, wherein such confidence and testing inferences are required. Examples include confidence intervals and tests (of equality to given constants) for the means or main effects* in a balanced one- or higher-way layout, for the parameters in an orthogonal polynomial regression or for

the independently estimated sets of parameters that arise in many experimental designs.

In fact, the validity of inferences based on (1) is not limited to independently estimated parameters. The validity of these inferences is extended to arbitrarily correlated estimators by the results of Sidak [10] and Jogdeo [5], which show that $P[(1)]$ is minimized in the case of independently estimated parameters. Hence the confidence intervals and tests based on (1) are conservative in cases of correlated parameter estimators. Note, however, that the power of this test is no longer $P(M^{\ddagger} > M^{\ddagger}_{\alpha;k,\nu})$ in correlated cases.

Due to Šidák's inequality [10], the Studentized maximum modulus distribution has many other potential uses in simultaneous inference with normally distributed estimators. In any such case, (1) gives conservative, level α simultaneous inference about any k parameters β_j on the basis of normal $N(\beta_j, d_j^2\sigma^2)$ estimators $\hat{\beta}_j$. For example, Hochberg [4] applies this fact to generalize Tukey's Studentized range* procedure for inference about contrasts to the case of an estimator with an arbitrary convariance matrix.

In multiple comparisons problems, the Studentized maximum distribution is the one-sided analog of the Studentized maximum modulus. That is, to achieve all the upper (or lower) bounds in (1) simultaneously and with probability $1 - \alpha$ merely requires replacing $M^{\ddagger}_{\alpha;k,\nu}$ with $M_{\alpha;k,\nu}$, the upper α point of the appropriate (central) Studentized maximum distribution. Again, these confidence intervals and tests are derived and are exact for independently estimated parameters, but they also can be applied conservatively in much more general cases. For example, Slepian's inequality [11] guarantees coverage by the one-sided, Studentized maximum analog to (1) with probability at least $1 - \alpha$ whenever all the estimators $\hat{\beta}_j$ are nonnegatively correlated.

Tables for the central Studentized maximal distributions (see Pillai and Ramachandran [8], Stoline and Ury [12], and Ury et al. [14]) and a Fortran algorithm [1] for the noncentral distributions through numerical integration are available. Exact tabulations for cases of equally correlated X_i are given by Dunn and Massey [2] and extended by Hahn and Hendrickson [3].

The idea of the Studentized maximum modulus as a tool in multiple comparisons is seen also in the Studentized augmented range.

References

[1] Bohrer, R., Schervish, M. J., and Sheft, J. (1983). *Appl. Statist.*, **32**, 309–317. (Fortran algorithm for noncentral Studentized maximal distributions and related probabilities.)

[2] Dunn, O. J. and Massey, F. J. (1965). *J. Amer. Statist. Ass.*, **60**, 573–583. (Tables for the Studentized maximum modulus with equally correlated ($\rho = 0(0.1)1.0$) variates; $k = 2, 6, 10, 20$; $\nu = 4, 10, 30, \infty$.)

[3] Hahn, G. J. and Hendrickson, R. W. (1971). *Biometrika*, **58**, 323–332. (Tables for the Studentized maximum modulus with equally correlated ($\rho = 0, 0.2, 0.4, 0.5$) variates; $k = 1(1)6(2)12, 15, 20$; $\nu = 3(1)12, 15(5)30, 40, 60$.)

[4] Hochberg, Y. (1974). *J. Multivariate Anal.*, **4**, 224–234.

[5] Jogdeo, K. (1970). *Ann. Math. Statist.*, **41**, 1357–1359. (Simple proof of Šidák's (1968) inequality.)

[6] Miller, R. G., Jr. (1966). *Simultaneous Statistical Inference*. McGraw-Hill, New York. (Comprehensive development of the subject up to 1966.)

[7] Miller, R. G., Jr. (1977). *J. Amer. Statist. Ass.*, **72**, 779–788. (Nicely annotated bibliography of developments of interest during the decade after publication of ref. 6; included in 1983 printing of ref. 6.)

[8] Pillai, K. C. S. and Ramachandran, K. V. (1954). *Ann. Math. Statist.*, **25**, 565–72. (Derivation and tabulation of 5, 95, and 99% points of Studentized maximal distributions:

Studentized Maximum Tables	Studentized Maximum Modulus Tables
$\alpha = 0.05$	$\alpha = 0.05$
$k = 1(1)8$	$k = 1(1)8$
$\nu = 3(1)10, 12, 14, 15, 16, 18,$	$\nu = 5(5)20, 24, 30,$
$20, 24, 30, 40, 60, 120$	$40, 60, 120, \infty$
$\alpha = 0.95, k = 1(1)10,$	
$\nu = 1(1)5, 10(5)20, 24, 30,$	
$40, 60, 120, \infty.)$	

[9] Roy, S. N. and Bose, R. C. (1953). *Ann. Math. Statist.*, **24**, 513–536. (Early use of Studentized maximal distributions.)

[10] Šidák, Z. (1967). *J. Amer. Statist. Ass.*, **62**, 626–633.

[11] Slepian, D. (1962), *Bell System Tech. J.*, **41**, 463–501.

[12] Stoline, M. and Ury, H. (1979). *Technometrics*, **21**, 87–93. (Tables for the Studentized maximum modulus: $\alpha = 0.2, 0.1, 0.05, 0.01$; $k = 3(1)20$; $\nu = 5, 7, 10, 12, 16, 20, 24, 30, 40, 60, 120, \infty$.)

[13] Tukey, J. W. (1953). "The Problem of Multiple Comparisons," unpublished manuscript. (Early use of Studentized maximal distributions.)

[14] Ury, H. K., Stoline, M., and Mitchell, B. T. (1980). *Commun. Statist. B*, **9**, 167–178. (Extension of SMM tables [12]: $\alpha = 0.2, 0.1, 0.05, 0.01$; $k = 20(2)48, 50(5)100$; $\nu = 20(1)39, 40(2)58, 60(5)120, 240, 480, \infty$.)

(MULTIPLE COMPARISONS
RANGES
STUDENTIZED RANGE TESTS)

ROBERT BOHRER

NONCENTRAL _t_-DISTRIBUTION

The noncentral _t_-distribution arises in the study of one-sided acceptance sampling plans, one-sided tolerance limits (*see* TOLERANCE REGIONS, STATISTICAL), the power of Student's _t_-test*, the distribution of the sample coefficient of variation*, confidence limits on one-sided quantiles* of the normal distribution*, one-sided tolerance limits for linear regression, etc. The *noncentral t-distribution* is defined to be the ratio of a standardized normal random variable plus a constant δ to the square root of a chi-square distribution divided by its degrees of freedom, i.e.,

$$T_f(\delta) = (Z + \delta)/\sqrt{Y/f},$$

where $T_f(\delta)$ is a noncentral _t_-variate with f degrees of freedom and noncentrality parameter δ, Z is a normal variate with mean zero and variance one, and Y is a chi-square variate based on f degrees of freedom.

The typical application of the noncentral _t_-distribution starts off with a random sample, $X_1, X_2, \ldots, X_n$ from a normal distribution with unknown mean μ and unknown variance σ^2. The usual estimators of these parameters are, respectively,

$$\hat{\mu} = \bar{x} = \frac{1}{n} \sum_{i=1}^{n} x_i,$$

$$\hat{\sigma}^2 = s^2 = \frac{1}{n-1} \sum_{i=1}^{n} (x_i - \bar{x})^2.$$

There are many representations of the noncentral t-distribution. Johnson and Welch [13] give the density and the cumulative distribution function (cdf) in terms of the $Hh_f(y)$ function studied by Fisher [7]. They also give the following expression, which has been found to be quite tractable for most calculations:

$$\Pr\{T_f(\delta) \leqslant t_0\}$$

$$= \sqrt{2\pi} \left[\Gamma(\tfrac{1}{2}f)\right]^{-1} 2^{(2-f)/2}$$

$$\times \int_0^\infty \Phi(f^{-1/2}t_0 U - \delta) U^{f-1} \phi(U)\, dU,$$

$$\phi(x) = (\sqrt{2\pi})^{-1} e^{-x^2/2}, \quad \Phi(x) = \int_\infty^x \phi(t)\, dt.$$

The distribution has the properties:

$$\Pr\{T_f(\delta) \leqslant t_0\} = 1 - \Pr\{T_f(-\delta) \leqslant t_0\},$$

$$\Pr\{T_f(\delta) \leqslant 0\} = \Phi(-\delta).$$

Note that if $\delta = 0$, $T_f(0)$ has Student's t-distribution* and the integration indicated may be completed in terms of a finite series of elementary functions. If $\delta \neq 0$, the integration may also be carried out to form a finite series, but the result involves $\Phi(\cdot)$, $\phi(\cdot)$ and $T(h, a)$, where the $T(h, a)$ function first arose in evaluating bivariate normal* cumulatives. See Owen [25, pp. 464–465] for these expressions. Young and Minder [32] give an algorithm for computing $T(h, a)$. Additional representations of the noncentral t-distribution are given by Amos [1, 2], Hawkins [9], and Owen [25, pp. 465–467].

The noncentral t-distribution plays a central role in MIL-STD-414 [23], which gives acceptance sampling* plans based on the assumption of a random sample from a normal distribution. For one-sided plans, one of the procedures is to accept lots for which $\bar{x} + ks \leqslant U$, where k is determined by the conditions imposed on the sampling plan and U is an upper specification limit for the product being inspected. The quantity k

is obtained from a percentage point of a noncentral t-distribution. In fact, $t_0 = k\sqrt{n}$, and the noncentrality parameter $\delta = K_p\sqrt{n}$, where K_p is the pth quantile of a standardized normal distribution. The procedure for two-sided plans given in MIL-STD-414 [23] is to estimate the proportions below a lower limit and above an upper limit. The acceptance criterion is to accept if the total proportion in both tails is below a critical value of the proportion computed using the noncentral t-distribution, as described for the one-sided procedure. See Wetherill and Köllerstrom [31] for other considerations of normal sampling inspection.

For tolerance limits, a limit is sought so that we can be $100\gamma\%$ sure that at least $100P\%$ of the population is below that limit. The limit is found to be $\bar{x} + ks$, where k is determined in the same manner as that used for one-sided sampling plans.

The form of tolerance limits on linear regression* is essentially the same except that $\bar{x}$ is placed by the estimated regression and s is replaced by the usual estimator of the standard deviation in this situation. See Kabe [15] for more on this.

For confidence limits on the proportion P in the tail of a normal distribution, the problem is turned around; we are given k and are asked to find P^* from $K_{P*}\sqrt{n} = \delta$, where this P^* is then an upper (or lower) confidence limit on P. Durant [6] gives a nomogram* for finding confidence limits on proportions in tails, and tables are given in Odeh and Owen [22, pp. 149–191].

The sample coefficient of variation is defined to be $s/\bar{x}$. Then

$$\Pr\{s/\bar{x} > c\} = \Pr\{0 \leqslant T_f(\delta) \leqslant \sqrt{n}/c\},$$

$$\Pr\{s/\bar{x} < -c\} = \Pr\{-\sqrt{n}/c \leqslant T_f(\delta) \leqslant 0\},$$

$$\Pr\{s/\bar{x} < c\} = \Pr\{T_f(\delta) \geqslant \sqrt{n}/c\}$$

$$+ \Pr\{T_f(\delta) < 0\},$$

$$\Pr\{s/\bar{x} > -c\} = \Pr\{T_f(\delta) \leqslant -\sqrt{n}/c\}$$

$$+ \Pr\{T_f(\delta) > 0\},$$

where $\delta = \mu\sqrt{n}\,/\sigma$, $f = n - 1$, and c is a positive constant. Warren [30] discusses the adequacy of an approximation to the distribution of $s/\bar{x}$.

The noncentral t-distribution probably arose when the power of Student's t-test was first considered. Consider a test of the null hypothesis that $\mu = \mu_0$ for a normal distribution against the alternative hypothesis that $\mu > \mu_0$. We reject the null hypothesis H_0 if

$$\frac{\bar{x} - \mu_0}{s}\sqrt{n} \geqslant t_{\alpha, n-1},$$

where $t_{\alpha, n-1}$ is an upper $(1 - \alpha)$th quantile of Student's t-distribution with $n - 1 = f$ degrees of freedom. We are interested in the probability of rejection given that the mean is actually $\mu_1 > \mu_0$. That is, we seek

$$\Pr\{\text{Rejecting } H_0\} = \Pr\{T_f(\delta) \geqslant t_{\alpha, n-1}\}$$

where $\delta = (\mu_1 - \mu_0)\sqrt{n}\,/\sigma$. Neyman and Tokarska [21] tabulated values of δ for a given value of power. Note that it depends on the unknown standard deviation, which is unsatisfactory if it is necessary to characterize the alternative in terms of a drift in the mean, since this cannot be done if σ is unknown. Stein [29] gives a double sampling* procedure that has power independent of σ; *see* FIXED-WIDTH AND BOUNDED-LENGTH CONFIDENCE INTERVALS.

Browne and Owen [3] divide the power into two parts, one due to small s and one due to $\bar{x}$ being too large or too small. There are many other applications of the noncentral t-distribution (e.g., see Guenther [8]).

Johnson and Welch [13] first developed viable tables for finding t_0 given δ and for finding δ given t_0 through the use of an auxiliary function. Resnikoff and Lieberman [28] prepared a table of the CDF and density of noncentral t for selected values of $\delta = K_p\sqrt{n}$. Hogben et al. [10] studied the moments of noncentral t. The mean is $c_{11}\delta$ and the variance is $c_{22}\delta^2 + c_{20}$, where

$$c_{11} = \sqrt{\frac{f}{2}}\,\Gamma\!\left(\frac{f-1}{2}\right)\!\Big/\Gamma\!\left(\frac{f}{2}\right),$$

$$c_{20} = \frac{f}{f-2},$$

$$c_{22} = c_{20} - c_{12}^2.$$

Odeh and Owen [22] tabulated many of the values of t_0 and δ required by the various applications. A detailed summary of properties and applications appear in Owen [25] and, even earlier in Johnson and Welch [13]. Both Odeh and Owen [22] and Owen [25] have long bibliographies of the many articles written on this distribution; Jílek [12] gives a bibliography for statistical tolerance regions.

If Y is noncentral chi-square with noncentrality parameter λ, then we have the *doubly noncentral t-distribution*

$$T_f'(\delta, \lambda) = (Z + \delta)/\sqrt{Y/f}\,.$$

This has been studied by Krishnan [17, 18] Bulgren and Amos [4], Bulgren [5], and Mudholkar and Chaubey [19].

A generalization to two dimensions is discussed in Ramig and Nelson [27] and extended to the equicorrelated multivariate case in Nelson [20]. See also Juritz and Troskie [14]. A special case of the *bivariate noncentral t-distribution* is discussed in Owen [24]. This deals with two-sided tolerance limits (sometimes referred to as strong tolerance limits) and sampling plans, which control both tails of the normal distribution.

There are also many approximations to the noncentral t-distribution. Jennett and Welch [11] give one assuming that $\bar{x} + ks$ is approximately normally distributed; this has been widely used and found quite accurate in many applications. Kraemer and Paik [16] and Warren [30] also discuss approximations. Mudholkar and Chaubey [19] give approximations to the double noncentral t-distribution, and Pearson et al. [26] compare quantiles of many distributions, including noncentral t.

References

[1] Amos, D. E. (1964). *Biometrika*, **51**, 451–458. (Gives several representations of the cumulative distribution function of noncentral t. Also discusses some representations of central t.)

[2] Amos, D. E. (1978). *SIAM Rev.*, **20**, 778–800. (Gives computational procedures to obtain the distribution function of noncentral t among several distributions. Emphasis is on a general method for avoiding truncation errors.)

[3] Browne, R. H. and Owen, D. B. (1978). *Commun. Statist. Simul. Comp.*, **7**, 605–617. (Partitions the power of Student's *t*-test into two rejection categories, one due to too small a sample standard deviation and the other due to the mean being too small or too large.)

[4] Bulgren, W. G. and Amos, D. E. (1968). *J. Amer. Statist. Ass.*, **63**, 1013–1019. (Gives series representations of the doubly noncentral *t*-distribution and considers computational aspects of these series.)

[5] Bulgren, W. G. (1974). In *Selected Tables in Mathematical Statistics, Vol. 2*. American Mathematical Society, Providence, RI.

[6] Durant, N. F. (1978). *J. Quality Tech.*, **10**, 155–158. (Gives a nomogram for finding confidence limits on quantiles.)

[7] Fisher, R. A. (1931). In *Introduction to British Association Mathematical Tables*, Vol. I.

[8] Guenther, W. C. (1975). *Amer. Statist.*, **29**, 120–121. (Gives a two-sample test of the hypothesis that two quantiles are equal based on the noncentral *t*-distribution and then gives formulas for approximate sample sizes from each population.)

[9] Hawkins, D. M. (1975). *Amer. Statist.*, **29**, 42–43. (Gives a representation of the noncentral *t* cumulative distribution function involving incomplete beta functions.)

[10] Hogben, D., Pinkham, R. S., and Wilk, M. B. (1961). *Biometrika*, **48**, 465–468. (Contains expressions for the first four moments of the noncentral *t*-distribution.)

[11] Jennett, W. J. and Welch, B. L. (1939). *J. R. Statist. Soc. Suppl.*, **6**, 80. (Includes an approximation to the noncentral *t*-distribution assuming a linear combination of the sample mean and sample standard deviation are approximately normally distributed.)

[12] Jílek, M. (1981). *Math. Operationsforsch. Statist. Ser. Statist.*, **12**, 441–456. (Has a bibliography on statistical tolerance regions.)

[13] Johnson, N. L. and Welch, B. L. (1940). *Biometrika*, **31**, 362–389. (An excellent summary paper of applications and properties of the noncentral *t*-distribution.)

[14] Juritz, J. M. and Troskie, C. G. (1976). *S. Afr. Statist. J.*, **10**, 1–8. (Provides a representation of a noncentral matrix *t*-distribution.)

[15] Kabe, G. (1976). *J. Amer. Statist. Ass.*, **71**, 417–419. (Includes confidence limits on a percentile in a regression situation.)

[16] Kraemer, H. C. and Paik, M. (1979). *Technometrics*, **21**, 357–360. (Gives a central *t*-distribution approximation to the noncentral *t*-distribution. Shows that this approximation is more accurate than the normal approximation for small values of the noncentrality parameter.)

[17] Krishnan, M. (1967). *J. Amer. Statist. Ass.*, **62**, 278–287. (Provides representations and recurrence relations for the first four moments of the doubly noncentral *t*-distribution and some numerical values.)

[18] Krishnan, M. (1968). *J. Amer. Statist. Ass.*, **63**, 1004–1012. (Contains analytic expressions for the distribution function of the doubly noncentral *t*-distribution and some numerical values.)

[19] Mudholkar, G. S. and Chaubey, Y. P. (1976). *Commun. Statist. Simul. Comp.*, **5**, 85–92. (Includes two approximations to the doubly noncentral *t*-distribution, one an Edgeworth expansion.)

[20] Nelson, P. R. (1981). *Commun. Statist. Simul. Comp.*, **10**, 41–50. (Provides an expression for the distribution function of an equicorrelated multivariate noncentral *t*-distribution.)

[21] Neyman, J. and Tokarska, B. (1936). *J. Amer. Statist. Ass.*, **31**, 318–326. (One of the first tables of the power for Student's *t*-test.)

[22] Odeh, R. E. and Owen, D. B. (1980). *Tables for Normal Tolerance Limits, Sampling Plans, and Screening*. Marcel Dekker, New York.

[23] Office of the Assistant Secretary of Defense (Supply and Logistics). (1957). "Sampling Procedures and Tables for Inspection by Variables for Percent defective." *MIL-STD-414*. U.S. GPO, Washington, DC.

[24] Owen, D. B. (1965). *Biometrika*, **52**, 437–446. (A special case of a bivariate noncentral *t* distribution is introduced and representations derived. The distribution is then applied to tolerance limits and sampling plans that control both tails of the normal distribution with parameters unknown.)

[25] Owen, D. B. (1968). *Technometrics*, **10**, 445–478. (Has a survey of properties and applications of the noncentral *t*-distribution and an extensive bibliography.)

[26] Pearson, E. S., Johnson, N. L., and Burr, I. W. (1979). *Commun. Statist. Simul. Comp.*, **8**, 191–229. (Compares quantiles from many distributions including the noncentral *t*-distribution.)

[27] Ramig, P. R. and Nelson, P. R. (1980). *Commun. Statist. Simul. Comp.*, **9**, 621–631. (Gives a representation for a bivariate noncentral *t*-distribution and discusses an application to the power of the analysis of means test.)

[28] Resnikoff, G. J. and Lieberman, G. J. (1957). *Tables of the Noncentral t-distribution*. Stanford University Press, Stanford, CA.

[29] Stein, C. M. (1945). *Ann. Math. Statist.*, **16**, 243–258. (Gives a two-sample test to replace the usual Student *t*-test, which in this case has power independent of the unknown population variance.)

[30] Warren, W. G. (1982). *Commun. Statist. Simul. Comp.*, **11**, 659–666. (Discusses the accuracy of a chi-square approximation to the distribution of the sample coefficient of variation.)

[31] Wetherill, G. B. and Köllerstrom, J. (1979). *J. R. Statist. Soc. A*, **142**, 1–32; errata **142**, 404. (Provides a normal approximation to the cumulative distribution function of noncentral *t* and also Cornish–Fisher expansions.)

[32] Young, J. C. and Minder, C. E. (1974). *Appl. Statist.*, **23**, 455–457; remarks and erratum **28**, 113. (Gives an algorithm for a function that arises in certain representations of the cumulative noncentral *t*-distribution.)

(BIVARIATE NORMAL DISTRIBUTION
STUDENT'S *t*-DISTRIBUTION
STUDENT'S *t*-TESTS
TOLERANCE REGIONS, STATISTICAL)

D. B. OWEN

NONCENTRAL WISHART DISTRIBUTION   *See* WISHART DISTRIBUTIONS

NONCONFORMITY

Nonconformity is a modern term for *defect*. The current official definition as given in ref. 1 defines nonconformity as "departure of a quality characteristic from its intended level or state that occurs with severity sufficient to cause an associated produce or service not to meet a specification requirement." The main difference between the terms *nonconformity* and *defect* is that the former refers to conformance to specifications, whereas the latter refers to customer usage.

Reference

[1] (1978). "Terms, Symbols, and Definitions for Acceptance Sampling Involving the Percent or Proportion of Variant Units in a Lot or Batch." *ANSI/ASQC Standard A2*. American Society for Quality Control (ASQC), Milwaukee, WI.

(ACCEPTANCE SAMPLING
QUALITY CONTROL, STATISTICAL)

NONEXPERIMENTAL INFERENCE

A designed experiment involves selection of the controllable covariates to maximize the expected value of the experimental observations. If the model is linear, these controllable covariates are not allowed to vary at all. A designed experiment also involves randomization* of the treatments, to minimize both the bias that would be present if unobserved covariates and treatments were correlated and the risk of substantial collinearity* between treatments and observed but uncontrollable covariates which, if present, would greatly reduce the information yielded by the experiment.

Nonexperimental inference must substitute method and metaphor for experimental design. In place of controls, the nonexperimental scientist observes as many covariates as seems "reasonable" and controls for their effects with multivariate techniques. In place of randomization, the nonexperimental scientist builds a secondary model that describes how the "treatments" and covariates are selected. Also in place of randomization, the nonexperimental scientist selects a subset of variables that are taken to be exogenous (*see* ECONOMETRICS). Metaphorically speaking, these are the randomized treatments.

Inference in the context of a model with nonrandom treatments is discussed in the entries SIMULTANEOUS EQUATIONS SYSTEMS and ECONOMETRICS.

Although these subjects are more complex mathematically than the theory of inference with experimental data, they rely implicitly on the experimental metaphor and therefore do not require new concepts of statistical inference. For example, phrases such as *sampling distribution* that have a relatively clear meaning in a context in which experiments can be repeated are used also in contexts such as the analysis of macroeconomic data in which the notion of repeating an experiment stretches the imagination. In such nonexperimental contexts it is probably better to reject as inappropriate the frequency interpretation* of probability and to adopt instead the personal or Bayesian viewpoint (*see* BAYESIAN INFERENCE) in which the metaphorical nature of probabilities is made more or less explicit.

Once metaphorical probabilities are given, nonexperimental inference proceeds exactly as does experimental inference. But what is special about the nonexperimental setting is that the metaphor of randomization is usually a subject of intense debate, both personally and publicly. The form of the debate is what Leamer [1] calls a *specification search* in which many alternative statistical models are used as a basis for drawing inferences from the same data set. A nonexperimental inference is credible only when it can be shown to be adequately insensitive to the form of the model. Even then an extra element of uncertainty must attach to the inference because of the risk that a slightly larger search would lead to an entirely different inference. Therefore the theory of nonexperimental inference ought to include methods to define and to control the ambiguity in the inferences that is a consequence of the doubt about the experimental metaphor.

The following fictitious example will illustrate concretely the problems and the procedures of the nonexperimental scientist. Using a sample of fires occurring in a large city in one year, a statistician discovered that the more firemen who were sent to the scene of a fire, the worse the resulting damage was. The inference that firemen cause damage would have been appropriate if the allocation of firemen to fires had been random, as it would be in a designed experiment. But because no formal randomization occurred and because the finding that firemen cause damage conflicts sharply with one's prior beliefs, most of us would interpret the positive correlation between firemen and damage as evidence that a specific nonrandom rule was used to allocate firemen to fires: More firemen were sent to the relatively severe blazes. We are thus led to reject the metaphor that firemen were assigned randomly to the fires.

The first metaphor rejected as inappropriate, the nonexperimental scientist seeks another. Possibly the next step would refer to a model such as the following: Let $D =$ property damage, $N =$ number of firemen, $P =$ potential property damage and $Z =$ the fire-fighting capacity of the district. A sensible causal model then consists of two equations: $D = P - cN$ and $N = fP + gZ$. The first equation asserts that the actual damage is equal to the potential damage offset by a function of the number of firemen. The second equation hypothesizes that the number of firemen who are dispatched to fight a fire depends on the potential severity of the fire and on the fire-fighting capacity of the district. This two-equation model is different from the simple one implicit in the original examination of the simple correlation between D (damage) and N (number of firemen) in two respects. First, the covariate P (potential damage) has been identified. Second, an equation that describes the nonrandom generation of the "treatment" N has been selected.

The more direct and common way to deal with the correlation between P and N is to find measures of the covariate P and to form an estimate of c by a multiple regression* of D on N and measures of P. If the covariate P is measured with sufficient accuracy the metaphor that N is a randomized treatment may be apt for most people. But because there are likely to be various ways that potential damage can be measured sensibly or proxied and because the estimate of c may change greatly depending on how P is measured, inference about c is likely to be ambiguous. Consequently, the nonexperimental scientist requires tools for identifying, controlling, and communicating that ambiguity.

Another approach is to treat P as an unobservable and to use a simultaneous equations method such as the one now to be described. This too leads to serious ambiguity in the inferences and requires the same kind of sensitivity analysis. The simultaneous equations analysis begins with the "reduced form" of the system, consisting of the following two equations: $D = -cgZ + (1 - cf)P$ and $N = gZ + fP$. In this form it appears that a regression of D on Z yields an estimate of $-cg$, a regression of N on Z yields an estimate of g, and their ratio allows us to recover $-c$. But do we really get estimates of cg and g from these regressions? To put this question more directly: Can we comfortably act as if Z were a randomized treatment in the sense of being independent of P? It is easy to think of reasons why we cannot. Suppose the capacity variable Z was a constant within the different districts of the city. It seems sensible to expect that districts which were subject to the greatest risk would have had the greatest fire-fighting capacity. This seems to suggest two new equations, the first determining Z and the second determining P, both as functions of the property value of the district. For this model, the property value variable is taken to be a randomized treatment and used as an *instrumental* variable*. But the next step in this intellectual game would be to think of reasons why property value of the district is not regarded credibly as a randomized treatment and to identify yet another instrumental variable.

This example illustrates two distinctive features of nonexperimental inference: (a) Credible inferences about the effects of "treatment" variables must often be made in the context of the simultaneous equations model. This requires explicit hypotheses about the form of the nonrandomness of the "treatments." (b) The same data set is analyzed with many different models, and sharply different inferences can result as the model is changed. The model finally selected as a basis for drawing inferences is a consequence of a delicate interplay between data and opinion.

Theory concerned with the choice of a model and the effect of that choice on the consequent inferences falls under the heading of *metastatistics*, which considers how motives and opinions affect research. Metastatistics includes the study of memory and computing failures and also social information networks for the transmission of information and opinions among individuals. The complex process by which a nonexperimental scientist chooses a model is what Leamer [1] calls "specification searching." The broadest possible viewpoint about specification searching considers a researcher as a member of society and studies how social reward structures (e.g., tenure at major universities) affect the choice of model. Leamer [1] generally discusses the narrower topic of personal inference and identifies six differ-

ent reasons why specification searches are used by individuals:

1. To select a "true" model from a list of candidates. The formal subject of hypothesis testing*.

2. To interpret multidimensional evidence, that is, to pool information in a given data set with more or less vaguely held prior opinions. Stepwise regression* and ridge regression* make use of implicit prior opinions. Bayesian regression makes use of explicit prior opinion.

3. To identify the most useful models.

4. To find a quantitative facsimile of a maintained theory. A topic under this heading is regression with variables measured with error and its generalization, the factor analytic model (*see* FACTOR ANALYSIS).

5. To select a data subset. Robust regression* methods that place relatively low weight on outliers* are an example.

6. To uncover an entirely new model. The subject of data-instigated hypothesis discovery is beyond the scope of traditional statistical theory, which is designed to characterize the uncertainty provided an intellectual horizon is firmly established and fully committed. Exploratory data analysis* (EDA) offers a collection of data displays intended to excite the creative spark necessary to push out the horizon. But EDA does not address the effects of a successful data exploration on the inferences properly drawn from the data used to instigate the hypotheses.

Reference

[1] Leamer, Edward E. (1978), *Specification Searches: Ad Hoc Inference with Nonexperimental Data*. Wiley, New York.

Bibliography

The first reference and [1] contain discussions of the estimation of simultaneous equations when the metaphor of randomization is accepted without question. The last two references discuss the consequences of doubt about the metaphor.

Blalock, Hubert M., Jr., ed. (1971). *Causal Models in the Social Sciences*. Aldine-Atherton, Chicago.

Goldberger, A. and Duncan, O. D., eds. (1973). *Structural Equation Models in the Social Sciences*. Seminar Press, New York.

Leamer, Edward E. (1983). *Amer. Econ. Rev.*, **73**(1), 31–43.

(CAUSATION
ECONOMETRICS
FOUNDATIONS OF PROBABILITY
LOGIC IN STATISTICAL REASONING)

EDWARD E. LEAMER

NONFORMATION

A concept introduced by Sprott [2] and independently by Barndorff-Nielsen [1] to denote a submodel (and the corresponding part of the data) containing no information with respect to the parameters of interest.

References

[1] Barndorff-Nielsen, O. (1976). *Biometrika*, **63**, 567–71.

[2] Sprott, D. A. (1975). *Biometrika*, **62**, 599–605.

(ANCILLARY STATISTICS
SUFFICIENT STATISTICS)

NONHIERARCHICAL CLUSTERING

In nonhierarchical clustering procedures, new clusters are obtained by both lumping and splitting old clusters, and, unlike the case of hierarchical clustering, the intermediate stages of clustering do not have the monotone increasing strength of clustering as one proceeds from one level to another. For more details see Hartigan [2] and Everitt [1].

References

[1] Everitt, B. (1979). *Cluster Analysis*. Heinemann, London.

[2] Hartigan, J. A. (1975). *Clustering Algorithms*. Wiley, New York.

(CLASSIFICATION
HIERARCHICAL CLUSTER ANALYSIS)

NONLINEAR MODELS

Statistical literature has concentrated almost exclusively on models that are linear in the parameters. A model is an equation or set of equations that describes the behavior of some system, for example, the working of a chemical reactor or the growth of an animal. As the power and speed of computers increase, it becomes more feasible to entertain a wider class of nonlinear models. Since much theory in the physical, chemical, biological, social, and engineering sciences is described most readily in terms of nonlinear equations, it is to be expected that statisticians in the future will be able to incorporate a greater amount of such theory in their models.

Not surprisingly, R. A. Fisher* did pioneering work with nonlinear models, as he did in so many other areas of statistics. In a paper in 1922, Fisher [14] studied the design of dilution experiments for estimating the concentration of small organisms in a liquid. A paper written with W. A. Mackenzie appearing a year later concerned the analysis of data on potatoes [15]. In it, the authors introduced the idea of the analysis of variance*, giving the now familiar linear model, but demonstrating, for the data they were analysing, that a nonlinear model gave better results.

Further developments in the area of nonlinear models had to await the widespread availability of computing power in the 1960s and 1970s. At this time, intensive investigation into the properties and use of nonlinear models could begin, spurred, to a great extent, by the work of G. E. P. Box and his co-workers.

Putting this discussion in mathematical terms, we can say that an important part of science is the description of observable phenomena in terms of equations of the form

$$\mathbf{y} = f(\boldsymbol{\theta}, \boldsymbol{\xi}) + \boldsymbol{\epsilon},$$

where $\mathbf{y}$ is the measured value of one or more responses (i.e., the dependent variables or outputs of the system under study), $\boldsymbol{\epsilon}$ is the noise* or experimental error associated with this measurement, and $f(\boldsymbol{\theta}, \boldsymbol{\xi})$ is a mathematical expression containing p parameters $\theta_1, \theta_2, \ldots, \theta_p$ (a set conveniently denoted by $\boldsymbol{\theta}$) and k variables $\xi_1, \xi_2, \ldots, \xi_k$ (a set conveniently denoted by $\boldsymbol{\xi}$).

If, for example, $\mathbf{y}$ is a single response, the concentration of the product of a first-order chemical reaction, the function f may be of the form

$$f(\boldsymbol{\theta}, \boldsymbol{\xi}) = \theta_1 [1 - \exp(-\theta_2 \xi)],$$

where ξ indicates the time since the reaction started [13]. As another example, the growth of plants or organisms is often modeled by a logistic growth model of the form

$$f(\boldsymbol{\theta}, \boldsymbol{\xi}) = \frac{\theta_1}{1 + \theta_2 \exp(-\theta_3 \xi)},$$

where ξ again represents the time since the organism or plant started growing. These two examples provide relatively simple mathematical expressions for the model function. In real applications, the model function is often much more complicated. Situations with seven or more parameters are not uncommon. Moreover, the function f may be defined implicitly by a partial differential equation that has no analytic solution.

We first consider the single-response case where $\mathbf{y}$ is one-dimensional so that it can be written as y. After n experiments, there will be a set of n responses $y_1, \ldots, y_n$ with the associated values of $\boldsymbol{\xi}_1, \ldots, \boldsymbol{\xi}_n$ to provide information about the parameters $\boldsymbol{\theta}$. If, in addition, assumptions are made on the statistical properties of the noise terms ϵ_t, $t = 1, \ldots, n$, estimates for the parameters can be derived and other inferences about them made.

The usual assumptions about the noise terms are that each ϵ_t, $t = 1, \ldots, n$ is normally distributed with mean zero and variance σ^2, that σ^2 is constant over t (although unknown), and that ϵ_t and ϵ_s are independent for $t \neq s$. The reasonableness of these assumptions can be checked, after obtaining parameter estimates, by examining plots of the residuals*

$$r_t = y_t - f(\boldsymbol{\theta}, \boldsymbol{\xi}_t)$$

vs. predicted values, the ξ_j's, $j = 1, \ldots, k$, the time order in which the data were collected, or by making other plots deemed of interest by the experimenter.

DEFINITION OF NONLINEARITY

An important distinction is that between models that are linear in the parameters $\boldsymbol{\theta}$ and those that are not. A *linear* model can be written as

$$f(\boldsymbol{\theta}, \boldsymbol{\xi}) = \sum_{i=1}^{p} \theta_i g_i(\boldsymbol{\xi}) \qquad (2)$$

for some functions g_i that depend only on the values of $\boldsymbol{\xi}$ but not on the values of $\boldsymbol{\theta}$. Models that cannot be written in this form are *nonlinear in the parameters* or, more simply, *nonlinear*. Note that the linearity or nonlinearity of the model is determined by the way in which the *parameters* enter the model and *not* by the way in which the *variables* $\xi_1, \ldots, \xi_k$ enter. Thus a quadratic equation in ξ_1

$$f(\boldsymbol{\theta}, \boldsymbol{\xi}) = \theta_1 + \theta_2 \xi_1 + \theta_3 \xi_1^2$$

is considered to be a linear model because it can be expressed in the form of equation (2) with $g_1(\xi) = 1$, $g_2(\xi) = \xi_1$, and $g_3(\xi) = \xi_1^2$. On the other hand, the model

$$f(\boldsymbol{\theta}, \boldsymbol{\xi}) = \theta_1 \exp(-\theta_2 \xi_1)$$

is nonlinear.

An easy way of checking whether a model is linear or nonlinear is to examine the derivatives of f with respect to each of the parameters θ_i. If $\partial f / \partial \theta_i$ does not depend on any of the elements of $\boldsymbol{\theta}$, the model is linear in θ_i; if it is linear in all p parameters $(\theta_1, \theta_2,$

..., θ_p), the model is said to be linear in the parameters or simply linear.

Some nonlinear models can be converted into a linear form by taking logarithms or reciprocals or by some other transformation. For example,

$$f(\boldsymbol{\theta}, \xi) = \theta_1 \exp(-\theta_2 \xi_1)$$

is mathematically equivalent to

$$\ln f(\boldsymbol{\theta}, \xi) = -\theta_2 \xi_1 + \ln \theta_1,$$

so that $\ln f$ represents a linear model in the parameters θ_2 and $\ln \theta_1$. This transformed model could be fitted to the responses $\ln y$. Such "linearizing" transformations, however, have the effect of transforming ϵ as well as f and altering the relationship between f and ϵ. The assumption that ϵ is a zero-mean, additive, constant variance noise component for the original model generally means that ϵ will not have these properties after transformation. The transformation could have the effect of bringing the behavior of the residual values closer to the assumed behavior of ϵ, but it could have the opposite effect. It is always important to examine the residuals after fitting a model to decide if the assumptions regarding the noise term are reasonable.

NONLINEAR REGRESSION

Perhaps after transformation, data for a single-response, nonlinear model with the usual assumptions on the noise term ϵ are most often analyzed using nonlinear regression*. As in linear regression*, parameter estimates are taken to be the values of $\boldsymbol{\theta}$, which minimize the residual sum of squares

$$S(\boldsymbol{\theta}) = \sum_{t=1}^{n} \left[y_t - f(\boldsymbol{\theta}, \xi_t) \right]^2.$$

Let $\hat{\boldsymbol{\theta}}$ denote these least-squares estimates. Unlike linear regression, where, in principle, the parameter estimates can be computed directly, sometimes even without the aid of a computer, nonlinear regression usually requires calculation by iterative computer programs.

An overview of the theory and practice of nonlinear regression is provided in Draper and Smith [13, Chap. 10] and in the introductory articles by Gallant [16] and Watts [26]. Some of the disadvantages of nonlinear relative to linear regression are the need to use iterative estimation techniques, which require initial estimates or "starting values" to get the iterations under way, difficulties in converging to the least-squares* estimates, and the lack of exact theoretical properties of the estimates. The advantage of nonlinear models is the flexibility that they provide. Not all of the model forms of interest to experimenters can be expressed in a linear fashion with an additive, constant-variance noise term. A nonlinear model is sometimes more theoretically sensible from the point of view of the experimenter, the model perhaps having been derived on the basis of knowledge in the substantive field. Frequently the resulting model will contain fewer parameters than a corresponding linear model, which is an advantage since it permits more precise predictions to be made.

Since exact theoretical properties of nonlinear regression estimates usually cannot be derived, inferences drawn from the model are most often based on a linear approximation to the model of the form

$$f(\boldsymbol{\theta}, \xi) \cong f(\boldsymbol{\theta}^0, \xi) + \sum_{i=1}^{p} \frac{\partial f}{\partial \theta_i} (\theta_i - \theta_i^0),$$

where the derivatives are evaluated at $\boldsymbol{\theta}^0$. Then the appropriate region or interval for this approximating linear model is used.

MULTIRESPONSE ESTIMATION

When r different responses are measured on each experimental run, the information from all of the responses can be combined to help estimate common parameters. In this case the responses are represented by a r-dimensional vector $\mathbf{y}_t$ for each $t = 1, \ldots, n$ and the model function incorporates r separate functions $f_i(\boldsymbol{\theta}, \xi)$, $i = 1, 2, \ldots, r$. The analysis of such data is complicated by possible correlation* in the noise terms associated with the different components of **y** for each experimental run. Box and Draper [2] demonstrated by a Bayesian argument that the appropriate criterion for parameter estimation from such data was to choose $\boldsymbol{\theta}$ so as to minimize

$$\det\left[(\mathbf{Y} - \mathbf{F}(\boldsymbol{\theta}))^T(\mathbf{Y} - \mathbf{F}(\boldsymbol{\theta}))\right], \quad (3)$$

where **Y** is the $n \times r$ matrix of measured responses, $\mathbf{F}(\boldsymbol{\theta})$ is the $n \times r$ matrix of predicted responses given $\boldsymbol{\theta}$, so the (i, j)th entry of $\mathbf{F}(\boldsymbol{\theta})$ is $f_j(\boldsymbol{\theta}, \xi_i)$, and the superscript T indicates the transpose of a matrix.

One difficulty with this criterion is that there may be dependencies among the predicted reponses that are also reflected in the observed responses. For example, if the r responses correspond to the concentrations of components in a mixture, the predicted concentrations could total to a constant. If the same relationship were used to derive one of the components of the measured responses **y** from the values of the other components, the matrix $\mathbf{Y} - \mathbf{F}(\boldsymbol{\theta})$ would be singular for each value of $\boldsymbol{\theta}$. Thus the determinant in expression (3) would always be zero in theory although the calculated value may differ from zero due to numerical roundoff.

To detect and correct for such dependencies, Box et al. [8] show that the eigenvalues of $\mathbf{Y}^T\mathbf{Y}$ and $\mathbf{F}(\boldsymbol{\theta})^T\mathbf{F}(\boldsymbol{\theta})$ should be examined. McLean et al. [22] show that it is advisable to examine also the eigenvalues of $(\mathbf{Y} - \mathbf{F}(\boldsymbol{\theta}))^T(\mathbf{Y} - \mathbf{F}(\boldsymbol{\theta}))$ to check for singularities. Automatic checks for such singularities can be incorporated into multiresponse estimation programs.

Methods of defining approximate confidence regions* for the parameters or subsets of the parameters are discussed in Box [1], Box and Draper [11], and Ziegel and Gorman [27]. The latter reference also gives practical examples where multiresponse estimation provides much more useful information about model parameters than can be obtained by examining responses in isolation.

With multiresponse data the methods of handling missing data become more compli-

cated than in the single-response case (*see* INCOMPLETE DATA). Stewart and Sorensen [25] derive a Bayesian criterion for the estimation of common parameters in the presence of missing data.

DESIGN OF EXPERIMENTS

As with the estimation of parameters, the design of experiments for nonlinear models is more complicated than design for linear models. Experimental design for nonlinear models is the selection of values of ξ_t for some or all of the $t = 1, \ldots, n$ so as to maximize some desirable criterion, usually a criterion related to the parameter estimates. Criteria that are often used are the precise estimation of all of the parameters (see Box and Lucas [9]), the precise estimation of a subset of the parameters (see Box [10]), or the ability to discriminate between two or more rival models (see Box and Hill [3]).

A difficulty is that the value of the criterion for a proposed design depends on the value of the parameters. Cochran [12] described the situation as the statistician proposing to the experimenter that "You tell me the value of $\boldsymbol{\theta}$ and I promise to design the best experiment for estimating $\boldsymbol{\theta}$." One way to avoid this circular argument is to design the experiments sequentially where the design for the next experiment or group of experiments is chosen using the parameter estimates from the current set of experiments that have been run (see Box and Hunter [6]). The initial experiments are usually undesigned "screening" runs used to check that the apparatus is working properly, or they may be designed on the basis of rough guesses of the parameter values. Hill et al. [18] offer a sequential criterion that combines properties of model discrimination and precise parameter estimation.

Other Difficulties

Residual plots from nonlinear models will sometimes indicate an inhomogeneity of variance that cannot be corrected through a simple transformation such as using the logarithm of the original response. Box and Hill [4] give a technique for using power transformations to correct this; this technique is further refined by Pritchard et al. [24].

The choice of an appropriate model can also cause difficulties. Box and Hunter [5] describe a method of iteratively building and criticizing models; it is illustrated in more detail in Box et al. [7, Chap. 16], along with a discussion of model testing and diagnostic parameters as proposed in Hunter and Mezaki [19].

A method of letting the data itself determine the model function is given by Lawton et al. [21] for a number of data sets with the same general model form but with different scaling of the model. This "self-modeling" approach uses spline* functions.

APPLICATIONS

Nonlinear models have been applied in the physical and engineering sciences, the biological and life sciences, and in the social sciences. One of the areas of widest application is chemical kinetics. Examples of such applications are given in the bibliography.

Some of the uses in biological sciences include the description of growth curves* and dose-response curves (see Draper and Smith [13, Chap. 10] as well as compartment models such as those used in pharmacokinetics (see Jennrich and Bright [20] or Metzler [23]). Some of the social sciences applications are in econometrics* (see Gallant and Holly [17]).

Acknowledgments

We would like to thank Norman Draper and David Hamilton for the use of their bibliographies on nonlinear models while collecting the bibliography that follows. We also thank Andy Jaworski for his helpful comments and additions to the annotations of the bibliography.

References

[1] Box, G. E. P. (1960). *Ann. N.Y. Acad. Sci.*, **86**, 792–816.

[2] Box, G. E. P. and Draper, N. R. (1965). *Biometrika*, **52**, 355–365.

[3] Box, G. E. P. and Hill, W. J. (1967). *Technometrics*, **9**, 57–71.

[4] Box, G. E. P. and Hill, W. J. (1974). *Technometrics*, **16**, 385–389.

[5] Box, G. E. P. and Hunter, W. G. (1962). *Technometrics*, **4**, 301–318.

[6] Box, G. E. P. and Hunter, W. G. (1965). *Technometrics*, **7**, 23–42.

[7] Box, G. E. P., Hunter, W. G., and Hunter, J. S. (1978). *Statistics for Experimenters*. Wiley, New York.

[8] Box, G. E. P., Hunter, W. G., MacGregor, J. F., and Erjavec, J. (1973). *Technometrics*, **15**, 33–51.

[9] Box, G. E. P. and Lucas, H. L. (1959). *Biometrika*, **46**, 77–90.

[10] Box, M. J. (1971). *Biometrika*, **58**, 149–153.

[11] Box, M. J. and Draper, N. R. (1972). *Appl. Statist.*, **21**, 13–24.

[12] Cochran, W. G. (1973). *J. Amer. Statist. Ass.*, **68**, 771–778.

[13] Draper, N. R. and Smith, H. (1981). *Applied Regression Analysis*, 2nd ed. Wiley, New York.

[14] Fisher, R. A. (1922). *Philos. Trans. R. Soc. Lond. A*, **222**, 309–328.

[15] Fisher, R. A. and Mackenzie, W. A. (1923). *J. Agric. Sci.*, **13**, 311–320.

[16] Gallant, A. R. (1975). *Amer. Statist.*, **29**(2), 73–81.

[17] Gallant, A. R. and Holly, A. (1980). *Econometrica*, **48**, 697–720.

[18] Hill, W. J., Hunter, W. G., and Wichern, D. W. (1968). *Technometrics*, **10**, 145–160.

[19] Hunter, W. G. and Mezaki, R. (1964). *Amer. Inst. Chem. Eng. J.*, **10**, 315–322.

[20] Jennrich, R. I. and Bright, P. B. (1976). *Technometrics*, **18**, 385–399.

[21] Lawton, W. H., Sylvestre, E. A., and Maggio, M. S. (1972). *Technometrics*, **14**, 513–532.

[22] McLean, D. D., Pritchard, D. J., Bacon, D. W., and Downie, J. (1979). *Technometrics*, **21**, 291–298.

[23] Metzler, C. M. (1981). In *Kinetic Data Analysis—Design and Analysis of Enzyme and Pharmacokinetic Experiments*, L. Endrenyi, ed. Plenum, New York, pp. 25–37.

[24] Pritchard, D. J., Downie, J., and Bacon, D. W. (1977). *Technometrics*, **19**, 227–236.

[25] Stewart, W. E. and Sorensen, J. P. (1981). *Technometrics*, **23**, 131–141.

[26] Watts, D. G. (1981). In *Kinetic Data Analysis—Design and Analysis of Enzyme and Pharmacokinetic Experiments*, L. Endrenyi, ed. Plenum, New York, pp. 1–24.

[27] Ziegel, E. R. and Gorman, J. W. (1980). *Technometrics*, **22**, 139–151.

Bibliography

Single Response Models—Estimation

Bard, Y. (1974). *Nonlinear Parameter Estimation*. Academic Press, New York and London. (An engineering orientation toward nonlinear models.)

Beck, J. V. and Arnold, K. J. (1977). *Parameter Estimation in Engineering and Science*. Wiley, New York. (Examples from heat transfer.)

Bliss, C. I. and James, A. T. (1966). *Biometrics*, **22**, 573–602. (Special techniques for a rectangular hyperbola model.)

Box, G. E. P. (1960). *Ann. N.Y. Acad. Sci.*, **86**, 792–816. (An early landmark paper in the area.)

Box, G. E. P. and Hunter, W. G. (1962). *Technometrics*, **4**, 301–318. (Methods of scientific model building through an iterative process.)

Box, G. E. P. and Hunter, W. G. (1965). *Technometrics*, **7**, 23–42. (Application of model building in chemical kinetics.)

Box, G. E. P., Hunter, W. G., and Hunter, J. S. (1978). *Statistics for Experimenters*. Wiley, New York. (A widely used text in statistical methods for science and engineering.)

Box, G. E. P. and Tiao, G. C. (1973). *Bayesian Inference in Statistical Analysis*. Addison-Wesley, Reading, MA.

Draper, N. R. and Smith, H. (1981). *Applied Regression Analysis*, 2nd ed. Wiley, New York. (A fundamental reference in linear and nonlinear regression.)

Gallant, A. R. (1975). *Amer. Statist.* **29**(2), 73–81. (Introductory article on nonlinear regression.)

Kittrell, J. R. (1970). *Adv. Chem. Eng.*, **8**, 97–183. (Nonlinear regression for chemical kinetics.)

Kittrell, J. R., Hunter, W. G., and Watson, C. C. (1965). *Amer. Inst. Chem. Eng. J.*, **11**, 1051–1057. (Nonlinear least squares with catalytic rate model example.)

Marquardt, D. W. (1970). *Technometrics*, **12**, 591.

Mezaki, R., Draper, N. R., and Johnson, R. A. (1973). *Ind. Eng. Chem. Fund.*, **12**, 251–254. (Warns of common misuses of regression through inappropriate model transformations.)

Mezaki, R. and Kittrell, J. R. (1968). *Amer. Inst. Chem. Eng. J.*, **14**, 513.

Peterson, T. I. and Lapidus, L. (1966). *Chem. Eng. Sci.*, **21**, 655–664.

Pritchard, D. J., Downie, J., and Bacon, D. W. (1977). *Technometrics*, **19**, 227–236.

Reilly, P. M. and Patino-Leal, H. (1981). *Technometrics*, **23**, 221–231. (Errors-in-variables models.)

Watts, D. G. (1981). In *Kinetic Data Analysis—Design and Analysis of Enzyme and Pharmacokinetic Experiments*, L. Endrenyi, ed. Plenum, New York, pp. 1–24. (Introduction to nonlinear regression for pharmacokinetics and other uses.)

Multiresponse Models—Estimation

Box, G. E. P. and Draper, N. R. (1965). *Biometrika*, **52**, 355–365. (Defines the commonly used Bayesian estimation criterion.)

Box, G. E. P., Draper, N. R., and Hunter, W. G. (1970). *Technometrics*, **12**, 613–620. (Describes missing observations in multiresponse models.)

Box, G. E. P., Hunter, W. G., MacGregor, J. F., and Erjavec, J. (1973). *Technometrics*, **15**, 33–51. (Methods of detecting singularities in multiresponse modeling.)

Box, M. J. and Draper, N. R. (1972). *Appl. Statist.*, **21**, 13–24. (Nonhomogeneous variance methods.)

Draper, N. R., Kanemasu, H., and Mezaki, R. (1969). *Ind. Eng. Chem. Fund.*, **8**, 423–427. (Applications in chemical kinetics.)

Hunter, W. G. (1967). *Ind. Eng. Chem. Fund.*, **8**, 423–427. (Comparison of different methods of multiresponse parameter estimation.)

McLean, D. D., Pritchard, D. J., Bacon, D. W., and Downie, J. (1979). *Technometrics*, **21**, 291–298. (Detection of singularities.)

Mezaki, R. and Butt, J. B. (1968). *Ind. Eng. Chem. Fund.*, **7**, 120–125.

Stewart, W. E. and Sorensen, J. P. (1981). *Technometrics*, **23**, 131–141. (Discusses multiresponse methods with missing data.)

Ziegel, E. R. and Gorman, J. W. (1980). *Technometrics*, **22**, 139–151. (Chemical kinetics modeling and model building.)

Experimental Design—Precise Parameter Estimation

Atkinson, A. C. and Hunter, W. G. (1968). *Technometrics*, **10**, 271–289.

Box, G. E. P. and Lucas, H. L. (1959). *Biometrika*, **46**, 77–90. (An early landmark paper in nonlinear experimental design.)

Box, M. J. (1968). *J. R. Statist. Soc. B*, **30**, 290–302. (The occurrence of replicate points in designs.)

Box, M. J. (1968). In *Conference on the Future of Statistics*, D. G. Watts, ed. Academic Press, New York, pp. 241–257.

Box, M. J. (1970). *Technometrics*, **12**, 569–589. (Experiences with design criteria.)

Box, M. J. (1971). *Biometrika*, **58**, 149–153. (Precise parameter estimation of a subset of the parameters.)

Box, M. J. (1971). *Technometrics*, **13**, 19–31.

Box, M. J. and Draper, N. R. (1971). *Technometrics*, **13**, 731–742. (Discussion of practical relevance of *D*-optimality.)

Chernoff, H. (1953). *Ann. Math. Statist.*, **24**, 586–602.

Cochran, W. G. (1973). *J. Amer. Statist. Ass.*, **68**, 771–778.

Currie, D. (1982). *Biometrics*, **38**, 907–919. (Designs for the Michaelis–Menton model.)

Draper, N. R. and Hunter, W. G. (1967). *Biometrika*, **54**, 147–153. (Use of prior distributions in *D*-optimal designs.)

Evans, J. W. (1979). *Technometrics*, **21**, 321–330. (Augmentation of designs.)

Graham, R. J. and Stevenson, F. D. (1972). *Ind. Eng. Chem. Process Des. Dev.*, **11**, 160–164. (Application of sequential designs.)

Herzberg, A. M. and Cox, D. R. (1969). *J. R. Statist. Soc. B*, **31**, 29–67. (Bibliography of experimental design literature.)

Hill, P. D. H. (1980). *Technometrics*, **22**, 275–276. (Elimination of conditionally linear parameters.)

Hill, W. J. and Hunter, W. G. (1974). *Technometrics*, **16**, 425–434. (Designs for subsets of parameters.)

Hill, W. J., Hunter, W. G., and Wichern, D. W. (1968). *Technometrics*, **10**, 145–160. (Joint design criterion for model discrimination and precise parameter estimation.)

Hunter, W. G. and Atkinson, A. C. (1966). *Chem. Eng.*, **73**, 159–164.

Hunter, W. G., Hill, W. J., and Henson, T. L. (1969). *Canad. J. Chem. Eng.*, **47**, 76–80. (Applications in chemical kinetics.)

Hunter, W. G., Kittrell, J. R., and Mezaki, R. (1967). *Trans. Inst. Chem. Eng.*, **45**, T146–T152.

Juusola, J. A., Bacon, D. W., and Downie, J. (1972). *Canad. J. Chem. Eng.*, **50**, 796–801. (Experimental strategy in kinetic studies.)

Kittrell, J. R., Hunter, W. G., and Watson, C. C. (1966). *Amer. Inst. Chem. Eng. J.*, **12**, 5–10. (Catalytic kinetics example.)

Pritchard, D. J. and Bacon, D. W. (1977). *Technometrics*, **19**, 109–115. (Accounts for nonhomogeneous variance in designs.)

Reilly, P. M., Bajramovic, R., Blau, G. E., Branson, D. R., and Sauerhoff, M. W. (1977). *Canad. J. Chem. Eng.*, **55**, 614–622.

St. John, R. C. and Draper, N. R. (1975). *Technometrics*, **17**, 15–23. (Review paper.)

Experimental Design—Model Discrimination

Atkinson, A. C. (1981). *Technometrics*, **23**, 301–305. (Comparison of criteria for model discrimination.)

Atkinson, A. C. and Cox, D. R. (1974). *J. R. Statist. Soc. B*, **36**, 321–334; discussion, 335–348.

Atkinson, A. C. and Fedorov, V. V. (1975). *Biometrika*, **62**, 57–70. (Discriminating between two models.)

Atkinson, A. C. and Fedorov, V. V. (1975). *Biometrika*, **62**, 289–304. (Discriminating between several models.)

Box, G. E. P. and Hill, W. J. (1967). *Technometrics*, **9**, 57–71. (Discrimination among many models.)

Froment, G. F. and Mezaki, R. (1970). *Chem. Eng. Sci.*, **25**, 293–301. (Sequential discrimination and estimation.)

Hill, P. D. H. (1978). *Technometrics*, **20**, 15–21.

Hill, W. J. and Hunter, W. G. (1969). *Technometrics*, **11**, 396–400.

Hill, W. J., Hunter, W. G., and Wichern, D. W. (1968). *Technometrics*, **10**, 145–160. (Joint design criterion for model discrimination and precise parameter estimation.)

Hunter, W. G. and Reiner, A. M. (1965). *Technometrics*, **7**, 307–323. (Simple criterion for discrimination between two models.)

Kittrell, J. R. and Mezaki, R. (1967). *Amer. Inst. Chem. Eng. J.*, **13**, 389–392. (Applications to Hougen–Watson kinetic models.)

Moeter, D., Pirie, W., and Blot, W. (1970). *Technometrics*, **12**, 457–470. (Comparison of design criteria.)

Pritchard, D. J. and Bacon, D. W. (1974). *Canad. J. Chem. Eng.*, **52**, 103–109. (Practical aspects of model discrimination.)

Reilly, P. M. (1970). *Canad. J. Chem. Eng.*, **48**, 168–173. (Introduction to model discrimination for chemical engineering.)

Experimental Design—Multiresponse Models

Box, M. J. and Draper, N. R. (1972). *Appl. Statist.*, **21**, 13–24. (Compensating for nonhomogeneous variance.)

Draper, N. R. and Hunter, W. G. (1966). *Biometrika*, **53**, 525–553.

Draper, N. R. and Hunter, W. G. (1967). *Biometrika*, **54**, 662–665. (Use of prior distributions in design.)

Heteroscedasticity

Box, G. E. P. and Hill, W. J. (1974). *Technometrics*, **16**, 385–389. (Power transformation weighting.)

Box, M. J. and Draper, N. R. (1972). *Appl. Statist.*, **21**, 13–24.

Pritchard, D. J. and Bacon, D. W. (1977). *Technometrics*, **19**, 109–115.

Pritchard, D. J., Downie, J., and Bacon, D. W. (1977). *Technometrics*, **19**, 227–236.

Diagnostic Parameters

Box, G. E. P. and Hunter, W. G. (1962). *Technometrics*, **4**, 301–318.

Hunter, W. G. and Mezaki, R. (1964). *Amer. Inst. Chem. Eng. J.*, **10**, 315–322.

Kittrell, J. R., Hunter, W. G., and Mezaki, R. (1966). *Amer. Inst. Chem. Eng. J.*, **12**, 1014–1017.

Nonlinear Least Squares—Computing

Barham, R. H. and Drane, W. (1972). *Technometrics*, **14**, 757–766. (Exploiting conditional linearity.)

Bates, D. M. and Watts, D. G. (1981). *Technometrics*, **23**, 179–183. (Geometry-based termination criterion.)

Chambers, J. M. (1973). *Biometrika*, **60**, 1–13. (Review paper.)

Dennis, J. E., Jr., Gay, D. M., and Welsch, R. E. (1981). *ACM Trans. Math. Software*, **7**, 348–368. (Adaptive methods for large-residual problems.)

Golub, G. H. and Pereyra, V. (1973). *J. SIAM*, **10**, 413–432. (Calculation of derivatives after elimination of conditionally linear parameters.)

Guttman, I., Pereyra, V., and Scolnik, H. D. (1973). *Technometrics*, **15**(2), 209–218.

Hartley, H. O. (1961). *Technometrics*, **3**, 269–280.

Harville, D. A. (1973). *Technometrics*, **15**, 509–515.

Hiebert, K. L. (1981). *ACM Trans. Math. Software*, **7**(1), 1–16. (Comparison of software for nonlinear regression.)

Jennrich, R. I. and Bright, P. B. (1976). *Technometrics*, **18**, 385–399. (Special methods for systems of linear differential equations.)

Jennrich, R. I. and Sampson, P. F. (1968). *Technometrics*, **10**, 63–72.

Lawton, W. H. and Sylvestre, E. A. (1971). *Technometrics*, **13**, 461–467. (Exploiting conditionally linear parameters.)

Levenberg, K. (1944). *Quart. Appl. Math.*, **2**, 164–168.

Marquardt, D. W. (1963). *J. SIAM*, **11**, 431–441. (Compromise between Gauss–Newton and steepest descent methods.)

Meyer, R. R. and Roth, P. M. (1972). *J. Inst. Math. Appl.*, **9**, 218.

Peduzzi, P. N., Hardy, R. J., and Holford, T. R. (1980). *Biometrics*, **36**, 511–516; **37**, 595–596. (Variable selection procedure.)

Pedersen, P. V. (1977). *J. Pharmacokin. Biopharm.*, **5**, 513. (Curve fitting in pharmacokinetics.)

Pedersen, P. V. (1978). *J. Pharmacokin. Biopharm.*, **6**, 447.

Ralston, M. L. and Jennrich, R. I. (1978). *Technometrics*, **20**, 7–14. (A derivative-free algorithm.)

Nonlinear Least Squares—Theory

Bates, D. M. and Watts, D. G. (1980). *J. R. Statist. Soc. B*, **42**, 1–16; discussion, 16–25. (Measures of nonlinearity.)

Bates, D. M. and Watts, D. G. (1981). *Ann. Statist.*, **9**, 1152–1167. (Effects of transformations of the parameters.)

Beale, E. M. L. (1960). *J. R. Statist. Soc. B*, **22**, 41–76; discussion, 76–88. (Measuring nonlinearity.)

Clarke, G. P. Y. (1980). *J. R. Statist. Soc. B*, **42**, 227–237. (Moments of the estimates from second and third derivatives of the model function.)

Gallant, A. R. (1975). *J. Amer. Statist. Ass.*, **70**, 198–203. (Power of likelihood ratio tests.)

Gallant, A. R. (1975). *J. Amer. Statist. Ass.*, **70**, 927–932. (Tests on a subset of the parameters.)

Gallant, A. R. (1977). *J. Amer. Statist. Ass.*, **72**, 523–529.

Guttman, I. and Meeter, D. A. (1965). *Technometrics*, **7**, 623–637. (Application of Beale's measures.)

Halperin, M. (1963). *J. R. Statist. Soc. B*, **25**, 330–333. (Methods of defining exact confidence regions for parameters.)

Hamilton, D. C., Watts, D. G., and Bates, D. M. (1982). *Ann. Statist.*, **10**, 386–393. (Compensating for intrinsic nonlinearity.)

Hartley, H. O. (1964). *Biometrika*, **51**, 347–353.

Hougaard, P. (1982). *J. R. Statist. Soc. B*, **44**, 244–252. (Parameter transformation in nonlinear models.)

Linssen, H. N. (1975). *Statist. Neerlandica*, **29**, 93–99.

Ross, G. J. S. (1970). *Appl. Statist.*, **19**, 205–221. (Suggested parameter transformations.)

Ross, G. J. S. (1978). In *COMPSTAT 78, Third Symposium on Computation*, L. Corstein and J. Hermans, eds. Physica-Verlag, Vienna.

Wilks, S. S. and Daly, J. F. (1939). *Ann. Math. Statist.*, **10**, 225–239.

Williams, E. J. (1962). *J. R. Statist. Soc. B*, **24**, 125–139.

Wu, C. F. (1981). *Ann. Statist.*, **9**, 501–513. (Asymptotic properties of estimates.)

Applications

Bacon, D. W. (1970). *Ind. Eng. Chem.*, **62**(7), 27–34.

Behnken, D. W. (1964). *J. Polymer Sci. A*, **2**, 645–668. (Copolymer reactivity ratios.)

Bliss, C. I. and James, A. T. (1966). *Biometrics*, **22**, 573–602.

Boag, I. F., Bacon, D. W., and Downie, J. (1975). *J. Catal.*, **38**, 375–384. (Analysis of oxidation data.)

Currie, D. (1982). *Biometrics*, **38**, 907–919. (Michaelis–Menten model of enzyme kinetics.)

Draper, N. R., Kanemasu, H., and Mezaki, R. (1969). *Ind. Eng. Chem. Fund.*, **8**, 423–427.

Fisher, R. A. (1939). *Ann. Eugen.*, **9**, 238–249.

Froment, G. F. and Mezaki, R. (1970). *Chem. Eng. Sci.*, **25**, 293–301.

Gallant, A. R. and Holly, A. (1980). *Econometrica*, **48**, 697–720.

Graham, R. J. and Stevenson, F. D. (1972). *Ind. Eng. Chem. Process Des. Dev.*, **11**, 160–164.

Hoffman, T. and Reilly, P. M. (1979). *Canad. J. Chem. Eng.*, **57**, 367–374.

Hsiang, T. and Reilly, P. M. (1971). *Canad. J. Chem. Eng.*, **49**, 865–871.

Hunter, W. G. (1967). *Ind. Eng. Chem. Fund.*, **8**, 423–427.

Hunter, W. G. and Atkinson, A. C. (1966). *Chem. Eng.*, **73**, 159–164.

Hunter, W. G., Hill, W. J., and Henson, T. L. (1969). *Canad. J. Chem. Eng.*, **47**, 76–80.

Hunter, W. G., Kittrell, J. R., and Mezaki, R. (1967). *Trans. Inst. Chem. Eng.*, **45**, T146–T152.

Hunter, W. G. and Mezaki, R. (1964). *Amer. Inst. Chem. Eng. J.*, **10**, 315–322.

Johnson, R. A., Standal, N. A., and Mezaki, R. (1968). *Ind. Eng. Chem. Fund.*, **7**, 181.

Juusola, J. A., Bacon, D. W., and Downie, J. (1972). *Canad. J. Chem. Eng.*, **50**, 796–801.

Kittrell, J. R. (1970). *Adv. Chem. Eng.*, **8**, 97–183.

Kittrell, J. R., Hunter, W. G., and Mezaki, R. (1966). *Amer. Inst. Chem. Eng. J.*, **12**, 1014–1017. (Diagnostic parameters.)

Kittrell, J. R., Hunter, W. G., and Watson, C. C. (1965). *Amer. Inst. Chem. Eng. J.*, **11**, 1051–1057. (Catalytic rate models.)

Kittrell, J. R., Hunter, W. G., and Watson, C. C. (1966). *Amer. Inst. Chem. Eng. J.*, **12**, 5–10.

Kittrell, J. R. and Mezaki, R. (1967). *Amer. Inst. Chem. Eng. J.*, **13**, 389–392. (Hougen–Watson models.)

Kittrell, J. R., Mezaki, R., and Watson, C. C. (1965). *Ind. Eng. Chem.*, **57**(12), 18–27.

Kittrell, J. R., Mezaki, R., and Watson, C. C. (1966). *Brit. Chem. Eng.*, **11**(1), 15–19.

Kittrell, J. R., Mezaki, R., and Watson, C. C. (1966). *Ind. Eng. Chem.*, **58**(5), 50–59. (Determination of reaction order.)

McLean, D. D., Bacon, D. W., and Downie, J. (1980). *Canad. J. Chem. Eng.*, **58**, 608–619.

Mezaki, R. and Butt, J. B. (1968). *Ind. Eng. Chem. Fund.*, **7**, 120–125.

Mezaki, R., Draper, N. R., and Johnson, R. A. (1973). *Ind. Eng. Chem. Fund.*, **12**, 251–254.

Mezaki, R. and Kittrell, J. R. (1966). *Canad. J. Chem. Eng.*, **44**, 285.

Mezaki, R. and Kittrell, J. R. (1967). *Ind. Eng. Chem.*, **59**(5), 63–69. (Parameter sensitivity.)

Mezaki, R. and Kittrell, J. R. (1968). *Amer. Inst. Chem. Eng. J.*, **14**, 513. (Model screening.)

Mezaki, R., Kittrell, J. R., and Hill, W. J. (1967). *Ind. Eng. Chem.*, **59**(1), 93–95.

Peterson, T. I. and Lapidus, L. (1966). *Chem. Eng. Sci.*, **21**, 655–664. (Kinetics of ethanol dehydrogenation.)

Podolski, W. F. and Kim, Y. G. (1974). *Ind. Eng. Chem. Process Des. Dev.*, **13**, 415–421. (Water–gas shift reaction.)

Pritchard, D. J. and Bacon, D. W. (1974). *Canad. J. Chem. Eng.*, **52**, 103–109.

Pritchard, D. J. and Bacon, D. W. (1975). *Chem. Eng. Sci.*, **30**, 567–574.

Pritchard, D. J., McLean, D. D., Bacon, D. W., and Downie, J. (1980). *J. Catal.*, **61**, 430–434.

Reilly, P. M. (1970). *Canad. J. Chem. Eng.*, **48**, 168–173.

Reilly, P. M. and Blau, G. E. (1974). *Canad. J. Chem. Eng.*, **52**, 289–299.

Reilly, P. M., Bajramovic, R., Blau, G. E., Branson, D. R., and Sauerhoff, M. W. (1977). *Canad. J. Chem. Eng.*, **55**, 614–622. (First-order kinetic models.)

Sutton, T. L. and MacGregor, J. F. (1977). *Canad. J. Chem. Eng.*, **55**, 602–608. (Design and estimation in vapor–liquid equilibrium models.)

(COMPUTERS AND STATISTICS
GENERAL LINEAR MODEL

ITERATIVELY REWEIGHTED LEAST SQUARES
LEAST SQUARES
NONLINEAR REGRESSION
STATISTICAL AND PROBABILISTIC MODELING)

D. M. Bates
W. G. Hunter

NONLINEAR PROGRAMMING

Nonlinear programming is the study of the problem of minimizing a function of several variables when the variables are constrained by functional equalities and inequalities. A standard formulation of the general nonlinear programming problem is:

Minimize	$f(\mathbf{x})$
Subject to	$g_i(\mathbf{x}) \leqslant 0, \quad i = 1, \ldots, m,$
and	$h_j(\mathbf{x}) = 0, \quad j = 1, \ldots, p,$

where the f, g_i, and h_j are continuous, smooth functions defined on n-dimensional Euclidean space.

This problem is a particular example of the more general optimization problem discussed in *mathematical programming**. In order to distinguish the preceding problem from the simpler *linear programming** problem, it is assumed that at least one of the functions occurring in the formulation is nonlinear.

There are three major topics to be considered in any treatise on nonlinear programming: theory, computation, and applications. The mathematical questions concerning the existence and characterization of solutions, the dependence of the solutions on the form and parameters of the functions that define the problem, and the formulation of equivalent problems are central to the theory of the subject. Computation refers to the development of (usually iterative) algorithms* for generating good approximate solutions to the problem. The applications in management science, engineering, and statistics provide special structure within the general form of the problem and hence guide the development of the theory and computation.

THEORY

The set of vectors that satisfy all of the equality and inequality constraints is called the *feasible* set. The continuity of the constraint functions g_i, $i = 1, \ldots, m$ and h_j, $j = 1, \ldots, p$ implies that the feasible set is closed. Hence the continuity of f guarantees that an optimal solution to the nonlinear problem exists provided that the feasible set is bounded or that f has appropriate growth properties.

One of the major problems inherent in the study of nonlinear programming is the difficulty in distinguishing *global* from *local* solutions. A local solution to the nonlinear program is a feasible vector x^* for which $f(x^*) \leqslant f(x)$ for all other feasible vectors in some neighborhood of x^*. x^* is a global solution if it is a feasible vector that satisfies $f(x^*) \leqslant f(x)$ for *all* feasible vectors x. Unless the problem has some special structure such as linearity or convexity, it is generally very difficult, and often impossible in practice, to determine if a local solution is a global solution. Consequently, most of the following theory pertains to local as well as global solutions.

An important objective in the study of an optimization problem is the characterization of the optimal points. For example, if f is a function of a single variable, then x^* is an (unconstrained) optimizer of f only if $f'(x^*) = 0$ and $f''(x^*) \geqslant 0$. Hence a minimizer of f is characterized by its inclusion in the set of x for which $f'(x) = 0$ and $f''(x) \geqslant 0$.

In order to characterize optimal points for the nonlinear programming problem, certain regularity conditions, often called *constraint qualifications*, must be imposed upon the constraint functions. A common qualification, which will be assumed henceforth, is that for each feasible x, the set of active constraint gradients at x,

$$\{\nabla g_i(x) : i \in I(x)\}$$

$$\cup \{\nabla h_j(x) : j = 1, \ldots, p\}$$

is a linearly independent set. Here $I(x) = \{1, \ldots, m\} \cap \{i : g_i(x) = 0\}$. For a discussion of other constraint qualifications and their importance, see Avriel [2].

Let λ and ω represent m- and p-dimensional vectors, respectively. The function of x, λ, ω defined by

$$L(x, \lambda, \omega) = f(x) + \sum_{i=1}^{m} \lambda_i g_i(x) + \sum_{j=1}^{p} \omega_j h_j(x)$$

is called the *Lagrangian** function and is crucial to the theory of the nonlinear optimization. Denoting by $L_x(x, \lambda, \omega)$, $L_\lambda(x, \lambda, \omega)$, and $L_\omega(x, \lambda, \omega)$ the gradients of L with respect to x, λ, and ω, respectively, and by $L_{xx}(x, \lambda, \omega)$ the $n \times n$ Hessian matrix of L with respect to x, the local optimal points of the nonlinear program can be characterized as follows.

Necessary Conditions

Let x^* be a local minimum point. Then there exist vectors λ^* and ω^* such that the following hold:

(1) $L_x(x^*, \lambda^*, \omega^*) = 0$.
(2) $L_\lambda(x^*, \lambda^*, \omega^*) \leqslant 0$.
(3) $L_\omega(x^*, \lambda^*, \omega^*) = 0$.
(4) $\lambda_i^* g_i(x^*) = 0$, $\quad i = 1, \ldots, m$.
(5) $\lambda_i^* \geqslant 0$, $\quad i = 1, \ldots, m$.
(6) For any nonzero n-vector d satisfying

$$\nabla g_i(x^*)^T d = 0, \quad i \in I(x^*),$$

$$\nabla h_j(x^*)^T d = 0, \quad j = 1, \ldots, p,$$

it is the case that

$$d^T L_{xx}(x^*, \lambda^*, \omega^*) d \geqslant 0.$$

Sufficient Conditions

Let x^* satisfy (1)–(5) for some λ^* and ω^* and also suppose that the following holds:

(7) For any nonzero n-vector d satisfying

$$\nabla g_i(x^*)^T d \leqslant 0, \quad i \in I(x^*),$$

$$\nabla g_i(x^*)^T d = 0, \quad i \in I(x^*) \text{ and } \lambda_i^* > 0,$$

$$\nabla h_j(x^*)^T d = 0, \quad j = 1, \ldots, p,$$

it is the case that

$$\mathbf{d}^T L_{\mathbf{xx}}(\mathbf{x}^*, \boldsymbol{\lambda}^*, \boldsymbol{\omega}^*)\mathbf{d} > 0.$$

Then $\mathbf{x}^*$ is an isolated local minimum.

$\boldsymbol{\lambda}^*$ and $\boldsymbol{\omega}^*$ are called the *multiplier vectors* for $\mathbf{x}^*$. The conditions (1)–(5) are called the first-order necessary conditions. Conditions (2) and (3) force $\mathbf{x}^*$ to be feasible while (1), (4), and (5) restrict the gradient of f at $\mathbf{x}^*$ so that f cannot decrease in a direction pointing into the linearization of the feasible set at $\mathbf{x}^*$. The second-order conditions (6) and (7) impose restrictions on the curvature of the level set of f at $\mathbf{x}^*$ (relative to the curvature of the feasible region) that constrain the possible directions in which f can decrease. In the unconstrained case the preceding sufficient conditions reduce to the well-known conditions that $\nabla f(\mathbf{x}^*) = 0$ and the Hessian of f is positive definite at $\mathbf{x}^*$.

It can be seen that $\mathbf{x}^*$ may satisfy the necessary conditions without being a local optimal solution or that $\mathbf{x}^*$ may be a local optimal solution without satisfying the sufficient conditions. For the wide and important class of convex programs, however, conditions (1)–(5) are both necessary and sufficient. The program is said to be a *convex program* if the functions f and g_i, $i = 1, \ldots, m$, are convex and the h_j, $j = 1, \ldots, p$, are affine (linear plus a constant).

The simplest example of a convex program with the corresponding optimality conditions is given by the following quadratic problem:

Minimize $\frac{1}{2}\mathbf{x}^T\mathbf{Q}\mathbf{x} + \mathbf{q}^T\mathbf{x}$

Subject to $\mathbf{A}\mathbf{x} - \mathbf{b} = 0,$

where $\mathbf{Q}$ is a positive definite $n \times n$ matrix, $\mathbf{q}$ is an n-vector, $\mathbf{A}$ is a $p \times n$ matrix of rank p, and $\mathbf{b}$ is a p-vector. Since $\mathbf{Q}$ is positive definite, this is a convex problem; hence the first-order necessary conditions are also sufficient. Conditions (1) and (3) are

$$\mathbf{Q}\mathbf{x} + \mathbf{A}^T\boldsymbol{\omega} = 0,$$
$$\mathbf{A}\mathbf{x} - \mathbf{b} = 0.$$

Since the coefficient matrix is nonsingular, the system can be solved directly to obtain the *global* optimal solution and its multiplier.

The history of the development of the optimality conditions covers a long period. The first-order necessary conditions for the case where only equality constraints are present were known to Lagrange (see LAGRANGE MULTIPLIERS*) while the necessary conditions for the inequality-constrained problem are less than a half-century old. These latter conditions are sometimes called the Karush–Kuhn–Tucker conditions. For further results on optimality conditions and a discussion of their development, the reader is referred to Auriel [2], Cottle and Lemke [4], and Hestenes [11].

A practical consideration for any user of optimization problems as models is the behavior of the solution and the optimal value when the parameters that define the constraint and objective functions are changed. One such situation, which also provides insight to the meaning of the optimal multipliers, occurs when the right-hand sides of the constraint relations are perturbed as in the following problem:

Minimize $f(\mathbf{x})$

Subject to $g_i(\mathbf{x}) \leqslant b_i, \qquad i = 1, \ldots, m,$

and $h_j(\mathbf{x}) = c_j, \qquad j = 1, \ldots, p.$

Here $\mathbf{b} = (b_1, \ldots, b_m)$ and $\mathbf{c} = (c_1, \ldots, c_p)$ are small vectors. If a local optimal solution to this perturbed problem is denoted by $\mathbf{x}^*(\mathbf{b}, \mathbf{c})$ and $\phi(\mathbf{b}, \mathbf{c}) = f(\mathbf{x}^*(\mathbf{b}, \mathbf{c}))$, the following fundamental result is obtained as a consequence of the implicit function theorem.

Basic Perturbation Theorem

Suppose $\mathbf{x}^*$ is a local solution to the unperturbed problem at which the sufficient conditions given earlier hold. In addition, suppose $\lambda_i^* > 0$ for $i \in I(\mathbf{x}^*)$. Then there exist continuously differentiable functions $\mathbf{x}^*(\mathbf{b}, \mathbf{c})$ and $\phi(\mathbf{b}, \mathbf{c}) = f(\mathbf{x}^*(\mathbf{b}, \mathbf{c}))$ defined in a neighborhood of $(\mathbf{b}, \mathbf{c}) = (\mathbf{0}, \mathbf{0})$ such that $\mathbf{x}^*(\mathbf{b}, \mathbf{c})$ is a local solution to the perturbed problem, $\mathbf{x}^*(\mathbf{0}, \mathbf{0}) = \mathbf{x}^*$, $\nabla_b\phi(\mathbf{0}, \mathbf{0}) = -\boldsymbol{\lambda}^*$, and $\nabla_c\phi(\mathbf{0}, \mathbf{0}) = -\boldsymbol{\omega}^*.$

In Avriel [2] and Fiacco [6], the reader can find a more complete development of duality theory and perturbation theory for nonlinear programming.

COMPUTATION

Finding a solution, even a local one, to the nonlinear programming problem is by no means a simple task. This is especially true when the number of variables is large or the functions are highly nonlinear. Only the advent of electronic computers has made the solution of the general nonlinear optimization problem practical.

The most common methods for solving nonlinear problems are iterative in nature. That is, a sequence of vectors, $\{\mathbf{x}^k\}$, is generated, each $\mathbf{x}^k$ being, in some sense, a better approximation to a solution than the preceding $\mathbf{x}^{k-1}$. For the unconstrained problem, the most common iterative methods are the descent methods. They can be generally described as follows: Given a current iterate $\mathbf{x}^k$, the next iterate is chosen by the equation

$$(8) \qquad \mathbf{x}^{k+1} = \mathbf{x}^k - \alpha^k \mathbf{d}^k,$$

where $\mathbf{d}^k$ is an n-vector and $\alpha^k > 0$ is a scalar, called the *step-length parameter*. The vector $\mathbf{d}^k$ is chosen so that

$$(9) \qquad \nabla f(\mathbf{x}^k)^T \mathbf{d}^k > 0,$$

which implies that $f(\mathbf{x}^{k+1}) < f(\mathbf{x}^k)$ provided that α^k is properly chosen. Details of these and other iterative schemes for solving the unconstrained problem can be found in Dennis and Schnabel [5], Fletcher [8], and Hestenes [11].

In trying to adapt descent methods to constrained problems, one is confronted immediately with the difficulty of deciding which function should be decreased. Given a current iterate $\mathbf{x}^k$ which is feasible, a step generated by (8) and (9) may decrease f but cause $\mathbf{x}^{k+1}$ to be infeasible. On the other hand, if $\mathbf{x}^k$ is not feasible, decreasing f from $\mathbf{x}^k$ may be inappropriate if $\mathbf{x}^{k+1}$ is not closer to feasibility.

One approach to solving the constrained problem is to "penalize" the objective func-

tion when x is infeasible. In this method, a *penalty* function such as

$$P(\mathbf{x}, \mu) = f(\mathbf{x})$$
$$+ \mu \left(\sum_{i=1}^{m} (|g_i(\mathbf{x})|_+)^2 + \sum_{j=1}^{p} |h_j(\mathbf{x})|^2 \right)$$

is minimized by unconstrained techniques. If μ (called the penalty parameter) is sufficiently large, then a minimizer of P is close to a solution of the constrained problem. In ref. 7, these methods are discussed in some detail.

Another method for solving the constrained problem is motivated by the simplex algorithm of *linear programming**. In this approach, the nonlinear constraints are linearized and the variables partitioned into basic and nonbasic sets. A descent direction is then determined in terms of the nonbasic variables. A refinement of this procedure, called the method of *reduced gradients*, and related techniques (e.g., the method of *projected gradients*) are very effective iteration schemes when the nonlinear program has linear or nearly linear constraints. See Avriel [2], Fletcher [9], and Hestenes [11] for details.

A more recent development uses quadratic approximations of the nonlinear program to generate iterates. Variants of this method have been shown to have properties similar to those of the secant methods for unconstrained optimization. Consequently, they hold promise of being among the most effective of algorithms for solving the general nonlinear program. For a thorough presentation of these ideas, see Bertsekas [3] and Fletcher [9].

The methods mentioned here do not, by any means, exhaust the list of algorithmic schemes that have been developed for solving the nonlinear program. In particular, there are many procedures that have been proposed for programs with special structure (e.g., convex and nondifferentiable programs, programs with large sparse data sets). The reader is referred to the references at the end of this article for discussions of these methods.

APPLICATIONS

A major source of nonlinear optimization models is the field of management science*. For many of the models found in *linear programming*, an effort to achieve more meaningful results will often require the incorporation of nonlinearities into the model. Another, less obvious, way in which a linear model can be transformed into a nonlinear model is by taking into account the stochastic character of the parameters in the problem. For example, if the constants a_j in the linear program

$$\text{Minimize} \quad \sum_{j=1}^{n} c_j x_j$$

$$\text{Subject to} \quad \sum_{j=1}^{n} a_j x_j \leqslant b$$

are assumed to be independently distributed random variables, say, $a_j \sim N(\bar{a}_j, \sigma_j^2)$, one can replace the constraint by the *chance constraint*.

$$(10) \qquad \Pr\left[\sum_{j=1}^{n} a_j x_j \leqslant b\right] \geqslant 1 - \beta,$$

where β is an appropriate small number. A relatively simple argument shows that (10) is equivalent to the nonlinear deterministic constraint

$$\sum_{j=1}^{n} \bar{a}_j x_j + F^{-1}(\beta)\left(\sum_{j=1}^{n} \sigma_j^2 x_j^2\right)^{1/2} \leqslant b,$$

where F is the distribution function for the standard normal*. The section on stochastic programming in MATHEMATICAL PROGRAMMING contains more details and references to chance-constrained optimization. For a survey of applications of nonlinear programming to operations research* and management science, see Lasdon and Waren [12].

A second application of constrained nonlinear optimization is to find approximate solutions to *optimal control* problems. For example, the problem of choosing a control function $u(t)$ such that

$$L(u) = \int_0^T [u(s)]^2 ds$$

is minimized over all square integrable func-

tions on $[0, T]$ such that

$$\dot{x}(t) = \psi(x(t), u(t)), \qquad t \in [0, T],$$

$$x(0) = x_0, \qquad x(T) = x_T,$$

and

$$|u(t)| \leqslant K, \qquad t \in [0, T],$$

might be considered a typical optimal control problem that can occur in engineering or management science. By partitioning $[0, T]$ such that $0 = t_0 < t_1 < \cdots < t_n = T$ and identifying the n-vector $\mathbf{u}$ with the piecewise constant function $u(t)$ according to

$$u(t) = u_j, \qquad t \in [t_{j-1}, t_j),$$

the optimal control problem can be approximated by the discrete problem

$$\text{Minimize} \quad \sum_{j=1}^{n} u_j^2$$

$$\text{Subject to} \quad (t_j - t_{j-1})\psi(x(t_{j-1}), u_j)$$

$$= x(t_j) - x(t_{j-1}),$$

$$j = 1, \ldots, n,$$

$$x(t_0) = x_0, \qquad x(t_n) = x_T$$

and $\quad |u_j| \leqslant K, \qquad j = 1, \ldots, n.$

Here the solution to the differential equation has been approximated by the Euler method. The piecewise constant function determined by the optimal solution to this nonlinear program will, under certain conditions, closely approximate an optimal solution to the control problem. Gruver and Sachs [10] provide many practical examples of this nature.

Another major use of constrained nonlinear optimization is in regression analysis or parameter estimation. In a typical example a set of observations is generated by an unknown function (or according to a probability distribution). The form of the function (or the probability distribution) with unknown parameters is hypothesized, and the parameters which minimize a predetermined error function (such as the least-squares error) are obtained by solving the nonlinear program. The constraints are bounds on the parameters and functional relations between them. An example of this type of application is found in MATHEMATICAL PROGRAMMING;

see also MAXIMUM LIKELIHOOD ESTIMATION*. Arthanari and Dodge [1] and Tapia and Thompson [13] contain expositions of the theory and many examples in this area.

References

[1] Arthanari, T. S. and Dodge, Y. (1981). *Mathematical Programming in Statistics*. Wiley, New York. (Emphasizes linear programming applications.)

[2] Avriel, M. (1976). *Nonlinear Programming: Analysis and Methods*. Prentice-Hall, Englewood Cliffs, NJ. (The most complete reference currently available on nonlinear programming.)

[3] Bertsekas, D. P. (1982). *Constrained Optimization and Lagrange Multiplier Methods*. Academic Press, New York. (An up-to-date description of computational methods.)

[4] Cottle, R. and Lemke, C., eds. (1976). *Nonlinear Programming*. American Mathematical Society, Providence, RI. (A collection of articles on the history and recent developments in the field.)

[5] Dennis, J. E., Jr. and Schnabel, R. B. (1983). *Numerical Methods for Nonlinear Equations and Unconstrained Optimization*. Prentice-Hall, Englewood Cliffs, NJ.

[6] Fiacco, A. V. (1983). *Introduction to Sensitivity and Stability Analysis in Nonlinear Programming*. Academic Press, New York. (The best reference text available on the perturbation of nonlinear programs.)

[7] Fiacco, A. V. and McCormick, G. (1968). *Nonlinear Programming: Sequential Unconstrained Minimization Techniques*. Wiley, New York. (A text on penalty function methods.)

[8] Fletcher, R. (1980). *Practical Methods of Optimization*, Vol. 1. Wiley, New York. (A text on computational schemes for unconstrained problems.)

[9] Fletcher, R. (1980). *Practical Methods of Optimization*, Vol. 2. Wiley, New York. (A companion piece to ref. 8 on constrained optimization.)

[10] Gruver, W. A. and Sachs, E. (1980). *Algorithmic Methods in Optimal Control*. Pitman, Boston.

[11] Hestenes, M. (1975). *Optimization Theory: The Finite Dimensional Case*. Wiley, New York.

[12] Lasdon, L. S. and Waren, A. D. (1980). *Operat. Res.*, **28**, 1029–1073. (A survey of nonlinear programming applications with an excellent bibliography.)

[13] Tapia, R. A. and Thompson, J. R. (1978). *Nonparametric Probability Density Estimation*. Johns Hopkins University Press, Baltimore.

(INTEGER PROGRAMMING
LAGRANGE MULTIPLIERS
LINEAR PROGRAMMING
MATHEMATICAL PROGRAMMING
MAXIMUM LIKELIHOOD ESTIMATION
OPTIMIZATION IN STATISTICS)

JON W. TOLLE

NONLINEAR REGRESSION

Nonlinear regression is used when one wishes to estimate parameters in a nonlinear model that relates a response Y to some control or predictor variables (x_j, $j = 1$, $2, \ldots, J$). A nonlinear model consists of an expectation function η, a mathematical function of known form that depends on the values of the vector $\mathbf{x}$ and the parameters $\boldsymbol{\theta}$, plus an additive disturbance. That is, the response on the tth experiment, $t = 1, 2, \ldots, n$, can be written

$$Y_t = \eta(\mathbf{x}_t, \boldsymbol{\theta}) + \epsilon_t,$$

where $\eta(\mathbf{x}_t, \boldsymbol{\theta})$ is the value of the expectation function on the tth experiment, $\mathbf{x}_t = (x_{t1}, x_{t2}, \ldots, x_{tj})^T$ is the setting of the control variables on the tth experiment, the T indicating the transpose of the vector, $\boldsymbol{\theta}$ is a K-dimensional vector of unknown coefficients or parameters, $\boldsymbol{\theta} = (\theta_1, \theta_2, \ldots, \theta_K,)^T$, and ϵ_t is a noise* or disturbance term, usually assumed to be normally distributed with mean 0 and variance σ^2 and independent of the disturbance on any other experiment.

Linearity or nonlinearity of a model depends on how the parameters occur in the expectation function, but not on how the predictor variables do. Thus the expectation function

$$\eta(\mathbf{x}, \boldsymbol{\theta}) = \theta_1 x + \theta_2 x^2$$

is linear because each parameter is multiplied by a quantity that depends only on $\mathbf{x}$, and the resulting products are then summed. Alternatively, the model is seen to be linear because the derivatives of η with respect to the parameters do not depend on the parameters; that is, $\partial\eta/\partial\theta_1 = x$ and $\partial\eta/\partial\theta_2 = x^2$.

Nonlinear models arise when an investigator has derived, through development of a theory or otherwise, a functional relationship in which the parameters do not occur linearly. An example is the Michaelis–Menten function [16], which relates the initial rate, or "velocity," of an enzymatic reaction to the concentration of the substrate x through $\eta = \theta_1 x / (\theta_2 + x)$. This model is nonlinear because the derivatives $\partial \eta / \partial \theta_1 = x / (\theta_2 + x)$, and $\partial \eta / \partial \theta_2 = -\theta_1 x / (\theta_2 + x)^2$ involve at least one of the parameters.

Some expectation functions are transformable to a linear form, for example, the reciprocal of the Michaelis–Menten expectation function can be written

$$1/\eta = 1/\theta_1 + (\theta_2/\theta_1)(1/x),$$

which is linear in the parameters $\beta_1 = (1/\theta_1)$ and $\beta_2 = (\theta_2/\theta_1)$. One could therefore estimate β_1 and β_2 using linear regression of reciprocal velocity data on reciprocal substrate concentration and then solve for θ. It must be remembered, however, that transformation of the data involves transformation of the disturbance term as well as the expectation function, and so the assumptions of constant variance and normality required for simple linear regression may no longer be valid. As a consequence, the estimates of β, and hence of θ, may be biased* or suffer from other deficiencies. Linearization should only be used when the analyst is certain that the *transformed* data is adequately described by a model with an additive normal disturbance term.

As in linear regression, the maximum likelihood* estimates of the parameters θ are the least-squares* values; that is, the values that minimize the sum of squares:

$$S(\theta) = \sum_1^n \{ y_t - \eta(\mathbf{x}_t, \theta) \}^2,$$

where y_t is the observed data value on the tth experiment. In contrast to linear models, however, nonlinear models can cause several difficulties. First, it is not possible to write down an explicit expression for the least-squares* estimates, as it is in the linear regression case, and second, it is not usually possible to derive the exact distributional properties of parameter estimators. This precludes simple summaries of confidence* or likelihood regions in the nonlinear case. These difficulties usually are avoided by employing a linear expansion to derive approximate inference regions and iterative techniques to obtain parameter estimates (see, e.g., Draper and Smith [9] and Bard [1]).

The stages of a nonlinear regression analysis consist of the following:

1. Use the data $\mathbf{y}$, the control settings $\mathbf{x}_t$, and the expectation function $\eta(\mathbf{x}, \theta)$ to obtain starting estimates θ^0 for the parameters.

2. Use the information from (1) in an iterative nonlinear estimation computer program to obtain the least-squares estimates $\hat{\theta}$ and to produce linear approximation summary statistics.

3. Investigate the fitted model for adequacy of fit and for sensibleness of the parameter estimates by examining the residuals and the parameter estimates as in linear regression.

4. Determine the adequacy of the approximation used for the summary statistics.

DETERMINING PARAMETER ESTIMATES

Nonlinear least squares is a nonlinear optimization problem and so standard nonlinear optimization* algorithms such as steepest descent*, quasi-Newton, or conjugate gradient could be used (see, e.g., Chambers [7, Chap. 6] or Kennedy and Gentle [13, Chap. 10]). However, these general methods do not exploit the particular structure of nonlinear least squares, so specialized methods are preferred. The most common of these are the Gauss–Newton algorithm [9] (*see* NEWTON ITERATION EXTENSIONS), which uses a linear expansion of the expectation function, and the Levenburg–Marquardt [14, 15] algorithm, which provides a compromise between steepest descent and the Gauss–Newton method. Both of these methods require derivatives of the expectation function with respect to the parameters. Since the determination and coding of these deriva-

Table 1 Michaelis–Menten Data and Calculations

Substrate x	Velocity y	Expected Value at $\boldsymbol{\theta}^0$ $\eta(\boldsymbol{\theta}^0)$	Residuals at $\boldsymbol{\theta}^0$, $e^0 = y - \eta(\boldsymbol{\theta}^0)$	Derivatives at $\boldsymbol{\theta}^0$ By θ_1, v_1^0	By θ_2, v_2^0
0.3330	3.636	3.490	0.146	0.943	$-\;9.89$
0.1670	3.636	3.304	0.332	0.893	$-\;17.67$
0.0833	3.236	2.984	0.252	0.806	$-\;28.88$
0.0416	2.666	2.499	0.167	0.675	$-\;40.56$
0.0208	2.114	1.886	0.228	0.510	$-\;46.23$
0.0104	1.466	1.266	0.200	0.342	$-\;41.64$
0.0052	0.866	0.763	0.103	0.206	$-\;30.30$

tives can be the most error-prone and time-consuming aspect of fitting nonlinear regression models, numerical derivatives are frequently used. Other specialized methods such as the DUD [17] algorithm restructure the iterative scheme to make derivatives unnecessary.

Among algorithms using numerical or analytical derivatives, the NL2SOL algorithm of Dennis et al. [8] creates an approximation to the Hessian* matrix, $\partial^2 S(\boldsymbol{\theta})/\partial\theta_i\,\partial\theta_j$, using a quasi-Newton update. This update is designed to provide a closer approximation to the Hessian than is available with either the Gauss–Newton or Levenburg–Marquardt algorithm, particularly when the residuals for the fitted model are large, which should expedite convergence. Hiebert [12] compares computer packages that implement some of the algorithms mentioned.

An Example Using the Gauss–Newton Algorithm. To help understand nonlinear estimation, it is instructive to follow a simple example. We use the Michaelis–Menten expectation function introduced earlier, and the data obtained by Michaelis and Menten in their original paper [16]. The data are reproduced in the first two columns of Table 1.

OBTAINING STARTING VALUES

Because the Michaelis–Menten expectation function is transformably linear, one could regress the reciprocal velocity data on the reciprocal substrate concentration to obtain starting values for $\boldsymbol{\theta}$. For functions that are not transformably linear, other approaches can be used, as illustrated below.

Consideration of the expectation function as a function of x reveals that as x increases, η approaches θ_1; that is, θ_1 is the *maximum velocity*. Thus we can use the maximum observed value as a starting value for θ_1, say, $\theta_1^0 = 3.7$. Further consideration reveals that the initial slope of the curve, that is, $\partial f/\partial x$ at $x = 0$ is θ_1/θ_2, or $\theta_2 = \theta_1/(\text{slope near } x = 0)$, and hence we could use $\theta_2^0 = 3.7/(0.866/0.0052) = 0.02$. Alternatively, θ_2 is the *half-velocity concentration*, that is, the value of x such that $\eta = \theta_1/2$. From the data we see that a velocity of $3.7/2 = 1.85$ would occur at a substrate concentration of about 0.015, and so we could use $\theta_2^0 = 0.015$. To illustrate the process, we choose the starting vector $\boldsymbol{\theta}^0 = (3.7, 0.02)^T$, which we can substitute in the expression for η. Then we evaluate that function at the particular x values to obtain the values shown in column 3 headed $\eta(\boldsymbol{\theta}^0)$. Note that we have dropped the explicit dependence of η on $\mathbf{x}$ because these control values are fixed and hence are simply treated as constants in what follows.

ITERATING

Using the starting values $\boldsymbol{\theta}^0$, we develop a linear Taylor series approximation for the expectation function as

$$\eta(\boldsymbol{\theta}) \simeq \eta(\boldsymbol{\theta}^0) + (\partial\eta/\partial\theta_1)(\theta_1 - \theta_1^0)$$
$$+ (\partial\eta/\partial\theta_2)(\theta_2 - \theta_2^0)$$
$$= \eta(\boldsymbol{\theta}^0) + v_1^0\delta_1 + v_2^0\delta_2,$$

using an obvious notation. The derivatives evaluated at the design points are shown in columns 5 and 6 of Table 1. We now test whether the point $\boldsymbol{\theta}^0$ is the least-squares point using the procedure described under the following section, Testing for Convergence. Assuming $\boldsymbol{\theta}^0$ is not the best parameter estimate, we proceed as follows.

The original model can now be written approximately as

$$Y_t - \eta_t(\boldsymbol{\theta}^0) \simeq v_{t1}^0\delta_1 + v_{t2}^0\delta_2 + \epsilon_t,$$

which looks linear in the "parameters" $\boldsymbol{\delta}$. We then use linear regression to obtain the apparent least-squares* increment

$$\hat{\boldsymbol{\delta}} = (\mathbf{V}_0^T\mathbf{V}_0)^{-1}\mathbf{V}_0^T\mathbf{e}^0,$$

where $\mathbf{e}^0 = \mathbf{y} - \boldsymbol{\eta}(\boldsymbol{\theta}^0)$ is the residual vector at $\boldsymbol{\theta}^0$ and is given in column 4 of Table 1. The matrix $\mathbf{V}_0 = (\mathbf{v}_1^0, \mathbf{v}_2^0)$ is the derivative matrix evaluated at $\boldsymbol{\theta}^0$ and consists of columns 5 and 6 of Table 1. Using the preceding equation, we find $\hat{\boldsymbol{\delta}} = (0.208, -0.002)^T$. We now test whether the point $\boldsymbol{\theta}^1 = \boldsymbol{\theta}^0 + \hat{\boldsymbol{\delta}} = (3.908, 0.018)^T$ is better than $\boldsymbol{\theta}^0$ by comparing the sum of squared residuals, $\mathbf{e}^{1T}\mathbf{e}^1$ at the new test point to the sum of squares $\mathbf{e}^{0T}\mathbf{e}^0$ at the starting point. If $S(\boldsymbol{\theta}^1) < S(\boldsymbol{\theta}^0)$, we assume that $\boldsymbol{\theta}^1$ is a better estimate than $\boldsymbol{\theta}^0$, and so we start another iteration by calculating new derivative vectors and test to see whether $\boldsymbol{\theta}^1$ is the least-squares point. If it is not, we complete the iteration, and continue iterating until we achieve convergence. In this example, the initial sum of squares was 0.325 while the new value was only 0.023. Hence we continue iterating.

Sometimes the sum of squared residuals at $\boldsymbol{\theta}^i$ on the ith iteration is greater than at $\boldsymbol{\theta}^{i-1}$, so we have "overshot" the least-squares point. In that case, as recommended by G. E. P. Box [6] and H. O. Hartley [11], a step equal to a fraction, say, $g < 1$, times the increment is used, and the sum of squared residuals at $\boldsymbol{\theta}^i = \boldsymbol{\theta}^{i-1} + g\hat{\boldsymbol{\delta}}$ is determined. If $S(\boldsymbol{\theta}^i) \geqslant S(\boldsymbol{\theta}^{i-1})$, the subprocedure of reducing the step size and testing for a reduction is repeated until the $S(\boldsymbol{\theta}^i) < S(\boldsymbol{\theta}^{i-1})$, whereupon the iteration may be continued. To prevent the step size from staying small, g is increased when a reduction in the sum of squares occurs.

When convergence has been declared, summary statistics are calculated and printed. These should include the least-squares parameter estimates $S(\hat{\boldsymbol{\theta}})$, the residual mean square s^2 and its degrees of freedom $(n - K)$, and the approximate parameter estimators variance matrix $(\hat{\mathbf{V}}^T\hat{\mathbf{V}})^{-1}s^2$, where $\hat{\mathbf{V}}$ is the derivative matrix evaluated at $\hat{\boldsymbol{\theta}}$. The variance matrix can be then be used to determine linear approximation joint and marginal parameter inference regions using the methods of linear regression. However, as discussed in the section Effects of Nonlinearity, these linear regions can be very poor approximations of the true region.

The fitted model should be criticized by looking at plots of the residuals* vs. the calculated expected values and the control variables, as in the linear regression case.

TESTING FOR CONVERGENCE

Most computer programs for nonlinear least squares require one or more convergence criteria to be met before stopping. Frequently these criteria are based on the relative change in $S(\boldsymbol{\theta})$ from one iteration to the next or on the relative change in the components of $\boldsymbol{\theta}$, and so they indicate whether the algorithm is succeeding in approaching a minimum. Although they are usually reliable, they are not unambiguous indicators of convergence.

It is possible, however, to have an unambiguous indicator of convergence, since true convergence is attained when the residual vector $\hat{\mathbf{e}} = (\mathbf{y} - \boldsymbol{\eta}(\hat{\boldsymbol{\theta}}))$ is orthogonal to the expectation surface $\boldsymbol{\eta}(\boldsymbol{\theta})$ evaluated at $\hat{\boldsymbol{\theta}}$, since then the sum of squared residuals $\hat{\mathbf{e}}^T\hat{\mathbf{e}}$

must be least. (See the material on geometry* in the following section.) In practice, exact orthogonality is not necessary; it is enough that the residual vector on the ith iteration, say, $\mathbf{e}^i$, has a sufficiently small component in the tangent plane at $\boldsymbol{\eta}(\boldsymbol{\theta}^i)$. Bates and Watts [3] have proposed such a convergence criterion based on the relative length of the residual vector in the tangent plane at $\boldsymbol{\eta}(\boldsymbol{\theta}^i)$ to the squared length of the residual vector $\mathbf{e}^i$. This has the interpretation of a relative offset of the center of the inference region to the radius of the inference disk on the approximating tangent plane. Hence a statistically meaningful tolerance level for the convergence criterion is obtained.

EFFECTS OF NONLINEARITY

Exact distributional properties of least-squares estimators generally are not available in the nonlinear regression case, so an investigator is often unsure of the adequacy of linear approximation inference regions in a given situation. To provide information on this, measures of nonlinearity have been developed by Beale [5], and Bates and Watts [2], using geometrical ideas.

The geometrical approach makes use of a sample or response space in which the data are represented as an n-dimensional vector $\mathbf{y}$, and the expected responses as another n-dimensional vector $\boldsymbol{\eta}$ with components $\eta(\mathbf{x}_t, \boldsymbol{\theta})$, $t = 1, 2, \ldots, n$. Since the values of $\mathbf{x}_t$ are fixed when a given set of data is being analyzed, the expectation vector $\boldsymbol{\eta}$ is written simply as $\boldsymbol{\eta}(\boldsymbol{\theta})$. The surface generated by vectors calculated using all possible values of $\boldsymbol{\theta}$ is called the *expectation surface*. The residual sum of squares at any value of $\boldsymbol{\theta}$ is then seen to be the squared length of the vector $\mathbf{e}$ joining $\boldsymbol{\eta}(\boldsymbol{\theta})$ to $\mathbf{y}$, so that

$$S(\boldsymbol{\theta}) = \|\mathbf{y} - \boldsymbol{\eta}(\boldsymbol{\theta})\|^2 = \|\mathbf{e}\|^2.$$

To illustrate the geometrical approach and important concepts of nonlinearity, we provide a simple example with a one-parameter model and some data obtained by Count

Rumford [18]. The experiment Rumford performed during his investigations on the nature of heat was to measure the temperature of a cannon heated by grinding it with a closely fitting metal bore. The ambient temperature was 60°F and the cannon was brought to a temperature of 130°F. Assuming Newton's law of cooling which states that the rate of change of temperature, τ, of an object is negatively proportional to the difference between τ and the temperature τ_0 of the surrounding medium, we have

$$d\tau/dt = -\theta(\tau - \tau_0).$$

Solving this differential equation and satisfying the initial and final conditions gives the expectation function

$$\eta = 60 + 70\exp(-\theta t).$$

We use only two design points, $t = 4$ minutes and $t = 41$ minutes and calculate, for any specified value of the rate parameter θ, the values of the expectation vector

$$\boldsymbol{\eta}(\theta) = (60 + 70\exp(-4\theta),$$

$$60 + 70\exp(-41\theta))^T.$$

For example,

at	$\theta = 0.00$	$\boldsymbol{\eta}(0) = (130, 130)$
at	$\theta = 0.01$	$\boldsymbol{\eta}(0.01) = (127, 106)$
at	$\theta = 0.10$	$\boldsymbol{\eta}(0.1) = (107, 61)$
at	$\theta = 1.00$	$\boldsymbol{\eta}(1) = (61, 60)$.

We then plot the coordinates of the vector $\boldsymbol{\eta}(\theta)$ in the response space to produce an expectation surface shown as the curved line in Fig. 1.

The first thing to note is that the expectation surface is curved. This is in marked contrast to the linear regression situation where the expectation surface is always planar. Because the expectation surface is curved, we say that it has *intrinsic curvature*, and consequently there is *intrinsic nonlinearity*. Note also that the expectation surface terminates at (60, 60), which corresponds to infinite θ. This is also in contrast to the linear situation, in which the expectation surface is always of infinite extent. The tick marks on the expectation surface indicate

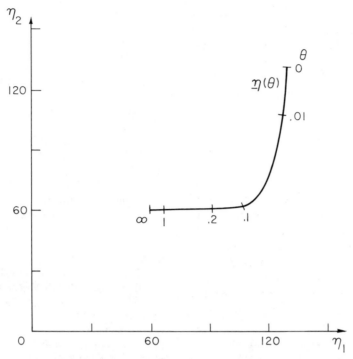

Figure 1   Expectation surface for the Count Rumford example.

values of θ, and it is seen that as θ goes from 0 to 0.1 and from 0.1 to 0.2 the distance traveled on the expectation surface is very different. This changing metric phenomenon is termed *parameter effects curvature* or *parameter effects nonlinearity*.

The term *intrinsic* is used because we could reformulate the expectation function in terms of, say, a time constant $\phi = 1/\theta$ rather than a rate constant, in which case the same expectation surface would be obtained but a different set of tick marks would occur (e.g., $\phi = 0$ would occur at (60, 60), and $\phi = \infty$ at (130, 130)). That is, the intrinsic expectation surface is completely specified by the form of the expectation function and the experimental design and cannot be changed once the experiment has been performed. The parameter effects nonlinearity, however, which depends on how parameters occur in the expectation function, can be changed. In this one-parameter example, the only parameter effect which can occur is that of a change of increment on the expectation surface for fixed changes

in the parameter value. When there are more parameters, other parameter effects can occur [4].

To determine the severity of any nonlinearities present, Beale [5] and Bates and Watts [2] investigated the behaviour of second-order Taylor series approximations to nonlinear models. These resulted in relative curvature measures of nonlinearity, the most useful of which appear to be the mean square intrinsic curvature and the mean square parameter curvature. Bates and Watts [4] have further shown that the parameter curvature under a new parameterization can be obtained simply and efficiently from the parameter curvature under the old parameterization.

Experience with measuring nonlinearity of different models and data sets has shown that in all the cases investigated the intrinsic curvature was smaller than the parameter curvature, and in almost all cases the intrinsic curvature was so small that it would not adversely affect a linear approximation inference region. In contrast, the parameter curvature was almost always large enough to adversely affect a linear approximation inference region. This suggests that most linear approximation inference regions are quite suspect, although likelihood regions are better approximated than confidence regions [10].

Summary

Nonlinear regression is a useful technique requiring considerable care in its application and in interpretation of results. Linear approximation marginal and joint parameter confidence regions can be especially misleading. If difficulties are encountered in obtaining convergence, the analyst should check the coding of the expectation function and its derivatives as well as the data being input to the program and make sure that the model is not overparameterized. It is also often advisable to use different starting values to see if the same parameter estimates are obtained. The final model should be criticized using residual plots and other techniques common to linear regression.

References

[1] Bard, Y. (1974). *Nonlinear Parameter Estimation.* Academic Press, New York, Chap. 5.

[2] Bates, D. M. and Watts, D. G. (1980). *J. R. Statist. Soc. B,* **42,** 1–25.

[3] Bates, D. M. and Watts, D. G. (1981). *Technometrics,* **23,** 179–183.

[4] Bates, D. M. and Watts, D. G. (1981). *Ann. Statist.,* **9,** 1152–1176.

[5] Beale, E. M. L. (1960). *J. R. Statist. Soc. B,* **22,** 41–76.

[6] Box, G. E. P. (1960). *Ann. N.Y. Acad. Sci.,* **86,** 792–816.

[7] Chambers, J. R. (1977). *Computational Methods for Data Analysis.* Wiley, New York, Chap. 6.

[8] Dennis, J. E., Jr., Gay, D. M., and Welsch, R. E. (1981). *ACM Trans. Math. Software,* **7,** 348–368.

[9] Draper, N. R. and Smith, H. (1981). *Applied Regression Analysis,* 2nd ed. Wiley, New York, Chap. 10.

[10] Hamilton, D. C., Watts, D. G. and Bates, D. M. (1982). *Ann. Statist.,* **10,** 386–393.

[11] Hartley, H. O. (1961). *Technometrics,* **3,** 269–280.

[12] Hiebert, K. L. (1981). *ACM Trans. Math. Software,* **7,** 1–16.

[13] Kennedy, W. J., Jr. and Gentle, J. E. (1980). *Statistical Computing.* Marcel Dekker, New York, Chap. 10.

[14] Levenburg, K. (1944). *Quart. Appl. Math.,* **2,** 164–168.

[15] Marquardt, D. W. (1963). *J. Soc. Ind. Appl. Math.,* **11,** 431–441.

[16] Michaelis, L. and Menten, M. L. (1913). *Biochemische Zeit.,* **49,** 333–339.

[17] Ralston, M. L. and Jennrich, R. I. (1978). *Technometrics,* **20,** 7–14.

[18] Roller, D. (1950). *The Early Development of the Concepts of Temperature and Heat.* Harvard University Press, Cambridge, MA, p. 71.

(BIAS
COMPUTERS AND STATISTICS
CONFIDENCE INTERVALS AND REGIONS
GEOMETRY IN STATISTICS
LEAST SQUARES
LIKELIHOOD
LINEAR REGRESSION
MAXIMUM LIKELIHOOD ESTIMATION
NONLINEAR MODELS
OPTIMIZATION IN STATISTICS
RESIDUALS)

D. G. WATTS
D. M. BATES

NONLINEAR RENEWAL THEORY

Let $X_1, X_2, \ldots$ be independent and identically distributed with positive mean μ and let ξ_n, $n \geq 1$, be any sequence of random variables for which $\xi_1, \ldots, \xi_n$ are independent of the sequence X_{n+k}, $k \geq 1$, for every $n \geq 1$. Let S_n, $n \geq 0$, denote the random walk*, $S_0 = 0$, and $S_n = X_1 + \cdots + X_n$, $n \geq 1$; let Z_n, $n \geq 1$, denote the perturbed random walk, $Z_0 = 0$ and

$$Z_n = S_n + \xi_n, \qquad n \geq 1. \qquad (1)$$

Renewal theory describes certain asymptotic properties of the random walk S_n, $n \geq 0$. See, for example, Feller [5, Chap. 11] and RENEWAL THEORY. Nonlinear renewal theory seeks to establish similar properties for the perturbed random walk.

A random variable X is said to be *arithmetic* iff there is an $h > 0$ for which

$$P\{X = \pm kh, \text{ for some } k = 0, 1, 2, \ldots\} = 1,$$

in which case the *span* of X is defined to be the largest such h.

For $a > 0$, let

$$t_a = \inf\{n \geq 1 : Z_n > a\},$$

$$R_a = Z_{t_a} - a,$$

where the infimum of the empty set is taken to be ∞ and R_a is defined only in the event that $t_a < \infty$. Thus t_a is the time at which the process Z_n, $n \geq 0$, first crosses the level a, and R_a is the overshoot or excess over the boundary. When $\xi_n = 0$ for all $n \geq 0$, a corollary to the renewal theorem asserts:

Corollary. *If X_1 is nonarithmetic and $\mu \geq 0$, then R_a has a limiting distribution G as $a \to \infty$ with density*

$$g(r) = P\{S_\tau > r\}/E(S_\tau), \qquad r > 0,$$

where $\tau = \inf\{n \geq 1 : S_n > 0\}$ is the time at which the random walk first becomes positive.

The nonlinear version of this result asserts that R_a has the same limiting distribution G, if X_1 is nonarithmetic, $\mu > 0$, and the following conditions are satisfied

(a) $(1/n)\max\{|\xi_1|, \ldots, |\xi_n|\} \to 0$ in probability as $n \to \infty$.

(b) for all $\epsilon > 0$, there is a $\delta = \delta(\epsilon) > 0$ for which

$$P\{\max_{k \leq n\delta}|\xi_{n+k} - \xi_n| \geq \epsilon\} \leq \epsilon$$

for all sufficiently large n.

If, in addition, X_1 has a finite positive variance σ^2 and $\xi_n/\sqrt{n} \to 0$ in probability, then $t_a^* = (t_a - a/\mu)/\sqrt{a/\mu}$ is asymptotically normal with mean 0 and variance $\mu^{-2}\sigma^2$ and t_a^* is asymptotically independent of R_a (*see* ASYMPTOTIC NORMALITY).

The renewal measure of the sequence Z_n, $n \geq 1$, is defined as

$$V\{J\} = \sum_{n=0}^{\infty} P\{Z_n \in J\}$$

for intervals $J \subset (-\infty, \infty)$, where $Z_0 = 0$. When $\xi_n = 0$ for all $n \geq 0$, the renewal theorem asserts:

Theorem. *If F is nonarithmetic and $\mu > 0$, then $\lim V\{(a, a+b]\} = b/\mu$ as $a \to \infty$ for all $b > 0$; and if X_1 is arithmetic with span $h > 0$ and $\mu > 0$, then $\lim V\{kh\} = h/\mu$ as $k \to \infty$.*

The nonlinear version of this theorem asserts the same limiting behavior if X_1 is nonarithmetic, $\mu > 0$, X_1 has higher moments, and ξ_n, $n \geq 0$, satisfy conditions slightly stronger than conditions (a) and (b).

Example 1. Let $Y_1, Y_2, \ldots$ be independent and identically distributed with finite mean θ; and let $\bar{Y}_n = (Y_1 + \cdots + Y_n)/n$ for $n \geq 1$. If u is a smooth function for which $u(\theta) > 0$, then $Z_n = nu(\bar{Y}_n)$, $n \geq 1$, may be written in the form (1) with $S_n = nu(\theta) + nu'(\theta)(\bar{Y}_n - \theta)$, $n \geq 1$, and $\xi_n = Z_n - S_n$, $n \geq 1$. Here ξ_n, $n \geq 1$, satisfy conditions (a) and (b) if Y_1 has a finite variance, and ξ_n, $n \geq 1$, satisfy the strengthened versions of conditions (a) and (b), if Y_1 has higher moments.

Example 2. Suppose $\theta > 0$; let $0 < \alpha < 1$ and $c > 0$. Then the random time

$$s = \inf\{n \geq 1 : Y_1 + \cdots + Y_n > cn^\alpha\}$$

at which $Y_1 + \cdots + Y_n$ first crosses the curved boundary cn^α is of the form t_a with $Z_n = n\max\{0, \bar{Y}_n\}^\beta$, $\beta = 1/(1 - \alpha)$, and $a = c^\beta$.

The renewal theorem was discovered by Blackwell [1, 2] and by Erdös et al. [4]. The nonlinear versions were formulated and established by Lai and Siegmund [8, 9], although the special case of Example 2 had appeared in Woodroofe [14]. Hagwood [6] and Lalley [10] considered the arithmetic case. Uniform integrability of R_a is considered in Hagwood and Woodroofe [7] and Lai and Siegmund [9], and uniform integrability of t_a^* is considered in Chow et al. [3]. Asymptotic expansions* are given in Takahashi and Woodroofe [13] for some special cases. Nonlinear renewal theory has been applied to approximate the properties of several sequential tests and estimates (see SEQUENTIAL ANALYSIS). The recent monograph by Woodroofe [15] and text by Siegmund [12] describe the development of nonlinear renewal theory and its applications to statistics. They include references to statistical applications. See also Lalley [11].

References

[1] Blackwell, D. (1948). *Duke Univ. Math. J.*, **15**, 145–150.

[2] Blackwell, D. (1953). *Pacific J. Math.*, **3**, 315–320.

[3] Chow, Y. S., Hsiung, C., and Lai, T. L. (1979). *Ann. Prob.*, **7**, 304–318.

[4] Erdös, P., Feller, W., and Pollard, H. (1949). *Bull. Amer. Math. Soc.*, **55**, 201–204.

[5] Feller, W. (1968). *An Introduction to Probability Theory and Its Applications*, Vol. 2. Wiley, New York.

[6] Hagwood, C. (1980). *Commun. Statist.*, **A9**, 1677–1698.

[7] Hagwood, C. and Woodroofe, M. (1982). *Ann. Prob.*, **10**, 844–848.

[8] Lai, T. L. and Siegmund, D. (1977). *Ann. Statist.*, **5**, 946–954.

[9] Lai, T. L. and Siegmund, D. (1979). *Ann. Statist.*, **7**, 60–76.

[10] Lalley, S. (1972). *Commun. Statist.*, **1**, 193–206.

[11] Lalley, S. (1983). *Zeit. Wahrscheinlichkeitsth.*, **63**, 293–321.

[12] Siegmund, D. (1984). *Sequential Analysis*. SIAM, Philadelphia, PA.

[13] Takahashi, H. and Woodroofe, M. (1981). *Commun. Statist.*, **A10**, 2113–2135.

[14] Woodroofe, M. (1976). *Ann. Prob.*, **4**, 67–80.

[15] Woodroofe, M. (1982). In *Sequential Analysis*, Regional Conference Series in Applied Mathematics No. 39. SIAM, Philadelphia.

(RANDOM WALK
RENEWAL THEORY
REPEATED SIGNIFICANCE TESTS)

MICHAEL B. WOODROOFE

NONMETRIC DATA ANALYSIS

Nonmetric data analysis in its broader sense refers to a set of models and techniques for analysis of nonmetric data. Nonmetric data here refer to nominal or ordinal data (see NOMINAL DATA *and* ORDINAL DATA) as opposed to metric data, which refer to interval or ratio data [17] (see MEASUREMENT STRUCTURES AND STATISTICS). Nonmetric data (sometimes called qualitative or categorical data) are obtained in a variety of ways. For example, in attitude surveys, the respondent may be asked to endorse attitude statements with which he or she agrees. In some mental tests, the examinee either passes or fails test items. In consumer research, the subject may be asked to rank-order food products according to preference. In multidimensional scaling*, stimulus confusion data which are used as (inverse) ordinal measures of subjective distances between the stimuli, may be taken. In some instances metric data may be "discretized" for the purpose of data analysis.

Methods to analyze nonmetric data may be classified into two major approaches. One is *quantitative analysis of qualitative data* [23], and the other is *parametric approaches to nonmetric scaling* [18–20]. The first approach is primarily descriptive but is more general in its applicability. Nonmetric data analysis, in its narrower sense, usually refers to this first approach. The second approach is less general but is more powerful in situations for which particular models are intended. For other approaches to nonmetric data analysis, see related entries listed after the references.

The essential idea behind the first approach is that nonmetric data are nonlinear transformations of metric data. Thus if an appropriate transformation is applied, the transformed data may be analyzed by a "quantitative" model. Unlike other methods that require data transformations, a specific transformation to be applied does not have to be predetermined in this approach. Both the best data transformation and the best parameter estimates of models are obtained on the basis of a single optimization criterion.

Let y_i denote the ith original observation. This y_i is assumed to be quantified a priori. For example, $y_i = 1$ or 0 depending on whether person i passes or fails a certain test item, or $y_1 = 2$, $y_2 = 3$, and $y_3 = 1$, if the y_i are rank-ordered and it is observed that $y_2 > y_1 > y_3$. The numbers are assigned and interpreted "nonmetrically." That is, for nominal data, only identity or nonidentity of the numbers (i.e., for any two numbers, $a = b$ or $a \neq b$) is meaningful, whereas for ordinal data, ordinal properties of the numbers (i.e., for $a \neq b$, either $a < b$ or $a > b$) are also meaningful. However, in either case neither the difference nor the ratio of two numbers is meaningful. The y_i is transformed by function f, and $f(y_i)$, the transformed data, is fitted by model $g(X_i, \alpha)$, where X_i is some auxiliary information about i (if there is any), and α is a vector of unknown parameters. Both f and g are real-valued functions, possibly defined only at discrete values of their arguments. The problem is to find f and g such that an overall discrepancy between $f(y_i)$ and $g(X_i, \alpha)$, $i = 1, \ldots, I$ is a minimum. More specifically, define a least-squares* loss function,

$$\text{Stress} = \sum_{i=1}^{I} \left(f(y_i) - g(X_i, \alpha) \right)^2.$$

This criterion is minimized with respect to both f and α under some appropriate normalization restriction.

General forms of f must be consistent with nonmetric properties of the data. That is, f must be such that the basic properties of nonmetric data are preserved through the transformation. (Such transformations are called admissible transformations.) This implies that f must be monotonic (order preserving) when the data are ordinal, and it must be one to one (identity preserving) when the data are nominal. Within the admissible types of transformations, a specific form of f is determined that minimizes Stress. For a given g the best monotonic transformation is obtained by Kruskal's [10] least-squares monotonic regression algorithm, and the best one-to-one transformation, by least-squares nominal transformation [5]. Since f is determined in such a way that it is closest to g among all admissible transformations, it may be considered to possess the same scale level as model g, provided that model g is appropriate for the

data. The scale level of a model is the type of admissible transformations by which defining properties of the model are not destroyed. For example, if g is a distance model, which is a ratio model since the defining properties of the distance (the metric axioms) are preserved by multiplying the distance by a positive constant, f is also considered ratio at least approximately. The nonmetric data are, so to speak, "scaled up" by f to g.

Similarly, specific models (g) to be fitted depend on the nature of the data. For example, if the data are similarity data (see MULTIDIMENSIONAL SCALING), a distance model may be employed. If the data are conjoint data (see MEASUREMENT STRUCTURES AND STATISTICS), an additive model may be appropriate. Other models that may be fitted include linear regression* models, bilinear models (principal components* and factor analysis* models), and a variety of distance models including the Minkowski and the weighted distance models [3, 9] and the unfolding model [4] (see MULTIVARIATE ANALYSIS and MULTIDIMENSIONAL SCALING). Whichever model is chosen, model parameters are determined in such a way that Stress is a minimum. For a given f, least-squares estimates of model parameters are obtained as if the current f were metric data.

To illustrate, consider the situation in which ordinal data are analyzed by the regression model. Such a situation arises, for example, when we wish to find out why some cars are regarded as more desirable than others, based on various attributes (e.g., gas mileage) of cars and a preference ranking among them. Let y_i be the ith observation on the dependent variable (the preference rank of the ith car) and X_i the corresponding observations on the independent variables (the values of the attributes). The dependent variable (y_i) is monotonically transformed (so that if $y_i > y_j$, then $f(y_i) \geq f(y_j)$), and the regression coefficients (α) are estimated in such a way that Stress is a minimum. Two algorithms are currently in use for minimizing Stress with respect to f and α. One is the steepest descent algorithm (see also OPTIMIZATION and SADDLE-POINT

APPROXIMATIONS) used originally by Kruskal [10] for his nonmetric multidimensional scaling. The other is the alternating least squares (ALS) algorithm developed by Young, de Leeuw, and Takane. (This work is summarized in Young [23].) In the steepest descent algorithm, f, which minimizes Stress for a fixed g, is expressed as a function of $g(\alpha)$ and then substituted in Stress. The Stress, which is now expressed as a function of α only, is then minimized with respect to α. In the ALS algorithm, LS estimates of f and g are obtained alternately with one of them fixed while the other is updated. This algorithm is monotonically convergent.

The origin of the quantitative analysis of qualitative data can be traced back to Guttman's scale analysis [8]. This method is still widely used and has regained considerable theoretical interest in recent years [6, 15] (see CORRESPONDENCE ANALYSIS). Coombs' unfolding analysis [4] is important in that it was the first to suggest the possibility of recovering metric information from nonmetric data. The current trend in the quantitative analysis of qualitative data began with Shepard's [16] and Kruskal's [9] landmark work on nonmetric multidimensional scaling. Following their work, it was soon realized that models other than the distance model could be fitted to nonmetric data in a similar manner, and several fitting procedures were developed along this line [9, 22]. More recently the ALS algorithm was proposed as a unified algorithmic framework for the quantitative analysis of qualitative data; this has considerably widened the scope of models that can be fitted [6, 23]. For a list of currently available procedures, see Young [23].

In the parametric approaches to nonmetric scaling, nonmetric data are viewed as incomplete data. That is, a complete metric process is supposed to underlie the nonmetric data generation process, but the metric information is assumed to be lost when the observations are made, leaving only ordinal or nominal information in the observed data. Thus, if this information reduction mechanism can be captured in a model, the metric information may be recovered from the nonmetric data by working backward from the data.

As an example, let us discuss Thurstone's [2, 21] classical pair comparison model. In a pair comparison experiment, stimuli are presented in pairs, and the subject is asked to choose one member of a pair according to some prescribed criterion. The data are a collection of partial rank orders. Suppose stimuli i and j are compared in a particular trial. It is hypothesized that each stimulus, upon presentation, generates a latent metric process that varies randomly from trial to trial. Let X_i and X_j denote the random variables for the latent processes of stimuli i and j, respectively. For simplicity let us assume that $X_i \sim N(\mu_i, \frac{1}{2})$ and $X_j \sim N(\mu_j, \frac{1}{2})$. (The μ_i and μ_j represent the mean subjective values of the two stimuli. The variances of X_i and X_j are assumed to be equal, but their size can be arbitrarily set.) It is assumed that stimulus i is chosen when $X_i > X_j$ and stimulus j is chosen when $X_i < X_j$. Under the distributional assumptions on X the probability (p_{ij}) of stimulus i over stimulus j can be stated as

$$p_{ij} = \phi(\mu_i - \mu_j),$$

where ϕ is the distribution function of the standard normal distribution*. The likelihood* of observed data is stated as a function of parameters in the latent processes. For computational convenience, ϕ may be replaced by the logistic distribution* [14]. In any case μ_i and μ_j may be estimated to maximize p_{ij} if in fact stimulus i is chosen over stimulus j.

This basic principle can be extended in various ways. Suppose that μ_i represents a combined effect of one or more factors. It may then be appropriate to characterize the μ_i by an additive function of these factors. Pair comparisons of such μ_i provide the data for additive conjoint analysis [19]. As another example, suppose two pairs of stimuli are presented and the subject is asked to choose a more similar pair (this method is called the method of tetrads, which involves pair comparisons of two similarities). Then stimulus (dis)similarities may be represented by a distance model, and then they are subject to pair comparisons. Nonmetric multidimensional scaling (in the sense of the second approach) is feasible with the pair comparison data [20]. As in the first approach, various other models may be fitted in a way that is consistent with the nature of the data.

Another line of extension is possible with regard to the kinds of judgments that are made. Stimuli may be rank-ordered. They may be rated on a categorical rating scale. A choice may be required among several comparison stimuli. In each case a specific model of information reduction mechanism (similar to that used in pair comparison situation) may be built into parameter estimation procedures. Then essentially the same analysis can be done as in the pair comparison case. Such procedures have been developed for similarity ratings [18], for similarity rankings [20], and for additivity analysis of rating and ranking data [19].

The history of the parametric approaches to nonmetric scaling is even older than the quantitative analysis of qualitative data. Thurstone's pair comparison model was originally proposed in the 1920s [21]. A similar model was developed in mental testing situations [13] in the early fifties. Around the same time, latent structure analysis* [12] was proposed, which accounts for observed response patterns to items by hypothesizing *latent structures*. (Again, conceptually, this is very similar to Thurstone's approach.) See Andersen [1], Bock and Jones [2], and Goodman [7] for recent developments in these models. More recently, Takane [18–20] has developed the conceptual framework for the parametric approaches to nonmetric scaling that is presented here.

References

[1] Andersen, E. B. (1980). *Discrete Statistical Models with Social Science Applications*. North-Holland, Amsterdam. (An excellent treatment of exponential family distributions for discrete data analysis.)

[2] Bock, R. D. and Jones, L. V. (1968). *The Measurement and Prediction of Judgment and Choice*.

Holden-Day, San Francisco. (A comprehensive statistical treatment of Thurstonian scaling.)

[3] Carroll, J. D. and Chang, J. J. (1970). *Psychometrika*, **35**, 283–319. (A proposal of individual differences model in MDS.)

[4] Coombs, C. H. (1964). *A Theory of Data*. Wiley, New York.

[5] De Leeuw, J., Young, F. W., and Takane, Y. (1976). *Psychometrika*, **41**, 471–503. (The first account of the ALS algorithm.)

[6] Gifi, A. (1981). *Non-linear Multivariate Analysis*. Department of Data Theory, University of Leiden, The Netherlands. (An account of Guttman's scale analysis by ALS.)

[7] Goodman, L. A. (1978). *Analyzing Qualitative/Categorical Data*. Abt Associates, Cambridge, MA. (Recent developments in latent structure analysis.)

[8] Guttman, L. (1941). In *The Prediction of Personal Adjustment*, P. Horst, ed. Social Science Research Council.

[9] Kruskal, J. B. (1964). *Psychometrika*, **29**, 1–27. (The first theoretically rigorous nonmetric MDS.)

[10] Kruskal, J. B. (1964). *Psychometrika*, **29**, 115–129.

[11] Kruskal, J. B. (1965). *J. R. Statist. Soc. B*, **27**, 251–265. (An application of the monotonic regression to additivity analysis.)

[12] Lazarsfeld, P. F. and Henry, N. (1968). *Latent Structure Analysis*. Houghton Mifflin, Boston.

[13] Lord, F. M. (1980). *Applications of Item Response Theory to Practical Testing Problems*. Earlbaum, Hillsdale, NJ. (An up-to-date illustration of latent trait test theory.)

[14] Luce, R. D. (1959). *Individual Choice Behavior*. Wiley, New York. (An axiomatic choice model and its mathematical properties.)

[15] Nishisato, S. (1980). *Analysis of Categorical Data*. University of Toronto Press. (The first English text on Guttman's scale analysis and its multidimensional extension, called dual scaling or correspondence analysis.)

[16] Shepard, R. N. (1962). *Psychometrika*, **27**, 125–140 and 219–246. (The first work on nonmetric MDS ever published.)

[17] Stevens, S. S. (1951). In *Handbook of Experimental Psychology*, S. Stevens, ed. Wiley, New York.

[18] Takane, Y. (1981). *Psychometrika*, **46**, 9–28.

[19] Takane, Y. (1982). *Psychometrika*, **47**, 225–241.

[20] Takane, Y. and Carroll, J. D. (1981). *Psychometrika*, **46**, 389–405. (References 18–20 describe parametric approaches to nonmetric scaling for different models and data.)

[21] Thurstone, L. L., (1959). *The Measurement of Values*. University of Chicago Press, Chicago. (A collection of his works.)

[22]   Young, F. W. (1972). In *Multidimensional Scaling*, Vol. 1, R. Shepard et al., eds. Seminar Press, New York. (Polynomial conjoint scaling. An extension of Kruskal's algorithm to other models.)

[23]   Young, F. W. (1981). *Psychometrika*, **46**, 357–388. (The most up-to-date account of the quantitative analysis of qualitative data. An excellent bibliography on this approach.)

(COMPONENT ANALYSIS
CORRESPONDENCE ANALYSIS
LATENT STRUCTURE ANALYSIS
MEASUREMENT STRUCTURES AND
    STATISTICS
MULTIDIMENSIONAL SCALING
NOMINAL DATA
OPTIMIZATION IN STATISTICS
ORDINAL DATA
REGRESSION (various entries))

YOSHIO TAKANE

NONOBSERVABLE ERRORS

In the general linear model*

$$Y = X\beta + \epsilon,$$

where **Y** is a $k \times 1$ vector of *observed* sample values the random component ϵ is often referred to as nonobservable errors.

(GENERAL LINEAR MODEL)

NONPARAMETRIC CLUSTERING TECHNIQUES

Write N for the number of clusters in a set of multivariate observations; given N, numerous clustering techniques estimate the cluster membership of each observation. Most of these techniques lack a statistical basis, making determination of N problematical.

One statistical formalization of the clustering problem assumes the data come from a mixture* of normal distributions. This assumption allows determination of N using a likelihood* or other statistical criterion since, under the assumption, N equals the number of component distributions. Several current clustering algorithms use this approach; see, e.g., Lennington and Rassbach [3]. The normality assumption is frequently violated, making interpretation of the resulting clusters difficult.

A generalization of the normal mixture model supposes the observations arise from a mixture of unspecified distributions [2, p. 205]. Based on this supposition, the clustering problem reduces to obtaining a nonparametric estimate of the underlying density function.

One nonparametric density estimate uses the equal cell histogram. Given a threshold, the clusters are the connected regions above the threshold level. No theoretically defined threshold currently exists, although some authors suggest the expected value of the density given a uniform distribution over the range of the observations. Goldberg and Shlien [1] apply this technique to obtain a preliminary clustering of LANDSAT data. Each observation consists of four measurements in the range from 0 to 127; the number of cells equals the number of possible combinations, 64^4, and the threshold value is the average number of observations per nonempty cell. All contiguous cells with density above the threshold are connected, and then all cells with density below the threshold are joined to the nearest connected set; N is the number of connected sets.

An improved estimate can be obtained by allowing the data to determine the cells, as in ref. 6, in which Wong partitions the data space into k regions, for k between N and the number of observations, obtaining a density estimate inversely proportional to the volume of the regions. The k regions are the partition of the data space minimizing the within region sum of squares of the observations and correspond to the clusters found by the k-means clustering algorithm. This set of estimates is then used to assign the observations in each region to the appropriate cluster.

An alternative approach to nonparametric density estimation and hence to the problem of estimating N and cluster assignment, uses the Parzen kernel density estimate. In this approach, the data is smoothed by averaging a set of normal densities with means corresponding to the observed values and common standard deviation h, giving an estimate of the underlying density; h is the window size. The estimate of N depends on h, with large h corresponding to small N and vice versa. For each N, a critical h exists such that the estimated density has N or more modes only for h less than the critical value. For each critical h, the significance of the associated N can be assessed by simulation using the estimated density, as in Silverman [5]. This approach has only been explored for one-dimensional data.

A different approach to determining N uses a likelihood criterion in the following way: The unique set of line segments connecting the observations and of minimum total length is calculated; this set is the minimum spanning tree. The value of N is then estimated by an iterative approach that considers each N from 1 to the number of observations; for a given N, the densities with N and $N + 1$ modes supported by the minimum spanning tree (*see* DENDRITES) and maximizing the likelihood are calculated. The N is increased until the difference in the log likelihoods no longer exceeds a threshold. This approach is suggested in ref. 2 and explored by Ramey [4] for the two-dimensional case. Not enough is known about the higher-dimensional properties of this technique to recommend it for application.

References

[1] Goldberg, M. and Shlien, S. (1978). *IEEE Trans. Syst. Man, Cybern.*, **8**, 86–92. (Describes histogram clustering scheme applied to multivariate LANDSAT data.)

[2] Hartigan, J. A. (1975). *Clustering Algorithms*. Wiley, New York. (Includes FORTRAN code implementing the algorithms along with many interesting examples.)

[3] Lennington, R. K. and Rassbach, M. E. (1978). *Proc. Tech. Sess. LACIE Symp.*, **2**, 671–689. NASA-JSC No. 16015, Houston, TX. (Describes normal mixtures model as implemented in CLASSY algorithm. Applied to LANDSAT data.)

[4] Ramey, D. B. (1982). "A Nonparametric Test of Bimodality with Applications to Cluster Analysis." Ph.D. dissertation, Yale University. (For references and discussion of significance problem in cluster analysis, see pp. 2–6.)

[5] Silverman, B. W. (1981). *J. R. Statist. Soc. B*, **43**, 97–99. (Kernel density estimation applied to clustering of chondrite data.)

[6] Wong, M. A. (1982). *J. Amer. Statist. Ass.*, **77**, 841–847. (Basic references on cluster analysis.)

Bibliography

Blashfield, R. K. and Aldenderfer, M. S. (1978). *Multivariate Behav. Res.*, **8**, 271–295. (Table 1 on p. 272, provides a comprehensive list of review articles on cluster analysis during the years 1965–1974, while Table 2, p. 273, lists texts on cluster analysis for the years 1960–1978.)

Cormack, R. M. (1971). *J. R. Statist. Soc. A*, **134**, 321–367. (Comprehensive list of references before 1971.)

Day, N. E. (1969). *Biometrika*, **56**, 463–474. (Proposes the application of normal mixtures to cluster analysis.)

Dubes, R. and Jain, A. K. (1979). *Pattern Recognition*, **11**, 235–254. (A semitutorial review of the problem of cluster significance from the practitioner's viewpoint.)

Everitt, B. S. (1979). *Biometrics*, **35**, 169–181. (Readable, with references to current literature on cluster analysis.)

Sneath, P. H. A. and Sokal, R. R. (1973). *Numerical Taxonomy*. W. H. Freeman, San Francisco. (A basic reference on clustering presented from a biologist's viewpoint with the required mathematics.)

(CLASSIFICATION
DENSITY ESTIMATION
GRAPH-THEORETIC CLUSTER ANALYSIS)

D. B. RAMEY

NONPARAMETRIC CONFIDENCE INTERVALS

The term *nonparametric* indicates that a statistical procedure can be used for large classes of distributions and that relevant probability statements are independent of the actual population distribution, at least if this distribution is continuous. Because of this latter property, the procedures are also referred to as *distribution-free** procedures. Examples of parameters for which nonparametric confidence intervals* are often found are the population median, the amount of shift of one population relative to another population, and the slope of a linear regression* line.

Another problem customarily discussed under the present heading is the problem of finding a confidence band* for an unknown (cumulative) distribution function. We shall use this problem to review basic ideas.

CONFIDENCE BAND*
FOR A DISTRIBUTION FUNCTION

Let X be a random variable with distribution function $F(x) = \Pr[X \leq x]$. Given a random sample $X_1, X_2, \ldots, X_N$ from $F(x)$, a point estimate of $F(x)$ at an arbitrary but fixed value x is given by $\#(X_k \leq x)/N$. A "point" estimate of the function $F(x)$ is given by the *empirical* or *sample* distribution function

$$\hat{F}(x) = \frac{1}{N} \#(X_k \leq x), \qquad -\infty < x < +\infty$$

a step function that increases by $1/N$ at each of the order statistics $X_{(1)} \leq X_{(2)} \leq \cdots \leq X_{(N)}$. The maximum distance between $\hat{F}(x)$ and $F(x)$ provides an indication of the accuracy of $\hat{F}(x)$ as an estimator of $F(x)$. More exactly, if we define the *Kolmogorov statistic* *

$$D = \sup_x |\hat{F}(x) - F(x)|$$

and denote its 100γ percentile by d_γ, $\Pr[D \leq d_\gamma] = \gamma$, then with probability γ, $\hat{F}(x)$ will not deviate anywhere from $F(x)$ by more than d_γ. Equivalently, with probability γ,

$$L(x) \leq F(x) \leq U(x),$$
$$-\infty < x < +\infty, \quad (1)$$

where $L(x) = \max[0, \hat{F}(x) - d_\gamma]$ and $U(x) = \min[1, \hat{F}(x) + d_\gamma]$. The interpretation of (1) requires some care. Under random sampling from $F(x)$, the probability is γ that over the whole range of x, $F(x)$ is contained between the *random* step functions $L(x)$ and $U(x)$. The functions $L(x)$ and $U(x)$ are called *lower* and *upper confidence contours* for $F(x)$; the region between $L(x)$ and $U(x)$ is called a *confidence band* for $F(x)$; the probability γ is called the *confidence coefficient*.

In view of a limit theorem by Kolmogorov (see, e.g., pages 221, 226–227, and Table A.24 of ref. 1), in large samples we can use the following asymptotic values for d_γ,

γ:	0.90	0.95	0.99
d_γ:	$1.22/\sqrt{N}$	$1.36/\sqrt{N}$	$1.63/\sqrt{N}$.

Relationship to Hypothesis Testing*

Let $F_0(x)$ be a completely specified distribution such as a normal distribution with prescribed mean and standard deviation. If $F(x)$ equals $F_0(x)$, with probability γ, $F_0(x)$ is completely contained in the confidence band (1). But with probability $1 - \gamma$, $F_0(x)$ crosses either the lower confidence contour $L(x)$ or the upper confidence contour $U(x)$. In either case, we have $D > d_\gamma$. Thus the test that rejects the hypothesis $F(x) = F_0(x)$ if

$$D = \sup_x |\hat{F}(x) - F_0(x)| > d_\gamma \quad (2)$$

has significance level $\alpha = 1 - \gamma$.

The Discontinuous Case

The Kolmogorov statistic* D is distribution-free only for a continuous distribution function $F(x)$. However, a very simple modification of the relevant probability statement validates the previous results in the discontinuous case. If $\Pr[D \leq d_\gamma] = \gamma$ in the continuous case, for discontinuous $F(x)$ we have $\Pr[D \leq d_\gamma] \geq \gamma$. For discontinuous distributions $F(x)$ both the confidence band and the test of the hypothesis $F(x) = F_0(x)$ are *conservative*: the true confidence coefficient associated with the confidence band (1) is *at least* γ and the true significance level associated with the test (2) is *at most* $\alpha = 1 - \gamma$.

CONFIDENCE INTERVALS FOR A SHIFT PARAMETER

Throughout the history of nonparametric statistics, hypothesis testing has received much greater emphasis than estimation. As a result, nonparametric confidence intervals usually are derived from existing nonparametric tests. The following example illustrates this approach. The most extensively investigated nonparametric problem is the *two-sample problem*. Given two independent random samples $X_1, X_2, \ldots, X_m$ and $Y_1, Y_2, \ldots, Y_n$ from populations with distribution functions $F(t)$ and $G(t)$, respectively, we want to test the hypothesis $G(t) = F(t)$. For many practical situations, the most satisfactory two-sample test is the Wilcoxon rank sum test* which uses as test statistic the sum of the ranks associated with either the X or the Y observations. If, as suggested by classical normal theory, alternatives are restricted to the *shift model*

$$G(t) = F(t - \Delta), \qquad (3)$$

it is natural to try to find a confidence interval for the shift parameter Δ. The confidence interval with confidence coefficient γ contains all values Δ such that the Wilcoxon test applied to observations $X_1, X_2, \ldots, X_m$ and $Y_1 - \Delta, Y_2 - \Delta, \ldots, Y_n - \Delta$ does not reject the two-sample hypothesis at significance level $\alpha = 1 - \gamma$. See, e.g., Hollander and Wolfe [1, Sect. 3 of Chap. 4].

For the present exposition, we prefer a more direct approach which has the additional advantage of easy generalization. For the two-sample shift model, we proceed as follows. Let X and Y be two random variables with distribution functions that satisfy the shift model (3). Then Y and $X + \Delta$ are distributed identically, so that any difference $Y - X$ provides an estimate of Δ. We shall refer to the set of mn sample differences $D_{ji} = Y_j - X_i$, $i = 1, 2, \ldots, m$; $j = 1, 2, \ldots, n$, as the set of *elementary estimates* for the shift parameter Δ. Since under the shift model (3) differences $Y - X$ are symmetrically distributed about Δ, it is intuitively

reasonable to assert that the true value Δ lies in an interval bounded by two sufficiently extreme elementary estimates. More precisely, if $D_{(1)} \leqslant D_{(2)} \leqslant \cdots \leqslant D^{(2)} \leqslant D^{(1)}$ are the mn elementary estimates $Y_j - X_j$ arranged according to size, we consider the set of confidence intervals

$$D_{(g)} \leqslant \Delta \leqslant D^{(g)}, \qquad g = 1, 2, \ldots . \quad (4)$$

For continuous distributions $F(t)$, these confidence intervals are distribution-free with confidence coefficient $\gamma = \gamma(g)$ depending only on g (and the sample sizes m and n). It is possible to find $\gamma(g)$ by simple enumeration over equally likely cases as illustrated in the example.

Example. For $m = 1$, $n = 2$, there are two elementary estimates $D_{(1)}$ and $D^{(1)}$. The following three possibilities are equally likely, (a) $D_{(1)}$ and $D^{(1)} < \Delta$, (b) $D_{(1)} \leqslant \Delta \leqslant D^{(1)}$, (c) $D_{(1)}$ and $D^{(1)} > \Delta$. The confidence interval (4) with $g = 1$ contains the true value Δ only if (b) occurs. Thus for $m = 1$, $n = 2$, we have $\gamma(1) = \frac{1}{3}$. For larger sample sizes m and n, the enumeration process is more laborious, but relatively straightforward.

There exist tables that list values of g corresponding to standard confidence levels 0.90, 0.95, and 0.99 [5, Table G]. Except for quite small sample sizes m and n, the following normal approximation is satisfactory for most practical purposes.

$$g \doteq \tfrac{1}{2} \left\{ mn + 1 - z \left[mn(m + n + 1)/3 \right]^{1/2} \right\},$$

where x is the appropriate normal deviate:

γ:	0.90	0.95	0.99
z:	1.645	1.960	2.576

For discontinuous distributions $F(t)$, the confidence intervals (4) are no longer distribution-free. But the true confidence coefficient associated with the closed interval (4) equals at least the tabulated value $\gamma(g)$ [4].

If we let g increase toward $(mn + 1)/2$ in (4), the confidence interval narrows down to a single point, the median* of the mn elementary estimates D_{ji}. This point furnishes an intuitively attractive point estimate for Δ, the Hodges–Lehmann estimate [1, Chap. 4, Sect. 2].

It is instructive to consider the test of the hypothesis $\Delta = \Delta_0$ which rejects the hypothesis if Δ_0 does not fall in the confidence interval (4). This happens if fewer than g of the elementary estimates are smaller than Δ_0 or greater than Δ_0. More formally, we can define two test statistics $T = \#(D_{ji} < \Delta_0)$ and $T' = \#(D_{ji} > \Delta_0)$ and reject the hypothesis $\Delta = \Delta_0$ at significance level $\alpha = 1 - \gamma(g)$ if the smaller of the two statistics T and T' is smaller than g. The statistics T and T' are the Mann–Whitney statistics* for the two-sample problem and linear functions of the Wilcoxon rank sum statistics.

THE METHOD OF ELEMENTARY ESTIMATES

The method of elementary estimates that we used for the two-sample shift case is easily generalized. We shall state the method in general terms and then apply it to several specific problems. Let θ be the parameter of interest. We define a set of elementary estimates E_k. (The choice will usually be quite obvious.) If $E_{(1)} \leqslant E_{(2)} \leqslant \cdots \leqslant E^{(2)} \leqslant E^{(1)}$ denote the ordered elementary estimates, we consider the set of confidence intervals I_g:

$$E_{(g)} \leqslant \theta \leqslant E^{(g)}, \qquad g = 1, 2, \ldots$$

with confidence coefficient $\gamma(g)$. As a point estimate of θ, we use

$$\hat{\theta} = \text{median } E_k.$$

A TWO-SAMPLE SCALE PROBLEM

Let X be a positive variable with distribution function $F(t)$. Let Y be distributed as τX, so that Y has distribution function $G(t) = F(t/\tau)$. As the set of elementary estimates for the scale parameter τ, we choose the set of mn sample ratios Y_j/X_i. The confidence

coefficient $\gamma(g)$ for the confidence interval I_g is the same as in the two-sample shift case.

THE ONE-SAMPLE PROBLEM

We are given a random sample $X_1, X_2, \ldots, X_N$ from a population with distribution function $F(x)$ and are interested in the population median η. The simplest estimates of η are the individual observations X_k themselves. For this set of elementary estimates, the confidence interval I_g is bounded by the gth smallest and largest order statistics in the sample. Tables of g values corresponding to customary confidence levels can be found in many statistics texts, e.g., Noether [5, Table E]. Unless N is quite small, the following normal approximation is usually satisfactory.

$$g \doteq \tfrac{1}{2}(N + 1 - zN^{1/2}),$$

where z is again the appropriate normal deviate at confidence level γ. The point estimate of η is the sample median and the test statistics T and T' are the *sign test** statistics.

Symmetric Populations

Let us add the assumption that $F(x)$ is symmetric about η. As our elementary estimates, we now take, in addition to the individual observations, all possible averages of the two observations: $(X_i + X_j)/2$, $1 \leqslant i \leqslant j \leqslant N$. Exact g-values are listed in Table F of Noether [5]. The normal approximation is given by

$$g \doteq \tfrac{1}{2}\Big[\tfrac{1}{2}N(N + 1) + 1 \\ - z\big\{ \tfrac{1}{18}N(N + 1)(2N + 1) \big\}^{1/2} \Big].$$

The corresponding test of the hypothesis $\eta = \eta_0$ is the *Wilcoxon signed rank test*.

LINEAR REGRESSION

For each of N distinct regression constants $x_1 < x_2 < \cdots < x_N$, we assume that we have independent random variables

$$Y_k = \alpha + \beta x_k + E_k, \qquad k = 1, 2, \ldots, N,$$

where the E_k are identically distributed "error" variables with median 0. Our primary interest is in the slope β. As the set of elementary estimates for β, we take the $N(N-1)/2$ sample slopes

$$S_{ji} = (Y_j - Y_i)/(x_j - x_i), \qquad 1 \leqslant i < j \leqslant N.$$

The following normal approximation for g is usually satisfactory:

$$g \doteq \tfrac{1}{2}\Big[\tfrac{1}{2}N(N-1) + 1$$
$$- z\big\{\tfrac{1}{18}N(N-1)(2N+5)\big\}^{1/2}\Big].$$

As point estimates of β and α, we use $\hat{\beta} = \text{median } S_{ji}$ and $\hat{\alpha} = \text{median } (Y_k - \hat{\beta}x_k)$. For testing the hypothesis $\beta = 0$, we compute the two test statistics $T = \#(S_{ji} < 0)$ and $T' = \#(S_{ji} > 0)$. The quantity $S = T' - T$ is known as *Kendall's S* and $2(T' - T)/N(N-1)$ is the *Kendall rank correlation coefficient**. The case when the individual x_k are not necessarily distinct is discussed in Sen [7].

INSENSITIVITY TO OUTLIERS

An important consideration for a practicing statistician is the effect of "outliers"* among the observations on the results of a statistical analysis. Since outliers among the observations can affect only a limited number of elementary estimates, in general, nonparametric confidence intervals are much less sensitive to outliers than are the customary normal theory confidence intervals.

EFFICIENCY OF NONPARAMETRIC CONFIDENCE INTERVALS

A natural question to ask is how nonparametric confidence intervals compare to standard normal theory confidence intervals. This question was investigated by Lehmann [2]. The answer essentially is that nonparametric intervals have the same efficiency relative to parametric intervals as do the associated nonparametric tests relative to parametric tests. In particular, the asymptotic relative efficiency of the nonparametric intervals for the center of symmetry η and the shift parameter Δ compared to the intervals based on one- and two-sample t statistics is $3/\pi = 0.955$ for normally distributed populations and is generally greater than 1 for distributions whose tails are longer than those of a normal population.

WEIGHTED ELEMENTARY ESTIMATES

We have exhibited the simplest and most common nonparametric confidence intervals in current use. More general confidence intervals are obtained by assigning weights w_k to the elementary estimates E_k [6]. If then $w_{(k)}$ is the weight assigned to the kth smallest elementary estimate $E_{(k)}$ and $w^{(k)}$, the weight assigned to the kth largest elementary estimate $E^{(k)}$, we can construct confidence intervals $E_{(g)} \leqslant \theta \leqslant E^{(g')}$, where g and g' are the smallest integers such that $w_{(1)} + w_{(2)} + \cdots + w_{(g)} > c$ and $w^{(1)} + w^{(2)} + \cdots + w^{(g')} > c'$, the constants c and c' having been determined in such a way that the confidence interval has prescribed confidence coefficient γ. The previously discussed confidence intervals correspond to weights $w_k \equiv 1$.

As an example of the use of weights, consider again the linear regression model. The distance $x_j - x_i$ suggests itself as an intuitively attractive weight for the elementary estimate $S_{ji} = (Y_j - Y_i)/(x_j - x_i)$ of β. For equally spaced regression constants, these weights are equivalent to weights $w_{ji} = j - i$ associated with the *Spearman rank correlation coefficient**.

SIMULTANEOUS CONFIDENCE INTERVALS. The reader interested in a nonparametric treatment of simultaneous confidence intervals is referred to Miller [3, Chap. 4].

References

[1] Hollander, M. and Wolfe, D. A. (1973). *Nonparametric Statistical Methods*. Wiley, New York.

[2] Lehmann, E. L. (1963). *Ann. Math. Statist.*, **34**, 1507–1512.

[3] Miller, R. G., Jr. (1966). *Simultaneous Statistical Inference.* McGraw-Hill, New York.

[4] Noether, G. E. (1967). *J. Amer. Statist. Ass.*, **62**, 184–188.

[5] Noether, G. E. (1976). *Introduction to Statistics: A Nonparametric Approach.* Houghton Mifflin, Boston.

[6] Noether, G. E. (1978). *Statist. Neerlandica*, **32**, 109–122.

[7] Sen, P. K. (1968). *J. Amer. Statist. Ass.*, **63**, 1379–1389.

Bibliography

Emerson, J. D. and Simon, G. A. (1979). *Amer. Statist.* **33**, 140–142. (The Encyclopedia article recommends that the closed confidence interval (4) be used both in the continuous and the discontinuous case. In contrast, most nonparametric tests require modifications in the discontinuous case to take care of possible ties. The paper shows that suitably modified sign tests for the population median may not correspond exactly to the confidence interval (4) based on sample order statistics.)

Maritz, J. S. (1981). *Distribution-Free Statistical Methods.* Chapman and Hall, London. (An intermediate-level theoretical text that pays more careful attention to nonparametric estimation than most comparable texts.)

Noether, G. E. (1972). *Amer. Statist.*, **26**(1), 39–41. (Survey paper.)

(CONFIDENCE INTERVALS
 AND REGIONS
DISTRIBUTION-FREE METHODS
MANN–WHITNEY–WILCOXON TEST
ORDER STATISTICS)

GOTTFRIED E. NOETHER

NONPARAMETRIC DISCRIMINATION

At its inception, discriminant analysis* was viewed as a tool for classifying an individual or object into one of a finite number (K) of groups (or populations) on the basis of a series of p observations (**X**) obtained on the individual or object. The scope of discriminant analysis has expanded over the years to address more than simply the issue of classification*. It now includes the study of group differences based on an analysis of variable characteristics associated with individuals assigned to each group. Klecka [12] has labeled this latter area the "interpretation" component of discriminant analysis. Nevertheless, when considering nonparametric methods, classification is still the major focus of discriminant analysis.

Classification problems can be subdivided into three major categories:

1. Those in which the underlying distributions are known and completely specified.

2. Those in which the distributions are known and specified except for one or more parameters.

3. Those in which the distributions are completely unknown.

KNOWN AND COMPLETELY SPECIFIED DISTRIBUTIONS

A theoretical probabilistic solution to the classification problem was introduced by Welch [19], who adapted the hypothesis testing concepts of Neyman and Pearson. Welch showed that for observations drawn from one of two populations, the optimal solution to the classification problem is based on the ratio of the known probability density functions, $f_1(\mathbf{X})/f_2(\mathbf{X})$, evaluated at the observation to be classified. The observation is classified into the first population if the ratio is greater than a constant κ and into the second population if the ratio is less than κ. When the ratio equals κ, the usual procedure is to assign the observation at random to one of the populations. Von Mises [18] extended Welch's work to include the problem of classifying an observation into one of K populations.

In this discussion we are assuming equal costs of misclassification and equal a priori probabilities of group membership. When the costs are equal and the a priori probabilities are equal, the constant κ is equal to 1. Were one to consider unequal costs of misclassification and/or unequal a priori probabilities of group membership, the constant κ for determining classification would not be equal to 1.

For multivariate normal* populations with equal variance–covariance matrices*, the ratio of the probability density functions* leads to a criterion identical to Fisher's linear discriminant function. Using an intuitive approach, Fisher [7] developed the linear discriminant function (LDF) which maximizes a function of the distance between the mean vectors of two samples. When the multivariate normal and equal variance–covariance matrices assumptions are satisfied, the LDF is the optimal classification procedure in the sense that it minimizes the overall probability of misclassification.

Gessaman and Gessaman [10] have shown that the LDF is not robust, which violates the assumption of equal variance–covariance matrices. If the variance–covariance matrices are not equal, the logarithm of the ratio of the multivariate normal density functions is a quadratic function, called the quadratic discriminant function (QDF).

KNOWN DISTRIBUTIONS
WITH AT LEAST ONE PARAMETER
UNSPECIFIED

When the underlying distributions are known and completely specified, the likelihood ratio procedure of Welch [19] is optimal. When the distributions are known and specified except for one or more parameters, sample data can be used to estimate the unknown parameters. In such instances, a common procedure is based on the ratio of the densities, $\hat{f}_1(\mathbf{X})/\hat{f}_2(\mathbf{X})$, with the sample estimates replacing the population parameters. Anderson [1] showed that when the densities are multivariate normal with unknown variance–covariance matrices and unknown mean vectors, the LDF using maximum likelihood sample estimates is a consistent procedure (i.e. asymptotically equiva-

lent to the optimum rule when the probability density functions are known).

COMPLETELY UNKNOWN
DISTRIBUTIONS

For many practical problems, however, it is theoretically impossible to specify the probability density function. When the form of the distribution is unknown and classification procedures must be developed solely on information derived from a sample, the procedures are referred to as nonparametric or distribution-free methods*. The two terms are often used interchangeably. Nonparametric procedures are employed when one is concerned with a wide class of distributions that cannot be expressed as a parametric family with a finite number of parameters [6]. Hand [11] placed nonparametric density function estimators into four categories described as (a) the histogram method, (b) the kernel method, (c) the nearest-neighbor method*, and (d) the series method. Das Gupta [6] divided the work in nonparametric or distribution-free methods into three main categories: (a) plug-in rules, (b) statistics employed in devising some nonparametric two-sample or K-sample tests, and (c) typical ad hoc methods for classification problems. Space will not permit a thorough description of all the procedures described in the literature. Nevertheless, a number of the more popular procedures based on Das Gupta's categorization will be presented.

PLUG-IN RULES

When the form of the probability density function is unknown, there are no parameters to estimate. Instead, the probability density function must be estimated. Classification rules are developed based on the ratio of the density estimates, $\hat{f}_1(\mathbf{X})/\hat{f}_2(\mathbf{X})$, evaluated at the observation to be classified. These rules are called nonparametric "plug-in" rules. In general, all nonparametric plug-in rules involve some sort of estimation of the probability density function. Fix and Hodges [8] showed that if $\hat{f}_1(\mathbf{X})$ and $\hat{f}_2(\mathbf{X})$ are consistent estimates of $f_1(\mathbf{X})$ and $f_2(\mathbf{X})$, respectively, for all $\mathbf{X}$, then the nonparametric plug-in rule is consistent with the optimal rule.

An example of the plug-in rule is the nearest-neighbor method*, first proposed by Fix and Hodges in 1951 [8]. Let $\mathbf{X}_{ij}$, $i = 1, 2, \ldots, n_j$, be a p-dimensional random sample from the jth population ($j = 1, 2$), $\mathbf{Z}$ be the p-dimensional observation to be classified, and $d(\mathbf{X}_{ij}, \mathbf{Z})$ be a distance function*. The estimates for $f_j(\mathbf{X})$ are determined by first pooling the samples and ordering the values of $d(\mathbf{X}_{ij}, \mathbf{Z})$. Choose a positive integer V that is large, but small compared to the size of the samples. Let L_j be the number of the V observations nearest to $\mathbf{Z}$ which are from the jth sample. Using the likelihood ratio rule, $\hat{f}_1(\mathbf{X})/\hat{f}_2(\mathbf{X})$ is replaced by $(L_1/n_1)/(L_2/n_2)$. If the ratio is greater than 1, $\mathbf{Z}$ is assigned to population I. If the ratio is less than 1, $\mathbf{Z}$ is assigned to population II. If the ratio equals 1, $\mathbf{Z}$ is assigned at random. Once a metric in the sample space has been specified, Fix and Hodges [8] showed that the nearest-neighbor method is a consistent procedure.

A classification procedure based on statistically equivalent blocks* has been proposed by Anderson [2] and Gessaman [9]. Assume n_1 objects are sampled from population I and n_2 objects from population II. Order the scores from population I on the basis of the first variable. Next partition the sample into m equal groups (blocks). The scores in each block are then ordered on the basis of the second variable and partitioned into m equal subblocks. Repeat the process over all p variables. It should be apparent that the sample will have to be large if the partitioning goes beyond $p = 3$ variables. Once all variables have been partitioned, calculate the number of observations in population II that lie within the boundaries for each block. Let L_{1i} represent the number of scores in the ith block from population I and L_{2i} be the

number of scores in the ith block from population II. Form the ratios $(L_{1i}/n_1)/(L_{2i}/n_2)$. When the ratio for the ith block is greater than 1, assign an observation within the boundaries of the ith block to population I. When the ratio is less than 1, assign to population II. If the ratio equals 1, assign at random.

Various other methods of estimating the densities have been proposed for both the univariate and multivariate situations. Among others, Parzen [16], Cacoullos [3], and Loftsgaarden and Quesenberry [15] have all developed consistent density estimators.

Gessaman and Gessaman [10] contrasted the effectiveness of the LDF to certain nonparametric density estimator procedures for data sampled from bivariate normal distributions. The LDF was compared with a nearest-neighbor procedure, equivalent blocks procedure, Parzen–Cacoullos density estimator, and the Loftsgaarden–Quesenberry density estimator for varied mean vectors, covariance matrices, and sample sizes. They showed that when the assumptions underlying the LDF were violated, the nonparametric procedures tended to have a lower proportion of misclassified observations than either the LDF or QDF. Koffler and Penfield [13] compared the same four nonparametric procedures with the LDF and QDF for normal as well as nonnormal distributions. They concluded that when observations were drawn from nonnormal distributions, the nonparametric procedures classified the observations more effectively than either the LDF or QDF.

STATISTICS INVOLVING STANDARD NONPARAMETRIC TESTS

Das Gupta's second category of nonparametric procedures (i.e., those based on customary statistics for two-sample and K-sample problems) can be exemplified by the work of Chanda and Lee [4], who developed a Wilcoxon-type statistic for purposes of classification. Assuming $n_1 = n_2 = n$, let

$$V = \left[\#(X_{i1}, X_{j2}) \text{ such that} \right.$$
$$\left. X_{i1} > X_{j2} \, (1 \leqslant i, j \leqslant n) \right] / n^2$$

and

$$W = \left\{ \left[\# X_{i1} \leqslant Z \, (1 \leqslant i \leqslant n) \right] \right.$$
$$\left. + \left[\# X_{j2} \leqslant Z \, (1 \leqslant j \leqslant n) \right] \right\} / n.$$

Z is classified into population I if $[V > \frac{1}{2}, W > 1]$ or $[V \leqslant \frac{1}{2}, W \leqslant 1]$. Z is classified into population II otherwise. Under certain conditions, the Chanda and Lee statistic is consistent with the optimal rule.

AD HOC METHODS

When the assumptions underlying the LDF or QDF are suspect, an alternative approach to the classification problem is to apply a transformation to the data. One such procedure suggested by Conover and Iman [5] requires ranking the data and then basing the classification on the ranked data. The transformed scores are then used to develop LDF or QDF methods. The procedure requires pooling the samples and ordering the scores for each variable. The observations for each variable are then replaced by their corresponding ranks. Tied observations may either be assigned at random or based on midranks. Values of the p variables associated with new observations are then replaced by scores obtained by linear interpolation between two adjacent ranks computed on the basis of the combined sample. Once the original data are ranked, the sample means and covariance matrix or matrices are computed on the ranked data. The LDF or QDF is calculated, and the new observation is classified according to the LDF or QDF rules. Conover and Iman [5] showed that for nonnormal distributions, the rank procedures consistently classified observations with greater accuracy than either the LDF or the QDF based on the original data. When comparisons were made against other nonparametric procedures, the rank proce-

dures, in most instances, produced smaller proportions of misclassified observations. Koffler and Penfield [14] showed that the normal scores* transformation worked as well as the rank transformation. Randles et al. [17] have proposed an alternative ranking procedure for discriminating between two populations.

References

[1] Anderson, T. W. (1951). *Psychometrika*, **16**, 31–50.

[2] Anderson, T. W. (1965). *Multivariate Anal. Proc. Int. Symp.*, Academic Press, New York, pp. 5–27.

[3] Cacoullos, T. (1966). *Ann. Inst. Statist. Math.*, **18**, 179–186.

[4] Chanda, K. C. and Lee, J. C. (1975). In *The Search for Oil—Some Statistical Methods and Techniques*, D. B. Owen, ed. Marcel Dekker, New York, pp. 83–119.

[5] Conover, W. J. and Iman, R. L. (1980). *Commun. Statist.*, **A9**, 465–487.

[6] Das Gupta, S. (1973). In *Discriminant Analysis and Applications*, T. Cacoullos, ed. Academic Press, New York, pp. 77–138.

[7] Fisher, R. A. (1936). *Ann. Eugen.*, **7**, 179–188. (LDF process is described.)

[8] Fix, E. and Hodges, J. L. (1951). *USAF School of Aviation Medicine*, Proj. 21-49-004, Rep. 4, Randolph Field, TX. (Describes consistent estimators and nearest-neighbor method.)

[9] Gessaman, M. P. (1970). *Ann. Math. Statist.*, **41**, 1344–1346.

[10] Gessaman, M. P. and Gessaman, P. H. (1972). *J. Amer. Statist. Ass.*, **67**, 468–472. (Compares LDF and QDF with nonparametric procedures.)

[11] Hand, D. J. (1981). *Discrimination and Classification*. Wiley, New York.

[12] Klecka, W. R. (1980). *Discriminant Analysis*. Sage Publications, Beverly Hills, CA.

[13] Koffler, S. L. and Penfield, D. A. (1979). *J. Statist. Comp. Simul.*, **8**, 281–299.

[14] Koffler, S. L. and Penfield, D. A. (1982). *J. Statist. Comp. Simul.*, **15**, 51–68.

[15] Loftsgaarden, D. O. and Quesenberry, C. P. (1965). *Ann. Math. Statist.*, **36**, 1049.

[16] Parzen, E. (1962). *Ann. Math. Statist.*, **33**, 1065–1076.

[17] Randles, R. H., Broffitt, J. D., Ramberg, J. S., and Hogg, R. V. (1978). *J. Amer. Statist. Ass.*, **73**, 379–384.

[18] Von Mises, R. (1945). *Ann. Math. Statist.*, **16**, 68–73.

[19] Welch, B. L. (1939). *Biometrika*, **31**, 218–220. (Presents derivation of classification procedures for known distributions.)

(CLASSIFICATION
DENSITY ESTIMATION
DISCRIMINANT ANALYSIS
DISTRIBUTION-FREE METHODS
LIKELIHOOD RATIO
MULTIVARIATE ANALYSIS
NEAREST-NEIGHBOR METHODS)

DOUGLAS A. PENFIELD
STEPHEN L. KOFFLER

NONPARAMETRIC ESTIMATION OF STANDARD ERRORS

Consider a statistic $\hat{T} = \hat{T}(x_1, x_2, \ldots, x_n)$, which is a function of random variables $X_1, \ldots, X_n$. The standard error of $\hat{T}$, $\sigma = [E\{\hat{T} - E(\hat{T})\}^2]^{1/2}$, is often estimated from the data based on parametric modeling of the distribution of the X_i's. Nonparametric estimates of this standard error may be of value if information about the distribution is unavailable or uncertain, if the estimation involves approximation, perhaps asymptotic assumptions that should be checked in finite samples, or for a variety of other reasons.

No one estimate of standard error is accepted as best in all situations. Four specific estimates will be discussed here in order to introduce some of the basic ideas that underlie such estimation. Some general indication of their applicability will be given.

METHODS

Efron [2] provides a very informative discussion of several nonparametric estimates of the standard error of a point estimate. Conceptually, the jackknife* estimate is perhaps the simplest. Let $\hat{T}_{-j} = \hat{T}(x_1, x_2, \ldots, x_{j-1}, x_{j+1}, \ldots, x_n)$ be the estimate of the same

form as $\hat{T}$ but calculated from the set of all x's except x_j. The *jackknife estimate* of the standard error of $\hat{T}$ is defined in terms of the $\hat{T}_{-j}$'s, $j = 1, 2, \ldots, n$, as

$$\hat{\sigma}_J = \left\{ \frac{n-1}{n} \sum_{j=1}^{n} \left(\hat{T}_{-j} - \frac{1}{n} \sum_{k=1}^{n} \hat{T}_{-k} \right)^2 \right\}^{1/2}.$$

A good review of the jackknife is given by Miller [9].

A procedure that is more generally applicable than the jackknife procedure is called *bootstrapping* [1]. The bootstrap approximates the sampling distribution of $\hat{T}$ in the following manner [1].

1. Construct the sample probability distribution $\hat{F}$ that puts mass $1/n$ at each point $x_1, x_2, \ldots, x_n$.

2. With $\hat{F}$ fixed, draw a random sample of size n from $\hat{F}$, which is denoted

$$X_i^* = x_i^* \quad X_i^* \overset{\text{ind}}{\sim} \hat{F}, \quad \text{for } i = 1, \ldots, n$$

and called the bootstrap sample. Note that **x*** is selected with repetition.

3. The sampling distribution of $\hat{T}(x_1, x_2, \ldots, x_n)$, which depends on F, is approximated by the sampling distribution of $\hat{T}^*(x_1^*, x_2^*, \ldots, x_n^*)$, which depends on $\hat{F}$. Although this can sometimes be calculated theoretically, the sampling distribution of $\hat{T}^*$ is commonly determined via Monte Carlo techniques.

If step (2) is repeated N times yielding N independent realizations of $\hat{T}^*$, say $\hat{T}_j^*$, $j = 1, \ldots, N$, then the bootstrap* estimate of the standard error of $\hat{T}$ is

$$\hat{\sigma}_B = \left\{ \frac{1}{N-1} \sum_{j=1}^{N} \left(\hat{T}_j^* - \frac{1}{n} \sum_{k=1}^{N} \hat{T}_k^* \right)^2 \right\}^{1/2},$$

which is just the sample standard deviation of the bootstrap sample.

Example 1. Assume that for data $x_1, x_2, \ldots, x_n$, a standard error is required for the sample standard deviation

$$\hat{T} = \left\{ \sum (x_i - \bar{x})^2 / (n-1) \right\}^{1/2}.$$

If $-4, -3, 1, 3$ represent the observed data, then $\hat{T} = 3.30$. If each observation is dropped from the set, the $\hat{T}_{-j}$'s are 3.06, 3.61, 3.79, and 3.00, respectively. This leads to an estimate $\hat{\sigma}_J = 0.59$.

To calculate $\hat{\sigma}_B$, samples of size four must be drawn from the set of data points, where the sampling is with replacement. Assume we draw four samples which are the sets $\{-3, 1, -4, 3\}$, $\{3, 1, -4, 1\}$, $\{-4, 3, -4, 1\}$, and $\{1, 3, -4, 3\}$. The statistics $\hat{T}_j^*$ based on these four samples are 3.30, 2.99, 3.56, and 3.30, respectively. From these four values, $\hat{\sigma}_B$ can be calculated to be 0.23. Of course, more than four samples would usually be drawn.

In some situations, the delta method ([10, p. 388]; *see also* STATISTICAL DIFFERENTIALS, METHOD OF) is useful in providing variance expressions. Assume the estimate $\hat{T}$ can be expressed as a function of k arguments $\hat{T}(\bar{S}_1, \ldots, \bar{S}_k)$, where each $\bar{S}_i$ is an observed average of n independent identically distributed random variables. Then it will often be reasonable to suppose that $\bar{\mathbf{S}} = (\bar{S}_1, \ldots, \bar{S}_k)$ is distributed approximately as k-variate normal. If the vector $\mathbf{S} = (S_1(x), \ldots, S_k(x))$, corresponding to the terms of each average based on one observation $X \sim F$, has mean $\boldsymbol{\mu}_F$ and covariance matrix $\boldsymbol{\Sigma}_F$, then the delta method gives an approximation to the standard error of $\hat{T}$ which is

$$\left\{ \frac{1}{n} (\nabla_F \boldsymbol{\Sigma}_F \nabla'_F) \right\}^{1/2},$$

where ∇_F is the gradient vector

$$(\partial \hat{T} / \partial S_r)|_{\mathbf{S} = \mu_F}.$$

The approximation derives from a first-order Taylor series expansion of $\hat{T}(\bar{\mathbf{S}})$ about $\hat{T}(\mu_F)$. A nonparametric estimate of the standard error of $\hat{T}$ is generated if $\hat{F}$ is substituted for F. This gives

$$\hat{\sigma}_D = \left\{ \frac{1}{n} (\nabla_{\hat{F}} \boldsymbol{\Sigma}_{\hat{F}} \nabla'_{\hat{F}}) \right\}^{1/2}.$$

In the sample survey* literature, where complex estimators are common, half-sampling* methods are used to generate nonparametric estimates of standard errors. The simplest and most commonly considered case is when the sample is naturally divided into strata of size two. Often the strata represent primary sampling units in a complex sampling plan. Frequently the estimate of interest will be a weighted average of information from the separate strata with weights related to the size of the populations from which the stratified sample is drawn. When the sampling plan does not allow a simple estimate of the variance of some statistic, then a replication of the sampling plan generates strata of size two and allows an estimate of variance. If $\hat{T}$ is the statistic based on the whole sample and $\hat{T}_1$ and $\hat{T}_2$ are the estimates based on the two replicates, then a simple standard error estimate for $\hat{T}$ is $\frac{1}{2} |\hat{T}_1 - \hat{T}_2|$.

More generally, based on any half-sample consisting of one observation from each stratum the estimate $\hat{T}^\S$ of the same form of $\hat{T}$ can be calculated. If there are M such half-samples possible, indexed by $1, \ldots, M$ then an improved estimated of the standard error of $\hat{T}$ is

$$\hat{\sigma}_{\mathrm{HS}} = \left\{ \frac{1}{M} \sum_{j=1}^{M} \left(\hat{T}_j^\S - \frac{1}{M} \sum_{k=1}^{M} \hat{T}_k^\S \right)^2 \right\}^{1/2}.$$

If the number of half-samples is very large, Monte Carlo techniques could be used to estimate $\hat{\sigma}_{\mathrm{HS}}$. In addition a variety of modifications of $\hat{\sigma}_{\mathrm{HS}}$ have been suggested, all based, however, on the use of half-samples.

Example 2. Assume that in the example discussed earlier the data were collected as two half-samples $(-4, 1)$ and $(-3, 3)$ where the observations -4 and -3 were from one stratum and 1 and 3 were from a second stratum. If we are interested in a standard error estimate for the sample standard deviation as before, then the two half-sample estimates are 3.54 and 4.24, respectively. A simple estimate of the standard error is then $\frac{1}{2} |3.54 - 4.24| = 0.35$.

All the possible half-samples are $(-4, 1)$, $(-4, 3)$, $(-3, 1)$ and $(-3, 3)$. The respective statistics $\hat{T}_j^\S$ are 3.54, 4.95, 2.83, and 4.24.

Based on these values $\hat{\sigma}_{HS}$ can be calculated to be 0.79.

McCarthy's [8] work on balanced repeated replication* is of particular value if $\hat{T}$ is an approximately linear statistic. By a careful choice of half-samples the number of estimates $T_j^\S$ in $\hat{\sigma}_{HS}$ can be reduced to approximately n, the number of observations, rather than $2^{n/2}$ in the situation with strata of size two. The variance of $\hat{\sigma}_{HS}$ is increased under this procedure, but often the increase is of no practical importance.

When no natural strata exist, artificial strata could be created or all possible half-samples could be considered. The performance of $\hat{\sigma}_{HS}$ in this situation has not been extensively studied, however. In addition, such a procedure ignores the fact that half-sample estimates initially were proposed to deal with situations when "exact" variance expressions were impossible to define. They will be most useful, therefore, in such situations.

DISCUSSION

The estimates $\hat{\sigma}_J, \hat{\sigma}_B, \hat{\sigma}_D, \hat{\sigma}_{HS}$ are four of the most common nonparametric estimates of the standard error of a statistic. All involve the concept of subsampling. Bootstrap techniques are being used increasingly in the statistical literature and initial investigations indicate that, in some circumstances, they can be markedly better than other alternatives. Efron [2], for example, compares different estimates of the standard error of the correlation coefficient*.

The estimate $\hat{\sigma}_J$ works well in many situations. The most frequently cited example of a breakdown in the jackknife methodology is where $\hat{T}$ is the sample median*, a single-order statistic*. Although not often of serious practical importance, it should be noted that $\hat{\sigma}_J$ tends to be biased upward [3].

Efron [1] argues that the bootstrap arises more naturally than the jackknife and that the jackknife is a linear approximation to the bootstrap. In many situations, however, the jackknife estimate will be comparable in performance to the bootstrap estimate and computationally it is much simpler. This computational advantage may frequently make the jackknife estimate more useful than the bootstrap estimate.

The estimate $\hat{\sigma}_D$ is closely related to the infinitesimal jackknife defined by Jaeckel [5]. In fact these two methods and the influence-function* approach of Hampel [4] are virtually identical and can be viewed as Taylor-series approximations to the bootstrap. Asymptotically these estimates should perform well. In small samples, comparisons by Efron and Stein [3] suggest that the bootstrap and jackknife estimates are superior.

Half-sampling methods have been widely useful in the context of sample surveys. Kish and Frankel [6] give simulation results suggesting that a particular half-sample estimate is, by some criteria, somewhat superior to a jackknifelike estimate. For the correlation coefficient, Efron [2] shows the opposite although since there is no natural stratification* in Efron's work, $\hat{\sigma}_{HS}$ may be at an unfair disadvantage. Krewski and Rao [7] give a theoretical discussion of the different methods for ratio estimation* in survey sampling. No method is shown or claimed to be the superior one.

It seems safe to conclude that for many problems adequate nonparametric estimates of standard errors can be produced. No method has yet been established as generally optimal. The characteristics of any particular application should continue to play a major role in the choice of an estimate.

References

[1]   Efron, B. (1979). *Ann. Statist.*, **7**, 1–26.

[2]   Efron, B. (1981). *Biometrika*, **68**, 589–599.

[3]   Efron, B. and Stein, C. (1981). *Ann. Statist.*, **9**, 586–596.

[4]   Hampel, F. R. (1974). *J. Amer. Statist. Ass.*, **69**, 383–393.

[5]   Jaeckel, L. (1972). *Bell Labs. Memo.* No. 72-1215–11.

[6]   Kish, L. and Frankel, M. (1974). *J. R. Statist. Soc.*, **B36**, 1–37.

[7] Krewski, D. and Rao, J. N. K. (1981). *Ann. Statist.*, **9**, 1010–1019.

[8] McCarthy, P. J. (1969). *Rev Int. Statist. Inst.*, **37**, 239–263.

[9] Miller, R. G. (1974). *Biometrika*, **61**, 1–16.

[10] Rao, C. R. (1973). *Linear Statistical Inference and Its Applications*, 2nd ed. Wiley, New York.

(BOOTSTRAP
HALF-SAMPLING TECHNIQUES
JACKKNIFE)

VERN FAREWELL

NONPARAMETRIC METHODS   *See*
DISTRIBUTION-FREE METHODS

NONPARAMETRIC REGRESSION   *See*
REGRESSION, DISTRIBUTION-FREE METHODS IN

NONPARAMETRIC TOLERANCE LIMITS

The following two types of problem have their solutions provided by nonparametric tolerance limits.

1. A given type of fuse is produced under conditions of quality control*. Of interest is the blowing time X associated with this type of fuse. On the basis of n observed values $x_1, x_2, \ldots, x_n$ of X, we want to determine a lower limit $L = L(x_1, x_2, \ldots, x_n)$ so that we can be reasonably sure (i.e., with specified probability γ close to 1) that at least $100P\%$ of all fuses produced have blowing times greater than L.

2. Physicians often quote "normal" lower and upper limits L and U for some physiological measurement X. The quoted limits are based on observed values $x_1, x_2, \ldots, x_n$ for n healthy people and are determined in such a way that the physician can be reasonably sure that at least $100P\%$ of healthy people have X values between L and U.

The intervals $x \geqslant L$ in 1 and $L \leqslant x \leqslant U$ in 2 are known as *100P percent tolerance*

intervals at probability level γ. Since the end points of a tolerance interval depend on observed values $x_1, x_2, \ldots, x_n$ of random variables $X_1, X_2, \ldots, X_n$, the actual *coverage* provided by the interval is random, and all we can assert is that unless an event of probability (at most) $1 - \gamma$ has occurred, the true coverage provided by the interval is at least $100P\%$.

ONE-DIMENSIONAL TOLERANCE INTERVALS

The problem of tolerance limits is nonparametric, since the limits are to depend only on the sample observations but not on the distribution function of the variable X. For simplicity of discussion, we assume that the variable X has (unknown) density function $f(x)$. The problem can then be stated as follows. On the basis of a random sample $X_1, X_2, \ldots, X_n$ from $f(x)$, we want to find two limits $L = L(X_1, X_2, \ldots, X_n)$ and $U = U(X_1, X_2, \ldots, X_n)$ such that

$$\Pr\left[\int_L^U f(x)\, dx \geqslant P \right] \geqslant \gamma \qquad (1)$$

for all $f(x)$. Since the limits L and U do not depend on $f(x)$, they are *distribution-free**.

Wilks [5, 6] showed that the problem can be solved by taking for L and U appropriately chosen order statistics*. Let $X_{(1)} < X_{(2)} < \cdots < X_{(n)}$ be the order statistics associated with the random sample $X_1, X_2, \ldots, X_n$ from $f(x)$ and set

$$C_k = \int_{X_{(k-1)}}^{X_{(k)}} f(x)\, dx, \qquad k = 1, 2, \ldots, n+1,$$

where $X_{(0)} = -\infty$ and $X_{(n+1)} = +\infty$. The C_k are known as *elementary coverages*. Elementary coverages play an important role in nonparametric statistics, since their distribution does not depend on the underlying density function $f(x)$. In fact, we can write $C_k = U_{(k)} - U_{(k-1)}$, where $U_{(1)} < U_{(2)} < \cdots < U_{(n)}$ are the order statistics in a random sample of size n from the uniform distribution on the interval $(0, 1)$ and $U_{(0)} = 0$, $U_{(n+1)} = 1$. It follows that the sum V_t of any t distinct elementary coverages is a random variable which has the beta distribution* with parameters t and $n + 1 - t$. With the help of tables of the beta distribution, it is then possible to determine the minimum value t such that $\Pr[V_t \geqslant P] \geqslant \gamma$. Any interval with $L = X_{(r)}$ and $U = X_{(r+t)}$, $r \geqslant 0$, $r + t \leqslant n + 1$, then satisfies requirement (1).

It is often more convenient to write $U = X^{(s)}$, where $X^{(s)}$ denotes the sth largest observation among $X_1, X_2, \ldots, X_n$. We find $s = n + 1 - (r + t) = m - r$, where $m = n + 1 - t = r + s$ equals the number of elementary coverages that have been omitted from the tolerance interval. Somerville [1] has tabulated the largest value m such that we may assert with probability at least γ that no less than $100P$ percent of a population lies between the limits $L = X_{(r)}$ and $U = X^{(s)}$, when $r + s = m$. For two-sided tolerance intervals, it usually is appropriate to choose r and s as nearly equal as possible subject to the requirement $r + s = m$. For one-sided limits, we choose $r = m$, $s = 0$, if a lower limit is required and $r = 0$, $s = m$, if an upper limit is required.

Example 1. For $n = 100$, $P = 0.90$, $\gamma = 0.95$, the Somerville table gives $m = 5$. Based on 100 observations, a 90% tolerance interval at probability level 0.95 extends from the second smallest to the third largest observations or from the third smallest to the second largest observations. The fifth smallest (largest) observation provides a lower (upper) bound for a one-sided interval.

Statisticians often use tolerance intervals with $r = s = 1$. Somerville's Table 2 lists probability levels γ at which we can assert that no less than $100P$ percent of the population is covered by the interval extending from the smallest to the largest observations in a sample of size n.

Example 2. For $n = 50$ and $P = 0.95$, we find $\gamma = 0.72$.

DISCONTINUOUS DISTRIBUTIONS

The preceding results are strictly true only for populations with continuous distribution functions. For such populations, the relevant probability statement remains unchanged, whether the tolerance interval is taken as an open or a closed interval. This is no longer true for discontinuous populations. Scheffé and Tukey [2] have shown that if the probability level of the tolerance interval bounded by the rth smallest and the sth largest observations equals γ in the continuous case, then for arbitrary distributions the probability level for the open interval is $\leqslant \gamma$; for the closed interval, it is $\geqslant \gamma$.

MULTIDIMENSIONAL TOLERANCE REGIONS

The concept of tolerance interval for a one-dimensional variable extends to the multivariate case. Generalizing an approach proposed by Wald [4], Tukey [3] with the help of a sequence of *ordering functions* defines $n + 1$ *statistically equivalent blocks* to take the place of the $n + 1$ elementary coverages of the one-dimensional case. The elimination of m blocks produces a tolerance region whose coverage obeys the same probability law as in the one-dimensional case.

References

[1] Somerville, P. N. (1958). *Ann. Math. Statist.*, **29**, 599–601.

[2] Scheffé, H. and Tukey, J. W. (1945). *Ann. Math. Statist.* **16**, 187–192.

[3] Tukey, J. W. (1947). *Ann. Math. Statist.*, **18**, 529–539.

[4] Wald, A. (1943). *Ann. Math. Statist.*, **14**, 45–55.

[5] Wilks, S. S. (1941). *Ann. Math. Statist.*, **12**, 91–96.

[6] Wilks, S. S. (1942). *Ann. Math. Statist.*, **13**, 400–409.

Further Reading

Pratt, J. W. and Gibbons, J. D. (1981). *Concepts of Nonparametric Theory*. Springer, New York, pp. 118–130. (Discusses possible difficulties associated with the interpretation of tolerance intervals.)

Wilks, S. S. (1948). *Bull. Amer. Math. Soc.*, **54**, 6–50. (Section 7 contains a very readable account of statistically equivalent blocks.)

Wilks, S. S. (1959). In *Probability and Statistics: The Harald Cramér Volume*. U. Grenander, ed. Almqvist & Wiksell, Stockholm, pp. 331–354.

These two Wilks papers and refs. 5 and 6 are reprinted in:

Anderson, T. W., ed. (1967). *S. S. Wilks: Collected Papers*. Wiley, New York.

(DISTRIBUTION-FREE METHODS
ORDER STATISTICS
TOLERANCE REGIONS, STATISTICAL)

GOTTFRIED E. NOETHER

NONRESPONSE (IN SAMPLE SURVEYS)

Nonresponse in sample surveys* may be defined as the failure to make measurements or obtain observations on some of the listing units selected for inclusion in a sample. Unfortunately, nonresponse is a problem that plagues virtually all sample surveys and, if it is extensive, may seriously compromise the validity and generalizability of any results.

The purpose of most sample surveys is to estimate, with the greatest possible precision, unknown population parameters such as means, totals, or proportions. Unbiased estimates of these parameters may be obtained using any one of many potential sampling strategies—provided a 100% response rate is attained. Unfortunately, nonresponse is always a problem and 100% response rates are never attained. The effect of nonresponse is to increase, perhaps dramatically, the bias of an estimate resulting from a particular sample survey.

To demonstrate the effects of nonresponse, it is convenient to think of the population as divided into two "strata."

Let

N = total number of enumeration units in the population.

N_1 = total number of potential responding enumeration units in the population.

N_2 = total number of potential nonresponding enumeration units in the population ($N_2 = N - N_1$).

If $\overline{X}_1$ = the mean level of a characteristic X among the N_1 potential responders and

if $\overline{X}_2$ = the mean level of characteristic X among the N_2 potential nonresponders, then $\overline{X} = (N_1\overline{X}_1 + N_2\overline{X}_2)/N$ = the mean level of X among the total population of N enumeration units.

Now if we take a simple random sample of n enumeration units, and no attempt is made to obtain data from the potential nonresponders, we are effectively estimating the mean level $\overline{X}_1$ rather than $\overline{X}$. If among the n enumeration units, n_1 respond, and if $\overline{x}$ denotes the mean level of X among these n_1 responders, then

$$E(\overline{x}) = \overline{X}_1$$

rather than $\overline{X}$. The bias* of $\overline{x}$ may be denoted as

$$\text{bias}(\overline{x}) = (N_2/N)(\overline{X}_1 - \overline{X}_2).$$

Clearly, the bias due to nonresponse is independent of n_1, the number of enumeration units actually sampled. It cannot be reduced by increasing n. The most effective way of reducing it is to decrease the proportion of potential nonresponders N_2/N since little can be done about the difference $\overline{X}_1 - \overline{X}_2$.

Before discussing methods of decreasing the size of N_2, it is necessary to understand that in any given survey, the potential nonresponders comprise a rather heterogeneous group. The methods used to encourage response should be carefully and specifically tailored to particular nonrespondents, and the type of survey being used.

One common problem in many surveys is that some targeted interviewees may be temporarily away from their homes or phones when the interviewer calls. In order to avoid the obvious bias these not-at-homes would cause, it is necessary to attempt making contact on other occasions. Each successive call increases the total number of respondents and, ultimately, decreases the overall bias. In personal interview surveys, provision should be made in the survey design to revisit households during the evening or on weekends.

In mail surveys nonresponse is a common problem, in part due to the high level of mobility of our modern population. Assuming the mailed survey reaches the intended responder, nonresponse is still a common problem since there are many individuals who refuse to cooperate and will not provide the desired information.

Numerous methods have been proposed for increasing response rates in mail surveys. For example, attractive packaging of the questionnaire is important as are carefully worded cover letters that specify the purpose of the survey and the organizations responsible for conducting the survey, and that assure the confidentiality of information provided by the respondent. Every attempt should be made to construct concise, clearly worded questionnaires that take less than 30 minutes to complete; longer forms run a much higher risk of refusal.

Telephone interviews have been shown to have higher response rates than mail surveys designed to collect the same information. Potential respondents find it more difficult to refuse telephone interviews than they do mailed questionnaires. Whether interviews are carried out over the telephone or in person, demographic and personal characteristics of the interviewer are extremely important factors in influencing response. It is particularly important for all interviewers to be provided with appropriate credentials.

In any type of survey, response rates can be increased if there is an effective publicity campaign in advance of the survey. Usually this is difficult to accomplish without great expense—especially in large metropolitan areas.

Appropriate use of endorsements can also reduce nonresponse dramatically. For instance, if an official agency or organization endorses the survey, this lends an air of importance that might persuade the respondent to complete the interview.

Incentives (rewards given to a respondent for participating in a survey) have been shown to be effective in increasing response rates (see, for example, refs. 2 and 11). A problem may arise, however, since cash incentives may be more likely to attract special subgroups of the population whereas nonmonetary incentives may attract other subgroups. Hence one should be sensitive to the issue of possible bias when such incentives are used.

Often respondents are *unable* to answer. This may occur for a myriad of reasons ranging from failure to understand what is being asked to the mental or physical inability to respond. These subjects do not refuse to be interviewed, they are simply incapable of providing the sought-after cooperation (compliance). Every attempt should be made in the survey plan to accommodate such people. For example, translations into other languages of the questionnaire should be available as should rewording of certain questions that makes it easier for the interviewee to respond.

Nonresponse may also occur as a result of failure to locate or to visit some units in the sample. This failure may be due to extreme hardship involved in reaching certain targeted individuals by virtue of inaccessibility, poor transportation, or weather conditions during the course of the survey. Again, failure to survey geographically extreme individuals can bias the results of the survey.

Finally, noncoverage is a potential source of nonresponse in surveys and may occur when an interviewer must find and list all households in a given area for subsequent sampling. If the resulting lists are incomplete, then the sample may not be representative.

A commonly used strategy to deal with nonrespondents is to sample in two stages. In stage 1, an attempt is made to collect information on n enumeration units (sampled from the N enumeration units in the population). Suppose that of these n initial contacts, n_1 respond and successfully provide the desired information. The remaining $n_2 = n - n_1$ enumeration units are nonrespondents. In stage 2, an intensive effort is made to collect information on a subset of the n_2 nonrespondents. This effort can simply be a followup questionnaire or phone calls or may involve an attempted personal interview. Letting n_2^* represent the number of the n_2 nonrespondents selected for the intensive followup effort and n_2' be the number of successful responses obtained from these n_2^* enumeration units, we take as an estimate of the population mean, $\overline{X}$,

$$\hat{\overline{X}} = \frac{n_1 \overline{x}_1 + n_2 \overline{x}_2}{n}$$

where $\overline{x}_1 = \sum_{i=1}^{n_1} x_i / n_1$ and $\overline{x}_2 = \sum_{i=1}^{n_2'} x_i' / n_2'$. If n_2' is close to n_2^*, then $\hat{\overline{X}}$ is a nearly unbiased estimator of the unknown population mean $\overline{X}$.

The decision as to how large n_2^* should be (relative to n_2) can be based on a strategy originally proposed by Hansen et al. [4]. This method takes into account the field costs as well as the expected nonresponse rate.

To determine n, the number of enumeration units to sample in stage 1, we must first determine the number of subjects necessary to meet specified requirements for precision. Let n' represent this required sample size, and let P_1 be an estimate of the response rate at stage 1 (i.e., n_1 / n). Then the required number of enumeration units to sample at stage 1 can be obtained by multiplying n' by a factor that takes nonresponse into consideration. A discussion and illustration of sample-size strategies for two-stage sampling designs can be found in Levy and Lemeshow [8, pp. 254–267]. Another approach to optimal sample size with nonresponse has been developed by Ericson [3]. Rubin [13] gives a method for estimating the effect of nonresponse in sample surveys. More recently, Ridley et al. [12] attempted to assess the extent and nature of nonresponse bias in a national survey of elderly women. Kalton et al. [6] have discussed such problems as household, person, and item nonresponse, as well as such methods of nonresponse adjustment as use of proxy reports and imputation procedures as they relate to the *Survey of Income and Program Participation* [9].

As a result of nonresponse and other problems in data collection, storage, and retrieval, missing values will inevitably result. Dealing with these missing values once the survey is complete is a serious and not easily resolved problem. Much has been written about data imputation methods (strategies for dealing with missing or clearly erroneous bits of information) and it is not the purpose of this entry to review them (*see* INCOMPLETE DATA). It is necessary to note, however, that nonresponse can destroy a sampling design that is otherwise self-weighting since the allocation of the actual sample has been altered from the one that was desired. Researchers deal with this problem in a number of ways. Some simply ignore the missing data and perform all statistical analyses on the actual collected data. This is very dangerous since the perhaps incorrect assumption is being made that the missing data are "similar" to the existing data. This will be true if observations are missed randomly, in which case no bias results. One should always compare the respondents and nonrespondents on whatever information is available to see if this assumption is valid. Failure to do anything with missing data through nonresponse is a kind of imputation. As a result, imputation is not only desirable but inevitable and the alternative of substituting "typical" values for missing observations using some acceptable imputation scheme has the advantage of preserving the self-weighting feature of many sampling designs. This substitution should be undertaken only if the assumption that the missing observations can be reasonably derived by the available ones can be justified.

Other discussions of nonresponse may be found in refs. 1, 5, 7, and 10.

References

[1] Cochran, W. G. (1977). *Sampling Techniques*, 3d ed. New York, Wiley.

[2] Erdos, P. (1970). *Professional Mail Surveys*. McGraw-Hill, New York.

[3] Ericson, W. A. (1967). *J. Amer. Statist. Ass.*, **62**, 63–78.

[4] Hansen, M. H., Hurwitz, W. N., and Madow, W. G. (1953). *Sample Survey Methods and Theory*, Vol. 1. Wiley, New York.

[5] Jessen, R. J. (1978). *Statistical Survey Techniques*. Wiley, New York.

[6] Kalton, G., Kaspryzyk, D., and Santos, R. (1980). *Proc. Sect. Surv. Res. Meth. Amer. Statist. Ass.*, **1980**, 501–506.

[7] Kish, L. (1965). *Survey Sampling*. Wiley, New York.

[8] Levy, P. S., and Lemeshow, S. (1980). *Sampling for Health Professionals*. Lifetime Learning Publications, Belmont, CA.

[9] Lininger, C. A. (1980). *Proc. Sect. Surv. Res. Meth. Amer. Statist. Ass.*, **1980**, 480–485.

[10] Moser, C. A., and Kalton, G. (1971). *Survey Methods in Social Investigations*. Heinemann, London.

[11] National Center for Health Statistics. (1975). "A Study of the Effect of Remuneration upon Response in the Health and Nutrition Examination Survey." *PHS Publication No. 1000, Series 2, No. 67*, U.S. GPO, Washington, DC.

[12] Ridley, J. C., Dawson, D. A., Tanfer, K., and Bachrach, C. A. (1979). *Proc. Sect. Surv. Res. Meth. Amer. Statist. Ass.*, **1979**, 353–358.

[13] Rubin, D. B. (1977). *J. Amer. Statist. Ass.*, **72**, 538–543.

(MEASUREMENT ERROR
STRATIFIED SAMPLING
SURVEY SAMPLING
TELEPHONE SURVEYS,
 COMPUTER-ASSISTED)

STANLEY LEMESHOW

NONSAMPLING ERRORS IN SURVEYS

Sampling errors in surveys arise from random variation caused by the selection of n randomly chosen sample units from a total population of N units. Nonsampling errors (often referred to as measurement errors) are those errors that are not associated with this inductive process of inference about the population.

The study of nonsampling errors has not been thoroughly discussed in sampling textbooks. Motivation for this work is difficult to develop in a few pages in a text since most of the examples that occur in practice are associated with complex survey designs. However, the concept of nonsampling errors was discussed as early as 1902 by K. Pearson [16], who demonstrated that even simple measurements can result in substantial measurement errors*.

Every step in the survey process, from development of survey specifications through sample selection, data collection, coding, editing*, summarization, and tabulation, is a potential source of nonsampling errors. Kish [10] gives a classification model of errors in surveys. The model classifies variable errors into sampling error and nonsampling errors. It also classifies bias into sampling, nonsampling, and constant statistical bias.

In this classification model, the variable sampling error is the standard (relative) error of an estimator. Examples of variable nonsampling errors include errors such as field error by interviewers and supervisors. Sampling bias includes frame bias caused by the possible duplication of units in the sampling frame, bias from consistent estimators, and any constant statistical bias, for example, any bias caused by using the median* to estimate a population mean for a skewed population. Nonsampling bias can include noncoverage of the population of interest caused by an inadequate sampling frame and nonresponse to the survey from sampled units.

A simple mathematical model to relate variable errors and bias to total survey error (TSE) is commonly used. For each unit in the population of interest, define a "true" value for the variable to be estimated. This true value is conceived of independent of the total response process and should be measurable under reasonable survey conditions. For some variables the true value is easy to define, e.g., the age or sex of the respondent, but oftentimes it may present a difficult concept, say a consumer preference or a respondent attitude toward government policy. Let this true value for each unit in the population of size N be denoted by Y_i, $i = 1, \ldots, N$. Then an estimate of the true population average, $\bar{Y} = N^{-1}\sum_{i=1}^{N} Y_i$, is given by $\bar{y} = n^{-1}\sum_{i=1}^{n} y_i$, where y_i is the observed value of the ith unit in a random sample of size n. The TSE is the expected squared deviation of the estimator $\bar{y}$ from $\overline{Y}$,

$$E\left[\bar{y} - \overline{Y}\right]^2 = E\left[\bar{y} - E(\bar{y})\right]^2$$
$$+ \left[E(\bar{y}) - \overline{Y}\right]^2, \quad (1)$$

where the expectation is taken over the distribution of all possible values of the estimator $\bar{y}$ based on samples of size n. The first component in (1) is the mean squared deviation of the variable errors around the expected value of $\bar{y}$ for the survey. The second component is the square of the deviation of the expected value from the true population value and is the bias* squared. Thus the square root of the TSE or the relative root mean square error of the survey can be expressed as

$$\text{TSE}^{1/2} = (\text{VE}^2 + B^2)^{1/2}, \quad (2)$$

where VE represents the variable errors from all sources and B represents the total of all biases associated with the survey. It is important to note that in this simple model the sampling error is but one component of all variable errors. Thus sampling error can contribute only a small part to the TSE. This model not only illustrates the two overall components of survey error but also provides a convenient breakout of the sources of error for further discussion.

MODELS

The purpose of this section is to illustrate the nature and measurement of variable nonsampling errors in surveys through the use of models. For simplicity it is assumed the survey is either a complete enumeration (census) or a sample survey in which all units have an equal probability of selection.

Hansen et al. [8] decompose the total variance of an estimator into components that reflect sampling and response variance as sources of survey estimator. Response variance reflects, but is not restricted to, the variability on the part of the respondent and/or interviewer to erroneously report the answer to a question in a survey. In this

model, a desired measure or true value is assumed, say a proportion of the population having some specified characteristic. That is, in a population of N units, each unit can be regarded as having some value Y, and the desired population proportion estimate is

$$\overline{Y} = N^{-1} \sum_{i=1}^{N} Y_i,$$

where Y_i has the value 1 if the ith unit has the specified characteristic and has the value zero otherwise.

In this model it is assumed the survey can be repeated under the same general survey conditions and that a particular survey is one trial from among all possible repetitions of that survey. Under this repeatability condition an observation on the ith unit is denoted by y_{it}, where y_{it} has the value 1 if the ith unit has the specified characteristic on the tth trial (survey) and has the value 0 otherwise. An estimate of $\overline{Y}$ from the survey is

$$P_t = n^{-1} \sum_{i=1}^{n} y_{it},$$

where n is the number of units in the sample. Let $Ep_t = P$ denote the expected value of p_t over all possible trials (surveys) taken under the same survey conditions. Also, let $E_i y_{it} = P_i$ denote the conditional expectation on the ith unit of the population, where the expectation is over all possible samples and trials under the same survey conditions. Then the difference between the observed value on the particular trial and the expected value for the unit,

$$d_{it} = y_{it} - P_i$$

is called the *response deviation.*

The total variance of the survey is

$$\sigma_{pt}^2 = E(p_t - p)^2 + 2E(p_t - p)(p - P)$$
$$+ E(p - P)^2, \quad (3)$$

where

$$p = n^{-1} \sum_{i=1}^{n} P_i$$

is the mean of the conditional expected values for those in the sample.

Therefore,

σ_{pt}^2 = response variance

+ 2 (covariance of response and

sampling deviations)

+ sampling variance.   (4)

The response variance is further decomposable into

$$\sigma_{\bar{d}_i}^2 = n^{-1}\sigma_d^2\left[1 + \rho(n - 1)\right],$$

where σ_d^2 is the variance of the individual response deviations over all possible trials, i.e., the simple response variance, and ρ is the intraclass correlation* among the response deviations in survey or trial. Note that for even a very small intraclass correlation, say, $\rho = 0.01$ with a sample size of $n = 1,000$, the intraclass correlation increases the response variance by a factor of $\rho(n - 1)$ $= 0.01(999) = 10$ or 1000%. Therefore, the correlated component can dominate the simple response variance and represent the largest contribution to the response variance.

The covariance of response and sampling deviations in equation (3) is zero in a complete enumeration of the entire population or in repetitions of a survey for a fixed sample of units. In other situations, it is generally assumed that this component is negligible. However, no research has thoroughly documented this assumption.

Battese et al. [4] suggest a component of variance model for an interview–reinterview situation to measure interviewer effects, sampling variance, and respondent response errors. Under the same survey conditions, the reinterview is obtained by a different interviewer. An additive model is used.

Under various distributional assumptions, 21 variables were examined for these errors in a survey of Iowa farm operators using this model. The estimated average respondent response errors were greater than the estimated standard errors, while the estimated interviewer variance contributed less than 0.10 of 1% of the total variance.

The preceding discussion does not include all of the models presented in the literature (see, e.g., refs. 3, 7, 9, 11, and 12) to estimate components of variance for variable nonsampling errors. However, they provide useful insight into the problem of variable nonsampling errors in surveys as well as providing the reader with the basic information on the complexity of assumptions that must be used and justified in order to estimate these components of error as one part of the total survey error.

VALIDITY STUDIES

In the preceding section several models were presented that illustrate the estimation of variable nonsampling errors. These models do not consider the problem of nonsampling bias. Validity studies use data external to the survey, either at the unit or aggregate level, to measure magnitude of nonsampling bias. The two major sources of nonsampling bias are noncoverage of the population of interest and nonresponse. Noncoverage of the population of interest is often caused by an inadequate sampling frame. List frames used to sample establishments and firms, for example, are incomplete and outdated as soon as they are built. National household surveys that use random digit dialing do not cover the entire population because some households do not have telephones. These biases may present more serious problems for certain domains in the population, say minority groups such as blacks and Hispanics, because they may have even fewer telephones.

The problem of nonresponse, either for entire sample units caused by refusals and inaccessibles or for missing items when a respondent fails to complete certain items, falls into a broad category of research called incomplete or missing data.

Several procedures for imputing for missing data have been presented in the literature. A detailed study of the problem of missing data has been undertaken by the National Academy of Science for the National Reasearch Council [14]. The theory of the current practice of incomplete data is studied in detail and contains the presentation of case studies on missing data procedures currently being used by survey organizations.

The main techniques used to estimate nonsampling bias are unit-by-unit validation and sample validation. For the former, the observed value for each unit in the sample is compared to a "true" value for that unit obtained from an external source. In the latter, the estimate obtained from the sample is compared to a "true" value obtained from a source external to the survey. Several studies have used these techniques to estimate nonsampling bias.

Anderson et al. [1] calculated the biases due to nonresponse errors, characteristics of respondent reporting, and processing errors resulting from the imputation of missing data for a national health survey. Validation data about the families' medical care and health insurance for the survey year for the area probability sample of the noninstitutionalized population of the United States was obtained from family physicians, clinics, hospitals, insuring organizations, and employers. It was found that the magnitude of the bias varied for different variables, e.g., being small for total expenditures for inpatient admissions and large for emergency and outpatient charges. It was noted that the main limitation in this study was the inability to identify false negatives, i.e., medical expenses not reported by the family.

Neter and Waksberg [15] find a difference in household expenditures depending on the length of the recall period used, and Arends et al. [2] find that farmers overreport "whole milk sold" to dairy plants.

Summary

Nonsampling errors can be classified into variable nonsampling errors and nonsampling bias. The former usually have been measured through the use of models while the magnitude of the latter are estimated by validation studies.

Because nonsampling errors can occur at every step in the survey (census) process it is impractical to list every source of error for this article. For example, Lessler [12] provides a comprehensive listing of the types of errors that can be associated with a sampling frame. However, for repetitive surveys, a description of each potential nonsampling error, along with any knowledge of the magnitude of the error, is a useful document not only aiding in understanding the errors in a given survey, but providing a working guide for the direction of research and improvements to ongoing surveys. Bailer and Brooks [6] have developed such an "error profile" for employment statistics from the Current Population Survey of the Bureau of the Census, U.S. Department of Commerce. A comprehensive list of the survey's operations was developed and the documentation of what is known about each survey operation as a potential source of nonsampling errors included sampling design, observational design, data preparation design, the estimation process, and analysis and publication. Beller [5] described an error profile for the multiple-frame cattle and hog surveys of the U.S. Department of Agriculture. Sources of error from each frame are discussed, and particular attention is paid to nonsampling errors that occur because the sampling frames used have common sampling units.

Since nonsampling errors can account for the larger part of the total survey error, even if the magnitude of these errors cannot always be measured, it is important for the survey designer to insure that survey procedures that control or reduce the magnitude of the errors are implemented. In fact, an important consideration should be the trade-off between the possible increase in sampling errors, caused, e.g., by the use of a less complex survey design, or an increase in the quality control aspects of a survey and the probable decrease in nonsampling errors.

References

[1] Anderson, R., Kasper, J., Frankel, M. R., and Associates. (1979) *Total Survey Error*. Jossey-Bass, San Francisco.

[2] Arends, W., Addison, R., Young, R., and Bosecker, R. (1973). "An Evaluation of Enumeration Techniques and Associated Response Errors

and Biases." *Statist. Rep. Serv. Staff Rep.* USDA, Washington, DC.

[3]   Bailar, B. A. and Dalenius, T. (1970). *Sankhyā Ser. B*, 341–360.

[4]   Battese, G. E., Fuller, W. A., and Hickman, R. D. (1976). *J. Indian Soc. Agric. Statist.*, **28**, 1–14.

[5]   Beller, N. D. (1979) "Error Profile for Multiple-Frame Surveys." *Econ. Statist. Coop. Serv. Rep. No. ESCS-63*, USDA, Washington, DC.

[6]   Brooks, C. A. and Bailar, B. A. (1978). "An Error Profile: Employment as Measured by the Current Population Survey." *Statist. Policy Work. Pap. No. 3*, U.S. Dept. Commerce. Washington DC.

[7]   Folsom, R., Jr. (1980). *Proc. Sect. Surv. Res. Amer. Statist. Ass.*, **1980**, 137–142.

[8]   Hansen, M. H., Hurwitz, W. N., and Bershad, M. A. (1961). *Bull. Int. Statist. Inst.*, **38**, 359–374.

[9]   Hartley, H. O. (1981). In *Current Topics in Survey Sampling*, D. Krewski, R. Platek, and J. N. K. Rao, eds. Academic Press, New York, pp. 31–46.

[10] Kish, L. (1965). *Survey Sampling.* Wiley, New York. pp. 509–573.

[11] Koch, G. G. (1973). *J. Amer. Statist. Ass.*, **68**, 906–913.

[12] Lessler, J. T. (1976). *Proc. Soc. Statist. Sect. Amer. Statist. Ass.*, **1976**, 520–525.

[13] Lessler, J. T. (1980). *Proc. Sect. Surv. Res. Amer. Statist. Ass.* **1980**, 125–130.

[14] National Research Council (to appear). *Panel on Incomplete Data, Theory and Bibliography.* Academic Press, New York.

[15] Neter, J. and Waksberg, J. (1964). *J. Amer. Statist. Ass.*, **59**, 18–55.

[16] Pearson, K. (1902). *Philos. Trans. R. Soc. Lond. A*, **198**, 235–299.

Bibliography

The following three articles present an author-alphabetized bibliography of non-sampling error research.

Dalenius, T. (1977) *Int. Statist. Rev.*, **45**, 71–89, 181–197, and 303–317.

(ACCURACY
CENSUS
EDITING STATISTICAL DATA
MEASUREMENT ERROR
NONRESPONSE IN SAMPLE SURVEYS
SURVEY SAMPLING
TELEPHONE SURVEYS,
    COMPUTER-ASSISTED)

ROBERT D. TORTORA

NONSENSE CORRELATION

When a significant correlation occurs between two variables that actually have no direct relation to one another, such a correlation may be referred to as a nonsense correlation, although a more common term is *spurious correlation**.

(CAUSATION
CORRELATION
SPURIOUS CORRELATION)

NONSINGULAR MATRIX

Let **A** be a square matrix in a *field K*. If there exists a matrix $\mathbf{A}^{-1}$ in K such that $\mathbf{A}\mathbf{A}^{-1} = \mathbf{A}^{-1}\mathbf{A} = \mathbf{I}$, where **I** is the unit (or identity matrix), then **A** is called nonsingular (or regular or invertible matrix). The matrix **A** is nonsingular if and only if the determinant of **A**, i.e., $|\mathbf{A}|$, is nonzero.

NONSTATIONARY TIME SERIES   *See* TIME SERIES, NONSTATIONARY

NORMAL APPROXIMATIONS TO THE POISSON*, BINOMIAL*, NEGATIVE BINOMIAL*, AND HYPERGEOMETRIC* DISTRIBUTIONS

Under suitable limiting conditions on the parameters, the widely used discrete distributions mentioned in the title tend to the normal distribution. This allows approximations to their cumulative probabilities, their quantiles, and the confidence bounds for their relevant parameters. By various modifications and transformations, the accuracy of such approximations can be improved. The results are then accurate up to a few digits for almost all cases (and thus useful for quick evaluation) and accurate up to many digits for cases near the limit (and thus useful for implementation on an electronic computer or programmable calculator for cases where a direct evaluation would become too costly or too inaccurate; see Ling [23]).

After an introduction, detailed recommendations for the approximation of the cumulative distribution function will be presented for each distribution separately. This is followed by a combined section with references on approximations to confidence bounds and quantiles* and a literature recommendation.

INTRODUCTION

The history of the normal approximation to the binomial distribution goes back to De Moivre* (*The Doctrine of Chances*, 1718) and Laplace* (*Théorie Analytique des Probabilités*, 1812). For other early references, *see* APPROXIMATIONS TO DISTRIBUTIONS. The problem of expressing one distribution in the form of a series expansion based on a second distribution (often the normal one) was successfully solved around 1900; *see* GRAM–CHARLIER SERIES, EDGEWORTH EXPANSIONS, and CHEBYSHEV–HERMITE POLYNOMIALS. With the Cornish–Fisher EXPANSIONS, published in the late thirties, normal approximations to quantiles and probabilities can be easily obtained from the moments of a distribution. As is documented in the Literature section of this entry and in remark (e) of the section on Poisson probabilities, the first two or three terms of such expansions are suboptimal as approximations because of slow convergence, but they are an important tool in comparing and improving other approximations.

The literature on this subject is vast and widespread; in the following only a selection will be presented. First, some general aspects are mentioned.

PRACTICALITY. Some authors concentrate on elegant mathematical results; others emphasize the applicability of their proposal.

ERROR MEASURE. In the order from the title, the distribution functions considered depend on two, three, three, and four arguments, respectively. One may consider absolute or relative errors per case or maximized across one or more arguments. Use of the maximum error in $\Pr[a \leqslant X \leqslant b]$ for fixed parameters, e.g., in Raff [34] and Gebhardt [16], stresses accuracy in the middle of the distribution. The use of the relative error in the smaller of $\Pr[X \leqslant \alpha]$ and $\Pr[X \geqslant \alpha]$, see Peizer and Pratt [32], stresses approximation of the extreme tails. Accuracy near the customary significance levels or their complements is emphasized in Molenaar [26]. Other criteria have been used as well. As the errors depend in a complicated way on all the arguments, simple rules of thumb or uniform bounds are hard to find.

COMPUTATIONAL EFFORT. Computing facilities have grown rapidly in the past few decades; people now tend to use approximations because they are cheaper and quicker rather than because direct evaluation is overwhelmingly difficult. It remains true, however, that potential users differ in the amount, costliness and nature of the computing aids that they have readily available (hardware, software, and statistical tables). In formulating recommendations, our position resembles that of a Consumers' Union comparing washers or motorcars: our users differ in available facilities, frequency and circumstances of use, and wishes as regards performance. An attempt is made to recommend the greatest accuracy for the least computational labor, but some subjective element cannot be avoided in judging either.

BOUNDS. As already stated, uniform bounds on the error are either too complicated or too coarse to be informative for a special application. In some rare cases, upper or lower bounds for the probability $\Pr[X \leqslant k]$ might be more useful than an approximation to it. For lack of space, we refer to Johnson and Kotz [21] and Patel and Read [31] for the many publications on such bounds based on approximations.

NOTATION. Throughout, X denotes a random variable having one of the four discrete distributions studied. In the next four sections $P = \Pr[X \leq k|\theta]$, where θ denotes parameter(s), is approximated by $\Phi(u)$, where u is a (preferably simple) function of k and θ and where Φ denotes the standard normal distribution function:

$$\Phi(u) = (2\pi)^{-1/2} \int_{-\infty}^{u} \exp\left(-\tfrac{1}{2}t^2\right) dt. \quad (1)$$

CUMULATIVE POISSON PROBABILITIES

Let X have a Poisson distribution* with expectation λ and cumulative probability

$$P = \Pr[X \leq k] = \sum_{j=0}^{k} \frac{e^{-\lambda}\lambda^j}{j!}. \quad (2)$$

Direct evaluation of P for a given a positive real λ and nonnegative integer k poses no problem when λ is small (say, $\lambda < 2$); only the first few terms contribute substantially to the sum. Normal approximations, with error tending to 0 for $\lambda \to \infty$, have been based on the following principles:

(a) X with Continuity Correction*. The classical approximation of P by $\Phi((k + \tfrac{1}{2} - \lambda)\lambda^{-1/2})$, follows from the central limit theorem* applied to the (infinitely divisible*) Poisson distribution. Even for moderately large λ, its accuracy is far from perfect: for $\lambda = 30$, $k = 17$, it gives 0.0113 for the exact value 0.0073, and near the median absolute errors exceed 0.01 for $\lambda = 30$. Asymptotic expansion and empirical evaluation of the difference $\Phi((k + c - \lambda)\lambda^{-1/2}) - P$ shows [26, p. 35] that the customary choice $c = \tfrac{1}{2}$ only improves upon $c = 0$ for roughly

$$0.06 < P < 0.95.$$

(b) Gamma* or Chi-square* Distribution. It follows from integrating the incomplete gamma integral by parts that (2) is exactly equal to $\Pr[C_{2k+2} > 2\lambda]$, where C_ν denotes a chi-square distributed random variable with ν degrees of freedom. Combining this with the asymptotic normality* of C_ν, however, leads to an error about twice that of (a); see Molenaar [26, p. 37].

(c) Square Root Transformation. From Curtiss [12] and Anscombe [3] it follows that $(X + c)^{1/2}$ for any constant c has a variance asymptotically independent of λ and a distribution closer to normality than that of X. The result is just as simple but about twice as accurate as (a) and therefore recommended:

$$\Pr[X \leq k] \approx \Phi\left((4k + 4)^{1/2} - (4\lambda)^{1/2}\right),$$

$$(3)$$

$$\Pr[X \leq k] \approx \Phi\left((4k + 3)^{1/2} - (4\lambda)^{1/2}\right).$$

$$(4)$$

Whereas (3) is especially accurate near the customary significance levels or their complements, (4) is a little better for $0.05 < P < 0.93$. See Bartlett [6] and, for expansions, Pratt [33] and Molenaar [26, p. 39]

(d) Other Power Transformations. All approximations (a)–(c) have an error with leading term of order $\lambda^{-1/2}$. From Anscombe [4] and later references, well summarized in Pratt [33], it follows that $(X + c)^{2/3}$ has skewness and error proportional to λ^{-1}. Various forms are compared in Molenaar [26, Sect. II.4] and Kao [22]. The same error order is achieved by combining (b) with the Wilson–Hilferty normal approximation to $C_\nu^{1/3}$. Transformations of this type are used in the numerical libraries NAG and IMSL. With (f), however, one obtains a still better result.

(e) Additional Terms. With $w = (k + \tfrac{1}{2} - \lambda)\lambda^{-1/2}$, one expands (see Cornish and Fisher [11]):

$$\Pr[X \leq k]$$
$$\approx \Phi(u)$$
$$= \Phi\left(w + \frac{1 - w^2}{6\sqrt{\lambda}} + \frac{5w^3 - 2w}{72\lambda}\right.$$
$$\left. + \frac{128 + 79w^2 - 249w^4}{6480\lambda\sqrt{\lambda}} + \cdots \right).$$

$$(5)$$

Surprisingly, the addition of subsequent

terms to w is far less effective than other approximations with the same error order, unless λ is so large that the additional terms are already superfluous. The same holds for addition to $\Phi(w)$ of suitable polynomials in w times the normal density in w. The secret is that a good approximation with a certain error order already eliminates most of the contribution of the next terms as well.

(f) **Very Accurate Transformation.** By an ingenious expansion aiming at accuracy both at the median and in the extreme tails, Peizer and Pratt obtained a very good approximation for a class of seven distributions related to the beta integral. For the Poisson case, it means

$$\Pr[X \leqslant k]$$
$$\approx \Phi\big(\{k + \tfrac{2}{3} - \lambda + a/(k+1)\}$$
$$\times \{1 + g[(k+\tfrac{1}{2})/\lambda]\}^{1/2}\lambda^{-1/2}\big),$$
$$(6)$$

where

$$g(z) = (1 - z^2 + 2z \log z)(1 - z)^{-2} \text{ and}$$
$$g(1) = 0. \qquad (7)$$

The constant a, only relevant for small λ, can be set to 0.02 [32] or for still more accuracy in the tails at 0.022 [26, p. 59]. This recommended approximation (error order $\lambda^{-3/2}$) is very accurate, uniformly in k, with some deterioration for the trivial case $k = 0$ for which $P = e^{-\lambda}$ does not require a normal approximation.

CUMULATIVE BINOMIAL PROBABILITIES

Let X have a binomial distribution* (n trials with success probability p) with cumulative probability

$$P = \Pr[X \leqslant k] = \sum_{j=0}^{k} \binom{n}{j} p^j q^{n-j}; \quad (8)$$

the notation $q = 1 - p$ and $\sigma^2 = npq$ is used. For some details in **(c)** and **(f)** below it is assumed that $p \leqslant \tfrac{1}{2}$ (if not, interchange the roles of success and failure). Approximations can be grouped as follows.

(a) **X with Continuity Correction.** The rule of thumb to use $\Phi((k + \tfrac{1}{2} - np)/\sigma)$ when $\min(np, nq) > 5$ is rather optimistic: for $n = 100$, $p = 0.05$, $\Pr[X \geqslant 11] = 0.0115$ it gives 0.0058; for $n = 10$, $p = 0.5$, $\Pr[X \geqslant 9] = 0.0107$ it gives 0.0080; and these are not the worst cases. Variations of the type

$$(k + c - np)/((n + b)pq + a)^{1/2}$$

are not helpful [26, pp. 75–78].

(b) **Beta* or F-Distribution*.** Integration by parts shows that (8) is exactly equal to $\Pr[B_{n-k,k+1} \leqslant q]$, where $B_{a,b}$ is a beta random variable with density proportional to $x^{a-1}(1 - x)^{b-1}$, in its turn related to the F-distribution. Combination with asymptotic normality of $B_{a,b}$ roughly doubles the error found in **(a)**; see Molenaar [26, p. 79].

(c) **Arcsine* and Square Root Transformations.** Variance stabilization by arc-sin$(\{(X + c)/(n + b)\}^{1/2})$ [3, 12] leads to approximations [15, 34, 19, 37]. After a small modification [15; 26, p. 87], the result is just as simple but about twice as accurate as **(a)**, and thus recommended:

$$\Pr[X \leqslant k] \approx \Phi\big((4k + 4)^{1/2}q^{1/2} \\ - (4n - 4k)^{1/2}p^{1/2}\big) \quad (9)$$
$$\Pr[X \leqslant k] \approx \Phi\big((4k + 3)^{1/2}q^{1/2} \\ - (4n - 4k - 1)^{1/2}p^{1/2}\big)$$
$$(10)$$

Near the customary significance levels Pinkham's (9) is best, and for $0.05 < P < 0.93$ Freeman and Tukey's (10). See refs. 15, 26 (p. 87), and 33.

(d) **Other Power Transformations.** As is well summarized in section 6 of ref. 33, the symmetrizing exponent $\tfrac{1}{3}$ for F leads to the Camp–Paulson [10, 21] approximation used in the IMSL and

NAG libraries. A related proposal is in Borges [9]. The error there is of order σ^{-2} (as in Ghosh [17]) while it was σ^{-1} for (a), (b), (c).

(e) **Additional Terms.** Just as in the Poisson case, the expansion [11] itself is not very effective.

(f) **Very Accurate Transformation.** Peizer–Pratt [32] use

$$\Pr[X \leqslant k] \approx \Phi(z), \qquad (11)$$

$$z = d \frac{\left\{ 1 + \frac{qg}{np}\left(k + \tfrac{1}{2}\right) + \frac{pg}{nq}\left(n - k - \tfrac{1}{2}\right) \right\}^{1/2}}{\left\{ \left(n + \tfrac{1}{6}\right)pq \right\}^{1/2}},$$

$$d = k + \tfrac{2}{3} - \left(n + \tfrac{1}{3}\right)p,$$

with g from (7). To the first factor in curly brackets, one may add a term

$$\frac{aq}{k+1} - \frac{ap}{n-k} + \frac{b\left(q - \tfrac{1}{2}\right)}{n+1} \qquad (12)$$

with $a = b = 0.02$ [32] or $a = 0.02$, $b = 0.13$ [26, p. 102]. The recommended (11) with error order σ^{-3} is uniformly in k very accurate; some deterioration occurs for the trivial cases $k = 0$ and $k = n - 1$, where $P = q^n$ and $P = 1 - p^n$ are readily obtained directly. Equally accurate but more laborious is the calculation in Bolshev et al. [8]; current work by Alfers and Dinges at Goethe Universität Frankfurt is not yet published.

(g) **Case $p = \tfrac{1}{2}$.** See refs. 17, and 26 (p. 101).

(h) **Small p or q.** For Poisson approximations, see, e.g., Bolshev et al. [8], Molenaar [24, 25], Raff [34], and Wise [39].

CUMULATIVE NEGATIVE BINOMIAL PROBABILITIES

As the events "the k-th success occurred at or before the t-th trial" and "the first t trials contain k or more successes" are logically equivalent, no new probabilities or approximations are required for the negative binomial*. The required substitutions are explicitly found in ref. 32 for (f) and in Bartko [5] and Johnson and Kotz [21] for the Camp–Paulson, and in ref. 31 for almost any approximation. As an example let Y be the number of failures preceding the sth success; then

$$P = \Pr[Y \leqslant y] = \sum_{j=0}^{y} \binom{s+y-1}{y} p^s q^y$$

is identical to $\Pr[X \geqslant s]$ where X has a binomial $(s + y, p)$ distribution. Thus one obtains from (9) that

$$\Pr[Y \leqslant y]$$
$$\approx \Phi\left((4y + 4)^{1/2}p^{1/2} - (4s)^{1/2}q^{1/2}\right).$$

CUMULATIVE HYPERGEOMETRIC PROBABILITIES

Let X have a hypergeometric* (n, r, N) distribution, with cumulative probability and 2×2 table

$$P = \Pr[X \leqslant k] = \sum_{j=0}^{k} \frac{\binom{n}{j}\binom{m}{r-j}}{\binom{N}{r}} \quad \text{and}$$

	X	$n - x$	n
	$r - X$	$m - r + X$	m
	r	s	N

respectively. By switching rows and/or columns if necessary it may—and will—be assumed that $n \leqslant r \leqslant \tfrac{1}{2} N$. Put

$$\mu = E[X] = nr/N,$$
$$\sigma^2 = \operatorname{var}(X) = mnrs / \left\{ N^2(N-1) \right\},$$
$$\tau = mnrsN^{-3}, \qquad w = \left(k + \tfrac{1}{2} - \mu\right)/\sigma,$$
$$\chi = \left(k + \tfrac{1}{2} - \mu\right)/\tau.$$

(a) **X and χ^2, Continuity Correction.** X is asymptotically normal for $N \to \infty$ if and only if $\mu \to \infty$ and $\tau \to \infty$ [36]; thus $\Phi(\chi)$ and $\Phi(w)$, which is slightly inferior [26, pp. 128–130; 29] can be shown to approximate P to order τ^{-1} (or τ^{-2} for $r = \tfrac{1}{2} N$); see Haagen and

Schweitzer [18] and Molenaar [26, Sect. IV.2], for proofs and expansions. Note that $\Phi(\chi)$ is equivalent to the well-known two-sided chi-square test, with a minor exception for $|k - \mu| < \frac{1}{2}$. The adequacy of the continuity correction is numerically evaluated in ref. 13 and asymptotically in ref. 26 (p. 130).

(b) **Square-Root Transform.** Molenaar [26, p. 125; 27] derives and recommends

$$\Pr[X \leqslant k]$$

$$\approx \Phi\big(2\{N - 1\}^{-1/2}$$

$$\times \big\{(k + 1)^{1/2}(N - n - r + k + 1)^{1/2}$$

$$- (n - k)^{1/2}(r - k)^{1/2}\big\}\big), \qquad (13)$$

$$\Pr[X \leqslant k]$$

$$\approx \Phi\big(2N^{-1/2}$$

$$\times \big\{(k + \tfrac{3}{4})^{1/2}(N - n - r + k + \tfrac{3}{4})^{1/2}$$

$$- (n - k - \tfrac{1}{4})^{1/2}(r - k - \tfrac{1}{4})^{1/2}\big\}\big), \qquad (14)$$

being both simple and roughly twice as accurate as (a), with (14) slightly better for $0.05 < P < 0.93$ and (13) otherwise.

(c) **More Accurate Normal Approximations.** There is no Peizer–Pratt proposal, as the hypergeometric probability cannot be reduced to a beta integral. Results from Nicholson [28], modified in ref. 26 (p. 133), are computationally unattractive; with exceptions for small tails in skew tables, Molenaar [26, p. 136] recommends the parsimonious order τ^{-2} approximation

$$\Pr[X \leqslant k]$$

$$\approx \Phi\bigg(\chi + \frac{(1 - \chi^2)(m - n)(s - r)}{6N^2\tau}$$

$$+ \frac{\chi(N^2 - 3mn)}{48N^2\tau^2}\bigg). \qquad (15)$$

See also Ling and Pratt [23a].

(d) **Binomial Approximation.** Neglecting the "without replacement" sampling, the binomial $(n, r/N)$ approximation is effective for small n/N. By taking an "average probability during sampling" [30, 40], it can be made accurate for all $n \leqslant r \leqslant \frac{1}{2}N$ (see refs 40, 29, 7, 35, and 26, sec. IV.4). The binomial probability of $X \leqslant k$ in n trials with success probability

$$p = \frac{2r - k}{2N - n + 1} - \frac{2n\big(k + \frac{1}{2} - nrN^{-1}\big)}{3(2N - n + 1)^2} \qquad (16)$$

can in turn be evaluated by some normal approximation. A similar refinement of the Poisson (nrN^{-1}) approximation is less effective. See ref. 26 (Sect. IV.3) for examples and expansions.

CONFIDENCE BOUNDS AND QUANTILES (ALL DISTRIBUTIONS)

In most cases the improved normal approximations to $\Pr[X \leqslant k \,|\, \theta] = P$ do not permit explicit solution for a parameter θ in terms of P and k (confidence bound) or for the argument k in terms of P and θ (quantiles). Normal approximations to confidence bounds, including accuracy assessments, are given in refs. 25 and 27 for the Poisson parameter λ, in refs. 1, 2, and 27, for the binomial parameter p (known n) and in ref. 27 for the hypergeometric parameter r (known n and N). For the negative binomial, use the binomial, as earlier.

Regarding quantiles, there are two closely related approximation methods. One is to take as many terms from the Cornish–Fisher expansions* as desired (the first four or five are given below). The other is to obtain an initial value k_0 from the first two or three terms, use one of the probability approximations recommended above to obtain $\Pr[X \leqslant k_0]$, and find the P-quantile by iteration as that value k for which $\Pr[X \leqslant k] - P$ equals zero. As noninteger values of k are not meaningful and the function is strictly increasing, this search process will converge in a few steps.

Let P be given and let z denote the standard normal P-quantile, thus $\Phi(z) = P$. Then the value of k for which $\Pr[X \leqslant k] = P$ is approximately given by

$$k \approx \lambda + z\lambda^{1/2} + (z^2 - 4)/6$$

$$+ (z^3 + 2z)/(72\lambda^{1/2})$$

$$+ (3z^4 + 7z^2 - 16)/(810\lambda)$$

in the Poisson case;

$$k \approx np + z\sigma - \tfrac{1}{2} + (q - p)(z^2 - 1)/6$$

$$- \left\{ z^3(1 + 2pq) + z(2 - 14pq) \right\}/(72\sigma)$$

in the binomial case, with $\sigma^2 = npq$ and $q = 1 - p$;

$$k \approx \mu + z\tau - \tfrac{1}{2} + (m - n)(s - r)(z^2 - 1)/(6N^2)$$

$$- (72\tau N^4)^{-1}\left\{ z^3(N^4 + 2mnN^2 + 2rsN^2 - 26mnrs) \right.$$

$$\left. + z(N^4 - 14mnN^2 - 14rsN^2 + 74mnrs) \right\}$$

in the hypergeometric case, in the notation explained earlier.

LITERATURE

The normal approximation to the binomial distribution goes back to DeMoivre* and Laplace*. Scanning the many subsequent publications, one may conclude that the best way, for both systematic comparison and improvement of approximations, is to use a combination of series expansions, as practiced, e.g., by Bolshev [7, 8], Cornish and Fisher [11], Curtiss [12], Feller [14 Chap. VII], Hill and Davis [20], and Wallace [38], and numerical comparisons, as carried out, e.g., Gebhardt [16], Ling [23], and Raff [34]. Such a combination is the strong point in the work of Peizer and Pratt [32, 33], which is still the key reference for normal approximations to Poisson, binomial, and negative binomial probabilities. The monograph by Molenaar [26], equally combining expansion and numerical work, also covers approximations to hypergeometric probabilities and Poisson approximations to binomial probabilities. An excellent guide to all references up to 1969, just before Peizer and Pratt and Molenaar, is the Johnson and Kotz [21] volume on discrete distributions. Patel and Read [31, Chap. 7] give a well-balanced and informative overview with a host of useful formulas and references.

References

[1] Anderson, T. W. and Burstein, H. (1967). *J. Amer. Statist. Ass.*, **62**, 857–862.

[2] Anderson, T. W. and Burstein, H. (1968). *J. Amer. Statist. Ass.*, **63**, 1413–1416.

[3] Anscombe, F. J. (1948). *Biometrika*, **35**, 246–254.

[4] Anscombe, F. J. (1953). *J. R. Statist. Soc. B*, **15**, 229–230.

[5] Bartko, J. J. (1966). *Biometrics*, **8**, 340–342.

[6] Bartlett, M. S. (1936). *J. R. Statist. Soc. Suppl.* **3**, 68–78.

[7] Bolshev, L. N. (1964). *Theory Prob. Appl.*, **9**, 619–624 (English translation from *Teory. Veroyatn. ee Primen.*).

[8] Bolshev, L. N., Gladkov, B. V., and Shcheglova, M. V. (1961). *Theory Prob. Appl.*, **6**, 410–419 (English translation from *Teory. Veroyatn. ee Primen.*).

[9] Borges, R. (1970). *Zeit. Wahrscheinlichkeitsth.*, **14**, 189–199 (in German).

[10] Camp, B. H. (1951). *Ann. Math. Statist.*, **22**, 130–131.

[11] Cornish, E. A. and Fisher, R. A. (1937). *Rev. Int. Statist. Inst.*, **5**, 307–320.

[12] Curtiss, J. H. (1941). *Ann. Math. Statist.*, **14**, 107–122.

[13] Doane, D. P. and Reese, R. M. (1977). *Proc. Statist. Comput. Sect., Amer. Statist. Ass.*, pp. 185–189.

[14] Feller, W. (1957). *An Introduction to Probability Theory and its Applications*, Vol. 1, 2nd ed. Wiley, New York.

[15] Freeman, M. F. and Tukey, J. W. (1950). *Ann. Math. Statist.*, **21**, 607–611.

[16] Gebhardt, F. (1969). *J. Amer. Statist. Ass.*, **64**, 1638–1646.

[17] Ghosh, B. K. (1980). *Commun. Statist. A*, **9**, 427–438.

[18] Haagen, K. and Schweitzer, W. (1975). *Statistische Hefte* **16**, 123–127 (in German).

[19] Hald, A. (1952). *Statistical Theory with Engineering Applications*. Wiley, New York.

[20] Hill, G. W. and Davis, A. W. (1968). *Ann. Math. Statist.*, **39**, 1264–1273.

[21] Johnson, N. L. and Kotz, S. (1969). *Distributions in Statistics: Discrete Distributions*. Wiley, New York.

[22] Kao, S. C. (1978). *Proc. Statist. Comput. Sect., Amer. Statist. Ass.*, pp. 264–267.

[23] Ling, R. F. (1978). *J. Amer. Statist. Ass.*, **73**, 274–283.

[23a] Ling, R. F. and Pratt, J. W. (1984). *J. Amer. Statist. Ass.*, **79**, 49–60.

[24] Molenaar, W. (1969). *Statist. Neerlandica*, **23**, 19–40, 241.

[25] Molenaar, W. (1970). In *Random Counts in Scientific Work*, Vol. 2, G. P. Patil, ed. Pennsylvania State University Press, PA, pp. 237–254.

[26] Molenaar, W. (1970). "Approximations to the Poisson, Binomial and Hypergeometric Distribution Functions." *Math. Centre Tracts No. 31*, Mathematisch Centrum, Amsterdam.

[27] Molenaar, W. (1973). *Biometrics*, **29**, 403–407.

[28] Nicholson, W. L. (1956). *Ann. Math. Statist.*, **27**, 471–483.

[29] Ord, J. K. (1968). *Biometrika*, **55**, 243–248.

[30] Overall, J. E. and Starbuck, R. R. (1983). *J. Educ. Statist.*, **8**, 59–73.

[31] Patel, J. K. and Read, C. B. (1982). *Handbook of the Normal Distribution*. Marcel Dekker, New York.

[32] Peizer, D. B. and Pratt, J. W. (1968). *J. Amer. Statist. Ass.*, **63**, 1416–1456.

[33] Pratt, J. W. (1968). *J. Amer. Statist. Ass.*, **63**, 1457–1483.

[34] Raff, M. S. (1956). *J. Amer. Statist. Ass.*, **51**, 293–303.

[35] Sandiford, P. J. (1960). *J. Amer. Statist. Ass.*, **55**, 718–722.

[36] Van Eeden, C. and Runnenburg, J. Th. (1960). *Statist. Neerlandica*, **14**, 111–126.

[37] Vijn, P. and Molenaar, I. W. (1981). *J. Educ. Statist.*, **6**, 205–235.

[38] Wallace, D. L. (1958). *Ann. Math. Statist.*, **29**, 635–654.

[39] Wise, M. E. (1950). *Biometrika*, **37**, 208–218.

[40] Wise, M. E. (1954). *Biometrika*, **41**, 317–329.

(APPROXIMATIONS TO DISTRIBUTIONS NORMAL DISTRIBUTION)

I. W. MOLENAAR

NORMAL DISTRIBUTION

Known also as the Gaussian distribution and as the bell-shaped curve, the normal distribution, denoted here as $N(\mu, \sigma^2)$, plays a key role in statistical theory and practice, as well as in the limit theorems* of probability theory. It is absolutely continuous with probability density function

$$f(x; \mu, \sigma) = \frac{1}{\sqrt{2\pi}\,\sigma} \exp\left[-\frac{(x-\mu)^2}{2\sigma^2} \right],$$
$$-\infty < x < \infty \quad (1)$$

and is completely determined by its expected value μ and variance σ^2, μ being also the median and mode. The distribution is symmetrical about μ, and the standardized form is given by

$$\phi(x) = f(x; 0, 1). \quad (2)$$

The higher central moments of (1) are

$$\mu_{2r+1} = 0,$$

$$\mu_{2r} = \frac{\sigma^{2r}(2r)!}{2^r r!}, \qquad r = 1, 2, \ldots .$$

The shape coefficients are thus $\sqrt{\beta_1} = 0$ and $\beta_2 = \mu_4/\mu_2^2 = 3$. The cumulants* κ_r are all zero if $r \geqslant 3$, the moment-generating function* being $\exp(\mu t + \frac{1}{2}\sigma^2 t^2)$, $t \neq 0$.

The normal distribution belongs to the exponential family (*see* NATURAL EXPONENTIAL FAMILIES), is stable (and strictly stable if $\mu = 0$), and infinitely divisible (*see* STABLE DISTRIBUTIONS and INFINITE DIVISIBILITY).

HISTORICAL BACKGROUND

The story of the emergence of the normal distribution is largely the story of the development of statistics as a science; it should be understood by students in statistics and historians of science alike. The entry LAWS OF ERROR and the biographical entries starred below should be consulted along with this section; also see Adams [2], Maistrov [24] and Patel and Read [30, Chap. 1], where further source references appear.

Charles Sanders Peirce* in 1873 may have been the first to label (1) the normal law [38], but the name did not catch on until after 1900. The distribution was also named after Laplace*, Maxwell*, and Quetelet*; Francis Galton* used a number of terms for it, including the *law of frequency of error* and *law of deviation from an average* (see Stigler [38] for a discussion of the nomenclature). It is surprising that (1) is not named after its discoverer Abraham de Moivre* [6], who obtained it in 1733 as an approximation to the probability that binomially distributed random variables lie between two quantities. Laplace* wove (1) into an intricate theory of mathematical statistics, including a more formal and general statement of de Moivre's result, a derivation of the minimum variance property of the least-squares estimator of linear regression under normality, and an early form of central limit theorem (*see* the Large-Sample Role section).

While de Moivre's investigation was prompted by a need to compute probabilities of winning in various games of chance* (*see* GAMBLING, STATISTICS IN), Gauss was motivated by problems of measurement in astronomy (*see* GAUSS, CARL FRIEDRICH). Galileo had already reasoned that errors of observation are distributed symmetrically and tend to cluster around their true value. In trying to estimate unknown quantities, Gauss [11] assumed them to take all values a priori with equal likelihood, and sought to find the underlying distribution which would lead to the realization of the *Principle of the Arithmetic Mean*; this asserts that the most probable value of an unknown quantity is the mean of all of its observed values (*see* the introduction to LAWS OF ERROR I). Gauss obtained the normal law (1).

In the middle of the nineteenth century, the distribution (1) came to be regarded as universally held by observations generally in nature. The work of Laplace and Gauss was responsible in part, but so was belief in the so-called *Hypothesis of Elementary Errors*, which was developed by several scientists in various forms. One of the most refined expressions was by the astronomer Bessel in 1838 [4]; for a more detailed discussion, *see*

LAWS OF ERROR II. Bessel assumed that each error of observation is the sum of a large number of independent component elementary errors of diverse origins, not necessarily identically distributed, but symmetrical about zero, each elementary error being negligible in comparison to the sum. Under this hypothesis, Bessel proved that the sum of elementary errors approximately follows a normal distribution, and (1) became known by such names as the *law of errors of observation*. Galton expressed the feeling of the age with the sense of wonder of Victorian romantic [10]:

> I know of scarcely anything so apt to impress the imagination as the wonderful form of cosmic order expressed by the "Law of Frequency of Error." The law would have been personified by the Greeks and deified, if they had known of it. It reigns with serenity and in complete self-effacement amidst the wildest confusion.

But Galton eventually questioned the universality of (1) through the principle of the arithmetic mean. In some frequency distributions, he observed, the mean of the logarithms of the observations appeared to better represent an unknown quantity of interest than did the mean of the observations themselves; thus was born the lognormal distribution*. The system of frequency curves* of Karl Pearson* was also based on the recognition that other laws exist in nature; (1) appears as a limiting case of all seven Pearson types. The publication in 1900 of Pearson's goodness-of-fit* test by chi-square [33] provided a broad scientific means of challenging the universality of (1). In view of a clear demonstration against this universality by Simeon-Denis Poisson* as early as 1824, it is surprising that the normal law held its place unchallenged by scientists in general for so many decades; *see* LAWS OF ERROR III.

TABULATION AND COMPUTATION

Tables of the distribution function Φ, tail probabilities $1 - \Phi(\cdot)$, density function ϕ, and quantiles* of (2) appear in Owen [28],

Pearson and Hartley [32], and in several other sources. There are also tables of values of $R(x) = [1 - \Phi(x)]/\phi(x)$ (*see* MILLS' RATIO), of derivatives of ϕ, and of quantile densities. See Section 13.1 of Johnson and Kotz [18] and Section 3.1 of Patel and Read [30] for details of sources. In addition, Odeh and Owen [27] have tabulated tolerance limits* for the normal distribution. The literature contains a variety of algorithms for evaluating these quantities, some suitable for use on high-speed computers, others that work very well on smaller desk calculators. They appear in the form of approximations, expansions, and inequalities, are of varying degrees of accuracy, and are scattered throughout the literature. We illustrate these with some that are both concise and reasonably accurate over a wide range. The examples that follow are not guaranteed to be "best" in any sense, and readers are directed to sources such as Abramowitz and Stegun [1], Johnson and Kotz [18, Chap. 13], and Patel and Read [30, Chap. 3] and to the tables cited therein. These references collect and present many of the algorithms, together with details of their coverage and accuracy. We also note some recent approximations; see Hawkes [15] who reviews these and adds some of his own.

Approximations to Mills' ratio provide further algorithms for normal tail probabilities

$$Q(x) = 1 - \Phi(x) = \Phi(-x).$$

See MILLS' RATIO, for example, for the excellent approximation of Patry and Keller [31].

1. A rational approximation by Hastings [14] for $\Phi(x)$:

$$2[1 - \Phi(x)]$$
$$= [1 + (0.196,854)x + (0.115,194)x^2$$
$$+ (0.000,344)x^3 + (0.019,527)x^4]^{-4}$$
$$+ 2\epsilon(x),$$
$$x \geq 0, \quad |\epsilon(x)| < 2.5 \times 10^{-4}.$$

Hastings gives several such approximations for Mills' ratio and for quantiles, in addition to those for $\Phi(x)$.

2.

$$\Phi(x) \simeq 1/(1 + e^{-2y}) = \tfrac{1}{2}(1 + \tanh y),$$
$$y = \left(\sqrt{2/\pi}\right)x\{1 + (0.044,715)x^2\}.$$

The maximum absolute error is 0.000179 [29]; and is reduced to 0.000140 if $y = (0.7988)x\{1 + (0.04417)x^2\}$. See also Hawkes [15].

3. Lew [21] gives the following approximation to normal tail probabilities $Q(x) = 1 - \Phi(x)$;

$$Q(x) \simeq \begin{cases} \tfrac{1}{2} - (2\pi)^{-1/2}(x - \tfrac{1}{7}x^3), \\ \hspace{3cm} 0 \leq x \leq 1, \\ (1 + x)\phi(x)/(1 + x + x^2), \quad x > 1, \end{cases}$$

with a maximum error of 0.00183; Hawkes [15] points out that the first form remains superior for $0 \leq x \leq 1.14$ and gives an improved approximation of the second form.

4. Derenzo [8] developed several approximations to $\Phi(x)$ and to quantiles, with integer coefficients. For example, let $\Phi(x_p) = 1 - p$ and $Y = -\log(2p)$. Then if $10^{-7} < p < 0.50$, or $0 < x_p < 5.2$, the quantile x_p of (2) is given by

$$x_p = \left[\frac{\{(4y + 100)y + 205\}y^2}{\{(2y + 56)y + 192\}y + 131} \right]^{1/2}$$
$$+ \epsilon(p),$$
$$|\epsilon(p)| < 1.3 \times 10^{-4}.$$

For an approximation that can be used in the extreme tails, that is, for $5.2 < x_p < 22.6$,

$$x_p = \left[\frac{\{(2y + 280)y + 572\}y}{(y + 144)y + 603} \right]^{1/2} + \epsilon(p),$$
$$|\epsilon(p)| < 4 \times 10^{-4}.$$

See Bailey [3] for a quantile approximation with a reduced error in the extreme tails.

LARGE-SAMPLE ROLE

Although the assumptions in the hypothesis of elementary errors are no longer regarded as universal in observations of data, the normal distribution (1) remains the most prominent in statistical inference. The most compelling reason for this is the approximate normality in large samples (and sometimes in samples of only moderate size) of many sampling statistics from other parent populations. A number of regularity conditions need to be, but frequently are satisfied; these usually include differentiability of the parent density function with respect to the parameter of interest and the existence of certain means and variances; *see* LIMIT THEOREM, CENTRAL and ASYMPTOTIC NORMALITY in conjunction with this section.

Chief among the sampling statistics that may asymptotically follow a normal law is the *sample mean* $\overline{X}_n$ of a random sample from a parent population with finite mean and variance. The asymptotic normality of the sample mean $\overline{X}_n$ of a random sample is expressed in the *central limit theorem**, which states in its simplest form:

Theorem 1. Let $X_1 X_2, \ldots$ be a sequence of independent and identically distributed random variables having common mean μ and finite variance σ^2. Then if $G_n(x) = \Pr[\sqrt{n}\,(\overline{X}_n - \mu)/\sigma \leqslant x]$,

$$\lim_{n \to \infty} G_n(x) = \Phi(x)$$

uniformly in x, where $\Phi(\cdot)$ is the cumulative distribution function (cdf) of a $N(0,1)$ variable.

As refined by Laplace, this was the result obtained by de Moivre in the special case of binomial probabilities, the X's being Bernoulli variables; *see also* MOIVRE–LAPLACE THEOREM. The hypothesis of elementary errors incorporated the genesis of a central limit theorem, but it was the St. Petersburg school in Russia that cast the theorem in a formal mathematical framework. Chebyshev* (or Tchebichef) [40] introduced the concept of *random variable** and *expected value** into the statement of (3) in 1890 and his pupil Liapunov* [22] generalized it in 1901 using characteristic functions*. The Lindeberg–Feller theorem* would give nec-

essary and sufficient conditions for (3) to hold.

The version of (3) given above is not only the simplest but also the most widely used in applications; it forms the basis for many parametric inference procedures about location parameters. When σ is unknown and is replaced by the sample standard deviation $[(n - 1)^{-1}\Sigma(X_i - \overline{X}_n)^2]^{1/2}$, procedures that treat $\sqrt{n}\,(\overline{X}_n - \mu)/S$ as having a Student t-distribution* turn out to be fairly robust against mild departures from normality.

When Theorem 1 is used, the question arises as to how large the sample size n needs to be before the normal approximation for $\overline{X}_n$ is reasonably good. For a parent distribution that is grossly skewed, a larger sample will be required than for a unimodal symmetric parent. Figure 1 illustrates the rate of convergence for random samples X_1, $X_2, \ldots$ from a Bernoulli distribution* that takes values 1 and 0 with probabilities 0.80 and 0.20, respectively. The parent distribution is thus highly skewed. Figure 1 demonstrates that the normal approximation here to the cdf of the binomial variable $S_n = X_1 + \cdots + X_n$ is unreliable for $n \leqslant 10$, is not very accurate if $n \leqslant 20$, and even for n as large as 100 noticeably overestimates $\Pr(a \leqslant S_n \leqslant b)$ if a and b are below the mean $(0.80)n$, while underestimating this quantity when a and b lie above $(0.80)n$.

The rate of convergence in (3) is formally expressed in the Berry–Esseen theorem, stated here in the context of Theorem 1:

Theorem 2

(a) Under the conditions of Theorem 1, let $\nu_3 = E(|X_1 - \mu|^3)$ exist. Then

$$\sqrt{n}\,|G_n(x) - \Phi(x)| \leqslant C\nu_3/\sigma^3 \qquad (4)$$

for all x, where $(\sqrt{10} + 3)/(6\sqrt{2\pi}) \leqslant C \leqslant 0.7882$.

See ASYMPTOTIC NORMALITY and Michel [25], who attributes the upper bound to Beek. The rate of convergence in Theorem 2a is uniform, but Michel also derived a pointwise

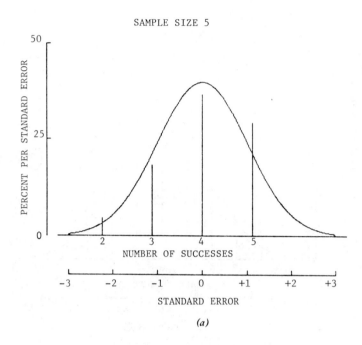

(a)

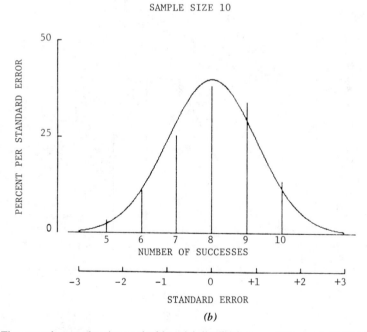

(b)

Figure 1 The normal approximation to the binomial distribution, $p = 0.80$; $n = 5, 10, 20, 50, 100, 1000$. The exact distribution is shown as a bar chart for $n = 5$, 10, and 20, (a)–(c), respectively and in histogram form for $n = 50$, 100, and 1000 (d)–(f), respectively.

SAMPLE SIZE 20

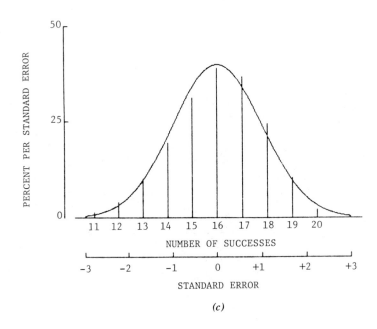

(c)

SAMPLE SIZE 50

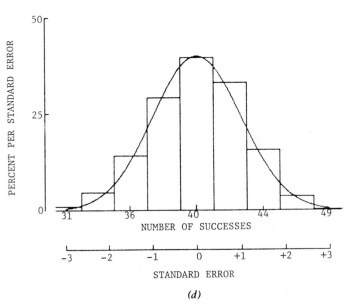

(d)

Figure 1 *(Continued)*

SAMPLE SIZE 100

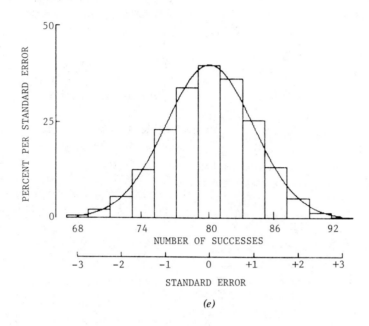

(e)

SAMPLE SIZE 1000

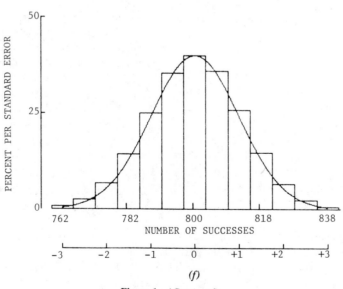

(f)

Figure 1 *(Continued)*

version:

(b) If, under the conditions of Theorems 1 and 2a,

$$F_n(x) = \Pr\left(\sqrt{n}\,(\bar{X}_n - \mu)/\sigma < x\right),$$

then

$$\sqrt{n}\,|F_n(x) - \Phi(x)| \leqslant c_1 \nu_3 \Big/ \big[\sigma^3(1 + |x|^3)\big],$$

where $c_1 \leqslant C + 8(1 + e) \leqslant 30.54$.

Expansions of $G_n(x) - \phi(x)$ in powers of $n^{-1/2}$ improve the accuracy of the approximation in Theorem 1; see ASYMPTOTIC EXPANSION *and* CORNISH–FISHER AND EDGEWORTH EXPANSIONS.

The moments of $Z_n = \sqrt{n}\,(\bar{X}_n - \mu)/\sigma$ do not necessarily converge to those of a $N(0, 1)$ variable. But the density of Z_n converges to $\phi(x)$ uniformly in x if the X's are absolutely continuous, and the following property is useful provided the density of the X's is known; it relates to *sample fractiles*. For $0 < p < 1$, the pth fractile statistic in a sample $X_1, \ldots, X_n$ is a value $\hat{x}_p$ such that the number of X_i's less than or equal to $\hat{x}_p$ and greater than or equal to $\hat{x}_p$ is at least $[np]$ and $[n(1 - p)]$, respectively, where $[r]$ is the largest integer less than or equal to r.

Theorem 3. (Rao [36]). Under the conditions of Theorem 1, let the X_i's be continuous with common density $g(x)$ and let x_p satisfy $\Pr(X_i \leqslant x_p) = p$ and $\Pr(X_i \geqslant x_p) = 1 - p$. Then $\sqrt{n}\,(\hat{x}_p - x_p)$ has an asymptotic normal distribution with mean zero and variance $p(1 - p)/\{g(x_p)\}^2$.

For further discussion of these and other central limit properties, see Patel and Read [30, Chap. 6].

Other quantities, such as U-statistics*, certain linear rank statistics*, and linear functions of order statistics* such as trimmed means and the sample median* are asymptotically normal. Since maximum likelihood estimators (MLEs) play an important role in estimation, it is worth noting that under cetain regularity conditions, MLEs are "best

asymptotically normal," being also consistent and asymptotically efficient (*see* LARGE-SAMPLE THEORY *and* MAXIMUM LIKELIHOOD ESTIMATION). Under certain conditions, *M*-estimators*, which seek to provide robustness against nonnormality, are asymptotically normal (*see* ROBUST ESTIMATION).

CHOICE OF A STANDARD NORMAL DENSITY

Tables of normal probabilities are based on the unit or standard $N(0, 1)$ variable. Thus if X has density (1), and if $Z = (x - \mu)/\sigma$, Z has density (2). While it is convenient to standardize in this way, Gauss [11] expressed the density function of (1) in the form

$$\phi\Delta = \frac{h}{\sqrt{\pi}}\, e^{-hh\Delta\Delta},$$

or $f(x) = (h/\sqrt{\pi})\exp[-h^2 x^2]$ in modern notation, where he called h the *precision* of measurement, so that $h = 1/(\sqrt{2}\,\sigma)$. (In his original discovery of (1) as the limiting distribution for a binomial variable with success probability $\frac{1}{2}$ in each trial, de Moivre [6] called 2σ "the *Modulus* by which we are to regulate our Eftimation").

Early work with (1) in the nineteenth century was in terms of the *error function*

$$\mathrm{erf}(x) = \int_{-x}^{x} \frac{1}{\sqrt{\pi}}\, e^{-u^2}\, du$$

$$= 2\Phi(\sqrt{2}\,x) - 1, \tag{5}$$

Φ being the $N(0, 1)$ cdf. The error function is still used in Europe; see, e.g., ref. 31. Stigler [39] has argued that the normal density

$$g(x) = e^{-\pi x^2}$$

with variance $1/(2\pi)$ be adopted as standard, in part because it is easier for less mathematically inclined students to deal with than $(\sqrt{2\pi})^{-1} e^{-x^2/2}$; it has standard deviation 0.40 and quartiles ± 0.25, approximately.

TRANSFORMATIONS
TO NORMALITY—APPROXIMATIONS

With the extensive available tabulation of $N(0, 1)$ probabilities and of $\mathrm{erf}(x)$, which goes back to de Morgan in 1837 [7], it has been convenient to seek transformations of other random variables to normality, or approximately so (*see* APPROXIMATIONS TO DISTRIBUTIONS). Certain variance-stabilizing transformations tend to normalize (*see* EQUALIZATION OF VARIANCE); these include the angular transformation* for binomial variables, the logarithmic transformation of chi-square, and the inverse hyperbolic tangent transformation of a bivariate normal correlation coefficient (*see* FISHER'S z-TRANSFORMATION). The Johnson system of distributions* incorporates three transformations to normality that combine to provide one distribution corresponding to every pair of values of the shape coefficients* $\sqrt{\beta_1}$ and β_2 (*see* FREQUENCY CURVES, SYSTEMS OF).

Various successful normalizing transformations arose by attempting to find that function of a random variable in some class of functions that has first, second, even third, and fourth moments equal or close to those of (2). A good example is the Wilson–Hilferty [42] approximation for chi-square; if Y has a chi-square distribution* with ν degrees of freedom, where ν is large, for what value of h does $(Y/\nu)^h$ minimize the skewness coefficient $\sqrt{\beta_1}(Y)$? The method of statistical differentials* yields $h \simeq \frac{1}{3}$, and an approximately standardized variable

$$\left\{\left(\frac{Y}{\nu}\right)^{1/3} - \left(1 - \frac{2}{9\nu}\right)\right\} \Big/ \left(\frac{2}{9\nu}\right)^{1/2}; \quad (6)$$

it is more nearly like (2) than $(Y - \nu)/\sqrt{2\nu}$, which is derived from Theorem 1, partly because the third and fourth moments of (6) are approximately equal to $8\sqrt{2}/(27\nu^{3/2})$ and $3 - 4/(9\nu)$, respectively, and tend as $\nu \to \infty$ to the $N(0, 1)$ values 0 and 3, respectively.

The central limit theorem is partly responsible for the fact that many statistical distributions can be approximated by the normal. The binomial*, Poisson*, gamma*, and chi-square distributions*, for example, can all be represented as sums of independent and identically distributed variables. Again, the binomial is a limiting case of the hypergeometric distribution*, which can therefore be approximated under suitable conditions by (1).

Normal approximations to several *discrete distributions* are presented by Molenaar [26]; *see also* NORMAL APPROXIMATIONS TO THE POISSON, BINOMIAL, NEGATIVE BINOMIAL, AND HYPERGEOMETRIC DISTRIBUTIONS. For approximations to discrete and continuous distributions, see particularly Peizer and Pratt [34], Patel and Read [30, Chap. 7] and Johnson and Kotz [17–19]. Peizer and Pratt use relations between the incomplete beta function and binomial tail probabilities to derive normal approximations to binomial, Pascal, negative binomial, beta, and F probabilities; and noting that "Poisson, chi-square and gamma tail probabilities are limiting cases of these," they treat normal approximations to these distributions as one problem, rather than eight.

We present a few concise yet reasonably accurate examples for *continuous distributions*.

1. Beta*. (Peizer and Pratt [34]). For a beta density

$$f(x; \alpha, \beta) = x^{\alpha-1}(1 - x)^{\beta-1}/B(\alpha, \beta),$$
$$0 < x < 1; \quad \alpha > 0, \quad \beta > 0,$$

and with cumulative distribution function (cdf) $F(x; \alpha, \beta)$,

$$F(x; \alpha, \beta) \simeq \Phi(u),$$

$$u = d\left[1 + xT\left\{\frac{\beta - 0.50}{(\alpha + \beta - 1)(1 - x)}\right\}\right.$$
$$\left. + (1 - x)T\left\{\frac{\alpha - 0.50}{(\alpha + \beta - 1)x}\right\}\right]^{1/2}$$
$$\times \left\{x(1 - x)\left(\alpha + \beta - \frac{5}{6}\right)\right\}^{-1/2},$$

$$T(y) = (1 - y^2 + 2y \log y)/(1 - y)^2, \quad y \neq 1,$$
$$T(1) = 0.$$

The accuracy of this approximation is within 0.001 if $\alpha \geqslant 2$ and $\beta \geqslant 2$, and within 0.01 if $\alpha \geqslant 1$, $\beta \geqslant 1$.

2. Chi-square. The approximation in (6) appears to be more accurate than that of Peizer and Pratt.

3. Chi-square Quantiles. Zar [43] compared 16 approximations, of which 12 are normal in some sense. If χ_p^2 is the desired quantile (for ν degrees of freedom) and z the corresponding standard normal quantile, then the two most acceptable and concise approximations are

(a) Wilson–Hilferty [43].

$$\chi_p^2 \simeq \nu \left\{ 1 - \frac{2}{9\nu} + z\sqrt{\frac{2}{9\nu}} \right\}^3.$$

This is recommended unless ν is "very small."

(b) A Cornish–Fisher expansion [13].

$$\chi_p^2 \simeq \nu + \nu^{1/2} x\sqrt{2} + \tfrac{2}{3}(x^2 - 1)$$

$$+ \nu^{-1/2}(x^3 - 7x)/(9\sqrt{2})$$

$$- \nu^{-1}(6x^4 + 14x^2 - 32)/405$$

$$+ \frac{\nu^{-3/2}(9x^5 + 256x^3 - 433x)}{4860\sqrt{2}}.$$

The first four terms alone improve in accuracy on the Wilson–Hilferty approximation. Zar [43] identified some longer algorithms that improve on both of these in the tails (beyond the ninety-ninth percentile).

4. Student's t. Let $F(t; \nu)$ be the cdf of a t-distribution with ν degrees of freedom. Then (Hill [16])

$$F(t; \nu) \simeq \Phi(u),$$

$$u = w + (w^3 + 3w)/b$$

$$- \frac{(4w^7 + 33w^5 + 240w^3 + 855w)}{10b(b + 0.80w^4 + 100)},$$

$$b = 48(\nu - 0.50)^2,$$

$$w = \left\{ (\nu - 0.50)\log(1 + t^2/\nu) \right\}^{1/2}, \quad t > 0.$$

This approximation is accurate within 10^{-1}, 10^{-3}, 10^{-5}, and 10^{-7} if $\nu \geqslant 1, 2, 4,$ and 6, respectively.

5. t Quantiles. Prescott [35] compared several approximations. If t_p is the desired quantile (for ν degrees of freedom) and z the corresponding standard normal quantile, then the best of those he compared is given by

$$t_p \simeq \sqrt{\nu} \left\{ \exp(z^2 b^2/\nu) - 1 \right\}^{1/2},$$

$$b = (8\nu + 3)/(8\nu + 1), \quad t_p > 0.$$

The scaling factor b aids in making this approximation good for small values of ν and large values of z.

6. F-distribution. This distribution has (m, n) degrees of freedom and cdf $F(x; m, n)$, say. The Peizer–Pratt approximation is accurate by less than 0.001 if $m \geqslant 4$, $n \geqslant 4$ and by less than 0.01 if $m \geqslant 2$, $n \geqslant 2$, i.e.,

$$F(x; m, n) \simeq \Phi(u),$$

$$u = \frac{d\left\{ 1 + qT\left(\dfrac{n-1}{p(m+n-2)} \right) + pT\left(\dfrac{m-1}{q(m+n-2)} \right) \right\}^{1/2}}{\left\{ pq(\tfrac{1}{2}m + \tfrac{1}{2}n - \tfrac{5}{6}) \right\}^{1/2}},$$

$$p = \frac{n}{mx + n}, \quad q = 1 - p,$$

$$d = \tfrac{1}{2}n - \tfrac{1}{3} - \left(\frac{m+n}{2} - \tfrac{2}{3} \right)p$$

$$+ \epsilon \left(\frac{q}{n} - \frac{p}{m} + \frac{q - \tfrac{1}{2}}{m + n} \right), \quad \epsilon = 0.04;$$

$$T(y) = (1 - y^2 + 2y \log y)/(1 - y)^2, \quad y \neq 1,$$

$$T(1) = 0.$$

A simpler computation with only a slight loss in accuracy occurs if $\epsilon = 0$.

When $m = n$, the approximation simplifies to

$$u = \pm \left(n - \tfrac{2}{3} + \frac{1}{10n} \right)$$

$$- \left[\frac{\log x}{(x+1)^2(n - \tfrac{5}{6})} \right]^{1/2},$$

where the sign agrees with that of 0.50 $- 1/(x + 1)$.

7. F Quantiles. Some of the best of these are based on Cornish–Fisher expansions and modifications thereof. We give that of Carter [5]; if x_p is the required quantile and z the corresponding standard normal quantile, and if $w_p = \frac{1}{2} \log x_p$, then

$$w_p \simeq zh^{-1}\sqrt{(h + \lambda)} - d\left[\lambda + \frac{5}{6} - \frac{2}{3h}\right],$$

$$\lambda = (z^2 - 3)/6,$$

$$h^{-1} = \frac{1}{2}\left\{(m - 1)^{-1} + (n - 1)^{-1}\right\},$$

$$d = (m - 1)^{-1} - (n - 1)^{-1}.$$

See also Sahai and Thompson [37] for some comparisons.

Johnson and Kotz [18, 19], Patel and Read [30, Chap. 7] list several other approximations to the cdf's and quantiles of these distributions, and also for noncentral χ^2, t, and F, and for the circular normal distribution* (see Upton [41] for several comparisons).

Efron [9] introduced a diagnostic function that measures, in a sense, the extent to which a monotone transformation $Y = g(X)$ deviates from normality for a parametric family of distributions of X, and he gives formulae for obtaining the form of $g(\cdot)$ when $g(X)$ is normal or nearly normal.

There are, finally, problems in which parameters appearing in a normalizing transformation have to be estimated from data. An example is the Box and Cox transformation*

$$Y = \begin{cases} (X - 1)/\lambda, & \lambda \neq 0 \\ \log X, & \lambda = 0, \end{cases}$$

where Y is to be assumed normal, and λ is to be estimated from observations $X_1, \ldots, X_n$.

ROLE IN SAMPLING THEORY

The Box–Cox transformation came about because in certain statistical problems, it was desirable that what are called "the usual assumptions" in the general linear model*

could be made; these assumptions include normally distributed errors as well as additivity of effects. This illustrates another reason for the key role of (1) in statistics (the first being the large-sample properties of $\bar{X}_n$ and other statistics); the mathematical elegance of normal sampling theory in the study of correlation, quadratic forms*, the general linear model, regression*, and the analysis of variance* and covariance*, as developed in particular by Friedrich Helmert*, Karl Pearson*, W. S. Gosset* ("Student"), R. A. Fisher*, John Wishart*, and Harold Hotelling*. Normal sampling theory led to the chi-square, t- and F-distributions for univariate samples, and to multivariate generalizations such as the Wishart distribution* and Hotelling's T^2*.

There are several properties of normal samples that contribute to the elegance and convenience of the theoretical results. These include:

(a)   In a sample of size n from a $N(\mu, \sigma^2)$ population, the sample mean $\bar{X}_n$ has a $N(\mu, \sigma^2/n)$ distribution.

(b)   Linear combinations $\sum a_i X_i$ of normal variables, whether these are independent or not, have a normal distribution. If $X_1, \ldots, X_n$ is a random sample, and $a_1, \ldots, a_n$ are all nonzero, this property characterizes the normal distribution; $\sum a_i X_i$ is normal if and only if the parent population is normal [23].

(c)   In a random sample from a normal population, the sample mean and sample variance are independent. This property also characterizes (1) [12]. Its importance for the role of (1) in statistical inference and particularly in regression analysis and analysis of variance and covariance under the assumption of normally distributed errors cannot be overstated; the structure of t-statistics and F-statistics relies upon it.

(d)   If the sample variance in (c) is given by $S^2 = (n - 1)^{-1}\sum(\bar{X}_i - X_n)^2$, then $(n - 1)S^2/\sigma^2$ has a chi-square distribution with $(n - 1)$ degrees of freedom.

Acknowledgment

The graphs for Fig. 1*a–f* were compiled by Michael Ames in the Statistical Laboratory, Southern Methodist University, Dallas, Texas.

References

[1] Abramowitz, M. and Stegun, I. A. (1964). *Handbook of Mathematical Functions*. National Bureau of Standards, Washington, DC.

[2] Adams, W. J. (1974). *The Life and Times of the Central Limit Theorem*. Caedmon, New York.

[3] Bailey, B. J. R. (1981). *Appl. Statist.*, **30**, 275–276.

[4] Bessel, F. W. (1838). *Astron. Nachr.*, **15**, 368–404.

[5] Carter, A. H. (1947). *Biometrika*, **34**, 352–358.

[6] De Moivre, A. (1733). "Approximatio ad Summan Terminorum Binomii $\overline{a + b}^n$ in Seriem expansi."

[7] De Morgan, A. (1837). *Encycl. Metropolitana* **2**, 359–468.

[8] Derenzo. S. E. (1977). *Math. Comp.*, **31**, 214–222.

[9] Efron, B. (1982). *Ann. Statist.*, **10**, 323–339.

[10] Galton, F. (1889). *Natural Inheritance*. Macmillan, London.

[11] Gauss, C. F. (1809). *Theoria Motus Corporum Coelestium*. Perthes and Besser, Hamburg. (See Sec. III of Liber 2).

[12] Geary, R. C. (1936). *J. R. Statist. Soc. B*, **3**, 178–184.

[13] Goldberg, G. and Levine, H. (1946). *Ann. Math. Statist.*, **17**, 216–225.

[14] Hastings, C. (1955). *Approximations for Digital Computers*. Princeton University Press, Princeton, NJ.

[15] Hawkes, A. G. (1982). *The Statistician*, **31**, 231–236.

[16] Hill, G. W. (1970). *Commun. Ass. Comput. Mach.*, **13**, 617–619.

[17] Johnson, N. L. and Kotz, S. (1969). *Distributions in Statistics: Discrete Distributions*. Wiley, New York.

[18] Johnson, N. L. and Kotz, S. (1970). In *Distributions in Statistics: Continuous Univariate Distributions*, Vol. 1. Wiley, New York. (Chapter 13 is devoted to the normal distribution.)

[19] Johnson, N. L. and Kotz, S. (1970). In *Distributions in Statistics: Continuous Univariate Distributions*, Vol. 2. Wiley, New York.

[20] Kruskal, W. (1978). *Amer. Sch.* **47**, 223–229.

[21] Lew, R. A. (1981). *Appl. Statist.*, **30**, 299–301.

[22] Liapunov, A. M. (1901). *Mém. Acad. Imp. Sci. St. Pétersbourg*, **12**, 1–24.

[23] Lukacs, E. (1956). *Proc. 3d Berkeley Symp. Math. Statist. Prob.*, **2**, 195–214.

[24] Maistrov, L. E. (1967). *Probability Theory: A Historical Sketch*. Academic Press, New York. (Translated into English in 1974 by S. Kotz.)

[25] Michel, R. (1981). *Zeit. Wahrscheinlichkeitsth. verw. Geb.*, **55**, 109–117.

[26] Molenaar, W. (1970). "Approximations to the Poisson, Binomial, and Hypergeometric Distribution Functions." *Math. Centre Tracts 31*. Mathematisch Centrum, Amsterdam.

[27] Odeh, R. E. and Owen, D. B. (1980). *Tables for Normal Tolerance Limits, Sampling Plans and Screening*. Marcel Dekker, New York.

[28] Owen, D. B. (1962). *Handbook of Statistical Tables*. Addison–Wesley, Reading, MA.

[29] Page, E. (1977). *Appl. Statist.*, **26**, 75–76.

[30] Patel, J. K. and Read, C. B. (1982). *Handbook of the Normal Distribution*. Marcel Dekker, New York.

[31] Patry, J. and Keller, J. (1964). *Numer. Math.*, **6**, 89–97 (in German).

[32] Pearson, E. S. and Hartley, H. O. (1966). *Biometrika Tables for Statisticians, Vol. 1*, (3d ed.) Cambridge University Press, London.

[33] Pearson, K. (1900). *Philos. Mag.*, *5th Ser.*, **50**, 157–175.

[34] Peizer, D. B. and Pratt, J. W. (1968). *J. Amer. Statist. Ass.*, **63**, 1416–1456.

[35] Prescott, P. (1974). *Biometrika*, **61**, 177–180.

[36] Rao, C. R. (1973). *Linear Statistical Inference and Its Applications*, 2nd ed. Wiley, New York.

[37] Sahai, H. and Thompson, W. O. (1974). *J. Statist. Comp. Simul.*, **3**, 81–93.

[38] Stigler, S. (1980). *Trans. N. Y. Acad. Sci. II*, **39**, 147–157.

[39] Stigler, S. (1982). *Amer. Statist.*, **36**, 137–138.

[40] Tchébichef, P. L. (1890). *Acta Math.*, **14**, 303–15. (In French. Reprinted in *Oeuvres*, Vol. 2 (1962). Chelsea, New York.)

[41] Upton, G. J. G. (1974). *Biometrika*, **61**, 369–373.

[42] Wilson, E. B. and Hilferty, M. M. (1931). *Proc. Natl. Acad. Sci.*, **17**, 684–688.

[43] Zar, J. H. (1978). *Appl. Statist.*, **27**, 280–290.

LAWS OF ERROR II
LIMIT THEOREM, CENTRAL
MILLS' RATIO
MOIVRE–LAPLACE THEOREM
MULTINORMAL DISTRIBUTION
NORMAL EXTREMES
TRANSFORMATIONS)

CAMPBELL B. READ

NORMAL DISTRIBUTION, BIVARIATE

See BIVARIATE NORMAL DISTRIBUTION

NORMAL EQUATIONS

These are the linear equations arising in obtaining least-squares* estimators of the coefficients in a linear model*. For example, the sum of squares

$$\sum_{j=1}^{n}\left(Y_j - \sum_{i=0}^{k}\beta_i X_{ij}\right)^2, \qquad k+1 \leqslant n$$

is minimized with respect to $\beta_0, \ldots, \beta_k$ by values satisfying the equations

$$\sum_{j=0}^{k}\beta_i\sum_{j=1}^{n}X_{hj}X_{ij} = \sum_{j=1}^{n}X_{hj}Y_j,$$

$$h = 0, 1, \ldots, k.$$

The equations obtained in weighted least squares* are also termed normal equations. For example, minimizing the weighted sum of squares

$$\sum_{j=1}^{n}w_j\left(Y_j - \sum_{i=0}^{k}\beta_i X_{ij}\right)^2$$

$$= (\mathbf{Y} - \mathbf{X}'\boldsymbol{\beta})'\mathbf{W}(\mathbf{Y} - \mathbf{X}'\boldsymbol{\beta}) \cdot$$

$(\mathbf{Y}' = (Y_2, \ldots, Y_n);$ $\boldsymbol{\beta}' = (\beta_0, \ldots, \beta_k);$ $\mathbf{X} = ((X_{ij}));$ $\mathbf{W} = \mathrm{diag}(w_1, \ldots, w_n))$ with respect to $\boldsymbol{\beta}$, we obtain the $(k+1)$ linear equations

$$\mathbf{XWX}'\boldsymbol{\beta} = \mathbf{XWY}.$$

Solutions of the normal equations may not be unique. (See IDENTIFIABILITY.)

The term *normal equations* is also used,

more generally, to denote the equations resulting from any least-squares estimation. When the model is not linear, the equations are no longer linear.

(GAUSS–MARKOV THEOREM
GENERALIZED INVERSE
GENERALIZED LINEAR MODEL
HAT MATRIX
LEAST SQUARES)

NORMAL EQUIVALENT DEVIATE

The *normal equivalent deviate* (sometimes called the *normit*) corresponding to a proportion P is the value u_P satisfying the equation

$$\Phi(u_P) = (2\pi)^{-1/2}\int_{-\infty}^{u_P}\exp(-u^2/2)\,du = P.$$

Transformation from an estimated value $\hat{P}$ of P to the corresponding $u_{\hat{P}}$ is used in the analysis of quantal response* data when the tolerance distribution* is assumed to be normal. It was once customary to use the *probit* (equal to the normal equivalent deviate plus five) instead, in order to avoid negative values.

(BIOASSAY, STATISTICAL METHODS IN QUANTAL RESPONSE ANALYSIS)

NORMAL EXTREMES

Let $X_1, X_2, \ldots X_n$ be a sequence of n independent and identically distributed random variables, each with a standard normal distribution* with CDF $\Phi(x) = \int_{-\infty}^{x}(2\pi)^{-1/2}\exp(-\frac{1}{2}\omega^2)\,d\omega$. In this article we shall summarize the main properties of the distribution of the largest variable $Y_n = \max\{X_i : 1 \leqslant i \leqslant n\}$—the "normal extreme." We shall be particularly interested in the limiting behavior of this distribution as $n \to \infty$ since this allows us to obtain useful and robust approximations to the CDF of

Y_n, $\Phi^n(x)$. For a discussion of the generalizations of this theory, *see* EXTREME VALUE DISTRIBUTIONS.

Tippett [15] considers the precise determination of the CDF $\Phi^n(x)$ for given $n \leqslant 1000$. He tabulates $\Phi^n(x)$ for several values of x and n, together with the 95 and 99% quantiles and first four (standardized) moments of the distribution for several values for n. Since Tippett's paper, several other authors have extended the tabulation (see, e.g., Gupta [10] and Pearson and Hartley [14]).

A method of obtaining fairly accurate approximating formulas for $\Phi^n(x)$ was developed by Hall [13]. Defining b_n to be the solution of

$$(2\pi)^{1/2}b_n\exp(\tfrac{1}{2}b_n^2) = n \qquad (1)$$

(some values of b_n are tabulated in Haldane [11] and Tippett [15]), Hall showed that for $x \geqslant b_n$,

$$Q_{1n}(x) = \exp\left\{-z_n(x)\left[1 - x^{-2} + 3x^{-4}\right.\right.$$
$$\left.\left. + z_n(x)/(2n-2)\right]\right\}$$
$$< \Phi^n(x) < Q_{2n}(x)$$
$$= \exp\left[-z_n(x)(1 - x^{-2})\right],$$

where $z_n(x) = (2\pi)^{-1/2}nx^{-1}\exp(-\tfrac{1}{2}x^2)$. $Q_{1n}(x)$ and $Q_{2n}(x)$ are useful lower and upper bounds for $\Phi^n(x)$ which give good estimates, especially when n and x are comparatively large. The same method can be used to find more precise bounds.

For large n (say $n > 1000$), estimates of $\Phi^n(x)$ can be obtained using the tools of extreme-value theory. As first shown by Fisher and Tippett [7], for many distribution functions F it is possible to choose normalizing constants α_n and β_n so that $F^n(\alpha_n x + \beta_n) \to G(x)$, where $G(x)$ is one of the three types of extreme-value distribution: type I, $\Lambda(x) = \exp(-e^{-x})$; type II, $\phi_k(x) = \exp(-x^{-k})$, $x, k > 0$; type III, $\Psi_k(x) = \exp(-(-x)^k)$, $x < 0$, $k > 0$. (For further details of this case, *see* EXTREME-VALUE DISTRIBUTIONS, Galambos [8], or Gumbel [9].) In particular, for normal extremes ($F = \Phi$), Fisher and

Tippett showed that as $n \to \infty$,

$$P((Y_n - \beta_n)/\alpha_n \leqslant x) = \Phi^n(\alpha_n x + \beta_n)$$
$$\to \Lambda(x) \qquad (2)$$

with $\alpha_n = b_n(1 + b_n^2)^{-1}$ and $\beta_n = b_n$ (defined at (1)) which corresponds to the approximation

$$\Phi^n(x) \approx \Lambda((x - \beta_n)/\alpha_n). \qquad (3)$$

As Fisher and Tippett observed by an empirical consideration of the distribution functions and their first four moments, the rate of convergence in (2) is fairly slow, so that the approximation (3) should not be used directly unless n is very large. However, in the corresponding statistical situation $\Lambda(Ax + B)$ (with A and B estimated from the data) will be a useful estimator of $P(Y_n \leqslant x)$ because of its robustness* to changes in the distributional assumptions.

More recently, several authors have considered the related problems of the estimation of the rate of convergence in (2) and the error in (3). The main results can be summarized in the following theorems, for which we define $a_n = b_n^{-1}$ and

$$S_n = \sup_x |\Phi^n(\alpha_n x + \beta_n) - \Lambda(x)|$$
$$= \sup_x |\Phi^n(x) - \Lambda((x - \beta_n)/\alpha_n)|.$$

Theorem 1. $S_n \leqslant 3/\log n$ if $\alpha_n = a_n$ and $\beta_n = b_n$.

Theorem 2. $S_n = O(1/\log n)$ if and only if $(\beta_n - b_n)/a_n = O(1/\log n)$, $(\alpha_n - a_n)/a_n = O(1/\log n)$.

Theorem 3. $S_n = o(1/\log n)$ cannot hold for any sequences α_n and β_n.

Theorems 1–3 are proven in Hall [12], Cohen [3] (theorem 4) and Anderson [1] (theorem 4.3.1), respectively. Essentially they show that with the optimal choice of sequences α_n and β_n, the convergence in (2) can be as fast as, but not faster than, order $1/\log n$.

Several authors [5, 11, 13] have investigated the possibility of improving the approximation (3) by applying a transformation $Y_n \to g(Y_n)$ to the extremes before applying the extreme-value approximation. This corresponds to the approximation

$$\Phi^n(x) \approx \Lambda((g(x) - d_n)/c_n), \qquad (4)$$

for $x \geqslant \lambda$ and suitable sequences $c_n > 0$ and d_n (assuming g is strictly increasing on $[\lambda, \infty)$). For example, if $g(x) = x^2$, then the error in (4) is uniformly (in $x \geqslant 0$) of order $(\log n)^{-2}$ if c_n and d_n are chosen optimally (see Hall [13]). However, the problem with this approach is that in practice it is impossible to choose precisely the right transformation for the actual distribution of the extremes.

A generally applicable (see Cohen [3]) approach to the problem of improving the approximation (3) while retaining the robustness property was introduced by Fisher and Tippett [7]. As Fisher and Tippett found empirically, $\Phi^n(x)$ may be more closely approximated by the type III extreme-value distribution $\Psi_{k_n}(A_n x + B_n)$ than by the (limiting) type I distribution $\Lambda((x - \beta_n)/\alpha_n)$. Cohen [2] made this result more precise by showing that the uniform error may be reduced to order $(\log n)^{-2}$ if the sequences k_n, A_n, and B_n are chosen optimally. (The main results are analogous to Theorems 1–3.)

Some related work on the distribution of Y_n is found in Dronkers [6] and Uzgören [16]. Also of interest is the recent and not fully developed work of Daniels ([5]—also see Daniels [4]) in which he investigates the variance-stabilizing* property of the transformation $g(Y_n)$, where

$$g(x) = \int^x \left[\operatorname{arccot}(\tfrac{1}{2}\omega) \right]^{-1} d\omega.$$

(See NORMAL EXTREMES, DANIELS' FORMULA.)

References

[1] Anderson, C. W. (1971). "Contributions to the Asymptotic Theory of Extreme Values." Ph.D. thesis, University of London, England.

[2] Cohen, J. P. (1982). Adv. Appl. Prob., 14, 324–339.

[3] Cohen, J. P. (1982). Adv. Appl. Prob., 14, 833–854.

[4] Daniels, H. E. (1942). Biometrika, 32, 194–195.

[5] Daniels, H. E. (1982). In Essays in Statistical Science, J. Gani and E. J. Hannan, eds. Applied Probability Trust, Sheffield, UK, pp. 201–206.

[6] Dronkers, J. J. (1958). Biometrika, 45, 447–470.

[7] Fisher, R. A. and Tippett, L. H. C. (1928). Proc. Camb. Philos. Soc., 24, 180–190.

[8] Galambos, J. (1978). The Asymptotic Theory of Extreme Order Statistics. Wiley, New York. (A wide range of theoretical problems are discussed. Contains a large bibliography.)

[9] Gumbel, E. J. (1960). Statistics of Extremes. Columbia University Press, New York, (Many practical applications are given.)

[10] Gupta, S. S. (1961). Ann. Math. Statist., 32, 888–893.

[11] Haldane, J. B. S. and Jayakar, S. D. (1963). Biometrika, 50, 89–94.

[12] Hall, P. (1979). J. Appl. Prob., 16, 433–439.

[13] Hall, P. (1980). Adv. Appl. Prob., 12, 491–500.

[14] Pearson, E. S. and Hartley, H. O. (1972). Biometrika Tables for Statisticians, Vol. II. Cambridge University Press, London, England.

[15] Tippett, L. H. C. (1925). Biometrika, 17, 364–387.

[16] Uzgören, N. T. (1954). In Studies in Mathematics and Mechanics Presented to Richard Von Mises. Academic Press, New York, pp. 346–353.

(EXTREME-VALUE DISTRIBUTIONS
NORMAL DISTRIBUTION
ORDER STATISTICS)

JONATHAN P. COHEN

NORMAL EXTREMES, DANIELS' FORMULA

Let $X_{(n)}$ be the the greatest observation in a random sample of size n from a standard normal distribution. Its expected value μ_n and standard deviation σ_n satisfy the approximate relation

$$\mu_n = 2 \cot \tfrac{1}{2} \pi \sigma_n$$

with reasonable accuracy over the whole range of values of $n = 1$ to ∞. As $n \to \infty$, the ratio of $2 \cot \tfrac{1}{2} \pi \sigma_n$ to μ_n tends to $4\sqrt{6}/\pi^2$ $= 0.9927$. The relation was observed empirically by Daniels [1] and further investigated by him in 1982 [2].

References

[1] Daniels, H. E. (1941). *Biometrika*, **32**, 194–195.

[2] Daniels, H. E. (1982). In *Essays in Statistical Sciences* (Papers in honor of P. A. P. Moran), J. Gani and E. J. Hannan, eds. *J. Appl. Prob.*, Special Vol. **19A**, 201–206.

(NORMAL EXTREMES
ORDER STATISTICS
RANGES)

NORMAL-GAMMA (PRIOR) DENSITY

Consider a normal random variable Y with mean μ and variance σ^2. A (natural) conjugate prior $g(\mu, \sigma)$ for (μ, σ) obtained by specifying a normal density for the conditional prior $g(\mu \mid \sigma)$ and an inverted gamma density for the marginal prior $g(\sigma)$ is known as a *normal-gamma prior*. Specifically,

$$
\begin{aligned}
g(\mu, \sigma) &= g(\mu \mid \sigma) \cdot g(\sigma) \\
&= \frac{\sqrt{\tau}}{\sqrt{2\pi}\, \sigma} \exp\left\{ -\frac{\tau}{2\sigma^2}(\mu - \mu_0)^2 \right\} \\
&\quad \times \frac{2}{\Gamma(v/2)} \left(\frac{vs^2}{2} \right)^{v/2} \\
&\quad \times \frac{1}{\sigma^{v+1}} \exp\left\{ -\frac{vs^2}{2\sigma^2} \right\}.
\end{aligned}
$$

Here τ, μ_0, v, and s^2 are the parameters of the prior distribution. In particular, $E(\mu \mid \sigma) = E(\mu) = \mu_0$ and $\operatorname{var}(\mu \mid \sigma) = \sigma^2 / \tau$.

Bibliography

Zellner, A. (1971). *An Introduction to Bayesian Inference in Econometrics*. Wiley, New York.

(CONJUGATE FAMILIES
   OF DISTRIBUTIONS
INVERTED GAMMA DISTRIBUTION
NORMAL DISTRIBUTION)

NORMALITY, TESTS OF *See* DEPARTURE FROM NORMALITY, TESTS FOR

NORMALIZED *T* SCORES

The transformation of raw test scores to standard scores is a common means of obtaining score comparability. Unfortunately, unless the distributions from which the scores are drawn have the same shape, standard scores will not be comparable across distributions. This problem will be most pronounced if distribution A is positively skewed and distribution B is negatively skewed. In this situation a Z score of 2.00 will represent a very different centile equivalent in the two distributions.

McCall [1] has suggested a procedure that normalizes a distribution and uses a standard score based on a conversion of Z scores to mean $(M) = 50$, standard deviation $(\sigma) = 10$. The result is called a McCall T score or a normalized T score.

The procedure requires finding the centile equivalent of each score, converting that centile to a Z score from an appropriate table of the unit normal distribution, and then converting the Z score to a standard score with $M = 50$, $\sigma = 10$. The resulting scores, if plotted against frequency, will be distributed normally regardless of the shape of the original distribution.

Reference

[1] McCall, W. A. (1939). *Measurement*. Macmillan, New York, pp. 505–508.

(PSYCHOLOGICAL TESTING THEORY)

HENRY E. KLUGH

NORMAL SCORES TESTS

Normal scores tests include techniques of estimation* and inference* that are based on linear rank statistics* whose constants are based on certain characteristics of the standard normal distribution* instead of ranks. These characteristics (called normal scores) are usually some function of either (a) expected values of order statistics* for a sample from the standard (or absolute) normal distribution or (b) quantiles* of the standard (or absolute) normal distribution*. These tests are distribution-free (*see* DISTRIBUTION-FREE METHODS) and have very good asymptotic relative efficiency*, especially for normal distributions. They are discussed in more detail in many nonparametric texts and reference books, including Bradley [3], Conover [6], Gibbons [12], Hájek [15], Hájek and Šidák [16], Lehmann [23], Marascuilo and McSweeney [25], Pratt and Gibbons [28], Randles and Wolfe [31], and van der Waerden and Nievergelt [37].

TWO-SAMPLE TESTS FOR LOCATION

For $X_1, X_2, \ldots, X_m$ and $Y_1, Y_2, \ldots, Y_n$ two mutually independent random samples of observations measured on at least an ordinal scale* and drawn from continuous populations, the null hypothesis* is that the populations are identical, or that the medians* M_X and M_Y satisfy $\theta = M_X - M_Y = 0$ under the shift model assumption that $X - \theta$ and Y have identical distributions. The normal scores test statistics are all of the form

$$\sum_{k=1}^{N} c_k I_k$$

where $I_k = \begin{cases} 1 & \text{if the } k\text{th order statistic of} \\ & \text{the pooled samples is an } X \\ 0 & \text{otherwise,} \end{cases}$

$$(1)$$

$c_1, c_2, \ldots, c_N$ are normal scores and $N = m + n$.

If $c_k = E(Z_{k:N})$ in (1), where $Z_{k:N}$ is the kth order statistic in a sample of N from the standard normal distribution, we have the test first proposed by Fisher and Yates [8] and then Hoeffding [19], but sometimes called the Fisher–Yates–Terry test or the Terry–Hoeffding test. These expected values are tabled in Fisher and Yates [8] to two decimal places and with greater precision in Teichroew [32], Owen [27], David et al. [7], and Harter [17, 18]. Terry [33] gives tables of the null distribution for $N \leq 10$ and dis-

cusses approximations for larger sample sizes; Klotz [22] gives critical values for $N \leq 20$ and discusses asymptotic approximations.

If $c_k = \Phi^{-1}[k/(N+1)]$, where $\Phi(x)$ denotes the standard normal CDF, we have the test proposed by van der Waerden [35, 36] for which no special tables are needed, although those provided in van der Waerden and Nievergelt [37] and Hájek [15] are useful.

These tests are asymptotically equivalent and differ little for moderate and even small sample sizes. For $m = n = 6$, for example, Pratt and Gibbons [28, pp. 267–268] show that the one-tailed probabilities are identical for all $P \leq 0.042$, and almost identical for P up to 0.053.

The asymptotic efficiency* of these normal scores tests relative to Student's t-test* is always at least one for all shift model distributions and equals one for normal distributions. Thus these tests for location are always at least as good as the uniformly most powerful* parametric procedure for normal distributions, and better for other distributions. Chernoff and Savage [5] proved this result, and Gastwirth and Wolff [10] simplified the proof.

Lehmann [23] suggests that the Mann–Whitney–Wilcoxon test* will tend to be more powerful than normal scores tests for large samples only when the distributions have heavy tails, e.g., a normal distribution contaminated by a large proportion of gross errors.

The small sample power of these two sample normal scores tests was investigated for $N \leq 10$ in Klotz [22] for normal shift alternatives and in Gibbons [11] for both normal shift alternatives and Lehmann-type alternatives* that do not specify the distributions; the latter paper compares the small sample power with that of other nonparametric tests. Ramsey [29] gives small sample power for $N \leq 10$ in double exponential* shift alternatives and makes comparisons with other nonparametric tests. All of these results conclude that two-sample normal scores tests for location have unusually good power even in small samples. However, Gordon [13] showed that the power function of both normal scores and Mann–Whitney–Wilcoxon tests* may decrease as sample size increases for a broad class of alternatives.

TWO-SAMPLE TESTS FOR SCALE

In the previous case of two mutually independent random samples, consider now the null hypothesis that the scale parameters σ_X and σ_Y satisfy $\theta = \sigma_X/\sigma_Y = 1$ under the scale model assumption that $(X - M_X)/\theta$ and $(Y - M_Y)$ have identical distributions. The Capon [4] normal scores test (*see* CAPON TEST) is of the form (1) with $c_k = E(Z_{k:N}^2)$, which is tabled in Teichroew [32]. The asymptotically equivalent Klotz test (*see* KLOTZ TEST) is of the form (1) with $c_k = \{\Phi^{-1}[k/(N + 1)]\}^2$; this latter reference tables the critical values for $N \leqslant 20$. The asymptotic efficiency* of these tests relative to the F test is one for normal distributions, but it can be smaller than one and in fact can range between 0 and infinity for other distributions. Klotz [20] gives some small sample power* calculations of his test and some other scale tests for $N \leqslant 10$ for normal scale alternatives.

ONE-SAMPLE AND
PAIRED-SAMPLE TESTS

For $X_1, X_2, \ldots, X_N$ a random sample of N observations measured on at least an ordinal scale and drawn from a population that is continuous and symmetric about its median* M, the null hypothesis is $M = M_0$. The normal scores test statistics for this problem are all of the form

$$\sum_{k=1}^{N} c_k I_k \qquad (2)$$

where $I_k = 1$ or 0 according as the value of X corresponding to the kth order statistic of $\{|X_i - M_0|; i = 1, \ldots, n\}$ is greater or less than M_0.

In particular, if $c_k = E(U_{k:N})$ in (2), where $U_{k:N}$ is the kth order statistic of a sample of N drawn from the absolute values of a standard normal distribution (chi distribution with one degree of freedom), we have the absolute normal scores test proposed by Fraser [9]. These expected values are tabled in Klotz [21] for $N \leqslant 10$ and Govindarajulu and Eisenstat [14] for $N \leqslant 100$. Critical values of the test statistic are tabled in Klotz [21] for $N \leqslant 10$ and extended to $N = 20$ in Thompson et al. [34]. If $c_k = G^{-1}[k/(N + 1)]$, where $G(x)$ denotes the CDF of the absolute value of a standard normal variable, we have a test of the van der Waerden type. This is the same as the van Eeden [38] test with $c_k = \Phi^{-1}\{\frac{1}{2} + \frac{1}{2}[k/(N + 1)]\}$ in (2); these constants represent only the positive half of the standard normal distribution.

Each of these tests can be used in a paired sample problem if X is interpreted to mean the difference between a pair of observations and M is interpreted as the median of these differences. These tests are all asymptotically equivalent. Their asymptotic efficiency relative to Student's t test* is the same here as it was for the two-sample normal scores tests for location. Pratt and Gibbons [28, p. 384] give a useful table of relative efficiencies of Student's t-test, normal scores tests, and three other nonparametric tests for various shift families of symmetric distributions. These same entries also hold for two-sample procedures.

The small sample power of the absolute normal scores test is given in Klotz [21] for normal alternatives when $N \leqslant 10$, and compared to that of the Wilcoxon signed rank test (*see* DISTRIBUTION-FREE METHODS). Arnold [1] studied small sample power for $N \leqslant 10$ for some other symmetric distributions.

NORMAL SCORES TESTS FOR OTHER
SAMPLING SITUATIONS AND/OR
OTHER HYPOTHESIS SITUATIONS

Procedures based on normal scores can be developed for virtually any ordinary nonparametric test procedure that is based solely on some function of the ranks of the observations in an array. The rank of each observation in an array of N observations is simply replaced by its expected normal order statistic or inverse normal score before calculating the ordinary test statistic. Some function of this statistic has the same asymptotic distribution as the ordinary test statistic.

For example, the ordinary Kruskal–Wallis one-way analysis of variance by ranks procedure (see DISTRIBUTION-FREE METHODS) for k mutually independent random samples of sizes $n_1, n_2, \ldots, n_k$ is to pool all the observations into an array and rank them from 1 to $N = n_1 + n_2 + \cdots + n_k$, while keeping track of which rank is for which sample. The test statistic is a linear function of $\sum_{j=1}^{k}(R_j^2/n_j)$, where R_j is the sum of the ranks in the jth sample. If we replace the observation with rank k by $E(Z_{k:N})$ and use these normal scores to compute R_j, we obtain a normal scores test of the Fisher–Yates–Terry–Hoeffding type; see FISHER–YATES TESTS. If we use $\Phi^{-1}[k/(N+1)]$, the normal scores test is of the van der Waerden type. The test statistic is

$$W = \frac{(N-1)\sum_{j=1}^{k}\left(U_j^2/n_j\right)}{\sum_{i=1}^{N}c_i^2},$$

where U_j is the sum of the normal scores in the jth sample and $\sum_{i=1}^{N}c_i^2$ is the sum of the squares of the N normal scores used in place of the ranks. This statistic is asymptotically chi-square distributed with $k-1$ degrees of freedom under the null hypothesis that the k samples come from identical continuous populations. This test was developed in McSweeney and Penfield [26], who also gave multiple comparisons* procedures to make pairwise comparisons among populations. Lu and Smith [24] give tables of the exact distribution of W for $U_j = E(Z_{k:N})$ for $k = 3$, $N \leqslant 15$, and selected significance levels.

Many numerical examples of these and other normal scores tests are given in Marascuilo and McSweeney [25] and in Conover [6].

OTHER NORMAL SCORES TESTS

The Bell and Doksum [2] distribution-free tests (see BELL–DOKSUM TESTS) might also be called normal scores tests because they use characteristics of normal deviates in place of the ranks. For example, in (1) for the two-sample location problem, $c_k = Z_{k:N}$, where $Z_1, Z_2, \ldots, Z_N$ are a sample of N selected from a table of random normal deviates, e.g., Rand Corporation [30]. Since these tests introduce an additional source of randomness, they may be less attractive to many applied researchers. The two-sample test for location is of theoretical interest since its asymptotic efficiency relative to Student's t-test is one for normal shift alternatives and is greater than one for other shift families.

Example. Hypoglycemia is a condition in which blood sugar is below normal limits. In order to compare two hypoglycemic compounds, X and Y, each one is applied to half of the diaphragm of each of nine rats in an experiment reported by Wilcoxon and Wilcox [39, p. 9]. Blood glucose uptake in milligrams per gram of tissue is measured for each half, producing the following data:

Rat	X	Y
1	9.4	8.4
2	8.5	8.7
3	4.7	4.1
4	3.9	3.6
5	4.7	5.1
6	5.2	5.2
7	10.2	10.0
8	3.3	4.6
9	7.0	6.1

We illustrate the use of a one-sample normal scores procedure to test $H_0 : M_D = 0$ vs. the one-sided alternative $H_1 : M_D > 0$ where M_D denotes the median of the population of differences $D = X - Y$.

The one difference that is equal to zero is discarded, leaving $N = 8$. The remaining differences, rearranged in order of absolute magnitude, are:

$-0.2, 0.2, 0.3, -0.4, 0.6, 0.9, 1.0, -1.3.$

For the Klotz paired sample test in this

example, the corresponding constants in (2) are

$$0.14, 0.29, 0.43, 0.61, 0.80, 1.02, 1.31, 1.78.$$

Note first that the two smallest differences are tied in absolute value. For each of these, we use $c_1 = c_2 = (0.14 + 0.29)/2 = 0.215$, the average of the normal scores they would be assigned if they were not tied. The value of the test statistic (2) for this example is

$$0.215 + 0.43 + 0.80 + 1.02 + 1.31 = 3.775.$$

From Klotz [21], the one-tailed critical value of this statistic, for $N = 8$, $\alpha = 0.05$, is 5.45. Thus we cannot reject the null hypothesis at the 0.05 level. For comparison purposes, the van der Waerden or van Eeden test statistic equals 3.59, and the Wilcoxon signed rank statistic is 22.5, each also nonsignificant at the 0.05 level.

References

[1] Arnold, H. J. (1965). *Ann. Math. Statist.*, **36**, 1767–1778.

[2] Bell, C. B. and Doksum, K. A. (1965). *Ann. Math. Statist.*, **36**, 203–214.

[3] Bradley, J. V. (1968). *Distribution-Free Statistical Tests*. Prentice-Hall, Englewood Cliffs, NJ. (Elementary; Chapter 6 is devoted to normal scores tests and references.)

[4] Capon, J. (1961). *Ann. Math. Statist.*, **32**, 88–100.

[5] Chernoff, H. and Savage, I. R. (1958). *Ann. Math. Statist.*, **29**, 972–994.

[6] Conover, W. J. (1980). *Practical Nonparametric Statistics*. Wiley, New York. (Elementary; normal scores tests are discussed in Section 5.10.)

[7] David, F. N., Barton, D. E., Ganeshalingham, S., Harter, H. L., Kim, P. J., and Merrington, M. (1968). *Normal Centroids, Medians, and Scores for Ordinal Data*. Cambridge University Press, London, England. (Tables.)

[8] Fisher, R. A. and Yates, F. (1938). *Statistical Tables for Biological Agricultural and Medical Research*. Oliver and Boyd, Edinburgh, Scotland. (Tables.)

[9] Fraser, D. A. S. (1957). *Ann. Math. Statist.*, **28**, 1040–1043.

[10] Gastwirth, J. L. and Wolff, S. (1968). *Ann. Math. Statist.*, **39**, 2128–2130.

[11] Gibbons, J. D. (1964). *J. R. Statist. Soc. B*, **26**, 293–304.

[12] Gibbons, J. D. (1971). *Nonparametric Statistical Inference*. McGraw-Hill, New York. (Intermediate level; mostly theory; no tables.)

[13] Gordon, R. D. (1978). *Commun. Statist. A*, **7**, 535–541.

[14] Govindarajulu, A. and Eisenstat, S. (1965). *Nippon Kagaku Gijutus.*, **12**, 149–164.

[15] Hájek, J. (1969). *A Course in Nonparametric Statistics*. Holden-Day, San Francisco. (Intermediate level; some tables of critical values given.)

[16] Hájek, J. and Šidak, Z. (1967). *Theory of Rank Tests*. Academic Press, New York. (Theoretical; no tables.)

[17] Harter, H. L. (1961). *Biometrika*, **48**, 151–165.

[18] Harter, H. L. (1969). *Order Statistics and Their Use in Testing and Estimation*, Vol. 2. U.S. GPO, Washington, DC.

[19] Hoeffding, W. (1951). In *Proceedings of the Second Berkeley Symposium*. University of California Press, Berkeley, CA, pp. 83–92.

[20] Klotz, J. (1962). *Ann. Math. Statist.*, **33**, 498–512.

[21] Klotz, J. (1963). *Ann. Math. Statist.*, **34**, 624–632.

[22] Klotz, J. (1964). *J. Amer. Statist. Ass.*, **59**, 652–664.

[23] Lehmann, E. L. (1975). *Nonparametrics: Statistical Methods Based on Ranks*. Holden-Day, San Francisco. (Intermediate.)

[24] Lu, H. T. and Smith, P. J. (1979). *J. Amer. Statist. Ass.*, **74**, 715–722.

[25] Marascuilo, L. A. and McSweeney, M. (1977). *Nonparametric and Distribution-Free Methods for the Social Sciences*. Brooks/Cole, Monterey, CA. (Elementary cookbook approach; many numerical examples; tables of normal scores given.)

[26] McSweeney, M. and Penfield, D. A. (1969). *Brit. J. Math. Statist. Psychol.*, **22**, 177–192.

[27] Owen, D. B. (1962). *Handbook of Statistical Tables*. Addison-Wesley, Reading, MA. (Tables.)

[28] Pratt, J. W. and Gibbons, J. D. (1981). *Concepts of Nonparametric Theory*. Springer-Verlag, New York. (Intermediate; conceptual approach to theory.)

[29] Ramsey, F. L. (1971). *J. Amer. Statist. Ass.*, **66**, 149–151.

[30] Rand Corporation (1955). *A Million Random Digits with 100,000 Normal Deviates*. Free Press, Glencoe, IL. (Tables.)

[31] Randles, R. H. and Wolfe, D. A. (1979). *Introduction to the Theory of Nonparametric Statistics*. Wiley, New York. (Intermediate; mostly theory; no tables.)

[32] Teichroew, D. (1956). *Ann. Math. Statist.*, **27**, 410–426.

[33] Terry, M. E. (1952). *Ann. Math. Statist.*, **23**, 346–366.

[34] Thompson, R., Govindarajulu, Z., and Doksum, K. A. (1967). *J. Amer. Statist. Ass.*, **62**, 966–975.

[35] Van der Waerden, B. L. (1952). *Proc. Kon. Ned. Akad. Wet. A*, **55**, 453–458.

[36] Van der Waerden, B. L. (1953). *Proc. Kon. Ned. Akad. Wet. A*, **56**, 201–207.

[37] Van der Waerden, B. L. and Nievergelt, E. (1956). *Tables for Comparing Two Samples by X-test and Sign Test*. Springer-Verlag, Berlin. (Tables.)

[38] Van Eeden, C. (1963). *Ann. Math. Statist.*, **34**, 1442–1451.

[39] Wilcoxon, F. and Wilcox, R. A. (1964). *Some Rapid Approximate Statistical Procedures*. Lederle Laboratories, Pearl River, NY.

(BELL–DOKSUM TESTS
CAPON TEST
DISTRIBUTION-FREE TESTS
FISHER–YATES TESTS
KLOTZ TEST
LINEAR RANK TEST
PURI'S EXPECTED NORMAL SCORES
 TEST
VAN DER WAERDEN TEST)

JEAN DICKINSON GIBBONS

NORMAL VARIABLES, RATIO OF BIVARIATE

If X_1, X_2 have a joint bivariate normal distribution* with expected values ξ_i $(i = 1, 2)$, standard deviation σ_i $(i = 1, 2)$ and correlation coefficient ρ, the probability density function (PDF) of $V = (X_1/X_2)$ is

$$f_V(v)$$

$$= \frac{hl}{\sqrt{2\pi}\,\sigma_1\sigma_2 g^3}\left\{2\Phi\left[\frac{h}{g\sqrt{(1-\rho^2)}}\right]-1\right\}$$

$$+ \frac{\sqrt{1-\rho^2}}{\pi\sigma_1\sigma_2 g^2}\exp\left\{\frac{-k}{2(1-\rho^2)}\right\},$$

$$g = \left(\frac{v^2}{\sigma_1^2} - \frac{2\rho v}{\sigma_1\sigma_2} + \frac{1}{\sigma_2^2}\right)^{1/2};$$

$$k = \frac{\xi_1^2}{\sigma_1^2} - \frac{2\rho\xi_1\xi_2}{\sigma_1\sigma_2} + \frac{\xi_2^2}{\sigma_2^2};$$

$$h = \left(\frac{\xi_1}{\sigma_1} - \frac{\rho\xi_2}{\sigma_2}\right)\frac{v}{\sigma_1} + \left(\frac{\xi_2}{\sigma_x} - \frac{\rho\xi_1}{\sigma_1}\right)\frac{1}{\sigma_2};$$

$$l = \exp\left[\frac{1}{2(1-\rho^2)}\left(\frac{h^2}{g^2} - k\right)\right].$$

Note that k is a constant independent of the variable v. (See also Nicholson [3].) The distribution of V approaches normality as the coefficient of variation* of the variable in the denominator $C_2 = \sigma_2/\mu_2$ tends to zero.

Shanmugalingam [4] carried out a Monte Carlo* study to ascertain the relation between the values of $C_i = \sigma_i/\mu_i$ $(i = 1, 2)$, for given ρ, for which the normal approximation is useful. The approach to normality is quite complicated.

References

[1] Fieller, E. C. (1932). *Biometrika*, **24**, 428–40.

[2] Hinkley, D. V. (1969). *Biometrika*, **56**, 635–39; (1970). erratum, **57**, 683.

[3] Nicholson, C. (1941). *Biometrika*, **32**, 16–28.

[4] Shanmugalingam, S. (1982). *The Statistician (Lond.)*, **31**, 251–258.

(BIVARIATE NORMAL DISTRIBUTION
FIELLER'S THEOREM
NORMAL DISTRIBUTION)

NORMIT   *See* NORMAL EQUIVALENT DEVIATE

NOTCHED BOX-AND-WHISKER PLOT

A notched box-and-whisker plot is a graphical display of univariate data that conveys basic information about its distribution. The display utilizes the lengths of the box, whiskers, and notch, and the width of the box to impart a quick overall assessment of the distribution (e.g., symmetry*, skewness*, outliers*). Such a display is far more informative than a look at raw numerical data and thus is useful in exploratory data analysis. For comparing several sets of data, notched box-and-whisker plots permit visual comparisons of relative spreads (via box lengths), pairwise tests of significance (via notches), and relative sample sizes (via box widths).

The basic concept was developed for either single or multiple batches of data by

Tukey [2, Vol. 1, Chap. 5]. Most commonly the displays are constructed according to the guidelines for schematic plots outlined in [3, Sect. 2E]. The additional refinements discussed here are useful for comparing several data sets and were suggested in an article by McGill et al. [1].

CONSTRUCTION OF THE PLOT

Given a set of data, $X_1, \ldots, X_n$, a notched box-and-whisker plot is constructed from the following calculations:

1. The median* X_m.
2. The lower and upper hinges X_L, X_U; i.e., the kth and $(n + 1 - k)$th order statistics, respectively, where $k = ([\frac{1}{2}(n + 1)] + 1)/2$, and $[\cdot]$ denotes the integer part of the argument. If k is not an integer, the mean of the adjacent order statistics is used.

3. The "step" $= 1.5(X_U - X_L)$ (i.e., $1\frac{1}{2}$ times the hinge spread).
4. The fences:
   (a) Inner fences: $f_1 = X_L -$ step; $f_2 = X_U +$ step.
   (b) Outer fences: $F_1 = X_L - (2$ steps); $F_2 = X_U + (2$ steps).
5. Special values:
   (a) Adjacent values: the data values that lie closest to, but just inside, the inner fences.
   (b) Outside values (outliers): the data values that lie between the inner and outer fences.
   (c) Far outside values (far outliers): the data values that lie outside the outer fences.
6. An interval about the median, $(X_m - W, X_m + W)$, where

$$W = 1.58(X_U - X_L)/\sqrt{n} .$$

The ends of this interval determine the

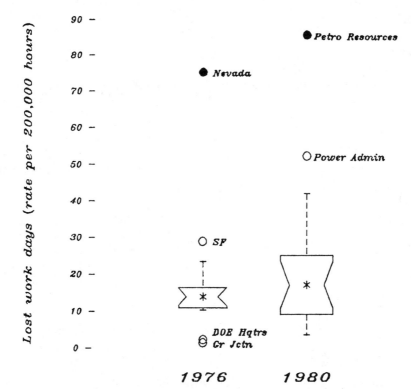

Figure 1   Notched box-and-whisker plots for the data in Table 1.

length of the notch. A pair of nonoverlapping notches indicates a significant difference in the population medians at the 5% level of significance. The factor 1.58 is a compromise between the case where the variances are equal and the case where they are grossly different (cf. McGill et al. [1, p. 16]). The resulting interval facilitates comparisons between samples from various underlying populations.

To actually construct the plot:

1. Place an asterisk or a bar at X_m.
2. Construct the ends of the box at X_L and X_U.
3. Complete the box by making its width proportional to $\sqrt{n}$.

Table 1 Lost Work Days at 15 Department of Energy Field Organizations in 1976 and 1980 (rate per 200,000 work hours)

(a) The Data		
	1976	1980
Schenectady N.R.	15.9	3.1
Savannah River	13.6	5.2
DOE Headquarters	1.9	6.3
Grand Junction	1.1	7.7
Albuquerque	13.8	9.9
Oak Ridge	23.0	13.1
San Francisco	28.6	15.3
Energy Tech Centers	15.0	16.8
Richland	10.5	22.2
Pittsburgh N.R.	12.0	22.3
Chicago	11.8	23.3
Idaho	9.9	25.9
Nevada	74.9	41.4
Power Admin.	—	51.8
Petroleum Resources	—	85.2

(b) The Calculations		
	1976	1980
Sorted values	$x(1), \ldots, x(13)$	$y(1), \ldots, y(15)$
Median	$x(7) = 13.6$	$y(8) = 16.8$
Lower hinge	$x(4) = 10.5$	$[y(4) + y(5)]/2 = 8.8$
Upper hinge	$x(10) = 15.9$	$[y(11) + y(12)]/2 = 24.6$
Hinge spread	$15.9 - 10.5 = 5.4$	$24.6 - 8.8 = 15.8$
One step	$1.5 \times 5.4 = 8.1$	$1.5 \times 15.8 = 23.7$
Inner fences	2.4, 24.0	$-14.9, 48.3$
Outer fences	$-5.7, 32.1$	$-38.6, 72.0$
Adjacent values	9.9, 23.0	3.1, 41.4
Outlier	1.1 (GJ), 1.9 (DOE), 28.6 (SF)	51.8 (Power Admin.)
Far outlier	74.9 (Nevada)	85.2 (Pet Resources)
Notch distance from median	$1.58 \times 5.4/\sqrt{13} = 2.4$	$1.58 \times 15.8/\sqrt{15} = 6.4$
Notches	(11.2, 16.0)	(10.4, 23.2)

Source. Injury and Property Damage Summary Report of the deputy assistant secretary for Environment, Safety, and Health (1976 and 1980).

4. Denote the locations of the adjacent values by dashes and connect them to the ends of the box with dashed lines ("whiskers").

5. Indicate outliers at the appropriate places with circles, far outliers with filled-in circles.

6. Form notches in the box, where the beginning and end of the notches are given by $X_m \pm W$. (The depth of the cuts into the box to form the notches has no statistical meaning.)

Example. Figure 1 illustrates notched box-and-whisker plots constructed from the data in Table 1. Notice that since the hinge spread is equal approximately to the interquartile range*, roughly 50% of the data lies within the box. For Gaussian data, one step is approximately 2σ, so observations that are more than 2σ outside the hinges would appear as outliers (roughly 2.7σ away from the median) and those that are more than 4σ outside the hinges would appear as far outliers. The notches in this figure overlap, so these data do not show evidence of a difference between the medians.

References

[1] McGill, R., Tukey, J. W., and Larsen, W. A. (1978). *Amer. Statist.*, **32**, 12–16.

[2] Tukey, J. W. (1970). *Exploratory Data Analysis*, limited preliminary ed. Addison-Wesley, Reading, MA.

[3] Tukey, J. W. (1977). *Exploratory Data Analysis*, Addison-Wesley, Reading, MA.

(EXPLORATORY DATA ANALYSIS
FIVE-NUMBER SUMMARIES
GRAPHICAL REPRESENTATION
  OF DATA
HYPOTHESIS TESTING
MEDIAN
ORDER STATISTICS
OUTLIERS)

KAREN KAFADAR

NP CHART *See* CONTROL CHARTS

_n_-POINT METHOD

This is a method of approximate evaluation of integrals of the form

$$\int_a^b f(x)g(x)\,dx \qquad (1)$$

by formulas of form

$$\{w_1 f(x_1) + \cdots + w_n f(x_n)\} \int_a^b g(x)\,dx. \qquad (2)$$

The values of $x_1, \ldots, x_n$ (the "n points") and of the weights $w_1, \ldots, w_n$ are chosen to get a good approximation. This is done by making them satisfy the $2n$ equations

$$\sum_{i=1}^n w_i x_i^j = m_j, \qquad j = 0, 1, \ldots, 2n - 1, \qquad (3)$$

where $m_j = \int_a^b x^j g(x)\,dx / \int_a^b g(x)\,dx$. The m's can be regarded as crude moments of a distribution over $[a, b]$ with density function $g(x)$. (Note that $m_0 = 1$.) Equations (3) require that $\sum w_i = 1$ and the first $(2n - 1)$ moments of the discrete distribution $\Pr[X = x_i] = w_i$ $(i = 1, \ldots, n)$ have the same values as those of $g(x)$.

If the Taylor series expansion* of $f(x)$ is valid, the remainder term of {formula (2) − formula (1)} is

$$\frac{1}{(2n)!} \left[f^{(2n)}(\xi_1) m_{2n} - f^{(2n)}(\xi_2) \sum_{i=1}^n w_i x_i^{2n} \right] \qquad (4)$$

$(a \leqslant (\xi_1, \xi_2) \leqslant b)$. If $f(x)$ is a polynomial of degree less than $2n$, $f^{(2n)}(x) \equiv 0$ and formula (2) is exactly equal to formula (1).

It is possible to evaluate the sum in (2) without calculating the w's and x's explicitly. We have

$$\begin{vmatrix} 1 & m_1 & \cdots & m_n \\ m_1 & m_2 & \cdots & m_{n+1} \\ \vdots & \vdots & & \vdots \\ m_n & m_{n+1} & \cdots & \sum_{i=1}^n w_i x_i^{2n} \end{vmatrix} = 0,$$

so that

$$\sum_{i=1}^{n} w_i x_i^{2n} = - \frac{\begin{vmatrix} 1 & m_1 & \cdots & m_n \\ m_1 & m_2 & \cdots & m_{n+1} \\ \vdots & \vdots & & \vdots \\ m_n & m_{n+1} & \cdots & 0 \end{vmatrix}}{\begin{vmatrix} 1 & m_1 & \cdots & m_{n-1} \\ m_1 & m_2 & \cdots & m_n \\ \vdots & \vdots & & \vdots \\ m_{n-1} & m_n & \cdots & m_{2n-2} \end{vmatrix}}$$

Perks [4] developed this method, with the restriction $w_1 = w_2 = \cdots = w_n = n^{-1}$, as a rationalization of an earlier empirical method (the *n-ages method*) of approximate evaluation of sums of products used to calculate certain actuarial functions. The constraint of equality on the *w*'s was removed by Jones [3], and the properties of the approximation were fully worked out by Beard [1]. Bivariate extensions to double summations or integrations are discussed in Perks [5].

Explicit solutions to (3) for $n = 2, 3, 4$ are given in Beard [1].

References

[1] Beard, R. E. (1947). *J. Inst. Actu.*, **73**, 356–403.

[2] Elderton, W. P. and Rowell, A. H. (1925). *J. Inst. Actu.*, **56**, 263–288.

[3] Jones, H. G. (1933). *J. Inst. Actu.*, **64**, 318–324.

[4] Perks, W. F. (1933). *J. Inst. Actu.*, **64**, 264–292 and 325–328.

[5] Perks, W. F. (1945). *J. Inst. Actu.*, **72**, 377–397.

(ACTUARIAL STATISTICS NUMERICAL INTEGRATION)

NUCLEAR MATERIAL SAFEGUARDS

The nuclear industry places great importance on the accountability of special nuclear materials (SNM), which occur in nuclear power work. The term SNM is defined to include plutonium-239, uranium-233, uranium enriched in the isotopes 235 or 233, or any substance containing the above. Such materials are not only expensive, but because of their role in weapons production, national security considerations are involved. The field of safeguards entails protecting SNM and monitoring its use to assess suspicions concerning possible loss of material.

Roughly speaking, safeguards can be thought of as the combination of two components: the first depends on physical security to monitor and control access to SNM, whereas the second involves accounting procedures that keep track of quantities and locations of SNM. An analogous situation exists in the banking industry, which also uses physical security (guards, cameras, locked vaults, etc.) together with accounting (or auditing) procedures as a matter of routine when doing business. The presence of measurement errors* is unique to problems involving bulk materials.

HISTORICAL BACKGROUND

The control of nuclear materials has been a major concern since the advent of the atomic era. The first major attempt to legislate controls was offered by Bernard Baruch, the U.S. delegate to the United Nations Atomic Energy Commission. The Baruch plan proposed an international nuclear industry operated under the auspices of an international governing body, but it was deemed unacceptable by the Soviet Union.

For several years much nuclear technology in the United States remained classified under the Atomic Energy Act of 1946. In 1953, President Eisenhower announced the Atoms for Peace program. Process details were provided to those nations that assured the information would be used strictly for peaceful purposes. The Atoms for Peace program was part of the Atomic Energy Act of 1954, which also authorized private ownership of nuclear facilities and the possession of SNM.

The establishment of the International Atomic Energy Agency (IAEA) in 1957 represented a landmark in international safeguards cooperation. The United States and the United Kingdom subsequently offered to allow their nuclear facilities to serve as safeguards demonstrations and to provide technical support for the nuclear activities of less developed countries. Partially as a result of this offer, the Nonproliferation Treaty of 1968 was ratified by more than 100 nations, who agreed to place their nuclear facilities under international safeguards. The IAEA, by international agreement, at present remains quite involved in verification efforts to monitor use of nuclear materials around the world. A lengthy account of the evolution of international safeguards is given by Willrich [7].

Recently there has been an increasing emphasis on domestic safeguards. In contrast to the international scene, where a primary concern is diversion of SNM by a nation through manipulation of its nuclear facilities, domestic safeguards are generally oriented toward detection and prevention of losses of SNM that might result from actions of individuals despite the best intentions of the facility operators involved. In the United States, the Nuclear Regulatory Commission and Department of Energy are the agencies most associated with these matters.

The prominence of safeguards issues has generated growing interest. Beyond the attention given the subject by the popular press (see Bibliography), many papers on statistical methods and their applications to safeguards problems have appeared in a number of journals, most notably the *Journal of the Institute of Nuclear Materials Management*. A literature survey concerning the major statistical efforts directed at important safeguards problems is included in the expository article of Goldman et al. [4].

STATISTICAL METHODS

As mentioned previously, "safeguards" is the use of physical security in conjunction with materials accounting methods. There are rel-

atively few statistical applications related to physical security, and these involve primarily discrimination problems that arise when mechanical devices (e.g., voice recognition equipment) are employed to control access to specified locations.

Statistical methods are at the core of proper interpretation of materials accounting information. Measurement of bulk quantities of SNM is nontrivial, and estimated values are often the combination of many individual observations. A good understanding of the instrumentation used is basic to evaluating the resulting data. Sher and Untermyer [6] provide an excellent review of this subject.

Estimated quantities of SNM are used in a variety of ways, one of which arises when material is received at a facility for processing. The amount of SNM is estimated upon arrival and compared to an invoice value provided by the shipper. A difference between these values is inevitable owing to measurement errors, if nothing else. The shipper–receiver difference must be reconciled and agreement reached regarding the amount involved. The same issue arises again when the material is processed and sent to its destination. Though seemingly straightforward, the treatment of shipper–receiver differences has had a colorful history [4].

A second major problem concerns monitoring SNM within distinct areas of a facility. Measurements are made at several locations to achieve localized accountability. At each area within the facility, the "balance" —or difference between beginning and ending inventories plus the difference between material transferred into and out of the area —should be zero. Measurement errors lead to nonzero observed values and induce correlations between estimated balances. Perlman [5] discusses some of the related multivariate testing issues.

A third subject presents itself in dealing with the temporal nature of accounting data. When convenient, materials balances can be closed frequently to provide timely information. To outline the statistical issues involved, consider a single balance established daily. Let the observed balance for the tth day be

$$\mathrm{MB}_t = I_t - I_{t-1} + T_t,$$

where I_t denotes the inventory at the end of day t and T_t is the difference between the measured amounts of SNM transferred into and out of the balance area during the day. Typically, MB_t is assumed normally distributed about its mean, which is zero if all material is properly handled. Because of the correlation between daily loss estimators, such as the individual MB_t or cumulative $\{\sum^t \mathrm{MB}_i\}$, analysis of the balance sequence can be difficult. Aspects of estimation have been addressed using standard time series* methodology (e.g., Downing et. al. [2, 3]) but the more important problems of testing have not yet been thoroughly examined.

A final topic of widespread concern is data verification. It should not be casually assumed that all measured values have been obtained in good faith, since a loss of material could be masked by falsified data. To counter this possibility, an inspector independently remeasures selected quantities of SNM. His or her measurements are then used to aid in detection of loss and/or falsification. Avenhaus [1] discusses many of the related questions.

Safeguards work is not confined to the subjects just described. Many interesting statistical problems in modeling*, calibration*, and variance estimation also arise, though such general matters are hardly unique to safeguards and need not be explicitly detailed here. The reference list [2–6] may be consulted for further information.

References

[1] Avenhaus, R. (1977). *Material Accountability: Theory, Verification, Applications*. Wiley, New York. (The leading reference on the verification problem.)

[2] Downing, D. J., Pike, D. H., and Morrison, G. W. (1978). *Nucl. Mater. Manag.*, **7**, 80–86.

[3] Downing, D. J., Pike, D. H., and Morrison, G. W. (1980). *Technometrics*, **22**, 17–22.

[4] Goldman, A. S., Picard, R. R., and Shipley, J. P. (1982). *Technometrics*, **24**, 267–275. [To date, the only comprehensive survey of statistical applications in the safeguards literature. The paper is followed (pp. 276–294) by comments from several discussants.]

[5] Perlman, M. D. (1969). *Ann. Math. Statist.*, **40**, 549–567. (A mathematically sophisticated discussion of one-sided testing problems in multivariate analysis.)

[6] Sher, R. and Untermyer, S. (1980). *The Detection of Fissionable Material by Nondestructive Means*. American Nuclear Society, LaGrange Park, IL.

[7] Willrich, M. (1973). *International Safeguards and Nuclear Industry*. Johns Hopkins University Press, Baltimore. (An interesting historical account of international safeguards.)

Bibliography

The following are recent nontechnical discussions of safeguards issues in the popular press that have received much attention.

Anderson, J. (1982). *Washington Post*, 28 March 1982, p. C7.

Emshwiller, J. R. and Brand, D. (1982). *Wall Street Journal*, 4 February 1982, p. 1.

Marshall, E. (1981). *Science*, **211**, 147–150.

Miller, J. (1981). *The New York Times*, 16 November 1981.

(CALIBRATION
CHEMISTRY, STATISTICS IN
DAM THEORY
EDITING STATISTICAL DATA
MEASUREMENT ERRORS
PRINCIPLES OF PROFESSIONAL
    STATISTICAL PRACTICE
QUALITY CONTROL, STATISTICAL
TIME SERIES)

R. R. PICARD
A. S. GOLDMAN
J. P. SHIPLEY

NUCLEOLUS OF GAME   *See* GAME THEORY

NUISANCE PARAMETERS

A nuisance parameter could be defined as a parameter* that is included in the probability model for the experiment at hand because it is necessary for the good fit of the model, but that is not of primary interest to the investigator. The appellation *nuisance* is generally appropriate for such a parameter because the statistical procedures available if its value were known are generally simpler and more powerful than those available when it is not. A simple explanation for this is that the statistical procedure in the unknown case must perform well across a larger class of probability models. For example, if in a normally distributed sample the mean is the parameter of interest, the result of not knowing the nuisance parameter σ^2 corresponds to the loss of power in going from the Z-statistic to the t statistic.

In the simplest setting for the problem, the probability model for the experiment has a parameter pair (θ, ϕ), where θ and ϕ are real- or real-vector valued, ϕ here denoting the nuisance parameter. However, richer problems can be included in the scheme by labeling all the unknown aspects of the distribution other than the parameter of interest as being the nuisance parameter. For example, in Cox's regression model*, where the hazard function of the survival time T is modeled as $\lambda_0(t)\exp(\boldsymbol{\beta}'\mathbf{x})$, the regression parameters $\boldsymbol{\beta}$ are generally of interest while the underlying hazard $\lambda_0(t)$ is an unknown nuisance function. A second important example is the location problem considered in the robustness* literature, which can be formulated as the problem of estimating θ in the distribution $F(x - \theta)$, where F is an unknown nuisance distribution, symmetric about zero.

EXACT METHODS

Sometimes a reasonable optimality criterion can be devised that results in a uniquely best procedure to use in the presence of the nuisance parameter. The classic text by Lehmann [15] extensively examines such methods in the theory of hypothesis testing*; with numerous applications in exponential family* models. Of particular import are the use of conditional tests and of invariant tests, as these methods show up repeatedly as ways of narrowing attention to conditional or marginal probability models where the nuisance parameter is absent or diluted in effect. Other strategies are available of course, as evidenced by the minimax* approach of Huber [12, Chap. 4] to the robust location problem.

ASYMPTOTIC METHODS

In the simplest version of asymptotic likelihood theory vector nuisance parameters are rather easily dealt with. A general textual treatment of such methods is found in Cox and Hinkley [19]. Generally speaking, given a random sample from the parametric model (θ, ϕ), one can estimate θ by $\hat{\theta}$, the first component of the maximum likelihood estimator* $(\hat{\theta}, \hat{\phi})$. Asymptotically, it will have a normal distribution with variance that can be found in the upper left corner of the inverse of the complete Fisher's information matrix* for (θ, ϕ). For testing, several methods are available, including the $C(\alpha)$ test* of Neyman [17] and the likelihood ratio test*, both of which have asymptotic distributions free of the nuisance parameter.

INFINITELY MANY NUISANCE PARAMETERS

It was Neyman and Scott [18] who dramatized the dangers of using the previously mentioned asymptotic theory in models with many nuisance parameters. We construct the following model. Let $X_1, X_2, \ldots, X_n, \ldots$ be a sequence of independent random variables, where X_i has parametric model (θ, ϕ_i), with ϕ_i allowed to depend on index i. For example, in the standard balanced one-way analysis of variance model, each element of the vector X_i of m independent observations would have mean ϕ_i and variance θ. In such a model, the standard asymptotic results valid when $m \to \infty$ may fail dismally when $n \to \infty$, a situation in which the number of nuisance parameters becomes infinite.

Consider the above example. In this case, the maximum likelihood estimator* of θ is

$$\hat{\theta} = \sum \left(X_{ij} - \bar{X}_{i.} \right)^2 / (mn),$$

which is consistent as $m \to \infty$ with n fixed, but is inconsistent as $n \to \infty$ with m fixed, as it then converges to $[m/(m-1)]\theta$. We can consider this example as illustrating the potentially dangerous bias in $\hat{\theta}$ when m is large relative to n. In a similar fashion, it can be shown that the likelihood ratio test of $H_0 : \theta = \theta_0$ against $H_a : \theta \neq \theta_0$ does not have a $\chi^2(1)$ distribution as $n \to \infty$ with m fixed, but rather diverges to ∞.

It is perhaps more surprising that the problem can be deeper than just bias*. A second Neyman and Scott example demonstrates that the maximum likelihood estimator of θ can be inefficient even when consistent*. That is, there are asymptotically normal estimators of θ with smaller asymptotic variance. The moral here is very similar to one associated with Stein shrinkage* estimators. In this latter case, a reduction in total mean square error* is obtained by treating the various parameters to be estimated as an aggregate, linked by being generated as observations from an unknown distribution. In the nuisance parameter case, there are also sometimes benefits involved in the estimation of θ by treating the collection of nuisance parameters as a sequence of observations from an unknown distribution.

A seminal role in this approach to the nuisance parameter problem is due to Kiefer and Wolfowitz [14]. Their paper treats the sequence of nuisance parameters ϕ_1, $\phi_2, \ldots, \phi_n, \ldots$ as independent and identically distributed observations from a completely unspecified distribution G, yielding a marginal distribution for X which is mixed on the parameter ϕ. Maximum likelihood estimation over the parameters (θ, G) gives an estimator for θ that is much more generally consistent than the maximum likelihood estimator in the nonmixed model of Ney-

man and Scott. A similar approach with a partially Bayesian spirit was suggested by Cox [8], who treated the distribution G as known to the investigator.

In some models there are alternative likelihood methods that reduce the impact of the nuisance parameters. They are linked in conception to the above-mentioned hypothesis testing methods. That is, instead of making θ inferences with the full likelihood, one uses a conditional or marginal likelihood whose dependence on the nuisance parameter is eliminated or reduced. The conditional likelihood is generated by conditioning on the sufficient statistics for the nuisance parameter. An invariance argument is commonly used to arrive at the marginal likelihood. In the one-way analysis of variance example, one can arrive at the marginal χ^2 distribution of $S^2/\sigma^2 = \sum_i \sum_j (X_{ij} - \bar{X}_{i.})^2/\sigma^2$ as the source of inference either by conditioning on the means $(\bar{X}_1, \ldots, \bar{X}_n.)$ or by citing the invariance of S^2 under location changes in each sample. This marginal distribution yields inferences correct as m or $n \to \infty$. Kalbfleisch and Sprott [13] have a general discussion of such methods; Andersen [1] deals extensively with the conditional approach in the infinite nuisance parameter model. Cox [7] has generalized this approach to dealing with nuisance parameters by considering partial likelihoods*, which are the products of conditional likelihoods. This results in particular in a likelihood for the Cox regression model* that is free of the nuisance hazard function* $\lambda_0(t)$.

INFORMATION

Even when there is but a single real-valued nuisance parameter ϕ in the model (θ, ϕ), the question of how to measure the information available concerning θ is difficult and controversial. An example with a long history concerns the log odds ratio* θ of a two-by-two table*. Here the particular question is whether the conditional distribution of the data given the marginal totals is completely informative as to the true value of θ. Fisher

[11] ignited the controversy in 1925 by suggesting that "if it be admitted that these marginal frequencies by themselves supply no information on the point at issue," then the correct inferential procedure is to condition on them, thereby treating them as ancillary statistics*. A whole series of authors with different philosophical and inferential points of view have considered this problem. Some key references are Barnard [2], Barndorff-Nielsen [3], Basu [4], Cox [6], Plackett [19], and Sprott [20].

One approach to the problem is to measure the information about θ through a generalization of Fisher's information suggested by Stein [21]. Efron [10] considers information in the Cox regression model within this framework. Lindsay [16] uses this approach in the (θ, G) model of Kiefer and Wolfowitz, with emphasis on when the conditional and partial likelihoods* are fully efficient. Bickel [5] provides a general discussion of when it is possible to estimate θ in any infinite-dimensional nuisance parameter model just as efficiently (asymptotically) when the nuisance parameter is unknown as when it is known.

References

[1] Andersen, E. B. (1973). *Conditional Inference and Models for Measuring*. Mentalhygiejnisk Forlag, Copenhagen.

[2] Barnard, G. A. (1963). *J. R. Statist. Soc. B*, **25**, 111–114.

[3] Barndorff-Nielsen, O. (1978). *Information and Exponential Families in Statistical Theory*. Wiley, New York. (Chapter 4 extensively discusses the information problem from a technical point of view.)

[4] Basu, D. (1977). *J. Amer. Statist. Ass.*, **72**, 355–367.

[5] Bickel, P. (1982). *Ann. Statist.*, **10**, 647–671.

[6] Cox, D. R. (1958). *Ann. Math. Statist.*, **29**, 357–372.

[7] Cox, D. R. (1975). *Biometrika* **62**, 269–276.

[8] Cox, D. R. (1975). *Biometrika* **62**, 651–654.

[9] Cox, D. R. and Hinkley, D. V. (1974). *Theoretical Statistics*. Chapman and Hall, London.

[10] Efron, B. (1977). *J. Amer. Statist. Ass.*, **72**, 557–565.

[11] Fisher, R. A. (1925). *Proc. Camb. Philos. Soc.*, **22**, 700–715.

[12] Huber, P. (1981). *Robust Statistics*. Wiley, New York.

[13] Kalbfleisch, J. D. and Sprott, D. A. (1970). *J. R. Statist. Soc. B*, **32**, 175–208.

[14] Kiefer, J. and Wolfowitz, J. (1956). *Ann. Math. Statist.*, **27**, 887–906.

[15] Lehmann, E. L. (1959). *Testing Statistical Hypotheses*. Wiley, New York.

[16] Lindsay, B. G. (1980). *Philos. Trans. R. Soc. Lond.*, **296A**, 639–665.

[17] Neyman, J. (1959). In *The Harold Cramér Volume*. Wiley, New York, pp. 213–234.

[18] Neyman, J. and Scott, E. L. (1948). *Econometrica*, **16**, 1–32.

[19] Plackett, R. L. (1977). *Biometrika*, **64**, 37–42.

[20] Sprott, D. A. (1975). *Biometrika*, **62**, 599–605.

[21] Stein, C. (1956). *Proc. 3d Berkeley Symp. Math. Statist. Prob.*, **1**, 187–195.

(ANCILLARY STATISTICS
EXPONENTIAL FAMILIES
FISHER INFORMATION
HYPOTHESIS TESTING
INFERENCE, STATISTICAL—I, II
INFORMATION THEORY
  AND CODING THEORY
INVARIANCE CONCEPTS IN STATISTICS
KULLBACK INFORMATION
LIKELIHOOD
MAXIMUM LIKELIHOOD
PARTIAL LIKELIHOOD
SUFFICIENCY)

BRUCE G. LINDSAY

NULL HYPOTHESIS

In hypothesis testing*, the hypothesis to be tested is frequently called the *null hypothesis*. The term *null* was coined by Fisher [1, Sect. 8]. While introducing the concept, he illustrated it with the well-known tea-tasting problem, in which a lady claims to be able to tell whether tea or milk has been added first to the cup from which she is drinking. The hypothesis to be tested is "that the judgments given are in no way influenced by the order in which the ingredients have been added."

In the analysis of data from experimental designs, the hypothesis being tested is usually one in which the term null is appropriate; a treatment has *no effect* or there are *no differences* between the effects of k treatments. It may also be appropriate in tests of data gathered from sample surveys*, in which the hypothesis being tested is that the percentage of voters in favor of certain legislation is *no different* from what is claimed by an advocate who belongs to one of the parties involved.

In other testing problems, however, the term null is not so clearly appropriate, and some writers omit it because they perceive it to be confusing; see Kendall and Stuart [2, Sect. 22.6, footnote], for example. Lindgren [3, Sect. 6.1] presents an example in which the use of the term is inappropriate and in which either one of two hypotheses could be labeled "the hypothesis being tested," the other being the alternative hypothesis*. In this example, archaeologists dig up some skulls that are believed to come from one of two tribes, A or B; the archaeologists are therefore interested in which of two corresponding hypothesis H_A or H_B is true.

The labels null hypothesis and alternative hypothesis may be assigned because a null effect is inherent in one of them, but the need to constrain the probability of making an erroneous decision is sometimes overriding. Since type I error probabilities are bounded by the size of the test in the classical Neyman–Pearson* approach, the null hypothesis is labeled accordingly. For example, in clinical trials*, it may be more serious to erroneously adopt a new drug B (believing it to perform better than a drug A which has a known 60% success rate in the past) than to erroneously rule out the use of B in favor of continuing to treat patients with A. If p_A and p_B are the success rates of the two drugs, the labels would then be:

Null hypothesis	$p_A \geqslant p_B$,
Alternative hypothesis	$p_A < p_B$,

The null effect when $p_A = p_B$ is then coincidental.

References

[1]　Fisher, R. A. (1951). *The Design of Experiments*, 6th ed. Oliver & Boyd, Edinburgh, Scotland.

[2]　Kendall, M. G. and Stuart, A. (1973). *The Advanced Theory of Statistics*, Vol. 2, 3d ed. Hafner, New York.

[3]　Lindgren, B. W. (1976). *Statistical Theory*, 3d ed. Macmillan, New York.

(ALTERNATIVE HYPOTHESIS
HYPOTHESIS TESTING
LEVEL OF SIGNIFICANCE
NEYMAN–PEARSON LEMMA
POWER)

NUMBER OF RUNS TEST　*See* RANDOMNESS TESTS

NUMERACY

WHAT IS NUMERACY?

Numeracy can be broadly defined as the ability to interpret mathematical and statistical evidence. It refers to the ability of an individual to understand numbers in terms of observation, measurement, evaluation, and verification.

The term *numeracy* was coined by Crowther in 1959 [19] to represent the understanding of the scientific approach in terms of numerical evidence. The word (which has not yet found its way into most standard dictionaries) is better explained than defined: What literacy is to words, numeracy is to numbers.

CAN NUMERACY BE LEARNED?

It is now generally agreed that being numerate is not the same as being mathematical [8, 10]. Ehrenberg [8] defines the objectives of numeracy as the ability to *understand* and *communicate* numerical information. Given this orientation, numeracy can be developed by following a few pragmatic "rules." Such rules include:

1. **Round all variable numbers to two significant digits.** For example, the statement that "Canada's population is 24 million" is easy to understand, easy to compare (i.e., it is about 10% of the U.S. population), and easy to remember. The statement that "Canada's population is 24,195,300," on the other hand, does not aid the understanding of numerical relationships.

2. **Use an index to facilitate comparisons.** It is easier to detect an exceptional observation in a numerical table when we compare each observation with an index, like the average. Comparing every number with every other number in the table, on the other hand, can be very confusing.

3. **Write down the figures to be compared in columns rather than in rows.** Unlike rows, columns are not interrupted by blank spaces. This makes the patterns and exceptions easier to detect when numbers are written columnwise than row-wise.

4. **Order the numbers by size** to facilitate easier comparisons.

5. **Use single spacing between rows.** Double spacing introduces blank spaces between numbers to be compared, thereby diluting the effect of writing numbers to be compared columnwise.

Theoretical and empirical support for "rules" **1–5** have come from different sources [3, 18]. There is considerable evidence to indicate [1, 9, 15, 20] that numeracy can be learned. The stumbling block to learning this skill can be related to the commonly held misconception that being numerate is the same as being mathematical.

NUMERACY AND APPLIED STATISTICS

The need for numeracy appears to be particularly important in understanding statistical evidence. While mathematics relies heavily on deductive reasoning, statistics involves the use of both deductive reasoning and probabilistic concepts. Applied statistics, in particular, calls for an interpretation of the calculated values. Inadequate levels of numeracy can and do lead to misleading interpretations of computed values [2, 11].

For instance, consider the results of a survey conducted in Toronto and Montreal in which consumers rated a new brand of detergent on a seven-point scale. The results are given in Table 1.

A mechanical application of statistics might have shown that the ratings given by Montreal consumers are significantly higher (compared to Toronto consumers) on some attributes, e.g., Convenient and Makes Clothes Bright and significantly lower on some others, e.g. Cleans Well and Removes Stains. A numerate, on the other hand, may realize quickly that these figures are perhaps not directly comparable: Consumers in Toronto tend toward the mean (4.0 ± 0.3) *no matter what the attribute is*; assuming that this is the case, any comparison between Montreal and Toronto consumers on specific attribute ratings is potentially misleading.

In another market survey, the correlations* in Table 2 were obtained among eleven ratings.

Such correlation matrices are often used in further analysis (e.g., factor analysis*) since it is not easy to see the underlying patterns in a correlation matrix produced by the computer. To see the patterns in this

Table 1   Ratings of a New Detergent

	Toronto	Montreal
Convenient	3.7	6.2
Economical	4.0	4.2
Safe	4.3	4.3
Cleans well	4.2	3.5
Softens clothes	3.8	3.6
Makes clothes bright	3.7	5.4
Removes stains	4.3	3.2
Average	4.0	4.3
Sample size	500	500

Table 2   Correlations among 11 Product Attributes

1.00000	− 0.10291	− 0.11784	0.30566	− 0.03699	− 0.02250	0.00763	0.04294	− 0.04760	0.22470	− 0.22749
− 0.10291	1.00000	0.27385	− 0.08736	0.27237	− 0.00497	− 0.02682	− 0.02401	− 0.13962	− 0.09211	0.13487
− 0.11784	0.27385	1.00000	− 0.10004	0.24169	0.13994	− 0.20327	− 0.12472	− 0.19723	− 0.03566	0.16144
0.30566	− 0.08736	− 0.10004	1.00000	− 0.13027	− 0.12490	0.20419	0.00755	0.01257	− 0.09832	− 0.13855
− 0.03699	0.27237	0.24169	− 0.13027	1.00000	0.09422	0.00266	− 0.00086	− 0.23673	0.09939	− 0.18951
− 0.02250	− 0.00497	0.13994	− 0.12490	0.09422	1.00000	− 0.02222	0.15800	0.06185	0.10322	0.09527
0.00763	− 0.02682	− 0.20327	0.20419	0.00266	− 0.02222	1.00000	0.16705	0.11872	0.02953	− 0.03452
0.04294	− 0.02401	− 0.12472	0.00755	− 0.00086	0.15800	0.16705	1.00000	0.20020	0.24826	0.01620
− 0.04760	− 0.13962	− 0.19723	0.01257	− 0.23673	0.06185	0.11872	0.20020	1.00000	0.04885	− 0.04185
0.22470	− 0.09211	− 0.03566	− 0.09832	0.09939	0.10322	0.02953	0.24826	0.04885	1.00000	− 0.11911
− 0.22749	0.13487	0.16144	− 0.13855	0.18951	0.90527	− 0.03452	0.01620	− 0.04185	− 0.11911	1.00000

Table 3   Correlations among Product Attributes (× 10, to nearest unit)

		AB	LM	LS	VR	LR	MU	HL	MD	SH	SF	MI
AB	Adds body to hair	—	− 1	− 1	3	0	0	0	0	0	2	− 2
LM	Leaves hair manageable	− 1	—	3	− 1	3	0	0	0	− 1	− 1	1
LS	Leaves hair soft	− 1	3	—	− 1	2	1	− 2	− 1	− 2	0	2
VR	Gives hair styling versatility	3	− 1	− 1	—	− 1	− 1	2	0	0	− 1	− 1
LR	Gives long-lasting results	0	3	2	− 1	—	1	0	0	− 2	1	2
MU	Is mild to use	0	0	1	− 1	1	—	0	2	1	1	1
HL	Leaves hair healthier looking	0	0	− 2	2	0	0	—	2	1	0	0
MD	Causes minimum hair damage	0	0	− 1	0	0	2	2	—	2	2	0
SH	Straightens hair well	0	− 1	− 2	0	− 2	1	1	2	—	0	0
SF	Safe for color-treated hair	2	− 1	0	− 1	1	1	0	2	0	—	− 1
MI	Causes minimum skin irritation	− 2	1	2	− 1	2	1	0	0	0	− 1	—

matrix, we may want to round and order the correlation matrix, as illustrated in Table 3. This revised matrix immediately shows that most correlations in fact range between 0 and 0.2 and no correlation is higher than 0.3. A numerate may thus conclude that no further analysis is warranted.

Statisticians who feel that statistical reasoning is at least as important as accurate computations have come up with several approaches that emphasize numeracy over mechanical calculations [2, 11, 16, 17, 22–24]. Although traditional courses still emphasize computational skills or the "learn now, apply later" [10] approach, there is a growing appreciation of the need for numeracy in interpreting statistical evidence.

NUMERACY AND GRAPHICAL PRESENTATION

There is a widespread belief that graphical presentations of numbers aid numeracy—it aids both the analyst and the end-user of numerical information.

To the analyst, simple graphic representation can be helpful in quickly spotting pat-

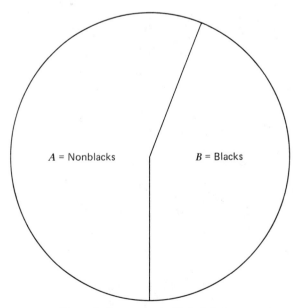

Figure 1   Poverty in the United States.

terns and exceptions in data. Tukey's *Exploratory Data Analysis** considers the use of graphical methods as an aid to numeracy. Tukey's methods, in fact, *facilitate numeracy by emphasizing the basic patterns in data.* The insights thus obtained can then be used in quantitative analysis.

The usefulness of graphical presentation to the end-user of numerical information is less clear. Although it seems to be universally assumed that graphic presentation is a good substitute for numeracy, i.e., even those who are "not good at numbers" can take in numerical information easily if it is presented graphically, recent findings [3, 7, 8] suggest that:

1. Graphical information is effective in conveying qualitative information, but *not quantitative* information. For example, in Fig. 1, it is obvious that *A* is bigger than *B* (qualitative information), but by how much? Five percent? Fifteen percent? This information can be effectively conveyed by numbers, not by graphs.

2. Plotting more than one (dependent) variable on a graph hinders rather than

helps understanding (although this may not be obvious to the reader). For example, consider the Defence and External Relations expenditures in Fig. 2. Compared to 1965, has the expenditure gone up or down in 1967? By how much? Has it decreased a lot as a proportion of the total expenditure? By how much? None of these questions can be answered reasonably well by simply looking at the graph.

It appears that *at best* graphical presentation can aid numeracy by highlighting important qualitative patterns in data; *at worst*, it can mislead the reader. (Graphic presentation can also be used to distort the data [11] but this weakness is shared by numeric presentation as well.) For a more complete discussion of the strengths and weaknesses of graphic presentation, please refer to other sources, e.g., refs. 3, 7, and 8.

NUMERACY AND BRAIN FUNCTIONS

"Rules" to aid the development of numeracy in an individual (such as the ones mentioned earlier) are concerned with the mechanics of numeracy and do not necessarily relate to the basis on which the ability develops. Although Piaget published his work on *The Child's Conception of Numbers* as early as 1941 [12] in French, it was not until the 1970s that interest began to develop in the processes by which we understand the information presented to us. This led to research on brain functions, particularly the specific contributions of the right and left hemispheres. (For a readable summary, see Ornstein [14] or Edwards [6].) Briefly, the left hemisphere of the brain is mainly responsible for verbal, analytic, symbolic, abstract, time-oriented, rational, classificatory, and logical abilities; the right hemisphere, on the other hand, takes the primary responsibility for nonverbal, integrative, analogic, non-time-oriented, spatial, intuitive, and holistic aspects [6]. Although we generally have access to both hemispheres, scientific method and deductive thinking, both of which

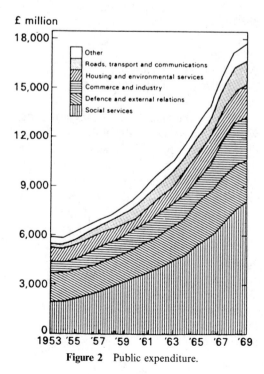

Figure 2   Public expenditure.

are encouraged in our culture, tend to draw upon the left hemisphere extensively. Such emphasis tends to inhibit the use of the right brain and most information is prone to be interpreted using the tools of the left hemisphere. Research based on these theories points to the fact that we may be using an inappropriate mode of evaluation to develop certain skills.

In this context, numeracy poses a special problem. *Digital* (classificatory) ability is a function of the left brain while *analogic* ability is a function of the right brain. The former relates to mathematical ability while the latter relates to numeracy. The traditional curriculum has emphasized the digital ability, treating the analogic ability as a byproduct. Such confusion between mathematical ability and numeracy may have discouraged even those who predominantly use their right brain from becoming numerate. Whether this is so or not is less relevant than the fact that mathematical ability and numeracy are two distinct skills requiring different approaches; success in teaching and learning numeracy is impeded by the prevailing confusion between the two skills.

PSEUDO-NUMERACY

A high degree of numeracy may lead a person to detect the less obvious relationships in numerical data. Many individuals have summarized such relationships in a way that would simplify and speed up calculations. Systems developed by Trachtenberg [21], mnemonic experts [13], and others fall into this category. While those who developed these systems might be numerate, the same cannot necessarily be said of the users. A user can quickly compute and remember figures without being numerate. Hence the term *pseudo-numeracy*. *Pseudo* is not used here pejoratively. Some of these systems are indeed highly useful and effective. Pseudo-numeracy simply refers to the condition in which one gives the appearance of being numerate when one is not really so.

TEACHING NUMERACY

Teaching mathematics through set-theoretical concepts was perhaps the first major attempt in which numeracy rather than the computational skill was emphasized. However, there appears to be a widespread disagreement as to the usefulness of this approach in developing the digital (computational) skills of the student. Obviously, numeracy cannot be substituted for computational skills. Currently many educators agree that numeracy, like literacy, should be a part of everyone's education; mathematical ability, on the other hand, is a specialized skill.

Considering the importance of numeracy, one might wonder why this has not been taught properly in an academic setting. A partial explanation may be obtained when one considers the traditional courses—they emphasize accuracy. Numeracy, on the other hand, is largely based on approximations. As John Craig [5] elegantly puts it: "Nor was he [Dr. Johnson] afraid of approximations—another sign of numeracy." The emphasis on accuracy may have contributed significantly to the neglect of numeracy.

PROBLEMS AND TRENDS IN TEACHING NUMERACY

Even when the aims of numeracy are understood by the educator, there are several practical problems in actually teaching it. First, there is a widespread "math anxiety" [20] even among well-educated individuals. Second, numeracy is a continuum with no agreed-upon stages of progression. Therefore, the subject matter to be taught depends more on the individual educator than on an agreed-on curriculum.

In recent years, there has been a growing interest in teaching numeracy, particularly in the United Kingdom [1, 8, 9, 10]. Given the current proliferation of computers that make complex number-crunching a simple task, numeracy may be expected to receive far more attention in the next few years.

References

[1] Buzan, T. (1977). *Make the Most of Your Mind.* Colt Books, London. (Contains a chapter on numeracy. A brief, nontechnical and nonrigorous view of the subject.)

[2] Campbell, S. K. (1974). *Flaws and Fallacies in Statistical Thinking.* Prentice-Hall, Englewood Cliffs, NJ. (A guide to properly interpreting statistical evidence. Nontechnical.)

[3] Chakrapani, C. and Ehrenberg, A. S. C. (1976). "Numerical Information Processing." Paper presented at the Poster Session, British Psychological Society, England. (Presents research evidence to support some of the "rules" designed to develop numeracy.)

[4] Chakrapani, C. (1982). "Data Analysis and Statistics." Paper presented at a symposium in Ottawa. (Statistics Canada and Professional Marketing Research Society.)

[5] Craig, J. (1979). *The Statistician,* **28**, 109–118. (A popular article on numeracy.)

[6] Edwards, B. (1979). *Drawing on the Right Side of the Brain.* Tarcher, Los Angeles. (Contains some lucid discussions on brain capabilities.)

[7] Ehrenberg, A. S. C. (1975). *Data Reduction.* Wiley, London and New York. (An introductory book on statistics with emphasis on numeracy.)

[8] Ehrenberg, A. S. C. (1977). *J. R. Statist. Soc. A,* **140**, 277–293. (A nontechnical explanation of numeracy and how it can be developed.)

[9] Ehrenberg, A. S. C. (1982). *A Primer in Data Reduction.* Wiley, New York and London. (An introductory text with emphasis on numeracy.)

[10] Glenn, J. A., ed. (1978). *The Third R—Towards a Numerate Society.* Harper & Row, London. (Discusses the objectives and methods of teaching numeracy from the educator's point of view.)

[11] Huff, D. (1954). *How to Lie With Statistics.* W. W. Norton, New York. (A popular introduction to the misuse of numerical data from a statistician's point of view.)

[12] Isaacs, N. (1972). *A Brief Introduction to Piaget.* Schocken Books, New York. (Contains a simple summary of Piaget's work on the development of number concepts in children.)

[13] Lorayne, H. (1980). *Develop Your Memory Power.* Coles Publishing Company, Toronto, Canada. (Contains materials that relate to carrying out mathematical manipulation using mnemonic techniques.)

[14] Ornstein, R. D. (1972). *The Psychology of Consciousness.* W. H. Freeman, San Francisco. (A popular introduction to the right- and left-brain activities. Written by a research psychologist.)

[15] Piaget, J. (1952). *The Child's Conception of Number.* Routledge and Kegan Paul, London. (An exploration into the child's mind and the development of numeracy, first published in French in 1941.)

[16] Roberts, H. V. (1974). *Conversational Statistics.* Hewlett-Packard Company, Cupertino, CA. (Computer-based interactive statistics with emphasis on interpretation. Introductory.)

[17] Runyon, R. P. (1977). *Winning with Statistics.* Addison-Wesley, Reading, MA. ("A painless look at numbers, ratios, percentages, means and inference.")

[18] Simon, H. A. (1969). *The Sciences of the Artificial.* MIT Press, Cambridge, MA. (Discusses short-term memory in relation to perception of numbers.)

[19] Stewart, W. A. C. (1977). In *The Harper Dictionary of Modern Thought,* A. Bullock and O. Stallybrass, eds. Harper & Row, New York. (A one-paragraph explanation of numeracy.)

[20] Tobias, S. (1978). *Overcoming Math Anxiety.* W. W. Norton, New York. (Techniques to overcome the fear of numbers and calculations.)

[21] Trachtenberg, J. (1973). *Speed System of Mathematics,* translated from German by A. Cutler and R. McShane. Pan Books, London. (Shortcuts to calculating quickly.)

[22] Tukey, J. W. (1977). *Exploratory Data Analysis.* Addison-Wesley, Reading, MA. (A novel approach to identifying the significance of statistical data. Introductory.)

[23] Wallis, W. A. and Roberts, H. V. (1956). *Statistics: A New Approach.* Free Press, Glencoe, IL. (An introduction to statistical reasoning with minimal emphasis on computations. Introductory.)

[24] Wallis, W. A. and Roberts, H. V. (1956). *The Nature of Statistics.* Free Press, Glencoe, IL. (Similar to [23]. Introductory.)

(CONSULTING, STATISTICAL
EDITING STATISTICAL DATA
EXPLORATORY DATA ANALYSIS
GRAPHICAL REPRESENTATION,
  COMPUTER-AIDED
GRAPHICAL REPRESENTATION
  OF DATA
STATISTICAL EDUCATION)

CHUCK CHAKRAPANI

NUMERICAL ANALYSIS

The name *numerical analysis* refers to the analysis and solution of mathematical problems by numerical means. It is probably as old as mathematics, since sophisticated numerical schemes are known to have been used by the ancient Egyptians, Babylonians, and Greeks. At one time, most mathematicians did some numerical analysis, just as they were also more involved in various areas of science. In the latter part of the 1800s, researchers began to specialize. Numerical analysis began to grow as a separate discipline, although most numerical methods were still developed in connection with a specific problem in applied mathematics or science.

The advent of computers in the 1940s caused major changes in applied mathematics and science, leading to a large increase in the use of numerical methods of solution. Computational methods no longer had to be laborious; the emphasis in applied mathematics changed from that of minimizing calculational effort to minimizing the need for time-consuming mathematical analyses carried out by humans. Because of this change in focus, numerical analysis today is almost entirely a product of the period since 1950. For an excellent history of numerical analysis up to 1900, see Goldstine [11].

COMMON PERSPECTIVES
IN NUMERICAL ANALYSIS

Numerical analysis is concerned with all aspects of the numerical solution of a problem, from the theoretical development and understanding of numerical methods to their practical implementation as reliable, efficient computer programs. Most numerical analysts specialize in smaller subareas, but they share some common threads of concerns, perspective, and mathematical methods of analysis. These include the following.

1. The solution of many problems is visualized in the following manner: since the given problem cannot be solved explicitly, it is approximated by a simpler problem that can be solved more easily. The hope is that the solution of the new problem will be close to that of the original problem. Examples include numerical integration* (replace the integrand by a nearby one whose integral can be easily evaluated, e.g., a polynomial) and finding the root of a function (replace the given function by a sequence of approximating functions whose roots are easily calculated, as in Newton's method* where tangent lines are used).

2. There is a wide use of the language and results of linear algebra, real analysis, and functional analysis (with its simplifying notation of norms, vector spaces, and operators); *see* Linz [17].

3. There is a fundamental concern with error, its size, and its analytic form. Error estimation arises from a knowledge of the form of the error, and it leads to extrapolation methods to accelerate the convergence of numerical methods; see Joyce [13].

4. Stability is a concept referring to the sensitivity of the solution of a problem to small changes in the data or the parameters of the problem. For example, the polynomial

$$p(x) = (x - 1)(x - 2)(x - 3)(x - 4)$$
$$\times (x - 5)(x - 6)(x - 7)$$
$$= x^7 - 28x^6 + 322x^5 - 1960x^4$$
$$+ 6769x^3 - 13{,}132x^2$$
$$+ 13{,}068x - 5040$$

has roots that are very sensitive to relatively small changes in the coefficients. If the coefficient of x^6 is changed to -28.002, then the original roots 5 and 6 are perturbed to $5.458676 \pm 0.5401258i$. Such a function $p(x)$ is called unstable or ill-conditioned with respect to the root-finding problem. In developing numerical methods for solving problems, they should be no more sensitive to changes in the data than the original problem to be solved. And generally, one tries to formulate the original problem as a stable or well-conditioned one.

5. Numerical analysts are very interested in the effects of using the finite precision arithmetic of digital computers. Many modern-day computers have poor error characteristics in their floating point arithmetic, and this needs to be taken into account when writing programs that are to be widely used.

6. One is almost always concerned with the cost of the numerical method in computing time and in the amount of needed memory storage. One also wants to write easily portable computer programs, ones that will run easily on computers with a variety of arithmetic characteristics. This promotes uniformity, and it avoids unnecessary duplication of programming time.

OUTLINE OF NUMERICAL ANALYSIS

The following gives a rough categorization of the areas within numerical analysis.

Approximation Theory

This covers the approximation of functions and methods based on those approximations.

(a) *Approximation of a function known analytically.* This includes the theory of best uniform approximations, needed in producing methods for evaluating functions on computers. For a discussion of obtaining such approximations in practice, see Hart et al. [12]; for a reference containing easy-to-use approximations for many of the standard functions, *see* Abramowitz and Stegun [1]. Other important topics include infinite series expansions using orthogonal polynomials* and trigonometric functions (leading to Fourier series), and approximation of functions by classes of functions other than polynomials.

(b) *Interpolation*. For discrete data, a function is to be found whose graph contains the given data points. Popular interpolation functions are polynomials, combinations of exponentials, trigonometric polynomials, and spline functions. Interpolation is used for several purposes. It extends tables of function values, as was done commonly in the past with tables of logarithms; also it is used to produce simple approximations to more complicated functions, for use in other problems such as integration or solving differential equations. Most introductory numerical analysis texts cover such interpolation adequately, e.g., Atkinson [2] and Stoer and Bulirsch [23]. Interpolation is also used to extend empirical data to a continuous function, usually with the intention of preserving the general geometric behavior of the data when graphed. The most popular methods for doing this are based on spline functions*; see de Boor [5].

(c) *Numerical integration and differentiation.* Most numerical methods for evaluating integrals and derivatives of a function f can be based on integrating or differentiating an approximation to f, often one based on polynomial interpolation. For numerical integration, this includes the trapezoidal and Simpson rules*, Gaussian quadrature, and other popular rules. Other rules are obtained by using extrapolation, based on a knowledge of the form of the error in a known rule. For one variable integration, see Davis and Rabinowitz [4]; and for multiple integration, see Stroud [25].

When doing differentiation of empirical data, there is an increase in the uncertainty in the derivatives as compared to that in the original data. One reasonable approach is to construct a least-squares* fit to the data using spline functions and then to differentiate this fit.

(d) *Least-squares* data fitting.* This is familiar to statisticians as part of regression analysis*; but often in statistics, the numerical problems of obtaining the least-squares fitting function are neglected. Use of a polynomial fitting function, written in a standard form as a sum of monomials, leads to a linear system of normal equations* which is very unstable to solve. To avoid this, other approaches are needed, usually based on numerical linear algebra methods or on writing the desired polynomial as an unknown combination of known polynomials that are nearly orthogonal over the domain of the independent variable. A reference from the numerical analysis perspective, including computer codes, is Lawson and Hanson [16].

Numerical Linear and Nonlinear Algebra

This refers to problems involving the solution of systems of linear and nonlinear equations, possibly of a large order. Most of the topics contained herein are ultimately reduced to the solution of nonsingular systems of linear equations, a subject that has been brought to a very high level in the two decades of 1960–1980, both in practice and in theory.

(a) *Linear systems of equations.* Denote such a system by $\mathbf{A}\mathbf{x} = \mathbf{b}$, with $\mathbf{A}$ a matrix of order n. When n is small to moderate in size, say $n \leqslant 100$, and when most of the elements of $\mathbf{A}$ are nonzero, the favorite method of solution is Gaussian elimination. This is now well understood, and very reliable and efficient computer codes are available [6]. One of the major problems that must be considered is the possible ill-conditioning of the matrix $\mathbf{A}$, along with efficient means of predicting the error. The least-squares fitting of data with a polynomial leads to some of the classic ill-conditioned linear systems. For a complete discussion of all aspects of the subject, see Rice [20].

For much larger linear systems, one must take advantage of the special structure of $\mathbf{A}$. Most such matrices are sparse*, which means that most of the elements of $\mathbf{A}$ are zero. For the systems arising from solving partial differential equations, iterative methods are often used; see Young [27]. For other techniques of solving a wider variety of sparse systems, see Duff [7].

(b) *Eigenvalue problems.* The standard matrix eigenvalue problem, $\mathbf{A}\mathbf{x} = \lambda\mathbf{x}$, has led to a large body of methods, most of them iterative. The major reference is Wilkinson [26], and a compendium of more recent results for the case of symmetric matrices is Parlett [18]. For excellent computer codes, see Smith et al. [22].

(c) *Systems of nonlinear equations.* One of the more popular methods is Newton's method. Let $\mathbf{f}(\mathbf{x}) = 0$ denote the system of nonlinear equations, with $\mathbf{x}$ and $\mathbf{f}(\mathbf{x})$ column vectors of order n. The Newton's method is

$$\mathbf{x}^{(m+1)} = \mathbf{x}^{(m)} - \left[\mathbf{f}'(\mathbf{x}^{(m)}) \right]^{-1} \mathbf{f}(\mathbf{x}^{(m)}),$$
$$m \geqslant 0,$$

where $\mathbf{f}'(\mathbf{x})$ denotes the order n Jacobian matrix of $\mathbf{f}(\mathbf{x})$. With a good initial guess $\mathbf{x}^{(0)}$, usually this is a rapidly convergent method. For a general survey of this and other methods, see Rheinboldt [19].

(d) *Optimization*.* Unconstrained optimization refers to finding the maximum or minimum of a function $f(\mathbf{x})$ of one or more variables. Constrained optimization adds constraints on the variables with the constraints given as inequalities or equalities. The simplest example is the linear programming* problem. For an extensive survey, see [8].

Differential and Integral Equations

These problems involve solving for a function f, where f is involved in differentiated or integrated form in an equation. Such equations include the models used in most fields of the physical sciences and engineering. The numerical methods involve discretizing the equations to obtain a finite linear or nonlinear system of equations. The discretization techniques used are from (a)–(c) in the first subsection. The systems are solved by methods related to those discussed in Numerical Linear and Nonlinear Algebra. A major consideration in the numerical analysis of these equations is whether the numerical method is stable or not, and usually the proof of convergence hinges on the question of numerical stability; *see* Linz [17]. These equations include:

Ordinary differential equations. For solving the initial value problem, see Gear [9] and Shampine and Gordon [21]; and for boundary value problems, see Keller [14].

Partial differential equations. For general surveys, see Gladwell and Wait [10] and Lapidus and Pinder [15]; for the finite element method, see Strang and Fix [24]. ***Integral equations*.*** See Baker [3].

MATHEMATICAL SOFTWARE

This is the name given to numerical analysis computer programs written for use by a general audience. Since about 1970, there has been a growing interest in producing programs that use the best possible methods, are efficient and reliable, are as convenient and flexible as possible, and that run on a wide variety of computers.

The first major project to produce such software was carried out at Argonne National Laboratory in Chicago, and it resulted in the popular matrix eigenvalue package EISPACK [22]. Such packages are also being prepared at several other centers, many associated with national research laboratories in the United States and other countries. In addition, there are two widely distributed general numerical analysis libraries, called the IMSL and NAG libraries; and information on them can be obtained through your computer center or through the references in Rice [20]. For new developments in mathematical software, see the algorithms section of the journal *ACM Transactions on Mathematical Software.* Also, see the *SIGNUM Newsletter* for discussions of various software projects. SIGNUM is a major organization for numerical analysts, and it is a subsidiary group of the Association for Computing Machinery (ACM).

References

[1] Abramowitz, M. and Stegun, I., eds. (1964). *Handbook of Mathematical Functions.* Dover, New York.

[2] Atkinson, K. (1978). *An Introduction to Numerical Analysis,* Wiley, New York.

[3] Baker, C. (1977). *The Numerical Treatment of Integral Equations,* Oxford University Press, New York.

[4] Davis, P. and Rabinowitz, P. (1975). *Methods of Numerical Integration.* Academic Press, New York.

[5] de Boor, C. (1978). *A Practical Guide to Splines,* Springer-Verlag, Berlin.

[6] Dongarra, J., Bunch, J., Moler, C., and Stewart, G. (1979). *LINPACK User's Guide.* SIAM, Philadelphia.

[7] Duff, I. (1977). *Proc. IEEE* **65**, 500–535.

[8] Fletcher, R. (1981). In *Practical Methods of Optimization,* Vol. I: *Unconstrained Optimization* and Vol II: *Constrained Optimization.* Wiley, New York.

[9] Gear, C. W. (1971). *Numerical Initial Value Problems in Ordinary Differential Equations.* Prentice-Hall, Englewood Cliffs, NJ.

[10] Gladwell, I. and Wait, R. eds. (1979). *A Survey of Numerical Methods for Partial Differential Equations.* Oxford University Press, New York.

[11] Goldstine, H. (1977). *A History of Numerical Analysis.* Springer-Verlag, Berlin.

[12] Hart, J., Cheney, E., Lawson, C., Maehly, H., Mesztenyi, C., Rice, J., Thacher, H., and Witzgall, C. (1968). *Computer Approximations.* Wiley, New York.

[13] Joyce, D. (1971). *SIAM Rev.,* **13**, 435–490.

[14] Keller, H. (1976). *Numerical Solution of Two-Point Boundary Value Problems.* SIAM, Philadelphia.

[15] Lapidus, L. and Pinder, G. (1982). *Numerical Solution of Partial Differential Equations in Science and Engineering.* Wiley, New York.

[16] Lawson, C. and Hanson, R. (1974). *Solving Least Squares Problems.* Prentice-Hall, Englewood Cliffs, NJ.

[17] Linz, P. (1979). *Theoretical Numerical Analysis: An Introduction to Advanced Techniques.* Wiley-Interscience, New York.

[18] Parlett, B. (1980). *The Symmetric Eigenvalue Problem.* Prentice-Hall, Englewood Cliffs, NJ.

[19] Rheinboldt, W. (1974). *Methods for Solving Systems of Nonlinear Equations.* SIAM, Philadelphia.

[20] Rice, J. (1981). *Matrix Computations and Mathematical Software.* McGraw-Hill, New York.

[21] Shampine, and Gordon, M. (1975). *Computer Solution to Ordinary Differential Equations: The Initial Value Problem.* W. H. Freeman, San Francisco.

[22] Smith, B., Boyle, J., Garbow, B., Ikebe, Y., Klema, V., and Moler, C. (1976). *Matrix Systems Routines–EISPACK Guide.* Springer-Verlag, Berlin.

[23] Stoer, J. and Bulirsch, R. (1980). *Introduction to Numerical Analysis.* Springer-Verlag, Berlin.

[24] Strang, G. and Fix, G. (1973). *An Analysis of the Finite Element Method*. Prentice-Hall, Englewood Cliffs, NJ.

[25] Stroud, A. (1971). *Approximate Calculation of Multiple Integrals*. Prentice-Hall, Englewood Cliffs, NJ.

[26] Wilkinson, J. (1965). *The Algebraic Eigenvalue Problem*. Oxford University Press, New York.

[27] Young, D. (1971). *Iterative Solution for Large Linear Systems*. Academic Press, New York.

(APPROXIMATIONS TO DISTRIBUTIONS
COMPUTERS AND STATISTICS
INTERPOLATION
MATHEMATICAL FUNCTIONS,
    APPROXIMATIONS TO
NEWTON–RAPHSON METHODS
NUMERICAL INTEGRATION)

K. Atkinson

NUMERICAL INTEGRATION

The Fundamental Theorem of Calculus states that

$$\int_a^b g(x)\,dx = G(b) - G(a),$$

where $G(x)$ is any function for which $G'(x) = g(x)$. This is the preferred method of evaluating a definite integral if $G(x)$ can be found. However, for many functions $g(x)$, $G(x)$ cannot be found in terms of "elementary" functions. For example, this is true for $g(x) = \exp(x^2)$. This means that numerical methods are needed for many problems. Numerical integration is one aspect of the topic of numerical analysis*.

A *numerical integration formula*, also called a *quadrature formula*, is the summation in an approximation of the form

$$\int_a^b w(x)f(x)\,dx \simeq \sum_{k=1}^{M} A_{M,k} f(x_{M,k}). \quad (1)$$

Here the $x_{M,k}, A_{M,k}, k = 1, \ldots, M$, are certain given constants; the $x_{m,k}$ are called the *points* or *nodes* in the formula; the $A_{M,k}$ are called the *coefficients* or *weights*; $w(x)$ is a given function called the *weight function*. In most cases the $x_{M,k}, A_{M,k}$ depend on M, on

$[a,b]$, and on $w(x)$, but not on $f(x)$. Various choices for the $x_{M,k}, A_{M,k}$ will be mentioned later. Usually one desires an approximation to a preassigned accuracy with M as small as possible.

We restrict this discussion to one-dimensional (univariate) integrals. Approximations analogous to (1) for higher dimensions (multiple integrals) come under the topic of n-dimensional quadrature*.

We say that approximation (1) has *algebraic degree* δ if it is an equality for all algebraic polynomials

$$P_m(x) = a_0 + a_1 x + a_2 x^2 + \cdots + a_m x^m$$

of degree $m \leqslant \delta$, and if (1) is not an equality for all such polynomials of degree $\delta + 1$. Also, (1) has *trigonometric degree* δ if it is an equality for all trigonometric polynomials of period $b - a$

$$\begin{aligned} T_m(x) = a_0 &+ a_1 \cos \lambda x + b_1 \sin \lambda x \\ &+ a_2 \cos 2\lambda x + b_2 \sin 2\lambda x \\ &+ \cdots + a_m \cos m\lambda x + b_m \sin m\lambda x, \\ &\lambda = 2\pi/(b - a), \end{aligned}$$

of degree $m \leqslant \delta$, and if (1) is not an equality for all such polynomials of degree $\delta + 1$.

Three well-known quadrature formulas for $w(x) = 1$ are:

1. *Trapezoidal formula*

$$\begin{aligned} \int_a^b f(x)\,dx \simeq \frac{h}{2}\, f(a) &+ hf(a + h) \\ &+ hf(a + 2h) \\ &+ \cdots + hf(a + (M - 2)h) \\ &+ \frac{h}{2}\, f(b), \quad (2) \\ h = (b - a)&/(M - 1); \end{aligned}$$

2. *Midpoint formula*

$$\int_a^b f(x)\,dx$$

$$\begin{aligned} \simeq h\bigg[f\bigg(a + \frac{h}{2}\bigg) &+ f\bigg(a + \frac{3h}{2}\bigg) + \cdots \\ &+ f\bigg(a + \bigg(\frac{2M - 1}{2}\bigg)h\bigg)\bigg], \quad (3) \\ h = (b - a)&/M; \end{aligned}$$

3. *Simpson's formula*

$$\int_a^b f(x)\,dx \simeq \frac{h}{3} f(a) + \frac{4h}{3} f(a+h)$$

$$+ \frac{2h}{3} f(a+2h) + \frac{4h}{3} f(a+3h)$$

$$+ \frac{2h}{3} f(a+4h) + \cdots$$

$$+ \frac{4h}{3} f(a+(M-2)h) + \frac{h}{3} f(b),$$

$$\tag{4}$$

$$h = (b-a)/(M-1),$$

M an odd integer $\geqslant 3$.

The algebraic degree of formulas (2), (3), and (4) are $\delta = 1$, 1, and 3, respectively. Their trigonometric degrees are $\delta = M - 2$, $M - 1$, and 0, respectively.

Standard references to numerical integration are the books by Davis and Rabinowitz [1], Engels [2], Krylov [5], and Stroud and Secrest [7].

GAUSS QUADRATURE

Assume that $w(x) > 0$ on $[a, b]$ and that M is any positive integer. Then there is a unique quadrature formula (1) of algebraic degree $2M - 1$. This is called the M-point (algebraic) *Gauss formula* for $w(x)$ and $[a, b]$. Gauss formulas have many important properties; some of these are:

The $x_{M,k}$ are the zeros of the Mth degree orthogonal polynomial for $w(x)$ and $[a, b]$. The $x_{M,k}$ are all distinct and all inside $[a, b]$.

All of the $A_{M,k}$ are positive. This is important for the following reason. If $f(x)$ is of one sign on $[a, b]$, then the approximation is assured of having the same sign as the integral.

The algebraic degree $2M - 1$ is the highest that can be obtained with M points.

The well-known Gauss formulas are named after the classical orthogonal polynomials. The names of these are listed in Table 1. A standard reference to orthogonal polynomials is Szegö [8].

The Gauss–Chebyshev formulas are known in closed form. For the interval $[-1, 1]$ these are as follows. For the first kind,

$$x_{M,k} = \cos \frac{(2k-1)\pi}{2M}, \qquad A_{M,k} = \frac{\pi}{M},$$

$$k = 1, 2, \ldots, M;$$

For the second kind,

$$x_{M,k} = \cos \frac{k\pi}{M+1},$$

$$A_{M,k} = \left(1 - x_{M,k}^2\right) \frac{\pi}{M+1},$$

$$k = 1, 2, \ldots, M.$$

Computer programs for generating the $x_{M,k}$, $A_{M,k}$ for general Gauss–Jacobi, generalized

Table 1   Names of the Well-Known Gauss Quadrature Formulas

Interval $[a, b]$	Weight Function $w(x)$	Name of the Quadrature Formula
$[-1, 1]$	constant = 1	Gauss–Legendre
$[-1, 1]$	$(1 - x^2)^{-1/2}$	Gauss–Chebyshev of the first kind
$[-1, 1]$	$(1 - x^2)^{1/2}$	Gauss–Chebyshev of the second kind
$[-1, 1]$	$(1 - x)^\alpha (1 + x)^\beta$ $\alpha > -1, \beta > -1$	general Gauss–Jacobi
$[0, \infty)$	e^{-x}	Gauss–Laguerre
$[0, \infty)$	$x^\alpha e^{-x}$ $\alpha > -1$	generalized Gauss–Laguerre
$(-\infty, \infty)$	e^{-x^2}	Gauss–Hermite

Gauss–Laguerre, and Gauss–Hermite formulas are given in [7, Chap. 2]. An efficient program for generating Gauss-Legendre formulas is given in [1 (p. 364)]. An excellent survey of Gauss formulas is given by Gautschi [3]. In particular, he lists all $w(x)$, $[a, b]$ for which numerical tables of the $x_{M,k}$, $A_{M,k}$ are available.

Trigonometric Gauss formulas also exist. An M-point formula (1) will be a trigonometric Gauss formula if it is exact for the $2M - 1$ trigonometric polynomials

$$1, \cos \lambda x, \sin \lambda x, \ldots, \cos(M - 1)\lambda x,$$

$$\sin(M - 1)\lambda x, \qquad \lambda = 2\pi/(b - a).$$

Therefore an M-point trigonometric Gauss formula has trigonometric degree $M - 1$. It is known that, with M points, one cannot obtain a formula exact for all $2M + 1$ linearly independent trigonometric polynomials of period $b - a$ of degree $\leqslant M$. It follows from this that the trapezoidal formula (2) and the midpoint formula (3) are trigonometric Gauss formulas for $w(x) = 1$. Note that if $f(x)$ has period $b - a$ then the two function values $f(a)$ and $f(b)$ in (2) should only be counted as one.

The idea of Gauss quadrature has been generalized to sets of functions other than algebraic and trigonometric polynomials. This theory is discussed by Karlin and Studden [4].

CONVERGENCE AND ERROR ESTIMATES

For short, let $Q_M[f]$ denote the summation in an approximation (1). Let us consider an infinite sequence of such sums

$$Q_{M_1}[f], Q_{M_2}[f], Q_{M_3}[f], \ldots \qquad (5)$$

with $M_1 < M_2 < M_3 < \cdots$. In a practical calculation, one should use a sequence (5) only if

$$\lim_{M \to \infty} Q_M[f] = \int_a^b w(x)f(x)\,dx. \qquad (6)$$

Since the integral in (6) can be defined as a limit of Riemann–Stieltjes sums, equation (6) will be true provided that each of the $Q_{M_i}[f]$ is a Riemann–Stieltjes sum [or a Rie-

mann sum, for the case $w(x) = 1$]. The most important formulas, including the Gauss formulas, do have this property. For Gauss formulas, a proof for the case $w(x) = 1$ is given in Stroud [6]; a proof in general is given in Szegö [8, p. 50].

A related question is that of bounding

$$\left| \int_a^b w(x)f(x)\,dx - Q_M[f] \right|.$$

Theoretical ways to do this are discussed in refs. 1, 2, 6, and 7. These bounds are mainly useful for comparing the relative merits of competing formulas; usually they are difficult or impossible to use in practical calculations. Often what is done in practice is to compute a sequence of approximations (5); if two of these, say $Q_{M_i}[f]$ and $Q_{M_{i+1}}[f]$, agree to k decimal places, then $Q_{M_{i+1}}[f]$ is assumed to be accurate to this number of places. In many cases this will give a reasonable estimate; however, in other cases it will not. One should use such an estimate with caution.

TRANSFORMATIONS

Various transformations can be used in conjunction with quadrature formulas. Linear transformations will be considered first. Assume that we start with a given formula

$$\int_a^b w(x)f(x)\,dx \simeq \sum_{k=1}^M A_{M,k} f(x_{M,k}). \qquad (7)$$

The linear transformation

$$x = \gamma u + \beta, \qquad u = (x - \beta)/\gamma,$$

$$\gamma = (b - a)/(d - c),$$

$$\beta = (ad - bc)/(d - c)$$

transforms the interval $a \leqslant x \leqslant b$ onto $c \leqslant u \leqslant d$ and transforms approximation (7) into

$$\int_c^d w^*(u)g(u)\,du \simeq \sum_{k=1}^M A_{M,k}^* g(u_{M,k}), \qquad (8)$$

where

$$u_{M,k} = (x_{M,k} - \beta)/\gamma, \qquad A_{M,k}^* = A_{M,k}/\gamma,$$

$$g(u) = f(\gamma u + \beta), \qquad w^*(u) = w(\gamma u + \beta).$$

If (7) has polynomial degree δ, then so does (8).

As an example, suppose we start with a Gauss–Hermite formula

$$\int_{-\infty}^{\infty} e^{-x^2} f(x)\, dx \simeq \sum_{k=1}^{M} A_{M,k}\, f(x_{M,k}). \quad (9)$$

The linear transformation

$$x = (u - \theta)/\sigma\sqrt{2}, \qquad u = x\sigma\sqrt{2} + \theta,$$
$$-\infty < \theta < \infty, \quad 0 < \sigma$$

transforms (9) into

$$\int_{-\infty}^{\infty} p(u\,|\,\theta,\sigma) g(u)\, du$$

$$\simeq \frac{1}{\sqrt{\pi}} \sum_{k=1}^{M} A_{M,k} g(u_{M,k}),$$

$$u_{M,k} = x_{M,k}\sigma\sqrt{2} + \theta, \quad k = 1, \ldots, M,$$

where the weight function

$$p(u\,|\,\theta,\sigma) = (\sigma\sqrt{2\pi})^{-1} \exp\left[-\tfrac{1}{2}(u - \theta)^2\sigma^{-2} \right]$$

is the normal probability density function.

Next we give two examples of nonlinear transformations. Assume we start with a generalized Gauss–Laguerre formula of polynomial degree δ

$$\int_{0}^{\infty} x^{\alpha} e^{-x} f(x)\, dx \simeq \sum_{k=1}^{M} A_{M,k}\, f(x_{M,k})$$

and apply the nonlinear transformation

$$x = 1/(\gamma y^2), \qquad y = (\gamma x)^{-1/2}, \qquad 0 < \gamma.$$

We obtain the following approximation for weight function the inverted gamma* probability density function

$$\frac{2}{\Gamma(a)} \int_{0}^{\infty} \frac{1}{\gamma^a y^{2a+1}} e^{-1/(\gamma y^2)} g(y)\, dy$$

$$\simeq \frac{2}{\Gamma(a)} \sum_{k=1}^{M} A_{M,k} g(y_{M,k}),$$

$$g(y) = f\big(1/(\gamma y^2)\big), \quad y_{M,k} = (\gamma x_{M,k})^{-1/2},$$
$$a = \alpha + 1 > 0.$$

This latter approximation will be exact whenever g(y) is a polynomial of degree ≤ δ in $1/(\gamma y^2)$.

For the final example, consider the integral

$$\int_{-\infty}^{\infty} p(u\,|\,\theta,h,\nu) g(u)\, du, \quad (10)$$

where $p(u\,|\,\theta,h,\nu)$ is the Student t^* PDF

$$\frac{1}{B(\tfrac{1}{2},\tfrac{1}{2}\nu)} \frac{(h\nu^{-1})^{1/2}}{\left\{1 + h\nu^{-1}(u - \theta)^2\right\}^{-(\nu+1)/2}}.$$

$$-\infty < \theta < \infty, \quad 0 < h, \quad 0 < \nu.$$

We can write (10) as the sum of two integrals

$$\int_{-\infty}^{\theta} p(u\,|\,\theta,h,\nu) g_-(u)\, du$$

$$+ \int_{\theta}^{\infty} p(u\,|\,\theta,h,\nu) g_+(u)\, du, \quad (11)$$

where $g_-(u) = g(u)$ for $u \leqslant \theta$, and $g_+(u) = g(u)$ for $\theta \leqslant u$. In (11), we make the transformation

$$x = \left[1 + h\nu^{-1}(u - \theta)^2\right]^{-1},$$

$$u = \theta \pm \left[\nu(1 - x)/(hx)\right]^{1/2},$$

where, in (12), the upper sign is used with g_+ and the lower sign with g_-. The sum (11) becomes

$$\left[2B\left(\frac{1}{2},\frac{\nu}{2}\right)\right]^{-1} \int_{0}^{1} x^{\nu/2-1}(1 - x)^{-1/2}$$

$$\times \left[g_-\left[\theta - \sqrt{\frac{\nu(1 - x)}{hx}}\right] \right.$$

$$\left. + g_+\left[\theta + \sqrt{\frac{\nu(1 - x)}{hx}}\right] \right] dx.$$

Since $x^{\nu/2-1}(1 - x)^{-1/2}$ is a Jacobi weight function for $0 \leqslant x \leqslant 1$, the latter integral can be approximated by a linearly transformed Gauss–Jacobi formula.

ADAPTIVE QUADRATURE

We assume the following:

$$w(x) = 1 \quad \text{and} \quad 0 < b - a < \infty,$$

To approximate

$$\int_{a}^{b} f(x)\, dx, \quad (13)$$

we use a finite sequence of formulas denoted by

$$Q_{M_i}[f; a, b], \qquad i = 1, 2, \ldots, s, \quad s \geqslant 2.$$

For $i = 1, 2, \ldots, s$, $Q_{M_i}[f; \alpha, \beta]$ denotes $Q_{M_i}[f; a, b]$ linearly transformed to $[\alpha, \beta]$. We have a criterion for deciding if each $Q_{M_i}[f; a, b]$, $i = 1, 2, \ldots$, is a sufficiently accurate approximation to (13).

The final approximation to be obtained for (13) will be denoted by Q.

We proceed as follows:

Step 1. Set $[\alpha, \beta] = [a, b]$; set $Q = 0$.

Step 2. Compute $Q_{M_i}[f; \alpha, \beta]$, $i = 1, 2, \ldots$. Here $[\alpha, \beta]$ is some subset of $[a, b]$. If for some i, $Q_{M_i}[f; \alpha, \beta]$ is a sufficiently accurate approximation to

$$\int_\alpha^\beta f(x)\, dx,$$

then replace Q by $Q + Q_{M_i}[f; \alpha, \beta]$ and record the fact that $[\alpha, \beta]$ does not have to be subdivided further; otherwise, go to Step 3.

Step 3. Subdivide $[\alpha, \beta]$ into $[\alpha, \frac{1}{2}(\alpha + \beta)]$ and $[\frac{1}{2}(\alpha + \beta), \beta]$. Go to Step 2, first with $[\alpha, \beta]$ replaced by $[\alpha, \frac{1}{2}(\alpha + \beta)]$ and then with $[\alpha, \beta]$ replaced by $[\frac{1}{2}(\alpha + \beta), \beta]$. In this way we recursively subdivide $[a, b]$ into smaller subintervals. This process is continued until we reach a partition of $[a, b]$, say,

$$[a, b] = [a_0, a_1] \cup [a_1, a_2]$$
$$\cup \cdots \cup [a_{m-1}, a_m],$$

where $a_0 = a$, $a_m = b$ and where each subinterval $[a_{j-1}, a_j]$ is such that either $a_j - a_{j-1}$ is less than some preassigned tolerance or for some i, the approximation $Q_{M_i}[f; a_{j-1}, a_j]$ is a sufficiently accurate approximation to

$$\int_{a_{j-1}}^{a_j} f(x)\, dx.$$

Such an algorithm is called an *adaptive quadrature algorithm*. In contrast with a quadra-

ture formula (1) in which the $x_{M,k}$ are independent of the integrand $f(x)$, in adaptive quadrature the points at which $f(x)$ is evaluated depend on $f(x)$.

What are the relative merits of adaptive quadrature compared with ordinary (nonadaptive) quadrature? One may expect adaptive quadrature to be well-suited for "poorly behaved" integrands; examples of poor behavior would be: $f(x)$ discontinuous; $f'(x)$ discontinuous; $f(x)$ with greatly varying behavior on different parts of $[a, b]$. For a well-behaved integrand, one may expect that the overhead involved in an adaptive algorithm will make it more expensive than a well-chosen nonadaptive formula.

Remarks. This article has not been meant as a discussion of specific integrals in statistics: it is intended only to convey some basic ideas about quadrature methods.

It should be noted that some integrals can be evaluated by other techniques. For example, the integral

$$\frac{1}{\sigma \sqrt{2\pi}} \int_x^\infty \exp\left[-\frac{1}{2\sigma^2}(u - \theta)^2 \right] du$$

has been studied as a function of x. In particular, polynomial and rational approximations are available for it. (See, for example, M. Abramowitz and I. A. Stegun, eds. (1964). *Handbook of Mathematical Functions*. National Bureau of Standards, Appl. Math. Series 55, Washington, DC. Chap. 26.) One of these approximations may be preferred over a quadrature formula. Such topics are beyond the scope of this article.

Computer programs for numerical integration are widely available and can probably be found at one's local computer center. Bibliographies of programs are given in ref. 1 and 2.

References

[1] Davis, P. J., and Rabinowitz, P. (1975). *Methods of Numerical Integration*, Academic Press, New York. (The best source for all aspects of numerical quadrature; includes a bibliography of computer programs and tables and a few program listings.)

[2] Engels, H. (1980). *Numerical Quadrature and Cubature*, Academic Press, New York. (This covers the

more recent developments in more depth than Davis and Rabinowitz and also includes a bibliography of programs and tables.)

[3] Gautschi, W. (1981). In *E. B. Christoffel, The Influence of His Work on Mathematics and the Physical Sciences, International Christoffel Symposium, A Collection of Articles in Honour of Christoffel on the 150th Anniversary of His Birth*, P. L. Butzer and F. Fehér, eds. Birkhäuser, Boston. (The best discussion of Gauss quadrature; it is particularly useful for its list of all published tables.)

[4] Karlin, S. and Studden, W. J. (1966). *Tchebycheff Systems: With Applications in Analysis and Statistics*, Interscience, New York. (This develops the theory of systems which are related to generalized Gauss quadrature formulas and to the theory of inequalities in statistical applications.)

[5] Krylov, V. I. (1962). *Approximate Calculation of Integrals*, Macmillan, New York. (The best discussion of the classical theory of quadrature.)

[6] Stroud, A. H. (1974). *Numerical Quadrature and Solution of Ordinary Differential Equations*, Springer-Verlag, New York, Sect. 3.13. (A textbook about one-third devoted to quadrature theory.)

[7] Stroud, A. H. and Secrest, D. (1966). *Gaussian Quadrature Formulas*, Prentice-Hall, Englewood Cliffs, NJ. (Tables of formulas with a discussion of their basic properties.)

[8] Szegö, G. (1959), *Orthogonal Polynomials*, rev. ed. American Mathematical Society, New York. (The standard reference to the classical theory of orthogonal polynomials.)

(FINITE DIFFERENCES, CALCULUS OF
INTERPOLATION
n-DIMENSIONAL QUADRATURE
NEWTON–COTES FORMULA
NUMERICAL ANALYSIS
SHOVELTON'S FORMULA
SIMPSON'S RULE
THREE-EIGHTHS FORMULA

TRAPEZOIDAL RULE
WEDDLE'S RULE)

A. H. STROUD

NUMEROSITY, SCALES OF

These were proposed by Tukey and Tukey [1], to describe the number of points in a data set.

Table 1, taken from Tukey and Tukey [1] describes these scales. The authors recommend reducing the number of points in a display to B, C, or possibly D. Of course, these are to be regarded as only very rough guides, not mandatory rules.

References

[1] Tukey, J. W. and Tukey, P. A. (1981). In *Interpreting Multivariate Data*, V. Barnett, ed. Wiley, New York, pp. 189–213.

(GRAPHICAL REPRESENTATION
OF DATA)

NYQUIST FREQUENCY

The Nyquist frequency is half the sampling frequency when a continuous time function is sampled at equally spaced time points. That is, the Nyquist frequency is π/Δ (in radians per unit time), where Δ is the time interval between two successive sampled

Table 1   Scales of Numerosity

Number of Data Points	Letter	Data Set	Data	Points
1–3				
4–8	A	miniscule	skinny	few
8–25	B	small		small number of
25–80	C	modest	moderate amount	modest number of
80–250	D	medium-sized	of	intermediate number of
250–800	E	substantial	copious	many
800–2500	F			
2500–8000	G	burdensome	extensive	crowds of
> 8000				

data. This article discusses the basic ideas of the Nyquist frequency and the relevant and very important sampling theorem.

In many applications involving processing a continuous-time signal, it is often preferable to convert the continuous-time signal to a discrete-time signal since discrete-time signal processing can be implemented with a digital computer. It is important to examine whether the discrete-time signal preserves all the information in the original continuous-time signal. We first consider the case that the signal $x(t)$ is a real-valued function. Assume that its Fourier transform* $X(\omega) = \int_{-\infty}^{\infty} x(t)\exp(-i\omega t)\,dt$ exists and that $x(t) = (1/2\pi)\int_{-\infty}^{\infty} X(\omega)\exp(i\omega t)\,d\omega$ for all t. The signal $x(t)$ is sampled at $t = n\Delta$, $n = \ldots, -1, 0, 1, \ldots$. We are interested in interpolating $x(t)$ from its samples $x(n\Delta)$. A natural question arises: Under what conditions can $x(t)$ be perfectly reconstructed from $x(n\Delta)$? The samples $x(n\Delta)$ can be related to $X(\omega)$ as follows:

$$x(n\Delta)$$
$$= \frac{1}{2\pi} \int_{-\infty}^{\infty} X(\omega)\exp(i\omega n\Delta)\,d\omega$$
$$= \frac{1}{2\pi} \int_{-\pi/\Delta}^{\pi/\Delta} \left[\sum_{k=-\infty}^{\infty} X(\omega + 2\pi k/\Delta) \right]$$
$$\times \exp(i\omega n\Delta)\,d\omega.$$

Therefore, $X_d(\omega) = \sum X(\omega + 2\pi k/\Delta)$ ($|\omega| < \pi/\Delta$) is the discrete Fourier transform of the sequence $x(n\Delta)$. Obviously, $X_d(\cdot)$ is obtained by folding $X(\cdot)$ every π/Δ radians per unit time. (Here we identify ω with $-\omega$.) This frequency π/Δ is the Nyquist frequency, also called the *folding frequency*. It is easy to see that $X(\cdot)$ is not uniquely determined by $X_d(\cdot)$. In other words, some sinusoidal components of different frequencies (e.g., $2\pi k/\Delta \pm \omega_0$, $k = \ldots, -1, 0, 1, \ldots$) in $x(t)$ cannot be distinguished from one another by the observations $x(n\Delta)$. This is called *aliasing*. Aliasing is the effect of undersampling. This effect is the principle on which the stroboscopic effect is based [6, Sect. 8.3].

When $x(t)$ is a band-limited signal with $X(\omega) = 0$ for $|\omega| > \omega_M$, $X_d(\omega)$ is identical to $X(\omega)$ (i.e., no aliasing) if $\omega_M < \pi/\Delta$. In other words, from the uniqueness property of Fourier transform, $x(t)$ is uniquely determined by its samples $x(n\Delta)$ under the condition that $X(\omega) = 0$ for $|\omega| > \pi/\Delta$. This is usually called the (Shannon) sampling theorem on information theory* [5]. From the sampling theorem, if we sample the signal $x(t)$ at a rate at least twice the highest frequency in $x(t)$, then $x(t)$ can be completely recovered from the samples. This sampling rate [twice the highest frequency in $x(t)$] is commonly referred to as the *Nyquist rate*. Actually $x(t)$ can be explicitly written, in terms of $x(n\Delta)$, as

$$x(t) = \sum_{n=-\infty}^{\infty} x(n\Delta) \frac{\sin \pi\{(t/\Delta) - n\}}{\pi\{(t/\Delta) - n\}}.$$

It should be noted that band-limited signals are generally not realizable physically, for $x(t) = (1/2\pi)\int_{-\Omega}^{\Omega} X(\omega)\exp(i\omega t)\,d\omega$ is analytic in t, as a complex variable, and therefore cannot vanish for all $t < -T$ for arbitrarily large T. Therefore, aliasing is inevitable in practice. A discussion on error bounds for aliasing can be found in Jerri [5].

In some applications, the signal $x(t)$ is assumed to be bandpass, i.e., there exist $0 \leq \omega_0 < \omega_1$ such that $X(\omega) = 0$ outside the intervals $[\omega_0, \omega_1]$ and $[-\omega_1, -\omega_0]$. The sampling theorem says that $x(t)$ can be recovered from equally spaced sampling at a rate of $2\omega_1$. Actually, this rate $2\omega_1$ is too conservative. It has been shown [4, Sect. 8.5] that a sampling rate of $2\omega_1/\nu$ is enough to recover $x(t)$ where ν is the largest integer not beyond $\omega_1/(\omega_1 - \omega_0)$.

The sampling theorem has been generalized to many situations such as random signals. When $x(t)$ ($-\infty < t < \infty$) is a wide-sense stationary stochastic process*, possessing a spectral density that vanishes outside the interval $[-\pi/\Delta, \pi/\Delta]$, Balakrishnan showed [1] that $x(t)$ has the representation

$$x(t) = \lim_{N \to \infty} \sum_{n=-N}^{N} x(n\Delta) \frac{\sin \pi\{(t/\Delta) - n\}}{\pi\{(t/\Delta) - n\}},$$

for every t, where lim stands for limit in the

mean square. Gardner [3] derived a similar result for nonstationary stochastic processes. Obviously, the Nyquist frequency and Nyquist rate can be defined similarly in the random signal case.

Blackman and Tukey [2, Sect. 12] provide good interpretations on aliasing. Jerri [5] gives an excellent review of the sampling theorem and its various extensions and applications. He discusses topics such as unequally spaced sampling, higher-dimensional functions, non-band-limited functions and error bounds for the truncation, aliasing, and jitter. A very exhaustive bibliography can be found therein.

References

[1] Balakrishnan, A. V. (1957). *IRE Trans. Inf. Theory*, **IT-3**, 143–146.

[2] Blackman, R. B. and Tukey, J. W. (1958). *The Measurement of Power Spectra*. Dover, New York.

[3] Garnder, W. A. (1972). *IEEE Trans. Inf. Theory*, **IT-18**, 808–809.

[4] Gregg, W. D. (1977). *Analog and Digital Communication*. Wiley, New York.

[5] Jerri, A. J. (1977). *Proc. IEEE*, **65**, 1565–1596.

[6] Oppenheim, A. V., Willsky, A. S., and Young, I. T. (1983). *Signals and Systems*. Prentice-Hall, Englewood Cliffs, NJ.

(COMMUNICATION THEORY,
    STATISTICAL
INFORMATION THEORY
    AND CODING THEORY
INTEGRAL TRANSFORMS
SAMPLING
SPECTRAL ANALYSIS
TIME SERIES)

YI-CHING YAO

O

OAKES'S TEST OF CONCORDANCE

For a random sample $\{(T_i^{(1)}, T_i^{(2)})\ i = 1, 2, \ldots n\}$ from an absolutely continuous joint distribution, Kendall's [5] coefficient of concordance* counts the proportion of concordant pairs, that is, pairs $\{(T_i^{(1)}, T_i^{(2)}), (T_j^{(1)}, T_j^{(2)}), i < j\}$ for which

$$\left(T_i^{(1)} - T_j^{(1)}\right)\left(T_i^{(2)} - T_j^{(2)}\right) > 0 \quad (1)$$

minus the proportion of discordant pairs, that is, pairs for which the reverse inequality holds. This coefficient gives a simple nonparametric test of independence.

The sample is said to be subject to censoring if there exist potential censoring times $(C_i^{(1)}, C_i^{(2)}; i = 1, 2, \ldots, n)$ such that only $X_i^{(l)} = \min(T_i^{(l)}, C_i^{(l)})$ and the indicator variable $\Delta_i^{(l)} = 1_i^{(l)}\{T_i^{(l)} \leq C_i^{(l)}\}$ are observed. The $\{C_i^{(l)}\}$ may be fixed or random, but if random must be independent of the $\{T_i^{(l)}\}$. The values of $C_i^{(l)}$ for which $T_i^{(l)} < C_i^{(l)}$ will not in general be known.

The simplest generalization of the concordance test statistic counts only the number N_c of definite concordances, that is, pairs for which the inequality (1) is known to hold, minus the number N_d of definite discor-

dances. Specifically, let

$$Y_i^{(l)} = \begin{cases} T_i^{(l)} & (\Delta_i^{(l)} = 1) \\ \infty & (\Delta_i^{(l)} = 0) \end{cases}$$

so that $Y_i^{(l)}$ may be thought of as the largest value of $T_i^{(l)}$ consistent with the observed data. Then on the basis of what is observed, $T_i^{(l)}$ is known to be less than or equal to $T_j^{(l)}$ if and only if $Y_i^{(l)} < X_j^{(l)}$ or equivalently $T_i^{(l)} \leq T_j^{(l)}$, $C_i^{(l)}$, and $C_j^{(l)}$. It can be shown [6] that, under the null hypothesis that $T^{(1)}$ and $T^{(2)}$ are independent,

$$\tau = \frac{N_c - N_d}{\binom{n}{2}}$$

has expectation zero. If the censoring times $(C_i^{(1)}, C_i^{(2)})$ have a common bivariate distribution then τ has variance

$$\frac{2\alpha}{n(n-1)} + \frac{4(n-2)\gamma}{n(n-1)},$$

where

$$\alpha = 4\Pr\left[Y_1^{(1)} \leq X_2^{(1)}; Y_1^{(2)} \leq X_2^{(2)}\right]$$

and

$$\gamma = \Pr\left[Y_1^{(1)} \leq X_2^{(1)}, X_3^{(1)}; Y_1^{(2)} \leq X_2^{(2)}, X_3^{(2)}\right]$$

are both easily estimated from the data. Asymptotic normality* follows from the results of Hoeffding [4].

For the case of censoring in one component only, this statistic was proposed by Brown et al. [1]. Concordance tests when both components of the bivariate sample are subject to censoring appear first to have been considered by Weier and Basu [8]. They considered a more complex statistic than that discussed here. Under the null hypothesis of independence, the Kaplan–Meier* estimated survivor functions of the two components $T^{(1)}, T^{(2)}$ can each be calculated in the usual way and these functions used to recover partial information from the indefinite pairs. The reference distribution for this statistic is obtained by permuting one component of the data, and the variance under this permutation distribution obtained from results of Daniels [3].

The censoring times $C_i^{(2)}$ must be subject to the same permutation as the $T_i^{(2)}$ under this approach, for otherwise the value of the test statistic could not be calculated. This is legitimate only when the two censoring mechanisms are independent, which is not required by Oakes's statistic.

In the absence of censoring, the sample coefficient of concordance is a consistent estimate of the population value, itself a useful descriptive measure of the association between $T^{(1)}$ and $T^{(2)}$. None of the statistics considered for censored data yields a consistent estimator of the corresponding population value for uncensored data from the same joint distribution.

However, for a special model proposed by Clayton [2] and considered further by Oakes [7], the coefficient of concordance does yield a useful method of estimation in the presence of censoring. In Clayton's model the joint density $f(t^{(1)}, t^{(2)})$ and survivor function $S(t^{(1)}, t^{(2)})$ of $T^{(1)}$ and $T^{(2)}$ satisfy

$$f(t^{(1)}, t^{(2)}) S(t^{(1)}, t^{(2)})$$
$$= \theta \int_{u=t^{(1)}}^{\infty} f(u, t^{(2)}) \, du \int_{v=t^{(2)}}^{\infty} f(t^{(1)}, v) \, dv$$

where θ ($\geqslant 1$) is a parameter governing the degree of association between $T^{(1)}$ and $T^{(2)}$.

For this model, the hazard functions for the conditional distributions of $T^{(1)}$ given $T^{(2)} = t$, and given $T^{(2)} \geqslant t$ have ratio θ.

It is easily shown that, given potential censoring times $\{C_i^{(l)}, C_j^{(l)};\ l = 1, 2\}$, the conditional probability that (i, j) is a definite concordance, given that it is either a definite concordance or a definite discordance equals $\theta/(\theta + 1)$. Under a random censorship model, the ratio of the number of definite concordances to the number of definite discordances consistently estimates θ.

References

[1] Brown, B. W., Hollander, M., and Korwar, R. M. (1974). In *Reliability and Biometry. Statistical Analysis of Life Length*. F. Proschan and R. J. Serfling, eds. SIAM, Philadelphia, pp. 327–353.

[2] Clayton, D. G. (1978). *Biometrika*, **65**, 141–151.

[3] Daniels, H. E. (1944). *Biometrika*, **32**, 129–135.

[4] Hoeffding, W. (1948). *Ann. Math. Statist.*, **19**, 293–325.

[5] Kendall, M. G. (1938). *Biometrika*, **30**, 81–93.

[6] Oakes, D. (1982). *Biometrics*, **38**, 451–455.

[7] Oakes, D. (1982). *J. R. Statist. Soc. B*, **44**, 414–422.

[8] Weier, D. R. and Basu, A. P. (1980). *J. Statist. Plan. Infer.* **4**, 381–390.

(CENSORED DATA
COEFFICIENT OF CONCORDANCE
SURVIVAL ANALYSIS)

DAVID OAKES

OBJECTIVE FUNCTION *See* LINEAR PROGRAMMING

OBJECTIVE PROBABILITY *See* AXIOMS OF PROBABILITY; FOUNDATIONS OF PROBABILITY

OBSERVATIONAL ERROR *See* MEASUREMENT ERROR; NONSAMPLING ERRORS IN SURVEYS

OBSERVATIONAL STUDIES

The term *observational* is employed to denote a type of investigation that can be described somewhat negatively as *not an experiment*. A general definition of such a study was presented by Wold, who, after setting down three criteria for a controlled experiment, defined observational data as those in which at least one of the three following conditions is violated:

(a) The replications of the experiment are made under similar conditions (so as to yield an internal measure of uncontrolled variation).

(b) The replications are mutually independent.

(c) The uncontrolled variation in the replication is subjected to randomization* in the sense of Fisher [16, p. 30].

Observational studies involving explanation rather than just description were defined as the collection of data in which the third criterion, randomization, is not possible.

> ... it is the absence of randomization that is crucial, for since this device is not available there is no clear-cut distinction between factors which are explicitly accounted for in the hypothetical model and disturbing factors which are summed up in the residual variation [16, p. 37].

Wold included in this class of studies all those concerned with the investigation of cause and effect and/or the development of predictive models.

In a seminal discussion of this type of investigation, Cochran [1] suggested two main distinguishing characteristics:

1. The objective is the investigation of possible cause-effect relationships.

2. This is implemented by the comparison of groups subject to different "treatments" that were preassigned in a nonrandom manner.

Most broadly, the term observational can apply to any investigation that is not an experiment, including descriptive surveys of populations, the essential characteristic being that the subjects or material under investigation are not manipulated in any way. Following the lead of Wold and Cochran, among others, however, the term has become most closely associated with that subset of studies investigating a hypothesized cause-effect relationship. This more narrow definition is assumed in the discussion presented here.

The two major defining characteristics of an observational study proposed by Cochran and quoted earlier are in a real sense in conflict. A cause-effect relationship can be established only under appropriately controlled experimental conditions that include randomization*. Without this control of variability, the inference that A causes B cannot be made and one is restricted to the weaker inference that A and B are associated. If this association includes a temporal sequence (e.g., A is always observed to occur before B), then this is suggestive of a cause-effect relationship. It is important to understand that data from an observational study cannot be used to demonstrate a cause-effect relationship—only to suggest one (for possible testing in an experiment). For further discussion of the unique inferential power of experiments, the reader is referred to articles by Kempthorne and McKinlay [5, 9] as well as to the appropriate entries in this encyclopedia.

Observational studies are variously designed and titled, but fall into either of two categories—prospective and retrospective. These categories and aspects of design associated with them are discussed in the following section.

PROSPECTIVE

This type of study is closest in design to an experiment. Groups are formed in a nonrandom (usually self-selected) manner according to categories or levels of a hypothesized cause (factor F) and subsequently observed with respect to the outcome of interest Y. The causal factor F can be referred to as the *design* or *independent* variable. The outcome

Y can also be termed the *dependent* variable. A well-known example is the prospective study of the association between smoking habit and lung cancer [4]. Groups of individuals who did and did not smoke (*F*) were identified and followed for a fixed period. All cases of lung cancer (*Y*) were recorded for each group in that time. The temporal sequence of *F* and *Y* suggested (but did not demonstrate) a cause-effect relationship. The equivalent experimental study would have required the identification of subjects who had never smoked to be randomly assigned to one of two groups, either to maintain their nonsmoking status or to start and continue to smoke over the observation period.

The major inferential weakness of this observational design is that subjects self-select themselves into groups. It is possible, for example, that the predisposition toward smoking in some individuals is also a causal factor for lung cancer and that the observed association between smoking and lung cancer is merely reflective of such a common cause. Other terms that have been used to describe prospective studies of this type are *cohort* and *quasi-experiment*.

The difference between the type of prospective, observational study described here and descriptive longitudinal* or *panel* studies should be noted. Panel studies (of opinions, health, etc.) describe changes in a population over time without predetermined groups for comparison.

The use of the term *prospective* to describe a type of study design must also not be confused with the type of data collection*. A prospective observational study may be designed entirely from preexisting records. For example, if sufficient information is available in a set of records (e.g., medical or occupational) to identify smokers and nonsmokers at the start of record keeping and to identify diagnosed lung cancer within a specified period, then a prospective study of the association between lung cancer and smoking can be designed from such records. Smokers and nonsmokers can be defined and their lung cancer experience observed prospectively over the period covered by the records. Investigations of associations of various prenatal factors with fetal outcome (e.g., McKinlay and McKinlay [8] and Neutra et al. [11]) illustrate this design.

RETROSPECTIVE

This type of design is particularly applicable in the epidemiology of chronic diseases or conditions, for which the period of observation between introduction of the hypothesized cause and appearance of the effect can be many years or even decades, and/or for which the effect is rare (as in many cancers).

The distinguishing characteristic of this design is that the groups for comparison are defined on an observed outcome and differences between the groups are sought on a hypothesized cause (or causes). For example, the initial investigations reporting an association between smoking and lung cancer in Britain [3] were comparing groups with and without lung cancer (the outcome or effect) with respect to potential causes. The roles of dependent and independent (design) variables are reversed. The dependent variable *Y* in a retrospective study is a hypothesized cause, while the independent or design variable *F* is the outcome or effect being investigated.

Although the term *retrospective* refers primarily to the design, it also frequently applies to the method of data collection, particularly when the cause(s) may have preceded the outcome chronologically, sometimes by many years. Heavy reliance on memory recall is a recurring disadvantage of this design, which must be weighed against the disadvantage of an alternative long and costly prospective study. A topical example that illustrates this design problem is the investigation of the role of diet in early adulthood on the subsequent development of selected cancers, requiring diet recall of 20 years or more.

The retrospective design does have a considerable advantage over equivalent prospective studies in terms of efficiency, particularly when the rate of occurence of the effect (outcome) is relatively low (e.g., endometrial cancer, cancer of the cervix, with rates under 5/1000). Numbers per group required for a prospective study may rapidly exceed 20 times the number required for the equivalent retrospective study. An excellent, detailed discussion of the advantages and disadvantages of the two designs is provided by Schlesselman [12].

Other terms for a retrospective observational study include *case-control* and *ex post facto*. The latter also has been used to denote a pair-matched design [14].

CONTROL OF VARIATION

A recurring issue in the design and analysis of observational studies (both retrospective and prospective) is the selective control of variation due to covariables ("intervening" or "confounding" variables).

Techniques that, in experimental design and analysis, are used to increase efficiency in the comparison, have the supplementary role of controlling for potential bias in observational comparisons—a role assumed by randomization* in experiments. The relative effectiveness of such techniques as pre- or post-stratification*, pair matching, and covariance adjustments has been investigated notably in the last two decades. Two comprehensive reviews provide full discussions of this issue [2, 6]. Results of this research show that pair matching is seldom more effective than analytic adjustments on independent samples. Moreover, pair matching may be very costly in terms of locating viable matches and/or discarding unmatchables [7], although it remains a popular technique in many fields.

ROLE OF OBSERVATIONAL STUDIES

This type of study has two major purposes: to establish the need for a subsequent experiment; and to provide optimal information in situations not amenable to the conduct of an experiment (for practical or ethical reasons). This design is not an alternative to a controlled, randomized experiment as it does not evaluate cause-effect relationships directly.

Retrospective studies, in particular, can be cost-effective preliminary investigations that provide the rationale for more costly experiments—especially when the observed associations are equivocal or contradictory. Recent retrospective and prospective studies of the association between cardiovascular disease and use of oral contraceptives [13, 15], for example, have stimulated the instigation of experiments to study the effect of this hormone combination on such cardiovascular disease risk factors as blood pressure and blood lipids [10].

The second important role for observational studies is as a less powerful substitute for experiments when the latter are not feasible. For example, it is not ethical to randomly assign subjects to smoking or nonsmoking groups in order to observe the relative lung cancer incidence. To investigate the impact of water composition on a community's health, it may not be feasible to randomly change water constituents in participating towns because of such constraints as variable water sources and the need for town council approval. In such a situation, the observational study may provide information quickly and cheaply concerning a potential cause-effect relationship. In some instances, this type of design—albeit inconclusive—may provide the only information on a cause-effect relationship.

References

[1] Cochran, W. G. (1965). *J. R. Statist. Soc. A*, **128**, 234–266.

[2] Cochran, W. G. and Rubin, D. B. (1973). *Sankhyā A*, **35**, 417–446. (Excellent review on controlling bias in observational studies.)

[3] Doll, R. and Hill, A. B. (1952). *Brit. Med. J.*, **2**, 1271–1286.

[4] Doll, R. and Hill, A. B. (1964). *Brit. Med. J.*, **1**, 1399–1410 and 1460–1467.

[5] Kempthorne, O. (1977). *J. Statist. Plan. Infer.*, **1**, 1–25.

[6] McKinlay, S. M. (1975). *J. Amer. Statist. Ass.*, **70**, 503–520. (Comprehensive review on design and analysis of observational studies.)

[7] McKinlay, S. M. (1977). *Biometrics*, **33**, 725–735.

[8] McKinlay, J. B. and McKinlay, S. M. (1979). *Epidemiol. Community Health*, **33**, 84–90.

[9] McKinlay, S. M. (1981), *Milbank Memorial Fund Quart. Health Soc.*, **59**, 308–323.

[10] National Institutes of Health, National Institute of Child Health and Human Development. *RFP No. NICHD-CE-82-4.* February 1, 1982.

[11] Neutra, R. R., Fienberg, S. E., Greenland, S., and Friedman, E. A., (1978). *N. Engl. J. Med.*, **299**, 324–326.

[12] Schlesselman, J. J. (1982). *Case-Control Studies: Design, Conduct, Analysis.* Oxford University Press, New York. (A comprehensive book on the retrospective design that provides an excellent discussion of its advantages relative to the prospective design.)

[13] Slone, D., Shapiro, S., Kaufman, D. W., Rosenberg, L., Miettinen, O. S., and Stolley, P. D. (1981). *N. Eng. J. Med.*, **305**, 420–424.

[14] Thistlethwaite, D. C. and Campbell, D. T. (1960). *J. Educ. Psychol.*, **51**, 309–317.

[15] Vessey, M. P. and Doll, R. (1968). *Brit. Med. J.*, **2**, 199–205.

[16] Wold, H. (1956). *J. R. Statist. Soc. A*, **119**, 28–60.

(BIOSTATISTICS
CAUSATION
CLINICAL TRIALS
DESIGN OF EXPERIMENTS
EPIDEMIOLOGICAL STATISTICS
PROSPECTIVE STUDIES
RETROSPECTIVE STUDIES
    (INCLUDING CASE-CONTROL))

S. M. McKINLAY

OBSERVATIONS, IMAGINARY

The "device of imaginary results" was suggested by Good [2, p. 35] as a method of obtaining order of magnitude quantifications of prior probabilities (*see* PRIOR PROBABILITIES) in situations where direct assessment appears difficult. The basic idea is that of a "thought experiment," exploiting the fact that the formal expression of Bayes' theorem* (*see* BAYES' THEOREM) enables one to infer values of prior probabilities from the direct specification of the various other probabilities appearing in the theorem and thus, in effect, inverting the "prior to posterior" sequence in which Bayes' theorem is most frequently applied and interpreted (*see* PRIOR PROBABILITIES, POSTERIOR PROBABILITIES, and BAYES' THEOREM).

Example. Suppose an experiment is to be conducted to investigate whether a subject has telepathic powers that enable him to give invariably correct answers to each of a sequence of yes-no questions (e.g., about hidden objects). It is assumed that either the subject has such powers—hypothesis H, corresponding to the certainty of a correct answer to each question—or is "just guessing" —hypothesis $\sim H$, corresponding to equal chances of correct or incorrect answers to each question, independently. If asked for an assessment of prior probability for H, $\Pr(H)$, you might well answer "rather small," but feel diffident about a precise quantification, e.g., 10^{-2} or 10^{-6}.

The device of imaginary observations proceeds by reformulating the question concerning $\Pr(H)$ in the following way. You are to imagine that the experiment has been conducted and that the subject has answered correctly all n questions put to him (where n might be 3, 10, 100, or whatever). Now ask yourself what the value of n would need to be in order that your revised (i.e., posterior) probability for H would have risen to $\frac{1}{2}$. If $n = N$ is your response, and D denotes the event that the subject answers the N questions correctly, then from Bayes' theorem we deduce immediately that, for you,

$$\tfrac{1}{2} = \Pr(H \mid D) = \frac{\Pr(D \mid H)\Pr(H)}{\Pr(D)},$$

so that

$$P(\sim H \mid D) = 1 - P(H \mid D) = \tfrac{1}{2}$$

and

$$1 = \frac{1/2}{1/2} = \frac{\Pr(H \mid D)}{\Pr(\sim H \mid D)}$$

$$= \frac{\Pr(D \mid H)\Pr(H)}{\Pr(D \mid \sim H)\Pr(\sim H)}$$

$$= \frac{\Pr(H)}{\left(\tfrac{1}{2}\right)^{N}(1 - \Pr(H))},$$

from which we deduce that $\Pr(H) = (\frac{1}{2})^N / [1 + (\frac{1}{2})^N]$.

The device of imaginary observations has been used most frequently for order of magnitude *numerical* assessment of small probabilities, as in the example. It is possible, however, that the idea will also prove useful in exploring the suitability of certain forms of *mathematical* assumptions commonly used in Bayesian procedures. Recent studies in this direction include investigations of the assignment of ratios of improper prior limits in model comparison using Bayes factors (*see* Spiegelhalter and Smith [3] and IMPROPER PRIOR DISTRIBUTIONS) and the assignment of priors over function spaces (see Diaconis and Freedman [1]).

References

[1] Diaconis, P. and Freedman, D. (1983). In *Statistical Inference, Data Analysis and Robustness*, G. E. P. Box, T. Leonard, and C. F. Wu, eds. Academic Press, New York, pp. 105–116. (Frequency properties of Bayes' rules.)

[2] Good, I. J. (1950). *Probability and the Weighing of Evidence*. Charles Griffin, London. (An early influential book on Bayesian theory and practice.)

[3] Spiegelhalter, D. J. and Smith, A. F. M. (1982). *J. R. Statist. Soc. B*, **44**, 377–387. (Bayes factors for linear and log-linear models with vague prior information.)

A. F. M. SMITH

OCCAM'S RAZOR *See* PARSIMONY, PRINCIPLE OF

OCCUPANCY PROBLEMS

CLASSICAL OCCUPANCY

There is a close relationship between urn models* and occupancy problems, because occupancy problems are described easily by urn models and many occupancy problems arise from urn models. Suppose we have a set of m urns and n balls. Balls are thrown, one at a time, in such a way that the proba-

Table 1 The Classical Occupancy Models

Models	Urns	Balls
(1)	DT	IDT
(2)	DT	DT
(3)	IDT	DT
(4)	IDT	IDT

bility of assignment to the jth urn is p_j for each ball. Clearly $p_1 + p_2 + \cdots + p_m = 1$. Let M_t be the number of urns containing exactly t balls ($t = 0, 1, \ldots, n$). Problems of occupancy arise when we are concerned with the distribution of M_t. This is called the classical occupancy model.

There are wide applications of occupancy distributions in many fields. For example, in computer theory [1], in cluster analysis, in the problem of predators and prey [12], and in theoretical physics. Johnson and Kotz [10] have written a book that gives a systematic and complete description of occupancy problems.

In many cases, each ball has an equal probability m^{-1} of falling into any specified one of the m urns, i.e., $p_1 = p_2 = \cdots = p_m = 1/m$. According to whether the urns and balls are distinguishable (DT) or indistinguishable (IDT), there are four possible models listed in Table 1. In theoretical physics, model (1) corresponds to the Bose–Einstein system and model (2) corresponds to the Maxwell–Boltzmann system (*see* FERMI–DIRAC STATISTICS). These models have been discussed systematically by Denning and Schwartz [1, Chap. 3].

DISTRIBUTIONS, MOMENTS* AND CHARACTERISTIC FUNCTIONS*

In many applied problems, people are interested in the numbers of urns containing specified numbers of balls, i.e., M_t ($t = 0, 1, \ldots, n$). For example, M_0 denotes the number of empty urns.

Let $T_{(i)}(n, m)$ denote the number of ways in which n balls are distributed among m urns under model (i). By elemental combinatorial techniques, we find (cf. ref. 10, p. 37):

$$T_{(1)} = \binom{n+m-1}{m-1}, \quad T_{(2)} = m^n,$$

$$T_{(3)} = \sum_{j=1}^{m} \frac{\Delta^j 0^n}{j!} \quad \text{and} \quad T_{(4)} = p_m(n), \tag{1}$$

where $\Delta f(x) = f(x+1) - f(x)$, $\Delta^n 0^m = \Delta^n x^m|_{x=0}$. There is no simple analytical expression for $p_m(n)$.

Denote by $Q_{(i)}(n, m, r, t)$ the number of ways of distributing the n balls among the m urns so that exactly r urns have exactly t balls under model (i). If we suppose that each way is equally likely, then for model (i) we have

$$\Pr(M_t = r) = Q_{(i)}(n, m, r, t) / T_{(i)}(n, m). \tag{2}$$

Let $x_j(w)$ denote the number of balls in the jth urn and F_j denote the event that there are exactly t balls in the jth urn, i.e., $F_j = \{w : X_j(w) = t\}$, $j = 1, \ldots, m$. By the inclusion-exclusion principle* (cf. ref. 10, p. 29), we have

$$\Pr(M_t = r) = \Pr(\text{exactly } r \text{ events among}$$

$$\text{the } m \text{ events } \{F_j\} \text{ occur})$$

$$= \sum_{j=r}^{m} (-1)^{j+r} \binom{j}{r} S_j, \tag{3}$$

where

$$S_j = \sum_{(a_1, \ldots, a_j)} \Pr\left(\bigcap_{b=1}^{j} F_{a_b}\right) \tag{4}$$

and the summation is over all possible $\binom{m}{j}$ sets $(a_1, \ldots, a_j)$. From the definition of the F_j's, we have, under model (1),

$$\Pr\left(\bigcap_{b=1}^{j} F_{a_b}\right) = \sum_{(a_1, \ldots, a_j)} \frac{T_{(1)}(n - jt, m - j)}{T_{(1)}(n, m)} \tag{5}$$

and under model (2)

$$\Pr\left(\bigcap_{b=1}^{j} F_{a_b}\right)$$

$$= \frac{n!}{(t!)^j (n - jt)!}$$

$$\times \sum_{(a_1, \ldots, a_j)} \frac{T_{(2)}(n - jt, m - j)}{T_{(2)}(n, m)}. \tag{6}$$

Combining (1)–(6), we have

$$\Pr(M_t = r)$$

$$= \frac{\binom{m}{r}}{\binom{n+m-1}{m-1}} \sum_{j=0}^{m-r} (-1)^j \binom{m-j}{j}$$

$$\times \binom{n+m-(r+j)(t+1)-1}{m-r-j-1} \tag{7}$$

under model (1) and

$$\Pr(M_t = r) = \frac{\binom{m}{r}}{m^n} \sum_{j=0}^{m-r} (-1)^j \binom{m-j}{j}$$

$$\times \frac{n!}{(t!)^{r+j} \{n - (r+j)t\}!}$$

$$\times (m - r - j)^{n-(r+j)t} \tag{8}$$

under model (2) (cf. ref. 10, p. 115; and ref. 3). They can be expressed in the following united form

$$\Pr(M_t = r)$$

$$= \binom{m}{r} \sum_{j=0}^{m-r} (-1)^j \binom{m-r}{j} g_i(n, t, j + r)$$

$$\times \frac{T_{(i)}(n - (r+j)t, m - r - j)}{T_{(i)}(n, m)} \tag{9}$$

under model (i), $i = 1, 2$, where

$$g_1(n, t, j) = 1 \tag{10}$$

$$g_2(n, t, j) = \frac{n!}{(t!)^j (n - jt)!}. \tag{11}$$

The sth descending factorial moment* of a random variable X is defined by $E(X^{(s)}) = E(X(X-1) \ldots (X - s + 1))$. We note that under model (i), $i = 1, 2$,

$$E(M_t^{(s)})$$

$$= m^{(s)} g_i(n, t, s) \frac{T_{(i)}(n - st, m - s)}{T_{(i)}(n, m)}. \tag{12}$$

We can express formula (9) as:

$$\Pr(M_t = r)$$

$$= \sum_{j=r}^{m} (-1)^{j+r} \binom{j}{r} \binom{m}{j}$$

$$\times \frac{T_{(i)}(n - jt, m - j) g_i(n, t, j)}{T_{(i)}(n, m)}.$$

We have

$$T_{(i)}(n,m)E\left(M_t^{(s)}\right)$$

$$= \sum_{r=s}^{m} r^{(s)} \sum_{j=r}^{m} (-1)^{j+r} \binom{j}{r}\binom{m}{j}$$

$$\times T_{(i)}(n-jt, m-j)\, g_i(n,t,j)$$

$$= \sum_{j=s}^{m} (-1)^{j} \binom{m}{j} T_{(i)}(n-jt, m-j)$$

$$\times g_i(n,t,j) \sum_{r=s}^{j} (-1)^{r} \binom{j}{r} r^{(s)}.$$

Note that

$$\sum_{r=s}^{j} (-1)^{r} \binom{j}{r} r^{(s)} = \begin{cases} r^{(s)}(-1)^{s} & \text{if } j = s, \\ 0 & \text{otherwise,} \end{cases}$$

and formula (12) follows.

If X is a discrete random variable with distribution

$$\Pr(X = j) = p_j, \qquad j = 0, 1, 2, \dots,$$

its generating function* is

$$P(v) = \sum_{j=0}^{\infty} p_j v^j$$

and its factorial moment generating function* is (if it exists)

$$G(v) = \sum_{j=0}^{\infty} E(X^{(j)}) v^j / j!.$$

It is a well-known fact [10, p. 61] that $P(v) = G(v-1)$ and the characteristic function* (cf) of X is $G(e^{iv} - 1)$ with $i = (-1)^{1/2}$. Now, applying the above formula for M_t we have

$$G(v) = \sum_{s=0}^{\infty} \frac{v^s}{s!} E\left(M_t^{(s)}\right)$$

$$= \sum_{s=0}^{\infty} \binom{m}{s} v^s g_k(n,t,s)$$

$$\times T_{(k)}(n-st, m-s) / T_{(k)}(n,m);$$

thus the cf of M_t is

$$\sum_{s=0}^{m} \binom{m}{s} (e^{iv} - 1) g_k(n,t,s)$$

$$\times T_{(k)}(n-st, m-s) / T_{(k)}(n,m). \quad (13)$$

RESTRICTED OCCUPANCY PROBLEM

The restricted occupancy problems are extensions of the classical occupancy problems and have been discussed by several authors. The following are some general models:

There are n balls and m urns. Each urn contains k cells. The n balls are assigned to the m urns in such a way that each cell contains at most one ball and the number of balls in the ith urn belongs to a given set I_i ($i = 1, 2, \dots, m$), where $I_1, \dots, I_m$ are subsets of the set of nonnegative integers.

According to whether urns, balls, and cells are DT or IDT, the five models listed in Table 2 are studied. When the cells are IDT, we always assume $k = \infty$, otherwise $k < \infty$.

These models include both those of classical occupancy models and those of restricted occupancy models. When $k = \infty$ and $I_1 = I_2 = \dots = I_m = I = \{0, 1, 2, \dots\}$, we have the classical occupancy models. If $n = m$ and $I_1 = \dots = I_m = \{0, 1, 2\}$, they reduce to Example 3.5 of [10], which arose in a chemical industry inquiry. Freund and Pozner [6] used model 1 to solve the problem of k judges rating a product. When $I_1 = \dots = I_m = I = \{0, 1, \dots, k\}$ or $I = \{1, 2, \dots, k\}$ the five models were considered by Fang [2]. Holst [9] discussed model 5 with $I_1 = I_2 = \dots = I_m = I$, and the more general models were studied by Fang and Niedzwiecki [4].

The methods used by the aforementioned were:

Elemental combinatorial method.

The inclusion-exclusion principle*.

Reduction to conditional distributions of independent random variables. (This is a more powerful technique than the others.)

Table 2 Occupancy Models

Models	Urns	Cells	Balls
1	DT	IDT	IDT
2	DT	DT	IDT
3	DT	DT	DT
4	DT	IDT	DT
5	IDT	IDT	DT

Let X_i and Z_i denote the number of balls in the ith urn ($i = 1, 2, \ldots, m$) in the restricted case and in the unrestricted case, respectively. The idea of the third method is to find some iid random variables $Y_1, \ldots, Y_m$ such that

$$\Pr(Z_i = x_i, i = 1, \ldots, m)$$
$$= \Pr(Y_i = x_i, i = 1, \ldots, m \mid$$
$$Y_1 + \cdots + Y_m = n), \quad (14)$$

where $x_1 + \cdots + x_m = n$. From this, we can obtain the number of ways of distributing the balls among the m urns under models i so that $X_j \in I_j, j = 1, \ldots, m$, for $i = 1, 2, 3, 4, 5$. Then as with the classical occupancy models (cf. the Distributions, Moments, and Characteristic Functions section) the distributions, moments and cf's of M_t follow.

RELATED OCCUPANCY PROBLEMS AND DISTRIBUTIONS

There are many variants of the classical occupancy problem arising both from practical requirements and mathematical interest. The following subsections discuss some of them.

Randomized Occupancy Models

n balls are randomly allocated to m urns with $p_1 = p_2 = \cdots = p_m = 1/m$ (cf. the Classical Occupancy Section). Each ball has probability p of staying in its urn and probability $1 - p$ of "falling through" or "leaking." We are concerned with the consequent occupancy distributions.

Waiting-Time Problems

In some applications people are interested in the number of balls needed to satisfy specified occupancy conditions for the urns, waiting-time problems arise. Sequential occupancy problems are a special class of them. We might, for example, consider the following conditions:

(a)   All urns to be occupied.

(b)   At least k urns to contain at least one ball each.

(c) k specified urns to contain at least one ball each.

(d)   The required number of empty urns to be achieved.

These problems are called sequential occupancy problems; they include the birthday problem, or the coupon collectors problem [10, Sects. 3.2.3 and 3.5].

Mixing Distribution Problems

There are m urns. Each ball is equally likely (with probability p) to be assigned to any one of b urns ($bp \leqslant 1$) that are called class I urns of the total population. The remaining $(m - b)$ urns are called class II urns. The probability of assignment to some one of the class II urns is $(1 - bp)$. Let M_t denote the number of class I urns containing exactly t balls. It is required to find the distribution of M_t.

Estimating the Number of Classes

There are equal numbers of balls of k different colors in an urn and sampling with replacement is continued until n balls have been drawn. If k is not known, we wish to estimate it [10, p. 137].

Multivariate Occupancy Distributions

So far we have only discussed univariate distributions. However, considering some multivariate distributions is required in some applications. Here are some interesting problems:

1.   The joint distribution of $M_0, M_1, \ldots, M_n$ [10, p. 115].

2.   Suppose that m urns are divided into s groups of $m_1, \ldots, m_s$ urns, respectively ($m_1 + m_2 + \cdots + m_s = m$). Let M_t^i de-

note the number of urns of the ith group containing exactly t balls ($1 \leqslant t \leqslant m_i$, $i = 1, \ldots, s$). We require the joint distribution of $M_{t_1}^1, \ldots, M_{t_s}^s$, in particular, the joint distribution of $M_0^1, M_0^2, \ldots, M_0^s$ [10, p. 147].

3. The distribution of $\max_{0 \leqslant j \leqslant n} M_j$ and the distribution of $\min_{0 \leqslant j \leqslant n} M_j$, or the joint distribution of them.

Committee Problems

A group contains n individuals, any w_i of whom can be selected at random to form the ith committee ($i = 1, 2, \ldots, r$). Find the probability that exactly m individuals will be committee members. This has been called the committee problem, which is an extension of the "chromosome problem" described by Feller [5]. This model has been studied by several authors including Mantel and Pasternack [11], Sprott [13], Gittelsohn [7], White [14], Johnson and Kotz [10], Holst [8], and Fang [3].

The committee problem is an excellent example of a problem that can be solved by several different methods. It can be solved by induction [11], by finite difference* operators [3, 14], by the inclusion-exclusion principle [13], by the method of moments* [7], and by the reduction to conditional distributions of independent random variables [8].

As the classical occupancy problem can be extended to the randomized occupancy models, so we can consider randomized committee problems (cf. Johnson and Kotz [10, Sect. 3.6.3]).

References

[1] Denning, P. J. and Schwartz, S. C. (1972). *Commun. ACM*, **15**, 191–198.

[2] Fang, K. T. (1982) *J. Appl. Prob.*, **19**, 707–711.

[3] Fang, K. T. (1982) "Some Further Applications of Finite Difference Operators." *Tech. Rep. No. 3, Contract DAAG29-82-K-0156*, Dept. of Statistics, Stanford University, Stanford, CA.

[4] Fang, K. T. and Niedzwiecki, D. (1983) In *Contributions to Statistics, Essays in Honor of Professor Norman Lloyd Johnson*, North Holland, pp. 147–158.

[5] Feller, W. (1957). *An Introduction to Probability Theory and Its Applications*, Vol. 1, 2nd ed. Wiley, New York.

[6] Freund, J. E. and Pozner, A. N. (1956). *Ann. Math. Statist.*, **27**, 537–540.

[7] Gittelsohn, A. M. (1969). *Amer. Statist.*, **23**(2), 11–12.

[8] Holst, L. (1980). *Scand. J. Statist.*, **7**, 139–146.

[9] Holst, L. (1981). "On Numbers Related to Partitions of Unlike Objects and Occupancy Problems." *Res. Rep. No. 121*, Stockholm University and Uppsala University, Sweden.

[10] Johnson, N. L. and Kotz, S. (1977). *Urn Models and Their Application*. Wiley, New York.

[11] Mantel, N. and Pasternack, B. S. (1968). *Amer. Statist.*, **22**(2), 23–24.

[12] Mertz, D. B. and Davies, R. B. (1968). *Biometrics*, **24**, 247–275.

[13] Sprott, D. A. (1969). *Amer. Statist.*, **23**(2), 12–13.

[14] White, C. (1971). *Amer. Statist.*, **25**(4), 25–26.

(BOOLE'S FORMULA
FINITE DIFFERENCES, CALCULUS OF
GENERALIZED HYPERGEOMETRIC
    DISTRIBUTIONS
HYPERGEOMETRIC DISTRIBUTION
INCLUSION-EXCLUSION PRINCIPLE
PROCESSES, DISCRETE
WARING'S FORMULA)

KAI-TAI FANG

OCCURRENCE RATE

This term is sometimes used for the parameter λ in the Poisson distribution* when given in the form

$$\Pr[X = x] = \frac{1}{x!} (\lambda t)^x \exp(-\lambda t),$$

$$x = 0, 1, \ldots,$$

where t is the length of a period of observation (see, e.g. Nelson [1]).

Reference

[1] Nelson, W. (1982). *Applied Life Data Analysis*. Wiley, New York.

(POISSON DISTRIBUTION)

OC CURVES   *See* ACCEPTANCE SAMPLING

OCTILES

Values of a variable dividing its distribution into eight equally probable portions. Generally any number x such that

$$\Pr[X \leqslant x] = j/8$$

is a *jth octile* of the distribution of X ($j = 1, \ldots, 7$). For discrete distributions, octiles (also medians, quartiles, etc.) are not uniquely defined.

For a sample of size n, the jth octile is the $\{\frac{1}{8}j(n - 1) + 1\}$th order statistic provided $\frac{1}{8}j(n - 1)$ is an integer. If this is not so, other more or less arbitrary definitions are used.

(DECILE
PERCENTILES, ESTIMATION OF
QUANTILES)

ODDS RATIO ESTIMATORS

The odds ratio is defined for a single 2×2 contingency table* or as a summary statistic for s 2×2 contingency tables where s represents the number of strata. A single 2×2 contingency table may be formed by two independent binomial* populations with parameters π_{11} and π_{21}. The odds in the first binomial population are $\pi_{11}/(1 - \pi_{11})$ and those in the second $\pi_{21}/(1 - \pi_{21})$. The odds ratio is then defined as the ratio of the two odds

$$\Psi = \frac{\pi_{11}\pi_{22}}{\pi_{21}\pi_{12}}$$

where

$$\pi_{22} = 1 - \pi_{21} \quad \text{and} \quad \pi_{12} = 1 - \pi_{11}.$$

A single 2×2 contingency table may also be formed by looking at a pair of random variables with dichotomous response defined over the same sample space*. Here π_{ij} represents the probability of an observation falling in cell (ij) for $i = 1, 2$; $j = 1, 2$. In this situation $\sum_{i=1}^{2}\sum_{j=1}^{2}\pi_{ij} = 1$, and the odds ratio Ψ, as defined above, is thought of as a measure of association* between the two

random variables defining the contingency table.

Under a logistic model (*see* LOGISTIC REGRESSION*) the odds ratio $\Psi = e^{\lambda}$ where

$$\lambda = \log(\pi_{11}/\pi_{12}) - \log(\pi_{21}/\pi_{22})$$

is the difference in logits. Cornfield [4] first used the odds ratio estimator as a measure of relative risk* for a retrospective study*. Mantel and Haenszel [19] continued in this vein with their classic paper, also dealing with retrospective studies. Because of its nature Fisher [7] and Mosteller [23] referred to the odds ratio as the cross-products ratio. Mosteller noted some of the properties of the odds ratio, including the fact it is invariant under row and column multiplication.

An estimate of the odds ratio for a single 2×2 table is

$$\hat{\Psi} = n_{11}n_{22}/n_{21}n_{12},$$

where n_{ij} represents the number of observations falling in cell (ij). This $\hat{\Psi}$ is the maximum likelihood* estimator for the likelihood resulting from the product of two binomials (see Gart [11]). The asymptotic variance of this estimator is

$$\Psi^2 \left(\frac{1}{n_{11}} + \frac{1}{n_{12}} + \frac{1}{n_{21}} + \frac{1}{n_{22}} \right)^{-1}.$$

Notice that $\hat{\Psi}$ and its variance may be undefined if any $n_{ij} = 0$. A method for avoiding this problem, suggested by Haldane [17], is to replace n_{ij} by $n'_{ij} = n_{ij} + \frac{1}{2}$. If a logistic model is assumed, then an estimate of λ is $\log(\hat{\Psi})$, which has asymptotic variance estimated by $(1/n_{11} + 1/n_{12} + 1/n_{21} + 1/n_{22})$. Birch [1] considered the maximum likelihood estimator of $\hat{\Psi}$ under the assumption of fixed marginals for the 2×2 table. This conditional maximum likelihood estimator is not easily found, requiring the solution of a polynomial equation of high degree. As Mantel and Hankey [20] point out, when all marginals are fixed, the preceding sample cross-product ratio is not the maximum likelihood estimate.

Exact confidence limit estimates for the odds ratio, for both the conditional and unconditional sample spaces, are available using computer programs developed by Thomas [25]. However, three approximate interval estimates are also available: those of Cornfield [5], Woolf [26], and Miettinen [22]. Brown [3] and Fleiss [9] both found the Cornfield approximation, with continuity correction*, preferable to the other two for the conditional sample space. Although an iterative computation more complicated than the others, the Cornfield estimate was more accurate than the estimators of Woolf and Miettinen for the various sample sizes and alternatives simulated [3]. A formula for computing the Cornfield estimate is given in Gart [11].

Gart and Thomas [12] compared the three methods when sampling from two independent binomials (the unconditional situation). They found the Cornfield method, without continuity correction, preferred over the other two methods. Thomas offers a computer program for the necessary calculations.

Difficulty in estimating the odds ratio arises when data from several 2×2 contingency tables are to be combined. Such a situation arises when attempting to control the effects of several covariables by stratifying the data on the values of the covariables (*see* MANTEL–HAENSZEL STATISTIC*). For this case let n_{ijk} be the number of observations in cell (ij) of the kth stratum, $k = 1, \ldots, s$ (as before $i = 1, 2; j = 1, 2$).

One of the earliest estimates of the combined odds ratio Ψ_s was given by Woolf [26]. This estimator, with the $\frac{1}{2}$ correction mentioned earlier and shown by Gart and Zweifel [13] to provide the least biased estimate, is

$$\hat{\Psi}_w$$

$$= \exp\left\{ \sum_{k=1}^{s} w_k \log \frac{(n_{11k} + \frac{1}{2})(n_{22k} + \frac{1}{2})}{(n_{21k} + \frac{1}{2})(n_{12k} + \frac{1}{2})} \bigg/ \sum_{k=1}^{s} w_k \right\},$$

where

$$w_k = \left(\frac{1}{n_{11k} + \frac{1}{2}} + \frac{1}{n_{12k} + \frac{1}{2}} \right.$$

$$\left. + \frac{1}{n_{21k} + \frac{1}{2}} + \frac{1}{n_{22k} + \frac{1}{2}} \right)^{-1}.$$

Mantel and Haenszel [19] offered five estimators of the combined odds ratio. The most well known of these estimators is

$$\hat{\Psi}_{mh} = \frac{\sum_{k=1}^{s}(n_{11k} n_{22k} / n_{..k})}{\sum_{k=1}^{s}(n_{21k} n_{12k} / n_{..k})},$$

which can be shown to be a weighted average of the individual stratum odds ratios. The . is used to indicate summation over a subscript; hence $n_{..k}$ is the total number of observations in the kth stratum.

Gart [10] considered the maximum likelihood estimator of Ψ_s when the marginals for each 2×2 table are *not* all assumed fixed. This unconditional maximum likelihood estimator involves an iterative solution of $(s - 1)$ third-degree equations for which a computer program by Thomas [25] is available. Gart also presented three noniterative estimators that are asymptotically as efficient as the proposed unconditional maximum likelihood estimator. Of these three, the best was that of Woolf, given earlier.

Birch [1] derived the maximum likelihood estimator of Ψ_s conditional on the marginals of each 2×2 table being assumed fixed. Such an approach required the solution of a high-degree equation which Birch approximated using a method proposed by Cox [6]. The resulting approximation

$$\hat{\Psi}_B = \exp\left\{ \frac{\sum_{k=1}^{s} n_{..k}^{-1}(n_{11k} n_{22k} - n_{21k} n_{12k})}{\sum_{k=1}^{s} n_{..k}^{-2}(n_{..k} - 1)^{-1} n_{.1k} n_{.2k} n_{1.k} n_{2.k}} \right\}$$

is appropriate when Ψ_s is in the neighborhood of one. Goodman [14] suggested a modification of $\hat{\Psi}_B$ that should be used when Ψ_s is not close to one, but involves substantially more computation.

McKinlay [21] compared the preceding estimators of Ψ_s using Monte Carlo methods and pointed out that all, with the possible exception of the Mantel–Haenszel estimator, require the assumption of a constant odds ratio across the strata. However, McKinlay does mention that when combining estimates from several 2×2 tables the variation in the individual table estimates (second-order interaction) should never involve reversals in direction. McKinlay concludes from the simulation study that the Mantel–

Haenszel estimator was preferable, even when the odds ratio was constant across all strata. Plackett [24] indicates how to test the hypothesis of a constant odds ratio across all strata.

Examination of the asymptotic properties of the preceding estimators [11, 21] has revealed that $\hat{\Psi}_w$ has the same asymptotic variance as the unconditional maximum likelihood estimator, which is also equivalent to the asymptotic variance of $\hat{\Psi}_B$. Hauck [18] has demonstrated that $\hat{\Psi}_{mh}$ is also efficient when the null hypothesis of no association is true.

Approximate confidence limits for the common odds ratio in the case of $k \geq 2$ strata can be obtained using an extension of Cornfield's single 2×2 table estimator. (See Gart [11].) The computation required for such intervals is complex, but, again, a computer program by Thomas [25] is available.

The preceding asymptotic results were derived under the situation where the number of tables remained fixed, but the number of observations within each table increases without bound. Breslow [2] considered the situation where the number of tables increases but the number of observations within a table is small. He concludes that $\hat{\Psi}_w$ and the unconditional maximum likelihood estimator of Ψ_s should *not* be used when there are small numbers of observations within a stratum, as, for example, in a matched pairs* design.

Example. This example illustrates several of the estimators using survey data from Pitt County, NC. This survey resulted in the following chart concerning raising the legal age for drinking alcoholic beverages from 18 to 21 years.

Age of Respondent	Sex of Respondent	Raise Legal Age		
		Yes	No	Total
< 21	Male	5	13	18
	Female	16	22	38
21–29	Male	12	25	37
	Female	28	29	57
≥ 30	Male	54	37	81
	Female	91	33	124

Woolf's estimator:

$$\hat{\Psi}_w = \exp(-10.73/19.53) = 0.577.$$

Mantel–Haenszel estimator:

$$\hat{\Psi}_{mh} = 14.35/27.58 = 0.520.$$

Birch's estimator:

$$\hat{\Psi}_B = \exp(-13.63/20.35) = 0.512.$$

As can be seen, the estimators are not substantially different.

There are numerous other measures of association for contingency tables, some of which appear in ASSOCIATION, MEASURES OF. Other older measures may be found in articles by Goodman and Kruskal [15, 16] and Mosteller [23].

For further readings on odds ratio estimators the previously mentioned papers of Breslow [2] and McKinlay [21] are recommended. Additional introductory readings on the subject can be found in the Mantel–Haenszel [19] paper and a book by Fleiss [8].

References

[1] Birch, M. W. (1964). *J. R. Statist. Soc. B*, **26**, 313–324.

[2] Breslow, N. (1981). *Biometrika*, **68**, 73–84.

[3] Brown, C. C. (1981). *Amer. J. Epidemiol.*, **113**, 226–235.

[4] Cornfield, J. (1951). *J. Natl. Cancer Inst.*, **11**, 1269–1275.

[5] Cornfield, J. (1956). *Proc. 3d Berkeley Symp. Math. Statist. Prob.*, **4**, 135–148.

[6] Cox, D. R. (1958). *J. R. Statist. Soc. B*, **20**, 215–242.

[7] Fisher, R. A. (1962). *Aust. J. Stat.*, **4**, 41.

[8] Fleiss, J. L. (1973). *Statistical Methods for Rates and Proportions*. Wiley, New York.

[9] Fleiss, J. (1979). *J. Chronic. Dis.*, **32**, 69–77.

[10] Gart, J. J. (1962). *Biometrics*, **18**, 601–610.

[11] Gart, J. J. (1971). *Int. Statist. Rev.*, **39**, 148–169.

[12] Gart, J. J. and Thomas, D. G. (1982). *Amer. J. Epidemiol.*, **115**, 453–469.

[13] Gart, J. J. and Zweifel, R. (1967). *Biometrika*, **54**, 181–187.

[14] Goodman, L. A. (1969). *J. R. Statist. Soc. B*, **31**, 486–498.

[15] Goodman, L. A. and Kruskal, W. H. (1954). *J. Amer. Statist. Ass.*, **49**, 723–764.

[16] Goodman, L. A. and Kruskal, W. H. (1959). *J. Amer. Statist. Ass.*, **54**, 123–163.

[17] Haldane, J. B. S. (1956). *Ann. Hum. Genet.*, **20**, 309–311.

[18] Hauck, W. W. (1979). *Biometrics*, **35**, 817–820.

[19] Mantel, N. and Haenszel, W. (1959). *J. Natl. Cancer Inst.*, **22**, 719–748.

[20] Mantel, N. and Hankey, B. F. (1975). *Proc. 40th Session Int. Statist. Inst.*, **46**, 128–136.

[21] McKinlay, S. M. (1978). *Biometrika*, **65**, 191–202.

[22] Miettinen, O. S. (1976). *Amer. J. Epidemiol.*, **103**, 226–235.

[23] Mosteller, F. (1968). *J. Amer. Statist. Ass.*, **63**, 1–28.

[24] Plackett, R. L. (1962). *J. R. Statist. Soc. B*, **24**, 162–166.

[25] Thomas, D. G. (1975). *Comp. Biomed. Res.*, **8**, 423–446.

[26] Woolf, B. (1955). *Ann. Hum. Genet.*, **19**, 251–253.

(ASSOCIATION, MEASURES OF
CHI-SQUARE TESTS
CONTINGENCY TABLES
LOGISTIC DISTRIBUTION
MANTEL–HAENSZEL STATISTIC
OBSERVATIONAL STUDIES
RELATIVE RISK)

GRANT W. SOMES
KEVIN F. O'BRIEN

OECD   *See* ORGANIZATION FOR ECONOMIC COOPERATION AND DEVELOPMENT (OECD)

OFFICE OF POPULATION CENSUSES AND SURVEYS, THE

The Office of Population Censuses and Surveys (OPCS) is the department of the U.K. government concerned with demographic matters broadly defined: the registration of births, marriages, and deaths, the collection of population statistics through censuses and surveys, and the demographic and epidemiological analysis of these and other sources. Most of its activities are limited to England and Wales, corresponding activities in Scotland being carried out by the General Register Office for Scotland and in Northern Ireland by the General Register Office for Northern Ireland. However OPCS coordinates population statistics for the United Kingdom as a whole.

The Director of OPCS is the Registrar General, an appointment made directly by Her Majesty the Queen. For his statistical functions, the Registrar General is subject to ministerial direction from the Secretary of State for Social Services. The OPCS currently employs a staff of some 2500 located in St. Catherine's House, Kingsway, London WC2, and in provincial locations in Hampshire and Merseyside.

HISTORY

OPCS was created on May 11, 1970 by merging two existing government departments: the General Register Office (GRO) for England and Wales and the Government Social Survey.

The title of Registrar General was created in 1836, the office of the Registrar General being named the General Register Office. The Registrar General's initial tasks were, first, to supervise the work of Superintendent Registrars appointed to deal with the preliminaries to all marriages outside the established church and to officiate at civil marriages, and the work of local registrars appointed to register all births and deaths and certain marriages; and second, to compile the central record of all births, marriages and deaths in England and Wales. (In the United Kingdom the primary function of the registration system has always been to provide a legal document recording the particular event; statistical analysis has been a secondary, though important, function.)

The first census* of population in England and Wales was taken in 1801; thereafter a census has been taken every 10 years with the exception of the war year 1941. From 1841 onward the task of taking the census fell on the Registrar General. In the 133 years of its separate existence, the General Register Office developed its demographic and epidemiological analysis based on material drawn from the registration of births, marriages and deaths, censuses of population, and other sources. The pioneering work of William Farr (1807–1883) is particularly remembered [1].

The Government Social Survey, the other partner in the merger, was a more recent creation. It had been developed during World War II to carry out sample surveys of persons on social and economic topics. Its achievements in the 30 years of its separate existence are particularly associated with the name of its first director, Louis Moss.

The idea of merging the GRO and the Government Social Survey was due to Professor Claus Moser (now Sir Claus Moser) when, in 1967, he became the Director of the Central Statistical Office. The aim was to bring together in one department the function of collecting population and related social statistics through the complementary instruments of registration, censuses and sample surveys, together with the associated analytic functions. It was part of a wider scheme for the development of the U.K. government's statistical service.

CENSUS OF POPULATION

The fieldwork in postwar censuses in the United Kingdom has followed conventional lines: the division of the country into "enumeration districts," currently numbering over 100,000, each containing some 150 households and to each of which an "enumerator" is appointed to list the addresses within it, to issue questionnaires to householders, and to collect questionnaires after householders have filled them in. Well over 99% response is obtained. Mail-out and mail-back methods have not been used.

Traditionally the census schedule in the United Kingdom has recorded the persons present at each address on census night rather than the persons *usually* resident there. The introduction into the 1931 and later censuses of a question on each person's usual address has also enabled statistics of the population usually resident in each area to be compiled (by transfer of the person's record from the area of enumeration to the

area of his usual residence). In addition, the 1961 and later censuses have asked for a return of persons absent from the enumerated address on census night though usually resident there; this offers a second approach to the measurement of the population usually resident in each area.

In 1961, sampling was introduced: every tenth questionnaire handed out was a "long form" containing more questions than the other nine questionnaires. But analysis of the results showed that some bias had crept into the distribution of the long forms. In 1966, a 10% *sample census* was taken—this has been the only mid-decade census in Britain. The sampling frame for 1966 was the list of addresses recorded in the 1961 census updated by reference to details of later housing developments, but, again, the sample introduced bias. In 1971 and 1981, sampling was confined to the processing stage; every householder received the same questionnaire but some of the topics that are expensive to code (e.g., occupation) were coded and analysed only for 10% of the households, the sample being chosen at headquarters after all the questionnaires had been received.

The list of questions in U.K. censuses is shorter than is customary in North America, reflecting opposition from some of the public to what may be seen as burdensome or intrusive questions. Questions on income and rentals have not been asked, nor have questions on race and ethnicity, though these are traditional in the United States. In an attempt to measure the numbers and characteristics of the ethnic groups that had recently arrived in Britain from the New Commonwealth and Pakistan, a question was asked in the 1971 census on the country of birth of the parents of each person; this question supplemented the traditional census question on the person's own country of birth. But following the controversy generated by a test in 1979 in Haringey (North London), in which a direct ethnic question was asked of some households and the question on parents' countries of birth was asked of other households, *both* questions were omitted from the 1981 census [2].

The results of censuses are now available on an extended geographical base. An array of statistics is published for each enumeration district, ward, local authority district and county, in 1971 and for selected areas in 1981 for each grid square, that is, the 1 square kilometer formed by the national grid lines, and in 1981 for each urban area with a population of over 2000 [3]. To help maintain confidentiality quasi-random errors of $+1$ and -1 are introduced into the counts that appear in a proportion of nonzero cells in statistics for small areas.

So far results of the census have not been released in the form of "public use tapes" (microdata).

The first interview survey to check on the quality of responses to the census took place in 1961 [4].

SOCIAL SURVEYS

The Social Survey Division of OPCS, successor to the separate Government Social Survey, carries out voluntary sample surveys on behalf of government departments, mostly by interview addressed to persons and private households. There are five main continuous surveys: the General Household Survey (GHS), the Family Expenditure Survey (FES), the National Food Survey, the International Passenger Survey (IPS), and the Labor Force Survey (LFS) [5].

The GHS is a multipurpose survey addressed to some 12,500 private households a year throughout Great Britain. It covers a wide range of topics—demographic, social, and economic—though its multipurpose nature restricts the depth of questioning on any one topic. Its value for surveying small groups of the population is limited by the relatively small sample size and the effect of geographical clustering in the sample design. Response rate is 80 to 85%. The GHS started in late 1970 [6] (*see* SURVEYS, HOUSEHOLD).

The Family Expenditure Survey collects information on the income and expenditure of private households as well as on the basic demographic and economic characteristics of the household. A detailed diary of the household's expenditure over a two-week period is a part of the information collected. The sample size is some 11,000 private households a year with a response rate of about 70%. The FES began in 1957 and is carried out on behalf of the Department of Employment; a main use is to provide weights for the index of retail prices [7].

The National Food Survey made on behalf of the Ministry of Agriculture is the oldest of all government sample surveys in the United Kingdom, having run since 1940. It collects detailed information from private households on the prices and quantities of food purchases. The sample size is about 13,500 private households a year, although the response rate is only just over a half.

The International Passenger Survey is very different in character. It is a survey of a sample of international passengers traveling through U.K. airports and seaports. In a short survey, taking on average about 4 minutes to complete, questions are asked about the origin and destination of the passenger, length of intended stay and past stay, place of birth, nationality, occupation, etc., and also about the expenditure abroad of British residents returning to the United Kingdom and about the expenditure in the United Kingdom of foreign residents leaving it. This information is the basis of estimates of overseas travel and its contribution to the U.K. balance of payments and of migration to and from the United Kingdom. Some 170,000 interviews take place annually, representing only 0.25% of the total number of international passenger movements. The IPS began in 1961 [8].

From 1973 through 1983, a biennial Labour Force Survey has been carried out on behalf of the Statistical Office of the European Communities (EEC). This is a survey aiming to collect comparable information about the demographic and economic characteristics of the population in each of the member countries of the EEC. In the United Kingdom some 100,000 private households —this is a sample of about $\frac{1}{2}$%—have been interviewed over a period of about six weeks in May and June, a response rate of about 85% being achieved. From 1984, the LFS has become a continuous survey, with interviews of 5000 households per month (20,000 households per month in the period March through May). The redesigned survey provides information on labor participation, broadly on the lines of the U.S. Current Population Survey and the Canadian Labour Force Survey [9].

Some dozen *ad hoc* social surveys are completed annually on behalf of other government departments (in addition to which departments themselves carry out some surveys using their own staff or employing contractors). Of particular interest to OPCS have been the surveys addressed to women on patterns of family building, the socioeconomic factors and attitudes affecting these, and on the use of family planning services [10].

DEMOGRAPHIC ANALYSIS

OPCS undertakes a wide range of analytic work on population and medical topics. The main data sources are:

1. Registrations of births, marriages, and deaths. The cause of death recorded on death certificates is a basic element in studies of mortality. Certain information is collected at the registration of a birth or death solely for statistical purposes: at a birth, the parents' dates of birth and date of marriage and the number of children previously born to the mother; at a death, the deceased person's marital status and, if married, the spouse's date of birth.

2. Censuses of population and surveys.

3. A variety of record systems maintained by OPCS or other organizations, including National Health Service registration, hospitals' records of their patients, medical practitioners' records of patients' treatment, central registrations of cancer cases, notifications of congenital malformations, notifications of infectious diseases, divorce registry, adoptions register, social security records,

electoral registration, and school enrollments.

Studies of marriage*, fertility*, mortality, morbidity*, and migration* (both internal to the United Kingdom and international) are central to the work of OPCS. Some of the topics studied would, in other countries, be of the kind carried out in the health department of government, e.g., in the National Center for Health Statistics* in the United States.

Census figures of population in each local authority district (a district has on average a population of 120,000) are updated annually on the basis of numbers of births and deaths and estimates of migration. The accuracy of these annual population estimates is limited by the lack of regular and reliable sources of information on migration at the local level [11].

Population projections*, analyzed by sex, age, and, to a limited extent, marital status, are prepared biennially at national and subnational levels. They extend up to 40 years into the future. Because of the particular difficulty in forecasting the birth rate, variant projections have been prepared to illustrate a range of possibilities [12].

Estimates are prepared of the population in Britain of New Commonwealth and Pakistani ethnic origin, which may be equated roughly with the coloured population. Projections to the end of the century have also been prepared [13].

An important technique, particularly in the work of the Medical Statistics Division of OPCS, is the linkage of two or more records of an identified person referring to different points of time. For example, death registrations are linked to birth registrations (in studies of infant mortality), to cancer registrations (in studies of cancer), and to records of persons who may have been exposed to particular risks (in studies of occupational mortality and of environmental hazards). Such linkage is facilitated by the use of the National Health Service Central Register; this is a clerically maintained set of records containing basic identifying and demographic information about virtually everyone in the country; it is maintained by OPCS primarily to service the operations of the National Health Service.

A longitudinal study* (LS) of a 1% sample of the population represents a significant application of record linkage techniques. In this study, the events routinely recorded by OPCS are collated for a 1% sample of the population to provide longitudinal information, i.e., statistics of flows and transition probabilities*. The initial sample was drawn from the records of the 1971 census of population by selecting everyone whose birthday fell on certain days of the year, and this sample has been updated by adding newly born children and immigrants with birthdays on the chosen days. The events included in the LS records are registration of a birth (to a mother in the sample), registration of a death, migration into or out of the United Kingdom, notification of cancer, and an entry in a census of population including some household details: the entries in the 1981 census of population in respect to members of the LS sample have already been linked into the study. Record linkage is facilitated by reference to the National Health Service Central Register using name and date of birth as the key; the success rate for linkage depends on the type of event, rates of over 90% being achieved in a number of cases. The longitudinal study's potentialities as a research tool are now being evaluated [14].

PUBLICATIONS

OPCS publishes, mainly through Her Majesty's Stationery Office, a wide variety of reports and statistics including annual volumes on such topics as marriage and divorce, births, mortality, migration, population estimates, and the General Household Survey [15]. Prior to 1974, much of this material had appeared in the *Registrar General's Statistical Review of England and Wales* (published annually in two or three parts). In 1978, an overview of population changes in Britain was published as *Demographic Review 1977* [16] and a new review of changes in the years to 1983 is in preparation. Research findings appear in the series *Studies on Medical and Population Subjects*.

References

Except where otherwise stated, publication is by Her Majesty's Stationery Office, London.

[1] (1975). *Vital statistics: A Memorial Volume of Selections from the Reports and Writings of William Farr*. The Scarecrow Press, Metuchen, NJ. (Republished with an introduction by Mervyn Susser and Abraham Adelstein.)

[2] Benjamin, Bernard. (1970). *The Population Census: An SSRC Review of Current Research*. Heinemann, London. [(A short review of censuses in Britain up to 1966.) The 1981 census is compared with previous censuses in Britain and with censuses in other countries in two articles by Philip Redfern that appeared in *Pop. Trends*, **23** and **24** (spring and summer 1981), and that were reproduced in OPCS *Occasional Paper 25*, published by OPCS, 1981.]

[3] (1980). *People in Britain: A Census Atlas*. (Statistics from the 1971 census for each square kilometer are mapped, with a commentary.)
(1984). *Census 1981: Key Statistics for Local Authorities, Great Britain.*
(1984). *Census 1981: Key Statistics Urban Areas, Great Britain.*

[4] (1985). An Enquiry into the Coverage and Quality of the 1981 Census in England and Wales.

[5] Barnes, B. (1979). *Popul. Trends*, **16**, 12.

[6] *General Household Survey*. (The annual reports give a description of the survey and a commentary on some of the results.)

[7] *Family Expenditure Survey*. (The annual reports issued by the Department of Employment describe the survey and results.)

[8] *International Migration 1975* (Series MN No. 2). (A description of the IPS with some of the results is given in the OPCS annual reference volume. Later volumes continue to report the results.)

[9] (1985). *Labour Force Survey 1983*. Results for all member countries of the EEC are given in the biennial *Labour Force Sample Survey*, which is accompanied by a separate volume on methodology (1977). *Labour Force Sample Survey: Methods and Definitions*. Statistical Office of the European Communities, Luxembourg.

[10] Dunnell, K. (1979). *Family Formation 1976* (ref. SS 1080).

[11] *Population Estimates*. (Series PP1). (The OPCS annual reference volume.)

[12] (1984). *Population Projections, 1981–2021*, Series PP2 no. 12. (The latest national projections with details of methods.)

[13] The latest estimates appear in: (1982). *Labour Force Survey 1981* (Chapter 5). (1985). *Labour Force Survey 1983*. Projections appeared in (1979) *Popul. Trends*, **16**, 22.

[14] (1973). *Cohort Studies: New Developments* (Studies on medical and population subjects No. 25.) (The aims of the LS are set out.) A series of volumes describing the results is in preparation, of which the first are (1982). *Longitudinal Study: Socio-demographic Mortality Differentials 1971–75*, Series LS No. 1 and (forthcoming) *Longitudinal Study: Social Class and Occupational Mobility. 1971–77*. Series LS No. 2.

[15] *Birth Statistics*, Series FM1, *Marriage and Divorce Statistics*, Series FM2, volumes on mortality statistics, Series DH, and volumes on morbidity, Series MB. (OPCS annual reference volumes not mentioned elsewhere in this reference list.)

[16] (1978). *Demographic Review 1977: A Report on Population in Great Britain*.

(BUREAU OF THE CENSUS, U.S.
CENSUS
DEMOGRAPHY
LONGITUDINAL STUDIES
SURVEY SAMPLING
SURVEYS, HOUSEHOLD)

P. Redfern

OFFSET NORMAL DISTRIBUTION
See DIRECTIONAL DISTRIBUTIONS

OGIVE

In statistics this name is mostly applied to graphs of empirical distribution functions (*see* EDF STATISTICS). If the proportion f_x of observed values in a sample that are less than or equal to x is plotted against x, a step function is obtained which, for large sample sizes, approximates an ogive.

For grouped distributions, plotting f_x against x at group boundaries and showing a smooth curve through the points so obtained usually produces an "ogival" curve—an "ogive."

Although the term might be applied more realistically to graphs of (population) cumulative distribution functions $F_X(x)$, against x, this is less common.

(FREQUENCY POLYGON
GRAPHICAL REPRESENTATION
OF DATA)

OLMSTEAD–TUKEY TEST FOR ASSOCIATION

This is also known as the *corner test for association*. The calculation of the test statistic is described in Olmstead and Tukey [1] as follows for the case of an odd number of observed pairs (x, y):

In the scatter diagram, draw two lines, $x = x_m$, $y = y_m$, where x_m is the median of the x-values without regard to the values of y, and y_m is the median of the y-values without regard to the values of x. Think of the four quadrants or corners thus formed as being labelled $+, -, +, -$, in order, so that the upper right and lower left quadrants are positive. Beginning at the right hand side of the diagram, count in (in order of abscissae) along the observations until forced to cross the horizontal median. Write down the number of observations met before this crossing, attaching the sign $+$ if they lay in the $+$ quadrant, and the sign $-$ if they lay in the $-$ quadrant. Repeat this process moving up from below, moving to the right from the left, and moving down from above. The quadrant sum is the algebraic sum of the four terms thus written down.

High (or low) values of the quadrant sum are regarded as significant. Table 1 is a "working table" of upper significance limits [1]. Note the limits vary very little with sample size.

To deal with ties, the practical rule:

When a tied group is reached, count the number in the tied group favorable to continuing and the number unfavorable. Treat the tied group as if the number of points

Table 1   Working Significance Levels for Magnitudes of Quadrant Sums

Significance Level (Conservative)	Magnitude of Quadrant Sum[a]
10%	9
5%	11
2%	13
1%	14–15
0.5%	15–17
0.2%	17–19
0.1%	18–21

[a]The smaller magnitude applies for large sample size, the larger magnitude for small sample size. Magnitudes equal to or greater than twice the sample size less six should not be used.

preceding the crossing of the median were

$$\frac{\text{number favorable}}{1 + \text{number unfavorable}}$$

is suggested.

Reference

[1] Olmstead, P. S. and Tukey, J. W. (1947). *Ann. Math. Statist.*, **18**, 495–513.

(ASSOCIATION, MEASURES OF
CORRELATION
TETRACHORIC CORRELATION)

OMEGA DISTRIBUTION

Let X be a random variable associated with the cumulative distribution function $F(x)$ and the probability density function $f(x)$ characterized by

$$F[x(p)] = p,$$

$$f[x(p)] = 1 - |2p - 1|^{v+1},$$

and

$$x(p) = \int_{\frac{1}{2}}^{p} \{f[x(z)]\}^{-1} dz,$$

where $v > -1$ and $0 < p < 1$. This symmetric distribution of X is parameterized by v (a shape parameter) and is termed the omega distribution, say Ω_v [1]. Special cases of Ω_v include (a) the double exponential (Laplace) distribution*, when $v = 0$, (b) the logistic

distribution*, when $v = 1$, and (c) the uniform distribution on $[-\frac{1}{2}, \frac{1}{2}]$ when $v \to \infty$. Specific probability density functions of Ω_v are illustrated in Fig. 1 for $v = -\frac{1}{2}, 0, 1, 3$, and ∞. Except for the three special cases when $v = 0$, 1, and ∞, moments of X expressed as explicit functions of parameter v are presently unknown.

While the term omega distribution was introduced in conjunction with an inference procedure for quantal response* assays (*see* QUANTIT ANALYSIS) [1, 2], the genesis of this distribution involves a specific class of two-sample rank tests that is correspondingly indexed by v. In particular, if $U_1, \ldots, U_m$ and $V_1, \ldots, V_n$ denote two independent random samples from two populations possessing continuous cumulative distribution functions, then this class of two-sample rank tests is based on the test statistic given by

$$C_v = m^{-1} \sum_{i=1}^{N} \psi_i Z_{Ni},$$

where $N = m + n$,

$$\psi_i = \begin{cases} |i - (N+1)/2|^v & \text{if } i > (N+1)/2, \\ 0 & \text{if } i = (N+1)/2, \\ -|i - (N+1)/2|^v & \text{if } i < (N+1)/2, \end{cases}$$

and Z_{Ni} is 1 or 0 if the ith smallest value in the pooled sample of U's and V's is a U or V, respectively (*see* SIGNED POWERS OF RANKS TESTS) [3, 4]. The asymptotic null distribution of C_v is normal if and only if

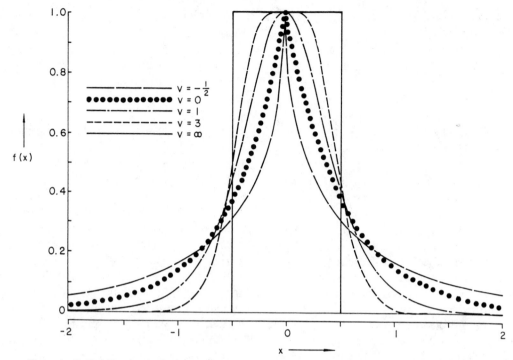

Figure 1 Probability density functions of omega distribution for selected values of v. Reproduced from ref. 1; reprinted with permission of *Biometrics*.

$v \geqslant -\frac{1}{2}$. The test based on C_v is (1) a locally most powerful rank test for detecting a location shift of Ω_v (i.e., $U = V + b$ where $b \neq 0$) when $v > -1$ and (2) is asymptotically most powerful for (Pitman-type sequences of) local (contiguous) location alternatives involving Ω_v if $v > -\frac{1}{2}$ [7]. Incidentally, the test based on C_v is the two-sample median test or the Wilcoxon–Mann–Whitney test* if $v = 0$ or 1, respectively. These same results also hold for an analogous class of one-sample (matched-pairs) rank tests [5, 6].

References

[1] Copenhaver, T. W. and Mielke, P. W. (1977). *Biometrics*, **33**, 175–186. (Introduces a quantal response analysis based on the omega distribution.)

[2] Magnus, A., Mielke, P. W., and Copenhaver, T. W. (1977). *Biometrics*, **33**, 221–223. (Provides a sum of an infinite-series application involving the omega distribution.)

[3] Mielke, P. W. (1972). *J. Amer. Statist. Ass.*, **67**, 850–854. (Introduces the omega distribution in conjunction with a specific class of two-sample rank tests.)

[4] Mielke, P. W. (1974). *Technometrics*, **16**, 13–16. (Provides further relations between two-sample rank tests and the omega distribution.)

[5] Mielke, P. W. and Berry, K. J. (1976). *Psychometrika*, **41**, 89–100. (Relates the omega distribution to a specific class of one-sample rank tests.)

[6] Mielke, P. W. and Berry, K. J. (1983). *Psychometrika*, **48**, 483–485. (Clarifies asymptotic properties between specific one-sample rank tests and the omega distribution.)

[7] Mielke, P. W. and Sen, P. K. (1981). *Commun. Statist. Theor. Meth.*, **A10**, 1079–1094. (Resolves nonparametric asymptotic theory questions motivated by the relation between rank tests and the omega distribution.)

(DISTRIBUTION-FREE METHODS
LAPLACE DISTRIBUTION
LOCALLY OPTIMAL STATISTICAL TESTS
LOGISTIC DISTRIBUTION
MANN–WHITNEY–WILCOXON
   STATISTIC
QUANTIT ANALYSIS
RANK TESTS
RECTANGULAR DISTRIBUTION)

PAUL W. MIELKE, JR.

OMEGA SQUARE TEST *See* CRAMÉR–
VON MISES TEST

OMNIBUS TESTS *See* PORTMANTEAU
TESTS

ONE- AND TWO-ARMED
BANDIT PROBLEMS

Suppose there are two available treatments
for a certain disease. Patients arrive one at a
time and one of the treatments must be used
on each. Information as to the effectiveness
of the treatments accrues as they are used.
The overall objective is to treat as many of
the patients as effectively as possible. This
seemingly innocent but important problem
is surprisingly difficult, even when the responses are dichotomous: success-failure. It
is a version of the two-armed bandit.

A bandit problem in statistical decision
theory* involves sequential selections from k
stochastic processes* (or "arms," machines,
treatments, etc.). Time may be discrete or
continuous, and the processes themselves
may be discrete or continuous. The processes are characterized by parameters that
are typically unknown. The process selected
for observation at any time depends on the
previous selections and results. A decision
procedure (or strategy) specifies which process to select at any time for every history of
previous selections and observations. A utility is defined on the space of all histories.
This provides a definition for the utility* of
a strategy in the usual way by averaging
over all possible histories resulting from that
strategy.

Most of the literature, and most of this
article, deals with discrete time. In this setting, each of the k arms generates an infinite
sequence of random variables. Making an
observation on a particular sequence is
called a *pull* of the corresponding arm. The
classical objective in bandit problems is to
maximize the expected value of the payoff
$\sum_1^\infty \alpha_i Z_i$, where Z_i is the variable observed at
stage i and the α_i are known nonnegative

numbers (usually $\alpha_i \geqslant \alpha_{i+1}$ is assumed) with
$\sum_1^\infty \alpha_i < \infty$; $(\alpha_1, \alpha_2, \dots)$ is called a *discount
sequence*. A strategy is *optimal* if it yields the
maximal expected payoff. An arm is optimal
if it is the first pull of some optimal strategy.

FINITE-HORIZON BERNOULLI BANDITS

Historically, the most important discount sequence has been "finite-horizon uniform":
$\alpha_1 = \cdots = \alpha_n = 1$, $\alpha_{n+1} = \cdots = 0$, and
most of the literature deals with Bernoulli
processes. The objective is then to maximize
the expected number of successes in the first
n trials.

The finite-horizon two-armed Bernoulli
bandit was first posed by Thompson [28]. It
received almost no attention until it was
studied by Robbins [22], and from a different point of view by Bradt et al. [10]. Robbins [22] suggested a selection strategy that
depends on the history only through the last
selection and the result of that selection (i.e.,
pull the same arm after a success and switch
after a failure), and compared its effectiveness with that of random selections. This
originated an approach called "finite memory" [12, 17, 26]. The decision maker's
choice at any stage can depend only on the
selections and results in the previous r
stages.

Bradt et al. [10] considered information
given as a joint probability distribution of
the Bernoulli parameters p_1 and p_2 (a so-
called Bayesian approach) and characterized
optimal strategies for the case in which one
parameter is known a priori. With such an
approach, a strategy requires (and "re-
members") only the sufficient statistics*: the
numbers of successes and failures on the
two arms. Most of the recent bandit literature (and the remainder of the article) takes
the Bayesian approach. It is not that most
researchers in bandit problems are "Bayesians"; rather, Bayes' theorem* provides a
convenient mathematical formalism that allows for adaptive learning, and so is an ideal
tool in sequential decision problems.

Myopic Strategies

Feldman [14] solved the Bernoulli two-armed bandit in the finite-horizon uniform setting for a deceptively simple initial distribution on the two parameters: both probabilities of success are known, but not which goes with which arm. Feldman showed that *myopic* strategies are optimal: at every stage, pull the arm with greater expected immediate gain (the unconditional probability of success with arm j is the prior mean of p_j). Feldman's result was extended in different directions by Fabius and van Zwet [13], Berry [2], Kelley [18], and Rodman [24].

It is important to recognize that myopic strategies are not optimal—or even good—in general. As a simple example, suppose p_1 is known to be $1/2$ (pulling arm 1 is like tossing a fair coin) and p_2 is either 1 or 0 (the other coin is either two-headed or two-tailed); let r be the initial probability that $p_2 = 1$. The fact that a single pull of arm 2 reveals complete information makes the analysis of this problem rather easy. If $r < 1/2$, then a myopic strategy indicates pulls of arm 1 indefinitely and has utility $n/2$. On the other hand, pulling arm 2 initially, and indefinitely if it is successful and never again if it is not, results in n successes with probability r and an average of $(n - 1)/2$ successes with probability $1 - r$. The advantage of this strategy over the myopic is

$$rn + \tfrac{1}{2}(1 - r)(n - 1) - \tfrac{1}{2}n$$
$$= \tfrac{1}{2}\{r(n + 1) - 1\}$$

which is positive for $r > 1/(n + 1)$. In this and other bandit problems, it may be wise to sacrifice some potential early payoff for the prospect of gaining information that will allow for more informed choices later. But the "information vs. immediate payoff" question is not usually as clear as it is in this example.

Stay with a Winner Rule

Partial characterizations of optimal strategies in the two-armed bandit with finite-horizon uniform discounting were given by Fabius and van Zwet [13]. Berry [2] gave additional characterizations when p_1 and p_2 are independent a priori. One such is the "stay-with-a-winner rule": if the arm pulled at any stage is optimal and yields a success, then it is optimal at the next stage as well. (Bradt et al. [10] give a counterexample to this result when the arms are dependent.) Nothing can be said in general about the arm to pull following a failure; the simplest example of staying with a loser involves an arm whose success rate is known, for then a failure contains the same information as a success.

The stay-with-a-winner rule, which characterizes optimal strategies, is to be contrasted with play-the-winner* rules [20, 27]. These are complete strategies and are based on "finite memory" as described earlier, where the memory length is one. That is, the same arm is pulled following a success and the other following a failure. Such a strategy is optimal only in very special circumstances; for example, $n = 2$ and the distribution of (p_1, p_2) is exchangeable*.

Solution by Dynamic Programming

Since the only uncertainty arises after a failure, one might say that the problem is half solved. But the stay-with-a-winner rule, while picturesque, is little help in finding optimal strategies. One must weigh the utility of all possible histories when deciding which arm to pull. The standard method of solution for such problems is dynamic programming* or backward induction. To determine optimal strategies, one first finds the maximal conditional expected payoff (together with the arm or arms that yield it) at the very last stage, for every possible $(n - 1)$-history (sequence of pulls and results), optimal and otherwise. Here, "conditional" refers to the particular history. Proceeding to the penultimate stage, one maximizes the conditional expected payoff from the last two observations for every possible $(n - 2)$-history. Continuing backward—remembering the optimal arms at each partial history—gives all optimal strategies. The problem is four-dimensional since that is the dimension of a minimal sufficient statistic. But a computer program requiring on the order of $n^3/6$ storage locations is possible.

GEOMETRIC DISCOUNTING

Much recent literature (notably refs. 15, 16, 19, 29, and 30) has dealt with geometric discounting: $\alpha_m = \beta^{m-1}$ where $\beta \geq 0$ and, usually, $\beta < 1$. In economic applications, for example, one may wish to assume that the rate of inflation does not change over time. Or, in a medical setting, a new and obviously better treatment may be discovered at each stage (or the disease could spontaneously disappear) with constant probability $1 - \beta$. Though it is quite special, geometric discounting is important, in part because, except for a multiplicative factor, it is invariant under a time shift. In addition, when the discount sequence is unknown, there are many instances in which the decision maker should act as though the discount sequence is geometric [4].

In k-armed bandits the attractiveness of an arm depends in general on the other arms available. But when the discount sequence is geometric (and the arms are independent), Gittins and Jones [16] showed that the arms can be evaluated separately by assigning a real number, a "dynamic allocation index" or "Gittins index," to each arm—any arm with the largest index is optimal. In view of this result, the desirability of a particular arm can be calculated by comparing the arm and various known arms; the Gittins index can be determined as the expectation of that known arm for which both arms are optimal in a two-armed bandit problem (this is quite different from the expectation of the unknown arm, cf. Berry and Fristedt [5]). In practice, it may not be possible to evaluate a Gittins index exactly, but it can be approxi-

mated by truncating and using backward induction.

Robinson [23] compares various strategies obtained for geometric discounting when the actual discount sequence is the finite-horizon uniform for large n. He finds that, for his criterion, a rule proposed by Bather [1] performs quite well.

ONE-ARMED BANDITS

When $k = 1$, the decision problem is trivial: the only available arm is pulled forever! The so-called "one-armed bandit" is formulated differently. There is a single process with unknown characteristics, and the decision maker has the option of stopping observation on the process at any time. Having this option makes the problem the same as a two-armed bandit, in which the second process is known with expectation 0 (or some number greater than 0 if there is a cost of observation). Stopping observation at stage i in the original formulation is analogous to choosing the known process at stages subsequent to i in the two-armed bandit. For this reason, two-armed bandits with one arm known are sometimes called one-armed bandits. Whether this is appropriate depends on the discount sequence.

Regular Discounting

When the characterisitics of one arm are known, Berry and Fristedt [5] showed that an optimal strategy is to pull the known arm indefinitely once it is pulled for the first time, provided the discount sequence is nonincreasing and *regular*:

$$\sum_{j}^{\infty} \alpha_i \sum_{j+2}^{\infty} \alpha_i \leq \left(\sum_{j+1}^{\infty} \alpha_i \right)^2$$

for $j = 1, 2, \ldots$. Moreover, if the discount sequence is not regular there is a distribution for the unknown arm such that all optimal strategies call for switches from the known arm [5]. Therefore, the term "one-armed bandit" is appropriate for this problem if and only if the discount sequence is regular.

Examples of regular discount sequences are $(1, 1, \ldots, 1, 0, 0, \ldots)$—finite-horizon, $(1, \beta, \beta^2, \ldots)$—geometric, and $(2, 1, 1, 0, 0, \ldots)$. Nonregular sequences include $(10, \beta, \beta^2, \ldots)$ for $0 < \beta < 1$, $(3, 1, 0, 0, \ldots)$, and $(2, 1, 1, 1, 0, 0, \ldots)$.

Chernoff's Conjecture

A conjecture of Chernoff [11] attempts to relate one- and two-armed bandits. For a given finite-horizon two-armed bandit, consider a modified problem in which arm 1 must be pulled forever if it is ever pulled once. If arm 1 is optimal in the modified problem, then Chernoff conjectured that it is also optimal in the unmodified problem. Since the modified problem is the equivalent of a one-armed bandit, this would imply that the solution of a two-armed bandit problem is partially determined by the solution of a corresponding one-armed bandit. The validity of the conjecture is easy to resolve for many regular discount sequences, but its truth remains an open question in the finite-horizon setting.

CONTINUOUS TIME

The preceding problems are set in discrete time. Chernoff [11] considered a continuous version in which the arms are time-continuous processes (in particular, independent Wiener processes with unknown means and known variances). An arm is observed, payoff accumulates equal to the value of the process, and information about the process accumulates continuously until a switch is made and the other process observed. Observation continues until some fixed time has elapsed. A less than helpful characteristic of every optimal strategy in this problem is that almost every switch is accompanied by an uncountable number of switches within every time interval of positive duration that includes the switch.

OTHER OBJECTIVES

The preceding discussion dealt with the objective of maximizing $\sum \alpha_i Z_i$. At least two different objectives have been considered in Bernoulli bandit problems.

Maximizing Successes before First Failure

Berry and Viscusi [9] consider k-armed Bernoulli bandits in which observation ceases with the first failure. The discount sequence may not be monotonic. This version is analytically simpler than classical bandits. Generalizing Berry and Viscusi [9], Berry and Fristedt [8] show that there is always an optimal strategy using a single arm indefinitely ("no switching") whenever the arms are independent and the discount sequence is superregular:

$$\alpha_{i+1}/\alpha_i \geqslant \alpha_{i+j+1}/\alpha_{i+j},$$

for all positive integers i and j. Superregular sequences are also regular; the geometric and finite-horizon uniform are both superregular.

When the discount sequence is superregular, the class of strategies to be considered is so restricted (to k in number) that an optimal strategy is easy to find. For example, suppose $\alpha_m > 0$ and $\alpha_i = 0$ for $i \neq m$. Then an arm is optimal if it has largest mth moment (average of p_j^m with respect to the prior distribution of the success rate p_j of arm j).

Reaching a Goal

Berry and Fristedt [6, 7] consider the problem of maximizing the probability of achieving n successes minus failures before m failures minus successes in a two-armed bandit (with no discounting). The problem is a sequential allocation version of "gambler's ruin" and is very difficult, with solutions having various nonintuitive characteristics. Speaking somewhat loosely, when one arm has a known success rate, there are circumstances in which it is optimal to use the known arm until "ruin" is approached, at which time the other, riskier arm is used [6].

Fields of Application

In addition to being classified in statistical decision theory, bandit problems are sometimes classified in the areas of learning processes, sequential stochastic design, control theory, on-line experimentation, and dynamic programming*. Since the decision problem typically involves weighing immediate gain against the possibility of obtaining information to increase the potential for future gain, they have attracted much interest in the fields of psychology [21], economics [25], biomedical trials [3], and engineering [31] among others.

As an example, consider the medical trial setting in the initial paragraph of this article. For the sake of simplicity, assume the length of the trial is known to be n and the responses are dichotomous with independent uniform priors on the two success rates. The following chart gives the expected proportion of successes for an optimal strategy for selected values of n:

n	1	5	10	50	100	∞
Maximum success proportion	0.500	0.578	0.602	0.640	0.649	0.667

Optimal strategies are difficult to find and specify. However, Berry [3] gives an easy-to-use strategy whose expected success proportion approximates that of an optimal strategy.

References

[1] Bather, J. A. (1981). *J. R. Statist. Soc. B*, **43**, 265–292.

[2] Berry, D. A. (1972). *Ann. Math. Statist.*, **43**, 871–897.

[3] Berry, D. A. (1978). *J. Amer. Statist. Ass.*, **73**, 339–345.

[4] Berry, D. A. (1983). In *Mathematical Learning Models—Theory and Algorithms*, U. Herkenrath et al., eds. Springer-Verlag, New York, pp. 12–25.

[5] Berry, D. A. and Fristedt, B. E. (1979). *Ann. Statist.*, **7**, 1086–1105.

[6] Berry, D. A. and Fristedt, B. E. (1980). *Adv. Appl. Prob.*, **12**, 775–798.

[7] Berry, D. A. and Fristedt, B. E. (1980). *Adv. Appl. Prob.*, **12**, 958–971.

[8] Berry, D. A. and Fristedt, B. E. (1983). *Stoch. Processes Appl.*, **15**, 317–325.

[9] Berry, D. A. and Viscusi, W. K. (1981). *Stoch. Processes Appl.*, **11**, 35–45.

[10] Bradt, R. N., Johnson, S. M., and Karlin, S. (1956). *Ann. Math. Statist.*, **27**, 1060–1070.

[11] Chernoff, H. (1968). *Sankhyā A*, **30**, 221–252.

[12] Cover, T. M. and Hellman, M. E. (1970). *IEEE Trans. Inf. Theory*, **16**, 185–195.

[13] Fabius, J. and van Zwet, W. R. (1970). *Ann. Math. Statist.*, **41**, 1906–1916.

[14] Feldman, D. (1962). *Ann. Math. Statist.*, **33**, 847–856.

[15] Gittins, J. C. (1979). *J. R. Statist. Soc. B*, **41**, 148–177.

[16] Gittins, J. C. and Jones, D. M. (1974). In *Progress in Statistics*, J. Gani et al., eds. North-Holland, Amsterdam, pp. 241–266.

[17] Isbell, J. R. (1959). *Ann. Math. Statist.*, **30**, 606–610.

[18] Kelley, T. A. (1974). *Ann. Statist.*, **2**, 1056–1062.

[19] Kumar, P. R. and Seidman, T. I. (1981). *IEEE Trans. Aut. Control*, **26**, 1176–1184.

[20] Nordbrock, E. (1976). *J. Amer. Statist. Ass.*, **71**, 137–139.

[21] Rapoport, A. and Wallsten, T. S. (1972). *Ann. Rev. Psychol.*, **23**, 131–176.

[22] Robbins, H. (1952). *Bull. Amer. Math. Soc.*, **58**, 527–536.

[23] Robinson, D. R. (1983). *Biometrika*, **70**, 492–495.

[24] Rodman, L. (1978). *Ann. Prob.*, **6**, 491–498.

[25] Rothschild, M. (1974). *J. Econ. Theory*, **9**, 185–202.

[26] Smith, C. V. and Pyke, R. (1965). *Ann. Math. Statist.*, **36**, 1375–1386.

[27] Sobel, M. and Weiss, G. H. (1970). *Biometrika*, **57**, 357–365.

[28] Thompson, W. R. (1933). *Biometrika*, **25**, 275–294.

[29] Whittle, P. (1980). *J. R. Statist. Soc. B*, **42**, 143–149.

[30] Whittle, P. (1982). *Optimization over Time, Dynamic Programming and Stochastic Control*. Vol. I. Wiley, New York, Chap. 14.

[31] Witten, I. H. (1976). *J. Franklin Inst.*, **301**, 161–189.

(CLINICAL TRIALS
GAMBLING, STATISTICS IN)

DONALD A. BERRY

ONE-SIDED TEST   *See* HYPOTHESIS TESTING

ONE-TAILED TEST   *See* HYPOTHESIS TESTING

ONE-THIRD SIGMA RULE

This is a rule for determining the number of decimal places to be retained in published values of statistics. It specifies that the last decimal place retained shall be the first significant place in the value of one-third of the standard deviation of the statistics.

For example, suppose the arithmetic mean of 60 independent values from a distribution with standard deviation 2 is

$$373/60 = 6\tfrac{13}{60}.$$

The standard deviation of the arithmetic mean is

$$2/\sqrt{60} = 0.258.$$

Since $0.258/3 = 0.086$, we retain up to the second decimal place for the published value of the arithmetic mean, giving

6.22.

This rule is proposed by Kelley [1]. It is not widely used, but it does provide one way of avoiding provision of excessive apparent accuracy.

Reference

[1] Kelley, T. L. (1947). *Fundamentals of Statistics*. Harvard University Press, Cambridge, MA, p. 223.

(ROUNDOFF ERROR)

ONE-WAY ANALYSIS OF VARIANCE

One-way analysis of variance is a method for comparing measurements obtained from two or more independent random samples. In the analysis, the total variability from one observation to another is divided into two parts, with one part reflecting differences among the means of the different random samples and the other part reflecting differences among observations within the individual random samples. It is the simplest of all analysis of variance techniques; its underlying model requires few assumptions and the total variability in the data is divided into only two parts rather than into several parts. A one-way analysis of variance is appropriate when the treatments are levels that involve only one factor. If, for example, six treatments consist of six types of fertilizer, a one-way analysis is used. If, on the other hand, the six treatments consist of three fertilizers, with each fertilizer used at low level and also at a high level, then a two-way analysis of variance is appropriate.

Example. Before discussing the model underlying the analysis, we give a small illustration of the method. In an experiment on corn yields, the objective might be to compare the yield of plots treated with several fertilizers. Yields from three plots using fertilizer containing $K_2O + P_2O_5$, three plots with $N + P_2O_5$, and five plots using no fertilizer might be as in Table 1. Here y_{ij}, $i = 1, 2$, and 3, $j = 1, 2, \ldots, n_i$, denotes the yield from the jth plot using the ith fertilizer, where $n_1 = 3$, $n_2 = 3$, $n_3 = 5$, and $n = 11$.

The total variability in the entire set of observations is measured by the sum of squared deviations from the overall mean $\bar{y}..$ of all $n = 11$ observations (here $\bar{y}.. = 79$). This sum is called the "total sum of squares," or

$$\sum_{i=1}^{3} \sum_{j=1}^{n_i} (y_{ij} - \bar{y}..)^2 = (99 - 79)^2 + (86 - 79)^2$$

$$+ \cdots + (78 - 79)^2$$

$$= 2482.$$

The total sum of squares is then divided into two parts. The first part is due to the difference in yield among the three fertilizer treatments. It is called the "due treatment," "among treatment," or "among fertilizer" sum of squares and is the sum of squared deviations of treatment means $\bar{y}_i.$ from the

Table 1   Illustrative Data Suitable for One-Way Analysis of Variance, Fixed-Effects Model

Fertilizer	Observation $= y_{ij}$	Sum $= \sum_{j=1}^{n_i} y_{ij}$	Mean $= \bar{y}_{i\cdot}$
$K_2O + P_2O_5$	99, 86, 88	273	91
$N + P_2O_5$	81, 90, 75	246	82
None	63, 92, 42, 75, 78	350	70

$$\sum_{i=1}^{3} \sum_{j=1}^{n_i} y_{ij} = 869, \text{ Overall mean} = \bar{y}_{\cdot\cdot} = 869/11 = 79$$

overall mean $\bar{y}_{\cdot\cdot}$ with each squared deviation weighted by the number of observations in the sample. Thus

$$\sum_{i=1}^{3} n_i (\bar{y}_{i\cdot} - \bar{y}_{\cdot\cdot})^2 = 3(91 - 79)^2 + 3(82 - 79)^2$$

$$+ 5(70 - 79)^2$$

$$= 864.$$

The second part of the variability is due to differences in yields in plots treated alike and called the "residual" or "within treatment" sum of squares. It is the sum of squared deviations of each observation from its own treatment mean. Here

$$\sum_{i=1}^{3} \sum_{j=1}^{n_i} (y_{ij} - \bar{y}_{i\cdot})^2$$

$$= (99 - 91)^2 + (86 - 91)^2$$

$$+ \cdots + (78 - 70)^2$$

$$= 1618.$$

These three sums of squares are entered into the second column of an analysis of variance table (see Table 2). The degrees of freedom associated with the three sums of squares are shown in column 3; they are

$a - 1$, $n - a$, and $n - 1$, where a is the number of independent samples, here 3. The mean squares in column 4 are the sums of squares from column 2 divided by the degrees of freedom from column 3. The within-treatment mean square s_e^2 is the usual pooled estimate of the variance of the yields of corn under a particular treatment. MS_a is also an estimate of this variance, provided the fertilizers have no effect on average corn yield. If fertilizers make a difference in mean corn yield, then MS_a reflects that difference and tends to be large. The ratio $MS_a/s_e^2 = 432/202 = 2.14$ (recorded in column 5) under $F_{\text{calc.}}$ should then be large also (*see F-TESTS*).

If, on the other hand, there is no difference among the three population mean yields, then the calculated ratio follows the F-distribution with $a - 1$ and $n - a$ degrees of freedom, provided that yields are approximately normally distributed and do not differ much in variability from one treatment to another. To make a test, with level of significance 0.05, of the null hypothesis that the three means are equal, one compares the calculated F (2.14 as recorded in column 5)

Table 2   Analysis of Variance Table

Source of Variation (1)	Sum of Squares (2)	Degrees of Freedom (3)	Mean Square (4)	$F_{\text{calc.}}$ (5)	F Tabled (0.05) (6)
Between treatments	864	$a - 1 = 2$	$MS_a = 432$	2.14	4.46
Within treatments	1618	$\sum_{i=1}^{a} n_i - a = 8$	$s_e^2 = 202$		
Total	2482	$\sum_{i=1}^{a} n_i - 1 = 10$			

with the ninety-fifth percentile of the F distribution with $a - 1$ and $n - a$ degrees of freedom (4.46 as recorded in column 6). With the rule that one rejects H_0 whenever the calculated F is larger than the tabled F, then if H_0 is actually true, one rejects H_0 5% of the time. If H_0 is false and the means are *not* all equal, then the MS_a and the calculated F tend to be large; thus it is reasonable to reject H_0 whenever $F_{calc} > F_{tabled}$. Since $2.14 < 4.46$, the null hypothesis of no difference among the population means (and thus among the three fertilizer treatments) is accepted; the data have not proved that the population mean yields differ. If the calculated F ratio had been, say, 6.14, then $6.14 > 4.46$ and one would conclude that differences exist among the three population mean yields. Comparisons among the various population means may also be made.

FIXED-EFFECTS MODEL

The model used in the example is called the fixed-effects* model; it is a reasonable model when the treatments studied have been chosen or fixed by the researcher. In the example, three fertilizer treatments were chosen because they were of interest to the experimenter.

In this model, each y_{ij} is expressed as the sum of three parts, $y_{ij} = \mu + \alpha_1 + \epsilon_{ij}$, $i = 1, \ldots, a$; $j = 1, \ldots, n_i$ where $\mu + \alpha_i$ is the population mean for the ith treatment, μ is the mean of the a population means, so that $\sum_{i=1}^{a} \alpha_i = 0$; the ϵ_{ij} are independent observations from a normal distribution with zero means and with variance σ^2. Thus for each treatment, the observations are from a normal distributions, and all a normal distributions have equal variances.

VARIABLE-EFFECT MODELS

In some situations in which one has an independent sample from each of a populations, it is more realistic to consider the treatment effects as variable (random or component-of-variance) effects. For example, the a samples might consist of test scores from a sample of graduating seniors at each of a high schools. In this case, the high schools might have been chosen at random from a population of high schools from a certain city. The data set is of the same form. A more appropriate model, however, if the population of high schools is very large, is as follows. Again the observations are expressed as $y_{ij} = \mu + \alpha_i + \epsilon_{ij}$, $i = 1, \ldots, a$; $j = 1, \ldots, n_i$ and again $\mu + \alpha_i$ is the population mean for the ith treatment, and the ϵ_{ij} are independently normally distributed with zero means and equal variances σ^2. The difference lies in the assumption that now the α_i have been drawn from a normal distribution with zero mean and variance σ_a^2; the α_i and ϵ_{ij} are also assumed to be distributed independently.

This variable-effects model is in appearance almost the same as that of the fixed-effects model. Now, however, μ is the population mean test score for the entire population of schools rather than the mean of just a schools. The one-way analysis of variance table and F-test are made exactly as with the fixed-effects model. The null hypothesis being tested is $H_0 : \sigma_a^2 = 0$, so that if H_0 is rejected, one concludes that mean test scores are not all the same for all the city's schools.

The essential difference between fixed- and variable-effects models is that for the fixed model, one has chosen to study a particular populations; in the variable model, the populations studied are considered to be a sample from a larger set of populations.

In the high school test score example, a somewhat different variable-effects model applies if the a high schools form a sizable proportion of the total number of high schools. One then thinks of sampling from a finite population of high school effects. For discussion of the various models, see Scheffé [5] (*see also* VARIANCE COMPONENTS).

THE ASSUMPTIONS

The assumptions underlying these models are thus normality, equality of variances, and independence. For a discussion of the robustness* of the analysis under violations of these assumptions, see Scheffé [5, 6].

In general, lack of independence can cause serious trouble in making inferences on the population means.

Departures from normality have little effect on the F-test and on comparisons among means under the fixed-effects model; under the variable-effects model, however, consequences can be serious.

Inequality of variances has little effect on inferences concerning means provided the sample sizes are equal; for unequal samples sizes, the effect can be serious.

Thus one is reasonably free to make inferences on means in a one-way analysis of variance provided one has independent observations, sample sizes approximately equal, and uses the fixed-effects model. In situations in which nonnormality and unequal variances may be present and the data set is too small to check these assumptions, then even though a variable-effects model is appropriate, it is still possible to analyze the data as a fixed-effects model provided the sample sizes are approximately equal. One then draws inferences concerning only the particular a populations. When the data set is larger or when one has some theoretical knowledge, nonnormality and inequality of variances can be reduced by transformations on the data. Bishop and Dudewicz [1] suggest a two-stage sampling procedure for use with the fixed-effects model when unequal variances are suspected. Their procedure yields an exact F-test.

Historical Background

Fisher [3, 4] suggested the basic ideas of the one-way analysis of variance using the test statistics $z = \frac{1}{2}\log F$. Snedecor (see [7]) introduced the technique into the United States, and named the F-test in honor of R. A. Fisher. For the historical development of variable-effects models, see Scheffé [5].

References

[1] Bishop, T. A. and Dudewicz, E. J. (1978). *Technometrics*, **20**, 419–30.

[2] Dunn, O. J. and Clark, V. (1974). *Applied Statistics: Analysis of Variance and Regression*. Wiley, New York.

[3] Fisher, R. A. (1925). *Statistical Methods for Research Workers*. Oliver and Boyd, Edinburgh, Scotland.

[4] Fisher, R. A. (1935). *The Design of Experiments*. Oliver and Boyd, Edinburgh, Scotland.

[5] Scheffé, H. (1956). *Ann. Math. Statist.*, **27**, 251–271.

[6] Scheffé, H. (1959). *The Analysis of Variance*. Wiley, New York.

[7] Snedecor, G. W. and Cochran, W. G. (1980). *Statistical Methods*, 7th ed. Iowa State University Press, Ames, IA.

(ANALYSIS OF VARIANCE
DESIGN OF EXPERIMENTS
FIXED-, RANDOM-, AND
  MIXED-EFFECTS MODELS
VARIANCE COMPONENTS)

OLIVE JEAN DUNN

ONE-WILD DISTRIBUTION

The One-Wild distribution refers to a collection of n (> 1) independent random variables of which $(n - 1)$ are from a standard normal population and one is from a normal population having the same mean but 100 times the variance. Strictly speaking, it is not a univariate distribution, and, since the random variables are not distributed identically, it is not a random sample. The One-Wild is used in simulations to represent a possible situation in which a data set contains an outlier*. For example, a decimal point may have been misplaced, or some other cause may have resulted in an observation whose precision is far less than that of the remainder of the sample observations.

As explained in Andrews et al. [1, p. 67], there is a correspondence between the One-Wild situation and the $(100/n)\%$ scale-contaminated normal distribution CN_p, which has density function

$$(1 - p)N(0, 1) + pN(0, 100),$$

and   (1)

$$N(\mu, \sigma^2) = \left(\sigma\sqrt{2\pi}\right)^{-1}\exp\left[-\tfrac{1}{2}(x - \mu)^2\sigma^{-2}\right]$$

where $p = 1/n$. The likelihood of a random sample generated by (1) may be expressed as

$$P\{\text{sample} \,|\, CN_p\} = \sum_{r=0}^{n} \binom{n}{r}(1-p)^{n-r}p^r$$
$$\times P\{\text{sample} \,|\, r \text{ wild}\}.$$

However, the likelihood for a One-Wild sample consists of only one of these terms, that corresponding to $r = 1$. It may be written

$$P\{\text{sample} \,|\, \text{One-Wild}\}$$
$$= n^{-1} \sum_{i=1}^{n} \prod_{j \neq i} \phi(X_j)$$

If the main part of the sample comes from $N(\mu, \sigma^2)$, and the contaminant from $N(\mu, 100\sigma^2)$, then the form for the One-Wild likelihood function makes it difficult to determine algebraically maximum likelihood estimates (MLEs) of μ and σ^2 from such a sample. Typically, however, the One-Wild situation is used in simulation studies as one of several extreme cases, where a procedure that compromises among all of the cases is the main objective. Thus an MLE for any one of the extreme cases, such as One-Wild, is not of interest.

Other values for the scale parameter of the contaminant have been used [e.g., $N(0, 9)$ to represent a mild outlier in ref. 1]. Simulation of the One-Wild distribution is straightforward: generate n pseudo-random normal deviates, generate one pseudo-random integer between 1 and n, say r, and multiply the rth normal deviate by 10.

Acknowledgment

The author wishes to thank J. W. Tukey for his comments on an earlier version of this article.

References

[1] Andrews, D. F., Bickel, P. J., Hampel, F. R., Huber, P. J., Rogers, W. H., Tukey, J. W. (1972). *Robust Estimates of Location: Survey and Advances.* Princeton University Press, Princeton, NJ.

[2] Gross, Alan M. (1976). *J. Amer. Statist. Ass.*, **71**, 409–417.

[3] Kafadar, K. (1982). *Commun. Stat. Theor. Meth.*, **11**, 1883–1901.

KAREN KAFADAR

O, o NOTATION

O means (loosely) "of order of"; o means (loosely) "of lower order than." The notation is used in connection with limiting processes, and these should be clearly identified. Thus

$a_n = O(1)$ as $n \to \infty$ means that $a_n \leqslant c$ for some c and all sufficiently large n.

$a_n = o(1)$ as $n \to \infty$ means that $a_n \to 0$ as $n \to \infty$.

$a_n = O(1/n)$ as $n \to \infty$ means that $a_n \leqslant c/n$ for some c and all sufficiently large n.

$a_n = o(1/n)$ as $n \to \infty$ means that $a_n n \to 0$ as $n \to \infty$.

"OPEN END" CLASS

A class with only one specific end-point, being unlimited at the other end. Either, or both, of the extreme classes of a grouped frequency distribution may have an open end. (For example, in a distribution of annual incomes, the top class might be "in excess of \$99,999.") In view of the indefiniteness involved, use of open end classes should be avoided if possible.

OPERATIONAL CHARACTERISTIC CURVE   *See* ACCEPTANCE SAMPLING

OPERATIONAL RESEARCH SOCIETY, JOURNAL OF THE

The Operational Research (O.R.) Club grew from a small group of World War II scientists who started to meet regularly in April 1948 to discuss their experiences and problems in developing the field of O.R. A restricted membership was established with confidential papers presented on various applications.

In March 1950 the Club launched its journal, the *Operational Research Quarterly*, under the joint editorship of Max Davies and R. T. Eddison, with a paper by Professor P. M. S. Blackett, FRS. This venture predated all other journals in the field by over

Figure 1   Contents of *Journal of the Operational Research Society*, Vol. 35, No. 9 (September 1984).

$2\frac{1}{2}$ years. In November 1953 the Club became a Society and the journal started to grow in size. With the ninth volume in 1958, there was a change of journal size and style and R. T. Eddison assumed sole editorial responsibility in 1959, which he continued to exercise until 1964. He was succeeded by Dr. J. R. Lawrence, who provided a number of changes, including the introduction of a fifth issue each year to this quarterly magazine. In 1969 Professor R. A. Cuninghame-Green assumed the role of editor and also effected a change in style and layout that was carried on from 1972 by Professor K. B. Haley, who reacted to the major publication pressures by dividing the quarterly issues into two parts in 1975 and then adopting a monthly journal in 1978. Peter Amiry assumed the mantle of editor in 1981 and, with the advent of microcomputers, started to introduce special camera-ready copy to provide a very rapid publication time for some articles.

The editors, although reflecting different demands and needs, have been able to publish an increasing number of papers. The average of two per issue in the first five years grew to seven in the 1960s. The last true quarterly in 1974 averaged 13 and the number then settled down to about 11 per issue or 130 full papers a year since 1977 (see Fig. 1).

Statistics have indicated a very steady growth in papers submitted to a situation where about 45% of the 300 received can be published. The division between content has also remained very steady in terms of the balance between Theoretical and Application. About 40% of all papers have always been concerned with the use in practice of O.R. and all editors have encouraged more contributions of this type. About half the articles are written by U.K. contributors and one quarter of the authors are industrially based.

The journal appears to have reached a stable position of about 1200 pages and 130 articles. It also includes book reviews and letters and, although there are now more than 20 other similar publications throughout the world, its tradition for practicality has not been surpassed.

K. B. HALEY

OPERATIONS RESEARCH

Operations Research (O.R.) is defined to be the application of scientific methods to provide a basis for executive decisions and actions. Variants on this definition are also extant. The importance of O.R. lies in the major improvements in payoff that can and often do result from its employment. Those who perform O.R. will be called here "operations analysts." The "operations" to which O.R. is applicable are usually complex organizations of people and equipments engaged in pursuit of one or more objectives.

The identifiable origins of O.R. were in military applications in the United Kingdom during the late 1930s. Military use of O.R. has subsequently expanded rather pervasively, and since World War II, O.R. has seen widespread use in nonmilitary governmental agencies and nongovernmental enterprises in many nations. (*See also* MILITARY STATISTICS.)

The relevance of the statistical sciences to O.R. is immediate. Evaluating the success of a course of action of an operation, aside from choosing among alternatives, is best based on empirical data, preferably interpreted by statistical inference*. The operations to which O.R. is applied are usually fraught with uncertainty, which is preferably accounted for through use of probability theory as a calculus of uncertainty. As an example of areas of O.R. in which the statistical sciences are especially prominent, *see* SEARCH THEORY*; *see also* DECISION THEORY*.

The rapid advances in computers and user-friendly software since the late 1960s have had an enormous impact on O.R. in simulation of an operation, in scientific computation associated with O.R. analysis, and in implementing O.R. results through computerized decision aids.

Optimization methods are natural mathematical tools in O.R. They include linear programming*, (nonlinear) mathematical programming*, dynamic programming*, integer programming*, combinatorial optimization, control theory, classical calculus of variations, game theory*, network analysis*, etc.

Synonyms for operations research include *operational research* (notably in the United Kingdom) and *operations analysis* (U.S. Air Force). A German language term for O.R. is the English *operations research*. *Management science** is the same as O.R., but those who use the term usually refer to nonmilitary applications, primarily to business problems. *Systems analysis* is close to O.R., but usually connotes emphasis on system design. *System engineering* usually goes further in the latter direction with more emphasis on systems of equipments. The methods employed under all these terms have a great deal in common.

In the definition given as our opening sentence, three words have significance worth noting.

O.R. merely provides a *basis* for decisions; decisions on courses of action are made by the executive in charge of an operation, who has the authority to take the actions and the responsibility for their consequences. An executive will often superimpose, on the results of O.R., knowledge and judgments outside the purview of an operations analyst, e.g., in morale areas. The *advisory* role of an operations analyst is not the same as the decision-making role of an executive, even when an executive puts an O.R. recommendation into action without change; *see* CONSULTING*, STATISTICAL.

O.R. is the *application* of scientific methods. Academic discourse on O.R. methodology can be highly useful preparation for participation in O.R., but the conduct of O.R. requires a user executive who has a problem amenable to O.R. and who can take action based on the O.R. results. This distinction is not merely semantic: The best learning of O.R. approaches is by doing, i.e., by application of scientific methods, learned academically or otherwise, to real problems of decision makers.

The methods of O.R. are *scientific*. To begin with, they place great demands on objectivity, and O.R. contributes much merely by enhancing objectivity in decision making. Scientists are not alone in their *desire* for truth, but as a disciplined search for truth, science has objectivity standards that are deeper, in general, than those of the executive decision makers. These standards are enhanced by empirical deduction and demonstration.

Moreover, the scientific aspect of O.R. is manifested in the *creativity* that is characteristic of the best in science. The techniques that are employed in O.R. are often deceptively elementary, but O.R. practice makes great demands on scientific insight to penetrate to the most important aspects of a decision problem and on the ability to create new techniques to solve new problems that are not found in textbooks.

Ensuing sections discuss in turn the conduct of an O.R. case, O.R. staffing and organization, organization of the O.R. profession, an O.R. case history, and the history of O.R. A general discussion of O.R. literature is given at the end.

CONDUCT OF O.R.

The conduct of an O.R. case usually begins with a problem discussion between a user executive and an operations analyst. It might be initiated by either. Presumably, the operation controlled by the executive appears to bear improvement in some respect. This might be perceived by the executive as a vaguely described conflict among competing influences.

Immediately the operations analyst can be useful in reducing the problem to quantitative statements that can be subsequently subjected to analysis. The operations analyst thus helps to *define* the problem, which is a significant advance toward its solution, although the problem might not really be completely understood until the investigation is well under way. At this problem definition stage, the operations analyst often helps to identify "payoff" parameters of most direct concern to the executive; such a parameter is usually called a *measure of effectiveness* (MOE). Examples of MOEs are stock-out probabilities in inventory control, market share in sales campaigns, detection or destruction probabilities in military operations, etc.

The most useful MOEs are quantities that vary in the same direction as ultimate payoff but are closer to measurable experience than ultimate payoff. An instructive example from World War II is *sweep width* [10, 13, 18] used to measure detection effectiveness of a sensor system against targets passing in relative motion. Let a target passing at distance to closest approach x have probability $f(x)$ of being detected at *some* point in its passage; $f(x)$ is determined empirically, theoretically, or both. The sweep width of the sensor system against this population of targets is $W = \int_{-\infty}^{\infty} f(x)\, dx$. Its dimension is distance. One may use this MOE to compare alternative sensors. One may also use it to plan search operations by laying search tracks W apart, or by estimating detection probability as W/L, where L is the frontage crossed by targets. If (relative) sensor speed V is variable, a better MOE is *sweep rate*, defined as WV, which gives swept area per unit time.

For O.R. to be useful, the potential improvement sought should be at least 10%–20% (this is expressed more strongly in the discussion of "hemibel thinking" in Morse and Kimball [13]), since O.R. conclusions usually do not attain accuracies to, say, 1%–2%. Some caveats to this traditional O.R. viewpoint should be noted: In business applications, large percentage changes in profit can be obtained by small percentage changes in revenue or expense. Also in some modern well-developed and well-defined areas of application, accuracies can be high and small percentage improvements can be important, e.g., mathematical programming* applications to certain problems in finance* and to oil transportation.

The approaches under discussion have parallels to those of laboratory science and abstract research, but they differ from those areas in that the MOEs are payoff parameters of most direct interest to the user executive and are usually not susceptible to laboratory measurement. This is one means of distinguishing O.R. from most other forms of applied science or engineering.

The operations analyst proceeds by acquiring a better understanding of the operation in question, probably through consultations with the executive's staff. Usually it is important to construct at this stage a quantitative description of the operation that is sensitive to the parameters that have the most significant influence on the MOE(s). Such a description is a *model* of the operation. The desired attributes of a model are that it be faithful to reality in that the significant influences on payoff are realistically reflected and that it be tractable to analysis, which usually requires that parameters of insignificant influence be suppressed. Often, perhaps usually, the model is stochastic in character. The ability to devise good models is one of the most important capabilities to be found in an operations analyst. It results from insightful talent and experience. It is difficult to learn or teach in a classroom, detached from real problems and their associated user executives.

Parameter estimation to complete the model usually requires empirical data reflecting past experience with the operation. Often it is necessary to conduct special trials by exercising the operation in question. These might lead to model reformulation. Such trials are expensive and deserve statistical design and analysis; good data thus acquired, usually in small samples, are much to be prized.

Given a problem statement and a reasonably faithful and tractable model of the operation, the operations analyst can proceed with analysis to determine recommended courses of action to be considered by the executive. Perhaps mathematical optimization methods are employed at this point. Perhaps the problem is to choose among a few alternatives, so that the difficulty is in making a realistic estimate of the outcome of each of them. Perhaps the model has been programmed as a computerized simulation, to be exercised repetitively with variations, possibly by Monte Carlo* computation. It is not uncommon for the analysis stage of an O.R. case to be relatively easy once the problem definition and modeling have been accomplished. On the other hand, deep abstract reasoning or expensive computation may be needed at this stage.

The results of the analysis might be subjected to verification through operational trials, again expensive. When the desired conclusions have been obtained, they are reported to the user executive, verbally and in writing, and probably the same is done at interim stages with tentative conclusions. The executive decides on a course of action and may utilize the operations analyst further in the implementation, for example, by preparing computerized decision aids for use by the executive's staff.

Indeed, it is important for the operations analyst to remain with the case during implementation, to recheck the assumptions, to react accordingly, to guard against misunderstandings between the recommending analysts and the implementing personnel, and to continue the education of the operations analyst. On this particular point, in-house O.R. personnel might have an advantage over outside consultants.

STAFF QUALIFICATIONS AND ORGANIZATION

What are the staff qualifications and organization desirable to provide O.R. support to one or more user executives? The principal staff qualification needed is excellence in a quantitative science. To succeed in O.R. practice, an operations analyst needs the ability to do creative independent research.

An equally important qualification is a sincere interest in applying one's scientific talents to serve the needs of user executives who are presumably not scientists themselves. The importance of the influence one's O.R. work exerts on "real life" affairs should be a major source of satisfaction to an operations analyst, in contrast somewhat to satisfaction one obtains from science for the sake of science and its elegance in abstraction.

An operations analyst must be able to communicate in both directions with nonscientist users, verbally and in writing.

It is important that an operations analyst have fairly direct access to the user executive whose decisions are being served by the O.R. work; see, e.g., Morse and Kimball [13].

The foregoing qualities should suffice to enable the conduct of worthwhile O.R. If an incipient operations analyst has had training in fundamentals of probability, statistics, optimization methods, and computing methods (and most modern training in quantitative science at least touches on most of these), so much the better. Otherwise, such fundamentals must be learned on the job. Also, for O.R. in business problems, a knowledge of accounting is useful. If one's scientific training is in O.R. per se, one will have more familiarity with O.R. approaches and techniques than otherwise, which is further desirable. However, the specialization in one's scientific training is not as important as demonstrated scientific excellence and motivation toward applications.

As to the composition of an O.R. group, much success has been obtained with "mixed teams." This refers to an aggregation of scientists from diverse disiplines brought together to do O.R. The mixed team was a point of emphasis in Blackett's early groups (see below) and has characterized most O.R. groups ever since. Some hold that a mixed team is *necessary* to do good O.R., considering that methods and viewpoints from separate disciplines reinforce each other when brought together. Other observers put the emphasis differently, i.e., as above, to the effect that background specialty is subordinated to excellence; while scientists of diverse backgrounds can be compatible with each other in the conduct of O.R., such diversity is not a necessity (see Wagner [19]).

An in-house O.R. team generally has an advantage over outside consultants in facility of remaining with the problem during the implementation of O.R. recommendations. On the other hand, outside consultants might be able to bring talents and expertise not available to an in-house group.

PROFESSIONAL ORGANIZATION

The worldwide scope of O.R. is indicated by the size of the International Federation of Operations Research Societies (IFORS), founded in 1959. In 1982 IFORS consisted of 37 component societies, including the principal O.R. societies in each of 32 separate nations. The Soviet Union is not represented.

IFORS publishes *International Abstracts in Operations Research* (IAOR), which presents abstracts (some by reviewers, but mostly by authors) of all of the articles in 24 O.R. periodicals and of selected articles in 50 supplemental periodicals that include O.R. among other areas of contribution. In 1981, IAOR had published 2360 abstracts. These are indexed by an extensive scheme of subjects. IFORS also holds triennial international conferences in O.R.

The oldest O.R. society is the United Kingdom's Operational Research Society formed in 1950. The largest are the Operations Research Society of America (ORSA) and The Institute for Management Sciences (TIMS), whose respective starts were in 1952 and 1953 and whose 1982 memberships were 6952 and 6760 (13,712 in combination). ORSA and TIMS share their national meetings and some other functions.

The O.R. periodicals of widest circulation are *Operations Research** (initiated 1952), published by ORSA, and *Management Science* (initiated 1954), published by TIMS. Prominent O.R. periodicals outside the United States include the *Journal of the Operational Research Society* (initiated 1950) and the *European Journal of Operational Research*. The latter is published by the Association of European Operational Research Societies, within IFORS, formed in 1975.

Excellent O.R. case histories in business applications are found in the prize papers from the Management Science Achievement Award Competition, held annually by the TIMS College on Practice of Management Science. These appear in the final issue each year of *Interfaces*, a TIMS–ORSA periodical. *Operations Research* frequently publishes feature articles, which are often surveys, as lead articles.

The profession's best known award is ORSA's annual Lanchester Prize, for the year's best English publication of O.R. work. This commemorates F. W. Lanchester (see below).

The Military Operations Research Society involves an additional considerable body of O.R. through semiannual symposia with published proceedings, access to which requires a U.S. Department of Defense security clearance.

Addresses of various O.R. societies and periodicals are given in IAOR.

ILLUSTRATIVE CASE HISTORY

Consider the first-prize paper [1], in the above-mentioned TIMS competition for 1981. Following is a somewhat simplifed summary of this O.R. work which was highly profitable to its users; a fuller account is given in Barker et al. [1].

The user management operates intercity trucking services, with nonspecialized cargo types, primarily less-than-truckload (LTL) shipments and secondarily full-truckload (TL) shipments that have lower tariff rates and lower profit margins. TL shipping is marketed in part to serve correction of routing imbalances in the LTL shipping. Transloading at "break-bulk" stations helps to blend LTL A-to-C freight with B-to-C freight, also A-to-B with A-to-C (A, B, and C are cities).

Business growth was straining the capacity of a major break-bulk station. Management assigned an in-house O.R. team to develop recommendations for location of a new break-bulk station. Management perceived that this problem should not be addressed by itself and that more general analyses of freight movement were needed.

The MOE is profit. Long-term impact on profit is recognized through service reputation and marketing considerations. It turned out to suffice to consider one-time and recurring contributions to profit separately.

As a first approach to modeling, an analytic model was developed that could have produced optimal guidance to load and route trucks. However, this approach lacked promise in timely computation and facility of implementation, so a simulation of LTL freight flow was developed and used instead. This LTL model utilized existing accounting, billing, and dispatch monitoring systems that were already in place, as natural sources of inputs. It also utilized an existing loading guide as a preliminary source of tactical advice and as a vehicle to improve future loadings. Outputs included load factor and line-haul costs to measure productivity, service statistics to afford trade-offs between service and cost, utilization of break-bulk stations, and total system costs.

An additional model was developed for TL traffic, complementing the LTL model. The TL model utilizes linear programming to determine whether TL traffic solves LTL imbalance, the acceptable quantity of loads on all lanes to achieve this, and the revenue of the least acceptable load.

By exercising the LTL and TL models interactively with operations and interim management reactions and conjectures, the O.R. team arrived at several useful recommendations for management. Evidently those were adopted, and the following results were reported, with identified profitability of several million dollars:

1. Contrary to a priori conjecture, a new break-bulk station was neither indicated nor created. In fact, much existing break-bulk activity was profitably distributed to origin/destination terminals. In particular, instances were identified

where it was advantageous to hold freight at origin for a day to ship direct and thus obviate handling and a day of delay at a break-bulk station.

2. Load-factors were improved without harm to service by using the LTL model to reroute trucks and freight and by increased loading discipline.

3. Adverse effects of a recession in business activity were withstood better than historically, as measured by empty miles. This was achieved by optimal routing of empties using the LTL model and identifying TL sources of empties using the TL model. Identification of TL-derived empties was incorporated advantageously into marketing.

4. More realistic accounting was afforded, particularly for TL operations.

5. Management effort in planning meetings to respond to changes in business activity was reduced by model assistance and was redirected to tactical improvements.

6. Fleet sizing was made more realistic.

7. Economies of scale were investigated as guidance to new acquisitions.

Barker et al. [1] emphasize the importance of management interest, support, and interaction to the success of their O.R. endeavors. Their report adds considerable elucidation and insights to the preceding summary.

O.R. HISTORY

Examples of executives, mostly military commanders, calling upon scientists for advice have occurred throughout history. Probably design of better weapons and other gadgetry was usually sought, but at times science was used to improve tactical employment of weapons.

In 1916, Lanchester [11] published theoretical analyses of military force levels by using differential equations to relate these levels to individual combat "exchange ra-

tios." Lanchester's work was extended and applied post-World War II, but did not see application in his era. As a consultant to the U.S. Secretary of the Navy in World War I, Edison (see Whitmore [21]) performed impressive O.R. investigations in antisubmarine problems. These had no affect on operations, probably because Edison lacked contact with uniformed commanders.

As an organized body of activity, O.R. had its origin in the planning, and later the execution, of the air defense of Britain in the late 1930s. A rather thorough account of the history of O.R. from that era through 1952 is given by Trefethen [17].

Radar developments in the 1930s gave the United Kingdom radically new capabilities in early warning of air attack. These capabilities raised numerous questions of operational employment, initially in terms of integration with existing warning-response organizations based on visual detections. The earliest study of such questions and accordingly the birth of organized O.R. is attributed to a group of scientists at Bawdsey Research Station under A. P. Rowe in 1937–39. Radar scientist R. Watson-Watt was an early instigator. This and ensuing O.R. work, in what was much later known collectively as communications, command, and control (C^3), was of enormous importance to the effectiveness of fighter and antiaircraft gun response to air attack during the Battle of Britain. O.R. spread extensively in all three of the U.K. armed services, notably in antisubmarine warfare, offensive air attack, and civil defense. The most prominent British O.R. pioneer was P. M. S. Blackett, later a Nobel laureate in physics. He was the first to staff O.R. work with scientists from diverse fields unrelated to the technology of the equipments employed. Early writings of his, reproduced in ref. 3, were influential in O.R. in the United States as well in the United Kingdom.

Literally the first U.S. O.R. organization was the Operational Research Group established at the Naval Ordnance Laboratory March 1, 1942. This arose from seminar activity led by E. A. Johnson. This group and its descendants addressed mining operations throughout the war.

The greatest thrust in U.S. World War II O.R. began with the recruitment April 1, 1942 of MIT physicist P. M. Morse to lead O.R. assistance to the antisubmarine command based at Boston. Morse had had experience with naval underwater acoustic devices. A month later, his group, the Anti-Submarine Warfare Operations Research Group (ASWORG), stood at seven scientists and had produced an impressive report on search analysis. After three months, ASWORG was transferred to the staff of Commander-in-Chief, U.S. Fleet in Washington, where it continued to make significant contributions to the antisubmarine thrust of the Battle of the Atlantic. From that position it extended its influence to most aspects of naval warfare. Hence in October 1944, ASWORG was renamed the Operations Research Group (ORG). ORG grew to about 75 scientists by the war's end. It had very direct involvement with fleet operations through rotation of its analysts on field assignment to combat commands.

In fall 1942, O.R. got underway in the U.S. Army Air Forces under impetus of its commanding generals in England and Washington, again inspired by U.K. experience in O.R. Most U.S.A.A.F major commands came to have O.R. groups. In contrast to the Navy's ORG, the Washington headquarters function was largely one of staffing the deployed groups.

For accounts of classic O.R. work and methods in World War II, see Morse and Kimball [13], Koopman [10], Blackett [3], Waddington [18], Crowther and Whiddington [5], Trefethen [17], and Brothers [2].

After World War II, military O.R. staffing in the United States and the United Kingdom temporarily declined but acquired permanence in all of the armed services and higher levels of these defense establishments. Subsequently, U.S. military O.R. involving private industry as well as government, has become a much larger community than in

World War II (and consequently its early elitism has been diluted), engaged in a wide variety of studies in current tactics, operational requirements for new systems, and strategic planning.

In the early 1960s, Secretary of Defense Robert McNamara and his Comptroller, C. J. Hitch, employed O.R. at the highest levels of U.S. defense planning. This employment appears to have had mixed success amid controversy over the extent of its contravention of professional judgments and its distance from measurable experience.

Development of nonmilitary O.R. began shortly after World War II. Most operations analysts returned to the academic scientific pursuits they had left to join war efforts. Many, however, turned eagerly to nonmilitary problems amenable to the O.R. methodologies and approaches that were developed in wartime.

In the United Kingdom, these efforts began with assistance to government-controlled nationalized industries. In the United States, the nonmilitary start was through leading management consulting firms, beginning with Arthur D. Little, Inc. Later most of the largest few hundred industrial and financial corporations came to have in-house O.R. groups. The oil industry has made particularly widespread use of O.R., e.g., linear programming to solve oil transport problems. Major accounting firms typically offer O.R. through their management consulting adjuncts. The growth of the professional societies formed in the 1950s, noted above, was much linked to progress in nonmilitary O.R. Propagation of O.R. in continental Europe was promoted by NATO in the late 1950s under the leadership of P. M. Morse.

After the mid-1970s a trend in corporate use of O.R. was discernible in that headquarters O.R. groups have often been dispersed to operational units. Intimacy with the realities of operations is advantageous, as was learned in World War II. However, an assignment that merges an operation analyst more or less *permanently* with an operating unit can have a disadvantage—loss of group scientific esprit. More importantly, it can lose freshness of approach that often comes from bringing to bear creative talent that has not become inbred with the operation.

Academic programs for graduate work in O.R. began soon after World War II. The first course offering was at MIT in 1948. Case Institute of Technology and the U.S. Naval Postgraduate School began the first degree programs in 1951. Subsequently most leading universities have offered doctorates in O.R., often in departments bearing that name.

In its fifth decade, the O.R. profession still has identification problems among the potential users of its work and problems in quality control of this work. Nevertheless, through a lengthy and extensive record of accomplishments, O.R. has a firmly established position as a means for science to expand its influence on the improvement of world affairs. With the increased complexity of world affairs and advances in computation tools, unbounded challenges and a wealth of interesting problems await the operations analysts of the future.

Literature

Some of the best sources in O.R. approaches and viewpoints are still to be found among accounts of classic O.R. work in World War II: Morse and Kimball [13], Koopman [10], Blackett [3], and Waddington [18] are particularly recommended. Subsequent advances in mathematical techniques and computation tools applicable to O.R. have led to literature much too prolific to be listed here. Good representation of such results is included in the following, which are among the leading texts used in academic instruction in O.R.: Hillier and Lieberman [8], H. Wagner [20], Eppen and Gould [6], and Starr [16]. Churchman et al. [4] is an earlier prominent text, emphasizing systematic examination of the relationships internal to an organization to which O.R. is applied.

A general survey article on O.R. is given by Miser [12]. Jewell [9] surveys analytic methods in O.R. For anatomy of O.R. cases, one may see Raisbeck [15]. A survey of the role of mathematicians in O.R. is given by D. Wagner [19]. Accounts of some theoretical modeling methods are given in Gass [7]. A guide to utilization of O.R. in developing countries is presented by a panel headed by Morse [14].

As noted earlier, the most comprehensive source of references to the O.R. literature, periodicals and texts, is IAOR.

References

[1] Barker, H. H., Sharon, E. M. and Sen, D. K. (1981). *Interfaces*, **11**, 4–20.

[2] Brothers, L. A. (1954). *J. Operat. Res. Soc. Amer.*, **2**, 1–16.

[3] Blackett, P. M. S. (1962). In *Studies of War*, Part II: *Operational Research*, Hill and Wang, New York.

[4] Churchman, C. W., Ackoff, R. L., and Arnoff, E. L. (1957). *Introduction to Operations Research*. Wiley, New York.

[5] Crowther, J. G. and Whiddington, R. (1948). *Science at War*. The Philosophical Library, New York.

[6] Eppen, G. D. and Gould, F. J. (1979). *Quantitative Concepts for Management*. Prentice-Hall, Englewood Cliffs, NJ.

[7] Gass, S. I., ed. (1981). "Operations Research: Mathematics and Models," Proceedings of Symposia in Applied Mathematics, Vol. 25, American Mathematical Society, Providence, RI.

[8] Hillier, F. S. and Lieberman, G. J. (1967). *Introduction to Operations Research*. Holden-Day, San Francisco.

[9] Jewell, W. S. (1977). *Philos. Trans. R. Soc. Lond. A*, **287**, 273–404.

[10] Koopman, B. O. (1946). "Search and Screening." *Operations Evaluation Group Rep. No. 56*. Office of the Chief of Naval Operations, Washington, DC.

[11] Lanchester, F. W. (1916). *Aircraft in Warfare: The Dawn of the Fourth Arm*. Constable, London.

[12] Miser, H. J. (1980). *Science*, **209**, 139–146.

[13] Morse, P. M. and Kimball, G. E. (1951). *Methods of Operations Research*. MIT Press, Cambridge, MA.

[14] Morse, P. M., chairman. (1976). *Systems Analysis and Operations Research: A Tool for Policy and Program Planning for Developing Countries*. National Academy of Sciences, Washington, DC.

(Panel on strengthening the capabilities of less developed countries in systems analysis.)

[15] Raisbeck, G. (1976). *Amer. Math. Monthly*, **83**, 681–701.

[16] Starr, M. K. (1971). *System Management of Operations*. Prentice-Hall, Englewood Cliffs, NJ.

[17] Trefethen, F. N. (1954). "A History of Operations Research," in *Operations Research for Management*, J. F. McCloskey and F. N. Trefethen, eds., Johns Hopkins University Press, Baltimore, MD, pp. 3–35.

[18] Waddington, G. C. (1973). *O.R. in World War 2, Operational Research Against the U-Boat*. Elek Science, London.

[19] Wagner, D. H. (1975). *Amer. Math. Monthly*, **82**, 895–905.

[20] Wagner, H. M. (1975). *Principles in Operations Research*. Prentice-Hall, Englewood Cliffs, NJ.

[21] Whitmore, W. F. (1953). *J. Operat. Res. Soc. Amer.*, **1**, 83–85.

Added in Proof:

Cunningham, W. P., Freeman, D., and McCloskey, J. F. (1984). *Oper. Res.*, **32**, 958–967. (Origins of O.R. in England, in the late 1930's.)

Acknowledgment

Grateful appreciation for critiques and assistance with source material, without, of course, attribution of responsibility, is expressed to Herbert F. Ayres, Henry H. Barker, David M. Boodman, Sidney W. Hess, John D. Kettelle, Philip M. Morse, William P. Pierskalla, Henry R. Richardson, Lawrence D. Stone, and Keith R. Tidman.

(CONSULTING, STATISTICAL
DECISION THEORY
DYNAMIC PROGRAMMING
ECONOMETRICS
LINEAR PROGRAMMING
MANAGEMENT SCIENCE
MATHEMATICAL PROGRAMMING
MILITARY STATISTICS
NONLINEAR PROGRAMMING
SEARCH THEORY)

DANIEL H. WAGNER

OPERATIONS RESEARCH

This journal is published by the Operations Research Society of America. Volume 32 is appearing in 1984. The editor is Dr. Thomas L. Magnanti, Sloan School of Management, M.I.T., Cambridge. There are area editors in the fields of:

Decision analysis, bargaining, and negotiation

Defense and international security

Distribution, networks and facility planning

Health care and service sector industries

Interfaces with computer science

Natural resource management, energy, and environment

Optimization

O.R. practice

Production and scheduling, inventory, and materials management

Simulation, implementation, and evaluation of stochastic models

Social systems and the public sector

Stochastic processes and their applications

An extract from the "Editorial Policy" of the journal reads:

Operations Research publishes quality operations research and management science work of interest to the OR practitioner and researcher in three substantive categories: operations research methods, data-based operational science, and the practice of OR. Included are papers reporting underlying data-based principles of operational science, observations of operating systems, contributions to the methods and models of OR, case histories of applications, review articles, and discussions of aspects of such subjects as the administrative environment, the history, policy, practice, future, or arenas of application of operations research.

Complete studies that contain data, computer experiments, etc. and integrate the theory, methods, and applications are of particular interest. Thus, we encourage case studies of lasting value. Contributors should submit informal descriptions of cases to the joint ORSA/TIMS publication *Interfaces*.

A "Technical Notes" section contains brief articles on all the topics mentioned. An

"OR Practice" section contains practitioner-oriented applications, tutorials, and surveys. Application papers whose utility is yet undemonstrated in practice or that are not tailored for practitioners should be submitted to the appropriate contextual Area Editor. An "OR Forum" section publishes papers on history, policy, analyses of current and future trends, and related subject matter including "Letters to the Editor." For more information on subject coverage and editorial policy, see editorials and Area Editor statements published in the January/February issues of both 1983 and 1984.

All papers published in *Operations Research* are refereed. Initial refereeing of clear, concise, well-written papers normally takes about four months for papers of average length, but generally a shorter time for notes.

OPHTHALMOLOGY, STATISTICS IN

Ophthalmology is the branch of medical science concerned with ocular phenomena. The types of research studies in ophthalmology are similar to those encountered in general biostatistical work, including epidemiologic studies*, clinical trials*, small laboratory studies based on either humans or animals, and genetic studies. For the biostatistician, the one important distinction between ophthalmologic data and data obtained in most other medical specialties is that information is usually collected on an *eye-specific* rather than a *person-specific* basis. This method of data collection has both advantages and disadvantages from a statistical point of view. We will discuss this in the context of (a) studies involving a comparison of two or more treatments such as in a clinical trial and (b) observational studies*.

CLINICAL TRIALS

Usually one wishes to compare two or more treatment groups where one (or more) of the treatment groups may serve as a control group(s). This is generally done in other medical specialties by either (a) randomly assigning treatments to different groups of individuals or (b) assigning both treatments to the same individual at different points in time using a crossover design and making intra-individual comparisons. If one wishes to compare treatment groups in (a), then one (1) relies on the randomization* to minimize differences between treatment groups on other covariates, (2) performs additional multivariate analyses* such as regression analyses* to adjust for differences in treatment groups that may emerge despite the randomization, or (3) matches individuals in different treatment groups on at least a limited number of independent variables at the design stage so as to minimize extraneous treatment group differences. One always has the problem that none of these methods may adequately control for confounding*, since one may not anticipate the appropriate potential confounding variables at the data collection phase. Design (b), if feasible, seemingly overcomes many of these problems since intra-individual rather than inter-individual comparisons are being made. However, in a crossover study, one must assume that the effect of one treatment is not carried over to the time of administration of the next treatment, an assumption that is not always easy to justify. In addition, a crossover study requires more commitment from the patient and the problem of loss to followup* becomes more important. In ophthalmologic work, two treatments can be assigned randomly to the left and right eyes, thus preserving the best features of the crossover design, that is, enabling intra-individual rather than inter-individual comparisons, while eliminating the undesirable feature of the carryover effect, since both treatments typically are administered simultaneously. The administration of the two treatments simultaneously also improves patient compliance over that of a typical crossover study. (*See also* CHANGEOVER DESIGNS.)

OBSERVATIONAL STUDIES*

In contrast to the clinical trial situation, the collection of data on an eye-specific basis has both advantages and disadvantages from the point of view of observational studies. An important advantage is that if responses on two eyes of an individual are treated as replicates, then one has an easily obtained independent estimate of intraindividual variability. Such estimates are more difficult to obtain when data is obtained on a person-specific basis. However, an important disadvantage is that collection of data on an eye-specific basis complicates even the most elementary of analyses such as the assessment of standard errors of means or proportions, and, in particular, poses challenging problems for conducting multivariate analyses. For this reason, many ophthalmologists often disregard this problem and present their results in terms of *distributions over eyes* rather than *distributions over individuals*. The observations in these eye-specific distributions are treated as independent random variables and standard statistical methods are used thereby. The problem with this formulation is that observations on two eyes of an individual are highly but not perfectly correlated random variables. Thus the assumption of independence results in standard errors that are generally underestimated and significance levels* that are too extreme, rendering some apparently statistically significant results actually not significant. A conservative approach that is sometimes used to avoid this problem is to select only one of the two eyes of each individual for analysis and then proceed with standard statistical methods. The "analysis" eye may be a randomly selected eye or may be an eye selected as having the better (or worse) visual function of the two eyes. This is a valid use of such data, but is possibly inefficient since two eyes contribute somewhat more information than one eye but not as much information as two independent observations. This problem has been discussed by Ederer [2]. One possible

solution for normally distributed outcome variables is offered by the following intraclass correlation* model [3]:

$$y_{ijk} = \mu + \alpha_i + \beta_{ij} + e_{ijk},$$

$$i = 1, \ldots, g, \qquad j = 1, \ldots, P_i, \qquad (1)$$

$$k = 1, \ldots, N_{ij},$$

where i denotes group, j denotes individual within group, k denotes eye within individual, α_i is a fixed effect, $\beta_{ij} \sim N(0, \sigma_\beta^2)$, $e_{ijk} \sim N(0, \sigma^2)$, and one permits a variable number of individuals per group (P_i) and a variable number of eyes available for analysis for each individual (N_{ij}). The data layout for the model in (1) is given in Table 1, where $\bar{y}_{ij\cdot} = \sum_k y_{ijk}/N_{ij}$, $\bar{y}_{i\cdot\cdot} = \sum_j \bar{y}_{ij\cdot}/P_i$, $\bar{y}_{\cdots} = \sum_i P_i \bar{y}_{i\cdot\cdot}/\sum_i P_i$, $P = \sum_i P_i$, $N = \sum_i \sum_j N_{ij}$. Typically, one would wish to test the hypothesis H_0: all $\alpha_i = 0$ vs. H_1: at least one $\alpha_i \neq 0$. An exact solution to this problem is difficult due to the unbalanced design whereby individuals can contribute either one or two eyes to the analysis. However, the method of unweighted means is a reasonable approximate method in this case since $\max N_{ij}^{-1/2}/\min N_{ij}^{-1/2} \leqslant 2$ for ophthalmologic work [5, p. 367]. Thus, one obtains the expected mean squares given in Table 1, where each individual is assumed to contribute $\bar{N} = P[\sum_i \sum_j (1/N_{ij})]^{-1}$ eyes to the analysis. It follows that an appropriate test statistic for the preceding hypotheses is given by $\lambda = \text{MSG}/\text{MSP} \sim F_{g-1, P-g}$ under H_0, where one would reject H_0 if $\lambda > F_{g-1, P-g, 1-\alpha}$. Specific groups ($i_1$ and i_2) can be compared by computing the test statistic

$$u_{i_1, i_2} = (\bar{y}_{i_1\cdot\cdot} - \bar{y}_{i_2\cdot\cdot})/\left[\text{MSP}(P_{i_1}^{-1} + P_{i_2}^{-1})\right]^{1/2}$$

and rejecting the null hypothesis that the group means are equal if $|u_{i_1, i_2}| > t_{P-g, 1-\alpha/2}$, $i_1, i_2 = 1, \ldots, g, i_1 \neq i_2$.

These methods are illustrated in the following data set obtained from an outpatient population of 218 persons aged 20–39 with retinitis pigmentosa (RP) who were seen at the Massachusetts Eye and Ear infirmary from 1970 to 1979 [1]. Patients were classi-

Table 1 Data Layout under the Intraclass Correlation Model

Source of Variation	Mean Square	d.f.	Expected Mean Square
Between groups	$\dfrac{\sum P_i(\bar{y}_{i..} - \bar{y}_{...})^2}{g - 1} \equiv \text{MSG}$	$g - 1$	$\dfrac{\sum_i P_i(\alpha_i - \bar{\alpha})^2}{g - 1} + \sigma_\beta^2 + (\sigma^2/\bar{N})$
Between persons within groups	$\dfrac{\sum\sum(\bar{y}_{ij.} - \bar{y}_{i..})^2}{P - g} \equiv \text{MSP}$	$P - g$	$\sigma_\beta^2 + (\sigma^2/\bar{N})$
Between eyes within persons	$\dfrac{\sum\sum\sum(y_{ijk} - \bar{y}_{ij.})^2}{N - P} \equiv \text{MSE}$	$N - P$	σ^2

Source. Reproduced from ref. 3 with permission.

Table 2 Comparison of Spherical Refractive Error of RP Patients by Genetic Type Using the Intraclass Correlation Model in (1)

(*a*) *Overall ANOVA Table*

Source of Variation	Sum of Squares	d.f.	Mean Square	F-statistic
Between groups	133.59	3	44.53	3.68 ($p = 0.013$)
Between persons within groups	2518.45	208	12.11	
Within persons	80.49	210	0.383	

(*b*) *Descriptive Statistics*: t-*Statistics and* p *Values for Comparisons of Specific Groups*

Group (i)	Mean ($\bar{y}_{i..}$)	Estimated Standard Error $(\text{MSP}/P_i)^{1/2}$	Number of Persons (P_i)
DOM	+ 0.127	0.658	28
AR	− 0.831	0.778	20
SL	− 3.299	0.820	18
ISO	− 0.842	0.288	146

	Comparison Group		
Group (i)	AR	SL	ISO
DOM	0.941[+] (NS)	3.259 ($p = 0.001$)	1.350 (NS)
AR	—	2.183 ($p = 0.030$)	0.013 (NS)
SL	—	—	− 2.826 ($p = 0.005$)

[+] t-statistic.

Source. Reproduced from ref. 3 with permission.

fied according to a detailed genetic pedigree into the genetic types of autosomal dominant RP(DOM), autosomal recessive RP(AR), sex-linked RP(SL), and isolate RP(ISO), the purpose being to compare patients in the four genetic types on the basis of selected ocular characteristics. For the sake of simplicity, only one person from the age group 20–39 was selected from each family; if more than one affected person in this age range was available in a given family, then a randomly selected affected person was chosen, thus yielding 218 individuals from 218 distinct families. In Table 2, we present an analysis comparing the four genetic types on spherical refractive error. The analysis is based on the subgroup of 212 individuals who had information on spherical refractive error in at least one eye, of whom 210 had information on both eyes and 2 had information on one eye. All refractive errors were determined with retinoscopy after cycloplegia.

There are overall significant differences between the four groups ($\lambda \equiv MSG/MSP = 3.68 \sim F_{3,208}$, $p = 0.013$). From the t-statistics in Table 2(b), we see that this overall difference is completely attributable to the lower spherical refractive error in the SL group as compared with the other three groups. The estimated intraclass correlation between eyes over all groups was 0.969!

For comparative purposes, a one-way ANOVA was also performed on the same data where the data from two eyes of an individual were treated as independent ran-

Table 3   Comparison of Spherical Refractive Error of RP Patients by Genetic Type Assuming Independence Between Eyes for an Individual

(*a*)   *Overall ANOVA Table*

Source of Variation	Sum of Squares	d.f.	Mean Square	F-statistic
Between groups	267.94	3	99.21	8.45 ($p = 0.00026$)
Within groups	4906.83	418	11.73	

(*b*)   *Descriptive Statistics: t-statistics, and p-values for Comparisons of Specific Groups*

Group (1)	Mean ($\bar{y}_{i..}$)	Estimated Standard Error	Number of Eyes (N_i)
DOM	+ 0.386	0.466	54
AR	− 0.831	0.542	40
SL	− 3.299	0.571	36
ISO	− 0.842	0.201	292

Group (1)	AR	SL	ISO
		Comparison Group	
DOM	1.704 (NS)	4.999 ($p < 0.001$)	2.421 ($p = 0.016$)
AR	—	3.135 ($p = 0.002$)	0.018 (NS)
SL	—	—	−4.059 ($p < 0.001$)

Source.   Reproduced from ref. 3 with permission.

dom variables, thus yielding a sample of 422 eyes. The results from this analysis are presented in Table 3.

All significant differences found in Table 2 (both overall and between specific groups) are much more significant in Table 3. The p value for the overall comparison between groups is 50 times larger in Table 2 than it is in Table 3 ($p = 0.013$ vs. $p = 0.00026$). In addition, the spherical refractive errors of the DOM and ISO group that were not significant in Table 2 become significant in Table 3 ($p = 0.016$). Clearly, the assumption of independence between eyes is inappropriate for this data set and has a major impact on the analysis.

A model similar to the one given in equation (1) has been developed for the case of a binomially distributed outcome variable [3]. These methods have been extended so that one can perform multiple linear regression and multiple logistic regression analyses in the context of ophthalmologic data [4].

References

[1] Berson, E. L., Rosner, B., and Simonoff, E. (1980). *Amer. J. Ophthalmology*, **89**, 763–775. (An outpatient population of retinitis pigmentosa and their normal relatives. Risk factors for genetic typing and detection derived from their ocular examinations.)

[2] Ederer, F. (1973). *Arch. Ophthalmol.*, **89**, 1–2. (Shall we count numbers of eyes or numbers of subjects?)

[3] Rosner, B. (1982). *Biometrics*, **38**, 105–114. (Statistical methods in ophthalmology with an adjustment for the intraclass correlation between eyes.)

[4] Rosner, B. (1985). *Biometrics*, **41** (in press).

[5] Searle, S. R. (1971). *Linear Methods*. Wiley, New York.

(ANALYSIS OF VARIANCE
BIOSTATISTICS
CLINICAL TRIALS
FOLLOWUP)

BERNARD ROSNER

OPINION POLL *See* PUBLIC OPINION POLL

O_p, o_p NOTATION

This extends the O, o notation* to random variables. $X_n = O_p(n^\alpha)$ means that for any $\delta > 0$, there is a constant $C(\delta)$, such that

$$\Pr\left[n^{-\alpha}|X_n| \leqslant C(\delta) \right] \geqslant 1 - \delta$$

for all sufficiently large n. Similarly $X_n = o_p(n^\alpha)$ means that for any $\epsilon > 0$

$$\lim_{n \to \infty} \Pr\left[n^{-\alpha}|X_n| \leqslant \epsilon \right] \to 0.$$

(CONVERGENCE OF SEQUENCES
OF RANDOM VARIABLES
LAW OF ITERATED LOGARITHM
LAWS OF LARGE NUMBERS)

OPTICAL ILLUSIONS IN CHARTS

Charts are used in descriptive statistics to provide quickly and easily assimilated summaries of data. It is therefore important to be on guard against the possible consequences of superficial judgment that may be affected by optical illusions in looking at charts. M. E. Spear [1] points out three major causes of confusion:

1. Insufficiently detailed grids.
2. Different types of shading in cross-hatched diagrams.
3. Misleading perspective effects.

Interesting examples are given in Spear [1].

Optical illusion effects should be distinguished from effects of *inaccurately constructed* charts. Optical illusion arises when the chart is accurate objectively, but gives, at least at first glance, a misleading impression.

Reference

[1] Spear, M. E. (1969). *Practical Charting Techniques*, McGraw-Hill, New York.

(GRAPHICAL REPRESENTATION
OF DATA)

OPTIMAL $C(\alpha)$-TESTS

In a general composite hypothesis-testing problem, the usual regularity conditions governing the existence and optimality of appropriate similar regions* may not hold. In such a case, an *asymptotic setup* (requiring the sample size to be indefinitely large) ensures the existence of optimal (or desirable) tests of composite hypotheses and provides good approximations for moderate sample sizes as well. The classical likelihood ratio tests* (LRT), under fairly general regularity conditions, are asymptotically optimal in a broad sense [5, 20, and 21]. However, they may appear to be computationally very cumbrous. Wald [20, 21] considered an alternative procedure (W-test) based on the maximum likelihood estimators* (MLEs) of the associated parameters and showed that, for local alternatives, his W-test shares the asymptotic (optimality) properties of the corresponding LRT. Some alternative tests based on the MLE and related scores have also been considered by Aitchison and Silvey [1], Silvey [19], Rao [17] and others. However, like the LRT, these tests are also quite involved computationally. Neyman [14] came up with a broad class of asymptotic tests of composite statistical hypotheses [called the $C(\alpha)$-tests or consistent level α-tests] and presented a general procedure for constructing optimal ones within this class that are termed the optimal $C(\alpha)$-tests.

Let $\{E_n\}$ be a sequence of sample points, $\{W_n\}$ the corresponding sequence of sample spaces and let $\{p_n(\,.\,;\theta,\gamma)\}$ be the associated sequence of densities, where θ and γ (possibly vector-valued) are unknown parameters, $\theta \in \Theta$, $\gamma \in \Gamma$. Consider the null hypothesis $H_0: \theta = \theta_0$ (specified) against an alternative θ different from θ_0, where γ is treated as a nuisance parameter. Note that w_n, a measurable subset of W_n, provides an *asymptotic test* of H_0 of size α ($0 < \alpha < 1$) if

$$\lim_{n \to \infty} P\{E_n \in w_n \mid H_0\} = \alpha, \qquad \forall \gamma \in \Gamma. \tag{1}$$

Consider a family ξ of alternatives θ^*, where $\theta^* = \{\theta_n^*\} \in \Theta$, such that $\theta_n^* \to \theta_0$ as $n \to \infty$; and let $K(\alpha)$ be a class of asymptotic size α tests of H_0. If there exists a sequence $\{w_n^0\} \in K(\alpha)$, such that for any other $\{w_n\} \in K(\alpha)$,

$$\liminf_{n \to \infty} \Big[P\{E_n \in w_n^0 \mid \theta_n^*, \gamma\} \\ - P\{E_n \in w_n \mid \theta_n^*, \gamma\} \Big] \geqslant 0, \tag{2}$$

for every (fixed) $\gamma \in \Gamma$ and $\theta^* \in \xi$, then with respect to the class ξ of (local) alternatives, $\{w_n^0\}$ is asymptotically optimal with the class $K(\alpha)$. With these definitions, consider first the case where θ is real and γ possibly vector-valued. Let $g(E_n, \gamma)$, defined on $W_n \times \Gamma$, be an arbitrary normalized function such that $g(\,.\,,\gamma)$ is differentiable (twice) with respect to the elements of γ and the usual Cramér [4] conditions hold on these derivatives. Let $\phi_\gamma(\,.\,,\theta,\nu) = (\partial/\partial\gamma)\log P_n(1;\theta,\gamma)$ and let $b_{\theta\gamma}^*$ be the vector that leads to a minimum value of $\{g(E_n,\gamma) - b'\phi_\gamma(E_n;\theta,\gamma)\}^2$, and let

$$\sigma_g^2(\theta,\gamma) = E\Big\{\big[g(E_n;\gamma) \\ - b_{\theta\gamma}^{*\prime}\phi_\gamma(E_n;\theta,\gamma)\big]^2 \,\big|\, \theta,\gamma\Big\}.$$

Then let

$$f(\,.\,;\theta,\gamma)$$
$$= \big[\sigma_g(\theta,\gamma)\big]^{-1}$$
$$\times \{ g(\,.\,,\gamma) - b_{\theta\gamma}^{*\prime}\phi_\gamma(\,.\,,\theta,\gamma)\}. \tag{3}$$

In the conventional case, where $E_n = (X_1, \ldots, X_n)$, $n \geqslant 1$ and the X_i are independent and identically distributed random elements, Neyman [14] considered an arbitrary sequence $\{\hat{\gamma}_n\}$ of square root n consistent estimators of γ [i.e., $n^{1/2}|\hat{\gamma}_n - \gamma| = O_p(1)$, i.e., bounded in probability], and defined

$$w_n = \{ E_n : f(E_n;\theta_0,\hat{\gamma}_n) \geqslant \tau_\alpha\}, \tag{4}$$

where τ_α is the upper $100\alpha\%$ point of the standard normal distribution. For one-sided alternatives: $\theta > \theta_0$, (4) leads to a class of $C(\alpha)$-tests. For two-sided alternatives, one may replace f and τ_α by $|f|$ and $\tau_{\alpha/2}$, respectively. Within this class of $C(\alpha)$-tests, an optimal one [in the sense of (2)] corresponds to the choice

$$g(E_n; \gamma) = (\partial/\partial\theta)\log p_n(E_n; \theta, \gamma)|_{(\theta_0, \gamma)},$$
$$\gamma \in \Gamma, \quad (5)$$

and the corresponding test shares the asymptotic optimality of the LRT and the W-test for the class of (local) alternatives ξ. Note that in (3)–(5) any square-root-n-consistent estimator $\hat{\gamma}_n$ may be used (instead of the MLE) and this makes the (optimal-) $C(\alpha)$-test computationally simpler and more flexible. But the convergence rate to the asymptotic theory may depend on the choice of such a sequence of estimators. Presumably, for the MLE (or some alternative efficient estimators), this rate may be faster. We may also remark that the asymptotic optimality of such a $C(\alpha)$-test is confined only to the chosen class of local alternatives ξ. For non-local alternatives, the LRT may behave much better than the optimal $C(\alpha)$-test (see Hoeffding [8]). We illustrate the theory with a couple of examples.

Example 1.   Consider a sample $X_1, \ldots, X_n$ from a population having the Cauchy density $(\gamma/\pi)\{\gamma^2 + (x - \theta)^2\}^{-1}$, where the null hypothesis H_0 states that the location parameter $\theta = 0$ against alternatives that $\theta \neq 0$, and the scale parameter γ is unkown. Uniformly most powerful (or UMP unbiased) tests for this hypothesis do not exist. The sample interquartile range* provides a square-root-n-consistent estimator $\hat{\gamma}_n$ of γ, and hence an optimal $C(\alpha)$-test may be based on $\sum_{i=1}^{n} X_i / (\hat{\gamma}_n^2 + X_i^2)$, which corresponds to (3) (before standardization), and this statistic is much simpler computationally than the LRT or the W-statistic based on the MLE of (θ, γ).

Example 2.   This example, due to Ray [18], corresponds to the 2×2 contingency table*, where p_{ij} denotes the probability that a subject chosen at random will fall into row i and column j, for $i, j = 1, 2$. The null hypothesis relates to the symmetry of this prob-

ability matrix, i.e., $H_0 : p_{12} = p_{21}$. If n_{ij}, $i, j = 1, 2$ denote the observed cell frequencies for this 2×2 table, the usual test for H_0 is based on $n_{12} - n_{21}$, conditional on the sum $S = n_{12} + n_{21}$ held fixed (*see* MCNEMAR STATISTICS).

This conditional test based on the simple binomial* law is, however, not the optimal one (in all cases), while the LRT or W-tests are more complicated. Ray [18] considered some optimal $C(\alpha)$-tests, which are always as efficient as the binomial test for symmetry and are much more efficient when the nuisance parameters are either close to 0 or 1.

For independent but not necessarily identically distributed random variables, optimal $C(\alpha)$-tests have been considered by Bartoo and Puri [2]. Also, Bühler and Puri [3] have considered the case of $C(\alpha)$-tests with multiple constraints on the parameters. Moran [13] has given a nice comparative picture of $C(\alpha)$-tests and the LRT and W-tests. For the case of vector-valued θ, $g(., \gamma)$ is also taken to be a vector of the same order, and, in (3), $b_{\theta\gamma}^*$ is then a matrix, while $\sigma_g(\theta, \gamma)$ is to be replaced by $D(\theta, \gamma)$ where $[D(\theta, \gamma)][D(\theta, \gamma)]' = $ dispersion matrix of the residual vector. Thus, in (3), $f(., \theta, \gamma)$ will be a vector and in (4), we can replace $f(.)$ by its Euclidean norm and τ_α by the upper $100\alpha\%$ point of an appropriate chi distribution*. Equation (5) extends naturally to this vector case. As a nice example of this vector-parameter case, we may refer to Johnson [10], who treated the $r \times c$ contingency tables and, for some specific "slippage" alternatives, provided some optimal $C(\alpha)$-tests.

In some problems, restrictions to invariance or conditionality may lead to some asymptotically optimal tests similar to such $C(\alpha)$-tests based on allied $g(., \gamma)$. We may refer to Hájek [6] for asymptotically optimal rank tests of this type (*see also* LOCALLY OPTIMAL STATISTICAL TESTS).

We conclude this article with the remark that for the likelihood estimating function, the LAN (local asymptotic normal) condition formulated in LeCam [12], Hájek [7], and Inagaki [9], among others, provides a very clear picture of the behavior of the likelihood function in the neighborhood of the true parameter point, which, in turn, provides a convenient way of choosing the appropriate $f(., \theta, \gamma)$ in (3) and thereby, the optimal $C(\alpha)$-tests.

References

[1]   Aitchison, J. and Silvey, S. D. (1960). *J. R. Statist. Soc. B*, **22**, 154–171.

[2]   Bartoo, J. B. and Puri, P. S. (1967). *Ann. Math. Statist.*, **38**, 1845–1852.

[3]   Bühler, W. J. and Puri, P. S. (1966). *Zeit. Wahrscheinlichkeitsth. verw. Geb.*, **5**, 71–88.

[4]   Cramér, H. (1946). *Mathematical Methods of Statistics*. Princeton University Press, Princeton, NJ.

[5]   Feder, P. I. (1968). *Ann. Math. Statist.*, **39**, 2044–2055.

[6]   Hájek, J. (1962). *Ann. Math. Statist.*, **33**, 1124–1147.

[7]   Hájek, J. (1970). *Zeit. Wahrscheinlichkeitsth. verw. Geb.*, **14**, 323–330.

[8]   Hoeffding, W. (1965). *Ann. Math. Statist.*, **36**, 369–408.

[9]   Inagaki, N. (1973). *Ann. Inst. Statist. Math.*, **25**, 1–26.

[10] Johnson, N. S. (1975). *J. Amer. Statist. Ass.*, **70**, 942–947.

[11] Klonecki, W. (1977). In *Symposium to Honour Jerzy Neyman*, R. Bartoszyński, E. Fidelis, and W. Klonecki, eds., PWN, Warsaw, Poland, pp. 161–175.

[12] LeCam, L. (1960). *Univ. California Publ. Statist.*, **3**, 37–98.

[13] Moran, P. A. P. (1970). *Biometrika*, **56**, 47–55.

[14] Neyman, J. (1959). In *Probability and Statistics*, the Harald Cramér volume, U. Grenander, ed. Wiley, New York.

[15] Neyman, J. (1979). *Sankhyā A*, **41**, 1–21.

[16] Neyman, J. and Scott, E. L. (1965). *J. Amer. Statist. Ass.*, **60**, 699–721.

[17] Rao, C. R. (1962). *J. R. Statist. Soc. B*, **24**, 46–72.

[18] Ray, R. M. (1976). *Commun. Statist.*, **A5**, 545–563.

[19] Silvey, S. D. (1959). *Ann. Math. Statist.*, **30**, 389–407.

[20] Wald, A. (1941). *Ann. Math. Statist.*, **12**, 1–19.

[21] Wald, A. (1943). *Trans. Amer. Math. Soc.*, **54**, 426–482.

[22] Wilks, S. S. (1938). *Ann. Math. Statist.*, **9**, 166–175.

(ASYMPTOTIC NORMALITY
ASYMPTOTIC EXPANSION
LIKELIHOOD RATIO TESTS
MAXIMUM LIKELIHOOD ESTIMATION
SIMILAR REGIONS)

P. K. Sen

OPTIMAL SAMPLE SIZE REQUIREMENTS

The determination of optimal sample sizes for experimental designs characterized by classical general linear hypothesis models (*see* GENERAL LINEAR MODEL) is discussed in a number of principal papers in the statistical literature [3, 4, 6–12, 14–16]. Answers to questions concerning adequate sample sizes generally depend on the number of categories to be compared, the level of risk an experimenter is willing to assume, and some knowledge of the noncentrality parameter, an unknown quantity that arises in the formulation of the test statistic under the alternative hypothesis. The experimenter, however, seldom has an appreciation of the meaning of a noncentrality parameter and may deal better intuitively with the standardized range* of the means. This quantity, which is the difference between the largest and smallest of a set of means divided by the standard deviation, may be expressed as an algebraic function of the noncentrality parameter.

Using this concept, extensive tables have been constructed that permit the experimenter to determine sample sizes without iteration under a variety of experimental conditions [2]. These tables present maximum values of the standardized range τ of a group of means for the single classification experiment in which K groups, each containing N observations, are to be compared at α and β levels of risk, and for the double classification experiment with K treatments,

B blocks, and N observations per cell. The standardized range is the difference between the largest and the smallest of K means divided by the standard deviation:

$$\tau = (\xi_{max} - \xi_{min})/\sigma.$$

α is the probability of a type I error or the risk an experimenter is willing to assume in rejecting the null hypothesis when it is true, and β is the probability of a type II error, or the risk an experimenter is willing to assume in accepting the null hypothesis when it is false. Values of τ are tabulated for $\alpha = 0.01, 0.05, 0.10, 0.20$; $\beta = 0.005, 0.01, 0.025, 0.05, 0.1, 0.2, 0.3, 0.4$, and $K = 2(1)11(2)$ $15(5)30(10)60$ in the single classification experiment for $N = 2(1)30(5)50(10)100(50)$ $200(100)500, 1000$, and in the double classification experiment for $B = 2(1)5$ blocks and $N = 1(1)5$ observations per cell.

SINGLE (ONE-WAY) CLASSIFICATION EXPERIMENT

Let $Z_{i1}, Z_{i2}, \ldots, Z_{in}$ $(i = 1, 2, \ldots, K)$ be independent, normal random variables with means ξ_i and variance σ^2. The usual test for $\xi_1 = \xi_2 = \cdots = \xi_K$ is based on the Snedecor F-statistic, (*see* F-DISTRIBUTION and NONCENTRAL F-DISTRIBUTION). For $\bar{\xi} = \sum_{i=1}^{K}\xi_i/K$ and $\delta_i = \xi_i - \bar{\xi}$, F has a noncentral F-distribution* with $f_1 = K - 1$ and $f_2 = K(N - 1)$ degrees of freedom, and noncentrality

Table 1   Single Classification
($K = 3$, $\alpha = 0.05$)

			β			
N	0.005	0.01	0.025	0.05	0.1	0.2
2	9.560	8.966	8.104	7.375	6.548	5.570
3	5.371	5.073	4.637	4.265	3.838	3.325
4	4.168	3.944	3.616	3.334	3.010	2.618
5	3.541	3.354	3.078	2.841	2.568	2.236
6	3.139	2.973	2.731	2.521	2.280	1.987
7	2.851	2.701	2.481	2.292	2.073	1.808
8	2.631	2.493	2.291	2.116	1.915	1.670
9	2.456	2.328	2.139	1.976	1.788	1.560
10	2.313	2.192	2.014	1.861	1.684	1.469
20	1.585	1.502	1.381	1.276	1.155	1.008
.						
.						
.						
26	1.381	1.309	1.203	1.112	1.007	0.878
27	1.354	1.284	1.180	1.091	0.987	0.861
.						
.						
.						
30	1.282	1.215	1.117	1.032	0.935	0.815
60	0.899	0.852	0.783	0.724	0.655	0.572
70	0.831	0.788	0.724	0.669	0.606	0.528
80	0.776	0.736	0.677	0.625	0.566	0.494
90	0.732	0.693	0.637	0.589	0.533	0.465
100	0.694	0.657	0.604	0.559	0.506	0.441
150	0.565	0.536	0.493	0.455	0.412	0.360
200	0.489	0.464	0.426	0.394	0.357	0.311
300	0.399	0.378	0.348	0.321	0.291	0.254
400	0.345	0.327	0.301	0.278	0.252	0.220
500	0.309	0.293	0.269	0.249	0.225	0.197
1000	0.218	0.207	0.190	0.176	0.159	0.139

parameter $\lambda = N \sum_{i=1}^{K} \delta_i^2 / 2\sigma^2$. Pearson and Hartley [11] show that the standardized range of the means, W/σ_Z in their notation, satisfies the inequality

$$W/\sigma_Z = \tau \leqslant 2\sqrt{\lambda/N} = \phi\sqrt{2K/N} \,,$$

where $\phi = \phi(f_1, f_2, \alpha, \beta)$ is Tang's [13] noncentrality parameter, with equality if and only if

$$\xi_i = (\xi_{max} + \xi_{min})/2$$

for all ξ_i other than ξ_{max} and ξ_{min}.

Values of $2\sqrt{\lambda/N}$ that give the maximum possible difference between any two standardized means are presented in Table 1 for $\alpha = 0.05$ and $K = 3$.

DOUBLE (CROSS-) CLASSIFICATION EXPERIMENT

The model for double classification with N observations per cell is

$$X_{ijl} = \xi + t_i + b_j + (tb)_{ij} + \epsilon_{ijl} \,,$$

where t_i is the ith treatment effect ($i = 1, 2, \ldots, K$), b_j is the jth block effect ($j = 1, 2, \ldots, B$), $(tb)_{ij}$ is the interaction* effect of treatment i with block j, and ϵ_{ijl} is the effect due to the lth observation on the ith treatment in block j ($l = 1, 2, \ldots, N$). The ϵ_{ijl} are assumed to be independent normal variables with mean zero and variance σ^2.

Degrees of freedom in the double classification experiment are $f_1 = (K - 1)$ for treatment and $f_2 = (K - 1)(B - 1)$ for error when the number of observations per cell is $N = 1$. For $N > 1$, $f_2 = KB(N - 1)$. The standardized maximum difference between any two treatment means is

$$\tau = (t_{max} - t_{min})/\sigma$$

$$\leqslant 2\sqrt{\lambda/(BN)} = \phi\sqrt{2K/(BN)} \,,$$

where $\phi = \phi(f_1, f_2, \alpha, \beta)$ is Tang's [13] noncentrality parameter, with equality if and only if $t_i = 0$, $\sum_i t_i = 0$ for all t_i other than t_{max} and t_{min}. Values of τ have been tabulated for $B = 2(1)5$ and $N = 1(1)5$. For

$N > 5$, sample sizes may be determined from corresponding τ values in the single classification tables by dividing the number of observations per group (N in Table 1) by the number of blocks (B). This procedure will never overestimate τ for the double classification experiment by more than 1%.

MULTIPLE (CROSS-) CLASSIFICATION EXPERIMENT

Let f_1 and f_2 be degrees of freedom in treatment and error, respectively. For prescribed α and β, the appropriate sample size N results from the equation $f_2 = K(N - 1)$, so that $N = (f_2/K) + 1$, where $K = f_1 + 1$. Find the value of τ that corresponds to N in the single classification table, and solve for $\tau_d = C_d \tau$ in Table 2, where C_d is a constant that depends on the experimental design and the treatment under consideration.

Example 1. The mean age at death of female mice of a particular strain is 900 days. Radiation of these animals at sublethal doses of X-rays above 100 roentgens is known to reduce their survival time. In fact, every increase of 100 roentgens of X radiation in the range of 100–400 roentgens appears to reduce longevity by approximately 40 days. A biologist wishes to know whether a corresponding reduction of longevity can be observed in the range 0–100 roentgens. He proposes to experiment with two groups of female animals. One group will be unirradiated; the other, irradiated with 100 roentgens of X-rays. How many animals should be placed in each of the two groups?

The number of groups $K = 2$, $\xi_{max} = 900$, $\xi_{min} = 860$ days. The levels of risk chosen by the experimenter are $\alpha = 0.05$ and $\beta = 0.10$, and the standard deviation, based on previous observations, is taken to be $\sigma = 200$. Thus $\tau = 40/200 = 0.20$, and the number of mice required in each of the two groups is about $N = 500$.

In the same situation, the experimenter might ask, How small a difference could I detect with 200 observations in each group?

Table 2   Values of C_d

Design	Treatment d.f. f_1	Error d.f. f_2	Constant C_d
Double classification K treatments in B blocks			
No interaction			
One observation per cell	$(K-1)$	$(K-1)(B-1)$	$(N/B)^{1/2}$
With interaction			
N_1 observations per cell			
Main effect	$(K-1)$	$BK(N_1-1)$	$(N/BN_1)^{1/2}$
Interaction	$(K-1)(B-1)$	$BK(N_1-1)$	$(N/N_1)^{1/2}$
Triple classification			
N_2 observations			
per cell			
Main effect	$(K-1)$	$KB_1B_2(N_2-1)$	$(N/B_1B_2N_2)^{1/2}$
Latin square			
K rows, K columns	$(K-1)$	$(K-1)(K-2)$	$(N/K)^{1/2}$
K treatments			

Here $\tau = 0.325$, where $\sigma = 200$ and $(\xi_{max} - \xi_{min})$ is unknown. Thus $\xi_{max} - \xi_{min} = 200(0.325) = 65$ days.

Alternatively, the experimenter might say, I can afford to run my experiment with 300 animals in each group. If my significance level is $\alpha = 0.05$, what is the probability that I will detect a reduction of 40 days in longevity at 100 roentgens of X-radiation? In this case, the probability of a type II error β is approximately 0.3; therefore, the probability of detecting a reduction of 40 days is $(1 - \beta) = 0.7$.

Example 2. Pearson and Hartley [11, p. 128] consider a double classification experiment with three treatments $(K = 3)$ and three blocks $(B = 3)$. In their notation, the standardized range of the means $\tau = W/\sigma_Z$. For levels of risk $\alpha = 0.05$ and $\beta = 0.1$, they recommend $N = 38$ replications when $W = 250$ and $\sigma_Z = 500$, and $N = 10$ replications when $W = \sigma_Z = 500$. The solutions provided by Pearson and Hartley result from a single iteration of their formula. Continued iteration or reference to Kastenbaum and Bowman tables would have led to values of $N = 34$ and $N = 9$, respectively (see Table 1).

References

[1] Bowman, K. O. (1972). *Biometrika*, **59**, 234.
[2] Bowman, K. O. and Kastenbaum, M. A. (1975). In *Selected Tables in Mathematical Statistics*, Vol. 3, H. L. Harter and D. B. Owen, eds. American Mathematical Society, Providence, RI, pp. 111–232.
[3] Bratcher, T. L., Moran, M. A., and Zimmer, W. J. (1970). *J. Qual. Tech.*, **2**, 156–164.
[4] Dasgupta, P. (1968). *Sankhyā B*, **30**, 73–82. (Tabulates λ for $\alpha = 0.01$, 0.05; $\beta = 0.1(0.1)0.9$; $f_1 = 1(1)10$, $f_2 = 10(5)50(10)100$, ∞.)
[5] David, H. A., Lachenbruch, P. A., and Brandis, H. P. (1972). *Biometrika*, **59**, 161–168. (Presents optimal sample sizes for the studentized range.)
[6] Fox, M. (1956). *Ann. Math. Statist.*, **27**, 484–497.
[7] Kastenbaum, M. A., Hoel, D. G., and Bowman, K. O. (1970). *Biometrika*, **57**, 421–430.
[8] Kastenbaum, M. A., Hoel, D. G., and Bowman, K. O. (1970). *Biometrika*, **57**, 573–577.
[9] Lehmer, E. (1944). *Ann. Math. Statist.*, **15**, 388–398. (Tabulates ϕ for $\alpha = 0.01$, 0.05; $\beta = 0.2$, 0.3; $f_1 = 1(1)10$, 12, 15, 20, 24, 30, 40, 60, 120, ∞; $f_2 = 2(2)20$, 24, 30, 40, 60, 80, 120, 240, ∞. For $f_1 = 1$, $f_2 = 2$, $\alpha = 0.05$, $\beta = 0.2$, Lehmer's value of ϕ is 3.898. It should be 3.998.)
[10] Odeh, R. E. and Fox, M. (1975). *Sample Size Choices: Charts for Experiments with Linear Models*. Marcel Dekker, New York. (The book has 107 charts that show contours of ϕ for values $(\alpha, 1 - \beta, f_1, f_2)$ with $f_1 \geq 1$, $f_2 \geq 4$ on a logarithmic scale, and $1 - \beta = 0.005$, 0.01, 0.025, 0.05,

0.1, 0.2, 0.3, 0.4, 0.5, 0.6, 0.7, 0.8, 0.9, 0.95, 0.975, 0.99, and 0.995; $\alpha = 0.001$, 0.005, 0.01, 0.025, 0.05, 0.1, 0.25, and 0.5 for the sets of values of $1 - \beta > \alpha$. The charts require iterations to find N for given α and power in experiments for which linear models are appropriate.)

[11] Pearson, E. S. and Hartley, H. O. (1951). *Biometrika*, **38**, 112–130.

[12] Pearson, E. S. and Hartley, H. O. (1972). *Biometrika Tables for Statisticians*, Vol. 2. Cambridge University Press, London.

[13] Tang, P. C. (1938). *Statistical Research Memoirs*, **2**, 126–149 and tables. (This seminal paper includes tabulations of β for $\alpha = 0.01$, 0.05; $f_1 = 1(1)8$; $f_2 = 2, 4, 6(1)30, 60, \infty$; $\phi = 1(0.5)3(1)8$.)

[14] Thompson, C. (1941). *Biometrika*, **32**, 151–181. (Tabulates the critical value of the F-distribution for $\alpha = 0.5$, 0.25, 0.10, 0.05, 0.025, and 0.005 for varying values of f_1 and f_2.)

[15] Tiku, M. L. (1967). *J. Amer. Statist. Ass.*, **62**, 525–539.

[16] Tiku, M. L. (1972). *J. Amer. Statist. Ass.*, **67**, 709–710. (These papers tabulate β for $\alpha = 0.005$, 0.01, 0.025, 0.05, 0.10; $f_1 = 1(1)10$, 12; $f_2 = 2(2)30$, 40, 60, 120, ∞; $\phi = 0.5$, 1.0(0.2)2.2(0.4)3.)

(DESIGN OF EXPERIMENTS
HYPOTHESIS TESTING
MULTIPLE COMPARISONS
POWER)

MARVIN A. KASTENBAUM
KIMIKO O. BOWMAN

OPTIMAL SAMPLING IN SELECTION PROBLEMS

A selection procedure typically consists of three ingredients: (1) a sampling rule, (2) a stopping rule, and (3) a decision rule, though these components usually are not explicitly labeled. Optimal sampling arises in different ways depending on the context. Broadly speaking, the problem of optimal (or optimum) sampling arises because of the need for balancing between the cost of sampling and the cost of making a wrong decision. Obviously, increasing the amount of sampling increases the former cost while decreasing the latter.

INDIFFERENCE ZONE FORMULATION

Suppose we have k independent populations $\pi_1, \pi_2, \ldots, \pi_k$, where the CDF of π_i is $F(x; \theta_i)$, where the parameter θ_i has an unknown value belonging to an interval Θ on the real line. Our goal is to select the population associated with the largest θ_i, which is called the best population. In the indifference zone formulation of Bechhofer [2], it is required that the selection rule guarantees with a probability at least equal to $P^*(1/k < P^* < 1)$ that the best population will be chosen whenever the true parametric configuration $\boldsymbol{\theta} = (\theta_1, \theta_2, \ldots, \theta_k)$ lies in a subset of the parametric space Ω_Δ characterizing the property that the distance between the best and the next best populations is at least Δ. The subset Ω_Δ is called the preference zone. The constants P^* and Δ are specified in advance by the experimenter. The probability guarantee requirement is referred to as the P^*-requirement.

Now, let us consider k independent normal populations $\pi_1, \pi_2, \ldots, \pi_k$ with unknown means $\mu_1, \mu_2, \ldots, \mu_k$, respectively, and common known variance σ^2. Based on samples of size n from each population, the single-stage procedure of Bechhofer [2] for selecting the population with the largest μ_i selects the population that yields the largest sample mean. Here the preference zone is defined by the relation $\mu_{[k]} - \mu_{[k-1]} \geqslant \Delta$, where $\mu_{[1]} \leqslant \cdots \leqslant \mu_{[k]}$ denote the ordered μ_i. The optimum sampling problem in this case is to determine the minimum sample size n subject to the P^*-requirement. The optimum value of n is given by the smallest integer n for which

$$\int_{-\infty}^{\infty} \Phi^{k-1}\left(x + \frac{\sqrt{n}\,\Delta}{\sigma}\right)\varphi(x)\,dx \geqslant P^*,$$

where Φ and φ denote the CDF and the density function of a standard normal* random variable.

Suppose that these normal distributions have unknown and possibly unequal variances. In this case, no single-stage procedure exists. Two-stage procedures have been studied in this situation by Bechhofer et al. [4], and Dudewicz and Dalal [9]. One may take a sample of size n_0 from each population at the first stage and on the basis of the information obtained from these samples, determine the sizes of additional samples to be taken from these populations. The selection rule is based on the total samples from all the populations. Even when the variances are known, one may use a two-stage procedure in which the first stage involves selection of a nonempty subset of random size with possible values $1, 2, \ldots,$ and k. If the first stage results in a subset of size larger than 1, then a second stage ensues with additional samples from those populations that still remain under consideration. Such procedures have been considered by Alam [1], Tamhane and Bechhofer [20, 21], and Gupta and Miescke [15] with some modifications. A problem of optimum sampling in these cases is to determine the optimal combination of the sample sizes in the two stages. This can be done, e.g. [20], by minimizing the maximum of the expected total sample size for the experiment over all parametric configurations subject to the P^*-requirement.

MINIMAX, GAMMA MINIMAX, AND BAYES TECHNIQUES

Consider again k normal populations π_1, $\pi_2, \ldots, \pi_k$ with unknown means $\mu_1, \mu_2, \ldots, \mu_k$ and common known variance σ^2. If the selection procedure is to take samples of size n from these populations and choose the population that yields the largest sample mean, one can consider a loss function* $L = c_1 n + \sum_{i=1}^{k} c_2 (\mu_{[k]} - \mu_i) I_i$, where c_1 is the sampling cost per observation, c_2 is a positive constant, and $I_i = 1$, if π_i is selected, and equals 0 otherwise. Optimum n can be obtained by minimizing the integrated risk assuming (known) prior distributions for μ_i's; see Dunnett [10]. One may also determine the optimum n by minimizing the maximum expected loss over all parametric configurations. However, the expected loss in

our case is unbounded above, and we can find a minimax* solution if we have prior information regarding the bounds on the differences $\mu_{[k]} - \mu_{[i]}$, $i = 1, \ldots, k - 1$.

Suppose we take a sample of size n_1 from each of k normal populations with unknown means $\mu_1, \mu_2, \ldots, \mu_k$, and common known variance σ^2. For a fixed t, $1 \leqslant t \leqslant k - 1$, we discard the populations that produced the t smallest sample means and take an additional sample of size n_2 from each of the remaining $k - t$ populations. We select as the best the population that entered the second stage and produced the largest sample mean based on all $n_1 + n_2$ observations. Given that the total sample size $T = kn_1 + (k - t)n_2$ is a constant, the problem is to determine the optimum allocation of (n_1, n_2) by minimizing the maximum expected loss, where the loss is $L = c_1 T + c_2 \sum_{i=1}^{k} (\mu_{[k]} - \mu_i) I_i$ as defined earlier. For details, see Sommerville [19] and Fairweather [11].

In these problems, we can also take the gamma-minimax approach and minimize the maximum expected risk over a specified class of prior distributions* for the parameters μ_i; see Gupta and Huang [14].

COMPARISON WITH A CONTROL

An optimal sampling problem can be, as we have seen, an optimal allocation problem. Such allocation problems are also meaningful when we compare several treatments with a control. Let $\pi_1, \pi_2, \ldots, \pi_k$ be k independent normal populations representing the experimental treatments and let π_0 be the control, which is also a normal population. Let π_i have unknown mean μ_i and known variance σ_i^2, $i = 0, 1, \ldots, k$. A multiple comparisons approach is to obtain one- and two-sided simultaneous confidence intervals for, say, $\mu_i - \mu_0$, $i = 1, 2, \ldots, k$. If n_i is the size of the sample from π_i, $i = 0, 1, \ldots, k$, such that $\sum_{i=0}^{k} n_i = N$, a fixed integer, then the problem is to determine the optimal allocation of the total sample size. The optimal allocation will depend on, besides other known quantities, a specified "yardstick" associated with the width of the interval. For details of these problems, see Bechhofer [3], Bechhofer and Nocturne [5], Bechhofer and Tamhane [6], and Bechhofer and Turnbull [7].

Instead of taking the preceding multiple comparisons* approach, one can use the formulation of partitioning the set of k experimental populations into two sets, one consisting of populations that are better than the control and the other consisting of the remaining (worse than the control). For a given total sample size, the problem is to determine the optimal allocation either by minimizing the expected number of populations misclassified or by maximizing the probability of a correct decision; for details, see Sobel and Tong [18].

SUBSET SELECTION APPROACH

As before, consider k independent populations $\pi_1, \pi_2, \ldots, \pi_k$, where π_i is characterized by the CDF $F(x; \theta_i)$, $i = 1, \ldots, k$. In the subset selection approach, we are interested in selecting a nonempty subset of the k populations so that the selected subset will contain the population associated with the largest θ_i with a guaranteed minimum probability P^*. The number of populations to be selected depends on the outcome of the experiment and is not fixed in advance as in the indifference zone approach.

Suppose we take a random sample of size n from each population. Let T_i, $i = 1, \ldots, k$, be suitably chosen statistics from these samples. In the case of location parameters, the procedure of Gupta [12] selects π_i if and only if $T_i \geqslant T_{\max} - D$, where $T_{\max} = \max(T_1, \ldots, T_k)$ and $D \geqslant 0$ is to be chosen such that the P^*-requirement is met. The constant D will depend on k, P^*, and n. In contrast to the indifference zone approach, one can obtain a selection rule satisfying the P^*-condition for any given n.

In the case of k normal populations with unknown means $\mu_1, \mu_2, \ldots, \mu_k$, and known common variance σ^2, the rule of Gupta [12] selects π_i if and only if $\overline{X}_i \geqslant \overline{X}_{\max} - d\sigma/\sqrt{n}$,

where $\overline{X}_i$ is the mean of a sample of size n from π_i, $i = 1, 2, \ldots, k$. The constant d is given by the equation

$$\int_{-\infty}^{\infty} \Phi^{k-1}(x + d)\varphi(x)\, dx = P^*.$$

The expected subset size, denoted by $E(S)$, is given by

$$E(S) = \sum_{i=1}^{k} \int_{-\infty}^{\infty} \prod_{j \neq i} \Phi\left\{x + d + \frac{\sqrt{n}}{\sigma}\left(\mu_{[i]}\right.\right.$$
$$\left.\left. - \mu_{[j]}\right)\right\} \varphi(x)\, dx,$$

where $\mu_{[1]} \leqslant \mu_{[2]} \leqslant \cdots \leqslant \mu_{[k]}$ denote the ordered μ_i. We can define the optimum sample size as the minimum sample size for which the expected subset size or, equivalently, the expected proportion of the populations selected, does not exceed a specified bound when the true parametric configuration is of a specified type. Relevant tables are available in Gupta [13] for the equidistant configuration given by $\mu_{[i+1]} - \mu_{[i]} = \delta$, $i = 1, 2, \ldots, k - 1$, and in Deely and Gupta [8] for the slippage configuration given by $\mu_{[1]} = \cdots = \mu_{[k-1]} = \mu_{[k]} - \delta$.

If we use the restricted subset selection approach in which the size of the selected subset is random subject to a specified upper bound, then the P^*-condition is met whenever the parametric configuration belongs to a preference zone, as in the case of Bechhofer's formulation. In this case, the minimum sample size (assuming common sample size) can be determined in a similar way [17].

In our discussion so far, the optimal sampling has been related to optimal sample sizes or optimal allocation under a given sampling scheme such as single-stage, two-stage, etc. One can also seek the optimal sampling scheme by comparing single-stage, multistage, and sequential procedures. Comparisons of different sampling schemes for several selection goals have been made and are available in the literature. In addition to the usual sampling schemes, inverse sampling* rules with different stopping rules and comparisons involving vector-at-a-time and play-the-winner* sampling schemes have been studied in the case of clinical trials involving dichotomous data. References to these and other problems discussed can easily be obtained from Gupta and Panchapakesan [16].

References

[1] Alam, K. (1970). *Ann. Inst. Statist. Math. Tokyo*, **22**, 127–136.

[2] Bechhofer, R. E. (1954). *Ann. Math. Statist.*, **25**, 16–39. (A pioneering paper introducing the indifference zone formulation.)

[3] Bechhofer, R. E. (1969). In *Multivariate Analysis II*, P. R. Krishnaiah, ed. Academic Press, New York, pp. 463–473.

[4] Bechhofer, R. E., Dunnett, C. W., and Sobel, M. (1954). *Biometrika*, **41**, 170–176.

[5] Bechhofer, R. E. and Nocturne, D. J. (1972). *Technometrics*, **14**, 423–436.

[6] Bechhofer, R. E. and Tamhane, A. C. (1983). *Technometrics*, **25**, 87–95.

[7] Bechhofer, R. E. and Turnbull, B. W. (1971). In *Statistical Decision Theory and Related Topics*, S. S. Gupta and J. Yackel, eds. Academic Press, New York, pp. 41–78.

[8] Deely, J. J. and Gupta, S. S. (1968). *Sankhyā A*, **30**, 37–50. (The first paper to consider a Bayesian approach to subset selection.)

[9] Dudewicz, E. J. and Dalal, S. R. (1975). *Sankhyā B*, **37**, 28–78.

[10] Dunnett, C. W. *J. R. Statist. Soc. B*, **22**, 1–40. (This is followed by a discussion by several statisticians.)

[11] Fairweather, W. R. (1968). *Biometrika*, **55**, 411–418.

[12] Gupta, S. S. (1956). "On a Decision Rule for a Problem in Ranking Means." *Mimeo. Ser. No. 150*, Institute of Statistics, University of North Carolina. Chapel Hill.

[13] Gupta, S. S. (1965). *Technometrics*, **7**, 225–245. (The first paper to present a general theory of subset selection.)

[14] Gupta, S. S. and Huang, D.-Y. (1977). In *The Theory and Applications of Reliability*, C. P. Tsokos and I. N. Shimi, eds. Academic Press, New York, pp. 495–505.

[15] Gupta, S. S. and Miescke, K.-J. (1983). In *Statistical Decision Theory and Related Topics III*, Vol. 1, S. S. Gupta and J. O. Berger, eds. Academic Press, New York, pp. 473–496.

[16] Gupta, S. S. and Panchapakesan, S. (1979). *Multiple Decision Procedures: Theory of Methodology of Selecting and Ranking Populations*. Wiley, New York. (A comprehensive survey of all aspects of selection and ranking problems with an extensive bibliography.)

[17] Gupta, S. S. and Santner, T. J. (1973). *Proc. 39th Sess. Int. Statist. Inst.*, **45**, Book 1, 409–417.

[18] Sobel, M. and Tong, Y. L. (1971). *Biometrika*, **58**, 171–181.

[19] Sommerville, P. N. (1954). *Biometrika*, **41**, 420–429.

[20] Tamhane, A. C. and Bechhofer, R. E. (1977). *Commun. Statist. Theor. Meth.*, **A6**, 1003–1033.

[21] Tamhane, A. C. and Bechhofer, R. E. (1979). *Commun. Statist. Theor. Meth.*, **A8**, 337–358.

(MULTIPLE COMPARISONS RANKING PROCEDURES SELECTION PROCEDURES)

SHANTI S. GUPTA

OPTIMAL SPACING PROBLEMS

Let $X_{1:1}, \ldots, X_{n:n}$ denote the order statistics* for a random sample from a distribution of the form $F[(x - \mu)/\sigma]$, where F is a known distributional form and μ and σ are, respectively, location and scale parameters. Often estimates of μ and/or σ are obtained by using linear functions of $k < n$ order statistics. Such estimators have received considerable attention in the statistical literature primarily due to their computational simplicity, high efficiency, and freqent robust behavior under departures from distributional assumptions. The loss in efficiency from using a subset of the order statistics is compensated in many instances by the decrease in time spent computing estimators and analyzing data. Moreover, these types of estimators are easy to use in censored samples where most other estimation techniques, such as maximum likelihood*, are of a less computationally tractable nature. One particularly simple estimator that is a linear combination of k sample quantiles (and hence of k sample order statistics) is the asymptotically best linear unbiased estimator* (ABLUE) developed by Ogawa [28]. We now discuss this estimator and the associated problem of optimal quantile (spacing) selection.

THE ABLUE

Let $Q(u) = \inf\{x : F(x) \geqslant u\}$, $0 < u < 1$, denote the *quantile function* for F and, assuming that F admits a continuous density $f = F'$, define the *density–quantile function* $fQ(u) = f(Q(u))$, $0 \leqslant u \leqslant 1$. Also define the *sample quantile function* by

$$Q_n(u) = X_{j\,:\,n}\,,$$

$$\frac{j-1}{n} < u \leqslant \frac{j}{n}\,, \quad j = 1, \ldots, n.$$

Given a *spacing* $U = \{u_1, \ldots, u_k\}$ (k real numbers satisfying $0 < u_1 < \cdots < u_k < 1$), it was shown by Mosteller [27] that, provided fQ is continuous and positive at each of the u_i, the corresponding sample quantiles, $Q_n(u_1), \ldots, Q_n(u_k)$, have a k-variate normal limiting distribution with means $\mu + \sigma Q(u_i)$, $i = 1, \ldots, k$, and variance–covariance matrix composed of the elements $\sigma^2 u_i(1 - u_j)/[nfQ(u_i)fQ(u_j)]$, $i \leqslant j$, $i, j = 1, \ldots, k$. Thus, asymptotically, the $Q_n(u_i)$ follow a linear model with known covariance structure so that ABLUEs of μ and σ may be constructed using generalized least squares*. Motivated by the work of Mosteller [27] and Yamanouchi [37], Ogawa [28] developed explicit formulas for these estimators and their asymptotic relative Fisher efficiencies (AREs) (see also Sarhan and Greenberg [33, pp. 47–55]). These formulas for the various estimation situations can be summarized as follows:

Case 1. σ known. The ABLUE of μ when σ is known is

$$\mu^*(U) = [X(U) - \sigma K_3(U)]/K_1(U) \quad (1)$$

where, taking $u_0 = 0$, $u_{k+1} = 1$ and assuming $f\,Q(0) = f\,Q(1) = f\,Q(0)Q(0) = f\,Q(1)Q(1) = 0$,

$$K_1(U) = \sum_{i=1}^{k+1} \frac{[fQ(u_i) - fQ(u_{i-1})]^2}{u_i - u_{i-1}}\,, \quad (2)$$

$K_3(U)$

$$= \sum_{i=1}^{k+1} (u_i - u_{i-1})^{-1}[\{fQ(u_i) - fQ(u_{i-1})\}$$

$$\times \{fQ(u_i)Q(u_i) - fQ(u_{i-1})Q(u_{i-1})\}], \quad (3)$$

and

$$X(U) = \sum_{i=1}^{k+1} (u_i - u_{i-1})^{-1}[\{fQ(u_i) - fQ(u_{i-1})\}$$

$$\times \{fQ(u_i)Q_n(u_i) - fQ(u_{i-1})Q_n(u_{i-1})\}]. \quad (4)$$

The ARE of $\mu^*(U)$ is given by

$$\mathrm{ARE}(\mu^*(U)) = K_1(U)/I_{\mu\mu}\,, \quad (5)$$

with $I_{\mu\mu}$ denoting the Fisher information* for location parameter estimation, i.e.,

$$I_{\mu\mu} = E\!\left[\left(\frac{f'(X)}{f(X)}\right)^2\right] = \int_0^1 [(fQ)'(u)]^2\,du.$$

Case 2. μ known. The ABLUE of σ when μ is known is

$$\sigma^*(U) = [Y(U) - \mu K_3(U)]/K_2(U), \quad (6)$$

where

$K_2(U)$

$$= \sum_{i=1}^{k+1} \frac{[fQ(u_i)Q(u_i) - fQ(u_{i-1})Q(u_{i-1})]^2}{u_i - u_{i-1}} \quad (7)$$

and

$Y(U)$

$$= \sum_{i=1}^{k+1} \Big[fQ(u_i)Q(u_i)$$

$$- fQ(u_{i-1})Q(u_{i-1}) \Big]$$

$$\cdot \frac{fQ(u_i)Q_n(u_i) - fQ(u_{i-1})Q_n(u_{i-1})}{u_i - u_{i-1}}. \quad (8)$$

The ARE of this estimator is given by

$$\mathrm{ARE}(\sigma^*(U)) = K_2(U)/I_{\sigma\sigma}\,, \quad (9)$$

with $I_{\sigma\sigma}$, the Fisher information for scale parameter estimation, defined by

$$I_{\sigma\sigma} = E\!\left[\left(\frac{Xf'(X)}{f(X)}\right)^2\right] - 1$$

$$= \int_0^1 [(fQ \cdot Q)'(u)]^2\,du;$$

$fQ \cdot Q$ denotes the product of fQ and Q.

Case 3. Both μ and σ unknown. The AB-LUEs for μ and σ are

$$\mu^*(U) = \frac{[K_2(U)X(U) - K_3(U)Y(U)]}{\Delta(U)},$$

(10)

$$\sigma^*(U) = \frac{[-K_3(U)X(U) + K_1(U)Y(U)]}{\Delta(U)},$$

(11)

where

$$\Delta(U) = K_1(U)K_2(U) - K_3(U)^2. \quad (12)$$

The joint ARE for these estimators is

$$\text{ARE}(\mu^*(U), \sigma^*(U)) = \Delta(U)/|I(\mu, \sigma)|$$

(13)

with $|I(\mu, \sigma)|$ denoting the determinant of the Fisher information matrix for location and scale parameter estimation. That is, $I(\mu, \sigma)$ is the 2×2 symmetric matrix with diagonal elements $I_{\mu\mu}$ and $I_{\sigma\sigma}$ and off-diagonal element

$$I_{\mu\sigma} = E\left[X\left(\frac{f'(X)}{f(X)} \right)^2 \right]$$

$$= \int_0^1 (fQ)'(u)(fQ \cdot Q)'(u) \, du.$$

Upon examination of the ARE expressions (5), (9), and (13) it is seen that they are all functions of the spacing U. Thus an estimator based on optimal quantiles can be obtained by choosing U to maximize the ARE for the estimator of interest. A spacing that maximizes one of (5), (9), or (13) [or, equivalently, (2), (7), or (12)] will be termed an *optimal spacing*. The determination of optimal spacings under various choices for F (e.g., the normal, Cauchy, etc.) has been a popular area of statistical research and is the subject of the next section.

OPTIMAL SPACING SELECTION

Early work on optimal spacing selection (although not explicitly discussed as such) dates back at least to papers by Sheppard

[36] and Pearson [29], who considered the use of certain simplified estimators (that were, in fact, ABLUEs) for the mean and standard deviation of a normal distribution. Both Sheppard and Pearson examined the problem of estimating μ or σ using estimators of the form $\tilde{\mu} = \frac{1}{2}[Q_n(1 - u_1) + Q_n(u_1)]$ and $\tilde{\sigma} = b[Q_n(1 - u_1) - Q_n(u_1)]$ for $u_1 \le 0.5$ and $b > 0$. They found the optimal value of u_1 to be approximately 0.27, for estimation of μ and took $u_1 \approx 0.069$ with $b \approx 0.34$ for the estimation of σ. These values are in agreement with those later obtained by Kulldorff [26] for optimal spacings for the normal distribution. Pearson [29] also considered the estimation of μ and σ using AB-LUEs based on four quantiles. The approximations he gave for the optimal spacings are quite close to the exact values given by Kulldorff.

The majority of progress on the optimal spacing problem, however, has been made by Ogawa [29] since the development of the ARE expressions (5), (9), and (13), which are applicable to general F. As these expressions are nonlinear functions of the u_i's, the computation of optimal spacings has also been facilitated in recent years by the advent of high-speed computers.

Spacings that satisfy a necessary condition for optimality may be obtained by differentiating expression (5), (9), or (13), with respect to the u_i's and equating the resulting expressions with zero. The usual approach to this problem has been to examine the solutions to these equations (there may be many solutions) for a particular probability law of interest. The solution that provides the highest ARE is then taken as the optimal spacing. For location parameter estimation, this procedure, for most distributions of practical interest, reduces to solving

$$2(fQ)'(u_i) - \frac{fQ(u_i) - fQ(u_{i-1})}{u_i - u_{i-1}}$$

$$- \frac{fQ(u_{i+1}) - fQ(u_i)}{u_{i+1} - u_i} = 0,$$

$$i = 1, \ldots, k, \quad (14)$$

and, for scale parameter estimation, a similar necessary condition is

$$2(fQ \cdot Q)'(u_i)$$

$$- \frac{fQ(u_i)Q(u_i) - fQ(u_{i-1})Q(u_{i-1})}{u_i - u_{i-1}}$$

$$- \frac{fQ(u_{i+1})Q(u_{i+1}) - fQ(u_i)Q(u_i)}{u_{i+1} - u_i} = 0,$$

$$i = 1, \ldots, k. \quad (15)$$

In some cases it is possible to show that (14) and (15) have unique solutions. However, even when this is not the case, their solutions provide a set of optimal spacing candidates that may be examined to locate a spacing providing high ARE. Important early references that utilize (14) and (15) for spacing selection are Higuchi [21], Saleh and Ali [32], and Chan and Kabir [8]. General methods for determining when (14) and (15) are necessary conditions for an optimal spacing can be found in Cheng [10] and Eubank et al. [18]. In particular, if $(fQ)''$ and $(fQ \cdot Q)''$ are continuous and positive on $[0, 1]$, (14) and (15) are satisfied by optimal spacings and if, in addition, $\log(fQ)''$ and $\log(fQ \cdot Q)''$ are concave on $(0, 1)$, these equation systems have unique solutions for each k (see Eubank et al. [18]). For simultaneous estimation of μ and σ, the maximization of (13) is usually mathematically, and frequently even numerically intractable. A notable exception is the Cauchy distribution, for which the optimal spacing consists of the uniformly spaced points $i/(k + 1)$, $i = 1, \ldots, k$.

For a discussion of some of the early work on optimal spacing selection, see Harter [20] and Johnson and Kotz [22, 23]. A bibliography containing many of the more recent references is provided in Eubank [16].

Approximate solutions to the problem of optimal spacing selection that are based on spacings generated by density functions on $[0, 1]$ have been considered by Särndal [35] and Eubank [15]. Assuming that $(fQ)''$ and $(fQ \cdot Q)''$ are continuous, let $\psi(u) = ((fQ)''(u), (fQ \cdot Q)''(u))'$; then Eubank [15] showed that asymptotically (as $k \to \infty$) optimal spacings are provided by the $(k + 1)$-

tiles of the densities

$$h(u) = \begin{cases} \dfrac{|(fQ)''(u)|^{2/3}}{\int_0^1 |(fQ)''(s)|^{2/3}\,ds}, & \sigma \text{ known,} \\[3ex] \dfrac{|(fQ \cdot Q)''(u)|^{2/3}}{\int_0^1 |(fQ \cdot Q)''(s)|^{2/3}\,ds}, & \mu \text{ known,} \\[3ex] \dfrac{\left[\psi(u)'I(\mu,\sigma)^{-1}\psi(u)\right]^{1/3}}{\int_0^1 \left[\psi(s)'I(\mu,\sigma)^{-1}\psi(s)\right]^{1/3}\,ds}, \\[2ex] \qquad\qquad \text{both } \mu \text{ and } \sigma \text{ unknown.} \end{cases}$$

$$(16)$$

Let H^{-1} denote the quantile function for h in the parameter estimation problem of interest. Then an approximate solution is provided by the spacing $U_k = \{H^{-1}[1/(k + 1)], \ldots, H^{-1}[k/(k + 1)]\}$. Examples of these solutions are given in Table 1. This approach provides spacings that are optimal in an asymptotic (as $k \to \infty$) sense. For instance, if $\{U_k\}$ denotes the sequence of spacings chosen from the density proportional to $|(fQ)''(u)|^{2/3}$ by successively increasing k, and $\{U_k^*\}$ is a corresponding sequence of optimal spacings, then

$$\lim_{k \to \infty} \frac{1 - \text{ARE}(\mu^*(U_k))}{1 - \text{ARE}(\mu^*(U_k^*))} = 1.$$

This can be interpreted to mean that the sequences $\{U_k\}$ and $\{U_k^*\}$ have identical asymptotic properties in terms of their ARE behavior, and it suggests that, for large k, U_k might be used as a computationally expedient alternative to the optimal spacing U_k^*. Similar results hold for the other estimation situations. The approximate solutions provided by the densities in (16) have been found to work surprisingly well even for k as small as 7 or 9. For example, when $k = 7$, the optimal spacing for the estimation of the scale parameter of the exponential distribution* is given by Sarhan et al. [34] as $U_7^* = \{0.3121, 0.5513, 0.7277, 0.8506, 0.9297, 0.9746, 0.9948\}$ with a corresponding ARE of 0.969. In contrast, the approximate solution U_7, obtained from Table 1, consists of

Table 1   Asymptotically Optimal Spacings for Various Distributions

Distribution	Unknown Parameter(s)	$H^{-1}(i/[k+1])$
Cauchy	μ and σ	$i/[k+1]$
Exponential	σ	$1-(1-i/[k+1])^3$
Extreme-Value (largest value)	μ	$(i/[k+1])^3$
Logistic	μ	$i/[k+1]$
Normal $(F=\Phi)$	μ	$\Phi(\sqrt{3}\,\Phi^{-1}(i/[k+1]))$
Pareto $[F(x)=1-(1+x)^{-v},x,v>0]$	σ	$1-(1-i/[k+1])^{3v/(2+v)}$

the points $1-(1-i/8)^3$ and provides an ARE of 0.958. Consequently, the loss in ARE from using the approximate solution (16) is, in this case, only 0.011. Such comparisons for the other laws in Table 1 provide analogous results. See Eubank [15] for further comparisons and details regarding the computation and computational savings available from the use of the densities in (16).

The spacing densities in (16), or, equivalently, their H^{-1} functions, may be viewed as describing (asymptotically) the areas of concentration of the optimal spacings for a distribution. Hence they are useful for the purpose of comparison between distributions. There are at least three common shapes for the H^{-1} functions:

Uniform: $H^{-1}(u)=u$, such as for the logistic* (σ known) and Pareto* (μ known) when $v=1$.
Skewed: $H^{-1}(u)$ frequently behaves like u^3 [e.g., the exponential (μ known) and extreme-value* (σ known)].
Symmetric: $H^{-1}(u)=H^{-1}(1-u)$, as illustrated by the normal distribution.

In contrast to uniform- or symmetric-shaped spacing densities, those that are skewed indicate that, for large k, estimators will be composed of quantiles corresponding to predominantly large or small percentage values. Thus, for instance, since $H^{-1}(u)=u^3$ for the extreme value, it follows that estimators of μ for the extreme-value distribution will be based predominantly on data values that are below the median for k sufficiently large.

For many problems, such as those arising from the study of survival data, the object of interest is the pth population percentile for some $0<p<1$. Under the assumed location and scale parameter model the pth percentile is $\mu+\sigma Q(p)$, for which an ABLUE is $\mu^*(U)+\sigma^*(U)Q(p)$. This estimator has asymptotic (as $n\to\infty$) variance

$$\sigma^2\big[K_2(U)+Q(p)^2K_1(U)$$
$$-2Q(p)K_3(U)\big]/n\Delta(U),$$

which may be minimized as a function of U to obtain spacings that are optimal for percentile estimation. Optimal spacing selection, in this setting, has been considered by Ali et al. [2] and Saleh [32] for the exponential and double exponential distributions. A general approximate solution similar to those discussed previously is given in Eubank [15].

Tests regarding certain hypotheses about μ and σ that are based on the ABLUEs have been developed also. When both μ and σ are unknown, a test of $H_0:\mu=\mu_0$ against $H_a:\mu\neq\mu_0$ can be conducted using the statistic

$$\sqrt{K_1(U)}\,(\mu^*(U)-\mu_0)\Big/\sqrt{\Omega_1(U)/(k-2)}$$

$$(17)$$

where, for a symmetric spacing (i.e., $u_{k-i+1}=1-u_i$),

$$\Omega_1(U)=S(U)-[K_1(U)+K_2(U)]\mu^*(U)^2$$

and

$$S(U)=\sum_{i=1}^{k+1}(u_i-u_{i-1})^{-1}$$
$$\times\{fQ(u_i)Q_n(u_i)-fQ(u_{i-1})Q_n(u_{i-1})\}^2$$

(cf. Ogawa [28] or Sarhan and Greenberg [34, pp. 291–299]). This statistic has a Student's t limiting distribution with $k - 2$ degrees of freedom, when H_0 is true and an asymptotic noncentral t-distribution* under H_a with noncentrality $\sqrt{K_1(U)}\,(\mu - \mu_0)/\sigma$. Thus the power of this test is an increasing function of $K_1(U)$ and, consequently, optimal spacings for testing purposes may be obtained by maximizing $K_1(U)$ over all symmetric spacings. In some (although not all) cases, this is equivalent to optimal spacing selection for $\mu^*(U)$. A test for $H_0 : \sigma = \sigma_0$ against $H_a : \sigma \neq \sigma_0$, when μ is known and, therefore, may be taken as zero, can be obtained from

$$\sqrt{K_2(U)}\,(\sigma^*(U) - \sigma_0)\Big/ \sqrt{\Omega_2(U)/(k - 1)}\,, \tag{18}$$

where $\Omega_2(U) = S(U) - K_2(U)\sigma^*(U)^2$ (cf. Sarhan and Greenberg [33, pp. 380–383]). This statistic also has an asymptotic t-distribution under H_0, with $k - 1$ degrees of freedom and (asymptotically) a noncentral t-distribution under H_a with noncentrality $\sqrt{K_2(U)}\,(\sigma - \sigma_0)/\sigma$. As the power of the test increases with $K_2(U)$, it follows that selecting optimal spacings for testing hypotheses about σ and for the estimation of σ are equivalent problems. The use of (17) and (18) with the Cauchy, logistic, and normal distributions has been considered by Chan and Cheng [5] and Chan et al. [7], who have also investigated the problem of spacing selection and some small-sample behavior. Similar results for (18) are given by Ogawa [28] (see also Sarhan and Greenberg [33, pp. 380–382]), Chan et al. [6], and Cheng [11] for the exponential, extreme-value, and Rayleigh distributions, respectively.

In the previous discussion it has been assumed that the quantiles utilized in estimation were selected from complete samples. However, the ABLUE is also readily adapted for use with censored samples. For instance, if the sample is censored from the left, with censoring proportion α, then this may be viewed as observing $Q_n(u)$ only on $[\alpha, 1]$. Consequently, the ABLUE may be computed as before, except now the spacing

must satisfy $\alpha \leqslant u_1 < u_2 < \cdots < u_k < 1$. Then optimal spacings are obtained by maximizing (5), (9), or (13) over spacings of this form. Similar comments hold for right and both left and right censored samples. Most of the previous results, such as the necessary conditions (14) and (15) and the approximate solutions (16) are found to hold for censored samples after appropriate modifications also (cf. Cheng [10] and Eubank [15]). References containing the optimal spacings for censored samples from various probability distributions can also be found in Eubank [16]. A closely related problem where the ABLUE is used to predict future observations in what may be viewed as right censored samples is discussed by Kaminsky and Nelson [24].

In practice, the distributional form F is now always known. Frequently our knowledge is sufficient only to restrict attention to several possible candidates for the underlying probability law. In such instances, it may be advantageous to use spacings that are robust relative to the various models being considered. One measure of robustness for a spacing is its guaranteed ARE (GARE), i.e., its minimum ARE over all the candidates for F. An approach to robust spacing selection for location parameter estimation that is based on GARE has been developed by Chan and Rhodin [9] (see also Eubank [17a]).

RELATED PROBLEMS

The problem of optimal spacing selection for the ABLUEs is closely related to a variety of other statistical problems. In the one-parameter case, the problem of optimal spacing selection is equivalent to (a) optimal grouping for maximum likelihood estimation of μ and σ from grouped data [25]; (b) optimal grouping for the asymptotically most powerful group rank tests for the two sample location and scale problems [19]; and (c) regression design selection for time-series models with regression function fQ or $fQ \cdot Q$ and Brownian bridge error [15]. In addition, there is a structural similarity between these problems and (a) optimal strata selection with proportional allocation* [13], (b) problems of optimal grouping considered by Cox [12], Rade [30], and Ekman [14]; and (c) certain problems of optimal grouping for chi-squared tests of homogeneity* and for multivariate distributions [3, 4, 21a].

The connections and relationships between these problems have been examined by Adatia and Chan [1] and Eubank [17]. Under appropriate restrictions all these problems are equivalent for normal* and gamma* distributions.

References

[1]   Adatia, A. and Chan, L. K. (1981). *Scand. Actuarial J.*, 193–202.

[2]   Ali, M. M., Umbach, D., and Hassanein, K. M. (1981). *Commun. Statist. Theor. Meth.*, **10**, 1921–1932.

[3]   Bofinger, E. (1970). *J. Amer. Statist. Ass.*, **65**, 1632–1638.

[4]   Bofinger, E. (1975). *J. Amer. Statist. Ass.*, **70**, 151–154.

[5]   Chan, L. K. and Cheng, S. W. H. (1971). *Technometrics*, **13**, 127–137.

[6]   Chan, L. K., Cheng, S. W. H., and Mead, E. R. (1972). *Naval Res. Logist. Quart.*, **19**, 715–723.

[7]   Chan, L. K., Cheng, S. W. H., Mead, E. R., and Panjer, H. H. (1973). *IEEE Trans. Rel.*, **R-22**, 82–87.

[8]   Chan, L. K. and Kabir, A. B. M. L. (1969). *Naval Res. Logist. Quart.*, **16**, 381–404.

[9]   Chan, L. K. and Rhodin, L. S. (1980). *Technometrics*, **22**, 225–237.

[10]   Cheng, S. W. (1975). *J. Amer. Statist. Ass.*, **70**, 155–159.

[11]   Cheng, S. W. (1980). *Tamkang J. Math.*, **11**, 11–17.

[12]   Cox, D. R. (1957). *J. Amer. Statist. Ass.*. **52**, 543–547.

[13]   Dalenius, T. (1950). *Skand. Aktuarietidskr.*, **33**, 203–213.

[14]   Ekman, G. (1969). *Rev. Int. Statist. Inst.*, **37**, 186–193.

[15]   Eubank, R. L. (1981). *Ann. Statist.*, **9**, 494–500.

[16]   Eubank, R. L. (1982). *Tech. Rep. No. 162*, Dept. of Statistics, Southern Methodist University, Dallas, TX.

[17]   Eubank, R. L. (1983). *Statist. Prob. Lett.*, **1**, 69–73.

[17a]   Eubank, R. L. (1983). *Commun. Statist. Theory Meth.*, **12**, 2483–2491.

[18]   Eubank, R. L., Smith, P. L., and Smith, P. W. (1982). *SIAM J. Sci. Statist. Comp.* **3**, 238–249.

[19]   Gastwirth, J. L. (1966). *J. Amer. Statist. Ass.*, **61**, 929–948.

[20]   Harter, H. L. (1971). In *Proceedings of the Symposium on Optimizing Methods in Statistics*, J. S. Rustagi, ed. Academic Press, New York, pp. 33–62.

[21]   Higuchi, I. (1954). *Ann. Inst. Statist. Math.*, **5**, 77–90.

[21a]   Hung, Y. and Kshirsagar, A. M. (1984). *Statist. Prob. Lett.*, **2**, 19–21.

[22]   Johnson, N. L. and Kotz, S. (1970). *Distributions in Statistics: Continuous Univariate Distributions —1*. Wiley, New York.

[23]   Johnson, N. L. and Kotz, S. (1970). *Distributions in Statistics: Continuous Univariate Distributions —2*. Wiley, New York.

[24]   Kaminsky, K. S. and Nelson, P. I. (1975). *J. Amer. Statist. Ass.*, **70**, 145–150.

[25]   Kulldorff, G. (1961). *Contributions to the Theory of Estimation from Grouped and Partially Grouped Samples*. Almqvist and Wiksell, Stockholm.

[26]   Kulldorff, G. (1963). *Skand. Aktuarietidskr.*, **46**, 143–156.

[27]   Mosteller, F. (1946). *Ann. Math. Statist.*, **17**, 377–408.

[28]   Ogawa, J. (1951). *Osaka Math. J.*, **3**, 175–213.

[29]   Pearson, K. (1920). *Biometrika*, **13**, 113–132.

[30]   Rade, L. (1963). *Skand. Aktuarietidskr.*, **46**, 56–69.

[31]   Saleh, A. K. Md. E. (1981). In *Statistics and Related Topics*, M. Csörgő, D. A. Dawson, J. N. K. Rao, and A. K. Md. E. Saleh, eds. North-Holland, Amsterdam, pp. 145–151.

[32]   Saleh, A. K. Md. E. and Ali, M. M. (1966). *Ann. Math. Statist.*, **37**, 143–151.

[33]   Sarhan, A. E. and Greenberg, B. G. (1962). *Contributions to Order Statistics*. Wiley, New York.

[34]   Sarhan, A. E., Greenberg, B. G., and Ogawa, J. (1963). *Ann. Math. Statist.*, **34**, 102–116.

[35]   Särndal, C. E. (1962). *Information from Censored Samples*. Almqvist and Wiksell, Stockholm.

[36]   Sheppard, W. F. (1899). *Philos. Trans. R. Soc. Lond. A*, **192**, 101–167.

[37]   Yamanouchi, Z. (1949). *Bull. Math. Statist.*, **3**, 52–57.

(ESTIMATION, POINT
ORDER STATISTICS)

RANDALL EUBANK

OPTIMAL STOCHASTIC CONTROL

Control charts*, cusums* and Shewart schemes are traditional methods used for industrial quality control. Although these techniques ascertain when a process is in statistical control, they do not provide a comprehensive strategy for controlling a process at a target value. In spite of this, they have enjoyed widespread use, due in part to the minimal knowledge required of the process and ease of implementation. With many products having more stringent quality requirements, more complex and theoretically based control strategies have been proposed. Stochastic control theory provides a unified framework for the design of controllers for industrial processes. By employing a model for the process dynamics and disturbances, very flexible control algorithms can be designed. The presence of serially correlated observations, inherent in many industrial processes, and delays associated with analytical measurements are incorporated in the control strategy. These algorithms are specifically designed to control the process variable at its target value and are implemented readily with microprocessors or nomographs. In addition to providing an appealing methodology for industrial process control, stochastic control theory has been applied to problems in econometrics, management science, and biological systems.

MATHEMATICAL DESCRIPTION OF THE PROCESS

The concepts of stochastic control theory are best introduced by considering a univariate process. Observations of a single process variable y are obtained at equispaced time intervals $t, t - T, t - 2T \ldots$. These observations are denoted by $y(k), y(k - 1), y(k - 2)$. Adjustments are made in a manipulated variable u (i.e., a flow rate, temperature, etc.) to affect y. The manipulated variable $u(k)$ is held constant over the interval $t, t - T$. Stochastic control theory differs

from traditional approaches to quality control in that a model of the process is required. The discrete time behavior of many industrial processes can often be described by a constant-coefficient linear difference equation* of the form

$$Y(k) = \delta_1 Y(k - 1) + \cdots + \delta_r Y(k - r)$$
$$+ \omega_0 U(k - b) - \cdots$$
$$- \omega_s U(k - b - s). \quad (1)$$

$Y(k)$ is the deviation of the process variable from its desired value. $U(k)$ is the deviation of the manipulated variable from the value required to keep the process variable on target. b is the delay in sampling periods between making a change in the manipulated variable and observing its initial effect on the process variable. By definition $b \geqslant 1$. The parameters $(\delta_1, \ldots \delta_r, \omega_0, \ldots \omega_s)$ model the time-dependent behavior of the process variable to changes in the manipulated variable. By an appropriate choice of the parameters, it is possible to model a damped sinusoidal response, an exponential rise to a new value, or an inverse response, where the process initially goes in the opposite direction to the final response. The process model (1) can be written in transfer function* notation as

$$Y(k) = \frac{\omega(z^{-1})}{\delta(z^{-1})} U(k - b)$$

$$= \frac{\omega_0 - \omega_1 z^{-1} - \cdots - \omega_s z^{-s}}{1 - \delta_1 z^{-1} - \cdots - \delta_r z^{-r}}$$

$$\times U(k - b). \quad (2)$$

z^{-1} is the backward shift operator defined by $z^{-1} Y(k) = Y(k - 1)$. Transfer functions* are used extensively in engineering to model the dynamic behavior of processes.

The process description (1) is incomplete. It predicts that if the manipulated variable is held at its equilibrium value, $Y(k)$ will return to and remain at its equilibrium value. Under such conditions, many industrial processes will tend to drift away from target value due to the presence of process disturbances. Deterministic disturbances can be modeled by steps, ramps, or sinusoidal functions. Many industrial disturbances are of a more random nature and can be represented mathematically by time-series* models [1, 2].

Box and Jenkins [2] and Astrom [1] characterize stochastic disturbances by mixed autoregressive–integrated moving average (ARIMA)*, time-series models. (*See* AUTO-REGRESSIVE–INTEGRATED MOVING AVERAGE (ARIMA) MODELS.) The general structure of these models is

$$N(k) = \frac{\theta(z^{-1})}{\phi(z^{-1})\nabla^d} a(k)$$

$$= \frac{1 - \theta_1 z^{-1} - \cdots - \theta_q z^{-q}}{1 - \phi_1 z^{-1} - \cdots - \phi_p z^{-p}}$$

$$\times (1 - z^{-1})^{-d} a(k), \quad (3)$$

where $\{a(k)\}$ is a sequence of independently distributed random variables with mean 0 and variance σ (i.e. a white noise sequence). ∇ is an abbreviation for $1 - z^{-1}$. There are q moving average* parameters and p autoregressive* parameters. When $d = 0$, the disturbance is mean stationary. If $d \geqslant 1$, it is possible to describe process disturbances that do not have a fixed mean. Equation (3) is known as an ARIMA (p, d, q) model.

The time-series model (3) can be considered as the output of a linear filter driven by white noise. Time series models are used extensively for filtering and prediction*.

Commonly encountered industrial disturbances are often described by ARIMA (0, 1, 1) models of the form

$$N(k) = N(k - 1) + a(k) - \theta a(k - 1). \quad (4)$$

When $\theta = 0$, the disturbance is known as a random walk. As θ increases, the nonstationary behavior of $N(k)$ is masked by short-term variability. It can be shown [2] that an ARIMA $(0, 1, 1)$ model is an aggregate of a random walk* on top of which is superimposed an uncorrelated white noise* measurement variability. The moving average parameter θ is proportional to the ratio of the measurement variability to the driving force of the random walk.

The process is now described by the superposition of the dynamics or deterministic component and disturbances (see Fig. 1).

$$Y(k) = \frac{\omega(z^{-1})}{\delta(z^{-1})} U(k - b) + N(k). \quad (5)$$

Low-order linear models with additive disturbances have been used extensively to model many complex industrial processes. When the underlying continuous process dynamics are described by a linear or linearized differential equation, there exists a unique difference equation representation. When the underlying continuous model is unknown or does not exist, commonly used identification techniques (i.e., maximum likelihood* or prediction error methods) can be used to estimate the structure and parameters of the model from data collected in a designed experiment [1, 2]. A comprehensive review of identification techniques appears in the 1981 special edition of *Automatica* on Identification Methods and Identification Applications. The papers on identification methods have been published separately by Isermann [10].

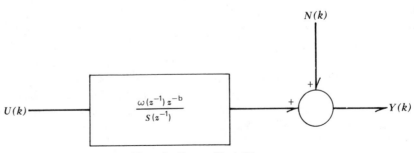

Figure 1 Process Block Diagram.

When special precautions are taken, the models can be identified from data collected in a closed-loop experiment [4, 12]. During such an experiment, the process value is kept close to its target value by manipulating U in some fashion. Superimposed on this control action is an extraneous control action, or "dither signal," that is uncorrelated with the primary control action. Data collected under closed-loop conditions minimize the main objections to plant experimental work —economic cost of off-grade product and safety considerations.

DESIGN OF FEEDBACK CONTROLLERS

Control schemes can now be devised using the mathematical description of the process dynamics and disturbances. A realistic objective from an economic and quality standpoint is to design a controller to minimize the deviations of the process variable from setpoint. This can be expressed mathematically as

$$J_1 = \lim_{N \to \infty} \frac{1}{N} E \left\{ \sum_1^\infty Y^2(k) \right\}$$

$$= \text{var } Y, \qquad (6)$$

where $E\{\ \}$ denotes the mathematical expectation. The cost of the control action is not considered in this design. The controller minimizing (6) is known as the unconstrained minimum variance controller. The application of an unconstrained minimum variance controller for monetary correction is discussed in Castrucci and Garcia [5].

In many process applications, the control action minimizing the variance of Y calls for excessive changes in the manipulated variable. To constrain the control action, the objective function is modified to include (6) a penalty on control

$$J_2 = \lim_{N \to \infty} \frac{1}{N} E \left\{ \sum_1^\infty Y^2(k) + \lambda \left(\nabla^d U(k) \right)^2 \right\}$$

$$= \text{var } Y + \lambda \text{ var } \nabla^d U. \qquad (7)$$

If the disturbances are nonstationary (i.e.,

$d \geqslant 1$), the variance of $\nabla^d U$ must be minimized since that of U is indefinitely large. The controller minimizing (6) or (7) is a linear combination of past deviations and control actions of the form

$$\nabla^d U(k) = a_0 Y(k) + a_1 Y(k-1) + \cdots$$

$$+ a_m Y(k-m) + b_1 \nabla^d U(k-1)$$

$$+ \cdots + b_n \nabla^d U(k-n). \qquad (8)$$

The coefficients $(a_0, \ldots a_m, b_1, \ldots b_n)$ are obtained by factorizing a covariance generating function. This requires finding the solution to a set of polynomial equations [13, 14]. The variance of Y and $\nabla^d U$ depend on the mean and variance of $\{a(k)\}$ and not on a particular realization of the white noise sequence. Thus it is possible to evaluate the variance of Y and $\nabla^d U$ numerically for different values of λ [13, 14]. The constraining factor is varied until the variance of Y and $\nabla^d U$ are jointly acceptable. In many cases a substantial reduction in the variance of the manipulated variable is achieved for a small increase in the variance of the process variable. This is especially true when the control interval T is small relative to the major process response time.

Controllers derived using objective functions (6) or (7) have some very desirable features. These controllers have reset or integral action if the process disturbances are nonstationary (i.e., $d \geqslant 1$). Consequently, the process variable will be kept on target in the presence of a persistent bias. These controllers also have dead-time compensation. That is, they allow for a rapid response to process disturbances without overcompensation or overcorrection resulting from time lag in the process. (Time lag arises many times from analytical delay.) As well, the controller minimizing (6) or (7) is independent of the distribution from which the [$a(k)$'s] originate. As a result, a controller designed to compensate for a random walk $N(k) = N(k-1) + a(k)$ will also guard against a disturbance in which the $\{a(k)$'s$\}$ are all zero except at one instant at which there is a nonzero value. The latter describes a step or load disturbance. The control schemes can be modified to allow for incorporation of exogeneous or feed-forward variables* [3]. Other properties of these controllers are discussed in Harris et al. [7] and Palmor et al. [11]. Optimal choice of the control interval is considered in MacGregor [8, 9]. Control algorithms with desirable features can also be designed by minimizing objective functions other than (6) or (7) [6].

The unconstrained minimum variance controller, $\lambda = 0$, can be derived explicitly. Ideally, one should choose the control action to cancel the process disturbance

$$U(k) = \frac{-\delta(z^{-1})}{\omega(z^{-1})} N(k + b). \qquad (9)$$

Due to the delay in the process ($b \geqslant 1$), this strategy requires that future values of the disturbance be known. Since this is impossible, the control action is chosen to cancel a prediction or forecast* of the disturbance

$$U(k) = \frac{-\delta(z^{-1})}{\omega(z^{-1})} \hat{N}\left(k + \frac{b}{k}\right) \qquad (10)$$

The forecast $\hat{N}(k + b/k)$ is restricted to date inclusive of time k. By selecting the forecast that has the smallest mean square prediction error, the control scheme (10) minimizes the variance of Y [1, 2]. The minimum variance forecast $\hat{N}(k + b/k)$ and forecast error $e(k + b)$ are determined uniquely by factoring the disturbance as follows [1, 2, 14]:

$$N(k + b) = \frac{\theta(z^{-1})}{\phi(z^{-1})\nabla^d} a(k + b)$$

$$= \psi(z^{-1})a(k + b)$$

$$+ \frac{T(z^{-1})}{\phi(z^{-1})\nabla^d} a(k)$$

$$= e(k + b) + \hat{N}(k + b/k). \qquad (11)$$

$\psi(z^{-1})$ is a polynomial of order $b - 1$. The term $\psi(z^{-1})a(k + b)$ is the prediction error. Substituting (11) into (10), the minimum

variance controller is given by

$$\nabla^d U(k) = \frac{-\delta(z^{-1})}{\omega(z^{-1})} \frac{T(z^{-1})}{\phi(z^{-1})} a(k). \qquad (12)$$

Under minimum variance control, the process deviation at time k will equal the prediction error b periods earlier

$$Y(k) = \psi(z^{-1})a(k). \qquad (13)$$

Using this, the control action can be expressed in terms of past control actions and output deviations as

$$\nabla^d U(k) = \frac{-\delta(z^{-1})}{\omega(z^{-1})} \frac{T(z^{-1})}{\phi(z^{-1})\psi(z^{-1})} Y(k). \qquad (14)$$

The process deviation is a moving average time-series model of order $b - 1$. For $b \geqslant 2$, the process output will not be a white noise sequence. Instead, adjacent values in time will be autocorrelated. When the delay exceeds one period, this autocorrelation will be present regardless of the control scheme implemented. Since the observations are not serially independent, care must be taken when applying conventional quality control tests to determine whether a process is in statistical control.

To illustrate the controller designs, consider the process described by

$$Y(k) = \frac{-0.076}{1 - 0.46z^{-1}} U(k - 2)$$

$$+ \frac{1 - 0.7z^{-1}}{1 - z^{-1}} a(k). \qquad (15)$$

Model (15) was identified from data collected in a designed experiment. The process disturbance can be factored in terms of its forecast error and forecast as

$$N(k + 2) = \frac{1 - 0.7z^{-1}}{1 - z^{-1}} a(k + 2)$$

$$= (1 + 0.3z^{-1})a(k + 2)$$

$$+ \frac{0.30}{1 - z^{-1}} a(k). \qquad (16)$$

The minimum variance controller is then

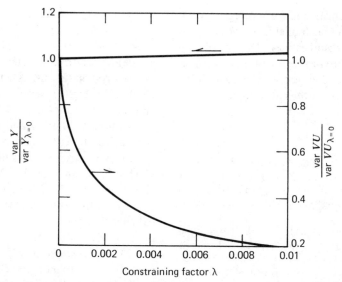

Figure 2   The effect of constraining on the variance Y and VU for the process described by (15).

given by

$$\nabla U(k) = \frac{0.300(1 - 0.46z^{-1})}{0.076(1 + 0.30z^{-1})} Y(k). \quad (17)$$

Using the definition of the backward shift operator, the controller can be expressed as

$$\nabla U(k) = 3.95 Y(k - 1) - 1.82 Y(k - 2)$$
$$- 0.30 \nabla U(k - 1). \quad (18)$$

The change in control action at time k is expressed as a linear combination of past deviations from setpoint and past control actions. The unconstrained controller requires large changes in the manipulated variable. By constraining the control action, an 80% reduction in the variance of ∇U is achieved at the expense of a 3% increase in the variance of Y (see Fig. 2). The coefficients of the unconstrained and constrained minimum variance controller for different values of the constraining factor are shown in Table 1. Constraining the control action in this fashion results in smooth changes in the manipulated variable. The performance of the constrained controller ($\lambda = 0.004$) was excellent.

Summary

Stochastic control theory provides an attractive approach for the control of industrial

Table 1   Coefficients of the Constrained Minimum Controller:

$$\nabla U(k) = a_0 Y(k) + a_1 Y(k - 1) + b_1 \nabla U(k - 1) + b_2 \nabla U(k - 2)$$

λ	a_0	a_1	b_1	b_2	$\dfrac{\text{Var } \nabla U}{\text{Var } \nabla U_{\lambda=0}}$	$\dfrac{\text{Var } Y}{\text{Var } Y_{\lambda=0}}$
0.000	3.95	-1.82	-0.30	0.00	1.00	1.000
0.002	2.85	-1.31	0.06	-0.06	0.44	1.006
0.004	2.44	-1.12	0.20	-0.08	0.32	1.012
0.006	2.19	-1.01	0.29	-0.10	0.26	1.018
0.008	2.00	-0.92	0.36	-0.16	0.22	1.022
0.010	1.87	-0.86	0.40	-0.13	0.20	1.026

processes. Practical control strategies result from employing a model for the process dynamics and disturbances. The control algorithms have desirable theoretical properties. In practice, they are very robust to the assumptions used in their development.

References

[1] Astrom, K. J. (1976). *Introduction to Stochastic Control Theory*. Academic Press, New York. (Procedures for identifying, fitting, and verifying dynamic and stochastic models are discussed in Chapter 6.)

[2] Box, G. E. P. and Jenkins, G. M. (1976). *Time Series Analysis: Forecasting and Control*, 2nd ed. Holden-Day, San Francisco. (Discusses the procedures mentioned in ref. 1 as well as properties of ARIMA models.)

[3] Box, G. E. P., Jenkins, G. M., and MacGregor, J. F. (1974). *Appl. Statist.*, **23**, 158–179.

[4] Box, G. E. P. and MacGregor, J. F. (1976). *Technometrics*, **18**, 371–380.

[5] Castrucci, P. B. L. and Garcia, J. D. G. (1981). *Automatica*, **17**, 221–229.

[6] Clarke, D. W. and Hasting-James, R. (1971). *Proc. IEE*, **118**, 1503–1506.

[7] Harris, T. J., MacGregor, J. F., and Wright, J. D. (1982). *Canad. J. Chem. Eng.*, **60**, 425–432.

[8] MacGregor, J. F., (1976). *Technometrics*, **18**, 151–160.

[9] MacGregor, J. F., (1977). *Technometrics*, **19**, 224.

[10] Isermann, R. ed. (1981). *System Identification*. Pergamon, New York.

[11] Palmor, Z. J. and Shinnar, R. (1979). *Ind. Eng. Chem. Process Des. Dev.*, **18**, 8–30.

[12] Soderstrom, T., Gustavsson, I., and Ljung, L. (1975). *Int. J. Control*, **21**, 243–255.

[13] Whittle, P. (1963). *Prediction and Regulation by Linear Least-Squares Methods*. D. Van Nostrand, Princeton, NJ.

[14] Wilson, G. T. (1970). "Modelling Linear Systems for Multivariable Control." Ph.D. thesis, University of Lancaster, England.

Further Reading

A very readable overview of the Box–Jenkins approach appears in *Time Series Analysis* and *Forecasting: The Box–Jenkins Approach* by O. D. Anderson (Butterworths, London, 1975). A comprehensive treatment of time series, and its relation to spectral analysis, is given in *Spectral Analysis and Time Series* by M. B. Priestley (Academic Press, New York, 1981). The application of an identification technique to a complex biological system is discussed in O. Brouko, D. M. Wiberg, L. Arena, and J. W. Bellville (1981) (*Automatica*, **17**, 213–220). The application of filtering and prediction and a comparison to a standard cusum method for inventory control appears in D. J. Downing, D. H. Pike, and G. W. Morrison, (1980) (*Technometrics*, **22**, 17–22).

An adaptive version of stochastic control, called self-tuning regulators* has been developed by K. J. Astrom and B. J. Wittenmark (1973). (*Automatica*, **9**, 185–199). An on-line identification scheme is combined with a feedback controller. There have been numerous industrial applications of this technique. A review of this approach is given in T. J. Harris, J. F. MacGregor, J. E. Wright, (1981) (*Technometrics*, **22**, 153–164).

The theory of multivariable stochastic control for continuous processes is covered in *Linear Estimation and Stochastic Control* by M. H. A. Davis (Methuen, London, 1977) and in ref. 1. Discrete processes are developed very nicely in Box and Jenkins [2]. Applications are fewer due to the complexities in modeling multivariable processes. Nonlinear stochastic control theory is, in most cases, mathematically intractable. In practice, a local linearization is made, and linear stochastic control theory applied. Stochastic control theory is but one of several more complex and theoretically based control strategies that have been used for industrial process control. Other approaches are critiqued by L. S. Kershenbaum and T. R. Forthesque, *Automatica*, **17**, 777–788.

Applications of stochastic control theory for the solution of economic and management problems are discussed in *Applied Stochastic Control in Econometrics and Management Science*, A. Bensoussan, P. Kleindorfer, and C. S. Tapiero, eds. (North-Holland, Amsterdam, 1980).

Articles on stochastic control-theory and applications appear regularly in *Automatica* and occasionally in *Technometrics*.

(AUTOREGRESSIVE–INTEGRATED
MOVING AVERAGE (ARIMA) MODELS
EVOLUTIONARY OPERATION (EVOP)
FEEDFORWARD FEEDBACK CONTROL
SCHEMES
FORECASTING
MAXIMUM LIKELIHOOD ESTIMATION
MOVING AVERAGE
QUALITY CONTROL, STATISTICS IN
SERIAL CORRELATION
TIME SERIES
TRANSFER FUNCTION)

THOMAS J. HARRIS

OPTIMAL STOPPING RULES

An optimal stopping problem is a particularly simple game of chance* in which a gambler is allowed to observe sequentially random variables $y_1, \ldots, y_n, \ldots$, having a known joint distribution and to decide at each stage to stop or continue the observation process. If after observing $y_1, \ldots, y_n$, the gambler decides to stop, he or she receives a reward r_n that is some known function of $y_1, \ldots, y_n$; and the game is finished. Otherwise the gambler observes y_{n+1} and again must decide whether to stop with the reward r_{n+1} or to continue for at least one more observation. The goal of the gambler is to find a stopping rule that maximizes the expected terminal fortune. Mathematically, a stopping rule is a random variable taking values (with probability one) in the set $\{1, 2, \ldots\}$ such that the event $\{t = n\}$ is defined entirely in terms of the available observations at time n, $y_1, \ldots, y_n$ and does not involve future observations. The gambler's terminal reward is r_t. The value V of the game is $\sup E(r_t)$, where the supremum is taken over all stopping rules t. The gambler's goal is to determine V and a stopping rule t^* such that $E(r_{t^*}) = V$.

Example. Suppose y_i is 0 or 1 according to whether the ith toss of a fair coin is tails or heads, and let $r_n = 2^n(y_1 y_2 \ldots y_n)$. This game is "fair" in the sense that $E(r_{n+1} \mid y_1, \ldots, y_n) = r_n$, i.e., r_n, $n = 1, 2, \ldots$ is a

martingale*. It is not surprising to learn that the gambler cannot do better in this game than always to stop with the first observation: For every stopping rule t, $Er_t \leqslant Er_1 = 1 = V$. The stopping problem defined by $\tilde{r}_n = -r_n$ is also fair with $E\tilde{r}_n = -1$, but in this case the stopping rule $t^* = $ first $n \geqslant 1$ such that $y_n = 0$ produces a terminal reward $\tilde{r}_{t^*}$ that is 0 with probability one, so t^* is obviously optimal.

The theory of optimal stopping had its birth in the context of sequential* statistical analysis, especially in the papers of Wald [12], Wald and Wolfowitz [13], and, most important, Arrow et al. [2], who were concerned with finding Bayes'* solutions to general sequential statistical decision* problems and specifically to the problem of testing a simple hypothesis against a simple alternative with data in the form of independent, identically distributed observations, each costing a unit amount. Abstract optimal stopping theory divorced from its statistical origins and with emphasis on its relation to martingale theory began in the paper of Snell [11] and received an important boost from Chow and Robbins [5], who described a class of interesting nonstatistical problems with elegant explicit solutions—the so-called monotone problems.

To understand the relation to statistical decision theory in its simplest form, assume that under the probability measure P_i ($i = 0$, 1), $y_1, y_2, \ldots$ are independent and identically distributed random variables that can be observed sequentially at unit cost per observation. A statistician does not know which of the probabilities P_i is the one generating the data, but there are a priori probabilities π that it is P_0 and $1 - \pi$ that it is P_1. Let $P = \pi P_0 + (1 - \pi)P_1$. Let w_i denote the cost incurred by the statistician in deciding that P_i is not the correct probability when in fact it is ($i = 0, 1$), and assume that no loss is incurred if the statistician makes the correct inference. The posterior risk to the statistician if he makes a decision after observing $y_1, \ldots, y_n$ is $\min\{w_0 \pi_n, w_1(1 - \pi_n)\}$, where π_n is the posterior probability that P_0 is the true underlying probability. The cost of observing $y_1, \ldots, y_n$ is n, so the total posterior expected *loss* associated with observation and an incorrect decision is $-r_n = \min\{w_0\pi_n, w_1(1 - \pi_n)\} + n$. An optimal Bayes' test is given by a stopping rule t, which maximizes Er_t. (E is expectation under $P = \pi P_0 + (1 - \pi)P_1$.) Wald and Wolfowitz [13] showed somewhat heuristically, and Arrow et al. [2] more precisely, that the optimal Bayes' test is a sequential probability ratio test*.

One general class of optimal stopping problems can be solved completely by a definite, although complicated algorithm. This is the finite case, in which there is a finite upper bound N on the number of random variables $y_1, y_2, \ldots, y_N$ that can be observed. The algorithm defines the so-called method of backward induction. Generalizations of it have been popularized in the literature of dynamic programming* under the name *principle of optimality*.

Let $R_n^{(N)}$ be defined for $n = N, N - 1, \ldots, 1$ by

$$R_N^{(N)} = r_N$$
$$R_n^{(N)} = \max\left(r_n, E\left(R_{n+1}^{(N)} \mid y_1, \ldots, y_n\right)\right), \tag{1}$$
$$1 \leqslant n \leqslant N - 1.$$

The random variable $R_n^{(N)}$ can be interpreted as the conditional expected reward of a gambler who has already observed $y_1, \ldots, y_n$ and proceeds optimally. Equation (1) states that if $n = N$, the gambler must stop and accept r_N. If $n < N$, the gambler can either stop and accept r_n or take the $(n + 1)$st observation and continue optimally thereafter. In the latter case, his or her conditional expected reward is $E(R_{n+1}^{(N)} \mid y_1, \ldots, y_n)$. The gambler can expect to achieve the maximum of these two quantities by stopping if

$$r_n \geqslant E\left(R_{n+1}^{(N)} \mid y_1, \ldots, y_n\right)$$

and continuing otherwise. (In the case of equality, the gambler is actually indifferent between stopping and continuing.) It may be

shown that an optimal rule is to stop sampling with

$$t^*(N) = \text{first } n \geqslant 1 \quad \text{such that} \quad r_n = R_n^{(N)} \tag{2}$$

and $V = ER_1^{(N)}$.

In general, there is no upper bound N on the number of available observations. Nevertheless, one can define a random variable R_n as the optimal conditional expected reward given that $y_1, y_2, \ldots, y_n$ have been observed. Analogously to (1) and (2), one expects to find that

$$R_n = \max(r_n, E(R_{n+1} \mid y_1, \ldots, y_n)),$$
$$n = 1, 2, \ldots \tag{3}$$

and that an optimal rule is defined by

$$t^* = \text{first } n \geqslant 1 \quad \text{such that} \quad r_n = R_n. \tag{4}$$

Although (3) is true, it does not define an algorithm because there is no starting value. Also, it may happen that $r_n < R_n$ for all n, so (4) does not define an actual stopping rule or it may happen that t^* in (4) is a stopping rule that is not optimal.

The most successful approach to studying the general optimal stopping problem has been to create a family of finite case problems by supposing that one must stop sampling after N observations. The finite case problems can be solved by (1) and (2), and one can attempt to show that for large N, this approximately solves the original problem in the sense that $R_n = \lim_{N \to \infty} R_n^{(N)}$ and $t^* = \lim_{N \to \infty} t^*(N)$ is optimal.

To see that the program sketched in the preceding paragraph contains numerous pitfalls, consider again the reward sequences r_n and $\tilde{r}_n = -r_n$ in the example. It is easy to see by the martingale property and (1) that in both of these problems an optimal stopping rule in the truncated game is to take exactly one observation. This is also optimal in the untruncated game for r_n, but not for $\tilde{r}_n$. In fact for $\tilde{r}_n$, the optimal expected reward in the truncated game is -1 no matter how large N is, and this by no means converges to 0, the optimal expected reward of the untruncated game.

Suppose now that $\tilde{\tilde{r}}_n = a_n r_n$, where a_n, $n = 1, 2, \ldots$ is any positive, strictly increasing, bounded sequence of real numbers, and r_n is as in the example. Now $E(\tilde{\tilde{r}}_n | y_1, \ldots, y_n) > \tilde{\tilde{r}}_n$ as long as $\tilde{\tilde{r}}_n$ is positive. Hence by (1), an optimal rule in the game truncated at N is to stop as soon as $\tilde{\tilde{r}}_n = 0$ or at the Nth observation. The expected reward for this rule is $E(\tilde{\tilde{r}}_N) = a_N$, which increases with N. However, the limit of these rules is to sample until $\tilde{\tilde{r}}_n = 0$, which, of course, is the worst possible stopping rule. It may be shown that there is no optimal stopping rule in this game.

A remarkable general result of the optimal stopping theory is the so-called *triple limit theorem* [6, p. 81; 10, p. 75]. It states that if r_n is replaced by $r_n(a, b) = \max(a, \min(r_n, b))$ for some $a < 0 < b$ and $R_n^{(N)}(a, b)$ is defined by (1) relative to $r_n(a, b)$, then, with probability one,

$$R_n = \lim_{b \to \infty} \lim_{a \to -\infty} \lim_{N \to \infty} R_n^{(N)}(a, b)$$

and

$$V = \lim_{b \to \infty} \lim_{a \to -\infty} \lim_{N \to \infty} ER_1^{(N)}(a, b).$$

Although one cannot actually use this algorithm for computational purposes, its qualitative implications are very important and allow development of a much simpler and more complete theory for optimal stopping than for general dynamic programming* or gambling* problems.

Optimal stopping theory as just described is a fairly well understood subject, the mathematical foundations of which are described in Chow et al. [6] and Shiryayev [10]. However, the number of cases in which one can compute an explicit optimal stopping rule with pencil and paper remains small. Often the computational problem is more manageable if the process evolves in continuous time. There the nonlinear integral equations* (1) can frequently be replaced by familiar (partial) differential equations, and the nonlinearity enters as a free boundary

(Stefan) condition. See Chernoff [3] and Shiryayev [10] for discussions of these ideas.

In addition to the general theory described, there are several concrete optimal stopping problems that, because of their importance, have generated their own literature. Some examples are (a) the secretary problem*, (b) the problem of disruption [6, p. 106; 10, pp. 193–207], and (c) Anscombe's model of sequential clinical trials* [1, 4, 9]. An important recent development due to Gittins [7, 8] (cf. also Whittle [14, 15]) shows that a difficult class of dynamic programming problems—the multiarmed bandit problems with discounting—can be reduced to optimal stopping problems, which, computationally, are much simpler to solve.

References

[1] Anscombe, F. J. (1963). *J. Amer. Statist. Ass.*, **58**, 365–383.

[2] Arrow, K. J., Blackwell, D., and Girshick, M. A. (1949). *Econometrica*, **17**, 213–244.

[3] Chernoff, H. (1972). *Sequential Analysis and Optimal Design*. SIAM, Philadelphia.

[4] Chernoff, H. and Petkau, J. (1981). *Biometrika*, **68**, 119–132.

[5] Chow, Y. S. and Robbins, H. (1961). *Proc. 4th Berkeley Symp. Math. Statist. Prob.*, **1**, 93–104.

[6] Chow, Y. S., Robbins, H., and Siegmund, D. (1971). *Great Expectations: The Theory of Optimal Stopping*, Houghton Mifflin, Boston.

[7] Gittins, J. C. (1979). *J. R. Statist. Soc. B*, **41**, 148–164.

[8] Gittins, J. C. and Jones, D. M. (1974). *Progress in Statistics*, J. Gani, ed. North-Holland, Amsterdam, pp. 241–266.

[9] Lai, T. L., Levin, B., Robbins, H., and Siegmund, D. (1980). *Proc. Natl. Acad. Sci. USA*, **77**, 3135–3138.

[10] Shiryayev, A. N. (1978). *Optimal Stopping Rules*. Springer-Verlag, New York.

[11] Snell, J. L. (1952). *Trans. Amer. Math. Soc.*, **73**, 293–312.

[12] Wald, A. (1947). *Econometrica*, **15**, 279–313.

[13] Wald, A. and Wolfowitz, J. (1948). *Ann. Math. Statist.*, **19**, 326–339.

[14] Whittle, P. (1980). *J. R. Statist. Soc. B*, **42**, 143–149.

[15] Whittle, P. (1982). *Optimization over Time*, Vols. I and II. Wiley, New York.

(DECISION THEORY
DYNAMIC PROGRAMMING
GAMBLING, STATISTICS IN
ONE- AND TWO-ARMED BANDIT
   PROBLEMS
SEQUENTIAL ANALYSIS
SEQUENTIAL SAMPLING)

D. Siegmund

OPTIMIZATION IN STATISTICS

Optimizing techniques are commonly applied to solve routine problems in many areas of science, business, industry and government. Topics in optimization have become important areas of study in disciplines such as operations research*, chemical engineering, electrical engineering, and economics. Optimizing methods play a central role in many statistical concepts, and many statistical techniques often turn out to be ordinary applications of optimization.

Mathematical techniques for optimization have been developed over the past several hundred years, and with the advent of modern computers, these techniques are making a significant impact in many areas of science and engineering. Optimizing techniques are used in a wide variety of applications. They include complicated engineering systems such as designing of nuclear reactors as well as involved management systems such as patient care in hospitals.

The very names of statistical procedures of estimation such as least squares*, minimum variance*, maximum likelihood*, least absolute deviation, and minimum chi-squared* indicate that they utilize techniques of optimization. In several other areas of statistics such as in regression analysis, testing hypotheses*, design of experiments*, decision theory*, sample survey*, and information theory*, optimization plays a major role.

A broad classification of optimizing methods can be made as follows.

1.   Classical optimizing methods.

2.   Numerical optimizing methods.

3.   Methods of mathematical programming*.

4.   Variational methods, including dynamic programming*.

Classical methods of optimization include those found in calculus and classical mathematical analysis. Necessary and sufficient conditions for an extremum of a function of several variables are available in textbooks. When solutions are not available in a closed form, practical application requires numerical approximations. The most elementary numerical optimization technique is by direct search, requiring the evaluation of the function to be optimized over the range of the variables leading to the optimizing value or values. Other commonly used numerical optimizing techniques include the methods of Gauss–Newton*; Newton–Raphson*; gradient methods, including methods of steepest ascent and descent; and iterative methods. Often the optimizing technique reduces to solutions of nonlinear equations and methods abound for solving them in numerical analysis*. For a recent survey, see Fletcher [5].

Applications of linear and nonlinear programming* occurs in problems of regression analysis where inequality constraints occur. They are used extensively in constrained estimation such as in Markov-chain* models, constrained regression, and optimal design of multifactor experiments. A survey of programming methods in statistics is given by Arthanari and Dodge [1].

Variational methods include those techniques used in optimizing functionals over function spaces. The classical methods of calculus of variations were developed to be used in engineering mechanics, and newer variational techniques, such as in Neyman–Pearson theory*, in testing hypotheses have led to the solution of many other statistical problems. Certain problems in control theory are variational, and efforts to solve them led to the development of Bellman's dynamic programming* and Pontryagin's maximum principle. Applications of these and many other variational methods to statistical problems is given by Rustagi [21].

Statistical examples are given here to illustrate the application of optimizing techniques. A wide variety of statistical applications of optimization may be found in the *Proceedings of the Conferences on Optimizing Methods in Statistics* [20, 23]. A special issue on optimization in statistics published by *Communications in Statistics** [22] is devoted to statistical studies using optimization. A collection of papers on optimization in statistics with special applications to management sciences* and operations research* has been recently published in a book edited by Zanakis and Rustagi [24].

APPLICATIONS OF CLASSICAL OPTIMIZING METHODS

Example 1. Let $p_1, p_2, \ldots, p_k, p_i \geq 0$, and $\sum_{i=1}^{k} p_i = 1$ be the probabilities of a trial ending in k possibilities. A sample of n trials leads to $x_1, x_2, \ldots, x_k$ occurrences of various possibilities. The maximum likelihood estimates of $p_1, p_2, \ldots, p_k$ are obtained by maximizing $L(p_1, p_2, \ldots, p_k)$ such that

$$L(p_1, p_2, \ldots, p_k) = p_1^{x_1} p_2^{x_2} \cdots p_k^{x_k},$$

with constraint, $p_1 + p_2 + \cdots + p_k = 1$. Lagrange's multiplier rule is used to obtain the solution.

Suppose that there are inequality restrictions such as

$$p_1 \leqslant p_2 \leqslant \cdots \leqslant p_k$$

on the p_i's. In that case, the estimates have to use the methods of mathematical programming*.

Example 2. Consider a normal p-variate population with mean μ and covariance matrix Σ. Let $\mathbf{x}_1, \ldots, \mathbf{x}_N$ be a random sample from the distribution. The logarithm of the likelihood is a constant multiple of $L(\mu, \Sigma)$ where

$$L(\mu, \Sigma) = -\log|\Sigma| - \text{tr}\,\Sigma^{-1}\mathbf{V}$$

with

$$\mathbf{V} = \frac{1}{N} \Sigma'(\mathbf{x}_\alpha - \bar{\mathbf{x}})(\mathbf{x}_\alpha - \bar{\mathbf{x}})'.$$

Differential calculus provides the maximum likelihood estimates of μ and Σ.

Example 3. Optimal Allocation in Survey Sampling. Suppose M is the number of total units to be divided among N clusters of size M_0 each. Let S_W^2 be within-cluster variance and S_B^2 be the between-cluster variance. Let the sample size selected be size n. Let the corresponding cost C_B associated with a cluster regardless of its size and C_W be the cost associated with each element regardless of cluster size. Then, for a fixed cost C, we have

$$C = nC_B + nM_0 C_W,$$

and we want to minimize the variance of the overall average $\bar{\bar{y}}$

$$V(\bar{\bar{y}}) = \left(1 - \frac{n}{N}\right)(S_W^2 + M_0 S_B^2)/(nM_0).$$

The optimal solution turns out to be

$$M_{\text{opt}} \propto \sqrt{C_B S_W^2 / C_W S_B^2}.$$

APPLICATIONS OF NUMERICAL OPTIMIZING METHODS

Example 4. Consider the problem of estimation of parameters of the gamma distribution*

$$f(x) = e^{-x/\beta} x^{\alpha-1} / \{\Gamma(\alpha)\beta^\alpha\}, \quad x > 0$$
$$= 0, \qquad\qquad\qquad \text{elsewhere.}$$

The maximum likelihood estimates are given by equating to zero the partial derivatives of $\log L$ with respect to α and β, where

$$\log L = -\Sigma x_i/\beta + (\alpha - 1)\Sigma \log x_i$$
$$- n \log \Gamma(\alpha) - n\alpha \log \beta.$$

The equations are

$$\bar{x} - \hat{\alpha}\hat{\beta} = 0$$

and

$$\Sigma \log x_i - n[\Gamma'(\hat{\alpha})/\Gamma(\hat{\alpha})] - n \log \hat{\beta} = 0.$$

These equations can be solved only through numerical methods although sometimes tables of the digamma function $\Gamma'(\alpha)/\Gamma(\alpha)$ [16] may facilitate the solution.

Example 5. Reliability. Realistic models in reliability theory* and survival analysis* require numerical evaluation frequently. Consider the three-parameter Weibull distribution* model for the time to failure. The probability density function is given by

$$f_T(t) = \begin{cases} \dfrac{\beta}{\delta}\left(\dfrac{t-\mu}{\delta}\right)^{\beta-1} \exp\left[-\left(\dfrac{t-\mu}{\delta}\right)^{\beta}\right], \\ \qquad\qquad\qquad\qquad t \geqslant \mu \\ 0, \quad t < \mu. \end{cases}$$

Here $\beta, \delta > 0$ and $\mu \geqslant 0$.

Suppose the experiment is conducted over the period $(0, t_0)$ and the times of failures of r_0 individuals out of r on test are given by $t_1, t_2, \ldots, t_{r_0}$. Then the likelihood of the sample is given by

$$L = \frac{r!}{(r-r_0)!\, r_0!}\left(\frac{\beta}{\delta}\right)^{r_0} \prod_{i=1}^{r_0}\left(\frac{t_i-\mu}{\delta}\right)^{\beta-1}$$

$$\cdot \exp\left(-\sum_{i=0}^{r_0}\left(\frac{t_i-\mu}{\delta}\right)^{\beta}\right)$$

$$\times \left\{1 - \exp\left[-\left(\frac{t_0-\mu}{\delta}\right)^{\beta}\right]\right\}^{r-r_0}.$$

The maximum likelihood estimates of μ, β, and δ can only be obtained numerically. Direct search procedures may be used. There are many approaches in the literature. Mann et al. [14] provide other models in reliability and survival analysis leading to numerical solutions.

Numerical methods are being used directly in statistical procedures. The methods of projection pursuits [1] utilize numerical optimization extensively. For recent applications to regression and density estimation*, see Friedman and Stuetzle [6] and Friedman et al. [8].

APPLICATIONS OF MATHEMATICAL PROGRAMMING

Example 6. Linear Regression. A common model of linear regression is

$$\mathbf{y} = \mathbf{X}\boldsymbol{\beta} + \boldsymbol{\epsilon},$$

where $\mathbf{y}$ is a vector of n dimensions, $\mathbf{X}$ is a $n \times p$ matrix of known constants, $\boldsymbol{\beta}$ is a p-dimensional vector of unknown parameters, and $\boldsymbol{\epsilon}$ is an n-vector of residuals. Suppose it is assumed that

$$\mathbf{C}\boldsymbol{\beta} \geqslant 0,$$

where $\mathbf{C}$ is some $g \times p$ known matrix. It is proposed to estimate $\boldsymbol{\beta}$ so as to minimize

$$(\mathbf{y} - \mathbf{X}\boldsymbol{\beta})'(\mathbf{y} - \mathbf{X}\boldsymbol{\beta})$$

subject to $\mathbf{C}\boldsymbol{\beta} \geqslant 0$. The problem reduces to the problem of quadratic programming. There are well-known algorithms to solve such problems, see Davis [4] and Judge and Takayama [12].

Example 7. Sampling. One of the common problems in sample survey is the estimation of the total y, of a finite population given by

$$y = \sum_{i=1}^{T} N_i y_i^*,$$

where $y_1^*, y_2^*, \ldots, y_T^*$ are the possible individual values of items in the population and N_i is the number of units in the population having values y_i^* so that under simple random sampling of size n, with $n = \sum n_i$, the maximum likelihood estimate of y is obtained by maximizing the likelihood

$$L = \prod_{i=1}^{T} \frac{\binom{N_i}{n_i}}{\binom{N}{n}}$$

with $n = \sum n_i$. The optimization in this case reduces to an integer programming* problem [9]. For other problems in survey sampling using mathematical programming, see Rao [17].

Example 8. Design of Experiments*. An important class of designs is concerned with factorial experiments. When the number of factors is large, all treatment combinations cannot be used in a block of ordinary size and hence fractional factorial* designs have been developed. A recent introduction in the study of fraction factorial is the concept of cost optimality. The problem of finding cost optimal fraction factorials naturally lead to programming problems [15].

Example 9. Estimation of Markov Chain Probabilities.

Consider the problem of estimating the transition probabilities $p_{ij}(t)$ of the Markov chain x_t, $t = 1, 2, \ldots, T$, and $i, j = 1, 2, \ldots, r$. Here

$$p_{ij}(t) = \Pr\{X_t = s_j \mid x_{t-1} = s_i\},$$

where s_i, $i = 1, 2, \ldots, r$ are the finite number of the states of the chain. Here

$$\sum_i \sum_j p_{ij}(t) = 1$$

and

$$0 \leqslant p_{ij} \leqslant t.$$

Suppose the chain is observed for $N(t)$ independent trials. Let $w_j(t)$ be the proportion of events that fall in the jth category. The likelihood* of the sample, then, can be obtained as follows

$$L = \prod_{t=1}^{T} \frac{N(t)!}{\prod_m (N(t)w_m(t))!}$$
$$\times \left(N(t) - \sum_k N(t)w_k(t) \right)!$$
$$\cdot \prod_j \left(w_i(t-1)p_{ij}(t) \right)^{N(t)w_j(t)}$$
$$\cdot \left(1 - \sum_k \sum_i w_i(t-1)p_{ij}(t) \right)^{N(t) - \sum_k N(t)w_k(t)}.$$

The problem of maximizing the likelihood subject to the preceding constraints is a nonlinear programming problem. This problem is studied by Lee et al. [13].

APPLICATION OF VARIATIONAL METHODS

Example 10. Order Statistics*.

Suppose $X_1 < X_2 < \cdots < X_n$ are the ordered statistics from a continuous distribution function $F(x)$. The expectation of the largest order statistic X_n is given by

$$L(F) = \int x \, dF^n(x).$$

An important problem in order statistics is to find upper and lower bounds of $L(F)$ when the mean and variance are given.

Similarly one may want to find the bounds of the expectation of the range $X_n - X_1$ of the sample. That is,

$$\min(\max) \int x \, d\left\{ 1 - F^n(x) - (1 - F(x))^n \right\}$$

subject to certain constraints.

Such problems occur in nonparametric statistical inference and have been discussed by Rustagi [18].

Example 11. Mann–Whitney–Wilcoxon* Statistic.

Suppose we are interested in the bounds of the variance of Mann–Whitney–Wilcoxon statistic. This is needed for applications such as in finding confidence intervals for $p = \Pr(X < Y)$. The integral we minimize (maximize) reduces to

$$I(F) = \int (F(x) - kx)^2 \, dx$$

subject to the condition

$$\int F(x) \, dx = 1 - p.$$

This is a variational problem and is treated by Rustagi [19].

Example 12. Efficiency of Tests.

Consider a random sample from a population having a continuous distribution function $F(x)$. Suppose we are interested in testing the hypothesis

$$H_0 : F(x) = G(x) \quad \text{vs.}$$
$$H_1 : G(x) = F(x - \theta)$$

with θ as some location parameter. The relative asymptotic efficiency of Wilcoxon test with respect to the t-test, which is used if F and G are normal distributions, is given in terms of the integral

$$I(f) = \int f^2(x) \, dx,$$

where $f(x)$ is the corresponding probability density function of X.

A problem of interest in nonparametric inference is to find bounds of $I(f)$ subject to side conditions such as

$$\int f(x)\,dx = 1, \qquad \int xf(x)\,dx = 0.$$

Hodges and Lehmann [10] provide further details.

Example 13. Robustness*. M-estimates* of a location parameter* θ for a probability density $f(x - \theta)$ and cumulative distribution function* $F(x - \theta)$ are defined by Huber [11]. M-estimate based on a random sample $X_1, X_2, \ldots, X_n$ from $f(x - \theta)$ is given by T_n if it maximizes, for some ρ,

$$\sum_{i=1}^{n} \rho(X_i - T_n).$$

A variational problem occurs if one finds a distribution function that minimizes the asymptotic variance of T_n.

Other variational problems on robustness have been discussed by Bickel [2] and Collins and Portnoy [3], for example.

References

[1]   Arthanari, T. S. and Dodge, Y. (1981). *Mathematical Programming in Statistics*. Wiley, New York.

[2]   Bickel, P. (1965). *Ann. Math. Statist.*, **36**, 847–858.

[3]   Collins, J. R. and Portnoy, S. L. (1981). *Ann. Statist.*, **9**, 567–577.

[4]   Davis, William W. (1978). *J. Amer. Statist. Ass.*, **73**, 575–579.

[5]   Fletcher, R. (1980). In *Unconstrained Optimization*, Vol. I: *Practical Methods of Optimization*. Wiley, New York.

[6]   Friedman, J. H. and Stuetzle, W. (1981). *J. Amer. Statist. Ass.*, **76**, 817–823.

[7]   Friedman, J. H. and Tukey, J. W. (1974). *J. IEEE Trans. Computers*, **C-23**, 881–890.

[8]   Friedman, J. H., Stuetzle, W., and Schroeder, A. (1981). *Technical Report*, Stanford University, Stanford, CA.

[9]   Hartley, H. O. and Rao, J. N. K. (1969). *New Developments in Survey Sampling*, N. L. Johnson and Harry Smith, eds. Wiley-Interscience, New York, pp. 147–169.

[10]   Hodges, J. L. and Lehmann, E. L. (1956). *Ann. Math. Statist.*, **27**, 324–355.

[11]   Huber, P. (1972). *Ann. Math. Statist.*, **43**, 1042–1067.

[12]   Judge, G. G. and Takayama, T. (1966). *J. Amer. Statist. Ass.*, **61**, 166–181.

[13]   Lee, T. C., Judge, G. G., and Zellner, A. (1958). *J. Amer. Statist. Ass.*, **63**, 1162–1179.

[14]   Mann, N. R., Schaffer, R. E., and Singpurwalla, N. D. (1974). *Methods for Statistical Analysis of Reliability and Lifedata*. Wiley, New York.

[15]   Neuhardt, J. and Mount-Campbell, C. A. (1978). *Commun. Statist. B*, **7**, 369–383.

[16]   Pearson, K. and Hartley, H. O. (1954). *Biometrika Tables for Statisticians*, Vols. I and II. Cambridge University Press, Cambridge, England.

[17]   Rao, J. N. K. (1979). *Optimizing Methods in Statistics*, J. S. Rustagi, ed. Academic Press, New York, pp. 419–434.

[18]   Rustagi, J. S. (1957). *Ann. Math. Statist.*, **28**, 309–328.

[19]   Rustagi, J. S. (1961). *Ann. Inst. Statist. Math. Tokyo*, **13**, 119–126.

[20]   Rustagi, J. S., ed. (1971). *Optimizing Methods in Statistics*. Academic Press, New York.

[21]   Rustagi, J. S. (1976). *Variational Methods in Statistics*. Academic Press, New York.

[22]   Rustagi, J. S., ed. (1978). *Commun. Statist. B*, **7**, 303–415.

[23]   Rustagi, J. S., ed. (1979). *Optimizing Methods in Statistics*. Academic Press, New York.

[24]   Zanakis, S. and Rustagi, J. S., eds. (1982). *Optimization in Statistics*. North-Holland, New York.

(MATHEMATICAL PROGRAMMING OPTIMIZATION, STATISTICS IN)

J. S. RUSTAGI

OPTIMIZATION, STATISTICS IN

The interface between the two fields of statistics and optimization forms a rich and diverse area of research that has received much research attention during the past decade. The variety of topics that belongs to this interface is considerable, thereby making it impossible for us to enumerate all the topics in this article. There is, however, one important example of the interplay between statistics and optimization that focuses on using statistical methods to estimate the best solution to various combinatorial and other difficult optimization problems. This article aims to review past and current research relating to this interesting area. In particular, we describe and review the results that enable one to obtain point estimates and confidence intervals for the optimal solution of difficult optimization problems based on sampling information.

A general optimization problem may be written as

$$\min f(x) \quad \text{such that} \quad x \in S \subseteq R^d, \quad (1)$$

where S is the set of all solutions to the problem and f is a known function generally called the objective function. Any solution x^* in S satisfying $f(x^*) \leqslant f(x)$ for all $x \in S$ is called an optimal solution and the corresponding objective function value $y^* = f(x^*)$ is called the optimal value of (1) or simply the optimum. Loosely speaking, one may distinguish between discrete and continuous optimization problems according to whether S is a discrete subset of R^d or not. In continuous problems, S is frequently a convex region or a suitable union of such regions. In combinatorial problems, on the other hand, S is a discrete set (e.g., all components of x may be restricted to have values 0 or 1). Usually continuous problems are solved by using some variant of the gradient to locate local minima. These problems become difficult to solve if either f or S is nonconvex since one would then be unable to distinguish between local and global optimal solutions. Most interesting combinatorial problems are difficult to solve optimally, due to the large number of candidate solutions (elements of S) that must be examined.

When faced with a difficult nonconvex or combinatorial optimization problem, an alternative to using deterministic algorithms for solving the problem is to sample solutions from the set S in some appropriate fashion. Suppose that $x_1, \ldots, x_n$ are n elements sampled randomly from the set S and that $y_i = f(x_i)$, $i = 1, \ldots, n$ are the corresponding values. The statistical approach involves using the values y_1 through y_n to obtain a point estimate and confidence interval for y^*. The key connection between statistics and optimization is provided by results due to Gnedenko [12] and Fisher and Tippett [11] on extreme-value distribu-

tions, that may be stated briefly as follows: Consider n independent random variables $Y_1, \ldots, Y_n$ bounded below by a with a common distribution function $F(\cdot)$ so that $F(a) = 0$ and $F(a + \epsilon) > 0$ for $\epsilon > 0$. If $Y_1' = \min\{Y_1, \ldots, Y_n\}$, then, under suitable assumptions on $F(\cdot)$, as n gets large, the distribution of Y_1' approaches a Weibull distribution* given by

$$G(y) = 1 - \exp\left[-\{(y - a)/b\}^c\right]$$
$$\text{for} \quad y \geqslant a \geqslant 0 \quad (b > 0, c > 0). \quad (2)$$

The quantities a, b, and c are called the location, scale, and shape parameters, respectively. We shall refer to this distribution as $W(a, b, c)$.

In the following section, we discuss how extreme-value distributions* may be used to provide statistical information about the optima of continuous nonconvex optimization problems. The following sections focus on applications of the statistical approach to combinatorial problems. In this setting, we shall see that the sampled minima (Y_1') correspond to heuristic solutions to the optimization problem. We shall also see how extreme-value results enable us to obtain impressively good estimates of optimal solution values for notoriously difficult combinatorial problems.

STATISTICAL ANALYSIS OF NONCONVEX OPTIMIZATION PROBLEMS

Consider the optimization problem in (1), where S is assumed to be bounded. Suppose that n elements of S are sampled randomly, resulting in the independent and identically distributed (iid) random variables $X_1, \ldots, X_n$ each uniformly distributed over S. Let $Y_i = f(X_i)$ for $i = 1, \ldots, n$ and denote the corresponding order statistics by $Y_1' \leqslant Y_2' \leqslant \cdots \leqslant Y_n'$. We know that Y_1' converges to y^* as n tends to infinity. We now ask how these observations can be used to provide statistical information about y^*.

One approach to this problem may be found in the work of de Haan [10] and Boender et al. [4] that we now review. Using extreme-value theory, de Haan shows that there exists a positive constant α such that the asymptotic CDF of αT approaches the exponential CDF $1 - e^{-t}$ as n gets large. Here

$$T = \ln\left(\frac{Y_2' - y^*}{Y_1' - y^*}\right). \tag{3}$$

If α is known, this fact can be used to construct a $(1 - p)$-level confidence interval for the optimal value y^* as specified in (4).

$$P\left[Y_1' - \beta_p(Y_2' - Y_1') \leqslant y^* \leqslant Y_1'\right] = 1 - p, \tag{4}$$

where $\beta_p^{-1} = (1 - p)^{-1/\alpha} - 1$. As we shall see, it is possible in some cases to establish the value of α theoretically. If, however, the value of α is unknown, then it can be estimated. It is shown by de Haan that the random variable

$$U = \ln\left(\frac{Y_{k(n)}' - Y_3'}{Y_3' - Y_2'}\right) \bigg/ \ln k(n)$$

converges to α^{-1} weakly as $n \to \infty$ provided that the sequence $\{k(n)\}$ is chosen to satisfy $k(n)/n \to 0$ as $n \to \infty$. This result implies that the random variable T/U retains the asymptotic CDF $1 - e^{-t}$ and hence a confidence interval for y^* similar to (3) can still be constructed.

It is well known that for the extreme-value limiting distributions to exist, certain conditions must be placed on the cumulative distribution function of the sampled results. In this case, the required condition relates to $F(\cdot)$, the distribution for Y_i, and may be stated as

$$\lim_{y \to 0^+} \frac{F(y^* + \sigma y)}{F(y^* + y)} = \sigma^\alpha$$
$$\text{for all} \quad \sigma > 0. \tag{5}$$

Since $F(y) = P[f(X) \leqslant y]$, where X is uniformly distributed over S, the condition in (5) may also be translated into a condition for f. Boender et al. show that a sufficient condition on f that ensures (5) is that x^* be a unique global minimum for f and that f be twice differentiable with a nonsingular Hessian at x^*. If this condition holds, then the α

occurring in (5) and the preceding discussion can be fixed at $d/2$, i.e., one-half the dimension of the feasible region S.

Patel and Smith [26] apply extreme-value theory to the optimization problem in (1) under the assumptions that f is concave and S is a compact polyhedral set. Their problem is thus a concave minimization problem subject to linear constraints. In these problems, local minima occur at extreme points of S but it is generally difficult to identify the global minimum solution x^*. By assuming that x^* is the unique global solution to (1), the authors can prove that condition (5) is satisfied with $\alpha = d$. Extreme-value theory then implies that the distribution of Y_1' is asymptotically Weibull $W(a, b, c)$, where $a = y^*$, $b = F^{-1}(1/n) - y^*$, and $c = \alpha = d$. Thus one already knows the shape parameter of the Weibull distribution and only needs to estimate $a = y^*$ and b. To do this, Patel and Smith randomly sample K sets of n observations each from S. If $Y_{1,k}'$ denotes the minimum objective value in the kth set of observations $(k = 1, \ldots, K)$, then for large n, $Y_{1,k}'$ is Weibull distributed, as discussed earlier. If one sets

$$Y_0 = \min\{Y_{1,k}'; k = 1, \ldots, K\},$$

then an unbiased estimate of y^* is

$$\hat{a} = \hat{y}^* = Y_0 - b(K + 1)^{-1/c}. \tag{6}$$

Solving for $\hat{y}^*$ provides a point estimate of the optimal solution value. Moreover, since Y_0 is the minimum of K iid Weibull random variables, $\tilde{W} = (1/b)(Y_0 - y^*)$ is distributed as $W(0, K^{-1/c}, c)$ with b and c as defined before. This allows us to construct confidence intervals for y^* in terms of the observed value of Y_0. Details may be found in Patel and Smith [26].

The importance of the preceding results lies in the theoretical justification they provide for using extreme-value distribution theory to obtain point estimates and confidence intervals for y^*. Moreover, we remarked that, under suitable assumptions, the value of α (corresponding to the shape parameter of the limiting distribution) may be fixed in terms of the dimension of the feasible region

S. Boender et al. have used the confidence interval in (4) in conjunction with a clustering approach to solve global nonconvex optimization problems by searching for all local minima of f over S. Their computational experience is reported in Boender et al. [4]. Patel and Smith [26] solved a number of concave minimization problems using their statistical approach. The dimension of S, that is, d, was 4 in their test problems, and n and K were set equal to 100. In all cases, their point estimate was within 2% of the true optimal solution value y^* and the width of the 99% confidence interval did not exceed 5% of y^*.

POINT ESTIMATION
FOR COMBINATORIAL PROBLEMS

Consider a combinatorial problem, such as the traveling salesman problem*, that is known to be difficult to solve optimally. Suppose that N samples consisting of n feasible solutions are each randomly sampled from the solution set S. Let Y^k be the minimum value obtained in the kth sample ($k = 1, \ldots, N$). By extreme-value theory, if n is large, each Y^k should be approximately Weibull. If the random variables $\{Y^k; k = 1, \ldots, N\}$ can be assumed to be independent, then N observations from the common distribution in (2) are available and may be used to estimate the parameters a, b, and c. Of these, the parameter a is of the greatest interest since it corresponds to the sought-after optimal value of the combinatorial problem.

McRoberts [24] applied this approach to combinatorial plant layout problems by taking as the Y^k's, the sequence of N intermediate solutions resulting from a single application of an iterative heuristic. These values were used to estimate the parameters of $W(a, b, c)$ with the help of Weibull probability paper.

Golden [13] significantly extended the work of McRoberts by applying this approach to the traveling salesman problem (TSP). The sampled minima (Y^k) in Gold-

en's approach correspond to heuristic solutions to the traveling salesman problem obtained by N "independent" runs of the heuristic. Golden argued that each heuristic solution may be viewed as the local minimum of a large number of possible solutions (traveling salesman tours). Thus the heuristic performs the task of implicitly sampling a large number of solutions and computing their minimum objective value. This allows one to invoke the Weibull distribution for each Y^k. To ensure that the Y^k's are independent, Golden used different runs of the heuristic starting from different initial solutions.

To estimate the parameters, Golden exploited the fact that if a is fixed, then least squares* can be applied to obtain estimates for α and β in the relation

$$\ln\left[-\ln\{ 1 - G(y)\} \right] = \alpha \ln(y - a) + \beta,$$

where $\alpha = c$ and $\beta = -c \ln b$. Thus once α and β are determined, b and c can be estimated also. This analysis is then repeated for different values of a and the best choice of a is determined by the largest correlation coefficient. The Weibull hypothesis was tested on networks with 70–130 nodes randomly generated over a square area. Computational results indicated that the Weibull hypothesis was justified statistically. The average point estimate was within 5% of the presumed optimal tour length for all networks tested.

To put these results in perspective, we may cite Golden's experiment with a 25-city TSP with a known optimal solution of 1711. In 25,000 random observations from the solution set S, the best value obtained was 3300, nearly twice the optimal solution value. This should underscore the significant advantages of estimating the optimal solution value from heuristics as opposed to random sampling of solutions. The ineffectiveness of the latter procedure is due to the extremely large number of feasible solutions in combinatorial problems such as the TSP. The poor performance of random sampling even when the sample size is in the thousands shows that good heuristic solutions, which generally come to within 10% of the optimum or closer, correspond more accurately to local minima of a truly astronomical number of solutions.

Golden's least-squares* approach in ref. 13 resulted in estimates that also satisfied the maximum likelihood* equations almost exactly while producing a correlation coefficient of close to one. In subsequent papers, Golden [14, 15] modified the least-squares approach in various ways. Over all the benchmark TSP's studied, the largest difference between the point estimate and the optimal solution was less than 3.0%.

King and Spachis [19] compared the maximum likelihood equation estimates to the least-squares estimates of Weibull parameters on six job-shop scheduling problems. The least-squares estimates showed an average deviation of 4.10% from the best known solution, with the greatest deviation being 11.11%. The maximum likelihood equation estimates showed an average deviation of 4.83% from the best known solution value, with the greatest deviation being 10.52%.

We remark briefly that Dannenbring [9] has also approached the point estimation problem, but from a distribution-free viewpoint.

INTERVAL ESTIMATION FOR COMBINATORIAL PROBLEMS

To the best of our knowledge, Clough [7] was the first to suggest and actually develop confidence intervals* for the optimal value of constrained optimization problems. His results pertain, however, only to the very restricted instance of the Weibull distribution where $c = 1$, that is, an exponential distribution*.

Golden and Alt [16] developed a straightforward interval estimation procedure for the Weibull distribution. From (2), we see that

$$G(a + b) = 1 - e^{-1}. \qquad (7)$$

Thus, if Y^k $(k = 1, \ldots, N)$ are iid, with $G(y)$ as their common distribution and Y_0

$= \min\{Y^k; 1 \leqslant k \leqslant N\}$, one obtains

$$P[Y_0 \leqslant a + b] = 1 - P[Y_0 > a + b]$$

$$= 1 - \prod_{k=1}^{N} P[Y^k > a + b]$$

$$= 1 - e^{-N}. \qquad (8)$$

Thus an approximate $100(1 - e^{-N})\%$ confidence interval for the optimal solution a is given by $[Y_0 - \hat{b}, Y_0]$, where $\hat{b}$ is estimated via the maximum likelihood approach.

Golden and Alt tested the Weibull hypotheses using the Kolmogorov–Smirnov goodness-of-fit* statistic and found that it could not be rejected. They proceeded to apply this confidence interval procedure to 50 two-opt solutions for each of the five 100-node benchmark problems presented by Krolak et al. [21] and to 50 three-opt solutions to the 318-node problem of Lin and Kernighan [22], derived by Padberg and Hong [25]. Computational results revealed that the width of the confidence interval was never more than 6.5% of the optimal solution and that the optimal solution was contained in every computed interval.

Los and Lardinois [23] extended the methodology of Golden and Alt by constructing a more general confidence interval that we now describe briefly.

Following the same line of thought used in the derivation of (8), one may write for any real number T:

$$G(a + bT^{-1}) = 1 - \exp(-T^{-c})$$

$$= P[Y^k \leqslant a + bT^{-1}]$$

so that

$$P[Y_0 \leqslant a + bT^{-1}]$$

$$= 1 - \prod_{k=1}^{N} P[Y^k > a + bT^{-1}]$$

$$= 1 - \exp(-N/T^c).$$

If we specify the level of significance by α, let $\hat{b}$ and $\hat{c}$ be the maximum likelihood estimates of b and c, and define T to be $[(-N/\ln \alpha)]^{1/\hat{c}}$, then an approximate $100(1 - \alpha)\%$ confidence interval for a is given by

$$P[Y_0 - \hat{b}T^{-1} \leqslant a \leqslant Y_0] = 1 - \alpha. \qquad (9)$$

In the second section, we noted that Patel and Smith [26] also derived confidence intervals based on the Weibull distribution using a different method.

DIRECT APPROACHES

Instead of appealing to extreme-value theory, several researchers have taken a more direct approach to estimating the cumulative distribution function of heuristic solution values that we denote by $F(y)$. Hartley and Pfaffenberger [18] assumed that $F(y)$ can be represented by a polynomial of specified degree. Chichinadze's approach [5, 6] was somewhat similar in that it involved a polynomial approximation to $F(y) = P[f(X) \leq y, X \in S]$ for the optimization problem in (1). Rubinstein [27] has discussed this area also.

Sielken and Monroe [28] recently were successful in extending such direct approaches to derive three procedures for determining confidence intervals on the lower bound based on sample proportions, smallest order statistics, and derivatives. Their Monte Carlo* study indicated that the confidence interval procedure based on smallest order statistics displayed the best empirical behavior.

In his dissertation, Klein [20] developed a direct approach based on an end-point approximation of the form

$$F(y) = \delta_1(y - \delta_2)^{\delta_3},$$

$$\text{where} \quad y \geq \delta_2 \quad \text{and} \quad \delta_1, \delta_3 \geq 0, \quad (10)$$

which is a three-parameter power function distribution* (a special case of the beta distribution*). Klein suggested a quick and straightforward procedure for estimating δ_1, δ_2, and δ_3 and applied the procedure to two 10-city traveling salesman problems* with mixed results.

Motivated by Klein's work, Ariyawansa and Templeton [1–3] showed that under certain conditions $F(y)$ behaves as a three-parameter power function near the lower truncation point. They also observed that $F(y)$ can be expressed as a Taylor–Riemann series. Finally, Ariyawansa and Templeton

performed computational experiments that indicated that the accuracy of their approach is about the same as for Klein's procedure.

Up to this point, all estimation procedures that we have discussed assumed an underlying continuous distribution governing the sampled values. In combinatorial problems, however, the solution set S in (1) is discrete and finite, thereby leading to a finite number of values for the objective function $f(x)$. Consequently, one might want to use a discrete distribution for such values. Golden and Wasil [17] explored this option to construct a point estimate for y^* using one of three discrete distributions: binomial*, negative binomial*, and Poisson*. Since each of these distributions refers to a random variable with a minimum value of 0, they must be translated to have a lower end point of y^*, i.e., the optimal value.

This estimation procedure was applied to 50 heuristic solutions to a TSP with 318 nodes [22] and also to 50 two-opt solutions to the 120 node TSP of Crowder and Padberg [8]. Five 100-node problems from ref. 21 were tested also. For each of these five problems, 200 two-opt solutions were generated and a translated distribution was fit for each sample of twenty-five heuristic solutions. The average deviation from optimality for all problems was less than 1.75%. Eighteen samples of heuristic solutions yielded estimates whose absolute deviation from optimality was less than 1.00%; fourteen samples yielded estimates with absolute deviations of from 1.00 to 1.99%, and 10 samples had estimates with absolute deviations of 2.00% or more.

CONCLUSION

In this article, we have focused on one important problem in the interface between statistics and optimization. In this particular problem, we seek to apply extreme-value theory and other statistical techniques in order to determine when a local minimum has an objective value that is "close enough" (as measured by point and interval estimates) to the optimal objective value. Such questions are extremely important in operations research and computer science, where so many practical optimization problems are nonconvex or combinatorial in nature, since, currently, efficient algorithms do not exist for obtaining optimal solutions to these problems.

In further research, a key issue will be the relationship between sample size and the degree to which some of the asymptotic results discussed in this article hold. Extensive computational testing will be required in this direction.

References

[1] Ariyawansa, K. A. and Templeton, J. G. C. (1980). "On Statistical Control of Optimization," *Working Paper No. 80-020*, Dept. of Industrial Engineering, University of Toronto, Canada.

[2] Ariyawansa, K. A. and Templeton, J. G. C. (1980). "Structural Estimation of Global Optima Using the Three-Parameter Power Function Distribution," *Working Paper No. 80-026*, Department of Industrial Engineering, University of Toronto, Canada.

[3] Ariyawansa, K. A. and Templeton, J. G. C. (1980). "Structural Inference for Parameters of a Power Function Distribution," *Working Paper No. 80-025*, Dept. of Industrial Engineering, University of Toronto, Canada.

[4] Boender, C., Rinnooy Kan, A. H. G., Timmer, G., and Stougie, L. (1982). *Math. Program.*, **22**, 125–140.

[5] Chichinadze, V. K. (1967). *Eng. Cybern.*, **1**, 115–123.

[6] Chichinadze, V. K. (1969). *Automatica*, **5**, 347–355.

[7] Clough, D. (1969). *CORS J.*, **7**, 102–115.

[8] Crowder, H. and Padberg, M. (1980). *Manag. Sci.*, **26**, 495–509.

[9] Dannenbring, D. (1977). *Manag. Sci.*, **23**, 1273–1283.

[10] de Haan, L. (1981). *J. Amer. Statist. Ass.*, **76**, 467–469.

[11] Fisher, R. and Tippett, L. (1928). *Proc. Camb. Philos. Soc.*, **24**, 180–190.

[12] Gnedenko, B. (1943). *Ann. Math.*, **44**, 423–453.

[13] Golden, B. (1977). *Networks*, **7**, 209–225.

[14] Golden, B. (1978). *Commun. Statist.*, **B7**, 361–367.

[15] Golden, B. (1978). In *Proceedings of the AIDS 1978 Annual Convention*, Vol. 1, R. Ebert, R. Monroe, and K. Roering, eds., St. Louis, pp. 255–257.

[16] Golden, B. and Alt, F. (1979). *Naval. Res. Logist. Quart.*, **26**, 69–77.

[17] Golden, B. and Wasil, E. (1981). In *Proceedings of the AIDS 1981 Annual Convention*, Vol. 2, R. Markland and T. Raker, eds., Boston, pp. 276–278.

[18] Hartley, H. O. and Pfaffenberger, R. (1971). In *Optimizing Methods in Statistics*, Jagdish S. Rustagi, ed., Academic Press, New York.

[19] King, J. R. and Spachis, A. S. (1980). *Omega*, **8**, 655–660.

[20] Klein, S. (1975). "Monte Carlo Estimation in Complex Optimization Problems," doctoral dissertation, The George Washington University, Washington, DC.

[21] Krolak, P., Felts, W., and Marble, G. (1971). *Commun. Ass. Comp. Mach. (ACM)*, **14**, 327–334.

[22] Lin, S. and Kernighan, B. (1973). *Operat. Res.*, **21**, 498–516.

[23] Los, M. and Lardinois, C. (1982). *Transport. Res.*, **16B**, 89–124.

[24] McRoberts, K. (1971). *Operat. Res.*, **19**, 1331–1349.

[25] Padberg, M. and Hong, S. (1980). *Math. Program. Stud.*, **12**, 78–107.

[26] Patel, N. and Smith, R. (1984). *Operat. Res.*, **31**, 789–794.

[27] Rubinstein, R. (1981). *Simulation and the Monte Carlo Method*. Wiley, New York.

[28] Sielken, R. L., Jr. and Monroe, H. M. (1984). *Amer. J. Math. Manag. Sci.*, **4**, 139–197.

(COMPUTERS IN STATISTICS
OPTIMIZATION IN STATISTICS)

A. A. Assad
B. L. Golden

OPTIMUM ALLOCATION *See* OPTIMUM STRATIFICATION

OPTIMUM DESIGN OF EXPERIMENTS *See* SUPPLEMENT

OPTIMUM ERROR RATE *See* DISCRIMINANT ANALYSIS

OPTIMUM STRATIFICATION

Certain characteristics of a population of elements—such as some means—must be estimated by a sample survey. Observational access to the population is provided by a frame, a set of sampling units. Stratified sampling calls for partitioning the frame into strata, selecting a sample of units from each stratum, and estimating the characteristics by (linear) combinations of estimates of the characteristics of the individual strata.

The use of stratified sampling actualizes four design problems, i.e., the choices of:

1. The stratification variable.
2. The number L of strata.
3. The mode of stratification, that is, how to partition a population into strata.
4. The sample sizes, that is, the sample allocation among the strata.

As a basis for the subsequent discussion, we introduce some notation. Thus the population numbers N elements. With each element, we associate the observable values of variables $Y, X, \ldots, Z$, with means $\overline{Y}, \overline{X}, \ldots, \overline{Z}$ and variances $\operatorname{var} Y, \operatorname{var} X, \ldots,$ $\operatorname{var} Z$. It is the objective of a sample survey to estimate the means.

We make the assumption that a list has been made of the N elements; this list will serve as the frame. The elements are partitioned into L disjoint groups, to be referred to as strata, with N_g elements in the gth stratum, $g = 1, \ldots, L$, means $\overline{Y}_g,$ $\overline{X}_g, \ldots, \overline{Z}_g$, and variances $\operatorname{var} Y_g, \operatorname{var} X_g,$ $\ldots, \operatorname{var} Z_g$. Hence

$$\overline{Y} = \sum_{g=1}^{L} W_g \overline{Y}_g$$

with $W_g = N_g / N$, and similarly for the other means. Also

$$\operatorname{var} Y = \sum_{g=1}^{L} W_g \operatorname{var} Y_g + \sum_{g=1}^{L} W_g \left(\overline{Y}_g - \overline{Y} \right)^2$$

with analogous expressions for $\operatorname{var} X, \ldots,$ $\operatorname{var} Z$.

Samples are selected from the L strata, with sample sizes equal to $n_g, g = 1, \ldots, L$. The mean $\overline{Y}$ is estimated by

$$\bar{y} = \sum_{g=1}^{L} W_g \bar{y}_g,$$

where $\bar{y}_g$ is an estimate of $\overline{Y}_g$, and similarly for the remaining means. This estimate has variance

$$\operatorname{var} \bar{y} = \sum_{g=1}^{L} W_g^2 \operatorname{var} \bar{y}_g.$$

SOME HISTORY

Stratified sampling has a long standing in the history of survey sampling. We will review briefly some early contributions to its methods and theory.

To begin with, the statisticians' interest focused on the problem of allocating a sample between the strata. In the years 1926–35, three papers appeared, which represented significant advances.

Thus in Bowley [1], which reflects the classical contributions to statistical theory associated with Poisson, theory was presented for the case when the size of the sample selected from a stratum is proportional to the total number of elements in that stratum. This allocation yields a *representative* sample, also referred to as a *miniature population*. In Neyman [26], it was shown how to allocate a sample among the strata in order to minimize the variance for a fixed total sample size. More important, this paper served to promote the use of probability sampling and introduced the notion of a confidence interval*. Finally, in Yates and Zacopanay [35], cost considerations were introduced, explicitly, which led to a theory for optimum allocation.

In the 1940s the problem of *how* to partition a population into strata was dealt with by various practitioners. Especially a *principle of equipartition*:

$$W_g \overline{Y}_g = \text{constant}, \quad \text{for} \quad g = 1, \ldots L$$

was used by the Indian Statistical Institute* [24] and the U.S. Bureau of the Census* [19]. Mention should also be made of Hagood and Bernert [17], who advocated the use of principal component analysis* as a technique for determining the strata boundaries.

Since around 1950, theories and methods have been developed dealing with the problems of choosing:

The sample sizes when more than one mean (or other population characteristics) is to be estimated.

The number of strata.

The mode of stratification.

In what follows we will give an overview of some of the theories that have been developed to determine the best mode of stratification. It is usual to refer to this mode as *optimum stratification*, although *minimum variance stratification* would in many instances be a more appropriate term. Special attention will be paid to two different cases, which for short will be referred to as *the uniparametric case* and *the multiparametric case*.

THE UNIPARAMETRIC CASE

The distinguishing feature of this case is that it deals with the problem of optimum stratification for the special case when a *single* population characteristic (usually the mean) is to be estimated. We will consider here in some detail one mathematical technique for tackling this problem.

This technique (the basic reference is Dalenius [5]) calls for replacing the population of N elements by a density $f(\cdot)$, distributed over the interval (a, b). The mean of this density—the parameter to estimate—is

$$\mu = \int_a^b yf(y)\,dy$$

and the variance is

$$\sigma^2 = \int_a^b y^2 f(y)\,dy - \mu^2.$$

The density is partitioned into L strata by points of stratification:

$$a = y_0 < y_1 < \cdots < y_g$$
$$< \cdots < y_{L-1} < y_L = b.$$

This amounts to using the variable, the mean of which is to be estimated, as the stratification variable.

For the gth stratum,

$$w_g = \int f(y)\,dy, \qquad w_g \mu_g = \int yf(y)\,dy,$$
$$w_g(\sigma_g^2 + \mu_g^2) = \int y^2 f(y)\,dy,$$

where the integrals are taken over (y_{g-1}, y_g).

A sample of size n is allocated to the L strata and observed with respect to y, yielding the estimates $\bar{y}_g$ of μ_g. These estimates are then combined to yield the estimate

$$\bar{y} = \sum w_g \bar{y}_g$$

of μ, the summation being from $g = 1$ to $g = L$. The variance of this estimate is

$$\sigma^2(\bar{y}) = \sum w_g^2 \sigma^2(\bar{y}_g).$$

For a given choice of L, the variance will depend on the sample allocation. For a given allocation, the points of stratification that minimize the variance are the solutions of the following equation:

$$\frac{\partial \sigma^2(\bar{y})}{\partial y_g} = 0.$$

The theory is summarized in Dalenius [6, Chap. 7]. Especially, it was shown that when n_g is chosen proportional to w_g, the optimum stratification is given by

$$y_g = \frac{\mu_g + \mu_{g+1}}{2},$$

and when n_g is chosen proportional to $w_g \sigma_g$, the condition for optimum stratification is given by

$$\frac{\sigma_g^2 + (\mu_g - y_g)^2}{\sigma_g} = \frac{\sigma_{g+1}^2 + (\mu_{g+1} - y_g)^2}{\sigma_{g+1}}$$

with an analogous condition for the case when n_g is chosen proportional to

$$w_g \sigma_g / \sqrt{C_g},$$

where C_g is the cost of observing an element. The equation

$$\frac{\partial \sigma^2(\bar{y})}{\partial y_g} = 0$$

given earlier is a *necessary but not sufficient* condition for the solutions to be the points of stratification, that minimize the variance.

For a discussion of the mathematical aspects of the technique, refer to Schneeberger [29]. In Eubank [12], both necessary and sufficient conditions, under proportional allocation*, are given.

For extension of the theory presented to the case when the stratification is carried out on a specific stratification variable, refer to Dalenius and Gurney [7]. And theory for the case when μ is estimated using auxiliary information is presented in, e.g., Singh and Sukhatme [34].

Approximate Optimum Stratification

The results just presented are computationally intractable. Fortunately, they lend themselves to approximations that are relatively easy to use. We will first consider approximations for the case when μ is estimated by $\bar{y}$ as already discussed.

In Dalenius and Gurney [7], it was conjectured that making

$$w_g \sigma_g = \text{constant},$$

for $g = 1, \ldots, L$, would serve well in the case of minimum variance allocation. For large L, this rule is equivalent to

$$w_g(y_g - y_{g-1}) = \text{constant},$$

as suggested by Ekman [10].

In Dalenius and Hodges [8], the *cum-root-rule* is suggested for the case when n_g is proportional to $w_g \sigma_g$ and L is large. The rule calls for constant-width stratification of the variable

$$y^* = G(x) = \int_a^x f(u)\, du,$$

that is, the choice of points y_g^* of stratification being the solutions of $G(y_g^*) = gK/L$, where $K = G(b)$.

Analogous approximations are available for the case when μ is estimated by a ratio or regression estimate. Especially Singh and Parkash [33] and Singh [32] provide rules for optimum stratification when the sample allocation is

$$n_1 = \cdots = n_g = \cdots = n_L.$$

The rules discussed earlier greatly simplify the computational work. In Sethi [31], the simplification is taken one step further. Thus the suggestion is made that the equations that determine the optimum points of stratification be solved and tabulated for a few standard densities $f_1, \ldots, f_i, \ldots, f_K$. Then, for a real population resembling one of these K densities, the corresponding points of stratification are taken from the table.

The exact (and also the approximate) solutions referred to make it possible to consider the choice of the optimum number of strata, as done in Dalenius [6, Chap. 8]. Serfling [30] shows how to tackle the problems of choosing L *and* the mode of stratification in an optimum way *simultaneously*.

Applications

In Cochran [3], the results of a comparison of the performances of four rules for determining the strata boundaries are presented, as applied to eight distributions reflecting real-life data. The four rules compared are

 I. The cum-root rule.

 II. $w_g \mu_g = \text{constant}$.

 III. The $w_g(y_g - y_{g-1}) = \text{constant}$ rule.

 IV. The cum $(r + f)$ rule suggested in Durbin [9].

Table 1 shows the ratios between the variances associated with the use of these rules and the corresponding minimum variances for $L = 2, 3, 4$.

The uniparametric approach has found applications in the design of large-scale sample surveys. For illustrations, reference is given to Hess et al. [20] and Kpedekpo [22]. It is characteristic of most applications that the theories—while formally providing exact solutions—play the role of providing a *guide* to the mode of stratification to adopt; typically this mode also reflects some *practical* considerations.

Finally, mention should be made of the fact that the use of the theories for optimum stratification is not restricted to the area of survey sampling as discussed here. These

Table 1   Ratios of the Variances Using the Four Rules to Solve for the Minimum Variances

L	I	II	III	IV	I	II	III	IV
	Agricultural Loans				Real Estate Loans			
2	1.03	1.36	1.03	1.03	1[a]	1.19	1.06	1.06
3	1.03	1.89	1.05	1[a]	1[a]	1.28	1[a]	1.07
4	1[a]	1.96	1[a]	1[a]	1[a]	1.39	1.05	1.10
	Industrial Loans				Bank Resources			
2	1[a]	1.07	1.07	1.07	1.03	1.01	1.03	1.03
3	1[a]	1.35	1.07	1.14	1[a]	1.07	1.04	1.12
4	1.01	1.42	1.01	1.21	1.09	1.05	1.09	1.22
	College Students				City Populations			
2	1.02	1.06	1.02	1.02	1[a]	1.03	1[a]	1[a]
3	1.08	1.18	1.00	1.08	1[a]	1[a]	1.01	1.16
4	1.08	1.16	1.02	1.10	1.03	1.02	1.00	1[a]
	No. of Farms per Sampling Unit				Gross Income			
2	1[a]	1[a]	1.16	1.16	1.25	1.70	1[a]	1[a]
3	1[a]	1.22	1[a]	1.45	1.02	1.81	1.08	2.76
4	1.02	1[a]	1[a]	1.39	1[a]	1.69	1.05	1.52

[a]Indicates that the boundaries given by the rule were those that make the variance a minimum.

theories have also been considered for use in the context of Monte Carlo experimentation, as discussed in Hammersley and Handscomb [18].

THE MULTIPARAMETRIC CASE

The objective of a sample survey typically is to estimate *several* means (or other population characteristics). In what follows, we will consider some theories that explicitly focus on the corresponding multiparametric problem.

The generalized variance* is but one of several possible criteria one could attempt to minimize by the design. This criterion is used in Ghosh [15] for a special case. A general discussion based on this criterion is given in Schneeberger [28]. Minimization of the generalized variance does not necessarily reflect the survey aims; for a discussion of this point, refer to Dalenius [6, Chap. 9].

Alternatively, techniques in the realm of cluster analysis may be applied, as done in Golder and Yeomans [16]. Again, the question may be raised how the result relates to the survey aims.

A computationally simple idea that circumvents the problem just mentioned is suggested in Jarque [21]. Let there be K variables and hence K parameters μ_i, $i = 1$, ..., K. Dealing with each variable separately as in the uniparametric case would result in stratifications S_i^* with the associated variances var*($\hat{\mu}_i$) of the estimates $\hat{\mu}_i$. If the stratification S is used for all K variables, the variances would be $\text{var}_S(\hat{\mu}_i) \geqslant$ var*($\hat{\mu}_i$). Define

$$d_i(S) = \frac{\text{var}_S(\hat{\mu}_i)}{\text{var}^*(\hat{\mu}_i)} .$$

Then select the stratification that minimizes

$$F(S) = \sum_i d_i(S).$$

A BROAD CLASS OF PARTITIONING PROBLEMS

The problem of optimum stratification may be viewed as a special case in a broad class of partitioning problems. Some other cases are

Grouping and spacing in the realm of order statistics* (see Kulldorff [23]).

Grouping in order to condense observations (see Cox [4] and Fisher [14]).

Mixing ore of varying quality to maximize the revenue as discussed in Ekman [11].

Combining parts to pairs in an industrial process, as discussed in Råde [27].

The solutions to the problems of determining the optimum partitioning in these and similar cases have a common mathematical structure, that allows a unified approach. Four pertinent references are Bühler and Deutler [2], Ekman [11], Eubank [13], and McClure [25].

For a related class of partitioning problems, refer to the vast literature on taxonomy* and classification*.

References

[1]   Bowley, A. L. (1926), *Bull. Int. Statist. Inst.*, **22**, 359–380.

[2]   Bühler, W. and Deutler, T. (1975). *Metrika*, **22**, 161–175. (Shows that dynamic programming may be viewed as a general-purpose tool to solve partitioning problems.)

[3]   Cochran, W. G. (1961). *Bull. Int. Statist. Inst.*, **38**, 345–358. (A careful comparison of the performance of four rules on real-life data.)

[4]   Cox, D. R. (1957). *J. Amer. Statist. Ass.*, **62**, 543–547. (Suggests a way to measure the loss of information due to grouping of data and how to minimize that loss.)

[5]   Dalenius, T. (1950). *Scand. Actuarial J.*, 203–213.

[6]   Dalenius, T. (1957). In *Contributions to the Methods and Theories of Sample Survey Practice*. Almqvist and Wiksell, Stockholm. (Presents a summary of the author's work up to 1957.)

[7]   Dalenius, T. and Gurney, M. (1951). *Scand. Actuarial J.*, 133–148. (Considers stratification on a specific stratification variable.)

[8]   Dalenius, T. and Hodges, J. L., Jr. (1957). *Scand. Actuarial J.*, 198–203.

[9]   Durbin, J. (1959). *J. R. Statist. Soc. A*, **122**, 246–248. (A review of ref. 6.)

[10] Ekman, G. (1959). *Ann. Math. Statist.* **30**, 219–229.

[11] Ekman, G. (1969). *Rev. Int. Statist. Inst.*, **37**, 186–193.

[12] Eubank, R. L. (1982). *Statist. Prob. Lett.*, **1**, 69–73.

[13] Eubank, R. L. (1982). *Technical Report No. 164*, Dept. of Statistics, Southern Methodist University, Dallas, TX.

[14] Fisher, W. D. (1958). *J. Amer. Statist. Ass.*, **53**, 789–798. (Addresses the same kind of problem that Cox [4] does).

[15] Ghosh, S. P. (1963). *Ann. Math. Statist.*, **34**, 866–872.

[16] Golder, P. A. and Yeomans, K. A. (1973). *J. R. Statist. Soc. C*, **22**, 213–219. (The pioneering paper on the use of cluster analysis.)

[17] Hagood, M. J. and Bernert, E. H. (1945). *J. Amer. Statist. Ass.*, **40**, 330–341.

[18] Hammersley, J. M. and Handscomb, D. C. (1964). *Monte Carlo Methods*. Methuen, London.

[19] Hansen, M. H., Hurwitz, W. G., and Madow, W. G. (1953). *Sample Survey Methods and Theory*, Vol. I. Wiley, New York.

[20] Hess, I., Sethi, V. K., and Balakrishnan, T. H. (1966). *J. Amer. Statist. Ass.*, **61**, 74–90. (A very illuminating discussion of the performance of various rules for stratification of real-life populations.)

[21] Jarque, C. M. (1981). *J. R. Statist. Soc. C*, **30**, 163–169. (Shows how the uniparametric approach may be used successfully in the multiparametric case.)

[22] Kpedekpo G. M. K. (1973). *Metrika*, **20**, 54–64.

[23] Kulldorff, G. (1961). *Contributions to the Theory of Estimation from Grouped and Partially Grouped Samples*. Almqvist & Wiksell, Stockholm.

[24] Mahalanobis, P. C. (1952). *Sankhyā*, **12**, 1–7.

[25] McClure, D. E. (1982). *Statist. Tidskr.*, **2**, 101–110 and **3**, 189–198. (A comprehensive and readable account of the mathematics of optimum partitioning.)

[26] Neyman, J. (1934). *J. R. Statist. Soc. A*, **109**, 558–606. (A landmark paper in the realm of probability sampling.)

[27] Råde, L. (1963). *Scand. Actuarial J.*, 56–69.

[28] Schneeberger, H. (1973). *Metrika*, **20**, 1–16.

[29] Schneeberger, H. (1979). *Sankhyā C*, **41**, 92–96.

[30] Serfling, R. J. (1968). *J. Amer. Statist. Ass.*, **63**, 1298–1309. (An ingenious extension of the cum-root-rule to take both variance and cost into account; most readable.)

[31] Sethi, V. K. (1963). *Austr. J. Statist.*, **5**, 20–33.

[32] Singh, R. (1977). *Austr. J. Statist.*, **19**, 96–104.

[33] Singh, R. and Parkash, D. (1975). *Ann. Inst. Statist. Math.*, **27**, 273–280.

[34] Singh, R. and Sukhatme, B. V. (1969). *Ann. Inst. Statist. Math.*, **21**, 515–528.

[35] Yates, F. and Zacopanay, (1935). *J. Agric. Sci.*, **25**, 545–577.

(NEYMAN ALLOCATION
PROPORTIONAL ALLOCATION
SURVEY SAMPLING)

T. Dalenius

OPTIONAL SAMPLING, OPTIONAL SAMPLING THEOREM

Suppose a gambler is playing a sequence of fair games (a martingale*) in which the expected size of his or her bank, given all that has been learned from previous games, is unchanged from one game to the next. In an effort to turn a profit, the gambler tries to use information from previous games to determine when an imminent game is unprofitable and then sits out such games. However, his or her efforts are doomed to failure. According to Doob's *optional sampling theorem*, no matter how devious a sampling scheme our gambler devises, the expected size of the bank (given the past) will remain unchanged after each game played.

Let us give a formal definition of optional sampling. Suppose $\{(S_n, \mathscr{F}_n), \, n \geq 1\}$ is a martingale. We can think of S_n as the size of the gambler's bank after the nth game if all games are played and the σ-field $\mathscr{F}_n$ as the collective experience acquired from the first n games. Let $\{m_j, \, j \geq 1\}$ be a finite or infinite sequence of positive integer-valued random variables, with the properties

$$1 \leq m_1 \leq m_2 \leq \cdots < \infty \qquad (1)$$

and

$$[m_j = n] \text{ is in } \mathscr{F}_n$$
$$\text{for all } n \geq 1 \text{ and } j \geq 1. \qquad (2)$$

The variables m_j are the sampling times (stopping times or optional times) selected by the gambler, and the last condition ensures that the gambler has not used any future clairvoyance to determine these times. With each variable m_j, we associate a σ-field $\mathscr{F}_{m_j}$, defined to be the set of all events E satisfying

$$E \cap [m_j = n] \text{ is in } \mathscr{F}_n \text{ for all } n \geq 1.$$

Doob's optional sampling theorem may be stated as follows (see Bauer [2, Theorem 11.2.4, p. 347], Chung [4, Theorem 9.3.5, p. 326] or Doob [5, Theorem 2.2, p. 302]).

Theorem. If $\{(S_n, \mathscr{F}_n), \, n \geq 1\}$ is a martingale, and $\{m_j, \, j \geq 1\}$ is a sequence of sampling times satisfying conditions (1), (2), and

$$E[\,|S_{m_j}|\,] < \infty \qquad \text{for each } j,$$

then $\{(S_{m_j}, \mathscr{F}_{m_j}), \, j \geq 1\}$ is also a martingale. This result continues to hold if the word *martingale* is replaced everywhere by supermartingale*.

Optional sampling differs from optional stopping in that the former involves observing the martingale at a sequence of stopping times, not necessarily consecutive, while the latter consists of observing the martingale at all points up to a single random time m, after which observation stops. Indeed, optional stopping is a special case of optional sampling although the former is usually studied in its own right; see e.g., Chow et al. [3].

The notion of optional sampling and the optional sampling theorem, may be generalized from the discrete parameter case to the case of a general index set; see Doob [5, p. 366], Kurtz [6], and Meyer [7, p. 98]. Aló et al. [1] describe an optional sampling theorem for convex set-valued martingales.

We now given a simple application of the optional sampling theorem, taken from Chung [4, p. 327]. Let S_n denote the position at time n of a simple, symmetric random walk* on the integers starting at $S_0 = 0$. If a, b are positive integers, eventually the walk will reach either $-a$ or b. Let m denote the first time that one or the other of these points is visited. Then m is a stopping time, and $E[|S_m|] < \infty$. The theorem implies that $\{S_1, S_m\}$ is a martingale, and so $E[S_m] = E[S_1] = 0$. That is,

$$-aP[S_m = -a] + bP[S_m = b] = 0,$$

so that the probability of reaching $-a$ before visiting b is $b/(a + b)$.

References

[1] Aló, R., de Korvin, A., and Roberts, C. (1979). *J. Reine angew. Math.*, **310**, 1–6.

[2] Bauer, H. (1981). *Probability Theory and Elements of Measure Theory*, 2nd English ed. Academic Press, New York. (Section 11.2 contains a graduate level introduction to stopping times and optional sampling.)

[3] Chow, Y. S., Robbins, H., and Siegmund, D. (1971). *Great Expectations: The Theory of Optimal Stopping*. Houghton Mifflin, Boston. (A graduate level introduction to stopping times and optional stopping.)

[4] Chung, K. L. (1974). *A Course in Probability Theory*. Academic Press, New York. (Chapters 8 and 9 contain an introduction to stopping times and optional sampling.)

[5] Doob, J. L. (1953). *Stochastic Processes*. Wiley, New York. (Pages 300 on describe optional sampling in the case of a discrete parameter martingale while pp. 366 ff. treat the case of a general index set.)

[6] Kurtz, T. G. (1980). *Ann. Prob.*, **8**, 675–681.

[7] Meyer, P. A. (1966). *Probability and Potentials*. Blaisdell, Waltham; Meyer, P. A. and Dellacherie, C. (1978). *Probabilities and Potentials*, 2nd ed. North-Holland, Amsterdam. (Section VI.2 is a graduate level introduction to optional sampling with continuous parameter martingales.)

(GAMBLING, STATISTICS IN
MARTINGALE
OPTIMAL STOPPING
RANDOM WALK
STOPPING RULES)

PETER HALL

ORBIT

The locus of all points in a sample space* on which a maximal invariant had a constant value is called an *orbit*.

If the invariance principle* is used in constructing a test of a hypothesis H_0, then all sample points on the same orbit must lead to the same conclusion about the validity of H_0.

ORD–CARVER SYSTEM OF DISTRIBUTIONS   *See* PEARSON SYSTEM OF DISTRIBUTIONS

ORDERED ALTERNATIVES, TESTS FOR   *See* MULTIDIMENSIONAL SCALING; PAGE TEST FOR ORDERED ALTERNATIVES; WILCOXON-TYPE TEST FOR ALTERNATIVES IN RANDOMIZED BLOCKS

ORDERING DISTRIBUTIONS BY DISPERSION

Often one encounters a pair of random variables X and Y with distribution functions F and G, respectively, where it is intuitively clear that one distribution, say G, is more dispersed than F. For example, in some situations in nonparametric inference*, one desires to state formally a one-sided alternative for the null hypothesis, which claims that F and G have the same dispersion. In queueing theory*, it can be expected that if the interarrival and service times of a queue become "more variable" then the waiting time would increase stochastically [22]; in such situations, again, one needs to define the meaning of *more variable*. Some well-known sets of data seem to be "more dispersed" than simple parametric fits of them (see Shaked [19] and references therein); again one may wish to describe exactly what *more dispersed* means.

Various partial orderings of distributions by dispersion have been introduced as results of efforts to formalize the intuition. Two main approaches have been used to define such orderings.

The first approach tries to order F and G according to the dispersion about some central point μ such as the mean, the median, or, when applicable, the center of symmetry.

Such orderings compare, stochastically, quantities such as $|X - \mu_X|$ and $|Y - \mu_Y|$ or other convex functions of $X - \mu_X$ and $Y - \mu_Y$ where μ_X and μ_Y are the appropriate central points of F and G, respectively.

In the second approach, F and G are ordered by comparing appropriate functions of pairs of quantiles* (percentiles*). For example, if F^{-1} and G^{-1} denote the quantile functions (inverses) of F and G, respectively, then pairs of quantities such as $F^{-1}(\beta) - F^{-1}(\alpha)$ and $G^{-1}(\beta) - G^{-1}(\alpha)$ or $F^{-1}(\beta)/F^{-1}(\alpha)$ and $G^{-1}(\beta)/G^{-1}(\alpha)$, are compared for all (α, β), which satisfy $0 < \alpha < \beta < 1$.

A third approach is to order F and G according to some properties of $G^{-1}F$ such as convexity, star-shapedness or superadditivity. The objectives of this approach are not necessarily to order F and G according to their dispersion, but according to the aging properties of the underlying random variables X and Y which, in this case, are thought of as life lengths. For example, if $G^{-1}F$ is convex, then F can be thought of as being more IFR (= increasing failure rate) than G, or, if $G^{-1}F$ is star-shaped then F can be thought of as being more IFRA (= IFR average) than G [see RELIABILITY entries]. However, as will be shown, these orderings can be used as dispersion orderings in some situations.

ORDERINGS BASED ON FUNCTIONS OF $X - \mu_X$ AND $Y - \mu_Y$

Birnbaum [4] called X *more peaked* about μ_X than Y about μ_Y if

$$P\{|Y - \mu_Y| \geq x\} \geq P\{|X - \mu_X| \geq x\}$$
$$\text{for all} \quad x \geq 0, \quad (1)$$

that is, $|Y - \mu_Y|$ is stochastically larger than $|X - \mu_X|$. Birnbaum restricted his discussion to random variables with symmetric densities about μ_X and μ_Y, respectively, and derived some closure properties of the peakedness orderings. Without loss of generality, he assumed that $\mu_X = \mu_Y = 0$. A typical result

of his says that if $X_1, \ldots, X_n$ are independent and identically distributed (iid), if $Y_1, \ldots, Y_n$ are iid and independent of the X_i's, and if Y_1 is more peaked about 0 than X_1 about 0, then, under some conditions, $\overline{Y} = n^{-1}\sum_{i=1}^{n} Y_i$ is more peaked about 0 than $\overline{X} = n^{-1}\sum_{i=1}^{n} X_i$ about 0. Proschan [14], Karlin [10, p. 326], and Bickel and Lehmann [2] have obtained related results. For example, if $X_1, \ldots, X_n$ are iid with a log concave density that is symmetric about zero, then $P(n^{-1}\sum_{i=1}^{n} X_i \leq t)$ is nondecreasing in n for all $t > 0$.

Note that (1) is equivalent to

$$E\big[g(|Y - \mu_Y|) \big] \geq E\big[g(|X - \mu_X|) \big] \quad (2)$$

for every nondecreasing function g such that the expectations in (2) exist. Thus, if (1) holds, then

$$E|Y - \mu_Y| \geq E|X - \mu_X|, \quad (3)$$

and if $\mu_X = E(X)$ and $\mu_Y = E(Y)$, then

$$\text{var}(Y) \geq \text{var}(X). \quad (4)$$

Inequalities such as (3) and (4) show that the ordering defined by (1) or, equivalently, by (2), has some desirable properties that every ordering by dispersion should have.

A related ordering is the dilation (or dilatation) ordering. A random variable Y (or its distribution G) is called a *dilation* of X (or of its distribution F) if

$$Eh(Y) \geq Eh(X) \quad (5)$$

for all convex functions h. Some reflection shows that relation (5) indeed formalizes the idea that Y is more dispersed than X. The ordering defined by (5) will be denoted by $Y \succ X$ (or $G \succ F$). See GEOMETRY IN STATISTICS: CONVEXITY for some discussion about the dilation ordering.

Note that if the expectations μ_X and μ_Y of X and Y exist, than (5) implies $\mu_X = \mu_Y$. Thus, to be able to order distributions with different expectations, it is better to consider the ordering defined by the inequalities

$$Eh(Y - \mu_Y) \geq Eh(X - \mu_X) \quad (6)$$

for all convex functions h for which the expectations in (6) exist, that is, $(Y - \mu_Y)$

$\succ (X - \mu_X)$. For example, if X and Y are normal random variables such that var (Y) $>$ var(X), then (6) holds.

Shaked [19] and Schweder [18] considered some conditions that arise in real life and that yield the inequalities (5) and (6). For a function $a(x)$ defined on a subset I of the real line, the number of sign changes of a in I is defined by

$$S^-(a) = \sup S^-[a(x_1), \ldots, a(x_m)], \quad (7)$$

where $S^-(y_1, \ldots, y_m)$ is the number of sign changes of the indicated sequence, zero terms being discarded, and the supremum in (7) is extended over all sets $x_1 < x_2 < \cdots < x_m$ ($x_i \in I$); $m < \infty$. Shaked [19] showed that if F and G have densities $f = F'$ and $g = G'$ on I and if $\mu_X = \mu_Y$, then

$$S^-(g - f) = 2 \qquad (8)$$

with sign sequence $+, -, +$ implies that

$$S^-(G - F) = 1 \qquad (9)$$

with sign sequence $+, -$, which, in turn, implies (5). Some real situations in which (8) and (9) hold have been discussed by Shaked [19]. Note that either (8) or (9) can be used to define a partial ordering by dispersion that is embedded in the dilation ordering.

Rolski [15], Whitt [22], and Brown [6], among others, studied the ordering defined by

$$Ek(X) \leqslant Ek(Y) \qquad (10)$$

for all nondecreasing convex functions k such that the expectations in (10) exist. Roughly speaking, if (10) holds, then Y is more dispersed or stochastically larger than X.

ORDERING BASED ON THE QUANTILE FUNCTIONS

Roughly speaking, the orderings of the first section take into account the dispersion relative to a fixed point such as the mean or the median. This section, considers orderings that take into account the spread of a random variable throughout its support.

In this section we will consider only distributions that have no atoms and whose supports are intervals. Such distributions have strictly increasing and continuous inverses on $(0, 1)$. Lewis and Thompson [11] have considered the more general case.

A distribution G is said to be *more dispersed* or *more spread out* than a distribution F if

$$G^{-1}(\beta) - G^{-1}(\alpha) \geqslant F^{-1}(\beta) - F^{-1}(\alpha),$$
$$0 < \alpha < \beta < 1, \quad (11)$$

where F^{-1} and G^{-1} are the inverses of F and G. Relation (11) means that the difference between any two percentiles of G is not smaller than the difference between the corresponding percentiles of F. Thus (11) indeed conveys the notion that G is more dispersed than F. When F and G satisfy (11), we write $G \overset{disp}{>} F$. It is easy to check, for example, that for every random variable X and any constant $a > 1$ (using obvious notation), $aX \overset{disp}{>} X$.

A concept that contains the essence of this definition was introduced by Brown and Tukey [5]. Further studies of the ordering $\overset{disp}{<}$ can be found in Fraser [8], Doksum [7], Yanagimoto and Sibuya [24], Saunders [16], Saunders and Moran [17], Bickel and Lehmann [3], Lewis and Thompson [11], Oja [13], Shaked [20], and Lynch and Proschan [12]. Proofs of the results, which are only mentioned in this article, can be found in these references.

Condition (11) may be compared to (1) which, for symmetric distributions, is equivalent to

$$G^{-1}(v) - G^{-1}(\tfrac{1}{2}) \gtrless F^{-1}(v) - F^{-1}(\tfrac{1}{2})$$
$$\text{as} \quad v \lessgtr \tfrac{1}{2}.$$

Condition (11) has many equivalent forms. For example, $F \overset{disp}{<} G$ if and only if

$$S^-(F(\cdot - c) - G(\cdot)) \leqslant 1$$
$$\text{for every} \quad c \in (-\infty, \infty) \quad (12)$$

and, in case of equality, the sign sequence is $-, +$. If F and G are differentiable on their support with densities $f = F'$ and $g = G'$, then $F \overset{\text{disp}}{<} G$ if and only if

$$g(G^{-1}(u)) \leqslant f(F^{-1}(u)), \qquad u \in (0, 1).$$

Let I_F and I_G be, respectively, the supports of F and G, then $F \overset{\text{disp}}{<} G$ if and only if there exists a function $\phi : I_F \to I_G$ such that (denoting equality in distribution by $\overset{\text{d}}{=}$)

$$Y \overset{\text{d}}{=} \phi(X) \tag{13a}$$

and

$$x < x' \Rightarrow \phi(x') - \phi(x) \geqslant x' - x. \tag{13b}$$

Here $\phi = G^{-1}F$. When ϕ is differentiable, (13b) is equivalent to $\phi' \geqslant 1$. Thus (13a) and (13b) agree with the idea that Y is more variable than X.

Another way of writing (13) is to require the existence of a function ψ such that

$$Y \overset{\text{d}}{=} X + \psi(X), \tag{14}$$

where ψ is a nondecreasing function. Here $\psi(x) = G^{-1}(F(x)) - x$.

We note that if $X \overset{\text{disp}}{<} Y$, then (3) and (4) hold.

If F and G have densities $f = F'$ and $g = G'$, then a sufficient condition for $F \overset{\text{disp}}{<} G$ is

$$S^-(f(\cdot - c) - g(\cdot)) \leqslant 2$$

$$\text{for every} \quad c \in (-\infty, \infty) \tag{15}$$

with the sign sequence being $-, +, -$ in case of equality in (15).

Shaked [20] studies the ordering $\overset{\text{disp}}{<}$ when the underlying random variables are non-negative.

If X and Y are random variables that are positive with probability one, then

$$\log Y \overset{\text{disp}}{>} \log X \tag{16}$$

if and only if

$$\frac{G^{-1}(\beta)}{G^{-1}(\alpha)} \geqslant \frac{F^{-1}(\beta)}{F^{-1}(\alpha)}, \qquad 0 < \alpha < \beta < 1, \tag{17}$$

so (17) is a relation that orders by dispersion pairs of logarithms of random variables.

Condition (17) has been studied by Gastwirth [9], Yanagimoto and Sibuya [14, 25], Saunders [16], Saunders and Moran [17], and Shaked [20], who obtained analogs of (12)–(15). Saunders and Moran [17] and Shaked [20] showed, for example, that every pair of gamma distributions satisfy (17). Further, Yanagimoto and Sibuya [24] have found some transformations of X and Y that preserve the ordering in (17).

The ordering defined by condition (17) is equivalent to the so-called star ordering* $F \overset{*}{<} G$ which means, for nonnegative random variables, that

$$\frac{G^{-1}(F(x))}{x} \text{ is nondecreasing in } x > 0$$

$$\tag{18}$$

(see, e.g., Doksum [7] and Barlow et al. [1]). It is easy to show that (18) implies (9). A related ordering is the convex ordering* $F \overset{c}{<} G$ of van Zwet [21], which means, for nonnegative random variables, that

$$G^{-1}F \text{ is convex on } (0, \infty). \tag{19}$$

Since $F \overset{c}{<} G$ implies $F \overset{*}{<} G$, it follows that (19) is a sufficient condition for (17). Thus the examples of van Zwet [21] satisfy (17).

APPLICATIONS

Many of the papers that deal with the dispersive ordering of the preceding sections were motivated by problems in nonparametric inference*. Doksum [7] and Gastwirth [9] used these orderings to obtain monotonicity of the asymptotic relative efficiencies and of the powers of nonparametric tests as functions of the underlying distributions. Fraser [8], Yanagimoto and Sibuya [24] and Bickel and Lehmann [2, 3] used these orderings to formalize statistical hypotheses that roughly claim that one distribution is more spread out than another.

For example, Yanagimoto and Sibuya [24] considered testing the hypothesis H_0: for some $a \in R$, $F(x) = G(x - a)$ for all $x \in R$ versus $H_1 : F \overset{\text{disp}}{>} G$, where G is a given distribution and F is the common distribution of the iid random variables $X_1, \ldots, X_n$. They considered test statistics that are functions of $\mathbf{V} = (X_{(1)} - \overline{X}, \ldots, X_{(n)} - \overline{X})$, where the $X_{(i)}$'s are the order statistics* of the sample and $\overline{X}$ is the sample mean. Note that $\mathbf{V}$ is a random point in $\Gamma \equiv \{\mathbf{x} : x_1 \leqslant x_2 \leqslant \cdots \leqslant x_n, \sum_{i=1}^{n} x_i = 0\}$. They proved that $\phi(\mathbf{V})$ is an unbiased test statistic whenever ϕ satisfies:

$$\phi(a + b) \geqslant \phi(a) \qquad \text{for all} \quad a, b \in \Gamma.$$

Other uses of the orderings of the preceding sections in other areas of probability and statistics include the following.

Birnbaum [4] used (1) to obtain various useful probability inequalities. Shaked [19], using (8) and (9), obtained various other inequalities concerning mixtures from exponential families and symmetric random walks*. Rolski [15] and Whitt [22] used (5) and (10) to obtain inequalities that are useful in queuing theory*. Saunders [16] and Saunders and Moran [17] have used (11) and (17) to identify distributions that are of importance in the study of the matching of optical receptors to the subjects they are designed to detect. Saunders and Moran [17] and Shaked [20] showed that the family of the gamma distributions* as well as other families of distributions are ordered by dispersion [(11) and (17)] using an appropriate parameter. Lewis and Thompson [11] and Lynch and Proschan [12] characterized unimodal distributions using ordering (11). Yanagimoto and Hoel [23] have used (17) to introduce measures of the heaviness of the tails of a distribution. Yanagimoto and Sibuya [25] have used (17) to examine models for estimating safe doses.

References

[1] Barlow, R. E., Bartholomew, D. J., Bremner, J. M., and Brunk, H. D. (1972). *Statistical Inference under Order Restrictions*. Wiley, New York.

[2] Bickel, P. J. and Lehmann, E. L. (1976). *Ann. Statist.*, **4**, 1139–1158.

[3] Bickel, P. J. and Lehmann, E. L. (1979). In *Contributions to Statistics, Jaroslaw Hajek Memorial Volume*, J. Jureckova, ed. Reidel, Dordrecht, pp. 33–40.

[4] Birnbaum, Z. W. (1948). *Ann. Math. Statist.*, **19**, 76–81.

[5] Brown, G., and Tukey, J. W. (1946). *Ann. Math. Statist.*, **7**, 1–12.

[6] Brown, M. (1981). *Ann. Prob.*, **9**, 891–895.

[7] Doksum, K. (1969). *Ann. Math. Statist.*, **40**, 1167–1176.

[8] Fraser, D. A. S. (1975). *Nonparametric Methods in Statistics*. Wiley, New York.

[9] Gastwirth, J. L., (1970). *Nonparametric Techniques in Statistical Inference*, M. L. Puri, ed. Cambridge University Press, Cambridge, pp. 89–101.

[10] Karlin, S. (1968). *Total Positivity*. Stanford University Press, Stanford, CA.

[11] Lewis, T. and Thompson, J. W. (1981). *J. Appl. Prob.*, **18** 76–90.

[12] Lynch, J. and Proschan, F. (1983). "Dispersive Ordering Results." *Tech. Rep. No. M651*, Dept. of Statistics, Florida State University.

[13] Oja, H. (1981). *Scand. J. Statist.*, **8**, 154–168.

[14] Proschan, F. (1965). *Ann. Math. Statist.*, **36**, 1703–1706.

[15] Rolski, T. (1976). "Order Relations in the Set of Probability Distribution Functions and Their Applications in Queueing Theory." *Dissertationes Mathematicae No. 82*, Polish Academy of Sciences, Warsaw.

[16] Saunders, I. W. (1978). *Adv. Appl. Prob.*, **10**, 587–612.

[17] Saunders, I. W., and Moran, P. A. P. (1978). *J. Appl. Prob.*, **15**, 426–432.

[18] Schweder, T. (1982) *Scand. J. Statist.*, **9**, 165–169.

[19] Shaked, M. (1980). *J. R. Statist. Soc. B.*, **42**, 192–198.

[20] Shaked, M. (1982). *J. Appl. Prob.*, **19**, 310–320.

[21] van Zwet, W. R. (1964). *Convex Transformation of Random Variables*. Mathematisch Centrum, Amsterdam.

[22] Whitt, W. (1980). *J. Appl. Prob.*, **17**, 1062–1071.

[23] Yanagimoto, T. and Hoel, D. G. (1980). *Ann. Inst. Statist. Math. B*, **32**, 465–480.

[24] Yanagimoto, T. and Sibuya, M. (1976). *Ann. Inst. Statist. Math Tokyo A*, **28**, 329–342.

[25] Yanagimoto, T. and Sibuya, M. (1980). *Ann. Inst. Statist. Math Tokyo A*, **32**, 325–340.

(GAMMA DISTRIBUTION
HAZARD RATE AND OTHER
  CLASSIFICATIONS OF DISTRIBUTIONS
QUANTILES
QUEUEING THEORY

RANDOM WALK
RELIABILITY, PROBABILISTIC)

MOSHE SHAKED

ORDERING OF DISTRIBUTIONS, PARTIAL

The following three different ways may be used to analyze systematically the properties of probability distributions.

1. A *partial ordering* "$<$" of distributions is defined corresponding to the property of interest. One then writes $F_X < F_Y$, or, alternatively, $X < Y$, if F_Y possesses the property under consideration more strongly than F_X does. The relation $<$ is thus required to be *reflexive* (i.e. $F_X < F_X$, and *transitive*, i.e., if $F_X < F_Y$ and $F_Y < F_Z$, then $F_X < F_Z$). All pairs of distributions need not be comparable, however.

2. The properties of distributions are measured *quantitatively*. Sometimes such measures are purely heuristic but a natural requirement is that they should be consistent with some partial ordering corresponding to the same property. The classical measures of location, scale, or scatter; skewness; and kurtosis* provide examples.

3. The distributions are *classified* according to the attributes under consideration. Univariate distributions, e.g., may be symmetric, skew to the right, skew to the left, etc.; lifetime distributions* can be increasing hazard rate (IHR), decreasing hazard rate (DHR), etc. Often the classification is based on partial ordering.

The idea of ordering distributions with respect to the considered property is not very old. In 1947 Mann and Whitney [16] introduced the notion of stochastic ordering and ordering for location. A year later Birnbaum [8] proposed a dispersion ordering. The skewness and kurtosis orderings (c and s ordering) by van Zwet [23] and the star-shaped ordering by Barlow and Proschan [2] are now used widely. In a series of papers

[5–7] Bickel and Lehmann thoroughly discussed problems concerning location and spread. These approaches to the univariate case were unified in Oja [17] by applying the notion of convexity of order k [11, p. 23] to the shift function $\Delta(x) = F_Y^{-1}(F_X(x)) - x$. Stochastic and other orderings of multivariate distributions have been studied since 1955 with early important contributions by Lehmann [14], Veinott [24], and Stoyan [21], among others.

In the following we list some generally used partial orderings.

PARTIAL ORDERING OF UNIVARIATE DISTRIBUTIONS

Location

Let X and Y be univariate random variables and F_X and F_Y the respective cumulative distribution functions. Then F_Y (or Y) is *stochastically larger* than F_X (or X) if

$$E(u(X)) \leqslant E(u(Y)) \qquad (1)$$

for all bounded increasing functions $u : \mathbb{R} \to \mathbb{R}$. This is denoted by $F_X \leqslant_{st} F_Y$ (or $X <_{st} Y$). The stochastic comparison of two given distributions F_X and F_Y can be done more easily by using the following facts: $F_X <_{st} F_Y$ if and only if $F_X(x) \geqslant F_Y(x)$ for all x. If the respective densities f_X and f_Y exist and the density curves cross each other exactly once, then F_X and F_Y are stochastically ordered. It is also enough to use continuous bounded increasing test functions u in the preceding definition.

An interval I is the *support* of F_X if it is the smallest interval with $\Pr(X \in I) = 1$. Now let F_X and F_Y be continuous and strictly increasing on their supports. Then the transformation $R(x) = F_Y^{-1}F_X(x)$ is such that the distribution function of $R(X)$ is F_Y. The function $\Delta(x) = R(x) - x$ usually is called the *shift function*, since X, when shifted by $\Delta(X)$, has the same distribution as Y. Now $X \leqslant_{st} Y \Leftrightarrow R(x) \geqslant x$ for all x in the support of $F_X \Leftrightarrow \Delta(x) \geqslant 0$ for all x in the support of F_X (i.e., Δ is convex of order 0).

Stochastic ordering has proved useful, e.g., in the following cases: It can be used in formulating hypotheses and constructing tests for nonparametric models. For example, the Mann–Whitney U-test is unbiased in testing the hypotheses $H_0: F_X = F_Y$ and $H_1: F_X \leqslant_{st} F_Y$. It is usually required [5] that a *measure of location* (or *location parameter*) $\mu(F_X)$ or $\mu(X)$ should be such that

1⁰.$\qquad X \leqslant_{st} Y \Rightarrow \mu(X) \leqslant \mu(Y)$

and

2⁰.$\qquad \mu(aX + b) = a\mu(X) + b$

$\qquad\qquad\qquad\qquad$ for all a and b.

Most of the generally used location parameters (mean, median, etc.) are measures of location in this sense. Two or more tests of significance for the same hypotheses can be compared if the respective P values* are stochastically ordered under the alternative.

Other orderings for considering location can be found by using convex increasing (or concave increasing) test functions in (1). (See Stoyan [21].) These orderings "combine" location and dispersion. (*See* ORDERING DISTRIBUTIONS BY DISPERSION.)

Scale or Dispersion

For symmetrical distributions, scale is usually understood as the nearness of the distribution to the symmetry center. Accordingly, one can define [6] that F_Y is *more dispersed* about ν than F_X about μ if

$$|X - \mu| \leqslant_{st} |Y - \nu|. \qquad (2)$$

For asymmmetrical distributions, F_Y is said to be *more spread out* than F_X if

$$F_X^{-1}(v) - F_X^{-1}(u) \leqslant F_Y^{-1}(v) - F_Y^{-1}(u) \qquad (3)$$

for all $0 < u < v < 1$ [7]. Any two fractiles of Y are then more apart than the respective fractiles of X. F_X and F_Y are *spread ordered* if and only if, for all a, the curves of $F_X(x + a)$ and $F_Y(x)$ cross each other once at most [17, 19]. Continuous and strictly increasing F_X and F_Y are spread ordered if $R'(x) \geqslant 1$ or $\Delta'(x) \geqslant 0$ (i.e., Δ is convex of order 1) for

all x in the support of F_X. Other scale orderings can be defined through (1) by using convex (or convex increasing) test functions u. (Again, *see* ORDERING DISTRIBUTIONS BY DISPERSION.)

Nonparametric scale tests for comparing two distributions have been constructed in this spirit. A *measure of spread* (or *scale parameter*) $\sigma(F_X)$ or $\sigma(X)$ is defined in Bickel and Lehmann [7] to be such that

1⁰.$\qquad\qquad \sigma(X) \leqslant \sigma(Y)$

whenever F_Y is more spread out than F_X and

2⁰.$\qquad\quad \sigma(aX + b) = |a|\sigma(X)$

$\qquad\qquad\qquad\qquad$ for all a and b.

The classical scale parameters (standard deviation, mean deviation, range, etc.) satisfy these requirements.

Other Partial Orderings

Now let F_X and F_Y be continuous and strictly increasing cumulative distribution functions. F_Y is then said to be *more skew to the right* than F_X if R (or Δ) is convex (of order 2) on the support of F_X (convex ordering, *c-ordering*, see van Zwet [23]). For symmetrical distributions F_X and F_Y, F_Y has *more kurtosis* than F_X if R (or Δ) is concave-convex (*s-ordering* [23]). Convexity of order 3 of Δ implies s ordering. F_X and F_Y are c ordered (Δ is s ordered) if for all $a > 0$ and b the curves of $F_X(x)$ and $F_Y(ax + b)$ cross each other at most twice (for Δ, three times). (See [17].) For nonnegative random variables the *star-shaped ordering*, weaker than the c ordering, is defined as follows [2]: F_X is star-shaped with respect to F_Y if $R(x)/x$ is increasing for $x > 0$ (*see* ORDERING, STAR-SHAPED).

Orderings of skewness are used widely in reliability theory [3]. The distribution F_T of a nonnegative random variable (a lifetime) T is *increasing hazard rate* if the exponential distribution is more skew to the right than F_T and *decreasing hazard rate* (DHR) if F_T is more skew to the right than the exponential distribution. (The concepts *increasing hazard rate average* and *decreasing hazard rate average* can be given similarly by using the star-shaped ordering). These notions are also intimately connected with stochastic ordering: If $F_{T,t}$ is the distribution of the residual life at time t, given that T exceeds t, then the distribution of T is IHR (DHR) if and only if $F_{T,t} \leqslant_{st} F_{T,s}$ ($F_{T,s} \leqslant_{st} F_{T,t}$) whenever $t \geqslant s$. (*See also* HAZARD RATE AND OTHER CLASSIFICATIONS OF DISTRIBUTIONS.)

It is natural to require that measures of skewness and measures of kurtosis should preserve the respective orderings. The classical $\sqrt{\beta_1}$ and β_2 are skewness and kurtosis measures in this sense, but the Pearson measure of skewness does not preserve the c ordering [17, 23]. Among the many applications of skewness and kurtosis orderings, we mention the selection of test hypotheses, the construction of tests for normality and exponentiality, the choice of suitable location parameters, the study of the efficiency* and the robustness* of tests and estimators.

The concept of uniform conditional stochastic order (UCSO) combines the notions of stochastic order and conditioning [25]. A distribution is less than or equal to another in the sense of UCSO if the usual stochastic ordering holds for each pair of conditional distributions obtained by conditioning on all events in an appropriate class. When densities of the distributions exist, a version of UCSO is equivalent to the MLR (monotone likelihood ratio $f_Y(x)/f_X(x)$) ordering [14] needed in the study of uniformly most powerful tests. Monotonicity of the ratios $F_Y(x)/F_X(x)$ and $[1 - F_Y(x)]/[1 - F_X(x)]$, also versions of UCSO, and other related orderings are studied in Keilson and Sumita [13].

PARTIAL ORDERING
OF MULTIVARIATE DISTRIBUTIONS

We briefly consider some partial orderings for bivariate or, more generally, for k-variate distributions.

Let $\mathbf{X} = (X_1, X_2)'$ and $\mathbf{Y} = (Y_1, Y_2)'$ be two bivariate random variables. Then $F_{\mathbf{Y}}$ (or

$\mathbf{Y}$) is *stochastically larger* than $F_{\mathbf{X}}$ (or $\mathbf{X}$) if

$$E(u(\mathbf{X})) < E(u(\mathbf{Y})) \qquad (4)$$

for all bounded coordinatewise increasing $u : \mathbb{R}^2 \to \mathbb{R}$ [4, 21]. The generalization of the definition to the k-variate case is obvious [10]. The stochastic ordering implies that $F_{X_1, X_2}(x_1, x_2) \geqslant F_{Y_1, Y_2}(x_1, x_2)$, but the converse is not true [21]. The stochastic ordering could be defined also by using bounded continuous coordinatewise increasing test functions u or by applying upper and lower sets [20]. As in the univariate case, other location orderings are obtained by using convex increasing (or concave increasing) test functions u in (4). (See Stoyan [21].)

The multivariate stochastic ordering arises naturally, e.g., when seeking for hypotheses corresponding to multivariate analogs of the Mann–Whitney U-test or when stating the properties of multivariate location parameters [18]. Multivariate increasing failure rate (MIHR) and multivariate decreasing failure rate (MDHR) distributions can also be defined by using multivariate stochastic ordering [1].

The notion of multivariate scale or scatter is not obvious. For symmetrical distributions, one can define, however, as in (2) that $F_{\mathbf{X}}$ is *more concentrated* about $\boldsymbol{\mu}$ than $F_{\mathbf{Y}}$ about $\boldsymbol{\nu}$ if

$$\Pr(\mathbf{X} - \boldsymbol{\mu} \in C) \geqslant \Pr(\mathbf{Y} - \boldsymbol{\nu} \in C) \qquad (5)$$

for all convex symmetrical sets C [9]. (C is symmetrical if $C = -C$.) One can also say [18] that F_Y is *more scattered* than F_X if there exists a function $\Phi : \mathbb{R}^2 \to \mathbb{R}^2$ such that

1^0. The distribution of $\Phi(\mathbf{X})$ is $F_{\mathbf{Y}}$.
2^0. $\Delta(\Phi(\mathbf{x}_1), \Phi(\mathbf{x}_2), \Phi(\mathbf{x}_3)) \geqslant \Delta(\mathbf{x}_1, \mathbf{x}_2, \mathbf{x}_3)$ for all $\mathbf{x}_1, \mathbf{x}_2, \mathbf{x}_3 \in \mathbb{R}^2$, where $\Delta(\mathbf{x}_1, \mathbf{x}_2, \mathbf{x}_3)$ is the area of the triangle defined by $\mathbf{x}_1$, $\mathbf{x}_2$, and $\mathbf{x}_3$.

This ordering generalizes the spread ordering and is preserved by the generalized variance* of Wilks. There are no partial orderings corresponding to multivariate skewness or kurtosis. Therefore, definitions of the respective measures are not clear.

Multivariate versions of UCSO and MLR orderings can be found as follows: A distribution F_X with density f_X is said [12] to be *multivariate totally positive of order 2* (MTP$_2$) if

$$f_X(x \vee y) f_X(x \wedge y) \geqslant f_X(x) f_X(y)$$

$$\text{for all } x, y, \quad (6)$$

where $\vee$ and $\wedge$ are the lattice operations, i.e., $x \vee y = (\max(x_1, y_1), \max(x_2, y_2))$ and $x \wedge y = (\min(x_1, y_1), \min(x_2, y_2))$. The marginals of X are then *associated* and *positively regression dependent*. A related ordering is the TP_2 *ordering* (*strong MLR ordering*) [12, 26]: $F_X \leqslant_{tp} F_Y$ if for the respective densities

$$f_X(y) f_Y(x) \leqslant f_X(x \wedge y) f_Y(x \vee y)$$

$$\text{for all } x, y. \quad (7)$$

This ordering implies the stochastic ordering. A weaker ordering (*weak MLR ordering*) is proposed in Whitt [26] by defining that $F_X \leqslant_r F_Y$ if

$$f_X(y) f_Y(x) \leqslant f_X(x) f_Y(y), \quad (8)$$

whenever $x \leqslant y$ (i.e., $x_1 \leqslant y_1$ and $x_2 \leqslant y_2$). A version of UCSO obtained by conditioning on all lattice subsets of $\mathbb{R}^2$ (A is a lattice if $x, y \in A \Rightarrow x \wedge y \in A$ and $x \vee y \in A$) is weaker than $\leqslant_{tp}$ but stronger than $\leqslant_r$. All these three orderings are equivalent if either F_X or F_Y is TP$_2$. (See Whitt [26] for this.)

Other qualitative concepts of bivariate and multivariate dependence are introduced in Lehmann [15] and an ordering of concordance, equivalent to the stochastic ordering, for bivariate discrete distributions is proposed in Tchen [22]. Some classical nonparametric measures of positive and negative dependence preserve this ordering. (*See*. DEPENDENCE, CONCEPTS OF.)

References

[1] Arjas, E. (1981). *Math. Operat. Res.*, **6**, 263–276. (Multivariate increasing (decreasing) failure rate distributions.)

[2] Barlow, R. E. and Proschan, F. (1966). *Ann. Math. Statist.*, **37**, 1593–1601.

[3] Barlow, R. E. and Proschan, F. (1975). *Statistical Theory of Reliability and Life Testing*. Holt, Rinehart and Winston, New York. (Partial orderings in reliability theory. Convex ordering and star-shaped ordering.)

[4] Bergman, R. (1978). *Math. Nachr.*, **82**, 103–114. (Partial orderings for bivariate distributions and distributions on general product spaces.)

[5] Bickel, P. J. and Lehmann, E. L. (1975). *Ann. Statist.* **3**, 1039–1069. (General reference providing measures of location.)

[6] Bickel, P. J. and Lehmann, E. L. (1976). *Ann. Statist.*, **4**, 1139–1158. (Scale or dispersion of symmetrical distributions.)

[7] Bickel, P. J. and Lehmann, E. L. (1979). In *Contributions to Statistics*, J. Jurečkova, ed. Academia, Prague, Czechoslovakia, pp. 33–40. (Scale in asymmetrical models.)

[8] Birnbaum, Z. W. (1948). *Ann. Math. Statist.*, **19**, 76–81. (A dispersion ordering.)

[9] Eaton, M. L. (1982). *Ann. Statist.*, **10**, 11–43. (A review of some multivariate probability inequalities.)

[10] Kamae, T., Krengel, U., and O'Brien, G. L. (1977). *Ann. Prob.*, **5**, 899–912. (Stochastic ordering on partially ordered spaces.)

[11] Karlin, S. (1968). *Total Positivity*. Stanford University Press, Stanford, CA.

[12] Karlin, S. and Rinott, Y. (1980). *J. Multivariate Anal.*, **10**, 467–498. (Multivariate total positivity (TP$_2$) and multivariate MLR.)

[13] Keilson, J. and Sumita, U. (1982). *Canad. J. Statist.*, **10**, 181–198. (The uniform orderings and the local ordering.)

[14] Lehmann, E. L. (1955). *Ann. Math. Statist.*, **26**, 399–419. (MLR ordering and stochastic ordering.)

[15] Lehmann, E. L. (1966). *Ann. Math. Statist.*, **37**, 1137–1153. (Concepts of positive and negative dependence.)

[16] Mann, H. B. and Whitney, D. R. (1947). *Ann. Math. Statist.*, **18**, 50–60. (Stochastic ordering.)

[17] Oja, H. (1981). *Scand. J. Statist.*, **8**, 154–168. (A review of the univariate case.)

[18] Oja, H. (1983). *Statist. Prob. Lett.*, **1**, 327–332. (Descriptive statistics for multivariate distributions.)

[19] Shaked, M. (1982). *J. Appl. Prob.*, **19**, 310–320. (Discussion on dispersion.)

[20] Simons, G. (1980). *Ann. Statist.*, **8**, 833–839. (Multivariate generalizations of the stochastic ordering.)

[21] Stoyan, D. (1977). *Qualitative Eigenschaften und Abschätzungen Stochastischer Modelle*. Akademie-Verlag, Berlin. (General reference on stochastic ordering and related orderings.)

[22] Tchen, A. H. (1980). *Ann. Prob.*, **8**, 814–827. (A partial ordering for the notion of concordance.)

[23] van Zwet, D. R. (1964). *Convex Transformation of Random Variables*. Mathematisch Centrum, Amsterdam. (Definitions of c and s orderings. Applications.)

[24] Veinott, R. (1965). *Operat. Res.*, **13**, 761–778. (A multivariate generalization of stochastic ordering.)

[25] Whitt, W. (1980). *J. Appl. Prob.*, **17**, 112–123. (Definition of UCSO.)

[26] Whitt, W. (1982). *J. Appl. Prob.*, **19**, 695–701. (Multivariate MLR ordering and UCSO.)

Acknowledgment

I am grateful to Professor Elja Arjas for many helpful discussions.

(DEPENDENCE, CONCEPTS OF
HAZARD RATE AND OTHER
    CLASSIFICATIONS OF DISTRIBUTIONS
ORDERING DISTRIBUTIONS
    BY DISPERSION
ORDERING, STAR-SHAPED
STOCHASTIC ORDERING
TOTAL POSITIVITY)

H. OJA

ORDERING PROCEDURES

Consider a set $A = \{A_1, \ldots, A_k\}$ of k alternatives where A_i for each i is characterized by a parameter set θ_i. For example, if A_i represents a normal population, then θ_i can be its mean and variance. Let $C_1, \ldots, C_m$ be m functions, called *criteria*, of θ_i. Then the m criteria induce a partial order B on A where $A_i B A_j$, (read A_i better than A_j) if and only if $C_l(\theta_i) \geq C_l(\theta_j)$ for all $l = 1, \ldots, m$. Formally, the input to an ordering problem is a set A of k alternatives, a set I of prior information on θ or on B, a goal G which is a partial order with k elements, and a loss function L. The output is a mapping from A to (the elements of) G. The loss function $L = L(M, B)$ is defined for every mapping M, and essentially measures how inconsistent M is with B. So if B were known, the loss would be minimized. In reality, B is unknown, and we have to collect data for making inferences on B.

An ordering procedure consists of a design D on data collection and a mapping $M_{I,D}$ from A to G. An ordering procedure is evaluated by the expected loss $E(L)$ as well as by the cost of collecting data. A standard approach is to bound the loss and minimize the cost.

Ranking and selection procedures are special cases of ordering procedures with special forms of G. However, ordered statistics is a completely different subject from our discussions.

GOALS OF ORDERING PROCEDURES

The goal of an ordering procedure is represented by the given partial order G. However, only partial orders of the following special form have been studied in the literature. Define a $\{k_1, \ldots, k_l\}$ complete lattice (CL) as a lattice of l levels where the ith level has k_i elements, and an element is better than all elements at the next level. Some special cases are:

1. Select the t best. G is a $(t, k - t)$ CL.
2. Rank the tth best. G is a $([t], k - t)$ $\equiv (1, \ldots 1, k - t)$ CL.
3. Select the tth best. G is a $(t - 1, 1, k - t)$ CL.

There are two basic types of data, absolute data, and comparison data. Absolute data are in real numbers where data collected for A_i provide estimates of θ_i, independent of data collected for other alternatives. A piece of comparison data is a partial order on a subset S of A (hence also a partial order on A). It rates each alternative in S with respect to other alternatives in S. The comparison data is called regular if the partial orders always have the same form. We call it $(c, s - c)$- or $([c], s - c)$-*regular* if the partial order on S is the $(c, s - c)$ or $([c], s - c)$ CL.

Note that comparison data provide information only on the ordering of θ_i but not its value. As θ_i cannot be estimated by data, it no longer serves as a useful medium to induce a partial order on A. We will replace it by something else, estimable from comparison data, that can also induce a partial order on A (possibly a different one). For example, if the comparison data is $(1, s)$-regular, then the parameter set of A_i is $\{P_i(S): S$ an s-subset of $A\}$ where $P_i(S)$ is the probability that A_i is the winner in a comparison involving S. The induced partial order is $A_i B A_j$ if and only if $P_i(S) \geqslant P_j(S')$ for all s subsets S and S' such that $S \cup \{A_j\} = S' \cup \{A_i\}$.

LOSS FUNCTION

The prevalent loss function used for ordering problems is the correct–incorrect (zero–one) function. Therefore, in most literature we talk about the probability of correct ordering (PCO) instead of the expected loss.

Since B is determined by $\{\theta_i\}$, we can also write $L(M, B)$ as $L(M, \{\theta_i\})$. A set $\{\theta_i\}$ is often referred to as a configuration. The fact that $L(M, \{\theta_i\})$ depends on the unknown true configuration introduces a problem in assessing the loss. The current wisdom is to take a conservative line by substituting $E[L(M, \{\theta_i^*\})]$ for $E[L(M, \{\theta_i\})]$, where $\{\theta_i^*\}$ maximizes the loss (called a *least favorable* configuration) over all configurations in a given zone Z of configurations. (If no such $\{\theta_i^*\}$ exist, use $\lim \sup E[L]$.) Usually Z is determined by practical considerations for a given problem.

THE INDIFFERENCE ZONE APPROACH

The indifference zone approach, first proposed by Bechhofer [3], partitions all configurations into two disjoint zones called the *preference zone* and the *indifference zone*, respectively. In principle, the preference zone consists of configurations for which the alternatives are far apart and a correct ordering is strongly preferable; the indifference zone consists of configurations for which the

alternatives are close to each other and we are, presumably, indifferent to the correctness of ordering.

While it is easy to define a distance function to measure the closeness of two alternatives, the partition of configurations into the preference zone and the indifference zone requires a global measure of closeness that is lacking. The classical approach is to substitute some local measure for the global measure. For example, for the goal of selecting the t best a configuration is in the indifference zone if the tth best and the $(t + 1)$st best alternative are close enough. However, since the tth best and the $(t + 2)$nd best alternative do not have to be close, we certainly are not indifferent to a selection of the $(t + 2)$nd best alternative. Such discrepancies cast doubt on the meaningfulness of the PCO in the classical approach.

A remedy has been proposed (see Chen [5] for references) that eliminates the somewhat artificial distinction between the preference zone and the indifference zone. The new concept is to label *indifferent* not the configurations, but the alternatives. If A_i and A_j are judged close enough, then we are indifferent to which one is selected. That is, A_i and A_j are interchangeable in the selection without affecting its correctness.

THE TYPES OF ORDERING PROCEDURES

An ordering procedure consists of two parts: a design part on collecting data, and a decision part on choosing the mapping. For the design part, ordering procedures can be classified into the following types:

1. **Nonadaptive.** Every piece of data is independently sampled. Sampling stops when the size reaches a prespecified number n.

2. **Sequential.** Data are sampled a piece at a time such that the information from early data can be used in later sampling. Sampling stops when the data obtained satisfy certain conditions.

3. Multistage. Each stage consists of a nonadaptive subprocedure but the information from earlier stages can be used in later stages.

Usually the decision on mapping is reached by first obtaining an estimator $\tilde{\theta}_i$ of θ_i and then choosing the mapping that minimizes $L(M, \tilde{B})$, where $\tilde{B}$ is induced from $\{\tilde{\theta}_i\}$. Maximum likelihood* estimators are typical choices of estimators, but they can be cumbersome to compute and their small-sample properties are unknown.

Another estimator often used in single-criterion cases due to its simplicity is the score estimator. The data are first normalized to eliminate bias due to any imbalance of the design. The score of an alternative is the sum of the normalized data values it receives. Then the alternatives are ordered linearly by their scores. Bühlmann and Huber gave conditions under which the score estimator maximizes the PCO for the goal of a $(1,1)$ CL. Huber [9] generalizes the result for the goal of a $([t], 0)$ CL.

Next we will discuss procedures with absolute data and procedures with comparison data separately. Since the former procedures are well covered by standard texts (see, e.g., Gibbons et al. [7]), we will emphasize our discussion on the latter procedures.

PROCEDURES WITH ABSOLUTE DATA

The k alternatives are usually k populations of the same family but with different parameters, and B is a linear order with respect to a parameter θ. In most cases, a sufficient estimator $\tilde{\theta}_i$ of θ_i can be obtained by a random sample on A_i. Then the decision part of any ordering procedure is to replace B by the ordering of $\tilde{\theta}_i$. So procedures differ only with respect to their designs. Typically, we look for a procedure minimizing the sample size but still meeting a prespecified PCO. The following types of procedures have been studied in the literature.

(a) Fixed Sample Size. Let n_i be the sample size of A_i with $\sum_{i=1}^{k} n_i = n$ fixed. A surprising phenomenon is that increasing n_i does not necessarily reduce the loss [13]. When the n_i's are all equal, Hall [8] showed that if $\tilde{\theta}_i$ has a monotone likelihood ratio* and θ_i is a location parameter, then the $\tilde{\theta}$ ordering minimizes the sample size for meeting a given PCO when the goal is a $(k_1, \ldots k_l)$ CL.

(b) Two-Stage. Usually used in the presence of nuisance parameters*. Data in the first stage are used to estimate the nuisance parameters. Then it is determined how much additional data are required for the second stage.

(c) Inverse Sampling* Procedure. The sample size of each alternative is increased one at a time. Sampling stops whenever the data obtained satisfy certain conditions.

(d) Play the Winner*. The alternatives are permuted randomly. Sampling starts from the first alternative and continues on the same alternative if the data obtained satisfy certain conditions. Otherwise, we sample the next alternative on line.

PROCEDURES WITH COMPARISON DATA

We will, as the literature almost exclusively does, cover comparison data only of the $(1, s - 1)$-regular CL type.

For $s = k$ there is only one s subset, which is A. Therefore, the data are just sample trials from a multinomial* distribution with k probabilities $P_1, \ldots, P_k$, where $P_i = P_i(S)$. Procedures for multinomial data are similar to those discussed in the preceding section, except that play-the-winner procedure does not exist and the inverse sampling procedure should be modified as follows:

(c') Inverse Sampling Procedure. Sampling stops whenever the counts of the alternatives in the sample satisfy certain conditions.

For $s < k$, there is a new aspect to the design problem, that is how to select the s subsets for comparisons. This problem was first studied in scheduling tournaments for sports events, which usually involve only two players in a comparison. Therefore, we shall describe the procedures assuming $s = 2$ (the paired-comparison case), although they are applicable for general s. We write $P_i(S) = P_{ij}$ for $S = \{i, j\}$.

In sports, the time constraint, which can be translated to constraints on sample size, usually is more severe than the need to meet a certain PCO. Therefore, the general approach is to consider procedures satisfactory from the sample-size viewpoint and then analyze their PCOs. The following types of procedures have been considered in the literature.

ROUND ROBIN. Every player plays one match against every other player. This is a nonadaptive procedure capable of ranking all players. Ford [6] gave the maximum likelihood estimator $\tilde{\theta}_i$ for θ_i when $P_i(S)$ is assumed to equal $\theta_i / (\theta_i + \theta_j)$. He also showed that the linear order induced by $\tilde{\theta}$ is the same as the score estimator when each match consists of the same number of comparisons.

KNOCKOUT*. This is a multistage procedure with approximately $\log_2 k$ stages for selecting the best player. At the first stage, the players are paired off for matches (adding a dummy player if necessary), and only the winners survive to the next stage. A similar procedure now applies to each stage recursively until only one player is left. This player is selected as the best.

BEAT THE WINNER. This is a sequential procedure for selecting the best player. The players are randomly permuted, and the first two players are paired off for a match. The winner then plays against the next player in the permutation and so on. The sampling stops and the last winner is selected as best when he or she has won at least t times or at least t times more than anyone else.

Example 1. Suppose that four players are engaged in a knockout tournament. Let P_{ij}

$= \pi_i / (\pi_i + \pi_j)$, where π_i can be interpreted as the strength of player i. Without loss of generality, assume $\pi_1 + \pi_2 + \pi_3 + \pi_4 = 1$. A configuration can then be represented by a set $\{\pi_1, \pi_2, \pi_3, \pi_4\}$ with sum one. Suppose that the zone we are interested in consists of all configurations in which the largest π_i is at least twice all other π_i's. Then the least favorable configuration is $(0.4, 0.2, 0.2, 0.2)$. Define $p = 0.4/(0.4 + 0.2) = \frac{2}{3}$. If each match consists of one game (one comparison), the PCO is $p^2 = \frac{4}{9}$. If the final match consists of three games, the PCO is $p \cdot (p^3 + 3p^2(1 - p)) = \frac{40}{81}$.

PROCEDURES* WITH DETERMINISTIC PAIRED COMPARISONS*

We consider data of the error-free paired-comparison type. That is, if A_i is better than A_j, then A_i beats A_j with certainty in every comparison (therefore, there is never a need to compare two alternatives more than once). Since there is no sampling error, the PCO can reach one. The only problem is minimizing the number of paired comparisons. A procedure is called *optimal* (*minimax*) if it minimizes the average (worst-case) number of comparisons. *Best* will be used in the sense of being currently best. The following ordering problems have been extensively studied by computer scientists (see Aigner [1] for an excellent survey). In general, minimax* or optimal procedures rarely are found except for small parameters. Often, procedures have to be compared by the leading terms of their asymptotic formulas.

Sorting

The goal is a $([k], 0)$ CL. When two disjoint linear orderings of m and $k - m$ alternatives are given as prior information on B, the problem is then called *merging*. Merging procedures are often building blocks for sorting procedures. The correct leading term for number of comparisons in sequential sorting is $k \log_2 k$ for both average case and worst case and is easily attainable (e.g., by recursive insertions). Thus the real competition is in the second-order term. Ford and Johnson have the best sequential procedure and for some time it was unknown whether this procedure was minimax*. Recently, Manacher answered this question in the negative by showing that for many values of k, the Ford and Johnson results can be improved by making use of a better version of Hwang and Lin's merging procedure. (See Christen [4] for references.) Minimax merging procedures are known only for small values of m. The cases of $m = 2$ and $m = 3$ have been published, while the cases of $m = 4$ and $m = 5$ are rumored to exist. Tanner [12] gave a merging algorithm whose average numbers of comparisons stay within 6% of the information-theoretic lower bound. The best nonadaptive merging procedure is due to Batcher and proved by Yao and Yao [14] to be optimal up to a factor of $2 + \epsilon$, where $\epsilon \rightarrow 0$ as the problem size gets large. The corresponding Batcher's nonadaptive sorting procedures requires $O(k \ln k)$ comparisons, when parallel processors are allowed. Ajtai et al. [2] gave a nonadaptive sorting procedure that completes the $O(k \ln k)$ comparisons in $O(\ln k)$ time.

Finding the *t*th Best

The goal is a $(t - 1, 1, k - t)$ CL. All procedures discussed here are sequential. For t fixed, the correct leading term for number of comparisons is k for both average- and worst-case procedures and is easily attainable. In fact, for the worst case the correct second-order term, $(t - 1)\log_2 k$, is also easily attainable. The best average-case procedure is due to Matula and has the correct second-order term $O(t \ln \ln k)$. For $t = k/2$ (or $(k - 1)/2$ and $(k + 1)/2$ if k is odd), the problem is known as the median problem. The best worst-case procedure is due to Schonhage, Paterson, and Pippenger and has leading term $3k$ (the correct leading term is known to be at least $1.75k$). The best average-case procedure is due to Floyd and Rivest and has leading term $1.5k$ (the cor-

rect leading term is known to be at least $1.375k$). See Yao and Yao [14] for references. Minimax procedures for $t \leqslant 3$ and similar results for finding the t best and the ordered t best can be found in Aigner [1].

Example 2. To find the second best among 16 players, we first run a knockout tournament to identify the best player in fifteen comparisons. There are four players who played against the best player and lost. All other players have lost to one of these four players and hence cannot be second best. We run a second knockout tournament on these four players, and the winner is the second best, determined in three more comparisons.

References

[1] Aigner, M. (1982). *Discrete Appl. Math.*, **4**, 247–267.

[2] Ajtai, M., Komlós, J., and Szemerédi, E. (1983). *Proc. 15th ACM Symp. Theory Computing.* Association for Computing Machinery, New York, pp. 1–9.

[3] Bechhofer, R. E. (1954). *Ann. Statist.*, **25**, 16–39.

[4] Christen, C. (1978). *Proceedings of the Nineteenth Annual IEEE Conference on the Foundations of Computer Science.* IEEE, Long Beach, CA, pp. 259–266.

[5] Chen, P. Y. (1982). "*An Alternative Definition of Correct Selection in Ranking and Selection Problems.*" Ph.D. dissertation, University of California, Santa Barbara, CA.

[6] Ford, L. R. (1957). *Amer. Math. Monthly*, **64**, 28–33.

[7] Gibbons, J. D., Olkin, I., and Sobel, M. (1977). *Selecting and Ordering Populations, A New Statistical Methodology.* Wiley, New York.

[8] Hall, W. J. (1959). *Ann. Statist.*, **30**, 964–969.

[9] Huber, P. J. (1963). *Ann. Statist.*, **34**, 511–520.

[10] Hwang, F. K. (1980). *SIAM J. Comp.*, **9**, 298–320.

[11] Knuth, D. E. (1973). *The Art of Computer Programming*, Vol. 3: *Sorting and Searching.* Addison-Wesley, Reading, MA.

[12] Tanner, R. M. (1978). *SIAM J. Comp.*, **7**, 18–38.

[13] Tong, Y. L. and Wetzel, D. E. (1979). *Biometrika*, **66**, 174–176.

[14] Yao, A. C. and Yao, F. F. (1978). *Proceedings of the Nineteenth Annual IEEE Conference on the Foundations of Computer Science.* IEEE, Long Beach, CA, pp. 280–289.

(KNOCKOUT TOURNAMENTS
RANKING PROCEDURES
SELECTION PROCEDURES
STOCHASTIC ORDERING)

F. K. HWANG

ORDERING, STARSHAPED

A real-valued function $\varphi : [0, \infty) \to R$ is starshaped if

$$\varphi(\alpha x) \leqslant \alpha\varphi(x)$$

for all $\alpha \in [0, 1]$ and all $x \geqslant 0$.

Equivalently, such a function φ is starshaped iff either

$$\varphi(0) \leqslant 0,$$

$$\varphi(x)/x \text{ is nondecreasing in } x > 0,$$

or

$$z \in \text{epigraph } \varphi \Rightarrow \alpha z \in \text{epigraph } \varphi$$

$$\text{for all} \quad \alpha \in [0, 1],$$

where epigraph φ is defined to be $\{(x, y); x \geqslant 0, \varphi(x) \leqslant y\}$.

A set is starshaped through the point a if a straight line connecting a and any point in the set lies entirely within the set. (Clearly a star as it is usually drawn is starshaped through its center, hence the term.) Thus a function is starshaped iff its epigraph is starshaped through the origin.

The property of being starshaped is weaker than convexity ($\varphi(\alpha x + (1 - \alpha)y) \leqslant \alpha\varphi(x) + (1 - \alpha)\varphi(y) \; \forall x, y, \; \alpha \in [0, 1]$) and stronger than superadditivity ($\varphi(x + y) \geqslant \varphi(x) + \varphi(y) \; \forall x, y$). In fact, Bruckner and Ostrow [6] have established a finer ordering as follows. Let the average function Φ of φ be given by

$$\Phi(x) = \frac{1}{x} \int_0^x \varphi(t) \, dt, \qquad (1)$$

and say that φ is convex on the average, starshaped on the average, or superadditive on the average if Φ is convex, starshaped, or superadditive, respectively. The following six conditions are then successively weaker (strictly):

φ is convex.

φ is convex on the average.

φ is starshaped.

φ is superadditive.

φ is starshaped on the average.

φ is superadditive on the average.

The starshaped property is involved in several statistical concepts. In reliability theory, a cumulative distribution function (cdf) F is said to be IFRA (increasing failure rate average) if its hazard function $-\ln(1 - F)$ is a starshaped function. The class of IFRA distributions is important as it is the smallest class of distributions (a) containing the exponential distributions, (b) closed under formation of coherent systems, and (c) closed under limits in distribution. (See Barlow and Proschan [3] for a discussion of coherent systems*.) It follows that if the lifetimes of independent components in a coherent system are IFRA, then the lifetime of the whole system is also IFRA.

The term *starshaped* is also applied to vectors [13]. In particular, the vector $\mu = (\mu_1, \ldots, \mu_n)$ is said to be upper (lower) starshaped with respect to the weights $w = (w_1, \ldots, w_n)$ if $0 \leqslant \bar{\mu}_1 \leqslant \bar{\mu}_2 \leqslant \cdots \leqslant \bar{\mu}_n$ ($\bar{\mu}_1 \geqslant \bar{\mu}_2 \geqslant \cdots \geqslant \bar{\mu}_n \geqslant 0$), where $\bar{\mu}_m = \sum_1^m \mu_i w_i / \sum_1^m w_i$.

Shaked has obtained elegant, closed forms for maximum likelihood* estimates (MLEs) of a vector of starshaped means for independent Poisson* and normal* populations.

Dykstra and Robertson [8, 9] have obtained exact distribution theory for likelihood ratio tests* when testing equal means versus starshaped means or when testing starshaped means against all alternatives in a normal distribution setting. They have also obtained MLEs for starshaped parameters in a multinomial setting and derived asymptotic theory for testing hypotheses concerning starshaped parameters in multinomial* and Poisson settings.

The concept of starshaped functions is

also used to induce a partial ordering among CDFs of nonnegative random variables. In particular, if $F(0) = G(0) = 0$, F is said to be *starshaped with respect to G* if the composite function $G^{-1}F(x)$ is starshaped on $[0, \infty)$. This concept, sometimes written $F \underset{*}{\leqslant} G$, yields a partial order among CDFs that has been studied by various authors. Marshall et al. [11] have given various characterizations of $F \underset{*}{\leqslant} G$ and used these to obtain several inequalities. For example, they show that if $F \underset{*}{\leqslant} G$, $X_1 \geqslant X_2 \geqslant \cdots \geqslant X_n$ are order statistics from F, and $Y_1 \geqslant Y_2 \geqslant \cdots \geqslant Y_n$ are order statistics from G, then $\sum_1^k X_i / \sum_1^n X_i$ is stochastically less than or equal to $\sum_1^k Y_i / \sum_1^n Y_i$ for $k = 1, \ldots, n$.

If one takes the CDF G to be the exponential distribution $G(x) = 1 - e^{-x}$ for $x \geqslant 0$, then $F \underset{*}{\leqslant} G$ is equivalent to F being IFRA. If $G(x) = x$, $0 \leqslant x \leqslant 1$, $F \underset{*}{\leqslant} G$ clearly requires that F be starshaped. MLEs for starshaped CDFs can be found (see Barlow et al. [1, p. 255], who attributes the original unpublished work to Marshall and Proschan). Surprisingly, however, these MLEs prove to be inconsistent. Similarly, the MLEs for IFRA CDFs are also inconsistent. Barlow and Scheuer [4] have proposed isotonic estimators for star-ordered families of distributions and established the consistency of these estimators. Barlow et al. [1] also discuss percentile estimators in the star-ordering case.

Star ordering among distributions is also important when testing for exponentiality. In particular, Barlow et al. [1] have shown that an appealing class of tests for testing exponentiality has isotonic power with respect to star ordering among the alternative distributions. Along similar lines, Doksum [7] relates star ordering to the heaviness of the tails of distributions and proves isotonic power with respect to star ordering for monotone rank tests.

References

[1] Barlow, R. E., Bartholomew, D. J., Bremner, J. M., and Brunk, H. D. (1972). *Statistical Infer-*
ence Under Order Restrictions. The Theory and Application of Isotonic Regression. Wiley, New York. (Standard comprehensive reference for order-restricted inferences prior to 1972.)

[2] Barlow, R. E., Marshall, A. W., and Proschan, F. (1969). *Pacific J. Math.*, **29**, 19–42. (Establishes some inequalities, primarily of the type where an integral sign and a function are interchanged under starshaped and convexity assumptions.)

[3] Barlow, R. E. and Proschan, F. (1981). *Statistical Theory of Reliability and Life Testing: Probability Models*. To Begin With, Silver Springs, MD. (A standard reference for material on reliability and life testing.)

[4] Barlow, R. E. and Scheuer, E. M. (1971). *Technometrics*, **13**, 145–159. (Discusses estimates of stochastically ordered IFRA distributions. Also establishes consistency of monotonic regression estimates for IFRA distributions.)

[5] Birnbaum, Z. W., Esary, J. D., and Marshall, A. W. (1966). *Ann. Math. Statist.*, **37**, 816–826. (Invokes a stochastic characterization of wear-out for components and systems that involves starshaped orderings.)

[6] Bruckner, A. M. and Ostrow, E. (1962). *Pacific J. Math.*, **12**, 1203–1215. (Establishes hierarchy of classes of functions involving convex, starshaped, and superadditive concepts.)

[7] Doksum, K. (1969). *Ann. Math. Statist.*, **40**, 1167–1176. (Establishes isotonic power for starshaped orderings of CDFs in certain classes of tests and hence is able to establish some simple optimality theory for rank tests.)

[8] Dykstra, R. L. and Robertson, T. (1982). *Ann. Statist.*, **19**, 1246–1252. (Finds MLEs for starshaped multinomial parameters. Also develops asymptotic theory for likelihood ratio tests.)

[9] Dykstra, R. L. and Robertson, T. (1983). *J. Amer. Statist. Ass.*, **78**, 342–350. (Develops exact distribution theory for various tests involving starshaped restrictions in a normal distribution setting.)

[10] Marshall, A. W. and Olkin, I. (1979). *Inequalities: Theory of Majorization and its Applications*. Academic Press, New York. (Recent comprehensive survey of inequalities pertaining to majorization.)

[11] Marshall, A. W., Olkin, I., and Proschan, F. (1967). *Inequalities*, O. Shisha, ed. Academic Press, New York, pp. 177–190. (Obtains various inequalities, some of which depend on the assumption of distributions being star-ordered.)

[12] Marshall, A. W. and Proschan, F. (1965). *J. Math. Anal. Appl.*, **12**, 87–90. (Proves an inequality for convex functions involving majorization.)

[13] Shaked, M. (1979). *Ann. Statist.*, **7**, 729–741. (Finds elegant, closed-form expressions for MLEs of starshaped mean vectors for normal and Poisson distributions.)

(ORDERING DISTRIBUTIONS
    BY DISPERSION
ORDERING OF DISTRIBUTIONS,
    PARTIAL)

RICHARD L. DYKSTRA

ORDERING, STOCHASTIC *See* STO-
CHASTIC ORDERING

ORDER OF STATIONARITY *See* STA-
TIONARITY, ORDER OF

ORDER-RESTRICTED INFERENCES

In many situations, it is desirable to make statistical inferences concerning several parameters when it is believed a priori that they satisfy certain order restrictions. Typically, inferences that take this information into account will perform better than those that do not. Further, if one can postulate a functional form, such as linear or exponential, so that the parameters can be expressed as a function of one or more independent variables with fewer unknown parameters, then the inferences can be improved even more. Order-restricted inferences, which make use of the ordering information without assuming a functional representation for the parameters, provide middle-of-the-road alternatives. They are more powerful than the omnibus procedures and provide protection against an incorrect assumption concerning the form of the function. The text by Barlow et al. [3] contains an excellent, detailed discussion of the origins, development, and applications of order-restricted inference.

Techniques for determining an unknown ordering on a set of parameters are discussed under RANKING and SELECTION PROCEDURES.

ISOTONIC INFERENCE*

Isotonic inferences form a large subclass of order-restricted inferences. If a parameter set inherits its order restrictions from a partial order on its index set, it is said to be *isotonic* (with respect to the partial order). Recall, a partial order $\ll$ on a set S is a binary relation satisfying

1. $a \ll a$ for all a in S.
2. If $a \ll b$ and $b \ll a$, then $a = b$.
3. If $a \ll b$ and $b \ll c$, then $a \ll c$.

A partial order allows for elements that are not comparable, but if for each a and b in S, $a \ll b$ or $b \ll a$, then $\ll$ is called a total order. Let $\ll$ be a partial order on $\{1, 2, \ldots, k\}$. If $\theta = (\theta_1, \theta_2, \ldots, \theta_k)$ is a k-dimensional parameter with $\theta_i \leqslant \theta_j$ whenever $i \ll j$, then θ as well as inferences concerning θ that incorporate this ordering information are said to be isotonic (with respect to $\ll$).

The following example, which is an extension of the standard one-sided comparison of two normal means illustrates the basic ideas of an order-restricted inference. Consider k increasing levels of a treatment and suppose that the responses are distributed normally with means μ_i and common variance. In many situations, it would be reasonable to assume that the mean response increases with the treatment level, and so we may assume that $H_1 : \mu_1 \leqslant \mu_2 \leqslant \cdots \leqslant \mu_k$ holds. To determine whether the treatment has an effect, one might test $H_0 : \mu_1 = \mu_2 = \cdots = \mu_k$ against $H_1 - H_0$, that is, H_1 holds with at least one strict inequality. Of course, this is the usual one-way analysis of variance* except that typically, the alternative hypothesis is $\sim H_0$, the complement of H_0. Brunk [6] obtained the maximum likelihood* estimates (MLEs) of the means subject to the restrictions in H_1 and Bartholomew [4] developed the likelihood ratio test (lrt) for H_0 vs. $H_1 - H_0$. To provide a test of the underlying assumption, Robertson and Wegman [23] developed the lrt for H_1 vs. $H_2 : \sim H_1$. Computation algorithms for these estimates and the form of the tests are discussed in ISOTONIC INFERENCE and in greater detail in Barlow et al. [3]. Similar results for the general linear model*, for parameters other than normal means, and nonparametric analogs are discussed there also.

OTHER ORDER RESTRICTIONS

The hypothesis $\theta_1 \leqslant \theta_2 \leqslant \cdots \leqslant \theta_k$ specifies in a strong way a trend among the parameters θ_i. Dykstra and Robertson [10, 11] studied the weaker notion, θ_i is nondecreasing on the average provided $\sum_{j=1}^{i} \theta_j / i$ is nondecreasing. They consider normal and Poisson means as well as multinomial parameters. The MLEs and the distribution theory needed for the tests are less complicated than in the case of the stronger trends.

The MLEs of nondecreasing means provide least-squares* estimates for a nondecreasing regression* function based on samples obtained at a finite number of values of the independent variable. (For details, see Brunk [7].) In certain applications, concave regression functions are of interest. If $\bar{y}_i$, $i = 1, 2, \ldots, k$, are the means of independent random samples of size n_i taken at observation points t_i, then a least-squares concave estimator minimizes $\sum_{i=1}^{k} n_i (\bar{y}_i - \theta(t_i))^2$ subject to $(\theta(t_i) - \theta(t_{i-1}))/(t_i - t_{i-1}) \geqslant (\theta(t_{i+1}) - \theta(t_i))/(t_{i+1} - t_i)$ for $i = 2, \ldots, k-1$. Hildreth [16] discussed the computation of such estimates and their application in econometrics. Hanson and Pledger [15] established their strong consistency.

Another interesting pattern for a parameter set is the following: $\theta_1 \leqslant \theta_2 \leqslant \cdots \leqslant \theta_l \geqslant \theta_{l+1} \geqslant \cdots \geqslant \theta_k$. With $l = k$ and 1, this includes nondecreasing and nonincreasing trends. Mack and Wolfe [19] considered sums of Mann–Whitney statistics to obtain nonparametric tests of homogeneity vs. these umbrella alternatives, in the cases l known and unknown. Using Monte Carlo techniques, they compared this test with Jonckheere's test* for trend.

CONTINGENCY TABLES*

Much of the literature concerning contingency tables with ordered categories does not involve order-restricted parameter sets. (Williams and Grizzle [30] consider tables in which all the variables have ordered responses, Simon [25] considers those in which at least one of the variables is ordinal level, and McCullagh [20] develops models for ordered square tables such as those occurring in paired sampling experiments. Agresti [1], Goodman [12], and Wahrendorf [28] discuss measures of association* and partial association for such tables.)

In the area of goodness-of-fit* tests for contingency tables*, order-restricted inferences for a multinomial parameter set p_1, $p_2, \ldots, p_k$ have been developed. Chacko [8], Weisberg [29], Lee [17], and Bergman [5] developed tests of homogeneity*, $p_1 = p_2 = \cdots = p_k = k^{-1}$, vs. the trend $p_1 \leqslant p_2 \leqslant \cdots \leqslant p_k$. Chacko derived the likelihood ratio test, Lee derived some maximin tests and Weisberg gave a Bayesian treatment. Robertson [22] considered the analogous situation when the p_i are assumed to obey a partial ordering.

Comparisons between multinomial distributions can be made via order restrictions. Alam and Mitra [2] developed a one-sample test for the equality of two multinomial parameter sets $p_1, p_2, \ldots, p_k$ and $q_1, q_2, \ldots, q_k$ with the alternative that one is more polarized than the other. If $\sum_{j=1}^{i} p_{(j)} \geqslant \sum_{j=1}^{i} q_{(j)}$ for $i = 1, 2, \ldots, k$, with $p_{(1)} \geqslant p_{(2)} \geqslant \cdots \geqslant p_{(k)}$ the ordered p_i, then p majorizes q (in the Schur sense), and p is said to be more polarized than q. Of course, $(1, 1, \ldots, 1)/k$ is the least polarized and a degenerate set of multinomial probabilities is the most polarized. They considered the test statistic $\sum_{i=1}^{k} X_i^2 / n$, where X_i is the number of occurrences of category i in n trials, obtaining its asymptotic distribution and investigating the rate of convergence to this limiting distribution.

If the cells of two multinomial distributions* are ordered, then the concept of stochastic ordering can be expressed as H_1: $\sum_{j=1}^{i} p_j \geqslant \sum_{j=1}^{i} q_j$ for $i = 1, 2, \ldots, k$. Robertson and Wright [24] developed one- and two-sample likelihood ratio tests for $H_0: p = q$ vs. $H_1 - H_0$ and for H_1 vs. $H_2: \sim H_1$. Dykstra et al. [9] derived a similar test of stochastic ordering but the underlying distributions were assumed to be continuous. There is extensive literature on nonparametric tests of the equality of several distributions with stochastic ordering alternatives. See Barlow et al. [3], Tryon and Hettmansperger [27], Govindarajulu and Haller (1974), Lee and Wolfe [18], Skillings [26], and references therein.

Independent random samples from c Bernoulli populations can be organized into a $2 \times c$ table. One may wish to test for a trend among the Bernoulli probabilities. Bartholomew's test, which was discussed in the section on isotonic inferences, can be modified for this purpose. For the details and a discussion of alternative tests, see Section 4.3 of Barlow et al. [3].

One approach to measuring association in a contingency table is to use the odds ratios*. If p_{ij} is the probability corresponding to the cell in the ith row and jth column for $i = 1, 2, \ldots, r$ and $j = 1, 2, \ldots, c$, then the odds ratios are defined by

$$\phi_{ij} = p_{ij}p_{i+1,j+1}/(p_{i,j+1}p_{i+1,j}),$$
$$1 \leqslant j \leqslant r-1, \quad 1 \leqslant j \leqslant c-1.$$

Patefield [21] discusses tests of no association, $\phi_{ij} = 1$ for each i, j versus tests of positive association, $\phi_{ij} \geqslant 1$ for all i, j with at least one strict inequality. Grove [14] restricted his attention to $2 \times c$ tables and considered two tests of no association versus positive association. One is based on the cross-product ratios, $(\sum_{j=1}^{q} p_{1j})(\sum_{j=q+1}^{c} p_{2j})$ $/(\sum_{j=1}^{q} p_{2j})(\sum_{j=q+1}^{c} p_{1j})$, which is equivalent to testing homogeneity vs. a stochastic ordering among two multinomial populations. The second is based on the odds ratios, ϕ_{1j}, which is equivalent to testing for homogeneity vs. a trend in independent Bernoulli parameters.

References

[1]   Agresti, A. (1977). *J. Amer. Statist. Ass.*, **72**, 37–45.

[2]   Alam, K. and Mitra, A. (1981). *J. Amer. Statist. Ass.*, **76**, 107–109.

[3]   Barlow, R. E., Bartholomew, D. J., Bremner, J. M., and Brunk, H. D. (1972). *Statistical Inference under Order Restrictions*. Wiley, New York.

[4]   Bartholomew, D. J. (1959). *Biometrika*, **46**, 36–48.

[5]   Bergman, B. (1981). *Scand. J. Statist.*, **8**, 218–227.

[6]   Brunk, H. D. (1955). *Ann. Math. Statist.*, **26**, 607–616.

[7]   Brunk, H. D. (1970). *Nonparametric Techniques in Statistical Inference*. Cambridge University Press, Cambridge, England, pp. 177–197.

[8]   Chacko, V. J. (1963). *Ann. Math. Statist.*, **34**, 945–956.

[9]   Dykstra, R. L., Madsen, R. W., and Fairbanks, K. (1983). *J. Statist. Comp. Simul.*, **18**, 247–264.

[10] Dykstra, R. L. and Robertson, T. (1983). *J. Amer. Statist. Ass.*, **78**, 342–350.

[11] Dykstra, R. L. and Robertson, T. (1982). *Ann. Statist.*, **10**, 1246–1252.

[12] Goodman, L. A. (1979). *J. Amer. Statist. Ass.*, **74**, 537–552.

[13] Govindarajulu, Z. and Haller, H. S. (1977). *Proc. Symp. J. Neyman*, 91–102.

[14] Grove, D. M. (1980). *J. Amer. Statist. Ass.*, **75**, 454–459.

[15] Hanson, D. L. and Pledger, G. (1976). *Ann. Statist.*, **4**, 1038–1050.

[16] Hildreth, C. (1954). *J. Amer. Statist. Ass.*, **49**, 598–619.

[17] Lee, Y. J. (1980). *J. Amer. Statist. Ass.*, **75**, 673–675.

[18] Lee, Y. J. and Wolfe, D. A. (1976). *J. Amer. Statist. Ass.*, **71**, 722–727.

[19] Mack, G. A. and Wolfe, D. A. (1981). *J. Amer. Statist. Ass.*, **76**, 175–181.

[20] McCullagh, P. (1978). *Biometrika*, **65**, 413–418.

[21] Patefield, W. M. (1982). *Appl. Statist.*, **31**, 32–43.

[22] Robertson, T. (1978). *J. Amer. Statist. Ass.*, **73**, 197–202.

[23] Robertson, T. and Wegman, E. J. (1978). *Ann. Statist.*, **6**, 485–505.

[24] Robertson, T. and Wright, F. T. (1981). *Ann. Statist.*, **9**, 1248–1257.

[25] Simon, Gary (1974). *J. Amer. Statist. Ass.*, **69**, 971–976.

[26] Skillings, J. H. (1980). *Technometrics*, **22**, 431–436.

[27] Tryon, P. V. and Hettmansperger, T. P. (1973). *Ann. Statist.*, **1**, 1061–1070.

[28] Wahrendorf, J. (1980). *Biometrika*, **67**, 15–21.

[29] Weisberg, H. (1972). *J. Amer. Statist. Ass.*, **67**, 884–890.

[30] Williams, O. D. and Grizzle, J. E. (1972). *J. Amer. Statist. Ass.*, **67**, 55–63.

(ISOTONIC INFERENCE)

F. T. WRIGHT

ORDER STATISTICS

Order statistics deals with the properties and applications of ordered random variables and functions involving them. Let the variates $X_i (i = 1, 2, \ldots, n)$ be arranged in ascending order and written as

$$X_{1:n} \leqslant X_{2:n} \leqslant \cdots \leqslant X_{n:n}.$$

Both the ordered variate $X_{r:n}$ $(r = 1, 2, \ldots, n)$ and the corresponding observed value $x_{r:n}$ are called the *rth order statistic*. Frequently the (unordered) X_i are taken to be statistically independent and identically distributed (iid).

Order statistics occur in a very natural manner when n items, e.g., electric light bulbs, are simultaneously put on test. As successive failures occur, such a life test generates in turn the ordered observations $x_{1:n}, x_{2:n}, \ldots, x_{n:n}$. Thus the variates $X_{1:n}$ and $X_{n:n}$ (the *extremes*) represent time to first failure and duration of this kind of test, respectively, and $X_{r:n}$ the duration of a test terminated or *censored* after occurrence of the *rth* failure. Closely related is the interpretation of $X_{1:n}$ and $X_{n:n}$ as the respective lifetimes of series and parallel systems of n components.

Usually the data analyst will have to do the ordering of the observations. Important early examples of functions of order statistics are the *median* M_n, defined as $X_{(n+1)/2:n}$ or $\frac{1}{2}(X_{n/2:n} + X_{n/2+1:n})$ according as n is odd or even, and the *range* $W_n = X_{n:n} - X_{1:n}$. If the X_i are independently drawn from a common population, then M_n and W_n are well-known measures of location* and dispersion*. In small samples, W_n is widely used in quality control. Slightly more elaborate measures of location are the *trimmed means*

$$T_n(j) = \sum_{i=j+1}^{n-j} \frac{X_{i:n}}{n-2j}, \qquad 0 < j < \tfrac{1}{2}n. \quad (1)$$

The basic properties of order statistics and some additional uses are outlined in the next two sections of this article. Further majoi applications are indicated in the final section.

DISTRIBUTION THEORY AND ESTIMATION OF PARAMETERS

Let $X_1, X_2, \ldots, X_n$ be iid with CDF $F_X(x)$. Then the CDF of $X_{r:n}$ is given by

$$\begin{aligned}
F_{X_{r:n}}(x) &= \Pr[X_{r:n} \leqslant x] \\
&= \Pr[\text{at least } r \text{ of the } X_i \text{ are} \\
&\qquad \text{less than or equal to } x] \\
&= \sum_{i=r}^{n} \binom{n}{i} F_X^i(x) [1 - F_X(x)]^{n-i}, \quad (2)
\end{aligned}$$

since the term in the summand is the binomial probability that exactly i of $X_1, X_2, \ldots, X_n$ are less than or equal to x.

Usually we assume that $F_X(x)$ is absolutely continuous. Note that this assumption is not required for the basic result (2). If X is a discrete variate, then the probability of ties among the observations is positive, which tends to complicate the distribution theory (see, e.g., David [7, p. 13]). However, the probability function $f_{X_{r:n}}(x)$ is easily found from (2). Thus if X can take only integral values, then

$$f_{X_{r:n}}(x) = F_{X_{r:n}}(x) - F_{X_{r:n}}(x-1).$$

If the probability density function (PDF) $f_X(x)$ exists, then differentiation of (2) gives

$$\begin{aligned}
f_{X_{r:n}}(x) &= \frac{n!}{(r-1)!(n-r)!} F_X^{r-1}(x) f_X(x) \\
&\quad \times [1 - F_X(x)]^{n-r}. \quad (3)
\end{aligned}$$

Also the joint PDF of $X_{r:n}$ and $X_{s:n}$ $(1 \leqslant r < s \leqslant n)$ is, for $x \leqslant y$ (e.g., see David [7, p. 10]),

$$\begin{aligned}
f_{X_{r:n}, X_{s:n}}(x, y) &= \frac{n!}{(r-1)!(s-r-1)!(n-s)!} \\
&\quad \times F_X^{r-1}(x) f_X(x) \\
&\quad \times [F_X(y) - F_X(x)]^{s-r-1} \\
&\quad \times f_X(y)[1 - F_X(y)]^{n-s}. \quad (4)
\end{aligned}$$

The distribution of many commonly used functions of order statistics can be derived from (4) (and from its generalization to more than two order statistics) by standard transformation of variables methods. For example, the PDF of $W_n = X_{n:n} - X_{1:n}$ is [39]

$$f_{W_n}(w) = n(n-1) \int_{-\infty}^{\infty} f_X(x)$$
$$\times \left[F_X(x+w) - F_X(x) \right]^{n-2}$$
$$\times f_X(x+w) \, dx,$$

giving, after integration, the useful formula

$$F_{W_n}(w) = n \int_{-\infty}^{\infty} f_X(x)$$
$$\times \left[F_X(x+w) - F_X(x) \right]^{n-1} dx.$$

When X is a normal $N(\mu, \sigma^2)$ variate, the CDF and percentage points of W_n/σ have been tabulated (see, e.g., Pearson and Hartley [31]), thus allowing an immediate simple test of the null hypothesis $\sigma = \sigma_0$. The well-known range estimate of σ is simply $\hat{\sigma}_w = W_n/d_n$, where the multiplier $1/d_n$ is widely available for small n; here $d_n = E(W_n/\sigma)$. Let $\hat{\sigma}_s$ denote the unbiased root-mean-square* estimator of σ. It can be shown that in small samples the efficiency of $\hat{\sigma}_w$, i.e., $\mathrm{var}\,\hat{\sigma}_s/\mathrm{var}\,\hat{\sigma}_w$, is quite high (e.g., 0.955 for $n = 5$); also $\hat{\sigma}_w$ is at least as robust as $\hat{\sigma}_s$ to departures from normality. For such investigations, knowledge of the mean and variance of W_n/σ under normality and under alternatives is needed. (*See also* RANGES.)

Proceeding from the range to more general linear functions of the order statistics $L_n = \sum_{i=1}^{n} a_{in} X_{i:n}$, we see the need for tables of expected values, variances, and covariances of order statistics for various important standard populations so that $E(L_n)$ and $\mathrm{var}(L_n)$ can be obtained. Extensive tables are available in the normal case [19, 38]. For other populations, see the listings in David [7, Appendix Section 3.1].

A major use of such tables arises in the estimation of location and scale parameters μ and σ (not necessarily mean and standard deviation) for distributions with PDF of the form

$$f_X(x) = \frac{1}{\sigma} g\left(\frac{x-\mu}{\sigma}\right), \qquad \sigma > 0. \quad (5)$$

Here we suppose that for the standardized

variate $Y = (X - \mu)/\sigma$, tables are available giving $\alpha_{r:n} = E(Y_{r:n})$ and $\beta_{rs:n} = \mathrm{cov}(Y_{r:n}, Y_{s:n})$ $(r,s = 1, 2, \ldots, n)$. It follows that $Y_{r:n} = (X_{r:n} - \mu)/\sigma$, so that

$$E(X_{r:n}) = \mu + \sigma\alpha_{r:n},$$
$$\mathrm{cov}(X_{r:n}, X_{s:n}) = \sigma^2 \beta_{rs:n}.$$

Thus $E(X_{r:n})$ is linear in the parameters μ and σ with known coefficients, and $\mathrm{cov}(X_{r:n}, X_{s:n})$ is known apart from σ^2. Therefore, the Gauss–Markov least-squares* theorem may be applied, in a slightly generalized form since the covariance matrix is not diagonal [25]. This gives the best linear unbiased estimators (BLUEs)

$$\mu^* = \sum_{i=1}^{n} a_{in} X_{i:n} \quad \text{and} \quad \sigma^* = \sum_{i=1}^{n} b_{in} X_{i:n},$$

where the coefficients a_{in} and b_{in}, which are functions of the $\alpha_{i:n}$ and $\beta_{ij:n}$, can be evaluated once and for all. In the normal case, Sarhan and Greenberg [34, pp. 218–251] provide, for $n \leqslant 20$, tables of coefficients covering also all forms of type II censoring with the lowest r_1 and the highest r_2 of the $X_{i:n}$ missing $(0 < r_1 + r_2 < n)$.

In cases where exact values of the β_{ij} are not available, approximations supported by asymptotic theory may be substituted [4]. This leads to *unbiased nearly best* linear estimators; even the $\alpha_{i:n}$ may be approximated to give *nearly unbiased nearly best* linear estimators. Note that all these estimators are linear functions of the order statistics (*L* statistics* or *L* estimators).

It is obvious that the results of this section depend on the underlying CDF $F_X(x)$. The subject of order statistics is therefore *not* a branch of nonparametric or distribution-free statistics. Nevertheless, there is some overlap, the most important instance occurring in the setting of confidence limits for the population median $\xi_{1/2}$ of a continuous population. Since, as is easily shown,

$$\Pr\left[X_{r:n} < \xi_{1/2} < X_{n-r+1:n} \right] = 2^{-n} \sum_{i=r}^{n-r} \binom{n}{i},$$

$$r \leqslant \tfrac{1}{2} n,$$

we see that the confidence interval $(X_{r:n}, X_{n-r+1:n})$ covers the median with a probability that does not involve $F_X(x)$. For generalizations of this result and also for distribution-free tolerance intervals, see, e.g., David [7, pp. 15–19].

Properties of order statistics can, in fact, characterize the underlying distribution. For example, if X_1 and X_2 are independent variates with common absolutely continuous CDF $F_X(x)$ and if $X_{1:2}$ and $X_{2:2} - X_{1:2}$ are independent, then X is exponentially distributed. Results of this type are mainly of theoretical interest. See Galambos and Kotz [12, esp. Chap. 3]. (*See also* NONPARAMETRIC CONFIDENCE INTERVALS *and* NONPARAMETRIC TOLERANCE LIMITS.)

ASYMPTOTIC THEORY

As we have seen, the practical use of order statistics in finite samples depends heavily on the availability of tables appropriate for the underlying population. Consequently, there is much motivation for asymptotic theory, which may be able to provide approximate results that are both simpler and more widely applicable.

First we consider the distribution of $X_{r:n}$, suitably normed, as $n \to \infty$. We assume that $X_1, X_2, \ldots, X_n$ are independent with common CDF $F_X(x)$. Fortunately it turns out that many "mild" kinds of departures from these assumptions do not disturb the form of the limiting distribution, a feature that adds greatly to the usefulness of the theory. If $r = [n\lambda] + 1$, where $[x]$ denotes the integral part of x, fundamentally different results are obtained according to whether: (a) $0 < \lambda < 1$ or (b) $\lambda = 0$ or 1.

Under (a), $X_{r:n}$ is said to be a *sample quantile** and, subject to minimal regularity conditions, it has an asymptotic normal distribution centered at the corresponding population quantile $\xi_\lambda = F_X^{-1}(\lambda)$. We state a multivariate version of the result, proved earlier under stronger conditions by Mosteller [27] in a form due to Ghosh [13].

Theorem 1. Let $n_j = [n\lambda_j] + 1$, with $j = 1, 2, \ldots, k$ (fixed) and $0 < \lambda_1 < \lambda_2 < \cdots < \lambda_k < 1$. If $0 < p(\xi_{\lambda_j}) < \infty$, $j = 1, 2, \ldots k$, then the asymptotic joint distribution of

$$n^{1/2}(X_{n_1:n} - \xi_{\lambda_1}), \ldots, n^{1/2}(X_{n_k:n} - \xi_{\lambda_k})$$

is k-dimensional normal with zero mean vector and covariance matrix

$$\frac{\lambda_j(1 - \lambda_{j'})}{f_X(\xi_{\lambda_j})f_X(\xi_{\lambda_{j'}})}, \quad j \leq j'.$$

The situation is more complex for (b), the extreme-value case, which has been discussed in detail by Galambos [11]. A famous result of Fisher and Tippett [10] is that any limiting distribution of $X_{n:n}$, suitably normed, must take one of three forms:

$$
\begin{aligned}
\Lambda_1(x) &= 0 & x &\leq 0 \\
&= \exp(-x^{-\alpha}) & x &> 0, \ \alpha > 0; \\
\Lambda_2(x) &= \exp\left[-(-x)^\alpha\right] & x &\leq 0, \ \alpha > 0, \\
&= 1 & x &> 0; \\
\Lambda_3(x) &= \exp(-e^{-x}) & -\infty &< x < \infty
\end{aligned}
$$

(6)

Gnedenko [15] provides necessary and sufficient conditions on $F_X(x)$ for the distribution of $(X_{n:n} - a_n)/b_n$ to tend to one of the preceding extreme-value distributions*, where a_n and $b_n > 0$ are norming constants. Alternative conditions on $F_X(x)$ have been given by de Haan [8].

The simplest of the extreme-value distributions, Λ_3, has been found very useful in describing the distribution of floods and of other extreme meteorological phenomena [17, 20]. Results similar to (6) hold for the distribution of the normed minimum. Most important in this case is $\Lambda_2^*(x)$, given by

$$
\begin{aligned}
\Lambda_2^*(x) &= 0, & x &\leq 0, \\
&= 1 - \exp(-x^{-\alpha}), & x &> 0, \alpha > 0.
\end{aligned}
$$

This is just the Weibull distribution* (in standardized form), which is well known to provide a good fit to the distribution of the strength of materials that break at their areas of minimum strength.

Linear Functions of Order Statistics (L-statistics)

From Theorem 1, it is clear that a linear function of a finite number of quantiles must be asymptotically normally distributed. On the other hand, limiting distributions of the extremes are nonnormal. In view of the importance of L-statistics as estimators of location and scale, the asymptotic behavior of

$$L_n = \sum_{i=1}^{n} c_{in} X_{i:n}$$

is of considerable interest. Asymptotic normality of L_n requires suitable conditions on both the c_{in} and the form of $F_X(x)$. Various sets of conditions have been developed, some severe on the c_{in} and weak on $F_X(x)$, others the reverse (e.g., Chernoff et al. [6], Stigler [37], Shorack [36], Ruymgaart and van Zuijlen [33], and Boos [5]). For an overview, see Serfling [35, Chap. 8].

It is convenient in this context to write L_n as

$$L_n = \frac{1}{n} \sum_{i=1}^{n} J\left(\frac{i}{n}\right) X_{i:n},$$

where the weight-generating function $J(u)$ is a function of u ($0 \leqslant u \leqslant 1$) such that $J(i/n) = nc_{in}$. We give a widely applicable result due to Stigler [37], who uses Hájek's projection lemma* to represent L_n as a linear combination of iid random variables plus an asymptotically negligible remainder term. Let

$$\sigma^2(J, F)$$

$$= 2 \iint\limits_{-\infty < x < y < \infty} J[F(x)]J[F(y)]$$

$$\times F(x)[1 - F(y)] \, dx \, dy.$$

Theorem 2. Assume that $E(X^2) < \infty$ and that $J(u)$ is bounded and continuous a.e. F^{-1}. Then

$$\lim_{n \to \infty} n\sigma^2(L_n) = \sigma^2(J, F).$$

Also, if $\sigma^2(J, F) > 0$, then

$$\lim_{n \to \infty} \left[L_n - E(L_n) \right] / \sigma(L_n) \xrightarrow{L} N(0, 1).$$

Here $\xrightarrow{L}$ means *tends in law*. Note that neither $J(u)$ nor $F_X(x)$ need be continuous. The condition $J(u) \ldots$ continuous a.e. F^{-1} means that $J(u)$ must not be discontinuous at a point u' if there is a discontinuity in F at a point x' given by $F_X(x') = u'$.

The question of how to choose the c_{in} so as to make L_n an efficient estimator, solved for finite n by Lloyd [25], has been considered by Bennett [3] and Chernoff et al. [6]. As an example of their results, consider X to be normal $N(\mu, \sigma^2)$. Then the asymptotically efficient estimator of σ is

$$\sum_{i=1}^{n} \Phi^{-1}\left(\frac{i}{n+1} \right) X_{i:n}.$$

FURTHER APPLICATIONS

Some major easily described applications of order statistics have been discussed at the beginning of this entry. We now draw attention to other areas of statistics in which order statistics play a significant role.

Treatment of Outliers*

The proper treatment of outlying observations has long been a subject of interest [21] and one in which order statistics enter naturally. For example, the standardized extreme deviate $B = (X_{n:n} - \overline{X})/\sigma$ is clearly sensitive to the presence of an observation having a larger mean than the remainder of the sample. When $X_1, X_2, \ldots, X_n$ are independent $N(\mu, \sigma^2)$ variates (the null case most studied), percentage points of B can be obtained [16]. If the observed value of B is large enough, the null hypothesis is rejected and $X_{n:n}$ becomes a candidate for further examination or rejection. If σ is not known, a suitably studentized form of $X_{n:n} - \overline{X}$ may be used and referred to appropriate tables (e.g., Pearson and Hartley [31]). Two-sided versions of these procedures are also available. The subject has many ramifications and has recently received book-length treatment by Barnett and Lewis [2]. See also the monograph by Hawkins [23].

Robust Estimation

Here one gives up the classical idea of using an estimator that is optimal under tightly specified conditions in favor of an estimator that retains reasonably good properties under a variety of departures from the hypothesized situation. Since the largest and smallest observations tend to be more influenced by such departures, including the presence of outliers, a good many robust estimators are based on the more central order statistics. A simple example is provided by the trimmed means of (1). These are among a large number of estimators of location, some quite complex and not necessarily involving order statistics, that are studied, mainly using Monte Carlo methods*, in Andrews et al. [1]. The emphasis is on the behavior of the estimators over a range of symmetrical parent distributions.

Simultaneous Inference and Multiple Decision Procedures

Order statistics play an important supporting role in many of the procedures falling under the preceding two related headings. A well-known example is provided by Tukey's [40] use of the studentized range* in setting simultaneous confidence intervals for the "treatment" means μ_i ($i = 1, 2, \ldots, n$) in a fixed-effects balanced one-way classification* of X_{ij} ($j = 1, 2, \ldots, m$). With the usual assumptions, let S_ν^2 denote the error mean square on ν degrees of freedom and $q_{n,\nu;\alpha}$ (q_α for short) the upper α significance point of the studentized range W_n/S_ν. Then the inequality

$$\sqrt{m}\,\text{range}(\overline{X}_i - \mu_i) < q_\alpha S_\nu$$

that holds with probability $1 - \alpha$, is equiva-

lent to

$$\sqrt{m}\,|\overline{X}_i - \overline{X}_{i'} - (\mu_i - \mu_{i'})| < q_\alpha S_\nu$$

$$\text{for all} \quad i, i' = 1, 2, \ldots, k.$$

This is in turn equivalent to the $\binom{n}{2}$ simultaneous confidence statements

$$\overline{X}_i - \overline{X}_{i'} - q_\alpha S_\nu/\sqrt{m}$$

$$< \mu_i - \mu_{i'} < \overline{X}_i - \overline{X}_{i'} + q_\alpha S_\nu/\sqrt{m},$$

which must also hold with probability $1 - \alpha$. In fact, since for $\sum c_i = 0$, $\sum |c_i| = 2$, we have

$$\left|\sum c_i \overline{X}_i\right| \leqslant \overline{X}_{(n)} - \overline{X}_{(1)},$$

it follows that the infinity of confidence statements

$$\sum c_i \overline{X}_i - q_\alpha S_\nu/\sqrt{m} < \sum c_i \mu_i$$

$$< \sum c_i \overline{X}_i + q_\alpha S_\nu/\sqrt{m}$$

hold simultaneously with probability $1 - \alpha$.

A general account of simultaneous inference is given in Miller [26]. (*See also* MULTIPLE COMPARISONS.)

If in the balanced one-way classification we wish to test the null hypothesis of equal μ's against the "slippage" alternative,

$$\mu_1 = \mu_2 = \cdots = \mu_{i-1} = \mu_{i+1} = \cdots = \mu_n$$

$$\text{and} \quad \mu_i = \mu_1 + \Delta,$$

where i and $\Delta > 0$ are unknown, we may take one of the $n + 1$ decisions, that is, D_0: H_0 holds or D_i: the ith treatment group has slipped to the right. Paulson [30] has shown that for this multiple decision procedure an optimal method for detecting slippage is: Let

$$R = m(\overline{X}_M - \overline{X})/(\text{TSS})^{1/2}.$$

If

$$\begin{cases} R < b_\alpha, & \text{choose } D_0; \\ R > b_\alpha, & \text{choose } D_M. \end{cases}$$

Here M is the subscript corresponding to the largest $\overline{X}_i$ ($i = 1, 2, \ldots, n$), TSS is the analysis of variance total sum of squares, and b_α is the upper α significance point of R (on H_0). It will be seen that R is essentially the extreme deviate, studentized in a special way; b_α is given in Table 26a of Pearson and Hartley [31].

For extensive treatments of multiple decision procedures see Gibbons et al. [14] and Gupta and Panchapakesan [18].

Data Compression and Optimal Spacing

Given a large random sample of size n from a population with PDF (5), we may wish to estimate μ or σ or both from a fixed small number k of order statistics. There are interesting possibilities for data compression here [9] since a large sample (e.g., of particle counts taken on a space craft) may be replaced by enough order statistics to allow (on the ground): (a) satisfactory estimation of parameters; (b) a test of the assumed underlying distribution by probability plotting or otherwise.

With the help of Theorem 1, Ogawa [29] has shown how to find the ranks $n_1, n_2, \ldots, n_k$ of the k order statistics that provide the most efficient estimator of μ (say) for the assumed population. In the normal case, the estimator corresponding to this optimal spacing is, for $k = 4$,

$$\mu^* = 0.1918\left(X_{(0.1068n)} + X_{(0.8932n)}\right)$$
$$+ 0.3082\left(X_{(0.3512n)} + X_{(0.6488n)}\right),$$

where $(0.1068n)$ is to be interpreted as the integral part of $0.1068n + 1$, etc. The estimator μ^* has asymptotic efficiency (A.E.) 0.920 and, since it does not involve the more extreme order statistics, is much more robust than $\bar{X}$. Likewise one has

$$\sigma^* = 0.116\left(X_{(0.9770n)} - X_{(0.0230n)}\right)$$
$$+ 0.236\left(X_{(0.8729n)} - X_{(0.1271n)}\right),$$

with A.E. 0.824. Estimators μ_0^* and σ_0^* corresponding to the *same* k order statistics can be found by minimizing var $\mu_0^* + c$ var σ_0^*, where c is a predetermined constant. Normal-theory tables for even $k \leqslant 20$ and $c = 1, 2, 3$ are given in ref. 9.

Probability Plotting*

This is a very useful graphical method typically applied to data assumed to follow a distribution depending only on location and scale parameters μ and σ. Plot the ordered observations $x_{i:n}$ ($i = 1, 2, \ldots, n$) against $v_i = G^{-1}(p_i)$, where G is the CDF of $Y = (X - \mu)/\sigma$ and the p_i are probability levels, such as $p_i = (i - 1/2)/n$. The plotting is facilitated if probability paper corresponding to G is available, but such paper is by no means essential. Now fit a straight line by eye through the points $(v_i, x_{i:n})$. If such a fit seems unreasonable, doubt is cast on the appropriateness of G. Often this simple procedure is adequate, but it may be followed up by a more formal test of goodness of fit*. Given a satisfactory straight-line fit, a second use may be made of the graph: σ may be estimated by the slope of the line and μ by its intercept on the vertical axis.

For a detailed account of probability plotting, and the related hazard plotting*, see Nelson [28].

Historical Remarks

Because of their simplicity, order statistics have a very long history. An extreme case is the sample range whose use has been traced back as far as Ptolemy (circa A.D. 150). However, a derivation of the *distribution* of the range in random samples had to wait until 1942 [22]. The history of the use and gradual theoretical development of the range and many other functions of order statistics is well charted in the annotated bibliography of Harter [21].

References

The number of references has been kept down, in view of ref. 7, which provides more details on most aspects of this entry and includes a guide to tables. For an earlier account, with tables, see ref. 34. References 31 and 32 contain many relevant tables and 19 consists of two volumes of tables confined to order statistics. Many aspects of order statistics referring to specific distributions are treated in Johnson and Kotz [24].

[1] Andrews, D. F., Bickel, P. J., Hampel, F. R., Huber, P. J., Rogers, W. H. and Tukey, J. W.

(1972). *Robust Estimates of Location.* Princeton University Press, Princeton, NJ.

[2] Barnett, V. and Lewis, T. (1978). *Outliers in Statistical Data.* Wiley, Chichester, England.

[3] Bennett, C. A. (1952). "Asymptotic Properties of Ideal Linear Estimators." Ph.D. thesis, University of Michigan, Ann Arbor, MI.

[4] Blom, G. (1958). *Statistical Estimates and Transformed Beta-Variables.* Almqvist and Wiksell, Uppsala, Sweden; Wiley, New York.

[5] Boos, D. D. (1979). *Ann. Statist.*, **7**, 955–959.

[6] Chernoff, H., Gastwirth, J. L. and Johns, M. V., Jr. (1967). *Ann. Math. Statist.*, **38**, 52–72.

[7] David, H. A. (1981). *Order Statistics*, 2nd ed. Wiley, New York.

[8] de Haan, L. (1970). *On Regular Variation and Its Application to the Weak Convergence of Sample Extremes*, Mathematical Centre Tracts Vol. **32**. Mathematisch Centrum, Amsterdam, Netherlands.

[9] Eisenberger, I. and Posner, E. C. (1965). *J. Amer. Statist. Ass.*, **60**, 97–133.

[10] Fisher, R. A. and Tippett, L. H. C. (1928). *Proc. Camb. Philos. Soc.*, **24**, 180–190.

[11] Galambos, J. (1978). *The Asymptotic Theory of Extreme Order Statistics.* Wiley, New York.

[12] Galambos, J. and Kotz, S. (1978). *Characterizations of Probability Distributions.* Lecture Notes in Mathematics No. 675. Springer, Berlin and New York.

[13] Ghosh, J. K. (1971). *Ann. Math Statist.*, **42**, 1957–1961.

[14] Gibbons, J. D., Olkin, I. and Sobel, M. (1977). *Selecting and Ordering Populations: A New Statistical Methodology.* Wiley, New York.

[15] Gnedenko, B. (1943). *Ann. Math.*, **44**, 423–453.

[16] Grubbs, F. E. (1950). *Ann. Math. Statist.*, **21**, 27–58.

[17] Gumbel, E. J. (1958). *Statistics of Extremes.* Columbia University Press, New York.

[18] Gupta, S. S. and Panchapakesan, S. (1979). *Multiple Decision Procedures: Theory and Methodology of Selecting and Ranking Populations.* Wiley, New York.

[19] Harter, H. L. (1970). *Order Statistics and Their Uses in Testing and Estimation*, Vols. 1 and 2. U.S. GPO, Washington, DC.

[20] Harter, H. L. (1978). *Int. Statist. Rev.*, **46**, 279–306.

[21] Harter, H. L. (1978). *A Chronological Annotated Bibliography on Order Statistics*, Vol. 1. U.S. GPO, Washington, DC. (This bibliography is pre-1950.)

[22] Hartley, H. O. (1942). *Biometrika*, **32**, 334–348.

[23] Hawkins, D. M. (1980). *Identification of Outliers.* Chapman and Hall, London.

[24] Johnson, N. L. and Kotz, S. (1970). *Distributions in Statistics: Continuous Univariate Distributions— 1, 2.* Wiley, New York.

[25] Lloyd, E. H. (1952). *Biometrika*, **39**, 88–95.

[26] Miller, R. G. (1981). *Simultaneous Statistical Inference*, 2nd ed. Springer, New York.

[27] Mosteller, F. (1946). *Ann. Math. Statist.*, **17**, 377–408.

[28] Nelson, W. (1982). *Applied Life Data Analysis.* Wiley, New York.

[29] Ogawa, J. (1951). *Osaka Math. J.*, **3**, 175–213.

[30] Paulson, E. (1952). *Ann. Math Statist.*, **23**, 610–616.

[31] Pearson, E. S. and Hartley, H. O. (1970). *Biometrika Tables for Statisticians*, Vol. I. 3rd ed. (with additions). Cambridge University Press, Cambridge, England.

[32] Pearson, E. S. and Hartley, H. O. (1972). *Biometrika Tables for Statisticians*, Vol. II. Cambridge University Press, Cambridge, England.

[33] Ruymgaart, F. H. and van Zuijlen, M. C. A. (1977). *Ned. Akad. Wet. Proc. A*, **80**, 432–447.

[34] Sarhan, A. E. and Greenberg, B. G., eds. (1962). *Contributions to Order Statistics.* Wiley, New York.

[35] Serfling, R. J. (1980). *Approximation Theorems in Mathematical Statistics.* Wiley, New York.

[36] Shorack, G. R. (1969). *Ann. Math. Statist.*, **40**, 2041–2050.

[37] Stigler, S. M. (1974). *Ann. Statist.*, **2**, 676–693; erratum, **7**, 466.

[38] Tietjen, G. L., Kahaner, D. K. and Beckman, R. J. (1977). *Select. Tables Math. Statist.*, **5**, 1–73.

[39] Tippett, L. H. C. (1925). *Biometrika*, **17**, 364–387.

[40] Tukey, J. W. (1953). "The Problem of Multiple Comparisons." Princeton University, Princeton, NJ (unpublished memorandum).

Acknowledgment

This work was supported by the U.S. Army Research Office.

(CENSORED DATA
DEPARTURES FROM NORMALITY,
  TESTS FOR
EXTREME-VALUE DISTRIBUTIONS
LINEAR ESTIMATORS, BAYES
L-STATISTICS
OUTLIERS
PROBABILITY PLOTTING
RANGES
ROBUST ESTIMATION
STUDENTIZED RANGE TESTS
TRIMMING AND WINSORIZATION)

H. A. DAVID

ORDER STATISTICS, CONCOMITANTS

OF   *See* MULTIVARIATE ORDER STATISTICS

ORDINAL DATA

An ordinal variable is one that has a natural ordering of its possible values, but for which the distances between the values are undefined. Ordinal variables usually have categorical scales. Examples are social class, which is often measured as upper, middle, or lower, and political philosophy, which might be measured as liberal, moderate, or conservative. Continuous variables that are measured using ranks are also treated as ordinal.

In this article we describe methods for analyzing only ordinal categorical variables. In particular, we summarize some association measures and models that are appropriate for the analysis of contingency tables* having at least one ordered classification. Methods for analyzing continuous observation variables are summarized in DISTRIBUTION-FREE STATISTICS and RANK TESTS.

ORDINAL MEASURES OF ASSOCIATION

We present three types of ordinal measures of association: measures based on the notions of concordance* and discordance, which utilize ordinal information only; correlation* and mean measures that require a user-supplied or data-generated scoring of ordered categories; sets of odds-ratio measures that contain as much information regarding association as the original cell counts. For a discussion of the rationale of measures of association, *see* ASSOCIATION, MEASURES OF.

Concordance–Discordance Measures

We discuss most of the methodology in this article in the context of a two-way contingency table*. Denote the cell counts of an $r \times c$ table by $\{n_{ij}\}$ and let $\{p_{ij} = n_{ij}/n\}$ be the corresponding cell proportions. Let X

denote the row variable and Y the column variable. For now, we suppose that the rows and columns are both ordered, with the first row and first column being the low ends of the two scales.

A pair of observations is *concordant* if the member that ranks higher on X also ranks higher on Y. A pair of observations is *discordant* if the member that ranks higher on X ranks lower on Y. The numbers of concordant and discordant pairs are

$$C = \sum_{i' > i} \sum_{j' > j} n_{ij} n_{i'j'} \quad \text{and}$$

$$D = \sum_{i' < i} \sum_{j' > j} n_{ij} n_{i'j'} .$$

Let $n_{i+} = \sum_j n_{ij}$ and $n_{+j} = \sum_i n_{ij}$. We can express the total number of pairs of observations as

$$n(n-1)/2 = C + D + T_X + T_Y - T_{XY},$$

where $T_X = \sum_i n_{i+}(n_{i+} - 1)/2$ is the number of pairs tied on X, $T_Y = \sum_j n_{+j}(n_{+j} - 1)/2$ is the number of pairs tied on Y, and $T_{XY} = \sum n_{ij}(n_{ij} - 1)/2$ is the number of pairs from a common cell (tied on X and Y).

Several measures of association* are based on the difference $C - D$. They are discrete generalizations of the Kendall's tau* measure for continuous variables. For each, the greater the relative number of concordant pairs, the more evidence there is of a positive association.

Of the untied pairs, $C/(C + D)$ is the proportion of concordant pairs and $D/(C + D)$ is the proportion of discordant pairs. The measure *gamma*, proposed by Goodman and Kruskal [9], is the difference between these proportions,

$$\hat{\gamma} = (C - D)/(C + D).$$

For 2×2 tables $\hat{\gamma}$ is also referred to as Yule's Q. In 1945, Kendall [14] proposed the related measure *tau-b* given by

$$\hat{\tau}_b = \frac{C - D}{\left[\left\{ \frac{1}{2} n(n-1) - T_X \right\} \left\{ \frac{1}{2} n(n-1) - T_Y \right\} \right]^{1/2}} .$$

For 2×2 tables $\hat{\tau}_b$ simplifies to the Pearson correlation obtained by assigning any scores to the rows and to the columns that reflect their orderings.

Gamma and tau-b assume the same values regardless of whether X or Y (or neither) is regarded as a response variable. In 1962, Somers proposed the asymmetric measure

$$d_{YX} = (C - D)/[n(n-1)/2 - T_X],$$

the difference between the proportions of concordant and discordant pairs, out of those pairs that are untied on X. For 2×2 tables d_{YX} simplifies to the difference of proportions $n_{11}/n_{1+} - n_{21}/n_{2+}$. For $2 \times c$ tables, it estimates $P(Y_2 > Y_1) - P(Y_1 > Y_2)$, where Y_1 and Y_2 are independent observations on the column variable in rows one and two of the tables, respectively.

All three of these ordinal measures are restricted to the range $[-1, +1]$. Independence implies that their population values equal zero, but the converse is not true. Note that $|\hat{\tau}_b| \leqslant |\hat{\gamma}|$ and $|d_{YX}| \leqslant |\hat{\gamma}|$, and $\hat{\tau}_b^2 = d_{YX} d_{XY}$, where d_{XY} has T_Y instead of T_X in its denominator. Tau-b may be interpreted as a Pearson correlation and Somers' d may be interpreted as a least-squares* slope for a linear regression* model defined using sign scores for pairs of observations.

Measures Based on Scores

Many methods for analyzing ordinal data require assigning scores to the levels of ordinal variables. To compute the Pearson correlation* between the row and column variables, e.g., one must assign fixed scores to the rows and to the columns. The canonical correlation is the maximum correlation obtained out of all possible choices of scores. The scores needed to achieve the maximum need not be monotone, however. Alternatively, one can generate monotone scores from the data. For example, one could use average cumulative probability scores, which for the column (Y) marginal distribution are

$$r_j = \sum_{i=1}^{j-1} p_{+i} + p_{+j}/2, \qquad j = 1, \ldots, c.$$

The correlation measure then obtained is a discrete analog of Spearman's rank correlation coefficient, referred to as $\hat{\rho}_b$ (see Kendall [14, p. 38]). Or one could use scores at

which a distribution function (such as the normal or logistic) takes on the $\{r_j\}$ values.

If X is nominal (unordered levels) and Y is ordinal, often it is useful to compute a mean score on Y within each level of X. The scores $\{r_j\}$ just defined are called *ridits* for the marginal distribution of Y. The measure $\bar{R}_i = \sum_j r_j(n_{ij}/n_{i+})$ is the sample mean ridit in row i. It estimates

$$P(Y_i > Y^*) + \tfrac{1}{2} P(Y_i = Y^*),$$

where Y_i and Y^* are categories of Y for observations randomly selected from row i and from the marginal distribution of Y, respectively. It is necessary that $\sum_j p_{+j} r_j = \sum_i p_{i+} \bar{R}_i = 0.5$. See Bross [4] for a discussion of ridit analysis*. For an example of a scaling method that assumes a particular form for an underlying continuous distribution, see Snell [21].

Odds-Ratio* Measures

The measures discussed summarize association by a single number. To avoid the loss of information we get by this condensation, we can describe the table through a set of $(r-1)(c-1)$ odds ratios. For ordinal variables, it is natural to form the *local odds ratios*

$$\hat{\theta}_{ij} = n_{ij} n_{i+1,j+1}/(n_{i,j+1} n_{i+1,j}),$$
$$i = 1, \ldots, r-1, \quad j = 1, \ldots, c-1.$$

Each $\hat{\theta}_{ij}$ describes the sample association in a restricted region of the table, with $\log \hat{\theta}_{ij}$ indicating whether the association is positive or negative in that region. Goodman [7] suggested log-linear models for analyzing the $\{\hat{\theta}_{ij}\}$. An alternative set of odds ratios is based on the $(r-1)(c-1)$ ways of collapsing the table into a 2×2 table.

ORDINAL MODELS

In recent years, much work has been devoted to formulating models for cross-classifications of ordinal variables. The models discussed here are directly related to standard log-linear and logit models (*see*

CONTINGENCY TABLES and MULTIDIMENSIONAL CONTINGENCY TABLES).

Log-linear Models

Suppose that $\{\rho_{ij}\}$ denotes the true cell proportions in an $r \times c$ contingency table, where $\sum \rho_{ij} = 1$. For a random sample of size n, the expected number of observations in a cell is $m_{ij} = n\rho_{ij}$. If the variables are independent, then $m_{ij} = m_{i+}m_{+j}$ for all i and j. There is a corresponding additive relationship for $\log m_{ij}$. That is, we can describe independence by the log-linear model $\log m_{ij} = \mu + \lambda_i^X + \lambda_j^Y$, where μ is the mean of the $\{\log m_{ij}\}$ and $\sum \lambda_i^X = \sum \lambda_j^Y = 0$. Haberman [12], Simon [20], and Goodman [7] have formulated more complex log-linear models for situations where at least one variable is ordinal and there is some association.

The log-linear models can be described in terms of properties of the local odds ratios $\{\theta_{ij} = (m_{ij}m_{i+1,j+1})/(m_{i,j+1}m_{i+1,j})\}$. A simple model has the form $\log \theta_{ij} = \beta$ for all i and j, whereby the local association is uniform throughout the table. A more general model is obtained by assigning monotone scores $\{u_i\}$ to the rows and $\{v_j\}$ to the columns and assuming that

$$\log m_{ij} = \mu + \lambda_i^X + \lambda_j^Y + \beta u_i v_j.$$

In this model

$$\log \theta_{ij} = \beta(u_{i+1} - u_i)(v_{j+1} - v_j).$$

When the $\{u_i\}$ are equally spaced and the $\{v_j\}$ are equally spaced, we obtain the uniform association model. When $\beta = 0$, we obtain the independence model. The goodness of fit of the uniform association model can be tested with a chi-squared statistic* having $rc - r - c$ degrees of freedom.

Goodman [7] discussed several other models that include the uniform association model as a special case. A row effects model has the property $\log \theta_{ij} = \alpha_i$ for all i and j. The row variable may be nominal for this model, which can be tested with a chi-squared statistic having $(r - 1)(c - 2)$ degrees of freedom. This model is itself a special case of two row and column effects mod-

els, one for which $\log \theta_{ij} = \alpha_i + \beta_j$ and the other of which has the multiplicative form $\log \theta_{ij} = \alpha_i \beta_j$. These models have $(r - 2)(c - 2)$ residual degrees of freedom. Analogous models can be formulated for multidimensional tables. See Clogg [5] for details.

These log-linear models treat the variables alike in the sense that no variable is identified as a response. Iterative methods are necessary to obtain maximum likelihood estimates of parameters and goodness-of-fit statistics for these models. See the sections on Estimation and Computer Packages.

Logit Models*

Suppose now that an ordinal variable Y is a response variable and let $\mathbf{X}$ denote explanatory variables. Let $\rho_i(\mathbf{x})$ denote the probability that Y falls in category i when $\mathbf{X} = \mathbf{x}$, where $\sum_{i=1}^c \rho_i(\mathbf{x}) = 1$. When $c = 2$, the logit transformation is $\log[\rho_2(\mathbf{x})/\rho_1(\mathbf{x})]$. The linear logit regression model

$$\log[\rho_2(\mathbf{x})/\rho_1(\mathbf{x})] = \alpha + \boldsymbol{\beta}'\mathbf{x}$$

is one that yields predicted values of $\rho_i(\mathbf{x})$ between 0 and 1, the relationship being S-shaped between $\rho_i(\mathbf{x})$ and each x_i (see LOGIT).

When there are $c > 2$ responses, there are several ways of forming logits that take the ordering of the categories into account. The *cumulative logits*

$$L_j = \log\left[\sum_{i>j} \rho_i(\mathbf{x}) \Big/ \sum_{i \leqslant j} \rho_i(\mathbf{x})\right],$$
$$j = 1, \ldots, c - 1,$$

are logits of distribution function values and lend themselves nicely to interpretation. Williams and Grizzle [22] and McCullagh [17] have suggested models for them.

We illustrate with a logit model for a two-way table having column variable Y as a response. The jth cumulative logit in row i is

$$L_{ij} = \log\left(\frac{\rho_{i,j+1} + \cdots + \rho_{ic}}{\rho_{i1} + \cdots + \rho_{ij}}\right),$$

$i = 1, \ldots, r, j = 1, \ldots, c - 1$. Suppose that X is also ordinal and that we assign scores

$\{u_i\}$ to its levels. A simple linear model is

$$L_{ij} = \alpha_j + \beta u_i,$$

$$i = 1, \ldots, r, \quad j = 1, \ldots, c - 1.$$

This model implies that the effect β of X on the logit for Y is the same for all cut points $j = 1, \ldots, c - 1$ for forming the logit. For the integer scores $\{u_i = i\}$, $L_{i+1,j} - L_{ij} = \beta$ for all i, j. Thus this logit model can also be regarded as a type of uniform association model. In this case, β is a log odds ratio formed using adjacent rows when the response is collapsed into two categories. Like the log-linear uniform association model, it has $rc - r - c$ residual degrees of freedom for testing goodness of fit*.

Logit models for multidimensional tables can be constructed like multiple regression models by including terms for qualitative and quantitative explanatory variables. Iterative methods are needed for maximum likelihood estimation of the models, as described in the sections on Estimation and Computer Packages.

Models for Square Tables

In some applications, each classification in a table has the same categories. This happens, for example, for matched-pairs data such as occur in social mobility tables. Cell probabilities in square tables often exhibit a type of symmetry relative to the main diagonal. Also, when the categories are ordered, it is often of interest to study whether one marginal distribution tends to have larger responses, in some sense, than the other.

An example of the type of model that has been proposed for $r \times r$ ordinal tables is Goodman's [8] diagonals-parameter symmetry model,

$$m_{ij} = m_{ji}\delta_{j-i}, \quad i < j.$$

The parameter δ_k, $k = 1, \ldots, r - 1$ is the odds that an observation falls in a cell k diagonals above the main one instead of in a corresponding cell k diagonals below the main one. For the special case $\delta_1 = \cdots = \delta_{r-1} = \delta$, this model exhibits the conditional symmetry $P(X = i, Y = j \mid X < Y) = P(X = j, Y = i \mid X > Y)$. The further spe-

cial case, in which all $\delta_k = 1$, gives the symmetry model $m_{ij} = m_{ji}$, $i \neq j$. Each of these models can be expressed as a log-linear model and tested using standard chi-squared statistics. Whether the delta parameters in these models exceed one or are less than one determine how the marginal distributions are stochastically ordered.

There are several other log-linear models in which the effect of a cell on the association depends on its distance from the main diagonal. Also, standard log-linear models for ordinal variables (e.g., uniform association model) often fit square tables well when the main diagonal is deleted. See Haberman [13, pp. 500–503] and Goodman [8] for examples; *see also* MARGINAL SYMMETRY.

Other Models

Several alternative ways have been proposed for modeling ordinal variables. Some of these assume an underlying continuous distribution of a certain form. McCullagh [17] discussed a "proportional hazards" model that utilizes the $\log(-\log)$ transformation of the complement of the distribution function of the response variable. He argued that it would be appropriate for underlying distributions of the types used in survival analysis.

If one feels justified in assigning scores to the levels of an ordinal response variable, then one can construct simple models for the mean response that are similar to analysis of variance and regression models for continuous variables. This approach is especially appealing if the categorical nature of the ordinal response is due to crude measurement of an inherently continuous variable. Grizzle et al. [11] gave a general weighted least-squares approach for fitting models of this type. Similar models have been constructed for mean ridits* (see Semenya and Koch [19]).

INFERENCE FOR ORDINAL VARIABLES

In this section we discuss estimation of ordinal measures of association and models and describe ways of using the estimates to test certain basic hypotheses. We assume that the sample was obtained by full multinomial sampling or else by independent multinomial sampling within combinations of levels of explanatory variables.

Estimation

Under these sampling models, the measures of association discussed in the first section are asymptotically normally distributed. Goodman and Kruskal [10] applied the delta method (*see* STATISTICAL DIFFERENTIALS) to obtain approximate standard errors for these measures. Hence one can form confidence intervals for them.

The ordinal log-linear and logit models can be fit using weighted least squares* (WLS) or maximum likelihood* (ML). The WLS estimate has a simple closed-form expression. See Williams and Grizzle [22], e.g., for WLS estimation of the cumulative logit model.

The ordinal log-linear models discussed in the Log-linear Models section are special cases of generalized linear models* proposed by Nelder and Wedderburn [18]. The ML estimates may be obtained using the iterative Newton–Raphson method described in their paper, which corresponds to an iterative use of WLS. ML estimates can also be obtained using an iterative scaling approach given by Darroch and Ratcliff [6] or by using a Newton unidimensional iterative procedure suggested by Goodman [7]. The latter approaches are simpler than Newton–Raphson*, but convergence is much slower. McCullagh [17] showed how to use the Newton–Raphson method to obtain ML estimates for a class of models that includes the cumulative logit models.

Testing Hypotheses

Basic hypotheses concerning independence, conditional independence, and higher-order interactions can be tested using estimates of measures of association or estimates of certain model parameters. For example, consider the null hypothesis of independence for the ordinal–ordinal table. Goodman and Kruskal [10] showed that a broad class of measures of association have asymptotic normal distributions for multinomial sampling. In particular, an ordinal measure such as gamma or tau-b divided by its standard error has an asymptotic standard normal null distribution. This statistic will (as $n \to \infty$) detect associations where the true value of the measure is nonzero. If the logit or log-linear uniform association model holds, then independence is equivalent to $\beta = 0$. The estimate of β divided by its standard error also has an asymptotic standard normal null distribution. Alternatively, the difference in values of the likelihood-ratio statistics for testing goodness of fit of the independence model and the uniform association model has an asymptotic chi-squared distribution* with a single degree of freedom.

Similar remarks apply to the two-way table with r unordered rows and c ordered columns. Independence can be tested using a discrete version of the Kruskal–Wallis test, which detects differences in true mean ridits. If log-linear or logit row effects models fit the data, it can also be tested using the difference in likelihood-ratio statistics between the independence model and the row effects model. Each of the approaches gives a statistic that has an asymptotic null chi-squared distribution with $r - 1$ degrees of freedom. Analogous tests can be formulated for multidimensional tables.

Computer Packages

Several computer packages can be used for the computational aspects of analyzing ordinal data. Some of these are large, general-purpose statistical packages that have components or options for categorical data*. For example, the widely available package BMDP has a program (4F) that, among other things, computes several measures of association and their asymptotic standard errors. The package GLIM* is particularly useful for fitting log-linear models, including the ordinal ones mentioned in the Log-linear Models section. Other programs have been designed specifically for categorical data* and can be used for certain ordinal methods. These include FREQ [13] for ML estimation of log-linear models, MULTIQUAL [3] for ML fitting of log-linear and logit models, and GENCAT [16], which can be used to fit a large variety of models using WLS (see also the FUNCAT program in the SAS package). (*See also* STATISTICAL SOFTWARE.)

Summary

More detailed surveys of methods for analyzing ordinal data are given by Semenya and Koch [19] and by Agresti [2]. Ordinal measures of association have been surveyed by Goodman and Kruskal [9, 10], Kruskal [15], and Kendall [14]. Summary discussions of methods for modeling ordinal variables were presented by Goodman [7], McCullagh [17], Clogg [5], and Agresti [1].

References

[1] Agresti, A. (1984). *J. Amer. Statist. Ass.*, **78**, 184–198.

[2] Agresti, A. (1984). *Analysis of Ordinal Categorical Data*. Wiley-Interscience, New York.

[3] Bock, R. D. and Yates, G. (1973). *Log-linear Analysis of Nominal or Ordinal Qualitative Data by the Method of Maximum Likelihood*. National Education Resources, Chicago.

[4] Bross, I. D. J. (1958). *Biometrics*, **14**, 18–38.

[5] Clogg, C. (1982). *J. Amer. Statist. Ass.*, **77**, 803–815.

[6] Darroch, J. N. and Ratcliff, D. (1972). *Ann. Math. Statist.*, **43**, 1470–1480.

[7] Goodman, L. A. (1979). *J. Amer. Statist. Ass.*, **74**, 537–552. (An easy-to-read development of log-linear models based on local odds ratios.)

[8] Goodman, L. A. (1979). *Biometrika*, **66**, 413–418.

[9] Goodman, L. A., and Kruskal, H. (1954). *J. Amer. Statist. Ass.*, **49**, 723–764. (A classic paper on measures of association for ordinal and nominal variables.)

[10] Goodman, L. A. and Kruskal, W. H. (1972). *J. Amer. Statist. Ass.*, **67**, 415–421.

[11] Grizzle, J. E., Starmer, C. F., and Koch, G. G., (1969). *Biometrics*, **25**, 489–504. (A good exposition of the use of weighted least squares for fitting a wide variety of models to categorical data.)

[12] Haberman, S. J. (1974). *Biometrics*, **30**, 589–600.

[13] Haberman, S. J. (1979). *Analysis of Qualitative Data*, Vol. 2: *New Developments*. Academic Press, New York. (One of the few categorical data books that devotes much space to models for ordinal variables, but not easy reading.)

[14] Kendall, M. G. (1970). *Rank Correlation Methods*, 4th ed. Charles Griffin, London.

[15] Kruskal, W. H. (1958). *J. Amer. Statist. Ass.*, **53**, 814–861.

[16] Landis, J. R., Stanish, W. M., Freeman, J. L., and Koch, G. G. (1976). *Computer Programs Biomed.*, **6**, 196–231.

[17] McCullagh, P. (1980). *J. R. Statist. Soc. B*, **42**, 109–42. (Discusses important issues to be considered in modeling ordinal response variables.)

[18] Nelder, J. A. and Wedderburn, R. W. M. (1972). *J. R. Statist. Soc. A*, **135**, 370–384.

[19] Semenya, K. and Koch, G. G. (1980). *Institute of Statistics Mimeo Series No. 1323*, University of North Carolina, Chapel Hill, NC. (A good survey of the use of weighted least squares for fitting various models to ordinal data.)

[20] Simon, G. (1974). *J. Amer. Statist. Ass.*, **69**, 971–976.

[21] Snell, E. J. (1964). *Biometrics*, **20**, 592–607.

[22] Williams, O. D. and Grizzle, J. E. (1972). *J. Amer. Statist. Ass.*, **67**, 55–63.

(ASSOCIATION, MEASURES OF
CONTINGENCY TABLES
GOODMAN–KRUSKAL TAU
    AND GAMMA
LOGIT
NOMINAL DATA
ODDS-RATIO ESTIMATORS
RANKING PROCEDURES
RANK TESTS
SCALE TESTS)

ALAN AGRESTI

ORDINARY LEAST SQUARES (OLS)

See LEAST SQUARES

ORGANIZATION FOR ECONOMIC COOPERATION AND DEVELOPMENT (OECD)

The OECD is the Paris-based international organization of the industrialized, market-economy countries. Its membership includes the countries of Western Europe, Canada and the United States, Japan, Australia, and

New Zealand. At OECD, representatives from member countries meet to exchange information, to compare experiences, and to attempt to harmonize policy. To help member country representatives in these tasks, the OECD Secretariat compiles internationally standarized and comparable economic statistics. Much of this statistical work is published and made available to the public through the Publications Shop at OECD headquarters in Paris, the OECD Publications and Information Centers in Bonn, Tokyo, and Washington, and the OECD sales agents in various countries throughout the world.

The following statistics are published on an ongoing basis.

MAIN ECONOMIC INDICATORS. Monthly indicators of GNP; industrial production; deliveries, stocks, and orders; construction; wholesale and retail sales; employment and wages; prices; finance; foreign trade; and balance of payments for all OECD countries. (GNP = gross national product.)

NATIONAL ACCOUNTS. A quarterly publication featuring GNP, gross capital formation, and private consumption expenditure tables for the United States and 11 other OECD countries. Annual data is published in two volumes each year. The main aggregates volume provides a 30-year run of data, and includes a set of comparative tables in U.S. dollars and in purchasing power parities. *The Detailed Tables* (Vol. II) includes the following tables for each country: main aggregates; GNP by activity; government and private final consumption expenditure according to purpose; distribution of national disposable income; financing of gross capital formation; income and outlay transactions of households, government, and business; and external transactions. Figures are in each country's national currency.

INDICATORS OF INDUSTRIAL ACTIVITY. Production, deliveries, orders, prices, and employment indicators on a quarterly basis for major industries in OECD countries.

LABOR FORCE STATISTICS. Quarterly and annual publications providing figures on the size of the labor force, unemployment, etc. in OECD countries.

OECD FINANCIAL STATISTICS. Includes, on a monthly basis, figures on interest rates, bond and security issues, and international bank loans; and on an annual basis, flow-of-funds and balance sheet accounts broken down by institutional sector and financial instrument, and nonfinancial enterprises' financial statements.

FINANCIAL MARKET TRENDS. Information on the international and major domestic financial markets of the OECD area, including commentary statistics and charts on current developments in Eurocredits and Eurobonds. Published three times a year.

STATISTICS OF FOREIGN TRADE*. The monthly bulletin features monthly and quarterly trade totals of each OECD country broken down by main commodity categories and by trading partner. The annual volume features the year's trade of each OECD country with every partner country broken down by SITC commodity code to the three-digit level. Information to the five-digit level is available on microfiche and magnetic tape.

REVENUE STATISTICS. Annual figures on national, state, local, and social security tax revenues in OECD countries.

DEVELOPMENT COOPERATION. Annual figures on foreign aid given by OECD Development Assistance Committee members.

GEOGRAPHICAL DISTRIBUTION OF FINANCIAL FLOWS TO DEVELOPING COUNTRIES. Shows aid and resource flows received by each developing country from individual, bilateral, and multilateral sources. Also gives external debt position of each country. Figures are issued annually. Annual reports featuring production and trade statistics, and in some cases prices and employment, are published for the following industries: chemical, engineering, fisheries, meat, dairy, leather and footwear, iron and steel, and nonferrous metals. The annual report on tourism policy and international tourism gives figures on tourism flows, international tourism receipts and expenditures, air traffic, occupancy rates, accommodation facilities, industry employment, and prices. The annual report on maritime transport presents statistics on fleets and freight carried.

Every two years OECD's Nuclear Energy Agency (NEA) publishes *Uranium Resources, Production, and Demand*, which provides detailed uranium statistics from 40 uranium-producing countries.

OECD's International Energy Agency (IEA) publishes oil statistics, on a quarterly and an annual basis and, on annual basis, energy statistics, energy balances (all figures are in million tons of oil equivalent), and crude oil import prices. In addition, the annual *Energy Policies and Programmes of IEA Countries* provides data on each member country's energy use.

Besides these statistical series, which are updated on a regular basis, OECD publishes a wide variety of monographs that present statistics on many subjects. The annual *OECD Economic Survey* of each member country provides a good statistical overview of each member country. The biannual *OECD Economic Outlook* provides the statistics on aspects of the economy that are most likely to be important in the next 6 to 18 months. Consult the *OECD Catalog of Publications* for a complete list of available titles.

Most OECD statistics that are produced on an ongoing basis, with the exception of the industry reports, are available on magnetic tape. All published OECD statistics are indexed in the *Index to International Statistics* published by the Congressional Information Service. Persons wishing to be kept informed about OECD publications can receive *OECD Recent Publications* each quarter by writing to:

OECD Publications and Information Center
1750 Pennsylvania Avenue, NW
Washington, DC 20006-4582.
Publications orders and information requests can also be sent to that office.

(FINANCE, STATISTICS IN
FOREIGN TRADE STATISTICS,
INTERNATIONAL)

H. De Vroom

ORNSTEIN–UHLENBECK PROCESS

The initial modeling of an Ornstein–Uhlenbeck process (O-U process) was developed by Uhlenbeck and Ornstein [22] in 1930. This modeling can be explained in the following way. If a Wiener process* represents the position of a Brownian particle, the derivative of the Wiener process should represent the particle's velocity. But the derivative of the Wiener process does not exist at any time interval. The O-U process is an alternative model that overcomes this defect by directly modeling the velocity of the Brownian particle as a function of time. Here, instead of the displacement $W(t)$ of the Wiener process, the velocity $X(t) = W'(t)$ at time t is considered. The equation of motion of a Brownian particle can be written as

$$dX(t) = -\beta X(t)\,dt + dB(t) \qquad (1)$$

with $X(0) = x_0$, where $-\beta X(t)$ represents the systematic part due to the resistance of the medium and $dB(t)$ represents the random component. It is assumed that these two parts are independent and that $B(t)$ is a Wiener process with drift $\mu = 0$ and variance parameter σ_0^2. Doob [7, 8] showed that a stationary Gaussian process* $\{X(t);\ t \geq 0\}$ is Markov if and only if its covariance function can be written as

$$R(t) = E\left\{\left[X(s+t) - EX(s+t)\right]\right.$$
$$\times\left[X(s) - EX(s)\right]\right\}$$
$$= \sigma^2 e^{-\beta|t|}, \qquad (2)$$

where σ^2 is the variance of $X(t)$ and $\infty > \beta > 0$. In general, this is served as the definition of an O-U process. It is easy to show that the O-U process $X(t)$ is a diffusion process and its transition probability density function (PDF) $p(x_0; x, t)$ satisfies the forward Kolmogorov equation

$$\frac{\partial p}{\partial t} = \beta \frac{\partial}{\partial x}(xp) + \frac{1}{2}\sigma_0^2 \frac{\partial^2 p}{\partial x^2} \quad (3)$$

and is given by

$$p(x_0; x, t)$$
$$= \frac{\partial}{\partial x} \Pr\left[X(t) \leqslant x \mid X(0) = x_0\right]$$
$$= \left[2\pi V(t)\right]^{-1/2}$$
$$\times \exp\left\{-\tfrac{1}{2}\left[x - x_0 e^{-\beta t}\right]^2 / V(t)\right\},$$
$$(4)$$

where $V(t) = \sigma^2[1 - \exp(-2\beta t)]$, $t > 0$. The solution of the stochastic differential equation* (1) is

$$X(t) = x_0 e^{\beta t} + \int_0^t e^{-\beta(t-s)} \, dW(t),$$

$$m_x(t) = E\left[X(t)\right] = x_0 e^{-\beta t},$$

and

$$V_x(t) = \mathrm{var}\left[X(t)\right] = \left(\tfrac{1}{2}\sigma_0^2 / \beta\right)(1 - e^{-2\beta t}),$$

which agrees with the transition PDF given in (4).

PROPERTIES

There are some interesting relationships between an O-U process $X(t)$ with no drift and a Wiener process $W(t)$:

$$X(t) = e^{-\beta t} W(e^{2\beta t}) \quad (5)$$

and

$$W(t) = \sqrt{t}\, X\left[(2\beta)^{-1}\log t\right], \quad (6)$$

where $W(t)$ is a Wiener process with drift $\mu = 0$ and variance parameter σ^2.

Doob [7] also showed the following properties: The sample function of an O-U process is continuous with probability 1 and

$$\limsup_{t\to 0} \frac{X(t) - X(0)}{\left[4\sigma^2 \beta t \log\log(1/t)\right]^{1/2}} = 1,$$

$$\limsup_{t\to\infty} X(t)\left(2\sigma^2 \log t\right)^{-1/2} = 1,$$

with probability 1. The properties expressed

in the equations are the counterparts of those well-known facts for a Wiener process.

APPLICATIONS

The O-U processes have been used widely in various fields. Some typical applications are listed as follows:

1. The velocity of a Brownian particle is modeled as an O-U process, as mentioned earlier.

2. In multirisk actuarial problems, an O-U process is a model for the deviations in investment performance, operating expenses, and lapse expenses (see Beekman [2]).

3. An O-U process is a limiting process of a discrete random process, such as a birth-and-death process* [14] or an Ehrenfest urn model [11].

4. Let $V(t, x)$ be the nerve membrane potential at a time t and a point x along the axis. Walsh [3] showed that $V(t, x)$ can be expressed as an infinite series of O-U processes. He then mentioned that the processes also arise in contexts such as heat conduction and electrical cables.

5. An observation process $Y(t)$ is expressed as $Y(t) = X(t) + f(t)$, where $f(t)$ is a deterministic function and $X(t)$ is an O-U process acting as an error fluctuation part in an estimation problem.

6. Others such as Tier and Hanson [21] and Prajneshu [18] have used the O-U process in demography*; Beekman and Fuelling [3] have used it as a risk model.

STATISTICAL INFERENCES

Related to application 5, statistical estimation and testing about parameters in an O-U process are given by Mann [12, 13] and Striebel [19]. The estimate of $2\beta\sigma^2$ is given by $D_n = (1/T)\sum_{i=1}^n (Y_i - Y_{i-1})^2$, where $Y_i = Y(iT/n)$, $i = 1, \ldots, n$, are observation points in $[0, T]$. Furthermore, $\sqrt{n}(D_n - 2\beta\sigma^2)$ has an asymptotic normal distribution with zero mean and variance $8\sigma^4\beta^2$ [12]. The maximum likelihood* estimate of β is

$$\hat{\beta} = \frac{A + \left\{ A^2 + 8k\int_0^T Y^2(t)\,dt \right\}^{1/2}}{4\int_0^T Y^2(t)\,dt},$$

where $A = KT - Y^2(0) - Y^2(T)$ and $K = 2\beta\sigma^2$. Therefore, estimating K by D_n, the asymptotic maximum likelihood estimate $\hat{\beta}$ of β can be obtained. Also based on n copies of short sequences of observations, an asymptotic maximum likelihood estimate of β and an asymptotic likelihood ratio test* statistic in the testing of independence against Markov dependency are given by Hsu and Park [10].

FIRST PASSAGE TIME PROBLEMS

Since the O-U process can be used to approximate more complicated or less tractable processes and also can be shown as a limiting process of the classical birth-and-death process, the properties of the first passage time to absorbing or reflecting barrier of an O-U process are useful. For example, in a meteorological problem, it may be desirable to estimate the probability of an event of a heat wave (i.e., outdoor air temperature) exceeding a certain level continuously for a few days. Denote the first passage time to $X(t) = a$ by

$$T_a = \inf_{t \geqslant 0} \left\{ t : X(t) \geqslant a \right\}, \qquad X(0) = x_0 < a.$$

or

$$T_a^* = \inf_{t \geqslant 0} \left\{ t : X(t) \leqslant a \right\}, \qquad X(0) = x_0 > a.$$

Darling and Siegert [5] presented the Laplace transform of the distribution of T_a involving the Weber function, which has not been inverted; Thomas [20] gave the mean and the variance of the first passage time and the polynomial approximations. Mehr and McFadden [15] and Wang and Uhlenbeck [24] solved the first passage time problem for $a = 0$ using equation (5) and properties of the Wiener process.

Breiman [4] obtained the approximate as-

ymptotic distribution of T_a^* when $t \to \infty$. Park and Schuurmann [17] derived a solution in the form of a functional equation that can be computed by numerical approximation. Beekman and Fuelling [3] applied Park and Schuurmann's result to compute actuality multirisk probabilities. Gringorten [9] took a completely different approach to give the Monte Carlo simulation of the first passage time density.

References

[1] Beekman, J. A. (1975). *J. Appl. Prob.*, **12**, 107–114.

[2] Beekman, J. A. (1976). *Scand. Actuarial J.*, 175–183.

[3] Beekman, J. A. and Fuelling, C. P. (1977). *Scand. Actuarial J.*, 175–183.

[4] Breiman, L. (1966). *Proc. 5th Berkeley Symp. Math. Statist. Prob.*, **2**, 9–16.

[5] Darling, D. A. and Siegert, A. J. F. (1953). *Ann. Math. Statist.*, **24**, 624–638.

[6] Dirkse, J. P. (1975). *J. Appl. Prob.*, **12**, 595–599.

[7] Doob, J. L. (1942). *Ann. Math.*, **43**, 351–369.

[8] Doob, J. L. (1953). *Stochastic Processes.* Wiley, New York.

[9] Gringorten, I. I. (1968). *J. Amer. Statist. Ass.*, **63**, 1517–1521.

[10] Hsu, Y. S. and Park, W. J. (1980). *Commun. Statist. A*, **9**, 529–540.

[11] Karlin, S. and Taylor, H. M. (1981). *A Second Course in Stochastic Processes.* Academic Press, New York.

[12] Mann, H. B. (1954). *Sankhyā*, **13**, 325–350.

[13] Mann, H. B. and Moranda, P. B. (1954). *Sankhyā*, **13**, 351–358.

[14] McNeil, D. R. and Weiss, G. H. (1977). *Biometrika*, **64**, 553–558.

[15] Mehr, C. B. and McFadden, J. A. (1965). *J. R. Statist. Soc. B*, **27**, 505–522.

[16] Park, C. and Schuurmann, F. J. (1976). *J. Appl. Prob.*, **13**, 267–275.

[17] Park, C. and Schuurmann, F. J. (1980). *J. Appl. Prob.*, **17**, 363–372.

[18] Prajneshu, (1980). *Stoch. Processes Appl.*, **10**, 87–99.

[19] Striebel, C. T. (1959). *Ann. Math. Statist.*, **30**, 559–567.

[20] Thomas, M. V. (1975). *J. Appl. Prob.*, **12**, 600–604.

[21] Tier, C. and Hanson, F. B. (1981). *Math. Biosci.*, **53**, 89–117.

[22] Uhlenbeck, G. E. and Ornstein, L. S. (1930). *Phys. Rev.*, **36**, 823–841.

[23] Walsh, J. B. (1981). *Adv. Appl. Prob.*, **13**, 231–281.

[24] Wang, M. C. and Uhlenbeck, G. E. (1945). *Rev. Mod. Phys.*, **17**, 323–342.

(BROWNIAN MOTION
GAUSSIAN PROCESSES
STOCHASTIC PROCESSES
WIENER PROCESS)

YU-SHENG HSU
WON JOON PARK

ORTHANT PROBABILITIES

An orthant probability is the probability that n random variables, $X_1, X_2, \ldots, X_n$, are all positive when the n variates have a joint multivariate normal distribution with all the means zero and all the variances one. Orthant probabilities have numerous applications. Among them are simultaneous statistical testing and confidence limits, pattern recognition*, classification* procedures, market forecasting, multiple drug interactions, etc. *See* MULTINORMAL DISTRIBUTIONS for additional information.

We define the correlation* between X_i and X_j to be ρ_{ij}, and we let E_i be the event that X_i is positive. Then

$$\Pr[E_i] = \Pr(X_i > 0) = \tfrac{1}{2},$$

and when $\rho_{ij} = 0$ for all i and j, then

$$\Pr[E_1 E_2 \ldots E_n] = 2^{-n}.$$

When $\rho_{ij} = \tfrac{1}{2}$ for all i and j, we have

$$\Pr[E_1 E_2 \ldots E_n] = 1/(n+1).$$

Sheppard [6] is usually credited with obtaining the result for two dimensions

$$\Pr[E_1 E_2] = \Pr(X_1 > 0, X_2 > 0)$$

$$= \tfrac{1}{4} + (2\pi)^{-1} \arcsin \rho_{12}.$$

This formula applies to any pair X_i, X_j for $i \neq j$ with ρ_{12} replaced with ρ_{ij}.

David [3] notes a result with several analogs in different branches of mathematics and develops the following formula when n is odd:

$$\Pr[E_1 E_2 \ldots E_n]$$

$$= \tfrac{1}{2}\Big\{ 1 - \sum_i \Pr[E_i] + \sum_{i<j} \Pr[E_i E_j]$$

$$- \sum_{i<j<l} \Pr[E_i E_j E_l] + \cdots, \text{etc.} \Big\}$$

The last term in this expression is the sum of n probabilities where all but one of the events $E_1 \ldots E_n$ is included in turn. Also the sign alternates with each summation. Hence for n odd, we can obtain any orthant probability if we know all orthant probabilities of lower dimensions.

If this is applied to the case $n = 3$,

$$P\{E_1 E_2 E_3\} = \tfrac{1}{8} + \frac{1}{4\pi}(\arcsin \rho_{12} + \arcsin \rho_{13}$$

$$+ \arcsin \rho_{23}).$$

For $n \geqslant 4$, there is no general closed representation of $P\{E_1 E_2 \ldots E_n\}$, but there have been many approximations proposed, usually for special cases (e.g., Steck [7] for the equicorrelated case).

On the other hand, if we had the orthant probabilities for $n = 4$, then we could use the earlier relationships given by David to obtain them for $n = 5$. Hence a great deal of effort has gone into the case $n = 4$.

Abrahamson [1] expresses the general quadrivariate case in terms of a linear combination of six orthoscheme probabilities, defined as

$$P_4(a, b, c)$$

$$= \frac{1}{16}\Big\{ 1 + \frac{2}{\pi}(\arcsin a + \arcsin b + \arcsin c)$$

$$+ \left(\frac{2}{\pi}\right)^2 \int_0^c \int_0^a \big[(1 - x^2)(1 - y^2)$$

$$- b^2\big]^{1/2} dx\, dy \Big\}.$$

The rules for computing a, b, and c in terms of ρ_{12}, ρ_{13}, ρ_{14}, ρ_{23}, ρ_{24}, and ρ_{34} and for expressing $P\{E_1 E_2 E_3 E_4\}$ as a sum of several $P_4(a, b, c)$ are given in Abrahamson [1, Sect. 4]; this reduces the original four-dimensional integral to several two-dimensional integrals. Abrahamson also gives a table of orthoscheme probabilities. Cheng [2] obtains a closed-form expression involving the dilogarithm function for some special cases of the correlation matrix based on Abrahamson's work.

Essentially the orthoscheme probabilities correspond to quadrivariate probabilities with $\rho_{13} = \rho_{14} = \rho_{34} = 0$. That is, the general quadrivariate orthant probability can be expressed as a linear combination of six orthoscheme probabilities that involve only three correlations.

Gehrlein [4] obtains a representation for the quadrivariate case that is a function of no more than three integrals over a single variable. When all correlations are equal to ρ, Gehrlein's result simplifies to

$$P\{E_1 E_2 E_3 E_4\}$$

$$= \frac{1}{16} + \frac{3}{4\pi} \arcsin \rho$$

$$+ \frac{3}{4\pi^2} \int_0^\rho \frac{\arcsin\{g(x;\rho)\}}{\sqrt{1-x^2}} \, dx,$$

where $g(x;\rho) = (\rho - x^2)/(1 + \rho - 2x^2)$. He gives similar forms for additional special cases and compares the computational times of his procedure and Abrahamson's. There are many variations of this result.

In n dimensions with all the correlations equal,

$$P\{E_1 E_2, \ldots, E_n\}$$

$$= \binom{n}{2} \frac{1}{2\pi} \int_0^\rho \sqrt{\frac{1+x}{x}}$$

$$\times \left\{ \int_{-\infty}^{+\infty} [\Phi(y)]^{n-2} \right.$$

$$\left. \times \phi\left(y\sqrt{\frac{1+x}{x}}\right) dy \right\} d(\arcsin x)$$

$$+ 2^{-n},$$

where $\Phi(z)$ is the cumulative distribution function and $\phi(z)$ is the density of the standardized univariate normal distribution at z.

Reductions of general multivariate normal distributions for cases where the means are not zero may be applied to the orthant prob-

abilities as special cases. See, e.g., Nelson [5], who gives an algorithm for computing multivariate normal distributions and multivariate t-distributions* for the case where $\rho_{ij} = \alpha_i \alpha_j$.

A complete listing of articles on orthant probabilities would be extensive. The following sources are either the most recent and hence contain references to the older literature or they are key papers that are mentioned in the article.

References

[1] Abrahamson, I. G. (1964). *Ann. Math. Statist.*, **35**, 1685–1703. (This important paper should be consulted in evaluating any four-dimensional orthant probability. It gives a reduction formula for evaluating the quadrivariate normal orthant probability with six parameters in terms of a three-parameter function that involves a double integral. There is a 24-item bibliography with many references not mentioned here.)

[2] Cheng, M. C. (1969). *Ann. Math. Statist.*, **40**, 152–161. (Gives quadrivariate orthant probabilities in terms of the dilogarithm functions when the correlation matrices are of specific forms. Also gives the orthant probability of five normal variates when the off-diagonal elements of the correlation matrix are equal. This paper is summarized in the Johnson and Kotz reference in the Bibliography.)

[3] David, F. N. (1953). *Biometrika*, **40**, 458–459. (Describes the state of the problem of evaluating orthant probabilities in 1953.)

[4] Gehrlein, W. V. (1979). *Commun. Statist. B*, **8**, 349–358. (Gives a representation of the orthant probability for $n = 4$ that is a function of no more than three integrals over a single variable. This article also has a bibliography of 27 references.)

[5] Nelson, P. R. (1982). *Commun. Statist. Simul. Comp.* **11**, 239–248. (Gives an algorithm for computing multivariate normal and t-distributions with $\rho_{ij} = \alpha_i \alpha_j$.)

[6] Sheppard, W. F. (1898). *Philos. Trans. Roy. Soc. London A*, **192**, 101–167.

[7] Steck, G. P. (1962). *Biometrika*, **49**, 433–445. (Considers orthant probabilities for the equicorrelated cases with tables for $n = 4$ and 5 and for general dimensions. This paper should be consulted whenever orthant probabilities are to be compared with equal correlations.)

Bibliography

Bacon, R. H. (1963). *Ann. Math. Statist.*, **34**, 191–198. (Gives intuitive approximations to orthant probabilities in n dimensions that are surprisingly accurate although in some specific cases other approximations are better.)

Choi, J. R. (1975). *Commun. Statist. A*, **4**, 1167–1175. (Gives an expression for the sum of two orthant probabilities with a general covariance matrix instead of the usual correlation matrix with variances one. The representation is in terms of a sum of products of two orthant probabilities of lower dimensions.)

Dutt, J. E. and Lin, T. K. (1975). *J. Statist. Comput. Simul.*, **4**, 95–120. (Gives a table of normal orthant probabilities for dimensions 4 and 5.)

Gupta, S. S. (1963). *Ann. Math. Statist.*, **34**, 792–828. (Gives a summary of known results to 1963 for probability integrals of multivariate normal and multivariate t. Contains some tables for the equicorrelated multivariate normal distribution function.)

Gupta, S. S. (1963). *Ann. Math. Statist.*, **34**, 829–838. (Gives a comprehensive bibliography on multivariate normal integrals to 1963.)

Horn, P. S. (1982). *Biometrika*, **69**, 681–682. (Shows that a t statistic based on two symmetric order statistics has a distribution that is expressible in terms of a sum of orthant probabilities.)

Johnson, N. L. and Kotz, S. (1972). *Distributions in Statistics*, Vol. 4: *Continuous Multivariate Distributions*. Wiley, New York, pp. 43–58. (Gives a summary of the evaluation of special cases of multinormal probabilities including orthant probabilities. This is a good source of information on the attempts to evaluate these functions in series form as well as other representations. Generally the expansions in series form are very slow to converge and as such are not easily evaluated numerically. The bibliography on pages 76–83 contains many references to work on orthant probabilities.)

Six, F. B. (1981). *Commun. Statist. A*, **10**, 1285–1295. (Gives a generalization of the special cases where $\rho_{ij} = \alpha_i \alpha_j$ and $\rho_{ij} = -\alpha_i \alpha_j$. The representation requires the evaluation of a single integral and is simpler to compute than previously available procedures.)

(MULTINORMAL DISTRIBUTIONS)

D. B. OWEN

ORTHOGONAL ARRAYS AND APPLICATIONS

INTRODUCTION: RELATION TO FACTORIAL EXPERIMENTS AND OPTIMUM PROPERTIES

Two mutually orthogonal Latin squares* (MOLs) of order 3, L_1 and L_2, are presented below, both written with the symbols 0, 1, and 2.

$$
L_1: \begin{array}{ccc} 0 & 1 & 2 \\ 1 & 2 & 0 \\ 2 & 0 & 1 \end{array} \qquad L_2: \begin{array}{ccc} 0 & 1 & 2 \\ 2 & 0 & 1 \\ 1 & 2 & 0 \end{array}
$$

Let the rows (also the columns) of a standard 3×3 square be numbered 0, 1, and 2, and the following array **A** of type 4×9 be constructed from the squares L_1 and L_2.

$$
A: \begin{array}{ccccccccc} 0 & 0 & 0 & 1 & 1 & 1 & 2 & 2 & 2 \\ 0 & 1 & 2 & 0 & 1 & 2 & 0 & 1 & 2 \\ 0 & 1 & 2 & 1 & 2 & 0 & 2 & 0 & 1 \\ 0 & 1 & 2 & 2 & 0 & 1 & 1 & 2 & 0 \end{array}
$$

The first two rows of **A** give the row and column numbers of the respective cells of a standard 3×3 square. In the $(k+2)$th row of **A** are written down the symbols occurring in the corresponding cells of L_k, $k = 1, 2$. **A** is observed to satisfy the property that any two-rowed submatrix of **A** contains each of the ordered pairs $(ij)'$ exactly once, $i, j = 0, 1, 2$. This specific property, as we shall see presently, makes **A** an orthogonal array of strength 2 and index 1. The method of construction illustrated is general, and an array of type **A** with s^2 columns and $m + 2$ rows can be constructed similarly if we can get hold of m MOLs of order s.

The idea of orthogonal Latin squares was extended to orthogonal Latin cubes and Latin hypercubes. Rao [13] generalized the concept to hypercubes of strength d and then brought these ideas together under the broad framework of an orthogonal array (OA), the definition of which follows.

Definition 1. An $r \times n$ matrix **A** (array) with entries from a set S of s ($\geqslant 2$) symbols is called an *orthogonal array* of size n, r constraints, s levels, strength d ($\geqslant 1$) and index λ ($\geqslant 1$), if every $d \times n$ submatrix of **A** contains all the s^d possible $d \times 1$ column vectors based on s symbols of S with the same frequency λ.

Such an array is denoted by OA (n, r, s, d). We refer to n, r, s, and d as the parameters of the array. Obviously for such an OA, $n = \lambda s^d$, where λ is the index of the array. In case λ is a power of s, the array reduces to a hypercube of strength d. Given an OA $(s^2, m + 2, s, 2)$, $m \geqslant 1$, we can construct m MOLs of order s, by utilizing any two rows of the array for coordinatization of a standard $s \times s$ square, and each of the remaining m rows of the array in constructing one Latin square of order s. Thus an OA $(s^2, m + 2, s, 2)$ implies and is implied by the existence of m MOLs of order s.

The principal importance of an OA lies in its use as an appropriate fractional factorial* experiment, the constraints representing the factors, and the columns representing treatment combinations included in the experiment. In order to accommodate appropriate fractions of asymmetrical factorial experiments into the system, OA has been extended to the situation where variable numbers of levels are permissible in different rows, i.e., constraints of the array [15]. To distinguish this case from an ordinary OA, the symbol that has been reserved for the symmetrical case, some authors prefer the use of the symbol OAVS to denote an OA permitting variable numbers of levels in the rows.

Definition 2. A $r \times n$ matrix $\mathbf{A}$ (array) whose ith row is written with the entries from a set S_i of s_i ($\geqslant 2$) symbols is called an *orthogonal array with variable numbers of symbols* (OAVS) of size n, r constraints, levels $s_1, s_2, \ldots, s_r$ for the r respective constraints and strength d if a $d \times n$ submatrix of A consisting of the rows, say, $i_1, i_2, \ldots, i_d$ where $\{i_1, i_2, \ldots, i_d\} \subset \{1, 2, \ldots, r\}$, contains with the same frequency each one of the $\prod_{j=1}^{d} s_{i_j}$ possible $d \times 1$ column vectors that can be constructed with jth symbol of the column derived from the set S_{i_j} of s_{i_j} symbols, $j = 1, 2, \ldots, d$, and the same property is satisfied for every choice of a $d \times n$ submatrix of $\mathbf{A}$.

Denote such an array by OAVS $(n, r, s_1, \ldots, s_r, d)$. Evidently, an OAVS $(n, r, s_1, \ldots, s_r, d)$ reduces to an OA (n, r, s, d) in the special case $s_1 = s_2 = \cdots s_r = s$.

While constructing fractions of factorial experiments, we normally require the fraction to satisfy certain desirable properties from the points of view of statistical inference. Some of these properties are reflected in the definition of the so-called orthogonal (g, t) plan, $1 \leqslant g \leqslant t$, also sometimes referred to as a resolution $(g + t + 1)$ plan. Consider a factorial experiment involving r factors, ith factor occurring at s_i levels, $i = 1, 2, \ldots, r$. Let the $v = s_1 \times s_2 \times \cdots s_r$ treatments represented by v r-tuples appropriately be written in a certain order as 1, 2, \ldots, v. Then a general factorial experiment $\mathbf{T}$ consists of a number of treatments, say m, where the ith treatment occurs r_i times with $r_i \geqslant 0$, $\forall_i$ and $\sum_{i=1}^{v} r_i = m$.

Definition 3. The factorial experiment $\mathbf{T}$ described in the preceding paragraph gives an orthogonal (g, t) plan, $1 \leqslant g \leqslant t \leqslant r$, provided

(a) All contrasts belonging to main effects* and interactions* involving g or fewer factors are estimable.

(b) All interactions involving more than t factors are assumed absent.

(c) The best linear unbiased estimator of a contrast belonging to a main effect or interaction involving g or fewer factors is uncorrelated with the best linear unbiased estimators of all the estimable contrasts belonging to all other main effects and interactions involving t or fewer factors.

Now $\mathbf{T}$ can be written down as an array with rows designating the respective factors and columns representing treatments included in the experiment. It has been proved by Mukherjee [8] that a necessary and sufficient condition for the array $\mathbf{T}$ to be an orthogonal (g, t) plan for $t \geqslant 2$ is that $\mathbf{T}$ constitutes an OAVS of strength $d = \min(g + t, r)$. A weaker condition, namely that of proportional frequency proposed by Addelman [1] is known to be both necessary and sufficient for an orthogonal $(1, 1)$ plan. Addelman's [1] orthogonality condition for main effect plans can be stated as follows:

In an array representing a factorial experiment with N treatments, let us select a $2 \times N$ submatrix, the two rows selected representing the factors A and B, say, in that specific order. Let the number of levels of A and B be s_1 and s_2, respectively, and suppose the ith level of A is denoted by α_i, $i = 1, 2, \ldots, s_1$ and the jth level of B by β_j, $j = 1, 2, \ldots, s_2$. Also, let N_{ij} represent the number of times the ordered pair $(\alpha_i \beta_j)'$ occurs in the $2 \times N$ submatrix chosen and $N_{i.} = \sum_{j=1}^{s_2} N_{ij}$, $N_{.j} = \sum_{i=1}^{s_1} N_{ij}$. Then the main effects of A and B are orthogonally estimable if and only if

$$N_{ij} = \frac{N_{i.} \times N_{.j}}{N},$$

for all i, j, $i = 1, 2, \ldots, s_1$ and

$$j = 1, 2, \ldots, s_2.$$

Obviously, an OAVS of strength 2 satisfies Addelman's orthogonality condition with respect to each pair of factors. Thus an OAVS of strength d necessarily provides us with an orthogonal resolution $(d + 1)$ plan for all $d \geqslant 2$. It also has the property that it ensures the orthogonal estimability of the general mean or intercept. These important properties of an OAVS naturally induce us to look for an appropriate fraction of a factorial experiment invariably within the class of OAs and OAVS's whenever they exist. The OAVS's of strength 2 are also known via Cheng [7] to satisfy a very general class of optimality criteria proposed by Kiefer with regard to each of the main effects.

Starting with an OA of strength 2 and the number of levels s, a prime power, Addelman [2] developed classes of what are called *compromise plans*, where all main effects and certain useful sets of two-factor interactions* are orthogonally estimable. Addelman [1] also gave a simple and ingenious method of collapsing the levels of factors in an OA of strength 2 to construct fractionally replicated orthogonal main effect plans in the asymmetrical case. The techniques have been successfully applied with and without modifications by several research workers and practitioners in constructing useful fractionally replicated plans.

CONSTRUCTION OF OAs AND BOUNDS ON THE NUMBER OF CONSTRAINTS

Construction of OAs of strength 2 and index unity from MOLs has been explained in the preceding section. Bush [6] gave a method of construction based on GF(s) for an OA $(s^d, s + 1, s, d)$, $s \geqslant d \geqslant 2$, for s a prime power. Perhaps the most widely used method of construction of OAs of strength 2 is the method of differences developed by Bose and Bush [5]. See Raghavarao [12] for a general description and a critical appreciation of these and other allied methods. These traditional techniques have been largely exploited, modified, and extended to construct various series of OAs of different strengths by later authors. An up-to-date list of references is available in Mukhopadhyay [11].

Finite or Galois geometries have provided a simple and elegant tool for the construction of some series of OAs. Let us have a close look at the construction of an OA $(s^t, (s^t - 1)/(s - 1), s, 2)$, with s a prime power and $t \geqslant 2$ from the geometrical point of view. Let the size be $n = s^t$, $t \geqslant 2$. The first t rows of the OA can be written as all the s^t possible t-tuples with entries from the Galois field* GF(s). We can denote these t rows by t linearly independent symbols ρ_1, $\rho_2, \ldots, \rho_t$. Then any row of the array would be constructed as a linear combination, $\beta_1 \rho_1 + \beta_2 \rho_2 + \cdots + \beta_t \rho_t$, where β_i's are elements of GF(s), $(\beta_1, \beta_2, \ldots, \beta_t) \neq (0, 0, \ldots, 0)$ and the first nonnull element of the β_i's is arbitrarily chosen to be the unit element of GF(s). Note that an array constructed in this manner is an OA of strength 2, and there is a one-to-one correspondence between the rows of the OA thus arrived at and the points of a PG $(t - 1, s)$ constructed by Bose [4] via GF(s). The number of rows in the array is evidently $(s^t - 1)/(s - 1)$ and it is saturated in the sense that no more rows can be added to it, preserving its orthogonality of strength 2. It is elementary to observe that the maximum number of rows of this array that can be selected to form an OA of strength d ($2 \leqslant d \leqslant t$) = the maximum number of points of the PG ($t - 1, s$), no d of which are linearly dependent, and this number is usually denoted by the symbol $m_d(t, s)$. The latter problem referred to is the well-known packing problem, which originated with Bose's [4] treatment of symmetrical factorial experiments.

The method of construction described can be sufficiently modified to obtain a number of saturated OAVS's of strength 2, accommodating factors with levels as different powers of s. For a detailed discussion of the problem and particularly the use of t-spreads of projective geometrics in the context, see Mukhopadhyay [10]. There are various other successful methods of construction of OAs and OAVS's, some being applicable in the cases where the numbers of levels are not prime powers. For a detailed reference, see Mukhopadhyay [11].

In general, finding the maximum number of constraints that can be accommodated in an OA of size λs^d, strength d, and levels s, denoted by the symbol $f(\lambda s^d, s, d)$, and, in the absence of an explicit expression, finding as sharp as possible bounds to it, both upper and lower, are of utmost importance in the study of OAs. Regarding bounds on k for an OA ($\lambda s^d, k, s, d$), $d \geqslant 2$, the first important result, obtained by Rao [14], was:

$$\lambda s^d - 1 \geqslant \binom{k}{1}(s - 1) + \cdots + \binom{k}{u}(s - 1)^u$$

$$\text{if} \quad d = 2u,$$

$$\lambda s^d - 1 \geqslant \binom{k}{1}(s - 1) + \cdots + \binom{k}{u}(s - 1)^u$$

$$+ \binom{k - 1}{u}(s - 1)^{u+1}$$

$$\text{if} \quad d = 2u + 1.$$

Bush [6] substantially improved these bounds for OAs of index unity. It is to be noted that $f(s^d, s, d) = d + 1$ for all $s \leqslant d$. Raghavarao [12] gives a good account of

Bush's bounds and the subsequent results by other authors on the subject. Addelman and Kempthorne (1962) constructed the saturated OA $(2s^t, 2(s^t - 1)/(s - 1) - 1, s, 2)$, $t \geqslant 2$ and s, a prime power. Not many series of saturated OAs are known. A Hadamard matrix* of order 4λ implies and is implied by a saturated OA $(4\lambda, 4\lambda - 1, 2, 2)$, which in turn is equivalent to a saturated OA $(8\lambda, 4\lambda, 2, 3)$, $\lambda \geqslant 1$. Several authors have handled the problem of these bounds, which boils down to the packing problem under suitable conditions from the geometrical point of view. For a detailed up-to-date information one can look up Mukhopadhyay [9] and Barlotti [3], and the references given therein.

There are many useful combinatorial arrangements related to OAs, specially of strength 2 and numerous problems of combinatorial importance connected with the construction of OAs. For general information to date on the subject, see Raghavarao [12] and Shrikhande [16].

EXTENSIONS OF OAs

The extension of OAs to OAVS's has been adequately dealt with. Because of the stringent conditions prevailing among the set of parameters, an OA cannot exist for all possible values of them. Hence weakening the conditions to an extent, new arrays have been defined that are also balanced in a broad mathematical and statistical sense. For a general discussion on them and a study of their properties, most of the references to date are available in Rao [15].

References

[1] Addelman, S. (1962). *Technometrics*, **4**, 21–46.

[2] Addelman, S. (1962). *Technometrics*, **4**, 47–58.

[3] Barlotti, A. (1980). *Combinatorial Mathematics, Optimal Designs, and Their Applications*, J. Srivastava, ed. North-Holland, Amsterdam, pp. 1–5.

[4] Bose, R. C. (1947). *Sankhyā*, **8**, 107–166.

[5] Bose, R. C. and Bush, K. A. (1952). *Ann. Math. Statist.*, **23**, 508–524.

[6] Bush, K. A. (1952). *Ann. Math. Statist.*, **23**, 426–434.

[7] Cheng, C. S. (1980). *Ann. Statist.*, **8**, 436–446.

[8] Mukherjee, R. (1980). *Calcutta Statist. Ass. Bull.*, **29**, 143–160.

[9] Mukhopadhyay A. C. (1978). *J. Comb. Theory A*, **25**, 1–13.

[10] Mukhopadhyay A. C. (1980). *Combinatorics and Graph Theory, Proceedings*, Springer-Verlag, New York, p. 885.

[11] Mukhopadhyay A. C. (1981). *Sankhyā B*, **43**, 81–92.

[12] Raghavarao, D. (1971). *Construction and Combinatorial Problems in Design of Experiments*. Wiley, New York.

[13] Rao, C. R. (1946), *Calcutta Math. Soc. Bull.*, **38**, 67–78.

[14] Rao, C. R. (1947), *J. R. Statist. Soc. B*, **9**, 128–139.

[15] Rao, C. R. (1973), In *A Survey of Combinatorial Theory*, J. N. Srivastava et al., eds. North-Holland, Amsterdam.

[16] Shrikhande, S. S. (1980). *Combinatorics and Graph Theory, Proceedings*. Springer-Verlag, New York, 885.

(DESIGN OF EXPERIMENTS
FACTORIAL EXPERIMENTS
FRACTIONAL FACTORIAL DESIGNS
GALOIS FIELDS
ORTHOGONAL DESIGNS)

A. C. MUKHOPADHYAY

ORTHOGONAL CHEBYSHEV POLYNOMIALS

For definitions and some applications, *see* CHEBYSHEV, PAFNUTY LVOVICH and CHEBYSHEV–HERMITE POLYNOMIALS.

Tables of these polynomials with a discussion of their applications to fitting polynomial regression* and, more generally, to least-squares estimation of parameters in linear models are presented, e.g., in Beyer [1].

Reference

[1] Beyer, W. H. (1971). *Basic Statistical Tables*, The Chemical Rubber Co., Cleveland, OH.

(CHEBYSHEV, PAFNUTY LVOVICH)

ORTHOGONAL DECOMPOSITION

Suppose $\mathbf{y}$ belongs to a vector space V, such as n-dimensional Euclidean space $\mathbb{R}^n$. Let V_1 be a vector subspace of V, e.g., $\{\mathbf{y} : \mathbf{y} \in V, \mathbf{A}\mathbf{y} = \mathbf{0}\}$, where $\mathbf{A}$ is an $m \times n$ matrix ($m < n$). Then $\mathbf{y}$ can be expressed *uniquely* in the form

$$\mathbf{y} = \mathbf{y}_1 + \mathbf{y}_2, \qquad (1)$$

where $\mathbf{y}_1 \in V_1$ and $\mathbf{y}_2 \in V_1^\perp \cap V = V_2$. Here V_2, the orthogonal complement of V_1 with respect to V, is the set of all vectors in V that are perpendicular to V_1. Since $\mathbf{y}_1$ is perpendicular to $\mathbf{y}_2$, (1) is called an *orthogonal decomposition*. Furthermore $\mathbf{y}_1'\mathbf{y}_2 = 0$ implies that $\mathbf{y}'\mathbf{y} = \mathbf{y}_1'\mathbf{y}_1 + \mathbf{y}_2'\mathbf{y}_2$ so that the square of the length of $\mathbf{y}$ is the sum of the squares of the lengths of $\mathbf{y}_1$ and $\mathbf{y}_2$—a generalization of Pythagoras' theorem. It can be shown that $\mathbf{y}_1$ is the orthogonal projection of $\mathbf{y}$ onto V_1 and is given by $\mathbf{y}_1 = \mathbf{P}_1\mathbf{y}$, where $\mathbf{P}_1$ is the *unique* symmetric idempotent matrix* (sometimes called a projection matrix) such that the range of $\mathbf{P}_1$ is V_1. Then

$$\mathbf{y}'\mathbf{y} = \mathbf{y}'\mathbf{P}_1\mathbf{y} + \mathbf{y}'\mathbf{P}_2\mathbf{y}, \qquad \mathbf{P}_2 = \mathbf{I} - \mathbf{P}_1.$$

The decomposition (1) can be extended to k components, that is, $\mathbf{y} = \mathbf{y}_1 + \mathbf{y}_2 + \cdots + \mathbf{y}_k$, where $\mathbf{y}_i'\mathbf{y}_j = 0$ ($i \neq j$). This leads to a partitioning of the sum of squares $\mathbf{y}'\mathbf{y}$ into

$$\mathbf{y}'\mathbf{y} = \mathbf{y}'\mathbf{P}_1\mathbf{y} + \mathbf{y}'\mathbf{P}_2\mathbf{y} + \cdots + \mathbf{y}'\mathbf{P}_k\mathbf{y}.$$

Such a partition forms the basis for an analysis of variance* and regression. By the application of projection matrices and Cochran's theorem on quadratic forms*, these sums of squares are related to mutually independent central or noncentral chi-squared distributions*. For further details and applications, see Seber [5] and Appendices A and B of Seber [4, 6].

These results can be extended in two ways. First, one can use a coordinate free approach using a general inner product $\langle \, , \, \rangle$. This includes special cases $\langle \mathbf{y}_1, \mathbf{y}_2 \rangle = \mathbf{y}_1'\mathbf{y}_2$ for $\mathbb{R}^n$ and $\langle \mathbf{y}_1, \mathbf{y}_2 \rangle = \mathbf{y}_1^*\mathbf{y}_2$ for complex n-dimensional space. Such an approach is demonstrated, e.g., by Drygas [2] and Kruskal [3]. Second, V can be a general infinite dimensional space. For example, y could represent the realization of a continuous time series (a point in a Hilbert space) or an infinite sequence of random variables (a point in a Banach space): see Doob [1, Chap. 4]. A Fourier series can then be regarded as a projection.

References

[1] Doob, J. L. (1953). *Stochastic Processes*. Wiley, New York.

[2] Drygas, H. (1970). *The Coordinate-Free Approach to Gauss–Markov Estimation*. Lecture Notes in Operations Research and Mathematical Systems, No. 4. Springer-Verlag, New York.

[3] Kruskal, W. (1960). *Proc. 4th Berkeley Symp. Math. Statist. Prob.*, **1**, 435–461.

[4] Seber, G. A. F. (1977). *Linear Regression Analysis*. Wiley, New York.

[5] Seber, G. A. F. (1980). *The Linear Hypothesis: A General Theory*, 2nd ed. Charles Griffin, London.

[6] Seber, G. A. F. (1984). *Multivariate Observations*. Wiley, New York.

(ANALYSIS OF VARIANCE
GEOMETRY IN STATISTICS
REGRESSION ANALYSIS
TIME SERIES)

G. A. F. Seber

ORTHOGONAL DESIGNS

The term *orthogonal design* is used widely and sometimes imprecisely. Broadly speaking, there are two types of experiment to which it may be applied. First, there are experiments in which a number of treatments (possibly with their own factorial structure) are to be compared and in which it is desirable to eliminate such systematic effects as rows, blocks, etc. Second, there are response surface* experiments in which the response is a function of one or more quantitative explanatory variables.

EXPERIMENTS FOR COMPARING TREATMENTS

In structured experiments to compare a number of treatments two factors are often said to be *orthogonal* if each level of the first factor occurs as frequently as each level of the second. The simplest example of this is a randomized block design*, where each of t treatments occurs just once in each of b blocks. Every possible block-treatment combination occurs once, and so we say that blocks and treatments are orthogonal. One result of this type of balance is that the least-squares estimate of any contrast* of treatment parameters (under the usual additive linear model with second-order assumptions) is simply the same contrast of treatment means. In addition, the sum of squares for testing equality of treatment parameters depends only on the treatment means and does not involve elimination of blocks.

These properties also hold in the next simplest type of orthogonal design, the Latin square*. In an $n \times n$ Latin square, each of n treatments occurs just once in each row and once in each column. Here treatments are orthogonal to both rows and columns. Indeed, rows and columns are themselves orthogonal. Strictly speaking, when we say that a Latin square is an orthogonal design, we mean that it is orthogonal for the estimation of row, column, and treatment effects in the usual additive linear model.

The idea of a Latin square can be generalized in a number of ways. First, there is the concept of a Graeco-Latin square*. Two $n \times n$ Latin squares are said to be orthogonal (or to form a Graeco-Latin square) if each of the n^2 ordered pairs of symbols from the two separate squares occurs just once when the squares are superimposed. If the treatments of the experiment are the symbols of the first square, then these are orthogonal to rows, columns, and the symbols of the second square. In fact, any pair of factors (rows, columns, first-square symbols, second-square symbols) are orthogonal and the least-squares estimate of any contrast, under the usual additive linear model, is the corresponding contrast of means.

Graeco-Latin squares can be generalized to the idea of a set of mutually orthogonal Latin squares. Any pair in such a set forms a Graeco-Latin square. Further generalization is provided by the concept of an orthogonal array*. An orthogonal array of size N, c constraints, level n, and strength s is a rectangular arrangement of n different symbols in N columns and c rows such that in any s rows the N columns contain all possible ordered sets of n^s symbols the same number λ of times, this number being called the *index* of the array. Thus a Latin square is equivalent to an orthogonal array of size n^2, three constraints, strength 2, and index 1. In general, to convert an orthogonal array to a design, each row of the array corresponds to a factor in the experiment, e.g., columns, blocks, and treatments. The least-squares properties outlined will then hold in any model involving up to $(s - 1)$th order interactions.

There are further, less widely used generalizations of Latin squares. For example an $n \times n \times n$ Latin cube consists of n symbols each repeated n^2 times, every symbol being present exactly n times in each of its planes. A useful reference for these and related designs is Vajda [5].

Mention should also be made here of orthogonal partitions of designs. For example, in a 4×4 Latin square with treatments A, B, C, and D arranged in rows and columns, we may wish to use two observers P and Q. To achieve a design in which observers are orthogonal to rows, columns and treatments, we find a Graeco-Latin square* orthogonal to the original one and use two of its symbols for P and two for Q. The result might be

$$
\begin{array}{cccc}
AP & BQ & CQ & DP \\
BQ & AP & DP & CQ \\
CP & DQ & AQ & BP \\
DQ & CP & BP & AQ
\end{array}
$$

It is perhaps useful to mention some designs that are not orthogonal. For example,

a balanced incomplete block design* is nonorthogonal because each treatment does not occur equally frequently in each block. Note that in this case the least-squares* estimates of treatment contrasts do involve quantities other than the corresponding treatment means. Similar remarks apply to other incomplete block designs. In general, unbalanced designs tend to be nonorthogonal although the least-squares properties discussed earlier do hold under slightly more general conditions than stated at the beginning of this section. For example, in a block design with the property that each treatment occurs a fixed proportion of times in each block, with possibly different proportions for different treatments, the least-squares estimates of treatment contrasts do depend solely on the treatment means.

RESPONSE SURFACE DESIGNS

In response surface* experiments, the term orthogonal design is used more precisely. We assume that the expected response to quantitative variables $x_1, \ldots, x_n$ is a linear function of p unknown parameters and that the usual second-order error assumptions apply. A design is said to be orthogonal if the p columns of the design matrix (or matrix of explanatory variable values) are mutually orthogonal. For example, if the expected response to explanatory variables x_1 and x_2 is of the form $\beta_0 + \beta_1 x_1 + \beta_2 x_2 + \beta_3 x_1 x_2$, where β_i ($i = 0, 1, 2, 3$) are unknown parameters, then the design with k (> 1) observations at each of the points $(-1, -1), (-1, 1)$, $(1, -1), (1, 1)$ is orthogonal. In an orthogonal response surface design, the least-squares estimate of any parameter, say β_j, will depend only on the data and the values in that column of the design matrix* corresponding to β_j. The same remark applies to the sum of squares for testing a linear hypothesis about β_j. In addition, the least-squares estimates of the unknown parameters are uncorrelated. When the treatments themselves have a factorial structure, it is often considered desirable to construct factorial designs arranged in blocks such that the usual least-squares estimates of estimable treatment contrasts belonging to different factorial effects are uncorrelated. This problem is considered by Mukerjee [4], who calls these designs *effect-wise orthogonal* and gives a unified theory of such designs together with a method of construction by means of Kronecker products.*

MATHEMATICAL ASPECTS OF ORTHOGONAL DESIGNS

The term orthogonal design has been used in a more precise sense by mathematicians. For example, Geramita and Seberry Wallis in their book [2] and in their paper [1] define an orthogonal design of order n and type $(s_1, \ldots, s_i)$, where the s_i's are positive integers, to be an $n \times n$ matrix $\mathbf{X}$ with entries from $\{0, \pm x_1, \ldots, \pm x_i\}$, the x_i's being commuting indeterminates, satisfying

$$\mathbf{X}\mathbf{X}^T = \left(\sum_{i=1}^{i} s_i x_i^2 \right) \mathbf{I}_n.$$

Here $\mathbf{I}_n$ is the $n \times n$ identity matrix. Each row of $\mathbf{X}$ has s_i entries of type $\pm x_i$, and the rows are orthogonal. For example, the Hadamard matrices with entries ± 1 such that $\mathbf{X}^T\mathbf{X} = \mathbf{X}\mathbf{X}^T = n\mathbf{I}_n$ are orthogonal of type $s_1 = n$. For a review, see Hedayat and Wallis [3]. Geramita and Wallis's work is concerned with the existence and construction of orthogonal designs.

References

[1] Geramita, A. V. and Seberry Wallis, J. (1975). *J. Comb. Theory A*, **19**, 66–83. (This paper is a source for further references.)

[2] Geramita, A. V. and Seberry Wallis, J. (1979). *Lecture Notes in Pure and Applied Mathematics*, Vol. 45: *Orthogonal Designs: Quadratic Forms and Hadamard Matrices*. Marcel Dekker, New York.

[3] Hedayat, A. and Wallis, W. D. (1978). *Ann. Statist.*, **6**, 1184–1238.

[4] Mukerjee, R. (1981), *J. Statist. Plan. Infer.*, **5**, 221–229.

[5] Vajda, S. (1967). *The Mathematics of Experimental Design*. Charles Griffin, London.

(ANALYSIS OF VARIANCE
BLOCKS, BALANCED INCOMPLETE
BLOCKS, RANDOMIZED COMPLETE
CONSTRAST
DESIGN OF EXPERIMENTS
GRAECO-LATIN SQUARES
INTERACTION
LATIN SQUARES
ORTHOGONAL ARRAYS
   AND APPLICATIONS
RESPONSE SURFACE DESIGNS)

L. V. WHITE

ORTHOGONAL EXPANSIONS

UNIVARIATE EXPANSIONS

Meixner* [12] studied a class of orthogonal polynomials on some of the common statistical distributions. These orthogonal polynomials and more general orthogonal functions can be used to construct expansions of probability distributions. Later workers, particularly Lancaster and Sarmanov, studied the structure of bivariate distributions through the use of orthonormal functions leading to the construction and characterization of classes of bivariate distributions with given marginals.

Let X be a random variable with distribution function $F(x)$. An *orthogonal sequence* of functions $\{\xi_i(X)\}$ on $F(x)$ is one such that

$$\mathscr{E}(\xi_j(X)\xi_k(X)) = \int_{-\infty}^{\infty} \xi_j(x)\xi_k(x)\,dF(x)$$
$$= 0 \quad \text{if} \quad j \neq k.$$

The sequence is *orthonormal* if

$$\mathscr{E}(\xi_j(X)\xi_k(X)) = \delta_{jk},$$

where δ_{jk} is the Kronecker delta. Conventionally, $\xi_0(X) \equiv 1$, and when this is so,

$$\mathscr{E}(\xi_j(X)) = 0, \quad j = 1, 2, \ldots .$$

An orthonormal sequence is complete if any function $g(X)$ with a finite variance has an expansion in mean square

$$g(X) = \sum_{i=0}^{\infty} a_i \xi_i(X), \tag{1}$$

where $a_i = \mathscr{E}(g(X)\xi_i(X))$. The expansion is in the sense that

$$\lim_{n \to \infty} \mathscr{E}(g(X) - S_n)^2 = 0,$$

where

$$S_n = \sum_{i=0}^{n} a_i \xi_i(X).$$

An equivalent definition is that

$$\text{var } g(X) = \sum_{i=1}^{\infty} a_i^2 .$$

If $\sum_0^\infty |a_i| < \infty$, the series converges with probability one. Classical sequences of orthogonal functions are the *orthogonal polynomials*, defined on a distribution with finite moments, and the trigonometric series

$$1, \sqrt{2} \sin(2\pi m X), \sqrt{2} \cos(2\pi m X),$$

$$m = 1, 2, \ldots$$

on the uniform distribution on $(0, 1)$. The Meixner class of orthogonal polynomials is of particular interest in statistics. When these polynomials $\{P_n(x)\}$ are normalized to have leading coefficients of unity, their generating function* is of the form

$$G(x; t) = \sum_{n=0}^{\infty} P_n(x) t^n / n!$$

$$= f(t)\exp(xu(t)),$$

where $f(t)$ and $u(t)$ are functions with power series expansions in t. This class includes the following distributions [1]:

The coefficients $\{b_n\}$ in a mean square expansion

$$g(X) = \sum_{n=0}^{\infty} b_n P_n(X)$$

can be conveniently found by noting that $b_n \mathscr{E}(P_n(X)^2)$ is the coefficient of $t^n / n!$ in

$$f(t)\mathscr{E}(g(X)\exp(Xu(t))).$$

Let X be a continuous random variable with a probability density function $h(x)$, and let $g(x)$ be another probability density function with an associated complete orthonormal sequence $\{\xi_n(X)\}$. $h(x)$ is said to be ϕ^2-*bounded* with respect to $g(x)$ if

$$\phi^2 + 1 = \int_{-\infty}^{\infty} \{h(x)/g(x)\}^2 g(x)\, dx < \infty.$$

Then there is an expansion in mean square:

$$h(x) = g(x)\left\{ 1 + \sum_{i=1}^{\infty} a_i \xi_i(x) \right\}, \qquad (2)$$

where

$$a_i = \int_{-\infty}^{\infty} \xi_i(x) h(x)\, dx \quad \text{and} \quad \phi^2 = \sum_{i=1}^{\infty} a_i^2 .$$

An analogous result holds for discrete and more general distributions. Lancaster [9] develops orthogonal function techniques extensively in his book, where a good bibliography is given. See also Lancaster [11]. If $g(x)$ in (1) belongs to the Meixner class and the

Distribution		$G(x; t)$
Normal	$\dfrac{1}{\sqrt{2\pi}\,\sigma} \exp\{-\tfrac{1}{2}x^2/\sigma^2\},$	$\exp\{xt - \tfrac{1}{2}\sigma^2 t^2\}$
	$-\infty < x < \infty, \sigma > 0$	
Gamma	$\dfrac{x^{\alpha-1}}{\Gamma(\alpha)} \exp(-x),$	$(1 + t)^{-\alpha}\exp\{xt/(1 + t)\}$
	$x > 0, \alpha > 0$	
Poisson	$\dfrac{\lambda^x}{x!} e^{-x},$	$(1 + t)^x e^{-\lambda t}$
	$x = 0, 1, \ldots ; \lambda > 0$	
Negative binomial	$(1 + \theta)^{-\alpha-x}\theta^x \binom{x + \alpha - 1}{x}$	$(1 + t\theta)^{-x-\alpha}(1 + t(1 + \theta))^x$
	$x = 0, 1, \ldots ; \theta > 0$	
Binomial	$\binom{n}{x} p^x (1-p)^{n-x},$	$(1 + (1-p)t)^x (1 - pt)^{n-x}$
	$x = 0, 1, \ldots, n$	

orthogonal functions are polynomials, then

$$h(x) = g(x)\left\{1 + \sum_{n=1}^{\infty} b_n P_n(x)\right\},$$

where $b_n \mathscr{E}(P_n^2(X))$ is the coefficient of $t^n/n!$ in $f(t)M(u(t))$ and $M(\theta)$ is the moment-generating function corresponding to $h(x)$.

BIVARIATE EXPANSIONS

Orthonormal expansions have been used very successfully in bivariate distribution theory. Let $f(x, y)$ be a bivariate probability density function of continuous random variables X, Y with $g(x), h(y)$ the respective marginals. If $\{\xi_i(X)\}, \{\eta_j(Y)\}$ are complete orthonormal sets of functions on the respective marginal distributions, then the product set $\{\xi_i(X)\eta_j(Y)\}$ is a complete orthonormal set on the joint independence distribution $g(x)h(y)$. An extension of (2) is

$$f(x, y) = g(x)h(y)$$
$$\times \left\{1 + \sum_{i=1}^{\infty} \sum_{j=1}^{\infty} \rho_{ij}\xi_i(x)\eta_j(y)\right\},$$
$$(3)$$

where $\rho_{ij} = \mathscr{E}(\xi_i(X)\eta_j(Y))$, $\mathscr{E}(\cdot)$ denoting expectation in the bivariate distribution $f(x, y)$. The expansion (3) holds provided

$$\phi^2 + 1 = \int_{-\infty}^{\infty}\int_{-\infty}^{\infty} \{f(x, y)/g(x)h(y)\}^2$$
$$\times g(x)h(y)\,dx\,dy < \infty,$$

and then

$$\phi^2 = \sum_{i=j}^{\infty} \sum_{j=1}^{\infty} \rho_{ij}^2.$$

Again the expansion holds for more general distributions, provided they are ϕ^2-bounded. X and Y are independent if and only if $\phi^2 = 0$. Expansion (3) is defined by Lancaster [9] to be in canonical form if it is diagonal. That is,

$$f(x, y) = g(x)h(y)\left\{1 + \sum_{i=1}^{\infty} \rho_i\xi_i(x)\eta_i(y)\right\},$$
$$(4)$$

where $1 \geqslant \rho_1 \geqslant \rho_2 \geqslant \cdots \geqslant 0$. The nonzero coefficients $\{\rho_i\}$ are called the *canonical correlations* and the pairs $\{(\xi_i, \eta_i)\}$ the *canonical variables* (*see* CANONICAL ANALYSIS).

Properties are

$$\mathscr{E}(\xi_i(X)\eta_j(Y)) = \rho_i\delta_{ij}, \qquad i, j = 1, 2, \ldots,$$
$$\mathscr{E}(\xi_i(X)\,|\,Y) = \rho_i\eta_i(Y), \quad i = 1, 2, \ldots,$$

and, similarly for X,

$$\rho_1 = \sup_{(u,v)} |\mathscr{E}(u(X)v(Y))|,$$

where u, v have zero means and unit variances: X and Y are independent if and only if $\rho_1 = 0$. An expansion (4) can always be found for a ϕ-bounded distribution. The expansion (4) has a regression significance of potential practical importance. If $g(x)$ is a function with zero mean and finite variance, then it has an expansion of the form (1) and

$$\mathscr{E}(g(x)\,|\,Y) = \sum_{i=1}^{\infty} a_i\rho_i\eta_i(Y).$$

Many of the classical one-dimensional distributions have bivariate analogs of the form (4). The Meixner class has been studied by Eagleson [1]. The individual distributions in this class are additive. If U, V, W are independent random variables with the same type of distribution,

$$X = U + V, \qquad Y = V + W$$

have a bivariate distribution whose canonical variables are the orthonormal polynomials. The canonical correlations are found using a generating function* technique. In particular, if the variances of U, V, and W are $\sigma_1^2, \sigma_2^2, \sigma_3^2$, the first pair of canonical variables are linear and

$$\rho_1 = \text{correlation } (X, Y)$$
$$= \sigma_2^2/\sqrt{\{(\sigma_1^2 + \sigma_2^2)(\sigma_2^2 + \sigma_3^2)\}}.$$

Lancaster [10] has studied a class of bivariate binomial distributions and their multivariate generalizations. Let $(U_1, V_1), \ldots,$ (U_n, V_n) be independent pairs of random variables such that marginally

$$P(U_j = 1) = P(V_j = 1) = p,$$
$$P(U_j = 0) = P(V_j = 0) = 1 - p,$$

and correlation $(U_j, V_j) = r_j, j = 1, \ldots, n$.

Let $X = U_1 + \cdots + U_n$, $Y = V_1 + \cdots + V_n$, then (X, Y) has a bivariate binomial distribution that has an expansion (4) where the canonical variables are the orthonormal polynomials, and

$$\rho_j = \binom{n}{j}^{-1} \sum r_{i_1} r_{i_2} \cdots r_{i_j},$$

summation being over distinct products.

A problem studied by a number of authors is to characterize the sequences $\{\rho_i\}$, given the marginals and that $\{\xi_i\}, \{\eta_j\}$ are orthonormal polynomials, for which $f(x, y)$ in (3) is a bivariate distribution. The condition $1 \geqslant \rho_1 \geqslant \cdots \geqslant 0$ is relaxed for the characterization. A general result is that if the marginal distributions are identical and have an infinite number of points of increase, then it is necessary that $\{1, \rho_1, \rho_2, \ldots\}$ is a moment sequence of a random variable on $[-1, 1]$. Tyan and Thomas [14] prove this and also examine the nonidentical marginals case. Of particular interest and a more difficult problem, is to determine which sequences are sufficient. Bivariate normal, gamma, Poisson, negative binomial, and binomial distributions exist when the marginals are identical and $\rho_i = \rho^i$. The bivariate normal so obtained is the classical one where $-1 \leqslant \rho \leqslant 1$, and for the gamma, Poisson, and negative binomial $0 \leqslant \rho \leqslant 1$. When considering mixtures over ρ, it is necessary and sufficient that $\{\rho_i\}$ be a moment sequence of a random variable on $[-1, 1]$ for the bivariate normal and a moment sequence on $[0, 1]$ for the other three distributions.

Sarmanov [13] initiated this line of research with the normal distribution. If $F(x, y; \rho)$ is the distribution function of the classical bivariate normal with zero means and unit variances, then Sarmanov's bivariate distribution has the form

$$\int_{-1}^{1} F(x, y; \rho) \, dK(\rho),$$

where $K(\cdot)$ is a distribution function on $[-1, 1]$. Since the binomial does not have an infinite number of points of increase it is not necessary that $\{\rho_i\}$ be a moment sequence.

Eagleson [2] characterizes which of these sequences give bivariate binomial distributions. In particular $\rho_i = \rho^i$ is only a sequence of canonical correlations if

$$\max\{-(1-p)/p, -p/(1-p)\} \leqslant \rho \leqslant 1.$$

MULTIVARIATE EXPANSIONS

There is an immediate extension of (2) to multivariate distributions, e.g., for three variables

$$f(x, y, z) = g(x)h(y)k(z) \times \{1 + S_{xy} + S_{yz} + S_{xy} + S_{xyz}\},$$

where $S_{xy} = \sum_{i,j \geqslant 1} \rho_{ij} \xi_i(x) \eta_j(y)$ and similarly for the other two pairs, and for

$$S_{xyz} = \sum_{i,j,k \geqslant 1} \rho_{ijk} \xi_i(x) \eta_j(y) \zeta_k(z),$$

provided the distribution is ϕ^2-bounded. The three sequences are orthonormal and complete on their respective marginals and

$$\rho_{ijk} = \mathcal{E}(\xi_i(X) \eta_j(Y) \zeta_k(Z)),$$

where expectation is taken with respect to f. The (X, Y) bivariate distribution is

$$g(x)h(y)(1 + S_{xy}),$$

and it is similar for the other two pairs.

$$\phi^2 = \phi_{XY}^2 + \phi_{YZ}^2 + \phi_{XZ}^2 + \phi_{XYZ}^2,$$

where the first three are the bivariate ϕ^2's and

$$\phi_{XYZ}^2 = \sum_{i,j,k \geqslant 1} \rho_{ijk}^2.$$

The dependence structure of (X, Y, Z) is characterized by which of the ϕ^2's are zero. It is possible to have all three pairs of random variables independent, but X, Y, Z not mutually independent when

$$\phi_{XY}^2 = \phi_{XZ}^2 = \phi_{YZ}^2 = 0, \quad \text{but} \quad \phi_{XYZ}^2 \neq 0.$$

An application is provided in three-way contingency tables. The total χ^2 for the table can be partitioned into independent parts to test which of the four ϕ^2's are zero.

Eagleson [3] proves several limit theorems for U-statistics* by using orthogonal expansions. Let $h(x, y)$ be a symmetric function and denote its associated U-statistic by

$$U_n = \binom{n}{2}^{-1} \sum_{1 \leq i < j \leq n} h(X_i, X_j),$$

where $X_1, X_2, \ldots, X_n$ is a random sample from a distribution function F. h is called *degenerate* if

$\int h(x, y) \, dF(y)$ is a constant for almost all x.

If

$$\int \int h^2(x, y) \, dF(x) \, dF(y) < \infty$$

and h is degenerate, then there is a complete orthonormal sequence $\{\psi_n(x)\}$ on F (with $\psi_0(x) \equiv 1$) such that

$$h(x, y) = \sum_{k=0}^{\infty} \lambda_k \psi_k(x) \psi_k(y)$$

and

$$\sum_{k=0}^{\infty} \lambda_k^2 < \infty.$$

For such a U-statistic, $n(U_n - \mathscr{E}(U_n))$ converges in distribution to

$$\sum_{k=1}^{\infty} \lambda_k (Z_k^2 - 1)$$

as $n \to \infty$, where $\{Z_k\}$ is a sequence of mutually independent normal random variables with zero means and unit variances.

The classical limit theorem's assertion that for nondegenerate U-statistics, $n^{1/2}(U_n - \mathscr{E}(U_n))$ converges in distribution to normality with zero mean and variance

$$4\mathscr{E}(h(X_1, X_2) h(X_1, X_3))$$

comes out simply from the limit theorem for degenerate U-statistics.

Griffiths [4] obtains a sequence of orthogonal polynomials on the multinomial and constructs a bivariate multinomial distribution in the following form. Let $\{p_{ij}; i = 1, \ldots, r, j = 1, \ldots, s\}$ be a bivariate probability distribution with, say, $r \leq s$ and an expansion of the form (3):

$$p_{ij} = p_i \cdot p_{\cdot j} \left\{ 1 + \sum_{m=1}^{r-1} \rho_m \xi_m(i) \eta_m(j) \right\}.$$

Now suppose a random sample of n pairs

(U_k, V_k) is taken from $\{p_{ij}\}$, and let

X_i = number of U_k equal to i,

Y_j = number of V_k equal to j.

Then $(\mathbf{X}, \mathbf{Y})$ has a bivariate multinomial distribution

$$P(\mathbf{x}) P(\mathbf{y}) \left\{ 1 + \sum_m \rho_1^{m_1} \cdots \rho_{r-1}^{m_{r-1}} \right.$$
$$\left. \times Q_{\mathbf{m}}(\mathbf{x}) Q_{\mathbf{m}}(\mathbf{y}) \right\},$$

where

$$P(\mathbf{x}) = \frac{n!}{x_1! \ldots x_r!} \, p_1^{x_1} \cdots p_r^{x_r},$$

there is a similar expression for $\mathbf{y}$, and $\{Q_{\mathbf{m}}\}$ is a set of multidimensional orthonormal polynomials.

Griffiths and Milne [5] obtain a characterization of a class of bivariate Poisson processes in terms of orthogonal functions. In its simplest form, let $\{N_1(t); t \geq 0\}$, $\{N_2(t); t \geq 0\}$ be two homogeneous Poisson processes* with rates 1 and $I_1, \ldots, I_k$ disjoint intervals of lengths $a_1, \ldots, a_k$. Denote $p(x; a) = e^{-a} a^x / x!$ and let $\{c_n(x; a)\}$ be the orthonormal Poisson–Charlier polynomial set on $p(x; a)$.

Denote

$$X_j = N_1(I_j), \qquad Y_j = N_2(I_j),$$
$$j = 1, \ldots, k,$$

$$p(\mathbf{x}; \mathbf{a}) = \prod_{j=1}^{k} p(x_j; a_j),$$

$$C_{\mathbf{n}}(\mathbf{x}, \mathbf{a}) = \prod_{j=1}^{k} c_{n_j}(x_j, a_j)$$

and set up a similar expression for y. Griffiths and Milne use these to obtain a characterization of the joint distribution

$$p(\mathbf{x}, \mathbf{y}) = p(\mathbf{x}; \mathbf{a}) p(\mathbf{y}; \mathbf{a})$$
$$\times \left\{ 1 + \sum_{\mathbf{n}} \rho_{\mathbf{n}} C_{\mathbf{n}}(\mathbf{x}; \mathbf{a}) C_{\mathbf{n}}(\mathbf{y}; \mathbf{a}) \right\}$$

in that the constants $\{\rho_{\mathbf{n}}\}$ be moments $\rho_{\mathbf{n}} = \mathscr{E}(Z_1^{n_1} \ldots Z_k^{n_k})$ of random variables on $[0, 1]$. The random variables have an interpretation as $Z_j = Z(I_j)$, where $\{Z(\cdot)\}$ is a stochastic process on [0, 1]. Series expansions of some other multivariate distributions are found in Johnson and Kotz [6].

Orthogonal expansions are of importance in stochastic processes*. Under regularity conditions the density of a reversible Markov process* $\{X(t); \ t \geqslant 0\}$ will have the form

$$f(x \mid x_0, t) = f(x)\left\{ 1 + \sum_{i=1}^{\infty} e^{-\lambda_i t} \xi_i(x) \xi_i(x_0) \right\},$$

$$(5)$$

where $f(x)$ is the stationary density and $X(0) = x_0$. Karlin and McGregor [8] derive an expansion in population genetics* where $X(t)$ is a gene frequency. This gives a class of conditional densities of the form (5), where

$$f(x) = \frac{\Gamma(\mu + \nu)}{\Gamma(\mu)\Gamma(\nu)} x^{\mu - 1}(1 - x)^{\nu - 1},$$

$$0 < x < 1, \quad \mu, \nu > 0,$$

$\{\xi_i(x)\}$ is the set of orthonormal polynomials on the beta distribution (easily found from the Jacobi polynomials), and $\lambda_i = \frac{1}{2} i(i + \mu + \nu - 1)t$. The backward diffusion equation of the process is

$$\frac{\partial f}{\partial t} = \frac{1}{2} x(1 - x) \frac{\partial^2 f}{\partial x^2}$$

$$- \left[\mu(1 - x) + \nu x \right] \frac{\partial f}{\partial x}.$$

The density is found as a limit from a discrete birth-and-death process*, whose distribution has a similar expansion in terms of the Hahn polynomials.

Karlin and McGregor [7] have constructed an orthogonal expansion of the transition probabilities for birth-and-death processes*. Let λ_i, μ_i (> 0) be constants and

$$P_{ij}(t) = P(X(t + s) = j \mid X(s) = i)$$

be such that

$$P_{ii+1}(t) = \lambda_i t + o(t)$$

$$P_{ii}(t) = 1 - (\lambda_i + \mu_i)t + o(t)$$

$$P_{ii-1}(t) = \mu_i t + o(t) \qquad \text{as} \quad t \to 0.$$

Denote

$$\pi_0 = 1,$$

$$\pi_n = (\lambda_0 \ldots \lambda_{n-1})/(\mu_1 \ldots \mu_n), \qquad n \geqslant 1.$$

Then there is a representation

$$P_{ij}(t) = \pi_j \int_0^\infty e^{-xt} Q_i(x) Q_j(x) \, d\psi(x),$$

where ψ is a distribution function. $\{Q_i(x)\}$ is an orthogonal polynomial set, defined by $Q_0(x) \equiv 1$, as

$$- xQ_0(x) = -(\lambda_0 + \mu_0)Q_0(x) + \lambda_0 Q_1(x),$$

$$- xQ_n(x) = \mu_n Q_{n-1}(x) - (\lambda_n + \mu_n)Q_n(x)$$

$$+ \lambda_n Q_{n+1}(x), \qquad n \geqslant 1.$$

There is a similar representation for random walks*.

References

[1] Eagleson, G. K. (1964). *Ann. Math. Statist.*, **35**, 1208–1215.

[2] Eagleson, G. K. (1969). *Aust. J. Statist.*, **11**, 29–38. (An application of orthogonal functions.)

[3] Eagleson, G. K. (1979). *Aust. J. Statist.*, **21**, 221–237.

[4] Griffiths, R. C. (1971). *Ann. Math. Statist.*, **13**, 27–35.

[5] Griffiths, R. C. and Milne, R. K. (1978). *J. Multivariate Anal.*, **8**, 380–395.

[6] Johnson, N. L. and Kotz, S. (1972). *Distributions in Statistics: Continuous Multivariate Distributions.* Wiley, New York.

[7] Karlin, S. and McGregor, J. (1957). *Trans. Amer. Math. Soc.*, **85**, 489–546. (An application of orthogonal functions.)

[8] Karlin, S. and McGregor, J. (1962). *Proc. Camb. Philos. Soc.*, **58**, 299–311. (An application of orthogonal functions.)

[9] Lancaster, H. O. (1969). *The Chi-squared Distribution.* Wiley, New York. (Includes an excellent bibliography and the application of orthogonal function techniques to bivariate distributions and χ^2.)

[10] Lancaster, H. O. (1974). In *Studies in Probability and Statistics* (In honor of E. J. G. Pitman), E. J. Williams, ed. North-Holland, Amsterdam, pp. 13–19.

[11] Lancaster, H. O. (1979). *Aust. J. Statist.*, **21**, 188–192. (This volume of *Aust. J. Statist.* is dedicated to H. O. Lancaster and has some interesting applications of orthogonal functions.)

[12] Meixner, J. (1934). *J. Lond. Math. Soc.*, **9**, 6–13. (Historical paper dealing with orthogonal polynomials on a class of statistical distributions.)

[13] Sarmanov, O. V. (1966). *Dokl. Akad. Nauk. SSSR*, **168**, 32–35.

[14] Tyan, S., Derin, T. and Thomas, J. B. (1976). *Ann. Statist.*, **4**, 216–222. (Includes references to papers dealing with the characterization of classes of bivariate distributions.)

(ASSOCIATION, MEASURES OF
CONTINGENCY TABLES
DEPENDENCE, CONCEPTS OF
GRAM–CHARLIER SERIES
JACOBI POLYNOMIALS
KRAWTCHOUK POLYNOMIALS
MEIXNER POLYNOMIALS
MULTIVARIATE DISTRIBUTIONS
U-STATISTICS)

R. C. GRIFFITHS

ORTHOGONALIZATION OF MATRICES *See* LINEAR ALGEBRA, COMPUTATIONAL

ORTHOGONAL LATIN SQUARES *See* GRAECO-LATIN SQUARES

ORTHOGONALLY INVARIANT DISTRIBUTIONS *See* ISOTROPIC DISTRIBUTIONS

ORTHOGONAL MATRIX

A real-valued square matrix Γ is called orthogonal if its rows considered as vectors are mutually perpendicular and have unit lengths. This implies that the columns are also mutually perpendicular and have orthogonal unit lengths. Alternatively, a matrix is orthogonal if its inverse equals its transpose. The totality of orthogonal matrices form a group with respect to matrix multiplication.

The determinant of an orthogonal matrix is either 1 or -1. If $|N| = 1$, it is called a *proper orthogonal matrix*. Orthogonal matrices are used extensively in multivariate statistical analysis.

(ORTHOGONAL TRANSFORMATION)

ORTHOGONAL POLYNOMIALS *See* CHEBYSHEV–HERMITE POLYNOMIALS; JACOBI POLYNOMIALS; LAGUERRE SERIES; ORTHOGONAL EXPANSIONS

ORTHOGONAL PROCESSES

Orthogonal processes are stochastic processes* with mean zero, whose increments over disjoint intervals are uncorrelated. They are also called *orthogonal increment processes*. Formally, a complex-valued process $Z(t)$, $-\infty < t < \infty$ is an orthogonal process if $Z(0) = 0$ and

$$E\left[(Z(t_2) - Z(t_1)) \overline{(Z(t_3) - Z(t_4))} \right] = 0$$

for all $t_1 < t_2 < t_3 < t_4$. The bar denotes the complex conjugate and can be ignored if $Z(t)$ is real-valued.

All finite variance Levy processes $Z(t)$ with their mean subtracted are real-valued orthogonal processes because their increments over disjoint intervals of the same length are iid. Examples include Brownian motion*, the Poisson process*, and the gamma process, each with their mean subtracted. If $h(t)$ is a monotone function, then $Z(h(t)) - EZ(h(t))$ is also an orthogonal process, but its increments may not be stationary anymore. *See* SELF-SIMILAR PROCESSES for examples of orthogonal processes whose increments are stationary but dependent.

Mean zero orthogonal processes are used to define stochastic integrals* and these in turn play a fundamental role in the spectral* representation of a weakly stationary process*. Let $X(t)$ be an orthogonal increment process with mean zero, and variance $F(t) = E|X(t)|^2$. Then the integral $Y(t) = \int_{-\infty}^{+\infty} g(t) \, dX(t)$ is well defined for all complex-valued functions g satisfying $\int_{-\infty}^{+\infty} |g(t)|^2 \, dF(t) < \infty$. The process $Y(t)$ is called a *general linear process*, and it can be interpreted as the output of a linear filter whose inputs are the increments of $X(t)$. One can also define the Wiener–Itô integrals

$$Y(t) = \int \cdots \int g(t_1, \ldots, t_n) \, dX(t_1) \ldots dX(t_n)$$

for all complex-valued functions $g(t_1, \ldots, t_n)$ satisfying

$$\int \cdots \int |g(t_1, \ldots, t_n)|^2 \, dF(t_1) \ldots dF(t_n) < \infty.$$

These $Y(t)$ can be interpreted as outputs of nonlinear filters.

Complex-valued orthogonal processes appear in the spectral representation of time series. Let $Y(t)$, $-\infty < t < \infty$, be a real-valued, mean zero, weakly stationary process with continuous covariances $R(t) = EY(s)Y(s + t)$. Then $Y(t)$ has the same covariances as its spectral representation $\int_{-\infty}^{+\infty} e^{itv} \, dZ(v)$, where $Z(v)$ is a complex-valued orthogonal process with variance $E|Z(v)|^2 = F(v)$. The function $F(v)$ is such that $R(t) = \int_{-\infty}^{+\infty} e^{itv} \, dF(v)$. The index $|v|$ can be interpreted as a frequency, and the function $F(v)$ as the integrated spectrum. If $Y(t)$ is a time series* defined on $t = 0, \pm 1, \pm 2, \ldots$, then the range of integration in the spectral representation becomes $-\pi$ to π instead of $-\infty$ to $+\infty$. *See also* PREDICTION and SPECTRAL ANALYSIS.

MURAD S. TAQQU

ORTHOGONAL SQUARES   *See* GRAECO-LATIN SQUARES; LATIN SQUARES

ORTHOGONAL TRANSFORMATION

In statistics, this term usually means a *linear* orthogonal transformation, that is, a linear transformation from a set of m variables $\mathbf{X}$ ($m \times 1$) to another set $\mathbf{Y}$ such that

$$\mathbf{Y} = \mathbf{\Gamma X},$$

where $\mathbf{\Gamma}$ is an $m \times m$ orthogonal matrix*. It is sometimes convenient to visualize this as a rotation of Cartesian axes in m-dimensional space. The orthogonality property ensures that:

1. The sums of squares (corresponding to squared distances of $(Y_1, \ldots, Y_m)$ and $(X_1, \ldots, X_m)$ from the origin) $\mathbf{Y'Y}$ and $\mathbf{X'X}$ are equal.

2. If $\mathbf{X}_1, \mathbf{X}_2$ are two $\mathbf{X}$ vectors and $\mathbf{Y}_1, \mathbf{Y}_2$ are their corresponding $\mathbf{Y}$ transforms, $\mathbf{X}_1'\mathbf{X}_2 = \mathbf{Y}_1'\mathbf{Y}_2$, i.e., the sums of their products are unchanged.

The Jacobian of an orthogonal transformation $|\mathbf{\Gamma}|$ is 1.

From property 1, it follows that if the joint density function of $X_1, \ldots, X_m$ is spherically symmetrical*, i.e., if $f_{\mathbf{X}}(\mathbf{x}) = g(\sum_{j=1}^{n} x_j^2)$, then that of $Y_1, \ldots, Y_m$ is also spherically symmetrical. As a special case, if $X_1, \ldots, X_m$ are independent normal variables each with zero mean and variance σ^2, so are $Y_1, \ldots, Y_m$.

(ORTHOGONAL EXPANSIONS
ORTHOGONAL MATRIX)

ORTHONORMAL EXPANSIONS   *See* ORTHOGONAL EXPANSIONS

OSCILLATORY PROCESS   *See* NARROWBAND PROCESS

OSCULATORY INTEGRATION

This is effected by replacing the integral by its estimate according to an osculatory interpolation* formula and integrating the resulting polynomial.

Bibliography

Salzar, H. E., Shoultz, D. C., and Thompson, E. P. (1960). *Tables of Osculatory Integration Coefficients.* Convair Aeronautics, San Diego, CA.

(NUMERICAL INTEGRATION)

OSCULATORY INTERPOLATION

Interpolation* is usually effected by fitting a polynomial to a number of successive available values of the function, say, u_x, to be estimated. Effectively, several polynomial arcs are fitted, giving rise to a continuous fitted function, but one with discontinuous slope (values of derivative du_x/dx). In *osculatory* interpolation it is further ensured that the first derivative, also, is continuous. (If second or higher derivatives u are made to be continuous, we have *hyperosculatory* interpolation.)

The simplest formula for osculatory interpolation between x and $x+1$ is

$$u_{x+1} = (1-t)u_x + tu_{x+1}$$
$$+ \tfrac{1}{4}t(t-1)(\delta^2 u_x + \delta^2 u_{x+1})$$
$$+ \tfrac{1}{2}t(t-\tfrac{1}{2})(t-1)\delta^3 u_{x+1/2}, \quad (1)$$

where δ is the central difference* operator $(\delta u_x = u_{x+1/2} - u_{x-1/2})$. In (1), the first three terms on the right-hand side are identical with the first three terms of Bessel's interpolation formula*.

Between $x+1$ and $x+2$, the formula will of course, be

$$u_{x+1+t} = (1-t)u_{x+1} + tu_{x+2}$$
$$+ \tfrac{1}{4}(t-1)(\delta^2 u_{x+1} + \delta^2 u_{x+2})$$
$$+ \tfrac{1}{2}t(t-\tfrac{1}{2})(t-1)\delta^3 u_{x+3/2}.$$

Letting $t \to 1$ in (1) and $t \to 0$ in (2), we get the common limit u_{x+1}.

An alternative formula for (1) is

$$u_{x+t} = (1-t)u_x + tu_{x+1} + \tfrac{1}{2}t(t-1)\Delta^2 u_x$$
$$+ \tfrac{1}{2}t(t-1)^2\Delta^3 u_{x-1},$$

where Δ is the forward difference* operator $(\Delta u_x = u_{x+1} - u_x)$.

If the values of derivative(s) du_x/dx $(d^2 u_x/dx^2, \dots)$ are known for the same values of x for which u_x is known, we may use special forms of *Hermite's general osculatory interpolation formula*:

$$u_{x+t} = \sum_{j=-[(n-1)/2]}^{[n/2]} \{L_{j,n}(t)\}^2 \left[\left\{ 1 - 2\frac{dL_{j,n}(t)}{dt} \right\} \right.$$
$$\left. - u_{x+j} + (t-j)\frac{du_{x+j}}{dx} \right],$$

where $[s]$ denotes the integer part of s and

$$L_{j,n} = \prod_{\substack{i=-[(n-1)/2] \\ i \neq j}}^{[n/2]} \left(\frac{t-i}{j-i} \right), \quad 0 < t < 1.$$

This provides osculatory interpolation between x and $x+1$. The formula is exact if u_x is a polynomial of degree less than $2n$. Tables in Salzer [2] assist in the use of this formula.

Recent developments are described in SPLINE FUNCTIONS. Formulas and tables have also been developed for bivariate osculatory interpolation (see Salzer and Kimbro [1]).

References

[1] Salzer, H. E. and Kimbro, G. M. (1958). *Tables for Bivariate Osculatory Interpolation over a Cartesian Grid*. Convair Aeronautics, San Diego, CA.

[2] Salzer, H. E. (1959). *Tables of Osculatory Interpolation Coefficients*. Natl. Bur. Stand. (U.S.) Appl. Math. Ser. **56** (Washington, DC).

(INTERPOLATION
SPLINE FUNCTIONS)

OSTROWSKI–REICH THEOREM   *See* GAUSS–SEIDEL ITERATION

OUTLIER-PRONE DISTRIBUTION

Statistical inference based on normal theory often treats outlying observations as aberrations to be detected and discarded. Based on their study of daily rainfall data, Neyman and Scott [5] suggested that some data might be drawn from a distribution likely to produce outliers—an *outlier-prone distribution*. Far from being abberrations, apparent outliers might be the most important observations.

Neyman and Scott made their idea precise by defining an *outlier-prone family of distributions*. Using Dixon's definition [1], they termed the largest observation $x_{n:n}$ a *k-outlier on the right* if its value exceeds that of $x_{n-1:n}$ by more than $k(x_{n-1:n} - x_{1:n})$. For particular values $n \geqslant 3$ and $k > 0$, a family of distributions $\mathscr{F}$ is termed (k, n)-*outlier-prone* if for each value of $p < 1$ there is an $F \in \mathscr{F}$ such that the probability of a k-outlier from a sample of size n is at least p.

If a family is (k, n)-outlier-prone for some $n \geqslant 3$ and $k > 0$, then it is (k, n)-outlier-prone for all such n and k [3], which Neyman and Scott referred to as *outlier-prone completely*. Families not outlier-prone are *outlier-resistant*. Examples of outlier-prone families are the gamma family and the log-normal family. Unfortunately, the condition that the probability of finding a k-outlier be arbitrarily high is too strong to make this definition useful in practice; no single distribution is outlier-prone in the sense of Neyman and Scott.

The idea of outlier-proneness is applied to individual distributions by weakening the definition. Green [4] considered the absolute and relative differences between the two largest values, $A(n) = x_{n:n} - x_{n-1:n}$ and $R(n) = x_{n:n}/x_{n-1:n}$, respectively, for large sample sizes. A distribution F is termed *absolutely outlier-prone* if there exist values $a > 0$ and $p > 0$ such that

$$\lim_{n \to \infty} P\{A(n) > a\} \geqslant p.$$

If, for all $a > 0$,

$$\lim_{n \to \infty} P\{A(n) > a\} = 0,$$

then F is *absolutely outlier-resistant*. *Relatively outlier-prone* and *outlier-resistant* are defined similarly for nonnegative distributions.

Using these definitions, the emphasis is on outlier resistance. Conditions for absolute and relative outlier resistance are equivalent to those given by Gnedenko [2] for the law of large numbers and relative stability of maxima, respectively. With these definitions, normal distributions are absolutely and relatively outlier-resistant, Cauchy distributions are absolutely and relatively outlier-prone, and gamma distributions are absolutely outlier prone, but relatively outlier-resistant.

The mathematical definitions of outlier-prone and outlier-resistant are of little practical importance, but they generate two important practical ideas. First, some distributions are likely to produce outliers, and this should be taken into account. Second, for outlier-resistant distributions, parameters may be estimated consistently using extremes. These estimates will be consistent even if the sample size is known only to order of magnitude, or if data is randomly censored so that extreme observations are more available than others.

References

[1] Dixon, W. J. (1950). *Ann. Math. Statist.*, **21**, 488–506.

[2] Gnedenko, B. (1943). *Ann. Math.*, **44**, 423–453. (An excellent paper, gives limit theorems for maxima analogous to those for sums. In French.)

[3] Green, R. F. (1974). *Ann. Statist.*, **2**, 1293–1295. (Shows that (k, n)-outlier-prone is equivalent to outlier-prone completely.)

[4] Green, R. F. (1976). *J. Amer. Statist. Ass.*, **71**, 502–505. (Defines outlier-prone and outlier-resistant distributions and shows the connection with extreme-value theory.)

[5] Neyman, J. and Scott, E. L. (1971). In *Optimizing Methods in Statistics*, J. Rustagi, ed. Academic Press, New York, pp. 413–430. (Introduces the idea of outlier-proneness and defines outlier-prone family of distributions.)

(ORDER STATISTICS OUTLIERS)

RICHARD F. GREEN

OUTLIERS

The intuitive definition of an outlier, which is adopted in this entry, is some observation whose discordancy from the majority of the sample is excessive in relation to the assumed distributional model for the sample, thereby leading to the suspicion that it is not generated by this model. Note the key fact that outliers are only in relation to a distributional model. If the model is changed, they may become concordant.

When a sample contains outliers, they may give rise to two superficially distinct problems. In *accommodation*, one wishes to use the sample to make model inferences that are minimally affected by the number, nature, or values of any observations—the outliers are regarded as a mere nuisance impeding such inferences. In *identification*, the object is to partition the sample into inliers and outliers, so that the latter, objects of interest in their own right, may be studied. The distinction is partially artificial in that good identification methods can provide accommodation and vice versa. As accommodation methods are covered separately under the headings of robust and nonparametric inference (see also Huber [4]), no more will be said about them here.

Identification of outliers can be put within the conventional frameworks of statistics very easily by the formulation of suitable models. A powerful class of such models is the mixture models class. Given a sample $X_1 \ldots X_n$, with probability $(1 - p)$, observation X_i comes from the "base" distribution $f_{1i}(\cdot \mid \theta)$, while with probability p, it comes from the contaminating distribution $f_{2i}(\cdot \mid \theta, \tau)$. In practice, f_{2i} has appreciable density where f_{1i} does not, generally in one or both tails, and $f_{1i}(\cdot \mid \theta) = f_{2i}(\cdot \mid \theta, \tau = 0)$. An example is $f_1 = N(\theta_1, \theta_2)$, $f_2 = N(\theta_1 + \tau, \theta_2)$. It does not follow that all contaminants will be outlying; however, likelihood-ratio considerations show that the best classification rule is to identify the furthest outlying observations as the contaminants.

The transition from the mixture model to the sample may be thought of, if appropriate, as occurring in two stages: in the first stage, Bernoulli trials with probability p take place, successes giving rise to contaminants or outliers; failures to inliers. The total number of contaminants, say, K, follows a binomial bi(n, p) distribution. Subsequently, X_i is drawn from f_{1i} if it is an inlier or f_{2i}, if a contaminant.

Looking at the intermediate stage, if K is zero, then there are no outliers. Otherwise condition on K. Let $j_1, j_2, \ldots, j_K$ be the indices of the X's that are outliers. A model in which $\{ j_1 \ldots j_K \}$ are known is termed *labeled*—for such models it is known which observations are outliers. If the $\{ j_i \}$ are not known, the model is termed *unlabeled*, and unlabeled models may be further partitioned into those in which K is known and those in which it is unknown.

For testing purposes, one wishes to know whether outliers are present. If a particular K is specified, then with a labeled model the test reduces to a test of the hypothesis $\tau = 0$. This is easily carried out using general Neyman–Pearson* theory, and, in particular, generalized likelihood ratio tests*. If the model is unlabeled but K is specified, then the test may be carried out in the same way by taking each set of possible labels $\{ j_1, \ldots, j_K \}$, evaluating the test statistics using these labels, and then making the identification and simultaneously carrying out the test by finding that set of labels for which this statistic is maximally discrepant from the null hypothesis. This generally applicable procedure produces many of the standard tests whose optimality has been proved directly, such as the maximum residual (studentized if the variance is unknown) for normal samples and the ratio of the largest or smallest x_i to the mean for gamma-distributed samples.

If the model is unlabeled and K is also unknown, then further options arise in that one may carry out *block* tests, in which blocks of suspect outliers of various sizes are tested, or *consecutive* tests. The latter set up a sequence of hypotheses of $0, 1, 2, \ldots, K_0$ outliers, and then test each as null against its successor alternative, proceeding either from left to right (forward testing) or from right to left (backward testing).

The block tests require some subsidiary rule to determine the number of outliers to test for. Although many such rules have been proposed (see the discussion following Beckman and Cook [2]), none is fully satisfactory, and thus block tests appear to be inferior to consecutive tests, though their power at the correct K is very much higher.

Obtaining the critical points for these outlier tests is, in general, extremely difficult. The best studied case is $K = 1$, testing for a single outlier against a no-outlier alternative. Here the two-stage model for the unlabeled procedure has a distribution whose tail is bounded and generally superbly approximated by the Bonferroni inequality*. Letting T_i denote the value of the two-sample test for the labeled model in which X_i is specified as the outlier,

$$\Pr\left[\bigcup_{i=1}^{n}(T_i \geqslant c_i)\right] \leqslant \sum_{i=1}^{n}\Pr[T_i \geqslant c_i].$$

Thus, for example, by choosing each c_i so that $\Pr[T_i \geqslant c_i] = \alpha/n$, an overall size not exceeding α is obtained.

While applicable as a bound for $k = 2$, 3, . . . outliers, the Bonferroni method for many (but not all) models and test statistics rapidly becomes hopelessly conservative. For example, consider a sample of size 10 from $N(\xi, \sigma^2)$ with both parameters unknown, but with a $\sigma^2\chi_5^2$ variate external to the sample providing additional information on σ^2. The F statistic to test for two outliers has a 5% point that is the $0.05/7.69$ point of the F-distribution* and not the $0.05/90$ implied by the Bonferroni inequality.

Particularly when more than one outlier is present in the sample, the problems of *masking* and *swamping** may occur. Masking arises when the configuration of the model or of other outliers makes an outlier appear inlying *to the particular method of identification being used.* Swamping is the inverse problem in which outliers, configuration, and method conspire to make inliers appear outlying. While both problems become critical with structured data (e.g., regressions), they can arise in even simple random samples and militate against the use of procedures subject to them. These include:

1. Consecutive testing in which suspected outliers are forward-tested (masking-prone).

2. Attempts to remove more than one outlier at a time using ordinary residuals (masking- and swamping-prone).

3. Block testing when the value of K is in doubt (masking- or swamping-prone, depending on the error in K).

4. Gap-based rules for selecting K (masking-prone).

The best-studied distribution in outlier theory is, not surprisingly, the normal. The mixture model here might be of a basic $N(\xi_i, \sigma^2)$ model with a contaminating $N(\xi_i + \delta_i, \sigma^2)$; or $N(\xi_i, \lambda^2\sigma^2)$ distributions (in the latter case, $\lambda^2 > 1$). The two models of location and scale contamination, respectively, often are used interchangeably or indiscriminately [and the first gives rise to the second if a $N(0, (\lambda^2 - 1)\sigma^2)$ distribution is put on δ_i], but in fact they are distinct logically in that in the latter model the outliers convey some information about ξ_i, whereas in the former they do not.

Special cases of this model are the simple random sample case, $\xi_i = \xi$, and linear regression $\xi_i = w_i\beta$. In principle, both can be solved easily by setting up the location contamination model and the resulting test statistic for K outliers is the F ratio (or any monotonic function thereof) obtained from the residual sum of squares of the entire sample and of the sample less the K suspected outliers. As noted above, in the unlabeled case, the test and labeling will be obtained by maximizing this F ratio over all possible sets of labels for the K outliers. In the regression model, it can also be helpful (and is mathematically equivalent) to write the linear model with possible outliers as

$$\mathbf{Y}_{n\times 1} = \mathbf{X}_{n\times p}\boldsymbol{\beta}_{p\times 1} + \mathbf{I}_n\boldsymbol{\delta}_{n\times 1} + \boldsymbol{\epsilon}_{n\times 1}, \quad (*)$$

where K elements of the regression parameter $\boldsymbol{\delta}$ will be nonzero, identifying the outliers. Thus the outlier identification and testing are formally equivalent to solving a subset regression.

This assumes that the δ_i are arbitrary. Another model holds that the nonzero δ_i are equal, in which case the F-test is replaced by a two-sample t-test contrasting the outliers with the inliers; this test statistic has an easy equivalent with regression data. As one seldom has a situation in which multiple outliers are known to be shifted in mean by the same amount, this model has had a very limited impact.

The sequential test for multiple outliers (K unknown) for the normal model reduces to the stepwise-regression t tests for successive introduction of parameters δ_i to the model $(*)$. To avoid problems of masking and swamping, the testing must be applied backward—i.e., by first eliminating all possible outliers and then testing for replacement in the sample one by one, stopping replacement at the first statistically significant test statistic. There is some empirical evidence that an overall size not exceeding α may be obtained by testing the ith deleted suspected outlier, applying the $0.75\alpha/(n - i + 1)$ fractile to the t value for the observation.

These univariate models all have multivariate analogs: For example, the univariate t-test becomes a Hotelling's T^2 test*, or, equivalently, Wilks' likelihood ratio (*see* LAMBDA CRITERION, WILKS'S). There are no difficulties of principle with these extensions to multivariate data, though, as is generally the case in multivariate analysis, dropping the usual requirement of invariance under full-rank transformation may give rise to special-purpose tests more powerful than Wilks'. If, for example, it is known that any outliers must take the form of a mean shift in the first component of $\mathbf{X}$, then the optimal test will look only at this component, using the others as covariates, and it is reduced to a univariate test for an outlier from the regression of X_1 on the remaining components of $\mathbf{X}$.

When multiple outliers are to be located, the only sure method is an evaluation of the test statistics for all $\binom{n}{k}$ labelings—a heavy computational task. There is considerable interest in the questions of whether multiple outliers may be located reliably without this combinatorial search. To date it has been shown that ordinary regression residuals are completely unable to detect multiple outliers reliably in one pass, and that stepwise operation of using residuals to locate one outlier, deleting it, recomputing residuals, and so on, is reliable with simple random samples but not necessarily with regression data.

Given the Bayesian flavor of the mixture model, it is not surprising that simultaneous accommodation/identification Bayesian procedures exist. There is a practical problem with these in that they involve the very substantial combinatorial calculations associated with all partitions of the data into inliers and outliers with varying numbers of the latter; the methods are, however, very elegant. A somewhat related approach is the application of the E-M algorithm (*see* MISSING INFORMATION PRINCIPLE) to the mixture model of the data, which gives accommodation in the form of robust estimates of the f_{1i} parameters and identification in the form of posterior probabilities of each observation's coming from the contaminating f_{2i} distribution.

While the normal distribution has received the most attention, the general two-stage procedure sketched in this article has been applied to the gamma distribution, to time series, and to contingency tables.

Apart from the conventional uses already sketched for the outlier model, it has intriguing possibilities as a parsimonious method of describing interactions in multiway contingency or ANOVA tables in terms of more additive models plus several identified discordant cells.

The literature on outliers is vast, but the best starting points for obtaining more detail on the problems and techniques are the two recent books by Barnett and Lewis [1] and Hawkins [3] and the review article by Beckman and Cook [2]. Barnett and Lewis provides excellent broad coverage of the different methods to be found in the literature up to the mid-1970s. Hawkins' coverage is limited generally to optimal tests and is more suitable for research or similar high-level work in the area. Beckman and Cook recover some of the ground in both books briefly, but also provide a crisp discussion of the post-1978 publications, a period covering, in particular, most of the work on structured data.

References

[1] Barnett, V. and Lewis, T. (1978). *Outliers in Statistical Data*. Wiley, New York.

[2] Beckman, R. J. and Cook, R. D. (1983). *Technometrics*, **25**, 119–149. (Discussion on pp. 150–163.)

[3] Hawkins, D. M. (1980). *Identification of Outliers*. Chapman and Hall, London.

[4] Huber, P. J. (1981), *Robust Statistics*. Wiley, New York.

(CHAUVENET'S CRITERION
MASKING AND SWAMPING
MIXTURE DISTRIBUTIONS
ORDER STATISTICS
OUTLIER-PRONE DISTRIBUTION
PEIRCE'S CRITERION
THOMPSON'S CRITERION)

DOUGLAS M. HAWKINS

OUTPUT INDEX *See* INDEX OF INDUS-
TRIAL PRODUCTION; PRODUCTIVITY MEASURE-
MENT

OVERALL AND WOODWARD TEST FOR HOMOGENEITY OF VARIANCES

The Overall and Woodward (O&W) test
was first defined in 1974 [7]. The original
authors have not made any further attempts
to expand on this article (as of May 1983).

The goal is to test for heterogeneity of
variances "in the analysis of factorial effects
on sample variances in more complex designs" [7, p. 311]. Given JK samples of sizes
n_{jk} from independent populations with unknown means and unknown variances, σ_{jk}^2,
we may define marginal variances using additive procedures typically used on means.
Then $\sigma_{j.}^2 = \sum^K \sigma_{jk}^2 / K$, $\sigma_{.k}^2 = \sum^J \sigma_{jk}^2 / J$, and $\sigma_{..}^2$
$= \sum\sum^{JK} \sigma_{jk}^2 / JK$. Interaction effects may be
defined as $(\alpha\beta)_{jk} = \sigma_{jk}^2 - \sigma_{j.}^2 - \sigma_{.k}^2 + \sigma_{..}^2$.
Then there will be three null hypotheses as
in the analysis of variance on means in a
factorial design. $H_{01}: \sigma_{j.}^2 = \sigma_{..}^2$, $H_{02}: \sigma_{.k}^2$
$= \sigma_{..}^2$, and $H_{03}: (\alpha\beta)_{jk} = 0$ for all j, k. Overall and Woodward define the statistic

$$Z_{jk} = \sqrt{\frac{c_{jk}(n_{jk} - 1)s_{jk}^2}{MS_W}} - \sqrt{c_{jk}(n_{jk} - 1) - 1},$$

$$(1)$$

where

$$c_{jk} = 2 + (1/n_{jk}),$$

$$MS_W = \frac{\sum_k^K \sum_j^J (n_{jk} - 1)s_{jk}^2}{\sum_k^K \sum_j^J (n_{jk} - 1)},$$

and s_{jk}^2 is the usual unbiased estimate of the
cell variance.

The statistic is an extension of the Fisher
and Yates [4] Z-score transformation for
chi-square statistics. Once the sample variances have been converted to this form, then
the Z_{jk} entries are submitted to a conventional analysis of variance, and the mean
square (MS) ratios computed for the row
effect (MS$_A$), the column effect (MS$_B$), and
the interactions (MS$_{AB}$). These MS's are
each tested against the expected theoretical
variance of 1.0. Thus MS$_A$/1.0 is compared
to a critical value of $F_{\alpha, J-1, \infty}$. If MS$_A$ is
larger than the critical value, H_{01} is rejected;
otherwise it is retained. Similar procedures
are used on the other two hypotheses.

When $n_{jk} = n$, then $c_{jk} = c$ is a constant
and

$$Z_{jk} = \sqrt{\frac{c(n - 1)s_{jk}^2}{MS_W}} - \sqrt{c(n - 1) - 1}$$

$$= a\sqrt{s_{jk}^2} - b,\qquad(2)$$

where a is a multiplicative constant and b an
additive constant. Thus for equal n's, we see
the O&W z_{jk} is merely a linear transformation of the cell standard deviation s_{jk}. The
goal of the linear transformation is to secure
entries that have a theoretical variance of
approximately one under the condition that
the raw score distributions of the cells are
normally distributed $[y_{ijk} \sim \text{IND}(\mu_{jk}, \sigma_{jk}^2)]$
and $n_{jk} \geq 10$.

Overall and Woodward properly empha-
size that their statistical test is greatly dependent on the assumption of normality, and
their 1974 paper includes a small Monte
Carlo study for one-factor designs showing
family-wise risk of type I errors (FWI) that
are very comparable to the classic Bartlett
[1] and Box [3] tests, under both equal n's
and unequal n's.

$E(s^2) = \sigma^2$ and $\sigma_{s^2} = \sigma^2[2/(n - 1) + (\gamma_2/n)]^{1/2}$, where γ_2 is the kurtosis* index of
the population. Note that the mean of the
sampling distribution of s^2, σ^2, is proportional to the standard deviation of s^2, σ_{s^2}.
When this is the case, then the logarith-

mic transformation is a variance-stabilizing transformation*. However, the O & W test uses the square-root transformation instead (*see* EQUALIZATION OF VARIANCES). Thus Games et al. [5] predicted that the Bartlett and Kendall [2] test would have greater power than the O & W test for the equal n case. This difference was confirmed in this Monte Carlo study; however, the differences were not massive (about 0.062 over four points of moderate to high power). This difference was also somewhat mitigated by the fact that the Bartlett and Kendall test had an average FWI of 0.064, slightly above the alpha of 0.05, while the O & W was considerably closer with an FWI of 0.052. Thus for the equal n case, the O & W test appears acceptable, given normality.

Overall and Woodward state that "the primary advantage of the Z-variance test appears to be the simple generalization to multi-factor designs" [7, p. 312]. The Bartlett and Kendall [2] test may also be extended to multifactor designs, but it is concerned with multiplicative models of variances. The O & W test may be considered for additive models of variances in multifactor designs (see Games and Wolfgang [6]). The Bartlett and Kendall test requires equal n's, while the O & W test does not. However, this author recommends striving for equal n's in factorial designs where the O & W test is going to be used.

For the balanced case of equal n's, the multifactor generalization is straightforward since there is only one way of defining mar-

ginal means, i.e., $\bar{Z}_{.k} = \sum^{J} Z_{jk}/J$. For unequal n's, there are several different ways of defining marginal means. As O & W state, "In a multiway classification with unequal n, the rationale would appear analogous to that of unweighted means ANOVA" [7, p. 313].

Example. Table 1 contains the data of a two-factor design with proportional n's. The first step in computing the O & W test would be to determine MS_W as the usual weighted average of the cell variances. In this case, $MS_W = 189.23780$. This value plus the cell s_{jk}^2 and n_{jk} yield the Z_{jk} values at the bottom of each cell in Table 1. These z_{jk} values would then be submitted to an analysis of variance yielding $MS_A = 20.81549$, $MS_B = 0.43339$, and $MS_{AB} = 2.00457$. Using $\alpha = 0.01$, the critical value for $MS_A/1$ is $F_{0.01,1,\infty} = 6.63$ so that we reject the H_{01}: $\sigma_{j.}^2 = \sigma_{..}^2$ and conclude that treatment A_2 yields larger variances than does treatment A_1. For the other two MS's, the critical value is $F_{0.01,2,\infty} = 4.61$ so that we retain H_{02} and H_{03}.

Summary

The O & W test is a valuable addition to the statistician's toolbox. It permits looking at additive models of variances in multifactor designs with independent groups. Striving for equal n's in such designs is still recommended though the O & W can be used with unequal n's in a procedure roughly comparable to an unweighted means analysis of cen-

Table 1 Sample Variances (s_{jk}^2), Sample Sizes (n_{jk}) and the Resulting Z_{jk} Values

		B_1	B_2	B_3	$\bar{Z}_{j.}$
	s_{jk}^2	83.	104.	96.	
A_1	n_{jk}	20	25	40	
	Z_{jk}	-2.02713	-1.73811	-2.50074	-2.08866
	s_{jk}^2	298.	218.	316.	
A_2	n_{jk}	20	25	40	
	Z_{jk}	$+1.67138$	$+0.58477$	$+2.65342$	$+1.63652$
$\bar{Z}_{.k}$		-0.17788	-0.57667	$+0.07634$	$\bar{Z}_{..} = -.22607$

tral tendency. The major limitation of the O&W test is that shared by all classical tests on variances: the FWI rate may greatly exceed alpha when the populations are leptokurtic. If you are confident your data are normally distributed or platykurtic, the O&W may be used. If leptokurtosis is suspected, one of the more robust tests is recommended. See Games and Wolfgang [6] for a review of these tests.

References

[1] Bartlett, M. S. (1937). *Proc. R. Soc. Lond.*, **160**, 268–282.

[2] Bartlett, M. S. and Kendall, D. G. (1946). *J. R. Statist. Soc. B*, **8**, 128–138.

[3] Box, G. E. P. (1949). *Biometrika*, **36**, 317–349.

[4] Fisher, R. A. and Yates, F. (1963). *Statistical Tables for Biological, Agricultural, and Medical Research*. Hafner, New York.

[5] Games, P. A., Keselman, H. J., and Clinch, J. J. (1979). *Psychol. Bull.*, **86**, 978–984.

[6] Games, P. A. and Wolfgang, G. S. (1983). *Comp. Statist. Data Anal.*, **1**, 41–52.

[7] Overall, J. E. and Woodward, J. A. (1974). *Psychometrika*, **39**, 311–318.

(ANALYSIS OF VARIANCE
BARTLETT'S TEST OF HOMOGENEITY OF
   VARIANCES
HOMOGENEITY
JACKKNIFE)

PAUL A. GAMES

OVERGRADUATION

This term is used to describe a situation in which the graduated values agree with the observed values too closely, reproducing to some extent accidental irregularities unlikely to correspond to population distributions. Assessment of overgraduation is largely subjective, but excessive overgraduation is easily recognized.

(GRADUATION
WHITTAKER–HENDERSON
   GRADUATION)

OVERIDENTIFICATION

Let Y be a random variable with distribution function $F(y, \Theta)$, where Θ belongs to a subset of R^m. Suppose that Θ is identified; that is, $F(y, \Theta_1) = F(y, \Theta_2)$ for all y implies $\Theta_1 = \Theta_2$. (*See* IDENTIFIABILITY and IDENTIFICATION PROBLEMS.) Then Θ is *overidentified* if it satisfies restrictions that effectively reduce the dimension of the parameter space and if it would be identified even without these restrictions. However, the use of the term usually is reserved for cases in which these restrictions arise from the choice of parameterization.

As an example, consider the multiple indicator-multiple cause model described by Goldberger [1]. We have a set of N regression equations

$$y_i = \beta_i X^* + \epsilon_i, \qquad i = 1, \ldots, N$$

relating observable indicators y_i to an unobservable X^*. We also have an equation relating the unobservable to K observable causes X_j:

$$X^* = \sum_{i=1}^{K} \alpha_j X_j + u.$$

In terms of the observables, this implies the set of N equations

$$y_i = \sum_{j=1}^{K} \pi_{ij} X_j + v_i,$$

where $\pi_{ij} = \beta_i \alpha_j$ (and $v_i = \epsilon_i + \beta_i u$).

The original set of parameters is of dimension $N + K - 1$: N of the β_i plus K of the α_j, minus one normalization to set the scale of the α's and β's. However, under reasonable assumptions about the errors and the observables, the likelihood function* will be "naturally" written in terms of the π_{ij}, a set of parameters of dimension NK. Either set of parameters is identified for all $N \geqslant 1$ and $K \geqslant 1$. However, if $N \geqslant 2$ and $K \geqslant 2$, the model is overidentified, since then $N + K - 1 < NK$; there are fewer parameters of intrinsic interest than there are parameters naturally appearing in the likelihood function. The restriction that reflects the presence of overidentification is that the rank of the matrix $[\pi_{ij}]$ is equal to one, which is indeed a restriction, unless N or K equals one.

Other examples that are very similar in nature could be cited; for example, the question of the identification of the structural parameters in a system of simultaneous linear equations (*see* ECONOMETRICS). Loosely speaking, such examples share the following common form. The distribution function (or likelihood function) depends naturally on a set of parameters Θ of dimension m. These are in turn determined by another set of parameters α, of dimension n, by some function, say $\Theta = h(\alpha)$, so that we could parametrize in terms of α. If $m = n$, so that the number of Θ's is the same as the number of α's, generally no interesting questions of identification arise. If $n > m$, α usually will not be identified, since the inverse function $\alpha = h^{-1}(\Theta)$ generally will not exist. However, if $n < m$, α typically will be overidentified, and this will be reflected by the implication of $(m - n)$ restrictions on Θ.

Reference

[1] Goldberger, A. S. (1974). In *Frontiers in Econometrics*, Paul Zarembka, ed. Academic Press, New York.

(ECONOMETRICS
IDENTIFIABILITY
IDENTIFICATION PROBLEMS
MULTICOLLINEARITY
MULTIPLE REGRESSION)

PETER SCHMIDT

OVERLAP DESIGN   *See* SERIAL DESIGN

OVERLAPPING COEFFICIENT

The overlapping coefficient (OVL) refers to the area under two probability (density) functions simultaneously. See the shaded area of Fig. 1. The overlapping coefficient is a measure of agreement between two distributions and ranges between zero and unity. If OVL = 0, then the distributions are nonoverlapping and if OVL = 1 then the distributions must be identical.

Let $f_1(\mathbf{x})$ and $f_2(\mathbf{x})$ be probability (density) functions defined on the n-dimensional real numbers R_n. OVL is formally defined in the continuous case by

$$OVL = \int_{R_n} \min\left[f_1(\mathbf{x}), f_2(\mathbf{x}) \right] d\mathbf{x}$$

and in the discrete case by

$$OVL = \sum_{\mathbf{x}} \min\left[f_1(\mathbf{x}), f_2(\mathbf{x}) \right].$$

HISTORICAL DEVELOPMENT

The overlapping coefficient was originally suggested by Weitzman [7] as a measure of agreement* when comparing distributions of income*. He applied the measure to census data on incomes of black and white families in the United States. Gastwirth [2] discussed the merits of using the OVL when comparing income distributions with particular reference to male versus female earnings distributions. Sneath [6] has proposed the overlap

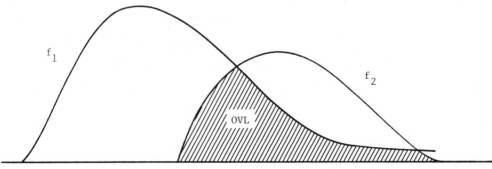

Figure 1   Graphical representation of the overlapping coefficient.

as a method to test for distinctness of two clusters in Euclidean space. He computes the coefficient for the distributions of the projections of the members of both clusters onto the line joining the cluster centroids. Bradley and Piantadosi [1] generalized the OVL to n dimensions and reported results for the case when both distributions are normal. Inman [3] developed both parametric and nonparametric estimates of the overlapping coefficient. He provides a thorough discussion of their sampling distributions and properties.

NORMAL DISTRIBUTION RESULTS

Consider the case of two normal distributions with expectations μ_i and standard deviations σ_i, $i = 1, 2$. Let $\Phi(\cdot)$ denote the CDF for a standard normal distribution.

If $\sigma_1 = \sigma_2$, then the PDFs cross at the point $x = (\mu_1 + \mu_2)/2$ and the overlapping coefficient is

$$OVL = 2\Phi(-|\delta|/2),$$

where $\delta^2 = (\mu_2 - \mu_1)^2/\sigma_1^2$ is the Mahalanobis [4] distance* measure.

If $\sigma_1 < \sigma_2$, then the PDFs cross at two distinct points which are given by the roots of a quadratic expression so that

$$(x_1, x_2) = (\sigma_2^2 - \sigma_1^2)^{-1}\Big[\mu_1\sigma_2^2 - \mu_2\sigma_1^2$$

$$\pm \sigma_1\sigma_2\big\{(\mu_1 - \mu_2)^2 + 2(\sigma_2^2 - \sigma_1^2)\log(\sigma_2/\sigma_1)\big\}^{1/2}\Big].$$

We take x_1 to be the smaller root. Define $z_{ij} = (x_i - \mu_j)/\sigma_j$. Then the overlapping coefficient is given by

$$OVL = 1 + \Phi(z_{11}) - \Phi(z_{12})$$

$$- \Phi(z_{21}) + \Phi(z_{22}).$$

For the case where the parameters are not known, we replace μ_i and σ_i^2 by the usual unbiased estimators, the sample mean $\bar{x}_i$, and sample variance s_i^2, respectively.

Comments

Gastwirth [2] considered a flaw in the OVL: its undue emphasis on the points of intersec-

tion of the two density functions. The OVL remains constant if the shape of the densities on either side of the intersection points are changed, as long as no further crossing points are produced. He then proposed a new measure, that does not possess this objection, which is the probability that a variate from one density exceeds a variate from the second. Gastwirth's measure, in the empirical case, is related to the Mann–Whitney* form of the Wilcoxon* test.

An attempt to compute a standard error and tests of significance for OVL has been reported by Marx [5]. He derived his results under the (erroneous) assumption that a standardized form of OVL has a Student's t-distribution.

References

[1] Bradley, E. L. and Piantadosi, S. (1982). "The Overlapping Coefficient as a Measure of Agreement Between Distributions." *Tech. Rep.*, Department of Biostatistics and Biomathematics, University of Alabama at Birmingham.

[2] Gastwirth, J. L. (1975). *Amer. Statist.*, **29**, 32–35.

[3] Inman, H. F. (1984). "Behavior and Properties of the Overlapping Coefficient as a Measure of Agreement Between Distributions," Ph.D. Dissertation, University of Alabama at Birmingham.

[4] Mahalanobis, P. C. (1936). *Proc. Natl. Inst. Sci. India*, **12**, 49–55.

[5] Marx, W. (1976). *Zeit. exper. angew. Psychol.*, **23**, 267–270. (In German.)

[6] Sneath, P. H. A. (1977). *Math. Geol.*, **9**, 123–143.

[7] Weitzman, M. S. (1970). "Measures of Overlap of Income Distributions of White and Negro Families in the U.S." *Tech. Rep.*, Bureau of the Census, U.S. GPO, Washington, DC.

(INCOME DISTRIBUTIONS
MAHALANOBIS D^2
MEASURES OF AGREEMENT
NORMAL DISTRIBUTION)

EDWIN L. BRADLEY

OVERREPRESENTATION *See* REPRESENTATIVE SAMPLING

P

PAASCHE–LASPEYRES INDEX NUMBERS

Paasche and Laspeyres index numbers* are generic labels attached by [11] to a class of indices widely used in practice and constantly debated in economics. The basic form of these indices is that of two price indices advocated in the 1870s by Laspeyres [8] and Paasche [9]. Let $\mathbf{p}^t \cdot \mathbf{q}^t = \sum_{i=1}^{n} p_{it} q_{it}$ be the inner product of the price and quantity vectors $\mathbf{p}$ and $\mathbf{q}$ of n commodities observed identically at time $t = 0, 1, \ldots, T$;

$$V_{rs} = \mathbf{p}^s \cdot \mathbf{q}^s / \mathbf{p}^r \cdot \mathbf{q}^r;$$

and

$$w_{ir} = p_{ir} q_{ir} \Big/ \sum_{i=1}^{n} p_{ir} q_{ir},$$

for any two periods r and s. The two-period or *binary* Laspeyres–Paasche indices are given by:

LASPEYRES PRICE INDEX:

$$P_{rs}(\mathbf{q}^r) = \mathbf{p}^s \cdot \mathbf{q}^r / \mathbf{p}^r \cdot \mathbf{q}^r. \tag{1}$$

LASPEYRES QUANTITY INDEX:

$$Q_{rs}(\mathbf{p}^r) = \mathbf{p}^r \cdot \mathbf{q}^s / \mathbf{p}^r \cdot \mathbf{q}^r. \tag{2}$$

PAASCHE PRICE INDEX:

$$P_{rs}(\mathbf{q}^s) = \mathbf{p}^s \cdot \mathbf{q}^s / \mathbf{p}^r \cdot \mathbf{q}^s. \tag{3}$$

PAASCHE QUANTITY INDEX:

$$Q_{rs}(\mathbf{p}^s) = \mathbf{p}^s \cdot \mathbf{q}^s / \mathbf{p}^s \cdot \mathbf{q}^r. \tag{4}$$

(1)–(4) have the same *base reference r*, usually set equal to 100. Laspeyres indices are, however, *base weighted*, while the Paasche indices are *current weighted*, and the Paasche price (quantity) index is the *implicit* index to match the corresponding Laspeyres quantity (price) index in the sense of the weak factor reversal test $[V_{rs} = (2) \times (3) = (1) \times (4)]$. The matching of (1) and (4) is quite common in practice because usually base weights are more readily available than current weights and the price index is more immediately required than the quantity index.

The *run* of a Laspeyres/Paasche index is a time series* of indices for more than two successive values of $t = 0, 1, \ldots, T$. Until fairly recently, as indicated in Allen [1], Craig [4], and Fowler [6], there has been little agreement on how to define the runs of Laspeyres and Paasche indices. But it is now generally accepted as pointed out in Allen [2, 3], that the run of Laspeyres index for

$t = 1, 2, \ldots, T$ has a *fixed* weight base at time 0 and a freely chosen reference base r as given by:

RUN OF LASPEYRES PRICE INDEX:

$$P_{rt}(\mathbf{q}^0) = \mathbf{p}^t \cdot \mathbf{q}^0 / \mathbf{p}^r \cdot \mathbf{q}^0$$
$$= P_{0t}(\mathbf{q}^0) / P_{0r}(\mathbf{q}^0). \qquad (5)$$

RUN OF LASPEYRES QUANTITY INDEX:

$$Q_{rt}(\mathbf{p}^0) = \mathbf{p}^0 \cdot \mathbf{q}^t / \mathbf{p}^0 \cdot \mathbf{q}^r$$
$$= Q_{0t}(\mathbf{q}^0) / Q_{0r}(\mathbf{q}^0). \qquad (6)$$

The run of Paasche index is *either* the current-weighted index *or* the implicit index corresponding to the run of Laspeyres. In binary comparisons, as was noted, there is only one Paasche form that is both current-weighted and implicit. There are, however, two runs of Paasche price indices given by the

RUN OF CURRENT-WEIGHTED PAASCHE PRICE INDEX:

$$P_{rt}(\mathbf{q}^t) = \mathbf{p}^t \cdot \mathbf{q}^t / \mathbf{p}^r \cdot \mathbf{q}^t. \qquad (7)$$

RUN OF IMPLICIT PAASCHE PRICE INDEX:

$$P_{rt}^*(\mathbf{q}^t) = V_{rt} / Q_{rt}(\mathbf{p}^0)$$
$$= P_{0t}(\mathbf{q}^t) / P_{0r}(\mathbf{q}^r). \qquad (8)$$

The corresponding two runs of Paasche quantity index are defined similarly.

PROPERTIES OF LASPEYRES–PAASCHE INDICES

Property 1. (1)–(4) can each be interpreted as weighted means [e.g., $P_{rs}(\mathbf{q}^r) = \sum_{i=1}^{n} w_{ir}(p_{is}/p_{ir})$]. Equation (7) retains this property but it does not correspond to (6) in the sense that (3) corresponds to (2), while (8) corresponds to (6), but it cannot be interpreted as a weighted average. This is the main difference between (7) and (8).

Property 2. Laspeyres–Paasche indices satisfy the time reversal test*; unlike Fisher's index and Vartia indices, they fail the factor reversal test; like Vartia's but unlike Fisher's, they are consistent in aggregation*; and like many other indices including Fisher's, Vartia's and the Törnquist index, they are not transitive. (*See* FISHER'S IDEAL INDEX NUMBER *and* INDEX NUMBERS.)

Property 3. The divergence between Laspeyres and Paasche indices can be expressed in terms of a well-known statistical relation originally due to Bortkiewicz (see e.g., Allen [1] and Jazairi [7]) as follows:

$$\frac{P_{rs}(\mathbf{q}^s)}{P_{rs}(\mathbf{q}^r)} = \frac{Q_{rs}(\mathbf{p}^s)}{Q_{rs}(\mathbf{p}^r)}$$
$$= 1 + \hat{\rho} \cdot \frac{\hat{\sigma}_p}{P_{rs}(\mathbf{q}^r)} \cdot \frac{\hat{\sigma}_q}{Q_{rs}(\mathbf{q}^r)} \qquad (9)$$

where $\hat{\rho}$ is the weighted sample correlation between (p_{is}/p_{ir}) and (q_{is}/q_{ir}), and $\hat{\sigma}_p$ and $\hat{\sigma}_p$ are the weighted standard deviations of (p_{is}/p_{ir}) and (q_{is}/q_{ir}), respectively. Thus Paasche index $\gtrless$ Laspeyres index if $\hat{\rho} \gtrless 0$.

Property 4. The Laspeyres–Paasche indices have a statistical interpretation that does not seem to have been noted in the literature. Suppose we want to estimate the mean β of the price ratios (p_{is}/p_{ir}) from

$$\frac{p_{is}}{p_{ir}} = \beta + \epsilon_i, \qquad (10)$$

and we assume that

$$E(\epsilon_i) = 0 \quad \text{and} \quad \text{var}(\epsilon_i) = \frac{\sigma^2}{w_{ir}}. \qquad (11)$$

Then the weighted least-squares estimator of β is

$$\hat{\beta} = \left[\sum_{i=1}^{n} \sqrt{w_{ir}}\,(p_{is}/p_{ir})\sqrt{w_{ir}} \right] \Big/ \sum_{i=1}^{n} \left(\sqrt{w_{ir}}\right)^2$$
$$= P_{rs}(\mathbf{q}^r). \qquad (12)$$

Thus if (11) is valid, then (1) is the best linear unbiased estimator of the mean of

(p_{is}/p_{ir}). If we further assume that $\epsilon_i \sim N(0, \sigma^2/w_{ir})$, then (1) is the maximum likelihood estimator of β. This interpretation of (1) makes it possible to estimate the variance of $\hat{\beta}$ and to apply confidence interval and hypotheses testing procedures. Many index number formulas, as well as the Laspeyres–Paasche forms, can be interpreted similarly under different specifications of the error term ϵ_i in (10). The problem, of course, is that the distribution of ϵ_i is unknown.

ECONOMIC INTERPRETATION OF LASPEYRES–PAASCHE INDICES

The most popular interpretation of Laspeyres–Paasche indices is that the Laspeyres (Paasche) price index measures the change in the cost of the base (current) period "basket" of goods and services, and the Laspeyres (Paasche) quantity index measures the change in the basket of goods and services valued at the base (current) period prices. These indices can, however, be given more formal economic interpretations under certain economic and technical assumptions. It can be shown that the Laspeyres–Paasche price indices may approximate or coincide with the *cost-of-living index* and that the (L-P) quantity indices may approximate or coincide with the *real-income index*. For example, if the consumer is utility maximizing and his utility function $u = f(\mathbf{q})$ is of the simple linear form $u = \mathbf{a}'\mathbf{q}$, where $\mathbf{a}$ is a vector of positive constants, then, as shown in Diewert [5], his real income index $f(\mathbf{q}^s)/f(\mathbf{q}^r)$ coincides with (2) and (4), and the corresponding implicit price indices (3) and (1) will also coincide with the cost-of-living index of this consumer. The economic theories of index numbers, including Laspeyres–Paasche indices, are surveyed in Diewert [5] and Samuelson and Swamy [10].

References

[1]   Allen, R. G. D. (1963). *Int. Statist. Rev.*, **31**, 281–301.

[2]   Allen, R. G. D. (1975). *Index Numbers in Theory and Practice*. Macmillan, London.

[3]   Allen, R. G. D. (1980). *An Introduction to National Accounts Statistics*. Macmillan, London.

[4]   Craig, J. (1969). *J. R. Statist. Soc. C*, **18**, 141–142.

[5]   Diewert, W. E. (1981). In *Essays in the Theory and Measurement of Consumer Behaviour*, A. S. Deaton, ed. Cambridge University Press, London, pp. 163–208.

[6]   Fowler, R. F. (1974). *J. R. Statist. Soc. A*, **137**, 75–88.

[7]   Jazairi, N. T. (1972). *Int. Statist. Rev.*, **40**, 47–51.

[8]   Laspeyres, E. (1871). *Jb. Natlekon. Statist.*, **16**, 296–314.

[9]   Paasche, H. (1874). *Jb. Natlekon. Statist.*, **23**, 168–178.

[10]   Samuelson, P. A. and Swamy, S. (1974). *Amer. Econ. Rev.*, **64**, 566–93.

[11]   Walsh, C. H. (1901). *The Measurement of General Exchange Value*. Macmillan, New York.

Nuri T. Jazairi

PACKED ADJACENCY MATRIX   *See* Graph-Theoretic Cluster Analysis

PADÉ AND STIELTJES TRANSFORMATIONS

Some basic aspects of summing divergent series are given in Levin's Summation Algorithm. The works of Padé [7] and Stieltjes [12] are nearly a century old and relate fundamentally to rational fraction approximants to series. Here, however, we shall confine our attention to the broad picture and how these studies relate to moment series in statistics.

THE TRANSFORMATION

Consider the series

$$g(n) \sim e_0 + e_1/n + \cdots \qquad (1)$$

relating to moment series for samples of size n or its traditional form $g(1/x) \sim e_0 + e_1 x + \cdots$. The Padé approximants

$$(s|r) = \frac{P_s(s)}{Q_r(x)} = \sum_{t=0}^{s} a_t x^t \Big/ \sum_{t=0}^{r} b_t x^t \qquad (2)$$

are determined from the linear equations

arising from the equivalence of the first $s + r + 1$ coefficients in (1) and those in (2) when expanded in a power series in x; the numerical implementation of this uses $P_s(x) \sim Q_r(x)g(1/x)$. In general $r = s$ or $r = s + 1$, $r = 0, 1, \ldots$, provide the most useful sequences of approximants. The Padé table consists of entries $(s \mid r)$ on a coordinate system $x = s$, $v = r$, and clearly

$$g(1/x)\, Q_r(x) - P_s(x) = (x^{s+r+1}), \quad (3)$$

where $(\cdot)$ indicates the order of the first surviving term in the remainder. Note that the notation $(s \mid r)$ refers to a power series in x.

Padé approximants received but one chapter of explanation in Wall's [13] treatise on analytic continued fractions. An explosion of interest occurred in subsequent years, mainly relating to advances in theoretical physics (see Baker and Gammel [2] and Baker [1]).

CONTINUED FRACTIONS AND STIELTJES

Why are rational fraction approximants powerful in summatory situations? The key lies in the works of Stieltjes and comments on his studies made by Émile Borel [3]—a captivating translation of his *Leçons* is given by Critchfield and Vakar (1975).

The bridge from series to rational fractions to regular functions is illustrated in the classical moment problem*. Given a set of moments $\{ \mu'_r \}$, is a unique distribution function determined by them? The Stieltjes moment problem is based on a semi-infinite interval and seeks a function $\sigma(\cdot)$ such that

$$\mu'_s = \int_0^\infty t^s \, d\sigma(t), \qquad s = 0, 1, \ldots . \quad (4)$$

Equations (4) are formally equivalent to

$$f(z) = \int_0^\infty \frac{d\sigma(t)}{t + z} \sim \frac{\mu'_0}{z} - \frac{\mu'_1}{z^2} + \cdots ,$$

Note that historically "moments" referred to the mechanical distribution of mass on a straight line (semi-infinite in the Stieltjes case); the center of gravity and radius of gyration relate to the first two moments in statistics. There could be point masses, continuous mass, or mixtures. What are the conditions for which a bounded nondecreasing function $\sigma(\cdot)$, with infinitely many points of increase, exists? Stieltjes found the answer in the continued fraction form

$$f(z) = \frac{1}{\alpha_1 z} + \frac{1}{\alpha_2} + \frac{1}{\alpha_3 z} + \frac{1}{\alpha_4} + \cdots \quad (6)$$

arising from rational fraction approximants to the series (5). There is a unique mass distribution provided $\alpha_1, \alpha_2, \ldots$ are positive, and $\sum \alpha_s$ diverges; the first condition relates to the existence of an orthogonal set developed from the moments. In this case, the integral in (5) defines a regular function equal to (6) for all z outside the slit $(-\infty, 0)$.

The generating series in (5) and likewise (6), may converge or diverge, convergent series can relate to divergent fractions and vice versa; in any event generally the domain of convergence of (6) far exceeds that of the series, but this should not be taken to imply rapid convergence (in passing it should be noted that remainder terms for rational fraction approximants are usually complicated). References are Shohat and Tamarkin [11], Perron [8], Graves-Morris [4], and Saff and Varga [9].

STIELTJES FRACTIONS
AND THE STANDARD DEVIATION
IN TYPE III SAMPLING

In Table 4 of METHOD OF MOMENTS, continued fraction approximants are given for the moments of $s(n) = \sqrt{(m_2/\mu_2)}$ from the population $(x^3/6)\exp(-x)$; the series for the mean (Table 1) has alternating signs and a divergence rate somewhere between the single and double factorial series. There are a limited number of anomalous signs in the continued fraction development corresponding to (6), but there is general agreement between the Levin, Padé, and simulation results for the sample sizes considered.

In Shenton et al. [10] we give the series for $E\sqrt{m_2}$ in exponential sampling. A Stieltjes form

$$E\left[\sqrt{m_2}\right] = 1 - \frac{1.5}{n} + \frac{p_1}{1} + \frac{q_1}{n} + \cdots \quad (7)$$

with partial numerators (p_s, q_s) taken as far as $s = 13$ is given: the largest of these positive monotone increasing coefficients does not exceed 2000, whereas the twenty-eighth coefficient in the series is 5.4×10^{74} approximately, showing how rational fractions absorb the shock of very large coefficients. The agreement between Padé, Levin, and simulation assessments improves to three or more significant digits for $n = 10$. There is the conjecture, in terms of a Stieltjes integral, that

$$E\left[\sqrt{m_2}\right] = 1 - \int_0^\infty \frac{d\sigma(t)}{n+t}, \quad n = 1, 2, \ldots$$

$$(8)$$

A similar conjecture holds for Student's t^*.

Further Comments

Our studies all relate to moments for which theoretically there is a corresponding uniquely defined multiple integral over the sample space, out of reach in general by quadrature techniques except for very small samples. Depending on the structures involved there may be available anything from 2 to 50 or so terms in a moment series depending on the dimensionality of the related Taylor series and also on whether an implicit function is involved. The coefficients may be subject to round-off errors*, especially in view of the magnitude involved.

Padé methods may well fail when the sign and magnitude patterns oscillate and also if divergence is extreme (in the region of the triple factorial series). In the latter case, the Borel dilution technique may be appropriate. On the positive side, the general usefulness of the approach is indicated by two conjectures of Baker [1]. We briefly state the first, referred to as the Baker–Gammel–Wills first conjecture: For a power series $P(z)$, regular for $z \leqslant 1$ (except for poles, and except for $z = 1$) there is at least one Padé subsequence $(s \mid s)$ that converges uniformly

to the function as $s \to \infty$ and is within the acceptable domain described.

With extended series of the future, this conjecture may or may not turn out to have a counterexample.

Another outstanding problem concerns the detection of singularities in moments of moments, an important aspect especially when poles or branch points are concerned. Are there singularities in the moments of the skewness and kurtosis in general sampling? The logarithmic approach [1, pp. 274–279] has been tried with strong divergent series with indifferent success. Mulholland [5, 6] gave the subject some thought, but his studies were cut short by his untimely death.

References

[1] Baker, George, A., Jr. (1975). *Essentials of Padé Approximants*. Academic Press, New York.

[2] Baker, G. A., Jr. and Gammel, J. L. (eds.) (1970). *The Padé Approximant in Theoretical Physics*. Academic Press, New York.

[3] Borel, E. (1928). *Leçons sur les Séries Divergentes*. Gauthier-Villars, Paris. (Translated by Charles L. Crutchfield and Ann Vakar (1975), Los Alamos Scientific Laboratory, Los Alamos, NM.)

[4] Graves-Morris, P. R., ed. (1973). *Padé Approximants and Their Applications*. Academic Press, New York.

[5] Mulholland, H. P. (1965). *Proc. Camb. Philos. Soc.*, **61**, 721–39.

[6] Mulholland, H. P. (1977). *Biometrika*, **64**, 401–409.

[7] Padé, H. (1892). *Ann. Sci. Éc. Norm. Super. Suppl.*, **9**(3), 1–93.

[8] Perron, O. (1954). *Die Lehre von den Kettenbruchen*, 3rd ed., Vols. 1 and 2. Teubner, Stuttgart.

[9] Saff, E. B. and Varga, R. S., eds. (1977). *Padé and Rational Approximation Theory and Applications*. Academic Press, New York.

[10] Shenton, L. R., Bowman, K. O., and Lam, H. K. (1979). *Proc. Statist. Comp. Sect. Amer. Statist. Ass.*, 20–29.

[11] Shohat, J. A., and Tamarkin, J. D. (1963). *The Problem of Moments*. American Mathematical Society, Providence, RI.

[12] Stieltjes, T. J. (1918). *Oeuvres Complètes*, Vol. 2. Noordhoff, Groningen, The Netherlands.

[13] Wall, H. S. (1948). *Analytic Theory of Continued Fractions*. Van Nostrand-Reinhold, Princeton, NJ.

Bibliography

See the following for more information on continued fractions.

Brezinski, C. (1977). *A Bibliography on Padé Approximations and Related Subjects*, Publications 18, 56, 96, and 118, Université des Sciences et Techniques de Lille.

K. O. Bowman
L. R. Shenton

PAGE TEST FOR ORDERED ALTERNATIVES

The Page test [5] is a nonparametric procedure for ordered alternatives in the complete randomized blocks model. Ordered alternatives are discussed in more detail elsewhere (*see* JONCKHEERE TESTS FOR ORDERED ALTERNATIVES).

The data consist of a collection $\{X_{ij} : j = 1, \ldots, k; \; i = 1, \ldots, m\}$ of independent random variables where X_{ij} has unknown CDF F_{ij}, index j identifies the k treatment levels of interest, and i identifies blocks, a nuisance effect. A minor generalization permits symmetric dependence within blocks, in which case the F_{ij} can be interpreted as marginal CDFs.

If attention is restricted to ordered location shift, then $F_{ij}(t) = F_i(t - \theta_j)$ and the problem can be expressed as

$$H_0 : \theta_1 = \theta_2 = \cdots = \theta_k$$

versus

$$H_1 : \theta_1 < \theta_2 < \cdots < \theta_k.$$

It is assumed that the treatment order under H_1 is a priori and that treatments are indexed in that order. As the notation implies, the distribution type F_i may differ among blocks.

Page's solution uses the method of *m*-rankings, introduced by Kendall and Babington Smith [4]. Observations are ranked separately within each block, letting r_{ij} be the rank of X_{ij} $(1 \leqslant r_{ij} \leqslant k)$. The ranks are summed within treatments,

$$R_j = \sum_{i=1}^{m} r_{ij},$$

and the test rejects for large values of

$$L = \sum_{j=1}^{k} j R_j.$$

Under the assumptions that F is a continuous CDF and that no ties exist within blocks, the test is distribution-free* and exact tables exist, for example in Hollander and Wolfe [3], for $k = 3$, $m = 2(1)20$, and $k = 4(1)8$, $m = 2(1)12$.

For problems including larger samples, noncontinuous distributions, or frequent ties within blocks (e.g., due to roundoff), an asymptotic $(m \to \infty)$ test rejects H_0 in favor of the a priori ordering when

$$Z_L = (L - \mu_L)/\sigma_L$$

exceeds the right-tail α-level critical value of the standard normal.

The exact null moments of L are given by

$$\mu_L = mk(k + 1)^2/4$$

and

$$\sigma_L^2 = m(k^3 - k)^2/\{144(k - 1)\}.$$

In the presence of ties within blocks, let $\{a_{ij}; j = 1, 2, \ldots, e \leqslant k\}$ represent the distinct values of the data in block i, with t_{ij} the number of occurrences of a_{ij}. Then the exact null variance of L becomes

$$\sigma_L^2 = k(k^2 - 1)$$
$$\times \frac{mk(k^2 - 1) - \sum_i \sum_j (t_{ij}^3 - t_{ij})}{144(k - 1)},$$

assuming the method of midranking is used to resolve ties.

Example.

Blocks (i)	Treatments		
	1 x_{i1}/r_{i1}	2 x_{i2}/r_{i2}	3 x_{i3}/r_{i3}
1	5.8/1	6.6/2	7.1/3
2	6.5/1	6.8/2	7.9/3
3	7.8/1	8.6/3	8.4/2
R_j	3	7	8

$L = 1(3) + 2(7) + 3(8) = 41$, $P[L \geqslant 41] \doteq 0.05$ (cf. Table A.16 in Hollander and Wolfe

[3]), indicating moderate support for the a priori ordering $\theta_1 < \theta_2 < \theta_3$.

The Page test can also be used in a two-sided form, testing if treatment effects $\{\theta_i\}$ either increase or decrease with the a priori order.

Page [5] notes that if each rank r_{ij} is paired with its predicted rank (treatment order), the ordinary product-moment correlation r for all such pairings is related to L by

$$r = 12L \Big/ \big[m(k^3 - k) \big] - 3(k+1)/(k-1).$$

It is trivial to show that r is also the average Spearman rank correlation* between the observations in each block and their predicted ranks.

Page's test is easily generalized to two-way layouts with arbitrary numbers of observations per cell $\{n_{ij}, \; i = 1, \ldots, m, \; j = 1, \ldots, k\}$, simply ranking all observations within each block. The assumption of no block-treatment interaction is necessary. For equal cell sizes, $n_{ij} \equiv n$, the large-sample version requires $R_j = \sum_{i=1}^{m} \sum_{l=1}^{n} r_{ijl}$, L and Z_L defined as before, and

$\mu_L = nmk(k+1)(nk+1)/4,$

$\sigma_L^2 = nk(k^2 - 1)$

$\quad \times \dfrac{nmk(n^2k^2 - 1) - \sum\sum(t_{ij}^3 - t_{ij})}{144(nk - 1)}.$

COMPETITORS AND COMPARISONS

Many other rank procedures have been investigated for ordered alternatives in the randomized blocks problem. In addition to the Jonckheere tests for ordered alternatives, Doksum [1] and Hollander [2] proposed very similar tests, based on the Wilcoxon signed

rank test*, Pirie and Hollander considered a normal scores version of Page's test [7] and a generalized sign test procedure [8], Sen [9] showed how aligned rank procedures can be applied, and Shorack [10] proposed a procedure that pools adjacent treatments when the corresponding rank sums violate the predicted order $(R_j \geqslant R_{j+1})$. The Page test is the most widely known and used, undoubtedly because it is one of the oldest and simplest to apply, and because it is the only one for which small-sample tables exist (cf. Table A.16 in Hollander and Wolfe [3]). In addition, the Page and aligned ranks tests are easiest to extend to multiple observations per cell.

Pitman asymptotic relative efficiencies* (AREs) have been evaluated for all of the preceding tests; the results are quite variable and complex.

The examples of ARE, fixed k, $m \to \infty$, in Table 1 are for one observation per cell and normal data, comparing the indicated rank test to the likelihood ratio, normal-theory test, when $\theta_j = a + j\theta$.

For distributions with heavier than normal tails, the corresponding efficiencies are typically greater than 1.0 except for the generalized sign test.

Efficiencies for fixed m, $k \to \infty$ have also been investigated by Pirie [6], who concludes that which rank test is favored varies considerably depending on F, k, and m.

For general location shift alternatives, where $H_1 : \theta_j \neq \theta_{j'}$, for at least one pair of treatments j and j', the Friedman test* is the rank test most commonly used. It, too, is based on the method of m rankings and the behavior of the rank sums R_j. For problems where the ordered alternative is appropriate, the Page test provides much higher power than the Friedman test.

Table 1   AREs of Ordered Alternatives Ranking Procedures: Randomized Blocks

TEST	Page or Shorack	Hollander	Aligned Ranks	General Sign Test
ARE($k = 3$)	0.716	0.963	0.966	0.286
ARE($k = \infty$)	0.955	0.989	0.955	0

As a demonstration, the value of the Friedman test statistic for the preceding sample data is $S = 4\frac{2}{3}$ and $P(S \geqslant 4\frac{2}{3}) = 0.194$ (cf. Table A.15 in ref. 3). Thus the Friedman test clearly fails to detect any differences, while the Page test moderately supports the a priori order.

References

[1] Doksum, K. (1967). *Ann. Math. Statist.*, **38**, 878–883.

[2] Hollander, M. (1967). *Ann. Math. Statist.*, **38**, 867–877.

[3] Hollander, M. and Wolfe, D. A. (1973). *Nonparametric Statistical Methods*. Wiley, New York.

[4] Kendall, M. G. and Smith, B. B. (1939). *Ann. Math. Statist.*, **10**, 275–287.

[5] Page, E. B. (1963). *J. Amer. Statist. Ass.*, **58**, 216–230.

[6] Pirie, W. R. (1974). *Ann. Statist.*, **2**, 374–381.

[7] Pirie, W. R. and Hollander, M. (1972). *J. Amer. Statist. Ass.*, **67**, 855–857.

[8] Pirie, W. R. and Hollander, M. (1975). *Ann. Inst. Statist. Math. Tokyo*, **27**, 521–523.

[9] Sen, P. K. (1968). *Ann. Math. Statist.*, **39**, 1115–1124.

[10] Shorack, G. R. (1967). *Ann. Math. Statist.*, **38**, 1740–1752.

(ISOTONIC INFERENCE
MULTIDIMENSIONAL SCALING
WILCOXON-TYPE TESTS
   FOR ORDERED ALTERNATIVES
   IN RANDOMIZED BLOCKS)

W. PIRIE

PAIRED COMPARISONS

Basic paired comparisons have t treatments, items, or individuals, $T_1, \ldots, T_t$, that are compared pairwise in an experiment or competition on one or more characteristics or attributes. Let $n_{ij} \geqslant 0$ be the number of comparisons of T_i with T_j, $n_{ji} = n_{ij}$, $i \neq j$, $i, j = 1, \ldots, t$. Some of the n_{ij} may be zero, but the design must be connected in the sense that it must not be possible to divide the treatments into two disjoint, exhaustive subsets such that no treatment in one subset is

compared with any treatment of the other subset.

Attention is limited to qualitative responses, indications that T_i is "preferred" to T_j, $T_i \rightarrow T_j$. The term preference is used generally to mean superiority of T_i over T_j on some defined attribute. For each comparison, the response is designated as one or zero, $a_{ij\alpha} = 1$ if $T_i \rightarrow T_j$, $a_{ij\alpha} = 0$ otherwise, $a_{ij\alpha} + a_{ji\alpha} = 1$, $i \neq j$, $\alpha = 1, \ldots, n_{ij}$. Define also $a_{ij} = \sum_{\alpha=1}^{n_{ij}} a_{ij\alpha}$ and $a_i = \sum_j a_{ij} - a_{ii}$, the total number of preferences for T_i. Paired comparisons may be considered an incomplete block design* and some nonparametric methods* apply. If a quantitative response in paired comparisons can be obtained, analysis of variance* of the data, transformed or otherwise, should be possible. Scheffé [37] considered such methods and provided for a possible effect of order of tasting in the comparison of food samples. The reader interested in this aspect of paired comparisons should *see also* BLOCKS, BALANCED INCOMPLETE; PARTIALLY BALANCED DESIGNS, and Clatworthy [11]. Interest in paired comparisons developed from the planning of tournaments, the desire to order and scale participants or items, and the design of comparative experiments in psychophysical testing when no natural measurement system is available. Basic stochastic models were proposed by Thurstone [38] and Zermelo [40], the former being concerned with psychophysical testing and the latter with the rating of chess players.

In this article, models for paired comparisons are discussed, some detail on statistical procedures is given for one model, model extensions are referenced, and some other approaches are noted. There is a rich literature of paired comparisons with some 400 references in the bibliography of Davidson and Farquhar [17].

MODELS FOR PAIRED COMPARISONS

It is assumed that the various comparisons of a paired comparisons experiment are stochastically independent.

Bradley and Terry [10], in a heuristic extension of the Bernoulli distribution*, specified treatment parameters, $\pi_1, \ldots, \pi_t$, $\pi_i \geqslant 0$, $\sum_{i=1}^{t} \pi_i = 1$, $i = 1, \ldots, t$, regarded as relative selection probabilities, for $T_1, \ldots, T_t$, respectively. The probability of selection of T_i when compared with T_j was written

$$P[T_i \rightarrow T_j] = \pi_i / (\pi_i + \pi_j),$$
$$i \neq j, \quad i, j = 1, \ldots, t. \quad (1)$$

This is the model of Zermelo, reinvented by Ford [22]. The n_{ij} comparisons of T_i and T_j were regarded as a set of independent Bernoulli trials leading to a likelihood function for the entire paired comparisons experiment in the form,

$$L = \prod_{i<j} \{\pi_i / (\pi_i + \pi_j)\}^{a_{ij}} \{\pi_j / (\pi_i + \pi_j)\}^{a_{ji}}$$

$$= \prod_i \pi_i^{a_i} \Big/ \prod_{i<j} (\pi_i + \pi_j)^{n_{ij}}. \quad (2)$$

Note that $a_1, \ldots, a_t$ constitute a set of sufficient statistics* for the estimation of $\pi_1, \ldots, \pi_t$.

Thurstone devised the concept of a subjective continuum, an inherent sensation scale, on which order, but not physical measurement, could be discerned. With suitable scaling, each treatment is assumed to have a location point on the continuum, say μ_i for T_i, $i = 1, \ldots, t$. A sensation X_i in response to T_i is received by an individual, perhaps in a taste-testing experiment, X_i assumed normally distributed about μ_i. The individual comparing T_i and T_j is assumed to report the order of sensations X_i and X_j, which may be correlated, $X_i > X_j$ associated with $T_i \rightarrow T_j$. Thurstone's Case V takes the variances of the X_i to be homogeneous and their correlations to be equal. Then the model is

$$P[T_i \rightarrow T_j] = P[X_i > X_j]$$

$$= \frac{1}{\sqrt{2\pi}} \int_{-(\mu_i - \mu_j)}^{\infty} \exp\left(-\frac{1}{2} y^2\right) dy. \quad (3)$$

Further detail is given in THURSTONE'S THE-ORY OF COMPARATIVE JUDGMENTS and by Mosteller [31].

More general models have been developed for paired comparisons. They have been reviewed by Bradley [7] under such classifications as linear models, the Lehmann model, psychophysical models, and models of choice and worth. David [12, Sect. 1.3] defined a "linear" model as one for which $P[T_i \rightarrow T_j] = H(V_i - V_j)$, where V_i is the "merit" of T_i on a merit scale and H is a symmetric distribution function, $H(-x) = 1 - H(x)$. The model (3) is clearly a linear model, as is (1), since

$$P[T_i \rightarrow T_j] = \frac{1}{4} \int_{-(\log \pi_i - \log \pi_j)}^{\infty} \operatorname{sech}^2(y/2) \, dy$$

$$= \pi_i / (\pi_i + \pi_j), \quad (4)$$

the distribution functions involved being those for the very similar normal and logistic* distributions, respectively. Comparison of (3) and (4) suggests similar roles for μ_i and $\log \pi_i$ as location parameters for T_i.

In the remainder of this article, attention is focused on model (1) and its extensions.

STATISTICAL PROCEDURES

Likelihood estimation and test procedures for model (1), based on the likelihood function L in (2), were developed by Bradley and Terry [10] and Dykstra [19]. If $\log L$ is maximized subject to the constraint $\sum_{i=1}^{t} \pi_i = 1$, after minor simplifications, the resulting estimation equations are

$$\frac{a_i}{p_i} - \sum_{j, j \neq i} \frac{n_{ij}}{p_i + p_j} = 0, \qquad 1, \ldots, t,$$
$$\quad (5)$$
$$\sum_{i=1}^{t} p_i = 1,$$

where p_i is the estimator of π_i, $i = 1, \ldots, t$. The equations are solved iteratively, an easy process with a computer. A first approximation $p_i^{(0)}$ is specified to start the process; one may take $p_i^{(0)} = 1/t$. The kth approximation is obtained from the preceding one through the following computations:

$$p_i^{*(k)} = a_i \Big/ \sum_{j, j \neq i} \left[n_{ij} / \left(p_i^{(k-1)} + p_j^{(k-1)} \right) \right],$$

$$p_i^{(k)} = p_i^{*(k)} \Big/ \sum_j p_j^{*(k)},$$

$$i = 1, \ldots, t, \quad k = 1, 2, \ldots .$$

The iterative process converges as shown by Ford [22], and convergence is sufficiently rapid for easy use.

Approximate large-sample theory associated with the method of maximum likelihood* may be used in inference procedures and for confidence regions—see Bradley [6] and Davidson and Bradley [16]. Let $\mu_{ij} = n_{ij}/N$, $N = \sum_{i<j} n_{ij}$. Then $\sqrt{N}(p_1 - \pi_1), \ldots, \sqrt{N}(p_t - \pi_t)$ have the singular multivariate normal distribution* of dimensionality $(t-1)$ in a space of t dimensions with zero mean vector and dispersion matrix $\Sigma = [\sigma_{ij}]$ such that

$$\sigma_{ij} = \text{cofactor of } \lambda_{ij} \text{ in } \begin{bmatrix} \Lambda & 1 \\ 1' & 0 \end{bmatrix} \Big/ \begin{vmatrix} \Lambda & 1 \\ 1' & 0 \end{vmatrix},$$

$$(6)$$

where $\Lambda = [\lambda_{ij}]$, $1'$ is the t-dimensional unit row vector, and

$$\lambda_{ii} = \frac{1}{\pi_i} \sum_{i, j \neq 1} \mu_{ij} \pi_j / (\pi_i + \pi_j)^2,$$

$$\lambda_{ij} = -\mu_{ij} / (\pi_i + \pi_j)^2,$$

$$(7)$$

$i \neq j$, $i, j = 1, \ldots, t$. In practice, σ_{ij} must be estimated; p_i is substituted for π_i in (7) to obtain the $\hat{\lambda}_{ij}$ with $\hat{\sigma}_{ij}$ resulting from the subsequent substitution of the $\hat{\lambda}_{ij}$ in (6). Standard multivariate normal procedures are used to obtain confidence regions* for subsets of the π_i. Some detail with examples is given by Bradley [6], who also considered confidence regions for the $\log \pi_i$.

A test of treatment equality or equal treatment preferences is available. The hypothesis $H_0: \pi_i = 1/t$, $i = 1, \ldots, t$, is tested against the alternative $H_a: \pi_i \neq \pi_j$ for some $i \neq j$, through the test statistic

$$-2 \log \lambda_1 = 2N \log 2 - 2B_1,$$

where $N = \sum_{i<j} n_{ij}$ and

$$B_1 = \sum_{i<j} n_{ij} \log(p_i + p_j) - \sum_i a_i \log p_i.$$

For large N, $-2 \log \lambda_1$ has the central chi-square distribution with $(t-1)$ degrees of freedom under H_0. The exact discrete distribution of B_1 is available for small, balanced, paired comparisons, each $n_{ij} = n$ (see Bradley and Terry [10] and Bradley [5]).

Dykstra [19] gave an example of unbalanced paired comparisons for a preference test on four variations of a product: $n_{12} = 140$, $n_{13} = 54$, $n_{14} = 57$, $n_{23} = 63$, $n_{24} = 58$, $n_{34} = 0$, $N = 372$; $a_1 = 66$, $a_2 = 205$, $a_3 = 56$, $a_4 = 45$, $p_1 = 0.1082$, $p_2 = 0.5193$, $p_3 = 0.2294$, $p_4 = 0.1431$; $B_1 = 206.3214$; $-2 \log \lambda_1 = 103.06$ with three degrees of freedom. These values permit a check on the computing procedures established.

Respondents in paired comparisons, e.g., in consumer testing, may be grouped by some demographic criteria into g groups. Group by treatment interaction or preference agreement may be tested. For group u, $u = 1, \ldots, g$, define the quantities n_{ij}^u, a_i^u, N^u, p_i^u, and B_1^u in accordance with the notation established. A separate analysis of the data from each group yields the values of p_i^u and B_1^u, $u = 1, \ldots, g$. An analysis for the pooled data (grouping ignored) yields values p_i and B_1 with $N = \sum_{u=1}^g N^u$. The statistic

$$-2 \log \lambda_2 = 2 \left(B_1 - \sum_{u=1}^g B_1^u \right),$$

has the central chi-square distribution with $(t-1)(g-1)$ degrees of freedom under the null hypothesis of no interaction $H_0: \pi_i^u = \pi_i$, $i = 1, \ldots, t, u = 1, \ldots, g$.

When a stochastic model is postulated for an experimental situation, it is incumbent on the practitioner to check the appropriateness of that model. This can be done for model (1) in terms of the frequencies a_{ij}. Given values of the p_i, a_{ij} may be estimated by $\hat{a}_{ij} = n_{ij} p_i / (p_i + p_j)$, $i \neq j$, $i, j = 1, \ldots, t$. Then the usual goodness-of-fit chi-square statistic* may be computed; in this case, we have

$$\chi^2 = \sum_{i \neq j} \frac{(a_{ij} - \hat{a}_{ij})^2}{\hat{a}_{ij}}$$

with the central chi-square distribution* with $\binom{t}{2} - t + 1$ degrees of freedom when the model is correct. A second related test statistic and examples are given by Bradley [4].

MODEL EXTENSIONS

Model (1) has been highlighted because a number of extensions are available. The major ones are noted below.

Davidson [13] permitted no-preference judgments:

$$P[T_i \rightarrow T_j] = \pi_i / (\pi_i + \pi_j + \nu\sqrt{\pi_i\pi_j}),$$

$$P[T_i = T_j] = \nu\sqrt{\pi_i\pi_j} / (\pi_i + \pi_j + \nu\sqrt{\pi_i\pi_j}),$$

$$i \neq j.$$

He developed associated test procedures. Rao and Kupper [34] have a somewhat different, but asymptotically equivalent, model and analysis.

Both Beaver and Gokhale [1] and Davidson and Beaver [14] allow for an effect of order of presentation in paired comparisons. The latter allow for both ties and order effect. For the order T_i, T_j, their model is

$$P[T_i \rightarrow T_j] = \pi_i / (\pi_i + \gamma\pi_j + \nu\sqrt{\pi_i\pi_j}),$$

$$P[T_j \rightarrow T_i] = \gamma\pi_j / (\pi_i + \gamma\pi_j + \nu\sqrt{\pi_i\pi_j}),$$

$$P[T_i = T_j] = \nu\sqrt{\pi_i\pi_j} / (\pi_i + \gamma\pi_j + \nu\sqrt{\pi_i\pi_j}),$$

where γ represents the order effect and ν is associated with the probability of a no-preference judgment.

Bradley and El-Helbawy [9] extended the flexibility of analyses of paired comparisons through provision of a general method of testing specified treatment contrasts, contrasts in terms of the $\log\pi_i$. This work permits consideration of factorial treatment combinations that are often used in practice in product development.

Davidson and Solomon [18] considered a Bayesian approach to parameter estimation. Pendergrass and Bradley [32] extended (1) to obtain a model for triple comparisons. Multivariate extensions of (1) are available (see Davidson and Bradley [15] and Kousgaard [29]).

Statistical properties of paired comparisons analyses have been investigated by Bradley [6], Davidson and Bradley [16], and El-Helbawy and Bradley [21]. Convergence properties of iterative estimation procedures have been considered by Ford [22], Davidson and Bradley [15], and El-Helbawy and Bradley [20].

OTHER METHODS
FOR PAIRED COMPARISONS

The Thurstone model (3) is very similar to (1) and yields similar results in applications. Harris [26] adjusted the model to allow for order effects and Glenn and David [23] allowed for no-preference judgments, while Sadasivan [36] permitted unequal numbers of judgments on pairs.

Kendall and Babington Smith [28] considered the count of circular triads as a measure of consistency of judgment, a circular triad resulting from preference statements, $T_i \rightarrow T_j$, $T_j \rightarrow T_k$, $T_k \rightarrow T_i$. They also defined a coefficient of concordance* as a measure of agreement of judgments by several respondents. Bezembinder [2], concerned with transitivity of scaling, considered measures of *external* consistency of paired comparison data with respect to a given linear order of items and also measures of *internal* consistency. In so doing, he examined the concordance and circular triad procedures, among others, and the alternative hypotheses that might be associated with them. Gokhale et al. [24] presented a model robust approach to analysis of paired comparisons and noted that it permits examination of the assumption of linearity of intrinsic treatment worths.

Guttman [25] developed a method of scaling treatments in paired comparisons and

Saaty [35] proposed a group consensus method to provide treatment scores on a ratio scale. Mehra [30] and Puri and Sen [33] extended the use of signed ranks to paired comparisons. Bliss et al. [3] used "rankits"* for paired comparisons. Wei [39] and Kendall [27] proposed an iterative scoring system based not only on direct comparisons, but also on roundabout comparisons involving other treatments.

References

[1] Beaver, R. J. and Gokhale, D. V. (1975). *Commun. Statist. A*, **4**, 923–939.

[2] Bezembinder, T. G. G. (1981). *Brit. J. Math. Statist. Psychol.*, **34**, 16–37.

[3] Bliss, C. I., Greenwood, M. L., and White, E. S. (1956). *Biometrics*, **12**, 381–403.

[4] Bradley, R. A. (1954). *Biometrics*, **10**, 375–390.

[5] Bradley, R. A. (1954). *Biometrika*, **41**, 502–537.

[6] Bradley, R. A. (1955). *Biometrika*, **42**, 450–470.

[7] Bradley, R. A. (1976). *Biometrics*, **32**, 213–232.

[8] Bradley, R. A. (1985). In *Handbook of Statistics*, Vol. 4: *Nonparametric Methods*, P. R. Krishnaiah and P. K. Sen, eds. North-Holland, Amsterdam, pp. 299–326.

[9] Bradley, R. A. and El-Helbawy, A. T. (1976). *Biometrika*, **63**, 255–262.

[10] Bradley, R. A. and Terry, M. E. (1952). *Biometrika*, **39**, 324–345.

[11] Clatworthy, W. H. (1955). *J. Res. Natl. Bur. Stand.*, **54**, 177–190.

[12] David, H. A. (1963). *The Method of Paired Comparisons*. Charles Griffin, London.

[13] Davidson, R. R. (1970). *J. Amer. Statist. Ass.*, **65**, 317–328.

[14] Davidson, R. R. and Beaver, R. J. (1977). *Biometrics*, **33**, 693–702.

[15] Davidson, R. R. and Bradley, R. A. (1969). *Biometrika*, **56**, 81–95.

[16] Davidson, R. R. and Bradley, R. A. (1970). In *Nonparametric Techniques in Statistical Inference*, M. L. Puri, ed. Cambridge University Press, Cambridge, pp. 111–125.

[17] Davidson, R. R. and Farquhar, P. H. (1976). *Biometrics*, **32**, 241–252.

[18] Davidson, R. R. and Solomon, D. L. (1973). *Biometrika*, **60**, 477–487.

[19] Dykstra, O. (1960). *Biometrics*, **16**, 176–188.

[20] El-Helbawy, A. T. and Bradley, R. A. (1977). *Commun. Statist. A*, **6**, 197–207.

[21] El-Helbawy, A. T. and Bradley, R. A. (1978). *J. Amer. Statist. Ass.*, **73**, 831–839.

[22] Ford, L. R., Jr. (1957). *Amer. Math. Monthly*, **64**, 28–33.

[23] Glenn, W. A. and David, H. A. (1960). *Biometrics*, **16**, 86–109.

[24] Gokhale, D. V., Beaver, R. J., and Sirotnik, B. W. (1983). *Commun. Statist. Theor. Meth.*, **12**, 25–36.

[25] Guttman, L. (1946). *Ann. Math. Statist.*, **17**, 144–163.

[26] Harris, W. P. (1957). *Psychometrika*, **22**, 189–198.

[27] Kendall, M. G. (1955). *Biometrics*, **11**, 43–62.

[28] Kendall, M. G. and Babington Smith, B. (1940). *Biometrika*, **31**, 324–345.

[29] Kousgaard, N. (1982). "Models for Multivariate Paired Comparisons Experiments," *Res. Rep. No. 80*, University Statistical Institute, Copenhagen.

[30] Mehra, K. L. (1964). *Ann. Math. Statist.*, **35**, 122–137.

[31] Mosteller, F. (1951). *Psychometrika*, **16**, 3–9.

[32] Pendergrass, R. N. and Bradley, R. A. (1960). In *Contributions to Probability and Statistics*, I. Olkin et al., eds. Stanford University Press, Stanford, CA, pp. 331–351.

[33] Puri, M. L. and Sen, P. K. (1969). *Ann. Inst. Statist. Math. Tokyo*, **21**, 163–173.

[34] Rao, P. V. and Kupper, L. L. (1967). *J. Amer. Statist. Ass.*, **62**, 194–204; erratum, **63**, 1550.

[35] Saaty, T. L. (1977). *J. Math. Psychol.*, **15**, 234–280.

[36] Sadasivan, G. (1982). *Commun. Statist. Theor. Meth.*, **11**, 821–833.

[37] Scheffé, H. (1952). *J. Amer. Statist. Ass.*, **47**, 381–400.

[38] Thurstone, L. L. (1927). *Amer. J. Psychol.*, **38**, 368–389.

[39] Wei, T. H. (1952). "The Algebraic Foundations of Ranking Theory," thesis, Cambridge University, Cambridge, England.

[40] Zermelo, E. (1929). *Math. Zeit.*, **29**, 436–460.

Further Reading

References [7] and [8] provide general discussions of paired comparisons, the first emphasizing model formulations and the second, additional information on topics treated or noted in this article. The bibliography of Davidson and Farquhar provides a guide to in-depth investigation of the subject generally.

(ANALYSIS OF VARIANCE
BERNOULLI DISTRIBUTION

BLOCKS, BALANCED INCOMPLETE
DESIGN OF EXPERIMENTS
DISTRIBUTION-FREE METHODS
MAXIMUM LIKELIHOOD ESTIMATION
PARTIALLY BALANCED DESIGNS
SELECTION PROCEDURES)

RALPH A. BRADLEY

PAIRS, MATCHED   *See* MATCHED PAIRS
TEST; MATCHED PAIRS *t*-TEST

PAIRWISE INDEPENDENCE

If each of the $\binom{k}{2}$ pairs of events (E_i, E_j) from
the set $(E_1, E_2, \ldots E_k)$ is mutually independent, the set is said to be *pairwise independent*.

A *mutually independent set* of events
$E_1, \ldots E_k$ must satisfy the conditions that
each of the $\frac{1}{2}(3^k - 2^{k+1} + 1)$ possible pairs
of *disjoint subsets* of the k events are mutually independent.

A mutually independent set is also pairwise independent, but the converse is not
true. Many examples have been given. One
of the simplest is the following.

Suppose there are four balls; one is colored white, one red, one blue, and one
white, red, and blue. One of the balls is
chosen at random (so that the probability
that any specific ball is chosen is $1/4$). Let
E_1, E_2, E_3 denote the events that the chosen
ball will have the color white, red, blue (respectively) on it. Any two of the events are
mutually independent $(\Pr[E_i] = \Pr[E_i \mid E_j]$
$= 1/2$, for any distinct pair $i, j = 1, 2, 3$), so
the set is pairwise independent, but

$$\Pr[E_1 \mid E_2 \cap E_3] = 1 \neq \Pr[E_1],$$

so the set is not mutually independent.

(DEPENDENCE, CONCEPTS OF
SETWISE DEPENDENCE)

PALM FUNCTIONS,
PALM–KHINCHIN EQUATIONS

For a (Borel) subset A of the real line $\mathbb{R}$ let
$N(A)$ denote the number of points in A of a
point process* on $\mathbb{R}$. Assume the following:

1. The point process is stationary, i.e., for
   each positive integer n and sets A_1,
   $\ldots, A_n$ the joint distribution function
   of the random vector $(N(a + A_i): 1 \leqslant i$
   $\leqslant n)$ is independent of $a \in \mathbb{R}$.

2. The point process is orderly in the sense
   that if $N_t = N((0, t])$, then

   $$\lim_{t \to 0+} t^{-1} P(N_t \geqslant 2) = 0.$$

It can then be shown that the *intensity*

$$\lambda = \lim_{t \to 0+} t^{-1} P(N_t \geqslant 1)$$

always exists [7]. If $\lambda < \infty$ then for $j = 0$,
$1, \ldots$, the functions, called *Palm functions*,

$$\phi_j(t) = \lim_{\tau \to 0+} P(N((\tau, \tau + t]) = j \mid N_\tau \geqslant 1)$$

can be defined. They can be interpreted as
the distribution of the number of points
occurring in an interval of length t measured
from an arbitrarily chosen point of the process.

Palm [10] first invoked this intuitive idea
to define $\phi_0(t)$ and used it in connection
with problems arising in queueing theory*.
Here points correspond to calls to a telephone exchange, and Palm showed that the
aggregate of calls is a Poisson process* if it
can be regarded as the superposition of
many sparse point processes emanating from
subscribers [7, p. 82]. Khinchin first showed
that all the $\phi_j(t)$ can be defined, and hence
Gnedenko and Kovalenko [5] call them
Palm–Khinchin functions.

Let $P_j(t) = P(N_t \leqslant j)$. Khinchin [7] established the formula $P_j(t) = \lambda \int_t^\infty \phi_j(u) \, du$, although the $j = 0$ case was known to Palm.
These formulas were named after Palm
by Khinchin, but now are called Palm–
Khinchin equations [1, 5]. They link the
distributions of numbers of points falling in
an interval measured from an arbitrary origin and numbers of points in an interval
measured from an arbitrary point of the
process.

For two sample applications of these formulas see Daley [2], where they are extended
slightly and used to derive an integral ex-

pression for var N_t, and Oakes [9], where they are used to elucidate properties of Poisson cluster processes.

Example. Let (N_t^0) denote an ordinary renewal* process with an interevent time distribution F having finite mean $\mu = \int_0^\infty tF(dt)$. The interpretation above implies that for the corresponding stationary, or equilibrium, renewal process $\phi_j(t) = P(N_t^0 = j) = F_j(t) - F_{j+1}(t)$, where F_j denotes the jth convolution power of F. Since $P_0(0+) = 1$ it follows that $\lambda = \mu^{-1}$ and also that the distribution function of the distance from an arbitrary origin to the next point on its right is $\mu^{-1}\int_0^t(1 - f(u))\,du$ [1, p. 30].

Generalizations of the Palm–Khinchin equations have been obtained for multivariate point processes [3] and also for nonstationary and nonorderly point processes [4]. In a more abstract direction, the Palm functions can be generalized by trying to define a conditional probability $P(B \mid N_{0+} > 0)$ where B belongs to the ring generated by sets of the form $\{N((\tau_i, \tau_i + t_i]) \leqslant j_i : 1 \leqslant i \leqslant n\}$. The measure so obtained is called a *Palm measure*, and this notion can be extended to point processes on quite general carrier spaces, and even to random measures [6, 8]. Also, the Palm–Khinchin equations can be extended to these abstract settings [8, §8.2].

References

[1] Cox, D. R. and Isham, V. (1980). *Point Processes.* Chapman and Hall, London.

[2] Daley, D. J. (1978). *Stoch. Proc. Appl.*, **7**, 255–264.

[3] Daley, D. J. and Milne, R. K. (1975). *J. Appl. Prob.*, **12**, 383–389.

[4] Fieger, W. (1965). *Math. Scand.*, **16**, 121–147.

[5] Gnedenko, B. V. and Kovalenko, I. N. (1968). *Introduction to Queueing Theory.* Israel Program for Scientific Translations, Jerusalem.

[6] Jagers, P. (1973). *Z. Wahrscheinlichkeitstheorie verw. Geb.*, **26**, 17–32. (Semiexpository account of Palm measures.)

[7] Khinchin, A. Ya. (1955). *Mathematical Methods in the Theory of Queueing.* Trudy Matem. Inst. Steklov, Vol. 49. (English translation Griffin, London, 1960.)

[8] Matthes, K., Kerstan, J., and Mecke, J. (1978). *Infinitely Divisible Point Processes.* Wiley, New York. (Detailed abstract treatment of Palm measures.)

[9] Oakes, D. (1975). *J. R. Statist. Soc. B*, **37**, 238–247.

[10] Palm, C. (1943). *Ericsson Technics*, **44**, 1–189.

(POINT PROCESSES; QUEUEING THEORY; RENEWAL THEORY; STATIONARITY)

ANTHONY G. PAKES

PANDIAGONAL SQUARE *See* MAGIC SQUARE DESIGNS

PANEL DATA

Panel or longitudinal data refers to data sets in which multiple observations are taken for each sampled unit (*see* LONGITUDINAL DATA ANALYSIS). The terms are nearly synonymous. Longitudinal studies are generally concerned with multiple observations on realizations of stochastic processes*, whereas panel studies include this possibility but generally involve an explicitly sampled panel and may address an essentially static behavioral model. In economics, common examples are generated by pooling time-series observations across a variety of cross-section units: e.g., countries, states, firms, or randomly sampled individuals. Without reference to time series* or cross-section variation, data with the same structure is produced by experimental designs* classified as one-way layouts (see Scheffé [10]).

A growing number of large panel data sets with an economic or social science background are available in the United States. Panels for samples of individuals include:

1. The Panel Study on Income Dynamics conducted by the Office of Economic Opportunity and the Survey Research Center of the University of Michigan.

2. The National Longitudinal Survey conducted by the Center for Human Resource Research at the Ohio State Uni-

versity (known as the NLS or Parnes Data).

3. Data from the negative income tax experiments conducted by the Institute for Research on Poverty at the University of Wisconsin.

These data sets are characterized by a large number of randomly selected individuals (N of the order of several thousand) and relatively few years (T no more than ten for yearly observations).

A longitudinal data set has a number of advantages over either a single cross-section or a single time series. In the first place, new sources of variation in the data are introduced. For example, in the early econometric demand studies of Stone [12], income and price elasticities had to be estimated separately from cross-section and time-series data, respectively, because of inadequate variation in both commodity prices over a single cross section and aggregate personal income over time. Panel data permits simultaneous estimation of income and price effects by pooling observations that vary over individuals with observations on the same individuals at different points in time.

A second advantage of longitudinal data is the ability to control for the presence of unobserved individual-specific or time-specific effects that would otherwise bias coefficient estimates in a linear regression* or analysis of variance* model. This is possible because transforming the data into deviations over time from individual means eliminates any individual-specific effect (measurable or not) and transforming into individual deviations from time averages eliminates all time-specific effects. Thus with panel data, one can test for such bias-inducing latent effects and construct consistent estimators that account for them (see Hausman and Taylor [5]).

Finally, a number of complex issues in dynamic behavior can be resolved if multiple observations are available for the same individual at different points in time. For example, with panel data it is possible to explain the fact that individuals experienc-

ing an event in the past—say unemployment—are more prone to experience it in the future and to distinguish two competing hypotheses: that people are altered by the event, or that people have different propensities to experience the event. See Heckman [6].

A rich set of models and estimators for use with longitudinal data have been developed. Since observations are often generated by an explicit sampling scheme, there is often interest in allowing parameters to be distributed randomly in the population. Although work has been done on random parameter models in general (see, e.g., Hsiao [7]), the principal distinction in the literature has been between fixed or random intercepts specific to individuals or time periods. Fixed-effects models* can be treated as ordinary linear regression models with intercepts specific to time periods or individuals; following Eisenhart [3], these are classified as Model I analysis of variance. Random effects models are also known as variance components models or Model II analysis of variance models (*see* MODELS I, II, AND III). For an individual-specific effect, estimation for a variance components linear model is easily outlined. Time-specific effects can be added with no conceptual difficulties, but nonlinear random effects models have a number of qualitative differences. See Chamberlain [1].

Suppose the data is organized first by individual and then by time. The model is

$$y_{it} = \mathbf{X}_{it}\boldsymbol{\beta} + \alpha_i + \epsilon_{it},$$

$$i = 1, \ldots, N; \quad t = 1, \ldots, T;$$

$$\mathbf{y} = \mathbf{X}\boldsymbol{\beta} + \boldsymbol{\alpha} + \boldsymbol{\epsilon},$$

where $\mathbf{X}_{it}$ is a vector of observations on k explanatory variables, $\boldsymbol{\beta}$ is a k vector of unknown coefficients, α_i is an unobserved random variable with mean zero and constant variance σ_α^2, and ϵ_{it} is a zero mean random disturbance with variance σ_ϵ^2. The preceding moments are conditional on the $\mathbf{X}_{it}$'s; and α_i and ϵ_{it} are assumed to be conditionally independent.

Least-squares* estimates of $\boldsymbol{\beta}$ are unbiased but inefficient, since $\text{cov}(\alpha_i + \epsilon_{it})$ is nonscalar. Let the $TN \times TN$ matrix **A** be the block diagonal matrix with $T \times T$ blocks of $1/T$'s on the diagonal. With data organized first by individual, then by time, premultiplication by **A** replaces a TN vector by its time averages, organized by individuals. Similarly if $\mathbf{B} = \mathbf{I}_{TN} - \mathbf{A}$, premultiplication by **B** replaces a vector with individual deviations from its time average, where $\mathbf{I}_{TN}$ is the order TN identity matrix. The columns of **A** and **B** span the eigenspaces of Ω corresponding to its two distinct eigenvalues. Transform the model into deviations from individual averages by premultiplying by **B** and estimate β by ordinary least squares. The resulting estimator

$$\hat{\beta}_W = (\mathbf{X}'\mathbf{B}\mathbf{X})^{-1}\mathbf{X}'\mathbf{B}\mathbf{y}$$

is termed the *within-groups estimator*. If the model is transformed by **A**,

$$\hat{\beta}_B = (\mathbf{X}'\mathbf{A}\mathbf{X})^{-1}\mathbf{X}'\mathbf{A}\mathbf{y}$$

is the *between-groups estimator*. Assuming the α_i are uncorrelated with the $\mathbf{X}_{it}$, both $\hat{\beta}_W$ and $\hat{\beta}_B$ are unbiased; the best linear unbiased estimator for β is a matrix-weighted average of the between- and within-groups estimators with weights depending upon $\text{cov}(\hat{\beta}_W)$ and $\text{cov}(\hat{\beta}_B)$. (See Maddala [8].) A computationally simpler method is to transform the columns of **y** and **X** as follows:

$$\tilde{y}_{it} = y_{it} - \theta y_i.\ ,$$

$$\tilde{\mathbf{X}}_{it} = \mathbf{X}_{it} - \theta \mathbf{X}_i.\ ,$$

where $\theta = 1 - [\sigma_\epsilon^2/(\sigma_\epsilon^2 + T\sigma_\alpha^2)]^{1/2}$ and $y_i. = (1/T)\sum_{t=1}^{T} y_{it}$. The best linear unbiased estimate of β is then produced by the least-squares regression of $\tilde{y}$ on $\tilde{\mathbf{X}}$ (see Hausman [4]).

Since $\mathbf{B}\alpha = \mathbf{0}$, $\hat{\beta}_W$ is unbiased for β even if the individual effects are correlated with the **X**'s. Hence a test for such correlations can be based on the length of the vector $\hat{\beta}_W - \hat{\beta}_B$ (see Hausman [4]). A drawback of the within-groups estimator is that efficiency is lost and coefficients of time-invariant variables are not estimable. Efficient estimation of all of the coefficients in this model is possible provided there is prior information that certain explanatory variables are uncorrelated with α_i. If there are as many time-varying variables uncorrelated with α_i as there are time-invariant variables correlated with α_i, the coefficients of the latter are identifiable and an efficient instrumental variable* estimator is constructed by Hausman and Taylor [5]. A feature of this approach is that the prior exogeneity assumptions are fully testable.

Examples of theoretical and applied econometric work with panel data are contained in the *Annales de l'INSEE* volume in reference 6. Chamberlain [2] is an idiosyncratic survey of the econometrics* literature; Maddala [9] is a textbook reference. For the statistics literature, see Searle [11]; the classic text is still Scheffé [10].

References

[1] Chamberlain, G. (1980). *Rev. Econ. Stud.*, **47**, 225–238. (Survey of methodology for panel studies with discrete data.)

[2] Chamberlain, G. (1983). In *Handbook of Econometrics*, M. L. Intriligator and Z. Griliches, eds. North-Holland, Amsterdam. (Survey of recent econometric literature for linear and nonlinear models.)

[3] Eisenhart, C. (1947). *Biometrics*, **3**, 1–21.

[4] Hausman, J. A. (1978). *Econometrica*, **46**, 1251–1272. (Theoretical development of a class of tests of model specification.)

[5] Hausman, J. A. and W. E. Taylor (1981). *Econometrica*, **49**, 1377–1398. (Panel models when latent individual effects are correlated with explanatory variables. An application to measuring the returns to education when ability is unobserved.)

[6] Heckman, J. J. (1978). *Ann. l'INSEE* **30/31**, 227–269. (Theoretical treatment of models for discrete panel data.)

[7] Hsiao, C. (1975). *Econometrica*, **43**, 305–325. (Technical treatment of the random parameters model for panel data.)

[8] Maddala, G. S. (1971). *Econometrica*, **39**, 341–358. (Technical derivation of the best linear unbiased estimator for the variance components model.)

[9] Maddala, G. S. (1977). *Econometrics*. McGraw-Hill, New York. (An elementary textbook containing an excellent treatment of the linear model with fixed and random effects.)

[10] Scheffé, H. (1959). *The Analysis of Variance*. Wiley, New York. (Still the classic textbook reference.)

[11] Searle, S. R. (1971). *Biometrics*, **27**, 1–76. (An excellent though dated survey of the statistical and biometric literature.)

[12] Stone, J. R. N. (1954). *The Measurement of Consumers' Expenditure and Behaviour in the United Kingdom, 1920–1938*, Vol. 1. Cambridge University Press, New York. (A milestone in applied economics.)

(ECONOMETRICS
FIXED, RANDOM- AND
    MIXED-EFFECTS MODELS
LONGITUDINAL DATA ANALYSIS
SURVEY SAMPLING)

WILLIAM E. TAYLOR

PAPADAKIS METHOD

In many agricultural experiments an experimental site is divided into approximately congruent plots, usually rectangular. The final data, assumed univariate here, consist of a measurement on each plot. Neighboring plots are essentially contiguous, and such microfactors as soil fertility and climate lead, in the absence of treatments, to the observations on plots close together tending to be more alike than those on plots well separated. Thus the assumption that the observations are independent is seldom justifiable.

R. A. Fisher*—showed that randomization*—choosing a design at random from a suitable set of designs—allows a valid estimate of variance to be attached to the ordinary least-squares* estimator. The efficiency of this estimator can be increased by a suitable choice of the design set, in particular, by blocking, i.e., forming the plots into blocks within which the plots should have very similar fertility. There are three points of dissatisfaction with randomization theory. First, it is often difficult to judge the best way to group plots into blocks, so that the design set chosen may be inefficient. Second, within a design set, the conditional efficiencies of designs vary when observa-

tions are correlated, so that the actual design selected may be inefficient. In some countries, efficient systematic designs* are still used. Third, the ordinary least-squares estimator may be inefficient unless the observations are essentially independent. This has led to methods that attempt to take the underlying correlation directly into account in the estimation and analysis. The first of these methods was proposed by J. S. Papadakis in 1937 and examined by Bartlett in 1938 [2]. Although mentioned again by D. R. Cox in 1952, the method suffered relative neglect until interest was revived in it in the 1970s by S. C. Pearce, at about the same time as interest was growing, with developments in modeling and computing, in methods that attempted to model the underlying correlation by a spatial stochastic process*.

The Papadakis method is an analysis of covariance* in which, unusually, the covariate is formed from the observations. This property has led to doubt about the validity of the suggested analysis. Various extensions have been suggested, including different definitions of the covariate, the use of more than one covariate, and an iterated analysis. The value of a covariate on plot i is usually formed by averaging deviations over a suitable set of neighboring plots Ni, where for each plot the deviation is formed by subtracting an estimate of the effect of the treatment from the observation. Note, however, that Kempton and Howes [4] have proposed that when there is any dependence between yields, as for disease trials or for when competition exists, the covariate be formed from the unadjusted observations. For plots forming a linear sequence, the simplest case is to form the covariate from adjoining plots, so that $N1 = \{2\}$, $Nn = \{n - 1\}$, and $Ni = \{i - 1, i + 1\}$ for $1 < i < n$, and to do a single analysis forming the deviations from the least-squares estimates. This is illustrated in the following example, which uses artificial data in a linear sequence with three treatments and only two replicates. The model only contains the treatment effects $\{\theta_k\}$.

Plot	i	1	2	3	4	5	6
Treatment	k	1	2	3	2	1	3
Observation	y_i	27	9	48	21	3	12
Deviation	d_i	12	-6	18	6	-12	-18
Covariate	x_i	-6	15	0	3	-6	-12

Here $\hat{\theta}_1 = \frac{1}{2}(27 + 3) = 15$, $\hat{\theta}_2 = 15$, $\hat{\theta}_3 = 30$, $d_1 = 27 - \hat{\theta}_1$, etc., and $x_1 = d_2$, $x_2 = \frac{1}{2}(12 + 18)$, etc. The analysis is then the usual analysis of covariance of y given x.

Sums of Squares and Products

	Corrected		Residual	
	y	x	y	x
y	1308	-6	1008	144
x	-6	444	144	144

Thus the estimate of the regression coefficient is $\hat{\beta} = 1$, and, since the treatment means for x, $\{\hat{\theta}_{x,k}\}$, are $-6, 9, -6$, respectively, the Papadakis estimates are $21, 6, 36$, respectively. The sums of squares (ss) in the unadjusted and adjusted analyses are:

Unadjusted			Adjusted		
Source	d.f.	SS	Source	d.f.	SS
			Regression	1	0.08
Treatments	2	300	Treatments | Regression	2	443.92
Residual	3	1008	Residual	2	864

The variances and covariances of the Papadakis estimators are estimated in the usual way. For an equireplicate design with r replicates, the estimated $\text{cov}(\hat{\theta}_{P,k}, \hat{\theta}_{P,l})$ is

$$\hat{\sigma}^2 \left\{ \frac{\delta_{k,l}}{r} + \frac{\hat{\theta}_{x,k}\hat{\theta}_{x,l}}{\text{residual sum of squares for } x} \right\},$$

where $\delta_{i,j}$ is the Kronecker delta. Hence in the example the dispersion matrix of $\hat{\theta}_P$ is estimated by

$$\begin{bmatrix} 324 & -162 & 108 \\ -162 & 459 & -162 \\ 108 & -162 & 324 \end{bmatrix}.$$

For linear sequences, the covariate could be extended to cover plots up to 2 away, so that $Ni = \{i - 2, i - 1, i + 1, i + 2\}$ for $2 < i < n - 1$, etc., or a second covariate could be

formed for plots two apart so that $N_2i = \{i - 2, i + 2\}$ for $2 < i < n - 1$, etc., in each case obvious adjustments being made for $i = 1, 2, n - 1, n$.

The covariate $\mathbf{x}$ depends on the initial estimate of $\boldsymbol{\theta}$, and not on $\hat{\boldsymbol{\theta}}_P$. In the iterated version, suggested by Bartlett [3], the covariate at iteration s is formed using deviations from the treatment estimates found at iteration $s - 1$, beginning with ordinary least squares, so that the first iteration is the original Papadakis analysis. Thus for the preceding example, the deviation and covariate in the second iteration are

Deviation	$d_i^{(2)}$	6	3	12	15	-18	-24
Covariate	$x_i^{(2)}$	3	9	9	-3	-4.5	-18

where the $d_i^{(2)}$ are the deviations formed by subtracting the initial Papadakis estimates from the data. The analysis then proceeds as before.

Sums of Squares and Products

	Corrected		Residual	
	y	x	y	x
y	1308	391.5	1008	504
x	391.5	520.875	504	464.625

Thus $\hat{\beta}^{(2)} \simeq 1.085$, $\hat{\theta}_x^{(2)} = (-0.75, 3, -4.5)'$, and the Papadakis estimates are $\hat{\theta}_P^{(2)} \simeq (15.81, 11.75, 34.88)'$. The sums of squares are:

Source	d.f.	SS
Regression	1	294.26
Treatments | Regression	2	552.45
Residual	2	461.29

A fixed number of iterations can be made, or iteration can continue until covergence, although convergence may be very slow. In some cases, the estimates will diverge or oscillate, as in the preceding example where everything else, including estimates of contrasts*, does converge.

The role of blocks in the analysis when the plots are contiguous is not clear. Papadakis originally proposed his method as a replacement of the use of blocks, but Bartlett [2] used both. The use of a step function to describe the trend in fertility, about which the errors are autocorrelated, is unsatisfactory. It seems preferable to fit a simple continuous surface or to remove the trend by differencing. However, the Papadakis method with block effects appears to fit smoothed step functions, but rarely seems to increase the precision, so that few would now recommend its use. Note that the covariate is formed from the deviation and not from the least-squares residual, and hence that in the uniterated version the covariate, and $\hat{\theta}_x$, are the same as without blocks, only $\hat{\beta}$, $\hat{\theta}_P$, and the sums of squares differing. The method with block effects is illustrated next using the preceding example with two blocks of size three.

Sums of Squares and Products
Treatment

	+ Residual		Residual	
	y	x	y	x
y	924	-198	624	-48
x	-198	348	-48	48

Thus $\hat{\beta} = -1$, and the Papadakis estimates are 9, 24, and 24 respectively. The sums of squares in the unadjusted and adjusted analyses are:

Unadjusted			Adjusted		
Source	d.f.	SS	Source	d.f.	SS
			Regression	1	0.08
Blocks	1	384	Blocks | Regression	1	496.57
Treatments	2	300	Treatments | Regression	2	235.34
Residual	2	624	Residual	1	576

For a second iteration, note that the deviations are:

Deviation $d_i^{(2)}$ 18 -15 24 -3 -6 -12

where $d_1^{(2)} = y_1 - 9$, etc.

In the spatial case there are no new principles involved, although there are more possibilities for different covariates. The simplest covariate uses the adjoining plots, so that $N(i, j) = \{(i - 1, j), (i + 1, j), (i, j - 1), (i, j + 1)\}$ for $1 < i < n_1$ and $1 < j < n_2$, $N(1, j) = \{(2, j), (1, j - 1), (1, j + 1)\}$ for $1 < j < n_2$, etc., and $N(1, 1) = \{(2, 1), (1, 2)\}$, etc. A numerical example is given in Pearce [8, pp. 51–54]. Other possibilities are considered by Pearce and Moore [9].

The first investigation of the validity of the analysis was by Bartlett [2], who suggested that for one covariate the residual degrees of freedom should be two less than in the uncorrected analysis and that the number of treatments should be large to give a better estimate of the variance and an approximately valid analysis. However, recent work by Wilkinson et al. [10] and others using simulations on uniformity trial data has shown that over randomization the uniterated Papadakis method leads to a slight upward bias in the residual mean square*, while the iterated Papadakis method gives more efficient estimators than the uniterated method, but the residual mean square is seriously downward biased. Theoretical work by Atkinson [1], and, more recently, by Bartlett [3] and Martin [5] has suggested that for a given set of covariates, the iterated Papadakis method essentially seems to fit a particular spatial correlation model to the data, the uniterated version being a first approximation. However, it does this inefficiently and often produces estimates of the spatial dependence parameters that lie outside their range. Hence some people believe (see the discussions in Bartlett [3] and Wilkinson et al. [10]) that there is much to be gained by making an explicit assumption about the structure of the errors and then fitting the model by exact maximum likelihood. Advantages are that it is clear what the model is, and other models can be compared for their fits. Simulations on uniformity trial data (see the comments of J. Besag in the discussion to [10]) suggest that this method, and the equivalent iterated Papadakis method, are as efficient as the best classical methods that use row and column blocking, and Martin [6] has shown some robustness to misspecification of the error model in the estimation of θ. The estimated standard errors of treatment differences do not appear as biased for the model-based approach as for the iterated Papadakis method.

Recently, the Papadakis method has been thought of as a means of rescuing classical experiments in which misjudgement of the fertility has led to an unfortunate choice of block and plot structure. However, if it is decided beforehand to use a spatial method of analysis, then the design could be chosen accordingly. It is shown in Martin [5, 6] that for some designs the Papadakis and generalized least-squares estimators are scarcely more efficient than the ordinary least-squares estimators and that the efficiency ordering of designs is highly dependent on how rapidly the correlations decline with distance and on the criterion by which the efficiency is measured. Wilkinson et al. [10] have suggested that designs with second-level partial nearest-neighbor* balance be used with their approximately valid nearest-neighbor analysis, these designs being highly efficient under their model and having approximately equal standard errors for estimated treatment differences. Several discussants to that paper have noted that the proposed estimators seem to be essentially the iterated Papadakis estimators or those for the equivalent spatial correlation model, with the regression parameters estimated by a different method.

Despite the current revival of interest in the Papadakis method, it seems possible that it will be remembered in the future only for providing a stimulus in the growth of spatial methods for the analysis of agricultural experiments.

It should be noted that Papadakis himself has developed his method in a series of papers unknown to those reexamining the original method mentioned here (see Papadakis [7]).

References

[1] Atkinson, A. C. (1969). *Biometrika*, **56**, 33–41. (An investigation of the theory of the method for some special one-dimensional designs.)

[2] Bartlett, M. S. (1938). *J. Agric. Sci.*, **28**, 418–427. (Contains a summary of Papadakis's 1937 paper. Gives two examples, but uses approximate results. An initial investigation of the theory.)

[3] Bartlett, M. S. (1978). *J. R. Statist. Soc. B*, **40**, 147–174. (A return to the theory 40 years on. A difficult paper to read, but a useful discussion. Gives four further examples. Unfortunately all analyses use blocks, are approximate, and appear to contain numerical errors.)

[4] Kempton, R. A. and Howes, C. W. (1981). *Appl. Statist.*, **30**, 59–70. (The iterated method on experimental data and a simulation investigation using uniformity trial data.)

[5] Martin, R. J. (1982). *Biometrika*, **69**, 597–612. (A theoretical investigation of the method.)

[6] Martin, R. J. (1983). *Res. Rep. No. RJM/2/83*, Dept. of Probability and Statistics, Sheffield University, England. (A theoretical investigation of design efficiency under generalized least squares.)

[7] Papadakis, J. (1984). *Proc. Acad. Athens*, **59** (forthcoming). (A summary of developments proposed by the originator of the method which have been overlooked in the English-speaking world.)

[8] Pearce, S. C. (1983). *The Agricultural Field Experiment: A Statistical Examination of Theory and Practice*. Wiley, Chichester, Sects. 2.4, 3.9, and 9.2. (Contains a recent more advanced introduction to the method, plus a worked example, 2C, pp. 51–54. However, some results quoted have been superseded. See the note on p. 56, relating to ref. 10.)

[9] Pearce, S. C. and Moore, C. S. (1976). *Exper. Agric.*, **12**, 267–272. (A practical investigation of different covariates on uniformity trial data.)

[10] Wilkinson, G. N., Eckert, S. R., Hancock, T. W., and Mayo, O. (1983). *J. R. Statist. Soc. B*, **45**, 151–210. (A simulation investigation using uniformity trial data. Notes errors in refs. 1, 3, and 4. Introduces an improved analysis of a similar type. The valuable discussion contains a review of recent research on spatial methods in agricultural experiments.)

Bibliography

Bartlett, M. S. (1981). *J. R. Statist. Soc. B*, **43**, 100–102. (Some further theoretical remarks.)

Pearce, S. C. (1978). *Trop. Agric.*, **55**, 97–106. (A practical comparison of the method with others using experimental data.)

Pearce, S. C. (1980). *Trop. Agric.*, **57**, 1–10. (A practical comparison of the iterated method with others on experimental data.)

Ripley, B. D. (1981). *Spatial Statistics*. Wiley, New York, Sect. 5.3. (Some theoretical comments. A useful book on all aspects of spatial analysis.)

(AGRICULTURE, STATISTICS IN
ANALYSIS OF COVARIANCE
ANALYSIS OF VARIANCE
RANDOMIZATION TESTS
SPATIAL DATA ANALYSIS)

R. J. MARTIN

PARADOX OF VOTING *See* VOTING PARADOX

PARALLELISM, TESTS FOR *See* HOLLANDER PARALLELISM TESTS

PARALLEL-REDUNDANT SYSTEM *See* EXTREME-VALUE DISTRIBUTIONS

PARAMETER

In everyday usage, *parameter* means a condition restricting possible action. In statistical theory and probability, the term has the more precise meaning of a quantity entering into the distribution of a statistic or a random variable.

Thus in the normal distribution with cumulative distribution function (CDF)

$$\Pr[X \leqslant x]$$
$$= \left(\sigma\sqrt{2\pi}\right)^{-1}$$
$$\times \int_{-\infty}^{x} \exp\left\{-\tfrac{1}{2}\sigma^{-2}(y - \xi)^2\right\} dy,$$

the quantities σ and ξ are parameters. The values of these parameters define the particular normal distribution* that is appropriate.

A parameter may be a scalar or a vector (corresponding to a set of quantities). In practical usage, the dimension of the vector is finite and usually quite small. It is possible to envision an infinite-dimensional vector parameter, but this usually is not necessary.

If a family of distributions has a CDF of form $F(x \mid \boldsymbol{\theta})$, where $F(x \mid \boldsymbol{\theta})$ is a known form, it is called a *parametric* family, and $\boldsymbol{\theta}$ is said to *index* the family.

Many problems in testing hypotheses* and estimation* relate to the values of the parameters.

PARAMETER-FREE INFERENCE *See* SUFFICIENT ESTIMATION AND PARAMETER-FREE INFERENCE

PARETO DISTRIBUTION

Although Pareto distributions are useful modeling and predicting tools in a wide variety of socioeconomic contexts, there is a definite advantage in focusing discussion on one specific field of application: the size distribution of income. It was in that context that Vilfredo Pareto [13] introduced the concept in his well-known economics text. Pareto observed that in many populations the number of individuals in the population whose income exceeded a given level x was well approximated by $Cx^{-\alpha}$ for some real C and some $\alpha > 0$. Subsequently, it became apparent that such an approximation was only acceptable for large values of x. Pareto asserted that some underlying law actually determined the form of income distributions*. He was not too precise in specifying how the law worked or, if you wish, what kind of stochastic mechanism might lead to such tail behavior (subsequently called Paretian tail behavior) of income distribution survival functions. Nor was it clear what constitutes a large value of x or whether α might or might not vary from population to population. His own view on this changed with time. The parameter α appeared immutable when introduced (i.e., "always about 1.5"), but later was admitted to depend on changes in the population and changes in the definition of income used in deriving the distribution (e.g., individual income, family income, income before taxes). It is from such imprecise but generally data-supported beginnings that lively controversies arise. The battle swayed back and forth for 40 to 50 years. Shirras [15], one of the harshest critics of Pareto's law, argued that when he plotted log income against the log of the survival function using Indian income data, "the points did not lie even roughly on a straight line" as predicted by the Pareto law. The configuration of points is indeed slightly parabolic, but the deviation from linearity is not as distressing to most as it was to Shirras. An excellent summary of the early life and times of the Pareto distribution is provided by Cirillo [3]. Cirillo also provides a translation of Pareto's original discussion of what was to become known as his law.

Despite the criticisms of doubters like Shirras, it became, chiefly on the basis of empirical evidence rather than on any theoretical grounds, generally accepted that most income distributions did indeed exhibit Paretian tail behavior. Indeed, it became, and to a great extent continues to be, standard procedure to go ahead and estimate the Paretian index α (as a measure of dispersion or inequality of income distribution) without much concern about how well the Pareto model fitted the actual data. When such a critical test is performed, it is remarkable how often the upper tail of the survival function does appear to be proportional to $x^{-\alpha}$. Reference may be made to Zipf [18], where a large number of social and economic phenomena are observed to have size distributions which, like income distributions, exhibit Paretian tail behavior.

To be sure, there have been attempts to describe stochastic models that might account for Paretian income distributions or at least Paretian tail behavior for such distributions. Some would argue that individual log incomes are basically random walks* and that a limiting stable distribution* for log income seems plausible. A popular specialization of this argument is used to buttress the view that income should be modeled by the lognormal distribution*. If the random walk model assumed for log income involves a reflecting barrier (or, equivalently, if a specific minimum income is built into the model), then Paretian tail behavior can be expected (see Ord [12] for a survey of competing income models described in terms of random walks*). Mandelbrot [10] argues that stable distributions* provide the appropriate model for income. The Paretian tail behavior observed in data sets reflects, in this view, the asymptotic regular variation of the stable survival functions. Arnold [1] provides a brief survey of these and other models. Few, if any, of the models proposed permit identification of the explanatory stochastic mechanism used with any economically meaningful parameters or processes. The bald fact remains that in the upper tail, income distributions are reasonably well approximated by Pareto distributions and, certainly for predictive purposes, the lack of a compelling explanatory model should not deter one from capitalizing on the appealing simplicity of Pareto's distribution.

Figure 1, based on data from the Republic of Botswana [14], illustrates the upper tail of the survival function of the income distribution in Botswana in 1974. The curve is plotted in a log vs. log scale, and perfect fit to a Pareto model would be evidenced by a linear plot. The barely perceptible parabolic form of the plot is not atypical. It formed the basis for Shirras' harsh criticism (alluded to earlier) and even earlier prompted Pareto to consider a slightly more general distribution for income, one whose density function has the form $kx^{-(\alpha+1)}e^{-\beta x}$ for large values of x. On the other hand, most authors are impressed by the apparent linearity of plots such as Fig. 1 and have been led to use Pareto's original simple model for prediction and fitting purposes.

Size distributions are of interest in many contexts besides that of income. Paretian tail behavior has been observed and studied in a wide variety of settings. Independent parallel development of related distributional concepts has occurred in many fields. Phenomena whose size distributions exhibit Paretian tail behavior include city sizes, sizes of busi-

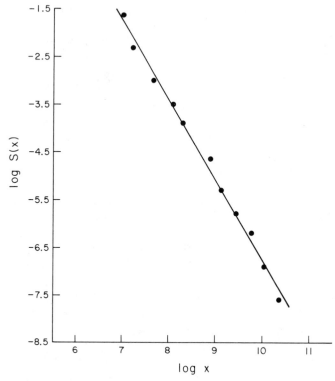

Figure 1 Upper tail of the survival function of income distribution, Botswana, 1974.

nesses, geological site sizes, and insurance claims. Some idea of this diversity of fields can be seen by perusing the Bibliography.

THE CLASSICAL PARETO DISTRIBUTION AND RELATED VARIANTS

The classical Pareto distribution has a survival function* of the form

$$S(x) = (x/\sigma)^{-\alpha}, \qquad x \geqslant \sigma. \qquad (1)$$

The scale parameter σ and the index of inequality α are both positive parameters. A common variant involves introduction of a new location parameter yielding a survival function of the form

$$S(x) = \left[1 + \left(\frac{x-\mu}{\sigma}\right)\right]^{-\alpha}, \qquad x \geqslant \mu. \qquad (2)$$

The additional flexibility acquired by using (2) instead of (1) comes at a high price. A sample from the classical Pareto distribution has a two-dimensional sufficient statistic,

$(X_{1:n}, \prod_{i=1}^{n} X_{i:n})$. No such reduction via sufficiency is possible for the family (2). In fact, if X has survival function (1), then $\log(X/\sigma)$ has an exponential (α) distribution. Thus if σ can be estimated with precision or is fixed by the sampling scheme, then analysis of classical Pareto samples is directly reducible to analysis of exponential samples. For added flexibility in fitting data, an additional parameter is introduced, transforming (2) to

$$S(x) = \left[1 + \left(\frac{x-\mu}{\sigma}\right)^{1/\gamma}\right]^{-\alpha}, \qquad x \geqslant \mu, \qquad (3)$$

sometimes referred to as a Burr distribution*. Perhaps the ultimate in generalized Pareto distributions is provided by random variables of the form $\mu + \sigma(Z_1/Z_2)^\gamma$, where $\mu \in \mathbb{R}$; $\sigma, \gamma \in \mathbb{R}^+$ and Z_1 and Z_2 are independent gamma random variables with unit scale parameter and possibly different shape parameters. Such a distribution may be dubbed Feller–Pareto (e.g., Arnold [1]) or generalized F (e.g., Kalbfleisch and Prentice [8]). It is easily verified that the Pareto distributions (1)–(3) are special cases of this generalized F-distribution.

DISTRIBUTIONAL PROPERTIES

Moments of generalized Pareto distributions are readily derived. The representation as a function of independent gamma random variables is often helpful in this regard. For example the δth moment of the classical Pareto distribution (1) is of the form $\alpha\sigma^\delta/(\alpha - \delta)$ (where $\delta < \alpha$ for convergence).

The distribution theory associated with samples from a generalized Pareto distribution [i.e., (3)] is generally complicated. It is not difficult to determine that convolutions* of such Pareto distributions exhibit Paretian tail behavior, but closed expressions for the convolved distribution usually are not available (for $n > 3$). Products of classical Pareto variables are well behaved, and distributional properties of classical Pareto order statistics* are tractable (and readily derived by the logarithmic transformation to exponentiality). For example the γth moment of the kth largest order statistic from a classical Pareto sample is

$$E(X^\gamma_{k\,:\,n})$$

$$= \sigma^\gamma \frac{n!}{(n-k)!} \frac{\Gamma(n-k+1-\gamma\alpha^{-1})}{\Gamma(n+1-\gamma\alpha^{-1})},$$

$$(4)$$

provided $\gamma < \alpha(n - k + 1)$. Similar expressions for moments of order statistics are not available for the generalized Pareto distribution (3) unless $k = 1$. However, it is a straightforward matter to generate tables of expectations of order statistics from the distribution (3) [2].

The translated Pareto, (2), has a representation as a gamma mixture* of exponential variables. The corresponding mean residual life function* is linear, and thus such distributions are sometimes considered alternatives to exponential distributions (which

have constant mean residual life). Suitably normalized minima of generalized Pareto (3) samples are asymptotically distributed as Weibull* variables.

One interesting feature of the family of classical Pareto distributions is that it is clearly closed under the operation of truncation from below. Thus if X has a classical Pareto distribution (1) with parameters σ and α, then X truncated from below at τ has a classical Pareto distribution with parameters τ and α. Numerous characterizations are based on this observation. Many are intimately related to lack of memory characterizations of the exponential distribution*. Basically, if truncation is equivalent to rescaling, then the distribution can be expected to be classical Pareto.

INFERENCE

Parameter Estimation

Maximum likelihood estimation of the parameters of generalized Pareto distributions is, in principle, straightforward. However, anomalous behavior of the likelihood surface can be encountered when sampling from the distribution (3). In the classical Pareto case closed-form expressions for the maximum likelihood* estimates are available:

$$\hat\sigma = X_{1\,:\,n},$$

$$\hat\alpha = \left[\frac{1}{n}\sum_{i=1}^{n}\log\left(\frac{X_{i\,:\,n}}{X_{1\,:\,n}}\right)\right]^{-1}. \tag{5}$$

Unbiased modifications of these estimates are readily derived if desired. They are of the form

$$\hat\alpha_u = (n-2)n^{-1}\hat\alpha,$$

$$\hat\sigma_u = \left[1 - (n-1)^{-1}\hat\alpha^{-1}\right]\hat\sigma. \tag{6}$$

Alternatively, one may choose to use estimates obtained by equating selected sample moments and/or sample quantiles* to their corresponding population values and solving for the unknown parameters. Consistency and asymptotic distribution results are relatively easily verified for moment, quantile, and maximum likelihood estimates.

If one is willing to assume that, in the distribution (2), the parameter α is known, then the family of distributions is a location and scale family*. One may then use available information about the means and variances of order statistics and the Gauss–Markov theorem* to construct minimum variance estimates of μ and σ among the class of all linear combinations of the order statistics (see Kulldorff and Vannman [9]). Instead, we may prespecify that the linear combination only involve a fixed number k of order statistics. The decision regarding which k order statistics to use is based on minimizing either the exact or asymptotic variances of the resulting estimates. Initial work in this area was due to Kulldorff and Vannman [9] and Vannman [17].

Testing the Adequacy of Pareto Models

In order to decide whether a generalized Pareto model adequately describes a given data set, generalized likelihood ratio tests* appear to be the most useful basic tool. An alternative approach focuses on characteristic properties of the distribution in question. Thus one might compare the sample survival function of the full data set with the rescaled sample survival function of the data truncated at some point in order to see if a classical Pareto model were plausible.

RELATED DISTRIBUTIONS

Multivariate classical Pareto distributions were introduced by Mardia [11]. After introducing location parameters, Mardia's k-variate Pareto distribution has joint survival function of the form

$$S_{\mathbf{X}}(\mathbf{x}) = \left[1 + \sum_{i=1}^{k} \left(\frac{x_i - \mu_i}{\sigma_i} \right) \right]^{-\alpha},$$

$$x_i > \mu_i, \quad i = 1, \ldots, k. \quad (7)$$

The marginal survival functions are of the form (2). The Mardia distribution (7) has the remarkable property that all its marginal

and conditional distributions are members of the same family. Some of the attractive distributional features of this distribution are attributable to the fact that it can be represented as a gamma mixture of independent exponential variables. Specifically, if W_1, $W_2, \ldots, W_k$ are independent identically distributed exponential variables and if Z is a gamma* random variable with shape parameter α and unit scale parameter, then, if we define

$$X_i = \mu_i + \sigma_i(W_i/Z), \quad i = 1, 2, \ldots, k$$

the random vector $\mathbf{X}$ has (7) as its joint survival function. An unfortunate feature of this Mardia multivariate Pareto distribution is that all its one-dimensional marginal distributions share a common value of α. More general multivariate Pareto distributions have been described [e.g., the marginals could be of form (3)]. Takahasi [16] introduced such a multivariate Burr distribution in the form of a gamma mixture of independent Weibull random variables. More recent contributors to this area are Cook and Johnson [4].

Hutchinson [6] describes four biometric settings in which a bivariate Pareto model is plausible. Bivariate or multivariate Pareto models might be expected to be useful in the analysis of incomes of related individuals (e.g., employed husband–wife pairs) or in the analysis of incomes of individuals measured at several different times.

A variety of distributions with Paretian tail behavior have been developed in the income distribution literature. Surveys may be found in Arnold [1], Hart [5] and Johnson and Kotz [7]. The discrete or quantized version of the classical Pareto distribution is known as the Zipf distribution (*see* ZIPF'S LAW).

References

[1] Arnold, B. C. (1983). *Pareto Distributions*. International Cooperative Publishing House, Fairland, MD.

[2] Arnold, B. C. and Laguna, L. (1977). "On Generalized Pareto Distributions with Applications to Income Data." *International Studies in Economics*

Monograph No. 10, Dept. of Economics, Iowa State University, Ames, IA.

[3] Cirillo, R. (1979). The Economics of Vilfredo Pareto. Frank Case, London.

[4] Cook, R. D. and Johnson, M. E. (1981). J. R. Statist. Soc., B43, 210–218.

[5] Hart, P. E. (1980). In The Statics and Dynamics of Income, N. A. Klevmarken and J. A. Lybeck, eds. Tieto Ltd., Clevedon, Avon, England.

[6] Hutchinson, T. P. (1979). Biometric J., 21, 553–563.

[7] Johnson, N. L. and Kotz, S. (1970). Distributions in Statistics: Continuous Univariate Distributions—1, 2. Wiley, New York.

[8] Kalbfleisch, J. D. and Prentice, R. L. (1980). The Statistical Analysis of Failure Time Data. Wiley, New York.

[9] Kulldorff, G. and Vannman, K. (1973). J. Amer. Statist. Ass., 68, 218–227. Corrigenda, 70, 494.

[10] Mandelbrot, B. (1960). Int. Econ. Rev., 1, 79–106.

[11] Mardia, K. V. (1962). Ann. Math. Statist., 33, 1008–1015.

[12] Ord, J. K. (1975). In Statistical Distributions in Scientific Work, Vol. 2, G. P. Patil et al., eds., pp. 151–158. D. Reidel, Dordrecht, Netherlands.

[13] Pareto, V. (1897). Cours d'Economie Politique, Vol. 2. F. Rouge, Lausanne, Switzerland.

[14] Republic of Botswana (1976). The Rural Income Distribution Survey in Botswana 1974–75. Central Statistics Office, Ministry of Finance and Planning, Republic of Botswana.

[15] Shirras, G. F. (1935). Econ. J., 45, 663–681.

[16] Takahasi, K. (1965). Ann. Inst. Statist. Math. Tokyo, 17, 257–260.

[17] Vannman, K. (1976). J. Amer. Statist. Ass., 71, 704–708.

[18] Zipf, G. (1949). Human Behavior and the Principle of Least Effort. Addison-Wesley, Reading, MA.

Bibliography

Aigner, D. J. and Goldberger, A. S. (1970). J. Amer. Statist. Ass., 65, 712–723. (Estimation from grouped data.)

Baxter, M. A. (1980). Metrika, 27, 133–138. (Minimum variance unbiased estimation.)

Berger, J. M. and Mandelbrot, B. (1963). IBM J. Res. Dev., 7, 224–236. (Modeling clustering in telephone circuits.)

Bhattacharya, N. (1963). Sankhyā B, 25, 195–196. (Characterization of the Pareto distribution using truncation from below.)

Blum, M. (1970). SIAM J. Appl. Math., 19, 191–198. (Convolutions of Pareto variables.)

Bowman, M. J. (1945). Amer. Econ. Rev., 35, 607–628. (A good survey of graphical techniques.)

Burr, I. W. (1942). Ann. Math. Statist., 13, 215–232. [A catalog of density functions including the equation (3) in this article.]

Champernowne, D. G. (1973). The Distribution of Income. Cambridge University Press, Cambridge, England.

Chan, L. K. and Cheng, S. W. (1973). Tamkang J. Math., 4, 9–21. (Optimal spacing for linear systematic estimates of Paretian scale parameters.)

Davis, H. T. and Feldstein, M. L. (1979). Biometrika, 66, 299–306. (Progressively censored survival data.)

Dubey, S. D. (1968). Naval Res. Logist. Quart., 15, 179–188. [A derivation of (3) as a compound Weibull distribution.]

DuMouchel, W. H. and Olshen, R. A. (1975). In Credibility: Theory and Applications. Academic Press, New York, pp. 23–50, 409–414. (Insurance claims costs.)

Dyer, D. (1981). In Statistical Distributions in Scientific Work, Vol. 6, G. P. Patil, ed. Reidel, Dordrecht and Boston, pp. 33–45. (Offshore oil lease bidding.)

Fisk, P. R. (1961). Econometrica, 29, 171–185. (Income modeling.)

Hagstroem, K. G. (1925). Skand. Aktuarietidskr., 8, 65–88. (Early characterization of the Pareto distribution.)

Harris, C. M. (1968). Operat. Res., 16, 307–313. (Application to queueing.)

Hartley, M. J. and Revankar, N. S. (1974). J. Econ., 2, 327–341. (Underreported income modeling.)

Hill, B. M. (1970). J. Amer. Statist. Ass., 65, 1220–1232. (Stochastic genesis of Zipf's law.)

Horvath, W. J. (1968). Behav. Sci., 13, 18–28. (Duration of wars and strikes.)

Irwin, J. O. (1968). J. R. Statist. Soc. A, 131, 205–225. (Accident theory and the generalized Waring distribution.)

Kaminsky, K. S. and Nelson, P. I. (1975). J. Amer. Statist. Ass., 70, 145–150. (Best linear unbiased prediction of Paretian order statistics.)

Koutrouvelis, I. A. (1981). Commun. Statist. Theory Meth., 10, 189–201. (Quantile estimation.)

Krishnaji, N. (1970). Econometrica, 38, 251–255. (Characterization based on underreported incomes.)

Leimkuhler, F. F. (1967). J. Doc., 23, 197–207. (Usage of library books.)

Likes, J. (1969). Statist. Hefte, 10, 104–110. (Minimum variance unbiased estimation.)

Lomax, K. S. (1954). J. Amer. Statist. Ass., 49, 847–852. (Business failures.)

Maguire, B. A., Pearson, E. S., and Wynn, A. H. A. (1952). Biometrika, 39, 168–180. (Times between accidents.)

Malik, H. J. (1966). *Skand. Aktuarietidskr.*, **49**, 144–157. (Moments of order statistics.)

Malik, H. J. (1970). *Metrika*, **15**, 19–22. (Products of Pareto variables.)

Malik, H. J. (1970). *Skand. Aktuarietidskr.*, **53**, 6–9. (Bayesian estimation.)

Mandelbrot, B. (1963). *J. Bus. Univ. Chicago*, **40**, 393–413. (Stock prices.)

Muniruzzaman, A. N. M. (1957). *Bull. Calcutta Statist. Ass.*, **7**, 115–123. (Estimation and hypothesis testing for classical Pareto.)

Quandt, R. E. (1966). *Metrika*, **10**, 55–82. (A survey of estimation techniques.)

Robertson, C. A. (1972). "Analyses of Forest Fire Data in California." *Tech. Rep. No. 11*, Dept. of Statistics, University of California, Riverside.

Sahin, I. and Hendrick, D. J. (1978). *Appl. Statist.*, **27**, 319–24. (Duration of strikes.)

Seal, H. L. (1952). *J. Inst. Actuaries*, **78**, 115–121. (Maximum likelihood for discrete Pareto.)

Seal, H. L. (1980). *ASTIN Bull.*, **11**, 61–71. (Insurance claims.)

Simon, H. A. and Bonini, C. P. (1958). *Amer. Econ. Rev.*, **48**, 607–617. (Size of business firms.)

Steindl, J. (1965). *Random Processes and the Growth of Firms: A Study of the Pareto Law*. Hafner, New York.

Talwalker, S. (1980). *Metrika*, **27**, 115–119. (Characterization by truncation from below.)

Targhetta, M. L. (1979). *Biometrika*, **66**, 687–688. (Confidence interval for α in multivariate Pareto.)

Thorin, O. (1977). *Scand. Actuarial J.*, 31–40. (Infinite divisibility.)

Turnbull, B. W., Brown, B. W., Jr., and Hu, M. (1974). *J. Amer. Statist. Ass.*, **69**, 74–80. (Heart transplant survival.)

Vartia, P. L. I. and Vartia, Y. O. (1980). In *The Statics and Dynamics of Income*, N. A. Klevmarken and J. A. Lybeck, eds. Tieto Ltd., Clevedon, Avon, England. (Scaled *F* distribution.)

(ECONOMETRICS
INCOME DISTRIBUTION MODELS
ZIPF'S LAW)

B. C. ARNOLD

PARETO OPTIMALITY

There are many situations in which a group of individuals has to choose one of two alternatives. There are also many situations where an individual must choose one of two alternatives, knowing that the choice will affect a group of individuals. In these situations, it is often possible for the choice to be made on the basis of comparisons that specify, for each alternative, whether or not it is at least as good for the group as the other alternative. For instance, if (according to the comparisons that have been made) each of the alternatives is at least as good as the other, then the two alternatives are equally good—and either one can be selected. On the other hand, if (according to the comparisons that have been made) one of the alternatives is at least as good for the group as the other, but the other alternative does not have this property (i.e., is not at least as good for the group as the first alternative), then the alternative that is at least as good as the other is the better alternative—and should, accordingly, be selected. The one exception is when (according to the comparisons that have been made) neither of the alternatives is at least as good for the group as the other. In this case, the comparisons that have been made don't provide a basis for making a choice. Rather, additional comparisons must be made before there is a basis for making a choice.

In the situations mentioned, comparisons can be made according to the following criterion: If one of the two alternatives is *preferred* to the other alternative by every individual in the group, then the preferred alternative is at least as good for the group as the other alternative. When these comparisons are the only ones that are made, they imply: If one of the two alternatives is preferred to the other alternative by every individual in the group, then the preferred alternative is the better alternative—and should be the one that is chosen. This implication has been variously labeled the *principle of unanimity*, the *weak Pareto principle*, and the *weak Pareto rule*. (As a historical note: The last two terms are in honor of Vilfredo Pareto, who was the first major scholar to use this principle as the basis for a systematic analysis of economic alternatives [6].)

A stronger criterion for making comparisons (in the sense that it more frequently specifies that one alternative is at least as good as the other) is: If one of the two alternatives is *at least as good* as the other alternative for each individual in the group, then it is at least as good for the group as the other alternative. When these comparisons are the only ones that are made, they imply two things. First: if, for each individual in the group, each of the alternatives is at least as good as the other, then the two alternatives are equally good—and either one can be chosen. This is called the *Pareto indifference rule*. Second: if one alternative is at least as good as the other alternative for each individual in the group *and* there is at least one individual who prefers it to the other alternative, then the one that is preferred is the better alternative and should be chosen. This part is known as the *strict Pareto rule*. Taken together, they constitute the *strong Pareto principle* (or the *strong Pareto rule*). If the intention is to make a choice that is beneficial for the individuals and the benefits are to be measured by their preferences (rather than according to, say, some paternalistic view of what's best for them) both the weak Pareto principle and the strong Pareto principle are compelling.

FORMAL DEFINITIONS

The verbal definitions already given can be restated in formal terms as follows. Let $\{a,b\}$ be the set of possible alternatives. Let $N = \{1, \ldots, n\}$ be an index set for the n individuals whose preferences are to be considered. For each $i \in N$, define the set $R_i \subseteq \{a,b\} \times \{a,b\}$ as follows. For each ordered pair, $(x, y) \in \{a,b\} \times \{a,b\}$, the ordered pair is an element of R_i if and only if x is at least as good as y for i. Since the set R_i is a binary relation on $\{a,b\}$, it is called "i's preference relation on $\{a,b\}$." P_i will denote the asymmetric part of R_i. That is, $P_i = \{(x, y) \in R_i : (y, x) \notin R_i\}$. In words: $(x, y) \in P_i$ if and only if i prefers x to y. I_i will denote the symmetric part of R_i. That is, $I_i = \{(x, y) \in R_i : (y, x) \in R_i\} = R_i - P_i$. In words: $(x, y) \in I_i$ if and only if i is indifferent between x and y. The set $R \subseteq \{a,b\} \times$

$\{a,b\}$ will be a set of ordered pairs that summarize the pairwise comparisons that a particular criterion or rule requires. As earlier, P and I will denote the asymmetric and symmetric parts of R. That is to say, $(x, y) \in P$ if and only if x is a better alternative than y; $(x, y) \in I$ if and only if x and y are equally good.

The set $R^w = \bigcap_{i \in N} P_i$ is called the *weak Pareto relation on the set* $\{a,b\}$. Clearly, the asymmetric part of R^w is all of R^w; that is, $P^w = R^w$ (and $I^w = \emptyset$). The weak Pareto rule is: For each pair $x, y \in \{a,b\}$,

$$(x, y) \in P^w \Rightarrow (x, y) \in P. \qquad (1)$$

The set $R^s = \bigcap_{i \in N} R_i$ is called the *strong Pareto relation on the set* $\{a,b\}$. Clearly, $R^w \subseteq R^s$. The symmetric part of R^s is $I^s = \bigcap_{i \in N} I_i$. The Pareto indifference rule is: For each pair $x, y \in \{a,b\}$,

$$(x, y) \in I^s \Rightarrow (x, y) \in I. \qquad (2)$$

The asymmetric part of R^s is

$$P^s = \bigcup_{i \in N} \left[P_i \cap \left[\bigcap_{j \in N-i} R_j \right] \right].$$

The strict Pareto rule is: For each pair $x, y \in \{a,b\}$,

$$(x, y) \in P^s \Rightarrow (x, y) \in P. \qquad (3)$$

(It should be noted that these definitions imply that we always have $P^w \subseteq P^s$.) Finally, the strong Pareto rule is: For each pair $x, y \in \{a,b\}$, (2) and (3) are satisfied.

MAKING CHOICES

As indicated earlier, the Pareto relations can sometimes be used as the basis for making a choice. This can be stated formally, as follows. Let the symbols $C(\{x, y\}) = A$ have the following meanings: If $A = \{x\}$ or $\{y\}$, then the choice from the set $\{x, y\}$ should be the element in A; if $A = \{x, y\}$, then the choice can be either x or y; if $A = \emptyset$, then there is no basis for choice. Let $x = a$ and $y = b$. For the weak Pareto relation R^w:

$$C(\{x, y\}) = \{x\} \Leftrightarrow (x, y) \in P^w,$$

$$C(\{x, y\}) = \{y\} \Leftrightarrow (y, x) \in P^w,$$

$$C(\{x, y\}) = \emptyset, \qquad \text{otherwise.}$$

For the strong Pareto relation R^s:

$$C(\{x, y\}) = \{x, y\} \Leftrightarrow (x, y) \in I^s,$$

$$C(\{x, y\}) = \{x\} \Leftrightarrow (x, y) \in P^s$$

$$C(\{x, y\}) = \{y\} \Leftrightarrow (y, x) \in P^s,$$

$$C(\{x, y\}) = \emptyset, \qquad \text{otherwise.}$$

THREE OR MORE ALTERNATIVES

There are also many situations in which a group has to choose one alternative out of a set that has three or more alternatives, and many situations where an individual has to choose one alternative out of a set that has three or more alternatives knowing that the choice will affect a group of individuals. In these situations, the Paretian criteria and rules often can be used to obtain comparisons for at least some of the possible pairs of elements from the set. Let X denote the set of possible alternatives. Then the definitions that are appropriate can be obtained directly from the definitions for two-alternative sets given in the Formal Definitions section. In particular, this can be accomplished by just replacing $\{a, b\}$ with X every time it appears in that section. By doing this, we specifically end up with Pareto relations and Pareto rules that are defined on X.

The following examples will illustrate the Pareto relations and Pareto rules that we can end up with when $X = \{a, b, c\}$ and $N = \{1, 2\}$. In them, the symbol Δ will be used to denote the set $\{(a, a), (b, b), (c, c)\}$.

Example 1. Full Agreement. $R_1 = R_2 = \{(a, b), (b, c), (a, c)\} \cup \Delta$.

To begin with, $P_1 = P_2 = \{(a, b), (b, c), (a, c)\}$ and $I_1 = I_2 = \Delta$. Since the two individuals are in complete agreement, the weak Pareto relation is $R^w = P_1 = P_2$. Therefore, the weak Pareto rule implies that $P = P_1 = P_2$.

The presence of complete agreement also implies that the strong Pareto relation is $R^s = R_1 = R_2$. Hence $I^s = I_1 = I_2$ and P^s

$= P_1 = P_2$. Therefore, the Pareto indifference rule implies that $I_1 = I_2 \subseteq I$. In addition, the strict Pareto rule implies that $P_1 = P_2 \subseteq P$. As a consequence, the strong Pareto rule implies $R = R_1 = R_2$.

Example 2. No Agreement on Pairs with Differing Elements. R_1 is the same as in Example 1, but $R_2 = \{(b, a), (c, b), (c, a)\} \cup \Delta$.

To begin with, $P_2 = \{(b, a), (c, b), (c, a)\}$ and $I_2 = \Delta$. In this example, the weak Pareto relation is $R^w = \emptyset$. Therefore, the weak Pareto rule has no implications for R.

The strong Pareto relation, on the other hand, is $R^s = I_1 = I_2$. Hence $I^s = I_1 = I_2$ and $P^s = \emptyset$. Therefore, the Pareto indifference rule implies that $I_1 = I_2 \subseteq I$. The strict Pareto rule, however, does not have any implications for R. As a consequence, the strong Pareto rule only implies $I_1 = I_2 \subseteq I$.

It should be noted that, unlike in Example 1, we have $R^w \neq P_1$, $R^w \neq P_2$, $P^s \neq P_1$, and $P^s \neq P_2$.

Example 3. Partial Agreement. $R_1 = \{(a, b), (b, a), (b, c), (a, c)\} \cup \Delta$ and $R_2 = \{(a, b), (b, c), (c, b), (a, c)\} \cup \Delta$.

To begin with, $P_1 = \{(b, c), (a, c)\}$ and $I_1 = \{(a, b), (b, a)\} \cup \Delta$; $P_2 = \{(a, b), (a, c)\}$ and $I_2 = \{(b, c), (c, b)\} \cup \Delta$.

In this example, the weak Pareto relation is $R^w = \{(a, c)\}$. Therefore, the only implication from the weak Pareto rule is: $(a, c) \in P$.

The strong Pareto relation, on the other hand, is $R^s = \{(a, b), (b, c), (a, c)\} \cup \Delta$. Hence $I^s = \Delta$ and $P^s = \{(a, b), (b, c), (a, c)\}$. Therefore, the Pareto indifference rule implies $\Delta \subseteq I$. In addition, the strict Pareto rule implies $\{(a, b), (b, c), (a, c)\} \subseteq P$. As a consequence, the strong Pareto rule implies $R = \{(a, b), (b, c), (a, c)\} \cup \Delta$.

Note that, unlike in Examples 1 and 2, $R^w \neq P^s$, $I^s \neq I_1$, and $I^s \neq I_2$.

Remark. In Examples 1–3, we have $I^s = \Delta$. This, however, is a special feature of these examples. For instance, suppose $R_1 = R_2 = X \times X$. Then $I^s = X \times X \neq \Delta$. For a more interesting example, suppose that R_1 is the same as in Example 3 and $R_2 = X \times X$. Then $I^s = \{(a,b),(b,a)\} \cup \Delta$.

CHOICES AND PARETO SETS

Let the symbols $C(X) = A$ have the following meanings: if $A \neq \emptyset$, then the choice can be any element in A; if $A = \emptyset$, then there is no basis for making a choice. For any given relation R on X, let

$$C(X) = \{x \in X : (x,y) \in R, \forall y \in X - x\}.$$

$C(X)$ is called the *choice set* for the relation R. For instance, in Example 1,

$$R = R^w \Rightarrow C(X) = \{a\},$$
$$R = R^s \Rightarrow C(X) = \{a\}.$$

In Example 2,

$$R = R^w \Rightarrow C(X) = \emptyset,$$
$$R = R^s \Rightarrow C(X) = \emptyset.$$

In Example 3,

$$R = R^w \Rightarrow C(X) = \emptyset,$$
$$R = R^s \Rightarrow C(X) = \{a\}.$$

In the first example used in the Remark,

$$R = R^w \Rightarrow C(X) = \emptyset,$$
$$R = R^s \Rightarrow C(X) = \{a,b,c\}.$$

In the second example used in the Remark,

$$R = R^w \Rightarrow C(X) = \emptyset,$$
$$R = R^s \Rightarrow C(X) = \{a,b\}.$$

It should be noted that since $R^w = P^w \subseteq P^s \subseteq R^s$, the choice set for R^w is always either the same as, or a proper subset of, the choice set for R^s—as in the examples.

The examples in this entry illustrate that, sometimes, $C(X) \neq \emptyset$ when $R = R^w \subseteq X \times X$ or $R = R^s \subseteq X \times X$. Therefore, sometimes, the weak or strong Pareto rule will provide a basis for making a choice from X. Frequently, however, $R = R^w$ or $R = R^s$ implies $C(X) = \emptyset$—especially when the preferences of a lot of different people are being taken into account (i.e., when n is large), since the potential for disagreement is then

(accordingly) also large. Nonetheless, the relations can still usually provide a basis for narrowing down the set of possible choices. In particular, for any given relation R, we can define the set of *maximal* elements, as

$$M(X) = \{ y \in X : \nexists x \in X$$
$$\text{such that } (x,y) \in P \}.$$

When $R = R^w \subseteq X \times X$, a particular $y \in X$ is in $M(X)$ if and only if there is *no* $x \in X$ that is preferred to y by each $i \in N$. An alternative that satisfies this requirement is called a *weakly Pareto optimal alternative*. The set of all such alternatives (i.e., $M(X)$ for R^w) is called the *weak Pareto set*. When $R = R^s \subseteq X \times X$, a particular $y \in X$ is in $M(X)$ if and only if there is *no* $x \in X$ that is at least as good as y for each individual and preferred to y by at least one individual. An alternative that satisfies this requirement is called a *strongly Pareto optimal alternative*. The set of all such alternatives (i.e., $M(X)$ for R^s) is called the *strong Pareto set*.

In Example 1, the weak Pareto set
= the strong Pareto set = $\{a\}$.
In Example 2, the weak Pareto set
= the strong Pareto set = X.
In Example 3, the weak Pareto set = $\{a,b\}$;
the strong Pareto set = $\{a\}$.

In the first example used in the Remark, the weak Pareto set = the strong Pareto set = X. In the second example used in the Remark, the weak Pareto set = X; the strong Pareto set = $\{a,b\}$.

It should be noted that since $P^w \subseteq P^s$, the strong Pareto set is always either the same as, or a proper subset of, the weak Pareto set —as in the examples. Thus, every strongly Pareto optimal alternative is also weakly Pareto optimal. At the same time, however, there are situations in which there are weakly Pareto optimal alternatives that are not strongly Pareto optimal (for instance, in Example 3). Hence saying "x is a strongly Pareto optimal alternative" is (as the terminology suggests) a stronger statement than saying "x is a weakly Pareto optimal alternative." Indeed, it is this particular relationship between the sets of maximal elements for R^w and R^s that has caused the terms *strongly* and *weakly* (and, in the earlier definitions, *strong* and *weak*) to be used in the way in which they have been.

Once one of the Pareto rules is accepted as being appropriate, it follows that the final choice should be limited to the corresponding Pareto set. Unfortunately, though, after the Pareto rule has been used to narrow down the original set of alternatives, X, to its set of Pareto optimal alternatives $M(X)$, the rule cannot narrow the set down any more. That is, $M(M(X)) = \{ y \in M(X) : \nexists x \in M(X) \text{ such that } (x, y) \in P \} = M(X)$ for the given rule. Therefore, the set of possible choices can be narrowed further only if additional rules or criteria are used.

For further discussion of the concepts described and analyses of the limitations on the possibility of using other rules and criteria at the same time, see Arrow [1] and/or Sen [7, 8]. For analyses of when outcomes from competitive markets are Pareto optimal, see Arrow and Hahn [2] and/or Debreu [4]. For analyses of when alternative schemes of taxation and government expenditure are Pareto optimal, see Atkinson and Stiglitz [3]. For uses of the concept of Pareto optimality in analyses of cooperative and noncooperative games, see Shubik [9]. For a discussion of how the concept of Pareto optimality can be thought of and analyzed as a generalization of the concept of a maximum for a real-valued function, see Smale [10]. For a discussion of the relation between the concept of Pareto optimality and the statistical concept of admissibility* and an analysis of the properties they have in common, see Kozek [5].

References

[1] Arrow, K. (1963). *Social Choice and Individual Values*, 2nd ed. Wiley, New York.

[2] Arrow, K. and Hahn, F. (1971). *General Competitive Analysis*. Holden-Day, San Francisco, CA.

[3] Atkinson, A. and Stiglitz, J. (1980). *Lectures on Public Economics*. McGraw-Hill, New York.

[4] Debreu, G. (1959). *Theory of Value*. Wiley, New York.

[5] Kozek, A. (1982). *Ann. Statist.*, **10**, 825–837.

[6] Pareto, V. (1909), *Manuel d'Économie Politique*. Giard, Paris.

[7] Sen, A. (1970), *Collective Choice and Social Welfare*. Holden-Day, San Francisco.

[8] Sen, A. (forthcoming). In *Handbook of Mathematical Economics*, Vol. III, K. Arrow and M. Intriligator, eds. North-Holland, Amsterdam.

[9] Shubik, M. (1982). *Game Theory in the Social Sciences*. MIT Press, Cambridge, MA.

[10] Smale, S. (1981). In *Handbook of Mathematical Economics*, Vol. I, K. Arrow and M. Intriligator, eds. North-Holland, Amsterdam.

(ADMISSIBILITY
DECISION THEORY
GAME THEORY
NASH AXIOMS
OPTIMIZATION IN STATISTICS
PAIRED COMPARISONS
UTILITY THEORY)

Peter Coughlin

PARSIMONY, PRINCIPLE OF

When fitting a model to a given set of observations, it is generally preferable to use the model with the smallest number of parameters that gives adequate ("reasonable") representation. (See, e.g., Box and Jenkins [1].)

The term *Occam's razor* is also used, named after William of Occam (or Ockham) (1280–1349), perhaps the most influential philosopher of the fourteenth century. He made frequent use of the medieval rule of economy, expressed in the dictum *Pluralitas non est ponenda sine necessitate*, literally, "Multiplicity ought not to be posited without necessity," or translated in the language of model fitting: "What can be accounted for by fewer assumptions is explained in vain by more"; see Moody [3].

Let E represent evidence or events observed in some experiment or experiments, and $\Pr(E \mid H)$ the probability of E, given a specified model or hypothesis H. Occam's razor states that if

$$\Pr(E \mid H_1) = \Pr(E \mid H_2) = \cdots = \Pr(E \mid H_k)$$
$$(1)$$

for models or hypotheses $H_1, \ldots, H_k$, then the simplest among $H_1, \ldots, H_k$ is to be preferred.

Good [2] presents a sharpened form of the razor for choosing between models when (1) is not satisfied. For an explanation of the concepts involved and for further discussion, see ref. 2.

References

[1] Box, G. E. P. and Jenkins, G. M. (1970). *Time Series Analysis Forecasting and Control.* Holden-Day, San Francisco.

[2] Good, I. J. (1968). *Brit. J. Philos. Sci.*, **19**, 123–143.

[3] Moody, E. A. (1974). In *Dictionary of Scientific Biography*, Vol. 10. Scribners, New York, pp. 171–175.

(DECISION THEORY
WEIGHT OF EVIDENCE)

PARTIAL CONFOUNDING   *See* CON-FOUNDING

PARTIAL CORRELATION   *See SUPPLEMENT*

PARTIAL EXCHANGEABILITY

A classical statistician would formulate many inference problems by assuming observations come from independent random variables $X_1, X_2, \ldots, X_N$ with some fixed but unknown distribution F. A Bayesian statistician, instead of regarding F as fixed but unknown, would regard F as chosen at random according to some prior. A pure subjectivist (e.g., de Finetti [5]) would argue that, instead of being assumed independent, the variables X_i should be assumed *exchangeable*:

$$(X_1, X_2, \ldots, X_N) \stackrel{D}{=} (X_{\pi_1}, X_{\pi_2}, \ldots, X_{\pi_N})$$

for each permutation $\pi = (\pi_i)$.

Here $\stackrel{D}{=}$ denotes equality in distribution. (*See* EXCHANGEABILITY and Kingman [7].) De Finetti's theorem asserts that if (X_i) is an infinite exchangeable sequence, then there exists a random distribution function F such that, conditional on F, the variables (X_i) are independent with common distribution F [more briefly, (X_i) is a mixture of iid sequences]. So in this setting the Bayesian and the pure subjectivist positions are mathematically equivalent. Aside from this philosophical interpretation, the theorem is mathematically interesting because it uses a hypothesis of invariance of distributions to obtain a conclusion of a certain dependence structure (in contrast to most theorems in probability theory, where the dependence structure is given in the hypotheses).

Partial exchangeability is a loose term covering various generalizations, specializations, and analogs of de Finetti's theorem. In practice, classical statisticians assume F belongs to some parametric family F_θ and make inferences about θ. In this setting, a Bayesian would regard θ as distributed according to some prior. Here the pure subjectivist would regard (X_i) as belonging to some subclass of exchangeable sequences. Thus it is natural to ask what subclasses correspond to what parametric families. On the other hand, not all observations would be modeled by a classical statistician as arising from iid observations; in some settings (X_i) might be modeled as, say, a Markov chain* with unknown transition matrix **P**. In this setting a Bayesian could regard **P** as random. What is the corresponding pure subjectivist model (i.e., what is an intrinsic description of the class of mixtures of Markov chains)? These kinds of topics constitute partial exchangeability. A few results are outlined below; see Aldous [1], Diaconis and Freedman [3], and Lauritzen [10] for more comprehensive treatments.

Perhaps the simplest topic concerns sampling from two populations. Suppose X_1, $X_2, \ldots$ ($Y_1, Y_2, \ldots$, respectively,) are the heights of a random sample of men (women). The classical statistician would model these variables as independent, with each X_i

(Y_i) having distribution $F(G)$, where F and G are unknown. The Bayesian would regard (F, G) as distributed according to some prior (on the space of pairs of distributions). A variant of de Finetti's theorem says this is equivalent (for infinite sequences) to the subjectivist's intrinsic condition

$$(X_1, X_2, \ldots; Y_1, Y_2, \ldots)$$
$$\stackrel{D}{=} (X_{\pi_1}, X_{\pi_2}, \ldots; Y_{\sigma_1}, Y_{\sigma_2}, \ldots)$$

for all finite permutations π, σ. This result goes back to de Finetti [4], who coined the term "partial exchangeability" for this case. Examples of Bayesian analysis in this setting are given by Dekker [6], Volpe di Prigno [in 9], and Wedlin [in 9].

A related result concerns infinite arrays of variables. Consider infinite arrays $(X_{i,j} : i, j \geqslant 1)$ of random variables. A natural subjectivist model would be to assume the sequence of rows to be exchangeable and similarly for columns:

$$(X_{i,j} : i, j \geqslant 1) \stackrel{D}{=} (X_{\pi_i}, Y_{\sigma_j})$$
$$\text{for finite permutations } \pi, \sigma. \quad (1)$$

(This is weaker than complete exchangeability of the array.) An example of an array with this property is

$$X_{i,j} = f(R_i, C_j, I_{i,j}), \quad (2)$$

where the R_i, C_i and $I_{i,j}$ are all iid and f is an arbitrary function. Thus the value $X_{i,j}$ is derived from a row effect R_i, a column effect C_j, and an individual effect $I_{i,j}$. Now (2) describes a parametric family of distributions of arrays where f is the parameter. A Bayesian could model an array as (2) with f chosen at random from some prior; it can be shown that the class of distributions thus obtained is precisely the class satisfying subjectivist's intrinsic condition (1).

The intuitive meaning of exchangeability is that the order of the observations does not matter. Let us treat one case where the order does matter. Let $\mathbf{x} = (x_1, \ldots, x_n)$ denote a sequence taking values in a finite set S. For each ordered pair (s_1, s_2) of elements of S, count the number of successive pairs (x_i, x_{i+1}) that equal (s_1, s_2). This set of numbers, as (s_1, s_2) varies, forms the *transition count* of $\mathbf{x}$. A possible subjectivist condition on a sequence $\mathbf{X} = (X_i)$ is for each n,

$$P(\mathbf{X} = \mathbf{x}) = P(\mathbf{X} = \mathbf{x}') \quad (3)$$

whenever $\mathbf{x}$ and $\mathbf{x}'$ have the same transition count and the same initial state. It can be shown that the class of (infinite, recurrent) processes satisfying (3) is precisely the Bayesian class, discussed earlier, of Markov chains in which the initial state and transition matrix are chosen at random from some prior.

Finally, we mention some specializations of de Finetti's theorem that characterize mixtures of iid sequences with distributions in a specified parametric family. Call an infinite sequence (X_i) *spherically symmetric* if for each n the random n-vector $(X_1, \ldots, X_n)$ has a distribution that is invariant under rotations in R^n. An independent spherically symmetric sequence must be iid $N(0, \sigma^2)$. Dropping the assumption of independence, spherical symmetry implies exchangeability, and then de Finetti's theorem and the independent result imply that the general spherically symmetric sequence is a mixture (over σ^2) of iid $N(0, \sigma^2)$ sequences.

There is a general technique, based on sufficient statistics*, for giving intrinsic conditions on a sequence that are equivalent to the sequence being a mixture from a specified family. Consider the $N(\mu, \sigma^2)$ family. The usual sufficient statistics are $(T_1, T_2) = (T_1(X_1, \ldots, X_n), T_2(X_1, \ldots, X_n)) = (\bar{X}, s_X^2)$. And conditional on $(T_1, T_2) = (t_1, t_2)$, the random vector

$(X_1, \ldots, X_n)$ is distributed uniformly

on the $(n-1)$ sphere

$$\{\mathbf{x} : (T_1(\mathbf{x}), T_2(\mathbf{x})) = (t_1, t_2)\}. \quad (4)$$

Now consider the class of infinite sequences (X_i) such that (4) holds for each n. It can be shown that this class is precisely the class of mixtures (over μ, σ^2) of iid $N(\mu, \sigma^2)$ sequences. This result and the analogous results for other classical families are treated by Diaconis and Freedman [3] and Lauritzen [10].

References

[1]   Aldous, D. J. (to appear). In *École d'Été St. Flour 1983*, Lecture Notes in Mathematics Series. Springer, New York. (A long survey of exchangeability, but emphasizes probabilistic rather than statistical topics.)

[2]   Dawid, A. P. (1982). In ref. 9, pp. 217–232. (A modern subjectivist's account of partial exchangeability.)

[3]   Diaconis, P. and Freedman, D. (1984). "Statistics: Applications and New Directions," *Proc. Indian Statist. Inst. Conf.*, pp. 205–236. (A concise survey of the "sufficient statistics" methods.)

[4]   de Finetti, B. (1938). *Acta Sci. Ind.*, **739**. (Translated in (1980). *Studies in Inductive Logic and Probability II*, R. C. Jeffrey, ed., University of California Press, Berkeley.)

[5]   de Finetti, B. (1972). *Probability, Induction and Statistics*. Wiley, New York. (The definitive classical account of subjectivism.)

[6]   Dekker, J. (1982). In *Test Equating*, P. W. Holland and D. B. Rudin, eds. Academic Press, New York, pp. 327–38.

[7]   Kingman, J. F. C. (1978). *Ann. Prob.*, **6**, 183–197. (A short elegant survey of exchangeability—the best place to start reading.)

[8]   Kingman, J. F. C. (1980). *Mathematics of Genetic Diversity*. (Contains some nice applications of exchangeability ideas in population genetics.)

[9]   Koch, G. and Spizzichino, F., eds. (1982). *Exchangeability in Probability and Statistics*. North-Holland, Amsterdam. (This proceedings of a conference to honor Professor de Finetti provides an overview of current research interests.)

[10] Lauritzen, S. L. (1982). *Statistical Models as Extremal Families*. Aalborg University Press, Aalborg, Denmark. (Contains a long, detailed account of the *sufficient statistics* methods.)

(BAYESIAN INFERENCE
EXCHANGEABILITY
STATISTICAL INFERENCE, I, II
SUFFICIENT STATISTICS)

DAVID J. ALDOUS

PARTIAL LEAST SQUARES

Path models with latent variables (PMLVs), an epoch-making innovation of the 1960s, combine econometric prediction with psychometric modeling of latent variables (LVs) that are indirectly observed by multiple manifest variables (MVs), called *indicators*. Figure 1 shows an array of path models, to the left with MVs, to the right with LVs. The numbering marks increasing levels of generality. Models III*–V* are territory opened up by PMLVs.

The conceptual graphic design by an arrow scheme is similar for PMLVs as first estimated by ML (Jöreskog's LISREL*, 1973–) and later by LS (my partial least squares, PLS, 1975–), but the "hard" distributional ML assumptions against the "soft" distribution-free PLS bring fundamental differences in technique and scope [21, 22].

First, ML–LISREL models the covariance matrices of MVs and LVs; PLS models and predicts MVs and LVs and in the process models their covariance matrices. Second, ML–LISREL assumes the MVs to be jointly ruled by a specified multivariate distribution which is subject to independent observations; in PLS the predictive relations are subject to *predictor specification* and are otherwise distribution-free, and independence of the observations is not stipulated. Third, in LISREL the parameter estimation is *consistent*, but the case values of the LVs are not estimated; in PLS the case values of all LVs are estimated as weighted averages of their block of indicators, giving estimates of parameters and LV case values that are inconsistent but are consistent in the limit, *consistency at large*, as the blocks of indicators increase indefinitely in size. Fourth, model evaluation in ML–LISREL uses hypothesis tests and standard errors computed by the classic asymptotic approaches, whereas PLS uses Stone–Geisser's test for predictive relevance and Tukey's jackknife*, both of which are distribution- and independence-free.

THE BASIC PLS DESIGN [21]

In the basic design LISREL and PLS models are linear, and the observations range over time or a cross section. The six-LV model IV* in Fig. 1 is used for illustration in the subsequent exposition of PLS.

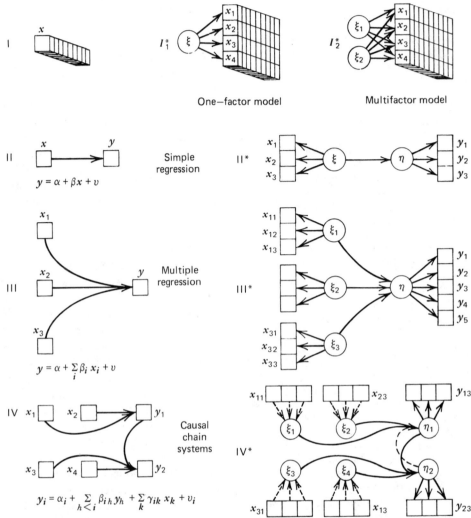

One–factor model   Multifactor model

Simple
regression

$y = \alpha + \beta x + v$

Multiple
regression

$y = \alpha + \sum_i \beta_i x_i + v$

Causal
chain
systems

$y_i = \alpha_i + \sum_{h < i} \beta_{ih} y_h + \sum_k \gamma_{ik} x_k + v_i$

Figure 1 *a–b   Graph of path models with (a) manifest variables, and (b) latent variables indirectly observed by multiple manifest indicators. MVs are illustrated by squares, LVs by circles. The cases observed and analyzed are illustrated in Models I, I_1^*, and I_2^*, but not in the other models. The arrow scheme in the right-hand column illustrates Model V* or Model IV* as the broken arrow is or is not included. (See ref. 22, Fig. 2.)*

The Arrow Scheme

The arrow scheme shows the theoretical-conceptual design of the model. Once the arrow scheme is specified, it is a direct matter to write the formal model and the estimation algorithm.

Each LV is indirectly observed by a block of manifest indicators. The LVs and the path of "inner" relations between the LVs are the core of the model. Two LVs are called *adjoint* if they are directly connected by an inner relation. In model IV*, for example, adjoint to ξ_1 is η_1; adjoint to η_1 are ξ_1, ξ_2, and η_2.

The arrow scheme is designed in accordance with the aims of the model. Using prior knowledge and intuition the investigator is free to specify the LVs, to design the inner relations, and to compile a selection of indicators for each LV.

Fundamental Principle for the Model Design: All information between the blocks of indicators is conveyed by the LVs via the inner relations.

For later use in the model construction the investigator should specify the following:

1. The expected sign of the correlation between each indicator and its LV.
2. The expected sign of the correlation between any two adjoint LVs.

Formal Definition of the Model

MVS AND LVS.   For simplicity in notation we renumber the LVs in one sequence: $\eta_1 = \xi_5$; $\eta_2 = \xi_6$. The indicators of ξ_j are denoted x_{jk}. In the present model we have

$$\xi_j, x_{jh}, \qquad j = 1,6; \quad h = 1, H_j. \quad (1)$$

Assuming that there are N cases under analysis, the case values of MVs and LVs are denoted

$$x_{jhn}, \xi_{jn}, \qquad n = 1, N \qquad (2)$$

where n ranges over time or a cross section. We shall often let the ranges of the subscripts j, h, n be tacitly understood.

INNER RELATIONS.   Linear relations in accordance with the arrow scheme:

$$\xi_5 = \beta_{50} + \beta_{51}\xi_1 + \beta_{52}\xi_2 + v_5, \qquad (3a)$$

$$\xi_6 = \beta_{60} + \beta_{63}\xi_3 + \beta_{64}\xi_4 + \beta_{65}\xi_5 + v_6. \quad (3b)$$

Predictor specification is assumed for the inner relations:

$$E(\xi_5 | \xi_1, \xi_2) = \beta_{50} + \beta_{51}\xi_1 + \beta_{52}\xi_2 \quad (4)$$

and similarly for (3b).

OUTER RELATIONS.   Each indicator is linear in its LV,

$$x_{jh} = \pi_{jh0} + \pi_{jh}\xi_j + \epsilon_{jh},$$

$$j = 1,6; \quad h = 1, H_j, \quad (5a)$$

and is subject to *predictor specification*:

$$E(x_{jh} | \xi_j) = \pi_{jh0} + \pi_{jh}\xi_j. \qquad (5b)$$

The coefficient π_{jh} is called the *loading* of x_{jh}, following psychometric usage for Models $I_1^*-I_2^*$.

Equations (5a) and (5b) imply

$$E(\xi_{jh}) = 0, \qquad r(\epsilon_{jh}, \xi_j) = 0,$$

$$j = 1,6; \quad h = 1, H_j. \quad (6)$$

Furthermore, on mild supplementary conditions ordinary least squares (OLS) regression gives consistent estimates of π_{jh0} and π_{jh}.

Both factors π_{jh} and ξ_j are unknown in (5a); hence we need a standardization of scales for unambiguity (SSU). PLS modeling achieves SSU by standardizing each LV to have unit variance:

$$\text{var}(\xi_j) = E\left\{ \left[\xi_j - E(\xi_j) \right]^2 \right\} = 1,$$

$$j = 1,6. \quad (7)$$

Part of the SSU problem is the specification of the signs of π_{jk} and ξ_j. The expected signs of the correlations between each indicator and its LV serve to meet this problem.

PREDICTION.   For the endogenous LVs and their indicators the model provides prediction by the inner and outer relations, respectively.

Substitutive prediction provides prediction by substitutive elimination of LVs from inner and outer relations.

1. *Prediction of LVs in Terms of LVs.*   In the present model prediction of $\eta_2 (= \xi_6)$:

$$\text{pred } \eta_2 = \beta_{60}^* + \beta_{63}\xi_3 + \beta_{64}\xi_4$$

$$+ \beta_{65}(\beta_{51}\xi_1 + \beta_{52}\xi_2) + v_6^* \quad (8a)$$

with location parameter and prediction error given by

$$\beta_{60}^* = \beta_{60} + \beta_{65}\beta_{50}, \qquad v_6^* = v_6 + \beta_{65}v_5.$$

$$(8b)$$

2. *Prediction of MVs in Terms of LVs.*   In the present model prediction of y_{1h} ($h = 1, H_5$), and similarly for y_{2h}:

$$\text{pred } y_{1h} = \pi_{5h0}^* + \pi_{5h1}(\beta_{51}\xi_1 + \beta_{52}\xi_2) + \epsilon_{5h0}^*$$

$$(9a)$$

with location parameters and prediction errors:

$$\pi_{5h0}^* = \pi_{5h0} + \pi_{5h1}\beta_{50},$$

$$\epsilon_{5h0}^* = \epsilon_{5h0} + \pi_{5h1}v_5. \qquad (9b)$$

3. *Repeated Substitution.* In pred y_{2h} as given by 2 above, substitution of ξ_5 gives pred y_{2h} in terms of $\xi_1, \xi_2, \xi_3, \xi_4$. The prediction (9a) does not allow such repeated substitution since in the present model there are only two endogenous LVs.

COMMENTS ON THE FORMAL MODEL

1. The formal specified above is distribution-free, except for the predictor specifications (4), (5b).

2. In real-world applications the observations x_{jhn} refer to a finite number of cases, denoted N. Formal PLS models need not specify N, nor the numbers of indicators, H_j.

3. The case values x_{jhn} are the *raw data* of the model. The *product data* are the means and the product moments:

$$\bar{x}_{jh}; \quad N^{-1}\Sigma_n(x_{ihn}x_{jkn}). \quad (10a\text{-}b)$$

4. Simulation experiments on our model generate data x_{jhn} from a multivariate distribution (a population), specified in accordance with the inner and outer relations. While the inner and outer relations and their parameters are *generative*, the relations of substitutive prediction in **1–3** above are *implicative*. Typically, the N cases of simulated data (2a) are generated as independent observations of the population. In real-world applications with nonexperimental data the assumption of independent observations of a specified population is more or less unrealistic.

5. The fundamental principle defined earlier and its implications

$$r(\epsilon_{ih}, \epsilon_{jk}) = r(\epsilon_{ih}, \xi_j) = 0 \quad (11)$$

with $i \neq j = 1, 6$; $h = 1, H_i$; $k = 1, H_j$ are in the nature of idealizations, and are not part of the formal model. Thus (11) is not in operative use in the subsequent PLS algorithm. The degree to which the estimated model honors (11) provides a partial test of the realism of the model.

The Model in Estimated Form

We shall set forth the estimated model in terms of *raw data*, (2). The estimates of parameters and LVs will be denoted by corresponding roman letters.

MVS AND LVS. Each LV is estimated as a weighted aggregate of its indicators, with weights w_{jh} that are auxiliary parameters:

$$X_{jn} = \text{est}(\xi_{jn}) = \Sigma(w_{jh}x_{jhn}). \quad (12)$$

INNER RELATIONS FOR X_5 AND X_6. To specify for X_6:

$$X_{6n} = b_{60} + b_{63}X_{3n} + b_{64}X_{4n} + b_{65}X_{5n} + u_{6n}. \quad (13)$$

OUTER RELATIONS. With $j = 1, 6$; $H = 1$, H_j; $n = 1, N$:

$$x_{jhn} = p_{jh0} + p_{jh}X_{jn} + e_{jhn}. \quad (14)$$

PREDICTION AND SUBSTITUTIVE PREDICTION. The same formulas as in the Prediction section are used, with roman letters. Furthermore, using (12) to substitute any predicting LV by MVs all through that section, we obtain the following.

PREDICTION OF LVS AND MVS IN TERMS OF MVS. For example, (9a) gives the following formula:

$$\text{pred } y_{1hn} = p_{5h0}^* + p_{5h1}\left[b_{51}\Sigma_h(w_{1h}x_{1hn}) + b_{52}\Sigma_h(w_{2h}x_{2hn}) \right] + e_{5h0}^*. \quad (15)$$

The PLS Estimation Algorithm

The data input for the algorithm is either raw data (2) or product data (10). We shall set forth the procedure with raw data input.

The PLS algorithm proceeds in three stages. The first gives estimates (12) of the case values of the LVs, the second the generative relations (13)–(14). The two first stages use data normalized to zero mean, giving

$$\bar{x}_{jh} = N^{-1}\Sigma_n(x_{jhn}) = 0,$$

$$j = 1,6; \quad H = 1, H_j \quad (16)$$

The third stage estimates the location parameters of the LVs and the generative relations.

Stage 1 is iterative: a sequence of OLS regressions, linear operations, and square root extractions. Stages 2–3 are noniterative OLS regressions.

FIRST STAGE OF THE PLS ALGORITHM. The first stage estimates each LV as a weighted aggregate of its indicators:

$$X_{jn} = \Sigma_h(w_{jh}x_{jhn}) = f_j\Sigma_h(v_{jh}x_{jhn}) \quad (16a)$$

using weights $w_{jh} = f_jv_{jh}$, where f_j is a scalar that gives X_{jn} unit variance over N,

$$f_j = \pm N^{1/2}\{\Sigma_n[\Sigma_h(v_{jh}x_{jhn})]^2\}^{-1/2} \quad (16b)$$

where the sign ambiguity is resolved by making

$$\text{signum } f_j = \text{signum}\{\Sigma_h[\text{signum } r(x_{jh}X_j)]\}. \quad (17)$$

Weight Relations. These determine the weights v_j. The weight relations for ξ_j involve a sign weighted sum, denoted SwS_j, of estimates for those ξ_k which are adjoint to ξ_j:

$$\text{SwS}_{jn} = \Sigma_k[(\pm)_{jk}X_{kn}] \quad (18a)$$

where the sign ambiguity is resolved by making

$$(\pm)_{jk} = \text{signum } r(X_j, X_k). \quad (18b)$$

Estimation Modes. For each ξ_j the investigator has the option to choose between two types of weight relations, called Modes A and B, which take the form of simple and multiple regressions over $n = 1, N$.

Mode A. For each h the simple OLS regression of x_{jh} on SwS_j:

$$x_{jhn} = v_{jh}\text{SwS}_{jn} + d_{jhn}, \quad h = 1, H_j. \quad (19)$$

Mode B. The multiple OLS regression of SwS_j on x_{jh} ($h = 1, H_j$):

$$\text{SwS}_{jn} = \Sigma_h(v_{jh}x_{jhn}) + d_{jn}. \quad (20)$$

In the present model, as seen from Fig. 1, Model IV*,

$$\text{SwS}_1 = \text{SwS}_2 = X_5; \quad (21a)$$

$$\text{SwS}_3 = \text{SwS}_4 = X_6; \quad (21b)$$

$$\text{SwS}_5 = (\pm)_{51}X_1 + (\pm)_{52}X_2 \\ + (\pm)_{56}X_6; \quad (21c)$$

$$\text{SwS}_6 = (\pm)_{63}X_3 + (\pm)_{64}X_4 \\ + (\pm)_{65}X_5. \quad (21d)$$

As is often appropriate, let us choose Mode A and Mode B, respectively, for the endogenous and exogenous LVs. Then in the present model the weight relations read

$$X_{5n} = \Sigma_h(v_{jh}x_{jhn}) + d_{jn}, \quad j = 1,2; \quad (22a)$$

$$X_{6n} = \Sigma_h(v_{jh}x_{jhn}) + d_{jn}, \quad j = 3,4; \quad (22b)$$

$$x_{5hn} = v_5h[(\pm)_{51}X_{1n} + (\pm)_{52}X_{2n} \\ + (\pm)_{56}X_{6n}] + d_{5hn}; \quad (22c)$$

$$x_{6hn} = v_{6h}[(\pm)_{63}X_{3n} + (\pm)_{64}X_{4n} \\ + (\pm)_{65}X_{5n}] + d_{6hn}. \quad (22d)$$

The PLS Algorithm. This alternates between (16) and (22), and proceeds in steps $s = 1, 2, \ldots,$ with substeps for the six LVS. As indicated in Table 1 the PLS algorithm in the step from s to $s + 1$ involves proxy estimates from step s marked ($'$) and from step $s + 1$ marked ($''$). Thus the first substep estimates ξ_{1n} by Mode B, using SwS'_{1n} from substep s, the multiple OLS regression (21a) giving v''_{1h} and then (16a–b) gives X''_{1n}. The third substep estimates ξ_{5n} by Mode A, using X''_{1n}, X''_{2n} from step $s + 1$ and X'_{6n} from step s to compute SwS_{5n}, the simple OLS regressions (22c) giving v''_{5n}, and then (16a–b) gives X''_{5n}.

The Starting Values. In step $s = 1$ the starting values are largely arbitrary, say $v'_{51} = v'_{61} = 1$ and $v'_{5h} = v'_{6k} = 0$ when $h, k > 1$.

Passage to the Limit. Continue the iterative procedure until each estimated weight, say

Table 1   The PLS Algorithm of Model IV*. The General Step from s to $s + 1$ and Its Six Substeps

(1) Substep	(2) LV	(3) Mode	(4) SwS	(5) (4) Computed From	(6) Ensuing Proxies	(7)
1	ξ_{1n}	B	SwS_1	X'_{5n}	v''_{1h}	X''_{1n}
2	ξ_{2n}	B	SwS_2	X'_{5n}	v''_{2h}	X''_{2n}
3	ξ_{5n}	A	SwS_5	$X''_{1n}, X''_{2n}, X'_{6n}$	v''_{5h}	X''_{5n}
4	ξ_{3n}	B	SwS_3	X'_{6n}	v''_{3h}	X''_{3n}
5	ξ_{4n}	B	SwS_4	X'_{6n}	v''_{4h}	X''_{4n}
6	ξ_{6n}	A	SwS_6	$X''_{3n}, X''_{4n}, X''_{5n}$	v''_{6h}	X''_{6n}

w_{jh}, converges according to a conventional stopping rule, say:

$$(w''_{jh} - w'_{jh})/w'_{jh} \leqslant 10^{-5}. \qquad (23)$$

SECOND STAGE OF THE PLS ALGORITHM. Using the LVs estimated in Stage 1, the noniterative second stage estimates the inner and outer relations by OLS regressions, without location parameters. The estimated relations can be written down directly from (3)–(5):

1. *Inner relations* for X_5 and X_6; to specify for X_5:

$$X_{5n} = b_{51}X_{1n} + b_{52}X_{2n} + u_{5n},$$
$$n = 1, N. \quad (24)$$

2. *Outer relations*:

$$x_{jhn} = p_{jh}X_{jn} + e_{jhn},$$
$$j = 1, 6; \quad h = 1, H_j. \quad (25)$$

3. *Prediction and substitutive prediction.* Same immediate operations as in the preceding sections on Prediction, Substitutive Prediction, and Prediction of LVs and MVs without location parameters, e.g., (15).

THIRD STAGE OF THE PLS ALGORITHM: ESTIMATION OF THE LOCATION PARAMETERS. This is immediate matter, as always in OLS regression. To spell out for the fifth LV, its inner and outer relations, and the repeated substi-

tutive prediction (15):

$$\bar{X}_5 = \Sigma_h(w_{5h}\bar{x}_{5h}),$$
$$b_{50} = \bar{X}_5 - b_{51}\bar{X}_1 - b_{52}\bar{X}_2,$$
$$p_{5h0} = \bar{x}_{5h} - p_{5h}\bar{X}_5, \qquad (26)$$
$$p^*_{5h0} = \bar{x}_{5h} - p_{5h1}b_{51}\Sigma_h(w_{1h}\bar{x}_{ih})$$
$$+ b_{52}\Sigma_h(w_{2h}\bar{x}_{2h}).$$

PRODUCT DATA INPUT. It is an immediate matter to carry over the PLS algorithm (16)–(26) with raw data input (2) to product data input (10). The ensuing parameter estimates are numerically the same, except for rounding errors. The difference is that for each LV the product data only give aggregate values over n, such as means, variances, and covariances of the LV estimates, whereas raw data input is needed to obtain estimates for the case values X_{jn} ($n = 1, N$) for each LV, and thereby the predictions for MVs and LVs in preceding sections on Prediction.

COMPUTER PROGRAMS. The programs of the PLS algorithm (manual and tapes) are available at nominal cost, and they cover both raw data input and product data input [10; also see ref. 21, p. 24].

SPECIAL CASES: MODELS I*–II* AND II–IV IN FIG. 1. Here PLS overlaps with well-known earlier models:

1. *The first principal component*, Model I*, as estimated by the classical algebraic

method in terms of eigenvalues and eigenvectors, and subject to an SSU different from (7), namely, $\Sigma_n(\pi_{jh}^2) = 1$, is numerically equivalent to the one-LV model estimated PLS Mode A [21, p. 19].

2. *The first canonical correlation*, Model II*, is numerically equivalent to a two-LV model estimated PLS Mode B [14, p. 20].

3. *Models II–IV in Fig.* 1. If each LV in Models II*–IV* has just one indicator, they reduce to Models II–IV. Then each LV estimate (16) reduces to its indicator, and the OLS estimates of Models II–IV will coincide with the PLS estimates.

Model Evaluation [22]

Maximum likelihood (ML) modeling has for disposal a highly developed framework for (a) hypothesis testing and (b) assessment of standard errors (SEs) of the estimated unknowns. For LS modeling a general counterpart to (a) and (b) has emerged, namely (a*) the Stone–Geisser (SG) test for predictive relevance, and (b*) assessment of SEs by Tukey's jackknife. A blindfolding device is a joint feature of (a*–b*). Jackknife SEs of all estimated unknowns are obtained as a byproduct of the SG test. The adaptation of (a*–b*) to PLS modeling is straightforward.

Comments:

1. In traditional model evaluation the inference is asymptotic and valid for large N. The ML test (a) asks whether the model is *true* (i.e., in accordance with the postulated multivariate distribution), and if N is allowed to increase indefinitely the model is rejected, sooner or later. The SG test (a*) explores the predictive relevance of the model; the answer is a matter of degree, and it is valid for any size of N.

2. Although the jackknife dates from 1958 and the SG test from 1974 it is only recently that these methods have gained momentum; see [2].

STONE-GEISSER (SG) TESTING FOR PREDICTIVE RELEVANCE [2, 21, 22]

1. The SG test criterion Q^2 is an R^2 evaluated without loss of degrees of freedom.

2. PLS models usually provide several modes of predictive inference, as noted in sections on Prediction and Substitutive Prediction.

3. The construction and evolution of a PLS model is very much a dialogue with the computer. Tentative improvements of the model are tested for predictive relevance.

ASSESSMENT OF SES BY TUKEY'S JACKKNIFE [2, 22]

1. The assumption of independent observations underlies the classical SEs in (b), but not so for jackknife SEs. From jackknife SEs based on real-world data it appears that the assumption of independent observations makes the classical SEs inconsistent, often an underestimation by 25%, 100% or more.

2. Twofold use of the blindfolding device provides a jackknife SE for the SG test criterion Q^2.

COMMENTS ON PLS MODEL EVALUATION

1. For model estimation and evaluation in general, including PLS modeling, it is a general requisite that the data are homogeneous and uniform. If subclasses of the data have different structure, the inference from the estimated model will not have homogeneous validity over the subclasses. Accordingly, classification is an important auxiliary tool in PLS and other model building. Principal components and other PLS models provide useful methods for classification [9, 24],

a topic that cannot be taken up here for considerations of space.

2. In PLS modeling both estimation and evaluation are distribution- and independence-free. This has the important advantage that neither the inconsistency of PLS as discussed in the introductory remarks nor other sources of PLS inaccuracy disturb the model evaluation; the SG test and the jackknife evaluate the estimated model as it emerges from the model, the inaccuracies in the model and the estimation being a *fait accompli*, part of the evaluated inference.

For example, PLS tends to over- and underestimate the inner and outer parameters, respectively; the ensuing balance is part of the model output evaluated by the SG test and the jackknife.

3. Since the SG test and the jackknife are distribution-free, they apply irrespective of the number of observations, N, even if N is quite small. For an extreme case with 28 observables and $N = 10$, see ref. 20.

EXTENSION OF THE BASIC PLS DESIGN IN THEORY AND PRACTICE

Some few items will be noted, with reference to typical applications. The references reflect both the recent origin of PLS modeling and its rapid evolution.

INDICATORS. The indicators of any LV can be replaced or supplemented by some functional transform [11, 21]. For example,

$$\log x_{jh} ; x_{jh}^2 ; x_{jh} x_{jk} ; \ldots . \quad (27)$$

INNER RELATIONS. In the inner relations one or more LVs may be replaced by MVs [11, 16].

CATEGORICAL VARIABLES AND CONTINGENCY TABLES [3, 12]. Primarily designed for scalar observations, the PLS algorithm carries over to categorical indicators and thereby to contingency tables. A contingency table with

just one categorical variable in each margin is the covariance matrix of the two variables.

HIGHER DIMENSIONS OF THE LVS [7; II, CH. 10, 11, 21; SECT. 9.2]. The basic PLS design estimates the first dimension of each LV. Higher dimensions of an LV can be estimated consecutively, using as data input the residuals of the outer relations.

Special cases include principal components and canonical correlations of higher "orders," in PLS called "dimensions."

HIERARCHIC STRUCTURE. In psychometric factor analysis*, hierarchic structure, pioneered by L. L. Thurstone (1935, 1947), allows extension to PLS modeling [21, Sect. 9.3]. In ref. 16 each hierarchic structure is treated as a single complex LV, the inner β coefficients of each hierarchic structure being equated to zero. *See* HIERARCHICAL CLASSIFICATION.

Extensions That Change the Basic PLS Algorithm

MODELS WITH FEEDBACKS OR INTERDEPENDENCE IN THE INNER RELATIONS. Models V and V* are obtained from Models IV and IV* by introducing interdependence between the endogenous variables. Models V can be estimated by the fix-point* method, Models V* by combining the fix-point and PLS algorithms. See ref. 7 (II, chap. 11) and ref. 21 (Sect. 9.5), with reference to B. S. Hui (1978).

TWO-WAY OBSERVATION OF THE INDICATORS [21, SECT. 9.4]. In the basic PLS design the indicators may be observed either over time or a cross section. For two-way observation of the indicators an ingenious PLS algorithm using Kronecker products* has been introduced by J.-B. Lohmöller [10, 11].

NONLINEARITIES IN THE INNER RELATIONS. There are many types of nonlinearity, and the ensuing estimation problems vary widely in difficulty. The transfer from MVs to LVs is rather straightforward in the subsequent case 1 [21, Sect. 9.7] but more sophisticated in 2 [23].

Case 1. In a two-LV model the inner relation is nonlinear of second order, say,

$$\eta_n = \beta_0 + \beta_1 \xi_n + \beta_2 \xi_n^2 + \omega_n. \qquad (28)$$

Case 2. The three LVs $\xi_{1t}, \xi_{2t}, \eta_t$ of the model (t ranges over time) form the first level of a hierarchic structure where at the second level they satisfy a third-degree equation:

$$\eta_t^3 + \xi_{1t}\eta_t + \xi_{2t} = 0 \qquad (29)$$

In this model η_t may involve discontinuities in the sense of René Thom's catastrophe theory*.

PLS Applications: Experiences and Outlook

Experiences

Initiated some ten years ago, PLS is now firmly consolidated and is rapidly gaining momentum. The comments that follow draw from the many-sided experiences of the accumulating material.

The broad scope and flexibility of the PLS approach is reflected in the diversity of PLS modeling. The data can be scalar, ordinal, or categorical; the inner relations can be linear or nonlinear; the purpose can be prediction, classification*, or causal analysis. The SIMCA program for classification [24] is based on disjoint principal component PLS models where the appropriate number of dimensions is assessed by Stone–Geisser's test for predictive relevance.

PLS was initially designed for research contexts that are simultaneously data-rich and theory-primitive [19, 21]. In a short time PLS has invaded chemistry, medicine, and other sciences with well established complex theories, PLS having become a useful tool at the research frontier [5, 7-II, 8, 9].

Ever larger PLS models are reported, and it is safe to say that PLS has its forte in the analysis of large data-rich systems.

In substantive research, PLS has inspired investigators to construct large models. For example, the educational system in ref. 16 is a synthesis of six recent models that dealt with specific aspects of educational systems.

Fornell (1982) in his review of LISREL and PLS states [5; I, Chap. 1, p. 19]: "As is readily observed LISREL and PLS are the most general and flexible methods [of multivariate analysis], both allowing a variety of different types of relationships." Fornell further states that as regards the information from the model PLS is definite, whereas LISREL is indeterminate. For one thing, and this is what Fornell has in mind, PLS but not LISREL provides explicit estimates of the case values of the LVs. For another thing, in interdependent systems (Models V and V*) PLS but not LISREL provides a definite causal-predictive direction for the structural relations.

To judge from accumulating experience, there is a division of labor between LISREL and PLS. LISREL is at a premium in small or smallish models where each parameter has operative significance, and accurate parameter estimation is essential. PLS comes to the fore in larger models, when the importance shifts from individual variables and parameters to packages of variables and aggregate parameters.

In contrast to PLS, the technical difficulty of LISREL modeling increases rapidly with the size of the model [22]. Jöreskog (personal communication, 1983) gives 30 MVs as the upper limit for current ML–LISREL modeling.

In contemporary statistics there is often a large distance and even some friction and tension between substantive research and theoretical statistics (see ref. 6, p. 42). This is only natural as substantive analysis, with its firm requirements for qualified theorizing and time-consuming data work, is very different from the sophisticated, specialized statistical theories in the ML mainstream. PLS modeling has drastically reduced the distance between substantive research and statistical method. To quote from model builders using PLS, several advantages combine in the process:

1. The broad scope and flexibility of the PLS approach in theory and practice.
2. The simplicity of the statistical implementation, the arrow scheme of a PLS model being a sufficient basis for the formal specification and for the PLS algorithm.
3. The easy and speedy computer work, which gives "instant estimation."
4. A PLS model develops by a dialogue between the investigator and the computer. Tentative improvements of the model—such as the introduction of a new LV, an indicator, or an inner relation, or the omission of such an element —are tested for predictive relevance by the SG test, discussed earlier, and the various pilot studies are a speedy and low-cost matter.

Outlook

Breaking away from the ML mainstream, and placing emphasis on applied work, PLS modeling has from the outset attracted active interest from subject matter researchers. PLS modeling combined with the SG and jackknife methods of model evaluation are now firmly established as a distribution-free approach of general scope for quantitative systems analysis. In this broad perspective PLS modeling is at an early stage of evolution. There is an abundance of potential PLS applications.

Model building proceeds from simple to more complex models (see Fig. 1). The step from intra- to interdisciplinary systems is a quantum jump in complexity, and is a big jump even when modeling with latent variables [1, 17, 18, 20]. In large, complex models with latent variables PLS is virtually without competition. The advent of PLS has drastically reduced the distance between subject matter analysis and statistical technique; accordingly, it can be expected that PLS will come to the fore in the modeling of complex systems in domains with access to a steady flow of reliable data. This trend

has already started in economics [8], education [16], political science* [1, 4, 13, 18], and chemistry* [9]. Examples of other domains with rich access to data are mortality, health, and migration*. The situation is less fortunate in sociology, psychiatry, and other sciences where most of the relevant data are obtained by surveys with limited range in time and/or space.

In contrast to the classic asymptotical methods of model evaluation the SG test and the jackknife have the advantage that they apply irrespective of the number of observations, N, and even for quite small N. In PLS and other modeling this feature is always important. For one thing, it matters when models based on two different surveys are tested for uniformity, as well as in model building in general when using classification to test for uniformity. Furthermore, the applicability when N is limited is important when the model, because of "structural change," is expected to pass the test only over a limited range in time or space [14, 23].

References

[1] Adelman, I., et al. (1980). In *Quantitative Economics and Development*, L. R. Klein, M. Nerlove, and S. C. Tsiang, eds. Academic Press, New York, Chapter 1.

[2] Bergström, R. and Wold, H. (1983). *Fix-Point Estimation in Theory and Practice*. Vandenhoeck & Ruprecht, Göttingen, W. Germany.

[3] Bertholet, J.-L. and Wold, H. (1984). In *Measuring the Unmeasurable*, P. Nijkamp, ed. Martinus Nijhoff, The Hague.

[4] Falter, J. W., et al. (1983). *Kölner Z. Sozwiss. Psych.*, **53**, 525–551.

[5] Fornell, C., ed. (1982). *A Second Generation of Multivariate Analysis*, Vols. I and II. Praeger, New York.

[6] Fox, K. A. (1980). *Philosophics*, **25**, 33–54.

[7] Jöreskog, K. G. and Wold, H., eds. (1982). *Systems under Indirect Observation*, Vols. I and II. North-Holland, Amsterdam.

[8] Knepel, H. (1981). *Sozialökonomische Indikatormodelle zur Arbeitsmarktanalyse*. Campus, Frankfurt/New York.

[9] Lindberg, W., Persson, J.-Å., and Wold, S. (1983). *Chemistry*, **55**, 643–648.

[10] Lohmöller, J.-B. (1981). *LVPLS Program Manual, Version 1.6: Latent Variables Path Analysis with*

Partial Least Squares Estimation. Hochschule der Bundeswehr, Munich. (2nd ed., 1984, Zentralarchiv für empirische Sozialforschung, Cologne.)

[11] Lohmöller, J.-B. (1983). Path Models with Latent Variables and Partial Least Squares Estimation. Doctoral dissertation, Hochschule der Bundeswehr, Munich, W. Germany. (To be published by Physika, Stuttgart.)

[12] Lohmöller, J.-B. and Wold, H. (1984). In *Cultural Indicators, An International Symposium*, G. Melischek et al., eds. Austrian Academy of Sciences, Vienna, pp. 501–519.

[13] Meissner, W. and Uhle-Fassing, M. (1983). *Weiche Modelle und Iterative Schätzung. Eine Anwendung auf Probleme der Neuen Politischen Ökonomie*. Campus, Frankfurt/New York.

[14] Mensch, G. O. (1982). *Proc. Tenth Int. Conf. Unity of the Sciences*. The International Cultural Foundation, New York, pp. 499–515.

[15] Nijkamp, P., ed. (1984). *Measuring the Unmeasurable*. Martinus Nijhoff, The Hague. (In press.)

[16] Noonan, R. and Wold, H. (1984). *Evaluating School Systems Using Partial Least Squares*, Pergamon, Oxford, England.

[17] Schneewind, K. A., Beckman, M., and Engfer, A. (1983). *Umwelteinflüsse auf das Familiäre Verhalten*. Kohlhammer, Stuttgart, W. Germany.

[18] Wilkenfeld, J., Hopple, G. W., Rossa, P. J., and Andriole, S. J. (1982). *Interstate Behavior Analysis*. Sage, Beverly Hills, CA.

[19] Wold, H., ed. (1975). Modelling in Complex Situations with Soft Information. Third World Congress of Econometric Society, August 21–26, Toronto, Canada.

[20] Wold, H. (1980). Factors Influencing the Outcome of Economic Sanctions: An Application of Soft Modeling. Fourth World Congress of Econometric Society, August 28–September 2, Aix-en-Provence, France.

[21] Wold, H. (1982). In *Systems under Indirect Observation*, Vols. I, II. K. G. Jöreskog and H. Wold, eds. North Holland, Amsterdam, II: 1–54.

[22] Wold, H. (1984). In *Measuring the Unmeasurable*, P. Nijkamp, ed. Martinus Nijhoff, The Hague.

[23] Wold, H. and Mensch, G. O. (1983). "Nonlinear Extensions in Soft Modeling." *Working Paper WSCM 83-017*, Case Western Reserve University, Cleveland, Ohio.

[24] Wold, S. (1978). *Technometrics*, **20**, 397–405.

(ECONOMETRICS
FIX-POINT METHOD
LEAST SQUARES
LISREL
PREDICTOR SPECIFICATION)

H. WOLD

PARTIAL LIKELIHOOD

Partial likelihood was introduced by Cox [3] as a technique for making inferences in the presence of many nuisance parameters*. The methods allow reduction in the dimensionality of certain problems.

Suppose Y is a random variable whose density $f(y; \theta, \beta)$ depends on the parameters θ and β, where β is of particular interest with θ a nuisance parameter. Suppose Y can be transformed into the sequence $(A_1, B_1, \ldots, A_m, B_m)$. The density $f(a^{(m)}, b^{(m)})$ can be written

$$\prod_{j=1}^{m} f\left(b_j \mid b^{(j-1)}, a^{(j-1)}\right) \prod_{j=1}^{m} f\left(a_j \mid b^{(j)}, a^{(j-1)}\right),$$

(1)

where $a^{(j)} = (a_i, \ldots, a_j)$ and $b^{(j)} = (b_1, \ldots, b_j)$ for $j = 1, \ldots, m$.

The second term is called the *partial likelihood* based on **A** in the sequence (A_j, B_j). This factorization will be useful when the partial likelihood is a function of β only. The partial likelihood in general cannot be interpreted in a simple way in terms of a probability statement of some derived experiment as is the case for marginal and conditional likelihood (*see* PSEUDO-LIKELIHOOD).

The marginal likelihood $f(a^{(m)})$ based on **A** can be written $\prod_{j=1}^{m} f(a_j \mid a^{(j-1)})$ while the conditional likelihood $f(a^{(m)} \mid b^{(m)})$ based on **A** given **B** is $\prod_{j=1}^{m} f(a_j \mid a^{(j-1)}, b^{(m)})$. The partial likelihood will be identical to the marginal likelihood only if the A's and the B's are independent, and if A_j is independent of $B_{j+1}, \ldots, B_m$ for each $j = 1, \ldots, m - 1$, the conditional and partial likelihoods coincide.

Example. The ideas of partial likelihood have grown out of regression* methods for the analysis of survival* data where the baseline survivor function is left unspecified or specified up to a finite number of parameters. Cox [2] proposes such a model in which the hazard function for an individual with explanatory variables $\mathbf{x} = (x_1, \ldots, x_p)$ is given by

$$\lambda(t; \mathbf{x}) = \lambda_0(t) e^{\beta' \mathbf{x}}, \qquad (2)$$

where $\boldsymbol{\beta} = (\beta_1, \ldots, \beta_p)'$ is a vector of regression coefficients* and $\lambda_0(t)$ is left unspecified. The parameters $\boldsymbol{\beta}$ are of particular interest, and $\lambda_0(t)$ can be considered as a nuisance function. Survival data usually will be recorded in a form that involves censorship. Assuming that the mechanism producing censoring* is unrelated to that associated with failure, the full likelihood can be expressed as

$$\prod_{i=1}^{n} f(t_i ; \mathbf{x}_i)^{\delta_i} \{1 - F(t_i ; \mathbf{x}_i)\}^{1-\delta_i}, \quad (3)$$

where for individual i, t_i is the (possibly censored) survival time, $\mathbf{x}_i$ is a vector of explanatory variables,

$$\delta_i = \begin{cases} 0 & \text{if } t_i \text{ is a censoring} \\ 1 & \text{if } t_i \text{ is a death} \end{cases}$$

and $f(\cdot)$ and $F(\cdot)$ denote densities and distribution functions, respectively.

This likelihood function will contain the function $\lambda_0(t)$. Note again that this is a function of t only. However, the preceding considerations can lead to a partial likelihood function for $\boldsymbol{\beta}$.

Let $t_{(1)} < \cdots < t_{(k)}$ be the ordered (distinct) uncensored survival times. Let $\mathbf{x}_{(j)}$ be the explanatory variables for the individual who fails at $t_{(j)}$ and let $R(t_{(j)})$ be the set of labels for those individuals with censored or uncensored survival times $\geq t_{(j)}$. Let B_j specify the censoring that occurs in $[t_{(j-1)}, t_{(j)})$ and the information that a failure occurs on some individual at $t_{(j)}$ and let A_j specify the individual who fails at $t_{(j)}$. Then

$$f(a_j \mid b^{(j)}, a^{(j-1)}) = \frac{\lambda(t_{(j)} ; x_{(j)})}{\sum_{i \in R(t_{(j)})} \lambda(t_{(j)} ; \mathbf{x}_i)}$$

$$= \frac{e^{\beta' \mathbf{x}_{(j)}}}{\sum_{i \in R(t_{(j)})} e^{\beta' \mathbf{x}_i}}. \quad (4)$$

The partial likelihood based on $\mathbf{A}$ in the sequence (A_j, B_j) is then

$$\prod_{j=1}^{k} \frac{e^{\beta' \mathbf{x}_{(j)}}}{\sum_{i \in R(t_{(j)})} e^{\beta' \mathbf{x}_i}}.$$

Note that this is independent of the baseline hazard $\lambda_0(\cdot)$.

ASYMPTOTIC PROPERTIES

Cox [3] indicates that the usual asymptotic properties for maximum likelihood estimators* hold for estimators obtained from maximization of partial likelihoods. Appropriate conditions for these results to hold have not been established except in special cases. Tsiatis [6] and Liu and Crowley [4] have investigated results associated with the preceding example. Cox [3] outlines the arguments involved in establishing general results as follows. Let

$$\mathbf{u}_j = \frac{\partial \log f(a_j \mid b^{(j)}, a^{(j-1)})}{\partial \boldsymbol{\beta}},$$

$$j = 1, \ldots, m.$$

Then $E\{\mathbf{U}_j \mid B^{(j)}, A^{(j-1)}\} = \mathbf{0}$ and

$$\text{var}\{\mathbf{U}_j \mid B^{(j)}, A^{(j-1)}\}$$

$$= E\left\{ -\frac{\partial^2 \log f(A_j \mid B^{(j)}, A^{(j-1)})}{\partial \boldsymbol{\beta}^2} \middle| B^{(j)}, A^{(j-1)} \right\}$$

$$= \mathbf{I}_j(\boldsymbol{\beta}) = \mathbf{I}_j.$$

It follows that $E\{\mathbf{U}_j\} = \mathbf{0}$ and

$$E\{\mathbf{U}_i \mathbf{U}_j'\} = E\left\{ E(\mathbf{U}_i \mathbf{U}_j' \mid B^{(j)}, A^{(j-1)}) \right\}$$

$$= E\left\{ \mathbf{U}_i E(\mathbf{U}_j' \mid \mathbf{B}^{(j)}, A^{(j-1)}) \right\} = \mathbf{0}$$

$$\text{for} \quad i < j$$

so that $\text{var}(\mathbf{U}_j) = E\{\mathbf{I}_j\}$.

Under mild conditions concerning the independence of the $\mathbf{U}$'s and that the sum $\sum_{j=1}^{m} E\{\mathbf{I}_j\}$ is not dominated by individual large terms, a central limit theorem* will apply to the score vector, and asymptotically, as $m \to \infty$,

$$\mathbf{U} \sim N\left(\mathbf{0}, \sum_{j=1}^{m} E\{\mathbf{I}_j\}\right).$$

In addition, $\mathscr{I}_j(\boldsymbol{\beta})/m$ and $\mathscr{I}_j(\hat{\boldsymbol{\beta}})/m$, where $\hat{\boldsymbol{\beta}}$ is the maximum partial likelihood estimate of $\boldsymbol{\beta}$ and

$$\mathscr{I}_j(\boldsymbol{\beta}) = -\frac{\partial^2 \log f(a_j \mid b^{(j)}, a^{(j-1)})}{\partial \boldsymbol{\beta}}$$

will be consistent estimators of $E(\mathbf{I}_j)/m$.

The asymptotic normality of $(\hat{\boldsymbol{\beta}} - \boldsymbol{\beta})$ $[\mathscr{I}(\hat{\boldsymbol{\beta}})]^{1/2}$, where $\mathscr{I}(\boldsymbol{\beta}) = \sum_{j=1}^{m} I_j(\boldsymbol{\beta})$, will then follow through a Taylor series expansion of $U(\beta)$ about $\beta = \hat{\beta}$.

Similar arguments will also justify test procedures and confidence intervals based on the large-sample chi-square distribution of transformed likelihood ratios.

Oakes [5] has provided an extensive and more advanced summary of the concept. In the discussion of that paper by Anderson and Gill it is proposed that the Cox model and the resulting partial likelihood can be considered within the framework of martingale* theory. Central limit theory for continuous-time martingales can then be used to derive the preceding asymptotic properties of $\hat{\beta}$. The details are given in Anderson and Gill [1].

References

[1] Anderson, P. K. and Gill, R. D. (1981). "Cox's Regression Model for Counting Processes: A Large Sample Study." *Res. Rep. No. 81/6*, Statistical Research Unit, Danish Medical and Social Science Research Councils.

[2] Cox, D. R. (1972). *J. R. Statist. Soc. B*, **34**, 187–220.

[3] Cox, D. R. (1975). *Biometrika*, **62**, 269–276.

[4] Liu, P.-Y. and Crowley, J. (1978). "Large Sample Theory of the MLE based on Cox's Regression Model for Survival Data." *Tech. Rep. No. 1*, Wisconsin Clinical Cancer Center.

[5] Oakes, D. (1981). *Int. Statist. Rev.* **49**, 235–264.

[6] Tsiatis, A. A. (1978). *Ann. Statist.*, **6**, 93–108.

(COX'S REGRESSION MODEL
LIKELIHOOD
MARTINGALES
SURVIVAL ANALYSIS)

RICHARD KAY

PARTIALLY BALANCED DESIGNS

This article should be read in conjunction with BLOCKS, BALANCED INCOMPLETE; BLOCKS, RANDOMIZED COMPLETE; GENERAL BALANCE; GROUP-DIVISIBLE DESIGNS; and INCOMPLETE BLOCK DESIGNS.

Balanced incomplete block designs (BIBDs) have many desirable qualities. They are easy to analyze; many families of them are easily constructed; the loss of information on treatment estimates (in the sense of increase in variance) due to blocking is as small as it can be for the given block size; and it is easy to compare them with other incomplete block designs, because usually the BIBDs are superior in every respect. However, for many combinations of block size, number of treatments and number of replications, there is no balanced incomplete block design. Bose and Nair [5] introduced partially balanced incomplete block designs in 1939 for use in such situations, hoping to retain many of the desirable properties of BIBDs. They succeeded in their aim to a certain extent: some of the new class of designs are very good, while others are undoubtedly useless for practical experimentation.

Unfortunately, any discussion of partial balance must necessarily be more technical than that of BIBDs, because there is an extra layer of complication. Moreover, many of the ideas have been developed and clarified by pure mathematicians, without ever being reexpressed in statistical language. Thus some sections of this article are unavoidably technical. It is hoped that the division into sections with titles will enable the nonmathematical reader to learn something from this article.

During the development of the subject there have been minor modifications to the definition of partial balance: the most generally accepted definition, which we use here, is from a review paper in 1963 by Bose [3].

The key idea is the combinatorial concept of *association scheme*, to which the first part of this article is devoted. The designs themselves are introduced in the second part of the article; the third part discusses generalizations of partial balance and related ideas.

ASSOCIATION SCHEMES

Suppose that there are t treatments. There are three equivalent ways of defining an *association scheme* on the treatments: as a partition of the pairs of treatments, as a set of matrices, and as a colored graph. According to the most common definition, an *association scheme with s associate classes* is a partition of the unordered pairs of treatments into s associate classes. If u and v are treatments and $\{u,v\}$ is in the ith associate class, then u and v are called *ith associates*. The partition must have the property that there are numbers p_{ij}^k, for $i, j, k = 1, 2, \ldots, s$, such that if u and v are *any* kth associates the number of treatments that are ith associates of u and jth associates of v is p_{ij}^k. (These numbers are not powers, so some authors write them as p_{ijk}.) It follows that each treatment has n_i ith associates, where

$$n_i = p_{i1}^k + p_{i2}^k + \cdots + p_{is}^k$$

for all $k \neq i$.

It is convenient to say that each treatment is its own zeroth associate. The preceding property still holds, with

$$p_{ij}^0 = p_{i0}^j = p_{0j}^i = 0, \quad \text{if } i \neq j;$$

$$p_{ii}^0 = n_i; \qquad p_{i0}^i = p_{0i}^i = 1.$$

An association scheme may be represented by a set of $t \times t$ matrices. For $i = 0, \ldots, s$ let $\mathbf{A}_i$ be the $t \times t$ matrix whose (u,v) entry is 1 if u and v are ith associates and 0 otherwise. The matrices $\mathbf{A}_0, \mathbf{A}_1, \ldots, \mathbf{A}_s$ are called *association matrices*, and satisfy the following conditions:

(a)   Each element of $\mathbf{A}_i$ is 0 or 1.
(b) $\mathbf{A}_i$ is symmetric.
(c) $\mathbf{A}_0 = \mathbf{I}$, the identity matrix.
(d) $\sum_{i=0}^{s} \mathbf{A}_i = \mathbf{J}$, the all-ones matrix.
(e) $\mathbf{A}_i \mathbf{A}_j = \sum_k p_{ij}^k \mathbf{A}_k$.

Conversely, any set of square matrices satisfying conditions **(a)**–**(e)** defines an association scheme.

An association scheme may also be represented by a colored graph. There is one vertex for each treatment, and there are s colors. The edge uv is colored with the ith color if u and v are ith associates. Now condition **(e)** may be interpreted as, e.g.: If

uv is a green edge then the number of green-red-blue triangles containing uv is independent of u and v. If $s = 2$, no information is lost by erasing the edges of the second color to obtain an ordinary graph. A graph obtained in this way is called a *strongly regular graph* [4]. Much literature on association schemes with two associate classes is to be found under this heading, with no explicit mention of association schemes. *See also* GRAPH THEORY.

The simplest association scheme has one associate class, and all treatments are first associates. Denote this scheme by $B(t)$, where t is the number of treatments. While not very interesting in its own right, $B(t)$ can be used to generate other association schemes.

Examples of Association Schemes with Two Associate Classes

GROUP-DIVISIBLE ASSOCIATION SCHEME. Suppose that $t = mn$. Partition the treatments into m sets of size n. The sets are traditionally called *groups* in this context, even though there is no connection with the technical meaning of *group* used later in this article. Two treatments are first associates if they are in the same group; second associates, otherwise. Call this scheme GD(m,n).

TRIANGULAR ASSOCIATION SCHEME. Suppose that $t = n(n-1)/2$. Put the treatments into the lower left and upper right triangles of an $n \times n$ square array in such a way that the array is symmetric about its main diagonal, which is left empty. Two treatments are first associates if they are in the same row or column; second associates, otherwise. Call this scheme $T(n)$. Then $T(4)$ is GD(3,2) with the classes interchanged. The square array for $T(5)$ is shown in Table 1. The

Table 1

*	A	B	C	D
A	*	E	F	G
B	E	*	H	I
C	F	H	*	J
D	G	I	J	*

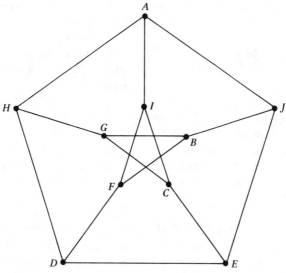

Figure 1

complement of the corresponding strongly regular graph is in Fig. 1, known as a *Petersen graph*.

In an alternative description of $T(n)$, the t treatments are identified with the unordered pairs from a set of size n. Then treatments are first associates if, as pairs, they have an element in common.

The association scheme $T(n)$ arises naturally in diallele cross experiments, where the treatments are all individuals (plants or animals) arising from crossing n genotypes. If there are no self-crosses, and if the gender of the parental lines is immaterial, then there are $\binom{n}{2}$ treatments and the first associates of any given treatment are those having one parental genotype in common with it.

LATIN SQUARE ASSOCIATION SCHEMES. Suppose that $t = n^2$, and arrange the treatments in an $n \times n$ square array. Two treatments are first associates in the association scheme $L_2(n)$ if they are in the same row or column. Let Λ be any $n \times n$ Latin square*. Then Λ defines an association scheme of type $L_3(n)$: two treatments are first associates if they are in the same row or column or letter of Λ. Similarly, if $\Lambda_1, \ldots, \Lambda_{r-2}$ are mutually orthogonal Latin squares ($3 \le r \le n + 1$), we may define an association scheme of type $L_r(n)$ by declaring any two treatments in the

same row, column, or letter of any of $\Lambda_1, \ldots, \Lambda_{r-2}$, to be first associates; any other pair, to be second associates. When $r = n + 1$, this scheme degenerates into $B(t)$.

CYCLIC ASSOCIATION SCHEMES. Identify the treatments with the integers modulo t. Suppose that D is a set of nonzero integers modulo t with the properties that if $d \in D$, then $t - d \in D$; and that the differences

$$d - e \text{ modulo } t$$

with d, e in D include each element of D exactly N times and each other nonzero integer modulo t exactly M times, for some fixed numbers N and M. In the cyclic association scheme $C(t, D)$ two treatments u and v are first associates if $u - v$ (modulo t) is in D; second associates, otherwise.

For example, if $t = 5$ we may take $D = \{1, 4\}$. Then $N = 0$ and $M = 1$. The association scheme $C(5, D)$ is shown in Table 2. Here

Table 2

	0	1	2	3	4
0	0	1	2	2	1
1	1	0	1	2	2
2	2	1	0	1	2
3	2	2	1	0	1
4	1	2	2	1	0

the (u, v) entry of the table is i if u and v are ith associates.

Combining Association Schemes

For $n = 1, 2$, let C_n be an association scheme with s_n associate classes for a set T_n of t_n treatments. There are two simple ways of combining these two schemes to obtain an association scheme for $t_1 t_2$ treatments. In both cases, we identify the new treatments with the set $T_1 \times T_2$ of ordered pairs (u_1, u_2) with $u_1 \in T_1$ and $u_2 \in T_2$ (*see also* NESTING AND CROSSING IN DESIGN).

We may *nest* C_2 in C_1 to obtain the association scheme C_1 / C_2 with $s_1 + s_2$ associate classes. Treatments (u_1, u_2) and (v_1, v_2) are ith associates if u_1 and v_1 are ith associates in C_1 (for $1 \leqslant i \leqslant s_1$) or if $u_1 = v_1$ and u_2 and v_2 are $(i - s_1)$th associates in C_2 (for $s_1 + 1 \leqslant i \leqslant s_2$). For example, $B(m) / B(n)$ is GD(m, n).

We may *cross* C_1 and C_2 to obtain the association scheme $C_1 \times C_2$ with $s_1 s_2 + s_1 + s_2$ associate classes. If the zeroth associate class is included in C_1, C_2 and $C_1 \times C_2$, then the associate classes of $C_1 \times C_2$ may be labeled by ij, for $0 \leqslant i \leqslant s_1$ and $0 \leqslant j \leqslant s_2$. Treatments (u_1, u_2) and (v_1, v_2) are ijth associates in $C_1 \times C_2$ if u_1 and v_1 are ith associates in C_1 and if u_2 and v_2 are jth associates in C_2. The crossed scheme $B(m) \times B(n)$ is called a *rectangular* association scheme $R(m, n)$, because its treatments may be arranged in an $m \times n$ array: two treatments are 01-th or 10-th associates if they are in the same row or the same column, respectively, and 11-th associates otherwise.

Factorial Association Schemes

There is a natural way of defining an association scheme on treatments with a factorial* structure. For example, suppose that treatments are all combinations of the t_1 levels of factor F_1 with the t_2 levels of factor F_2. If F_2 is nested in F_1, then the association scheme $B(t_1) / B(t_2)$ [i.e., GD(t_1, t_2)] corresponds naturally to the breakdown of treatment ef-

fects into the two factorial effects: F_1 and F_2-within-F_1. If there is no nesting, then the association scheme $B(t_1) \times B(t_2)$ [i.e., $R(t_1, t_2)$] corresponds to the three factorial effects: main effects* F_1 and F_2, and interaction* $F_1 F_2$.

More generally, if there are n treatment factors $F_1, F_2, \ldots, F_n$ with $t_1, t_2, \ldots, t_n$ levels, respectively, a factorial association scheme may be built up from $B(t_1), B(t_2), \ldots, B(t_n)$ by repeated use of crossing and/or nesting. Roy [49] described repeated nesting, and Hinkelmann [19], repeated crossing, but the two operations may be combined. One of the simplest crossed-and-nested schemes is $(B(t_1) \times B(t_2)) / B(t_3)$, which is also called a *generalized right angular scheme*. A somewhat more general construction, depending only on the nesting relations among the factors, is given by Speed and Bailey [54].

Example. An experiment was conducted to find out how the ultrasonic fat probe measurement of a sow was affected by the particular sow, by the site on the sow where the measurement was taken, and by the operator who did the measuring. Each of eight operators measured each of four sows at each of two sites on the body (technically known as $P2$ and H). The operators consisted of two batches of four: those in the first batch were experienced users of the small portable measuring instruments being used, while those in the second were novices. The 64 treatments (all combinations of sow, site, and operator), therefore, had a $4 \times 2 \times (2/4)$ factorial structure. The corresponding association scheme is shown in Table 3. Here the 64 treatments are shown in an 8×8 array: the entry for treatment u is i if u is an ith associate of the treatment in the top left-hand corner.

An unstructured set of treatments may have a dummy factorial structure imposed on it to help with the construction of a design or the subsequent analysis. Any corresponding factorial association scheme is called *pseudofactorial*.

Table 3

Sow	Site	Operators							
		Experienced				Novices			
		1	2	3	4	5	6	7	8
1	1	0	1	1	1	2	2	2	2
1	2	3	4	4	4	5	5	5	5
2	1	6	7	7	7	8	8	8	8
2	2	9	10	10	10	11	11	11	11
3	1	6	7	7	7	8	8	8	8
3	2	9	10	10	10	11	11	11	11
4	1	6	7	7	7	8	8	8	8
4	2	9	10	10	10	11	11	11	11

Other Association Schemes

We have space to mention only a few of the many other families of association schemes. The d-dimensional triangular association scheme $T(d, n)$ has $t = \binom{n}{d}$ and $s = \min(d, n - d)$. The treatments are identified with subsets of size d from a set of size n. Two treatments are ith associates if, as subsets, they have $d - i$ elements in common. Thus $T(2, n) = T(n)$. These schemes are also called *Johnson schemes*. Conceptually similar is the d-dimensional *lattice association scheme*, also called the *Hamming association scheme* $H(d, n)$, which has $t = n^d$ and $s = d$. The treatments are identified with ordered d-tuples of the numbers $1, 2, \ldots, n$, and two treatments are ith associates if their coordinates are the same in $d - i$ positions.

The Latin square* association scheme $L_r(n)$ also has generalizations. The scheme $L_r'(n)$ differs from $L_r(n)$ in that it has $r + 1$ associate classes. Treatments are first, second, third, $\ldots$, rth associates if they are in the same row, column, letter of $\Lambda_1, \ldots$, letter of Λ_{r-2}. Any way of amalgamating associate classes in $L_r'(n)$ produces another association scheme (this is *not* true for association schemes in general): in particular, amalgamation of the first r classes produces $L_r(n)$.

The cyclic scheme $C(t, D)$ is formed by amalgamating classes of the general cyclic association scheme $C(t)$, which has $(t - 1)/2$ associate classes if t is odd, $t/2$ if t is even. Treatments u and v are ith associates if $u - v = \pm i$ modulo t. Both $C(t, D)$ and $C(t)$ may be further generalized by replacing the integers modulo t by any Abelian group (in the algebraic sense of "group").

There are also many other individual association schemes that do not fit into any general families.

Algebra

Conditions (a)–(e) for an association scheme have algebraic consequences that are important for partially balanced designs. Let $\mathscr{A}$ be the set of linear combinations of the matrices $\mathbf{A}_0, \mathbf{A}_1, \ldots, \mathbf{A}_s$, which is called the *Bose–Mesner algebra* of the association scheme. Every matrix $\mathbf{M}$ in $\mathscr{A}$ is "patterned" according to the association scheme; that is, the entry M_{uv} depends only on the associate class containing $\{u, v\}$.

If $\mathbf{M}$ and $\mathbf{N}$ are in $\mathscr{A}$, so are $\mathbf{M} + \mathbf{N}$, $\mathbf{MN}$ and $\alpha \mathbf{M}$, for every real number α; moreover, $\mathbf{MN} = \mathbf{NM}$. It follows that the matrices in $\mathscr{A}$ are simultaneously diagonalizable; i.e., the t-dimensional vector space corresponding to the treatments is a direct sum of spaces $W_0, W_1, \ldots, W_s$, each of which is an eigenspace of every matrix in $\mathscr{A}$. Hence if $\mathbf{M}$ is invertible and $\mathbf{M}$ is in $\mathscr{A}$, then $\mathbf{M}^{-1}$ is in $\mathscr{A}$.

Usually the eigenspaces $W_0, W_1, \ldots, W_s$

are easy to calculate. We may always take W_0 to be the one-dimensional space spanned by the vector $(1, 1, \ldots, 1)$. At worst, the problem of finding $W_0, W_1, \ldots, W_s$ can be reduced to a similar problem for $(s + 1) \times (s + 1)$ matrices. See Bose and Mesner [6], Cameron and van Lint [9], Delsarte [16], and Dembowski [17] for further details.

The trivial association scheme $B(t)$ has two eigenspaces, W_0 and W_1. The space W_1 consists of all contrasts, i.e., all vectors whose entries sum to zero.

For the group-divisible association scheme $GD(m, n)$ the spaces W_1 and W_2 consist of all *between-groups* contrasts* and all *within-groups* contrasts, respectively. That is, a vector is in W_1 if its entries are constant on each group and sum to zero overall; a vector is in W_2 if its entries sum to zero on each group. The rectangular association scheme has four eigenspaces: W_1 and W_2 consist of between-rows contrasts and between-columns contrasts respectively, while W_3 contains all vectors that are orthogonal to W_0, W_1 and W_2. In general, the eigenspaces of every factorial association scheme are the spaces corresponding to main effects and interactions in the factorial sense.

If t is odd, every eigenspace (except W_0) of the cyclic association scheme $C(t)$ has dimension 2. If the treatments are written in the order $0, 1, 2, \ldots, t - 1$, a basis for W_i consists of

$$(1, \cos(2\pi i / t), \cos(4\pi i / t), \ldots,$$
$$\cos(2\pi(t - 1)i / t));$$
$$(0, \sin(2\pi i / t), \sin(4\pi i / t), \ldots,$$
$$\sin(2\pi(t - 1)i / t)),$$

for each i with $1 \leqslant i \leqslant (t - 1)/2$. If t is even, there is an additional eigenspace $W_{t/2}$ spanned by $(1, -1, 1, \ldots, -1)$.

For every association scheme, there are real numbers e_{ij} $(0 \leqslant i, j \leqslant s)$ such that every vector in W_i is an eigenvector of $\mathbf{A}_j$ with eigenvalue e_{ij}. The array $[e_{ij}]$ is sometimes called the *character table* of the association scheme. Table 4 shows the character tables of some association schemes.

Denote by $\mathbf{S}_i$ the matrix representing orthogonal projection onto W_i. The eigenpro-

Table 4[a]

Association Scheme	Eigenspace W_i	Dimension of W_i	Associate Classes A_j			
			A_0	A_1		
$B(t)$	W_0	1	1	$t - 1$		
	$W_1 = $ contrasts	$t - 1$	1	-1		
			A_0	A_1	A_2	
$GD(m, n)$	W_0	1	1	$m - 1$	$m(n - 1)$	
	$W_1 = $ between groups	$m - 1$	1	$m - 1$	$-m$	
	$W_2 = $ within groups	$m(n - 1)$	1	-1	0	
			A_{00}	A_{01}	A_{10}	A_{11}
$B(m) \times B(n)$	W_0	1	1	$m - 1$	$n - 1$	$(m-1)(n - 1)$
	$W_1 = $ between rows	$m - 1$	1	$m - 1$	-1	$-(m - 1)$
	$W_2 = $ between columns	$n - 1$	1	-1	$n - 1$	$-(n - 1)$
	W_3	$(m - 1)(n - 1)$	1	-1	-1	1
			A_0	A_1	A_2	
$C(5)$	W_0	1	1	2	2	
	W_1	2	1	0.618	-1.618	
	W_2	2	1	-1.618	0.618	

[a] The entry in row W_i and column A_j is the character e_{ij}.

jections are symmetric, idempotent, and mutually orthogonal; i.e., $S_i' = S_i$, $S_i^2 = S_i$, and $S_i S_j = 0$ if $i \neq j$. Moreover, $\Sigma_i S_i = I$. For every j,

$$A_j = \sum_i e_{ij} S_i .$$

The $(s + 1) \times (s + 1)$ matrix $[e_{ij}]$ is invertible, with inverse $[f_{ij}]$, say. Then

$$S_i = \sum_j f_{ji} A_j$$

for each i. (See refs. 6, 9, and 16 for more details.)

PARTIALLY BALANCED INCOMPLETE BLOCK DESIGNS

Let Δ be an incomplete block design* for t treatments, each replicated r times, in b blocks each of size k. Suppose that no treatment occurs more than once in any block: such a design is said to be *binary*. For treatments u and v, denote by λ_{uv} the number of blocks in which u and v both occur, that is, the *concurrence* of u and v. The design Δ is said to be *partially balanced* with respect to an association scheme C if λ_{uv} depends only on the associate class of C that contains $\{u, v\}$. Usually we abbreviate this and simply say that Δ is a partially balanced (incomplete block) design, or PBIBD. A design may be partially balanced with respect to more than one association scheme; usually, the simplest possible association scheme is assumed.

The balanced incomplete block design* (BIBD) is just a special case of a PBIBD, for a BIBD with t treatments is partially balanced with respect to the association scheme

Table 5

A	A	A	D
B	B	B	E
C	C	C	F
D	D	G	G
E	E	H	H
F	F	I	I
G	J	J	J
H	K	K	K
I	L	L	L

$B(t)$. A less trivial example is shown in Table 5. Here $t = 12$, $r = 3$, $b = 4$, and $k = 9$; treatments are denoted by capital letters and blocks are columns. This design is partially balanced with respect to GD(4, 3): the "groups" of treatments are $\{A, B, C\}$, $\{D, E, F\}$, $\{G, H, I\}$, and $\{J, K, L\}$.

Partially balanced designs often inherit the name of the appropriate association scheme. Thus a design that is partially balanced with respect to a group-divisible association scheme is called a *group-divisible design**. Partial balance with respect to a factorial association scheme is known as *factorial balance*. Although PBIBDs were introduced only in 1939, the importance of factorially balanced designs had been recognized previously by Yates [56].

Construction and Catalogs

There are two elementary methods of constructing a PBIBD for a given association scheme. Each method has a block $B(u, i)$ for each treatment u and a fixed i with $1 \leqslant i \leqslant s$. In the first construction $B(u, i)$ consists of all ith associates of u; in the second, u is also in $B(u, i)$.

If Δ_1 is partially balanced with respect to an association scheme C for t_1 treatments, we may obtain a new design Δ for $t_1 t_2$ treatments by replacing each treatment of Δ_1 by t_2 new treatments. Then Δ is partially balanced with respect to $C / B(t_2)$. In particular, if Δ_1 is a BIBD, then Δ is group divisible. The design in Table 5 was constructed by this method with $t_1 = 4$ and $t_2 = 3$. For further constructions of group divisible designs, *see* GROUP DIVISIBLE DESIGNS.

The lattice designs* for n^2 treatments are partially balanced with respect to Latin square* association schemes $L_r(n)$ for various r: for a simple lattice, $r = 2$; for a triple lattice, $r = 3$. The scheme $L_2(n)$ is also the two-dimensional Hamming scheme $H(2, n)$. Designs partially balanced with respect to $H(d, n)$ for $d = 3, 4, 5$ are sometimes called cubic, quartic, or quintic, respectively. A simple construction of such a PBIBD gives a block $B(i, j)$ for $1 \leqslant i \leqslant d$ and $1 \leqslant j \leqslant n$:

Table 6

123	124	125	134	135
A	A	A	B	B
B	C	D	C	D
E	F	G	H	I

145	234	235	245	345
C	E	E	F	H
D	F	G	G	I
J	H	I	J	J

the treatments in $B(i, j)$ are all those d-tuples whose ith coordinate is equal to j.

The method of *triads* gives a PBIBD for the triangular association scheme $T(n)$. There are $\binom{n}{3}$ blocks of size 3, one for each subset of size 3 of the original set of size n. The block corresponding to the set $\{\alpha, \beta, \gamma\}$ contains the treatments corresponding to the pairs $\{\alpha, \beta\}$, $\{\alpha, \gamma\}$, and $\{\beta, \gamma\}$. The triad design for $T(5)$ is shown in Table 6. A PBIBD with $\binom{n}{m}$ blocks of size $\binom{m}{2}$ may be constructed in a similar manner. A simpler PBIBD for $T(n)$ has n blocks of size $n - 1$: the blocks are the columns of the square array that defines the association scheme.

Most factorial designs in common use have factorial balance. Some constructions may be found in CONFOUNDING and FACTORIAL EXPERIMENTS.

CYCLIC DESIGNS*. These designs are partially balanced with respect to cyclic association schemes. An *initial block* B_0 is chosen. A second block B_1 is generated by adding 1 (modulo t) to every element of B_0. Further blocks $B_2, B_3, \ldots$ are generated from B_1, $B_2, \ldots$ in the same way. The process stops *either* with B_{t-1} *or* with B_{u-1}, where B_u is the first block identical to B_0. The cyclic design for five treatments in blocks of size 3 with initial block $\{1, 2, 4\}$ is shown in Table 7. It is also possible to have more than one initial block. Essentially the same method

Table 7

1	2	3	4	0
2	3	4	0	1
4	0	1	2	3

gives PBIBDs with association scheme based on any Abelian group.

Although there are many fascinating methods of constructing PBIBDs, the practitioner who needs a single PBIBD should consult a catalog. Clatworthy [14] gives an extensive catalog of PBIBDs with two associate classes. John et al. [26] list cyclic designs with high efficiency factors (see below).

Randomization

The randomization* of a PBIBD is in two parts. First, the blocks of the design should be allocated randomly to the actual blocks. Second, in each block independently the treatments prescribed for that block should be randomly allocated to the plots. (If the design is resolvable*, and the replicates are intended to match features of the experimental material or its management, the blocks should be randomized only within replicates.)

There is no need for any randomization of treatments to the letters or numbers used in the design. Any such randomization would be positively harmful if the association scheme has been chosen with reference to the specific treatments, as, e.g., in a factorial design. If a series of similar experiments is to be conducted at different sites, any randomization of treatments at individual sites will hamper, or even prevent, the subsequent analysis across all sites. The first two stages of randomization should, of course, be done, independently, at all sites.

Efficiency Factors

The standard analysis of an incomplete block design* uses a generalized inverse* of the $t \times t$ matrix $\Theta = \mathbf{I} - r^{-1}k^{-1}\Lambda$, where Λ is the matrix whose diagonal elements are equal to r, and whose (u, v) element is equal to λ_{uv} if $u \neq v$. Thus $\Theta = \sum_i \theta_i \mathbf{A}_i$, where $\theta_0 = 1 - k^{-1}$ and $\theta_i = -\lambda_i/rk$ otherwise, where λ_i is the common value of the concurrence of pairs of treatments which are ith associates. Hence $\Theta = \sum_j E_j \mathbf{S}_j$, where E_j

$= \sum_i e_{ji} \theta_i$ and the e_{ji} are the characters defined in the subsection "Algebra" of the section "Association Schemes". A generalized inverse $\mathbf{\Phi}$ of $\mathbf{\Theta}$ is $\sum_j E_j^{-1} \mathbf{S}_j$, the sum being restricted to those j for which $E_j \neq 0$. Moreover, $\mathbf{\Phi}$ has the same "pattern" as $\mathbf{\Theta}$, for $\mathbf{\Phi} = \sum_i \phi_i \mathbf{A}_i$, where $\phi_i = \sum_j f_{ij} E_j^{-1}$.

Denote by τ_u the effect of treatment u. If $\mathbf{x}$ is any contrast vector then $\mathbf{x} \cdot \boldsymbol{\tau} = \sum_u x_u \tau_u$ is a linear combination of the treatment effects, and the variance of the intrablock* estimate of $\mathbf{x} \cdot \boldsymbol{\tau}$ is $\mathbf{x}\mathbf{\Phi}\mathbf{x}'\sigma^2/r$, where σ^2 is the intrablock variance (*see* INCOMPLETE BLOCK DESIGNS). Thus, if $\mathbf{x}$ is a vector in W_i, then the variance of $\mathbf{x} \cdot \boldsymbol{\tau}$ is $(\mathbf{x} \cdot \mathbf{x})\sigma^2/(rE_i)$ and hence the *efficiency factor* (*see* BLOCKS, BALANCED INCOMPLETE) for $\mathbf{x} \cdot \boldsymbol{\tau}$ is E_i. The E_i are called the *canonical efficiency factors* of the design. They lie between 0 and 1, and E_0 is zero. If no other canonical efficiency factor is zero the design is said to be *connected*: in this case, there is an intrablock estimate of $\mathbf{x} \cdot \boldsymbol{\tau}$ for every contrast $\mathbf{x}$. The efficiency factor for such an $\mathbf{x} \cdot \boldsymbol{\tau}$ may be calculated in terms of the E_i, and it lies between the extreme values of the E_i with $i \neq 0$. In particular, the efficiency factor for $\tau_u - \tau_v$ is $1/(\phi_0 - \phi_i)$ if u and v are ith associates.

For example, consider the design given at the top of Table 8. It is a group-divisible design with $t = b = 4$, $r = k = 2$, $\lambda_1 = 0$,

$\lambda_2 = 1$. The groups are $\{A, C\}$ and $\{B, D\}$. The contrast eigenspaces are the between-groups eigenspace W_1, with basis $\mathbf{x}^{(1)} = (1, -1, 1, -1)$, and the within-groups eigenspace W_2, with basis consisting of $\mathbf{x}^{(2)} = (1, 0, -1, 0)$ and $\mathbf{x}^{(3)} = (0, 1, 0, -1)$. The canonical efficiency factors are given by

$$E_i = e_{i0}(1 - \tfrac{1}{2}) - e_{i2}(\tfrac{1}{4}).$$

The character table $[e_{ij}]$ is the second part of Table 4. Thus

$$[e_{ij}] = \begin{bmatrix} 1 & 1 & 2 \\ 1 & 1 & -2 \\ 1 & -1 & 0 \end{bmatrix}$$

and we have

$$E_0 = 1 \times \tfrac{1}{2} - 2 \times \tfrac{1}{4} = 0,$$

$$E_1 = 1 \times \tfrac{1}{2} + 2 \times \tfrac{1}{4} = 1,$$

$$E_2 = 1 \times \tfrac{1}{2} + 0 \times \tfrac{1}{4} = \tfrac{1}{2}.$$

Moreover, $[f_{ij}] = [e_{ij}]^{-1} =$

$$\frac{1}{4}\begin{bmatrix} 1 & 1 & 2 \\ 1 & 1 & -2 \\ 1 & -1 & 0 \end{bmatrix}$$

and so

$$\phi_0 = \tfrac{1}{4}E_1^{-1} + \tfrac{1}{2}E_2^{-1} = \tfrac{5}{4},$$

$$\phi_1 = \tfrac{1}{4}E_1^{-1} - \tfrac{1}{2}E_2^{-1} = -\tfrac{3}{4},$$

$$\phi_2 = -\tfrac{1}{4}E_1^{-1} = -\tfrac{1}{4}.$$

Table 8

Block	1	1	2	2	3	3	4	4
Treatment	A	B	B	C	C	D	D	A
Yield y	4	2	6	-11	-7	5	10	1
$\mathbf{x}_D^{(1)}$	1	-1	-1	1	1	-1	-1	1
$\mathbf{x}_B^{(1)}$	0	0	0	0	0	0	0	0
$\mathbf{x}_P^{(1)}$	1	-1	-1	1	1	-1	-1	1
$\mathbf{x}_D^{(2)}$	1	0	0	-1	-1	0	0	1
$\mathbf{x}_B^{(2)}$	$\tfrac{1}{2}$	$\tfrac{1}{2}$	$-\tfrac{1}{2}$	$-\tfrac{1}{2}$	$-\tfrac{1}{2}$	$-\tfrac{1}{2}$	$\tfrac{1}{2}$	$\tfrac{1}{2}$
$\mathbf{x}_P^{(2)}$	$\tfrac{1}{2}$	$-\tfrac{1}{2}$	$\tfrac{1}{2}$	$-\tfrac{1}{2}$	$-\tfrac{1}{2}$	$\tfrac{1}{2}$	$-\tfrac{1}{2}$	$\tfrac{1}{2}$
$\mathbf{x}_D^{(3)}$	0	1	1	0	0	-1	-1	0
$\mathbf{x}_B^{(3)}$	$\tfrac{1}{2}$	$\tfrac{1}{2}$	$\tfrac{1}{2}$	$\tfrac{1}{2}$	$-\tfrac{1}{2}$	$-\tfrac{1}{2}$	$-\tfrac{1}{2}$	$-\tfrac{1}{2}$
$\mathbf{x}_P^{(3)}$	$-\tfrac{1}{2}$	$\tfrac{1}{2}$	$\tfrac{1}{2}$	$-\tfrac{1}{2}$	$\tfrac{1}{2}$	$-\tfrac{1}{2}$	$-\tfrac{1}{2}$	$\tfrac{1}{2}$

Thus we have

$$\Theta = \frac{1}{4} \begin{bmatrix} 2 & -1 & 0 & -1 \\ -1 & 2 & -1 & 0 \\ 0 & -1 & 2 & -1 \\ -1 & 0 & -1 & 2 \end{bmatrix},$$

$$\Phi = \frac{1}{4} \begin{bmatrix} 5 & -1 & -3 & -1 \\ -1 & 5 & -1 & -3 \\ -3 & -1 & 5 & -1 \\ -1 & -3 & -1 & 5 \end{bmatrix},$$

and direct calculation shows that $\Phi\Theta = \Theta\Phi = I - \frac{1}{4}J$, so that Φ is a generalized inverse for Θ. Thus the variance–covariance matrix for the intrablock estimates of treatment effects is $\Phi\sigma^2/r$. Furthermore, $\Phi\mathbf{x}^{(1)} = \mathbf{x}^{(1)}$, and so the variance of the intrablock estimate of $\mathbf{x}^{(1)} \cdot \boldsymbol{\tau}$ is $\mathbf{x}^{(1)} \cdot \mathbf{x}^{(1)}\sigma^2/r$, just as it would be in a complete block design with the same variance, so the efficiency factor for this contrast is 1. On the other hand, $\Phi\mathbf{x}^{(2)} = 2\mathbf{x}^{(2)}$ and $\Phi\mathbf{x}^{(3)} = 2\mathbf{x}^{(3)}$ so the variance of the intrablock estimates of the within-groups contrasts $\mathbf{x}^{(2)} \cdot \boldsymbol{\tau}$ and $\mathbf{x}^{(3)} \cdot \boldsymbol{\tau}$ is $2\mathbf{x}^{(2)} \cdot \mathbf{x}^{(2)}\sigma^2/r$, which is twice the variance that would be achieved from a complete block design (assuming that the larger blocks could be found with the *same* intrablock variance), and so the efficiency factor for within-groups contrasts is $\frac{1}{2}$. In fact, $\mathbf{x}^{(2)}$ is the contrast vector for $\tau_A - \tau_C$, and the efficiency factor $\frac{1}{2}$ is equal to $\{\frac{5}{4} - (-\frac{3}{4})\}^{-1}$ $= (\phi_0 - \phi_1)^{-1}$, in agreement with theory, since A and C are first associates. In the same way, if $\mathbf{z} = (1, -1, 0, 0)$, then the efficiency factor for $\tau_A - \tau_B$ is

$$\mathbf{z} \cdot \mathbf{z}/\mathbf{z}\Phi\mathbf{z}' = 2/3$$
$$= 1/\{\tfrac{5}{4} - (-\tfrac{1}{4})\}$$
$$= 1/(\phi_0 - \phi_2),$$

and A and B are second associates.

Analysis

The algebraic structure of a PBIBD gives an alternative method of performing the calculations required to estimate means and variances that does not explicitly use matrix inverses. Choose an orthogonal basis for

each eigenspace W_i. If $\mathbf{x}$ is a basis vector for W_i, calculate the following *rt*-vectors:

$\mathbf{x}_D$, which has entry x_u on each plot
        that receives treatment u;

$\mathbf{x}_B$, in which each entry of $\mathbf{x}_D$ is replaced
        by the mean of the entries
        in the same block;

$$\mathbf{x}_P = \mathbf{x}_D - \mathbf{x}_B.$$

Let $\mathbf{y}$ be the *rt*-vector of yields. The intrablock estimate of $\mathbf{x} \cdot \boldsymbol{\tau}$ is $(\mathbf{x}_P \cdot \mathbf{y})/(rE_i)$. Such estimates are found for each of the chosen basis vectors $\mathbf{x}$. Intrablock estimates of any other contrasts are obtained as linear combinations of these. The contribution of $\mathbf{x}$ to the intrablock sum of squares is

$$(\mathbf{x}_P \cdot \mathbf{y})^2/(\mathbf{x}_P \cdot \mathbf{x}_P) = (\mathbf{x}_P \cdot \mathbf{y})^2/(rE_i(\mathbf{x} \cdot \mathbf{x})).$$

The residual intrablock sum of squares is obtained by subtraction. Interblock* estimates and sums of squares are obtained similarly, using $\mathbf{x}_B$ in place of $\mathbf{x}_P$ and $1 - E_i$ in place of E_i. Sometimes it is desirable to combine the inter- and intrablock information. (See Brown and Cohen [7], Nelder [35], Sprott [55], and Yates [57, 58].)

We demonstrate the calculations on some fictitious data on eight plots using the design given at the top of Table 8. Although this design is so small that it is unlikely to be of much practical use and the calculations are not difficult to perform by other means, it serves to demonstrate the method, which is no more difficult for larger designs. We may use the basis vectors $\mathbf{x}^{(1)}, \mathbf{x}^{(2)}, \mathbf{x}^{(3)}$ given in the preceding section, where we calculated the canonical efficiency factors to be $E_0 = 0$, $E_1 = 1$, $E_2 = \frac{1}{2}$. The calculations of effects and sums of squares are shown in Table 9, and the analysis of variance* in Table 10.

Pros and Cons of Partially Balanced Designs

The advantage of partially balanced designs is the great simplification that may be achieved in the calculations needed to compare potential designs for an experiment and to analyze the results of the experiment. Incomplete block designs are usually as-

Table 9

	$\mathbf{x}^{(1)}$	$\mathbf{x}^{(2)}$	$\mathbf{x}^{(3)}$
$\mathbf{x}_P \cdot \mathbf{y}$	-36	11	-3
rE_i	2	1	1
Intrablock estimate	-18	11	-3
Intrablock sum of squares	162	60.5	4.5
$\mathbf{x}_B \cdot \mathbf{y}$	0	12	-4
$r(1 - E_i)$	0	1	1
Interblock estimate	—	12	-4
Interblock sum of squares	0	72	8

Table 10

Stratum	Source	d.f.	SS
Blocks	W_2	2	80
	Residual	1	0.5
	Total	3	80.5
Plots	W_1	1	162
	W_2	2	65
	Residual	1	32
	Total	4	259

sessed by the values of their canonical efficiency factors, all of which should be as large as possible. Some contrasts are more important to the experimenter than others, so it is relatively more important for the efficiency factors for these contrasts to be large.

Many catalogs of designs give a single summary value for the canonical efficiency factors, such as their harmonic mean*, geometric mean*, minimum, or maximum. Like the canonical efficiency factors themselves, these summaries can in general be calculated only by diagonalizing a $t \times t$ matrix for each design. However, for any given association scheme, the eigenspaces W_i, eigenprojections $\mathbf{S}_i$, and characters e_{ij} may be calculated by diagonalizing a few $s \times s$ matrices, and the canonical efficiency factors for *every* PBIBD for this association scheme may be calculated directly from the characters and the concurrences, with no further matrix calculations. Thus it is extremely easy to compare the canonical efficiency factors of two of

these PBIBDs, and the comparison is particularly relevant because the efficiency factors apply to the *same* eigenspaces.

The story is the same for the analysis of PBIBD experiments. In general, each incomplete block design requires the inversion of a $t \times t$ matrix for the analysis of experiments with that design. We have shown that PBIBD experiments may be analyzed without using this inverse. Even if the standard method of analysis is preferred, the inverse can be calculated directly from the canonical efficiency factors and the matrix $[f_{ij}]$, which is obtained by inverting the $(s + 1) \times (s + 1)$ matrix $[e_{ij}]$. Moreover, this *single* matrix inversion is all that is needed for *all* PBIBDs with a given association scheme.

In the days before electronic computers, these advantages of PBIBDs were overwhelming. Even today, routine use of PBIBDs permits considerable saving of time and effort and more meaningful comparison of different designs than is generally possible. When the association scheme makes sense in terms of the treatment structure, as with factorial designs, the case for using PBIBDs is still very strong.

If the treatment structure does not, or cannot, bear any sensible relationship to an association scheme, the efficiency factors of most interest to the experimenter will not be the canoncial efficiency factors of any PBIBD, so the ease of the preceding calculations does not seem very helpful (see Pearce [42]). Moreover, for an arbitrary treatment structure, the best designs (according to some efficiency criterion) may not be PBIBDs. Thus other designs should certainly be examined if the necessary computing facilities are available. However, PBIBDs still have a role, and if one is found with a very small range of nonzero canonical efficiency factors, then the efficiency factors that interest the experimenter will also be in that range, and the design is probably close

to optimal according to many optimality criteria.

GENERALIZATIONS OF PBIBDs AND RELATED IDEAS

Generalizations of Association Schemes

Various authors have tried to extend the definition of PBIBD by weakening some of the conditions (**a**)–(**e**) for an association scheme without losing too much of the algebra. Shah [52] proposed weakening condition (**e**) to

(**e**)′ For all i and j, there are numbers q_{ij}^k such that $\mathbf{A}_i\mathbf{A}_j + \mathbf{A}_j\mathbf{A}_i = \sum_k q_{ij}^k \mathbf{A}_k$.

This condition is sufficient to ensure that

Table 11

		(*a*)	**Association Scheme (Shah Type)**				
		A	*B*	*C*	*D*	*E*	*F*
A		0	1	1	2	3	4
B		1	0	1	4	2	3
C		1	1	0	3	4	2
D		2	4	3	0	1	1
E		3	2	4	1	0	1
F		4	3	2	1	1	0

	(*b*)	**Shah's Design**						
A	*A*	*B*	*B*	*C*	*C*	*A*	*B*	*C*
D	*D*	*E*	*E*	*F*	*F*	*E*	*F*	*D*

(*c*)	**Other Design**				
A	*B*	*C*	*D*	*E*	*F*
B	*C*	*A*	*F*	*D*	*E*
D	*E*	*F*	*A*	*B*	*C*

		(*d*)	**Association Scheme (Nair Type)**				
		A	*B*	*C*	*D*	*E*	*F*
A		0	1	5	2	3	4
B		5	0	1	4	2	3
C		1	5	0	3	4	2
D		2	4	3	0	5	1
E		3	2	4	1	0	5
F		4	3	2	5	1	0

every invertible matrix in $\mathscr{A}$ still has its inverse in $\mathscr{A}$, and that every diagonalizable matrix in $\mathscr{A}$ has a generalized inverse in $\mathscr{A}$, and so any one inverse can be calculated from s linear equations. Thus Shah's designs retain the property that the variance of the estimate of the elementary contrast $\tau_u - \tau_v$ depends only on what type of associates u and v are. However, the property of simultaneous eigenspaces W_i for *every* PBIBD with a given association scheme is lost, so the calculation of inverses must be done afresh for each design. Moreover, there is no easy method of calculating efficiency factors.

As an example, Shah gave the "generalized association scheme" in Table 11a. (This is obtained in the same way as a cyclic association scheme, but using the non-Abelian group of order 6.) The same convention is used as in Table 2. Parts b and c of Table 11 show incomplete block designs that are "partially balanced" with respect to this generalized association scheme. (Design b was given by Shah [52] and also discussed by Preece [43]. Design c is formed from its initial block by a construction akin to the cyclic construction.) The concurrence matrices of both designs have W_0 and W_1 as eigenspaces, where W_1 is spanned by $(1, 1, 1, -1, -1, -1)$. However, the remaining eigenspaces for design b are W_2, spanned by $(2, -1, -1, -\sqrt{3}, 0, \sqrt{3})$ and $(0, -\sqrt{3}, \sqrt{3}, -1, 2, 1)$, and W_3, spanned by $(2, -1, -1, \sqrt{3}, 0, -\sqrt{3})$ and $(0, \sqrt{3}, -\sqrt{3}, -1, 2, -1)$. The remaining eigenspaces for design c are V_2, spanned by $(2, -1, -1, 1, -2, 1)$ and $(0, 1, -1, 1, 0, -1)$, and V_3, spanned by $(2, -1, -1, -1, 2, -1)$ and $(0, 1, -1, -1, 0, 1)$.

Nair [32] suggested retaining condition (e) while weakening condition (b) by allowing nonsymmetric association relations, with the proviso

(b)′   If A_i is an association matrix, then so is A_i'.

Then it remains true that $\mathscr{A}$ is an *algebra* in the sense of containing the sum and product of any two of its elements, and the scalar multiples of any one of its elements; more-over, $\mathscr{A}$ still contains generalized inverses of its symmetric elements. Thus the Nair schemes have the good properties of the Shah schemes.

However, for a Nair scheme, the algebra $\mathscr{A}$ may or may not be *commutative* in the sense that $\mathbf{MN} = \mathbf{NM}$ for all $\mathbf{M}$ and $\mathbf{N}$ in $\mathscr{A}$. All the examples given in Nair's paper did have commutative algebras, and Nair wrongly concluded that this is always the case. A counterexample is provided by Table 11d, which is a Nair association scheme with respect to which the designs in Tables 11b and 11c are partially balanced. If $\mathscr{A}$ is commutative, then the Nair scheme has all the good properties of genuine association schemes. However, since concurrence and covariance are symmetric functions of their two arguments, one may as well replace A_i by $A_i + A_i'$ if A_i is nonsymmetric, and obtain a genuine association scheme, because the commutativity of $\mathscr{A}$ ensures that condition (e) will still hold. If $\mathscr{A}$ is *not* commutative then the Nair scheme has the same disadvantages as the general Shah schemes. In fact, most Shah schemes arise by taking a Nair scheme and fusing nonsymmetric associate classes, as in the preceding example.

Another possibility is to allow self-associates to form more than one class: thus condition (c) becomes

(c)′   The nonzero elements of A_i are either all on, or all off, the diagonal.

This was first suggested implicitly in the statistical literature when Pearce [38] introduced *supplemented balance* as the judicious addition of a control treatment to a BIBD. There are two classes of self-associates: the control, and all other treatments. There are also two other associate classes: control with new treatment, and new treatment with another new treatment. More general designs satisfying (c)′ rather than (c) were given by Rao [48] and Nigam [36]. For these designs $\mathscr{A}$ cannot be commutative, because the association matrices for self-associates do not commute with $\mathbf{J}$, and so the useful property of common eigenspaces is lost. If the association relations must be symmetric, then condition (e)' is satisfied while (e) is not, whereas if the pairs (control, new) and (new, control) are allocated to different classes then (b)' and (e) are satisfied. In either case, $\mathscr{A}$ contains a generalized inverse of each of its symmetric elements, and so the pattern of variances of elementary contrasts matches the pattern of concurrences.

Bearing in mind the importance of condition (e) for properties of $\mathscr{A}$, Higman [18] defined a generalization of association schemes called *coherent configurations*. These satisfy conditions (a), (b)', (c)', (d), and (e). If they also satisfy (c), they are called *homogeneous*. Homogeneous coherent configurations arise naturally in the context of *transitive permutation groups*, and so they are intimately related to recent ideas of Sinha [53], who defined an incomplete block design to be *simple* if there is a group G of permutations of the treatments with the properties that

(i) if the rows and columns of the concurrence matrix Λ are both permuted by the same element of G, then the matrix is unchanged as a whole;

(ii) given any two treatments, there is a permutation in G that carries one into the other.

Such designs are indeed simple in the sense that once their structure has been analyzed with respect to a single treatment, everything is known about their structure. This fact is often used implicitly by other authors, and the idea of its simplicity seems worth pursuing. As a whole, coherent configurations are of considerable mathematical interest and have many potential applications to the design of statistical experiments, but they cannot be discussed in any further detail here.

Approximations to Balance*
in Terms of Concurrence

In a review paper on types of "balance" in experimental design, Pearce [39] defined an incomplete block design to be partially balanced if there are real numbers $\theta_0, \theta_1, \ldots, \theta_s, n_1, \ldots, n_s$ such that every diagonal element of the matrix Θ is equal to θ_0 and every row of Θ contains exactly n_i elements equal to θ_i. Thus he effectively removed condition (e) (and even (e)') from the definition of partial balance. Although Pearce [42] has argued strongly that condition (e) is entirely abstract and artificial and has no place in practical experimentation, without condition (e) all the algebraic theory collapses, and general designs that are partially balanced in Pearce's sense have none of the good properties we have described.

It is unfortunate that Pearce retained the term *partial balance*, because his paper was influential and caused other authors to use the term in his way, which may account for a certain amount of confusion about the meaning of the term. Sinha [53] called this property the *extended Latin Square property* (ELSP) and attempted to relate ELSP, partial balance, and simplicity. Unfortunately, some of the specific results in his paper are incorrect. Jarrett [24] used the term *s-concurrence design* for Pearce's PBIBD and clarified the situation somewhat. However, he insisted that $\theta_i \neq \theta_j$ unless $i = j$. Although *s*-concurrence designs have an obvious attraction, in general, Θ^{-1} does not have the same pattern as Θ, and little can be predicted about the efficiency factors.

For a 2-concurrence design, the matrices A_1 and A_2 define a regular graph and its complement. If balance is impossible, the nearest approximation seems, intuitively, to be a 2-concurrence design whose concur-

Table 12

E	A	B	C	D	A	B	C	D
F	B	C	D	A	B	C	D	A
G	E	F	G	H	C	D	A	B
H	F	G	H	E	G	H	E	F
I	H	E	F	G	I	I	I	I

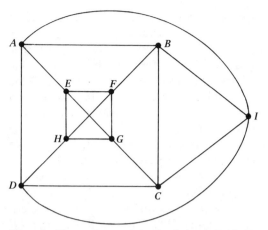

Figure 2 The regular graph for the design in Table 12.

Approximations to Balance in Terms of Variance or Efficiency

In general, if one allows the incomplete block design Δ to have unequal replication, unequal block sizes, or more than one occurrence of a treatment per block, then the information matrix $r\Theta$ must be replaced by the matrix $\mathbf{L} = \mathbf{R}^2 - \mathbf{NK}^{-1}\mathbf{N}'$, where $\mathbf{R}$ is the $t \times t$ diagonal matrix whose (u,u) entry is $\sqrt{r_u}$ if treatment u is replicated r_u times; $\mathbf{K}$ is the $b \times b$ diagonal matrix whose (c,c) entry is the size of block c; and $\mathbf{N}$ is the $t \times b$ matrix whose (u,c) entry is the number of times that treatment u occurs in block c.

Various authors have extended the definition of partial balance to cover designs that have one or more of the following conditions relaxed: equireplicate, equal block sizes, binary. Unequal block sizes make very little difference to the theory discussed so far; while for nonbinary designs equal replication is no longer equivalent to $\mathbf{L}$ having equal diagonal elements. Mehta et al. [30] defined Δ to be partially balanced if it has equal replication, equal block sizes, and $\mathbf{L}$ is in the algebra $\mathcal{A}$ of an association scheme; while Kageyama [29] relaxed the complementary conditions and defined Δ to be partially balanced if it is binary and $\mathbf{L}$ is in the algebra $\mathcal{A}$ of an association scheme. Note that in both cases the diagonal elements of $\mathbf{L}$ are all the same. The supplemented balance designs described earlier are similar in spirit, except that $\mathbf{L}$ belongs to the algebra $\mathcal{A}$ of an inhomogeneous coherent configuration, whose classes of self-associates consist of all treatments with the same replication.

For all these extended definitions, $\mathbf{L}$ has a generalized inverse $\mathbf{M}$ in $\mathcal{A}$. Since the variance of the estimate $\mathbf{x} \cdot \tau$ is $\mathbf{xMx}'$, the variance of the elementary contrast $\tau_u - \tau_v$ depends only on the associate class containing the pair (u,v). Thus it would be correct to call these designs *partially variance-balanced*. Nigam's [36] designs have only two variances: he called his designs *nearly balanced* if the two variances are within 10% of each other, but this is not related to NBIBDs.

However, if treatments are not equally

rences differ by 1. Such designs were called *regular graph designs** (RGDs) by John and Mitchell [25]. An example is given in Table 12, with one of its corresponding regular graphs in Fig. 2. (Thus a regular graph design is partially balanced if and only if its graph is strongly regular.) Intuition is not quite correct, however, and some RGDs are poor in terms of efficiency factors. Mitchell and John [31] gave a catalog of RGDs that are *E*-optimal in the sense of having their smallest efficiency factor as large as possible, and they conjectured [25] that, for each fixed value of *t*, *b*, and *k*, if there exists any RGD then there exist RGDs that are *A*-optimal, *D*-optimal, and *E*-optimal respectively. Cheng [10–12] and Jacroux [21] have shown that this is true in several special cases; the optimal designs are often simultaneously RGDs and PBIBDs.

Cheng and Wu [13] allowed for the case where the number of plots is not divisible by *t* and defined a *nearly balanced incomplete block design* (NBIBD) to have treatment replications differing by at most 1, and, for all treatments *u*, *v*, and *w*, concurrences λ_{uv} and λ_{uw} differing by at most 1. This concept is so far (algebraically) from partial balance that there seems little hope of any good general theory about NBIBDs. Nevertheless, Cheng and Wu were able to show that the best NBIBDs do have high efficiency factors (*see also* NEARLY BALANCED DESIGNS).

replicated then treatment contrasts would not have equal variances even in an unblocked design, and it may be unreasonable to expect them to have equal variances in the incomplete block design. The efficiency factor takes account of the unequal replication by comparing the variance in the blocked and unblocked designs, and so it is a more useful measure of information loss than variance. James and Wilkinson [22] showed that the efficiency factors may be obtained from the matrix $\mathbf{R}^{-1}\mathbf{L}\mathbf{R}^{-1}$. This matrix is symmetric, so it has a complete eigenvector basis, and its eigenvalues are the canonical efficiency factors: if there are m nonzero eigenvalues, then James and Wilkinson define Δ to have *mth order balance*. As Houtman and Speed [20] pointed out, this is just a special case of the *general balance** introduced by Nelder [34].

Following a similar, but independent, line of thought, Caliński [8] showed that the canonical efficiency factors are the eigenvalues of $\mathbf{R}^{-2}\mathbf{L}$. Since $\mathbf{R}^{-2}\mathbf{L}$ and $\mathbf{R}^{-1}\mathbf{L}\mathbf{R}^{-1}$ are similar ($\mathbf{R}^{-2}\mathbf{L} = \mathbf{R}^{-1}(\mathbf{R}^{-1}\mathbf{L}\mathbf{R}^{-1})\mathbf{R}$) they have the same eigenvalues, and the eigenspaces of one can be obtained easily from those of the other. In particular, $\mathbf{R}^{-2}\mathbf{L}$ also has a complete eigenvector basis. However, Caliński did not realize this, and Puri and Nigam [44] defined Δ to be *partially efficiency balanced* if $\mathbf{R}^{-2}\mathbf{L}$ has a complete eigenvector basis. Since all incomplete block designs are partially efficiency balanced, this definition seems unnecessary.

A stream of papers followed the Caliński/Puri/Nigam school (e.g., Puri and Nigam [46]) and another stream followed the Nelder/James/Wilkinson school (e.g., Jarrett [23]); there has been little cross-reference between the two. However, Pal [37] has recently pointed out that all designs are partially efficiency balanced, and it is to be hoped that these separate strands will be unified and simplified.

More Complicated Block Structures

The definition of partial balance can be extended to more complicated block struc-

tures obtained from several block factors by crossing and nesting* (e.g., the *simple orthogonal block structures* of Nelder [33]). The treatments must form a PBIBD with each system of blocks separately, ignoring all the other block systems, and the association scheme must be the same throughout. The results described in this article extend to these structures with no difficulty. See Houtman and Speed [20] for details.

It is also possible to use partially balanced designs when time is a block factor, or there is some other definite order on the experimental units. Blaisdell and Raghavarao [2] have used partial balance in changeover* designs.

Literature

There is an enormous statistical and mathematical literature on association schemes and partial balance, and it is impossible to cite here all who have made important contributions. The statistical reader with some mathematical background will find comprehensive accounts of partial balance in the textbooks by John [27, 28] and Raghavarao [47]; the references in Bose [3], Clatworthy [14], Pearce [42], and Preece [43] are also useful sources of further reading.

A more detailed account of the practical role of the eigenspaces in some particular cases is in Corsten [15]. A full account of the role of the eigenspaces is given by Houtman and Speed [20], James [22], and Nelder [34], while papers such as refs. 40 and 41 concentrate on just the eigenvalues (i.e., the canonical efficiency factors). References 44 and 46 explain the eigenspace analysis more specifically in the context presented in this article.

For a mathematical treatment of association schemes and strongly regular graphs, see refs. 1, 9, 16, 17, 50, and 51.

References

[1] Biggs, N. L. and White, A. T. (1979). "Permutation Groups and Combinatorial Structures." *Lond. Math. Soc. Lect. Notes Ser.*, **33**, Cambridge University Press, Cambridge, England. (Strictly

for mathematicians; includes a section on strongly regular graphs.)

[2] Blaisdell, E. A. and Raghavarao, D. (1980). *J. R. Statist. Soc. B*, **42**, 334–338. (PBIBDs as change-over designs.)

[3] Bose, R. C. (1963). *Sankhyā*, **25**, 109–136. (A review article on partial balance.)

[4] Bose, R. C. (1963). *Pacific J. Math.*, **13**, 389–419. (Introduction of strongly regular graphs.)

[5] Bose, R. C. and Nair, K. R. (1939). *Sankhyā*, **4**, 337–372. (Introduction of partial balance.)

[6] Bose, R. C. and Mesner, D. M. (1959). *Ann. Math. Statist.*, **30**, 21–38. (The major step in the algebra of partial balance.)

[7] Brown, L. D. and Cohen, A. (1974). *Ann. Statist.*, **2**, 963–976.

[8] Caliński, T. (1971). *Biometrics*, **27**, 275–292.

[9] Cameron, P. J. and van Lint, J. H. (1980). "Graphs, Codes and Designs." *Lond. Math. Soc. Lect. Notes Ser.*, **43**, Cambridge University Press, Cambridge. (Strictly for mathematicians; includes very clear sections on strongly regular graphs and on association schemes.)

[10] Cheng, C.-S. (1978). *Commun. Statist.*, **A7**, 1327–1339.

[11] Cheng, C.-S. (1980). *J. R. Statist. Soc. B*, **42**, 199–204.

[12] Cheng, C.-S. (1981). *Ann. Inst. Statist. Math. Tokyo*, **33**, 155–164.

[13] Cheng, C.-S. and Wu, C.-F. (1981). *Biometrika*, **68**, 493–500.

[14] Clatworthy, W. H. (1973). "Tables of Two-Associate Class Partially Balanced Designs." *Natl. Bur. Stand. (U.S.) Appl. Math. Ser. 63* (Washington, DC). (A large catalog; very useful.)

[15] Corsten, L. C. A. (1976). In *Essays in Probability and Statistics*, S. Ikeda, ed. Shinko Tsusho, Tokyo, 125–154.

[16] Delsarte, P. (1973). "An Algebraic Approach to the Association Schemes of Coding Theory." Thesis, Université Catholique de Louvain (appeared as Philips Research Reports Supplement, 1973, No. 10). (For mathematicians, a splendid account of association schemes and their relation to other mathematical objects.)

[17] Dembowski, P. (1968). *Finite Geometries*. Springer-Verlag, Berlin. (Strictly for mathematicians; includes an appendix on association schemes and partial designs.)

[18] Higman, D. G. (1975). *Geom. Dedicata*, **4**, 1–32.

[19] Hinkelmann, K. (1964). *Ann. Math. Statist.*, **35**, 681–695.

[20] Houtman, A. M. and Speed, T. P. (1983). *Ann. Statist.*, **11**, 1069–1085. (Contains the natural extension of PBIBDs to more complicated block structures; also a useful bibliography on the use of efficiency factors and eigenspaces in analysis.)

[21] Jacroux, M. (1980). *J. R. Statist. Soc. B*, **42**, 205–209.

[22] James, A. T. and Wilkinson, G. N. (1971). *Biometrika*, **58**, 279–294. (A lucid account of the use of efficiency factors and eigenspaces in analysis—for those who are familiar with higher-dimensional geometry and algebra.)

[23] Jarrett, R. G. (1977). *Biometrika*, **64**, 67–72.

[24] Jarrett, R. G. (1983). *J. R. Statist. Soc. B*, **45**, 1–10. (A clear separation of designs with equal concurrence patterns from PBIBDs.)

[25] John, J. A. and Mitchell, T. (1977). *J. R. Statist. Soc. B*, **39**, 39–43. (Introduction of regular graph designs.)

[26] John, J. A., Wolock, F. W., and David, H. A. (1972). "Cyclic Designs." *Natl. Bur. Stand. (U.S.) Appl. Math. Ser. 62* (Washington DC). (A catalog of designs with high efficiency factors.)

[27] John, P. W. M. (1971). *Statistical Design and Analysis of Experiments*. Macmillan, New York. (A textbook for statisticians, with a comprehensive section on PBIBDs.)

[28] John, P. W. M. (1980). *Incomplete Block Designs*. Marcel Dekker, New York.

[29] Kageyama, S. (1974). *Hiroshima Math. J.*, **4**, 527–618. (Includes a useful review of different types of association schemes.)

[30] Mehta, S. K., Agarwal, S. K., and Nigam, A. K. (1975). *Sankhyā*, **37**, 211–219.

[31] Mitchell, T. and John, J. A. (1976). "Optimal Incomplete Block Designs." *Oak Ridge Nat. Lab. Rep. No. ORNL/CSD-8*. Oak Ridge, TN. (A catalog of regular graph designs.)

[32] Nair, C. R. (1964). *J. Amer. Statist. Ass.*, **59**, 817–833.

[33] Nelder, J. A. (1965). *Proc. R. Soc. Lond. A*, **283**, 147–162.

[34] Nelder, J. A. (1965). *Proc. R. Soc. Lond. A*, **283**, 163–178.

[35] Nelder, J. A. (1968). *J. R. Statist. Soc. B*, **30**, 303–311.

[36] Nigam, A. K. (1976). *Sankhyā*, **38**, 195–198.

[37] Pal, S. (1980). *Calcutta Statist. Ass. Bull.*, **29**, 185–190.

[38] Pearce, S. C. (1960). *Biometrika*, **47**, 263–271. (Introduction of supplemented balance.)

[39] Pearce, S. C. (1963). *J. R. Statist. Soc. A*, **126**, 353–377.

[40] Pearce, S. C. (1968). *Biometrika*, **55**, 251–253.

[41] Pearce, S. C. (1970). *Biometrika*, **57**, 339–346.

[42] Pearce, S. C. (1983). *The Agricultural Field Experiment*. Wiley, Chichester, UK. (This text, full of good advice on practical experimentation, dis-

cusses the place of various kinds of theoretical balance in the practical context.)

[43] Preece, D. A. (1982). *Utilitas Math.*, **21C**, 85–186. (Survey of different meanings of "balance" in experimental design; extensive bibliography.)

[44] Puri, P. D. and Nigam, A. K. (1977). *Commun. Statist.*, **A6**, 753–771.

[45] Puri, P. D. and Nigam, A. K. (1977). *Commun. Statist.*, **A6**, 1171–1179. (Relates efficiency balance, pairwise balance, and variance balance.)

[46] Puri, P. D. and Nigam, A. K. (1982). *Commun. Statist. Theory Meth.*, **11**, 2817–2830.

[47] Raghavarao, D. (1971). *Constructions and Combinatorial Problems in Design of Experiments*. Wiley, New York. (This remains the best reference work on construction of incomplete block designs; it requires some mathematical background.)

[48] Rao, M. B. (1966). *J. Indian Statist. Ass.*, **4**, 1–9.

[49] Roy, P. M. (1953). *Science and Culture*, **19**, 210–211.

[50] Seidel, J. J. (1979). In "Surveys in Combinatorics," B. Bollobas, ed. *Lond. Math Soc. Lect. Notes Ser.*, **38**. Cambridge University Press, Cambridge, England, pp. 157–180. (Survey of strongly regular graphs.)

[51] Seidel, J. J. (1983). In *Proceedings of the Symposium on Graph Theory*, Prague, 1982 (to appear).

[52] Shah, B. V. (1959). *Ann. Math. Statist.*, **30**, 1041–1050.

[53] Sinha, B. K. (1982). *J. Statist. Plan. Infer.*, **6**, 165–172.

[54] Speed, T. P. and Bailey, R. A. (1982). In *Algebraic Structures and Applications*, P. Schultz, C. E. Praeger, and R. P. Sullivan, eds. Marcel Dekker, New York, pp. 55–74. (Gives the eigenprojection matrices for a class of association schemes.)

[55] Sprott, D. A. (1956). *Ann. Math. Statist.*, **27**, 633–641.

[56] Yates, F. (1935). *Suppl. J. R. Statist. Soc.*, **2**, 181–247.

[57] Yates, F. (1939). *Ann. Eugen. (Lond.)* **9**, 136–156.

[58] Yates, F. (1940). *Ann. Eugen. (Lond.)* **10**, 317–325.

ROSEMARY A. BAILEY

PARTIALLY SUFFICIENT *See* CONDITIONAL INFERENCE

PARTIALLY TRUNCATED DISTRIBUTIONS

This term applies to distributions formed by discarding a proportion (chosen at random) of all values of a random variable (X) that fall into certain interval(s). For simplicity, we will suppose that the region is $(X > x)$ and denote the proportion discarded by e. If the cumulative distribution function (CDF) of the original distribution is $F_X(x)$ and the probability density function (PDF) is $f_X(x)$, then the PDF of the partially truncated distribution is

$$\begin{cases} \theta f_X(x) & \text{for } x \leqslant a, \\ \theta(1-e)f_X(x) & \text{for } x > a, \end{cases}$$

where $\theta = \{1 - e(1 - F_X(a))\}^{-1}$. The CDF is

$$\begin{cases} \theta F_X(x) & \text{for } x \leqslant a, \\ \theta\{F_X(x) - e(F_X(x) - F_X(a))\} \\ \qquad\qquad \text{for } x > a. \end{cases}$$

If $e = 1$, we have a (fully) truncated distribution; if $e = 0$, there is no truncation.

Partially truncated distributions of this type can arise in quality control* when it is desired to exclude high values $(> a)$ of some characteristic, but the control applied is not fully effective. In these circumstances, a is termed the *action level* and e is called the *enforcement level*. An example, in which X is the proportion of mercury in seafood is given in Lee and Krutchkoff [1]. (In this paper there are detailed calculations for the case when the original distribution is lognormal*.)

There are, of course, natural generalizations introducing different *enforcement levels* for different sets of values of X. These are all included in the general class of *weighted distributions**, but the term *partially truncated* is commonly used only for the simplest case, as described.

Multivariate generalization is straightforward.

Reference

[1] Lee, L. and Krutchkoff, R. G. (1980). *Biometrics*, **36**, 531–536.

(TRUNCATION
WEIGHTED DISTRIBUTIONS)

PARTIAL ORDER SCALOGRAM ANALYSIS

Partial Order Scalogram Analysis (POSA) is a graphic technique designed to analyze a scalogram (a collection of empirical profiles as in the Example below) as for a parsimonious description of the context behind the variables under study. POSA maps out the partial order configuration that best accommodates the data and yet has a relatively simple structure in order to interpret directions in profile spaces. Once directions in the partial order configuration space have been determined, individuals may be scored with respect to each of them. Given the configuration, original profiles may be reproduced from these new scores, and inasmuch as the partial order dimensionality is substantially smaller than the number of observed items, a considerable parsimony in data presentation is attained. Furthermore, the contents

attributable to those directions are likely to point out more fundamental notions than those represented by the specific items. Often, the empirical profiles can be characterized by only two scores.

POSA was introduced by Louis Guttman and has been employed mainly in social studies as an extension of the one-dimensional Guttman scale. Given n variables (e.g., questions in a questionnaire), whose response range has *common* meaning, the empirical responses to them by a certain population may be ordered so as to form a one-dimensional Guttman scale with possible implications for the content investigated. (For a basic reference on scales, see, e.g., Guttman [2] and Stouffer et al. [11].) Thus data material that a priori requires n-dimensional portrayal to describe its order relations may be reduced to a one-dimensional scale that reproduces the empirical data. In other words, in a Guttman scale, a single new score corresponds to each of the original n-score profiles and yet the order relations among them are preserved.

POSA is an extension of the idea, where the reduction in dimensionality is from the original n (n being the number of variables) to any smaller number m (not necessarily 1 as in Guttman scale), while preserving all (and stating only) order relations in the empirical data profiles. Thus m scores are sufficient to reproduce the original responses. The $m = 2$ scores used later in examples are called *joint* and *lateral*, respectively. For example, on social status studies, the joint score implies the *level* of status and the lateral score implies the *kind* of status. Let us now present the mathematical formulation of partial order scalograms.

Suppose that each of N subjects, p_1, $p_2, \ldots, p_N$, in a population P receives a score in each test item $v_1, v_2, \ldots, v_N$, where the range of item v_i is $A_i = 0, 1, 2, \ldots, \alpha_i$ for all $i = 1, 2, \ldots, n$. Let us regard A_i as a set ordered by the relation $<$ and consider the Cartesian set $A = A_1 A_2 \ldots A_n$. Each component set A_i is called a *facet*, and each element of A, $a \in A$, where $a = a_1 a_2 \ldots a_n$, $a_i \in A_i$, is a *profile* in the n items. We shall say that $a' < a$ if $a_i' \leqslant a_i$ for all $i = 1, 2,$ $\ldots, n$ and if there exists i', $1 \leqslant i' \leqslant n$, with $a_{i'}' < a_{i'}$ where, of course, $a = a_1 a_2 \ldots a_n$ and $a' = a_1' a_2' \ldots a_n'$. Here a and a' are said to be *comparable* (denote: $a' \gtrless a$) if and only if $a' < a$ or $a < a'$ or $a' = a$ and otherwise are *incomparable* ($a' \not\gtrless a$). The score $S(a)$ of a profile a is defined by

$$S(a) = \sum_{i=1}^{n} a_i.$$

A *scalogram* is a mapping $P \rightarrow A$ from a population P to the Cartesian set A. The subset of profiles A' in A ($A' \subset A$) to which elements of P are mapped is called the *scalogram range* or, briefly, the *scalogram*. If every two profiles in the range A' of a scalogram are comparable, the scalogram is called a *scale*. In general, however, A is a partially ordered set with respect to the relation $\leqslant$; hence A', which is a subset of A, is also a partially ordered set with respect to the same relation (see, e.g., Birkhoff [1]).

In Shye [6], we find the following definition for scalogram dimensionality: The *partial order dimensionality* of a scalogram A' is the smallest integer m ($m \leqslant n$) for which there exists an m-faceted Cartesian set $X = X_1 X_2 \ldots X_m$, $X_i' = 1, 2, \ldots, \xi_i$, $i = 1,$ $2, \ldots, m$, and there exists a one-to-one mapping, $Q: X' \rightarrow A'$ from a subset X' of X onto A' so that if $Q(x') = a'$ and $Q(x) = a$, then $a' < a$ if and only if $x' < x$; that is, both relations, of comparability and incomparability, are preserved under mapping Q.

Some results follow from the foregoing definition of partial order dimensionality:

1. The partial order dimensionality of a perfect scale is one.
2. The appending of a new profile (row in the data matrix) to a given scalogram results in the new scalogram having a dimensionality no smaller than that of the original scalogram. In particular, with a complete scale, namely, a scale with $\sum_{i=1}^{n}(\alpha_i + 1)$ different profiles, the appending of any additional profile would increase the dimensionality.

3. Adding a new item to a scalogram (a column to the data matrix) may decrease its dimensionality.

Example. For illustration, let us look at a hypothetical example of the two-dimensional scalogram configuration taken from Shye [6]. Consider a population of the parents of four children in a community where all children of the appropriate ages attend high school. The number of items $n = 4$ and all are dichotomized, that is, $\alpha_i = 1$ for all i. Let each parent in this population be scored on four items, as follows:

$$\text{Parent } (p)\text{'s } \left\{ \begin{matrix} \text{first} \\ \text{second} \\ \text{third} \\ \text{fourth} \end{matrix} \right\} \text{child}$$

$$\text{attends high school now} \rightarrow \left\{ \begin{matrix} \text{yes} = 1 \\ \text{no} = 0 \end{matrix} \right\}.$$

A set of possible profiles is given for the four dichotomized items by the symmetric diamond scalogram whose partial order configuration is presented in Fig. 1a. In Fig. 1b, the rightmost path of profiles in Fig. 1a is exhibited as one of many possible perfect scales.

This pattern is a symmetric one. An essential feature of the figure representing the partial order relationships of a scalogram is that all and only comparable profiles are connected by downward lines. The possibilities for complex patterns increase with the number of items. Each profile has two scores (coordinates); the joint score $S_n(a) = \sum_{i=1}^{n} a_i$ counts the number of children, out of four, who attend high school at the time of the investigation. The lateral score of a profile $a = a_1 a_2 \ldots a_n$ is given by the formula

$$T_n(a) = \{ S_n(a) \}^{-1} \sum_{i=1}^{n} [2i - (n+1)] a_i + n.$$

In scalograms having a dimensionality of two or more, several profiles can occur with an identical joint score. Differentiation among them is facilitated by the lateral directions, each of which orders profiles belonging to a specific joint-score level. In the

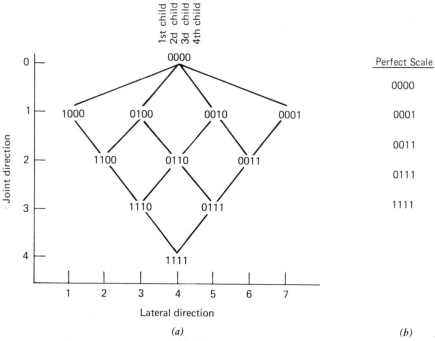

Figure 1 (*a*) Diamond scalogram configurations; (*b*) An example of a Guttman scale.

present example, we can interpret the lateral score as something like "the average seniority" of the four children relative to high school ages. Profiles having the same joint score may be of a different *kind* depending on their lateral score. In the foregoing example, a priori $2^4 = 16$ different profiles are possible in principle located in four-dimensional space. The restricted structures of scale and POSA (both perfect) allow only 5 and 11 different profiles in one- and two-dimensional spaces, respectively.

POSA by examples are given in Yalan et al. [9], Shye [5, 6], and Levy [3]. Let us present a short empirical example of partial-order stratification of American cities by crime rates. The data came from the 1970 *U.S. Statistical Abstract* and are analyzed in more detail in Levy [3]. The data in Table 1 present original crime rates and their transformation into four ranks for each crime separately for 16 cities. The seven kinds of crime are: murder, rape, assault, robbery, burglary, larceny, and auto theft. The analysis, therefore, will pertain only to these cities,

with no necessary implications for the remaining cities in the United States. The stratification of these 16 cities among themselves, is, of course, automatically a part of the stratification of all the cities, since the partial order of a subset does not depend on the partial order of the total set; the internal ordering holds regardless of the traits of cities not in the sample.

From the definitions of the seven crimes, it is possible to classify them into three distinct subsets: crimes against person (murder, rape, assault), and crimes against property (larceny, auto theft), with robbery and burglary in between. To anticipate, this threefold classification will relate to a lateral direction of the partial order derived from the crime items.

Each city was given one of four ranks on each crime item: 1 indicates the lowest rate, while 4 indicates the highest crime rate of that item. Thus each city was assigned a structuple composed of seven structs, one for each crime item. The lowest structuple rank on crime is 1111111, and characterizes the city of Hartford. The highest ranking

Table 1 Crime Rates and Their Ranking for 16 American Cities

City	Murder Manslaughter [a]	[b]	Rape [a]	[b]	Assault [a]	[b]	Robbery [a]	[b]	Burglary [a]	[b]	Larceny [a]	[b]	Auto Theft [a]	[b]
Atlanta	16.5	4	24.8	2	147	2	106	1	1112	2	905	2	494	1
Boston	4.2	1	13.3	1	90	1	122	1	982	1	669	1	954	4
Chicago	11.6	3	24.7	2	242	3	340	3	808	1	609	1	645	2
Dallas	18.1	4	34.2	3	293	4	184	2	1668	4	901	2	602	2
Denver	6.9	2	41.5	4	191	2	173	2	1534	3	1368	4	780	3
Detroit	13.0	3	35.7	3	220	3	477	4	1566	3	1183	3	788	3
Hartford	2.5	1	8.8	1	103	1	68	1	1017	1	724	1	468	1
Honolulu	3.6	1	12.7	1	28	1	42	1	1457	3	1102	3	637	2
Houston	16.8	4	26.6	2	186	2	289	3	1509	3	787	1	697	2
Kansas City	10.8	3	43.2	4	226	3	255	2	1494	3	955	2	765	3
Los Angeles	9.7	2	51.8	4	355	4	286	3	1902	4	1386	4	862	4
New Orleans	10.3	2	39.7	3	283	4	266	2	1056	1	1036	3	776	3
New York	9.4	2	19.4	1	267	3	522	4	1674	4	1392	4	848	4
Portland	5.0	1	23.0	2	144	2	157	1	1530	3	1281	4	488	1
Tucson	5.1	1	22.9	2	148	2	85	1	1206	2	756	1	483	1
Washington	12.5	3	27.6	2	217	3	524	4	1496	3	1003	3	739	3

[a] Rate per 100,000 population (from the 1970 *U.S. Statistical Abstract*).
[b] Rank in this sample of cities (1 = lowest; 4 = highest).

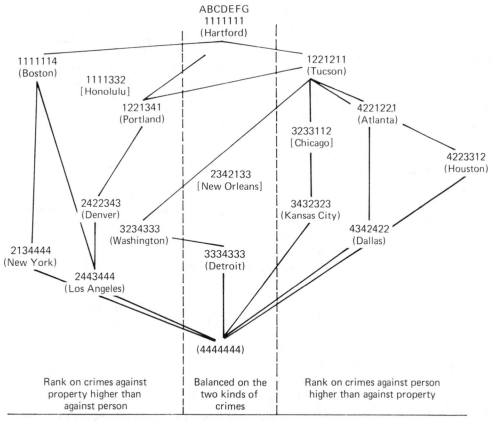

ABCDEFG
1111111
(Hartford)

1111114
(Boston)

1111332
[Honolulu]

1221341
(Portland)

1221211
(Tucson)

4221221
(Atlanta)

3233112
[Chicago]

4223312
(Houston)

2342133
[New Orleans]

2422343
(Denver)

3234333
(Washington)

3432323
(Kansas City)

4342422
(Dallas)

2134444
(New York)

3334333
(Detroit)

2443444
(Los Angeles)

(4444444)

Rank on crimes against
property higher than
against person

Balanced on the
two kinds of
crimes

Rank on crimes against person
higher than against property

Figure 2   The partial order of 16 American cities according to their rates on seven crimes. (A = murder; B = rape; C = assault; D = robbery; E = burglary; F = larceny; G = auto theft.)

structuple is 4444444, but this did not occur empirically in this sample of cities. All other structuples are partly ordered between these two extreme structuples. The POSA of this listing reveals that the 16 cities are stratified by their crime rates in essentially $M = 2$ dimensions. The partial order is presented in Fig. 2. Thirteen of the 16 cities (81%) fit perfectly into the two-space. The remaining three have their structuples enclosed in square brackets.

According to the joint direction of *level* of crime, Hartford and Boston are at the top, Detroit, New York, and Los Angeles at the bottom. As for any POSA, the partly ordered crime stratification shows not only the level of crime rate but also the *kind* of crime. It is interesting that the lateral direction differentiates among cities according to the contrast between the respective ranks on crimes against person and crimes against

property. To the left of the diagram—regardless of level of crime—are located cities that rank higher on the property-oriented crime rate than the person-oriented crime rate. Among these are Boston, Denver, New York, and Los Angeles. Conversely, to the right of the diagram are located cities that rank higher on person-oriented crime and lower on property-oriented crime, again, regardless of level of crime. Among these are Tucson, Atlanta, Houston, and Dallas. In the middle band are located the cites with balanced ranks on both kinds of crimes, Hartford having the balance between the low ranks, and Detroit having the balance between high ranks. Information of this same kind for both preceding and later years could be subjected to POSA in order to ascertain trends in both the level and kind of crime in American cities.

Two assumptions are made in using

POSA to process empirical data. The first is that a substantive rationale has guided the selection of the variables to be processed together, indicating that they sample, or cover, the different aspects of a well-defined content universe. The second is that the range of each variable is ordered and that the order is uniform in its direction and general meaning for all variables included in the analysis.

Although POSA may be regarded as an extension of the Guttman scale to profile configurations of dimensionality higher than one, it is a special case of Guttman's multidimensional scalogram analysis (MSA). In using MSA, we do not need the second assumption stated in the preceding paragraph, and even the first assumption is necessary only in a very general sense. The advantage offered by POSA in cases where these assumptions are made is that the interpretation of one direction in the profile configuration space is given, and the task of interpreting other directions promises to be easier than it is in MSA. For more details on MSA, see Zvulun [13] and its references. A new computer program has been developed at the Israel Institute of Applied Social Research (see Shye and Amar [10]). For small numbers of items, fitting by hand can sometimes be accomplished.

References

[1] Birkhoff, G. (1948). *Lattice Theory*, rev. ed. American Mathematical Society, Colloquium Publications, New York.

[2] Guttman, L. (1950). In S. A. Stouffer et al., *Measurement and Prediction*. Princeton University Press, Princeton, NJ.

[3] Levy, S. (1984). "Partly-Ordered Social Stratification." *Social Indicators Research*, **15**. Institute of Applied Social Research, Jerusalem, Israel.

[4] Nathan, M. and Guttman, R. (1984). *Acta Genet.*, **33**, 213–218.

[5] Shye, S. (1976). "Partial Order Scalogram Analysis of Profiles and Its Relationship to Smallest Space Analysis of the Variables." *Technical monograph*. Institute of Applied Social Research, Jerusalem, Israel.

[6] Shye, S. (1978). In S. Shye, ed., *Theory of Construction and Data Analysis in the Behavioral Sciences*, Jossey-Bass, San Francisco, pp. 265–279.

[7] Shye, S. and Elizur, D. (1976). *Hum. Relat.*, **29**, 63–71.

[8] Shye, S. (in press). *Multiple Scaling Theory*, Elsevier, Amsterdam.

[9] Shye, S. (1984). In *Facet Theory: Approaches to Social Research*, D. Cantor, ed. Springer-Verlag, New York, pp. 97–148.

[10] Shye, S. and Amar, R. (1984). *Ibid.*, pp. 277–298.

[11] Stouffer, S. A., Guttman, L., Suchman, E. A., Lazarsfeld, P. F., Star, S. A. and Clausen, J. A. (1950). *Measurement and Prediction*. Princeton University Press, Princeton, NJ.

[12] Yalan, E., Finkel, C., Guttman, L., and Jacobsen, C. (1972). *The Modernization of Traditional Agricultural Villages: Minority Villages in Israel*. Settlement Research Center, Rehovot, Israel.

[13] Zvulun, E. (1978). In S. Shye, ed. *Theory Construction and Data Analysis, In The Behavioral Sciences*, Jossey-Bass, San Francisco, pp. 237–264.

(COMPONENT ANALYSIS CORRESPONDENCE ANALYSIS MULTIDIMENSIONAL SCALING)

ADI RAVEH

PARTIAL REGRESSION

The word *partial*, when used as a modifier of regression*, can refer to several concepts. We address first the meaning of a partial regression coefficient in the context of a multivariate normal* model. Following somewhat the notation of Graybill [3], let $\mathbf{Y}^{*\prime} = [Y_0, Y_1, \ldots, Y_p]$ be a random vector of $p + 1$ components. Suppose $\mathbf{Y}^*$ is distributed as a multivariate normal with expectation vector $\boldsymbol{\mu}^*$ and covariance matrix $\boldsymbol{\Sigma}^*$. Let $\mathbf{Y}^*$, $\boldsymbol{\mu}^*$, and $\boldsymbol{\Sigma}^*$ be partitioned as follows:

$$\mathbf{Y}^* = \begin{bmatrix} Y_0 \\ \mathbf{Y} \end{bmatrix}, \qquad \boldsymbol{\mu}^* = \begin{bmatrix} \mu_0 \\ \boldsymbol{\mu} \end{bmatrix},$$

$$\boldsymbol{\Sigma}^* = \begin{bmatrix} \sigma_{00} & \boldsymbol{\sigma}_{01} \\ \boldsymbol{\sigma}_{10} & \boldsymbol{\Sigma} \end{bmatrix}.$$

Denote the conditional random variable Y_0 given $\mathbf{Y} = \mathbf{X}$, by the symbol $Y_{\mathbf{X}}$. $Y_{\mathbf{X}}$ is distributed normally with expectation $\mu_0 - \boldsymbol{\sigma}_{01}\boldsymbol{\Sigma}^{-1}\boldsymbol{\mu} + \boldsymbol{\sigma}_{01}\boldsymbol{\Sigma}^{-1}\mathbf{X}$ and variance $\sigma^2 = \sigma_{00} - \boldsymbol{\sigma}_{01}\boldsymbol{\Sigma}^{-1}\boldsymbol{\sigma}_{10}$. Let $\beta_0 = \mu_0 - \boldsymbol{\sigma}_{01}\boldsymbol{\Sigma}^{-1}\boldsymbol{\mu}$, $\boldsymbol{\beta}^\prime = \boldsymbol{\sigma}_{01}\boldsymbol{\Sigma}^{-1}$ and $E = Y_{\mathbf{X}} - \beta_0 - \boldsymbol{\beta}^\prime\mathbf{X}$. Then

$Y_X = \beta_0 + \boldsymbol{\beta}'\mathbf{X} + E$, where E is distributed normally with mean zero and variance σ^2.

Since $\boldsymbol{\beta}'\mathbf{X} = \sum_{i=1}^{p} \beta_i X_i$, we could also write $\epsilon(Y_x) = \beta_0 + \sum_{i=1}^{p} \beta_i X_i$. The β_i, $i = 1$, $2, \ldots, p$ are defined as the *partial regression coefficients*. β_j is the amount that the variable Y_0 is expected to change if Y_j is changed one unit while the other Y_i's are held fixed.

If we denote conditional variances and covariances by $\sigma^2_{j \mid 1,2,\ldots,j-1,\ldots,p}$ and $\sigma_{0j \mid 1,2,\ldots,j-1,j+1,\ldots,p}$, then partial regression coefficients can be expressed in terms of these conditional variances and covariances:

$$\beta_j = \frac{\sigma_{0j \mid 1,2,\ldots,j-1,j+1,\ldots,p}}{\sigma^2_{j \mid 1,2,\ldots,j-1,j+1,\ldots,p}} .$$

A more precise notation for the partial regression coefficient would be $\beta_{0j \mid 1,2,\ldots,j-1,j+1,\ldots,p}$.

The partial regression coefficient β_j is closely related to the partial correlation coefficient*

$$\rho_{0j \mid 1,2,\ldots,j-1,j+1,\ldots,p}$$
$$= \beta_j \frac{\sigma_{j \mid 1,\ldots,j-1,j+1,\ldots,p}}{\sigma_{0 \mid 1,\ldots,j-1,j+1,\ldots,p}} .$$

A partial correlation coefficient can be thought of as a standardized partial regression coefficient.

The case where there are four random variables is a general enough case to demonstrate the essence of partial regression concepts in the multivariate normal context. Suppose Y_0, Y_1, Y_2, and Y_3 are multivariate normal with mean vector $\boldsymbol{\mu}^*$ and covariance matrix $\boldsymbol{\Sigma}^*$. The conditional random variable $Y_X = \{Y_0 \mid Y_i = X_i\}$ has expected value $\epsilon(Y_X) = \beta_0 + \beta_1 X_1 + \beta_2 X_2 + \beta_3 X_3$, where, e.g.,

$$\beta_1 = \frac{\mathrm{cov}[Y_0, Y_1 \mid Y_2 = X_2, Y_3 = X_3]}{\mathrm{var}[Y_1 \mid Y_2 = X_2, Y_3 = X_3]} .$$

Furthermore, $\rho_{01 \mid 23} = \beta_1 \sigma_{1 \mid 23} / \sigma_{0 \mid 23}$.

The square of $\rho_{01 \mid 23}$ is called the *coefficient of partial determination*. It can be shown that $\rho^2_{01 \mid 23} = (\sigma^2_{0 \mid 23} - \sigma^2_{0 \mid 123}) / \sigma^2_{0 \mid 23}$. The coefficient of partial determination measures the relative reduction in conditional variability when one moves from conditioning on Y_2 and Y_3 to conditioning on Y_1, Y_2, and Y_3.

Because of the relationship between a partial regression coefficient and a corresponding partial correlation coefficient, inference statements concerning one are closely related to inference statements concerning the other. In particular, testing that $\rho_{0j} = 0$ is equivalent to testing $\beta_{0j} = 0$.

The word *partial* is sometimes used to modify the term coefficient in a situation where a linear (fixed predictors) model is used. Presumably the term is borrowed from mathematics where an apparently similar concept is termed a *partial derivative*.

Consider a multiple regression* situation with three predictors, namely, X_1, X_2, and X_3 (see MULTIPLE REGRESSION). Suppose that $\epsilon(Y) = f(X_1, X_2, X_3) = \beta_0 + \beta_1 X_1 + \beta_2 X_2 + \beta_3 X_3$. The regression coefficient β_1 is the amount that Y is expected to change if X_1 is changed one unit while X_2 and X_3 are held fixed. If X_2 and X_3 are functionally unrelated to X_1, then the partial derivative $\partial f / \partial X_1 = \beta_1$. Therefore, if X_2 and X_3 are functionally unrelated to X_1, then it is natural and proper to call β_1 a partial regression coefficient. Suppose, however, that X_3 is functionally related to X_1 and X_2. Suppose, in fact, that $X_3 = X_1 X_2$ so that $\epsilon(Y) = g(X_1, X_2) = \gamma_0 + \gamma_1 X_1 + \gamma_2 X_2 + \gamma_3 X_1 X_2$ with $\gamma_3 \neq 0$. The partial derivative $\partial g / \partial X_1 = \gamma_1 + \gamma_3 X_2$. Only when $X_2 = 0$ is γ_1 a partial derivative. γ_1 is the change in $\epsilon(Y)$ when X_1 is changed one unit and $X_2 = 0$. Note that as X_1 changes, both X_2 and $X_3 = X_1 X_2$ cannot be held fixed unless $X_2 = 0$. γ_1 is called on occasions a partial regression coefficient, but it has the same meaning as β_1 only when $X_2 = 0$.

There exists another aspect of the word *partial* when employed in conjunction with estimation of regression coefficients*. Consider the problem of fitting a curve or a surface for a data set involving a response and with k predictor variables. The numerical value of, the meaning of, and the inference given to, say, $b_1 X_1$, depend, of course, on the data, but they also depend very heavily on the model chosen. For example, the partial regression coefficient b_1 plays an entirely different role in the least-squares* linear fit $\hat{Y} = b_0 + b_1 X_1$ than it does in the least-squares quadratic fit $\hat{Y} = b_0 + b_1 X_1 + b_2 X_1^2$. In summary, partial regression coefficients are model dependent. Consequently, any inference statement made relative to partial regression coefficients must be made relative to the model chosen.

The third use of the word *partial* in regression is in conjunction with regression analysis of variance* sums of squares. Consider, for illustrative purposes, a response variable Y and p predictors $X_1, X_2, \ldots, X_p$. Consider the matrix formulation of regression where $Y = \mathbf{X}\boldsymbol{\beta} + E$, but allow $\mathbf{X}$ to represent the model matrix for each of the following regressions. In each case SSR($\cdot$) denotes the sum of squares due to regression, i.e., $\mathbf{Y}'\mathbf{X}(\mathbf{X}'\mathbf{X})^{-1}\mathbf{X}'\mathbf{Y}$.

Situation	SS Due to Regression
Regress Y on X_i only	SSR(X_i) $i = 1, 2, \ldots, p$
Regress Y on X_i and X_j	SSR(X_i, X_j)
	$i, j = 1, 2, \ldots, i \neq j$
Regress Y on X_i, X_j, X_k	SSR(X_i, X_j, X_k)
	$i, j, k = 1, 2, \ldots, p,$
	$i \neq j, \ i \neq k, \ j \neq k$

Partial sums of squares are then defined to be

$$SSR(X_j \mid X_i) = SSR(X_i, X_j) - SSR(X_i),$$

$$i, j = 1, 2, \ldots, p_j, \quad i \neq j$$

$$SSR(X_k \mid X_i, X_j)$$
$$= SSR(X_i, X_j, X_k) - SSR(X_i, X_j),$$

$$i, j, k = 1, 2, \ldots p_j, \quad i \neq j, \ i \neq k, \ j \neq k$$

$$SSR(X_h \mid X_i, \ldots, X_{h-1}, X_{h+1}, \ldots, X_p)$$
$$= SSR(X_1, \ldots, X_p)$$
$$- SSR(X_1, \ldots, X_{h-1}, X_{h+1}, \ldots, X_p).$$

A partial sum of squares is related to a partial regression coefficient when the standard regression assumptions are applied. In particular, under $H_0 : \beta_h = 0$

$$\frac{SSR(X_h \mid X_1, \ldots, X_{h-1}, X_{h+1}, \ldots, X_p)}{MSE}$$

$$\sim F(1, n - p) \quad \text{and} \quad \frac{b_h}{S_{b_h}} \sim t(n - p)$$

Hence,

$$SSR(X_h \mid X_1, \ldots, X_{h-1}, X_{h+1}, \ldots, X_p)$$

$$= \frac{MSE\, b_h^2}{s_{b_h}^2} = b_h^2 f(X_1, \ldots, X_p),$$

where $f(X_1, \ldots, X_p)$ is a function of the values of the predictor variables in the data. The coefficient of partial determination for predictor X_h is defined to be

$$\frac{SSR(X_h \mid X_1, \ldots, X_{h-1}, X_{h+1}, \ldots, X_p)}{SSE(X_1, \ldots, X_{h-1}, X_{h+1}, \ldots, X_p)},$$

where SSE is the sum of squares for residual in a regression of Y on all predictors except X_h. If the multivariate normal model is assumed, then the coefficient of partial determination is equal to the square of the sample estimate $r_{0h \mid 1,2, \ldots, h-1, h+1, \ldots, p}$ of the partial correlation coefficient $\rho_{0h \mid 1,2, \ldots, h-1, h+1, \ldots, p}$.

Bibliography

Allen, D. M. and Cady, F. B. (1982). *Analyzing Experimental Data by Regression*. Wadsworth, Belmont, CA.

Draper, N. and Smith, H. (1981). *Applied Regression Analysis*, 2nd ed. Wiley, New York.

Graybill, F. A. (1976). *Theory and Application of the Linear Model*. Duxbury, Belmont, CA.

Gunst, R. F. and Mason, R. L. (1980). *Regression Analysis and Its Application*. Marcel Dekker, New York.

Kleinbaum, D. G. and Kupper, L. L. (1978). *Applied Regression Analysis and Other Multivariate Methods*. Duxbury, Belmont, CA.

Mendenhall, W. (1968). *Introduction to Linear Models and the Design and Analysis of Experiments*. Wadsworth, Belmont, CA.

Mosteller, F. and Tukey, J. W. (1977). *Data Analysis and Regression*. Addison-Wesley, Reading, MA.

Neter, J. and Wasserman, W. (1974). *Applied Linear Statistical Models*. Richard D. Irwin, Georgetown, Ontario, Canada.

Searle, S. R. (1971). *Linear Models*. Wiley, New York.

Younger, M. S. (1979). *A Handbook for Linear Regression*. Duxbury, North Scituate, MA.

(ANALYSIS OF COVARIANCE
CORRELATION
REGRESSION (Various Entries))

ROBERT HULTQUIST

PARTICLE-SIZE STATISTICS

The study of small particulate matter is important in many different fields. A characterization of particles is essential to powder technologists for improving process behavior, to ceramists for making better materials, to environmentalists in determining better air quality, and to geologists for examining rock structure. Histologists trying to detect diseases such as cancer or diabetes need to study the distribution of blood-cell sizes. Unfortunately, the fact that particle-size statistics is useful in such varied disciplines has worked to the disadvantage of the researcher. It becomes tedious to find relevant articles because publications are so widespread in the literature. In addition, the lack of a unified effort has caused considerable overlap as well as an absence of consistent notation.

Important papers by Hatch and Choate [13] and Hatch [12] gave formulas for lognormal distributions* and log-normal transformations that enabled experimenters to conveniently display size distributions and to estimate parameters. The assumption of spherical particles permitted *sizes* to be defined as diameters. In turn, sizes were assumed to follow a lognormal distribution, which simplified procedures for transforming a weight fraction probability measure to a "standard" frequency density. This proved to be of such great practical utility in the analysis of particle-size data that Hatch's work can be considered a classic paper—one of the first to utilize applied statistics in this field.

Another classical paper related to the lognormal distribution is attributed to Kolmogorov [17]. His work substantiated Hatch's empirical results by demonstrating that logarithms of sizes obtained from successive random fragmentations of a particle will, under mild regularity conditions, approach the normal distribution. Halmos [11] gives similar results in a completely different setting. A fixed amount of gold dust is distributed sequentially among a countably infinite number of beggars. Each beggar, in turn, receives a random amount of gold x_i. The problem is to find the distribution of x_i, where $i = 1, 2, \ldots$. Halmos presented results compatible with Kolmogorov's. In addition, he proved that it is almost certain that all of the gold dust will be disseminated. O. Barndorff-Nielsen [3] extended these results by introducing the hyperbola as an approximation function for the distribution of log sizes of mixtures of normal distributions. (*See* HYPERBOLIC DISTRIBUTIONS.)

Much of the early experimental work was done by elementary methods such as the use of sieves, but after 1945 an ever-increasing number of electronic sizing devices resulted in a need for more sophisticated statistical analysis. This demand for technology has arisen from a substantial growth of interest among scientists, whose applications are also increasing at a fast rate. One of the many contributors is Allen [1], whose references include over 150 different sizing devices and nearly 700 articles and books.

Associated with defining the size of a particle is its *shape*. The size of a spherical particle is readily defined as its diameter (or radius). Analogously a cube size may be defined as a length of its edge. Sizes of irregularly shaped particles are defined in various ways. Equivalent spherical sizes can be obtained by relating measurements to a range of mesh sizes in sieves, replacing irregularly shaped photographed data with circles of equal area, or using formulas that associate terminal velocity of particles settling in a liquid to spheres of equal volume (e.g., Stokes' diameter). The assumption of equivalent particle size is directly related to the type of instruments. For example, it is a poor assumption to choose equivalent sphere sizes of mica when using screening techniques because of large resultant errors. *Feret's diameter* is obtained by taking the average of distances between pairs of parallel tangents to the projected ($2d$) outline of the particle. *Martin's diameter* is computed as the average chord length of the projected outline. Unfortunately, these diameters and similar empirical measurements do not define particle shape uniquely. The sizes mentioned are linear; however the dimension may be nonlinear. Instruments such as the Quantimet and the Coulter counter measure frequencies of surface area and volume, respectively. These variables are appropriately defined as size. Many authors define a weight fraction (or weight) probability measure vs. size as a "*weight distribution*." This is wrong. Frequencies or relative frequencies of weight would imply a distribution of weights. A detailed explanation, unified notation, and proper classification of sizes is given in Chapter 1 of Fayde and Otten [9].

Solutions to problems of describing particles become almost impossible when one attempts to use just one or two numbers to quantify shape. A new field called *particle morphology* discusses methods of describing shape (see Beddow and Meloy [4]). Meloy [21] compares the problem of defining a shape factor with a man buying a tailor-made suit but giving the tailor only one measurement. On the other hand, shapes can be defined qualitatively. British Standard 2995 gives the following breakdown: acicular—needle-shaped; angular—sharp-edged or having roughly polyhedral shape; crystalline—freely developed in a fluid medium of geometric shape; dentritic—having a branched crystalline shape; flaky—platelike; granular—having approximately an equidimensional irregular shape; irregular—lacking any symmetry; modular—having rounded, irregular shape; and spherical—global shape.

Shape factor is generally defined as a value that relates the average volume (or surface area) to an average particle size. In the equation

$$\bar{v} = \alpha_v d_p^3,$$

where $\bar{v}$ denotes the average volume, d_p is a projected diameter (i.e., a diameter of sphere having an equal volume), and α_v is a volume shape factor. If d_p was considered to be an "equivalent" sphere diameter, then $\alpha_v = \pi/6$. Heywood [15] has arrived at classical formulas for a pair of shape factors that correct for both differences in dimensionality of the original measurements and shape

irregularities. His work is summarized by Herdan [14]. A more recent approach has made use of automatic scanners that provide (x, y) coordinates of particle silhouettes. The center of gravity is determined and random distances to the edge of the particle and angles from a horizontal line computed. This data is used to construct a Fourier series whose coefficients relate to shape. Theoretical and experimental results may be found in Beddow [4], Beddow and Meloy [5, 6], or in numerous articles that have appeared in *Powder Technology* (a convenient list may be found under "Particle Shape" in a 25-year index [22]). Other sources are Ehrlich et al. [7] and Ehrlich and Weinberg [8]. A discussion of instrumentation for morphological analysis is available from Lenth et al. [18].

An experimenter's choice from the numerous sizing devices available depends on the type of particulate being investigated as well as the size range of the material. Output from any of the instruments rarely lends itself to a routine statistical analysis. Some of the problems are: tails of a distribution are apt to be biased because small particles are difficult to measure precisely and large particles often have small probabilities of being detected because of a nonhomogeneous mix; agglomeration of particles makes determination of size difficult; data displayed in continuous form are dependent and transforming to discrete form becomes troublesome; the data may not be in a form usable to the investigator; some data may be noisy and need to go through some form of filtering; and analysis is made difficult as a result of instrumental measurement errors* and reading errors. Lewis and Goldman [19] have categorized data and summarized methods that partially answer these problems.

A difficult problem inherent in particle-size analysis is to obtain a representative sample of particulates. Consider a container filled with sand and a marble considerably larger than a grain of sand. If the container is tapped at the bottom, the marble will rise to the top. This example illustrates that too much shaking will not result in a homogeneous mix. Powders will segregate when moved in containers. Small particles will cluster in the center when powder is poured in a heap. Material tends to agglomerate; for example, salt will form lumps in damp weather. Collected samples are only a very small subset of the tremendous volume of airborne particles. Fortunately, appropriate sampling methods are available for solving most of these problems (e.g., see the first three chapters of Allen [1]).

Packing density of material is defined as the fraction or percentage of material that occupies a container. In many fields it is important to obtain the maximum packing of particulates. This is particularly relevant to the pharmaceutical industry* (e.g., when manufacturing capsules). Stronger metals often result from the fabrication of mixtures of powders having a higher packing fraction than the individual powders. The basic as yet unsolved problem can be stated quite simply: Given a container with a known "random" distribution of solid spherical particles, what is its packing density? As an elementary step toward a solution, packing of stacked arrays of monosized spheres are computed:

Packing	Description	Points of Contact	Packing Fraction
(Lines connecting centers)			
Cubic	Stacked directly on top	6	52.36%
Ortho- rhombic	As cubic except shift	8	60.46%
Hexagonal (loose pack)	Directly on top	10	69.81%
Hexagonal (close pack)	Nested in pockets	12	74.05%

The chart shows that the maximum packing of any stacked array is hexagonal close pack with a packing fraction of 74.05%. Scott and Kilgour [24] conducted an experiment that used a random close pack distribution of spheres that gave a packing density of 63.66%. Results of this experiment have yet

to be duplicated either theoretically or by simulation. Several computer simulations have met with limited success achieving results in the vicinity of 58–60%. Powell [23] used the following algorithm: Conveniently pack one layer of monosized hard spheres. Form subsequent layers by concentrating on different regions, sequentially dropping spheres into the deepest pocket to attain a limiting packing fraction. Visscher and Bolsterli [25] added spheres sequentially until each sphere attained a stable three-point contact. A résumé of important packing and simulation experiments and packing fractions may be found in Hoare [16] or Powell [23]. Another approach to the packing problem is to determine what percentages of coarse and fine powders should be mixed together in order to "improve" packing density. Results by Lewis and Goldman [19] indicate under certain conditions what percentages of powders are needed to attain maximum packing. These results are a function of the coefficient of variation*.

A contrived packing arrangement was devised by Wise [28] that has interesting theoretical aspects but little practical value. This research assumed a packed array of random-sized spheres in which all touch in such a way that tetrahedrons can be obtained by connecting lines to centers of touching spheres. Packing fractions were obtained from various size distributions. As a by-product, Wise discovered some relevant mensuration formulas for tetrahedra.

Methods of stereology* (see discussion given in that entry) are needed for sizing devices that render two-dimensional output. For example, thin slices of skin tissue are observed through a microscope to determine cell sizes. A photograph will give a two-dimensional display of "equivalent" circles to be analyzed. The fundamental problem is to then determine the three-dimensional size distribution. The procedure for doing this involves solving an Abel integral equation* that was first done in a classical paper by Wicksell [27]. This paper stimulated much research. Andersen and Jakeman [2] reviewed the literature and an update was furnished by Visscher and Goldman [26]. A rigorous explanation of mathematical development and solutions may be found in Golberg [10].

References

[1] Allen, T. (1975). *Particle Size Measurement*. Chapman and Hall, London.

[2] Andersen, R. S. and Jakeman, A. J. (1975). In the Fourth International Congress of Stereology. National Bureau of Standards, Gaithersburg, MD.

[3] Barndorff-Nielsen, O. (1977). *Proc. R. Soc. Lond. A*, **353**, 401–419.

[4] Beddow, J. K. (1980). *Particulate Science and Technology*, Chemical Publishing Company, New York.

[5] Beddow, J. K. and Meloy, T. P. eds. (1980). *Advanced Particulate Morphology*. CRC Press, Cleveland, OH.

[6] Beddow, J. K. and Meloy, T. P. (1980). *Testing and Characterization of Powders and Fine Particles*. Hayden, London.

[7] Ehrlich, R., Orzeck, J. and Weinberg, B. (1974). *J. Sedimentol Petrol.*, **44**, 145–150.

[8] Ehrlich, R. and Weinberg, B. (1970). *J. Sedimentol. Petrol.*, **40**, 205–212.

[9] Fayed, M. D. and Otten, L. eds. (1984). In *The Handbook of Power Science and Technology*. Van Nostrand Reinhold, New York, Chap. 2.

[10] Golberg, M., ed. (1979). *Solution Methods for Integral Equations*. Plenum, New York, Chap. 6.

[11] Halmos, P. (1944). *Ann. Math. Statist.*, **15**, 182–189.

[12] Hatch, T. (1933). *J. Franklin Inst.*, **215**, 27–37.

[13] Hatch, T. and Choate, S. (1929). *J. Franklin Inst.*, **207**, 369–387.

[14] Herdan, G. (1960). *Small Particle Statistics*. Academic Press, New York.

[15] Heywood, H. (1947). *Inst. Chem. Eng. Suppl.*, **25**, 14.

[16] Hoare, M. R. (1978). *J. Non-Crystalline Solids*, **31**, 157–179.

[17] Kolmogorov, A. N. (1941). *Dokl. Akad. Nauk. SSSR*, **31**, 99–101.

[18] Lenth, R., Chang, C. R., Beddow, J. K., and Vetter, A. F. (1980). "Particle Image Analyzing System." Materials Engineering Division, University of Iowa, Iowa City, IA.

[19] Lewis, H. D. and Goldman, A. S. (1966). *J. Amer. Ceram. Soc.*, **49**, 323–327.

[20] Lewis, H. D. and Goldman, A. S. (1968). "Theoretical Small-Particle Statistics: A Summary of Techniques for Data Analysis with Recent Developments in Data Comparison, Notation, and Mixture Theory." *Rep. No. LA-3656*, Los Alamos Scientific Laboratory, Los Alamos, NM.

[21] Meloy, T. (1977). *Powder Tech.*, **16**, 233–253.

[22] (1980). *Powder Tech.*, index ed.

[23] Powell, M. J. (1980). *Powder Tech.*, **25**, 45–52.

[24] Scott, G. D. and Kilgour, D. M. (1969). *Brit. J. Appl. Phys. (J. Phys. D)*, **2**, 863–866.

[25] Visscher, W. M. and Bolsterli, M. (1972). *Nature*, **239**, 504–509.

[26] Visscher, W. M. and Goldman, A. S. (1983). *SIAM J. Sci. Statist. Comp.*, **4**, 280–290.

[27] Wicksell, S. D. (1925). *Biometrika*, **17**, 84–89.

[28] Wise, M. E. (1952). *Philips Res. Rep.*, **7**, 321–343.

(BIOSTATISTICS
GEOLOGY, STATISTICS IN
LOGNORMAL DISTRIBUTION
MEASUREMENT ERROR
STEREOLOGY)

AARON S. GOLDMAN

PARTITION OF CHI-SQUARE

Consider a hierarchy of models*

$$H_{01} \subset H_1 \subset H_2 \subset \cdots \subset H_k, \qquad (1)$$

in which H_{01} and H_1 are *inner models*, H_k is an *outer model* (usually a general class of alternatives), and $H_{01} = H_0 \cap H_1$ is the model that includes H_0 and H_1. The notation H_i may also denote the hypothesis that model H_i holds.

If H_i is "accepted" under some hypothesis testing* procedure, then logically all outer models containing H_i should also be acceptable. A natural framework is that of likelihood ratio (LR) tests*, which we adopt here (see Hogg [3], who developed the rationale following). Suppose that $\mathbf{x} = (x_1, x_2, \ldots, x_n)$ denotes data sampled from the population of interest, that $\Lambda_{i:j}$ is the LR criterion for testing H_i against $H_j - H_i$ ($H_i \subset H_j$), where

$$\Lambda_{i:j} = L(\mathbf{x} \mid H_i) / L(\mathbf{x} \mid H_j),$$

$L(\mathbf{x} \mid H_i)$ being the maximum likelihood of

the data **x** under model H_i. Then

$$\Lambda_{01;2} = \frac{L(\mathbf{x}\,|\,H_{01})}{L(\mathbf{x}\,|\,H_1)} \cdot \frac{L(\mathbf{x}\,|\,H_1)}{L(\mathbf{x}\,|\,H_2)} = \Lambda_{01;1} \cdot \Lambda_{1;2}$$

$$= \Lambda_{0|1} \cdot \Lambda_{1;2}, \qquad (2)$$

the notation $\Lambda_{0|1} = \Lambda_{01;1}$ indicating that a test of H_{01} vs. H_1 is essentially a test of H_0 against alternatives in which H_1 is given to hold, i.e., of what we denote by $H_{0|1}$; and $\Lambda_{0|1}$ is a *conditional* LR test criterion. Logically, for a test of H_{01} vs. H_1 to be meaningful, H_1 should first be acceptable. In testing the hierarchy $H_{01} \subset H_1 \subset H_2$, then, these steps should be followed:

1. Test H_1 vs. $H_2 - H_1$ with the criterion $\Lambda_{1;2}$.
2a. If H_1 is rejected, there is no purpose in testing H_{01} vs. $H_1 - H_{01}$.
2b. If H_1 is acceptable, test H_{01} vs. $H_1 - H_{01}$ with the criterion $\Lambda_{0|1}$.

Similar logic dictates that, with the extended hierarchy (1), the outer models are tested first, H_{k-1} vs. $H_k - H_{k-1}$, then H_{k-2} vs. $H_{k-1} - H_{k-2}$, and so on, until some model H_i is rejected.

If $\lambda_{i;j} = -2\log\Lambda_{i;j}$, then (2) becomes

$$\lambda_{01;2} = \lambda_{1;2} + \lambda_{0|1}, \qquad (3)$$

and from the hierarchy (1) we derive the equation

$$\lambda_{01;k} = \lambda_{k-1;k} + \lambda_{k-2|k-1}$$
$$+ \cdots + \lambda_{1|2} + \lambda_{0|1}. \qquad (4)$$

If H_{01} holds, then each $\lambda_{i;j}$ (or $\lambda_{i|j}$) in (4) has an approximate chi-square distribution* when n is large, the degrees of freedom (d.f.) partitioning additively (*see* CHI-SQUARE TESTS). Equations (3) and (4) present the *partition of chi-square*, known in generalized linear models* as the *analysis of deviance*.

Suppose that $H_{01} \subset H_0 \subset H_2$ and $H_{01} \subset H_1 \subset H_2$ are both hierarchies of interest. If in (3), $\lambda_{0|1} = \lambda_{0;2}$ so that

$$\lambda_{01;2} = \lambda_{0;2} + \lambda_{1;2}, \qquad (5)$$

then $\lambda_{1|0} = \lambda_{1;2}$ and H_0 and H_1 are called

independent models with respect to H_2 (see Darroch and Silvey [1]) in the sense that both hierarchies lead to the same partition of chi-square; the LR test of H_0 (H_1, respectively) against H_2 is unaffected by whether or not H_1 (H_0, respectively) is known to hold.

Hogg [3] showed that if the $\Lambda_{i;i+1}$ are each functions of complete sufficient statistics of nuisance parameters not specified in H_1, then the $\lambda_{i;i+1}$'s are statistically independent (*see* BASU THEOREMS).

Example 1. (See Hogg [3].)   Let the density function of an exponential distribution be given by

$$f(x;\theta,\rho) = \rho^{-1}\exp[-(x-\theta)/\rho],$$
$$x > \theta, \qquad (6)$$

zero elsewhere, where $\rho > 0$. Define

H_2 :   (6) holds for any θ, any $\rho > 0$;

H_1 :   (6) holds for any θ, $\rho = \rho_1$;

H_0 :   (6) holds for $\theta = \theta_0$, any $\rho > 0$.

From the hierarchy $H_{01} \subset H_1 \subset H_2$,

$$\Lambda_{1;2} = \left[\sum_{i=1}^{n}(X_i - X_1)/(n\rho_1)\right]^n$$

$$\times \exp\left[-\left\{\sum_{i=1}^{n}(X_i - X_1)/n\right\} + n\right],$$

$$\Lambda_{0|1} = \exp[-n(X_1 - \theta_0)/\rho_0],$$

where $X_1 \leqslant X_2 \leqslant \cdots \leqslant X_n$ are the order statistics* of a random sample of size n. The partition of chi-square appears in Table 1. For a discussion of the degrees of freedom of $\lambda_{0|1}$, see Hogg [2]; under H_{01}, $\lambda_{0|1}$ has an exact chi-square distribution with 2 d.f. Further, $\lambda_{0|1}$ and $\lambda_{1;2}$ are independent.

Example 2. (Multidimensional contingency tables*.)   This example is discussed along with others in Read [5].

Suppose that frequencies n_{ijk} appear in a three-dimensional $r \times s \times c$ contingency table $(i = 1, \ldots, r; \; j = 1, \ldots, s; \; k = 1, \ldots, c)$, where each $n_{i..} = \sum_j \sum_k n_{ijk}$ is fixed and the i subscript values represent each of r

Table 1   Partition of Chi-square (Example 1)

Source	Test Statistic	d.f.
H_1 vs. H_2	$-2n\log\left[\sum_{i=1}^{n}(X_i - X_1)/(n\rho_1)\right]$	1
	$-2n + 2\sum_{i=1}^{n}(X_i - X_1) = \lambda_{1;2}$	
H_0 vs. H_1	$2n(X_1 - \theta_0)/\rho_0 = \lambda_{0\mid 1}$	2
H_{01} vs. H_2	$\lambda_{01;2}$	3

populations and $n_1..\,, n_2..\,, \ldots, n_r..$ units are sampled from these populations, respectively. Each level of i is an $s \times c$ contingency table with multinomial sampling. For discussion of LR tests and hierarchical models in this kind of setting *see* CONTINGENCY TABLES and MULTIDIMENSIONAL CONTINGENCY TABLES. In other notation, $n_{ij.} = \sum_k n_{ijk}$, p_{ijk} is the probability that a unit sampled from the ith population will be observed in cell (j,k) of the $s \times c$ table for that population, so that $p_{i..} = \sum_j\sum_k p_{ijk} = 1$ for $i = 1, \ldots, r$. Marginal probabilities are $p_{ij.} = \sum_k p_{ijk}$ and $p_{i.k} = \sum_j p_{ijk}$.

Models or hypotheses of interest, along with maximum likelihood* estimators $\hat{p}_{ij.}$, etc., might include those following, where i, j and k are understood to run through all their values.

H_0. $p_{ijk} = p_{ij.} \cdot p_{i.k}$; $\hat{p}_{ij.} = n_{ij.}/n_{i..}$, $\hat{p}_{i.k} = n_{i.k}/n_{i..}$ (independence of rows and columns in each layer or population).

H_1. $p_{ijk} = p_{(\cdot)jk}$; $\hat{p}_{(\cdot)jk} = n_{.jk}/N$ (homogeneity of populations), where $p_{(\cdot)jk}$ denotes that p_{ijk} is the same for each population.

$H_{01} = H_0 \cap H_1$. $p_{ijk} = p_{(\cdot)j} \cdot p_{(\cdot)\cdot k}$;

$\hat{p}_{(\cdot)j.} = n_{.j.}/N$, $\hat{p}_{(\cdot)\cdot k} = n_{..k}/N$.

H_2. General alternatives with multinomial sampling in each population.

The Pearson chi-square statistics for testing these models do not partition additively. The partitions of LR chi-square appearing in Ta-

ble 2 use the lnn function, defined for vectors $\mathbf{a} = (a_1, \ldots, a_m)$, $\mathbf{b} = (b_1, \ldots, b_l)$, etc. by

$$\mathrm{lnn}\left[\frac{\mathbf{a}}{\mathbf{b}}\right] = 2\sum_{\alpha=1}^{m} a_\alpha \log a_\alpha - 2\sum_{\beta=1}^{l} b_\beta \log b_\beta,$$

since LR statistics for such models are frequently sums and differences of quantities of the form $2n\log n$. Then

$$\mathrm{lnn}\left[\frac{\mathbf{a},\mathbf{b}}{\mathbf{c},\mathbf{d}}\right] + \mathrm{lnn}\left[\frac{\mathbf{e},\mathbf{f}}{\mathbf{b},\mathbf{g}}\right] = \mathrm{lnn}\left[\frac{\mathbf{a},\mathbf{b},\mathbf{e},\mathbf{f}}{\mathbf{c},\mathbf{d},\mathbf{b},\mathbf{g}}\right]$$

$$= \mathrm{lnn}\left[\frac{\mathbf{a},\mathbf{e},\mathbf{f}}{\mathbf{c},\mathbf{d},\mathbf{g}}\right].$$

This property renders the statistics in Table 2 concise and easy to add; $\mathbf{n}_{ijk}$ is an extended vector of all frequencies n_{ijk}; $\mathbf{n}_{ij.}$ is similarly defined over all values of i and j, and so on.

Unlike other hierarchies of models presented in Read [5, 6], H_0 and H_1 are not independent in the sense of Darroch and Silvey [1], and hence the partition based on $H_{01} \subset H_0 \subset H_2$ is different from that based on $H_{01} \subset H_1 \subset H_2$. The test statistics in the Chi-square column in Table 2 are those used for testing for independence or homogeneity in certain two-dimensional contingency tables. $\lambda_{1;2}$ in Table 2(b) is based on a $r \times sc$ table of the entire data with i representing stretched-out rows, say, of sc elements each (sc being the number of columns), and $\lambda_{0\mid 1}$ is based on a $s \times c$ table in which the frequencies are the marginal totals $n_{.jk}$. Thus $\lambda_{0\mid 1}$ is a conditional chi-square statistic in that, given that the r populations are homogeneous, it makes sense to pool them when testing for independence of the j and k variables.

Table 2   Partition of Chi-square (Example 2)

(a) $H_{01} \subset H_0 \subset H_2$

Hypothesis	Chi-square	d.f.
Independence, factor level 1	$\ln n \left[\dfrac{n_1.., \mathbf{n}_{1jk}}{\mathbf{n}_{1j}., \mathbf{n}_{1 \cdot k}} \right]$	$(s-1)(c-1)$
$\vdots$	$\vdots$	$\vdots$
Independence, factor level r	$\ln n \left[\dfrac{n_r.., \mathbf{n}_{rjk}}{\mathbf{n}_{rj}., \mathbf{n}_{r \cdot k}} \right]$	$(s-1)(c-1)$
H_0	$\ln n \left[\dfrac{\mathbf{n}_i.., \mathbf{n}_{ijk}}{\mathbf{n}_{ij}., \mathbf{n}_{i \cdot k}} \right]$	$r(s-1)(c-1)$
$H_{1\mid 0}$	$\ln n \left[\dfrac{N, N, \mathbf{n}_{ij}., \mathbf{n}_{i \cdot k}}{\mathbf{n}_i.., \mathbf{n}_i.., \mathbf{n}_{\cdot j}., \mathbf{n}_{\cdot \cdot k}} \right]$	$(r-1)(s+c-2)$
H_{01}	$\ln n \left[\dfrac{N, N, \mathbf{n}_{ijk}}{\mathbf{n}_i.., \mathbf{n}_{\cdot j}., \mathbf{n}_{\cdot \cdot k}} \right]$	$rcs - r - s - c + 2$

(b) $H_{01} \subset H_1 \subset H_2$

Hypothesis	Chi-square	d.f.
H_1	$\ln n \left[\dfrac{N, \mathbf{n}_{ijk}}{\mathbf{n}_i.., \mathbf{n}_{\cdot jk}} \right]$	$(r-1)(sc-1)$
$H_{0\mid 1}$	$\ln n \left[\dfrac{N, \mathbf{n}_{\cdot jk}}{\mathbf{n}_{\cdot j}., \mathbf{n}_{\cdot \cdot k}} \right]$	$(s-1)(c-1)$
H_{01}	$\ln n \left[\dfrac{N, N, \mathbf{n}_{ijk}}{\mathbf{n}_i.., \mathbf{n}_{\cdot j}., \mathbf{n}_{\cdot \cdot k}} \right]$	$rcs - r - s - c + 2$

The statistic $\lambda_{01;2}$ at the foot of Table 2 is identical to that for testing for independence of rows, columns, and layers in a $r \times s \times c$ table of three responses from one population; see Read [5, 6] for partitions of chi-square in the latter model and others in two-dimensional tables. Interpreting the conditional chi-square statistics such as $\lambda_{0\mid 1}$ and $\lambda_{1\mid 0}$ in such examples is important, and they are omitted in some textbooks; an early exception is Kullback [4], who first displayed the partition of chi-square in contingency table analysis.

The computer program TRICHI presents estimates and partitions of chi-square under fundamental models of independence and homogeneity in three-dimensional contingency tables; it can be obtained from the

Statistical Laboratory, Southern Methodist University, Dallas, Texas 75275.

References

[1] Darroch, J. N. and Silvey, S. D. (1963). *Ann. Math. Statist.*, **34**, 555–567.

[2] Hogg, R. V. (1956). *Ann. Math. Statist.*, **27**, 529–532.

[3] Hogg, R. V. (1961). *J. Amer. Statist. Ass.*, **56**, 978–989.

[4] Kullback, S. (1959). *Information Theory*. Wiley, New York.

[5] Read, C. B. (1977). *Commun. Statist. A*, **6**, 553–562.

[6] Read, C. B. (1978). *Psychometrika*, **43**, 409–420.

(CATEGORICAL DATA
CHI-SQUARE TESTS

CONTINGENCY TABLES
LIKELIHOOD RATIO TESTS
MULTIDIMENSIONAL CONTINGENCY
  TABLES)

CAMPBELL B. READ

PARTITIONS   See HIERARCHICAL CLUSTER
ANALYSIS

PARTY TRANSFER PROBABILITY   See
ELECTION FORECASTING; TRANSFERABLE VOTE
SYSTEM

PASCAL, BLAISE

> **Born:** June 19, 1623 in Clermont-
> Ferrand, France.
> **Died:** August 19, 1662 in Paris,
> France.
> **Contributed to:** mathematics, com-
> puting machines, physics, philoso-
> phy.

The history of probability is traditionally
taken to begin in 1654, the year of the
correspondence between Pascal and Pierre
Fermat*. Pascal's central notion was that
of mathematical expectation*, and he developed decision-theoretic* ideas based
thereon. He did not use the term *probability*
[12]. Nevertheless, there are narrowly probabilistic ideas in Pascal's writings: the notions
of equally likely events, of the random
walk*, and the frequency interpretation of
probability. Through this last Pascal seems
to have had influence on the first substantial
landmark of probability, Bernoulli's theorem.

The de Méré–Pascal double-six situation
[7] is encountered at the beginning of
courses in probability: this concerns the
minimum number of throws of a pair of fair
dice to ensure at least one double six with a
better than even chance. The number of
tosses until a double-six occurs has a geometric distribution, which may explain why
the geometric distribution* and, by exten-

sion, the negative binomial distribution*, are
sometimes named in his honor. In regard to
probability theory, however, Pascal's direct
concern was with the equitable division of
total stake when a game is interrupted before its agreed completion (*problème de par-
tis*, or the Problem of Points*). One problem
of this kind solved by Pascal and Fermat (by
correspondence and by different methods),
although for some particular cases, marks an
epoch in probability theory because of difficult structural features. At each of a number
of trials, each of two players has probability
1/2 of winning the trial. It is agreed that the
first player with n wins gains the total stake.
The game is interrupted when player A
needs a trial wins to gain the stake and
player B needs b. How should the stake be
divided? The obvious sample space*, of sequences leading to game termination, consists of sample sequences of unequal length
and therefore unequal probabilities. Pascal
expands the sample space to sequences each
of length $(a + b - 1)$, which are equiprobable, and the probability of a win by A, and
hence proportion of stake presently due to
him,

$$\sum_{r=a}^{a+b-1} \binom{a+b-1}{r}\left(\frac{1}{2}\right)^{a+b-1},$$

is then easily evident, and can easily be
obtained from "Pascal's Triangle"*. We
may identify the complicating factor in this
situation as the stopping time defined on an
infinite random sequence. Pascal's study of
the rule of formation of "Pascal's Triangle"
(which he did not discover) helped him to
discover and understand the principle of
Mathematical Induction ("reasoning by re-
currences").

Pascal discovered the principle of a calculating machine when he was about 20, and
in physics his name occurs in Pascal's Principle (or Law of Pressure).

However, he is most widely known for his
philosophical and theological later writings.
He was influenced by Jansenism, a religious
movement associated with Catholicism and
distinguished by its piety; the night of November 23–24, 1654 [his famous (second)

"conversion"] resulted in his joining the lay community associated with the movement at Port-Royal des Champs. Becoming interested in pedagogic research, he contributed to a "Logic," which appeared anonymously in 1662 entitled *La Logique ou L'Art de Penser*, often referred to as the *Logique de Port-Royal*. This is now attributed, in a recent reprinting, entirely to Arnauld and Nicole [1], and Pascal's part in it remains relatively obscure. In this 1970 reprint, the fourth part, entitled *De la méthode*, likely due to Pascal, contains elements of probabilistic thinking—for the first time in contexts other than games of chance*—that illustrate most clearly his decision-theoretic ideas; and, according to Jacob Bernoulli*, partly motivated Bernoulli's Weak Law of Large Numbers. In a passage of his *Ars Conjectandi*, reproduced in Uspensky [13], Bernoulli [2] calls the unknown author "magni acuminis et ingenii vir" (a man of great acumen and ingenuity). In particular, as regards decision theory, this section of the *Logique* contains allusions to Pascal's famous passage from his *Pensées: De la nécéssité du pari* or, as it is generally known in English, *The Wager*. Essentially a philosophical argument, this can be expressed in terms of a 2×2 loss table with a prior distribution* over the states of nature and puts forward that decision as solution which minimizes expected loss [5, 10].

During the Port-Royal period also, Pascal, in correspondence to Fermat, proposed the "gambler's ruin" problem in disguised form; in a random walk on the integers between two absorbing barriers, starting in the middle, with unequal probabilities of movement a unit left or right (and a positive probability of no movement at all from any intermediate position), the probability of ruin is sought [7].

The usual assessments of Pascal's probabilistic contributions have been based on the Pascal–Fermat correspondence, and on his *Traité du Triangle Arithmétique*, discovered after his death. This treats extensively the formulation and solution of the equitable division of stakes problem in association with Pascal's triangle*. One of the first substantial mathematical treatments of such problems arising out of Pascal's work, particularly the Problem of Points*, is due to Montmort*, an extensive summary of whose book is given as Chapter 8 of [12]. Realization of the probabilistic significance of the philosophical later writings (not referred to in historical treatments of probability such as David [4] and Todhunter [12]) is of recent origin [3, 5, 7].

Pascal's mother died when he was three; his father took charge of his education. In 1631 the family moved to Paris, where one sister, Jacqueline, entered the convent of Port-Royal in 1652, an event that seems to have had substantial significance on Pascal's intense religious development, of which some mention has already been made. At this distance in time it is as difficult to assess the validity of Pascal's scientific standing as it is to assess his personality from the extant fragments of his writings. We are inclined to agree with the conclusions of F. N. David ([4], p. 97); French writers incline to a more sympathetic view (e.g., [11], p. 337). General references on his life may be found in [6], Section 5.8.

References

[1] Arnauld, A. and Nicole, P. (1970). *La Logique ou L'Art de Penser*. Georg Olms, Hildesheim, W. Germany. (Reproduction of first anonymous Paris edition of 1662. An edition was also published in 1970 by Flammarion, Paris. There is an English translation in its fifth edition by P. and J. Dickoff, New York, 1964.)

[2] Bernoulli, J. (1713). *Ars Conjectandi*. Basileae Impensis Thurnisiorum Fratrum, Basel.

[3] Coumet, E. (1970). *Ann. Econ. Soc. Civilis.*, **5**, 574–598.

[4] David, F. N. (1962). *Games, Gods and Gambling: The Origins and History of Probability and Statistical Ideas from the Earliest Times to the Newtonian Era*. Charles Griffin, London. (Contains an entertaining assessment of the Pascal–Fermat correspondence and Pascal's scientific standing and personality.)

[5] Hacking, I. (1975). *The Emergence of Probability*. Cambridge University Press, London. (Focuses

on philosophical aspects. Chapter 8 deals with "The Wager" as a decision-theoretic problem.)

[6]   Heyde, C. C. and Seneta, E. (1977). *I. J. Bienaymé: Statistical Theory Anticipated*. Springer, New York. (Section 5.8 gives a biographical account in the context of probability and statistics.)

[7]   Ore, O. (1960). *Amer. Math. Monthly*, **67**, 409–419. (The first more-or-less complete account of Pascal's contributions to probability and decision theory.)

[8]   Pascal, B. (1904–1925). *Oeuvres*, Vols. 1–14, L. Brunschvieg, P. Boutroux, and F. Gazier, eds., "Les Grands Ecrivains de France", Hachette, Paris. (Collected works with editorial comments: a standard version.)

[9]   Pascal, B. (1970). *Oeuvres Complètes*, Vol. 2, J. Mesnard, ed. Desclée de Brouwer, Paris. (From a more recent collection than ref. 8. Possibly the best edition to date of complete writings with extensive editorial comments. Volume 1 was published in 1964. Volume 2 contains Pascal's work on the equitable division of stakes.)

[10] Seneta, E. (1979). In *Interactive Statistics*, D. McNeil, ed. North-Holland, Amsterdam. (Pascal's mathematical and philosophical writings pertaining to aspects of probability and decision theory are sketched and considered from a modern viewpoint.)

[11] Taton, R. (1974). In *Dictionary of Scientific Biography*, Vol. **10**, C. C. Gillispie, ed. pp. 330–342. (Complete survey of Blaise Pascal by an eminent French historian of science.)

[12] Todhunter, I. (1865). *A History of the Mathematical Theory of Probability from the Time of Pascal to that of Laplace*. Cambridge University Press, London and Cambridge. (Reprinted in 1949 and 1961 by Chelsea, New York. Standard reference on the early history.)

[13] Uspensky, J. V. (1937). *Introduction to Mathematical Probability*. McGraw-Hill, New York, pp. 105–106.

(BERNOULLIS, THE
CHANCE
DECISION THEORY
EXPECTED VALUE
FERMAT, PIERRE DE
FREQUENCY INTERPRETATION
    OF PROBABILITY
GAMBLING, STATISTICS IN
LAWS OF LARGE NUMBERS
PASCAL'S TRIANGLE
RANDOM WALK)

E. SENETA

PASCAL DISTRIBUTION   *See* NEGATIVE BINOMIAL DISTRIBUTION

PASCAL'S TRIANGLE

In his *Traité du Triangle Arithmétique* (written in 1654 and published posthumously in 1665) Blaise Pascal* defined the numbers in the "arithmetical triangle" (see Fig. 1): "The number in each cell is equal to that in the preceding cell in the same column plus that in the preceding cell in the same row." He placed an arbitrary number in the first cell (in the right angle of the triangle) and regarded the construction of the first row and column as special "because their cells do not have any cells preceding them."

The arbitrary number in the first cell is an uninteresting generalization, and we shall assume it to be unity, as did Pascal in all his applications. Then we have for the number in the ith row and jth column

$$f_{i,j} = f_{i-1,j} + f_{i,j-1}, \quad i, j = 2, 3, 4, \ldots$$
$$f_{i,1} = f_{1,j} = 1, \qquad i, j = 1, 2, 3, \ldots . \tag{1}$$

The numbers thus defined have a triple origin; as "figurate numbers," they have Pythagorean roots, as "combinatorial numbers," they have Hindu and Hebrew roots; and as "binomial numbers," they have Chinese and Arabic roots.

The successive rows (Fig. 1) form the figurate numbers; thus the third row lists the "triangular numbers," or numbers that can be represented by triangular arrays of points, and the fourth row the similar "tetrahedral numbers." The latter, consisting as they do of layers of triangular numbers, reflect the fact that each is the sum of the

```
1    1    1    1    1    1 . .
1    2    3    4    5    . . .
1    3    6   10    .    .  . .
1    4   10    .    .    .  . .
1    5    .    .    .    .  . .
1    .    .    .    .    .  . .
```

Figure 1   Figurate Numbers.

triangular numbers up to and including the one in the same column. This, indeed, is the general defining property of all the figurate numbers, easily derived from (1), and allows them to be extended to more than the three dimensions considered by the Pythagoreans.

The binomial numbers, the coefficients of a binomial expansion, are found in the diagonals of the arithmetical triangle (from upper right to lower left), their identity with the figurate numbers having been recognized in Persia and China as early as the eleventh century and in Europe in the sixteenth. Relation (1) is immediately apparent on considering the expansion of both sides of

$$(x + y)^n \equiv (x + y)(x + y)^{n-1}. \quad (2)$$

The fact that the coefficient of $x^r y^{n-r}$ in $(x + y)^n$ may be expressed as

$$\frac{n(n - 1)(n - 2) \ldots (n - r + 1)}{1 \cdot 2 \cdot 3 \ldots r} = \binom{n}{r}$$
$$(3)$$

was known to the Arabs in the thirteenth century and to the Renaissance mathematician Cardano in 1570. It provides the solution to (1) in closed form:

$$f_{i,j} = \binom{i + j - 2}{i - 1}. \quad (4)$$

In statistics, however, the occurrence of the numbers in the arithmetical triangle is due to their being the combinatorial numbers. If ^nC_r is the number of combinations of n different things taken r at a time, then

$$^nC_r = \binom{n}{r}, \quad (5)$$

an identity known in India in the ninth century, to Hebrew writers in the fourteenth century, and to Cardano in 1550. [Note that ^nC_r and $\binom{n}{r}$ are not merely alternative modern notations for the same thing; equation (5) requires proof.] The identity of the combinatorial and figurate numbers was known even earlier in India, being embodied in the *Meru Prastara* of Pingala (ca. 200 B.C.). By the twelfth century in India, it was appreciated that ^nC_r was also the number of arrangements of r things of one kind and

$(n - r)$ of another. The connection with the binomial coefficients is obvious.

The manifold properties, mutual relations, and applications of the numbers defined by (1) fascinated Pascal, although his only original contributions were to prove (3) (the first proof of the binomial theorem for integral index) and to note that if (1) is written in terms of the combinatorial numbers, it becomes

$$^{n+1}C_{r+1} = {}^nC_r + {}^nC_{r+1}, \quad (6)$$

which has a direct proof: Considering any particular one of the $(n + 1)$ things, ^nC_r gives the number of combinations that contain it while $^nC_{r+1}$ gives the number that excludes it, the two numbers together thus giving the total.

Since Pascal's time, endless further properties of the numbers have been discovered. In statistics they crop up wherever there is a connection with the binomial distribution, and thus they pervade the theory of discrete random processes. Knowledge of their properties enabled Pascal to solve the Problem of Points* and thus, with Fermat*, to give the theory of probability its initial impetus. In analysis, they led to the Bernoulli numbers*, to Leibniz' discovery of the calculus, and to Newton's discovery of the general binomial theorem. They are fundamental to the theory of finite differences* and in combinatorics* itself they are, of course, central.

Finally, we may note that the common modern form of the arithmetical triangle shown in Fig. 2 was not actually used by Pascal, though it is itself very old in both combinatorial and binomial applications. Since it was popularized by Cardano's con-

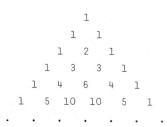

Figure 2 Pascal's Triangle.

temporary Tartaglia in 1556, Italians refer to Tartaglia's Triangle rather than Pascal's, though Tartaglia made no special study of it.

Literature

For the properties of the numbers of Pascal's triangle see D. E. Knuth, *The Art of Computer Programming*, 2nd ed., Vol. I: *Fundamental Algorithms* (Addison-Wesley, Reading, MA. 1978). For a historical account, see A. W. F. Edwards, *Pascal's Arithmetical Triangle* (Griffin, High Wycombe, England, in press).

(COMBINATORICS
PASCAL, BLAISE)

A. W. F. EDWARDS

PASCUAL'S ESTIMATOR

Pascual [1] proposes the approximately unbiased ratio estimator* for the ratio of mean values of variables X, Y:

$$t_p = r + \frac{1}{(n-1)\overline{X}}(\overline{y} - \overline{r}\overline{x})$$

with

$$r = \frac{\overline{y}}{\overline{x}} \quad \text{and} \quad \overline{r} = \sum_{i=1}^{n} \frac{y_i}{nx_i},$$

where $\overline{x}$ and $\overline{y}$ are sample means for variables x and y for a sample of n observations $(x_1, y_1), \ldots, (x_n, y_n)$ and $\overline{X}$ is the population mean of the x's.

He derives the asymptotic theory of t_p assuming that the higher-order population moments of $\delta x_i = (x_i - \overline{X})/\overline{X}$ and $\delta y_i = (y_i - \overline{Y})/\overline{Y}$ are negligible and that $\delta x_i < 1$. He concludes that for large n, t_p and r are about equally efficient.

Rao [2] gives the mean square error of t_p, if $y_i = \alpha + \beta x_i + \mu_i$, where x_i/n has a gamma distribution* with parameter h so that $\overline{x}$ has a gamma distribution with param-

eter $m = nh$. Then

$$\mathrm{MSE}(t_p)$$

$$= \alpha^2 \left\{ \frac{1}{(m-1)^2(m-2)} \right.$$

$$+ \frac{(2n-2)(m-2n) + mn - m + n}{mn^3(m-n)(m-2n)(n-1)}$$

$$+ \frac{(2n-1)(m-n)^2 + m^2(n-1)^2(n-2)}{m^2n^3(m-n)^2(n-1)^2}$$

$$\left. + \frac{(n-1)^4 - (m-1)^4}{m^2(n-1)^2(m-1)^2(m-n)^2} \right\}$$

$$+ \delta \left\{ \frac{1}{(m-1)(m-2)} - \frac{2}{m(m-1)(m-n)} \right.$$

$$- \frac{m+n}{m^2n^2(m-n)}$$

$$\left. + \frac{mn - m + n}{mn^2(m-n)(m-2n)} \right\}$$

for $m > 2n$ and δ a constant of order n^{-1}.

Comparisons of estimators t_p with other ratio estimators are given by Rao and Beegle [3] (*see* QUENOUILLE ESTIMATOR). In those comparisons, t_p is moderately efficient with no striking advantages or disadvantages.

References

[1] Pascual, J. N. (1961). *J. Amer. Statist. Ass.*, **56**, 70–87.

[2] Rao, J. N. K. (1967). *Biometrika*, **54**, 321–324.

[3] Rao, J. N. K. and Beegle, L. D. (1967). *Sankhyā B*, **29**, 47–56.

(RATIO ESTIMATORS
STATISTICAL DIFFERENTIALS
SURVEY SAMPLING)

HANS T. SCHREUDER

PASSAGE TIMES

The theory of passage times, or as more commonly referred to, first passage times, seeks to evaluate the rate at which suitably defined events can occur together with probabilities that may be associated with specific models. The simplest examples of this class of problems can be traced to gambling problems studied in the pioneering phase of probability theory [17]. In particular, the so-called *gambler's ruin problem* can illustrate some of the questions analyzed in the framework of first passage time problems [13]. A statement of this problem is as follows: A gambler with initial fortune N plays a game in which, at each turn he wins an additional unit of money with probability a or loses a unit with probability $b = 1 - a$. The first passage time would be the time at which the gambler's fortune first reaches a specified level, say, zero. A second aspect sometimes analyzed is the probability that the gambler's fortune reaches one level before reaching another. One context in which this arises is in a study of two gamblers playing against one another in which the probability of one gambler bankrupting the second, rather than being bankrupted himself, becomes a quantity of interest.

Passage time problems play a fundamental role in many areas of applied probability* and indeed the earliest studies of such problems appeared in the context of physics (*see* PHYSICS, STATISTICS IN, EARLY HISTORY) and chemical physics [21, 24]. The theory of sequential analysis* leans heavily on the formulation of first passage time theory [28], and Wald's identity* (*see* FUNDAMENTAL IDENTITY OF SEQUENTIAL ANALYSIS) was developed as a tool for calculating average sample numbers for such statistical designs [27]. The analysis of Kolmogorov–Smirnov statistics* reduces to the study of a particular set of passage-time problems [11, 12]. Passage time problems arise quite naturally in engineering where one is often interested in the statistics of the time for a signal to exceed a specified level [5] or in the time to system failure in the study of models of reliability* [2]. Other applications will be mentioned later, but this brief catalog should suffice to indicate the breadth of use of passage time models in statistics and applied probability. Because of this large number of applications, interest in the theory is high, and a considerable number of publica-

tions appear yearly on the subject in theoretical journals exemplified by the *Annals of Probability*, as well as in many more specialized journals.

A formal definition of the first passage time starts by defining a state space S that is partitioned into two disjoint subspaces S_1 and S_2. Let $x(t)$ be a random variable, depending on time, which takes values on S, and let $x(0) = x_0$ initially be in S_1. The first passage time to S_2 from $x(0)$ is a random variable defined as the earliest time at which $x(t)$ reaches S_2. In the terminology of random walks, S_2 can be characterized as an absorbing space. If S_2 is the boundary of S_1 and $x(t)$ is a diffusion process*, S_2 is sometimes referred to as an *exit boundary*, and the first passage time is sometimes referred to as the *killing time*. The formal definition of the first passage time is

$$T(x_0) = \inf\{t \mid x(t) \in S_2, x(0) = x_0 \in S_1\}.$$

(1)

When the absorbing space S_2 is composed of disjoint distinguishable subspaces

$$S_2 = S_{2,1} \cup S_{2,2} \cup S_{2,3} \cup \cdots,$$

and absorption is irreversible, then a second set of quantities of interest consists of the probabilities of absorption by the $S_{2,j}$, $j = 1, 2, \ldots$:

$$P_j(x_0) = \Pr\{x(\infty) \in S_{2,j} \mid x(0) = x_0 \in S_1\}.$$

(2)

A more rigorous formulation of some of these definitions is found in Belyayev [3]. Passage time problems involve the calculation of statistical properties of $T(x_0)$ and/or the $P_j(x_0)$. Other quantities occasionally are of interest. As an example, one might wish to calculate statistical properties of the first passage time conditional on termination in a subspace $S_{2,j}$.

To show how one defines these spaces in a particularly simple example, consider the gambler's ruin problem mentioned earlier. If the gambler plays against an infinitely rich opponent, the total space S is defined as $S = \{0, 1, 2, 3, \ldots\}$, the absorbing space is the state of bankruptcy, or equivalently, $S_2 = \{0\}$, and $S_1 = S - S_2 = \{1, 2, \ldots\}$. When the gambler plays against an opponent who himself has a finite fortune M the game will end when either fortune is exhausted. In this case $S = \{0, 1, 2, \ldots, M + N\}$, $S_{2,1} = \{0\}$, $S_{2,2} = \{M + N\}$, and $S_1 = S - S_{2,1} - S_{2,2} = \{1, 2, 3, \ldots, M + N - 1\}$. The quantities of interest for this system are the statistical properties of the time to completion of the series of games and the respective probabilities of winning. Other applications of probability theory may require the study of first passage time problems without explicit use of the terminology. For example, in the context of queuing theory*, if one is interested in finding a maximum length of a queue during a fixed time period T then one defines an absorbing state $S_2 = \{n\}$ and calculates the probability that the queue length will remain less than n during the interval T. If this probability is $p_n(T)$ the probability that the maximum queue length is equal to n in the interval T is $p_{n+1}(T) - p_n(T)$. However, $p_n(T)$ can also be interpreted as the probability that the first passage time to n is greater than T, thus establishing the relation between the two ways of looking at the problem. Many similar applications of this reasoning occur in probability theory.

Formalisms for the solution of a wide variety of first passage time problems are available, but generally useful computational methods for implementing these methods depend on simplifying features being present such as a one-dimensional geometry or special symmetries. The random variable $x(t)$ can be either discrete or continuous, depending on the application, and the time can likewise be measured in discrete or continuous units. Thus the evolution of $x(t)$ will be described by an equation that can be either a difference, a partial differential, integral, or integrodifferential equation. All types are met in practice. However, many techniques for finding formal solutions to first passage time problems can be dealt with in a unified way.

For concreteness let us consider the case of continuous x and continuous t. The evolution of x can be described in terms of the probability density function $p(x, t \mid x_0, 0)$, which will be assumed to obey a linear equation of the form

$$\frac{\partial p}{\partial t} = L(x) p, \tag{3}$$

where the operator L is generally independent of time. The probability that $T(x_0)$ will exceed t is

$$F(t; x_0) \equiv \Pr\{T(x_0) > t\}$$
$$= \int_{S_1} p(x, t_0 \mid x_0, 0) \, dx. \tag{4}$$

Only a small class of operators $L(x)$ lead to solvable forms for (3) so that (4) is not often a practical prescription for calculating $F(t; x_0)$. However, if one can find a related, or adjoint operator to $L(x)$, the calculation of moments of $F(t; x_0)$ leads to a somewhat simpler mathematical problem. Specifically, suppose that the adjoint operator $L^+(x_0)$ depends only on x_0, and that in terms of $L^+(x_0)$, p satisfies

$$\frac{\partial p}{\partial t} = L^+(x_0) p. \tag{5}$$

To derive an equation directly for $F(t; x_0)$ all we need do is integrate (5) with respect to x. This leads to

$$\frac{\partial F(t; x_0)}{\partial t} = L^+(x_0) F(t; x_0). \tag{6}$$

In particular, the mean first passage time

$$\mu_1(x_0) = \int_0^\infty F(t; x_0) \, dt \tag{7}$$

can be found by integrating (6) over t. Therefore, it satisfies

$$L^+(x_0) \mu_1(x_0) = -1. \tag{8}$$

Similar recursive equations can be developed for higher-order moments. Equation (8) is a much simpler way to calculate $\mu_1(x_0)$ than starting from (3) and (4) followed by the integration in (7). This argument was first developed for the Fokker–Planck equation* in [20], but it is valid more generally.

A simple illustration of the preceding formalism can be given by the gambler's ruin problem. Let $p(r, n \mid N, 0)$ be the probability that the gambler's fortune is r at the nth play given an initial fortune of N. Then the discrete analog of (3) is

$$p(r, n + 1 \mid N, 0) = bp(r + 1, n \mid N, 0)$$
$$+ ap(r - 1, n \mid N, 0) \quad (9)$$

so that the operator L is a matrix in this case. The adjoint is then the transpose of L so that the analog of (5) is

$$p(r, n \mid N, 0) = ap(r, n - 1 \mid N - 1, 0)$$
$$+ bp(r, n - 1 \mid N + 1, 0)$$
$$(10)$$

A more sophisticated and useful application of extension of ideas inherent in the gambler's ruin problem is developed as the theory of risk (see RISK THEORY), which is most often applied by insurance companies, where premiums must be calculated to minimize the risk of being wiped out by catastrophic claims [22]. The most useful application of the formalism in (3)–(8) is in the analysis of systems whose evolution is described by a Fokker–Planck or forward Kolmogorov equation [8]. In this case, the operator $L(\mathbf{x})$ is

$$L(\mathbf{x}) = \sum_i \sum_j \frac{\partial^2}{\partial x_i \, \partial x_j} \left(a_{ij}(\mathbf{x}) \right)$$
$$- \sum_i \frac{\partial}{\partial x_i} \left(b_i(\mathbf{x}) \right) \quad (11)$$

in any number of dimensions, where the a_{ij} and b_i are derivable from the infinitesimal moments of the underlying stochastic process. The adjoint operator $L^+(\mathbf{x}_0)$ is given by Cox and Miller [8]:

$$L^+(\mathbf{x}_0) = \sum_i \sum_j a_{ij}(\mathbf{x}_0) \frac{\partial^2}{\partial x_{0i} \, \partial x_{0j}}$$
$$+ \sum_i b_i(\mathbf{x}_0) \frac{\partial}{\partial x_{0i}}. \quad (12)$$

While the preceding formalism is independent of dimension, in fact the most generally useful results have only been found for one-dimensional problems, because of the necessarily simple geometry. In one dimension, e.g., when $L^+(x_0)$ is given in (12), the equation for the mean first passage time in (8)

can be solved explicitly. A number of first passage problems on the line, useful in statistical testing, are worked out in detail in Darling and Siegert [10]. Related material is presented in Siegert [23].

A general result, often rediscovered for different processes, is that of the asymptotic exponentiality of the first passage time distribution to infrequently visited states. An early example of this occurs in the theory of the maxima of one-dimensional diffusion processes* [19], but the same phenomenon appears in the analysis of Markov chains and other models related to reliability studies [15, 25]. A fruitful application of these ideas is to the study of the occurrence of high levels in random noise processes [5]. A second generally useful result is Wald's identity, whose principal utility is for first passage-time problems on a bounded line [26]. Using Wald's identity, one can derive exact relations between moments of the passage time to absorption at either end of the line and moments of single-step displacements.

APPLICATIONS

The problem and variety of applications of first passage time concepts is so great that no pretense of completeness can be made for the few that we now enumerate.

We list just three statistical applications. In sequential analysis, one studies the (generally two-dimensional) random walk* defined by a sequential probability ratio test*, whose argument is determined by experiments performed up to the nth step [28]. Boundaries for terminating the experiment are calculated so as to satisfy requirements set on absorption probabilities. It then becomes of interest to determine statistical properties of the number of samples required to terminate the experiment as well as the probability of termination at specific parts of the boundary. This information is used for experimental design. A second type of sequential testing is that of inverse sampling (see INVERSE SAMPLING), in which one waits until a random event of interest occurs for the nth time. Thus one is interested in statistical properties of the passage time till that occurrence. Similar ideas occur in renewal theory* [7]. In the analysis of Kolmogorov–Smirnov tests (*see* KOLMOGOROV–SMIRNOV STATISTICS; KOLMOGOROV–SMIRNOV TYPE TESTS OF FIT) one is interested in the construction and evaluation of tests for testing hypotheses about the fit of empirical distributions to known ones [12]. If $F_n(x)$ is the proportion of the values $x_1, x_2, \ldots, x_n < x$, then various statistics such as $D_n = \sup_{-\infty < x < \infty} |F_n(x) - F(x)|$ can be related to statistical properties of the crossing by $F_n(t) - t$ of boundaries that may be curvilinear. Such first passage time problems give rise to questions of considerable mathematical complexity and have been studied mainly for Brownian motion* [4, 11, 14].

Applications of passage time methodology abound in engineering and the physical sciences. A problem that arises in many guises in engineering concerns the statistics of the crossing of different levels by signals contaminated by noise* [5] as well as the intervals between successive level crossings. These are directly describable in terms of passage problems and make use of methodology in this mathematical area. Many reliability problems involve calculation of the first time to system failure under various programs for equipment maintenance [2]. The theory of rate processes and reaction times in chemical physics is often phrased in terms of first passage times [29]. Passage time problems for diffusing particles were among the earliest applications of these methodologies in any discipline [21, 24]. In astronomy, the rates of escape of stars from clusters constitute an important use of these ideas [6].

In biology, the study of the ultimate fixation of mutant genes constitutes an important part of theoretical population genetics [9]. Questions related to maximum epidemic size [1] and other problems arising in population theory can be discussed naturally in the framework of passage time methodology [16]. Interesting recent applications are to the statistics of nerve spike trains [26] and to the estimation of the maxima of carbon monoxide concentration in the lungs of smokers [18].

References

[1] Bailey, N. T. J. (1976). *The Mathematical Theory of Infectious Diseases and Its Applications*. Hafner, New York.

[2] Barlow, R., and Proschan, F. (1975). *Statistical Theory of Reliability and Life Testing: Probability Models*. Wiley-Interscience. New York.

[3] Belyayev, Yu. K. (1972). In *Proc. Sixth Berkeley Symp.*, *Math. Statist. Prob.*, Vol. 3, University of California Press, Berkeley, CA, pp. 1–17. (A rigorous definition of first passage problems in the context of point processes.)

[4] Berman, S. M. (1982). *Ann. Prob.*, **10**, 1–46. (A good review of a large body of work on sojourns and extremes of stationary processes.)

[5] Blake, I. F. and Lindsey, W. C. (1971). *IEEE Trans. Inf. Theory*, 295–315. (A useful introduction to the engineering literature on level-crossing problems up to 1971. The bibliography has 133 references.)

[6] Chandrasekhar, S. (1960). *Principles of Stellar Dynamics*. Dover, New York.

[7] Cox, D. R. (1962). *Renewal Theory*. Wiley, New York.

[8] Cox, D. R. and Miller, H. D. (1965). *The Theory of Stochastic Processes*. Wiley, New York.

[9] Crow, J. F., and Kimura, M. (1970). *An Introduction to Population Genetics Theory*. Burgess, Minneapolis, MN.

[10] Darling, D. A. and Siegert, A. J. F. (1953). *Ann. Math. Statist.* **24**, 624–639. (Several one-dimensional problems arising in statistics are solved in detail essentially using the formalism discussed in this article.)

[11] Durbin, J. (1971). *J. Appl. Prob.*, **8**, 431–453. (Studies of boundary-crossing probabilities related to the theory of Kolmogorov–Smirnov tests.)

[12] Durbin, J. (1973). *Distribution Theory for Tests Based on the Sample Distribution Function*. SIAM, Philadelphia.

[13] Feller, W. (1968). *An Introduction to Probability Theory and Its Applications*, Vol. I. Wiley, New York.

[14] Ferebee, B. (1983). *Z. Wahrscheinlichkeitsth. verw. Geb.*, **63**, 1–15. (A systematic expansion of the density function for the first passage time of a Brownian motion over a moving boundary of general form.)

[15] Keilson, J. (1979). *Markov Chain Models—Rarity and Exponentiality*. Springer-Verlag, New York.

(The theory of Markov chains with nearest-neighbor transitions. A good deal of this monograph is devoted to first passage time problems.)

[16] Keyfitz, N. (1977). *Applied Mathematical Demography*. Wiley-Interscience. New York.

[17] Maistrov, L. E. (1974). *Probability Theory. A Historical Sketch*, translated by S. Kotz. Academic Press, New York.

[18] Marcus, A. H. and Czajkowski, S. (1979). *Biometrics*, **35**, 539–548.

[19] Newell, G. F. (1962). *J. Math. Mech.*, **11**, 481–496. (Extreme-value theory for one-dimensional diffusion processes.)

[20] Pontryagin, L., Andronow, A., and Witt, A. (1933). *Zh. Eksp. i Teor. Fiz.*, **3**, 177–189.

[21] Schrödinger, E. (1915). *Phys. Zeit.*, **16**, 289–293.

[22] Seal, H. (1978). *Survival Probabilities: The Goal of Risk Theory*. Wiley-Interscience, New York.

[23] Siegert, A. J. F. (1951). *Phys. Rev.*, **81**, 617–623.

[24] Smoluchowski, M. (1915). *S. B. Akad. Wiss. Wien*, **2a**, 124–150 and 339–364.

[25] Soloviev, A. D. (1972). In *Proc. Sixth Berkeley Symp. Math. Statist. Prob.*, Vol. 3, University of California Press, Berkeley, CA, pp. 17–86. (Asymptotic theory of the first passage time for the crossing of a high level by a birth-and-death process with jumps allowed to nearest neighbors only.)

[26] Tuckwell, H. S. (1976). *J. Appl. Prob.*, **13**, 39–48.

[27] Wald, A. (1944). *Ann. Math. Statist.*, **15**, 283–294.

[28] Wald, A. (1947). *Sequential Analysis*. Wiley, New York.

[29] Weiss, G. H. (1967). *Adv. Chem. Phys.*, **13**, 1–18. (Discusses several problems in chemical physics in the framework of first passage times.)

Bibliography

Karlin, S. and Taylor, H. M. (1981). *A Second Course in Stochastic Processes*. Academic Press. New York. (Contains a sophisticated treatment of diffusion processes as well as several interesting applications of first passage times.)

Kemperman, J. H. B. (1961). *The Passage Problem for a Stationary Markov Chain*. University of Chicago Press, Chicago. (A very detailed study of first passage times for lattice models. Many exact results are given that would be difficult to apply in practice.)

Mandl, P. (1968). *Analytical Treatment of One-Dimensional Markov Processes*. Springer-Verlag, New York. (Contains an excellent and exhaustive treatment of first passage times for one-dimensional diffusion processes.)

(BROWNIAN MOTION
CUMULATIVE SUM CONTROL CHARTS
DAM THEORY
DIFFERENCE EQUATIONS
FOKKER–PLANCK EQUATION
FUNDAMENTAL IDENTITY
 OF SEQUENTIAL ANALYSIS
INVENTORY THEORY
INVERSE SAMPLING
KOLMOGOROV–SMIRNOV STATISTICS
KOLMOGOROV–SMIRNOV-TYPE
 TESTS OF FIT
MARKOV PROCESSES
QUEUEING THEORY
RANDOM WALKS
RENEWAL THEORY
RISK THEORY
SEQUENTIAL ANALYSIS
STOCHASTIC DIFFERENTIAL
 EQUATIONS)

GEORGE H. WEISS

PATH ANALYSIS

This statistical technique has, until recent years when attempts at mathematical formalization have been made, been confined to applied contexts such as population genetics and the social sciences; consequently, its specific structure has remained unclear although related work has been done in econometrics* in the guise of structural equation models*. It appears to be, at least in its sociological manifestations, a technique for measuring the strength of postulated causal relationships and for substantiating (or rejecting) the internal "causal" consistency of a network of such relationships.

The basic object of interest is a set of random variables $\mathbf{X}' = \{X_1, \ldots, X_{m+n}\}$, $m \geqslant 0$, $n \geqslant 1$, $m + n \geqslant 2$. Associated with each $j > m$ is a nonempty subset

$$S_j = \left\{ X_{c_j(1)}, X_{c_j(2)}, \ldots, X_{c_j(v_j)} \right\}$$

of $\{X_1, \ldots, X_{m+n}\} - \{X_j\}$: the $X_{c_j(i)}$ are said to be the *direct causes* of X_j, and we write for each i, $X_{c_j(i)} \to X_j$. If $m \geqslant 1$, $X_1, \ldots, X_m$ are said to be the *exogenous* (or basic) variables of the system $\mathbf{X}'$, i.e., have no direct causes. The variables $X_{m+1}, \ldots, X_{m+n}$ are said to be *endogenous* (*see also* ECONOMETRICS). If for some $X_{i_1}, X_{i_2}, \ldots, X_{i_k}$

in $\mathbf{X}'$, X_{i_r} is a direct cause of $X_{i_{r+1}}$ for $r = 1, 2, \ldots, k - 1$ for $k > 2$, then X_{i_1} is said to be an indirect cause of X_{i_k} and X_{i_k} an indirect effect of X_{i_1}. By *cause* (and similarly, by *effect*), a direct or indirect cause is meant; a cause may be simultaneously direct and indirect for some effect. If X_i is an indirect cause of itself for some i, the system is said to be *nonrecursive*; otherwise it is said to be *recursive*. The direct causal relationships are assigned (and hence the exogenous and endogenous variables defined) in the social sciences by the statistical investigator to reflect the causal relationship between the real quantities they represent (from the viewpoint of mathematical analysis the assignment can be made more or less arbitrarily); the assignment may be changed after initial analysis. Generally, in population genetics, the initial path diagram is part of the underlying and immutable causal information. A useful feature for further development and intuitive grasp is a pictorial representation called a *path diagram*; the initial diagram is obtained by drawing an arrow from each direct cause to each endogenous variable. In addition, for each endogenous variable X_j, a *residual* random variable U_j is added to the diagram, this being treated as a direct cause of X_j also.

If $\mathbf{X}' = (\mathbf{Z}', \mathbf{Y}')$, where $\mathbf{Z}'$ is the original set of exogenous variables, a linear relationship

$$\mathbf{Y} = \mathbf{BY} + \mathbf{\Gamma Z} + \mathbf{U} \qquad (1)$$

is postulated, where the nonzero entries of the matrices $\mathbf{B}$ and $\mathbf{\Gamma}$ correspond to arrows in the path diagram to each of the endogenous variables in the set $\mathbf{Y}'$; and $\mathbf{U} = \{U_j\}$. For a recursive system $\mathbf{I} - \mathbf{B}$ is nonsingular, so

$$\mathbf{Y} = (\mathbf{I} - \mathbf{B})^{-1}\mathbf{\Gamma Z} + \mathbf{V} \qquad (2)$$

where $\mathbf{V} = (\mathbf{I} - \mathbf{B})^{-1}\mathbf{U}$. The numerical values of the nonzero entries of $\mathbf{B}$ and $\mathbf{\Gamma}$, which are also entered on the corresponding arrows in the path diagram, are called *path coefficients*, and they are obtained from a *pairwise correlation table* of the whole set of the random variables $\mathbf{X}'$; that is, effectively $(n + m)(n + m - 1)/2$ pairwise correlation

coefficients. The path diagram and correlation table are the essential ingredients of a path analysis. The path coefficients are interpreted as absolute measures of direct causal influence on each endogenous variable by its direct causes.

Conventions differ on how the numerical values of the nonzero entries of $\mathbf{B}$ and $\mathbf{\Gamma}$ are to be obtained from the correlation table. One possibility is to regard U_j as the residual of a best linear prediction* in the least-squares* sense for an endogenous random variable X_j in terms of its direct causes, assuming all variables in the system $\mathbf{X}'$ have been standardized. Then (as is well-known) U_j has zero mean and is uncorrelated with the direct causes of X_j; but in general will be correlated with other residuals and with exogenous random variables other than the direct causes of X_j. This is an approach directly via equation (1), which defines the residual variable U_j and which may also be given a path coefficient to accord with its standardization to variance unity. This approach is found to give a mathematically consistent structure with all coefficients uniquely identified by ordinary least squares and all information in the correlation table used [5]. It may be, however, that the investigator wishes to insist at the outset that $\mathrm{Corr}(X_i, U_j) = 0$ for each nonresidual exogenous variable X_i and each *residual* U_j; then the role of specifying the path coefficients is taken over in part by such a requirement. Superposition of the best linear predictor approach often leads to problems of overidentification* of coefficients as a manifestation of internal mathematical inconsistency. Another approach is to define the elements of $\mathbf{V}$ in (2) as the residuals of a best linear prediction for $\mathbf{Y}$ in terms of $\mathbf{Z}$. This will specify the entries of the matrix $\mathbf{\Pi} = (\mathbf{I} - \mathbf{B})^{-1}\mathbf{\Gamma}$ which will sometimes be adequate to yield in turn all entries of $\mathbf{B}$ and $\mathbf{\Gamma}$ uniquely (the system is just identified); however, sometimes there will be many solutions (underidentified), and sometimes none (overidentified). This is in contrast to the best linear predictor approach directly via (1), which always gives a just identified system.

At least one of the aims of path analysis is

to attain a path diagram that is internally consistent from the point of view of causal interpretation in that it represents, in a plausible mathematical sense, the actual cause-and-effect network. That is, for every pair (X_i, X_j) of $\mathbf{X}'$ where X_i is an indirect cause of X_j, the path coefficient which would have been attached to the path $X_i \to X_j$ is negligible. If one adopts the best linear predictor approach, the problem may be considered in terms of partial correlation coefficients: If in the original formulation X_j has direct causes $X_{c_j(2)}, \ldots, X_{c_j(v_j)}$ and X_i is an indirect cause of X_j, by setting $X_i = X_{c_j(1)}$ and computing the partial correlation* between X_j and X_i when the linear effect of each of $X_{c_j(2)}, \ldots, X_{c_j(v_j)}$ has been removed, we can check (by nonnegligibility or otherwise of the partial correlation coefficient) whether X_i, if included as a direct cause of X_j, would indeed give a nonnegligible path coefficient. This is the mathematical formalization of the Simon–Blalock procedure in the social sciences. If the system turns out to be internally inconsistent, the system will need to be recalculated.

If the best linear predictor approach to the definition of the residuals is adopted, path analysis emerges as a second-order linear (correlational) analysis of interlocking predictor systems.

Another historical aspect of path analysis, dating to the original work of Wright [9, 10] is the decomposition of correlation* according to contributions from various sources. Components are sometimes classified as being due to direct effect, indirect effect, and spurious effect. Path-tracing rules of Wright, applied to an augmented path diagram, enable correlations between pairs of endogenous random variables in $\mathbf{X}'$ to be obtained in terms of path coefficients and correlations between an augmented set of exogenous random variables, although equivalent analytical procedures are safer.

Additional problems that have been considered include the treatment of "unobserved" random variables, problems of estimation (we have assumed the correlation table to be known) and the interrelation between the structure of a correlation table

and a corresponding parsimonious internally consistent path diagram.

The traditional beginnings of path analysis lie in the 1920s with the geneticist Sewall Wright and the genetics association has been continued in his writings and those of his disciples.

For a detailed treatment of several numerical examples, see Kang and Seneta [5].

References

[1] Blalock, H. M., ed. (1971). *Causal Models in the Social Sciences*. Macmillan, London.

[2] Duncan, O. D. (1975). *Introduction to Structural Equation Models*. Academic Press, New York. (The approach from the standpoint of the social sciences. Extensive bibliography.)

[3] Goldberger, A. S. and Duncan, O. D., eds. (1973). *Structural Equation Models in the Social Sciences*. Seminar Press, New York.

[4] Heise, D. R. (1975). *Causal Analysis*. Wiley, New York.

[5] Kang, K. M. and Seneta, E. (1980). In *Developments in Statistics*, Vol. 3, P. R. Krishnaiah, ed. Academic Press, New York, pp. 217–246. (Uses the best linear predictor approach and views path analysis from a correlational viewpoint. A modern attempt to put path analysis on a proper statistical footing.)

[6] Li, C. C. (1975). *Path Analysis: A Primer*. Boxwood Press, Pacific Grove, CA. (Elementary account of usual treatment of examples, including population genetics. Useful bibliography.)

[7] Moran, P. A. P. (1961). *Aust. J. Statist.*, **3**, 87–93. (An early attempt at rigorous probabilistic formulation of path analysis.)

[8] Wermuth, N. (1980). *J. Amer. Statist. Ass.*, **75**, 963–972. (A modern attempt to put path analysis on a proper statistical footing.)

[9] Wright, S. (1921). *J. Agric. Res.*, **20**, 557–585. (Generally acknowledged as the founding paper.)

[10] Wright, S. (1934). *Ann. Math. Statist.*, **5**, 161–215. (Wright's major exposition.)

(CAUSATION
CORRELATION
ECONOMETRICS
IDENTIFICATION PROBLEMS
PARTIAL CORRELATION
STRUCTURAL EQUATION MODELS)

E. SENETA

PATH DIAGRAMS   *See* PATH ANALYSIS

PATTERNED COVARIANCES

Patterned covariance matrices arise from multivariate* data with externally imposed structure. Let $\mathbf{X}$ ($p \times 1$) be a random p-variate column vector with mean vector $\boldsymbol{\mu}$ ($p \times 1$) and covariance* matrix $\boldsymbol{\Sigma}$ ($p \times p$). In general, $\boldsymbol{\Sigma}$ must be positive definite. A patterned covariance matrix has additional constraints on its elements that give the matrix a structured form. (See the following examples.) Although explicit forms for maximum likelihood estimates* (MLE) and likelihood ratio tests* exist in some cases, hypothesis testing* and estimation* of parameters can be problematic with this additional structure.

Several well-known patterned matrices are motivated by structured data as described in the following examples.

Example 1. Consider the scores obtained by an individual on p mathematics tests and arrayed in a $p \times 1$ vector $\mathbf{X}$. Assuming that the test means are equal, their variances are equal, and that the covariances among the p tests are equal, yields the complete symmetry pattern studied by Wilks [21]. With $p = 3$, we have

$$\boldsymbol{\mu} = \begin{bmatrix} \mu_1 \\ \mu_1 \\ \mu_1 \end{bmatrix}, \qquad \boldsymbol{\Sigma} = \begin{bmatrix} a & b & b \\ b & a & b \\ b & b & a \end{bmatrix}.$$

Assuming that $\mathbf{X}$ came from a multivariate normal distribution*, $\mathbf{X} \sim N_p(\boldsymbol{\mu}, \boldsymbol{\Sigma})$, the complete symmetry pattern arises if the distribution of $\mathbf{X}$ is assumed to be invariant under permutations of elements of $\mathbf{X}$.

Block complete symmetry is a generalization of complete symmetry motivated by mean and covariance patterns generated by the assumption that the mean and/or covariance matrices remain invariant under permutations of subvectors of $\mathbf{X}$.

Example 2. If $\mathbf{X}$ (6×1) consists of three (2×1) subvectors of scores, each subvector

comprising a mathematics and a verbal score from one of three parallel tests so that $\mathbf{X}' = (\mathbf{x}'_1, \mathbf{x}'_2, \mathbf{x}'_3)$, then we have the block complete symmetry pattern:

$$\boldsymbol{\mu} = \begin{bmatrix} \boldsymbol{\mu}_1 \\ \boldsymbol{\mu}_1 \\ \boldsymbol{\mu}_1 \end{bmatrix}, \qquad \boldsymbol{\Sigma} = \begin{bmatrix} \mathbf{A} & \mathbf{B} & \mathbf{B} \\ \mathbf{B} & \mathbf{A} & \mathbf{B} \\ \mathbf{B} & \mathbf{B} & \mathbf{A} \end{bmatrix},$$

with $\boldsymbol{\mu}_1$ (2×1) and $\mathbf{A}$ and $\mathbf{B}$ (2×2).

Votaw [20] generalized the concept of complete symmetry to compound symmetry, which involves patterns that arise not only from the invariance of means and covariances under exchangeability within subsets of elements of $\mathbf{X}$, but also between subsets.

Example 3. Let $p = 4$ with X_1 the score on an outside criterion, and X_2, X_3, and X_4 scores on three parallel mathematics tests. Then, under the invariance assumption not only of exchangeability* among X_2, X_3, and X_4, but also of each with respect to the outside criterion X_1, the compound symmetry pattern is

$$\boldsymbol{\mu} = \begin{bmatrix} \mu_1 \\ \cdots \\ \mu_2 \\ \mu_2 \\ \mu_2 \end{bmatrix},$$

$$\boldsymbol{\Sigma} = \begin{bmatrix} a & \vdots & b & b & b \\ \cdots & & \cdots & & \\ b & \vdots & c & d & d \\ b & \vdots & d & c & d \\ b & \vdots & d & d & c \end{bmatrix}. \tag{1}$$

Olkin and Press [9] examine circular symmetry patterns. Such a pattern arises if there are six radio receivers located symmetrically on a circle with a radio transmitter in the center.

Example 4. If $\mathbf{X}$ (6×1) consists of a simultaneous measurement of a transmitted signal at the six receivers, we might assume that the correlation of the measurements between the six receivers is a function of the distance between receivers on the circle, generating a symmetry pattern of the form

$$\Sigma = \begin{bmatrix} a & b & c & d & \vdots & c & b \\ b & a & b & c & \vdots & d & c \\ c & b & a & b & \vdots & c & d \\ d & c & b & a & \vdots & b & c \\ \cdots & \cdots & \cdots & \cdots & \vdots & \cdots & \cdots \\ c & d & c & b & \vdots & a & b \\ b & c & d & c & \vdots & b & a \end{bmatrix} \quad (2)$$

The stationary symmetry pattern that arises in time-series* models as discussed by Anderson [1] is illustrated in (2) by the upper left 4×4 matrix.

LINEAR COVARIANCE PATTERNS

A special family of patterned covariance matrices can be expressed as a linear pattern [1, 2]. Let $X_1, \ldots, X_n$ be independent and identically distributed (iid) random p-variate column vectors from a multivariate normal distribution, $N_p(\mu, \Sigma)$, where $\Sigma(\sigma) = \sum_1^m \sigma_g G_g$, the G's are known symmetric linearly independent $p \times p$ matrices, and the σ's are unknown scalars with the property that there exists at least one value of σ for which Σ is positive definite. This model includes all the examples discussed so far. The MLEs of these patterned covariances have explicit representations, i.e., are known linear combinations of elements of the sample covariance matrix, if and only if Σ^{-1} has the same pattern as Σ [3, 16]. Szatrowski [14] shows that naive averaging of subsets of elements of the sample covariance matrix will not generally yield the MLEs and will not necessarily even yield a positive definite estimate of the patterned covariance matrix. Explicit MLEs exist for all the specific examples given except for Example 4. In Examples 1–3 the explicit MLE can be obtained by "averaging."

For example, suppose we observed the random sample $x_1, \ldots, x_n$, p-variate column vectors with common distribution $N_p(\mu, \Sigma)$. The multivariate sample mean $\bar{x}$

$= (\sum_1^n x_i)/n$ and the multivariate cross-product sum of squares matrix $A = \sum_1^n (x_i - \bar{x})(x_i - \bar{x})'$ are sufficient statistics*. Let C be defined by $nC = A + n(\bar{x} - \hat{\mu})(\bar{x} - \hat{\mu})'$, where $\hat{\mu}$ is the MLE of μ. To illustrate averaging to find the MLE, consider Example 3 of compound symmetry with $p = 4$. We first find the MLE of μ by averaging to get $\hat{\mu}_1 = \bar{x}_1$ and $\hat{\mu}_2 = (\sum_2^4 \bar{x}_i)/3$. We then use this value of $\hat{\mu}$ in the formula for C. The MLE of Σ is obtained by averaging to obtain the MLE of a, b, c, and d, yielding $\hat{a} = c_{11}$, $\hat{b} = (\sum_2^4 c_{1j})/3$, $\hat{c} = (\sum_2^4 c_{ii})/3$, and $\hat{d} = (c_{23} + c_{24} + c_{34})/3$, where c_{ij} is the ijth element of C. The explicit MLE for complete symmetry, block complete symmetry, and circular symmetry can all be obtained similarly.

Iterative algorithms, including the Newton–Raphson* method of scoring* and E-M algorithms for finding the MLE, are used either when explicit solutions are not available or to obtain the explicit solution (see ITERATED MAXIMUM LIKELIHOOD ESTIMATES). They have been derived even in cases of missing data (e.g., Szatrowski [18]). The likelihood equations for the unknown parameter σ based on the random sample $x_1, \ldots, x_n$ with common distribution $N_p(\mu, \Sigma)$ are given by

$$\hat{\sigma} = \left[\text{tr} \hat{\Sigma}^{-1} G_g \hat{\Sigma}^{-1} G_h \right]^{-1} (\text{tr} \hat{\Sigma}^{-1} G_g \hat{\Sigma}^{-1} C),$$

$$(3)$$

where $[\cdot]$ is an $m \times m$ matrix with element gh given inside $[\cdot]$, $(\cdot)$ is an $m \times 1$ vector with element g given inside $(\cdot)$, and C is as defined. These likelihood equations suggest the following iterative procedure: Given an estimate of σ, form $\hat{\Sigma} = \hat{\Sigma}(\sigma)$, which, when substituted into the right-hand side of (3), yields a new estimate of σ, where, for this example, we assume either μ is known (and use $\hat{\mu} = \mu$) or μ is unpatterned (and use $\hat{\mu} = \bar{x}$). This iterative procedure is the method of scoring; it converges in one iteration from any positive definite starting point for patterns with explicit MLE in the nonmissing data case [16].

From (2) we observe that some patterns

without explicit MLEs can be viewed as submatrices of larger patterned matrices with explicit MLEs. Rubin and Szatrowski [12] show that this allows one to have a closed-form solution in the M step of the E-M algorithm. In general, convergence is not guaranteed with any of these algorithms for patterns without explicit MLEs.

Likelihood ratio tests (LRT) are used to compare two nested patterns, with the null hypothesis being the more restricted pattern. The LRT λ when sampling from one population is of the form $\lambda^{2/n} = |\hat{\Sigma}_1|/|\hat{\Sigma}_0|$, where $\hat{\Sigma}_0$ and $\hat{\Sigma}_1$ are the MLEs of Σ under the null and alternative hypotheses, respectively. When the null hypothesis is true, $-2(\ln\lambda)$ has an asymptotic chi-squared distribution* with degrees of freedom equal to the difference in the number of parameters under the null and alternative hypotheses. We reject the null hypothesis if $-2(\ln\lambda)$ is too large. Approximate and exact distributions of the LRT statistics have been obtained for special patterns (e.g., Olkin and Press [9], Rogers and Young [11], Votaw [20], and Wilks [21]). For the general linear pattern, only asymptotic null and nonnull distributions have been obtained for MLE and likelihood ratio statistics (LRS) [15, 18]. For special cases, asymptotic distributions of the LRS for contiguous alternatives have been obtained (e.g., Fujikoshi [5], Nagao [8], and Sugiura [13]). (Contiguous alternatives form a sequence of alternatives which, as a function of increasing sample size, approaches the null hypothesis parameterization at such a rate that the type I* and type II* error probabilities converge to constants between 0 and 1.) LRTs are also used to determine if k sampled populations have the same values of μ and Σ for given patterns of μ and Σ [15, 17, 20].

LINEAR PATTERNED CORRELATIONS

Another general class of covariance patterns is parameterized as $\Sigma = D(\sigma)P(\rho)D(\sigma)$, where $D(\sigma) = \sum_1^t \sigma_r \Lambda_r$ is a diagonal matrix,

the Λ's are known, linearly independent $p \times p$ diagonal matrices and the σ's are unknown scalars with the property that there exists at least one value of σ for which $D(\sigma)$ is positive definite; and where $P(\rho) = P^* + \sum_1^m \rho_g G_g$, where P^* is a known $p \times p$ symmetric matrix with diagonal elements equal to one, the G's are known linearly independent $p \times p$ symmetric matrices with diagonal entries equal to zero, and the ρ's are unknown scalars with the property that there exists at least one value of ρ such that $P(\rho)$ is positive definite and the diagonal entries are equal to one. Typically, $P^* = I$. Testing and estimation problems for this pattern have been studied by McDonald [7] and Browne [4]. Szatrowski [19] gives iterative algorithms for finding the MLEs, LRTs, and the asymptotic null and nonnull distributions of the LRS for these patterns even in the presence of missing data.

OTHER PATTERNS

Jöreskog [6] considers a generalization of the patterned mean* model of Potthoff and Roy [10] to a patterned mean and covariance model. The patterned covariance is of the form $\Sigma = B(\Lambda\Phi\Lambda' + \Psi^2) + \Omega^2$, where all matrices are parameter matrices with Φ symmetric and Ψ and Ω diagonal. Jöreskog studies estimation and testing problems involving further restrictions on the matrices to eliminate identifiability* problems among the parameters. These restrictions include fixing some parameters and constraining subsets of parameters to be equal.

References

[1] Anderson, T. W. (1970). In *Essays in Probability and Statistics*, R. C. Bose, I. M. Chakravarti, P. C. Mahalanobis, C. R. Rao, and K. J. C. Smith, eds. University of North Carolina Press, Chapel Hill, pp. 1–24. (Gives the Newton–Raphson algorithm for finding the MLEs of patterned covariances.)

[2] Anderson, T. W. (1973). *Ann. Statist.*, **1**, 135–141. (Gives the method of scoring algorithm for finding the MLEs of patterned covariances.)

[3] Andersson, S. A. (1975). *Ann. Statist.*, **3**, 132–154. (Uses group theory to derive properties of invariant normal models and patterned covariances arising from symmetries. Highly mathematical.)

[4] Browne, M. W. (1977). *Brit. J. Math. Statist. Psychol.*, **30**, 113–124. (Uses generalized least squares to estimate linear patterned correlation matrices. Contains examples.)

[5] Fujikoshi, Y. (1970). *J. Sci. Hiroshima Univ. Ser. A-1*, **34**, 73–144. (Inverts characteristic functions to obtain asymptotic nonnull distributions of test statistics in terms of hypergeometric functions with matrix arguments. Highly technical.)

[6] Jöreskog, K. G. (1970). *Biometrika*, **57**, 239–251. (Contains both techniques for and examples of estimating some general covariance patterns.)

[7] McDonald, R. P. (1975). *Psychometrika*, **40**, 253–255. (Uses the Newton–Raphson algorithm to obtain estimates of patterned correlation matrices. Includes an example.)

[8] Nagao, H. (1974). *J. Multivariate Anal.*, **4**, 409–418. (Uses noncentral chi-square variable expansions to obtain nonnull distributions of certain covariance hypothesis test criteria. Highly technical.)

[9] Olkin, I. and Press, S. J. (1969). *Ann. Math. Statist.*, **40**, 1358–1373. (Extensive details on estimation, testing, and distribution results for the circular symmetry covariance pattern.)

[10] Potthoff, R. F. and Roy, S. N. (1964). *Biometrika*, **51**, 313–326. (Generalizes the MANOVA model for growth curves. Contains examples.)

[11] Rogers, G. S. and Young, D. L. (1978). *J. Amer. Statist. Ass.*, **73**, 203–207. (Extensive estimation and distribution details on testing a multivariate linear hypothesis when the patterned covariance has explicit MLEs.)

[12] Rubin, D. B. and Szatrowski, T. H. (1982). *Biometrika*, **69**, 657–660. (MLEs for patterned covariances using the E-M algorithm obtained by posing a complete data nonexplicit *M*-step patterned covariance as a missing data explicit MLE *M*-step patterned covariance. Contains examples.)

[13] Sugiura, N. (1973). *Ann. Statist.*, **1**, 718–728. (Asymptotic nonnull distributions of LRS under local alternatives in terms of hypergeometric functions. Highly technical.)

[14] Szatrowski, T. H. (1978). *Ann. Inst. Statist. Math. A*, **30**, 81–88. (Presents sufficient conditions for explicit MLEs. Contains examples.)

[15] Szatrowski, T. H. (1979). *Ann. Statist.*, **7**, 823–837. (Asymptotic nonnull distributions of LRS for fixed alternatives for nested patterned means and covariances are obtained using the delta method.)

[16] Szatrowski, T. H. (1980). *Ann. Statist.*, **8**, 802–810. (Necessary and sufficient conditions for explicit MLE for linear patterned means and covariances are given.)

[17] Szatrowski, T. H. (1982). *J. Educ. Statist.*, **7**, 3–18. (Details on estimation and testing for block compound symmetry, with examples.)

[18] Szatrowski, T. H. (1983). *Ann. Statist.*, **11**, 947–958. (Generalizes ref. 15 in the presence of missing data.)

[19] Szatrowski, T. H. (1985). *Lin. Algebra Applic.* (to appear).

[20] Votaw, D. F. (1948). *Ann. Math. Statist.*, **19**, 447–473. (Compound symmetry model.)

[21] Wilks, S. S. (1946). *Ann. Math. Statist.*, **17**, 257–281. (Complete symmetry model.)

(COMMUNICATION THEORY CONTIGUITY)

TED H. SZATROWSKI

PATTERNED MEANS

Patterned mean vectors arise from data with externally imposed structure. Let $X_1, \ldots, X_n$ be independent random ($p \times 1$) column vectors with common covariance matrix Σ. Let X ($n \times p$) be the random matrix with ith row given by X_i'. Patterns arise in different parameterizations of $E(X)$. Several examples with linear patterns follow, all assuming sampling from a multivariate normal distribution.

Potthoff and Roy [4] study estimation* and hypothesis testing* for the mean pattern $E(X) = A \Xi P$ with A ($n \times g$) and P ($h \times p$) being known matrices of rank g and h, respectively, ($g \leqslant n$, $h \leqslant p$) and Ξ ($g \times h$) being a matrix of unknown parameters. They derive estimates for the unknown parameters and develop hypothesis tests for several nested mean patterns. Their model generalizes the MANOVA* model and yields applications to growth analysis. Jöreskog [2] investigates estimation and hypothesis testing problems using the Potthoff and Roy patterned mean and a patterned covariance*.

In the MANOVA model, $E(X_i) = \mu_i = B z_i$ with z_i ($g \times 1$) a vector of known elements and B ($p \times g$) a matrix of unknown parameters, $i = 1, \ldots, n$ [1]. If we let $\mathbf{Z} = (\mathbf{z}_1, \ldots, \mathbf{z}_n)$, $(g \times n)$, then $E(\mathbf{X}) = \mathbf{Z}'\mathbf{B}'$, a special case of the Potthoff and Roy model with $\mathbf{A} = \mathbf{Z}'$, $\mathbf{\Xi} = \mathbf{B}'$, and $\mathbf{P} = \mathbf{I}$.

Another linear mean pattern takes the form $E(\mathbf{X}_i) = \boldsymbol{\mu} = \mathbf{Z}\boldsymbol{\beta}$, where $\mathbf{Z}$ $(p \times h)$ is a known matrix of rank $h \leqslant p$ and $\boldsymbol{\beta}$ $(h \times 1)$ is a vector of unknown parameters. With $\mathbf{A} = (1, \ldots, 1)'$, $(n \times 1)$, $\mathbf{\Xi} = \boldsymbol{\beta}'$, and $\mathbf{P} = \mathbf{Z}'$, we see that this is another special case of the Potthoff and Roy model. Estimation and hypothesis testing problems involving this pattern often involve a patterned covariance matrix [5].

An ANOVA model can be obtained as a special case of the MANOVA model with $p = 1$, where we have assumed a fixed-effects* model. If we wish to investigate the mixed model of ANOVA, a model with both fixed and random effects, we use $p = 1$ and assume that the mean of $\mathbf{X}$ is $\mathbf{Z}\boldsymbol{\beta}$ and the covariance of $\mathbf{X}$ (not of the rows of $\mathbf{X}$) is $\boldsymbol{\Sigma}$. $\boldsymbol{\Sigma}$ is a patterned covariance matrix with nonnegativity constraints on some of its elements that impose further restrictions beyond positive definiteness on $\boldsymbol{\Sigma}$ [3, 6].

References

[1] Anderson, T. W. (1958). *An Introduction to Multivariate Statistical Analysis.* Wiley, New York.

[2] Jöreskog, K. G. (1970). *Biometrika*, **57**, 239–251.

[3] Miller, J. J. (1977). *Ann. Statist.*, **5**, 746–762. (Asymptotic properties of MLE in the mixed model of ANOVA.)

[4] Potthoff, R. F. and Roy, S. N. (1964). *Biometrika*, **51**, 313–326.

[5] Szatrowski, T. H. (1983). *Ann. Statist.*, **11**, 947–958.

[6] Szatrowski, T. H. and Miller, J. J. (1980). *Ann. Statist.*, **8**, 811–819.

(PATTERNED COVARIANCES)

Ted H. Szatrowski

PATTERN RECOGNITION

Pattern recognition is a branch of science concerned with the development of systems capable of performing various tasks of recognition. The term *pattern recognition* refers to identification of patterns by automatic means. Although pattern recognition in the human is an interesting aspect of psychology, the phrase pattern recognition generally is used to refer to automatic identification without active human participation in the process.

The automatic recognition is done by a machine system. In this sense, pattern recognition can be thought of as "machine perception." Usually a computer is an integral part of such a system; in this respect pattern recognition is viewed appropriately as a branch of computer science as well as statistics.

The patterns to be recognized may be physical shapes, forms, or configurations, but they can just as well be electrical patterns or sounds.

A relatively early application of pattern recognition was the computer recognition and classification* of chromosomes with a view toward finding abnormalities. Also, in the biomedical area, there is the task of automatic blood-cell sorting. An early success of pattern recognition was the classification of white blood cells; nucleus and cytoplasm area and color were used as features to obtain adequate separation into the standard categories for classification of white blood cells. Automated detection of cancer in gynecologic cytology specimens has been studied.

The systematic study of fingerprints goes back to Galton [7]. An early reference on matching fingerprints by computer is Wegstein and Rafferty [23]. Fingerprint classification involves categorizing each of the ten fingers as either an arch, loop, or whorl. The focus in Osterburg et al. [16] is not on pattern recognition, but rather on the development of a mathematical formula for fingerprint probabilities based on individual characteristics; however, in that article the definitions of the details (ridge endings, forks, etc.) and types (loops, whorls, and arches) are reviewed. Automated approaches to fingerprint classification involve making line drawings from the ridges. The characteristics noted are various types of line endings and intersections. Computer classification of single fingerprints into types (subdivisions of arch, loop, and whorl) by a syntactic approach (see below) has been achieved; see, e.g., Rao and Balck [17].

The field of pattern recognition has experienced extraordinary growth in recent years. Tasks that can now be performed by pattern recognition systems include the automated reading of the "zebra code" (Universal Product Code) on boxes and cans at checkout counters in stores, the recognition of typed, printed, or even handwritten characters, the detection of patterns of received radar signals (radar "signatures"), and the analysis of electrocardiograms (EKGs) and electroencephalograms (EEGs). Pattern recognition is involved in diagnosing the faults of a nuclear reactor by means of findings produced by its monitoring systems. It plays a role in the development of "expert systems," i.e., sets of automatic procedures for performing higher-level tasks such as computer-assisted medical diagnosis*, oil exploration, and robotics. For example, robotics involves computer vision, which in turn involves pattern recognition. These applications cannot be discussed here in any detail. Rather, some simpler and shorter examples will be given, along with the presentation of some of the statistical theory relevant to pattern recognition. Many references on various applications can be obtained, e.g., from the applications review articles in Krishnaiah and Kanal [15].

MATHEMATICAL FORMULATION

Transducers sense the physical world and transform the sensed signals into numerical data. These data may be considered a vector, which is the numerical *pattern*. The pattern recognition algorithm may utilize these raw data, but often they are transformed or combined mathematically. The measurements so derived are then the measurements on which the pattern recognition is based.

These measurements will be denoted by the vector **X**. It is to be assigned to one of a prescribed set of classes, i.e., it is to be "recognized."

Statistics plays a major role in the mathematical description (model) of pattern recognition tasks because usually that description is probabilistic: The vector **X** is treated as having a different probability distribution in each class. Then statistics plays an essential role in the development of an appropriate method for the problem. Although successful pattern recognition algorithms are sometimes obtained by heuristic means, from the statistician's point of view the design of the pattern recognition algorithm should follow from the probabilistic model and the corresponding statistical method.

CLASSIFICATION*

From the introduction, it is clear that *classification* forms the theoretical basis for pattern recognition. Here we mean classification in the sense of placing an object into one of a given set of categories. Another meaning of *classification* is the grouping together of similar things (*see* CLASSIFICATION), i.e., classes are *defined* by the grouping procedure. However, here the term is used with its other, more specific and technical meaning, i.e., the assignment of an object to one of a *prescribed* set of classes. (Synonyms for this meaning of classification are *assignment* and *allocation*.)

According to the classification model there are classes $\Pi_1, \Pi_2, \ldots, \Pi_k$ with prior probabilities π_c, $c = 1, 2, \ldots, k$. The conditional probability density function of **X**, given class c, will be denoted by $f_c(\mathbf{x})$. This is called the *class-conditional* density of **X**.

Maximum Posterior Probability Classification

The posterior probability* that x arose from class c is the conditional probability of class c, given **x**, or

$$\Pr(c \mid \mathbf{x}) = \pi_c f_c(\mathbf{x}) / f(\mathbf{x}).$$

[In fact $f(\mathbf{x}) = \sum_c \pi_c f_c(\mathbf{x})$.] The maximum posterior probability classification rule is to classify $\mathbf{x}$ as having arisen from class d if $\Pr(d \mid \mathbf{x})$ is maximal among $\{\Pr(c \mid \mathbf{x}), c = 1, 2, \ldots, k)\}$, i.e., if setting $c = d$ maximizes $\Pr(c \mid \mathbf{x})$. Since the denominator $f(\mathbf{x})$ of $\Pr(c \mid \mathbf{x})$ does not depend on c, this is the same as: Classify $\mathbf{x}$ into class d if setting $c = d$ maximizes $\pi_c f_c(\mathbf{x})$.

In the parametric case there is a function, say $h(\mathbf{x}; \beta)$, with a parameter β such that $\beta = \beta_c$ in the cth class, i.e.,

$$f_c(x) = h(\mathbf{x}; \beta_c).$$

In the multivariate normal* case, for example, β consists of the mean vector and covariance matrix*. In this case, we are led to discriminant analysis* with linear or quadratic discriminant functions.

Example 1. Suppose we had to classify an adult human as male or female, given only the person's height. It is easy to classify a six-foot person as a male and a five-foot person as a female. Suppose we use the model that male heights are normally distributed with a mean of 68 inches and a standard deviation of 3 inches and female heights are normally distributed with a mean of 63 inches and a standard deviation of 3 inches. Then with equal prior probabilities one classifies the person into that sex for which the probability density at the person's height x is higher. This is equivalent to classifying the person as a male if and only if x exceeds $(63 + 68)/2 = 65.5$. But now suppose that it is given that the person is from a population that is 90 percent female. Now we would be even more sure about classifying a five-foot person as a female. But perhaps even a six-foot person should be classified as a female. The maximum posterior probability classification rule is to classify the person as a male if and only if

$$x > \left[3^2 / (68 - 63) \right] \log_e(0.9/0.1) + 65.5$$
$$= 69.46.$$

If the standard deviations in the male and female height distributions were unequal, the classification rule would involve x^2 as well as x and would reduce to the form: Classify the person as a male if $x > b$ or $x < a$. The condition $x < a$ seems nonsensical; however, it would come into play only with small probability. In this example the observation was a scalar, height; usually it will be a vector of several, or a large number of variables.

Minimum Posterior Expected Loss Classification

Given a loss function $L(c \mid d) = $ loss of classifying into class c when the true class is d, the posterior expected loss of classifying into class c

$$\sum_d L(c \mid d)\Pr(d \mid x) = M(c \mid \mathbf{x}),$$

say. The minimum posterior expected loss rule is to assign $\mathbf{x}$ to class d if setting $c = d$ minimizes $M(c \mid \mathbf{x})$. Note that if $L(c \mid d) = 0$ if $c = d$ and $= 1$ if $c \neq d$, then $M(c \mid \mathbf{x}) = \sum_{d \neq c} \Pr(d \mid \mathbf{x}) = 1 - \Pr(c \mid \mathbf{x})$ so that minimizing $M(c \mid \mathbf{x})$ is equivalent to maximizing $\Pr(c \mid \mathbf{x})$, i.e., minimum posterior loss classification reduces to maximum posterior probability classification in this case. (More generally, this is true when the loss is of the form $L(c \mid d) = A$ if $c = d$ and $= B$ if $c \neq d$.)

MODELING

Class-Conditional Probability Functions

The preceding section assumes that there is a stochastic model for the observation $\mathbf{X}$. The class-conditional probability functions usually must be estimated. Often parametric forms are specified. Then usually the parameters must be estimated. This involves a training phase. For example, in constructing a device to recognize the spoken integers $0, 1, 2, \ldots, 9$, a speaker or speakers would speak the word *zero* into the device a number of times. The results of these voicings then are prototypes for zero. Relevant statistics, such as Fourier coefficients at important frequencies, would be extracted for later matching to input words. The device would be similarly trained for *one, two, three,* etc.

The development of an appropriate stochastic model in a given contex is an important part of pattern recognition. Sometimes the modeling of $\mathbf{X}$ involves use of stochastic processes* such as Markov chains or random fields*. This will be discussed to some degree later in the article.

Example 2. Consider a dot matrix representation of letters. If everything goes right an E will look like this:

$$
\begin{array}{ccc}
\bullet & \bullet & \bullet \\
\bullet & & \\
\bullet & \bullet & \bullet \\
\bullet & & \\
\bullet & \bullet & \bullet
\end{array}
\qquad (1)
$$

But suppose that sometimes some of the dots do not light up. Then what is supposed to look like diagram (1) might look like this:

$$
\begin{array}{ccc}
\bullet & \bullet & \bullet \\
\bullet & & \\
\bullet & \bullet & \\
\bullet & & \\
\bullet & \bullet & \bullet
\end{array}
\qquad (2)
$$

Or it might come out like this, which is an appropriate signal for a C.

$$
\begin{array}{ccc}
\bullet & \bullet & \bullet \\
\bullet & & \\
\bullet & & \\
\bullet & & \\
\bullet & \bullet & \bullet
\end{array}
\qquad (3)
$$

The probability of the letter E, given the observation x, is

$$
\Pr(E \mid \mathbf{x}) = \Pr(\mathbf{x} \mid E)\Pr(E)/P(\mathbf{x}).
$$

The observation will be classified as E if $\Pr(E \mid \mathbf{x})$ is maximal in the set $\{\Pr(d \mid \mathbf{x}),$ where d is any member of the alphabet$\}$. This will be the case if $\Pr(\mathbf{x} \mid E)\Pr(E)$ is maximal among such terms. This product involves the prior probability $\Pr(E)$ and the conditional probability $\Pr(\mathbf{x} \mid E)$ of the observation $\mathbf{x}$, given that the true letter is in fact E. This probability is the probability that a true E will show up as $\mathbf{x}$. For $\mathbf{x}$ different from (1), this may be viewed as the probability that E as represented in (1) is *distorted* to $\mathbf{x}$. It is necessary to have estimates of these probabilities in order to do the classification. One might obtain these experimentally, transmitting an E a number of times and tabulating the results. Then by studying these results one might find some

model for the distortion. One such model would be that there is a small probability, say* 0.01, that any one of the dots fails to light up, and these failures occur independently. Then the probability of (2) given E is $0.01(0.99)^{10} \simeq 0.000914$ and the probability of (3) given E is $0.01^2(0.99)^9 \simeq 0.0000914$. The probability of the correct transmission (1) is the probability that all 11 dots light up, or $0.99^{11} \simeq 0.895$. Note that under this model, the probability of (3) given C is $0.99^9 \simeq 0.914$, so with $\mathbf{x} = (3)$, $\Pr(\mathbf{x} \mid C)\Pr(C)$ is larger than $\Pr(\mathbf{x} \mid E)\Pr(E)$ unless $\Pr(E)/\Pr(C)$ exceeds 10,000. Now E is much more probable than C, but unless this ratio is as large as 10,000, the signal (3) will be appropriately classified as C, not E. In those rare instances where (3) resulted from E, not C, one would be out of luck unless a more complicated model were used, such as one taking account of sequences of letters.

Temporal or Spatial Dependence

Usually patterns occur in time or space, so there is temporal or spatial dependence among the elements of the pattern. The simplest such cases are the ones in which there is linear structure, as in a time series*. Then ordinary Markov models, with a one-dimensional indexing parameter, can be applied.

For example, in English, the letter q is almost always followed by the letter u; this information can be built into the classifier by making the probability of a q-to-u transition high. In fact, higher-order transitions are needed in analysis of the English language. To illustrate the use of these higher-order transitions, suppose, for example, that the word PRIbR, where b denotes a letter that was not successfully received, is to be recognized. Using a table of transition probabilities for trigrams (three-letter sequences), one might find that PRI is most likely to be followed by OR, giving the word PRIOR.

The Labeling Model

It is useful to consider classification and pattern recognition problems in terms of a *labeling* model. With each observation X_t, $t = 1, 2, \ldots, n$, where $n =$ the number of observations, is associated an unobservable label Z_t, where the possible values of Z_t may be identified with $1, 2, \ldots, k$, and k is the number of classes. Then $f_c(x_t)$ is the probability density function for X_t, given that $Z_t = c$, $c = 1, 2, \ldots, k$. When X_t is discrete,

$$f_c(x_t) = \Pr(X_t = x_t \mid Z_t = c).$$

Temporal or spatial correlation can be modeled by treating the process $\{Z_t\}$ as a Markov chain.

STATISTICAL AND SYNTACTIC PATTERN RECOGNITION

The emphasis in this article has been on *statistical* pattern recognition. Some other approaches will be mentioned briefly; also they are discussed in some of the entries in the references.

In the *structural* approach, primitive elements are defined from which patterns are built up. For example, in describing various items of furniture such as chairs and tables, the primitives might include legs and planes. A table top is a plane, as is the seat of a chair. Both tables and chairs have legs. Object description involves specification of the spatial relationships among the primitives.

The *syntactic* approach [5, 6] uses the analogy between the structure of patterns and that of language. It depends on the definition of certain elements as primitives, the building blocks out of which the patterns are constructed. Just as sentences are formed by the syntactical combination of words and phrases according to the rules of grammar, so are the patterns constructed from the primitives. In handwriting, for example, the primitives are loops and certain strokes. These can follow one another in some orders, but not in others. The primitives are like parts of speech, subjects, and objects; the feasible temporal and spatial relationships embody restrictions like those of a grammar.

The statistical approach can be combined with the structural/syntactic approach. For example, probabilities can be used for adjacency information. This was alluded to earlier in the context of English orthography. Similarly, in analysis, one can incorporate information such as the fact that a bird is very likely in a tree; a table often has chairs near it, etc. It is possible that some of the major new advances in pattern recognition may result from a combination of the statistical and structural/semantic approaches.

Literature

There is a large literature on pattern recognition. In the bibliography only a few texts are listed. Among books reporting recent advances are those in the series Kittler [13]. Some particularly interesting recent developments are reported in Bogner [2], Das Gupta [3], Gelsema and Kanal [8], Kittler and Devijver [14], Tarter [20], Tsai and Fu [21], Turner and Tsokos [22], and Yakowitz [24]. The reader can follow developments in the field through journals and through the proceedings of various conferences, such as the International Joint Conference on Pattern Recognition, sponsored by the International Association for Pattern Recognition (a federation of pattern recognition societies), the Institute of Electrical and Electronics Engineers (IEEE) Computer Society Conference on Pattern Recognition and Image Processing, and various NATO Advanced Study Institutes.

Some journals particularly relevant to pattern recognition are *Pattern Recognition, Artificial Intelligence, IEEE Transactions on Pattern Analysis and Machine Intelligence, IEEE Transactions on Biomedical Engineering, IEEE Transactions on Information Theory, IEEE Transactions on Systems, Man and Cybernetics*, and *Computer Vision, Graphics, and Image Processing*.

References

[1] Andrews, H. (1972). *Introduction to Mathematical Techniques in Pattern Recognition*. Wiley-Interscience, New York. (Relatively short and elementary introduction to mathematical pattern recognition.)

[2] Bogner, R. E. (1981). *IEEE Trans. Pattern Anal. and Mach. Intell.*, **3**, 128–133.

[3] Das Gupta, S. (1977). In *Multivariate Analysis IV*, P. R. Krishnaiah, ed. North-Holland/American Elsevier, Amsterdam/New York, pp. 457–472.

[4] Duda, R. O. and Hart, P. E. (1972). *Pattern Classification and Scene Analysis.* Wiley-Interscience, New York. (A unified, comprehensive treatment of both statistical and descriptive methods for pattern recognition.)

[5] Fu, K. S. (1974). *Syntactic Methods in Pattern Recognition.* Academic Press, New York.

[6] Fu, K. S. (1982). *Syntactic Pattern Recognition and Applications.* Prentice-Hall, Englewood Cliffs, NJ.

[7] Galton, F. (1892). *Finger Prints.* Macmillan, London; rpt. 1965, DaCapo Press, New York.

[8] Gelsema, E. S. and Kanal, L. N. (1980). *Pattern Recognition in Practice.* North Holland/American Elsevier, Amsterdam/New York.

[9] Grenander, U. (1976). *Pattern Synthesis: Lectures in Pattern Theory*, Vol. 1. Springer-Verlag, New York.

[10] Grenander, U. (1978). *Pattern Analysis: Lectures in Pattern Theory*, Vol. 2. Springer-Verlag, New York.

[11] Grenander, U. (1981). *Regular Structures: Lectures in Pattern Theory*, Vol. 3. Springer-Verlag, New York. (Introductory; can be read before Volumes 1 and 2.)

[12] Kanal, L. N. and Rosenfeld, A., eds. (1982). *Progress in Pattern Recognition*, Vol. 1. Elsevier, New York.

[13] Kittler, J., ed. (1984). *Pattern Recognition and Image Processing Research Studies Series.* Research Studies Press (Wiley), New York. (An ongoing series.)

[14] Kittler, J. and Devijver, P. A. (1981). *Pattern Recognition*, **13**, 245–249.

[15] Krishnaiah, P. R. and Kanal, L. N. eds. (1982). *Handbook of Statistics*, Vol. 2: *Classification, Pattern Recognition, and Reduction of Dimensionality.* North-Holland, New York. (Includes articles on statistical methodology as well as review articles on applications such as EKGs, speech, radar, white blood cells, remote sending, optical character recognition, and oil spill identification.)

[16] Osterburg, J. W., Parthasarathy, T., Raghavan, T. E. S., and Sclove, S. L. (1977). *J. Amer. Statist. Ass.*, **72**, 772–778.

[17] Rao, K. and Balck, K. (1980). *IEEE Trans. Pattern Anal. Mach. Intell.*, **2**, 223–231.

[18] Rosenfeld, A. (1969). *Picture Processing by Computer.* Academic Press, New York. (One of the early works on *image processing*, which has an extensive literature of its own.)

[19] Sebestyen, G. S. (1962). *Decision Making Processes in Pattern Recognition.* Macmillan, New York. (An early text relating to statistical pattern recognition.)

[20] Tarter, M. E. (1981). In *Proceedings of the Thirteenth Symposium on Interface: Computer Science and Statistics*, W. F. Eddy, ed., pp. 105–110.

[21] Tsai, W. H. and Fu, K. S. (1980). *IEEE Trans. Syst., Man, Cybern.*, **10**, 873–875.

[22] Turner, J. C. and Tsokos, C. P. (1977). *Metron*, **35**, 89–104.

[23] Wegstein, J. H. and Rafferty, J. F. (1968). *NBS Tech. Note No. 466*, U.S. GPO, Washington, DC.

[24] Yakowitz, S. (1981). In *Proceedings of the Thirteenth Symposium on Interface: Computer Science and Statistics*, W. F. Eddy, ed., pp. 85–91.

(COMMUNICATION THEORY
DISCRIMINANT ANALYSIS
INFORMATION AND CODING THEORY
LINGUISTICS, STATISTICS IN
LITERATURE, STATISTICS IN)

STANLEY L. SCLOVE

PATTERN RECOGNITION LETTERS

This is an official publication of the International Association for Pattern Recognition. Volume I appeared in 1982. Papers are limited to six pages in length. The subject matter can be "theory, methodology, empirical studies or applications both in image processing and in pattern recognition." The editors are E. Becker, Delft University of Technology, and E. S. Gelsema, Free University, Amsterdam, The Netherlands. The journal is published by Elsevier Science Publishers.

PATTERN RECOGNITION SOCIETY

The Pattern Recognition Society is a society formed "to fill a need for information exchange among research workers in the pattern-recognition field." Its address is

c/o National Biomedical Research Foundation
Georgetown University Medical Center
3900 Reservoir Road, NW
Washington, DC 20007 (United States)

PATTERN RECOGNITION SOCIETY, JOURNAL OF THE

Pattern Recognition is the official journal of the Pattern Recognition Society*. It is intended to expedite communication among research workers interested in pattern recognition in various fields, including "high energy physics, target recognition, biological taxonomy, meteorology, space science, oceanography, character recognition, optical instrumentation, industrial applications, neuron physiology, and many others." Special issues, devoted to specific topics, are organized from time to time.

The editor-in-chief of the journal (1984) is Robert S. Ledley, National Biomedical Research Foundation, Georgetown Medical Research Center, Washington, DC. The journal is published by Pergamon Press. Volume 17 appeared in 1984.

PAULSON APPROXIMATION

This is an approximation to the distribution of F obtained by combining the Wilson–Hilferty* approximation to the chi-squared distribution* with Fieller's approximation to the distribution of the ratio of two normal variables*.

Define

$$F_{\nu_1, \nu_2} = \left(\nu_1^{-1} X_1 \right) / \left(\nu_2^{-1} X_2 \right),$$

where X_j denotes a variable having a chi-squared distribution with ν_j degrees of freedom ($j = 1, 2$), and X_1 and X_2 are mutually independent. Using the Wilson–Hilferty approximation for each X_1 and X_2, we find

$$F_{\nu_1, \nu_2}^{1/3} \doteq \left\{ 1 - \frac{2}{9\nu_2} + U_1 \sqrt{\frac{2}{9\nu_2}} \right\}$$

$$\times \left\{ 1 - \frac{2}{9\nu_2} + U_2 \sqrt{\frac{2}{9\nu_2}} \right\}^{-1},$$

where U_1, U_2 are mutually independent unit normal variables*. Now applying Fieller's

approximation, we obtain the result that

$$\left\{ \left(1 - \frac{2}{9\nu_2} \right) F_{\nu_1, \nu_2}^{1/3} - \left(1 - \frac{2}{9\nu_1} \right) \right\}$$

$$\times \left(\frac{2}{9\nu_2} F_{\nu_1, \nu_2}^{2/3} + \frac{2}{9\nu_1} F_{\nu_1, \nu_2}^{2/3} \right)^{-1/2}$$

has, approximately, a unit normal distribution. This approximation is remarkably accurate for $\nu_2 \geqslant 10$. See also Severo and Zelen [2].

References

[1] Paulson, E. (1942). *Ann. Math. Statist.*, **13**, 233–235.

[2] Severo, N. S. and Zelen, M. (1960). *Biometrika*, **47**, 411–416.

(ANALYSIS OF VARIANCE
CHI-SQUARED DISTRIBUTION
F-DISTRIBUTION
RATIO OF BIVARIATE NORMAL
 VARIABLES
WILSON–HILFERTY TRANSFORMATION)

PAYOFF TABLE   *See* DECISION THEORY; GAME THEORY; HUYGENS, CHRISTIAAN

P CHART   *See* CONTROL CHARTS

PEAKEDNESS   *See* KURTOSIS

PEAK OF AN UMBRELLA   *See* UMBRELLA ALTERNATIVES

PEAK TEST, THE

The peak test is a nonparametric test of the null hypothesis of homoscedasticity in the linear regression model. The model is

$$y_t = \beta' x_t + u_t \qquad (1)$$

with $E(u_t) = 0$ and $E(u_t u_\tau) = 0$ for $t \neq \tau$. The null hypothesis is stated as $H_0 : E(u_t^2) = \sigma^2$ (homoscedasticity) and the alternative as $E(u_t^2) = \sigma_t^2$, $\sigma_t^2 \neq \sigma_\tau^2$ for $t \neq \tau$ (heteroscedasticity*). Detecting violations of the null hypothesis is important because ordinary least-squares estimates of (1) are not efficient in the presence of heteroscedasticity.

If the distribution of u_t is known (e.g., if

it is Gaussian), parametric tests of H_0 are possible [7]. These tests, however, are not appropriate if the distributional assumptions about u_t do not hold. The peak test, related to earlier work on a similar topic by Foster and Stuart [3], is designed to cope with this problem (see also Goldfeld and Quandt [6, 7]). Since it utilizes the absolute values of residuals*, it is also related by Glejser [5] to tests in which the absolute values of residuals (or their squares) are regressed on an exogenous variable (*see* ECONOMETRICS), the values of which are believed to be monotone in error variance.

Assume that under H_1 the variance of u_t is believed to be increasing with the values of a variable z_t (which may be one of the regressors or the time index t). Since ordinary least-squares residuals are correlated even if $E(u_t u_\tau) = 0$, $(t \neq \tau)$, we require a residual vector with scalar covariance matrix. Such a vector is provided by recursive residuals [2] or by BLUS residuals (for a particular application using the latter see Hedayat et al. [8]). Assume that such uncorrelated residuals have been computed, and denote their absolute values by $|\hat{u}_t|$. If H_1 is true, large values of $|\hat{u}_t|$ will tend to be paired with large values of z_t. Reorder the variable z_t so that its values form an ascending sequence and correspondingly, reorder $|\hat{u}_t|$ so that, for $|\hat{u}_t|$ and $|\hat{u}_\tau|$, $t < \tau$ if and only if $z_t < z_\tau$. A peak occurs in the list of ordered residuals at t if $|\hat{u}_t| \geq |\hat{u}_\tau|$ for all $\tau < t$. The first residual, by convention, is taken not to represent a peak. (Note that if the number of observations is T and the number of regressors p, there exist only $n = T - p$ residuals in all.) The probability of k peaks in a sequence of n residuals can then be obtained as follows.

Define $N(n, k)$ as the number of ways in which n residuals can yield k peaks and take $N(1, 0) = 1$. Then

$$N(n, n - 1) = 1$$
$$\vdots$$
$$N(n, k) = (n - 1)N(n - 1, k) + N(n - 1, k - 1)$$
$$\vdots$$
$$N(n, 0) = (n - 1)N(n - 1, 0),$$

whence the required probabilities $P(n, k)$ are given by $N(n, k)/n$. (For tables, see Goldfeld and Quandt [7].)

As an illustration, consider the regression* $y_i = \beta_0 + \beta_1 x_i + (\sqrt{i})u_i$, where the x_i are integers and u_i are iid $N(0, 1)$. For 22 observations $(x_i = 1, 2, \ldots, 22)$, $y_i = (1.314, 4.013, 1.643, -0.277, 3.756, 2.251, 6.017, 2.206, 7.222, 2.994, 5.691, 3.487, 9.887, 9.982, 4.390, 10.945, 4.570, 9.731, 5.985, 7.979, 2.676, 14.487)$, 20 recursive residuals can be computed according to Brown et al. [2], and the absolute values of these recursive residuals are $(2.069, 1.605, 2.447, 0.012, 2.709, 1.743, 2.699, 2.096, 0.537, 1.797, 3.912, 2.566, 3.408, 2.768, 3.939, 1.118, 2.862, 0.870, 5.863, 5.650)$, showing five peaks. The probability of more than four peaks in a sample of 20 under H_0 is 0.094, just permitting H_0 to be rejected at the 0.1 level of significance.

An alternative nonparametric test for heteroscedasticity is to obtain the rank correlation* between z_t and $|\hat{u}_t|$. Some other tests [4] also use the ranks of the residuals, but require an assumption about the distribution generating errors and also specify the general form of heteroscedasticity as $\log \sigma_t^2 = \alpha_0 + \alpha_1 z_t$. Under these circumstances, the Giaccotto and Ali test is more powerful than the peak test. Finally, Bickel [1] proposes a robust test of heteroscedasticity.

References

[1] Bickel, P. J. (1978). Using residuals robustly. I: Tests for heteroscedasticity, nonlinearity. *Ann. Statist.*, **6**, 266–291.

[2] Brown, R. L., Durbin, J. and Evans, J. M. (1975). Techniques for testing the constancy of regression relationships over time. *J. R. Statist. Soc. B*, **37**, 149–192.

[3] Foster, F. G. and Stuart, A. (1954). Distribution-free tests in time series based on breaking records. *J. R. Statist. Soc. B*, **16**, 1–22.

[4] Giaccotto, C. and Ali, M. M. (1982). Optimum distribution-free tests and further evidence of heteroscedasticity in the market model. *J. Finance*, **37**, 1247–1257.

[5] Glejser, H. (1968). A new test for heteroscedasticity. *J. Amer. Statist. Ass.*, **64**, 316–323.

[6] Goldfeld, S. M. and Quandt, R. E. (1965). Some tests for homoscedasticity. *J. Amer. Statist. Ass.*, **60**, 539–547.

[7] Goldfeld, S. M. and Quandt, R. E. (1972). *Nonlinear Methods in Econometrics*. North-Holland, Amsterdam.

[8] Hedayat, A., Raktoe, B. L. and Talwar, P. P. (1977). Examination and analysis of residuals. *Commun. Statist.*, **A6**, 497–506.

(BARTLETT'S TEST OF HOMOGENEITY
OF VARIANCES
HETEROSCEDASTICITY)

RICHARD E. QUANDT

PEARL–REED CURVE

Pearl–Reed curve is an alternative name for logistic curve*.

PEARSON, EGON SHARPE

Born: August 11, 1895, in Hampstead (London), England.

Died: June 12, 1980, in Midhurst, Sussex, England.

Contributed to: applications of statistical techniques, statistical theory, quality control, operations research, statistical education.

Egon Sharpe Pearson (E. S. P.), the only son of the British statistician Karl Pearson*, was born in 1895, a time when the latter's interest in statistical methods, kindled by the work of Francis Galton*, was growing rapidly. Following school education at the Dragon School, Cambridge (1907–9) and Winchester (1909–14), he entered Trinity College, Cambridge to read mathematics in 1914. His undergraduate studies were interrupted first by a severe bout of influenza (August–December 1914) and then by war work (1915–18). After finally obtaining his first degree in 1919, he continued with graduate studies in astronomy, attending, inter alia, lectures by A. S. Eddington on the theory of errors and F. J. M. Stratton on combinations of observations. He also worked with F. L. Engledow and G. U. Yule at this time.

In the fall of 1921, he took a position in the Department of Applied Statistics at University College, London. This department, the first of its kind, had been established through the efforts of Karl Pearson in 1911 and was still very much under his control. The next five years were an apprenticeship; Egon was not required (or permitted) to teach until 1926, when his father was suffering from cataract. By this time, he felt the need to work out his own philosophy on the use of statistical inference and scientific method generally if he were to develop an independent career. He was not fully satisfied with the methods of the Pearsonian school, with their considerable reliance on a wide variety of "indices," and he was ripe to receive fresh external influences. In fact, two such influences of some importance entered his life about this time. While they exerted considerable interplay on each other (and were not, of course, exhaustive), it is convenient to regard them as representing the more theoretical and practical interests in his statistical work.

The first, and perhaps better known, was the collaboration with Jerzy Neyman*, whom he met when the latter was a visitor in the department in 1925–26. The second was a sustained correspondence with W. S. Gosset ("Student")* which started in 1926. Both episodes lasted about the same time, till Neyman's departure for the United States in 1938 and Gosset's death in 1937, respectively.

As already noted, this dichotomy, although a convenient simplification, omits much that is relevant and important. Egon Pearson never regarded the corpus of methods associated with the name Neyman–Pearson as a *theory*, but rather as a collection of principles representing an approach to statistical—or more broadly scientific—problems in general. The essentials of this approach are consideration of the results to be expected from applying statistical procedures, usually expressed in probabilistic terms. The approach is not limited to the traditional fields of hypothesis testing and estimation, but has been applied in other contexts, notably, in discriminant analysis (wherein a property of major interest is probability of correct classification).

One of the basic concepts in the Neyman–Pearson approach—that of an alternative hypothesis*, which might be valid if a hypothesis under test were not valid—was, in fact mentioned by Gosset in a letter to Egon Pearson in 1926. In November of that year, the latter wrote to Neyman, now returned to Warsaw, mentioning this concept. At first Neyman suggested that some form of inverse probability argument might provide an appropriate technical basis but it was soon recognized that sufficient knowledge of prior probabilities* would be available but rarely. A period of intensive, though mostly long-range, discussions ensued. The remarkable story of the development of this collaboration, whose members were usually some thousand miles apart until 1934, is told in some detail in David [1] and Reid [23]. Progress was accelerated by occasional meetings, starting with ten days in Paris in April 1927 and continuing in Warsaw in the summer of 1929 and January 1932. Consequent on the first meeting, the first general outline of the approach appeared in two long papers in 1928 [14, 15]. From the 1929 discussions in Warsaw came the basic papers [16, 17] on two- and k-sample problems, published in 1930–31; in 1933 there appeared what may be regarded as a definitive summary [18] of this phase of the Neyman–Pearson* approach to hypothesis testing*.

We now return to the other major influence and describe some contemporaneous developments associated with the correspondence with Gosset. At an early stage, the latter had drawn Pearson's attention to the important topic of robustness*—lack of sensitivity of statistical procedures to departures from assumptions (e.g., of independence, normality, etc.) on which their "standard" properties are assessed. The tables of random numbers* produced by L. H. C. Tippett while working in the department in University College (1925–27) made it possible to investigate robustness by means of sampling experiments (later termed *Monte Carlo methods* and currently and more generally, *simulation**). The results of some of this work

were published in 1928–31 [3, 4, 11, 12]. (The first published experimental estimate of power* of a test, an important concept in the Neyman–Pearson approach, appeared in Pearson and Adyanthaya [12].) The topic of robustness remained of perennial interest and was reflected in a publication as late as 1975 [22]. A little later in the correspondence, Gosset drew attention to the application of statistical methods in quality control* of industrial production, then beginning to flourish in the United States under the enthusiastic guidance of W. A. Shewhart and F. W. Winters. Pearson visited the United States in 1931 and met Shewhart and a number of other prominent American statisticians, giving an address on Probability and Ways of Thinking at the annual meeting of the American Statistical Association. In this talk, he set out some basic ideas of the then very novel Neyman–Pearson approach.

Later, he played an active role in the formation of the Industrial and Agricultural Research Section of the Royal Statistical Society* in 1932–33 and its initial progress. (Its later development into the Research Section, and its organ's development from the *Supplement* into *Series B* of the *Journal of the Royal Statistical Society** was accompanied by a marked increase in emphasis on theory. The regional and specialized sections of the Royal Statistical Society, and *Applied Statistics* or *Section C* of the *Journal* have greater claims to be regarded as true heirs of the original organization.)

At this time there also began a long association with the British Standards Institution. In 1936, the monograph in ref. 5 appeared. It was the first official British standard on the use of statistical methods in industrial quality control. The association continued for some 40 years, and Pearson was attending meetings of committees of the International Standards Organization as late as 1976.

On Karl Pearson's retirement in 1933, his department at University College, London was split into the Department of Eugenics and the Department of Applied Statistics. Egon was appointed head of the latter. He soon found it possible (in 1934) to invite

Neyman to take a position on the staff—at first temporary but, before long, permanent. It might have been expected, and must have been hoped, that this would lead to more rapid further developments in the Neyman–Pearson approach to the philosophy of statistical procedures. In fact, this did not transpire, for several reasons. There were new preoccupations—Pearson now had new administrative responsibilities and had married in 1934. On his father's death in 1936, he became managing editor of *Biometrika**, a post he held, creating a growing reputation for care and thoroughness, both for the journal and himself, for the next 30 years. During this time also he began his work on successive revisions of Karl Pearson's *Tables for Statisticians and Biometricians* in collaboration first with L. J. Comrie and later with H. O. Hartley*. He was personally concerned in calculating many new tables, notably those of distribution of range in normal samples, percentage points of Pearson curves, and distribution of skewness and kurtosis coefficients. Although further joint work [19–21] with Neyman was published in the short-lived but influential departmentally produced journal *Statistical Research Memoirs* in 1936 and 1938, by 1936 Pearson felt he had gone as far as he wished to go along the path of further elaboration of the Neyman–Pearson approach. In his later work, there is, in general, more emphasis on specific applications.

This shift in emphasis received powerful support with the beginning of World War II in 1939. Pearson was head of a group of statisticians working on weapons assessment with the Ordnance Board. There is an interesting general account of some phases of this work in ref. 8. During this period there was a very rapid increase in the use of statistical methods in quality control*—already one of Perason's major interests. There were related developments in formation of the new discipline of *operations research**—Pearson was a founding member of the (British) Operational Research Club (later Society) in 1948.

This very full life continued until Pearson retired as head of the department at University College in 1960. Thereafter, Pearson's activity decreased only gradually. With the relaxation of administrative burdens, time was again found for more scholarly work—especially in the field of frequency curves*, in which his father was pioneering at the time of his birth. His last statistical paper [13] included wide-ranging comparisons among no fewer than eight systems of frequency curves.

His last major work was the fulfillment of a promise made to his father—production of a scholarly annotated version [10] of Karl Pearson's lectures on early history of statistics and University College in 1921–33 based on Pearson's own lecture notes, but considerably enriched by illuminating editorial comments.

References[1]

NOTE

1. References 14–21 are included in *Joint Statistical Papers of J. Neyman and E. S. Pearson* (University of California Press, Berkeley, 1967). References 3, 4, 6–8, 11 and 12 are in *The Selected Papers of E. S. Pearson* (University of California Press, 1966).

[1] David, F. N., ed. (1966). In *Research Papers in Statistics. Festschrift for J. Neyman*. Wiley, New York, pp. 1–23.

[2] Moore, P. G. (1975). *J. R. Statist. Soc.*, **138A**, 129–130.

[3] Pearson, E. S. (1929). *Biometrika*, **21**, 337–360.

[4] Pearson, E. S. (1931). *Biometrika*, **23**, 114–133.

[5] Pearson, E. S. (1936). *The Application of Statistical Methods to Industrial Standardization and Quality Control*, Brit. Stand. 600. British Standards Institution, London.

[6] Pearson, E. S. (1947). *Biometrika*, **34**, 139–167.

[7] Pearson, E. S. (1956). *J. R. Statist. Soc.*, **119A**, 125–146.

[8] Pearson, E. S. (1963). *Proc. 8th Conf. Des. Exper. Army Res. Dev.*, 1–15.

[9] Pearson, E. S. (1968). *Biometrika*, **55**, 445–457.

[10] Pearson, E. S., ed. (1979). *The History of Statistics in the Seventeenth and Eighteenth Centuries*. Macmillan, New York. (Lectures by Karl Pearson given at University College, London, during the academic sessions 1921–1933.)

[11] Pearson, E. S. and Adyanthaya, N. K. (1928). *Biometrika*, **20A**, 356–60. (**21**, 259–86 (1929)).

[12] Pearson, E. S. and Adyanthaya, N. K. (1928). *Biometrika*, **20A**, 259–286.

[13] Pearson, E. S., Johnson, N. L., and Burr, I. W. (1979). *Commun. Statist. B*, **8**, 191–229.

[14] Pearson, E. S. and Neyman, J. (1928). *Biometrika*, **20A**, 175–240.

[15] Pearson, E. S. and Neyman, J. (1928). *Biometrika*, **20A**, 263–294.

[16] Pearson, E. S. and Neyman, J. (1930). *Bull. Acad. Pol. Sci.*, 73–96.

[17] Pearson, E. S. and Neyman, J. (1931). *Bull. Acad. Pol. Sci.*, 460–81.

[18] Pearson, E. S. and Neyman, J. (1933). *Philos. Trans. R. Soc. Lond.*, **231A**, 289–37.

[19] Pearson, E. S. and Neyman, J. (1936). *Statist. Res. Memo.*, **1**, 1–37.

[20] Pearson, E. S. and Neyman, J. (1936). *Statist. Res. Memo.*, **2**, 25–57.

[21] Pearson, E. S. and Neyman, J. (1936). *Statist. Res. Memo.*, **1**, 113–137.

[22] Pearson, E. S. and Please, N. W. (1975). *Biometrika*, **62**, 223–241.

[23] Reid, C. (1982). *Neyman from Life*. Springer, New York.

(FREQUENCY CURVES, SYSTEMS OF
GOSSET, WILLIAM SEALY ("STUDENT")
NEYMAN, JERZY
NEYMAN–PEARSON LEMMA)

PEARSON, KARL

Born: March 27, 1857, in London, England.

Died: April 27, 1936, in Coldharbour, Surrey, England.

Contributed to: anthropology, biometry, eugenics, scientific method, statistical theory.

Karl Pearson was educated at University College School and privately, and at Kings College, Cambridge, where he was a scholar. He took his degree in 1879, being the Third Wrangler in the Mathematical Tripos. In 1885 he was appointed to the Chair of Applied Mathematics at University College, London, where he stayed for the whole of his working life, moving to the newly instituted Chair of Eugenics in 1911. Both his parents were of Yorkshire stock, and he thought of himself as a Yorkshireman. He was married twice, and there were three children, Sigrid, Egon, and Helga, by the first marriage.

A man of great intellect, Pearson, when young, had many and varied interests. His ability as a mathematician was unquestioned, but he also studied physics and metaphysics, law, folklore, and the history of religion and social life in Germany. The echoes of this last are found among his first publications: *The Trinity: A Nineteenth Century Passion Play* (1882), *Die Fronica* (1887), and the *Ethic of Freethought* (1888). Throughout the intellectual ferment of these early years, however, his interest in the sciences persisted and emerged in the happy mingling of science, history, and mathematics in his completion (1886) of Todhunter's *History of the Theory of Elasticity* and his own production *The Grammar of Science* (1892).

The appointment to the Chair of Applied Mathematics (1885) did not check the outpourings of this fertile mind, although it inevitably gave them some direction. His principal professorial duty was teaching mathematics to engineering students, which influenced the way in which he subsequently thought about observational problems, but the real impetus in his thinking, both then and for many years, came from his friendship with Francis Galton* and with Weldon. The latter came to University College in 1890. He was interested in Galton's *Natural Inheritance* (1889) and had been collecting data on shrimps and shore crabs and appealed to Pearson for the mathematical tools to carry out his analyses with the idea of studying evolution through morphological measurements. Pearson clearly found his lifework and in the process of answering his friends' pleas for help laid the foundations of modern mathematical statistics. For the fact that foundations must be buried does not mean that they are not there. In an astonishing output of original papers, we get the now familiar apparatus of moments and correlation*, the system of frequency curves*, the probable errors of moments and product-moments reached by maximizing the likelihood function*, the χ^2 goodness-of-fit* test, all in the space of 15 years when he was also lecturing many hours a week. It was not until 1903 that a grant from the Draper's Company enabled "biometric" research to be carried on with Pearson under less pressure. He was elected to the Royal Society in 1896 and was the Darwin Medallist in 1898. Among the students during these years were G. U. Yule and L. N. G. Filon.

The concept of *ancestral inheritance* put forward by Pearson and Weldon was not received kindly by many biologists—chief among them, William Bateson—and the inevitable controversies ensued. From these quarrels, arising from critical prepublication reviews, sprang the statistical journal *Biometrika**, which Pearson edited from its inception in 1901 until his death in 1936. The effect of the controversies on Pearson was to make him cut himself off from participation in meetings of learned societies. Thus, for example, he was never a Fellow of the Royal Statistical Society. He must have felt even more isolated by the movement of Weldon to a chair at Oxford and by Weldon's death in 1906.

Francis Galton was supportive of Pearson both in friendship and financial aid throughout the period 1885 until his death in 1911. He helped with the founding of *Biometrika*. When eventually the Biometrics Laboratory was started, he helped with the funding, and in his will money was left for a Chair of Eugenics, which Pearson was the first to occupy. The Biometrics Laboratory and Galton's Eugenics Laboratory both became part of the new Department of Applied Statistics with Pearson at the head.

During the years after Weldon left for Oxford, there was a lull in Pearson's mathematical-statistical activity. Endless data of all kinds were collected and analyzed, many controversies—mostly in the eugenics field—were pursued, many graduate students were instructed in the new biometric field, and it was from one of these that the next stage in statistical methods originated. W. S. Gosset* ("Student"), was a student in 1906, bringing with him the small-sample problem. This was to occupy the attention of statisticians for many years, although it did not affect the course of Pearson's work unduly. He was always concerned with the collection and analysis of large quantities of data, with the mathematics very definitely ancillary to the main purpose—a tool to aid the understanding and not an end in itself. This led him in later years into many mathematical controversies in which he was wrong in the exact sense, but close enough in the practical sense.

But if at this time and in the later years the research seems to have been directed toward the working out of the ideas, both mathematical and practical, of a former era, the teaching side flourished. In 1915 the University of London instituted an undergraduate honors degree in statistics, and after a brief interlude of war service, courses were drawn up and lectures given. In addition to the students, professors from all over the world came to these classes and went back after their sabbaticals were over to teach and research in their turn. These postwar years, remembered by many for the bitter controversies with R. A. Fisher*, are more important historically for this dissemination by teaching of how to analyze data and for the elevation of statistics both in England and abroad to the status of a respectable field of university study.

Pearson resigned in 1933. His department was split into two, one part as the Department of Eugenics with R. A. Fisher as Galton Professor, and the other as the Department of Statistics under E. S. Pearson*, with the latter responsible for the undergraduate teaching. Pearson moved to rooms in the Department of Zoology where he continued to edit *Biometrika* until his death. The goal for the statistician, which he always taught his students, may be summed up in some words which he wrote in 1922: "The imagination of man has always run riot, but to imagine a thing is not meritorious unless we demonstrate its reasonableness by the laborious process of studying how it fits experience."

(BIOSTATISTICS
CORRELATION
ENGLISH SCHOOL OF BIOMETRY

F. N. DAVID

PEARSON'S CHI-SQUARE   *See* CHI-SQUARE TESTS

PEARSON'S COEFFICIENT OF CONTINGENCY

This is a measure of association* computed from the frequencies $\{n_{ij}\}$ in an $(r \times c)$ contingency table*, by the formula

$$P = \left(\frac{X^2}{n + X^2} \right)^{1/2},$$

where $n = \sum_{i=1}^{r}\sum_{j=1}^{c}n_{ij}$ is the total number of observations and

$$X^2 = \sum_{i=1}^{r} \sum_{j=1}^{c} \left(\frac{n_i.n._j}{n} \right)^{-1} \left(n_{ij} - \frac{n_i.n._j}{n} \right)^2$$

$(n_i. = \sum_{j=1}^{c}n_{ij};\ n._j = \sum_{i=1}^{r}n_{ij})$ is the *mean square contingency*. It can be shown that if the contingency table is formed from a bivariate normal distribution with correlation coefficient ρ then, as $n, r, c \to \infty$, P^2 tends to ρ^2.

Related coefficients are *Chuprov's* coefficient*

$$T = \left[\frac{X^2}{n\{(r-1)(c-1)\}^{1/2}} \right]^{1/2}$$

and *Cramer's coefficient*

$$K = \left\{ \frac{X^2}{n \min(r-1, c-1)} \right\}^{1/2}.$$

The upper bounds of P^2, T^2, and K^2 are

$$\frac{\min(r-1, c-1)}{1 + \min(r-1, c-1)},$$

$$\frac{\min(r-1, c-1)}{\max(r-1, c-1)}$$

and 1, respectively. (They are attained when $n_{ij} = 0$ for $i \neq j$.)

PEARSON SYSTEM OF DISTRIBUTIONS

In his work in the 1890s on evolution, Karl Pearson* came across large data sets which often exhibited considerable skewness* and other systematic departures from normality. These discrepancies led to the development of the Pearson system of frequency curves [18, 19]. In addition to data fitting, the system has proved useful in modern statistics as a source of approximations* to sampling distributions when only the first few moments are available [16, 21] and in providing a class of "typical" nonnormal forms which may be used to examine the robustness* of standard procedures, as in Posten [20]. A valuable feature of the Pearson system is that it contains many of the best known continuous univariate distributions, as will be shown.

In his original development, Pearson noted that the probabilities (p_r) for the hypergeometric* distribution satisfied the difference equation

$$p_r - p_{r-1} = \frac{(r-a)p_r}{b_0 + b_1 r + b_2 r^2}$$

for values of r inside the range. A limiting argument suggests a comparable differential equation for the probability density function (PDF):

$$p'(x) = \frac{d}{dx}\, p(x) = \frac{(x-a)p(x)}{b_0 + b_1 x + b_2 x^2}.$$

This is Pearson's basic equation, and it is satisfied by the normal PDF when $b_1 = b_2 = 0$. It follows that the distributions are unimodal, although they may have maxima at the end(s) of their range. The equation has been criticized because of its ad hoc nature, but it is interesting to note that if we consider a continuous-time, continuous-state birth-and-death process* in equilibrium (see Cox and Miller [5, pp. 213–215]), we obtain the limiting differential equation

$$\frac{1}{2} \frac{d^2}{dx^2} \left[b(x)p(x) \right] = \frac{d}{dx} \left[a(x)p(x) \right],$$

where $a(x)$ and $b(x)$ denote the rates of

change in the mean and variance, respectively. When $a(x)$ is linear in x and $b(x)$ is quadratic in x, we return to Pearson's equation. Thus a stochastic model does indeed underlie the system.

DISTRIBUTIONS IN THE SYSTEM

There are three main distributions in the system, designated types I, VI, and IV by Pearson, generated by the roots of the quadratic in the denominator (real and opposite signs, real and same sign, and complex, respectively). Ten more "transition" types follow as special cases. The distributions possess the density functions shown in Table 1, where g and h are functions of the skewness and kurtosis coefficients defined later. (The constants of integration have been omitted, and we may replace x by $(y - \mu)/\sigma$ in each case.) With changes in location where necessary, we may identify several of these as standard distributions: beta* of first kind (I), F or beta of the second kind (VI), Student's t* (VII), chi-square or gamma* (III), reciprocal gamma (V), exponential* (X), and Pareto* (XI). For further discussion of individual members of the system and various properties, see Elderton and Johnson [8], Ord [14], and Dagum [6].

FITTING A DISTRIBUTION

A key feature of the Pearson system is that the first four moments (when they exist) may be expressed explicitly in terms of the four parameters (a, b_0, b_1, and b_2). In turn, the two moment ratios

Skewness $\quad \beta_1 = \mu_3^2/\mu_2^3$

Kurtosis $\quad \beta_2 = \mu_4/\mu_2^2$

provide a complete taxonomy of the system as shown in Fig. 1 (some of the more specialized types are omitted since they correspond to a single point in the plane). The (β_1, β_2) values giving rise to U- or J-shaped distributions are indicated separately. *See also* MOMENT-RATIO DIAGRAMS.

Given the β_1, β_2 values (or estimates thereof), a distribution may be selected and then fitted. Pearson suggested the method of moments* and later clashed with Fisher over the latter's claim that maximum likelihood* was to be preferred. Modern practice would seem to favor the use of the (β_1, β_2) chart for selection and then fitting the selected model by maximum likelihood.

Percentage points of the Pearson distribution for select values of β_1 and β_2 are given in E. S. Pearson and Hartley [17] and Bowman and Shenton [2, 3].

Various extensions to the Pearson system have been proposed, generally using higher-order polynomials. A recent development of interest is due to Dunning and Hanson [7], who consider the differential equation

$$p'(x)/p(x) = R_m(x)/S_n(x),$$

where R_m and S_n are polynomials of order m and n, respectively, in x. A generalized Pearson curve is fitted to a histogram* using nonlinear programming* with a mean square error objective function. Linear constraints, such as fixing the first k moments, are readily incorporated into this formulation.

A restricted scheme of interest is to set $b_0 = 0$ and consider the three-parameter family defined on $x \geqslant 0$ or a known lower end point. This includes versions of all the distributions except types IV, VII, and the

Table 1 Pearson Distributions

Type	Density	Support*
(I)	$(1 + x)^{m_1}(1 - x)^{m_2}$	$(-1 \leqslant x \leqslant 1)$
(VI)	$x^{m_2}(1 + x)^{-m_1}$	$(0 \leqslant x < \infty)$
(IV)	$(1 + x^2)^{-m}$	
	$\times \exp\{-v \tan^{-1}(x)\}$	$(-\infty < x < \infty)$
Normal	$\exp(-\frac{1}{2}x^2)$	$(-\infty < x < \infty)$
(II)	$(1 - x^2)^m$	$(-1 \leqslant x \leqslant 1)$
(VII)	$(1 + x^2)^{-m}$	$(-\infty < x < \infty)$
(III)	$x^m \exp(-x)$	$(0 \leqslant x < \infty)$
(V)	$x^{-m} \exp(-x^{-1})$	$(0 \leqslant x < \infty)$
(VIII)	$(1 + x)^{-m}$	$(0 \leqslant x \leqslant 1)$
(IX)	$(1 + x)^m$	$(0 \leqslant x \leqslant 1)$
(X)	e^{-x}	$(0 \leqslant x < \infty)$
(XI)	x^{-m}	$(1 \leqslant x < \infty)$
(XII)	$[(g + x)/(g - x)]^h$	$(-g \leqslant x \leqslant g)$

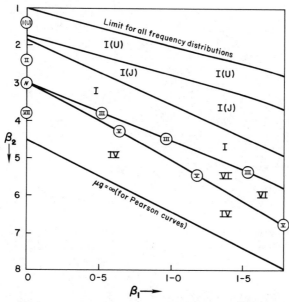

Figure 1 The β_1, β_2 chart for the Pearson curves. Equations of bounding curves as follows:
Upper limit for all frequency distributions: $\beta_2 - \beta_1 - 1 = 0$.
Boundary of $I(J)$ area: $4(4\beta_2 - 3\beta_1)(5\beta_2 - 6\beta_1 - 9)^2 = \beta_1(\beta_2 + 3)^2(8\beta_2 - 9\beta_1 - 12)$.
Type II line $2\beta_2 - 3\beta_1 - 6 = 0$.
Type V line $\beta_1(\beta_2 + 3)^2 = 4(4\beta_2 - 3\beta_1)(2\beta_2 - 3\beta_1 - 6)$. (From Pearson and Hartley [17] with permission of the authors.)

normal, and the appropriate type may be selected using only the first three moments [12; 14, p. 4].

A variety of approximations has been developed for particular Pearson curves; for further details, *see* APPROXIMATIONS TO DISTRIBUTIONS [14, pp. 182–183; 23, for type IV].

Several attempts were made by Pearson and his co-workers to develop a bivariate system of curves based on partial differential equations such as

$$\frac{\partial}{\partial x} \log p(x, y) = \frac{\text{cubic in } (x, y)}{\text{quartic in } (x, y)} \; ;$$

details are given in Ord [14, Chap. 3]. Usually, these systems proved too cumbersome for general use, but the simpler scheme developed by Steyn [22] with a linear numerator and quadratic denominator yields a system with linear regression functions. Other multivariate extensions are reviewed by Kotz [11]. *See also* FREQUENCY SURFACES, SYSTEMS OF.

Parrish and Bargmann [15] have devel-

oped high-precision quadrature techniques for the evaluation of the distribution function in the bivariate case.

Other continuous systems that cover most (β_1, β_2) values are the Johnson* system, based on the normal, and the Burr* system, based on the logistic.

THE DISCRETE SYSTEM

Returning to the difference equation from which Pearson began, we may develop an analogous system of discrete distributions. Rather iniquitously, this is sometimes known as the Ord–Carver system [4, 13]. The resulting distributions and their properties are described in Ord [14, Chap. 5]. The main types are hypergeometric* series distributions, first examined systematically by Kemp and Kemp [10]. The binomial*, negative binomial*, and Poisson* arise as special cases when $b_2 = 0$. The (β_1, β_2) chart is not very useful in the discrete case, but when attention is restricted to nonnegative variates, the

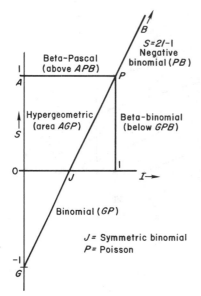

Figure 2 The I, S chart for the discrete Pearson system. Adapted from Ord [13] with permission.

ratios

$$I = \mu_2/\mu_1', \qquad S = \mu_3/\mu_2$$

provide a convenient description; see Fig. 2. The principal hypergeometric series distributions included are as in Table 2 ($0 < M < N$ in all cases). The binomial-beta reduces to the negative hypergeometric when all its parameters are integers. Indeed, all these distributions may be generated from urn schemes* when the parameters are integer. For estimation procedures, see Kemp and Kemp [10].

When $b_2 = 0$, the difference equation* may be rewritten as

$$u_r = rp_r/p_{r-1} = c_0 + c_1 r;$$

c_1 is $>$, $=$, or $<$ zero for the negative

Table 2   Hypergeometric Series Distributions

Hypergeometric*	$\binom{M}{r}\binom{N-M}{n-r} \Big/ \binom{N}{n}$
Negative hypergeometric*	$\binom{k+r-1}{r}\binom{N-k-r}{M-r} \Big/ \binom{N}{M}$
Binomial-beta	$\binom{n}{r}B(r+a, n+b-r)/B(a,b)$
Pascal-beta	$\binom{k+r-1}{r}B(k+a, r+b)/B(a,b)$

binomial, Poisson, and binomial distributions, respectively, so that the sample versions of the u_r ratios may be used, together with the sample (I, S) values, to select a distribution from the system.

For a discussion of systems of discrete multivariate distributions, see Steyn [22], Ord [14, Chap. 7], and Johnson and Kotz [9].

RELATED COMPUTER PROGRAMS

Bouver and Bargmann [1] provide an algorithm for fitting Pearson curves. Several useful algorithms have appeared in the Algorithms section of *Applied Statistics** for the evaluation of certain distributions and their percentage points. These are as follows (with reference numbers):

Distribution	Algorithms
Normal	2, 24(p), 66, 70(p), 111(p)
Student's t	3, 27
Gamma	32, 91(p), 147
Beta	63, 64(p)
Hypergeometric	59, 152
Pearson curve	192(p)

where p denotes the evaluation of percentage points.

References

[1] Bouver, M. and Bargmann, R. E. (1977). *Amer. Statist. Ass. Proc. Statist. Computing*, 127–132. (Computer program for fitting Pearson curves.)

[2] Bowman, K. O. and Shenton, L. R. (1979). *Biometrika*, **66**, 147–152.

[3] Bowman, K. O. and Shenton, L. R. (1979). *Commun. Statist. B*, **8**, 231–244. (Provides extended tables of percentage points.)

[4] Carver, H. C. (1919). *Proc. Casualty Actuarial Soc. Amer.*, **6**, 52–72. (First use of discrete Pearson curves for smoothing data.)

[5] Cox, D. R. and Miller, H. D. (1965). *The Theory of Stochastic Processes*. Methuen, London.

[6] Dagum, C. (1981). *J. Inter-Amer. Statist. Inst.*, **35**, 143–183.

[7] Dunning, K. A. and Hanson, J. N. (1978). *J. Statist. Comp. Simul.*, **6**, 115–121.

[8] Elderton, W. P. and Johnson, N. L. (1969). *Systems of Frequency Curves*. Cambridge University Press, Cambridge, England. (A comprehensive account of the Pearson system and the individual curves.)

[9] Johnson, N. L. and Kotz, S. (1982). *Int. Statist. Rev.*, **50**, 71–101. (Review of recent developments on systems of discrete distributions, including over 700 references.)

[10] Kemp, C. D. and Kemp, A. W. (1956). *J. R. Statist. Soc. B*, **18**, 202–211. (Describes the different hypergeometric series distributions and their properties.)

[11] Kotz, S. (1975). In *Statistical Distributions in Scientific Work*, Vol. 1, G. P. Patil, S. Kotz, and J. K. Ord, eds. D. Reidel, Boston and Dordrecht, pp. 247–270.

[12] Müller, P. H. and Vahl, H. (1976). *Biometrika*, **63**, 191–194.

[13] Ord, J. K. (1967). *Biometrika*, **54**, 649–656.

[14] Ord, J. K. (1972). *Families of Frequency Distributions*. Charles Griffin, London. (Gives an account of various systems of continuous and discrete distributions, plus bibliography.)

[15] Parrish, R. S. and Bargmann, R. E. (1981). In *Statistical Distributions in Scientific Work*, Vol. 5, C. Taillie, G. P. Patil, and B. Baldessari, eds. D. Reidel, Boston and Dordrecht, pp. 241–257.

[16] Pearson, E. S. (1963). *Biometrika*, **50**, 95–112. (Use of Pearson curves to approximate the noncentral chi-square.)

[17] Pearson, E. S. and Hartley, H. O. (1966, 1972). *Biometrika Tables for Statisticians*, Vols. I and II. Cambridge University Press, Cambridge, England. (Gives tables of percentage points for Pearson curves.)

[18] Pearson, K. (1894, 1895). *Philos. Trans. R. Soc. Lond.*, **A185**, 719–810; **A186**, 343–414.

[19] Pearson, K. (1948). *Early Statistical Papers*. Cambridge University Press, Cambridge, England. (The two original papers [18] are reproduced in this volume and show the development of Pearson's ideas.)

[20] Posten, H. O. (1978, 1979). *J. Statist. Comp. Simul.*, **6**, 295–310; **9**, 133–150. (Robustness of *t*-tests over the Pearson system.)

[21] Solomon, H. and Stephens, M. A. (1977). *J. Amer. Statist. Ass.*, **72**, 881–885. (Use of Pearson system to approximate the distribution of a sum of weighted chi-square variables.)

[22] Steyn, H. S. (1955, 1957, 1960). *Kon. Ned. Akad. Wet. Proc.*, **A58**, 588–595; **A60**, 119–127; **A63**, 302–311. (The first two papers deal with discrete multivariate systems of the hypergeometric type

and the third with a multivariate continuous Pearson system.)

[23] Woodward, W. A. (1976). *J. Amer. Statist. Ass.*, **71**, 513–514.

Acknowledgment

We are grateful to the *Biometrika* trustees for permission to reproduce Figs. 1 and 2.

(APPROXIMATIONS TO DISTRIBUTIONS
BETA DISTRIBUTION
BINOMIAL DISTRIBUTION
BIRTH-AND-DEATH PROCESSES
EXPONENTIAL DISTRIBUTION
FREQUENCY CURVES, SYSTEMS OF
GAMMA DISTRIBUTION
HYPERGEOMETRIC DISTRIBUTION
MAXIMUM LIKELIHOOD ESTIMATION
METHOD OF MOMENTS
NEGATIVE BINOMIAL DISTRIBUTION
PARETO DISTRIBUTION
POISSON DISTRIBUTIONS
STUDENT'S *t*-DISTRIBUTION
URN MODELS)

J. K. ORD

PECKING ORDER PROBLEM

The term *pecking order problem* describes the situation where a ranking of individuals $i = 1, \ldots, r$ is to be established on the basis of some pairwise comparisons of them. The problem occurs in consumer preference tests, psychological studies, and sporting contests (see, e.g., refs. 4, p. 101, 6, and 7). Such a ranking is equivalent to a set of numerical ratings or values $\{\rho_i : i = 1, \ldots, r\}$ for the individuals with the property that $\rho_i > \rho_j$ if and only if individual i is ranked higher than j. In general these values $\{\rho_i\}$ will not be uniquely determined: even their existence entails the implicit assumption of a one-dimensional ordering which, for example, might otherwise be regarded as a dominant factor in a principal component analysis*.

The situation is simplest when the number of pairwise comparisons is the same for all $r(r-1)/2$ distinct pairs, as occurs for example in a round-robin tournament*. Suppose that the random variable W_{ij} measures the preference for i rather than j, and that for some function $g(\cdot, \cdot)$,

$$E\left[W_{ij}\right] = g(\rho_i, \rho_j),$$

where $g(x, y)$ increases in x and the sum $g(x, y) + g(y, x)$ is constant for all x and y. Then the quantities $W_{i\cdot} = \sum_k W_{ik}$ have the property that $E[W_{i\cdot}] > E[W_{j\cdot}]$ if and only if $\rho_i > \rho_j$. Consequently, summing the observed preference measures w_{ik} over k for each i and using the method of moments* to estimate $\{\rho_i\}$ yields the same pecking order as using the sums $\{w_{i\cdot}\}$.

A similar approach can be used when there are N_{ik} comparisons of i and k with N_{ik} no longer necessarily constant for all pairs. Supposing that w_{ik} denotes the total observed preference measure from these N_{ik} comparisons, the $\{\rho_i\}$ can be estimated from the equations

$$w_{i\cdot} = \sum_k N_{ik} g(\rho_i, \rho_k).$$

An altogether different modelling approach, though it is immediately applicable only to comparisons that yield $W_{ik} = 0$ or 1 when $N_{ik} = 1$, is to interpret ρ_i as a handicap or stake value that i loses to any opponent k defeating him. Writing $\pi_{ik} = \Pr[W_{ik} = 1] = 1 - \pi_{ki}$, a fair or equitable allocation of the handicaps $\{\rho_i\}$ is to demand that the expected gain for each i in a round-robin tournament should be zero, i.e., $0 = E[\sum_k (\rho_k W_{ik} - \rho_i W_{ki})]$, equivalently, $(r - 1)\rho_i = \sum_k (\rho_i + \rho_k)\pi_{ik}$. This approach, of appealing to fair handicaps or bets, admits other variants (see ref. 5 for discussion and references), but estimation of the probabilities π_{ij} is a serious practical difficulty. Related Markov chain modelling approaches are indicated in ref. 3. As in the first approach, the conditions $\pi_{ik} \geqslant \pi_{jk}$ for all $k \neq i$ and j and $\pi_{ij} \geqslant 0.5$ imply $\rho_i \geqslant \rho_j$.

The models most discussed in the literature have $g(x, y) = F(x - y)$ for some distribution function F. Taking F to be Gaussian yields the Thurstone Case V model that is much discussed in the literature of psychology*, while letting F be the logistic distribution* is equivalent to the Bradley–Terry model*, in which $g(x, y) = x/(x + y)$. This model has been discussed extensively in the statistical literature on paired comparisons (see, e.g., refs. 2 and 4).

The preceding discussion envisages a fixed set of individuals with constant values $\{\rho_i\}$. The so-called Elo rating method, devised originally for chess players [6], allows for the incorporation of newcomers (i.e., expansion of the population of individuals) and regards the rating values $\{\rho_i\}$ as being functions that may change (not too rapidly) in time. The system is pragmatic and involves various approximations, some of which have been detailed in a more mathematical account [1].

References

[1] Batchelder, W. H. and Bershad, N. J. (1979). *J. Math. Psychol.*, **19**, 39–60.

[2] Bradley, R. A. (1976). *Biometrics*, **32**, 213–239.

[3] Daley, D. J. (1979). In *Interactive Statistics*, D. R. McNeil, ed. North-Holland, Amsterdam, pp. 247–254.

[4] David, H. A. (1963). *The Method of Paired Comparisons*. Griffin, London.

[5] David, H. A. (1971). *Rev. Internat. Statist. Inst.*, **39**, 137–147.

[6] Elo, A. E. (1978). *The Rating of Chess Players, Past and Present*. Batsford, London.

[7] Stefani, R. G. (1980). *IEEE Trans. Systems Man Cybernet*, **SMC-10**, 116–123.

(PAIRED COMPARISONS
RANKING PROCEDURES)

D. J. DALEY

PEELING DATA

Peeling methods are designed to provide a rudimentary form of rank for multivariate data. Rank is naturally determined by numerical order for univariate data, but in a higher number of dimensions order is not unique [1]; peeling can be used to define

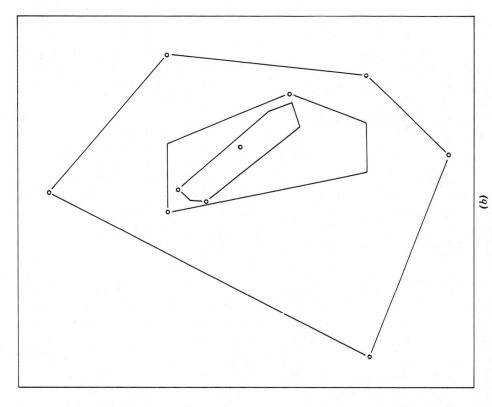

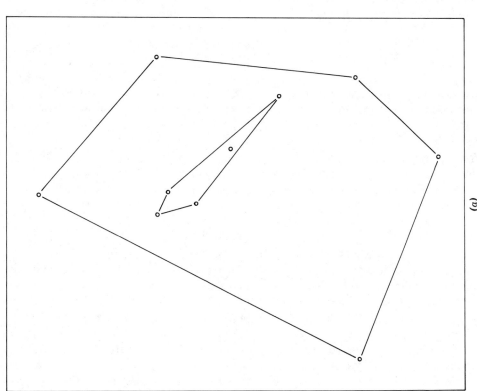

Figure 1 Two methods of peeling an artificial data set comprising ten points: (*a*) convex hull peeling; (*b*) Tukey peeling. The points are shown as circles; the polygons defining the peels are given by the straight lines.

instead a ranking by extremity, measuring the depth of each point in a multivariate sample, its distance from the "outside" of the data set. Crudely, peeling methods proceed by successive deletion of points around the periphery of the data until none is left: the term peeling is thus used as applied to an onion, not an orange.

Several different ways of peeling a data set have been suggested, corresponding to different notions of "outside," and leading to different resulting depths. The definitions that follow apply to bivariate data where they are readily visualized: translation to a higher number of dimensions is straightforward.

These peeling methods are more fully described, along with several others, in a review article by the present author [5].

An entirely different use of the term "peeling" deserves brief mention. In the algebraic computation of genetic information expressed as probability functions defined on pedigrees [3], the basic procedure consists of "peeling off" individuals in the pedigree one by one.

CONVEX HULL PEELING

John Tukey is credited [7, 1] with first proposing the idea of peeling as a higher-dimensional analogue of trimming*, in which the outermost peel comprises those points forming the extreme vertices of the convex hull of the data points. (The convex hull of a set is the smallest convex set containing that set: for a finite set of points in the plane, it is a convex polygon whose vertices are points in that set.) These points are assigned depth one and then completely discarded. The process is then repeated, the extreme points at the next stage being assigned depth two, and so on. The result is a nested sequence of polygons whose vertices together comprise the original data points (see Fig. 1a). Programs for the construction of these peels are available [6]. As with the two alternative proposals below, application to univariate data coincides with trimming.

ELLIPTICAL PEELING

Such a recursive method of peeling is readily modified by replacing the convex hull by some other set of specified shape, and possibly orientation or location. For example, use of the ellipse of smallest area containing the point set has been proposed [9] in a paper describing an efficient algorithm for the purpose. Necessarily, three, four, or five points of the set lie on the perimeter of the ellipse: these are assigned depth one and discarded, and the process repeated.

TUKEY PEELING

Another definition of peeling that is not simply recursive, and bearing perhaps a closer connection with univariate ordering, is again due to Tukey [11] (see also Eddy [4], where this is called convex hull peeling). For each integer d, we simultaneously slice off all open half-planes containing fewer than d points. There remains a convex polygon C_d: it is in fact the intersection of the convex hulls of all subsamples of $(n + 1 - d)$ points chosen from the original sample of n. Any data points in C_d but not C_{d+1} are assigned depth d: they are necessarily vertices of C_d. This system of nested polygons again assigns a depth to each data point, which is in fact the minimum rank of that point obtained when ordering all possible one-dimensional projections of the data set. The construction is illustrated in Fig. 1b for the same set of points as Fig. 1a.

PROPERTIES

Theoretical properties of the methods are difficult to obtain for randomly distributed data: It is, however, clear that the three methods described in the preceding section possess affine invariance, in the sense that the peels of an affinely transformed data set may be obtained by applying the same transformation to the peels of the original data. (Affine transformations are those ob-

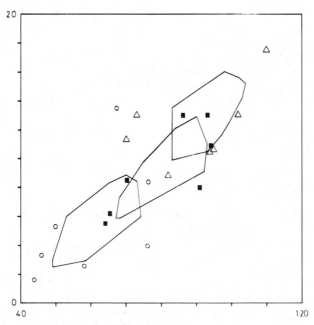

Figure 2 A decorated scatter plot of weight (in kilograms; vertical axis) versus height (in centimeters; horizontal axis) for samples from three different age groups of Nepalese children: 0–1, 2–3, and 4–5 years. The symbols (circles, squares, and triangles) mark the outermost convex hull peels for these age groups respectively. The polygons display the second peels. The sample sizes are 213, 206, and 128 respectively.

tained from successive rotations, reflections, and shears. Invariance follows since under such transformations, straight lines remain straight and ellipses elliptical, polygons continue to include their interiors, and areas are scaled proportionately.)

It is also apparent that they differ in degrees of coarseness: Tukey and elliptical peeling assign fewer points to each peel than convex hull peeling, for most distributions. Asymptotic distribution theory for Tukey peeling is given by Eddy [4].

APPLICATIONS

Applications of peeling, which are also discussed in ref. 5, range considerably in degrees of formality. Use of the peel depths in determining weights for robust*/resistant* statistical procedures has been proposed, although the performance of such methods is difficult to assess except by simulation. At a simpler level, one or more peels may be simply discarded as potential outliers* [2,

10] by analogy with univariate trimming. Use of peel-based ranks permits concepts such as median and inter-quartile range* to be extended to multivariate data [8]. Finally, and most simply, the peels may be highlighted to enhance the visual impact of a graphical display. As an illustration, Fig. 2 depicts such a decorated scatter plot*, displaying three bivariate samples.

References

[1] Barnett, V. (1976). *J. R. Statist. Soc. A*, **139**, 318–354. (A paper, with discussion, covering the whole field of multivariate order.)

[2] Bebbington, A. C. (1978). *Appl. Statist.*, **27**, 221–226. (The correlation coefficient and convex-hull trimming.)

[3] Cannings, C., Thompson, E. A., and Skolnick, M. H. (1978). *Adv. Appl. Prob.*, **10**, 26–61.

[4] Eddy, W. F. (1982). *Compstat 1982*. Physica-Verlag, Vienna, pp. 42–47.

[5] Green, P. J. (1981). In *Interpreting Multivariate Data*, V. D. Barnett, ed. Wiley, Chichester, England, pp. 3–19. (A review of peeling methods for bivariate data. Many more references.)

[6] Green, P. J. and Silverman, B. W. (1979). *Computer J.*, **22**, 262–266. (A comparative study of algorithms for convex hulls of bivariate data sets, with applications.)

[7] Huber, P. J. (1972). *Ann. Math. Statist.*, **43**, 1041–1067.

[8] Seheult, A. H., Diggle, P. J., and Evans, D. A. (1976). In ref. 1.

[9] Silverman, B. W. and Titterington, D. M. (1980). *SIAM J. Sci. Statist. Computing*, **1**, 401–409. (The minimum ellipse algorithm.)

[10] Titterington, D. M. (1978). *Appl. Statist.*, **27**, 227–234. (The correlation coefficient and elliptical trimming.)

[11] Tukey, J. W. (1975). *Proc. Int. Conf. Math., Vancouver, 1974*, **2**, 523–531. (A typically original contribution on "picturing data.")

(EXPLORATORY DATA ANALYSIS
GRAPHICAL REPRESENTATION OF DATA
ORDER STATISTICS
OUTLIERS
TRIMMING AND WINSORIZATION)

P. J. GREEN

PEIRCE'S CRITERION

This is a criterion for rejection of outlying observed values, proposed by B. Peirce [3] in 1852. It is based on the principle that "the proposed observations should be rejected when the probability of the system of errors obtained by retaining them is less than that of the system of errors obtained by their rejection, multiplied by the probability of making so many, and no more, abnormal observations."

The principle appears to be quite arbitrary, and the criterion is not in current use. In application "probability" is replaced by what is now called "likelihood"*.

Suppose that "errors," Y, exceeding Δ in absolute value will be rejected. If n errors $Y_1, Y_2, \ldots Y_n$ are available and of these $|Y_1|, \ldots, |Y_s|$ exceed Δ, with common probability density function $f(y)$, then the first probability is

$$\prod_{j=1}^{n} f(Y_j)$$

and the second is

$$\left\{ \prod_{j=s+1}^{n} f(Y_j) \right\} \binom{n}{s} \{ P(\Delta) \}^{n-s} \{ 1 - P(\Delta) \}^{s},$$

where

$$P(\Delta) = \Pr[\,|Y| < \Delta\,] = \int_{-\Delta}^{\Delta} f(y)\,dy.$$

So Peirce's rule requires that

$$\prod_{j=1}^{s} f(Y_j) < \binom{n}{s} \{ P(\Delta) \}^{n-s} \{ 1 - P(\Delta) \}^{s}$$

If $f(y)$ involves unknown parameters they are estimated from the $(n - s)$ values less than Δ.

The rule can be applied sequentially starting with the largest $|Y_j|$ and including further possible outliers one by one. Chauvenet [1] describes further details, involving estimation of nuisance parameters*.

An application of the method is set out by Gould [2], who assumed a normal distribution of errors, with zero expected value.

References

[1] Chauvenet, W. (1864). *A Manual of Spherical and Practical Astronomy*, Lippincott, Philadelphia.

[2] Gould, B. A. (1855). *Astron. J.*, **4**, 81–87.

[3] Peirce, B. (1852). *Astron. J.*, **2**, 161–168.

(CHAUVENET'S CRITERION
OUTLIERS
THOMPSON'S CRITERION)

PENALIZED LIKELIHOOD ESTIMA-TOR   *See* DENSITY ESTIMATION; PENALIZED MAXIMUM LIKELIHOOD ESTIMATION

PENALIZED MAXIMUM LIKELIHOOD ESTIMATION

It often happens when estimating a curve or function nonparametrically that the method of maximum likelihood* either cannot be applied at all or else leads to unsatisfactory results. These difficulties are illustrated below by considering density estimation* and regression* respectively. Penalized maximum likelihood estimates (PMLEs) provide a unified approach to curve estimation avoiding some of the problems raised by ordinary maximum likelihood.

The basic idea behind PMLEs is the notion that there are two aims in curve estimation; one is to maximize fidelity to the data, as measured by the log likelihood $l(g)$, while the other is to avoid curves which exhibit too much roughness or rapid variation. Roughness of a curve g can be measured by a *roughness functional* $R(g)$ in various ways. A typical choice of $R(g)$ is $\int g''(t)^2 \, dt$, which will have a high value if g exhibits a large amount of local curvature and the value zero if g is a straight line. The possibly conflicting aims of obtaining a high value of $l(g)$ while guarding against excessive values of $R(g)$ are reconciled by subtracting from the log likelihood a multiple of $R(g)$, called a *roughness penalty*, to obtain

$$l_p(g) = l(g) - \alpha R(g).$$

A penalized maximum likelihood estimate is then a curve g which maximizes the penalized log likelihood over the class of all curves satisfying sufficient regularity conditions for $R(g)$ to be defined. The *smoothing parameter* α controls the trade-off between high likelihood and smoothness and hence determines implicitly how much the data are smoothed to produce the estimate. The degree of smoothing increases as α increases.

The idea of penalizing a measure of goodness of fit* by a term based on roughness was suggested, in the regression context, by Whittaker [13]. PMLE regression estimates have been widely considered in the last 20 years by numerical analysts interested in splines* (see, for example, Chapter 14 of ref. 4) as well as by statisticians (see ref. 11 and the references mentioned there). The use of PMLEs for density estimation was suggested by Good [6] and has been discussed by several authors since (see refs. 10 and 9 for detailed historical discussion and references). PMLEs have also been used for a variety of other curve and function estimation problems, for example the estimation of discriminants [8], logistic regression [1], and hazard rate estimation [2]. Wahba and co-

workers [3, 11, 12] have contributed widely to the theoretical and practical aspects of the application of the PMLE approach to the estimation of functions f where one observes

$$X_i = (Af)(t_i) + \epsilon_i$$

for some linear operator A; the nonparametric regression problem is a special case of this.

The philosophical justification of the PMLE approach was originally as an *ad hoc* device but the method can be given a Bayesian* interpretation [6]; the PMLE is the posterior mean assuming a certain (improper) prior distribution* over function space, and the smoothing parameter is then a parameter in the prior specification.

DENSITY ESTIMATION

Suppose $x_1, \ldots, x_n$ are observations of independent identically distributed random variables whose underlying PDF f, on an interval $[a, b]$, is of interest. Given any PDF g, the log likelihood of g as an estimate of f is

$$l(g) = \sum \log g(x_i).$$

Since l can be made arbitrarily large [6] essentially by choosing g to be a sum of narrow peaks centred at the observations, there is no maximum likelihood estimate of f in the class of all PDFs.

If the roughness functional $\int g''(t)^2 \, dt$ is used then the penalized log likelihood will be

$$l_p(g) = \sum_i \log g(X_i) - \alpha \int_a^b g''(t)^2 \, dt,$$

which will be maximized subject to the constraint that g is a PDF to obtain the PMLE. Another possible roughness functional [6] for density estimation is $\int g'^2/g$, which may also be written $4\int(\sqrt{g})'^2$, and which may be viewed both as a measure of the ease of detecting small shifts in g and as a measure of variability in $\sqrt{g}$. Yet another possibility [9] is $\int (\log g)'''^2$, a measure of high order curvature in $\log g$ which is zero if and only if g is a Normal density. References 6 and 9 contain further discussion on the form of the roughness functional.

Under suitable conditions, for certain roughness penalties, PMLEs of density functions can be shown to exist and to be unique, and to be consistent in various senses. Those results that have been obtained on exact rates of consistency indicate that PMLEs achieve the same accuracy of estimation as other methods of density estimation, under comparable assumptions [9]. Establishing theoretical properties of PMLEs of densities can involve difficult technical argument, because of the implicit definition of the estimators, the nonlinearity inherent in the form of l_p, and the constraint that the estimate is a PDF.

The computation of PMLEs of densities involves methods for constrained nonlinear optimization in function space. Methods that have been used include [6] the Rayleigh–Ritz method where g is expanded as an orthogonal series; the maximization is then performed by re-expressing $l_p(g)$ in terms of the coefficients of g in the series expansion. If the series is chosen suitably, l_p will then be in a more amenable form for maximization. Another suggestion [10, Chap. 5] is a discretization procedure where the roughness penalty is replaced by a sum of squared second differences and the values of the PMLE on a fine grid are then found by solving a finite-dimensional optimization problem. For the particular roughness functional $\int g'^2/g$, the PMLE is a piecewise exponential curve, the various parameters of which can be found by a fairly straightforward iterative scheme; see ref. 5 for details.

Published practical applications of PMLEs of densities are given in refs. 10 and 7, which also include suggestions for methods of choosing the smoothing parameter.

NONPARAMETRIC REGRESSION

Given paired observations (t_i, X_i) suppose it is of interest to fit a model of the form

$$X_i = g(t_i) + \epsilon_i,$$

where the ϵ_i are independent Normal errors with equal variance σ^2 and g is a curve to be estimated. If the variance is assumed known, the log likelihood of g, up to a constant, is $-\sum(X_i - g(t_i))^2/(2\sigma^2)$, by standard properties of least squares estimation*. This will be maximized by any g which actually interpolates the data points. Such a curve would, in general, exhibit rapid local variation and would be dismissed as over-fitting the data; thus PMLEs are appropriate. Because of the form of $l(g)$, finding the PMLE will correspond to minimizing

$$\sum(X_i - g(t_i))^2 + \lambda \int g''(t)^2 \, dt$$

over the class of suitable functions g. Here the usual roughness functional is used, and the smoothing parameter α is replaced by $\lambda = 2\alpha\sigma^2$.

The solution to this minimization problem is a cubic spline, i.e., a piecewise cubic polynomial with continuous second derivative. The parameter λ controls the amount of smoothing; as $\lambda \to 0$ the PMLE approaches a curve which interpolates the data, while the limit as $\lambda \to \infty$ is the linear regression* fit. The smoothing parameter can be chosen automatically by a method called cross-validation [3]. Because they are the solution of an unconstrained quadratic minimization problem, nonparametric regression PMLEs are more easily dealt with, both computationally and theoretically, than PMLEs of densities, and are more widely used in practice. Numerical algorithms and a historical discussion are given in Chapter 14 of De Boor [4]. The techniques for computing the estimates all use the spline nature of the solution; the various polynomial coefficients are found by solving large systems of linear equations. The special properties of these equations make it possible to solve them in ways which are computationally fast and numerically stable.

OTHER APPLICATIONS

The PMLE approach can easily be generalized to deal with any curve estimation problem where the likelihood or partial likelihood* of a curve can be found. An example is given in ref. 8 where the estimation of the log density ratio $g = \log(f_1/f_0)$ is considered, given independent samples $X_1, \ldots, X_n$ from f_0 and $Y_1, \ldots, Y_m$ from f_1. The (partial) log likelihood of g is given by

$$l(g) = \sum \psi_0(g(X_i)) + \sum \psi_1(g(Y_j))$$

where ψ_0 and ψ_1 are known functions. The form of this likelihood is such that the maximum would occur if $g(X_i) = -\infty$ for all i and $g(Y_j) = +\infty$ for all j. Introducing a roughness penalty term leads to an estimate which is a cubic spline. A nonlinear system has to be solved to find the polynomial coefficients.

Another application [1] is to hazard rate* estimation. In the particular problem considered, the model (*see* COX'S REGRESSION MODEL)

$$\lambda(t, \mathbf{z}) = \lambda_0(t)\exp(\boldsymbol{\beta}^T\mathbf{z})$$

is used for the hazard rate at time t given covariates $\mathbf{z}$. The vector $\boldsymbol{\beta}$ is estimated by maximum likelihood and the log-likelihood of the curve λ_0 can then be written in a fairly complicated form $l(\lambda_0)$ which involves values and integrals of λ_0. The log-likelihood l is unbounded above, but a suitable PMLE, using the roughness functional $\int \lambda'(t)^2\,dt$, gives a smooth nonparametric estimate of λ_0.

PMLEs can be used to solve problems of deconvolution. Consider [11] the approximate solution of the integral equation*

$$\int K(t, s)g(s)\,ds = f(t)$$

from observations X_i, with error, of $f(t_i)$ at various points t_i. The case where the integral operator is the identity transformation is precisely the nonparametric regression problem considered above. As before, maximizing the log-likelihood of g corresponds to minimizing

$$S(g) = \sum \left(X_i - \int K(t_i, s)g(s)\,ds\right)^2$$

over the class of suitable functions g. Minimizing S directly involves the solution of an ill-conditioned linear system, and so the usual statistical drawbacks of maximum like-

lihood curve estimation are compounded with numerical difficulties. Adding a roughness penalty term to S resolves these problems.

References

[1] Anderson, J. A. and Blair, V. (1982). *Biometrika*, **69**, 123–136.

[2] Anderson, J. A. and Senthilselvan, A. (1980). *J. R. Statist. Soc. B.*, **42**, 322–327.

[3] Craven, P. and Wahba, G. (1979). *Numer. Math.*, **31**, 377–403.

[4] De Boor, C. (1978). *A Practical Guide to Splines*. Springer-Verlag, New York. (A good introduction, with emphasis on numerical rather than statistical aspects. Contains FORTRAN programs. Mainly Chapter 14 is relevant.)

[5] Ghorai, J. and Rubin, H. (1979). *J. Statist. Comp. Simul.*, **10**, 65–78.

[6] Good, I. J. and Gaskins, R. A. (1971). *Biometrika*, **58**, 255–277.

[7] Good, I. J. and Gaskins, R. A. (1980). *J. Amer. Statist. Ass.*, **75**, 42–73. (Includes discussion.)

[8] Silverman, B. W. (1978). *Appl. Statist.*, **27**, 26–33.

[9] Silverman, B. W. (1982). *Ann. Statist.*, **10**, 795–810. (Contains further references on theory of PMLEs of density estimates and regression functions.)

[10] Tapia, R. A. and Thompson, J. R. (1978). *Nonparametric Probability Density Estimation*. Johns Hopkins University Press, Baltimore. (An interesting treatment which lays emphasis on the PMLE approach.)

[11] Wahba, G. (1977). *SIAM J. Numer. Anal.*, **14**, 651–667.

[12] Wahba, G. and Wendelberger, J. (1980). *Monthly Weather Rev.*, **108**, 1122–1143. (An application that uses PMLEs generalized to the multivariate case.)

[13] Whittaker, E. (1923). *Proc. Edinburgh Math. Soc.*, **41**, 63–75.

(BAYESIAN INFERENCE
DENSITY ESTIMATION
ESTIMATION, POINT
LEAST SQUARES
MAXIMUM LIKELIHOOD ESTIMATION
NEAR NEIGHBOR ESTIMATOR
PARTIAL LIKELIHOOD
SPLINE FUNCTIONS
WHITTAKER METHOD
  OF GRADUATION)

B. W. SILVERMAN

PENTAGAMMA FUNCTION

The pentagamma function is the fourth derivative of the digamma or psi function* $(\Psi(x) = d \log \Gamma(x)/dx)$. It is customarily denoted by

$$\Psi'''(x) = \frac{d^4 \log \Gamma(x)}{dx^4}.$$

PENTAMEAN

This statistic, sometimes used as a measure of location* in exploratory data analysis*, is

$\frac{1}{10}$ {(sum of greatest and least observations)

+ median

+ (sum of upper and lower quartiles*))}

or in "exploratese,"

$\frac{1}{10}$ {(sum of extremes)

+ median + (sum of hinges)}.

PERCENTAGE

Percentage = (Proportion) × 100.

The symbol % is used conventionally to denote a percentage. If 75 individuals among 250 possess a property E, say, then the proportion possessing E is $75/250 = 0.3$ and the percentage possessing E is $0.3 \times 100\% = 30\%$.

Percentages are widely used in descriptive statistics to assist comparison of data from different sources. Such comparisons are easily made; a possible drawback is that the amounts of data on which the percentages are based may not be shown.

PERCENTILES, ESTIMATION OF

Percentile estimation has frequently been considered in the context of reliability, especially for the exponential* and Weibull* dis-

tributions and is addressed in refs. 34, 6, 11, and 30. Further applications of percentile estimation are found throughout the literature in such diverse areas as mortality trials [25], bioassay [10], material fatigue of rotor blades [14], mineral resources [18], aerial radiometric data [7], space shuttle ground operations [46], and clinical ophthalmology [17]. In general, the 100εth percentile of a random variable X is denoted as X_ϵ and defined as

$$X_\epsilon = \inf\{ x : F(x) \geqslant \epsilon \}$$

where F is the CDF of X and $0 \leqslant \epsilon \leqslant 1$.

LOCATION AND SCALE PARAMETER DISTRIBUTIONS*

Suppose X has a location, μ, and scale, σ, parameter distribution with parameter space $\Omega = \{(\mu, \sigma) : -\infty < \mu < \infty, \ \sigma > 0\}$. The CDF of X is given by

$$F(x; \mu, \sigma) = G((x - \mu)/\sigma) \qquad (1)$$

where G is a parameter-free CDF. Under certain regularity conditions on G (i.e., continuity and with a unique inverse)

$$X_\epsilon = \mu + G^{-1}(\epsilon)\sigma. \qquad (2)$$

Many prominent statistical distributions fall into this class directly (e.g., normal*, exponential, uniform*, Laplace*, Cauchy*, logistic*, and Gumbel) whereas other families of distributions can be affiliated with this class by transformation (e.g., lognormal*, inverse Gaussian*, Pareto*, loguniform, and Weibull).

Estimation by Order Statistics

Let $X_{1:n}, \ldots, X_{n:n}$ be the n order statistics from a random sample of size n from (1). $Y_{i:n} = (X_{i:n} - \mu)/\sigma$ has a parameter-free distribution for $i = 1, \ldots, n$ with the result that

$$E(X_{i:n}) = \mu + \sigma E(Y_{i:n}),$$
$$\text{cov}(X_{i:n}, X_{j:n}) = \sigma^2 \text{cov}(Y_{i:n}, Y_{j:n}) \qquad (3)$$

for $i, j = 1, \ldots, n$. Define $\mathbf{X}' = (X_{1:n} \cdots X_{n:n})$ and $\mathbf{Y}' = (Y_{1:n} \cdots Y_{n:n})$. Then it follows that $E(\mathbf{Y}) = \boldsymbol{\alpha}$ and $\text{cov}(\mathbf{Y}) = \mathbf{B}$ where $\boldsymbol{\alpha}$ is an $n \times 1$ column vector of known constants and $\mathbf{B}$ is an $n \times n$ matrix of known constants. This formulation gives rise to the general linear model*

$$\mathbf{X} = \mathbf{C}\boldsymbol{\beta} + \boldsymbol{\delta}, \qquad (4)$$

where $E(\boldsymbol{\delta}) = \mathbf{0}$, $\text{cov}(\boldsymbol{\delta}) = \sigma^2 \mathbf{B}$, $\mathbf{C} = (\mathbf{1}\,\boldsymbol{\alpha})$, $\mathbf{1}$ is an $n \times 1$ column vector of ones and $\boldsymbol{\beta}' = (\mu\,\sigma)$.

For our purpose of estimation of X_ϵ, define $\mathbf{L}$ as

$$\mathbf{L} = \begin{bmatrix} 1 & G^{-1}(\epsilon) \\ 0 & 1 \end{bmatrix}.$$

Consequently it follows that

$$\mathbf{X} = \mathbf{CL}^{-1}\mathbf{L}\boldsymbol{\beta} + \boldsymbol{\delta} = \mathbf{C}_0\boldsymbol{\beta}_0 + \boldsymbol{\delta}, \qquad (5)$$

where $\mathbf{C}_0 = \mathbf{CL}^{-1}$ and $\boldsymbol{\beta}_0 = \mathbf{L}\boldsymbol{\beta} = (X_\epsilon\,\sigma)'$. From the generalized Gauss-Markov theorem* the best linear unbiased estimator* (BLUE) of $\boldsymbol{\beta}_0$ is

$$\hat{\boldsymbol{\beta}}_0 = \begin{pmatrix} X_\epsilon^* \\ \sigma^* \end{pmatrix} = (\mathbf{C}_0'\mathbf{B}^{-1}\mathbf{C}_0)^{-1}\mathbf{C}_0'\mathbf{B}^{-1}\mathbf{X}. \qquad (6)$$

Although we employ both estimates X_ϵ^* and σ^*, X_ϵ^* can be expressed separately by partitioning $\mathbf{C}_0'\mathbf{B}^{-1}\mathbf{C}_0$. Hence

$$X_\epsilon^* = \boldsymbol{\alpha}_0'\boldsymbol{\Gamma}\mathbf{X}, \qquad (7)$$

where $\boldsymbol{\Gamma} = \mathbf{B}^{-1}(\boldsymbol{\alpha}_0\mathbf{1}' - \mathbf{1}\boldsymbol{\alpha}_0')\mathbf{B}^{-1}/\Delta$, $\Delta = |\mathbf{C}_0'\mathbf{B}^{-1}\mathbf{C}_0|$ and $\boldsymbol{\alpha}_0 = \boldsymbol{\alpha} - G^{-1}(\epsilon)\mathbf{1}$. Moreover, the variance of X_ϵ^* is

$$\text{var}(X_\epsilon^*) = (\boldsymbol{\alpha}_0'\mathbf{B}^{-1}\boldsymbol{\alpha}_0)\sigma^2/\Delta. \qquad (8)$$

X_ϵ^* is simply the appropriate linear combination of the BLUE's of μ and σ respectively.

Following the pioneering work of Lloyd [32], Sarhan and Greenberg [42] were instrumental in collecting and documenting formulas for $\boldsymbol{\alpha}$ and $\mathbf{B}$ for the prominent location and scale parameter distributions. Moreover, the calculation of X_ϵ^* and σ^* is readily implemented in APL.

Mann [33], seeking an improvement on the BLUE, derived the unique minimum mean squared error* linear estimator of X_ϵ as

$$\tilde{X}_\epsilon = X_\epsilon^* - \{e/(1 + d)\}\sigma^*, \qquad (9)$$

where $\text{var}(\sigma^*) = d\sigma^2$ and $\text{cov}(X_\epsilon^*, \sigma^*) = e\sigma^2$. $\tilde{X}_\epsilon$ has become popularly known in the literature as the best linear invariant estimator (BLIE) of X_ϵ and the mean squared error of $\tilde{X}_\epsilon$ is given by

$$\text{mse}(\tilde{X}_\epsilon) = \{f - e^2/(1 + d)\}\sigma^2, \qquad (10)$$

where $f = (\boldsymbol{\alpha}_0'\mathbf{B}^{-1}\boldsymbol{\alpha}_0)/\Delta$. Although the appeal of the BLIE is clear mathematically in reducing the risk under a quadratic loss function, one should not presume the general acceptability of $\tilde{X}_\epsilon$ over X_ϵ^* until the findings of Rao [37] are considered. Rao provides a variety of examples in which shrinking unbiased estimators to minimum mean squared error estimators does not improve the intrinsic property of Pitman Nearness. *See also* SHRINKAGE ESTIMATORS.

The preceding estimates of X_ϵ can be modified so as to contain any specified r of the n order statistics*. Of course $\boldsymbol{\alpha}$ and $\mathbf{B}$ must be modified accordingly. Such modifications allow for the treatment of Type II censored samples, trimming the first r_1 and the last r_2 order statistics so as to obtain a robust estimate of X_ϵ [2] and the optimal selection of two or three order statistics [1]. Although the BLUE and BLIE are appealing for their adaptability under various sampling situations and providing good estimates for families which do not admit complete sufficient statistics, these estimators, obtained for the location and scale parameter family, are often simply transformed for estimates of X_ϵ in the affiliated families. Unfortunately, in this process the property of unbiasedness* and minimum mean squared error of X_ϵ^* and $\tilde{X}_\epsilon$, respectively, is lost.

Equivariant Estimators

For a random sample of size n with common continuous CDF given by (1), consider estimators of μ and σ, denoted by $\hat{\mu}(\mathbf{X}; n)$ and $\hat{\sigma}(\mathbf{X}; n)$, which are equivariant estimators*

[47]. The location and scale invariant nature of these estimators makes them natural candidates for estimators of μ and σ. It is well known that $(\hat{\mu} - \mu)/\hat{\sigma}, \hat{\sigma}/\sigma,$ and $Z_\epsilon = (X_\epsilon - \hat{\mu})/\hat{\sigma}$ are pivotal quantities* for μ, σ, and X_ϵ, respectively. Thus

$$E(\hat{\mu}) = \mu + a\sigma \quad \text{and} \quad E(\hat{\sigma}) = b\sigma. \quad (11)$$

$\hat{\mu}$ and $\hat{\sigma}$ can be assumed to be uncorrelated [note that if we define $\hat{\tilde{\mu}} = \hat{\mu} - \beta\hat{\sigma}$ where $\beta = \text{Cov}(\hat{\mu}, \hat{\sigma})/\text{Var}(\hat{\sigma})$ then $(\hat{\sigma}, \hat{\tilde{\mu}})$ are uncorrelated].

The choice of an equivariant pair $(\hat{\mu}, \hat{\sigma})$ is not unique so that the importance of a class $\{\hat{X}_\epsilon : \hat{\mu} + \gamma\hat{\sigma}, \gamma = \gamma(\epsilon, n)\}$ is limited unless of course $(\hat{\mu}, \hat{\sigma})$ is jointly sufficient for (μ, σ) so that the choice is not arbitrary but rather optimal. The maximum likelihood estimators (MLE) and BLUEs of μ and σ are equivariant [30].

A mean unbiased estimator of X_ϵ based on $\hat{\mu}$ and $\hat{\sigma}$ is given by

$$\hat{X}_\epsilon(\text{MN}) = \hat{\mu} + \gamma'\hat{\sigma}, \quad (12)$$

where $\gamma' = (G^{-1}(\epsilon) - a)/b$. A minimum mean squared error estimator can be found in the same class as

$$\hat{X}_\epsilon(\text{SE}) = \hat{\mu} + \gamma^{\cdot}\hat{\sigma}, \quad (13)$$

where

$$\gamma^{\cdot} = \left(G^{-1}(\epsilon) - a\right)b/c$$

and

$$c = E\left((\hat{\sigma}/\sigma)^2\right).$$

However, Zidek [48] and Rukhin and Strawderman [41] exemplify the inadmissibility of the best equivariant estimator of X_ϵ in the normal and exponential cases, respectively. Rukhin and Strawderman develop a class of minimax* estimators based on a technique in ref. 8.

Keating [23], for the same class, demonstrates the existence of optimal estimators predicated on absolute loss. The median unbiased estimator* [31] is given as

$$\hat{X}_\epsilon(\text{MD}) = \hat{\mu} + \gamma''\hat{\sigma},$$

where γ'' is the median* of Z_ϵ. Moreover,

the minimum mean absolute error estimator is given as

$$\hat{X}_\epsilon(\text{AE}) = \hat{\mu} + \gamma^*\hat{\sigma}, \quad (15)$$

where γ^* is the median of a special noncentral distribution. In the normal distribution, Dyer and Keating [13] derive (12) through (15) and compare them on the basis of absolute risk and Pitman Nearness [37]. In Refs. 13 and 23, results can be found which support Rao's [37] findings.

Parzen [36] gives a concise description of a quantile estimation technique based on reproducing kernels in Hilbert spaces. The resulting estimators are computationally simple and asymptotically efficient. A detailed discussion of Parzen's quantile estimation technique is contained in QUANTILES.

UNIVERSAL TECHNIQUES

Parametric Methods

Several techniques exist for estimating percentiles in an arbitrary distribution. MLEs are especially appealing because of their invariance property, their ability to handle progressively censored* samples (Bain [6] gives a complete treatment of likelihood* estimation under various censoring criteria) and their well-known asymptotic properties [30].

A novel technique of estimating percentiles based on error in the predicted distribution function has been employed in refs. 39, 43, 4, and 3. In this procedure $\hat{X}_\epsilon$ is chosen within a restricted class so as to minimize $E\{[F(\hat{X}_\epsilon) - \epsilon]^2\}$. Another technique involves the use of the 50% conditional (on the values of ancillary statistics*) confidence interval* on X_ϵ. The resultant estimator has some appealing properties among which is conditional median unbiasedness. The conditional confidence interval has been advocated in refs. 15, 26 to 29, and 40 and reduced to a point estimation procedure in ref. 24 with application to the extreme-value distribution*.

Nonparametric Methods

In nonparametric methods the quantile function, $Q(\epsilon) = X_\epsilon$, is more prevalent and will be adopted for this section. Parzen [36] proposes estimators based on quantiles of a random sample selected from a continuous population:

$$\hat{Q}_1(\epsilon) = X_{j:n}$$

$$\text{where} \quad (j-1)/n < \epsilon \leqslant j/n. \quad (16)$$

For $\epsilon = 0$, define $\hat{Q}_1(\epsilon) = X_{0:n}$, which is $X_{1:n}$ or a natural minimum (e.g., $X_{0:n} = 0$ when X is nonnegative). Since $\hat{Q}_1(\epsilon)$ is a piecewise step function, smoother functions obtained in a piecewise linear fashion are appealing (e.g.,

$$\hat{Q}_2(\epsilon) = n(j/n - \epsilon)X_{j-1:n}$$
$$+ n\{\epsilon - (j-1)/n\}X_{j:n} \quad (17)$$

whenever $(j-1)/n \leqslant \epsilon \leqslant j/n$, for $j = 1, \ldots, n$). For samples from symmetric densities, the following estimator may behave better:

$$\hat{Q}_3(\epsilon) = n\{(2j+1)/n - \epsilon\}X_{j:n}$$
$$+ n\{\epsilon - (2j-1)/n\}X_{j+1:n} \quad (18)$$

whenever $(2j-1)/n \leqslant \epsilon \leqslant (2j+1)/n$, for $j = 1, \ldots, n-1$. $\hat{Q}_3(\epsilon)$ remains undefined for $\epsilon < 1/(2n)$ or $\epsilon > 1 - 1/(2n)$.

$\hat{Q}_1(\epsilon)$ is efficient among the class of translation equivariant and uniformly asymptotically median unbiased estimators. However, Harrell and Davis [19] point out the many drawbacks associated with $\hat{Q}_1(\epsilon)$ and we note that the asymptotic properties of $\hat{Q}_2(\epsilon)$ and $\hat{Q}_3(\epsilon)$ have not been investigated.

The generalized sample quantile estimators developed by Kaigh and Lachenbruch [22] and Kaigh [21] are obtained by averaging an appropriate subsample quantile over all possible subsamples of a fixed size selected from the complete sample from a continuous population. With the previous definition of order statistics for a sample of size n in mind, let $X_{1:k;n}, \ldots, X_{k:k;n}$ denote the order statistics of a subsample of size k selected from the complete sample,

$k = 1, \ldots, n$. For subsampling without replacement

$$\Pr(X_{r:k;n} = X_{j:n}) = \binom{j-1}{r-1}\binom{n-j}{k-r} \Big/ \binom{n}{k},$$

where $r \leqslant j \leqslant r + n - k$. By averaging over all possible subsamples of size k, define a collection of generalized order statistics $K_{1:k;n}, \ldots, K_{k:k;n}$ as

$$K_{r:k;n}$$
$$= \sum_{j=r}^{r+n-k} \left\{ \binom{j-1}{r-1}\binom{n-j}{k-r} \Big/ \binom{n}{k} \right\} X_{j;n}$$

for $r = 1, \ldots, k$. These are indeed generalized order statistics since for $k = n$, $K_{r:n;n} = X_{r:n}$. Then the generalized sample quantile estimator of X_ϵ is

$$\hat{Q}_4(\epsilon; k, n) = K_{[(k+1)\epsilon]:k;n}. \quad (19)$$

If the subsampling is done with replacement [21], then

$$\Pr(X_{r:k;n} = X_{j:n}) = \int_{(j-1)/n}^{j/n} m_{r,k}(x)\,dx,$$

where $m_{r,k}(x)$ is the PDF of a beta distribution with parameters r and $k - r + 1$. An alternative set of generalized order statistics $L_{1:k;n}, \ldots, L_{k:k;n}$ can be defined as

$$L_{r:k;n} = \sum_{j=1}^{n} \left\{ \int_{(j-1)/n}^{j/n} m_{r,k}(x)\,dx \right\} X_{j:n}$$

for $r = 1, \ldots, k$. Thus a generalized sample quantile estimator of X_ϵ is given by

$$\hat{Q}_5(\epsilon; k, n) = L_{[(k+1)\epsilon]:k;n}. \quad (20)$$

Harrell and Davis [19] consider the X_ϵ bootstrap* estimator which can be written as $L_{[(n+1)\epsilon]:n;n}$. $\hat{Q}_4$ and $\hat{Q}_5$ provide the sample mean when $\epsilon = 1/2$ and $k = 1$ and they are both translation and scale invariant. Since the extreme order statistics are ignored, $\hat{Q}_4$ is more robust and can be used with censored data* whereas $\hat{Q}_5$ can provide greater precision for samples from light to medium tailed distributions. Based on analytical and simulation results, $\hat{Q}_4$ and $\hat{Q}_5$ usually provide smaller mse than $\hat{Q}_1$. Jackknifing* these generalized estimators, Kaigh [21] obtained consistent estimators of their variances.

The procedure of estimating X_ϵ by kernel methods has previously been touched on for location and scale parameter families. In general for a random sample of size n, $\hat{f}(x)$ is a kernel estimator of the PDF f provided

$$\hat{f}(x) = (1/(nb)) \sum_{i=1}^{n} w((x - X_i)/b), \quad (21)$$

where the kernel function w is some even bounded PDF and $b > 0$. A commonly used kernel is the standard normal PDF and the choice of the kernel is not considered as important as the choice of the smoothing parameter b (see refs. 5, 16, 20, and 45 for considerations in the choice of b). Let $\hat{F}(x)$ be an estimate of the CDF formed by integrating $\hat{f}(x)$ in (21). Quantile estimators of $X_\epsilon = Q(\epsilon)$ are obtained by inverting the estimated CDF.

$$\hat{Q}(\epsilon) = \hat{F}^{-1}(\epsilon). \quad (22)$$

Under some regularity conditions, Nadaraya [35] observed that the asymptotic distribution of $\hat{Q}(\epsilon)$ is the same as that of $\hat{Q}_1(\epsilon)$.

Miscellaneous Methods

Quasiquantiles and adaptive quasiquantiles were investigated by Reiss [38] with the result that the relative performance of these estimators (for increasing sample size) to suitably defined adaptive estimators* was inferior. Schmeiser and Deutsch [44] studied percentile estimation for grouped data based on the cell midpoint. They concluded that larger cell widths resulted in smaller bias and reduced variance of their quantile estimator. Their conclusions on the effect of increasing sample size ran counter to the results of David and Mishriky [12] that the effects of grouping may generally be ignored. For some asymptotic results on sample quantiles based on dependent observations from linear processes and for some related references see Chanda [9].

References

The number of references is not small and is intended to represent a survey of past and current work in the field. References 34, 30, and 36 are excellent resources for the parametric approaches and are a rich source of further reading. References 21, 22, and 36 provide broad coverage of the modern nonparametric approaches to percentile estimation. Likewise, an attempt is made to acquaint the reader with some of the lesser known but nonetheless interesting techniques in percentile estimation.

[1] Ali, M. M., Umbach, D., Saleh, A., and Hassanein, K. M. (1983). *Commun. Statist.—Theory Meth.*, **12**, 2261–2271. (This article is an excellent reference for quantile estimators based on two or three optimally chosen order statistics.)

[2] Andrews, D. F., Bickel, P. J., Hampel, F. R., Huber, P. J., Rogers, W. H., and Tukey, J. W. (1972). *Robust Estimates of Location*. Princeton University Press, Princeton, N.J. (This seminal article on robust estimation provides the general methodology for obtaining robust quantile estimators based on order statistics.)

[3] Angus, J. E. (1983). *Commun. Statist.*, **12**, 1345–1358. (Angus provides a well written discussion of quantile estimation based on error in the predicted distribution function with an application to the normal distribution.)

[4] Angus, J. E. and Schafer, R. E. (1979). *Comm. Statist. A*, **8**, 1271–1284.

[5] Azzalini, A. (1981). *Biometrika*, **68**, 326–328. (This article presents a clear and brief discussion of estimating quantiles by a kernel method. The properties of such estimates are also investigated.)

[6] Bain, L. J. (1978). *Statistical Analysis of Reliability and Life-Testing Models*. Marcel Dekker, New York. (Bain's text provides a rich source of quantile estimation procedures in reliability especially for circumstances involving censored data. The text also has a comprehensive list of references of the important work done by Bain and his coworkers in quantile estimation.)

[7] Bement, T. R. and Pirkle, F. L. (1981). *Math. Geol.*, **13**, 429–442.

[8] Brewster, J. F. and Zidek, J. V. (1974). *Ann. Statist.*, **2**, 21–38.

[9] Chanda, K. C. (1976). *Commun. Statist. A*, **5**, 1385–1392. (Sample quantiles for dependent observations are examined along with their asymptotic properties.)

[10] Chmiel, J. J. (1976). *Biometrika*, **63**, 621–626.

[11] David, H. A. (1981). *Order Statistics*, 2nd. ed. Wiley, New York. (David's text provides an authoritative treatment of estimation procedures based on order statistics.)

[12] David, H. A. and Mishriky, R. S. (1980). *J. Amer. Statist. Ass.*, **63**, 1390–1398.

[13] Dyer, D. D. and Keating, J. P. (1979). *Commun. Statist. A*, **8**, 1–16. (This article provides methods for obtaining optimal estimators of percentiles

based on equivariant statistics. The article provides a comparison of these optimal estimators for the normal distribution.)

[14] Dyer, D. D., Keating, J. P., and Hensley, O. L. (1977). *Commun. Statist. B*, **6**, 269–284.

[15] Fraser, D. A. S. (1976). *J. Amer. Statist. Ass.*, **71**, 99–111.

[16] Fryer, M. J. (1976). *J. Inst. Math. Applic.*, **18**, 371–380.

[17] Garsd, A., Ford, G. E., Waring, G. O., and Rosenblatt, L. S. (1983). *Biometrics*, **39**, 385–394.

[18] Good, I. J. (1979). *J. Statist. Comp. Simul.*, **9**, 77–79.

[19] Harrell, F. E. and Davis, C. E. (1982). *Biometrika*, **69**, 635–640.

[20] Hill, P. D. (1982). *Commun. Statist.—Theory Meth.*, **11**, 2343–2356. (Various methods of estimating quantiles such as those based on sample quantiles and kernel estimates of the density function are presented. The resulting estimators are compared with respect to bias, variance and mean square error based on simulation.)

[21] Kaigh, W. D. (1983). *Commun. Statist.—Theory Meth.*, **12**, 2427–2443. (Consistent estimates of the variance of the generalized sample quantile estimators introduced in [22] are obtained by jackknifing. These results are utilized for giving confidence intervals for the population quantiles.)

[22] Kaigh, W. D. and Lachenbruch, P. (1982). *Commun. Statist.—Theory Meth.*, **11**, 2217–2238. (An alternative estimator, called a "generalized sample quantile," obtained by averaging an appropriate subsample quantile over all possible subsamples of a fixed size is introduced. Properties of such estimators are also examined. An essential study.)

[23] Keating, J. P. (1983). *Commun. Statist.—Theory Meth.*, **12**, 441–447. (Methodology is given for constructing optimal quantile estimators based on absolute error loss.)

[24] Keating, J. P. (1984). *Statist. Prob. Lett.*, **2**, 143–146.

[25] Lampkin, H. and Ogawa, J. (1976). *Canad. J. Statist.*, **4**, 65–94.

[26] Lawless, J. F. (1973). *Technometrics*, **15**, 857–865.

[27] Lawless, J. F. (1973). *J. Amer. Statist. Ass.* **68**, 665–669.

[28] Lawless, J. F. (1978). *Technometrics*, **20**, 355–364. (An excellent expository treatment of the conditional confidence interval approach for percentiles is given.)

[29] Lawless, J. F. (1980). *Technometrics*, **22**, 409–419.

[30] Lawless, J. F. (1982). *Statistical Models and Methods for Lifetime Data*. Wiley, New York. (Lawless' text is an excellent reference for quantile estimation in location and scale parameter families.

lies. Thorough discussions are given on equivariant estimators, conditional confidence intervals and maximum likelihood estimators.)

[31] Lehmann, E. L. (1959). *Testing Statistical Hypotheses*. Wiley, New York.

[32] Lloyd, E. H. (1952). *Biometrika*, **39**, 88–95. (This article opened up the field of statistical inference based on order statistics.)

[33] Mann, N. R. (1969). *Ann. Math. Statist.*, **40**, 2149–2155. (Mann provides the methodology and motivation for constructing BLIE's of percentiles.)

[34] Mann, N. R., Schafer, R. E., and Singpurwalla, N. D. (1974). *Methods for Statistical Analysis of Reliability and Lifetime Data*. Wiley, New York. (The text provides an expository treatment of BLUEs, BLIEs, and MLEs of percentiles in families of life distributions important in reliability.)

[35] Nadaraya, E. A. (1964). *Theor. Prob. Appl.*, **15**, 497–500.

[36] Parzen, E. (1979). *J. Amer. Statist. Ass.*, **74**, 105–121. (A very good discussion of the properties of quantile functions, density quantile functions, and how they behave under transformation. Some properties of sample quantile functions are also presented. A rich source of bibliography.)

[37] Rao, C. R. (1981). *Statistics and Related Topics*. North Holland, Amsterdam. (Rao provides an interesting contrast of the roles of measures of closeness and measures of risk in point estimation.)

[38] Reiss, R. D. (1980). *Ann. Statist.*, **8**, 87–105.

[39] Robertson, C. A. (1977). *J. Amer. Statist. Ass.*, **72**, 162–164. (This article is the pioneering paper in the estimation of percentiles based on error in the predicted distribution function.)

[40] Robinson, G. K. (1975). *Biometrika*, **62**, 155–161.

[41] Rukhin, A. L. and Strawderman, W. E. (1982). *J. Amer. Statist. Ass.*, **77**, 159–162.

[42] Sarhan, A. E. and Greenberg, B. G. (1962). *Contributions to Order Statistics*. Wiley, New York. (This text collected many papers for constructing BLUEs based on Lloyd's work.)

[43] Schafer, R. E. and Angus, J. E. (1979). *Technometrics*, **21**, 367–370.

[44] Schmeiser, B. W. and Deutsch, S. J. (1977). *Commun. Statist. B*, **6**, 221–234. (Properties of the cell midpoint estimator of the population ϵth quantile based on grouped data are examined.)

[45] Wegman, E. J. (1972). *Technometrics*, **14**, 533–546.

[46] Wilson, J. R., Vaughan, D. K., Naylor, E., and Voss, R. G. (1982). *Simulation*, **38**, 187–203.

[47] Zacks, S. (1971). *The Theory of Statistical Inference*. Wiley, New York. (Zacks provides a theoretical treatment of location and scale parameter

distributions, equivariant estimation and other procedures in statistical inference.)

[48] Zidek, J. V. (1971). *Ann. Math. Statist.*, **42**, 1444–1447.

(CLOSENESS OF ESTIMATORS
DENSITY ESTIMATION
ESTIMATION POINT
ORDER STATISTICS
QUANTILE ESTIMATION
QUANTILES)

J. P. KEATING
R. C. TRIPATHI

PERCOLATION THEORY

Percolation models were introduced by Broadbent and Hammersley [2] to model fluid flow in a medium, where fluid and medium may be broadly interpreted. In the

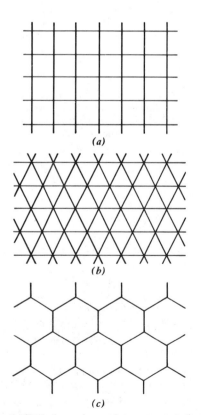

Figure 1 Typical graphs used to represent the medium: (*a*) square lattice, (*b*) triangular lattice, and (*c*) hexagonal lattice.

percolation approach, fluid flow is deterministic in a randomly structured medium, in contrast to the diffusion approach where fluid flow is viewed as random movement in a structureless medium. In a percolation model, fluid is imagined to flow through the network* of sites (vertices) and bonds (edges) of a graph which represents the medium (Fig. 1).

In a Bernoulli percolation model, sites and bonds are randomly passable or impassable by the fluid. The primary concern is the extent of flow of fluid, particularly in the probability that there exists an infinite connected cluster of passable sites and bonds. Typical applications include ferromagnetism and electrical conductivity in physics, where cluster properties play an important role. The critical probability, a threshold proportion of passable sites or bonds which distinguishes between local and infinite flow, is of interest to students of critical phenomena (see Essam [5]). There is a rapidly growing physics literature on Bernoulli percolation, containing numerous conjectures, Monte Carlo* studies, plausibility arguments, and applications. Important tools used in Bernoulli percolation are planar graph duality and correlation inequalities.

In a first-passage percolation model, for each bond there is a random variable which represents the time required for fluid to pass through the bond. Interest focuses on the rate of spread of fluid. Applications include traffic flow, tumor growth, and spread of epidemics or rumors. Topics of interest are the asymptotic velocity of spread, properties of optimal routes, and the asymptotic shape of the wetted region. The main probabilistic tool is subadditive process theory.

BERNOULLI PERCOLATION

Bond and Site Models

In a bond percolation model on a graph G, each bond is open (passable) with probability p, $0 \leqslant p \leqslant 1$, and closed (impassable) with probability $1 - p$, independently of all other bonds (see Fig. 2*a*). A path or circuit

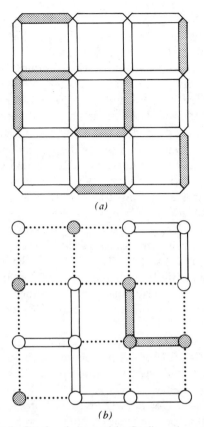

(a)

(b)

Figure 2 Bond and site models. Configurations of (*a*) the bond model and (*b*) the site model on a portion of the square lattice. Open bonds and sites are shown in white; closed bonds and sites are in solid black; bonds of the underlying graph which are neither open or closed are indicated by dotted lines in the site model.

in G is open (closed) if all its bonds are open (closed). In the site percolation model on G, sites are open with probability p, $0 \leqslant p \leqslant 1$ and closed with probability $1 - p$, independently, and a bond is open (closed) if and only if both its endpoints are open (closed) (see Fig. 2*b*). A path or circuit in G is open (closed) if all its sites are open (closed). In either model, let Pr_p and E_p denote the probability measure and expectation operator parameterized by p.

The covering graph G^c of a graph G is constructed by the "bond-to-site" transformation: Place a site of G^c at the midpoint of each bond of G, then insert a bond of G^c between each pair of sites of G^c for which the corresponding bonds of G share a common endpoint. The site percolation model

on G^c is equivalent to the bond percolation model on G. There is no "site-to-bond" transformation allowing a site model to be converted into an equivalent bond model, so the class of site models is more general than the class of bond models.

Critical Probability Definitions

The critical probability separates intervals in the parameter space corresponding to local flow and extensive flow of fluid. Several alternative definitions exist. The open cluster containing site v, denoted C_v, is the set of all vertices which are connnected to v by an open path. The cluster size $\# C_v$ is the number of sites in C_v.

For a fixed site v in an infinite graph G, the percolation probability is $\theta_v(p) = \mathrm{Pr}_p(\# C_v = \infty)$. The cluster size critical probability $p_H(G) = \inf\{ p : \theta_v(p) > 0 \}$ is independent of v if G is connected.

The mean cluster size critical probability is $p_T(G) = \inf\{ p : E_p(\# C_v) = \infty \}$. For any connected G, $p_T(G) \leqslant p_H(G)$.

For a graph imbedded in $\mathbb{R}^d$, let $T(n, i) = \{(x_1, \ldots, x_d) \in \mathbb{R}^d : 0 \leqslant x_i \leqslant n, \ 0 \leqslant x_j \leqslant 3n \text{ for } j \neq i\}$. Define $\tau(n, p; i) = \mathrm{Pr}_p\{\exists$ an open path in $T(n, i)$ from a site in $x_i \leqslant 0$ to a site in $x_i \geqslant n\}$. The sponge crossing critical probability

$$p_S(G)$$

$$= \inf\left\{ p : \limsup_{n \to \infty} \ \max_{1 \leqslant i \leqslant d} \tau(n, p; i) > 0 \right\}$$

was defined by Kesten [8], who modified slightly the definition of Seymour and Welsh [9]. Arising less naturally than the other versions, p_S is the key to rigorous evaluation of critical probabilities.

In a two-dimensional graph G, the number of open clusters per site in a rectangle R converges to a limiting function $\lambda(p)$ as R expands to the entire graph. Sykes and Essam [11] conjectured that λ has a unique singularity, denoted $p_E(G)$.

Dual and Matching Graphs

For bond percolation on a planar graph G, G^* denotes the dual graph, which has a site in each face of G and a bond crossing each bond of G. A bond of G^* is open if and only if it crosses an open bond in G. An open cluster in G is finite if and only if it is enclosed by a closed circuit in G^*.

Let M be a planar graph and $\mathcal{F}$ be a set of faces in M. Construct a graph G by inserting a bond between each pair of non-adjacent vertices of F, for each $F \in \mathcal{F}$. For site percolation on G, G^* denotes the matching graph, which is constructed by inserting such bonds in all faces not in $\mathcal{F}$. In site models, the matching graph plays the role of the dual graph in bond models. In fact, the covering graphs of a dual pair of graphs form a matching pair.

Critical Probability Values

Regularity conditions are imposed on G to obtain rigorous results. A graph G in $\mathbb{R}^d$ is periodic if its edge and vertex sets are invariant under translation by each of a set of d linearly independent vectors. Let Σ denote the class of graphs G satisfying: (a) G is connected and periodic in $\mathbb{R}^d$, $d \geqslant 2$; (b) every site of G is an endpoint of at most z bonds, $z < \infty$; and (c) each compact subset of $\mathbb{R}^d$ contains at most finitely many vertices of G.

For fixed $v \in G$, let f_n denote the number of n-step self-avoiding paths starting from site v. A measure of the complexity of the path structure of G is the connectivity constant $\lambda = \limsup_{n \to \infty} f_n^{1/n}$, which provides the lower bound $p_H \geqslant p_T \geqslant 1/\lambda$. If G satisfies (b) above, then $\lambda \leqslant z - 1$, so $p_H \geqslant p_T \geqslant 1/(z - 1)$.

Let $p_H^{(S)}$ and $p_H^{(B)}$ denote the cluster size critical probabilities for the site and bond models respectively: for any G, $p_H^{(B)}(G) \leqslant p_H^{(S)}(G)$. If G is a tree in which each site is an endpoint of z bonds, then $p_H^{(S)}(G) = p_T^{(S)}(G) = p_T^{(B)} = p_H^{(B)} = 1/(z - 1)$.

Exact critical probabilities are known for few graphs other than trees. A heuristic argument of Sykes and Essam [11] suggests that $p_E(G) + p_E(G^*) = 1$ for site models on matching lattices. This suggests that $p_E = \frac{1}{2}$ for self-matching site models, providing val-

ues for site models on fully triangulated graphs and the square lattice bond model. Using self-duality and a method of constructing open circuits which required two symmetry axes, Seymour and Welsh [9] showed $p_T = p_S$ and $p_T + p_H = 1$ for the square lattice bond model. Kesten [8] proved that $p_T = p_H = p_S = \frac{1}{2}$ for the square lattice bond model, and also for the site model on any fully triangulated $G \in \Sigma$ with one symmetry axis. However, Van den Berg [12] constructed a fully triangulated graph (not in Σ) for which $p_T = p_H = 1$ in the site model. Using the star-triangle transformation, Sykes and Essam derived the values $2 \sin(\pi/18)$ and $1 - 2 \sin(\pi/18)$ for the bond model critical probability p_E on the triangular and hexagonal lattices respectively. These were verified by Wierman [13] for P_H, P_T, and P_S.

Kesten [8] showed that $p_S(G) = p_T(G)$ for any $G \in \Sigma$. For site percolation on matching graphs in Σ with two nonperpendicular axes of symmetry, $p_H(G) + p_H(G^*) = 1$ and $p_H = p_T = p_S$ for both G and G^*. For site percolation on the square lattice, $p_H = p_T = p_S > \frac{1}{2}$. For the most general formulation of critical probability results see Chapter 3 in Kesten [8].

No exact critical probabilities are known for models in $\mathbb{R}^d$, $d \geqslant 3$. Knowledge of such models is based on numerical evidence.

Consider the site model on the square or triangular lattice or the bond model on the square, triangular, or hexagonal lattice. Let C_v^* denote the open cluster in G^* containing a fixed site v. The cluster sizes on G and G^* are related as follows: For $p < p_H(G)$, $E_p(\# C_v) < \infty$, $\mathrm{Pr}_{1-p}(\# C_v^* = \infty) > 0$, and $\mathrm{Pr}_p(\exists$ a unique infinite closed cluster in $G^*) = 1$. For $p = p_H(G)$, $\mathrm{Pr}_p(\# C_v = \infty) = \mathrm{Pr}_{1-p}(\# C_v^* = \infty) = 0$, $E_p(\# C_v) = E_p(\# C_v^*) = \infty$, and for each N, $\mathrm{Pr}_p(\exists$ an open circuit in G surrounding $[-N, N] \times [-N, N]$ and a closed circuit in G^* surrounding $[-N, N] \times [-N, N]) = 1$. For $p > p_H(G)$, $\mathrm{Pr}_p(\# C_v = \infty) > 0$, $E_{1-p}(\# C_v^*) < \infty$ and $\mathrm{Pr}_p(\exists$ a unique infinite open cluster in $G) = 1$.

For any $G \in \Sigma$, $E_{p_T}(\# C_v) = \infty$.

Percolation Probability Function

For site or bond percolation on the square lattice, $\theta(p)$ is continuous on $[0, 1]$ and is infinitely differentiable except at p_H. It is not known if $\theta(p)$ is analytic for $p > p_H$. Kesten [8] established bounds for the percolation function and expected cluster size: There exist constants $0 < k_i, \beta_i < \infty$, $1 \leq i \leq 6$, such that

$$k_1(p - p_H)^{\beta_1} \leq \theta(p)$$

$$\leq k_2(p - p_H)^{\beta_2}$$

$$\text{for } p > p_H,$$

$$k_3(p_H - p)^{-\beta_3} \leq E_p(\# C_v)$$

$$\leq k_4(p_H - p)^{-\beta_4}$$

$$\text{for } p < p_H,$$

and

$$k_5(p - p_H)^{-\beta_5} \leq E_p(\# C_v; \# C_v < \infty)$$

$$\leq k_6(P - p_H)^{-\beta_6}$$

$$\text{for } p > p_H.$$

Numerical evidence suggests that each of these functions behaves approximately as $|p - p_H|^\beta$ as $p \to p_H$, where β depends only on the dimension of the graph.

The proof of these bounds uses intricate bounds on the rate of decay of the cluster size distribution: For site or bond percolation on the square lattice, there exist $0 < k_i, \gamma_i < \infty$, $i = 1, 2$, such that $k_1 n^{\gamma_1 - 1} \leq \text{Pr}_{p_H}(\# C_v \geq n) \leq k_2 n^{-\gamma_2}$ $\forall n$. For $G \in \Sigma$ and $p < p_T(G)$, there exists a universal constant k_3 and $\gamma_3 = \gamma_3(p, G)$ such that $\text{Pr}_p(\# C_v \geq n) \leq k_3 e^{-\gamma_3 n}$ $\forall n$. Aizenman et al. [1] showed that for d-dimensional $G \in \Sigma$ and $p > p_H$,

$$\text{Pr}_p(n \leq \# C_v < \infty) \geq k_4 e^{-\gamma_4 n^{(d-1)/d}} \qquad \forall n,$$

where $0 < k_4, \gamma_4 < \infty$. For $d = 2$,

$$\text{Pr}_p(n \leq \# C_v < \infty) \leq k_5 e^{-\gamma_5 n^{1/2}}$$

$$\forall n \text{ for } p > p_H$$

where $0 < k_5, \gamma_5 < \infty$.

For site or bond percolation on the square lattice, $\lambda(p)$ is a real analytic function on $[0, 1] - p_H$, and is twice continuously differentiable on $[0, 1]$. It is not known if λ has a singularity at p_H.

Variations of the Model

Oriented (or directed) percolation models, where certain bonds are passable in only one direction, are increasingly popular in the physics literature. In multiparameter models, subsets of graph elements may have different probabilities of being open. Kesten [8, Chap. 3] reports progress on conjectures of Sykes and Essam concerning multiparameter asymmetric bond models on the square and triangular lattices. Kesten [8, Chap. 11] investigates the resistance of random electrical networks on the square lattice as a variant of the percolation model.

FIRST-PASSAGE PERCOLATION

First-passage percolation was introduced by Hammersley and Welsh [6]. Let $\{X_e, e \in E\}$, where E denotes the edge set of the square lattice, be i.i.d. nonnegative random variables with distribution function F. X_e represents the time required for fluid to flow through e in either direction. The travel time of a path $r = \{v_0, e_1, \ldots, e_n, v_{n+1}\}$ is $t_r = \sum_{i=1}^n X_{e_i}$. The first-passage time of a set of paths R is $t_R = \inf\{t_r : r \in R\}$. If there exists a path $u \in R$ such that $t_u = t_R$, then u is a route for t_R. Note that routes need not exist nor be unique.

First-Passage Time Processes

For $m, n \in Z$, let $R(m, n)$ $(\tilde{R}(m, n))$ be the set of paths from $(m, 0)$ to $(n, 0)$ (the line $x = n$). The unrestricted point-to-point and point-to-line first-passage times are defined by $a_{mn} = t_{R(m, n)}$ and $b_{mn} = t_{\tilde{R}(m, n)}$ respectively, for $m, n \in Z$. Using Bernoulli percolation results on the existence of circuits, Wierman and Reh [15] proved that routes exist for all a_{mn} and b_{mn} almost surely.

If $m \leq n \leq p$, then $a_{mn} + a_{np} \geq a_{mp}$. If also $\int_0^\infty x \, dF(x) < \infty$, then $\{a_{mn}\}$ is a subadditive process. The limiting behavior is given by $\lim_{n\to\infty} a_{0n}/n = \mu(F)$ a.s., where $\mu(F) = \lim_{n\to\infty} E(a_{0n}/n) = \inf_n E(a_{0n}/n)$ is the time constant of the process. This basic convergence result was shown to hold under weaker conditions by Cox and Durrett [3], who showed that if $F(x) = 0$ for $x < 0$, then there exists a finite constant $\mu(F)$ such that

$$\liminf_{n\to\infty} a_{0n}/n = \lim_{n\to\infty} b_{0n}/n = \mu(F) \qquad \text{a.s.}$$

A weak renewal theorem holds for the associated reach processes $x_t^u = \sup\{n : a_{0n} \leq t\}$ and $y_t^u = \sup\{n : b_{0n} \leq t\}$, stating that

$$\lim_{t\to\infty} \frac{x_t^u}{t} = \lim_{t\to\infty} \frac{y_t^u}{t} = \frac{1}{\mu(F)} \qquad \text{a.s.}$$

If $F(0) < \frac{1}{2}$, the convergence holds in L^p for all $p > 0$ (see Chap. 6 of Smythe and Wierman [10]). Thus the time constant is the reciprocal of the asymptotic velocity of spread of fluid.

Time Constant

A major goal of the theory is evaluation of the time constant. Hammersley and Welsh [6] showed that $\mu(F) \leq \int_0^\infty x\, dF(x)$, with equality if and only if F is degenerate. Wierman [10] proved that

$$\mu(F) \leq \left[\mu(F(0))/(1 - F(0)) \right] \int_0^\infty x\, dF(x),$$

where $\mu(p)$ denotes the time constant of the Bernoulli distribution with mean p. Smythe and Wierman [10] proved that $\mu(F) = 0$ if $F(0) \geq \frac{1}{2}$, and Kesten [7] proved that $\mu(F) > 0$ if $F(0) < \frac{1}{2}$ by deriving an exponential bound on the tail of the reach distribution.

A monotonicity property holds: If $F_1(x) \leq F_2(x)\ \forall x$, then $\mu(F_1) \geq \mu(F_2)$. Continuity theorems of Hammersley and Welsh [6] and Smythe and Wierman [10] were improved by Cox and Kesten [4] to show that if $F_n \xrightarrow{w} F$, then $\mu(F_n) \to \mu(F)$ as $n \to \infty$, where $\xrightarrow{w}$ denotes weak convergence of measures.

Route Length

Knowledge of the asymptotic behavior of the length of an optimal route provides information about the functional dependence

of $\mu(F)$ on F. Let $F \oplus r$ denote the distribution of the shifted travel times $\{X_e + r : e \in E\}$. Almost sure convergence of a_{0n}/n and b_{0n}/n to a time constant $\mu(F \oplus r)$ holds for $r \in (\beta, \infty)$ for some $\beta < 0$ if $F(0) < \frac{1}{2}$. $\mu(F \oplus r)$ is a concave function of $r \in (\beta, \infty)$, and thus the right and left derivatives, μ^+ and μ^- respectively, exist on (β, ∞). Smythe and Wierman [10] proved that if $F(0) < \frac{1}{2}$, then

$$\mu^+(0) \leq \liminf_{n\to\infty} \frac{N_n}{n}$$

$$\leq \limsup_{n\to\infty} \frac{N_n}{n} \leq \mu^-(0) \qquad \text{a.s.,}$$

where N_n is the minimum number of bonds in a route for a_{0n} or b_{0n}.

Asymptotic Shape

Cox and Durrett [3] described the asymptotic shape of the wetted region at time t. Let $a(0, x)$ denote the first passage time from the origin to $x \in Z^2$. For $x \in R^2$, let $a(0, x)$ be the passage time to the site in Z^2 nearest to x. Then

$$\lim_{n\to\infty} \frac{a(0, nx)}{n} = \psi(x) < \infty$$

in probability. Let $A_t = \{y : a(0, y) \leq t\}$. Cox and Durrett show that for $\epsilon > 0$,

$$\Pr\big(\{x : \psi(x) \leq 1 - \epsilon\} \subseteq t^{-1}A_t$$

$$\subseteq \{x : \psi(x) \leq 1 + \epsilon\}\ \forall\ t \text{ suff. large}\big) = 1$$

if and only if $\mu(F) > 0$ and

$$E\left[\min_{i=1,\ldots,r} X_{e_i}^2 \right] < \infty.$$

Without assumptions on F,

$$P\big(t^{-1}A_t \subseteq \{x : \psi(x) < 1 + \epsilon\}$$

$$\forall\ t \text{ suff. large}\big) = 1$$

and if $\mu(F) > 0$,

$$P\big(|\{x : \psi(x) \leq 1\} - t^{-1}A_t| < \epsilon$$

$$\forall\ t \text{ suff. large}\big) = 1$$

where $|\cdot|$ denotes Lebesgue measure. Thus, $t^{-1}A_t$ grows like $\{x : \psi(x) \leq 1\}$ and covers most, but not all, of the interior.

References

[1] Aizenman, M., Deylon, F., and Souillard, B. (1980). *J. Statist. Phys.*, **23**, 267–280.

[2] Broadbent, S. R. and Hammersley, J. M. (1957). *Proc. Camb. Phil. Soc.*, **53**, 629–641, 642–645. (The original papers on percolation.)

[3] Cox, J. T. and Durrett, R. (1981). *Ann. Prob.*, **9**, 583–603.

[4] Cox, J. T. and Kesten, H. (1981). *J. Appl. Prob.*, **18**, 809–819.

[5] Essam, J. W. (1972). In *Phase Transitions and Critical Phenomena*, Vol. 2, C. Domb and M. S. Green, eds. Academic, New York, pp. 197–270. (Surveys connections between percolation and statistical mechanics.)

[6] Hammersley, J. M. and Welsh, D. J. A. (1963). In *Bernoulli-Bayes-Laplace Anniversary Volume*, J. Neyman and L. M. LeCam, eds. Springer-Verlag, Berlin, pp. 61–110. (The seminal paper on first-passage percolation.)

[7] Kesten, H. (1980). *Adv. Appl. Prob.* **12**, 848–863. (Characterizes distributions with time constant equal to zero.)

[8] Kesten, H. (1982). *Percolation Theory for Mathematicians*. Birkhäuser, Boston. (Research-level monograph containing significant theoretical advances with intricate proofs.)

[9] Seymour, P. D. and Welsh, D. J. A. (1978). *Ann. Discrete Math.*, **3**, 227–245. (Initial treatment of sponge-crossing critical probability.)

[10] Smythe, R. T. and Wierman, J. C. (1978). *First-Passage Percolation on the Square Lattice*, Lecture Notes in Mathematics, Vol. 671. Springer-Verlag, Berlin. (Self-contained presentation of first-passage percolation theory.)

[11] Sykes, M. F. and Essam, J. W. (1964). *J. Math. Phys.*, **5**, 1117–1127. (Classic paper providing a heuristic method of determining critical probabilities.)

[12] Van den Berg, J. (1981). *J. Math. Phys.*, **22**, 152–157. (Provides a counterexample to Sykes and Essam critical probability values.)

[13] Wierman, J. C. (1981). *Adv. Appl. Prob.*, **13**, 298–313. (Critical probabilities for bond models on triangular and hexagonal lattices.)

[14] Wierman, J. C. (1982). *Ann. Prob.*, **10**, 509–524. (A survey of both Bernoulli and first-passage percolation.)

[15] Wierman, J. C. and Reh, W. (1978) *Ann. Prob.*, **6**, 388–397. (Proves existence of optimal routes.)

(SIMULATION
STOCHASTIC MECHANICS)

JOHN C. WIERMAN

PERIODICITY   *See* PERIODOGRAM ANALYSIS; TIME SERIES

PERIODIC VARIATION   *See* PERIODOGRAM ANALYSIS; TIME SERIES

PERIODOGRAM ANALYSIS

The statistical analysis of time series* may take place in either the time domain or the frequency domain. For a stationary time series $\{X_t\}$, with $E(X_t) = \mu$, $V(X_t) = \gamma_0$, analysis in the time domain may be based upon the autocorrelations

$$\rho_j = \gamma_j / \gamma_0, \qquad j = 1, 2, \ldots,$$

where $\gamma_j = \gamma_{-j} = E\{(X_t - \mu)(X_{t+j} - \mu)\}$ denote the autocovariances.

The sample autocorrelations, also known as serial correlations*, are given by

$$r_j = c_j / c_0, \qquad j = 1, 2, \ldots,$$

where $c_j = \sum_{t=1}^{n-j}(x_t - \bar{x})(x_{t+j} - \bar{x})/(n-1)$. Some writers use $(n - j - 1)$ rather than $(n - 1)$ in the denominator of c_j [6, pp. 443–445]. The plot of r_j against the lag j is known as the *correlogram*.

In the frequency domain, it is assumed that the series may be represented by a weighted average of sine waves. In his initial development in 1898, Schuster [11] assumed that the time series was made up of a small number of sine waves with unknown frequencies, leading to the search for "hidden periodicities." Now it is generally assumed that the time series may contain all possible frequencies $0 < \omega < \pi$ such that

$$X_t = \int_0^\pi \left[\cos \omega t \, dU(\omega) + \sin \omega t \, dV(\omega) \right],$$

where $dU(\omega)$ and $dV(\omega)$ are uncorrelated random incremental processes. This leads to the decomposition

$$\sigma_x^2 = \int_0^\pi f(\omega) \, d\omega,$$

where $f(\omega)$ is the population spectral density function.

Useful general references on the frequency domain approach are Bloomfield [2],

Brillinger [3], Kendall et al. [6], and Priestley [10].

From the theoretical perspective, the two descriptions of a time series are equivalent in that the set of all autocorrelations and the spectrum are a Fourier transform pair. However, when analyzing a sample series, each representation may be useful in highlighting different features of the data.

THE FORM OF THE PERIODOGRAM

Given T observations, x_t ($t = 1, \ldots, T$) equally spaced in time, we may write (for T odd, $T = 2M + 1$, say)

$$x_t = A(0) + \sum \left[A(k)\cos \omega_k t \right.$$

$$\left. + B(k)\sin \omega_k t \right], \quad (1)$$

where the sum is taken over $k = 1, \ldots, M$ and $\omega_k = 2\pi k / T$. When T ($= 2M$) is even, we add the term $A(M)\cos \pi t$. These equations contain T data values and T unknowns; their general solution is

$$A(k) = (2/T)\sum x_t \cos \omega_k t, \quad (2)$$

$$B(k) = (2/T)\sum x_t \sin \omega_k t, \quad (3)$$

$A(0) = \bar{x}$. Also, $A(M) = (1/T)\sum x_t (-1)^t$ when T is even. If we write

$$S^2(k) = A^2(k) + B^2(k),$$

it follows that

$$T \sum_{k=1}^{M} S^2(k) = \sum_{t=1}^{T} (x_t - \bar{x})^2, \quad (4)$$

a sample version of Parseval's identity. Expression (4) reveals that the terms $TS^2(k)$ represent a partition of the variance into M components. The plot of $TS^2(k)$ against the wavelength of its sine curve, $\lambda_k = T/k$, is the periodogram. Figure 1 shows the periodogram for the trend-free series on wheat prices developed by Lord Beveridge in 1921 for the years 1500–1869; the data are given in Sec. 47.7 of Kendall et al. [6]. The increasing spacing between successive λ_k values for small k is apparent. For this reason, it is now common practice to plot $TS^2(k)$ against ω_k. This plot represents the unsmoothed sample spectrum, although the term periodogram is used for this plot also. Figure 2 shows this plot for the wheat price series.

The sampling properties of the periodogram are extensively discussed by Priestley [10, Sec. 6.1].

TESTS FOR PERIODICITIES

Initially, attempts were made to ascribe meaning to every major peak in the periodogram, a daunting task as Figs. 1 and 2 illustrate. If only a small number of the possible frequencies were known to be of

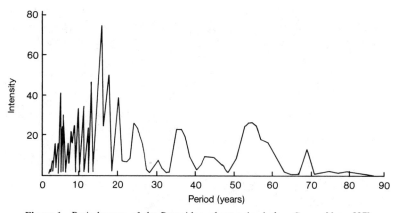

Figure 1   Periodogram of the Beveridge wheat price index. *Source* [6, p. 587].

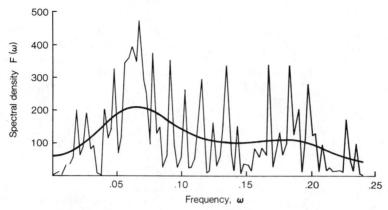

Figure 2 Spectrum of the Beveridge wheat price index. Only frequencies up to $\omega = 0.24$ are shown; the remainder of the spectrum is negligibly small. The curve is a smoothed estimate of the spectrum using a Parzen window. *Source* [6, p. 581].

interest (e.g., seasonal values), it would be feasible to drop the other frequencies and reformulate (1) as a regression* model. The estimates are still given by (2) and (3). Inferential methods for this model are described in Sec. 4.3 of Anderson [1]. As a null hypothesis, we may assume that the $\{X_t\}$ are independent and identically distributed normal variates. It follows that each $S^2(k)$, $k \leq M - 1$ is independently distributed as chi-square* with 2 degrees of freedom. Thus a test for peaks in the periodogram may be based upon the quantities

$$u_j = \sum_{k=1}^{j} S^2(k) \Big/ \sum_{k=1}^{M-1} S^2(k),$$

which behave like the order statistics of a random sample of $(M - 1)$ observations from a uniform distribution on $[0, 1]$. For example, the Kolmogorov-Smirnov test* may be used (a proposal due to M. S. Bartlett) or the various tests based on spacings*. Knoke [7] concludes that Bartlett's suggestions give a test with good power characteristics against general alternatives of autocorrelation being present. Much earlier in 1914, Walker [13] noted that

$$P\left[\max_{1 \leq k \leq m} S^2(k) \leq 4c\sigma^2/T\right] = (1 - e^{-c})^m,$$

so that the largest ordinate of the periodogram may be tested. Since σ^2 is unknown, Fisher [5] extended the test procedure by

using the sample variance in place of σ^2. Tables of critical points for the test were developed by Davis [4]. Anderson [1, Sec. 4.3.4] showed that Fisher's test is uniformly the most powerful invariant against single periodicities. Siegel [12] extended the test to cover multiple periodicities. Priestley [9] has developed a group periodogram statistic which allows single peaks to be tested within a range of adjacent harmonics. MacNeill [8] gives a test of whether several series share common periodicities. Priestley [10, Chap. 8] discusses the advantages and drawbacks of several of these tests and proposes a further test (*see* PRIESTLEY'S TEST FOR HARMONIC COMPONENTS*).

Since the presence of a harmonic component (or a narrow frequency band with large power) may seriously bias subsequent estimates of the spectrum derived by smoothing, it would appear that more routine use of such tests is desirable.

ESTIMATION OF THE SPECTRUM

When no known periodicities are present, we may view the time series as containing all possible frequencies. Reformulated as a plot against frequencies, the periodogram may then be used to estimate $f(\omega)$ by smoothing to ensure consistency. For example, smoothing may be carried out by taking weighted averages of adjacent periodogram ordinates (*see* SPECTRAL ANALYSIS).

THE BUYS-BALLOT TABLE

Suppose that the series comprises $N = rs$ terms and the observations are arrayed as

$$
\begin{array}{cccc}
x_1 & x_2 & \cdots & x_s \\
x_{s+1} & x_{s+2} & \cdots & x_{2s} \\
\vdots & \vdots & & \vdots \\
x_{(r-1)s+1} & x_{(r-1)s+2} & \cdots & x_{rs}
\end{array}
$$

with column totals

$$
m_1 \quad m_2 \quad \cdots \quad m_s .
$$

The components of the periodogram may be written as

$$
A(k) = (2/rs) \sum_{k=1}^{s} m_k \cos(2\pi k/s),
$$

$$
B(k) = (2/rs) \sum_{k=1}^{s} m_k \sin(2\pi k/s),
$$

enabling the periodogram to be computed more rapidly. This simplification, developed by Buys-Ballot in 1847, is the basis of the fast Fourier transform method now widely used to compute spectral estimates.

References

[1] Anderson, T. W. (1971). *The Statistical Analysis of Time Series*. Wiley, New York. (Chapter 4 includes an extensive development of tests and sampling properties.)

[2] Bloomfield, P. (1976). *The Fourier Analysis of Time Series: An Introduction*. Wiley, New York. (An excellent intermediate level introduction to the frequency domain.)

[3] Brillinger, D. R. (1975). *Time Series: Data Analysis and Theory*. Holt, Rinehart & Winston, New York. (An excellent introduction with major emphasis on the frequency domain.)

[4] Davis, H. T. (1941). *Cowles Comm. Res. Econ., Monagr. No. 6*. Bloomington, Indiana.

[5] Fisher, R. A. (1929). *Proc. R. Soc.* **A125**, 54–59.

[6] Kendall, M. G., Stuart, A., and Ord, J. K. (1983). *The Advanced Theory of Statistics*, 4th ed. Vol. 3. Griffin, London, and Macmillan, New York. (Chapter 49 discusses the frequency approach; extensive bibliography.)

[7] Knoke, J. D. (1977). *Biometrika*, **64**, 523–530.

[8] MacNeill, I. B. (1977). *Biometrika*, **64**, 495–530.

[9] Priestley, M. B. (1962). *J. R. Statist. Soc.*, **B24**, 511–529.

[10] Priestley, M. B. (1981). *Spectral Analysis and Time Series*. Academic, New York, 2 Vols. (Chapters 6 and 8 give extensive treatment of sampling properties and tests, respectively; extensive bibliography.)

[11] Schuster, A. (1898). *Terr. Mag. Atmos. Electr.*, **3**, 13–41.

[12] Siegel, A. F. (1980). *J. Amer. Statist. Ass.*, **75**, 345–348.

[13] Walker, G. T. (1914). *Indian Meteorol. Dept. (Simla) Mem.*, **21**(9), 22.

(SEASONALITY
SPECTRAL ANALYSIS
TIME SERIES)

J. KEITH ORD

PERKS' DISTRIBUTIONS

Perks' distributions constitute a class of distributions with PDFs of form

$$
f_X(x) = \left\{ \sum_{j=0}^{m_1} a_{1j} e^{-j\theta x} \right\} \Big/ \left\{ \sum_{j=0}^{m_2} a_{2j} e^{-j\theta x} \right\}
$$
$$(\theta > 0). \quad (1)$$

The coefficients $\{a_{ij}\}$ have to be chosen so that the conditions

$$f_x(x) \geqslant 0 \quad \text{for all } x \quad \text{and}$$

$$\int_{\infty}^{\infty} f_x(x)\, dx = 1 \quad \text{are satisfied.}$$

For practical use, both m_1 and m_2 should not be large. The logistic distribution* is a special case (with $m_1 = 1$, $m_2 = 2$, $a_{10} = 0$, $a_{20} = a_{22} = \frac{1}{2} a_{21}$). Another special case [4] is the *hyperbolic secant distribution* (with $m_1 = 1$, $m_2 = 2$, $a_{10} = 0$, $a_{11} = 2\theta\pi^{-1}$; $a_{20} = a_{22} = 1$, $a_{21} = 0$).

$$f_X(x) = \theta\pi^{-1} \cdot \frac{2}{e^{\theta x} + e^{-\theta x}} = \theta\pi^{-1} \operatorname{sech} \theta x.$$

The distribution of the sum (or arithmetic mean) of a number of independent random variables each having the same hyperbolic secant distribution (i.e., common value of θ)

—derived by Baten [1]—has also been called a "Perks' distribution."

In fact Perks [2], a British actuary concerned with graduation* of life tables*, introduced two classes of distributions (a) with *hazard rates* (force of mortality*) of form (1) and (b) with the function $q_x/(1 - q_x)$ ("odds ratio") of form (1), where $q_x = 1 - F_X(x + 1)/F_X(x)$, $F_X(x)$ being the CDF corresponding to (1). He also used only small values of m_1 ($\leqslant 1$) and m_2 ($\leqslant 2$). Rich [3] gave some theoretical reasons for using a formula of type (1) for the force of mortality, with $a_{2j} = (j + 1)^{-1}a_{1j}$, though Perks (in the discussion of ref. 3) pointed out that it could not adequately represent the whole range of life for human population data.

References

[1] Baten, W. D. (1934). *Bull. Amer. Math. Soc.*, **40**, 284–290.

[2] Perks, W. F. (1932). *J. Inst. Actu. Lond.*, **58**, 12–57.

[3] Rich, C. D. (1939). *J. Inst. Actu. Lond.*, **65**, 314–363. (Discussion 364–379.)

[4] Talacko, J. (1956). *Trab. Estadíst.*, 7, 159–174.

(GOMPERTZ DISTRIBUTION
LIFE TABLES
LIFE TIME DISTRIBUTIONS
LOGISTIC DISTRIBUTION)

PERMUTATIONAL CENTRAL LIMIT THEOREMS

In simple random sampling (without replacement) from a finite population as well as in nonparametric distribution theory (under the null hypothesis of permutational invariance), sampling distributions of various statistics are typically generated by the *equally likely permutations* of the elements of a finite set. Under such a permutational probability model, the associated random variables (or vectors) are, generally, not independent but interchangeable. Nevertheless, parallel to the case of linear functions of independent random variables, asymptotic normality of various permutation statistics holds under quite general regularity conditions. Permutational central limit theorems (PCLTs) relate to this general topic.

Consider a set $A_N = (a_{N1}, \ldots, a_{NN})$ of N elements and define a random vector $X_N = (X_{N1}, \ldots, X_{NN})$ which takes on each permutation of $a_{N1}, \ldots, a_{NN}$ with the equal probability $(N!)^{-1}$. This (discrete) uniform probability law (over the $N!$ permutations of the elements of A_N) is denoted by $\mathscr{P}_N$. For the time being, we assume that the a_{Ni} are all real valued, and define a *linear permutation statistic* $\mathscr{L}_N$ as

$$\mathscr{L}_N = c_{N1}X_{N1} + \cdots + c_{NN}X_{NN}, \quad (1)$$

where $C_N = (c_{N1}, \ldots, c_{NN})$ is a suitable vector of real numbers (not all equal). Then let $\bar{a}_N = N^{-1}\sum_{i=1}^{N}a_{Ni}$, $\bar{c}_N = N^{-1}\sum_{i=1}^{N}c_{Ni}$, $V_N^2 = \sum_{i=1}^{N}(a_{Ni} - \bar{a}_N)^2$, and $W_N^2 = \sum_{i=1}^{N}(c_{Ni} - \bar{c}_N)^2$. We thus have

$$E_{\mathscr{P}_N}\mathscr{L}_N = N\bar{a}_N\bar{c}_N \quad \text{and}$$

$$E_{\mathscr{P}_N}(\mathscr{L}_N - N\bar{a}_N\bar{c}_N)^2 \quad (2)$$

$$= (N - 1)^{-1}V_N^2W_N^2,$$

and consider the standardized form

$$Z_N = (N - 1)^{1/2}(\mathscr{L}_N - N\bar{a}_N\bar{c}_N)/(V_NW_N). \quad (3)$$

The exact distribution $H_N(x)$ ($= P\{Z_N \leqslant x \mid \mathscr{P}_N\}$, x real) of Z_N (under $\mathscr{P}_N$) can be obtained by enumeration (of all possible $N!$ realizations), when N is not very large. As N becomes large, the task becomes prohibitively laborious, and hence one seeks to provide suitable approximations. In this setup one conceives of sequences $\{A_N\}$ and $\{C_N\}$, and, under appropriate regularity conditions on them, seeks to approximate $H_N(x)$ by $\Phi(x)$, the standard normal distribution; this relates to the PCLT.

During the past four decades, PCLTs have been established in increasing generality by a host of research workers. The first systematic approach is due to Wald and Wolfowitz [16]. Their regularity conditions were subsequently relaxed by Noether [8] (*see* NOETHER AND RELATED CONDITIONS) and Hoeffding [4], among others. This PCLT may be presented as follows:

If $\{C_N\}$ satisfies the (Wald–Wolfowitz) condition that for every r ($= 3, 4, \ldots$), $N^{(1/2)r-1}\sum_{i=1}^{N}(c_{Ni} - \bar{c}_N)^r / W_N^r = O(1)$, while $\{A_N\}$ satisfies the (Noether) condition: if $V_N^{-1}\{\max_{1\leqslant i\leqslant N}|a_{Ni} - \bar{a}_N|\} \to 0$ as $N \to \infty$, then $H_N(x) \to \Phi(x)$, $\forall$ real x.

Note that $\mathscr{L}_N$ in (1) may be written as $c_{N1}a_{NR_1} + \cdots + c_{NN}a_{NR_N}$ where $(R_1, \ldots, R_N)$ takes on each permutation of $(1, \ldots, N)$ with the equal probability $(N!)^{-1}$. If we denote the antiranks by $S_1, \ldots, S_N$ (i.e., $R_{S_i} = S_{R_i} = i$, $i = 1, \ldots, N$), then we have $\mathscr{L}_N = c_{NS_1}a_{N1} + \cdots + c_{NS_N}a_{NN} = a_{N1}Y_{N1} + \cdots + a_{NN}Y_{NN}$, where $\mathscr{Y}_N = (Y_{N1}, \ldots, Y_{NN})$ takes on each permutation of $c_{N1}, \ldots, c_{NN}$ with the equal probability $(N!)^{-1}$. Therefore, in the PCLT presented above, the role of $\{\mathscr{A}_N\}$ and $\{\mathscr{C}_N\}$ may be interchanged. Also, as has been pointed out by Hoeffding [4], the Noether condition is equivalent to either of the following: for some $r > 2$, $\sum_{i=1}^{N}|a_{Ni} - \bar{a}_N|^r / V_N^r = O(1)$ or $V_N^{-1}\{\max_{1\leqslant i<j\leqslant N}|a_{Ni} - a_{Nj}|\} \to 0$, as $N \to \infty$. The proof of this theorem rests on the convergence of the moments of Z_N (under $\mathscr{P}_N$) to the corresponding ones of $\Phi(x)$, and is quite (combinatorially) involved. Since the convergence of moments is not necessary for the convergence of the distribution, we may remark that the regularity conditions (on $\{A_N\}$ and $\{C_N\}$) are sufficient, but not necessary. Madow [5] stressed the role of this PCLT in simple random sampling (without replacement) from a finite population. For example, if we let $c_{N1} = \cdots = c_{Nn} = 1/n$ and $c_{Ni} = 0$, $n + 1 \leqslant i \leqslant N$, then $\mathscr{L}_N$ in (1) reduces to the sample mean, where the Wald–Wolfowitz condition holds when n/N is bounded away from 0 and 1. Hence under the Noether condition on the a_{Ni}, the asymptotic normality* (of $\sqrt{n}(\bar{X}_{Nn} - \bar{a}_N)$) in simple random sampling holds. Extension of this result for general U-statistics* has been considered by Nandi and Sen [7]. Their regularity conditions are the same.

A somewhat more general PCLT is due to Hoeffding [4] and extended further by Motoo [6]. Let $B_N = \{b_N(i, j); 1 \leqslant i, j \leqslant N\}$ be a (square-) block of real elements and define a *bilinear permutation statistic* by $T_N = \sum_{i=1}^{N} b_N(i, R_i)$, where $\mathbf{R}_N = (R_1, \ldots, R_N)$ takes on each permutation of $(1, \ldots, N)$ with the equal probability $(N!)^{-1}$; here also, $\mathscr{P}_N$ stands for this permutational probability law. Then let

$$\xi_N = N^{-2} \sum_{i=1}^{N} \sum_{j=1}^{N} b_N(i, j),$$

$$D_N = \left[d_N(i, j) = b_N(i, j) \right.$$
$$\left. - N^{-1} \sum_{r=1}^{N} \{b_N(i, r) + b_N(r, j)\} + \xi_N \right],$$

$1 \leqslant i, j \leqslant N$, and

$$d_N^2 = N^{-1} \sum_{i=1}^{N} \sum_{j=1}^{N} d_N^2(i, j).$$

Suppose that D_N satisfies the Lindeberg condition that for every $\epsilon > 0$, as $N \to \infty$,

$$N^{-1} \sum_{i=1}^{N} \sum_{j=1}^{N} \left(d_N^2(i, j)/d_N^2 \right)$$
$$\times I\left(|d_N(i, j)| > \epsilon d_N \right) \to 0, \qquad (4)$$

then $(T_N - N\xi_N)/d_N$ has asymptotically (under $\mathscr{P}_N$) the standard normal distribution.

Like the preceding theorem, the proof of this theorem also rests on the convergence of the moments. A very significant contribution to PCLT is due to Hájek [3]. He was able to replace the "method of moments*" approach of the earlier workers by a very elegant "quadratic mean approximation" approach, which ties the classical central limit theorems (CLT) and PCLT in a very coherent manner. Without any loss of generality, we may set $a_{N1} \leqslant \cdots \leqslant a_{NN}$. Let then $a_N^0(u) = a_{Ni}$ for $(i - 1)/N < u \leqslant i/N$, $1 \leqslant i \leqslant N$ and let $U_1, \ldots, U_N$ be independent and identically distributed random variables having the uniform $(0, 1)$ distribution. Define then

$$S_N = \sum_{i=1}^{N} (c_{Ni} - \bar{c}_N)a_N^0(U_i) + N\bar{a}_N\bar{c}_N. \quad (5)$$

Hájek [3] showed that

$$E_{\mathscr{P}_N}(S_N - \mathscr{L}_N)^2 / \{(N-1)^{-1}A_N^2 C_N^2\} \to 0$$

(which ensures that

$$(N-1)^{1/2}|S_N - \mathscr{L}_N|/(A_N C_N) \to 0,$$

in probability), while the classical CLT is applicable on S_N. Hence the asymptotic normality of $\mathscr{L}_N$ can be obtained from that of S_N. Specifically, we have the following:

If both $\{A_N\}$ and $\{C_N\}$ satisfy the Noether condition, then for Z_N in (3), $H_N(x) \to \Phi(x)$, for all x, if and only if the Lindeberg condition in (4) is satisfied for $\{B_N\}$, where $B_N = \{b_N(i,j) = c_{Ni} a_{Nj}; \, 1 \leq i, j \leq N\}$.

Hájek [3] has also considered some variant forms of the theorem where the roles of A_N and C_N need not be symmetric. Further, in the context of rank tests*, the a_{Ni} depend on some underlying "score function" $\phi = \{\phi(u), 0 < u < 1\}$, and simplified conditions on ϕ for which (4) holds were considered by him. Finally he also considered the case where the a_{Ni} are p-vectors (i.e., $\mathbf{a}_{Ni} = (a_{Ni}^{(1)}, \ldots, a_{Ni}^{(p)})'$ for some $p \geq 1$) and/or the c_{Ni} are q-vectors for some $q \geq 1$. In such a case, to consider the joint (asymptotic) normality of the vector of statistics, it suffices to show that an arbitrary linear combination of them has the asymptotic normality under the same permutation law. Now, such an arbitrary linear combination is itself expressible as an $\mathscr{L}_N$ with appropriate $\{\mathscr{A}_N\}$ and $\{\mathscr{C}_N\}$, and hence the previous theorem remains applicable in such a case too.

PCLTs play a vital role in *nonparametric multivariate analysis*, where the rank statistics are usually only conditionally (permutationally) distribution-free. A variety of such multivariate tests has been discussed in Puri and Sen [10, 11]. Let us point out the role of the PCLT in this context in a brief manner. Let $\mathbf{X}_i = (X_{i1}, \ldots, X_{ip})', i = 1, \ldots, N$ be independent random vectors ($p \geq 1$) with continuous distributions $F_1, \ldots, F_N$, respectively, and we want to test for the equality of $F_1, \ldots, F_N$ against alternatives relating to shifts in location (vectors) or, in general,

regression models (including the one-way MANOVA* model as a special case). For each j ($= 1, \ldots, p$), let $R_{1j}, \ldots, R_{Nj}$ be the vector of ranks of $X_{1j}, \ldots, X_{Nj}$, among themselves, and let $\mathbf{R}_N = ((R_{ij}))$ be the $p \times N$ *rank collection matrix*. Also, for each j ($= 1, \ldots, p$), consider a set of scores $\{a_{Nj}(1), \ldots, a_{Nj}(N)\}$ and define

$$\mathscr{L}_{Njk} = \sum_{i=1}^{N} (c_{ik} - \bar{c}_{Nk}) a_{Nj}(R_{ij}),$$

$$\bar{c}_{Nk} = N^{-1} \sum_{i=1}^{N} c_{ik}, \quad (6)$$

for $j = 1, \ldots, p$, $k = 1, \ldots, q$, where the $\mathbf{c}_i = (c_{i1}, \ldots, c_{iq})', i \geq 1$ are given q-vectors. Note that for $p = 1$, under the null hypothesis, $\mathbf{R}_N$ has the uniform permutation distribution over the set of $N!$ permutations of $(1, \ldots, N)$. But for $p \geq 2$, $\mathbf{R}_N$ has $(N!)^p$ possible realizations and the distribution of $\mathbf{R}_N$ over this set depends on the underlying F, even when the null hypothesis holds (except when the coordinates of the $\mathbf{X}_i$ are all stochastically independent). Thus in general, the distribution of the $\mathscr{L}_{Njk}$ depends on the underlying F (under the null hypothesis), and hence a test based on these statistics is not generally genuinely distribution-free. To overcome this problem, Chatterjee and Sen [1] introduced the permutationally (conditionally) distribution-free* tests based on the following reduction. If we permute the columns of $\mathbf{R}_N$ in such a way that the first row is in the natural order $(1, \ldots, N)$ and denote the resulting matrix by $\mathbf{R}_N^*$ (the *reduced rank collection matrix*) then under the null hypothesis, given $\mathbf{R}_N^*$, $\mathbf{R}_N$ has a (discrete) uniform distribution* over the $N!$ possible column permutations of $\mathbf{R}_N^*$. We denote this conditional (permutational) probability law by $\mathscr{P}_N$. For $p = 1$, this conditional law agrees with the unconditional one, while for $p \geq 2$, the unconditional distribution of $\mathbf{R}_N$ is defined on a different set and is different from this conditional one. Since the conditional distribution does not depend on the underlying F, it may be used to provide conditionally distribution-free tests. If we let $\mathscr{L}_N = ((\mathscr{L}_{Njk}))$ then, by the same permutational arguments as leading to (2), we obtain that

$$E_{\mathscr{P}_N} \mathscr{L}_N = N\left((\bar{a}_{Nj}\bar{c}_{Nk})\right) \quad \text{and}$$
$$\text{Var}_{\mathscr{P}_N} \mathscr{L}_N = \mathbf{V}_N \otimes \mathbf{W}_N \tag{7}$$

where

$$\mathbf{W}_N = \sum_{i=1}^{N} (\mathbf{c}_i - \bar{\mathbf{c}}_N)(\mathbf{c}_i - \bar{\mathbf{c}}_N)',$$
$$\bar{\mathbf{c}}_N = (\bar{c}_{N1}, \dots, \bar{c}_{Nq})', \tag{8}$$
$$\mathbf{V}_N = (N-1)^{-1} \sum_{i=1}^{N} \left(\mathbf{a}_N^{(i)} - \bar{\mathbf{a}}_N\right)\left(\mathbf{a}_n^{(i)} - \bar{\mathbf{a}}_N\right)',$$
$$\bar{\mathbf{a}}_N = (\bar{a}_{N1}, \dots, \bar{a}_{Np})' \tag{9}$$

and $\mathbf{a}_N^{(i)} = (a_{N1}(R_{i1}), \dots, a_{Np}(R_{ip}))'$, for $i = 1, \dots, N$. Note that $\mathbf{V}_N$ is generally a stochastic matrix. If we roll out $\mathscr{L}_N$ into a pq-vector (and denote it by $\mathscr{L}_N^0$), then as a test statistic we may use $Q_N = (\mathscr{L}_N^0)'[\mathbf{V}_N \otimes \mathbf{W}_N]^{-1}(\mathscr{L}_N^0)$. For small values of N, the exact permutational distribution of Q_N (under $\mathscr{P}_N$) can be obtained by direct enumeration. This task becomes quite laborious as N becomes large. The PCLT can play a vital role in this context: If the permutation distribution of $(\mathbf{V}_N \otimes \mathbf{W}_N)^{-1/2}\mathscr{L}_N^0 \; (= \ell_N^*$, say) is asymptotically (in probability) multinormal with null mean vector and dispersion matrix $\mathbf{I}_{pq}$, then Q_N is asymptotically chi square with pq degrees of freedom. In passing, we may remark that the mean square equivalence of S_N and $\mathscr{L}_N$ (after ref. 5) may easily be extended to the multivariate case. However that would not be of much use in the context of the permutational distribution of ℓ_N^*, though the unconditional distribution theory may be studied with the aid of this result. For the permutational distribution theory of ℓ_N^*, we may either use the Hoeffding–Motoo version of the PCLT in (4) or use some martingale* approach. Toward this we define the (generalized) *Noether condition* as

$$\zeta_N = \max_{1 \leqslant i \leqslant N} \left\{ (\mathbf{c}_i - \bar{\mathbf{c}}_N)' \mathbf{W}_N^{-1}(\mathbf{c}_i - \bar{\mathbf{c}}_N) \right\} \to 0$$
$$\text{as} \quad N \to \infty, \tag{10}$$

and also let

$$\gamma_N = \max_{1 \leqslant i \leqslant N} \left\{ \left(\mathbf{a}_N^{(i)} - \bar{\mathbf{a}}_N\right)' \mathbf{V}_N^{-1}\left(\mathbf{a}_N^{(i)} - \bar{\mathbf{a}}_N\right) \right\}. \tag{11}$$

Note that ζ_N is nonstochastic while γ_N is stochastic in nature. Then, whenever $\zeta_N \gamma_N \to 0$, in probability, the permutation distribution of ℓ_N^* is asymptotically normal in probability. This result, due to Sen [14], exploits the Hoeffding–Motoo theorem. Note that the condition that $\zeta_N \gamma_N \xrightarrow{p} 0$ is more stringent than (10). If the scores $a_{Nj}(i)$, $i = 1, \dots, N$ are generated by a score generating function $\phi_j(u) : 0 < u < 1$, where ϕ_j is expressible as a difference of two nondecreasing and absolutely continuous square integrable functions, for each $j = 1, \dots, p$, then the PCLT in the multivariate case holds under the sole condition (10). This result is also due to Sen [14] and is based on a martingale characterization of progressively censored rank statistics, due to Chatterjee and Sen [2]. This permutational martingale characterization yields results deeper than the asymptotic normality. Indeed, permutational invariance principles relating to weak convergence to one- or two-dimensional Wiener processes based on this characterization have been studied by Sen and Ghosh [15], Sen [12], and Chatterjee and Sen [2], among others. Thus as in the case of the CLTs, the PCLTs have also been strengthened to appropriate invariance principles [13].

References

[1] Chatterjee, S. K. and Sen, P. K. (1964). *Calcutta Statist. Ass. Bull.*, **13**, 18–58.

[2] Chatterjee, S. K. and Sen, P. K. (1973). *Calcutta Statist. Ass. Bull.*, **22**, 13–50.

[3] Hájek, J. (1961). *Ann. Math. Statist.*, **32**, 506–523.

[4] Hoeffding, W. (1951). *Ann. Math. Statist.*, **22**, 558–566.

[5] Madow, W. G. (1948). *Ann. Math. Statist.*, **19**, 535–545.

[6] Motoo, M. (1957). *Ann. Inst. Statist. Math.*, **8**, 145–156.

[7] Nandi, H. K. and Sen, P. K. (1963). *Calcutta Statist. Ass. Bull.*, **12**, 125–143.

[8] Noether, G. E. (1949). *Ann. Math. Statist.*, **20**, 455–458.

[9] Pitman, E. J. G. (1937). *Suppl. J. R. Statist. Soc.*, **4**, 119–130.

[10] Puri, M. L. and Sen, P. K. (1969). *Ann. Math. Statist.*, **40**, 1325–1343.

[11] Puri, M. L. and Sen, P. K. (1971). *Nonparametric Methods in Multivariate Analysis*. Wiley, New York.

[12] Sen, P. K. (1976). *Ann. Prob.*, **4**, 13–26.

[13] Sen, P. K. (1981). *Sequential Nonparametrics*. Wiley, New York.

[14] Sen, P. K. (1983). *Sankhyā, Ser. A*, **45**, 141–149.

[15] Sen, P. K. and Ghosh, M. (1972). *Sankhyā, Ser. A*, **34**, 335–348.

[16] Wald, A. and Wolfowitz, J. (1944). *Ann. Math. Statist.*, **15**, 358–372.

(ASYMPTOTIC NORMALITY
FINITE POPULATIONS, SAMPLING FROM
LIMIT THEOREM, CENTRAL
LIMIT THEOREMS
LINDEBERG–FELLER THEOREM
NOETHER AND RELATED CONDITIONS)

P. K. SEN

PERMUTATION MATRIX

A permutation matrix is a square $m \times m$ matrix in which each column contains $(m - 1)$ zeros and one 1. If $\mathbf{P}$ is such a permutation matrix, and $\mathbf{x}$ is an m-rowed vector, then $\mathbf{Px}$ is a vector with components which are a permutation of $\mathbf{x}$.

PERMUTATION MODELS

Until recently the term "permutation distribution" referred explicitly to the probability distribution function of certain nonparametric test statistics. Now, however, this term can also be applied to special models which are developed for ranking* experiments and trials, such as horse races and preference testing, where the ordering of the sample is vital. Such permutation distributions are more appropriately referred to as permutation models and they are the subject of this article.

Early developments in this area concern the theory of paired comparisons*. This topic was initiated in the late 1920s and has received a great deal of attention since 1950 (see Davidson and Farquhar [4] for a comprehensive bibliography).

In basic form a paired comparison experiment has $t \geqslant 2$ treatments, $T_1, T_2, \ldots, T_t$, considered in pairs. There are n_{ij} independent selection decisions on the comparison of T_i with T_j. Let $T_i \to T_j$ represent "T_i is selected over T_j"; then a primitive model consists of $\binom{t}{2}$ functionally independent parameters, $0 \leqslant \Pi_{ij} \leqslant 1$, such that

$$P(T_i \to T_j) = \Pi_{ij}, \qquad \Pi_{ij} + \Pi_{ji} = 1,$$
$$i \neq j, \quad i, j = 1, 2, \ldots, t.$$

Some structure is introduced by the Bradley–Terry model*, which assumes treatment parameters $\Pi_1, \Pi_2, \ldots, \Pi_t$ that are associated with $T_1, T_2, \ldots, T_t$ and are such that Π_i is the relative selection probability of T_i. In these terms $\Pi_{ij} = \Pi_i / (\Pi_i + \Pi_j)$. The Π_i can be estimated by maximum likelihood* iteration, and large sample tests of hypotheses can be conducted in the usual way. The model has been extended to triple comparisons, to multivariate paired comparisons and in other directions.

Models which assume background random response variables to the T_i, X_i have also been proposed. In this context $T_i \to T_j$ if and only if $X_i > X_j$ and the resulting probability structure depends on the distributional assumptions concerning the X_i. For a full review of the paired comparison methodology outlined above see Bradley [1].

Plackett [7] discussed the problem of constructing place probabilities for a horse race with n runners. He used the horses' individual win probabilities, $p_1, p_2, \ldots, p_n$, to argue that under suitable conditions the probability of the order $ijk \ldots l$ is

$$p_i \{ p_j (1 - p_i)^{-1} \} \{ p_k (1 - p_i - p_j)^{-1} \} \ldots 1.$$

$$(1)$$

Writing $p_{ijk \ldots l}$ as the probability of the ordering $ijk \ldots l$, then $p_i = \sum_{jk \ldots l} p_{ijk \ldots l}$ where the sum is over all permutations with i in the first position. Thus the elementary model (1), which is equivalent to the Bradley–Terry models, specifies $n! - 1$ functionally independent probabilities by means of $n - 1$ parameters.

Plackett developed a logistic type structure which adds parameters in stages. At the first stage (1) there are $n - 1$ parameters and at the qth stage $n(n - 1) \ldots (n - q + 1) - 1$ parameters. Thus for $q = n$ the model is saturated. It is a feature of this model that parameters at the qth stage only appear when those of the $(q + 1)$th stage have been set to zero. The reader is referred to the original paper for further details where the theory is illustrated with a numerical example.

Simplistic models paralleling those of the analysis of variance* have been proposed by Tallis and Dansie [8]. These are easy to illustrate for $n = 3$. In fact, the saturated model is

$$p_{ijk} = \tfrac{1}{6} + \alpha_i/2 + \beta_{ij}, \qquad i \neq j,$$

$$\beta_{ij} = \beta_{ik} = 0, \qquad j, k \neq i, \quad i = 1, 2, 3, \quad (2)$$

$$\alpha_1 + \alpha_2 + \alpha_3 = 0.$$

The set of parameters $\{\alpha_i\}$ relate to the probability that i is ranked first. The second order parameters $\{\beta_{ij}\}$ represent the association between the rankings of elements i and j. Model (2) is easily extended to arbitrary n and maximum likelihood estimates of the parameters are obtained directly from the appropriate relative frequencies. Moreover (2) allows for the easy specification and testing of subhypotheses. For example, if it is thought that $\beta_{12} = \beta_{21} = \beta$, $\beta_{31} = 0$ then

$$p_{123} = 1/6 + \alpha_1/2 + \beta,$$
$$p_{132} = 1/6 + \alpha_1/2 - \beta$$
$$p_{213} = 1/6 + \alpha_2/2 + \beta, \qquad (3)$$
$$p_{231} = 1/6 + \alpha_2/2 - \beta$$
$$p_{312} = p_{321} = 1/6 + \alpha_3/2.$$

The three parameters α_1, α_2, and β can be estimated and the adequacy of (3) tested against (2) by standard likelihood ratio* methods.

Models which have background variables $X_1, X_2, \ldots, X_n$ assume that the permutation $ijk \ldots l$ has a probability given by

$$p_{ijk \ldots l} = P(X_i < X_j < X_k < \cdots < X_l). \qquad (4)$$

In simple cases the X_i are assumed independent but belong to the same family of PDFs.

For example, suppose X_i has PDF $\alpha_i e^{-\alpha_i x}$, then it is readily shown that

$$P_{ijk \ldots l} = \frac{\alpha_i}{A} \cdot \frac{\alpha_j}{A_i} \cdot \frac{\alpha_k}{A_{ij}} \cdots 1 \qquad (5)$$

where $A = \sum_1^n \alpha_i$, $A_i = A - \alpha_i$, $A_{ij} = A - \alpha_i - \alpha_j$. Actually, $P(X_i < \text{all } X_j, \ j \neq i) = \alpha_i/A$ and, identifying this quantity with p_i of Plackett's model, it is clear that (5) is equivalent to stage 1. In fact the equivalence of Plackett's hierarchical system to a hierarchical system of independent exponential variables was recently established by Dansie [3].

It is clear that (4) is invariant under an arbitrary monotone transformation of X_i. Therefore (5) is not as restrictive as might first appear. As for interpretation, X_i could be the time taken by the ith horse to run a race, in which case (5) gives the probability of the particular outcome $ijk \ldots l$. The assumption of stochastic independence requires no interaction between horses while the parameters α_i relate to the horse's potential speed over the course; innate ability, handicapping, the jockey, and other factors having been taken into account.

A logical extension is to let X_i have a PDF, $f(x)$, which is determined up to a shift parameter θ while maintaining the assumption of stochastic independence (Henery [5]). It is easy to show that, under these conditions, $\theta_1 < \theta_2 < \cdots < \theta_n$ implies $p_1 > p_2 > \cdots > p_n$ as required.

Let Π be the particular permutation 1, 2, $\ldots$, n. Henery considers $f(x) = \phi(x)$, the standardized normal PDF. He finds the approximate probability of Π, $P(\Pi)$, by expanding $P(\Pi)$ in a Taylor series around $\theta_i = 0$, $i = 1, 2, \ldots, n$, showing that

$$P(\Pi) \simeq \frac{1}{n!} + \sum_1^n \frac{\theta_i \mu_{i \cdot n}}{n!} \qquad (6)$$

where $\mu_{i \cdot n}$ is the expectation of the ith order statistic from a sample of size n from $\phi(x)$. Equation (5) has the same form as the generalization of (3) to an arbitrary number of linear terms. However, the interpretation of (6) is quite specific. Henery also presents similar approximations for the joint probability that X_1 is smallest, X_2 is second, and so on.

An obvious extension to the above models is the introduction of covariates. For example, a horse may have some average potential to win a race but on a particular day this could be modified by factors such as the jockey, the barrier position, weather conditions, and so on. For the sake of illustration assume background, independent exponential random variables $X_1, X_2, \ldots, X_n$ where X_i has PDF $\alpha_i e^{-\alpha_i x}$. It is notationally convenient to label the X_i so that $p_{n,n-1,\ldots,1}$ is the relevant permutation probability. Then (5) can be written

$$p_{n,n-1,\ldots,1} = \frac{\alpha_2}{(\alpha_1 + \alpha_2)} \cdot \frac{\alpha_3}{(\alpha_1 + \alpha_2 + \alpha_3)}$$

$$\cdots \frac{\alpha_n}{(\alpha_1 + \alpha_2 + \cdots + \alpha_n)}$$

(7)

If it is assumed that a row vector of covariates for the ith horse, $\mathbf{x}_i$, may modify the parameter α_i, a reasonable form for α_i allowing for dependence on $\mathbf{x}_i$ is $\alpha_i(\mathbf{x}_i) = e^{\mathbf{x}_i \boldsymbol{\beta}}$, where $\boldsymbol{\beta}$ is a vector of coefficients. Now let $\mathbf{w}_i = \mathbf{x}_i - \sum_{j=1}^i \mathbf{x}_j/i$, $\mathbf{X}_i' = [\mathbf{x}_1', \mathbf{x}_2', \ldots, \mathbf{x}_i']$, $\overline{S}_i = \mathbf{X}_i' \mathbf{X}_i/i$, then it can be shown that

$$p_{n,n-1,\ldots,1} \simeq \frac{1}{n!} \exp\left\{ \mathbf{W}_n \boldsymbol{\beta} - \tfrac{1}{2} \boldsymbol{\beta}' \mathscr{S}_n \boldsymbol{\beta} \right\}$$

(8)

where $W_n = \sum_2^n w_i$ and $\mathscr{S}_n = \sum_2^n \overline{S}_i$. Equation (8) should, in most cases, provide a serviceable covariate representation of $p_{n,n-1,\ldots,1}$, allowing a first estimate of $\boldsymbol{\beta}$ to be found easily by maximum likelihood from an arbitrary number of independent trials. This estimate can then be used in (7) as the initial value of $\boldsymbol{\beta}$ in a more refined analysis.

Pettitt [6] considers a similar structure and assumes $X_1, X_2, \ldots, X_n$, after an arbitrary monotone transformation, are independent and that X_i is normally distributed with mean $\mathbf{x}_i \boldsymbol{\beta}$ and unit variance, $\mathbf{x}_i$ being a row vector of covariates. He showed that

$$p_{n,n-1,\ldots,1} \simeq \frac{1}{n!} \exp\left\{ \mathbf{a}' \mathbf{X} \boldsymbol{\beta} - \tfrac{1}{2} \boldsymbol{\beta}' \mathbf{X}' \mathbf{C} \mathbf{X} \boldsymbol{\beta} \right\}$$

(9)

where $\mathbf{X}' = [x_1', x_2', \ldots, x_n']$ and, if $X_1' < X_2' < \cdots < X_n'$ are the order statistics of a sample of size n from $\phi(x)$ and $\mathbf{Z}' = [X_n', X_{n-1}', \ldots, X_1']$, $\mathbf{a} = E[\mathbf{Z}]$, $\mathbf{C} = \mathbf{I} - V[\mathbf{Z}]$. Equation (9) can be compared with (8), which is of a similar but simpler form. Pettitt considers points of inference for $\boldsymbol{\beta}$, the problem of ties, prediction, and some interesting applications. He also extends his results to nonnormal parent PDFs.

In principle, a covariate structure can also be imposed upon some of the models discussed earlier. However, an effort must be made to contain the computations to manageable proportions.

Another direction of generalization is to allow the X_i to be dependent, although care must be exercised to ensure that parameters are identifiable (see Daganzo [2] for details and applications to demand forecasting). For example, if $n = 3$ and X_1, X_2, X_3 have a trivariate normal PDF, there is a total of nine parameters to be fitted to five functionally independent probabilities. In fact,

$$p_{123} = P(X_1 < X_2 < X_3)$$

$$= P(Y_1 - Y_2 < 0, Y_2 < 0)$$

where $Y_1 = X_1 - X_3$, $Y_2 = X_2 - X_3$, which shows that the model is essentially two-dimensional. Moreover Y_2 can be standardized to unit variance and there are in fact only four identifiable parameters. The model does not allow for saturation.

It is evident from the bibliography that research into permutation models is contemporary. Applications abound and include horse racing, preference testing, and demand forecasting. The literature should be monitored for further useful and interesting results in this area.

References

[1] Bradley, R. A. (1976). *Biometrics*, **32**, 213–232.

[2] Daganzo, C. (1979). *Multinomial Probit: The Theory and Its Application to Demand Forecasting*. Academic, New York.

[3] Dansie, B. R. (1983). *J. R. Statist. Soc. B*, **45**, 22–24.

[4] Davidson, R. R. and Farquhar, P. H. (1976). *Biometrics*, **32**, 241–252.

[5] Henery, R. J. (1981). *J. R. Statist. Soc. B*, **43**, 86–91.

[6] Pettitt, A. N. (1982). *J. R. Statist. Soc. B*, **44**, 234–243.

[7] Plackett, R. L. (1975). *Appl. Statist.*, **24**, 193–202.

[8] Tallis, G. M. and Dansie, B. R. (1983). *Appl. Statist.* **32**, 110–114.

(CONCOMITANT VARIABLES
PAIRED COMPARISONS
PERMUTATION TESTS
RANDOMIZATION, CONSTRAINED
RANDOMIZATION TESTS
RANK PROCEDURES IN EXPERIMENTAL
   DESIGN)

G. M. TALLIS

PERMUTATION TESTS

Permutation tests are a special kind of randomization test*. Randomization tests include all tests based on the results of random assignment of a fixed set of observations into groups or categories, say, plus and minus. A value for the test statistic is computed for each of the resulting equally probable assignments and these values and their corresponding probabilities are called the randomization distribution of the test statistic. Hypothesis tests, *P*-values*, and confidence intervals* are based on this randomization distribution. The random assignment is most frequently generated by some kind of permutation of the fixed set of observations. When this is the case, the procedure might be called a permutation test. Pitman tests*, Brown-Mood median test*, and Fisher's exact test* are special kinds of permutation tests.

(RANDOMIZATION TESTS)

J. D. GIBBONS

PERSSON–ROOTZÉN ESTIMATOR

This is an explicit and rather simple estimator for the parameters in a type I censored normal sample (*see* CENSORED DATA), with

statistical properties which are virtually the same as for the maximum likelihood* (ML) estimator. In this situation the ML-equations cannot be solved explicitly, and numerical solution is sometimes problematic. Persson and Rootzén [7] consider single censoring, while Persson [6] and Schneider [9] propose related estimators for doubly censored samples.

A random sample is type I singly (left) censored if all observations below some fixed level c are removed, so that the remaining information consists of the values $x_1, \ldots, x_k$ which fell above c, and the number $n - k$ of those which fell below. Here either n or k or both may be random variables, and the distribution of the uncensored observations is assumed to be normal with mean μ and standard deviation σ. The Persson–Rootzén estimator for (μ, σ) is

$$\mu^* = \frac{1}{k} \sum_{i=1}^{k} x_i - \alpha \sigma_{\mathrm{RML}},$$

$$\sigma^* = \left[\frac{1}{k} \sum_{i=1}^{k} x_i^2 - \left(\frac{1}{k} \sum_{i=1}^{k} x_i \right)^2 \right.$$
$$\left. - \left\{ \alpha \lambda_{k/n} - \alpha^2 \right\} \sigma_{\mathrm{RML}}^2 \right]^{1/2},$$

where σ_{RML} and α are given by

$$\sigma_{\mathrm{RML}} = \frac{1}{2} \left[\lambda_{k/n} \frac{1}{k} \sum_{i=1}^{k} (x_i - c_i) \right.$$
$$+ \left\{ \left(\lambda_{k/n} \frac{1}{k} \sum_{i=1}^{k} (x_i - c_i) \right)^2 \right.$$
$$\left. + \frac{4}{k} \sum_{i=1}^{k} (x_i - c_i)^2 \right\}^{1/2} \right],$$

$$\alpha = \frac{n}{k} \frac{1}{\sqrt{2\pi}} e^{-\lambda_{k/n}^2/2},$$

and where $\lambda_{k/n}$ is the upper (k/n)th quantile of the standard normal distribution (which can easily be computed using, e.g., formula

26.2.23 of ref. 1). The idea behind the estimator is to use an approximative ML estimator to correct for the bias caused by the censoring in the sample mean and standard deviation. Bias and small sample and asymptotic variances of the estimators are computed in ref. 7. It may be noted that, for very small samples and severe censoring, both the Persson–Rootzén estimator and the ML estimator may be rather biased, and that, in contrast to the uncensored case, the estimates of μ and σ are correlated. In ref. 6 the effect of correlation in the sample is studied, the main conclusion being that the estimator seems reasonably robust against moderate correlations.

Cohen [2], Hald [3], and Halperin [4] construct diagrams and tables to reduce the amount of work involved in calculating the ML estimates. However, due to limited resolution in tables and diagrams, the simplified estimator (μ^*, σ^*) often gives at least as accurate approximations to the ML estimators. Harter and Moore [5] consider methods for numerical solution of the ML-equations, and Saw [8] proposes a simple linear estimator for (μ, σ) which, however, is less efficient.

References

[1] Abramovitz, M. and Stegun, I. A. (1964). *Handbook of Mathematical Functions*. Dover, New York.

[2] Cohen, A. C. (1961). *Technometrics*, **3**, 535–541.

[3] Hald, A. (1949). *Skand. Aktuarietidskr.*, **32**, 119–134.

[4] Halperin, M. (1952). *J. Amer. Statist. Ass.*, **47**, 457–465.

[5] Harter, H. L. and Moore, A. H. (1966). *Biometrika*, **53**, 205–213.

[6] Persson, T. (1983). Ph.D. dissertation, University of Lund, Sweden.

[7] Persson, T. and Rootzén, H. (1977). *Biometrika*, **64**, 123–128.

[8] Saw, J. G. (1961). *Biometrika*, **48**, 367–377.

[9] Schneider, H. (1984). *Biometrika*, **71**, 412–414.

(CENSORED DATA
LOCATION PARAMETER, ESTIMATION OF
MAXIMUM LIKELIHOOD ESTIMATION)

H. ROOTZÉN

PERT

The Program Evaluation and Review Technique (PERT) was developed in the late 1950s to help accelerate completion of the Polaris ballistic missile (for historical details, see Malcolm et al. [1]). It is commonly used today for planning and controlling major projects in diverse areas such as research and development, production, and construction. These projects are typically comprised of numerous jobs or activities that are governed by certain precedence relationships. That is, while certain activities may be carried out in parallel, others may begin only upon the completion of a set of predecessor activities. A project's duration is the shortest time required to complete all activities subject to the given precedence constraints. Due to the uncertainties involved in most project activities, individual activity times must be modeled stochastically. The project duration is consequently a random variable whose distribution and moments* are of major interest.

A PERT network is a directed acyclic graph* that depicts each activity, its precedence relations, and information regarding its time requirements. The arcs in the network represent activities and the nodes denote events corresponding to various stages in the project's duration. In a PERT network with n nodes, nodes 1 and n refer to the commencement and termination of the overall project. The network is constructed so that all activities, represented by arcs, leading into a given node must precede the activities or arcs emanating from that node.

Example 1. Consider the PERT network in Fig. 1 with a total of seven activities. Activity $(1, 2)$ must be finished before either $(2, 4)$ or $(2, 5)$ is initiated, and similarly, activities $(2, 4)$ and $(3, 4)$ must precede $(4, 5)$. Node 5 is the terminal event denoting completion of the project. If the durations of all activities were specified and attached as arc lengths to the PERT network, then each path from node 1 to node n would have a length equal to the sum of the arc lengths on that path.

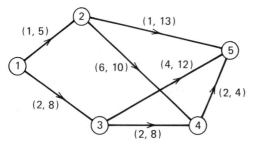

Figure 1   A stochastic network.

Since the activities on any path must be performed consecutively, each path length would yield a lower bound on the overall project duration. The project duration T would be the length of the longest path from node 1 to n. The longest path is called the *critical path* (CP).

In Fig. 1, the four paths from 1 to 5 are as follows: (P1) 1-2-5; (P2) 1-2-4-5; (P3) 1-3-5; (P4) 1-3-4-5. If we take the arc length to be the average of the two numbers given next to each arc of Fig. 1, then the above four paths have lengths 10, 14, 13, and 13 respectively. The critical path is, therefore, P2.

Since arc lengths are random variables, the CP will depend on the particular *realization* of the stochastic network where fixed durations are sampled from the activity time distributions. In this case the CP length T will itself be a random variable whose expectation $E(T)$ and variance $\text{Var}(T)$ will be of principal interest.

Example 2.   Consider again the PERT network of Fig. 1 and assume that each activity time may take on one of the two values given in parentheses with equal probability. There would then be $2^7 = 128$ realizations of this stochastic network. For example, in the case where all arcs require their maximum time, paths P1 through P4 would have lengths 18, 19, 20, and 20 respectively showing P3 and P4 to be critical. $E(T)$ may be found by averaging the CP lengths over the 128 possible realizations.

FINDING THE EXPECTED CRITICAL PATH LENGTH

Since PERT networks may easily contain in excess of 2000 arcs, the exact distribution of

the CP length would be impractical to obtain using the technique suggested in Example 2. Instead various procedures have been devised to obtain approximations to $E(T)$ and $\text{Var}(T)$ as described below.

Assume without loss of generality that the n nodes of a PERT network are numbered such that each arc $(i, j) \in A$ (A is the set of all arcs) is constructed with $i < j$. Let E_j be the set of arcs terminating at node j and let B_j be the initial nodes of these arcs. (In Fig. 1, $B_1 = \emptyset$, $B_2 = B_3 = \{1\}$, $B_4 = \{2, 3\}$, and $B_5 = \{2, 3, 4\}$.) Let t_{ij} be the random length (duration) of arc (i, j) with expected value $\bar{t}_{ij}$. We may collect all such lengths into a vector $\mathbf{t}$ with joint density $p(\mathbf{t})$. We denote the set of all arcs in the subnetwork composed of nodes $\{1, 2, \ldots, j\}$ only by D_j. For any subset X of arcs, $\mathbf{t}_X$ denotes the components of $\mathbf{t}$ referring to arcs in X alone. (Obviously, $\mathbf{t}_A = \mathbf{t}$.) Finally, let the random variable l_j be the length of the critical (longest) path from node 1 to node j. The project duration T then equals l_n. For the particular realization $\mathbf{t}_X^r$ of $\mathbf{t}_X$ where $X \supseteq D_j$, $l_j(\mathbf{t}_X^r)$ signifies the longest path length to j. We point out that since all arcs (i, j) have $i < j$, only realizations of arcs contained in D_j have an effect on l_j. It is usually assumed that the activity durations t_{ij} are independent random variables, allowing us to factor the joint density as

$$p(\mathbf{t}) = p(\mathbf{t}_{E_2}) \cdots p(\mathbf{t}_{E_n}). \tag{1}$$

The exact expected CP length to node j may be expressed as

$$e_j = E(l_j) = \sum_{\mathbf{t}^r} p(\mathbf{t}^r) l_j(\mathbf{t}^r) \tag{2}$$

$$= \sum_{\mathbf{t}_{j-1}^r} \sum_{\mathbf{t}_{E_j}^r} p(\mathbf{t}_{j-1}^r) p(\mathbf{t}_{E_j}^r) l_j(\mathbf{t}^r) \tag{3}$$

$$= \sum_{\mathbf{t}_{j-1}^r} p(\mathbf{t}_{j-1}^r) \sum_{\mathbf{t}_{E_j}^r} p(\mathbf{t}_{E_j}^r) \underset{i \in B_j}{\text{Max}} \{ l_i(\mathbf{t}_i^r) + t_{ij}^r \} \tag{4}$$

where $\mathbf{t}_i \overset{\Delta}{=} \mathbf{t}_{D_i}$. In going from (2) to (3) we exploit the independence assumption and the fact that $D_{j-1} \cup E_j = D_j$. The last equality follows from the relation $l_j = \text{Max}_{i \in B_j} \{ l_i + t_{ij} \}$. Various PERT techniques attempt to approximate e_j by assigning la-

Table 1   Expected CP Lengths Along with Approximations

Node j	1	2	3	4	5
g_j	0	3	5	11	14
f_j	0	3	5	12	$16\frac{1}{4}$
e_j	0	3	5	$12\frac{5}{8}$	$17\frac{5}{64}$

bels recursively to nodes $j = 1, \ldots, n$. The simplest approximation which is the one most frequently used, unfortunately, is not always very accurate. In this procedure all arc durations are replaced by their expected values. The node numbers g_j are then defined by the recursion

$$g_1 = 0, \qquad g_j = \operatorname*{Max}_{i \in B_j} \left\{ g_i + \bar{t}_{ij} \right\}. \qquad (5)$$

The values of g_i for Example 2 are listed in Table 1. Fulkerson [7] suggests a refinement of this procedure by retaining the stochastic nature of arcs in E_j, but approximating the CP lengths l_i by deterministic numbers f_i for $i \in B_j$. His recursion is

$$f_1 = 0, \qquad f_j = \sum_{t'_{E_j}} p(t'_{E_j}) \operatorname*{Max}_{i \in B_j} \left\{ f_i + t_{ij}^r \right\}. \qquad (6)$$

For instance, in Example 2, given that f_2 and f_3 equal 3 and 5, respectively, f_4 is computed as

$$f_4 = \tfrac{1}{4}(9 + 13 + 13 + 13) = 12,$$

where the numbers in parentheses list the possible values of $\operatorname{Max}\{3 + t_{24}, 5 + t_{34}\}$ as t_{24} and t_{34} vary over their allowed ranges.

Comparing the values of e_j, f_j, and g_j displayed in Table 1 (for Example 2), we see that

$$g_j \leqslant f_j \leqslant e_j \qquad \text{for all } j. \qquad (7)$$

Relationship (7) holds in general and demonstrates that the usual PERT procedure and Fulkerson's method each provide lower bounds on the true expected CP length. The PERT literature contains other refinements of these estimates. Clingen [5] provides a more computationally tractable version of the recursion in (6) that applies to both discrete and continuous activity time distributions. Elmaghraby [6] proposes two meth-

ods for calculating node numbers h_j that satisfy $f_j \leqslant h_j \leqslant e_j$ and, hence lead to tighter lower bounds. Robillard and Trahan [10] show that these bounds may be derived from Jensen's inequality* and, in addition, they provide yet sharper bounds. One criticism of the above procedures is that only lower bounds are derived. To remedy this, Kleindorfer [8] and, later, Shogan [11] have focused on deriving both lower and upper bounds on the cumulative distribution of l_j, that is, functions $\underline{F}_j$ and $\bar{F}_j$ satisfying

$$\underline{F}_j(t) \leqslant \Pr(l_j \leqslant t) \leqslant \bar{F}_j(t).$$

These may then be used to provide lower and upper bounds on e_j.

Lindsey [9] suggests a complicated estimation procedure that is based on certain normal approximations in order to obtain node numbers k_j satisfying $k_j \geqslant f_j$. As his node numbers may actually overestimate e_j, they are point estimates rather than lower bounds.

If all activity time distributions are normal, then it is possible to use the results of Clark [4] on computing the moments of the maximum of correlated normal random variables as illustrated in the next example.

Example 3.   Consider the network of Fig. 2 with arcs labeled a through e whose lengths t_a through t_e are taken to be independent and normal. Then $l_2 = t_b$, $l_3 = \operatorname{Max}\{t_a, t_b + t_c\}$, and $l_4 = \operatorname{Max}\{l_2 + t_d, l_3 + t_e\}$. Note that l_3 is the maximum of independent normal variables and its mean and variance can be computed analytically using Clark's formulas (which we do not present here). Now if we approximate l_3 by a normal variable l_3' with the same mean and variance, then l_4

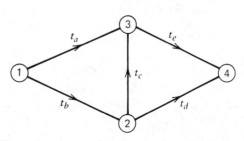

Figure 2   Stochastic network with normal activity times.

may be approximated by

$$l_4' = \text{Max}\{l_2 + t_d, l_3' + t_e\}.$$

Again l_4' is the maximum of normal random variables; however, this time they are correlated (through t_b). This correlation may also be computed using Clark's formulas. In this fashion, one may approximate $E(T)$ and $\text{Var}(T)$. For normal t_{ij}'s, Clark's procedure usually yields very accurate results.

Given $E(T)$ and $\text{Var}(T)$, and assuming the approximate normality of T by appealing to the Central Limit Theorem*, one may obtain probability statements of the form $\text{Pr}(T_1 \leqslant T \leqslant T_2)$ that are helpful in evaluating the likelihood of completing a project within certain time limits. In the usual PERT procedure, $E(T)$ is approximated by g as computed in (5), and $\text{Var}(T)$ is estimated by summing the individual variances of the activities on the CP implied by g_n. While one knows that $g_n \leqslant E(T)$, no similar bounding results hold for the estimate of $\text{Var}(T)$ derived through the procedure just outlined. However, the results from the usual PERT approach are reasonably accurate if in the network with arc lengths $\bar{t}_{ij}$ the CP length is much larger than that of the second longest path.

SIMULATION* AND STATISTICAL COMPUTATIONS

Obtaining the exact CP length distribution involves multivariate integrations that are computationally prohibitive for large networks. However, Monte Carlo simulation* may be employed to provide information about this distribution. Under this approach, individual activity time distributions are sampled to obtain realizations of the network on which the CP may be calculated deterministically. The result is a sampling distribution for the CP length. Early and important work describing this procedure is contained in a paper by Van Slyke [21]. The simulation output includes: (a) the expected CP length, (b) its variance, and (c) a criticality index for each activity, which estimates

the probability of that activity being on the CP. The sample size required to achieve the desired accuracy may be determined through standard confidence intervals. The simulation results may also be used to evaluate the accuracy of the approximations described in the previous section.

For large PERT networks, "straightforward" simulation may still be computationally too demanding. The following procedures have, therefore, been proposed to increase the accuracy of simulation results and to decrease the computational burden:

1. **Selective Sampling.** Arcs with zero or negligible criticality need not be sampled. Such arcs may be identified by analytic procedures (see MacCrimmon and Ryavec [29]), or by performing a trial simulation run to determine the arcs that are very rarely critical (see Van Slyke [21]).

2. **Network Reduction.** Portions of the stochastic network may be collapsed into single arcs whose distributions are computable analytically. For example, two arcs in series of lengths t_1 and t_2 can be combined into a single arc with a length of $t_1 + t_2$, whose distribution may be obtained by a convolution*. These reductions are similar to series-parallel reductions used in circuit theory. Martin [18], Hartley and Wortham [16], and Ringer [19, 20] give procedures for computations associated with such reductions.

3. **Conditional Sampling.** One may condition the lengths of paths from 1 to n on those activities which such paths share in common. This will "decouple" the paths and allow one to express the conditional distribution of the CP analytically. The sampling may then be limited to the conditioned activities alone. Burt and Garman [14] and Garman [15] discuss this procedure and report an improvement in the accuracy of the simulation as a result.

4. **Variance Reduction Techniques.** Standard simulation techniques for improv-

ing the accuracy (antithetic variables, stratified sampling*, etc.) as applied to PERT are described by Burt and Garman [13].

ESTIMATION OF INDIVIDUAL ACTIVITY TIMES

PERT procedures require the individual activity time distribution (or its moments) as input. The procedure used in practice consists of several steps. First, for each activity obtain optimistic, pessimistic, and most likely estimates of its duration t, denoted by a, b, and m respectively. Estimate the mean and variance of the duration by

$$E(t) = \frac{a + 4m + b}{6}, \qquad (8)$$

$$\text{Var}(t) = \sigma^2(t) = \frac{(b - a)^2}{36}. \qquad (9)$$

Swanson and Pazar [33] attribute the popularity of PERT, at least in part, to the fact that equations (8) and (9) allow for easy conversion from readily obtained estimates a, b, and m to the required activity time means and variances.

These approximations are based on the assumption that t has a beta distribution* with range $[a, b]$ and mode m. If we consider a general beta density over $[a, b]$,

$$f(t) = \frac{1}{B(p, q)(b - a)^{p + q - 1}}$$

$$\times (t - a)^{p - 1}(b - t)^{q - 1}, \quad (10)$$

where $B(p, q)$ is the complete beta function, then

$$E(t) = \frac{a + b + (p + q - 2)m}{p + q} \qquad (11)$$

and

$$\text{Var}(t) = \frac{pq}{(p + q)^2(p + q + 1)}(b - a)^2. \qquad (12)$$

This shows that (8) and (9) are only approximations to the exact relations (11) and (12). A number of papers have focused on the

errors resulting from this approximation. Indeed (8) and (9) place constraints on possible choices of the parameters p and q and yield exact results only if

$$p = 3 \pm \sqrt{2} \quad \text{and} \quad q = 3 \mp \sqrt{2},$$

$$\text{or} \quad p = q = 4.$$

(Note that in the latter case, $m = (a + b)/2$ due to symmetry.) Moreover, the skewness of a beta distribution satisfying (8) and (9) can only assume the values $\pm 1/\sqrt{2}$ or 0 (see Donaldson [23] and Grubbs [24]). Donaldson shows that if an estimate of the mean, rather than the mode, is used to estimate p and q, and if tangency of the beta density to the x-axis is assumed at endpoints a and b, then the skewness can vary over the range $-\sqrt{2}$ to $+\sqrt{2}$. However, Coon [22] remarks that Donaldson's estimating procedure still implies setting either p or q to 2.

McBride and McClelland [30] search for a choice of p and q that renders the variance estimate in (9) exact, i.e., they require that $pq/\{(p + q)^2(p + q + 1)\} = 1/36$. It can be shown that the error resulting from using (8) as opposed to the exact relation (11) for $E(t)$ will not exceed 19%.

In view of the difficulties associated with the beta assumption and the restrictive nature of relations (8) and (9), a number of researchers have proposed alternative approaches for estimating $E(t)$ and Var(t):

1. **Other Distributional Forms.** Since there is no a priori reason to postulate a beta fit, MacCrimmon and Ryavec [29] suggest using triangular distributions* whereas Kotiah and Wallace [28] propose a truncated normal distribution* based on a maximum entropy* approach subject to constraints imposed by a priori information.

2. **Use of Percentiles.** Moder and Rodgers [31] advocate using 100α and $100(1 - \alpha)$ percentiles $q_{100\alpha}, q_{100(1 - \alpha)}$ instead of a and b as input data into formulas analogous to (8) and (9). The main advantage of this approach is that

the resulting errors are much less sensitive to the exact form of the underlying distribution of t. Perry and Greig [32] provide the following relatively distribution-free formulas to estimate $E(t)$ and $\text{Var}(t)$:

$$E(t) = \frac{q_5 + 0.95m + q_{95}}{2.95} \qquad (13)$$

and

$$\text{Var}(t) = \sigma^2(t) = \left(\frac{q_{95} - q_5}{3.25} \right)^2. \qquad (14)$$

It must be noted that for any estimation procedure to be valuable and of practical use, it should be reasonably straightforward and should avoid extensive computations.

A body of research, typified by the studies of King et al. ([25, 26, 27]), investigates the subjective estimating behavior of the sources consulted for activity time data. The studies are based on experience from real project networks and include simple adjustment models for modifying subjective estimates based on the historical performance of the sources.

CONCLUSIONS

Despite the limitations of using the beta distribution equations (8) and (9), in the more than two decades since its inception, PERT has developed into a well known and widely applied management tool.

As pointed out in the preceding section of this article, theoretical problems with approximations (8) and (9) have motivated and continue to motivate researchers to seek more accurate activity duration distributions. In practice, activity durations are rarely independent of one another. As most researchers to date have assumed independence, this seems like a direction especially worth pursuing. In terms of project duration, recent work on bounding distributions (see Shogan [11]) demonstrates that PERT still poses interesting statistical questions.

The subject of controlling the project duration (say, by expediting certain activities at an appropriate cost) involves optimization* issues that, while reasonably understood in the deterministic case, still remain unexplored in stochastic networks (see Charnes et al. [34] and Jewell [35]). Future research may reveal important insights and connections between statistics and optimization.

References

GENERAL

[1] Malcolm, D., Roseboom, J., Clark, C., and Fazar, W. (1959). *Opns. Res.*, **7**, 646–669.

[2] Moder, J. and Phillips, C. (1970). *Project Management with CPM and PERT*, 2nd ed. Van Nostrand Reinhold, New York.

[3] Wiest, J. and Levy, F. (1977). *A Management Guide to PERT/CPM*, 2nd ed. Prentice-Hall, Englewood Cliffs, N.J.

CRITCAL PATH APPROXIMATIONS

[4] Clark, C. (1961). *Opns. Res.*, **9**, 145–162.

[5] Clingen, C. (1964). *Opns. Res.*, **12**, 629–632.

[6] Elmaghraby, S. (1967). *Manag. Sci.*, **13**, 299–306.

[7] Fulkerson, D. (1962). *Opns. Res.*, **10**, 808–817.

[8] Kleindorfer, G. (1971). *Opns. Res.*, **19**, 1586–1601.

[9] Lindsey, J. (1972). *Opns. Res.*, **20**, 800–812.

[10] Robillard, P. and Trahan, M. (1976). *Opns. Res.*, **24**, 177–182.

[11] Shogan, A. (1977). *Networks*, **7**, 359–381.

[12] Welsh, D. (1965). *Opns. Res.*, **13**, 141–143.

SIMULATION AND STATISTICAL COMPUTATIONS

[13] Burt, J. and Garman, M. (1971a). *INFOR*, **9**, 248–262.

[14] Burt, J. and Garman, M. (1971b). *Manag. Sci.*, **18**, 207–217.

[15] Garman, M. (1972). *Manag. Sci.*, **19**, 90–95.

[16] Hartley, H. O. and Wortham, A. (1966). *Manag. Sci.*, **12**, B469–481.

[17] Klingel, A. (1966). *Manag. Sci.*, **13**, B194–201.

[18] Martin, J. (1965). *Opns. Res.*, **13**, 46–66.

[19] Ringer, L. (1969). *Manag. Sci.*, **16**, B136–143.

[20] Ringer, L. (1971). *Manag. Sci.*, **17**, 717–723.

[21] Van Slyke, R. (1963). *Opns. Res.*, **11**, 839–860.

ESTIMATION OF ACTIVITY TIMES

[22]   Coon, H. (1965). *Opns. Res.*, **13**, 386–387.

[23]   Donaldson, W. (1965). *Opns. Res.*, **13**, 382–385.

[24]   Grubbs, F. (1962). *Opns. Res.*, **10**, 912–915.

[25]   King, W. and Lukas, P. (1973). *Manag. Sci.*, **19**, 1423–1432.

[26]   King, W. and Wilson, T. (1967). *Manag. Sci.*, **13**, 307–320.

[27]   King, W., Wittevrongel, D., and Hezel, K. (1967). *Manag. Sci.*, **14**, 79–84.

[28]   Kotiah, T. and Wallace, N. (1973). *Manag. Sci.*, **20**, 44–49.

[29]   MacCrimmon, K. and Ryavec, C. (1964). *Opns. Res.*, **12**, 16–37.

[30]   McBride, W. and McClelland, C. (1967). *IEEE Trans. Eng. Manag.*, **EM-14**, 166–169.

[31]   Moder, J. and Rodgers, E. (1968). *Manag. Sci.*, **15**, B76–83.

[32]   Perry, C. and Greig, I. (1975). *Manag. Sci.*, **21**, 1477–1480.

[33]   Swanson, L. and Pazar, H. (1971). *Decision Sci.*, **2**, 461–480.

RELATED RESEARCH

[34]   Charnes, A., Cooper, W., and Thompson, G. (1964). *Opns. Res.*, **12**, 460–470.

[35]   Jewell, W. (1965). *Manag. Sci.*, **11**, 438–443.

(NETWORK ANALYSIS
OPERATIONS RESEARCH)

A. A. ASSAD
B. L. GOLDEN

PESOTCHINSKY DESIGN *See* RESPONSE SURFACE DESIGNS

***P*-ESTIMATORS** *See* PITMAN ESTIMATORS

PETTY, SIR WILLIAM *See* BIOSTATISTICS

PHARMACEUTICAL INDUSTRY, STATISTICS IN

Pharmaceuticals are the medicines used in humans and animals for the treatment or prevention of disease. Many are "ethical" products in that throughout their manufacture and marketing standards are maintained consistent with the ethics of clinical and veterinary practice. In spite of the great variety of substances involved, both with regard to their origin and to the formulation in which they are presented, they possess two types of property in common. These are (a) the dependency of their effect upon the dose at which they are administered, and (b) the liability to induce harmful toxic effects at higher dose levels. Thus the dose-response relationship is of fundamental importance to the industry.

THE CLINICAL RESPONSE TO TREATMENT

When a treatment is prescribed it is with the intention of achieving one or more specific clinical objectives. If these be couched in purely qualitative terms (allowing time for the treatment to act), the outcome might be recorded as success or failure with respect to a particular objective. Replicated observations of this all-or-none type may provide a so-called quantal measure of response by noting the relative frequency of successes.

Commonly the clinical objectives will be capable of expression in quantitative terms and the degree of success achieved by the treatment may then be assessed by appropriate measurement. This may involve some physical or chemical determination, or it may be a subjective rating by the clinician or even self-rating by the patient.

Clinical experiments have shown that even a dummy treatment may at times induce effects simulating drug action—the "placebo" response.

If the medicine be given at too low a dose, either no response will be seen or it will be insufficient to meet the required criterion. Too high a dose may induce too profound an effect with its attendant dangers, or it may lead to other undesirable responses which reflect the toxicity of the substance. The various types of response which are associated with the complete dose range make up an "activity profile" for the preparation, from a knowledge of which it is possible to judge the margin of safety attaching to its use.

The Aims of Drug Research

Just as one drug may elicit several types of response, depending on its dosage, so is it that any required response may be produced by a number of quite distinct chemical compounds. Research effort is therefore deployed in attempts to discover substances which will produce the desired effect but with a greater freedom from side effects and a greater margin of safety. Another area for research is concerned with the duration of action of substances—in some cases to produce longer or, in others, shorter action. There will obviously also be a call to devise remedies for conditions where no satisfactory treatment at present exists.

Although the principles of drug design have been the subject of many recent publications, the medicinal chemist cannot predict the activity of a novel compound with anything approaching the level of assurance required for it to be tested in man. The only option which permits that step to be taken is to use experimental animals for preclinical testing.

Much preliminary work can be carried out in vitro on isolated tissues, their extracts, or on cultures of cells or parasites. Among the advantages of this approach is the humane aspect of reducing the numbers of conscious animals used.

Among mammalian species there is a broad similarity in the pattern of responses at different dose levels which provides the basis for preliminary studies in man. Unpredictable species differences occur, however, which call for exteme caution. Dosage is commonly adjusted on the basis of body weight, but in certain cases the two-thirds power of bodyweight appears preferable.

General theoretical arguments, based on the bioavailability of active constituent(s) of a drug, lead to the following formula for

response (R):

$$R = \frac{EQ \text{ dose}}{1 + Q \text{ dose}} \tag{1}$$

in which E is the maximum response or "efficacy," and Q is a potency factor. Since the equation does not indicate the direction of E relative to control, this must be specified by terms such as "stimulant" or "depressant." A similar, but more physiological approach, yields as the equation for the observed variable

$$V = \frac{C + LQ \text{ dose}}{1 + Q \text{ dose}} \tag{2}$$

where C is the control value, L is the asymptotic value, and Q the potency factor. In both equations the effect will be at the midpoint of its range when dose $= 1/Q$. This dose may conveniently be described as the $D50$ but care should be taken to distinguish it from the $ED50$ relating to quantal response* data.

Rearranging the terms of (1) and (2), taking logarithms, and inserting a slope constant b gives the logit* of effect:

$$Y = \ln\left(\frac{R}{E - R}\right) = \ln\left(\frac{V - C}{L - V}\right)$$
$$= \ln(Q) + b\ln(\text{dose}). \tag{3}$$

This is the equation of the Hill plot [2]. For data that conform to the simple model, the Hill plot will be linear with unit slope ($b = 1$).

For many practical purposes it is unnecessary to go beyond the simple linear relationship between the measurement of effect and the logarithm of dose. Over the full range of effect the regression* will be sigmoid, but between about 20 and 80% of the range of deviations from linearity can usually be ignored. In this usage, effect can be scaled in any way to suit the circumstances transformed as necessary to achieve homoscedasticity.

QUANTAL RESPONSES

The initial response to treatment will be in the form of quantitative changes in one or more of the response systems. A qualitative criterion of effect might be expected to correspond to some required threshold of response being achieved in these systems. Such a mechanistic approach to modelling the quantal response situation has not yet been fully developed and empirical statistical procedures dominate the scene.

In a very few instances it is possible to use drug infusions to measure directly the amount of drug required to produce a given effect. This would be defined as the threshold dose, or tolerance, for the effect. Tolerances so measured appear to be distributed either in a normal fashion or, more commonly, lognormally*. The quantal response, being the proportionate incidence of effect at a given dose, is therefore seen as corresponding to an estimate on the cumulative distribution of tolerances. Probit analysis permits the estimation of the parameters of the tolerance (or log-tolerance) distribution, but by convention the median rather than the mean is quoted. This is symbolized as the $ED50$.

The logit transformation of proportionate response, $\ln[p/(1-p)]$, provides a very different approach since it differs from the simple models for quantitative drug action solely by the slope parameter b, given in (3).

For practical purposes there is little to choose between the results of probit and logit analyses.

Results of titrations of infective materials such as bacteria or viruses may be treated as ordinary quantal responses, but Peto [3] advanced the theoretical relationship

$$\log S = -pn,$$

where S is the proportion of test subjects failing to develop symptoms, n is the dose in infective units, and p the probability of symptoms developing from each such unit.

DRUG INTERACTIONS

Various types of interaction between drugs can be identified. In the simplest case of additive effect, one drug will substitute for another when the two are given together, the substitution being in inverse proportion to their potencies. Drugs which act in the same mode, but with different efficacies, will display antagonism of one another over part or all of the response range. When they act in different modes, the result may either be one of noncompetitive antagonism or of potentiation.

The nature of the interaction will determine the form of analysis to be employed, but for a visual assessment the isobologram can be useful. An *isobole* is a contour line for a given level of effect drawn on a graph with axes corresponding to the scales of dose for the two drugs. The shape of the isobole, particularly with respect to its intercepts (if any), assists in the interpretation of the data [1].

Screening Drugs for Activity

Of the very large numbers of novel compounds synthesized, very few will qualify for eventual clinical study. The tests to which they are subjected are organized as a sequential screening process, some being designed to investigate the desirable properties of the compounds, others to assess side effects and toxicity. In the earlier stages of the screening, the tests are designed to be capable of a high throughput at minimal cost, usually with preset rules for the rejection of unsuitable compounds. Thereafter, as the numbers passing through diminish, so the nature of the testing becomes more detailed and specific. The emphasis shifts from merely detecting quantitative activity toward identifying the mechanism by which it is attained.

On the few compounds which pass through the whole screening process a large amount of information will have been assembled concerning their actions and acute toxicity. It is then a matter of economic necessity to select only one or two for the final stage of preclinical testing—the toxicity study. The selection process will often be extremely difficult since the advantages and disadvantages of each of the contenders must be considered from the multidimensional data, with the relative clinical importance of each action being taken into account.

The formal study of toxicity involves daily administration of the drug at two or more dose levels to groups of animals of both sexes drawn from two, three, or more species. At the higher levels the purpose is to identify the nature and incidence of pathological changes associated with near lethal dosage, while at the lowest level, chosen to represent a possible clinical dosage, it is to test for freedom from such changes. The duration of the study may range from one month to six months, one year or even longer, depending on the intended clinical use of the drug.

Even more sophisticated tests may be required to examine for embryo-toxicity or for mutagenic effects which might adversely influence reproduction.

BIOLOGICAL STANDARDIZATION (BIOASSAY*)

Chemical and physical analytical methods may be inadequate to estimate the biological activity of pharmaceutical products of natural origin. The standardization of such products must then be achieved by biological testing. Since any unit of activity based on response to a given dose will vary with the sensitivity of the animals employed, standard preparations are maintained against which samples from each batch of product may be directly compared. Potency is then quoted in standard units instead of animal units.

The statistical methods developed for the analysis of bioassays lend themselves to research applications where variation of animal sensitivity may also confuse the interpretation of various tests of activity. The place of the standard is taken by a reference compound, when possible selected from established drugs whose general profiles of activity are widely recognized. Since the substances to be compared in this type of test are not chemically identical, Finney proposed the term *comparative assay* to distinguish it from the analytical assay upon which the statistical methods were originally based.

STABILITY OF PRODUCTS

Many medicines, particularly the synthetic chemical compounds, remain substantially unchanged over long periods of storage provided extreme conditions are avoided. Others, however, will deteriorate on keeping and for them it is necessary to specify an expiry date together with the conditions of storage which will ensure that prior to that date the change of activity will be of little clinical significance. For therapeutic drugs a loss of about 10% may be tolerated, while for vaccines losses up to 50% may be allowed. To establish the conditions of storage and the shelf life will, in the first instance, require predictive testing. This is achieved by estimating the rate of loss at a series of elevated temperatures. The temperature coefficient of decay then allows, by extrapolation, the rates to be estimated for lower temperatures, which then become the basis for prediction.

Vaccines, and possibly other products of natural origin, may lose activity in a multistage process. This factor, combined with the larger errors of biological assay by which their potency must be tested, leads to data poorly represented by the simpler models of chemical decay. It is customary therefore to make retrospective checks between the predictions made and the potency found after prolonged storage of samples at lower temperatures. The shelf life of further batches of the same material can then be adjusted if necessary.

QUALITY CONTROL*

The pharmaceutical industry is not alone in seeking to ensure the quality of its products by tests carried out both during manufacture and on the finished products. In spite of the esoteric nature of many of the tests involved, the statistical problems which they present are common to many sampling procedures. These concern the establishment of rules governing the type of test, the size of a sample (often related to batch size), and of criteria leading to acceptance or rejection of the batch or to its retesting.

To avoid intolerable expense these rules and criteria have been established by the common consent of the industry and the statutory bodies charged with licensing its products, taking into account the accumulated data from past experience. If the precious element of trust which this implies is to be preserved, constant vigilance is required to guard against exceptional circumstances failing to be recognized.

EXPERIMENT DESIGN*

The importance of design derives from a desire (a) to maximize the information from a given set of resources, and (b) to minimize the costs of achieving requisite information; costs in terms of materials, labor, time and, where appropriate, animals.

The choice of design will often be determined by how much is known about the drug to be tested, how well understood is the method of testing, and on the nature of the test system. In clinical work crossover designs are commonly used, but obviously would be pointless if the effect of a treatment were to bring about a cure of the diseased state. In analytical assays, crossover designs, including Latin squares*, would often be the most economical. In preclinical studies the type of design is more determined by the test system. When intact animals are used, parallel groups are by far the most common. These will sometimes be controlled by a group receiving dummy treatment, sometimes by pretreatment measurements, and occasionally both. Isolated tissues and animals under anesthetic will commonly be subjected to sequential testing, although sometimes a form of crossover might be attempted.

For obvious reasons the incomplete block design* is least common in such work and is limited to those areas where the maximum of information must be extracted from a system constrained in some way.

NONPARAMETRIC METHODS

The traditional objective of drug research is to characterize the dose–response relationship for as many response systems as may be relevant. With so much emphasis on estimation, nonparametric methods have little to offer. They are welcomed, however, when experimental results do not lend themselves to valid analysis by parametric methods and there are numerous types of localized study in which evaluation is achieved by significance testing based on nonparametric techniques. These can be regarded as contributing qualitative statements to points in a broadly quantitative structure. The major problem is how to reconcile statistical significance with clinical importance in a highly multivariate system.

One must reserve a special place among the nonparametric methods for the double-dichotomy test (*see* TWO-BY-TWO TABLES). Its unpretentious form gives it a justifiable appeal to biologists and clinicians alike through all stages of drug investigation.

References

[1] De Jongh, S. E. (1961). In *Quantitative Methods in Pharmacology*, H. De Jonge, ed. North-Holland, Amsterdam.

[2] Hill, A. V. (1910). *J. Physiology*, **40** 4–7.

[3] Peto, S. (1953). *Biometrics*, **9**, 320–335.

Bibliography

Boyd, E. M. (1972). *Predictive Toxicometrics*. Scientechnica, Bristol, England.

Buncher, C. R. and Tsay, J. Y., eds. (1981). *Statistics in the Pharmaceutical Industry*. Dekker, New York. (Emphasis on background information with excellent reviews on selected topics. A few chapters deal with statistical methods, especially stability of drugs. Many references.)

Delaunois, A. L. ed., (1973). *International Encyclopedia of Pharmacology and Therapeutics*, Sec. 7, *Biostatistics in Pharmacology*. Pergamon, Oxford, England. 2 vols.

Finney, D. J. (1978). *Statistical Methods in Biological Assay*, 3rd ed. Griffin, London. (An invaluable vade mecum for workers in the field.)

O'Flaherty, E. J. (1981). *Toxicants and Drugs: Kinetics and Dynamics*. Wiley, New York. (Written for biologists, this book examines many of the quantitative relationships encountered in drug research with examples and references.)

Hewlett, P. S. and Plackett, R. L. (1979). *The Interpretation of Quantal Responses in Biology*. Edward Arnold, London.

Lamble, J. W., ed. (1981). *Towards Understanding Receptors*. Elsevier, Oxford, England. (A collection of published articles, selected to present a variety of modern concepts in molecular pharmacology.)

Laurence, D. R. and Black, J. W. (1978). *The Medicine You Take*. Croom Helm, London. (A sober assessment of contemporary medicine, written for the layman.)

Tallarida, R. J. and Jacob, L. S. (1979). *The Dose-Response Relation in Pharmacology*. Springer-Verlag, New York.

(BIOASSAY, STATISTICAL METHODS IN
CLINICAL TRIALS
DESIGN OF EXPERIMENTS
FOLLOW UP
QUALITY CONTROL, STATISTICAL
QUANTAL RESPONSE ANALYSIS
SURVIVAL ANALYSIS
TOLERANCE DISTRIBUTION)

P. A. YOUNG

PHASE FREQUENCY TEST   *See* WALLIS
AND MOORE'S PHASE FREQUENCY TEST

PHASE PROBLEM OF X-RAY CRYSTALLOGRAPHY, PROBABILISTIC METHODS IN THE

THE PHASE PROBLEM

The atomic arrangement in the unit cell of a crystal, i.e., the crystal structure, is determined once the intensities of a sufficient number of x-ray diffraction maxima have been measured. The number of these intensities usually exceeds by far the number of parameters required to describe the structure. From these intensities a set of num-

bers $|E_H|$ can be derived, one corresponding to each intensity. However the elucidation of the crystal structure requires also a knowledge of the complex numbers $E_H = |E_H|\exp(i\phi_H)$, of which only the magnitudes $|E_H|$ can be determined from experiment. Thus a "phase" ϕ_H must be assigned to each $|E_H|$, and the problem of determining the phases when only the magnitudes $|E_H|$ are known is called "the phase problem." Owing to the known atomicity of crystal structures and the redundancy of observed magnitudes $|E_H|$, the phase problem is solvable in principle.

It is the redundancy of the system of equations relating the magnitudes $|E|$ with the desired phases ϕ, as well as errors in the observed $|E|$'s, which makes possible, even indispensable, the use of probabilistic techniques in the solution of the phase problem.

THE NORMALIZED STRUCTURE FACTORS E

The relationship between the (complex) normalized structure factors E and the crystal structure is given by the pair of equations

$$E_H = |E_H|\exp(i\phi_H)$$

$$= \frac{1}{\sigma_2^{1/2}} \sum_{j=1}^{N} Z_j\exp(2\pi iH \cdot r_j), \quad (1)$$

$$\langle E_H\exp(-2\pi iH \cdot r)\rangle_H \left. \begin{array}{ll} = \dfrac{Z_j}{\sigma_2^{1/2}} & \text{if } r = r_j \\[2mm] = 0 & \text{if } r \neq r_j \end{array} \right\}$$

$$(2)$$

where H is an arbitrary reciprocal lattice vector, Z_j is the atomic number and r_j is the position vector of the atom labeled j, N is the number of atoms in the unit cell, and

$$\sigma_2 = \sum_{j=1}^{N} Z_j^2. \quad (3)$$

Clearly (2) shows that the crystal structure (i.e., the position vectors r_j) is determined in terms of the normalized structure factors E_H. However, it turns out that although the magnitudes $|E_\mathbf{H}|$ may be determined, at least approximately, from experiment, the phases $\phi_\mathbf{H}$, which are also needed if the crystal structure is to be found via (2), cannot be determined experimentally. Nevertheless, because the number of equations (1) usually exceeds by far the number of unknowns $\mathbf{r}_j$, the available data, i.e., the known $|E_\mathbf{H}|$'s, are in general more than sufficient to determine crystal structures uniquely. In fact, from (1) one naturally formulates the problem as the determination of the N position vectors $\mathbf{r}_j$ which minimize the weighted sum of squares

$$\sum_\mathbf{H} W_j \left[|E_\mathbf{H}| - \frac{1}{\sigma_2^{1/2}} \left| \sum_{j=1}^N Z_j \exp(2\pi i \mathbf{H} \cdot \mathbf{r}_j) \right| \right]^2,$$

(4)

in which the sum is taken over all reciprocal lattice vectors $\mathbf{H}$ for which magnitudes $|E_\mathbf{H}|$ are available, and the $W_\mathbf{H}$'s are a suitably chosen set of weights. This formulation clearly calls for a probabilistic approach. In practice the problem of finding the global minimum of (4) is too intractable to be solved *ab initio*. Instead the unknown position vectors $\mathbf{r}_j$ are eliminated from the system (1) to yield relationships among the $E_\mathbf{H}$'s having probabilistic validity. These in turn lead to approximate values of the unknown phases $\phi_\mathbf{H}$ which can then be used in (2) to determine a trial structure, i.e., approximate values of the unknowns $\mathbf{r}_j$. Employing standard iterative techniques (4), or something similar, is then used to obtain refined values for the $\mathbf{r}_j$. The techniques which employ the phases to determine crystal structures are known as direct methods, since the phases $\phi_\mathbf{H}$ are determined directly from the observed magnitudes $|E_\mathbf{H}|$ [rather than from a presumed known structure via (1)].

THE STRUCTURE INVARIANTS

Equation (2) implies that the normalized structure factors $E_\mathbf{H}$ determine the crystal structure. However (1) does not imply that, conversely, the crystal structure determines the values of the normalized structure fac-

tors $E_\mathbf{H}$ since the position vectors $\mathbf{r}_j$ depend not only on the structure but on the choice of origin as well. It turns out nevertheless that the magnitudes $|E_\mathbf{H}|$ of the normalized structure factors are in fact uniquely determined by the crystal structure and are independent of the choice of origin, but that the values of the phases $\phi_\mathbf{H}$ depend also on the choice of origin. Although the values of the individual phases depend on the structure and the choice of origin, there exist certain linear combinations of the phases, the so-called *structure invariants*, whose values are determined by the structure alone and are independent of the choice of origin.

The most important structure invariants are the linear combinations of three phases (triplets):

$$\phi_\mathbf{H} + \phi_\mathbf{K} + \phi_\mathbf{L},$$ (5)

where

$$\mathbf{H} + \mathbf{K} + \mathbf{L} = 0;$$ (6)

and the linear combination of four phases (quartets):

$$\phi_\mathbf{H} + \phi_\mathbf{K} + \phi_\mathbf{L} + \phi_\mathbf{M},$$ (7)

where

$$\mathbf{H} + \mathbf{K} + \mathbf{L} + \mathbf{M} = 0;$$ (8)

etc.

THE FUNDAMENTAL PRINCIPLE OF DIRECT METHODS

Two structures related by reflection through a point are said to be *enantiomorphs* of each other. The x-ray diffraction experiment is not capable of distinguishing the enantiomorphs when they are distinct. For this reason the fundamental principle of direct methods is formulated as follows. For fixed enantiomorph the observed magnitudes $|E|$ determine, in general, unique values for all the structure invariants. The latter, as certain well defined linear combinations of the phases, lead in turn to unique values for the phases ϕ. In short, the structure invariants serve to link the observed magnitudes $|E|$ with the desired phases ϕ (the fundamental principle of direct methods).

THE NEIGHBORHOOD PRINCIPLE

It has been seen that for fixed enantiomorph the values of the observed magnitudes $|E|$ determine the values of all the structure invariants. A major recent insight is that, for fixed enantiomorph, the value of any structure invariant T is primarily determined, in favorable cases, by the values of one or more small sets of magnitudes $|E|$, the neighborhoods of T, and is relatively insensitive to the values of the great bulk of remaining magnitudes $|E|$ (the neighborhood principle). The conditional probability distribution of T, assuming as known the magnitudes $|E|$ in any of its neighborhoods, yields an estimate for T which is particularly good in the favorable case that the variance of the distribution happens to be small.

THE NEIGHBORHOODS OF THE STRUCTURE INVARIANTS

The first neighborhood of the triplet (5), where (6) holds, consists of the three magnitudes

tudes

$$|E_\mathbf{H}|, |E_\mathbf{K}|, |E_\mathbf{L}|. \tag{9}$$

The first neighborhood of the quartet (7), where (8) holds, consists of the four magnitudes

$$|E_\mathbf{H}|, |E_\mathbf{K}|, |E_\mathbf{L}|, |E_\mathbf{M}|. \tag{10}$$

The second neighborhood of the quartet consists of the four magnitudes (10) in the first neighborhood plus the three additional magnitudes

$$|E_{\mathbf{H}+\mathbf{K}}|, |E_{\mathbf{K}+\mathbf{L}}|, |E_{\mathbf{L}+\mathbf{H}}|, \tag{11}$$

i.e., seven magnitudes $|E|$ in all.

CONDITIONAL PROBABILITY DISTRIBUTIONS OF THE STRUCTURE INVARIANTS

Triplets

Suppose that a crystal structure consisting of N atoms per unit cell is fixed. Denote by W the collection of all reciprocal lattice vectors $\mathbf{H}$ and by $\phi_\mathbf{H}$ the phase of the normalized

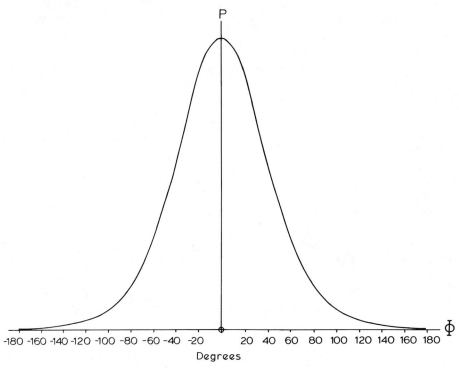

Figure 1 The distribution $P_{1|3}$, (16), for $A = 2.316$.

structure factor $E_{\mathbf{H}}$. Assume also that R_1, R_2, and R_3 are fixed nonnegative numbers. Suppose finally that the primitive random variable (vector) is the ordered triple $(\mathbf{H}, \mathbf{K}, \mathbf{L})$ of reciprocal vectors $\mathbf{H}, \mathbf{K}, \mathbf{L}$ which is assumed to be uniformly distributed over the subset of the threefold Cartesian product $W \times W \times W$ defined by

$$|E_{\mathbf{H}}| = R_1, \qquad |E_{\mathbf{K}}| = R_2, \qquad |E_{\mathbf{L}}| = R_3 \tag{12}$$

and

$$\mathbf{H} + \mathbf{K} + \mathbf{L} = \mathbf{0}. \tag{13}$$

Then the structure invariant

$$\phi_3 = \phi_{\mathbf{H}} + \phi_{\mathbf{K}} + \phi_{\mathbf{L}} \tag{14}$$

is a function of the primitive random variables $\mathbf{H}, \mathbf{K}, \mathbf{L}$ and therefore is itself a random variable. Denote by

$$P_{1|3} = P(\Phi \mid R_1, R_2, R_3) \tag{15}$$

the conditional probability distribution of ϕ_3, given the three magnitudes (12), the first neighborhood of ϕ_3. Then [1, 3]

$$P_{1|3} \approx \frac{1}{2\pi I_0(A)} \exp(A \cos \Phi), \tag{16}$$

where I_0 is the modified Bessel function*, A is defined by

$$A = \frac{2\sigma_3}{\sigma_2^{3/2}} R_1 R_2 R_3, \tag{17}$$

and

$$\sigma_n = \sum_{j=1}^{N} Z_j^n. \tag{18}$$

Graphs of the distribution (16) for $A = 2.316$ and $A = 0.731$ are shown in Figs. 1 and 2. Clearly this distribution has a unique maximum at $\Phi = 0$ in the interval $(-\pi, \pi)$ so that the most probable value of ϕ_3 is zero. The larger the value of A the smaller is the variance of the distribution and the more reliable is the estimate of ϕ_3, zero in this case.

Quartets

As before, suppose that a crystal structure consisting of N atoms per unit cell is fixed. Two distributions will be described. The first, in strict analogy with the preceding section ("Triplets"), is the conditional proba-

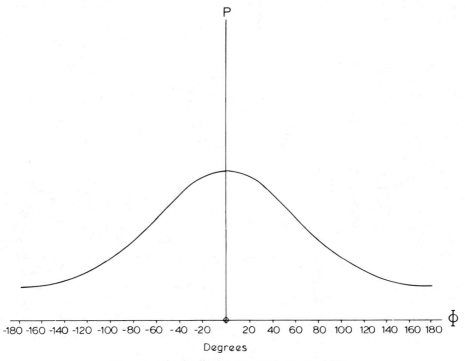

Figure 2   The distribution $P_{1|3}$, (16), for $A = 0.731$.

bility distribution of the quartet, given the four magnitudes in its first neighborhood; the second is the conditional probability distribution of the quartet, assuming that the seven magnitudes in its second neighborhood are known.

THE FOUR-MAGNITUDE DISTRIBUTION

Assume that R_1, R_2, R_3, and R_4 are fixed nonnegative numbers. Next suppose that the primitive random variable (vector) is the ordered quadruple $(\mathbf{H}, \mathbf{K}, \mathbf{L}, \mathbf{M})$ of reciprocal vectors $\mathbf{H}, \mathbf{K}, \mathbf{L}, \mathbf{M}$ which is assumed to be uniformly distributed over the subset of the fourfold Cartesian product $W \times W \times W \times W$ defined by

$$|E_{\mathbf{H}}| = R_1, \qquad |E_{\mathbf{K}}| = R_2,$$
$$|E_{\mathbf{L}}| = R_3, \qquad |E_{\mathbf{M}}| = R_4 \qquad (19)$$

and

$$\mathbf{H} + \mathbf{K} + \mathbf{L} + \mathbf{M} = 0. \qquad (20)$$

In view of (20), the linear function of four phases

$$\phi_4 = \phi_{\mathbf{H}} + \phi_{\mathbf{K}} + \phi_{\mathbf{L}} + \phi_{\mathbf{M}} \qquad (21)$$

is a structure invariant which, as a function of the primitive random variables $\mathbf{H}, \mathbf{K}, \mathbf{L}$, $\mathbf{M}$, is itself a random variable. Denote by

$$P_{1|4} = P(\Phi \,|\, R_1, R_2, R_3, R_4) \qquad (22)$$

the conditional probability distribution of ϕ_4, given the four magnitudes in its first neighborhood (19). Then [3]

$$P_{1|4} \approx \frac{1}{2\pi i_0(B)} \exp(B \cos \Phi), \qquad (23)$$

where B is defined by

$$B = \frac{2\sigma_4}{\sigma_2^2} R_1 R_2 R_3 R_4 \qquad (24)$$

and σ_n by (18). Thus $P_{1|4}$ is identical with $P_{1|3}$, but B replaces A. Hence similar remarks apply to $P_{1|4}$. In particular (23) always has a unique maximum at $\Phi = 0$, so that the most probable value of the structure invariant (21) is zero, and the larger the value of B the more likely it is that $\phi_4 \approx 0$. Since B values, of order $1/N$, tend to be less than A values, or order $1/N^{1/2}$, the estimate

(zero) of the quartet (21) is in general less reliable than the estimate (zero) of the triplet (14). Hence (23) is no improvement over (16) and the goal of obtaining a reliable nonzero estimate for a structure invariant is not realized by (23). The decisive step in this direction is made in the next section.

THE SEVEN-MAGNITUDE DISTRIBUTION

If one assumes as known not only the four magnitudes (19), but the additional three magnitudes $|E_{\mathbf{H}+\mathbf{K}}|$, $|E_{\mathbf{K}+\mathbf{L}}|$, and $|E_{\mathbf{L}+\mathbf{H}}|$ then, in favorable cases, one obtains a more reliable estimate for the quartet (21), and, furthermore, the estimate may lie anywhere in the interval 0 to π.

Assume that the seven nonnegative numbers $R_1, R_2, R_3, R_4, R_{12}, R_{23}, R_{31}$ are fixed. Suppose next that the ordered quadruple of reciprocal vectors $(\mathbf{H}, \mathbf{K}, \mathbf{L}, \mathbf{M})$ is a random variable which is uniformly distributed over the subset of the fourfold Cartesian product $W \times W \times W \times W$ defined by

$$|E_{\mathbf{H}}| = R_1, \qquad |E_{\mathbf{K}}| = R_2,$$
$$|E_{\mathbf{L}}| = R_3, \qquad |E_{\mathbf{M}}| = R_4; \qquad (25)$$
$$|E_{\mathbf{H}+\mathbf{K}}| = R_{12}, \qquad |E_{\mathbf{K}+\mathbf{L}}| = R_{23},$$
$$|E_{\mathbf{L}+\mathbf{H}}| = R_{31}; \qquad (26)$$

and

$$\mathbf{H} + \mathbf{K} + \mathbf{L} + \mathbf{M} = 0. \qquad (27)$$

Then the quartet (21) is a structure invariant which, as a function of the primitive random variable $(\mathbf{H}, \mathbf{K}, \mathbf{L}, \mathbf{M})$ is itself a random variable. Denote by

$$P_{1|7} = P(\Phi \,|\, R_1, R_2, R_3, R_4, R_{12}, R_{23}, R_{31})$$
$$(28)$$

the conditional probability distribution of the quartet (21), given the seven magnitudes in its second neighborhood, (25) and (26). The explicit formula for $P_{1|7}$ has been found [2, 3].

Figures 3 to 5 show the distribution (28) (solid line ——) for typical values of the seven parameters (25) and (26). For comparison the distribution (23) (broken line ———)) is also shown. Since the magnitudes $|E|$ have

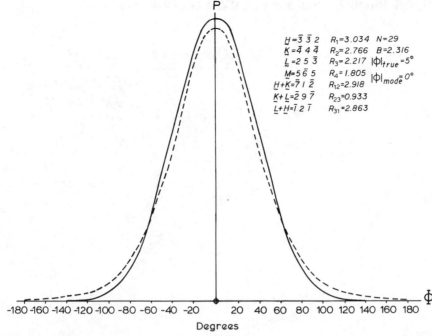

Figure 3 The distributions (28) (——) and (23) (– – –) for the values of the seven parameters (25) and (26) shown. The mode of (28) is 0, of (23) always 0.

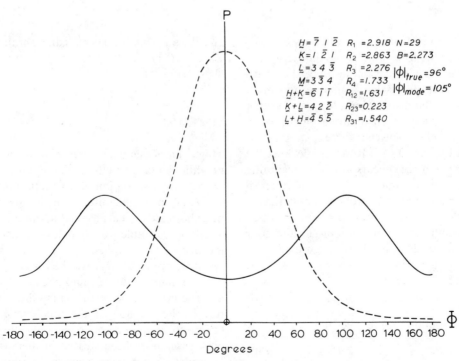

Figure 4 The distributions (28) (——) and (23) (– – –) for the values of the seven parameters (25) and (26) shown. The mode of (28) is 105°, of (23) always 0.

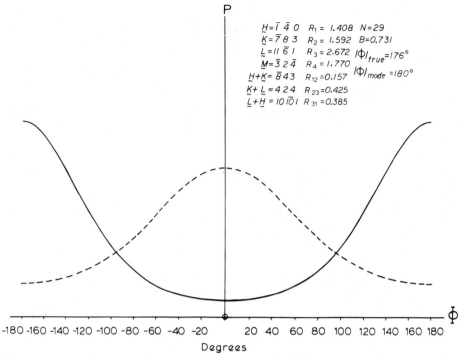

Figure 5 The distributions (28) (——) and (23) (– – –) for the values of the seven parameters (25) and (26) shown. The mode of (28) is 180°, of (23) always 0.

been obtained from a real structure, comparison with the true value of the quartet is also possible. As already emphasized, the distribution (23) always has a unique maximum at $\Phi = 0$. The distribution (28), on the other hand, may have a maximum at $\Phi = 0$, or π, or any value between these extremes, as shown by Figs. 3 to 5. Roughly speaking the maximum of (28) occurs at 0 or π according as the three parameters R_{12}, R_{23}, R_{31} are all large or all small, respectively. These figures also clearly show the improvement which may result when, in addition to the four magnitudes (25), the three magnitudes (26) are also assumed to be known. Finally, in the special case that

$$R_{12} \approx R_{23} \approx R_{31} \approx 0 \qquad (29)$$

the distribution (28) reduces to

$$P_{1|7} \approx \frac{1}{L} \exp(-2B'\cos\Phi), \qquad (30)$$

where

$$B' = \frac{1}{\sigma_2^3}\left(3\sigma_3^2 - \sigma_2\sigma_4\right)R_1 R_2 R_3 R_4, \qquad (31)$$

and L is a suitable normalizing parameter. Clearly (31) has a unique maximum at $\Phi = \pi$ (Fig. 5).

SUMMARY

Major emphasis has been placed on the neighborhood principle. The conditional probability distribution of a structure invariant T, given the magnitudes $|E|$ in any of its neighborhoods, yields a reliable estimate for T in the favorable case that the variance of the distribution happens to be small. Since the structure invariants are the essential link between magnitudes $|E|$ and phases ϕ, probabilistic methods are seen to play the central role in the solution of the phase problem. Owing to limitations of space, only the simplest cases have been treated in this article. However a much larger class of distributions is presently available. Not only have these distributions already proven to be of great value in the applications, particularly for complex structures when data sets have been limited in number and quality, but preliminary calculations strongly suggest that the distributions of the higher order structure invariants, particularly if one takes into account whatever elements of symmetry may be present, will play a vital role in future applications.

References

[1] Cochran, W. (1955). *Acta Crystallogr.*, **8**, 473–478.

[2] Hauptman, H. (1975). *Acta Crystallogr.*, **A31**, 680–687.

[3] Hauptman, H. (1976). *Acta Crystallogr.*, **A32**, 877–882.

Acknowledgements

This work was supported by Grant No. CHE-8203930 from the National Science Foundation and a grant from the James H. Cummings Foundation, Inc.

(PERIODOGRAM ANALYSIS
STATISTICS IN CRYSTALLOGRAPHY
TIME-SERIES)

H. HAUPTMAN

PHASE TYPE DISTRIBUTIONS

A nonnegative random variable T (or its distribution function) is said to be of phase type (PH) if T is the time until absorption in a finite-state continuous-time Markov chain. PH distributions can be thought of as a generalization of Erlang* distributions, and they are useful because of their versatility and the relative ease of their numerical implementations. An excellent review of PH distributions can be found in Neuts [21, Chap. 2] where some of their history is also given. Various applications of PH distributions in probability and statistics (e.g., queueing* theory, reliability* theory, and branching processes*) have been found recently.

The class of the univariate PH distributions and its properties will be described in the following section. Most of these results and their proof can be found in Neuts [21]

and in references there. Most of the references which are mentioned in this article are recent papers which were written or published after the book of Neuts [21] appeared. Other references can be found in that book.

The following sections consist of a survey of PH-renewal processes, a recent study dealing with multivariate PH distributions, and some applications of the PH distributions in stochastic models.

UNIVARIATE PH DISTRIBUTIONS

Consider an absorbing continuous-time Markov chain $\{X(t), t \geqslant 0\}$ with state space $\{1, 2, \ldots, m, \Delta\}$. We will assume that $X(t)$ is right-continuous. Assume that states 1, 2, ..., m are transient and that state Δ is absorbing. Thus, the infinitesimal generator, representing the rates of transitions among the states of this process, is of the form

$$\begin{bmatrix} \mathbf{A} & -\mathbf{Ae} \\ \mathbf{0} & 0 \end{bmatrix}$$

where $\mathbf{A} = \{a_{ij}\}$ is an $m \times m$ matrix with negative diagonal elements and nonnegative off-diagonal elements and $\mathbf{Ae} \leqslant \mathbf{0}$. Here $\mathbf{e}$ denotes the m-dimensional vector $(1, 1, \ldots, 1)$ and $\mathbf{0}$ denotes the m-dimensional vector of zeros. In the sequel it will usually be possible to determine the dimensions of $\mathbf{e}$, $\mathbf{0}$ and of other vectors and matrices from the equation in which they appear. Also it will be easy to see whether a given vector is a row or a column vector, although the same notation will be used in both cases.

The structure of $\mathbf{A}$ can be expressed as

$$\begin{aligned} a_{ij} &= -\lambda_i && \text{if } i = j \\ &= \lambda_i p_{ij} && \text{if } i \neq j \end{aligned}$$

where $\lambda_i > 0$ is the parameter of the exponential holding time in state i and p_{ij} is the transition probability from state i to state j. Absorption into Δ from any initial state is certain if and only if $\mathbf{A}$ is nonsingular.

Let $(\boldsymbol{\alpha}, \alpha_\Delta)$ be an initial probability vector [i.e., $\alpha_i = P(X(0) = i)$, $i = 1, \ldots, m$, and $\alpha_\Delta = P(X(0) = \Delta)$] and let $T = \inf\{t : X(t) = \Delta\}$ denote the time until absorption. Then the distribution function of T is

$$F(t) = 1 - \alpha \exp\{At\}e, \qquad t \geq 0. \quad (1)$$

Note that if $\alpha_\Delta > 0$, then F has an atom at 0 $\{P(T = 0) = \alpha_\Delta\}$ and is absolutely continuous on $(0, \infty)$. F can be computed by

$$F(t) = 1 - \nu(t)e, \qquad t \geq 0, \quad (2)$$

where $\nu(t)$ is the solution to the system of differential equations

$$\nu'(t) = \nu(t)A, \qquad t \geq 0, \quad (3)$$

with the initial condition $\nu(0) = \alpha$ (see, e.g., Neuts [21]). Alternatively one can choose a finite $\lambda \geq \max_{1 \leq i \leq m}\{\lambda_i\}$ and compute F by

$$F(t) = \sum_{n=0}^{\infty}\left(\sum_{r=0}^{n} g(r)\right)\frac{e^{-\lambda t}(\lambda t)^n}{n!}, \qquad t \geq 0, \quad (4)$$

where $g(0) = \alpha_\Delta$, $g(n) = \alpha R^{n-1}r_0$, $r_0 = -Ae/\lambda$ and the elements of the $m \times m$ matrix R are

$$r_{ij} = a_{ij}/\lambda \qquad \text{if} \quad i \neq j$$
$$= 1 - \lambda_i/\lambda \qquad \text{if} \quad i = j$$

(see, e.g., Shanthikumar [30]).

For simplicity we assume henceforth that $\alpha_\Delta = 0$. This prevents the process from starting at the absorbing state Δ, thus guaranteeing that $P(T > 0) = 1$ and simplifying some of the formulas below (for more detailed formulas see Neuts [21]).

A random variable with distribution function F of the form (1) is said to be *phase type* (PH) with representation (α, A). Usually a PH random variable has more than one representation.

The survival function $\bar{F}$ of T, its density f and its Laplace transform

$$\phi(s) = E[\exp\{-sT\}]$$

are

$$\bar{F}(t) = \alpha \exp\{At\}e, \qquad t \geq 0,$$
$$f(t) = -\alpha \exp\{At\}Ae, \qquad t \geq 0,$$
$$\phi(s) = \alpha[sI - A]^{-1}e, \qquad s \geq 0.$$

The Laplace transform ϕ is rational and it is actually defined also on some interval of the

form $(\sigma, 0)$ for some $\sigma < 0$ (see, e.g., Takahashi [32, p. 622]). Neuts [23] showed that if the representation (α, A) is irreducible then $-\sigma$ is equal to the eigenvalue of maximum real part of the matrix A. It follows that the moments of T are

$$ET^k = (-1)^k \alpha A^{-k}e, \qquad k = 0, 1, \ldots$$

Example 1. The Erlang distribution (i.e, the sum of i.i.d. exponential* random variables) with rate λ is PH with representation $\alpha = (1, 0, \ldots, 0)$ and

$$A = \begin{bmatrix} -\lambda & \lambda & 0 & \cdots & \cdots & \cdots & 0 \\ 0 & -\lambda & \lambda & 0 & \cdots & \cdots & 0 \\ \vdots & & & & & & \vdots \\ 0 & \cdots & \cdots & \cdots & 0 & -\lambda & \lambda \\ 0 & \cdots & \cdots & \cdots & \cdots & 0 & -\lambda \end{bmatrix}.$$

Example 2. The mixture of m exponential* distributions (i.e., hyperexponential) $F(t) = \sum_{i=1}^{m} p_i(1 - \exp\{-\lambda_i t\})$, $\lambda_i > 0$, $p_i \geq 0$, $i = 1, \ldots, m$, $\sum_{i=1}^{m} p_i = 1$, is PH and has the representation $\alpha = (p_1, \ldots, p_m)$, $A = -\text{diag}(\lambda_1, \ldots, \lambda_m)$.

Example 3. The generalized Erlang distribution (i.e., the sum of independent, but not necessarily identically distributed, exponential* random variables) is PH and has the representation $\alpha = (1, 0, \ldots, 0)$ and

$$A = \begin{bmatrix} -\lambda_1 & \lambda_1 & 0 & \cdots & & \cdots & 0 \\ 0 & -\lambda_2 & \lambda_2 & \cdots & & \cdots & 0 \\ \vdots & & & & & & \vdots \\ 0 & \cdots & \cdots & \cdots & -\lambda_{m-1} & \lambda_{m-1} \\ 0 & \cdots & \cdots & \cdots & 0 & -\lambda_m \end{bmatrix}$$

That these three examples are PH follows also from the fact that every exponential* random variable is PH, combined with some of the following closure properties.

Property 1. A sum of n independent PH random variables is PH.

In particular, if X and Y are independent PH random variables with representations

$(\alpha, \mathbf{A})$ and $(\beta, \mathbf{B})$ then $X + Y$ is a PH random variable with representation $(\alpha, 0)$ and

$$\begin{bmatrix} \mathbf{A} & -\mathbf{A}\mathbf{e}\beta \\ \mathbf{0} & \mathbf{B} \end{bmatrix},$$

where $\mathbf{e}\beta$ is the matrix with ijth element being β_j.

Property 2. A finite mixture, $F = \sum_{i=1}^{k} p_i F_i$, of PH distributions is PH. If F_i has the representation $\alpha^{(i)}, \mathbf{A}^{(i)}$, $i = 1, \ldots, k$, then F has the representation $(p_1 \alpha^{(1)}, \ldots, p_k \alpha^{(k)})$ and

$$\begin{bmatrix} \mathbf{A}^{(1)} & \mathbf{0} & \cdots & \cdots & \mathbf{0} \\ \mathbf{0} & \mathbf{A}^{(2)} & \mathbf{0} & & \mathbf{0} \\ \vdots & & & & \vdots \\ \mathbf{0} & \mathbf{0} & \cdots & \mathbf{0} & \mathbf{A}^{(k)} \end{bmatrix}.$$

Neuts [21] noted that if X and Y are independent PH random variables then $\min(X, Y)$ and $\max(X, Y)$ are PH. A more general result has been obtained by Assaf and Levikson [3].

Property 3. If $X_1, \ldots, X_k$ are independent nonnegative PH random variables then for every coherent* life function τ of order k, $Y = \tau(X_1, \ldots, X_k)$ is PH.

Here the representation of Y depends on τ. The representations of parallel and series systems of order 2 are given in Neuts [21, pp. 60–61].

Property 4. The class of PH distributions is (weakly) dense in the set of all distributions on $[0, \infty)$.

The reader should be warned, however, that the value of the approximation of Property 4 is limited. See Neuts [21, pp. 78–79] for a general discussion. Further it should be pointed out that since the Laplace transforms of PH distributions are rational, probability densities like

$$K e^{-\lambda t} [1 - \cos \mu t] \text{ and}$$

$$\sum_{n=0}^{\infty} \frac{e^{-\mu} \mu^{n+1}}{(n+1)!} \times \frac{e^{-\lambda t} (\lambda t)^n}{n!} \lambda$$

cannot be represented in the PH form.

So far we have discussed the continuous (except perhaps at 0) PH distributions on $[0, \infty)$. Similar definition and properties apply also to discrete distributions on the nonnegative integers [16]. We will mention here two closure results which involve both types of PH distributions.

Property 5. Let $X_1, X_2, \ldots$ be i.i.d. PH random variables and let N be a discrete PH random variable independent of $X_1, X_2, \ldots$. Then $Y = X_1 + X_2 + \cdots + X_N$ has a PH distribution.

Property 6. A mixture* of Poisson distributions with a continuous PH mixing distribution is a discrete PH distribution.

The representations of the resulting PH distributions in Properties 5 and 6 are given in Neuts [21, p. 54 and 59].

The class of PH distributions is convex. Assaf and Langberg [2] showed that this class does not have extreme points. Shanthikumar [30] and Syski [31] discuss some extensions of the PH class to those with infinite state space.

RENEWAL PROCESSES OF PHASE TYPE

In this section we consider renewal* processes with phase type interrenewal times. Let F be a PH distribution with representation $\alpha = (\alpha_1, \ldots \alpha_m)$ and an $m \times m$ generator $\mathbf{A}$ associated with the absorbing finite-state Markov chain $\{X(t), t \geqslant 0\}$.

If F is the common distribution of the independent interrenewal times then the renewal process is actually a Markov process* $\{Y(t), t \geqslant 0\}$ as follows: The transient states of X are m of the $m + 1$ states of Y. The state Δ is now an instantaneous state at which the process immediately jumps to state i with probability α_i, $i = 1, \ldots, m$.

The generator of this process is $\mathbf{A}^* = \mathbf{A} - \mathbf{Ae\alpha}$. The renewal density is $\theta(t) = -\alpha \exp\{\mathbf{A}^*t\}\mathbf{Ae}$ and, of course, $\lim_{t \to \infty} \theta(t) = \mu^{-1}$ where μ is the expected value of F. The limiting distribution of the age, $F^*(t) = \mu^{-1}\int_0^t(1 - F(u))\,du$, is also a PH distribution. It has the representation $\boldsymbol{\pi} = (\pi_1, \ldots, \pi_m)$ and $\mathbf{A}$, where $\boldsymbol{\pi}$ is the unique probability vector satisfying $\boldsymbol{\pi}\mathbf{A}^* = 0$ and $\boldsymbol{\pi}\mathbf{e} = 1$, i.e., $\boldsymbol{\pi}$ is the stationary probability vector of $\mathbf{A}^*$. The expected number of renewals $H(t)$ in $(0, t)$ is given by $\mu^{-1}t + 2^{-1}\mu^{-2}(\sigma^2 + \mu^2) + \mu^{-1}\boldsymbol{v}(t)\mathbf{A}^{-1}\mathbf{e}$ where σ^2 is the variance of F, $\boldsymbol{v}(t) = \alpha(\Pi - \exp\{\mathbf{A}^*t\})$ and $\Pi = \text{diag}(\pi_1, \ldots, \pi_m)$. $H(t)$ can be evaluated easily since $\boldsymbol{v}(t)$ is the solution of the system of differential equations

$$\boldsymbol{v}'(t) = \boldsymbol{v}(t)\mathbf{A}^*$$

with the boundary condition

$$\boldsymbol{v}(0) = \boldsymbol{\pi} - \boldsymbol{\alpha}.$$

Further results and expressions for other probabilistic quantities of interest can be found in Neuts [17; 21, pp. 63–70].

Neuts [18] and Latouche [13] studied related Markovian and semi-Markovian point processes. Neuts [21, pp. 256 and 262] and Baxter [5] discussed an alternating renewal process of phase type in the setting of queueing* theory and reliability* theory, respectively.

MULTIVARIATE PH DISTRIBUTIONS

Motivated by applications in reliability theory, Assaf et al. [4] have introduced some classes of multivariate PH distributions. Here we will describe only the most important one. As in the univariate case, the advantage of a multivariate PH distribution is that it can be written in a closed form and thus various probabilistic quantities of interest and various reliability criteria can be evaluated with a relative ease. Furthermore

multivariate PH distributions are useful in modeling real situations in which multivariate distributions with support in $[0, \infty)^n$ are involved. Only the bivariate case will be discussed in some detail.

As in the univariate case, we consider a continuous time Markov chain $\{X(t), t \geqslant 0\}$ on a finite state space E. Let Γ_1 and Γ_2 be two stochastically closed nonempty subsets of E such that $\Gamma_1 \cap \Gamma_2$ is a proper subset of E ($\Gamma \subset E$ is said to be *stochastically closed* if once X enters Γ, it never leaves). We assume that absorption into $\Gamma_1 \cap \Gamma_2$ is certain. Without loss of generality $\Gamma_1 \cap \Gamma_2$ consists of one state denoted by Δ. Thus without loss of generality $E = \{1, 2, \ldots, m, \Delta\}$ for some $m \geqslant 1$.

Let $\boldsymbol{\beta}$ be an initial probability vector on E. Define $T_k = \inf\{t : X(t) \in \Gamma_k\}$, $k = 1, 2$. We will call the joint distribution of T_1 and T_2 a *bivariate phase type* (BPH) distribution and (T_1, T_2) will be called a BPH random vector. To simplify the following discussion assume that $\boldsymbol{\beta}$ puts mass 0 at Δ. Write $\boldsymbol{\beta} = (\boldsymbol{\alpha}, 0)$ where $\boldsymbol{\alpha} = (\alpha_1, \ldots, \alpha_m)$. Furthermore to ensure $P(T_1 > 0, T_2 > 0) = 1$ we shall also assume that $\alpha_i \neq 0$ only if $i \in \Gamma_1^c \cap \Gamma_2^c$. Denote the infinitesimal generator by

$$\begin{bmatrix} \mathbf{A} & -\mathbf{Ae} \\ \mathbf{0} & 0 \end{bmatrix}.$$

The Marshall and Olkin* [14] exponential distribution, the Freund [10] distribution, and the Becker and Roux [6] extension of the gamma* distribution are all BPH distributions.

Assaf et al. [4] obtained explicit expressions for various probabilistic quantities of interest. The joint survival function of T_1 and T_2, for $t_1 \geqslant t_2 \geqslant 0$, is

$$\bar{F}(t_1, t_2)$$
$$= P(T_1 > t_1, T_2 > t_2)$$
$$= \boldsymbol{\alpha} \exp\{\mathbf{A}t_2\}\mathbf{g}_2\exp\{\mathbf{A}(t_1 - t_2)\}\mathbf{g}_1\mathbf{e},$$

where, for $k = 1, 2$, $\mathbf{g}_k$ denotes the diagonal matrix with entries $g_k(i) = 1$ if $i \in \Gamma_k^c$ and 0 otherwise. When $t_2 \geqslant t_1 \geqslant 0$, a similar expression can be obtained. The joint distribution, F, of T_1 and T_2 can be obtained from $\bar{F}$ in a routine manner.

Notice that, in general, F will have a singular component on $\{(t_1, t_2): t_1 = t_2\}$. Assaf et al. [4] obtained explicit expressions for the density of the absolutely continuous component of F, for $P(T_1 = T_2)$, for the moments $ET_1^i T_2^j$, for the Laplace transform $E[\exp\{-s_1 T_1 - s_2 T_2\}]$ and so on.

The ideas of the bivariate case extend naturally to the multivariate case. We omit the details.

This class of multivariate phase type (MPH) is closed under conjunctions, finite mixtures, and formation of coherent life functions. Also all the lower dimensional marginals of an MPH distribution are MPH.

APPLICATIONS OF PH DISTRIBUTIONS

The closure, computational, and weak denseness properties of the PH distributions have lent themselves to the efficient use of PH distributions in stochastic modeling. In general one can identify three kinds of use for PH distributions. They are as follows.

Closure of PH in Stochastic Models

In several stochastic models, the use of PH distributions for some or all of the input random variables results in a PH distribution for the output random variable of interest. In such a case one only needs to obtain an explicit representation of the resulting PH distribution. To begin with, Neuts [21, p. 57] has shown that the waiting time in an $M/PH/1$ queue has a PH distribution and he gives the explicit representation of this distribution. Neuts and Bhattacharjee [24] have shown that the time to system failure in some shock models* (such as the models of Esary et al. [9]) with PH intershock times has a PH distribution. Further applications of PH distributions of a similar nature can be found in Neuts and Meier [26] and Chakravarthy [8]. Analogous applications of

MPH distributions in the theory of the multivariate cumulative damage* shock models of Marshall and Shaked [15] are given by Assaf et al. [4].

Computational Results for Models with PH Distributions

One of the strengths of PH distributions is that the use of it in stochastic models leads to efficient algorithmic solutions to the performance measures of the system being modeled. Neuts [21] lists and discusses many examples of such nature. For instance, efficient algorithmic solutions for a single server queueing system are obtained through a PH/PH/1 queueing model [21] and through a GI/PH/1 queue [19]. For more applications of PH distributions of similar nature in queueing theory see Geiszler [11] and Neuts and Kumar [25]. It is worth noting that the use of PH distributions sometimes leads to semiexplicit results for the performance measures of stochastic systems (see, e.g., Neuts [22] and Gillent and Latouche [12]). The computation of the distribution of the number of component failures during a random time interval can be carried out using efficient numerical algorithms* if one assumes a PH distribution for this time interval [20, 29].

Qualitative Properties and Approximations for Stochastic Systems

Use of PH distributions in stochastic models may allow one to develop the qualitative properties of the system being modeled. For example, Takahashi [32] has proved the tail exponentiality of the waiting time distribution in a PH/PH/c queueing system (see also Neuts and Takahashi [27]). Similarly, Burman and Smith [7] consider an $M/PH/c$ queue and show that when the customer arrival rate goes to zero, the probability of delay of a customer depends only on the mean of the service time. Other insensitivity properties of this $M/PH/c$ queue are also given in Burman and Smith [7]. Thus because of Property **4** and the continuity property of the system being modeled, one may expect the above results to be true for the more general cases. In particular, the results of Takahashi [32] can be postulated for the $GI/G/c$ queue and the result of Burman and Smith [7] can be postulated for the $M/G/c$ queue. However, these propositions have not yet been proved. Tweedie [33], on the other hand, has successfully shown that the $GI/G/1$ has an operator geometric solution; a result conjectured by Neuts [21, p. 157] as a consequence of the analysis of a $GI/PH/1$ queue. Furthermore the use of PH distributions may lead to qualitative properties that may not be true in the general case. For example, Assaf [1], using PH distributions in a restricted way, obtains the characteristics of the optimal policy for a replacement problem with n possible types of replacements. In particular he shows that when the lifetimes of all the alternative replacements have a *common matrix* of PH distributions $F_i = 1 - \alpha^{(i)}\exp\{\mathbf{A}t\}\mathbf{e}$, $i = 1, \ldots, n$, such that $\mathrm{Span}\{\alpha^{(1)}\mathbf{A}, \ldots, \alpha^{(n)}\mathbf{A}\} \subset \mathrm{Span}\{\alpha^{(1)}, \ldots, \alpha^{(n)}\}$, then, under mild conditions, an interval policy is optimal. Under an interval policy there exist $0 \equiv t_0 \leqslant t_1 \leqslant \cdots \leqslant t_{n-1} \leqslant t_n \equiv \infty$ such that for some permutation β of $\{1, 2, \ldots, n\}$, type $\beta(i)$ replacement should be used during the time interval $(t_{i-1}, t_i]$, $i = 1, \ldots, n$. It should be noted that when the life distributions of the alternative replacements are general an interval policy need not be optimal. Ramaswami and Neuts [28], using PH distributions for interarrival and service times, develop a duality theory for a single server queue. Suppose $\mathbf{n} = ((n, 1), \ldots, (n, m))$ be a state descriptor with (n, j) representing the state of n customers in the system and the service [respectively, arrival] is in phase j at a customer arrival (respectively, service completion) epoch in a $GI/PH/1$ (respectively, $PH/G/1$) queue. Then Ramaswami and Neuts [28] show that the first passage time from $\mathbf{n}$ to $(\mathbf{n} + 1)$ in the $GI/PH/1$ queue converges in distribution, as $n \to \infty$, to the distribution of the first passage time from **1** and **0** in an appropriately defined (called dual) $PH/G/1$ queue.

References

[1] Assaf, D. (1982). Renewal decisions when category life distributions are of phase-type. *Math. Operat. Res.*, **7**, 557–567.

[2] Assaf, D. and Langberg, N. A. (1985). On the extreme points of the class of phase type distributions. *J. Appl. Prob.*, **22** (to appear).

[3] Assaf, D. and Levikson, B. (1982). Closure of phase type distributions under operations arising in reliability theory. *Ann. Prob.*, **10**, 265–269.

[4] Assaf, D., Langberg, N. A., Savits, T. H., and Shaked, M. (1984). Multivariate phase type distributions. *Operat. Res.*, **32**, 688–702. (This paper introduces bivariate and multivariate PH distributions.)

[5] Baxter, L. A. (1983). The moments of the forward recurrence times of an alternating renewal process. *Eur. J. Opnl. Res.*, **12**, 205–207.

[6] Becker, P. J. and Roux, J. J. (1981). A bivariate extension of the gamma distribution. *S. Afr. Statist. J.*, **15**, 1–12.

[7] Burman, D. Y. and Smith, D. R. (1983). A light-traffic theorem for multi-server queues. *Math. Operat. Res.*, **8**, 15–25.

[8] Chakravarthy, S. (1983). Reliability analysis of a parallel system with exponential life times and phase type repairs. *OR Spektrum*, **5**, 25–32.

[9] Esary, J. D., Marshall, A. W., and Proschan, F. (1973). Shock models and wear processes. *Ann. Prob.*, **1**, 627–643.

[10] Freund, J. E. (1961). A bivariate extension of the exponential distribution. *J. Amer. Statist. Ass.*, **56**, 971–977.

[11] Geiszler, C. (1981). A numerical procedure for the selection of the constant interarrival time to a single server queue. *Comput. Math. Appl.*, **7**, 537–546.

[12] Gillent, F. and Latouche, G. (1983). Semi-explicit solutions for $M/PH/1$-like queueing systems. *Eur. J. Operat. Res.*, **13**, 151–160.

[13] Latouche, G. (1982). A phase-type semi-Markov point process. *SIAM J. Alg. Discrete. Meth.*, **3**, 77–90.

[14] Marshall, A. W. and Olkin, I. (1967). A multivariate exponential distribution. *J. Amer. Statist. Ass.*, **62**, 30–44.

[15] Marshall, A. W. and Shaked, M. (1979). Multivariate shock models for distributions with increasing hazard rate average. *Ann. Prob.*, **7**, 343–358.

[16] Neuts, M. F. (1975). Probability Distributions of Phase Type. *Liber Amicorum Prof. Emeritus H. Florin*, Dept. of Mathematics, University of Louvain, Belgium, 173–206. (An excellent introduction to discrete PH distributions.)

[17] Neuts, M. F. (1978). Renewal processes of phase type. *Naval Res. Logist. Quart.*, **25**, 445–454.

[18] Neuts, M. F. (1979). A versatile Markovian point process. *J. Appl. Prob*, **16**, 764–779.

[19] Neuts, M. F. (1981a). Stationary waiting time distributions in the GI/PH/1 queue. *J. Appl. Prob.*, **18**, 901–912.

[20] Neuts, M. F. (1981b). An illustrative problem in computational probability. *Opsearch*, **18**, 171–177.

[21] Neuts, M. F. (1981c). *Matrix-Geometric Solutions in Stochastic Models: An Algorithmic Approach.* Johns Hopkins University Press, Baltimore, MD. (Chapter 2 is an excellent review on PH distributions.)

[22] Neuts, M. F. (1982). Explicit steady-state solutions to some elementary queueing models. *Operat. Res.*, **30**, 480–489.

[23] Neuts, M. F. (1984). The abscissa of convergence of the Laplace–Stieltjes transform of a PH distribution. *Commun. Statist.—Simul. Comp.*, **13**, 367–373.

[24] Neuts, M. F. and Bhattacharjee, M. C. (1981). Shock models with phase type survival and shock resistance. *Naval Res. Logist. Quart.*, **28**, 213–219.

[25] Neuts, M. F. and Kumar, S. (1982). Algorithmic solution of some queues with oberflows. *Manag. Sci.*, **28**, 925–935.

[26] Neuts, M. F. and Meier, K. S. (1981). On the use of phase type distributions in reliability modelling of systems with two components. *OR Spektrum*, **2**, 227–234.

[27] Neuts, M. F. and Takahashi, Y. (1981). Asymptotic behavior of the stationary distributions in the GI/PH/c queue with heterogeneous servers. *Z. Wahr. Geb.*, **57**, 441–452.

[28] Ramaswami, V. and Neuts, M. F. (1980). A duality theorem for phase type queues. *Ann. Prob.*, **8**, 974–985.

[29] Shanthikumar, J. G. (1982). A recursive algorithm to generate joint probability distribution of arrivals from exponential sources during a random time interval. *Inf. Processing Lett.*, **14**, 214–217.

[30] Shanthikumar, J. G. (1985). Bilateral phase-type distributions. *Nav. Res. Logist. Quart.*, **32**, 119–132.

[31] Syski, R. (1982). Phase-type distributions and perturbation model. *Zast. Mat.*, **17**, 377–399.

[32] Takahashi, Y. (1981). Asymptotic exponentiality of the tail of the waiting-time distribution in a PH/PH/c queue. *Adv. Appl. Prob.*, **13**, 619–630.

[33] Tweedie, R. L. (1982). Operator-geometric stationary distributions for Markov chains with application to queueing models. *Adv. Appl. Prob.*, **14**, 368–391.

(COHERENCE
CUMULATIVE DAMAGE MODELS
ERLANG DISTRIBUTION
EXPONENTIAL DISTRIBUTION
GAMMA DISTRIBUTION
INTEGRAL TRANSFORMS
MARKOV PROCESSES
MULTISERVER QUEUES
QUEUEING THEORY
RELIABILITY (PROBABILISTIC)
RENEWAL THEORY
SHOCK MODELS)

M. Shaked
J. G. Shanthikumar

PHI-COEFFICIENT

The phi-coefficient is actually a product–moment coefficient of correlation* and is a variation of Pearson's definition of r when the two states of each variable are given values of 0 and 1 respectively.

The phi-coefficient was designed for the comparison of truly dichotomous distributions, i.e., distributions that have only two points on their scale which indicate some unmeasurable attribute. Attributes such as living or dead, black or white, accept or reject, and success or failure are examples. It is also sometimes known as the Yule ϕ [1].

If certain allowances are made for continuity, the technique may be applied to observations grouped into two arbitrary, but clearly defined, divisions. It is clear that in many such divisions point attributes are achieved by a binary decision process employing demarcation lines through "grey" or "ill-defined" regions.

The phi-coefficient relates to the 2×2 table*:

| | Attribute 1 | |
Attribute 2	Yes	No
Yes	a	b
No	c	d

If a, b, c, and d represent the frequencies of observation, then ϕ is determined by the relationship

$$\phi = \frac{ad - bc}{\sqrt{\{(a + b)(c + d)(a + c)(b + d)\}}}$$

The phi-coefficient is particularly used in psychological and educational testing where the imposing of a dichotomy on a continuous variable is a frequent occurrence; variables where pass/fail categories are obtained in relation to a threshold score are typical (*see* PSYCHOLOGICAL TESTING THEORY).

It bears a relationship to χ^2, where

$$\phi^2 = \frac{\chi^2}{N} \quad \text{or} \quad \chi^2 = N\phi^2$$

and $N = a + b + c + d$ (*see* CHI-SQUARE TESTS).

The significance of ϕ may be tested by determining the value of χ^2 from the above relationship and testing in the usual way.

As an example, 43 persons were asked if they believed that there was any truth in horoscopes or in the existence of UFOs. The results gave

| | Horoscopes | |
UFO's	Some truth	No truth
Might exist	14	10
Don't exist	6	13

Applying the above formula, $\phi = 0.266$.

This value of ϕ corresponds to a value of χ^2 of $43 \times (0.266)^2 = 3.04$. This may then be tested against the relevant value of χ^2 for 1 degree of freedom.

An alternative significance test (rarely used) may be performed by considering the standard error of ϕ. Calculation of this is laborious but if N is not too small, then $1/\sqrt{N}$ approximates to it [2].

References

[1] Yule, G. U. (1912). *J. R. Statist. Soc.*, **75**, 576–642. (On the methods of measuring the association between two variables. The first identification of the phi-coefficient.)

[2] McNemar, Q. (1962). *Psychological Statistics*. Wiley, New York. (Justification after use of approximation.)

(BISERIAL CORRELATION
CORRELATION
PHI-MAX COEFFICIENT
TWO-BY-TWO TABLES)

O. B. Chedzoy

PHI-DEVIATION

Frequency distributions relating to continuous variables give rise to the determination of the formal parameters of the mean and standard deviation as measures of central value and dispersion. For data which is normally distributed, approximately 68% of all observations will fall within one standard deviation of the mean value.

Measurements which have been classified into suitable intervals and counted can be represented by a cumulative frequency diagram. In certain applications, for instance, in sedimentology, aggregate totals are obtained by weights of sediment passing through sieves and counting of grains is inappropriate.

If the cumulative frequencies are converted to proportions of the total observed frequency and expressed as percentages then the diagram becomes the basis for determining the values of the variable under consideration corresponding to various percentile points. These percentile points are referred to as ϕ_a, where the subscript a represents the percentile concerned. Thus ϕ_{50} is the median* value; ϕ_{25} and ϕ_{75} are the quartiles; and ϕ_{10} and ϕ_{90} are the 10- and 90-percentiles.

The establishing of a diagram enables all percentile values to be easily determined, in particular values for ϕ_{16} and ϕ_{84}. The observations which occur between ϕ_{16} and ϕ_{84} are the 68% which form the central group. It follows that it will normally be expected that these will represent the values at approximately one standard deviation from the mean.

Thus the empirical measure phi-deviation is defined as

$$\text{phi-deviation} = \sigma_\phi = \frac{\phi_{84} - \phi_{16}}{2},$$

which will be an estimate of the standard deviation of a set of normally distributed data.

The symmetry of these percentiles suggests a further definition of a central value

$$\text{phi-mean} = m_\phi = \frac{\phi_{84} + \phi_{16}}{2}.$$

These two measures may also be used, in connection with other percentile values, to give approximations for skewness and kurtosis:

$$\text{phi-skewness} = \frac{m_\phi - \phi_{50}}{\sigma_\phi};$$

$$\text{phi-kurtosis} = \frac{\frac{1}{2}(\phi_{95} - \phi_5) - \sigma_\phi}{\sigma_\phi}.$$

For example, suppose the weights of 100 marrows grown by a market gardener varied from 2 to 12 lb. The frequencies in the intervals 2–2.99, 3–3.99, etc., up to 11–11.99 were 2, 5, 5, 15, 26, 18, 12, 11, 3, and 3. These data may be alternatively presented as the total numbers less than 2 lb, less than 3 lb, etc., up to less than 12 lb; these cumulative frequencies are 0, 2, 7, 12, 27, 53, 71, 83, 94, 97, and 100. It is these which are plotted on the graph (Fig. 1).

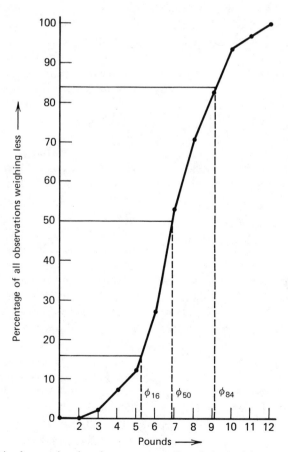

Figure 1   Cumulative frequencies plotted as percentage of total observed frequency against variable.

The values obtained from the graph are

$$\phi_{16} = 5.3; \qquad \phi_{50} = 6.9; \qquad \phi_{84} = 9.1;$$

$$\text{phi-mean} = 7.2 \text{ lb};$$

$$\text{phi-deviation} = 1.9 \text{ lb}.$$

Calculation by conventional methods gives the mean as 7.04 lb and the standard deviation as 1.93 lb.

(FREQUENCY POLYGON
OGIVE)

O. B. CHEDZOY

PHI-MAX COEFFICIENT

The phi-coefficient considered as a product–moment correlation coefficient* has absolute maximum and minimum values of $+1$ and -1 respectively. Given the marginal totals for any set of data, these theoretical extremes will in general have more severe limitations.

If, in the example given for the phi-coefficient*, the same marginal totals are assumed then the extreme values of correlation [1] will be given when one of the elements is zero.

20	4	24		1	23	24
0	19	19		19	0	19
20	23	43		20	23	43

$$\phi = +0.83 \qquad\qquad \phi = -0.95$$
(maximum value)   (minimum value)

It will be seen that the zero elements can only possibly occur in the row or column where the marginal total is lowest. It follows that the maximum and minimum possible values of the phi-coefficient (i.e., $+1$ and -1) can be attained only when the four marginal totals are equal.

It is generally considered that any determination of the phi-coefficient should be accompanied by the appropriate phi-max coefficient.

References

[1] Ferguson, G. A. (1941). *Psychometrika*, **6**, 323–333. (A factorial interpretation of test difficulty, incorporating a formula for calculating ϕ.)

(PHI-COEFFICIENT)

O. B. CHEDZOY

PHI-MEAN   *See* PHI-DEVIATION

PHI-SKEWNESS   *See* PHI-DEVIATION

PHYSICS, STATISTICS IN (EARLY HISTORY)

In retrospect one can say that statistical methods were already used in physics at a time when physicists would have referred to probability theory as the source of their methodological approach. This has to do with the fact that modern statistics as a mathematical discipline in its own right was established only in the twentieth century. As a consequence physicists did not distinguish between probability theory and statistics before the rise of quantum mechanics* and, even more importantly, the use of statistical methods in physics had to be justified in terms of probability theory. Witnesses for this situation are David Hilbert [14] and Emanuel Czuber [10].

All the physical problems that were treated by physicists with the methods of probability theory have in common that they are mass phenomena, or at least they were considered as such.

Among the early applications we can distinguish the following main problem groups: kinetic theory of gases, being the main root of statistical mechanics; heat- and light-radiation together with spectral analysis; fluctuation phenomena like Brownian motion*; and radioactivity. Nearly all these problems are now to be subsumed under statistical mechanics.

In addition to these we would need to include the use of error theory, as developed by Gauss* and Laplace*, in experimental physics. From the point of view of statistics, this application is not of special interest since it does not differ from error theory in geodetics or in astronomy while from the point of view of physics, error theory was very important in the late nineteenth century, e.g., in the measurement of black-body radiation where errors were considerable and the question of reliability of a measured intensity distribution was a serious one.

KINETIC THEORY OF GASES

The kinetic theory of gases tries to derive relations between macroscopic physical variables, such as volume, pressure, and temperature, from the motions of microscopic particles interacting through perfectly elastic collisions. Models of this kind were already used by Daniel Bernoulli* (1700–1782) in the eighteenth century and by John Herapath (1790–1868), J. J. Waterston (1811–1883), and others in the first half of the nineteenth century. All these attempts remained almost completely ignored due, first of all, to resistance to their implied atomism. Only in the 1850s when conservation of energy, especially the mutual transformation of heat into mechanical energy, became generally accepted could the kinetic theory be revived.

The pioneer of this revival was Rudolph Clausius (1822–1888), who justified the use of his kinetic model by probabilistic arguments. In his first paper in 1857 on the kinetic theory of gases [8] probability theory functioned only as a means to explain the transition from the seemingly chaotic movement of the individual microscopic particles to an ordered movement of groups of particles. He treated average movements of subsets of particles according to the model of Adolphe Quetelet*, who had described mass phenomena by averages taken from a great number of individual measurements.

The first kinetic models were still fairly crude and considered collisions only with the walls of the vessel containing the gas. Due to the relatively small diffusion velocity of a gas, collisions between particles had to be taken into account. For this Clausius introduced the concept of mean free path [9] which not only opened the way for the first real application of probability theory to the kinetic theory of gases but also became the starting point of J. C. Maxwell's (1831–1879) and Ludwig Boltzmann's (1844–1906) work in this field. Clausius divided the space of the gas into equally thick layers perpendicular to the direction of the motion of a single molecule. He found the probability that this molecule would pass a layer of thickness x as a^x, or $e^{-\alpha x}$, where $a = e^{-\alpha}$. This probability he interpreted as the fraction of n molecules all moving in the same direction that would pass a layer of thickness x without collision. Clausius then obtained the density function, i.e., the number of molecules that would collide with other molecules after passing exactly a distance x before collision. If we take this distance x as the variate, the expectation of x will give the mean free path as calculated by Clausius. The statistical character of his approach consists in a softening of the classical deterministic view, i.e., Clausius admitted that the gas laws deduced from the kinetic theory are only valid with a high probability.

Deviations from the laws are possible although with very small probabilities. Whether his insistence on complete irregularity of the individual molecular motions should be understood as indeterminism or as lack of information about determined motions is not clear. Clausius had thought of velocity distributions of the molecules in fluids and gases but based his calculation on a constant velocity, the root mean square velocity of the molecules involved.

It remained for Maxwell* to introduce explicitly a velocity distribution of the molecules. The background of Maxwell's interest in kinetic theory of gases was threefold: acquaintance with the literature on probability theory by Laplace, Boole, and especially by Quetelet [21, 13]; work on the rings of

Saturn; and familiarity with Clausius' publications of 1857 and 1858. It seems that Maxwell based his deduction of the velocity distribution less on specific assumptions about his mechanical model than on an a priori conviction that the normal curve of errors would apply. In fact, he used the same functional equation that John Herschel had used in order to get the error-distribution [18].

For the velocity distribution he found the density

$$\frac{4}{\alpha^3\sqrt{\pi}} v^2 e^{-v^2/\alpha^2}.$$

From these results, and using the concept of mean free path introduced by Clausius, Maxwell calculated transport properties of gases, such as the coefficients of viscosity, diffusion, and heat conduction. In 1867, Maxwell would deduce the velocity distribution on different assumptions [19]. He then showed that this distribution is the only stable one in the sense that collisions between molecules would not change it.

Maxwell's later reflections on the probabilistic character of the gas laws which can be explained by the kinetic theory of gases resemble and make more precise statements made earlier by Clausius.

The development of the kinetic theory of gases following Maxwell's contributions of 1860 and 1867 was dominated by Boltzmann's publications. In 1868 he extended Maxwell's results to the case in which the system is exposed to external forces [2]. The density of the velocity distribution assumed the form

$$f(v) = c \cdot \exp\left[-h\left(\tfrac{1}{2}mv^2 + V(x)\right)\right],$$

where $V(x)$ is a potential function describing the action of the external force depending on the position x. This density, now called the *Boltzmann factor*, became a basic concept in statistical mechanics. Here Boltzmann* used already two different meanings for the corresponding distribution function. It would give the fraction of any suitably long time interval during which the velocity of any molecule would vary within

given limits, or the fraction of the total number of molecules which would have velocities within those limits at any one moment, i.e., time average against microcanonical phase average.

In 1871 [3] Boltzmann felt compelled to introduce the hypothesis that the coordinates and velocities of the atoms or molecules would assume all possible values compatible with the total energy of the gas, or in Gibbsian terminology, that a closed mechanical system will pass through all possible points of the phase space lying on a given energy surface. This hypothesis became the target of many attacks and the starting point of the ergodic theory.

In his paper of 1872 [4] Boltzmann tackled the problem of finding the density of the energy distribution $f(x, t)$, which gives the number of molecules that possess the energy x at time t. This $f(x, t)$ satisfies the "Boltzmann transport equation" in its original form. Since the density corresponding to Maxwell's velocity distribution makes $\partial f(x, t)/\partial t = 0$, Maxwell's result that collisions would not alter the state of a gas that has already attained the Maxwell distribution could be demonstrated. Boltzmann would go even further and show that a distribution initially different from Maxwell's would become Maxwellian in the limit. For this he defined the function

$$E\left[f(x,t)\right]$$
$$= \int_0^\infty f(x,t)\left\{\log\left[\frac{f(x,t)}{\sqrt{x}}\right] - 1\right\} dx$$

which is monotically decreasing with time, $\partial E/\partial t \leqslant 0$.

It is unchanging, i.e., $\partial E/\partial t = 0$, only for the Maxwell–Boltzmann distribution. Boltzmann remarked that $-E$ is proportional to entropy* in the equilibrium state, where $\partial E/\partial t = 0$. The E-function, or later H-function, thus would extend the thermodynamical concept of entropy to nonequilibrium states. In 1876, when Loschmidt confronted Boltzmann with the apparent contradiction between the reversibility of the presupposed mechanical system of the gas molecules and the irreversible monotonic behavior of E, Boltzmann ultimately responded with a new statistical approach to the problem. In complete accordance with what Maxwell had tried to illustrate by his famous demon, Boltzmann now would try to clarify "how intimately the second law is related to the theory of probability." The new method, published in 1877 [5], was a combinatorial approach. Boltzmann first counted the number P of ways in which a fixed number n of molecules with constant total energy $\lambda\epsilon$ (λ integral) could be distributed over $(p + 1)$ energy levels—$0, \epsilon, 2\epsilon, \ldots, p\epsilon$—where w_r molecules have the energy $r\epsilon$ ($\sum_{r=0}^{p} w_r = n$).

Since this method failed to give the expected result, Boltzmann counted the number P of distributions of n molecules over $(p + 1)$ *velocity* levels under the same conditions. This number $P = n!/\prod_{r=0}^{p} w_r!$, called the *permutability measure*, is proportional to the probability of the given distribution, the proportionality factor being the total number J of possible permutability measures, $J = \binom{m+\lambda-1}{\lambda}$. Consequently, the distribution for maximum P is most probable. By the limit $\epsilon \to 0$, Boltzmann could show that the Maxwell–Boltzmann distribution is not only the unique stationary distribution but also by far the most probable distribution. In this case entropy is proportional to the logarithm of P and so of the probability of the distribution. Boltzmann would later try to settle the recurrence paradox brought forward by Poincaré and Zermelo by insisting on the probabilistic or statistical character of the second law of thermodynamics. With his 1877 publication Boltzmann had achieved the transition from a kinetic view determined by collision processes to a combinatorial view. The advantages of this equilibrium statistical mechanics became clear on the more abstract level of the phase space description created by Gibbs in 1902 [12].

RADIATION PHENOMENA

Max Planck, who belonged originally to Boltzmann's critics, adopted in 1900 [20] the original combinatorial approach for energy levels which Boltzmann had attempted to use in 1877. Planck's translation of the problem of black-body radiation into a combinatorial problem required the number of all possible distributions of n resonators of a given frequency over the $(p + 1)$ discrete energy levels, $0, \epsilon, \ldots, p\epsilon$, such that the total energy is $p\epsilon$. This approach permitted the deduction of the density of the energy distribution in radiation. Planck's radiation law provided the root from which quantum theory grew. The predecessor of Planck's law, Wien's radiation law, had been deduced on the assumption that the radiation function should be found by analogy to Maxwell's statistical basis for the velocity distribution.

A more explicit use of probability theory can be observed in light-radiation. Already in 1849 [7] Clausius had tested different physical hypotheses accounting for the atmospherical dispersion of light. One of the hypotheses would lead to the following problem: n spheres are randomly distributed in n pipes; what is the mean number of pipes that contain exactly k, $k \leqslant n$, spheres in the limit $n \to \infty$? Clausius showed that the mean value of the number of pipes containing k spheres is asymptotically equal to $1/(e \cdot k!)$, i.e., Poisson-distributed with parameter 1.

Another case of hypothesis testing* is offered in spectroscopy. In 1893, C. Runge [15] had decomposed the line spectrum of each of the elements, tin, lead, arsenic, antimony, and bismuth, into sets of repeating series. Each series repeated itself in the sense that the difference in wavelength between successive lines was the same. Runge had no explanation for this phenomenon but he was convinced that there must exist a still unknown law that would explain it. Therefore, like eighteenth century astronomers before him, he tested the hypothesis that the phenomenon was due to chance. For this he had to calculate, or at least estimate, the probability that a sample of m randomly chosen elements from $n > m$ natural numbers, 1, 2, \ldots, n contains r pairs of numbers, $2r \leqslant m$, that have the same difference. For the spectrum of tin, which contained at least two repeating series, each consisting of 13 lines, this probability was less than 10^{-7}, small enough to discard the hypothesis that the effect was due to chance.

Except for Runge, whose case of statistical inference was completely ignored, the combinatorial approaches of Clausius to atmospherical optics, Boltzmann to the second law of thermodynamics, and Planck to black-body radiation could be subsumed under the occupancy problem as was done, e.g., by von Mises in 1931 [27]. The occupancy problem* covers also quantum-statistical distribution like those of Bose–Einstein and Fermi–Dirac, and applies in part to fluctuation problems.

FLUCTUATION PROBLEMS

The most prominent fluctuation problem was offered by Brownian motion. Einstein had seen a possibility to explain the chaotic movement of microscopic particles in liquid —known as Brownian motion ever since Brown in 1827 had observed the phenomenon—as the effect of the collisions between the microscopic particles and submicroscopic molecules. Since the theory of Brownian motion as developed by Einstein [11] and Smoluchowski [25] accounted completely for the observed effects, it helped to justify the kinetic gas model and especially the statistical interpretation of the second law of thermodynamics. It also provided a means to get at properties of the unobservable molecules.

Einstein defined the density function $f(x,t)$ for the displacement x within time t and showed that t has the density of the normal distribution,

$$f(x,t) = \frac{1}{\sqrt{4\pi Dt}} \exp\left(-\frac{x^2}{4Dt}\right),$$

where D is the diffusion coefficient, and $\mathrm{var}(x(t)) = 2Dt$. He assumed that t is large in comparison with the time interval between successive collisions, each of which contributes to the increment x. Since D is a function of the absolute temperature, the

number of molecules, and the size of the microscopical particles, Einstein could predict the standard deviation of x, which eventually was observed and in turn was used to calculate Avogadro's number. Smoluchowski achieved the same result with a combinatorial approach. He had applied the theory of fluctuations, for which he had laid the foundations already in 1904, to the variation of the density of a gas. In 1908, he combined this approach with the observable phenomenon of opalescence of a gas in a critical state.

The theory of fluctuations also applied to radioactivity. Already in 1900 Rutherford had measured the decay curve of radioactive material. He described this curve by the formula $n(t) = n_0 \exp(-\lambda t)$ and in 1903 declared that the equivalent differential expression $dn/dt = -\lambda_n(t)$, where $n(t)$ is the number of radioactive atoms at time t, holds for any radioactive substance. In 1904 Rutherford explained the constant λ as "the fraction of atoms disintegrating per second" and called its inverse λ^{-1} the "average life of the corresponding radioactive body." In 1905 the physicist Egon von Schweidler [28] showed that the exponential decay of α-radiation as found by Rutherford could be interpreted by probability theory. He identified $\exp(-\lambda t)$ with the probability that a radioactive atom survives the time interval t. Since in reality $n(t)$ is always finite and integral, Schweidler was concerned about fluctuations, i.e., the differences between the experimentally observed and the theoretical values. He could show that for very large n the fluctuations of the counted scintillations in different time intervals are of the order $\sqrt{n}$, the variance of the distribution.

Following Schweidler there appeared numerous publications on radioactivity which employed statistical methods. The most important results achieved before the rise of quantum mechanics* are summarized in Bortkiewicz' book of 1913 [26].

Bortkiewicz distinguished two approaches: one is concerned with the distribution of time intervals between two successive scintillations and the other with the frequency of scintillations in given intervals of time. He found e^{-kt} to be the distribution function for the time elapsed until the next scintillation, where k is the reciprocal value of the expectation of t. In the second approach Bortkiewicz stated that the probability of r scintillations in a time interval of length t is Poisson-distributed, i.e., $(kt)^r \cdot e^{-kt}/r!$.

Even the proof that the dispersion of the numbers of scintillations is normal would still not establish, according to Bortkiewicz, the mutual independence of the radioactive atoms. So in 1913 neither Bortkiewicz nor Rutherford—who thought it important to settle "whether the alpha-particles are emitted at random in time and space" [22]—would surrender himself to indeterminism in radioactive decay.

It remained for the pioneers of quantum mechanics to assert that radical position.

The early history of statistics in physics as described in this article mainly served as an encouragement to develop quantum statistics of which classical statistical mechanics is a special case.

References

[1] Amaldi, E. (1979). In *Rendiconti S.I.F. 72*, North-Holland, Amsterdam.

[2] Boltzmann, L. (1868). *Sitzungsber.-Berichte Akad. Wiss. Wien*, **58**, 517–560.

[3] Boltzmann, L. (1871). *Sitzungsber.-Berichte Akad. Wiss. Wien*, **63**, 679–711.

[4] Boltzmann, L. (1872). *Sitzungsber.-Berichte Akad. Wiss. Wien*, **66**, 275–370.

[5] Boltzmann, L. (1877). *Sitzungsber.-Berichte Akad. Wiss. Wien*, **76**, 373–435.

[6] Brush, S. G. (1976). *The Kind of Motion We Call Heat*. North-Holland, Amsterdam, 2 vols. (Best and most comprehensive account of the development of nineteenth century kinetic theory.)

[7] Clausius, R. (1849). *Ann. Phys. Chem.*, **76**, 165f.

[8] Clausius, R. (1857). *Ann. Phys. Chem.*, **100**, 353–380.

[9] Clausius, R. (1858). *Ann. Phys. Chem.*, **105**, 239–258.

[10] Czuber, E. (1921). *Die Statistischen Forschungsmethoden*. Seidel, Wien. (These "statistical research methods" influenced by Yule's *Introduction to the Theory of Statistics* do not hint in a single instance to physics as a field of application.)

[11] Einstein, A. (1905). *Ann. Phys.*, 4th Ser. **17**, 549–560.

[12] Gibbs, J. W. (1902). *Elementary Principles in Statistical Mechanics*. Yale University Press, New Haven, Conn.

[13] Herschel, J. (1850). *Edinburgh Rev.*, **92**, 1–57. (Review of Quetelet's book.)

[14] Hilbert, D. (1900). *Göttinger Nachr.*, **1900**, 253–297. (The famous paper on 23, at the time unsolved, mathematical problems which considerably influenced the development of mathematics in our century. Number six postulates the axiomatization of *physical* disciplines like probability theory!)

[15] Kayser, H. and Runge, C. (1893). *Ann. Phys. Chem.*, 2nd Ser., **52**, 92–113.

[16] Klein, M. (1973). In *The Boltzmann Equation Theory and Applications*, E. G. D. Cohen and W. Thirring, eds. Springer, Wien, pp. 53–106.

[17] Kuhn, T. S. (1978). *Black-Body Theory and the Quantum Discontinuity 1894–1912*. Clarendon, Oxford, England.

[18] Maxwell, J. C. (1860). *Philos. Mag.*, 4th Ser., **19**, 19–32 and **20**, 21–37.

[19] Maxwell, J. C. (1867). *Philos. Trans. R. Soc. Lond.*, **157**, 49–88.

[20] Planck, M. (1901). *Ann. Phys.*, 4th Ser., **4**, 553–563. (The paper which started quantum physics.)

[21] Quetelet, A. (1849). *Letters . . . On the Theory of Probability as Applied to the Moral and Political Sciences*. Cayton, London. (Most popular account of the time about probability theory and representative of Quetelet's dogma that distributions in nature are normal.)

[22] Rutherford, E. (1913). *Radioactive Substances and Their Radiations*. Cambridge University Press, Cambridge, England.

[23] Schneider, I. (1974). *Arch. History Ex. Sci.*, **14**, 143–158.

[24] Schneider, I. (1975). *Arch. History Ex. Sci.*, **15**, 237–261.

[25] Smoluchowski, M. (1906). *Ann. Phys.*, 4th Ser., **21**, 756–780.

[26] von Bortkiewicz, L. (1913). *Die Radioaktive Strahlung als Gegenstand Wahrscheinlichkeitstheoretischer Untersuchungen*. Springer, Berlin.

[27] von Mises, R. (1931). *Vorlesungen aus dem Gebiete der Angewandten Mathematik*, Vol. 1, *Wahrscheinlichkeitsrechnung und ihre Anwendung in der Statistik und Theoretischen Physik*, Deuticke, Leipzig-Wien.

[28] von Schweidler, E. (1906). *Compt. Rend. Premier Congr. Int. Étude Radiologie Ionisation 1905*, Brussels.

(BOLTZMANN, LUDWIG EDWARD FERMI–DIRAC STATISTICS (WITH

MAXWELL–BOLTZMANN AND
BOSE–EINSTEIN STATISTICS)
MAXWELL, JAMES CLERK
QUANTUM MECHANICS:
  PROBABILISTIC OVERVIEW
QUANTUM MECHANICS:
  THEORETICAL ISSUES
QUANTUM PHYSICS:
  FUNCTIONAL ANALYSIS)

I. Schneider

PICTOGRAM *See* Graphical representation of data

**PIECEWISE EXPONENTIAL SURVIVAL
ANALYSIS** *See* poisson regression

PIE CHART

This is a way of representing the proportions
in which an entity, Ω for instance, is divided
into classes $\alpha_1, \alpha_2, \ldots, \alpha_k$, say. The chart
consists of a circle divided into k sectors
("slices of the pie") with areas (or equivalently, angles subtended at the center) pro-
portional to $\alpha_1 : \alpha_2 : \ldots : \alpha_k$. (It is also
called a *circle chart*.) The sectors are labeled
appropriately. Often they are colored distinctively to assist quick comprehension.

As an example suppose that we have
$k = 5$ and $\alpha_1 = 2,000$; $\alpha_2 = 5,000$; $\alpha_3 = 15,000$; $\alpha_4 = 1,000$; $\alpha_5 = 7,000$ so that $\alpha_1 + \alpha_2 + \alpha_3 + \alpha_4 + \alpha_5 = 30,000$. These classes
are in proportions $(2,000/30,000) = \frac{1}{15} : \frac{1}{6} : \frac{1}{2} : \frac{1}{30} : \frac{7}{30}$. Figure 1 is the corresponding pie
chart. The angles subtended at the center are

$$\alpha_1 :\!- 360 \times \tfrac{1}{15} = 24°$$
$$\alpha_2 :\!- 360 \times \tfrac{1}{6} = 60°$$
$$\alpha_3 :\!- 360 \times \tfrac{1}{2} = 180°$$
$$\alpha_4 :\!- 360 \times \tfrac{7}{30} = 84°$$
$$\alpha_5 :\!- 360 \times \tfrac{1}{30} = 12°$$

Such charts can be used, e.g., to represent

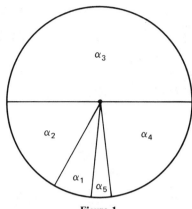

Figure 1

the distribution of a company's investments,
sources of income and expenditure, or distributions of religious or political affilia-
tions. They are a popular tool for business
and economical applications, especially in
weekly and monthly periodicals of wide circulation. They are not commonly used to
represent ordered variables (such as income,
weight, or tensile strength). Histograms* are
more commonly used for this purpose.

Pie charts may be used for comparative
purposes to exhibit proportions over different time periods or in different spatial areas
(e.g., nations or countries). When used in
this way it usually is very desirable to indicate changes in absolute amounts by corresponding changes in the total areas of the
corresponding circles.

Pie charts are competitive alternatives to
bar charts*. For single diagrams pie charts
may be preferable, but for comparisons over
different time or space units the change in
size of the pie charts (radius proportional to
the square root of total amount) may be less
easily comprehended than the change in
length of the bar charts (directly proportional to total amount). For a more detailed
discussion and comparison between pie
charts and other methods of graphical representation see Peterson and Schramm [1] and
an earlier pioneering paper by Eels [2].

Kruskal [3] gives a fascinating discussion
of Eels's work and also provides a comprehensive bibliography on statistical graphics.

References

[1] Peterson, L. V. and Schramm, W. (1954). How accurately are different kinds of graphs read? *Audio-Visual Comm. Rev.*, **2**, 178–189.

[2] Eels, W. (1926). *J. Amer. Statist. Ass.*, **21**, 119–132.

[3] Kruskal, W. H. (1982). *Utilitas Math.*, **21B**, 283–310.

(BAR CHART
GRAPHICAL REPRESENTATION OF DATA
HISTOGRAM
SEMI-PIE DIAGRAM)

PILLAI'S TRACE

Pillai's trace has been proposed (see Pillai, 1954, 1955) for tests of the same three hypotheses as stated in the *Introduction* of HOTELLING'S TRACE, namely: (*a*) equality of covariance matrices of two *p*-variate normal populations; (*b*) equality of *p*-dimensional mean vectors of *l* *p*-variate normal* populations having a common unknown covariance matrix, known as MANOVA (alternately general linear* hypotheses); and (*c*) independence between a *p*-set and a *q*-set ($p \leqslant q$) in a ($p + q$)-variate normal population. In the context of hypothesis (*b*), Pillai's trace is a generalization of $T^2/(v + T^2)$ as well as $cD^2/(v + cD^2)$. (*See* HOTELLING'S T^2 and MAHALANOBIS D^2 for T^2, D^2, c, and v.) Alternately,

$$V^{(s)} = \sum_{h=1}^{l} n_h (\bar{\mathbf{x}}_h - \bar{\mathbf{x}})' \mathbf{S}_0^{-1} (\bar{\mathbf{x}}_h - \bar{\mathbf{x}}) = D_0^2(N - 1),$$

where D_0^2 is a generalized D^2. Here $\bar{\mathbf{x}}_h$ is the mean vector of the *h*th sample of size n_h ($h = 1, \ldots, l$), $\bar{\mathbf{x}}$, the general mean vector, $\mathbf{S}_0$, the total sums and products (SP) matrix with $N - 1$ d.f., where $N = \sum_{h=1}^{l} n_h$.

As discussed in HOTELLING'S TRACE, tests proposed for these three hypotheses are generally invariant tests which, under the null hypotheses, depend only on the characteristic (ch.) roots of matrices based on samples. Pillai's trace test (Pillai, 1954, 1955) is such a test and the statistic is defined by

$$V^{(s)} = \sum_{i=1}^{s} b_i,$$

where the ch. roots, $b_1, \ldots, b_s$, of each of the three matrices defined below have the same form of the joint density given by

$$f(b_1, \ldots, b_s) = C(s, m, n) \prod_{i=1}^{s} \left\{ b_i^m (1 - b_i)^n \right\}$$

$$\times \prod_{i>j} (b_i - b_j),$$

$$0 < b_1 < \cdots < b_s < 1.$$

[See Pillai (1960, 1976) for $C(s, m, n)$.] Here s, m, and n are to be understood differently for the three hypotheses (*see below as well as* HOTELLING'S TRACE).

For (*a*), b_i's are the ch. roots of $\mathbf{S}_1(\mathbf{S}_1 + \mathbf{S}_2)^{-1}$ where $\mathbf{S}_1$ and $\mathbf{S}_2$ are sums of products (SP) matrices with n_1 and n_2 d.f.; for (*b*), they are those of $\mathbf{S}^*(\mathbf{S} + \mathbf{S}^*)^{-1}$, where $\mathbf{S}^*$ is the between SP matrix and $\mathbf{S}$, the within SP matrix with $l - 1$ and $N - l$ d.f., respectively, and N is the total of l sample sizes; for (*c*), those of $\mathbf{S}_{11}^{-1}\mathbf{S}_{12}\mathbf{S}_{22}^{-1}\mathbf{S}_{12}'$, where $\mathbf{S}_{ij}$ ($i, j = 1, 2$) is the SP matrix of the *i*th set with the *j*th set (1, denoting *p*-set and 2 denoting *q*-set). $\mathbf{S}_{12}\mathbf{S}_{22}^{-1}\mathbf{S}_{12}'$ and $\mathbf{S}_{11} - \mathbf{S}_{12}\mathbf{S}_{22}^{-1}\mathbf{S}_{12}'$ have q and $n' - 1 - q$ d.f. respectively where n' is the sample size. If v_1 and v_2 denote the two d.f. in each case in the order given, then $m = \frac{1}{2}(|v_1 - p| - 1)$ and $n = \frac{1}{2}(v_2 - p - 1)$. In (*a*) if $p \leqslant n_1, n_2$, then $s = p$; in (*b*), $s = \min(p, l - 1)$ and in (*c*) if $p + q < n'$ then $s = p$ (*see also* HOTELLING'S TRACE in this connection). Hence if $\mathbf{B}$ denotes in turn the matrix defined in (*a*), (*b*), or (*c*) then $V^{(s)} = \operatorname{tr} \mathbf{B}$. Further, $b_i = f_i/(1 + f_i)$, $i = 1, \ldots, s$, where $0 < f_1 < \cdots < f_s < \infty$ are the ch. roots defined in HOTELLING'S TRACE.

The trace test proposed by Pillai in 1954 for tests of the three hypotheses above has been considered by other authors in limited contexts [see Bartlett (1939); Hotelling (1947); and Nanda (1950)].

DISTRIBUTION

The null and nonnull distribution problems may be considered separately.

Null Distribution

Pillai (1954, 1956) has obtained a recurrence relation concerning the moment generating function* (mgf) of $V^{(s)}$ in terms of functions of mgf of $V^{(s-2)}$, using which he studied the first four moments of $V^{(s)}$ and suggested a beta function* approximation to the null distribution of $V^{(s)}$ (see Pillai, 1954, 1955, 1957, 1960), resulting in an F-statistic* of the form

$$F_{f_1, f_2} = \frac{(2n + s + 1)V^{(s)}}{(2m + s + 1)(s - V^{(s)})}$$

with $f_1 = s(2m + s + 1)$ and $f_2 = s(2n + s + 1)$ d.f. The approximation is recommended for $m + n > 30$. Pillai and Mijares (1959) have evolved an alternate method of expressing the moments of $V^{(s)}$ and have given a general expression for the fourth moment which was obtained earlier by Pillai only for $s = 2$, 3, and 4. The first four moments of $V^{(s)}$, in fact, the moment quotients, have been used to obtain approximate upper 5 and 1% points of $V^{(s)}$ (Pillai, 1957, 1960) for $s = 2(1)8$ and various values of m and n. Further, Mijares (1964) has tabulated such approximate upper and lower 5 and 1% points of $V^{(s)}$ for $s = 2(1)50$. [The method of using moment quotients to obtain percentage points is available in Pearson and Hartley (1956).] Further, James (1964) has given the mgf of $V^{(p)}$ by employing the multivariate beta distribution* given by

$$K(p, \nu_1, \nu_2)|\mathbf{B}|^{(1/2)(\nu_1 - p - 1)}$$
$$\times |\mathbf{I} - \mathbf{B}|^{(1/2)(\nu_2 - p - 1)}, \qquad 0 < \mathbf{B} < \mathbf{I},$$

i.e., $\mathbf{B}$ and $\mathbf{I} - \mathbf{B}$ are both positive definite. [See James (1964) for $K(p, \nu_1, \nu_2)$ and Srivastava and Khatri [9] for the multivariate beta distribution when $\nu_1 < p$.] In view of this approach

$$E(e^{t \, \mathrm{tr} \, \mathbf{B}}) = \sum_{k=0}^{\infty} \frac{t^k}{k!} \sum_{\kappa} \frac{(\frac{1}{2}\nu_1)_\kappa C_\kappa(\mathbf{I})}{(\frac{1}{2}(\nu_1 + \nu_2))_\kappa},$$

where the coefficient $(a)_\kappa$ and the zonal polynomial* $C_\kappa(\mathbf{I})$ corresponding to the partition κ of k, are given in James (1964). Also $E(V^{(p)})^k$ may be seen to be the co-

efficient of $t^k/k!$ in the above series which is a special case, $_1F_1(a; b; \mathbf{S})$, of the $_pF_q(a_1, \ldots, a_p; b_1, \ldots, b_q; \mathbf{S})$ hypergeometric function of a matrix variate, defined in James (1964).

No general forms for the exact null distribution have yet been obtained. See Nanda (1950) for the cdf of $V^{(s)}$ for $m = 0$ and $s = 2$ and 3 through the mgf and inversion and Pillai and Jayachandran (1970) for extension of Nanda's method for the mgf when $m > 0$, explicit expressions for the cdf for $s = 3$, $m = 0, 1, 2, 3$, and $s = 4$, $m = 0, 1$ and exact upper percentage points for these values of m and s, and selected values of n. Also see Mikhail (1965) for the exact density of $V^{(2)}$ and Davis (1970) for the distribution of $V^{(s)}$ satisfying a differential equation (d.e.), and the results showing that Pillai's approximate percentage points of $V^{(s)}$ (Pillai, 1957, 1960) are accurate to four decimal places except when ν_1 and ν_2 are very small. Other related references are Krishnaiah and Chang (1972) and Schuurmann et al. (1973) which includes the inversion of Laplace transform obtained through the Pfaffian method, the density of $V^{(s)}$ explicitly for $s = 2(1)6$ and several values of ν_1 and ν_2, and exact upper percentage points for the same values of s, $\alpha = 0.01, 0.025, 0.05$, and 0.10, and m and $n = 0(1)10(5)25$ (confirming the comments of Davis concerning Pillai's approximate percentage points).

Pillai (1964) has obtained the following lemma which enables one to obtain the moments of Hotelling's trace $U^{(s)}$ from that of $V^{(s)}$ and vice versa:

Lemma. Let $V_{i,m,n}^{(s)}$ and $U_{i,m,n}^{(s)}$ denote the ith elementary symmetric functions* in the s b_i's and s f_i's discussed in the preceding section; then $E(U_{i,m,n}^{(s)})^k$ is derivable from $E(V_{i,m,n}^{(s)})^k$ by making the following changes in the expression of the latter: (a) Multiply by -1 all terms except the term in n in each linear factor involving n and (b) change n to $m + n + s + 1$ after performing (a).

Davis (1970) has also obtained a d.e. for $V^{(s)}$ from the d.e. for $U^{(s)}$ [Davis (1968)] by making the following simple transformation: $U^{(s)} \to -V^{(s)}$, $\nu_2 \to s - \nu_1 - \nu_2 + 1$. The transformation brings out the relationship between moments of $V^{(s)}$ and $U^{(s)}$. Further, Davis has shown that the density $f_{\nu_1,\nu_2}(V^{(s)}) = f_{\nu_2,\nu_1}(s - V^{(s)})$, which is useful for obtaining lower percentage points.

In regard to the asymptotic distributions of $V^{(s)}$, Muirhead (1970, 1978) has obtained an asymptotic expansion up to terms of order ν_2^{-2} for the cdf of $V^{(s)}$, involving chi-square cdfs, the first term being a chi-square cdf with $p\nu_1$ d.f., and Davis (1970) has derived an asymptotic expansion of the percentile of $V^{(s)}$ up to terms of order ν_2^{-3} in terms of chi-square* percentiles.

Nonnull Distribution

The exact distribution of $V^{(s)}$ has not yet been obtained. See Pillai and Jayachandran (1967, 1968) for the distribution in the two-roots case for (a), (b), and (c), using up to sixth degree zonal polynomials, defined in James (1964), and power tabulations. Also see Khatri and Pillai (1968) for the density of $V^{(p)}$ for (b) as a zonal polynomial series which is convergent only for $|V^{(p)}| < 1$, and Pillai (1964, 1966) and Khatri and Pillai (1965, 1968) for moments of $V^{(p)}$ in the linear case, i.e., when there is only one nonzero deviation parameter, based on the idea of independent beta variables. Khatri and Pillai (1967) explore the method of derivation of the moments of $V^{(p)}$ from moments of $V^{(r)}$ for (b), where $r \leqslant p$ is the rank of the noncentrality matrix Ω, and the derivation of the first two moments of $V^{(2)}$, while Pillai (1968) gives the mgf of $V^{(p)}$ for (a), (b), and (c) in zonal polynomial series, and Khatri (1967) gives the mgf of (a) with a slight error which was corrected by Pillai (1968). Refer to Pillai and Sudjana (1975) for an extension of the work of Pillai and Jayachandran in the two-roots case (1967, 1968) and for robustness studies of (a) and (b), and Pillai and Hsu (1979) for that of (c) using Pillai's distribution of the characteristic roots of $S_1 S_2^{-1}$ under violations (1975).

In regard to asymptotic distributions,

Fujikoshi (1970) has used Pillai's mgf (1968) to derive an asymptotic expansion of chi-square terms of the distribution of $V^{(p)}$ for (b) up to terms of order ν_2^{-2} and has computed approximate upper 5 and 1% points comparing with those of Pillai and Jayachandran (1967) and powers for the three-roots case. Similarly Fujikoshi (1970) has obtained an asymptotic expansion up to terms of order ν_2^{-2} for (c) involving noncentral chi-squares*. [For other asymptotic expansions, see Lee (1971) for expansions for (b) up to terms of order ν_2^{-3}, extension to (c) of that for (b), and for power computations; Sugiura and Nagao (1973) for (c) up to terms of order $(\nu_1 + \nu_2)^{-1}$ in terms of normal cdf and its derivatives using the method of differential operators on symmetric matrices following Siotani (1957), Ito (1960) and others; and Fujikoshi (1972) for (b) to terms of order ν_1^{-2} employing a method similar to that of Sugiura and Nagao (1971) in terms of chi-squares.]

Optimum Properties

For (a), let $\lambda_1, \ldots, \lambda_p$ be the ch. roots of $\Sigma_1 \Sigma_2^{-1}$ where $\Sigma_h(p \times p)$ is the covariance matrix of the hth population, $h = 1, 2$. Similarly for (b), let $\omega_1, \ldots, \omega_p$ be the ch. roots of Ω and for (c) let $\rho_1^2, \ldots, \rho_p^2$ be the ch. roots of $\Sigma_{11}^{-1}\Sigma_{12}\Sigma_{22}^{-1}\Sigma_{12}'$, where Σ_{ij} is the covariance matrix of the ith set with the jth $(i, j = 1, 2)$. It has been shown by Anderson and Das Gupta (1964) that the power of the $V^{(p)}$-test for (a) increases monotonically in each nonzero population ch. root, i.e., in each λ_i, $i = 1, \ldots, p$, for one-sided alternatives: $H_1: \lambda_i \geqslant 1$, $\sum_{i=1}^{p} \lambda_i > p$. Perlman (1974) has shown such a monotonicity property for the $V^{(p)}$-test for (b) and (c) under the condition,

$$\text{upper } (1 - \alpha) \text{ percentile} \leqslant \max(1, p - \nu_2).$$

Perlman and Olkin (1980) have demonstrated the unbiasedness* of all invariant tests with acceptance regions monotone in the maximal invariant statistic (characteristic roots) which include Pillai's trace test. Kiefer and Schwartz (1965) have shown that the $V^{(p)}$-test is admissible Bayes, fully invariant, similar, and unbiased. Giri (1968) has shown that it is locally best invariant for (a) against one-sided alternatives. Pillai and Jayachandran (1967, 1968) have made exact numerical power* comparisons in the two-roots case for each of the three hypotheses, for (a) against one-sided alternatives as above, for tests based on the four statistics, $U^{(2)}$, $V^{(2)}$, $W^{(2)} = \prod_{i=1}^{2}(1 - b_i)$ and the largest root. While the first three tests compare favorably and behave somewhat in the same manner in regard to the three hypotheses, the largest root has generally lower power than the other three when the number of nonzero deviation parameters is greater than one. When the sum of the two population roots is constant, the $V^{(2)}$-test has been observed to have largest power compared to others when the two roots are close. Schatzoff (1966) has made a Monte Carlo study for (b), also for larger numbers of roots and his findings are similar. Again for (b), Fujikoshi (1970) has computed some approximate powers for $U^{(3)}$, $V^{(3)}$, and $W^{(3)}$, and Lee (1971) some approximate powers for $s = 3$ and 4. For the two-sided alternatives for (a), Chu and Pillai [1] have observed in the two-roots case that all the four tests are biased but with the largest root showing generally least bias. They have also made comparisons of the four tests under locally unbiased conditions.

As regards robustness*, Mardia [2] has shown, based on certain permutation distributions, that the $V^{(s)}$-test for (b) is robust against moderate nonnormality but for test of (a) it may be sensitive. Olson [4] has made a Monte Carlo study concerning robustness of six MANOVA tests, including the four discussed above. For general protection against departures from normality* and from homogeneity of covariance matrices in the fixed effects model, he has recommended the $V^{(s)}$-test as the most robust of the MANOVA tests with adequate power against a variety of alternatives. Further Pillai and Sudjana (1975) have carried out some exact robustness studies for the two-roots case for (a) against nonnormality

and for (b) against covariance heterogeneity, based on Pillai's distribution of the ch. roots of $S_1 S_2^{-1}$ under violations (1975), and Pillai and Hsu [8] have made a similar study for (c) against nonnormality. Based on numerical values of the ratio $e = (p_1 - p_0)/(p_0 - \alpha)$, where p_1 = power under violations, p_0 = power without violation and $\alpha = 0.05$, the $V^{(s)}$-test has been observed to be the most robust in all cases.

APPLICATION

In order to illustrate the test procedure for (b) one may refer to the numerical example given in HOTELLING'S TRACE. The between and within SP matrices are available there, computed from 1. height (inches), 2. weight (pounds), 3. chest (inches), and 4. waist (inches) of 25 male reserve officers in civilian status of the Armed Forces of the Philippines hailing from six different regions. The null hypothesis is $H_0: \mu_1 = \mu_2 = \cdots = \mu_6$ versus H_1: not all μ_h's equal in $N_4(\mu_h, \Sigma)$, $h = 1, \ldots, 6$. Now

$$V^{(4)} = \text{tr } S^*(S^* + S)^{-1} = 0.1799$$

$$< V_{0.95}^{(4)}\left[m = \tfrac{1}{2}(|l - 1 - p| - 1) = 0,\right.$$

$$\left. n = \tfrac{1}{2}(N - l - p - 1) = 69.5 \right],$$

since $N = 150$ (see Pillai (1960)). Hence do not reject H_0. This agrees with the outcome of the Hotelling's trace test.

References

(For references not listed here see ref. 7. References 6 and 7 are annotated in HOTELLING'S TRACE.)

[1] Chu, S. S. and Pillai, K. C. S. (1979). *Ann. Inst. Statist. Math.*, **31**, 185–205.

[2] Mardia, K. V. (1971). *Biometrika*, **58**, 105–127.

[3] Muirhead, R. J. (1978). *Ann. Statist.*, **6**, 5–33.

[4] Olson, C. L. (1974). *J. Amer. Statist. Ass.*, **69**, 894–908.

[5] Perlman, M. D. and Olkin, I. (1980). *Ann. Statist.*, **8**, 1326–1341.

[6] Pillai, K. C. S. (1976). *Can. J. Statist.*, **4**, 157–183.

[7] Pillai, K. C. S. (1977). *Can. J. Statist.*, **5**, 1–62.

[8] Pillai, K. C. S. and Hsu, Y. S. (1979). *Ann. Inst. Statist. Math.*, **31**, 85–101.

[9] Srivastava, M. S. and Khatri, C. G. (1979). *An Introduction to Multivariate Statistics*, North-Holland, Amsterdam.

(GENERAL LINEAR MODEL
HOTELLING'S T^2
HOTELLING'S TRACE
LAMBDA CRITERION, WILKS'S
MAHALANOBIS D^2
MULTIVARIATE ANALYSIS
MULTIVARIATE ANALYSIS OF VARIANCE
ZONAL POLYNOMIALS)

K. C. S. PILLAI

PITMAN, E. J. G.

Edwin James George Pitman was born in Melbourne, Australia on October 29, 1897. He was educated at the University of Melbourne, graduating with B.Sc. and M.A. degrees. He was acting professor of mathematics at the University of New Zealand 1922–1923, then tutor in mathematics and physics at residential colleges of the University of Melbourne 1924–1925. In 1926 he was appointed professor of mathematics at the University of Tasmania, a position he held until his retirement at the end of 1962.

As part of his duties at the University of Tasmania, Pitman was required to offer a course in statistics. He was also consulted by experimentalists of the Tasmanian Department of Agriculture about the analysis of their data. These circumstances led to his studying the then new statistical methods and ideas being developed by R. A. Fisher*. As Fisher's ideas were not then widely understood, Pitman set about mastering his work, in particular, *Statistical Methods for Research Workers*.

Pitman became interested in testing hypotheses about means without any accompanying specification of form of population distributions. He examined the nonnull distribution of the test statistic for what would now be called the permutation test, first for the test of equality of two means, and then for tests of several means (analysis of variance) and of correlation. The results are given in three papers [2, 3].

This work on distribution-free methods* culminated in the lecture notes [6] developed during a visit to the United States in 1948 and 1949. These notes, produced for the lectures given at the University of North Carolina, were never published but were widely circulated in mimeographed form. They were frequently cited and undoubtedly provided a starting point for much subsequent work in this field.

However, Pitman's first published papers dealt with basic theoretical questions arising from the work of R. A. Fisher. His first paper [1] not only discussed the applicability of the concept of intrinsic accuracy but also established the result that families of distributions admitting a nontrivial sufficient statistic* are of exponential type.

In two major papers [4, 5] on inference about location and scale parameters, Pitman systematically developed the ideas of Fisher [14] on estimation conditional on a sample "configuration." He applied the concept of invariance*, previously used by Hotelling [15], essentially as a practical restriction on the class of estimators to be considered. The restrictions on simultaneous estimation of location and scale parameters were clearly spelled out. The powerful methods developed were applied to show the unbiasedness of Bartlett's test* for homogeneity of variances, as against Neyman and Pearson's related test.

Concern that much of the basic mathematics of statistical inference was unnecessarily unattractive led Pitman to reexamine some of the theory of the subject; the result was the monograph [11]. As he states in the preface: "The book is an attempt to present some of the basic mathematical results required for statistical inference with some elegance as well as precision . . . ."

In his later years, Pitman turned his attention to topics in probability theory. The most important results are those [8, 9] on the behavior of the characteristic function* of a probability distribution in the neighborhood of the origin; this behavior determines the nature of the limiting distribution of sums.

His most recent contribution [12] was to subexponential distributions, for which he gave new results and methods.

Pitman was remarkable as a statistician in that, though he worked for many years in virtual isolation, he was able to contribute to the solution of central problems of statistical inference and to originate some important fundamental ideas. His first contact with statisticians outside Australia was in 1948–1949, when he was invited to visit Columbia University, the University of North Carolina, and Princeton University to lecture on nonparametric inference.

Pitman contributed several important concepts in statistics. Perhaps the most useful and widely known is that of asymptotic relative efficiency* [13, pp. 337–338], which provides a means of comparing alternative tests of a hypothesis.

Pitman was much in demand as a visiting lecturer. He was visiting professor of statistics at Stanford in 1957. After his retirement from the University of Tasmania he visited Berkeley, Johns Hopkins, Adelaide, Melbourne, Chicago, and Dundee.

He was honored by election as a Fellow of the Australian Academy of Science (FAA) in 1954, a Fellow of the Institute of Mathematical Statistics in 1948, President of the Australian Mathematical Society in 1958 and 1959, and a Vice-President in 1960 of the International Statistical Institute (to membership of which he was elected in 1956). He was elected an Honorary Fellow of the Royal Statistical Society in 1965, an Honorary Life Member of the Statistical Society of Australia in 1966, and an Honorary Life Member of the Australian Mathematical Society in 1968.

In honor of his seventy-fifth birthday a collection of essays by colleagues and former students [16] was published. In honor of his eightieth birthday in 1977, and in recognition of his outstanding contributions to scholarship, the University of Tasmania conferred on him the Honorary Degree of Doctor of Science. In 1977 the Statistical Society of Australia instituted the Pitman medal, a gold medal to be awarded to a member of the Society for high distinction in statistics; the first such medal was awarded to Pitman himself.

In 1981 the American Statistical Association invited Pitman, along with other leading statisticians, to record a lecture on videotape for historical purposes.

At his home in Hobart, Tasmania, Edwin Pitman and his wife Elinor provided warm-hearted hospitality and kept "open house" over the past 40 years to visiting statisticians, mathematicians, and others of the academic community.

Selected References of E. J. G. Pitman

[1]   (1936). *Proc. Camb. Phil. Soc.*, **32**, 567–579.

[2]   (1937). *J. R. Statist. Soc. Suppl.*, **4**, 119–130; 225–232.

[3]   (1938). *Biometrika*, **29**, 322–335.

[4]   (1939a). *Biometrika*, **30**, 391–421.

[5]   (1939b). *Biometrika*, **31**, 200–215.

[6]   (1949a). *Lecture Notes on Non-Parametric Statistical Inference (unpublished)*.

[7]   (1949b—with Herbert Robbins). *Ann. Math. Statist.*, **20**, 552–560.

[8]   (1960). *Proc. 4th Berkeley Symp. Math. Statist. Prob.*, **2**, University of California Press, Berkeley, CA, pp. 393–402.

[9]   (1968). *J. Aust. Math. Soc.*, **8**, 423–443.

[10]   (1978). *Aust. J. Statist.*, **20**, 60–74.

[11]   (1979). *Some Basic Theory for Statistical Inference.* Chapman & Hall, London.

[12]   (1980). *J. Aust. Math. Soc.*, **A29**, 337–347.

Other References

[13]   Cox, D. R. and Hinkley, D. V. (1974). *Theoretical Statistics.* Chapman & Hall, London.

[14]   Fisher, R. A. (1934). *Proc. Roy. Soc.*, **A144**, 285–307.

[15]   Hotelling, H. (1936). *Biometrika*, **28**, 321–377.

[16]   Williams, E. J. (1974). *Studies in Probability and Statistics: Papers in Honour of Edwin J. G. Pitman.* North-Holland, Amsterdam.

(DISTRIBUTION-FREE METHODS
PITMAN EFFICIENCY
PITMAN ESTIMATORS
PITMAN TESTS)

E. J. WILLIAMS

PITMAN EFFICIENCY

SOME HISTORICAL COMMENTS

The Pitman efficiency is an index of the relative performance in large samples of two sequences of test statistics or estimators. This index of relative efficiency, called the *Pitman asymptotic relative efficiency* (A.R.E.), was apparently introduced by Pitman in 1948 [7]. Properties of this index were discussed in the articles of Noether [6] and of VanEeden [10]. A more general approach to the definition of the concept of asymptotic relative efficiency was proposed by Hoeffding and Rosenblatt [2]. Their definition yields the Pitman A.R.E. as a special case. More recently, Weiand [11] established conditions under which the Pitman efficiency coincides with another measure of relative efficiency called the Bahadur efficiency*. Rothe [9] generalized the definition of the Pitman efficiency and provided conditions under which the general definition is reduced to the conventional one. Rothe illustrated also how the relative efficiency can be derived in nontrivial cases. A nice exposition of the Pitman efficiency is given in the recent book of Lehmann [5] on statistical estimation. A more advanced discussion is given in the monograph of Pitman, which was published only in 1979. Applications of the Pitman efficiency can be found in numerous papers on both parametric and nonparametric estimation and testing. The well known measure of efficiency of unbiased estimators, which compares the Cramér–Rao lower bound* of variances to the variance of a given estimator, can also be considered as a special case of the Pitman efficiency. Thus the applicability of the Pitman efficiency index is very wide. The present article provides an introductory treatment of the subject, stressing the main ideas and basic properties. The article does not restrict attention to testing only but considers estimation problems too.

INTRODUCTORY EXAMPLE

In order to introduce the more formal discussion several examples are provided.

Consider first the following testing problem. Let $X_1, \ldots, X_n$ be independent and identically distributed (i.i.d.) random variables (a random sample) having a normal distribution $N(\theta, 1)$. Consider the problem of testing the null hypothesis $H_0 : \theta \leqslant 0$ versus the alternative hypothesis $H_1 : \theta > 0$ at level of significance α. From the theory of hypotheses testing [3], it is well known that the uniformly most powerful (UMP) test* is

$$\phi_n^* : \text{reject } H_0 \text{ if } \overline{X}_n \geqslant Z_{1-\alpha}/\sqrt{n},$$

where $\overline{X}_n = (1/n)\sum_{i=1}^{n} X_i$ is the sample mean, $Z_{1-\alpha}$ is the $(1 - \alpha)$ fractile of the standard normal distribution, i.e., $Z_{1-\alpha} = \Phi^{-1}(1 - \alpha)$, and $\Phi(Z)$ is the standard normal integral. The power of this UMP test when $\theta = \theta_1$, $\theta_1 > 0$, is

$$\psi_n(\theta_1) = \Phi\left(\sqrt{n}\,\theta_1 - Z_{1-\alpha}\right)$$

and the number of observations required so that $\psi_n(\theta_1) \geqslant \psi^*$ is $N(\alpha, \theta_1, \psi^*) = $ *least integer n greater than or equal to* $(Z_{\psi^*} + Z_{1-\alpha})^2/\theta_1^2$. Notice that

(i) $\lim_{n \to \infty} \psi_n(\theta_1) = 1$ for each θ_1 (fixed) and

(ii) if $\delta > 0$ and $\theta_1 = \delta/\sqrt{n}$ then

$$\lim_{n \to \infty} \psi_n(\delta/\sqrt{n}) = \Phi(\delta - Z_{1-\alpha}) = \psi_\infty,$$

where $\alpha < \psi_\infty < 1$.

If there is some doubt whether the above normal model is valid it may be decided to consider a more general model that $X_1, \ldots, X_n$ are i.i.d. having a location parameter distribution, with a PDF $f(x; \theta)$ $= g(x - \theta)$, $-\infty < \theta < \infty$. We assume also, without loss of generality, that the variance of X_i $(i = 1, \dots, n)$ is 1. It is not clear now whether the test of H_0 versus H_1 should be based on the sample mean $\overline{X}_n$ or on the sample median, M_n. Recall that if $n = 2m$ then $M_n = (X_{(m)} + X_{(m+1)})/2$ and if $n = 2m + 1$ then $M_n = X_{(m+1)}$, where $X_{(1)} \leqslant \dots \leqslant X_{(n)}$ is the order statistic. One could make, however, the following analysis of what happens in large samples.

We recall first that, under the above assumptions, the Central Limit Theorem* holds, and the asymptotic distribution of $\overline{X}_n$ is normal with mean θ and variance $1/n$. Moreover, the asymptotic distribution of M_n [4, p. 354] is also normal with mean θ, but with variance $1/\{4ng^2(0)\}$. Accordingly, one can consider the following two sequences of tests:

$$T_n^{(1)} : \text{reject } H_0 \text{ if } \overline{X}_n \geqslant a_n/\sqrt{n},$$

and

$$T_n^{(2)} : \text{reject } H_0 \text{ if } M_n \geqslant a_n/\{2g(0)\sqrt{n}\},$$

where $a_n \downarrow Z_{1-\alpha}$ as $n \to \infty$. Both tests will have asymptotically a level of significance (size) α. The asymptotic power functions of these tests are, however,

$$\psi_n^{(1)}(\theta) \approx \Phi(\theta\sqrt{n} - Z_{1-\alpha}) \qquad \theta > 0$$

and

$$\psi_n^{(2)}(\theta) \approx \Phi(2\theta g(0)\sqrt{n} - Z_{1-\alpha}), \qquad \theta > 0.$$

Consider now the sequence of power functions $\{\psi_n^{(1)}(\theta_n); n \geqslant 1\}$ evaluated at $\theta_n = \delta/\sqrt{n}$, with $\delta > 0$. Obviously,

$$\lim_{n\to\infty} \psi_n^{(1)}(\theta_n) = \Phi(\delta - Z_{1-\alpha}) = \psi^*,$$

where $0 < \psi^* < 1$. Similarly, the sequence $\{\psi_{n'(n)}^{(2)}(\theta_n); n \geqslant 1\}$ with $\theta_n = \delta/\sqrt{n}$ and the sample size $n'(n) = n/\{4g^2(0)\}$, converges also to ψ^*. The Pitman A.R.E. is defined as the limit of $n/n'(n)$ as $n \to \infty$. This is the asymptotic ratio of sample sizes required to attain the same asymptotic power over a proper sequence of parameters which converge to $\theta_0 = 0$. In the above sample, the

Pitman A.R.E. is $\lim_{n\to\infty} n/n'(n) = 4g^2(0)$. Notice that if the original model of normally distributed random variables is correct then $g(0) = (2\pi)^{-1/2}$ and the Pitman A.R.E. of M_n with respect to $\overline{X}_n$ is $4g^2(0) = 0.637$. On the other hand, if the distribution of X_i $(i = 1, \dots, n)$ has a standard PDF $g(x)$ $= (1/\sqrt{2})\exp(-\sqrt{2}|x|)$, $-\infty < x < \infty$, i.e., the double exponential (Laplace) distribution*, then the Pitman A.R.E. is 2.

THE THEORETICAL FRAMEWORK FOR TESTING HYPOTHESES

Let $X_1, X_2, \dots$ be a sequence of random variables whose distributions depend on a real parameter $\theta \in \Omega$. Let T_n $(n \geqslant 1)$ be a real valued statistic, depending on $X_1, \dots, X_n$. Suppose that there exist functions $\mu(\theta)$ and $\sigma(\theta)$ such that for each θ in Ω, the asymptotic distribution of $Z_n = (T_n - \mu(\theta))/\sigma_n(\theta)$ is the standard normal distribution $N(0, 1)$. Generally one can express the scaling factor $\sigma_n(\theta)$ as $\sigma_n(\theta) = c(\theta)w(n)$, where $c(\theta) > 0$, $w(n) > 0$. It is assumed that $w(n)$ $\downarrow 0$ as $n \to \infty$. Typically $w(n) = n^{-\alpha}$ or $(n\log n)^{-\alpha}$ for some $\alpha > 0$. Consider the problem of testing the hypothesis $H_0 : \theta \leqslant \theta_0$ versus $H_1 : \theta > \theta_0$ at level of significance $\alpha_n \to \alpha$ as $n \to \infty$. Furthermore, consider testing procedures which reject H_0 whenever $(T_n - \mu(\theta_0))/\sigma_n(\theta_0) \geqslant k_n$, where $k_n \to Z_{1-\alpha}$ as $n \to \infty$. The power functions corresponding to this sequence of tests are given for large values of n by

$$\psi_n(\theta; T_n) \approx \Phi\left(\frac{\mu(\theta) - \mu(\theta_0)}{w(n)c(\theta_0)} \cdot \frac{c(\theta_0)}{c(\theta)} \right.$$
$$\left. - Z_{1-\alpha}\frac{c(\theta_0)}{c(\theta)} \right). \qquad (1)$$

We assume that

A.1 The derivative of $\mu(\theta)$, $\mu'(\theta)$ is continuous in a neighborhood of θ_0 and $\mu'(\theta_0) > 0$;

A.2 $c(\theta)$ is continuous in a neighborhood of θ_0, and $c(\theta_0) > 0$.

Under these assumptions, if $\theta_n = \theta_0 + \delta w(n)$, $n \geqslant 1$, with $\delta > 0$, then

$$\lim_{n \to \infty} \psi_n(\theta_n; T_n) = \Phi\left(\frac{\delta\mu'(\theta_0)}{c(\theta_0)} - Z_{1-\alpha} \right)$$

$$= \psi^*, \qquad (2)$$

where $\alpha < \psi^* < 1$. The function

$$J(\theta; T) = \frac{(\mu'(\theta))^2}{c^2(\theta)} \qquad (3)$$

is called the *asymptotic efficacy** of T_n [8, p. 351]. Let V_n be an alternative statistic such that $W_n = (V_n - \eta(\theta))/\{v(\theta)w(n)\}$ is asymptotically $N(0,1)$. The asymptotic efficacy of the sequence of tests based on V_n is $J(\theta; V) = (\eta'(\theta))^2/v^2(\theta)$. We discuss now the case of $w(n) = n^{-1/2}$. This is the common rate of approach to normality of $n^{-1/2}$-consistent statistics. Accordingly, the asymptotic power of the sequence of tests based on T_n will be examined on the sequence $\theta_n = \theta_0 + \sigma/\sqrt{n}$. Let $\psi_n(\theta_n; V_n)$ denote the sequence of power functions of the test based on $V_{n'}$, at the parameter values $\theta_n = \theta_0 + \delta/\sqrt{n}$, and sample size n'. In order that $\psi_n(\theta_n; V_{n'})$ would converge to the same limit as $\psi_n(\theta_n; T_n)$ we should require that n' will depend on n according to

$$n'(n) = nJ(\theta_0; T)/J(\theta_0; V). \qquad (4)$$

Thus

$$\lim_{n \to \infty} \frac{n}{n'(n)} = \frac{J(\theta_0; V)}{J(\theta_0; T)}$$

$$= \left[\frac{n'(\theta_0)}{\mu'(\theta)} \right]^2 \cdot \frac{c^2(\theta_0)}{v^2(\theta_0)} . \qquad (5)$$

The r.h.s. of (5) is called the *Pitman A.R.E.* of $\{V_n\}$ w.r.t. $\{T_n\}$. We denote this index of asymptotic relative efficiency by eff.$(\theta_0; V_n, T_n)$.

COMMENTS

(i) The derivation of the Pitman A.R.E. was performed under the assumption that the asymptotic distributions of $\{T_n\}$ and of $\{V_n\}$ are normal. This assumption can be

generalized to any asymptotic distribution, provided the two standardized sequences, $\{Z_n\}$ and $\{W_n\}$, converge in law to the same distribution [8, Chap. 7, 9]. If $\{Z_n\}$ and $\{W_n\}$ converge in law to two different distributions, the Pitman A.R.E. is *not* defined.

(ii) If $w(n) = n^{-\alpha}$, $0 < \alpha$, the Pitman A.R.E. is

$$\text{eff.}(\theta_0; V_n, T_n)$$

$$= (J(\theta_0; V)/J(\theta_0; T))^{1/(2\alpha)}. \qquad (7)$$

ESTIMATION

Let $\mathscr{F}$ be a family of distributions depending on a real parameter θ. Consider the problem of estimating a parametric function $w(\theta)$. Let $X_1, X_2, \ldots, X_n$ be a sample of i.i.d. random variables having a density $f(x; \theta)$. Let $\hat{w}_1(\mathbf{X}_n)$ and $\hat{w}_2(\mathbf{X}_n)$ be two estimators of $w(\theta)$, where $\mathbf{X}_n = (X_1, \ldots, X_n)$. We assume that

$$E_\theta\{\hat{w}_i(\mathbf{X}_n)\} = w_i(\theta), \qquad i = 1, 2$$

and

$$V_\theta\{\hat{w}_i(\mathbf{X}_n)\} = \sigma_{i,n}^2(\theta), \qquad i = 1, 2.$$

Often, $\sigma_{i,n}^2(\theta) = \sigma_i^2(\theta)/n + o(1/n)$, as $n \to \infty$. The efficacy of $\hat{w}_i(\mathbf{X}_n)$ is defined as

$$J(\theta; \hat{w}_i) = (w_i'(\theta))^2/\sigma_{i,n}^2(\theta), \qquad i = 1, 2$$

$$(8)$$

and the *Pitman relative efficiency* of $\hat{w}_2$ compared to $\hat{w}_1$ is defined [1] as

$$\text{eff.}(\theta; \hat{w}_2, \hat{w}_1) = J(\theta; \hat{w}_2)/J(\theta; \hat{w}_1). \qquad (9)$$

Under some regularity conditions on the family of p.d.f.'s, $f(x; \theta)$, one can obtain an upper bound to the efficacy of estimators of θ [8]. Such an upper bound can be substituted in (9) for $J(\theta; \hat{w}_1)$ to obtain an index of relative efficiency related to the Pitman relative efficiency.

It is common to replace $J(\theta; \hat{w}_1)$ by the inverse of the Cramér–Rao lower bound for the variances of unbiased estimators of $w(\theta)$, when proper regularity conditions are satisfied [12, p. 244].

MULTIPARAMETER DISTRIBUTIONS

Suppose that the family of distributions $\mathscr{F}$ depends on k real parameters. The density functions are $f(x; \boldsymbol{\theta})$, where $\boldsymbol{\theta} = (\theta_1, \ldots, \theta_k)'$. Consider r parametric functions

$$g_1(\theta_1, \ldots, \theta_k), \ldots, g_r(\theta_1, \ldots, \theta_k),$$

where $1 \leqslant r \leqslant k$. Assume that the partial derivatives $\partial g_i(\theta)/\partial \theta_j$, $i = 1, \ldots, r$; $j = 1, \ldots, k$, exist for all $\boldsymbol{\theta}$ in the parameter space. Let $\mathbf{D}_g(\boldsymbol{\theta}) = (\partial g_i(\theta)/\partial \theta_j)$, be an $r \times k$ matrix of the partial derivatives. Let $\mathbf{g}(\boldsymbol{\theta})$ designate the vector of the parametric functions. Suppose that $\hat{\mathbf{g}}(\mathbf{X})$ is an unbiased estimator of $\mathbf{g}(\boldsymbol{\theta})$, based on a random sample of n i.i.d. random variables $X_1, \ldots, X_n$; with a nonsingular covariance matrix $\boldsymbol{\Sigma}_n(\mathbf{g})$. Assume that $\boldsymbol{\Sigma}_n(\hat{\mathbf{g}}) \approx (1/n)\mathbf{V}(\boldsymbol{\theta})$ as $n \to \infty$, where $\mathbf{V}(\boldsymbol{\theta})$ is a positive definite $r \times r$ matrix. The *efficacy* of the estimator $\hat{\mathbf{g}}(\mathbf{X})$ is

$$J(\boldsymbol{\theta}; \hat{\mathbf{g}}) = |\mathbf{D}_g(\boldsymbol{\theta})\mathbf{D}_g'(\boldsymbol{\theta})|/|\mathbf{V}(\boldsymbol{\theta})|. \quad (10)$$

Let $\hat{\mathbf{h}}(\mathbf{X})$ be an unbiased estimator of $h(\boldsymbol{\theta}) = (h_i((\boldsymbol{\theta}), \ldots, h_r(\boldsymbol{\theta}))'$, and $\mathbf{D}_h(\boldsymbol{\theta})$ the corresponding matrix of partial derivatives $\partial h_i(\theta)/\partial \theta_j$. Assume that the asymptotic variance of $\hat{\mathbf{h}}$ is $\boldsymbol{\Sigma}_n(\hat{\mathbf{h}}) \approx (1/n)\mathbf{V}^*(\boldsymbol{\theta})$, and that there exist nonsingular matrices $\mathbf{C}_1(\boldsymbol{\theta})$ and $\mathbf{C}_2(\boldsymbol{\theta})$ such that $\sqrt{n}\,\mathbf{C}_1(\boldsymbol{\theta})(\hat{\mathbf{g}} - \mathbf{g})$ and $\sqrt{n}\,\mathbf{C}_2(\boldsymbol{\theta})(\hat{\mathbf{h}} - \mathbf{h})$ converge to the same multivariate distribution; then the Pitman relative efficiency of $\hat{\mathbf{h}}$ w.r.t. $\hat{\mathbf{g}}$, at $\boldsymbol{\theta}$, is

$$\text{eff.}\,(\boldsymbol{\theta}; \hat{\mathbf{h}}, \hat{\mathbf{g}}) = \frac{|\mathbf{D}_h(\boldsymbol{\theta})\mathbf{D}_h'(\boldsymbol{\theta})|}{|\mathbf{D}_g(\boldsymbol{\theta})\mathbf{D}_g'(\boldsymbol{\theta})|} \cdot \frac{|\mathbf{V}(\boldsymbol{\theta})|}{|\mathbf{V}^*(\boldsymbol{\theta})|}.$$

$$(11)$$

If $\hat{\mathbf{g}}$ and $\hat{\mathbf{h}}$ do not converge to the same type of asymptotic multivariate distribution, the A.R.E. is not defined.

Example 1. As shown by Lehmann [4, p. 376], the Pitman A.R.E. of the Wilcoxon rank-sum test* w.r.t. the Students t-test, in the two-sample location problem is

$$\text{eff.} = 12 \left(\int \phi^2(x)\,dx \right)^2,$$

where $\phi(x)$ is the standardized density of X.

In the normal case, this Pitman efficiency is $3/\pi$.

Example 2. Let $X_1, X_2, \ldots, X_n$ be i.i.d. Poisson random variables, with mean λ, $0 < \lambda < \infty$. Let $w(\lambda) = \exp(-\lambda)$. The maximum likelihood estimator (MLE) of $w(\lambda)$ is $\hat{w}_1 = \exp\{-\overline{X}_n\}$, where $\overline{X}_n = (1/n)\sum X_i$. Let $\hat{w}_2 = (1/n)\sum_{i=1}^n I\{X_i = 0\}$, where

$$I\{x_i = 0\} = \begin{array}{ll} 1, & \text{if } x_i = 0 \\ 0, & \text{if } x_i > 0. \end{array}$$

The asymptotic distribution of $\hat{w}_1$ is $N(w(\lambda), \lambda w^2(\lambda)/n)$. The asymptotic distribution of $\hat{w}_2$ is $N(w(\lambda), w(\lambda)(1 - w(\lambda))/n)$. Hence the asymptotic relative efficiency of $\hat{w}_2$ compared to $\hat{w}_1$ is $\text{eff.}(\lambda; \hat{w}_2, \hat{w}_1) = \lambda w(\lambda)/(1 - w(\lambda))$.

Example 3. Let $X_1, X_2, \ldots, X_n$ be i.i.d. random variables having a lognormal distribution* $\text{LN}(\mu, \sigma^2)$, i.e., $Y = \log X$ has a normal distribution, $N(\mu, \sigma^2)$. It is well known [12, p. 220] that the expected value of X is $\xi(\mu, \sigma^2) = \exp(\mu + \sigma^2/2)$. The MLE of ξ is $\hat{\xi}_n = \exp(\overline{Y}_n + \hat{\sigma}^2/2)$, where $\overline{Y}_n = (1/n)\sum_{i=1}^n Y_i$ and $\hat{\sigma}_n^2 = (1/n)\sum_{i=1}^n (Y_i - \overline{Y}_n)^2$. On the other hand, the sample mean $\overline{X}_n$ is an unbiased estimator of ξ. Both $\hat{\xi}_n$ and $\overline{X}_n$ are asymptotically normal. Moreover, the asymptotic variance of $\hat{\xi}_n$ is $\xi^2\sigma^2(1 + \frac{1}{2}\sigma^2)/n$, while that of $\overline{X}_n$ is $\xi^2(e^{\sigma^2} - 1)/n$. Thus the Pitman efficiency of $\overline{X}_n$ w.r.t. $\hat{\xi}_n$ is

$$\text{eff.}\,(\mu, \sigma^2; X_n, \xi_n) = \sigma^2(1 + \sigma^2/2)/(e^{\sigma^2} - 1).$$

We notice that the Pitman A.R.E. of $\overline{X}_n$ approaches 0 as $\sigma^2 \to \infty$.

Example 4. Let $\mathscr{F}$ be the family of rectangular distributions* on $(0, \theta)$, $0 < \theta < \infty$. Given a sample of n i.i.d. random variables, the complete sufficient statistic is $X_{(n)} = \max_{1 < i < n} X_i$. The uniformly minimum variance unbiased estimator of θ is $\hat{\theta}_n = [(n + 1)/n]X_{(n)}$. The asymptotic distribution of θ_n is, however, *not* normal. Indeed $P_\theta\{\hat{\theta}_n - \theta > 1 - \delta\} \approx 1 - \exp(-n\delta)$ as $n \to \infty$.

On the other hand, $\hat{\hat{\theta}}_n = 2\overline{X}_n$ where $\overline{X}_n$ is the sample mean is asymptotically distributed like $N(\theta, \theta^2/(4n))$. The Pitman efficiency of $\hat{\hat{\theta}}_n$ w.r.t. $\hat{\theta}_n$ is undefined.

References

[1] DeGroot, M. H. and Raghavachari, M. (1970). Relations between the Pitman efficiency and Fisher information, *Sankhyā*, **32**, 314–324.

[2] Hoeffding, W. and J. R. Rosenblatt (1955). The efficiency of tests, *Ann. Math. Statist.*, **26**, 52–63.

[3] Lehmann, E. L. (1959). *Testing Statistical Hypotheses*. Wiley, New York.

[4] Lehmann, E. L. (1975). *Nonparametrics: Statistical Methods Based on Ranks*. Holden-Day, San Francisco.

[5] Lehmann, E. L. (1983). *Theory of Point Estimation*. Wiley, New York. (A new comprehensive book on the theory of statistical estimation. Advanced graduate level.)

[6] Noether, G. F. (1955). On a theorem of Pitman, *Ann. Math. Statist.*, **26**, 64–68.

[7] Pitman, E. J. G. (1948). *Notes on Nonparametric Statistical Inference*, Institute of Statistics, University of North Carolina, Chapel Hill, N.C. (unpublished).

[8] Pitman, E. J. G. (1979). *Some Basic Theory for Statistical Inference*. Chapman & Hall, London.

[9] Rothe, G. (1981). Some properties of the asymptotic relative Pitman efficiency, *Ann. Statist.*, **9**, 663–669.

[10] VanEeden, C. (1963) The relation between Pitman's asymptotic relative efficiency of two tests and the correlation coefficient between the test statistics, *Ann. Math. Statist.*, **34**, 1442–1451.

[11] Weiand, H. S. (1976) A condition under which the Pitman and Bahadur approaches to efficiency coincide, *Ann. Statist.*, **4**, 1003–1011.

[12] Zacks, S. (1980). *Parametric Statistical Inference: Basic Theory and Modern Approaches*, Pergamon, Oxford, England. (A graduate level textbook in the theory of statistical inference, with many worked out examples and problems.)

(BAHADUR EFFICIENCY)

S. ZACKS

PITMAN ESTIMATORS

SMALL SAMPLE THEORY

The Concept

Pitman estimators have been introduced by Pitman [29] in order to solve the *point esti-mation** problem for location parameter models. In the following let $X_1 \ldots X_n$ be independent, identically distributed random vectors with values in $\mathbb{R}^k$ such that the common distribution of the deviations $X_i - \theta$ does not depend on θ. If the distribution has the density f and a finite first moment, then the Pitman estimator of θ is

$$\mu_P(x_1 \ldots x_n)$$
$$= \frac{\int \theta f(x_1 - \theta) \ldots f(x_n - \theta) \, d\theta}{\int f(x_1 - \theta) \ldots f(x_n - \theta) \, d\theta}.$$

Pitman estimators have a couple of remarkable properties which make them interesting for theoretical and practical purposes.

First, Pitman estimators are *equivariant**, i.e.,

$$\mu_P(x_1 + a, \ldots, x_n + a)$$
$$= \mu_P(x_1 \ldots x_n) + a.$$

They are distinguished in that they have minimal risk with respect to quadratic loss among equivariant estimators. This property has been the starting point of the definition of Pitman estimators in the work of Pitman [29].

Further, Pitman estimators μ_P are unbiased estimators of θ. They need not be of uniformly minimum variance since such unbiased estimators do not exist in general. However, if there exists an unbiased estimator of uniformly minimum variance, then it is identical with the Pitman estimator μ_P. Moreover, in this case the Pitman estimator is a necessary and sufficient statistic for the estimation of θ provided that the moments of f do not increase too fast [40].

Another important property of Pitman estimators is the minimax* property with respect to quadratic loss. It says that for any other estimator δ

$$\sup_\theta E_\theta\left[(\delta - \theta)^2\right] \geq \sup_\theta E_\theta\left[(\mu_P - \theta)^2\right].$$

This was proved by Girshick and Savage [12]. Under mild conditions the Pitman estimator is even admissible if $\theta \in \mathbb{R}$ [18, 31].

If the density f is of the form

$$f(x) = \frac{1}{\sqrt{2\pi}} e^{-x^2/2},$$

then the underlying location parameter model is called a *Gaussian shift*. For a Gaussian shift the Pitman estimator coincides with the sample mean $\bar{x}$. A partial converse of this fact is valid [7, 10]. The well-known minimax property and, if $\theta \in \mathbb{R}$, also the admissibility* of the sample mean in case of a Gaussian shift is mainly due to the fact that the sample mean is the Pitman estimator. This becomes very clear in Blyth [6], where admissibility of the sample mean was proved for the first time.

If the observations $X_1 \ldots X_n$ are not governed by a Gaussian shift, then, as a rule, the Pitman estimator is different from the sample mean $\bar{x}$ and even from the maximum likelihood* (ML) estimator $\hat{\theta}$. However in contrast to its theoretical superiority the Pitman estimator is often difficult to compute, which is a practical disadvantage.

Extensions

Define the risk of an estimator δ by $E_\theta(W(\delta - \theta))$ where W is an arbitrary loss function. Then one may again consider the problem of determining the equivariant estimator with minimal risk. Let us call this estimator the Pitman estimator for W and denote it by μ_W.

The form of the Pitman estimator μ_P admits an interesting interpretation. For a given sample $(x_1 \ldots x_n)$ let us call the distribution on $\mathbb{R}^k$ with density

$$g(\theta) = \frac{f(x_1 - \theta) \ldots f(x_n - \theta)}{\int f(x_1 - \theta') \ldots f(x_n - \theta') d\theta'}$$

the *posterior distribution* with respect to the uniform weight function. Then μ_P appears as the mean of this distribution or, in other words, as the *Bayes estimator* for quadratic loss (and the uniform weight function).

This interpretation is of general importance. The Pitman estimator for an arbitrary loss function W is always the Bayes estimator for W and the uniform weight function. If the Bayes solution is not uniquely determined one has to choose an equivariant one.

In case of a Gaussian shift the sample mean coincides with the Pitman estimator μ_W for every quasiconvex and centrally symmetric loss function W. This is due to the Bayes representation of the Pitman estimator and Anderson's lemma [3]. From the theorem of Hunt and Stein (*see* HUNT–STEIN THEOREM and Wesler [42]) it follows that μ_W is minimax*, which implies in particular that the sample mean is minimax for Gaussian shifts and the above mentioned class of loss functions.

If $\theta \in \mathbb{R}$ and in case of a Gaussian shift the sample mean is even admissible for every symmetric and quasiconvex loss function [6]. For more general location parameter models the problem of admissibility is more difficult to deal with.

In the following examples let $k = 1$.

Example 1. Take the loss function $W(x) = |x|$. Then the Pitman estimator equals the median $\mu_{0.5}$ of the posterior distribution. This Pitman estimator is median unbiased*, i.e.,

$$P_\theta\{\mu_{0.5} \geqslant \theta\} = \tfrac{1}{2}, \qquad P_\theta\{\mu_{0.5} \leqslant \theta\} \geqslant \tfrac{1}{2}.$$

Moreover, it is an admissible estimator of θ [11].

Example 2. Take the loss function $W(x) = ax$ if $x > 0$ and $W(x) = -bx$ if $x < 0$, where $a > 0$, $b > 0$. Then the Pitman estimator equals the α-quantile* μ_α of the posterior distribution*, for $\alpha = a/(a + b)$. These Pitman estimators satisfy

$$P_\theta\{\mu_\alpha \geqslant \theta\} = \alpha, \qquad P_\theta\{\mu_\alpha \leqslant \theta\} \geqslant 1 - \alpha,$$

which corresponds to the notion of unbiasedness defined by the loss function W. Moreover, these estimators are admissible for the estimation of θ [11]. The unbiasedness of μ_α can be used for testing the hypothesis $H = \{\theta \leqslant \vartheta\}$ against the alternative $K = \{\theta > \vartheta\}$ for any fixed ϑ. If

$$U = \frac{\int_{-\infty}^{\vartheta} f(x_1 - \theta) \ldots f(x_n - \theta) d\theta}{\int_{-\infty}^{+\infty} f(x_1 - \theta) \ldots f(x_n - \theta) d\theta},$$

then we have $U < \alpha$ iff $\mu_\alpha > \vartheta$ and hence $\{U < \alpha\}$ is a critical region* for testing H against K which is unbiased of level α. The test statistic U has the advantage of coinciding with the critical value of the class of tests defined by U.

Example 3. Take the loss function $W(x) = 1$ for $|x| \geqslant c$, and $W(x) = 0$ for $|x| < c$. The appertaining Pitman estimator is the maximum probability estimator* defined by Weiss and Wolfowitz [41]. The problem of admissibility of such an estimator depends on the uniqueness of the associated Bayesian estimation problem [9].

The problem of optimal equivariant estimation can also be treated within the framework of general statistical invariance theory. Already Pitman [29] considered location and scale parameter models, i.e., models which are generated by the group of location and scale transformations. His work has been continued by Hora and Buehler [16]. Another possibility consists of dropping the assumption of identically distributed observations. This leads to general linear models which are generated by linear subspaces of $\mathbb{R}^{k \cdot n}$. Particular cases of practical interest are the two-sample problem, analysis of variance* and regression* models. In such cases the problem of optimal equivariant estimation can be solved similarly to the most simple case of estimating a single location parameter. Again the solution is called *Pitman estimator*. If the underlying density f is Gaussian, then the Pitman estimators are exactly those projections which arise as least squares* estimators in the theory of linear models.

LARGE SAMPLE THEORY

Fix a loss function W and let μ_n be the Pitman estimator for the sample size n. In the following the loss function is specified if necessary.

Consistency

As far as consistency is concerned, Pitman estimators do not differ from Bayes estimators for arbitrary continuous weight functions. For example, Pitman estimators are consistent whenever ML estimators are consistent [36]. Moreover, Pitman estimators are consistent with the correct rate of convergence [23, 37, 38].

First Order Properties

If the density f is differentiable and has finite Fisher's information I, then the statistical model which governs the observations $X_1 \ldots X_n$ can be approximated by a Gaussian shift (for details see LeCam [20] and Hajek [13, 14]). In such cases the first order properties (asymptotic efficiency, asymptotic minimaxity, asymptotic admissibility) of Pitman estimators do not differ from those of arbitrary Bayes estimators or ML estimators. This was proved for the first time by LeCam [19]. If W is symmetric, then this follows from the fact that

$$\sqrt{n}\,(\hat{\theta}_n - \mu_n) \to 0,$$

where $\hat{\theta}_n$ denotes the ML estimator for the sample size n (see also Chao [8] and Bickel and Yahav [5]).

To illustrate the case of nonsymmetric W let us consider Example 2. In this case the Pitman estimators $\mu_{n,\alpha}$ are the α-quantiles of the posterior distribution and satisfy

$$\sqrt{n}\,(\hat{\theta}_n - \mu_{n,\alpha}) \to N_\alpha / \sqrt{I}\,,$$

where N_α is the α-quantile of the standard Gaussian distribution. This implies that the tests based on the statistic U are of asymptotic efficiency 1.

Higher Order Properties

If the density f is sufficiently smooth and possesses moments of sufficiently high order, then the asymptotic theory is able to distinguish between various estimators that appear equivalent as long as only first order properties are considered.

If the loss function W is symmetric, then the asymptotic equivalence of arbitrary Bayes estimators and ML estimators is exactly of the order

$$\sqrt{n}\,(\hat{\theta}_n - \mu_n) = O\!\left(\frac{1}{\sqrt{n}}\right)$$

[32, 33]. The difference between these estimators is not negligible if multiplied by the sample size n.

The classification of estimators in terms of

their higher order properties is the work of Pfanzagl [26] and Pfanzagl and Wefelmeyer [28]. By means of this theory it can be shown that the Pitman estimators for $W(x) = |x|$, i.e., the medians $\mu_{n,0.5}$ of the posterior distributions, are asymptotically optimal of order $o(n^{-1})$ within the class of all estimators which are median unbiased of order $o(n^{-1})$ [35]. This property distinguishes the Pitman estimators from ML estimators and other Bayes estimators which have to be improved by a bias-correction to obtain this optimality. Some further information concerning higher order properties of Pitman estimators can be found in Akahira and Takeuchi [2].

The role of the tests based on the quantiles $\mu_{n,\alpha}$ can also be clarified by their higher order properties. The suitable notion for this purpose is the concept of deficiency introduced by Hodges and Lehmann [15] and pursued by Pfanzagl [26]. Given a level α, for every $\beta \geqslant \alpha$ there exist tests $\varphi_{n,\beta}$ which keep the level α with an accuracy $o(n^{-1})$ and need a minimal number n of observations to reject relevant alternatives with power β (for details see Pfanzagl [26]). For a given test φ_n which keeps the level α with $o(n^{-1})$ the deficiency $d(\varphi_n, \beta)$ is the additional number of observations which is needed compared with $\varphi_{n,\beta}$ to reject relevant alternatives with power β. The point is that, in general, there do not exist tests with deficiency zero for all $\beta \geqslant \alpha$ [26].

The deficiency of the tests based on the Pitman estimators $\mu_{n,\alpha}$ is $d(\mu_{n,\alpha}, \beta) = D \cdot N_\beta^2$, where $D \geqslant 0$ is a number measuring the deviation of the density f from the Gaussian case, and N_β is the β-quantile of the Gaussian distribution [34]. It follows that the tests based on $\mu_{n,\alpha}$ have deficiency zero if $\beta = 0.5$, i.e., they need a minimal number of observations to reject relevant alternatives with power $\beta = 1/2$.

Nonregular Cases

Classical asymptotics of first order consider exclusively such cases where the density of the observations $X_1 \ldots X_n$ is smooth and has finite Fisher's information*. However,

this need not always be the case. Cases where such regularity conditions are not satisfied were considered by Prakasa Rao [30], Akahira [1], and Ibragimov and Has'minskii [17].

Considering such nonregular cases it is still possible to approximate the original statistical model by a simpler one [21]. Let us call this approximation the limiting model since it comes up letting $n \to \infty$. The extension of classical asymptotics [19, 14] by LeCam [21, 25] shows that the analysis of the finite sample case can be simplified by analyzing only the limiting model. Now in the present context it is of particular interest that limiting models are typically translation invariant [22]. The type of translation invariance proved by LeCam is of a more general nature than the invariance property of being a location parameter model. But by means of the theory of statistical experiments [24], it can be shown that these invariance properties are essentially equivalent [4]. Thus Pitman estimators are the general solution of the estimation problem for limiting models (for the minimax property see Strasser [39] and for admissibility see Becker and Strasser [4]).

In the light of the preceding remarks the importance of Pitman estimators goes far beyond the finite sample case of smooth location parameter models.

References

[1] Akahira, M. (1979). *Asymptotic Theory of Statistical Estimation in Non-Regular Cases*. Statistical Laboratory, Department of Mathematics, University of Electro-Communications, Chofu, Tokyo 182, Japan.

[2] Akahira, M. and Takeuchi, K. (1981). Asymptotic Efficiency of Statistical Estimators: Concepts and Higher Order Asymptotic Efficiency. *Lect. Notes Statist.*, **7**, Springer, New York.

[3] Anderson, T. W. (1955). *Proc. Amer. Math. Soc.*, **6**, 170–176.

[4] Becker, C. and Strasser, H. (1986). *Statistics and Decisions*, **3** (to appear).

[5] Bickel, P. J. and Yahav, J. A. (1969). *Z. Wahrscheinlichkeitsth. verwend. Geb.*, **11**, 257–276.

[6] Blyth, C. R. (1951). *Ann. Math. Statist.*, **22**, 22–42.

[7]   Bondesson, E. L. (1974). *Sankhyā*, **A36**, 321–324.

[8]   Chao, M. T. (1970). *Ann. Math. Statist.*, **41**, 601–608.

[9]   Farrell, R. H. (1964). *Ann. Math. Statist.*, **35**, 949–998.

[10] Fieger, W. (1976). *Sankhyā*, **A38**, 394–396.

[11] Fox, M. and Rubin, H. (1964). *Ann. Math. Statist.*, **35**, 1019–1031.

[12] Girshick, M. A. and Savage, L. G. (1951). *Proc. 2nd Berkeley Symp. Math. Statist. Prob.*, **1**, University of California Press, Berkeley, CA, pp. 53–74.

[13] Hajek, J. (1970). *Z. Wahrscheinlichkeitsth. verwend. Geb.*, **14**, 323–330.

[14] Hajek, J. (1972). *Proc. 6th Berkeley Symp. Math. Statist. Prob.*, **1**, University of California Press, Berkeley, CA, pp. 175–194.

[15] Hodges, J. L. and Lehmann, E. L. (1970). *Ann. Math. Statist.*, **41**, 783–801.

[16] Hora, R. B. and Buehler, R. J. (1966). *Ann. Math. Statist.*, **37**, 643–656.

[17] Ibragimov, I. A. and Has'minskii, R. Z. (1981). *Statistical Estimation—Asymptotic Theory* (trans. S. Kotz). Springer, New York.

[18] Karlin, S. (1958). *Ann. Math. Statist.*, **29**, 406–436.

[19] LeCam, L. (1953). *Univ. Calif. Publ. Statist.*, **1**, 277–330.

[20] LeCam, L. (1960). *Univ. Calif. Publ. Statist.*, **3**, 37–98.

[21] LeCam, L. (1972). *Proc. 6th Berkeley Symp. Math. Statist. Prob.*, **1**, University of California Press, Berkeley, CA, pp. 245–261.

[22] LeCam, L. (1973). "Sur les contraintes imposées par les passages à la limite usuels en statistique." I.S.I. Congress, Vienna, Austria.

[23] LeCam, L. (1973). *Ann. Statist.*, **1**, 38–53.

[24] LeCam, L. (1974). *Notes on Asymptotic Methods in Statistical Decision Theory*. I. Publ. du Centre de Recherches Mathématiques, Université de Montréal, Québec, Canada.

[25] LeCam, L. (1979). In *Contributions to Statistics—Hájek Memorial Volume*, J. Jureckova, ed. Reidel, Dordrecht, Netherlands, pp. 119–135.

[26] Pfanzagl, J. (1974). *Preprint in Statistics*, Vol. 8, University of Cologne, W. Germany.

[27] Pfanzagl, J. (1975). In *Statistical Inference and Related Topics*, Vol. 2, *Proc. Summer Res. Inst. Statist. Inference Stochastic Processes*. Academic Press, New York, pp. 1–43.

[28] Pfanzagl, J. and Wefelmeyer, W. (1978). *J. Multiv. Anal.*, **8**, 1–29.

[29] Pitman, E. J. G. (1939). *Biometrika*, **30**, 391–421.

[30] Prakasa Rao, B. L. S. (1968). *Ann. Math. Statist.*, **39**, 76–87.

[31] Stein, C. (1959). *Ann. Math. Statist.*, **30**, 970–979.

[32] Strasser, H. (1975). *J. Multiv. Anal.*, **5**, 206–226.

[33] Strasser, H. (1977). *Theor. Prob. Appl.*, **22**, 349–361.

[34] Strasser, H. (1977). *Recent Developments in Statistics*, J. R. Barra et al., eds. North-Holland, Amsterdam, pp. 9–35.

[35] Strasser, H. (1978). *Ann. Statist.*, **6**, 867–881.

[36] Strasser, H. (1981). *Ann. Statist.*, **9**, 1107–1113.

[37] Strasser, H. (1981). *J. Multiv. Anal.*, **11**, 127–151.

[38] Strasser, H. (1981). *J. Multiv. Anal.*, **11**, 152–172.

[39] Strasser, H. (1982). *Z. Wahrscheinlichkeitsth. verwend. Geb.*, **60**, 223–247.

[40] Strasser, H. (1985). *Metrika*, **43** (to appear).

[41] Weiss, L. and Wolfowitz, J. (1967). *Ann. Inst. Statist. Math.*, **19**, 193–206.

[42] Wesler, O. (1959). *Ann. Math. Statist.*, **30**, 1–20.

(ADMISSIBILITY
EFFICIENCY, SECOND-ORDER
EQUIVARIANT ESTIMATORS
ESTIMATION, POINT
INVARIANCE CONCEPTS IN STATISTICS
MAXIMUM LIKELIHOOD ESTIMATION)

H. STRASSER

PITMAN–MORGAN TEST

This is a test of equality of variances of two random variables X, Y having a joint bivariate normal distribution*, given n pairs of values (X_i, Y_i) from a random sample.

The method is based on noting that the variables $U_i = X_i - Y_i$ and $V_i = X_i + Y_i$ have a joint bivariate normal distribution with covariance $\mathrm{cov}(U, V) = \mathrm{var}(X) - \mathrm{var}(Y)$ so testing the hypothesis $\mathrm{var}(X) = \mathrm{var}(Y)$ is equivalent to testing the hypothesis that the correlation between U and V is zero.

The appropriate criterion for the latter test is the sample correlation* between U and V, r_{UV}. Significance limits for r_{UV} (corresponding to zero population correlation, i.e., equality of variances of X and Y) are obtained by referring $r_{UV}\sqrt{(n-2)} \big/ \sqrt{\left(1 - r_{UV}^2\right)}$ to a t distribution* with $(n-2)$ degrees of freedom.

The test was proposed simultaneously by

Morgan [2] and Pitman [3]. C. T. Hsu [1] extended the method, applying it to test several other hypotheses about parameters of bivariate normal distributions.

The test is easily adapted to test the hypothesis $\mathrm{var}(X) = \theta^2 \mathrm{var}(Y)$ by replacing Y by $Y' = \theta Y$ and testing the hypothesis $\mathrm{var}(X) = \mathrm{var}(Y')$.

References

[1] Hsu, C. T. (1940). *Ann. Math. Statist.*, **11**, 410–426.

[2] Morgan, W. A. (1939). *Biometrika*, **31**, 13–19.

[3] Pitman, E. J. G. (1939). *Biometrika*, **31**, 9–12.

(CORRELATION)

PITMAN TESTS

Pitman tests are tests of significance* that are valid without any assumptions about the population sampled and also without any assumptions about random sampling to obtain the observations or measurements. These tests are based on the randomization distribution of the observations and hence are members of the class of randomization tests*; they are also special kinds of distribution-free* or nonparametric tests. Pitman [15, 16, 17] proposed these significance tests for the two-sample location problem, the bivariate correlation problem, and the analysis of variance* problem, but the basic idea can be extended to testing in almost any kind of problem. The critical region* for any of these test statistics depends on the specific observations or measurements in the data set in the sense that for a level α test, the conditional probability of rejecting the null hypothesis H_0, given the data, is equal to α when H_0 is true. Thus these tests might also be called conditional tests.

Pitman tests are sometimes called *Fisher–Pitman* tests, because the principle of randomization is discussed at length in both of Fisher's first editions (1970 [5], first edition 1924 and 1966 [4], first edition 1935) and Fisher's Exact Test* is of the same type. They might also be called permutation tests* because they are based on a randomization distribution generated by permutations.

Pitman tests are sometimes included among the topics covered in books on distribution-free or nonparametric statistics. Books that have at least some significant discussion are Bradley [1, Chap. 4], Conover [2, pp. 327–334], Fraser [6], Lehmann [11], Marascuilo and McSweeney [13], Mosteller and Rourke [14, pp. 12–22], Pratt and Gibbons [18, Chaps. 4 and 6], Randles and Wolfe [19], and Siegel [20]. Edgington [3] gives a complete coverage of Pitman tests and other randomization tests, but his entry on RANDOMIZATION TESTS does not discuss Pitman tests in much detail.

TWO-SAMPLE PITMAN TEST

The first test of significance proposed by Pitman [15] applies to the situation where $X_1, X_2, \ldots, X_m$ is a sample from a continuous distribution and $Y_1, Y_2, \ldots, Y_n$ is a sample from another continuous distribution and all observations are mutually independent. We want to test the null hypothesis that the two distributions are identical with a procedure that is particularly sensitive to differences in locations, or we assume the location model.

Under the null hypothesis, conditional upon the $m + n$ observations, each of the $\binom{m+n}{m}$ separations of the actual combined sample data into m X's and n Y's is equally likely. Thus for any specific test statistic that is a function of these separations, we can compute the statistic's value for each possible separation and find the critical value for a test at level α from this enumeration. The value of the test statistic for the separation observed can then be compared with this critical value to make a decision. A natural test statistic here is simply $\overline{X} - \overline{Y}$ although several other equivalent forms could be used (including the usual two-sample t statistic). A numerical example is the easiest way to explain the procedure fully. The data, percent change in retail drug sales in Alabama over a one-year period, for three randomly

Table 1   Generation of Randomization Distribution of $\overline{X} - \overline{Y}$

X Set	$\sum X$	$\overline{X} - \overline{Y}$	X Set	$\sum X$	$\overline{X} - \overline{Y}$
$-0.2, 0.5, 0.9$	1.2	-8.2	$0.5, 0.9, 14.3$	15.7	0.3
$-0.2, 0.5, 2.0$	2.3	-7.5	$-0.2, 2.0, 14.3$	16.1	0.5
$-0.2, 0.9, 2.0$	2.7	-7.3	$0.5, 2.0, 14.3$	16.8	0.9
$0.5, 0.9, 2.0$	3.4	-6.9	$0.9, 2.0, 14.3$	17.2	1.2
$-0.2, 0.5, 6.5$	6.8	-4.9	$-0.2, 6.5, 11.5$	17.8	1.5
$-0.2, 0.9, 6.5$	7.2	-4.7	$0.5, 6.5, 11.5$	18.5	1.9
$0.5, 0.9, 6.5$	7.9	-4.3	$0.9, 6.5, 11.5$	18.9	2.2
$-0.2, 2.0, 6.5$	8.3	-4.0	$2.0, 6.5, 11.5$	20.0	2.8
$0.5, 2.0, 6.5$	9.0	-3.6	$-0.2, 6.5, 14.3$	20.6	3.1
$0.9, 2.0, 6.5$	9.4	-3.4	$0.5, 6.5, 14.3$	21.3	3.6
$-0.2, 0.5, 11.5$	11.8	-2.0	$0.9, 6.5, 14.3$	21.7	3.8
$-0.2, 0.9, 11.5$	12.2	-1.8	$2.0, 6.5, 14.3$	22.8	4.4
$0.5, 0.9, 11.5$	12.9	-1.4	$-0.2, 11.5, 14.3$	25.6	6.1
$-0.2, 2.0, 11.5$	13.3	-1.1	$0.5, 11.5, 14.3$	26.3	6.5
$0.5, 2.0, 11.5$	14.0	-0.7	$0.9, 11.5, 14.3$	26.7	6.7
$0.9, 2.0, 11.5$	14.4	-0.5	$2.0, 11.5, 14.3$	27.8	7.3
$-0.2, 0.5, 14.3$	14.6	-0.4	$6.5, 11.5, 14.3$	32.3	10.2
$-0.2, 0.9, 14.3$	15.0	-0.1			

chosen standard metropolitan statistical area (SMSA) counties (X) and four nonmetropolitan counties (Y) are taken from Pratt and Gibbons [18, p. 299]:

$$X: -0.2, 0.9, 2.0; \quad Y: 0.5, 6.5, 11.5, 14.3.$$

Since $m = 3$, $n = 4$, there are $\binom{7}{3} = 35$ separations of the 7 observations into a set of 3 X's and 4 Y's. Each X separation is listed in Table 1 and $\overline{X} - \overline{Y}$ is calculated for each using the constant relationship $\overline{X} - \overline{Y} = (7\sum X - 106.5)/12$ that holds for these data because $\sum(X + Y) = 35.5$ for any separation. This gives the randomization distribution of $\overline{X} - \overline{Y}$ conditional upon the observations. Since each possible value of $\overline{X} - \overline{Y}$ is different in this case, each has probability $1/35$. For example, Table 1 shows that $\Pr(\overline{X} - \overline{Y} \geqslant 6.06) = 5/35$. These results can be used to test H_0 at a given level or to find a P-value*. The observed separation gives $\overline{X} = 0.9$, $\overline{Y} = 8.2$ and $\overline{X} - \overline{Y} = 7.3$, and its one-tailed P-value from Table 1 is $3/35$ since only 3 of the separations have an $\overline{X} - \overline{Y}$ value as small as that observed.

Pitman [15] also gives a corresponding trial and error method for interval estimation of the difference in location of the X and Y populations and gives approximations to the null distribution of a function of $\overline{X} - \overline{Y}$ using the beta distribution*.

BIVARIATE CORRELATION PITMAN TEST

Pitman [16] proposes a test of significance of the correlation coefficient in a sample $(X_1, Y_1), (X_2, Y_2), \ldots, (X_n, Y_n)$ from a continuous bivariate distribution. The randomization distribution is generated here by enumerating all $n!$ possible pairings of the X's and Y's. Pitman uses the test statistic $\sum XY - n\overline{X}\overline{Y}$, which is a function of the ordinary product moment correlation coefficient r.

Pitman [16] finds a beta approximation to the distribution of r^2 and compares his test with tests based on normal distribution assumptions.

ANALYSIS OF VARIANCE PITMAN TEST

Pitman [17] proposes a test of significance of the treatment effects in a randomized complete block design*. Suppose there are n treatments and k blocks (or populations), where each block consists of n experimental units which are alike in some important way. We randomly assign each of the n units to a separate treatment group and observe the treatment effect. The null hypothesis is that the n treatment effects are all equal. The n observations in any one block can be arranged in $n!$ ways; since there are k blocks, $(n!)^k$ permutations of the observations are possible. But $n!$ of these are the same so there are only $(n!)^{k-1}$ different ways of arranging all the observations, and all are equally likely under H_0, as are the resulting values of any test statistic that is a function of these arrangements. Pitman's proposed test statistic is $W = S_T/(S - S_B)$, where S_T is the sum of squares due to treatments, S_B is the sum of squares due to blocks, and S is the total sum of squares. To carry out the test W is calculated for each arrangement of the observations, and each result has probability $1/(n!)^{k-1}$.

Pitman [17] finds a beta approximation to the distribution of W.

DISCUSSION

These Pitman tests can be applied to samples from any population and their power*, under normal distributions, is asymptotically equal to the corresponding normal theory test (see Lehmann and Stein [12] and Hoeffding [10]). Other tests based on the principle of randomization have been developed in the literature.

A disadvantage of Pitman tests is the necessity of determining a different critical region or significance probability for each set of observations. These determinations are quite tedious, except for very small sample sizes. With modern computers this disadvantage is lessened. Green [7] describes a practical computer program for finding P-values* with the two-sample Pitman test. Heller [8, 9] gives computer programs for exact P-values with the two-sample and analysis of variance Pitman tests. Edgington [3] gives computer programs for generation of randomization distributions. A great deal of research has been devoted to the asymptotic

distributions of the Pitman and other randomization test statistics. However, the accuracy of these asymptotic distributions cannot be measured in general because it depends on the observations.

Rank tests, normal scores tests*, and almost all distribution-free tests* are also based on the principle of randomization, and their exact distributions are known and apply to any set of data. These procedures are therefore used more frequently than the Pitman and other randomization tests.

References

[1] Bradley, J. V. (1968). *Distribution-Free Statistical Tests*. Prentice-Hall, Englewood Cliffs, N.J. (Elementary; references; Chap. 4 covers Pitman tests.)

[2] Conover, W. J. (1980). *Practical Nonparametric Statistics*. Wiley, New York. (Elementary; references; Pitman tests are covered on pp. 327–334.)

[3] Edgington, E. S. (1980). *Randomization Tests*. Dekker, New York. (Practical guide for development and application of randomization tests; contains computer programs for generation of randomization distributions.)

[4] Fisher, R. A. (1966). *The Design of Experiments*. Oliver and Boyd, Edinburgh, Scotland. (Primary discussion of the principle of randomization.)

[5] Fisher, R. A. (1970). *Statistical Methods for Research Workers*. Oliver and Boyd, Edinburgh, Scotland. (Extensive discussion of the principle of randomization.)

[6] Fraser, D. A. S. (1957). *Nonparametric Methods in Statistics*. Wiley, New York. (Theoretical; advanced.)

[7] Green, B. F. (1977). *Amer. Statist.*, **31**, 37–39.

[8] Heller, R. (1981a). *EDV Med. Biol.*, **12**, 62–64.

[9] Heller, R. (1981b). *EDV Med. Biol.*, **12**, 81–82.

[10] Hoeffding, W. (1952). *Ann. Math. Statist.*, **23**, 169–192. (Investigation of asymptotic power of Pitman tests.)

[11] Lehmann, E. L. (1975). *Nonparametrics: Statistical Methods Based on Ranks*. Holden-Day, San Francisco. (Intermediate; very brief mention of Pitman tests on pp. 43 and 106.)

[12] Lehmann, E. L. and Stein, C. (1949). *Ann. Math. Statist.*, **20**, 28–45.

[13] Marascuilo, L. A. and McSweeney, M. (1977). *Nonparametric and Distribution-Free Methods for the Social Sciences*. Brooks/Cole, Monterey, Calif. (Elementary; cookbook approach.)

[14] Mosteller, F. and Rourke, R. E. K. (1973). *Sturdy Statistics*. Addison-Wesley, Reading, Mass. (Elementary; Pitman tests are covered on pp. 12–22.)

[15] Pitman, E. J. G. (1937a). *J. R. Statist. Soc., B*, **4**, 119–130. (Original reference that develops the theory for the two-sample Pitman test for location.)

[16] Pitman, E. J. G. (1937b). *J. R. Statist. Soc. B*, **4**, 225–232. (Original reference that develops the theory for the Pitman bivariate correlation test.)

[17] Pitman, E. J. G. (1938). *Biometrika*, **29**, 322–335. (Original reference that develops the theory for the Pitman analysis of variance test.)

[18] Pratt, J. W. and Gibbons, J. D. (1981). *Concepts of Nonparametric Theory*. Springer-Verlag, New York. (Intermediate; theoretical; extensive treatment of Pitman tests in Chaps. 4 and 6.)

[19] Randles, R. H. and D. A. Wolfe (1979). *Introduction to the Theory of Nonparametric Statistics*. Wiley, New York. (Intermediate; theoretical; randomization tests are discussed on pp. 344–355.)

[20] Siegel, S. (1956). *Nonparametric Statistics for the Behavioral Sciences*. McGraw-Hill, New York. (Elementary; randomization tests are covered on pp. 88–92 and 152–156.)

(DISTRIBUTION-FREE TESTS
FISHER'S EXACT TEST
HYPOTHESIS TESTING
PITMAN ESTIMATORS
RANDOMIZATION TESTS)

J. D. GIBBONS

PIVOTAL INFERENCE

The term "pivotal" was introduced into statistics by R. A. Fisher*, who used it to denote quantities such as $(\bar{x} - \mu)\sqrt{n}/s_x$, Student's t, where $\bar{x}$ denotes the sample mean and s_x the sample standard deviation of a sample of size n from a normal distribution with unknown mean μ and unknown standard deviation σ. It is a function of observations and parameter(s) whose distribution is known—in the case of t, this distribution is Student's with $n - 1$ degrees of freedom. Fisher appears to have confined his use of the term to cases where the statistics entering into the pivotal are sufficient* for the unknown parameters, and for given values of these statistics the relationship between values of the pivotal and values of the

parameter is 1 to 1. In pivotal inference the term is extended to cover any function of observations and/or parameters whose distribution is known.

Any statistical model involving continuous distributions (and some involving discrete distributions) can be specified by giving the joint distribution of a set of pivotals, collectively termed the "basic pivotal" of the model. For example, to say that x_i ($i = 1, 2, \ldots, n$) is a sample from a normal distribution with unknown mean μ and unknown standard deviation σ is equivalent to saying that the vector $\mathbf{p}$ with ith component $p_i = (x_i - \mu)/\sigma$ has the standard normal distribution in n dimensions. *Known* functions of the basic pivotal will also have known distributions, and the general term "pivotal" is confined to such functions. The distribution of the basic pivotal may be known only approximately, and then the distribution of any pivotal will be known to a corresponding approximation. Much of the logic of pivotal inference turns on the pivotal, or nonpivotal, status of specific functions of observations and parameters. Insofar as this status does not depend on the specific distribution of the basic pivotal, this mode of inference has an important "built-in robustness*" property not shared, e.g., by inference procedures using concepts such as "sufficiency*" as is usually understood.

A pivotal which is a function only of observables is called an *ancillary*. When the observations are known, the value of an ancillary is known. The first step in a pivotal inference is to condition the distribution of the remaining pivotals—those involving the unknown parameters—on the known values of the ancillaries. Because all operations are performed on pivotals, this conditioning is unique.

The simplest case arises with a single unknown location parameter μ, when the basic pivotal $\mathbf{p}$ has ith component $p_i = x_i - \mu$, the x_i, $i = 1, 2, \ldots, n$ being observables. Making the 1–1 transformation $p_i = \bar{p} + c_i$, to $(\bar{p}, c_1, \ldots, c_{n-1})$, where $\bar{p} = \sum p_i/n$, we see that $\bar{p} = \bar{x} - \mu$ while $c_i = x_i - \bar{x}$. Thus the c_i are ancillary. The Jacobian of the transformation is n, so if the density for the basic pivotal is $\phi(p_1, \ldots, p_n)$ the joint density of $(\bar{p}, c_1, \ldots, c_{n-1})$ is $n\phi(\bar{p} + c_1, \ldots, \bar{p} + c_n)$. If c_{io} denotes the observed value of $x_i - \bar{x}$, the conditional density of $\bar{p}$ is $K\phi(\bar{p} + c_{1o}, \ldots, \bar{p} + c_{no}) = \psi(\bar{p} \mid c_o)$, say, and the information about μ given by the n observations x_i, with joint density specified by $\phi(p_1, \ldots, p_n)$, is equivalent to that given by the single observation $\bar{x}$ with density specified by $\psi(\bar{p} \mid c_o)$. For example, if the x_i are independent observations of μ subject to error uniformly distributed between $-\frac{1}{2}$ and $+\frac{1}{2}$, $\phi(p_1, \ldots, p_n) = 1$ if all the p_i are between $-\frac{1}{2}$ and $+\frac{1}{2}$, and $\phi = 0$ otherwise. If $c_u = \mathrm{Max}(c_1, \ldots, c_n)$ and $c_l = \mathrm{Min}(c_1, \ldots, c_n)$, the conditional density of $\bar{p}$ will be zero unless $-\frac{1}{2} < \bar{p} + c_l < \bar{p} + c_u < +\frac{1}{2}$, and $\bar{p}$ will be uniformly distributed between $-\frac{1}{2} - c_l$ and $+\frac{1}{2} - c_u$. Thus $\bar{x}$ can be regarded as an observation on μ with error uniformly distributed betweeen these limits.

If, for example, one found $\bar{x} = 0.09$ with $c_l = -0.45$ and $c_u = +0.40$ the information would be equivalent to that given by an observation $\mu = 0.09$ with error $\bar{x} - \mu$ uniform between -0.05 and $+0.10$. It is, however, customary to think of errors as symmetrically distributed around zero, when possible; and so a better formulation would be to write $\mu = 0.065 \pm 0.15u$. This equation is an "abuse of language," understood to mean that $(m - \mu)/s = u$ has the standard uniform distribution between $-\frac{1}{2}$ and $+\frac{1}{2}$, and the observed $m = 0.065$ and the observed $s = 0.15$. A user of this information with a special interest in the question whether μ might be negative could deduce that to suppose $\mu < 0$ would entail that an event of probability less than $1/15$ had occurred, making the supposition $\mu < 0$ somewhat implausible, though by no means impossible.

If, on the other hand, the basic pivotal were supposed to have the standard spherical normal distribution, we would find that the conditional distribution of $\bar{p}$, given $\mathbf{c}$, was normal with zero mean and variance $1/n$, no matter what the values of the ancillaries $\mathbf{c}$. The n observations x_i on μ would

then be equivalent to a single observation $\bar{x}$ on μ, with error variance $1/n$ instead of 1. This, of course, is a standard result of classical statistical inference*, though the latter relates to the marginal distribution of $\bar{x}$ rather than the conditional distribution. The fact that here the values of the c_i do not matter corresponds to the fact, in the classical theory, that $\bar{x}$ is here sufficient for μ.

The major difference between pivotal inference and inference of the classical Neyman–Pearson type is that pivotal inference is required to be conditional on ancillaries, while Neyman–Pearson* theory uses marginal distributions. It happens that, with normally distributed observations, marginal and conditional distributions agree, and it was concentration on normal observations in early years that led to neglect of the importance of conditioning.

When both location and scale are unknown, the basic pivotal takes the form $\mathbf{p} = (\mathbf{x} - \mu.\mathbf{1})/\sigma$, using $\mathbf{1}$ to denote a column of 1's and $\mathbf{x}$ to denote the column vector of observations. If we make the 1–1 transformation $\mathbf{p} = s_p(t.\mathbf{1} + \mathbf{c})$, subject to $\mathbf{1}'\mathbf{c} = 0$ and $\mathbf{c}'\mathbf{c} = n(n-1)$ (where $'$ denotes transpose), we find that the ith component of $\mathbf{c}$ is $(x_i - \bar{x})\sqrt{n}/s_x$, so that $\mathbf{c}$ is ancillary, and the remaining pivotals are $s_p = s_x/(\sigma\sqrt{n})$ and $t = (\bar{x} - \mu)\sqrt{n}/s_x$. If the basic pivotal is standard normally distributed the c_i do not appear in the joint conditional distribution of $(s_p, t \mid \mathbf{c})$, but for nonnormal distributions the c_i do appear. If $\psi(s_p, t \mid \mathbf{c})$ denotes the conditional density, given $\mathbf{c} = \mathbf{c}_o$ we can find a set Γ such that $\mathrm{Pr}\{(s_p, t) \in \Gamma \mid \mathbf{c}_o\} = \gamma$ and then, by the standard argument the set $C = \{(\mu, \sigma) : (s_{po}, t_o) \in \Gamma\}$ is a γ-confidence set for (μ, σ). In general this will be a conditional confidence set—its coverage frequency will be γ in repeated samples having the same $\mathbf{c}_o$ as that observed; but if the $\mathbf{p}$ is standard normal, C will be a γ-confidence in the full Neyman–Pearson sense—its overall coverage frequency in repeated samples will be γ. Here we denote by t_o the result of substituting the observed $\bar{x}_o$ and s_{xo} for $\bar{x}$ and s_x in the expression for t; and similarly for s_{po}.

If interest is concentrated on the location

parameter μ, a variety of situations may present themselves. σ may be entirely unknown, apart from the observed data, and in that case it is correct to take the marginal distribution of t, obtained by integrating s_p out from the joint conditional distribution, as conveying the information about μ, along with the observed $\bar{x}_o$ and s_{xo}. But we may have information to the effect that $a < \sigma < b$, where a and b do not differ greatly. In such a case s_p will be "approximately ancillary," since we know that it lies between $s_x/(b\sqrt{n})$ and $s_x/(a\sqrt{n})$, assuming our model to be correct. We can then find limits to the probability distribution of t by conditioning in turn on $s_p = s_x/(b\sqrt{n})$ and on $s_p = s_x/(a\sqrt{n})$. In the limiting case when $a = b$, of course, we would be required to condition on the known value of s_p, and the result would agree with that obtained by taking the basic pivotal to have ith coordinate $p_i = (x_i - \mu)/a$. When b/a is noticeably larger than 1, it may be that the corresponding values of s_p lie in a region of low probability density for s_p; in this case the model must become suspect. But if the corresponding values of s_p lie in a region of high probability density, then integrating s_p out over this range will give a result not very different from that obtained when σ is taken as wholly unknown. In practice, judgment will be called for in presenting the final conclusion.

The acceptance of this possibility of a restricted range of one or more of the parameters constitutes one respect in which pivotal inference differs from the structural inference* of Fraser. In structural inference considerable stress is placed on the requirement that the parameters should impose a group structure on the inference problem. So long as σ is taken to range over the whole positive real axis, its space is that of a group; but the group property is lost as soon as the range is restricted below this. Both structural and pivotal inference draw their inspiration from R. A. Fisher's 1934 paper [6] in which conditional inference for location and scale parameters was first introduced. But notions of group invariance have no fundamental

role in the theory of pivotal inference. In so far as structural inference imposes group restrictions on its models, it allows greater rigor and associated formality in its reasoning. Applications of judgment of the kind indicated in the preceding paragraph are not called for. Correspondingly, however, the domain of application would seem to be narrower than that of pivotal inference.

An example of a pivotal model which appears to have no structural counterpart is provided by Darwin's growth rate experiment discussed by Fisher [7], (see also Herzberg and Andrews [8]). If y_i denotes the height of a cross-fertilized plant and x_i that of its self-fertilized pair, one of Darwin's models corresponds to assuming that there is a number θ such that $(y_i - \theta x_i)/\sigma = p_i$ is approximately standard normal, with p_i independent of p_j for $i \neq j$. Darwin was interested in evaluating θ. If we here make the transformation $\mathbf{p} = s_p(t.\mathbf{1} + \mathbf{c})$ we find

$$ t = (\bar{y} - \theta\bar{x})\sqrt{n}/\sqrt{(s_y^2 + \theta^2 s_x^2 - 2\theta r_{xy}s_x s_y)}, $$

$$ s_p = \sqrt{(s_y^2 + \theta^2 s_x^2 - 2\theta r_{xy}s_x s_y)}/(\sigma\sqrt{n}) $$

while

$$ c_i = \{(y_i - \bar{y}) - \theta(x_i - \bar{x})\}/(\sigma s_p). $$

On the assumption of approximate normality, t will have Student's distribution on $(n-1)$ degrees of freedom, $(n-1)ns_p^2$ will have a χ^2 distribution on $(n-1)$ degrees of freedom, and the vector c will be uniformly distributed over an $(n-2)$-dimensional sphere. In the absence of knowledge of σ, s_p can range over its whole distribution for any fixed θ, and so cannot provide usable information about θ. It appears that $\mathbf{c}$ is likewise uninformative. So the available information about θ is contained in the pivotal t, with Student's distribution, together with the observed values $\bar{x}_o$, $\bar{y}_o$, s_{yo}, s_{xo}, and r_{xyo}. Because t_o is not a monotone function of θ, care needs to be taken, in general, in interpreting this information. But in the case of Darwin's actual data, $t_o = (65.53 - 44.60\theta)$ $\sqrt{15}/\sqrt{(837.27 + 317.97\theta + 269.40\theta^2)}$ and over the range from $t = -3.22$ to $t = +3.69$

we have, to good approximation, $t_o = 15.9 - 13.8\theta$. Ignoring the possibility (probability 0.005) that t should fall outside the range $(-3.22, 3.69)$, therefore, the information in the data is well represented by saying $\theta = 1.15 \pm 0.072 t_{14}$, meaning that the information is approximately equivalent to an observation 1.15 made on θ with an instrument with error distributed like 0.072 times Student's t with 14 degrees of freedom.

If the ith component of the basic pivotal is $p_i(\mathbf{x}, \boldsymbol{\theta})$, $i = 1, 2, \ldots, n$, and if the parameter $\boldsymbol{\theta}$ has jth component θ_j, $j = 1, 2, \ldots, k$, then the maximal ancillary will be found as the general solution to the k partial differential equations $\sum_i (\partial A / \partial p_i)(\partial p_1 / \partial \theta_j) = 0$, $j = 1, 2, \ldots, k$. If it is required to make inferences about a given function $\lambda(\boldsymbol{\theta})$ of the unknown parameters, without regard to the rest of $\boldsymbol{\theta}$, a corresponding set of partial differential equations must be solved. Solutions of the required form may not exist. Questions of this kind have been discussed by Barnard and Sprott [1]. In the absence of exact solutions approximate solutions may be useful. As is usual in probabilistic problems, "approximate" here typically means a small error with high probability.

In models of classical Neyman–Pearson type the parameters are regarded as unknown constants. The corresponding pivotal models may admit of only limited ancillaries, or perhaps none at all. In a fully Bayesian model all the parameters are thought of as random variables, all the observations then have a known marginal distribution, and so all the observations are ancillaries. From the pivotal point of view intermediate cases may arise in which some, but not all parameters appear as pivotals, giving "partially Bayesian" modes of inference.

The observed values of ancillaries may fall into improbable regions of their known distributions, perhaps giving rise to the "suspicious coincidences" which form the basis of significance tests. In this way the ancillaries may be used for "model criticism" in the sense of Box and Tiao [3]. In

the case of the rectangular distribution considered at the beginning of this article, for example, if $c_u - c_l$ were greater than 1, this would mean that the model assumed must be wrong.

In addition to its close relationship to Fraser's structural model pivotal inference has close connections with the "functional model" of H. Bunke and O. Bunke [4, 5] and with Fisher's original fiducial inference*. All of them involve expressing the information about an unknown parameter θ in terms of a quantity $p(t, \theta)$, where t is a function of observations, and the distribution of $p(t, \theta)$ in repeated experiments of the kind considered is taken as known. Differences between the approaches arise in relation to what is involved in substituting the observed t_o for t in the function $p(t, \theta)$, as well as in what is assumed about the models themselves. Fisher originally appeared to assume that provided $p(t, \theta)$ was fully informative, in a certain sense, concerning θ, $p(t_o, \theta)$ could generate a probability distribution of θ itself, but toward the end of his life he came to doubt this [2, p. 321, footnote].

References

[1]   Barnard, G. A. and Sprott, D. A. (1983). *Ann. Statist.*, **11**, 104–113.

[2]   Bennett, J. H., ed. (1973). *Collected Works of R. A. Fisher*, **3**, University of Adelaide, Australia.

[3]   Box, G. E. P. and Tiao, G. C. (1973) *Bayesian Inference in Statistical Analysis*, Addison-Wesley, Reading, MA.

[4]   Bunke, H. (1975). *Math. Opforsch. Statist.*, **6**, 667–676.

[5]   Bunke, O. (1976). *Math. Opforsch. Statist.*, **7**, 673–678.

[6]   Fisher, R. A. (1934). *Proc. R. Soc. Lond.*, **A144**, 285–307.

[7]   Fisher, R. A. (1960). *J. Oper. Res. Soc. Japan*, **3**, 1–10.

[8]   Herzberg, A. and Andrews, D. F. (1985). *Data*, Springer-Verlag, New York.

(ANCILLARY STATISTICS
CONFIDENCE INTERVALS
  AND REGIONS
FIDUCIAL INFERENCE

HYPOTHESIS TESTING
PIVOTAL QUANTITIES
STATISTICAL INFERENCE
STRUCTURAL INFERENCE
SUFFICIENT STATISTICS)

G. A. BARNARD

PIVOTAL QUANTITIES

DEFINITION AND NOTATION

Let $x_1, x_2, \ldots, x_n$ be a random sample from a distribution $F(x; \theta, \phi)$. Then a function $U(x_1, \ldots, x_n, \theta)$ whose distribution does not depend on any unknown parameters is known as a *pivot function* for θ. Perhaps the best known pivot function occurs when one samples from a normal with unknown μ and σ and forms the pivot function

$$t = \sqrt{n} \, (\bar{x} - \mu)/s$$

for inference about μ. The distribution of this pivot function was first given by William S. Gosset in 1908 [15]. His development of the distribution was less than rigorous but he did obtain the correct result.

The name pivotal quantity no doubt arises from their use in constructing confidence intervals*. Clearly constants u_1 and u_2 exist which do not depend on unknown parameters and which satisfy $P[u_1 < u(X_1, \ldots, X_n; \theta) < u_2] = \gamma$ for any specified γ. The pivotal quantity method is then to pivot or invert the inequalities to obtain a confidence interval of the form $[t_1(X_1, \ldots, X_n) < \theta < t_2(X_1, \ldots, X_n)]$. Note, however, that it may not always be possible to invert such inequalities in this manner.

Mood et al. [14] point out that a pivotal quantity for a continuous random variable with a single unknown parameter always exists, since

$$-2 \sum_{i=1}^{n} \ln F(X_i; \theta) \sim \chi^2(2n).$$

LOCATION AND SCALE PARAMETER DISTRIBUTIONS*

Let $x_1, x_2, \ldots, x_n$ be a random sample from a distribution $F_X(x; a, b)$ for which a density function is given by

$$f_X(x; a, b) = \frac{1}{b} \, g\left(\frac{x - a}{b}\right),$$

$$-\infty < x < \infty$$

(note that $x \geq a$ is permissible if $g(z) = 0$ for $z < 0$). Then $F_X(x; a, b)$ is a location and scale parameter distribution. Also let $R_X(x; a, b) = 1 - F_X(x; a, b)$ denote the reliability function. Let $\hat{a}$ and $\hat{b}$ be the maximum likelihood estimators of a and b. It can be shown that the distributions of the quantities

$$\sqrt{n} \, (\hat{a} - a)/b, \quad \sqrt{n} \, (\hat{a} - a)/\hat{b}, \quad \hat{b}/b$$

do not depend on a and b. Thus these are pivot functions for a with b known, a with b unknown, and b with a unknown, respectively (see, e.g., Fisher [8] or Antle and Bain [1]). This represents a very important generalization of the corresponding well-known results for the normal distribution. Clearly tests or confidence intervals for the parameters can be easily constructed from these quantities for any location-scale parameter distribution, if the necessary percentage points of the distributions of the pivotal quantities are available. With the advent of high speed computers one can easily generate by simulation* any percentage point not readily available. Also tables of the percentage points have been tabulated for many of the important distributions such as the Type 1 extreme value distribution*, and thus the related Weibull distribution* (Thoman et al. [16], Billman et al. [5], Lawless [11]), the logistic* (Antle et al. [2]), and the Cauchy* (Haas et al. [9]).

There are many useful extensions of the basic results given here. For example, with two location and scale parameter densities, the functions

$$(\hat{a}_1 - \hat{a}_2)/b, \quad \hat{b}_1/\hat{b}_2, \quad (\hat{a}_1 - \hat{a}_2)/\hat{b}$$

are useful pivot functions for inference

about $(a_1 - a_2)$ when b_1 and b_2 are known and are equal, (b_1/b_2), and $(a_1 - a_2)$ when b_1 and b_2 are unknown but are assumed to be equal. It seems that many of the usual well-known results for normal distributions can be obtained in general for any location and scale parameter family.

Another important property of location-scale parameter distributions is that $F(x_i; \hat{a}, \hat{b})$ is a pivotal quantity. For example, tests based on the empirical distribution function such as the Kolomogorov–Smirnov and Cramér–Von Mises goodness-of-fit* tests can be extended to the parameters-unknown case for these models. Somewhat similarly, for fixed x^* the distribution of $F_X(x^*; \hat{a}, \hat{b})$ or $R_X(x^*; \hat{a}, \hat{b})$ depends only on the reliability $R_X(x^*; a, b)$, and does not depend otherwise upon x^*, a, or b. Thus tests of hypotheses about the reliability at time x^* may be easily carried out and confidence intervals may be obtained by using the general method for confidence limits.

Along the same lines ratios of maximized likelihood functions of location-scale parameter distributions are pivotal quantities, say

$$RL = \frac{\Pi f_0(x_i; \hat{a}_0, \hat{b}_0)}{\Pi f_1(x_i; \hat{a}_1, \hat{b}_1)}.$$

Thus this is a suitable test statistic for selecting between two models when the parameters are unknown (see Dumonceaux et al. [6]).

Quite often useful test statistics may asymptotically be pivotal quantities and thus provide "approximate" pivotal quantities for fixed sample sizes. For example, it is true in general for the likelihood ratio* test statistic λ, that asymptotically $-2\ln\lambda$ follows a chi-square distribution [14].

APPLICATION TO THE TWO PARAMETER WEIBULL

The foregoing remarks also hold for distributions which may be transformed to location-scale distributions. For example, let

$$T \sim f_T(t; b, c) = c(t/b)^{c-1}e^{-(t/b)^c}$$

$$\text{for} \quad t \geqslant 0.$$

Then T is said to have a Weibull distribution* with scale parameter b and shape parameter c [3]. It is clear that if we let $X = \ln T$, then

$$X \sim f_X(x; b, c) = ce^{c(x - \ln b) - e^{c(x - \ln b)}}$$

$$\text{for all } x.$$

Thus X has a location and scale parameter density with parameters $\ln b$ and c^{-1} respectively (i.e., a Type 1 extreme value* for minimum density). Thus from the fundamental result for location and scale parameter densities we can conclude that

$$\hat{c}/c, \ c(\ln\hat{b} - \ln b) \text{ and } \hat{c}(\ln\hat{b} - \ln b)$$

are suitable pivot functions for inferences about c, b with c known and b with c unknown (see Thoman et al. [16]). These results for Weibull distributions may be extended to questions regarding inferences for reliability, percentile points, and tolerance intervals* (see Billman et al. [5] and Bain [3]).

CENSORED* SAMPLES

Whenever censoring may be regarded as censoring on order statistics*, all of the above results continue to hold with the modifications required by the censoring. For example, we would now say that the density of $\hat{b}/b$ depends only on n and the pattern of censoring. To censor other than on order statistics* invalidates all the results to some extent. However, it would seem that if one stopped an experiment after a fixed time determined in advance and 20 failures for 40 items (or some such numbers) had been observed, little harm would be done in assuming that the intent was to stop after the 20th failure and proceed accordingly.

THREE PARAMETER MODELS

Some results can also be stated for location-scale parameter models which also have a third parameter present. Suppose

$$f_X(x; a, b, c) = \frac{1}{b} g\left(\frac{x-a}{b}, c\right);$$

then the maximum likelihood* estimators have the property that

$$\hat{c}, \hat{b}/b, (\hat{a}-a)/b \text{ and } (\hat{a}-a)/\hat{b}$$

are distributed independently of a and b [7]. The distributions of these quantities will in general depend on c. However, for example, tests for c based on $\hat{c}$ may always be obtained with a and b unknown.

These results are also pertinent to problems such as model selection for models such as the Weibull and gamma where the parameter may not be location-scale (see, e.g., Bain and Engelhardt [4]).

PIVOTAL QUANTITIES FROM OTHER ESTIMATORS

Lawless [11] has generalized the fundamental invariant property ML estimators have in the location and scale parameter setting and uses the term "equivarent" to describe these properties. Clearly minimum variance linear unbiased estimators and other simple estimators can be shown to be equivarent. It follows that for these other methods of estimation one can develop all of the corresponding pivot functions suggested in the above discussion.

FIDUCIAL INFERENCE AND ANCILLARITY*

Pivot functions may also be used to generate fiducial distributions* and fiducial intervals. Wilk [17] gives a brief discussion of this application. The area of ancillary statistics is also concerned with pivotal quantities. Wilkinson [18] and discussants provide a recent review of fiducial concepts including references to pivot function and ancillary methods. Hinkley [10] discusses the relationship of a normalized likelihood function to the distribution of a pivot function conditioned on an ancillary statistic.

References

[1] Antle, C. E. and Bain, L. J. (1969). *SIAM Rev.*, **11**, 251–253.

[2] Antle, C. E., Harkness, W. L., and Klimko, K. A. (1970). *Biometrika*, **57**, 397–402.

[3] Bain, L. J. (1978). *Statistical Analysis of Reliability and Life Testing Models*. Dekker, New York.

[4] Bain, L. J. and Engelhardt, M. (1980). *Commun. Statist.*, **A9**, 375–387.

[5] Billman, B., Antle, C. E., and Bain, L. J. (1972). *Technometrics*, **14**, 831–840.

[6] Dumonceaux, R., Antle, C. E., and Haas, G. (1973). *Technometrics*, **15**, 19–27.

[7] Eastman, J. and Bain, L. J. (1973). *Commun. Statist.*, **2**, 23–28.

[8] Fisher, R. A. (1934). *Proc. R. Soc. Lond. A*, **144**, 285–307.

[9] Haas, G., Bain, L. J., and Antle, C. E. (1970). *Biometrika*, **57**, 403–408.

[10] Hinkley, D. V. (1980). *Biometrika*, **67**, 287–92.

[11] Lawless, J. F. (1978). *Technometrics*, **20**, 355–364.

[12] Lawless, J. F. (1982). *Statistical Models and Methods for Lifetime Data*. Wiley, New York.

[13] Mann, N. R., Schafer, R. E., and Singpurwalla, N. D. (1974). *Methods for Statistical Analysis of Reliability and Lifetime Data*. Wiley, New York.

[14] Mood, A. M., Graybill, F. A., and Boes, D. C. (1974). *Introduction to the Theory of Statistics*. McGraw-Hill, New York.

[15] 'Student'. (1908). *Biometrika*, **6**, 1–24.

[16] Thoman, D. R., Bain, L. J., and Antle, C. E. (1969). *Technometrics*, **11**, 445–460.

[17] Wilks, S. S. (1962). *Mathematical Statistics*. Wiley, New York.

[18] Wilkinson, G. N. (1977). *J. R. Statist. Soc. B*, **39**, 144–171.

(ANCILLARY STATISTICS
CONFIDENCE INTERVALS
  AND REGIONS
HYPOTHESIS TESTING
PIVOTAL INFERENCE)

C. E. Antle
L. J. Bain

PIVOT AND BI-PIVOT t-STATISTICS

The pivot and bi-pivot t-statistics are t-statistics based on two or four order statistics (*see* ORDER STATISTICS). If we consider the n order statistics, $X_{1:n} \leqslant X_{2:n} \leqslant \cdots \leqslant X_{n:n}$, then the pivot t-statistic is equal to $\frac{1}{2}(X_{i:n} + X_{n-i+1:n})/(X_{n-i+1:n} - X_{i:n})$, $X_{i:n} \neq X_{n-i+1:n}$, where i is equal to the pivot-depth (*see* PIVOT-DEPTH). Thus the pivot t-statistic has the usual form of a t-statistic (location estimate/spread estimate), where the location estimate is the average of two symmetric order statistics and the spread estimate is the difference between the same two order statistics [1].

Just as the pivot t-statistic is based on two order statistics using the pivot-depth, so is the bi-pivot t-statistic based on four order statistics using the bi-pivot-depth (*see* PIVOT-DEPTH). If the bi-pivot-depth is the half-integer $j + 0.5$, say, then the bi-pivot t-statistic is equal to

$$\frac{X_{j+0.5:n} + X_{n-(j+0.5)+1:n}}{2(X_{n-(j+0.5)+1:n} - X_{j+0.5:n})},$$

where $X_{k+0.5:n}$ is equal to

$$\tfrac{1}{2}(X_{k:n} + X_{k+1:n}),$$

for the integer k. Thus the bi-pivot t-statistic is equal to

$$\frac{X_{j:n} + X_{j+1:n} + X_{n-j+1:n} + X_{n-j:n}}{2(X_{n-j+1:n} + X_{n-j:n} - X_{j:n} - X_{j+1:n})}.$$

We see that the bi-pivot t-statistic has the same form as the pivot t-statistic, except that instead of two order statistics being used to estimate location and spread, two averages of two adjacent order statistics are so used [1].

As an example, suppose we wish to test $H : \mu = 0$ vs. $A : \mu > 0$ (*see* HYPOTHESIS TESTING) at the 0.05 level, where the population in question is Gaussian with unknown variance. We wish to test the above hypothesis based on the following seven (ordered) observations: $-2.1, 0.1, 0.2, 0.3, 0.6, 0.9,$ and 1.8. Since n is equal to 7, the pivot-depth is

2, and the sample pivot t-statistic is $(x_{2:7} + x_{6:7})/2(x_{6:7} - x_{2:7}) = (0.1 + 0.9)/2(0.9 - 0.1) = 0.625$. Using computational techniques such as those discussed in ref. 1, we find that the critical value is equal to 0.550. Since 0.550 is less than the value of the test statistic, 0.625, we reject the null hypothesis.

Similarly, the bi-pivot depth is 2.5, giving a sample bi-pivot t-statistic equal to

$$\frac{x_{2.5:7} + x_{5.5:7}}{2(x_{5.5:7} - x_{2.5:7})}$$

$$= \frac{\tfrac{1}{2}(0.1 + 0.2) + \tfrac{1}{2}(0.6 + 0.9)}{2\left[\tfrac{1}{2}(0.6 + 0.9) - \tfrac{1}{2}(0.1 + 0.2)\right]} = 0.750.$$

Using the same techniques as above we find that the critical value is equal to 0.745. Since 0.745 is less than the value of the test statistic, 0.750, we reject the null hypothesis.

We note that the pivot and bi-pivot t-statistics are more resistant to outliers* in the sample than is the traditional Student's t-statistic*. From the sample of seven points above, we compute the value of Student's t-statistic to be equal to 0.572. Since this number is less than the critical value of 1.943 (the 95% point of a Student's t-distribution with 6 degrees of freedom) we do not reject the null hypothesis. However if we change the minimum value of the sample from -2.1 to -0.3, we compute Student's t-statistic to be equal to 1.996. Since 1.996 exceeds the critical value of 1.943, we would now reject the null hypothesis. Thus Student's t-statistic is sensitive to outliers.

Note that the change of the value of the minimum observation has no effect on the pivot and bi-pivot t-statistics: the null hypothesis is rejected in either case. The resistant property of the pivot and bi-pivot t-statistics is an asset when the data being analyzed are not Gaussian. When the underlying distribution of the data is heavier-tailed than the Gaussian distribution, then points that would be considered outliers from a Gaussian distribution will be more prevalent. The reader is referred to ref. 1 where it is shown that when the Gaussian assumption is dropped, the pivot and bi-pivot t-statistics perform well, while the sensitivity of Student's t-statistic causes it to perform poorly.

Reference

[1] Horn, P. S. (1983). *J. Amer. Statist. Ass.*, **78**, 930–936.

(HYPOTHESIS TESTING
ORDER STATISTICS
OUTLIERS
PIVOT-DEPTH
STUDENT'S t-TESTS)

P. S. HORN

PIVOT-DEPTH

If we consider the n order statistics*, $X_{1:n} \leqslant X_{2:n} \leqslant \cdots \leqslant X_{n:n}$, the depth of the order statistic $X_{j:n}$ is defined as j if $j < (n+1)/2$, and $n - j + 1$ otherwise. Thus the order statistics $X_{m:n}$ and $X_{n-m+1:n}$ both have depth equal to m [2, p. 30].

The pivot-depth of a sample of size n is defined as that depth whose value is $[(n+1)/2]/2$ or $([(n+1)/2]+1)/2$, whichever is an integer, where $[\cdot]$ rounds down to the nearest integer. Since the pivot-depth is effectively equal to $[(n+3)/4]$, we see that for large n it is essentially $n/4$. Thus the order statistics at the pivot-depth are approximately quartiles for large n. The order statistics defined by the pivot-depth are preferable to quartiles for small samples. This is because the quartile is really a large sample notion and open to interpretation for small samples [1].

For example, let us consider the ordered data points, $-1, 2, 4, 12, 30$. Since n is 5 the pivot-depth is 2. The two order statistics whose depth is 2 are $X_{2:5}$ and $X_{4:5}$, here taking the value 2 and 12 respectively.

Similar to the pivot-depth, let us define the bi-pivot-depth to be equal to $[(n+1)/2]$ or $([(n+1)/2]+1)/2$ whichever is *not* an integer. Thus the bi-pivot-depth will equal $j + 0.5$, where j is an integer. If we define $X_{j+0.5:n}$ to be equal to $(X_{j:n} + X_{j+1:n})/2$ and treat the bi-pivot-depth as we did

the pivot-depth, then the bi-pivot-depth defines two averages of two pairs of adjacent order statistics, namely, $X_{j+0.5:n}$ and $X_{n-(j+0.5)+1:n}$.

Returning to the above example, with n equal to 5, the bi-pivot-depth is equal to 1.5. Thus the bi-pivot-depth refers to the two averages, $X_{1.5:5}$ and $X_{4.5:5}$. By definition $X_{1.5:5}$ is $(X_{1:5} + X_{2:5})/2$, here taking the value $(-1 + 2)/2 = 0.5$, and $X_{4.5:5}$ is $(X_{4:5} + X_{5:5})/2$, here taking the value $(12 + 30)/2 = 21$.

Asymptotically the bi-pivot-depth is equal to $n/4$, as is the pivot-depth. Thus the functions of the order statistics defined by the bi-pivot-depth are also equivalent to the quartiles for large n.

References

[1] Horn, P. S. (1983). *J. Amer. Statist. Ass.*, **78**, 930–936.

[2] Tukey, J. W. (1977). *Exploratory Data Analysis*. Addison-Wesley, Reading, Mass. (An important introductory text.)

(ORDER STATISTICS
PERCENTILES
PIVOT AND BI-PIVOT t-STATISTICS)

P. S. HORN

PIVOTING   *See* GAUSS–JORDAN ELIMINATION

PIXEL   *See* GEOGRAPHY, STATISTICS IN; PROPORTION ESTIMATION IN CROP SURVEYS

PLACEMENT STATISTICS

To compare statistically two or more populations with minimal assumptions about the forms of the probability distributions governing the measurement(s) of interest, statisticians have developed a body of knowledge generally known as nonparametric statistics*. Most nonparametric statistical test procedures are based on some type of ranking* among the collected observations. Much of the early work in this field was based directly on the combined samples ranks of the data. However, some of the important nonparametric statistics are more naturally viewed as functions of measures called *two-sample placements*, which are defined as follows.

Let $X_1, \ldots, X_m$ and $Y_1, \ldots, Y_n$ be independent random samples from populations with distribution functions $F(\cdot)$ and $G(\cdot)$, respectively. Define the random vector $\mathbf{U} = (U_1, \ldots, U_n)$ by $U_i = F_m(Y_i)$, $i = 1, \ldots, n$, where $F_m(\cdot)$ is the empirical distribution function (*see* EDF STATISTICS) for the X's. The quantity

$$P_i = mU_i = [\text{number of } X\text{'s} \leqslant Y_i]$$

is called the *two-sample placement of* Y_i *among the* X's. Since the combined samples rank, R_i, of Y_i can be written as

$$R_i = [(\text{number of } X\text{'s} \leqslant Y_i)$$
$$+ (\text{number of } Y\text{'s} \leqslant Y_i)],$$

we see that the placement, say $P_{(i)}$, of the ith order statistic*, $Y_{(i)}$, for the Y sample is related to the rank, say $R_{(i)}$, of $Y_{(i)}$ through the equation

$$R_{(i)} = P_{(i)} + i, \qquad i = 1, \ldots, n.$$

Hence there is a one-to-one relationship between the set of ordered placements $\{P_{(1)}, \ldots, P_{(n)}\}$ and the set of ordered ranks $\{R_{(1)}, \ldots, R_{(n)}\}$. As a result every rank statistic that is a function of the ordered Y ranks only (which is the case for any reasonable rank statistic) could also be written as a function of the ordered Y placements only and vice-versa. Thus for this fixed sample size case the class of useful rank statistics is equivalent to the class of useful placement statistics. However, certain statistics lend themselves more easily to interpretation and study when they are represented in terms of the sample placements.

The basic concept behind two-sample placements was introduced by Fligner and Wolfe [6] in their study of sample analogues to the probability integral transformation* and certain coverage properties. However, the term placement was first applied to the P_i's by Orban and Wolfe [12] when they proposed and studied a sizable class of test

statistics designated as linear placement statistics. Although similar in form to another large class of statistics known as linear rank* statistics, the only statistic belonging to both classes is the well-known Mann–Whitney–Wilcoxon [8, 16] statistic*. A member of the general class of linear placement statistics is of the form

$$S_{n,m} = \sum_{i=1}^{n} \phi_m(U_i),$$

for some real-valued Lebesgue measurable function $\phi_m(\cdot)$ defined on $[0, 1]$. The Mann–Whitney statistic is the member of this class corresponding to $\phi_m(t) = t$, while the corresponding, statistically equivalent, Wilcoxon rank sum statistic is a member of the class of linear rank statistics.

Although the term placement and this general class of linear placement statistics are relatively new to the statistical literature, the underlying idea of basing test statistics on the information contained in two-sample placements had previously been utilized in a variety of settings.

Many of these early usages of placements involved what are commonly called *exceedance* statistics (see Sec. 11.4 of Randles and Wolfe [13]). Included in this category are the Mathisen [10] median* statistic $T_1 = \text{median}_{1 \leqslant j \leqslant n} P_j$, when n is odd, and the Rosenbaum [15] statistic $T_2 = [m - \text{maximum}_{1 \leqslant j \leqslant n} P_j]$, both used to test for location differences in the X and Y populations, and a second Rosenbaum [14] statistic

$$T_3 = m - \max_{1 \leqslant j \leqslant n} P_j + \min_{1 \leqslant k \leqslant n} P_k$$

used to test for scale differences in the X and Y populations.

Two-sample placements have been used to construct hypothesis tests for general alternatives as well. One such statistic, equivalent to one proposed by Dixon [4] and later investigated by Blum and Weiss [1] and Blumenthal [2], is based on the expected values of the ordered placements under the null hypothesis of no differences between the X and Y populations. Letting $P_{(1)} \leqslant \cdots \leqslant P_{(n)}$ denote the ordered placements of the Y's, define the differences Q_i

$= [P_{(i)} - P_{(i-1)}]$, $i = 2, \ldots, n$, of adjacent $P_{(i)}$'s. Setting $Q_1 = P_{(1)}$ and $Q_{n+1} = [m - P_{(n)}]$, it can be shown that $E_0[Q_i] = (m/n + 1)$, for $i = 1, \ldots, n + 1$, where $E_0[Q_i]$ indicates the expected value of Q_i taken under the null hypothesis of no differences between the X and Y populations. The test studied by Dixon, Blum, Weiss, and Blumenthal is equivalent to rejecting the null hypothesis of equal X and Y populations for large values of the chi-square* type statistic

$$V = \sum_{i=1}^{n+1} \left[Q_i - E_0(Q_i) \right]^2$$
$$= \sum_{i=1}^{n+1} \left[Q_i - \frac{m}{n+1} \right]^2.$$

Some of the general null distributional properties of placements and ordered placements are presented in Fligner and Wolfe [6], including means, variances, and covariances. They also establish sample analogues based on placements of some well-known results about coverages*.

Properties of two-sample placements also find application in the area of nonparametric prediction intervals. They can be used to establish the crucial probability expression in Danziger and Davis [3] that leads to prediction intervals containing at least k of n future observations. Fligner and Wolfe [7] also demonstrate that distribution-free prediction intervals for a future sample median follow at once from these basic placement properties.

Fligner and Policello [5] have used placements in their approach to providing nonparametric tests for the famous Behrens–Fisher problem*. In particular they make use of placements in developing distribution-free estimators for the variance of the Mann–Whitney statistic.

Finally, Orban and Wolfe [11] adapt the idea of placement statistics to the setting where one of the samples is fixed size in nature and the second sample is obtained sequentially. In such partially sequential settings, where $X_1, \ldots, X_m$ is a random sample of fixed size m and the Y's are obtained sequentially, the sequential *placements* of the Y's are much more natural than the sequential *ranks* of the Y's, since at each collection stage the latter involves reranking all of the previously obtained Y's, which is not necessary in order to compute the sequential placement at any collection time. Orban and Wolfe study some of the properties associated with linear placement statistics applied to partially sequential data and find that the potential for sample size reduction is substantial in many cases.

The extension of this idea of two-sample placements to the k-sample, $k > 2$, setting has not yet appeared in the literature. However, it would seem to be a very natural and desirable extension, especially to the setting of many treatments and a control.

Example 1. To illustrate both the placements themselves and their part in the calculation of several useful test statistics, consider the following subset of the data obtained by March et al. [9] in a study dealing with differences between healthy and lead-poisoned (as a result of ingesting hunters' pellets from the bottom of ponds) Canadian geese. The data listed are plasma glucose (in mg/100 ml plasma) values for eight healthy and seven lead-poisoned Canadian geese.

Labeling the lead-poisoned geese as the Y-sample, we see that the placements of the Y's among the eight X observations are: $P_1 = 4$, $P_2 = 4$, $P_3 = 8$, $P_4 = 8$, $P_5 = 8$, $P_6 = 8$, and $P_7 = 5$. The associated Mann–Whitney statistic for these data is

$$W = \sum_{i=1}^{7} P_i = 45,$$

Table 1   Plasma Glucose Values

Healthy Geese	Lead-Poisoned Geese
297	293
340	291
325	370
227	430
277	510
337	353
250	318
290	

while the Mathisen median statistic is

$$T_1 = \underset{1 \leqslant j \leqslant 7}{\text{median}} P_j = 8.$$

The Rosenbaum location and scale statistics are

$$T_2 = m - \underset{1 \leqslant j \leqslant 7}{\text{maximum}} P_j$$

$$= 8 - 8 = 0$$

and

$$T_3 = m - \underset{1 \leqslant j \leqslant 7}{\max} P_j + \underset{1 \leqslant k \leqslant 7}{\min} P_k$$

$$= 8 - 8 + 4 = 4,$$

respectively.

References

[1]   Blum, J. R. and Weiss, L. (1957). *Ann. Math. Statist.*, **28**, 242–246. (Studies properties of the Dixon chi-square statistic.)

[2]   Blumenthal, S. (1963). *Ann. Math. Statist.*, **34**, 1513–1523. (Studies properties of the Dixon chi-square statistic.)

[3]   Danziger, L. and Davis, S. A. (1964). *Ann. Math. Statist.*, **35**, 1361–1365. (Prediction intervals for k out of n.)

[4]   Dixon, W. J. (1940). *Ann. Math. Statist.*, **11**, 199–204. (Proposed a two-sample chi-square statistic based on null hypothesis expected values of placements.)

[5]   Fligner, M. A. and Policello, G. E., II (1981). *J. Amer. Statist. Ass.*, **76**, 162–168. (Behrens–Fisher problem.)

[6]   Fligner, M. A. and Wolfe, D. A. (1976). *Amer. Statist.*, **30**, 78–85. (Null distributional properties of placements. Includes general discussion of a variety of applications.)

[7]   Fligner, M. A. and Wolfe, D. A. (1979). *J. Amer. Statist. Ass.*, **74**, 453–456. (Prediction intervals for future sample median.)

[8]   Mann, H. B. and Whitney, D. R. (1947). *Ann. Math. Statist.*, **18**, 50–60.

[9]   March, G. L., John, T. M., McKeown, B. A., Sileo, L., and George, J. C. (1976). *J. Wildl. Diseases*, **12**, 14–18.

[10] Mathisen, H. C. (1943). *Ann. Math. Statist.*, **14**, 188–194. (Two-sample median test.)

[11] Orban, J. and Wolfe, D. A. (1980). *Comm. Statist.*, **A9**, 883–904. (Application of linear placement statistics to setting where data are partially sequential in nature.)

[12] Orban, J. and Wolfe, D. A. (1982). *J. Amer. Statist. Ass.*, **77**, 666–672. (General class of linear placement statistics.)

[13] Randles, R. H. and Wolfe, D. A. (1979). *Introduction to the Theory of Nonparametric Statistics*. Wiley, New York. (Secs. 11.4 and 11.5 contain discussions on some of the properties and applications of two-sample placements.)

[14] Rosenbaum, S. (1953). *Ann. Math. Statist.*, **24**, 663–668. (Dispersion test and tables.)

[15] Rosenbaum, S. (1954). *Ann. Math. Statist.*, **25**, 146–150. (Location test and tables.)

[16] Wilcoxon, F. (1945). *Biometrics*, **1**, 80–83.

(BEHRENS–FISHER PROBLEM
COVERAGE
DISTRIBUTION-FREE METHODS
EDF STATISTICS
LINEAR RANK TESTS
MATHISEN'S MEDIAN TEST
ORDER STATISTICS
RANK ORDER STATISTICS
RANK PROCEDURES IN EXPERIMENTAL
   DESIGN
RANKED SET SAMPLING)

D. A. WOLFE

PLACKETT AND BURMAN DESIGNS

The term "Plackett and Burman design" has its origin in a paper written by R. L. Plackett and J. P. Burman in 1946 [9]. It is generally used to indicate certain two-level fractional factorial* designs which allow efficient estimation of the main effects of all the factors being explored, assuming that all interactions between the factors can, tentatively, be ignored. (However, the paper [9] also discusses designs with *more* than two levels, and provides some specific three-, five-, and seven-level designs with similar characteristics.)

Suppose we wish to examine how k factors (e.g., temperature, pressure, concentration of chemical A, flow rate of mix) affect a response variable (e.g., percentage yield). The more experimental runs* that can be performed, the more information can be discovered about the various main effects*, two-factor interactions*, three-factor interactions, and so on. If, however, we wish to perform only a very few runs, the possible information is limited. Nevertheless, if we feel able to make strong assumptions about the form of the response such as, e.g., compared with main effects, all interactions between two or more factors can be ignored, at least initially, we can then consider how to pick the "best" set of runs in our assumed circumstances. "Best" is typically taken to imply that we wish to obtain main-effect estimates with variances as small as possible. To estimate k main effects and the overall mean response level, a minimum of $(k + 1)$ runs is essential.

Specifically, consider the problem of estimating the main effects of seven $(k = 7)$ factors at two levels (coded to -1 and 1, say) in eight runs with minimum variance. This problem, it can be shown, reduces to finding seven orthogonal columns of length eight containing $+1$'s and -1's, which we simplify to $+$ and $-$ below. Plackett and Burman offer this solution:

Row	1	2	3	4	5	6	7
1	+	+	+	−	+	−	−
2	−	+	+	+	−	+	−
3	−	−	+	+	+	−	+
4	+	−	−	+	+	+	−
5	−	+	−	−	+	+	+
6	+	−	+	−	−	+	+
7	+	+	−	+	−	−	+
8	−	−	−	−	−	−	−

The pattern displayed requires $N = 8$ experiments (rows) on $k = 7$ variables or factors (columns) and is obtained by cyclic permutation of the top row (which Plackett and Burman provide) followed by the addition of a row of minus signs. Note that *any* two columns consist of the four runs $(- -)$, $(+ -), (- +), (+ +)$, namely a 2^2 factorial design, twice over. All Plackett and Burman designs have this basic pattern of signs repeated $N/4$ times.

The analysis of such a design is straightforward. We simply calculate and compare

means at upper and lower levels of each variable. Thus, for example, if the eight observations obtained are denoted $y_1, y_2, \ldots, y_8$, in row order, the estimator of main effect of factor 1 is

$$L_1 = \tfrac{1}{4}(y_1 + y_4 + y_6 + y_7)$$
$$- \tfrac{1}{4}(y_2 + y_3 + y_5 + y_8).$$

(Compare the form of L_1 with the signs in column 1.) The seven main effects plus the overall mean value $\bar{y}$ replace the eight observations by eight other quantities which can be directly interpreted in terms of the factors. Each main effect estimator has the same variance $V(L_j) = 4\sigma^2/N$, where N is the total number of observations (here, 8), where $\sigma^2 = V(y_i)$, and provided the y_i are uncorrelated.

The cyclic generation pattern applies to designs for other values of N also. For example, Plackett and Burman provide the following initial rows:

$N = 12$	+ + − + + + − − − + −
$N = 16$	+ + + + − + − + + − − + − − −
$N = 20$	+ + − − + + + + − + − + − − − − + + −
$N = 24$	+ + + + + − + − + + − − + + − − + − + − − − −

A final line of minus signs provides the Nth row in all cases. Other constructions are slightly different. The design provided by Plackett and Burman for $N = 28$ consists of three 9×9 blocks of signs A, B, and C, say, which, set side by side, form nine rows of width $k = 27$. Cyclic permutation to BCA and then to CAB gives a total of 27 rows, and a final row of minuses completes the $N = 28$ experiments. In all, Plackett and Burman provide designs for N equal to a multiple of four and for all such $N \leqslant 100$ (except 92). For details see ref. 9, pp. 323–324. Alternatively, for selected designs see ref. 2, pp. 333–334. The missing $N = 92$ case was later given by Baumert et al. [1]. These designs are *saturated* in that they examine $N - 1$ factors, the maximum possible, in N runs. To examine fewer than $N - 1$ factors, we simply allocate the factors to *some* of the columns and then utilize the unallocated columns to estimate error. (Actually they estimate combinations of interactions between the factors that are assumed to be errors. See the following paragraphs on alias relationships.) Geometrically the rows of a saturated design define N points which are the coordinates of a simplex in $N - 1$ dimensional space.

ALIAS RELATIONSHIPS

All Plackett–Burman two-level designs are fractions of (i.e., subsets of runs selected from) full 2^k factorial designs in which k factors are examined at every possible combination of levels. A full 2^k factorial design will provide estimates of the overall mean, k main effects, $\frac{1}{2} k(k - 1)$ two-factor interactions, $\frac{1}{6} k(k - 1)(k - 2)$ three-factor interactions, ..., and one k-factor interaction; there are 2^k estimates in all. Whenever only a fraction of the 2^k is performed, fewer than 2^k estimates are available and each column of signs in the fractional design actually estimates a linear combination of main effects and interactions. For Plackett and Burman designs in which N is a multiple of four but *not* a power of two, these linear combinations are *extremely* complicated; for $N = 12$, for example, see Draper and Stoneman [5] or Margolin [8, p. 571]. When $N = 2^{k-p}$, simpler combinations occur (see Box and Hunter [2] and Box et al. [3]). Main effects and interactions that are inextricably linked in estimable combinations in factorial designs are called *aliases* of one another (*see also* CONFOUNDING). The separation of aliased effects can be achieved by augmenting an original design with carefully chosen runs (see [2], [3], and Daniel [4]).

It follows that, because the basic Plackett and Burman designs are main-effect-only designs, their usefulness is conditional on (*a*) the interactions between factors being small or negligible (relative to the sizes of some or all main effects), *or* (*b*) the existence of relatively few "important" factors. Unless these circumstances exist, the results from the Plackett and Burman designs may be confusing, and the designs will need to be augmented. One standard type of augmentation is to follow up the original Plackett and Burman design with a "folded" portion in which all the signs are reversed. The combined design will always have main effects (plus their higher order aliases) unaliased with two-factor interactions (plus their higher order aliases) and will comprise a "resolution IV design" (explained in the 2^{k-p} section which follows).

In the field of response surface* methodology, the Plackett and Burman designs have application as first order orthogonal designs* suitable for fitting a planar regression* model. The essential use is identical, but the context and interpretation are different.

We now discuss the relationship between Plackett and Burman designs, Hadamard matrices*, and the 2^{k-p} fractional factorial designs*.

HADAMARD MATRICES*

If (e.g., in the first display table) we add a column of plus signs to a Plackett and Burman design we obtain an $N \times N$ matrix (**M**, say) with orthogonal columns such that $\mathbf{M'M} = N\mathbf{I}$. Such a matrix is called a Hadamard matrix and it occurs in many areas of experimental design. For an extensive discussion, see Hedayat and Wallis [6].

PLACKETT AND BURMAN DESIGNS CONSIDERED AS TWO-LEVEL FRACTIONAL FACTORIALS*

As already mentioned, the Plackett and Burman two-level designs are all fractions of full 2^k factorial designs. Let us reconsider the $N = 8$ Plackett and Burman design described earlier (see first display) and rearrange its rows into the pattern shown in the second display. Not that the original row numbering has been retained to make clear the nature of the rearrangement. The *divisor* row is simply the number of + signs in a column, namely $\frac{1}{2} N$.

Row	1	2	3	4	5	6	7
8	−	−	−	−	−	−	−
4	+	−	−	+	+	+	−
5	−	+	−	−	+	+	+
7	+	+	−	+	−	−	+
3	−	−	+	+	+	−	+
6	+	−	+	−	−	+	+
2	−	+	+	+	−	+	−
1	+	+	+	−	+	−	−
Divisor	4	4	4	4	4	4	4

In this rearrangement, the design can be immediately recognized [2, 3] as a 2_{III}^{7-4} design, read as a "two to the seven minus four resolution three fractional factorial design," i.e., a $1/2^4$ fraction of a full 2^7 factorial. (Resolution will be explained shortly.) It is defined by the column properties that $4 = -13$, $5 = 123$, $6 = -12$, $7 = -23$, or in words, each sign in the 4 column is minus the product of the signs in the 1 and 3 columns, row by row, and so on. It is conventional to rewrite these properties as $I = -134 = 1235 = -126 = -237$. The symbol I denotes a column of + signs and $I = -134$ means that (row by row) "minus the product of the signs in the 1 and 3 and 4 columns will be a plus." The four statements $I = -134$, etc., constitute *generators* for the design. They are not unique; e.g., if $I = -134$ and $I = 1235$, then $I = (-134)(1235) = -1^2 23^2 45 = -245$ because, if we square the elements in any column, the result will be all plusses. This new generator could replace either in the product which formed it. In general, a set of generators must be independent, i.e., no generator can be the product of other generators in the set.

Whenever N is a power of two, the Plackett and Burman designs are always 2_R^{k-p} fractional factorials, with $R = III$. R denotes the *resolution* of the design, and is the length of the smallest word in the *defining relation* (which is obtained as the result of multiplying together the generators in all possible ways and writing down all the *words*, i.e., groups of numbers, that result). A resolution *III* design does *not* enable separate estimation of main effects and two-factor (and higher) order interactions and thus only if two-factor (and higher) order interactions

are zero are we able to estimate all main effects in an unaliased manner. A *family* of designs consists of all the 2^p designs of type 2^{k-p} obtainable through all possible choices of signs for a set of generators.

Plackett and Burman provide one design choice for $N = 8, 16, 32, 64$, for examining $k = N - 1$ factors. For additional details on 2^{k-p} designs, see Box and Hunter [2], and/or Box et al. [3].

When N is a multiple of four but *not* a power of two, the Plackett and Burman designs are still two-level fractional factorials but not 2^{-p} fractions.

COMMENT

In written and conversational usage, the term "Plackett and Burman designs" is most often restricted to two-level main effect designs for which N is a multiple of four but *not* a power of two. When $N = 2^k$ the designs are more usually called two-level fractional factorials. In fact, both terms *could* be used for either type. Strictly speaking we should attach the name "Plackett and Burman designs" only to those designs given in the original Plackett and Burman paper [9]. In practice the name has become almost generic.

HISTORICAL NOTE

Yates [10, p. 210] considered the problem of how best to weigh objects on a biased scale requiring a zero correction and pointed out that greater precision for individual weights can be achieved by weighing the objects in groups rather than singly. Such groups can be symbolized by the rows of a two-level design matrix where each column represents one object. A plus means the corresponding object is in the (row) group, a minus that it is not. (Alternatively, the meanings of the signs can be reversed.) As Plackett and Burman point out [9, p. 325], their two-level designs "provide what is effectively a complete solution" to this problem. For other related references, consult Kempthorne [7, p. 423].

References

[1] Baumert, L., Golomb, S. W., and Hall, M. (1962). Discovery of an Hadamard matrix of order 92. *Amer. Math. Soc. Bull.*, **68**, 237–238.

[2] Box, G. E. P. and Hunter, J. S. (1961). The 2^{k-p} fractional factorial designs, Part I. *Technometrics*, **3**, 311–351.

[3] Box, G. E. P., Hunter, W. G., and Hunter, J. S. (1978). *Statistics for Experimenters. An Introduction to Design, Data Analysis, and Model Building.* Wiley, New York.

[4] Daniel, C. (1962). Sequences of fractional replicates in the 2^{p-q} series. *J. Amer. Statist. Ass.*, **57**, 403–429.

[5] Draper, N. R. and Stoneman, D. M. (1966). Alias relationships for two-level Plackett and Burman designs. *Tech. Rep. No. 96*, University of Wisconsin, Madison, WI.

[6] Hedayat, A. and Wallis, W. D. (1978). Hadamard matrices and their applications. *Ann. Statist.*, **6**, 1184–1238.

[7] Kempthorne, O. (1952). *The Design and Analysis of Experiments.* Wiley, New York.

[8] Margolin, B. H. (1968). Orthogonal main effect $2^n 3^m$ designs and two-factor interaction aliasing. *Technometrics*, **10**, 559–573.

[9] Plackett, R. L. and Burman, J. P. (1946). The design of optimum multifactorial experiments. *Biometrika*, **33**, 305–325 and 328–332.

[10] Yates, F. (1935). Complex experiments. *Suppl. J. R. Statist. Soc.*, **2**, 181–223; discussion, 223–247.

Acknowledgments

Helpful comments and suggestions from A. C. Atkinson, J. P. Burman, D. R. Cox, C. Daniel, A. M. Dean, A. M. Herzberg, J. S. Hunter, W. G. Hunter, W. E. Lawrence, T. J. Mitchell, R. L. Plackett, R. C. St. John, and the editors are gratefully acknowledged.

(FACTORIAL EXPERIMENTS
FRACTIONAL FACTORIAL DESIGNS
HADAMARD MATRICES)

N. R. DRAPER